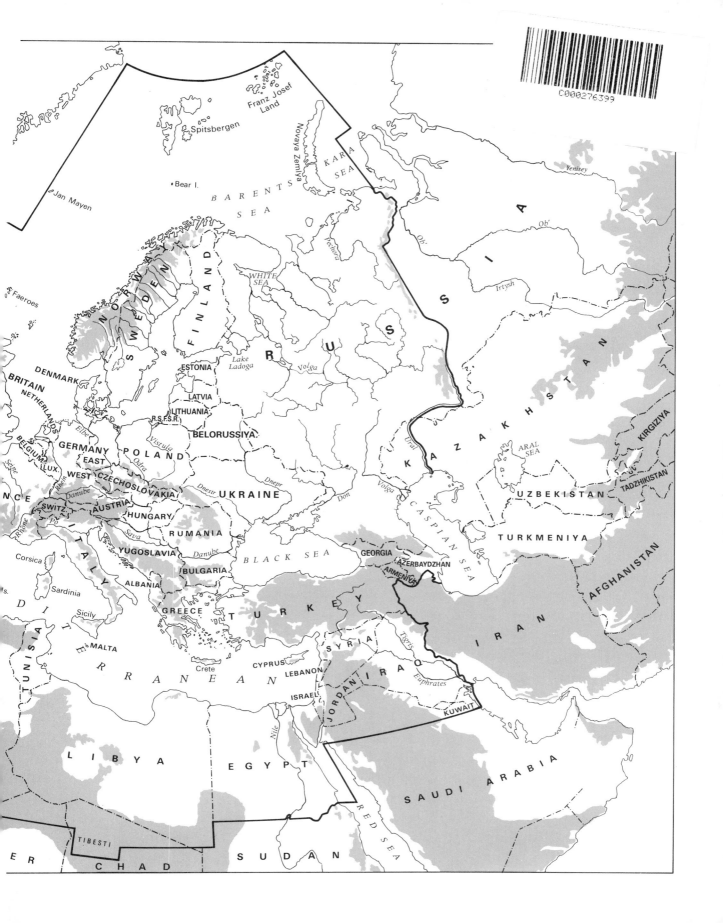

Handbook of the
Birds of Europe
the Middle East and
North Africa

The Birds of the
Western Palearctic

Volume IX

Handbook of the
Birds of Europe
the Middle East and
North Africa

The Birds of the Western Palearctic

Volume IX · Buntings and New World Warblers

†Stanley Cramp
C M Perrins   *Senior Editor*
Duncan J Brooks   *Executive Editor*
Euan Dunn   Robert Gillmor   Joan Hall-Craggs
Brian Hillcoat   P A D Hollom   E M Nicholson
C S Roselaar   W T C Seale   P J Sellar
K E L Simmons   D W Snow   Dorothy Vincent
K H Voous   D I M Wallace   M G Wilson

OXFORD   NEW YORK
OXFORD UNIVERSITY PRESS · 1994

DEDICATED TO THE MEMORY OF

H F WITHERBY

(1873–1943)

EDITOR OF *THE HANDBOOK OF BRITISH BIRDS*
(1938–41)

Oxford University Press, Walton Street, Oxford OX2 6DP
Oxford New York Toronto
Delhi Bombay Calcutta Madras Karachi
Kuala Lumpur Singapore Hong Kong Tokyo
Nairobi Dar es Salaam Cape Town
Melbourne Auckland Madrid
and associated companies in
Berlin Ibadan

Oxford is a trade mark of Oxford University Press

Published in the United States by
Oxford University Press Inc., New York

A catalogue record for this book is available from the British Library

Library of Congress Cataloging in Publication Data
Data available

ISBN 0–19–854843–5

Typeset by Latimer Trend Ltd., Plymouth
Printed in Hong Kong

CONTENTS

Contents

FOREWORD
by Max Nicholson

The concept of a team effort by leading ornithologists to produce comprehensive and definitive reviews of our current knowledge of the avifauna was thoroughly developed by H F Witherby in a succession of works, using the base which he had established with the monthly journal *British Birds*.

As early as 1913, he started on the new model *Practical Handbook of British Birds*, although, owing to the long interruption of World War I, Volume I could only appear in 1920, and the ensuing final volume in 1924. According to practice at that time, each subspecies had its separate heading, making a total of 493, but otherwise the layout and content were surprisingly modern, summarizing distribution abroad and field characters.

In 1934, that work went out of print and, instead of preparing a new edition, Witherby bravely decided to replace it by an even more ambitious one. Despite its five heavy volumes, he insisted on entitling this *The Handbook of British Birds*. While three of its four editors had served on its predecessor, many new helpers were brought in, including P A D Hollom and myself, who have carried the torch on to the present, even more monumental work.

Among the important improvements adopted for Witherby's *Handbook* was the widening of the colour plate series to include all the main plumages by a Continental bird artist and the enlargement of the 'Distribution Abroad' sections to outline the entire range of each species. Text on field characters was expanded and a new habitat section was introduced, as well as one on voice. Despite having to appear during the run-up to World War II (ending while it was actually being fought) the work was an outstanding success. This, however, proved an embarrassment since it stimulated so much intensive effort in filling gaps in knowledge, and exploring the remaining fields, that it quickly ceased adequately to cover its subject. Demands for revision raised great difficulties, not least because the whole concept of 'British Birds' was coming to be seen as outdated, both through wider views of ornithology and its role in the biological sciences, and through much increased travel and accompanying interest in the birds of other countries. The early death in 1950 of B Tucker, the sole surviving key editor, made the break even more final. Having had, at short notice and quite unprepared, to take on the Senior Editorship of *British Birds*, I found myself, as his and Witherby's heir, confronted with the question of what, if anything, was to be done by way of follow-up.

Further difficulties had to be faced, first over resolving the unhappy legacy of the false Hastings Rarities records, which Witherby had been prevented from tackling through threats of legal action, and which underlined the artificiality of the special status accorded to the 'British List', in so far as it diverted attention from aspects of greater scientific significance. A second more practical difficulty was that, in the new conditions of post-war publishing, and deprived of Witherby's unique capabilities, the Witherby firm, which had published the *Handbook*, no longer felt able to finance or manage such a large undertaking as any successor work must be. Indeed, much the same applied to almost any commercial publishing firm in existence.

While the material was there in embarrassing abundance, and ornithology could hope to muster a new team able to marshal and present it, the task had assumed such a scale in terms of management, financing, production, and marketing as to appear verging on the impracticable. Indeed, subsequent history has shown that to be true.

It cannot be denied that consultations on the subject in the early 1960s were marked by a general flinching from commitment. As pressure continued, and the need for a successor to the *Handbook* became more obvious, procrastination grew more difficult. How matters came to a head and action was eventually taken is not entirely clear, but in retrospect I think it was because I was able to devote more time to this project. *British Birds* was by now strong enough for me to pass on the editorship to Hollom, after a wide reorganization of management and production as well as of the editorial team, from the start of 1960. The success of this exercise, and the emergence of S Cramp as a prospective new leader, soon able to take on full-time responsibilities, cleared away some of the obstacles.

Planning for the great work then began in earnest and in the conviction that it must depart from a solely British basis and move to one that was European and West Palearctic. This had become easier thanks to the new outlook and network created by Hollom and his collaborators on *A Field Guide to the Birds of Britain and Europe*. Aided by the proven merits of the *Handbook* model, there was relatively little difficulty in finding an acceptable framework, and not much more in identifying most of the international ornithological team members required to cover the much expanded scope. At this stage, I still had to bear the central responsibility as initial Chairman and Chief Editor, but any possibility of my continuing in an executive role was ended when, as a result of the Duke of Edinburgh's Countryside in 1970 Conference, I became involved in the great pathfinding problems of environmental consultancy, which simply would not wait.

In order to get our show on the road, however, it was essential to find a publisher big enough in every sense to take on the daunting responsibility. Having some influence at Oxford, since I took part in founding the University's Edward Grey Institute of Field Ornithology, I approached the Oxford University Press. They, to my relief and indeed joy, allowed themselves to be persuaded to assume the burden, including the provision of much of the start-up finance, of which we had almost none. During the many and almost unending stresses and crises of the ensuing three decades, I have often been troubled by feelings of guilt over what I had let them in for—burdens undreamt of by anyone at the outset, which they have courageously borne through far more years than they had been led to expect.

Not long afterwards, Cramp was able to take early retirement from the Civil Service, and to assume both the Chairman's and the Chief Editor's role for the ensuing seventeen years. Unlike Witherby, the proprietor of a flourishing family printing and publishing firm with a track record of serious ornithological explorations and as an authority on species, Cramp came as a dark horse, little known before his forties beyond narrow circles of local and urban ornithology, and relatively little equipped with obvious qualifications for the immense charge which he was now to perform with such success. Essentially a field ornithologist, he was, less obviously, equipped by his degree in public administration and his career in Customs and Excise to handle all kinds of background problems in a deceptively quiet but most efficient way. Ornithologists tend not to want to know what their fellows do when birds are not on the agenda. It came as some surprise to find that the expert on urban birds was also an experienced man of business, able and willing to cope with an increasing range of problems which a mere ornithologist would have viewed with horror. The onset of runaway inflation simultaneous with the launching of the project, the inexorably expanding demand for paid staff to keep abreast of the torrent of new literature, and the difficulty of processing so much material within the bounds of space and time up to the searching standards required, taxed even Cramp's powers. Added to these management burdens were those of diplomacy in holding together and keeping up the morale of an international team which in total exceeded a hundred co-workers.

Early in the planning stage, it had reluctantly to be accepted that, although the scope would extend across from the Urals to the Azores and from Spitsbergen to the Sahel, there could be no practical alternative to control by the British board of directors of West Palearctic Birds Ltd—surely one of the most peculiar titles on the UK Register of Companies. The one shining exception, through the helpfulness at the outset of Prof. K H Voous of Leiden, was the delegation to a small expert team in the Netherlands of the sections on Plumages, Bare Parts, Moults, Measurements, Weights, Structure, and Geo-graphical Variation. The decline in expertise in Britain in these areas since the days of the *Handbook* made the assumption of this responsibility by our Dutch colleagues a vital contribution to the balance and authority of the work, and also relieved it of becoming a narrowly British project.

The necessary committee work, especially for editorial progress, presented a further challenge to Cramp's tact and firmness as Chairman of meetings. These had to settle many difficult issues, and meet the demands of the time-table for publication, and deal with laggards. Specialists with unique knowledge could be an intractable source of delay, as could some foreign correspondents, although Cramp had a way with these, and had particularly friendly relations with the Russian contributors.

An important aspect of the scheme was the inclusion of good colour pictures of every significant plumage of every species, a requirement which involved endless critical assessment and reference back to artists, more or less cheerfully accepted by them as they learnt to work to deadlines and standards of precision which were not always familiar. R Gillmor, as their unique ringmaster, was faced here with tasks no less demanding than those confronting Cramp, and he came through with a set of illustrations probably unmatched.

Another highly innovative and demanding area was that of Voice. I myself was initially responsible for this, and with J Hall-Craggs outlined the treatment in a twelve-page section of the Introduction to Volume I; but it soon proved necessary to transfer the main ongoing responsibility, the brunt of which was subsequently borne by the Cambridge specialists at Madingley, under the direction of the late Prof. W H Thorpe, and by the indefatigable P J Sellar, who sought out material and prepared the tapes, while J Hall-Craggs applied rare skills to preparing the sonagrams.

A broader special feature was the much greater attention given to the living bird, which was prominent in Field Characters, Habitat, and most of all in Social Pattern and Behaviour. Knowledge in this last area expanded so fast, even during the course of publication, as to amount to a major work in itself, under a succession of editors starting with the pioneer Dr K E L Simmons. Both Field Characters and Habitat were newly designed on fresh frameworks to extend the reach of these branches of ornithological expertise. One needs only to look at the row of volumes on a shelf to see at once how the scale of treatment has had to be increased with advancing knowledge. From the 720 pages of Volume I to the 1060 pages of Volume V (plus 78 pages of plates), after which what should have been Volume VI had to be divided into two on account of its excessive size. This progressive expansion naturally slowed the rate of publication and greatly increased costs in a period of mounting inflation, adding to the stress on all concerned, most of all the publishers but also the subscribers and other readers.

The great and prolonged burden of producing this

immense work was undertaken in the first place by a volunteer team consisting of Dr K E L Simmons (Social Pattern and Behaviour and general back-up to the Chief Editor), I J Ferguson-Lees (Field Characters), E M Nicholson (Habitat and Voice), R Hudson (Movements), P J S Olney (Food), Dr M A Ogilvie (Breeding), and Dr J Wattel (Plumages, Bare Parts, Moults, Measurements, Weights, Structure, and Geographical Variation). Gillmor assumed responsibility for the illustrations, and Cramp himself took care of Distribution and Population.

After Volume I appeared in 1977, changing circumstances dictated some adjustments—D I M Wallace took over Field Characters, and Simmons shared with Nicholson responsibility for Voice, while C S Roselaar joined Wattel for the 'taxonomic' group of text sections. D J Brooks took over most, and latterly all, of the editing formerly done by Simmons and Cramp. Dr N J Collar provided vital support at staff level for Volume II in 1980. Further changes occurred before Volume III appeared in 1983. In particular, Nicholson passed over his Voice responsibilities to Collar and Simmons, and from Volume IV onwards the work for this and for Social Pattern and Behaviour fell to Dr E K Dunn and M G Wilson with the continuing support of Hall-Craggs and Sellar. D F Vincent took over the responsibility for Movements from Volume VI, and B Hillcoat that for Food and Breeding from Volume VII.

Naturally, the stress was greatest for the Chief Editor and Chairman, and by 1985 it was becoming obvious that Cramp was suffering from it to a disturbing degree. Efforts to relieve him of part of the burden were, however, frustrated by his intense dedication and refusal to contemplate compromise or partnership. His colleagues had no choice but to see his health steadily worsening, not helped by his persistent chain-smoking, and the conduct of business and day-to-day organization deteriorated, with increasing delays and problems. Finally in August 1987, a stroke and pneumonia put him in hospital and on the 20th he died, having brought the great work within sight of completion.

The shock and sadness of losing him was compounded by the difficulty of providing for the carrying on of the enterprise. Fortunately in Sir William Wilkinson the Board acquired a new Chairman of rare quality and resolution. Like Cramp, he is both an ornithologist and a man of business, but in his case a senior merchant banker, whose talents had led to his appointment by the Government to the Chairmanship of the Nature Conservancy Council. He took all the time and trouble necessary to sort out the many urgent problems, not only internally but with the Oxford University Press, who were confronted with immense difficulties in carrying on. But for Wilkinson, it is highly questionable whether these could have been surmounted, and tribute must also be paid here to the exceptional and costly co-operation with which the publishers responded to the various crises. Dr D W Snow, at much personal inconvenience, took on the position of Senior Editor, his great international reputation going far to damp down alarm about the work's completion. He held the fort until Dr (later Prof.) C M Perrins, head of the Edward Grey Institute at Oxford, where much of the text was being written, most obligingly took over for the last lap.

Thus against all odds *The Birds of the Western Palearctic* is complete. Despite all the troubles it has entailed, the warmth and respect with which ornithologists have welcomed it has made the enterprise eminently worthwhile.

E M Nicholson
March 1993

INTRODUCTION

Scientific nomenclature in this final volume deviates from the *List of Recent Holarctic Bird Species* (Voous 1977) in using the names *Passerella iliaca* (Fox Sparrow) rather than *Zonotrichia iliaca*, *Melospiza melodia* (Song Sparrow) rather than *Zonotrichia melodia*, and *Euphagus carolinus* (Rusty Blackbird) rather than *E. carolinensis*.

For a full introduction, detailing the scope of the work as a whole and of the individual sections (including glossaries of terminology), see Volume V.

The editors with special responsibility for the various sections of the species accounts in this volume are as follows:

Field Characters   D I M Wallace
Habitat   E M Nicholson
Distribution and *Population*   Dr D W Snow
Movements   D F Vincent
Food   B Hillcoat
Social Pattern and Behaviour   Dr E K Dunn and M G Wilson
Voice   Dr E K Dunn, M G Wilson, J Hall-Craggs, W T C Seale, and P J Sellar
Breeding   B Hillcoat
Plumages, *Bare Parts*, *Moults*, *Measurements*, *Weights*, *Structure*, and *Geographical Variation*   Drs C S Roselaar
Family Treatments   Dr K E L Simmons and Drs C S Roselaar

D J Brooks is responsible for the editing of the entire volume.

The paintings are the work of Norman Arlott, Trevor Boyer, Hilary Burn, Dr P J K Burton, Ian Lewington, Darren Rees, Chris Rose, and C E Talbot-Kelly; their initials appear at the end of the caption for each plate. Euan Dunn, Robert Gillmor, Julian Hough, David Nurney, and David Quinn have prepared the line drawings for the Social Pattern and Behaviour sections. Robert Gillmor is the editor with general responsibility for artwork. R J Connor and A C Parker have given most generously of their time and expertise in providing photographs of eggs in the collection of the Natural History Museum at Tring where they were joined in this work by M P Walters of the Museum staff.

For purely practical reasons it has not always been possible to incorporate in our treatment the many recent changes in political boundaries and names of states both within and outside the west Palearctic. In general, the European part of the former USSR has been treated as an entity for distributional purposes, as in previous volumes. Other data from the former USSR have been linked to the various new republics individually wherever possible, but there are many instances where this could not be done conveniently and in such cases the term 'USSR' has been used for simplicity. The names Czechoslovakia and Yugoslavia have been retained throughout, and it has sometimes been necessary to refer to West and East Germany.

We have included a Combined Index to all the species accounts in the hope that this will make it easier for readers to locate the species to which they wish to refer. In the individual volumes there are indexes to the scientific, English, French, and German names of the birds. We have added indexes to the Spanish and Italian names to the Combined Index for the convenience of our many readers in those countries. We are grateful to Ms Beatriz Arroyo and Dr Alberto Masi for help with the preparation of these.

Kees Roselaar's contributions to BWP have been enormous. As we go to press we are pleased to note that his researches for this work have uncovered two new subspecies (*Dutch Birding* 15: 258-62, December 1993). The Greenfinch he names *Carduelis chloris voousi* after Prof. K H Voous, a member of the editorial team of BWP and a supporter of the project from its inception. The Fan-tailed Raven he names *Corvus rhipidurus stanleyi* with the note that it is named 'in honour of Stanley Cramp, OBE (1913-87), initiator of BWP. Stanley had a strong interest in both crows and the Middle East. Without his stimulating correspondence throughout the first five volumes, the handbook would never have reached its present status.'

CONCLUSION

It is an indication of the scale of the undertaking that the nine volumes of *The Birds of the Western Palearctic* have taken some three decades to complete. Indeed this final volume is appearing seventeen years after the publication of Volume I. It has required an enormous team effort to produce this outstanding work, which is a pre-eminent contribution to both European and World ornithology.

Max Nicholson, in his fascinating Foreword, pays tribute to the contribution of most of the leading players. Characteristically, he understates his own role. Without his vision and drive I doubt whether the work would have seen the light of day. Although no longer a Board member of West Palearctic Birds Ltd in recent years, his influence was always there, and his own submissions to the text were always the first to arrive. I would also like to emphasize again the part played by Phil Hollom—ever helpful, always astute in his judgements, and, of course, profound in his ornithological knowledge and experience. Peter Oliver, in recent years, has kept firm control of the finances besides providing a balanced and knowledgeable input to our Board's discussions. Similarly of late, Dr Jeremy Greenwood has supplied additional ornithological knowledge at Board level.

Production of this work has not been easy. Transitions in editorship, for a variety of reasons natural and personal, made continuity a particularly difficult problem. Fortunately, the vessel came safely to port under the mastership of Christopher Perrins with his energy and organizational skills, and Christopher has seen the work through to a triumphant close, despite his heavy departmental duties at Oxford University.

Max has paid tribute to the leaders of the work-force in the early days, and I gladly acknowledge and thank those who have carried the torch across the finishing line. Still with the team are Ian Wallace, Ken Simmons, Joan Hall-Craggs, and Patrick Sellar, while the outstanding contribution of Kees Roselaar in Amsterdam deserves special recognition. Latterly based at Tring, Ruth Wootton has compiled the references and foreign names and typed practically the entire work—much of it two or three times in pre-computer days—and Duncan Brooks, who started as assistant to Stanley in 1977, took over all the editorial reins and much of the administration during the time of Stanley's final illness and thereafter. Also at Tring, Dr David Snow continued Stanley's work on the maps and the texts for Distribution and Population.

The team based at the Edward Grey Institute in Oxford consists of Dr Euan Dunn (joined 1980), Michael Wilson (translator of German and Russian from 1971, and a full-time editor from 1980), Dorothy Vincent (1985), and Brian Hillcoat (1989). Euan helped with many parts of the book, particularly in the later volumes, but from the outset of his involvement wrote the texts, together with Mike Wilson, for Social Pattern and Behaviour, and Voice. Dorothy was responsible for the Movements texts, and Brian for the accounts on Breeding and Food. In addition to their ornithological knowledge, the team brought essential skills in languages, including fluency in French, German, and Russian and an ability to handle most others. We were particularly fortunate to have had Mike, who not only did the translations from Russian, essential for his own sections, but translated all the Russian literature necessary for other authors as well; without his expertise, coverage of the literature would have been far less thorough.

The production of the sonagrams deserves special credit. Working with the recordings of countless sound-recordists, Pat Sellar has expertly, and enthusiastically, compiled the tapes from which these are derived. Working from the start of BWP on her own, but in the last four volumes with help from William Seale, Joan Hall-Craggs has continued indefatigably to produce sonagrams and add the results of her analyses to the Voice accounts. Production of sonagrams for those skilful songsters, the passerines, has been particularly challenging and demanding, and Dr Peter McGregor (Nottingham University) helped by making the sonagrams for a number of species in Volumes VII, VIII, and IX. Overall, there can be no set of sonagrams more complete for any region in the world.

We were fortunate in the final volume to have help from a number of experts on various species; in particular, Dr A J Erskine undertook the Movements accounts for all the Nearctic vagrants, and Peter Colston produced the information on the occurrences of these in Europe. Tim Inskipp produced most of the information on changes to the west Palearctic list since the relevant *BWP* volumes had been produced. Dr Christopher Feare organized or wrote much of the text for the Sturnidae, Dr Denis Summers-Smith for the Passeridae, Dr David Snow for some of the Fringillidae and Emberizidae, Dr David Holyoak for a number of the Corvidae and Emberizidae, and Dr Jan Ekman for the Siberian Jay. Other accounts were produced, or co-authored, by Prof. T. R. Birkhead, Dr M Eens, Prof. P R Evans, Dr J R Harpum, Dr Nee Kang, Melanie Kershaw, Dr P K McGregor, C J Mead, Geoff Morgan, Prof. S J Peris, P J Sellar, and Prof. P J B Slater. Many others, too numerous to mention here, but who are included in the Acknowledgements, have contributed to the completion of this mammoth task.

At the Oxford University Press, John Manger and Bruce Wilcock have been patient yet persistent advocates of this great work, and, despite all the delays and frustrations, never lost their sense of vision as to its importance. The financial support of the Press, too, has been essential and the whole world of ornithology owes it its gratitude. Many other individuals and organizations have helped financially, and thanks are due to all of them, listed under Acknowledgements. Among these, the support of the Pilgrim Trust, the World Wide Fund for Nature, and the Royal Society for the Protection of Birds has been outstanding. Indeed, the RSPB's offer of a special edition for their members, at a difficult time in the work's pro-duction, may have tipped the scales decisively in favour of publication.

I said earlier that it was a team effort and so it was. So, thanks to the efforts of so many people, a real service to ornithology has been completed. I am sure that its exellence will stand the test of time.

William Wilkinson
Chairman, West Palearctic Birds Ltd
March 1993

ACKNOWLEDGEMENTS

In the preparation of this volume the continued substantial financial support of the Delegates of the Oxford University Press has been fundamental. Their generous help and understanding, together with the patience of other benefactors, particularly the Royal Society for the Protection of Birds, the Pilgrim Trust, and the World Wide Fund for Nature, who advanced loans long ago, are gratefully acknowledged.

Time has been generously given by many ornithologists and others throughout the world. For facilitating their labours, the Editorial Board are grateful to the Institute of Taxonomic Zoology (University of Amsterdam), and to the Edward Grey Institute of Field Ornithology (University of Oxford) where Dr L Birch of the Alexander Library has provided invaluable back-up over many years.

At the University of Cambridge the Directors of the Sub-Department of Animal Behaviour—Prof. P P G Bateson and his successor Dr E B Keverne—and Prof. G Horn, Head of the Department of Zoology, have afforded long-term and generous support through providing a working environment and all of the apparatus needed for preparation of the sonagrams and analysis of these and many other recordings. J Hall-Craggs, who has enjoyed free access to these facilities throughout the life of the project, is joined in this work by W T C Seale who has been solely responsible for voice illustration of twenty species. Dr P K McGregor and L M McGregor of the Behaviour and Ecology Research Group of the Department of Zoology, University of Nottingham, were responsible for the preparation of sonagrams for many species, and Prof. P J Slater for two. P J Sellar continued to give his time and expertise to amassing, indexing, and re-recording the quantities of material required for each volume, along with the use of his studio and much of his own equipment. R Ranft, Curator of the British Library of Wildlife Sounds, has been always ready and willing to give help and to seek out the best examples of items needed. Gratitude and keen appreciation are due to the members of the Oxford University Press who patiently guided the many sonagrams through the processing that transforms them into intelligible figures. The Voice text for many species has benefited from comments by L Svensson.

Where recordings used for sonagrams are available as published gramophone records or cassettes, references are given in the captions as follows:

Ferdinand, L (1991) *Bird voices in the North Atlantic*. Tórshavn.

Mild, K (1987) *Soviet bird songs*. Stockholm.

Mild, K (1990) *Bird songs of Israel and the Middle East*. Stockholm.

Roché, J-C (1964) *Guide sonore des oiseaux d'Europe* 1; (1966) 2; (1970) 3. Institut Echo, Aubenas-les-Alpes, France.

Roché, J-C (1968) *A sound guide to the birds of north-west Africa*. Institut Echo, Aubenas-les-Alpes, France.

Roché, J-C (1990) *All the bird songs of Europe*. Sittelle, La Mure.

Sveriges Radio (1972–80) *A field guide to the bird songs of Britain and Europe* by S Palmér and J Boswall 1–15 (discs).

Swedish Radio Company (1981) *A field guide to the bird songs of Britain and Europe* by S Palmér and J Boswall 1–16 (cassettes).

For recordings which have not been published commercially, assistance in contacting the original recordists may often be obtained from the Curator, British Library of Wildlife Sounds, National Sound Archive, 29 Exhibition Road, London SW7 2AS.

A vital part was played by the correspondents who provided much of the basic data on status, distribution, and populations for species occurring in the following regions:

ALBANIA Dr E Nowak
ALGERIA E D H Johnson
AUSTRIA Dr H Schifter
AZORES Dr G Le Grand
BRITAIN R Spencer
BULGARIA T Michev
CAPE VERDE ISLANDS C J Hazevoet
CHAD C Erard
CYPRUS P R Flint, P F Stewart
CZECHOSLOVAKIA Dr K Hudec
DENMARK U Gjøl Sørensen
EGYPT P L Meininger, W C Mullié, S M Goodman
FAEROES Dr D Bloch, S Sørensen
FINLAND Dr O Hildén
FRANCE R Cruon
GERMANY A Hill, S Schnabel
GREECE G I Handrinos
HUNGARY G Magyar
ICELAND Dr Æ Petersen
IRAQ H Y Siman
IRELAND C D Hutchinson
ISRAEL H Shirihai
ITALY P Brichetti, B Massa
JORDAN I Andrews, P A D Hollom, D I M Wallace
KUWAIT Prof. C W T Pilcher
LEBANON Lt-Col A M Macfarlane
LIBYA G Bundy
MADEIRA P A Zino, G Maul
MALI Dr J M Thiollay, B Lamarche
MALTA J Sultana, C Gauci
MAURITANIA R A Williams, J Trotignon, B Lamarche

MOROCCO Dr M Thévenot, J D R Vernon
NETHERLANDS Drs C S Roselaar
NIGER Dr J M Thiollay
NORWAY V Ree
POLAND Dr A Dyrcz, Dr L Tomiałojć
PORTUGAL R Rufino, G A Vowles
RUMANIA P Weber
SAUDI ARABIA M C Jennings, *Atlas of the Breeding Birds of Arabia* project
SPAIN A Noval
SWEDEN L Risberg
SWITZERLAND R Winkler
SYRIA Lt-Col A M Macfarlane
TURKEY R P Martins
USSR Prof. L S Stepanyan, Prof. A F Kovshar', V Konstantinov
YUGOSLAVIA V F Vasić

We also wish to thank all those who made available photographs, sketches, and published material on which the drawings illustrating the Social Pattern and Behaviour section were based; their names are given at the end of the relevant accounts.

F E Warr carried out much of the essential basic literature research, and L Cruickshank gave valuable assistance with typing. For help with translations we are grateful to Dr P Ahlberg, J E Arévalo, B Arroyo, M Cellier, Dr S Haywood, N Hillcoat-Kayser, Dr V Karpov, J King, T Köhler, M Kohlhaas, Pak Fook Chinese Restaurant (Oxford), E R Potapov, L Rode, P Stephenson, Dr E Syroechkovski, Dr T Székely, J C Yoo, L Zadorina, and M Zernicka-Goetz.

Finally, we are greatly indebted to the following, who assisted in many ways too diverse to specify in detail though credits are given in the text where appropriate:

G H Acklam, Dr V C Ambedkar, I J Andrews, Dr J S Ash, G Åström, P Baldwin, L Batt, W R R de Batz, M Beaman, Dr G S Bel'skaya, Dr C W Benkman, P Bennett, Prof. H-H Bergmann, H Biebach, H van den Bijtel, J Bjørn Andersen, B Bjørnsen, W J Bock, J H R Boswall, J Bowler, P Brichetti, B de Bruin, C de Bruyn, Dr I D Bullock, Dr I Byrkjedal, M Cellier, A E Chapman, K Colcomb-Heiliger, Dr N J Collar, P R Colston, P J Conder, Dr C de la Cruz Solis, A Dawson, Dr R W R J Dekker, H Delin, P von Dom, P J Dubois, Dr A Dyrcz, E B Ebels, S Eck, Dr C C Eley, Dr S Elliott, J Elmberg, Dr R van den Elzen, R E Emmett, P Enggist-Düblin, Dr A Evans, J Evans, Dr P G H Evans, Dr C J Feare, G D Field, Prof. V E Flint, Free University of Amsterdam, R J Fuller, D Goodwin, D Gosney, H Göttgens, A Grabher Meyer, M Grahn, G Le Grand, A Gretton, G I Handrinos, P S Hansen, Dr D G C Harper, S Harrap, Dr I R Hartley, C J Hazevoet, M Herremans, I and M Hills, Dr O Hogstad, Dr G Högstedt, D J Holman, P R Holmes, Dr D T Holyoak, P D Housley, Prof. V D Ilyichev, M P S Irwin, Dr V V Ivanitski, Dr B Ivanov, J Jackson, H Jännes, D C Jardine, Dr P J Jones, J Jukema, E I Khlebosolov, the late B King, J King, J L Kitwood, Dr A G Knox, Dr A J Knystautas, J Langer, P A Lassey, P G Lee, Dr G Le Grand, L R Lewis, C M Liebregts-Hooker, M Limbert, F Lindgren, M L Long, F de Lope Rebollo, Dr V M Loskot, F Lovaty, S C Madge, Dr W Mann, Dr A Martín, R P Martins, J A McGeoch, D McGinn, Dr G Mauersberger, C J Mead, B S Meadows, Dr G F Mees, Prof. H Mendelssohn, T Michev, Dr P Mierauskas, Dr A Mikkonen, K Mild, Dr A P Møller, N C Moore, M-Y Morel, A Motis, National University of Singapore, E Nemeth, Dr I A Neufeldt, Dr I Newton, Dr B Nicolai, Dr E Nieboer, Dr I C T Nisbet, the late M E W North, T B Oatley, E Olafsson, Dr G Olioso, U Olsson, Dr S L C O'Malley, Prof. D F Owen, J Palfery, Dr E N Panov, Dr I J Patterson, J Paul, Dr S J Peris, Dr Æ Petersen, G Pétursson, F Pieters, R Pinxten, E R Potapov, Prof. R L Potapov, Dr G R Potts, T Prins, L Profirov, J A Ramos, R Reijnders, A Renard, Dr G Rheinwald, S Rick, Dr D Robel, T J Roberts, V A D Sales, Dr D A Scott, R Scudamore, Y M Shchadilov, B C Sheldon, M Shepherd, H Shirihai, R D Smith, Dr T Stawarczyk, Prof. L S Stepanyan, Prof. B Stephan, J D Summers-Smith, L Svensson, Dr T Székely, P Tatner, C Thomas, J Tigner, L M Tuck, M Ullman, the late Prof. B N Veprintsev, B D Waite, Dr K Walasz, L K Wang, J M Warnes, Dr A Watson, F E Warr, R B Warren, Dr J Wattel, T C White, D Yakutiel, Dr Y Yom-Tov, Dr V A Zubakin, R L Zusi, E Zwart.

CITATION

The editors recommend that for references to this volume in scientific publications the following citation should be used: Cramp, S and Perrins, C M (eds) (1994) *The Birds of the Western Palearctic* Vol. IX.

Order PASSERIFORMES (continued)

Family PARULIDAE New World wood-warblers

Small 9-primaried oscine passerines (suborder Passeres); though unrelated to Old World warblers (Sylviidae), whose niche many of them occupy in New World, parulids as a group also known just as 'warblers' in North America—though a few species bear other names (e.g. 'chat', 'redstart'). Highly active; often arboreal though some (including Ovenbird *Seiurus aurocapillus* and Northern Waterthrush *S. noveboracensis*) are terrestrial. Feed mainly on insects, gleaned from foliage and (in a number of species) by aerial flycatching; fruit (berries), seeds, and nectar also taken by some species. Usually solitary but some form flocks outside breeding season or on migration. Except as vagrants, occur only in New World (from Alaska to southern South America); northerly species migratory. 126 species in 29 genera of which 19 species of 8 genera accidental in west Palearctic.

Sexes virtually identical in size. Bill straight, slender, and pointed in many parulids (with inconspicuous rictal bristles) but wider and flatter (with well-developed rictal bristles) in more specialized flycatching ones. Wing often fairly long and pointed in migratory species, shorter and rounder in non-migratory; 9 primaries (p10 minute and hidden). Tail of medium size or fairly long; often slightly rounded. Leg and foot rather short and slender. Parulids unusual in that head-scratching method varies even within same genus, some species (probably majority) using indirect method only, others direct method only, and 1 species (at least) both methods (see especially Burt and Hailman 1978). Bathing apparently by in-out manner (Slessors 1970) like babblers (Timaliidae). When sunning, use lateral and spreadeagle postures typical of most passerines (see Simmons 1986a). Direct (active) anting reported in at least 2 species (Whitaker 1957; Potter 1970; Potter and Hauser 1974).

Plumages typically yellow, olive, or blue-grey, with patches or streaks of contrasting yellow, orange, red, black, or white, especially on head, rump, and chest; wing often with contrasting white, yellow, or red bars; tail often with white, yellow, or red spots. ♂ brighter than ♀ in most North American species; seasonal difference sometimes marked.

Close affinity to tanagers (Thraupidae) and buntings (Emberizidae)—but not to vireos (Vireonidae)—is indicated by DNA evidence (Sibley and Ahlquist 1990), even to the extent that a boundary between members of Parulidae and Thraupidae is hard to draw. See Fringillidae, Thraupidae, and Emberizidae for further discussion of relations within 9-primaried assemblage.

Mniotilta varia **Black-and-white Warbler**

PLATE 1
[facing page 64]

Du. Bonte Zanger   Fr. Paruline noire et blanche   Ge. Kletterwaldsänger
Ru. Пегий певун   Sp. Reinita trepadora   Sw. Svartvit skogssångare

Motacilla varia Linnaeus, 1766

Monotypic

Field characters. 11·5–13 cm; wing-span 20·5–22·5 cm. Slightly smaller than Blackpoll Warbler *Dendroica striata*, with subtly different structure: rather longer bill with noticeably sharp culmen, shorter square tail, and long toes. Rather small but lithe, bark-creeping Nearctic warbler, with black upperparts striped white, and white underparts streaked black; white central crown-stripe diagnostic. Sexes dissimilar, especially when breeding; slight seasonal variation in ♂. 1st-winter separable.

ADULT MALE. Moults: July–August (complete); February–April (head, body, forewing). At all seasons, broad band of black across upper rear ear-coverts and jet-black lines and spots from breast along flank to under tail-coverts contrast dramatically with white ground-colour. Throat black in breeding plumage, joining breast-lines; dark-spotted or white in winter. Bill and legs blackish. ADULT FEMALE. Black on ear-coverts confined to narrow stripe behind eye. Throat always white, and dusky lines and spots on underparts less contrasting than ♂. Fore-supercilium and ground of rear flank, vent, and under tail-coverts tinged warm buff to brown. Lower mandible pale brownish-horn; legs dusky. FIRST-WINTER. Resembles winter adult. At close range, dull brownish tone of carpal area and flight-feathers may show (see also Plumages).

Unmistakable among known transatlantic vagrants. Breeding ♂ *D. striata* somewhat similar but with wholly black crown lacking white central stripe, unstriped mantle, and bright brown legs. *M. varia* has unique habit among Parulidae of persistently creeping up and down tree trunks and along branches, with actions recalling both nuthatch *Sitta* and treecreeper *Certhia*; unlike latter, does not depress tail for support. Hops and runs. Often confiding.

Calls include quite loud, hard 'chick' or 'tik', weak thin 'tsip' or 'tzit', and hiss in alarm.

Habitat. Breeds across cool to warm temperate Nearctic lowlands, on hillsides or ravines in all woodland types from mature deciduous or mixed stands to (more locally) northern conifer forests and also second growth. Forages on main branches or trunks of trees, rather than in foliage. (Pough 1949.) On Great Plains, generally found in semi-open upland stands of deciduous or coniferous forest, such as oaks *Quercus* and junipers *Juniperus* of scrubby character. Also lives in riverside forests by grasslands, and in hillside or ravine groves with thin understorey. (Johnsgard 1979.) Found in Canada in moist woodlands, and in willow *Salix* or alder *Alnus* shrub growth on edges of coniferous woodland, foraging in lower or middle branches (Godfrey 1979).

Distribution. Breeds in North America east of Rocky Mountains, from north-east British Columbia to northern Newfoundland, south to central Texas and central Georgia. Winters from northern Mexico and south-east USA south through Central America and West Indies to northern South America.

Accidental. Iceland, Faeroes, Britain, Ireland.

Movements. Migratory, with small overlap of breeding and winter ranges in southern Texas. Probably the most widely common parulid warbler across its winter range. (Keast and Morton 1980; Rappole *et al.* 1983; Pashley and Martin 1988.)

Migration begins early, in both seasons. Birds appear outside southern breeding areas from July (Oberholser 1974; Kale and Maehr 1990), with some records as far as West Indies (Brudenell-Bruce 1975; Bond 1985). Migration front moves rapidly south in August, reaching Honduras and Colombia (Monroe 1968; Hilty and Brown 1986), but northern populations are far behind. In Canada and northern USA, main passage begins only in late August (Stewart and Robbins 1958; Tufts 1986; Janssen 1987); west of Ontario, most have left by end of September (Houston and Street 1959; Sadler and Myres 1976), but migrants linger through October all over north-east (Bent 1953; Sprague and Weir 1984; Tufts 1986). Migration continues through Central and South America in October (e.g. Rogers *et al.* 1986a). Southward movement extends from Rockies to Atlantic, and even to Bermuda, continuing south through Mexico and Central America as well as across Gulf of Mexico and through Florida (Bent 1953; Peterson and Chalif 1973). Winter site-fidelity widely reported from ringing, e.g. in southern Mexico, Panama, and Puerto Rico, with recoveries up to 4 years later (Ely *et al.* 1977; Rappole *et al.* 1983; Faaborg and Arendt 1984).

Spring migration begins early March, with first records outside wintering areas (Sutton 1967; Potter *et al.* 1980; Alcorn 1988). By early April, South American winter range is largely vacated (Ridgely 1981; Hilty and Brown 1986), with arrivals north to Kentucky and West Virginia (Mengel 1965; Hall 1983). Early migrants before mid-April are mostly west of Appalachians, from Colorado to Ontario (Bailey and Niedrach 1967; Quilliam 1973), but are also reported occasionally from Nova Scotia with 'overshoots' by other parulids (Tufts 1986). First arrivals by end of April in Canada, from New Brunswick to Manitoba (Bent 1953; Squires 1976), and migration front expands northward to reach northern breeding areas mid-to late May (Munro and Cowan 1947; Peters and Burleigh 1951; Todd 1963). Spring migration is reverse of that in autumn, including both trans- and circum-Gulf movements, on broad front (Bent 1953; Stevenson 1957).

Birds stray widely west of Rockies from California to British Columbia (Garrett and Dunn 1981; Cannings *et al.* 1987) almost annually in both seasons, but these are few in number so are evidently wanderers. Those that reached arctic Alaska in October (Kessel and Gibson 1978) and Churchill (Manitoba) in June (Godfrey 1986) were certainly lost.

Rare autumn vagrant to Atlantic seaboard of west Palearctic, mainly Britain and Ireland where 11 records up to 1989, majority September–October, but noted also March and December (Dymond *et al.* 1989).   AJE, PRC

Voice. See Field Characters.

Plumages. ADULT MALE BREEDING. Broad black stripe at each side of forehead, crown, and nape, separated by white median stripe over crown and nape, narrowing on forehead to thin line. Discrete white and black stripes on nape merge into somewhat more irregular and more closely spaced black-and-white streaks on mantle, scapulars, and back; part of streaks on upper mantle and back sometimes dark grey. Rump either uniform black or streaked black, dark grey, and white; upper tail-coverts black, lateral ones with broad white outer fringe. Broad white supercilium extends from nostril to upper side of neck. Eye-ring conspicuous, white, broken by black at front and rear. Lore, upper cheek, and ear-coverts uniform black, bordered below by white stripe running from base of lower mandible to rear of lower cheek. Side of neck finely streaked black and white. Upper chin white, lower chin and all throat black, feathers narrowly fringed white when plumage fresh, showing as fine streaking or mottling; upper chest and side of breast black with more distinct white fringing, pattern merging into that of flank and under tail-coverts, which are white with contrasting black streaks *c.* 4 mm wide (on flank) or black spots (on coverts). Mid-belly and vent white; thigh black and white. Central pair of tail-feathers (t1) black with grey fringe, t2–t6 dull black with grey fringe along outer web (wider and paler grey towards base, narrow and white on t6); inner web of t5 with large white spot on tip, inner web of t6 with white tip *c.* 1·5–2·5 cm long. Flight-feathers greyish-black; narrow and sharp fringes along outer webs of primaries grey-white, distal 30–50% of outer webs of secondaries fringed pale ash-grey. Tertials deep black, 2 longer ones broadly fringed white on outer web (covering almost entire outer web in longest feather). All upper wing-coverts and bastard wing deep black, median with broad white tips, forming wing-bar *c.* 5 mm wide, greater with slightly narrower white tips, forming bar 2–3 mm wide, innermost greater (tertial coverts) with broad white outer fringe; longest feather of bastard wing with white inner border, middle feather with white tip which shows black centre. Under wing-coverts, axillaries, and broad basal inner borders of flight-feathers white, dark grey bases of some coverts visible. When plumage fresh, all black of plumage (except of tail and flight-feathers) slightly glossed blue; when worn, some dark grey of feather-bases visible, especially on upperparts, disrupting regular black and white streaking. ADULT FEMALE BREEDING. Upperparts, tail, and wing as adult ♂ breeding, but white streaking on mantle, scapulars, and (especially) back and rump tinged buff; white fringes on tertials and on inner web of longest feather of bastard wing narrower; white on tips of outer tail-feathers, on median and greater coverts, and on middle feather of bastard wing slightly less extensive (wing-bar on median coverts *c.* 3 mm wide, on greater *c.* 2 mm). Side of head and all underparts quite different, without solid black spots except for under tail-coverts. Supercilium and complete eye-ring pale buff or yellow-white. Lore grey. Upper cheek and ear-coverts mottled grey and pale buff, upper and rear border of ear-coverts sometimes uniform dusky grey. Side of neck off-white with dull black-and-grey mottling. Entire underparts buff (when fresh) to pale isabelline (when worn), grading to pale buff or off-white on central throat and on mid-breast, belly, and vent; side of breast with some dull black spots, chest with indistinct grey or olive-grey specks, sometimes forming faint gorget; flank with ill-defined grey or olive-grey streaks, becoming wider and more well-defined olive-brown on lower flank; under tail-coverts with bold dull black central spot. Under wing-coverts and axillaries tinged buff. ADULT MALE NON-BREEDING. Like adult ♂ breeding, but chin, throat, and mid-chest white, sometimes with a little black of feather-bases visible; unlike adult ♀, lore, upper cheek, and ear-coverts black, side of breast and flank broadly streaked black and white; white ground-colour of underparts slightly tinged pale pink-buff when plumage quite fresh. ADULT FEMALE NON-BREEDING. Like adult ♀ breeding, but ground-colour of side of head and all underparts rather more intensely buff when plumage fresh, partial dark surround to ear-coverts less clear or absent, gorget (if any) and streaks on flank less dark and even less sharply defined. JUVENILE. Like adult ♀ non-breeding, but streaks of upperparts (including crown-stripes) less sharply defined, black-brown and buff instead of black and off-white or isabelline; underparts greyish-buff with faint grey-brown streaks or almost unstreaked; for tail and wing, see First Adult Non-breeding. FIRST ADULT NON-BREEDING. Like adult non-breeding (sexes thus differ in ground-colour of side of head and underparts and in extent and contrast of streaking), but tail, flight-feathers, tertials, greater upper primary coverts, bastard wing, and sometimes a number of outer greater upper wing-coverts still juvenile, markedly browner and more worn than adult feathers at same time of year; in particular, tertials, primary coverts, bastard wing, and (if any) outer greater coverts markedly browner than neighbouring 1st adult median and inner greater coverts; also, tail-feather tips more pointed (less truncate than in adult), white on juvenile tertials and bastard wing less extensive than in adult (especially in ♂), white on outer tail-tips sometimes more restricted. In ♂, lore and upper ear-coverts dark grey or dull black, lower ear-coverts and upper cheek off-white or pale grey (in adult ♂, uniform black); mantle, scapulars, back, and rump less regularly streaked, black spots and streaks on side of breast and flank slightly narrower and shorter, not reaching to feather-tip. In ♀, black of crown-stripes, mantle, and back partly tinged brown or grey; lore, upper cheek, and ear-coverts buffish-grey, side of neck buff-white with faint darker specks; streaks and spots on side of breast and flank short, rather narrow, grey (less dark and olive than adult ♀ non-breeding). FIRST ADULT BREEDING. Like adult breeding, but tail, flight-feathers, primary coverts, and variable number of feathers of bastard wing and tertials still juvenile, brown and rather worn (as in 1st adult non-breeding); in adult, retained non-breeding flight-feathers, tertials, primary coverts, and bastard wing often slightly browner-black than newer tertials and coverts of breeding plumage, but difference less marked than in 1st adult.

Bare parts. ADULT. Iris brown to dark brown. Bill dark plumbeous, blue-black, or brown-black, culmen darkest, cutting edges paler blue-grey; in adult ♂ breeding, black. Leg and foot dusky green, dark brown with green shade, or dark brown-green, almost black in adult ♂ breeding. JUVENILE, FIRST ADULT. Iris dark brown. Bill dark grey, paler grey on cutting edges, basal 70% of lower mandible pale greyish-flesh, flesh-brown, or light brown. Leg and foot olive-green or brownish-green. Adult colours obtained during winter and spring.

Moults. ADULT POST-BREEDING. Complete; primaries descendent. On breeding grounds, July–August (Pyle *et al.* 1987). Migrant birds in Mexico, mid-August, had moult completed (RMNH). ADULT PRE-BREEDING. Partial: head, body, lesser and median upper wing-coverts, and variable number of tertials. In winter quarters, February–April. POST-JUVENILE. Partial: head, body, and lesser, median, and variable number of greater upper wing-coverts. July–August, on breeding grounds. (Dwight 1900; Pyle *et al.* 1987; RMNH, ZMA.) In birds examined, several from late July and early August had moult already completed (RMNH, ZMA). FIRST PRE-BREEDING. Like adult pre-breeding.

Measurements. ADULT, FIRST ADULT. Whole geographical range, all year; skins (RMNH, ZMA). Bill (S) to skull, bill (N) to distal corner of nostril; exposed culmen on average 3·4 shorter than bill (S).

WING	♂	71·7 (2·02; 37) 68–79	♀	68·8 (1·89; 15)	66–72
TAIL		48·7 (1·84; 20) 46–52		47·4 (1·33; 15)	45–50
BILL (S)		14·7 (0·73; 33) 13·5–16·0		15·0 (1·10; 13)	13·5–16·3
BILL (N)		9·2 (0·71; 20) 8·2–10·5		9·2 (0·69; 14)	8·2–10·3
TARSUS		17·1 (0·36; 19) 16·6–17·8		17·6 (0·55; 15)	16·8–18·5

Sex differences significant, except for bill. Ages combined, though retained juvenile wing and tail of 1st adult on average shorter than in adult. Thus, in ♂: wing, adult 72·1 (1·40; 19) 70–75, 1st adult 71·3 (2·51; 18) 68–79 (mainly 69–74, only once 79); tail, adult 49·5 (1·58; 12) 47–52, 1st adult 47·4 (1·52; 8) 46–50 (RMNH, ZMA). See also Woodford and Lovesy (1958).

Weights. Pennsylvania (USA): (1) July–September, (2) April–June (Clench and Lebermann 1978). (3) Georgia (USA), early October (Johnston and Haines 1957).

(1) AD	♂	11·8 (2·00; 9) 8·8–15·2	♀	10·5 (0·67; 25) 9·3–12·2
1ST AD		10·3 (0·81; 45) 7·9–11·8		10·3 (1·02; 34) 8·3–14·0
(2)		10·6 (0·88; 15) 9·1–12·2		10·6 (0·88; 19) 9·0–12·7

(3)   11·2 (— ; 10) 9·8–12·3   11·5 (— ; 14) 10·0–14·1

Pelee Island (Lake Erie, Canada), May: ♂♂ 10·1, 12·6; ♀♀ 10·5, 12·6 (Woodford and Lovesy 1958). Averages, Illinois (USA), September: adult ♂ 10·7 (n=2), adult ♀ 10·2 (n=4), 1st adult ♂ 12·2 (n=2) (Graber and Graber 1962). Georgia and South Carolina (USA), mainly August: ♂ 13·8 (n=5), ♀ 10·9 (n=1) (Norris and Johnston 1958). Ohio (USA): May, ♂ 10·8; August ♂ 11·2 (n=2), ♀ 9·2 (Baldwin and Kendeigh 1938). Kentucky (USA), April–September: ♂ 10·8 (3) 10·4–11·0, ♀♀ 11·2, 11·8 (Mengel 1965).

Mexico: October, ♂♂ 9·3, 9·5; April, ♂ 11·0, ♀ 10·0 (RMNH). Belize: October, ♂ 9·4 (ZMA); February, ♀ 9·1 (Russell 1964); autumn, 9·3 (1·0; 100) (Mills and Rogers 1990). Average on Jamaica approximately constant September–April, c. 9–10, range c. 8·1–11·5; heavier in May (Diamond et al. 1977, which see for details). Exhausted ♂ on ship off Florida, April: 6·9 (ZMA). For weight gain in birds trapped twice on East Ship Island (southeast Mississippi, USA), spring, see Kuenzi et al. (1991).

For fat content in autumn and spring, USA, see Caldwell et al. (1963).

Structure. Wing relatively long, broad at base, tip bluntly pointed. 9 primaries: p7–p8 longest, p9 0·5–2·5 shorter, p6 1–3, p5 5–9, p4 8–13, p3 10–15, p2 12–17, p10 15–19. Of 10 birds examined, none had any trace of even a tiny reduced p10. Outer web of p6–p8 emarginated (sometimes faint on p6), inner web of (p7–)p8 with fairly distinct notch, sometimes faintly on p6 and p9 also. Tip of longest tertial reaches tip of p1–p2 in closed wing. Tail of average length, tip square; 12 feathers. Bill long, visible culmen almost equal to head length; straight, fine, acutely pointed. Ridge of culmen rather sharp, middle and tip of bill compressed laterally. Nostril small, oval, partly protected by membrane above. Many short and soft bristles at base of both mandibles. Leg short, toes long; strong. Middle toe with claw 16·1 (12) 15·5–16·7; outer toe with claw c. 76% of middle with claw, inner c. 69%, hind c. 96% (markedly long); bases of front toes partly fused.

Geographical variation. Perhaps slight variation in size: wing of ♂♂ breeding east coast of USA (Carolina to Massachusetts) and of wintering ♂♂ from Florida and Cuba 70·8 (13) 68–74, of ♂♂ breeding Great Lakes area and of ♂♂ wintering Guatemala to Colombia 72·3 (17) 69–75 (once 79) (C S Roselaar).

For relationships with other Parulidae, see Parkes (1973).

CSR

Vermivora chrysoptera **Golden-winged Warbler**   PLATE 1
[facing page 64]

Du. Geelvleugelzanger   Fr. Paruline à ailes dorées   Ge. Goldflügel-Waldsänger
Ru. Золотокрылый пеночковый певун   Sp. Reinita alidorada   Sw. Gulvingad skogssångare

Motacilla chrysoptera Linnaeus, 1766

Monotypic

Field characters. 11·5–12 cm; wing-span 19–20·5 cm. Noticeably smaller than Blackpoll Warbler *Dendroica striata*, with wings and tail proportionately 5–10% shorter. Rather small, quite tit-like Nearctic warbler, with mainly blue-grey upperparts strikingly marked by yellow forecrown and wing-blaze; bold dark face-mask and bib; white panels on outer tail-feathers. Sexes dissimilar; no obvious seasonal variation. 1st-winter difficult to separate (see Plumages).

ADULT MALE. Moults: July–August (complete). Striking jet-black mask (from bill through eye to rear ear-coverts) and deep bib; feathers of bib tipped grey when fresh. Large brilliant golden-yellow patch on forecrown, bordered below by white supercilium. Fringes of tertials and inner secondaries blue-grey to olive, forming distinct panel when fresh but looking faded and browner when worn. Bill rather long, fine, and tapering; black when breeding, fading to greyish-horn. Legs mainly grey. ADULT FEMALE. Noticeably duller than ♂, with dusky mask and bib, greenish-yellow forecrown, black division of yellow wing-blaze on median coverts, and greener panel on inner flight-feathers. Bill paler horn on cutting edges than ♂.

Unmistakable. Behaviour recalls tit *Parus*; will forage in outer foliage, hanging downwards to look for invertebrates.

Call a short 'chip', not noted as distinctive.

Habitat. Breeds in eastern Nearctic lowland woods of temperate zone. Prefers edges of mature deciduous woods of tall oaks *Quercus*, and maples *Acer* shading meadows grown up with briars, tall herbage, bushes and grass, ideally by clear flowing stream (Forbush and May 1955). Sometimes breeds in overgrown clear-felled woodland patches or abandoned pastures in moist places. Nests on or near ground but lives at all levels up to treetops (Pough 1949). In Great Plains, occupies both upland forest such as aspen *Populus* and lowland edge habitats and openings, such as overgrown pastures or brushy fields with ferns and other moisture-loving plants (Johnsgard 1979). In South American winter quarters occupies equivalent habitats but favours higher altitudes in rain and cloud forests at 1000–3000 m, and dwarf forest in the higher region, foraging at low levels (Schauensee and Phelps 1978).

Distribution. Breeds in North America from south-east Manitoba and southern Ontario south to eastern Tennessee, northern Georgia, and south-east Pennsylvania. Winters in Central and South America, from Yucatán peninsula south to central Colombia and northern Venezuela.

Accidental. Britain: ♂, Maidstone (Kent), 24 January to 10 April 1989 (Doherty 1992).

Movements. Migratory.

Movements in late summer and autumn are inconspicuous. Some birds leave breeding areas in late July or early August (Bailey 1955; Stewart and Robbins 1958; Janssen 1987). By mid-September, most have left Ontario and northern states (Bull 1974; Sprague and Weir 1984; Janssen 1987), and some are arriving in wintering areas south of Caribbean (Monroe 1968; Hilty and Brown 1986; Stiles and Skutch 1989). Most migration through southern USA is in September, and few remain into October except in Louisiana (Bent 1953; Lowery 1974; Imhof 1976; Potter *et al.* 1980). Migration is mostly east of Great Plains and west of Appalachians, continuing across Gulf of Mexico; rare in Mexico and West Indies (Peterson and Chalif 1973; Rappole *et al.* 1979; American Ornithologists' Union 1983).

Spring route is reverse of autumn, but includes more movement through eastern Mexico (Stevenson 1957; Rappole *et al.* 1979). Migrants leave wintering areas late March to April (Schauensee and Phelps 1978; Hilty and Brown 1986; Stiles and Skutch 1989). Earliest birds reach Texas and east to Carolinas by mid-April (Oberholser 1974; Imhof 1976; Potter *et al.* 1980), and appear in southern Canada and New England by end of April (Bailey 1955; Quilliam 1973; Janssen 1987). Most reach northern breeding areas early to mid-May (Stewart and Robbins 1958; Sprague and Weir 1984; Bohlen 1989).

Rare vagrant north-west to southern Saskatchewan and north-east to New Brunswick and Nova Scotia (Squires 1976; Godfrey 1986; Tufts 1986). For vagrancy to west Palearctic, see Distribution. AJE

Voice. See Field Characters.

Plumages. ADULT MALE. Forehead to centre of crown bright yellow, bordered at side by broad white supercilium from nostril to above middle of ear-coverts. Hind-crown, upper side of neck, and all remainder of upperparts uniform medium or dark bluish-grey, sharply demarcated from yellow of mid-crown. Lore, upper cheek, and ear-coverts black, forming broad and distinctly defined mask, bordered below by white streak over lower cheek, narrowest at base of lower mandible, wider towards rear. Chin and throat black, forming distinct triangular bib. Lower side of neck, side of breast, flank, and thigh light or medium ash-grey; remainder of underparts white, sometimes suffused grey on chest.

Tail dark bluish-grey, outer web of outermost pair (t6) and inner web of t2–t6 sooty-black; small white subterminal patch sometimes on inner web of t3, a larger patch on t4 and patch increasing in size outwards: 15–20 mm of inner web of t4 white, 20–25 mm of inner web of t6; white patch bordered by black rim *c.* 1 mm wide on tip of t4, 3–10 mm on tip of t6, but inner webs of t4–t5 frequently have all-white tip. Flight-feathers, tertials, and greater upper primary coverts greyish-black, outer webs of secondaries fringed medium blue-grey, those of primaries pale blue-grey, those of primary coverts less sharply defined medium blue-grey; outer webs of tertials extensively washed medium blue-grey, sometimes partly washed green. Lesser upper wing-coverts medium blue-grey, median and greater contrastingly bright yellow, forming large patch on wing. Under wing-coverts and axillaries white. *In fresh plumage* (early autumn), side and rear of crown may show some narrow green feather-tips, black bib with faint and narrow white feather-tips. *In worn plumage* (spring and early summer), upperparts duller grey, less bluish; underparts mixed grey and dirty white; tertials almost uniform dusky grey; some black of bases of median and greater upper wing-coverts sometimes visible. ADULT FEMALE. In fresh plumage (autumn), forehead and crown yellowish-green, pure yellow of feather-bases of forehead and forecrown concealed; remainder of upperparts medium blue-grey, feather-tips washed green (hardly so on upper tail-coverts), partly concealing grey. Supercilium narrow, especially in front, where partly tinged yellow, tinged grey at rear, where gradually merging into grey side of neck. Lore, upper cheek, and ear-coverts dark grey or plumbeous, mask contrasting less with side of neck than in adult ♂. Rather broad but ill-defined dirty white stripe from base of lower mandible to rear of lower cheek. Chin and throat medium blue-grey, slightly darker than light ash-grey side of breast, flank, and thigh. Chest to under tail-coverts white with slight yellow suffusion. Tail and wing as adult ♂, but yellow on tips of greater and median upper wing-coverts more restricted, on each median covert appearing as large yellow triangle with contrasting black base; on each greater covert confined to tip of outer web, with base of outer web green and entire inner web black; outer webs of tertials greener. *In worn plumage* (spring), forehead and forecrown yellow, grading into green on hindcrown; remainder of upperparts blue-grey with traces of green (latter mainly on lower mantle, back, and tertials); supercilium and pale stripe on lower cheek more distinct; bib mottled paler and darker grey, remainder of underparts mottled grey and off-white, purer white restricted to lower belly and under tail-coverts. JUVENILE. Upperparts uniform greyish-olive or brownish-olive, more olive-green on rump; side of head and neck light greyish-olive with off-white stripes above lore and over lower cheek; off-white eye-ring; chin white, throat greyish-olive, remainder of underparts pale yellow; median and greater upper wing-coverts grey with pale yellow or off-white tips; remainder of wing and tail as 1st adult. FIRST ADULT MALE. Like adult ♂, but tips of retained juvenile tail-feathers distinctly pointed, less broad and truncate; retained juvenile secondaries and primary coverts have outer fringes partly green; outer webs of tertials strongly washed green. Yellow feathers of forehead and forecrown narrowly tipped green, those of side and rear of crown extensively green, blue-grey of hindneck to back with green feather-tips, partly concealing blue-grey; upperparts of some birds rather similar to adult ♀, but mask and bib black, not plumbeous-grey. Upper chin white, lower chin and throat black with narrow white feather-tips; remainder of underparts as adult ♂, but white parts washed pale yellow. Greater and median upper wing-coverts as adult ♂, but black at bases of median coverts and on inner webs of greater coverts slightly more extensive, sometimes partly visible; lesser coverts partly tipped green.

In worn plumage, similar to worn adult ♂, but rear of yellow on crown partly green, yellow not sharply demarcated from blue-grey hindneck; traces of green on lower mantle, back, and tertials; chin partly white; traces of yellow on chest and side of breast; tips of tail-feathers more pointed; secondaries and lesser coverts with traces of green. FIRST ADULT FEMALE. Like adult ♀, but flight-feathers, primary coverts, tertials, and tail still juvenile, as in 1st adult ♂. Upperparts (except upper tail-coverts), side of neck, and lesser upper wing-coverts extensively washed green (on average, less blue-grey visible than in adult ♀); part of chin whitish; chest to vent washed yellow (brighter and more extensively than in 1st adult ♂); some green-yellow feather-tips in grey of throat and side of breast, less uniform than in adult ♀. *In worn plumage*, yellow of forehead and crown mottled green, white of underparts mottled yellow, otherwise as in fresh plumage.

Bare parts. ADULT, FIRST ADULT. Iris brown or dark brown. Bill greyish-horn or grey-brown, tinged flesh-grey at base, culmen dark blue-grey, slate-grey, or horn-black. Leg and foot brown-green, dark grey with green cast, or dark horn-grey, soles and rear of tarsus paler yellow-horn or flesh-grey; in breeding ♂, bill black with dark horn cutting edges, leg and foot blackish grey-horn, foot slightly paler plumbeous-grey. JUVENILE. Iris brown. Bill dusky grey, base of lower mandible flesh-grey. Leg and foot dark green-grey. (BMNH, RMNH, ZMA.)

Moults. ADULT POST-BREEDING. Complete; primaries descendent. On breeding grounds, late June and July (Dwight 1900) or July–August (Pyle *et al.* 1987). In Michigan, one examined had moult already completed 20 July, another 6 August (RMNH; ZMA). No pre-breeding moult according to Pyle *et al.* (1987), but head and body of some spring adults appear newer and less abraded than wing, tertials, and tail, perhaps pointing to partial pre-breeding moult. POST-JUVENILE. Partial: head, body, and lesser, median, and greater upper wing-coverts. On breeding grounds, July–August (Pyle *et al.* 1987) or from early July, acquired *c.* 1 month after fledging (Dwight 1900). Birds examined from Wisconsin and Michigan had moult completed 15 and 22 August (BMNH, RMNH). No pre-breeding moult (Pyle *et al.* 1987), but one undated winter bird from Venezuela had belly in moult (SMTD).

Measurements. ADULT, FIRST ADULT. Whole geographical range, all year (mainly Great Lakes area, May–August); skins (BMNH, RMNH, SMTD, ZFMK, ZMA). Bill (S) to skull, bill (N) to distal corner of nostril; exposed culmen on average 3·3 shorter than bill (S).

		♂			♀	
WING		64·9 (1·84; 27)	61–68		60·2 (1·27; 13)	58–63
TAIL		47·2 (1·15; 11)	45–49		43·8 (1·15; 13)	42–45
BILL (S)		14·5 (0·46; 12)	13·9–15·1		14·3 (0·56; 13)	13·6–15·3
BILL (N)		8·8 (0·37; 12)	8·2–9·3		8·9 (0·35; 13)	8·2–9·2
TARSUS		17·4 (0·62; 12)	16·6–18·2		17·2 (0·42; 13)	16·5–17·8

Sex differences significant for wing and tail. For influence of abrasion on wing length, see Francis and Wood (1989).

Weights. Pennsylvania: (1) July–September, (2) April–June (Clench and Leberman 1978, which see for details).

			♂		♀	
(1) AD	♂	8·8 (0·66; 19)	7·6–9·8	♀	9·0 (0·67; 11)	7·8–10·4
	1ST AD	9·1 (0·65; 15)	8·0–10·4		8·5 (0·89; 17)	7·0–10·3
(2)		8·7 (0·55; 145)	7·2–11·2		8·9 (0·94; 79)	7·5–11·8

New York (USA), May: ♂ 8·8 (ZMA). Kentucky (USA): May, ♂ 8·8, ♀ 10·5; September, ♂ 9·4 (3) 8·9–9·7, ♀ 9·2 (Mengel 1964). Georgia and South Carolina (USA), early September: ♂♂ 8·2, 10·2 (Norris and Johnston 1958). Belize, February: ♀ 9·1 (Russell 1964).

Structure. Wing rather short, broad at base, tip fairly rounded. 10 primaries: p7–p8 longest, p9 0·5–2·5 shorter, p6 1–2, p5 4–7, p4 7–11, p3 9–13, p2 11–15, p1 12–17. P10 extremely reduced, a narrow tiny pin (*c.* 4 mm long, 0·5 mm wide), hidden below almost equally reduced outermost greater upper primary covert; 41–47 shorter than p7–p8, 5–7 shorter than longest upper primary covert. Outer web of p6–p8 emarginated (sometimes faint on p6), inner web of p7–p9 with notch (sometimes distinct on p8 only). Tip of longest tertial reaches to tip of p1–p2 in closed wing. Tail rather short, tip square; 12 feathers. Bill rather long, visible culmen *c.* 67% of head length; similar to bill of Tennessee Warbler *V. peregrina*, but relatively longer, deeper at base, and tapering more evenly towards tip. Tarsus rather short, toes rather long, slender. Middle toe with claw 13·6 (5) 13–14; outer toe with claw *c.* 80% of middle with claw, inner *c.* 75%, hind 83%.

Geographical variation. None.

Closely related to Blue-winged Warbler *V. pinus*, but the two do not form a superspecies, as ranges presently overlap widely. Though *V. pinus* superficially quite different (head and body of ♂ yellow, except for narrow black eye-stripe; wing blue-grey; median and greater coverts with rather narrow white tips), hybrids are frequent. First generation hybrid most frequently encountered is 'Brewster's Warbler *V. leucobronchialis*', which has yellow cap and mainly blue-grey upperparts of *V. chrysoptera*, but underparts pale yellow with grey-white flanks, without black on throat (nearer *V. pinus*), narrow black eye-stripe (as *V. pinus*), wing blue-grey with 2 rather broad yellow wing-bars, sometimes coalescing (intermediate). A rare backcross hybrid is 'Lawrence's Warbler *V. lawrencii*', which combines yellow head and body of *V. pinus* with broad black mask and throat-patch of *V. chrysoptera*; wing blue-grey with 2 rather narrow yellow wing-bars. Other backcrosses are nearer either parent in plumage, but may retain some characters of the other species. For genetics, morphology, and hybrid zones, see Parkes (1951) and American Ornithologists' Union (1983).   CSR

Vermivora peregrina **Tennessee Warbler**

PLATE 1
[facing page 64]

Du. Tennessee-zanger   Fr. Paruline obscure   Ge. Brauenwaldsänger
Ru. Зеленый пеночковый певун   Sp. Reinita peregrina   Sw. Tennesseesångare

Sylvia peregrina Wilson, 1811

Monotypic

Field characters. 11–12 cm; wing-span 18·5–20·5 cm. Noticeably smaller than Blackpoll Warbler *Dendroica striata*, with proportionately 5% shorter wings and 10% shorter tail; bill distinctive, rather short with sharply pointed tip. Rather small, usually dumpy but sharp-billed Nearctic wood warbler, with adult plumage suggesting small vireo *Vireo* and 1st-winter inviting confusion with other yellow Parulidae and *Phylloscopus* warblers. Yellowish lime-green above, with bright white vent and under tail-coverts. Bill pattern and call can be distinctive. Sexes closely similar; some seasonal variation. 1st-winter separable.

ADULT MALE. Moults: July–August (complete). Crown, hindneck, and ear-coverts dusky to bluish-grey, often sullied with green when fresh; rather short supercilium, lower eye-crescent, and throat almost white. Remainder of upperparts yellowish lime- to olive-green, brightest on rump. Underbody white, washed yellowish-green on side of breast and grey along flank. In winter, many dull ♂♂ indistinguishable from brighter ♀♀. Bill blackish with base of lower mandible rather paler. Legs mainly grey. ADULT FEMALE. As ♂, but grey of head less strong (more greenish) and not spreading to hindneck. Supercilium, rear face, and flank all more heavily marked yellow or buff. In winter, many ♀♀ even greener on head and then indistinguishable from 1st-winter. FIRST-WINTER. As adult ♀ but brighter and more yellowish above, and yellow below except for narrow whitish vent and under tail-coverts. At close range, shows yellow supercilium, bright whitish to yellow lower eye-crescent, obscure green eye-stripe, long diffuse yellowish bar across tips of greater coverts, indistinct dull yellowish bar on tips of median coverts, and (when fresh) narrow, almost white tips to primaries. Bill sometimes has pinkish or yellowish base to lower mandible.

1st-winter traditionally confused with Arctic Warbler *Phylloscopus borealis* (much larger, with strong bill with bright flesh lower mandible, long rakish supercilium, dark eye-stripe, whitish wing-bars and underparts, and straw legs: Volume VI, p. 536), but can also be confused with Philadelphia Vireo *Vireo philadelphicus* (stubby bill, grey always restricted to crown, no distinct wing-bars, more yellowish vent and under tail-coverts, blue-grey legs), Yellow Warbler *Dendroica petechia* (less pointed bill, no supercilium, wing-bar obvious only on inner greater coverts, deep yellow underparts), and other *Phylloscopus* warblers, particularly yellower, wing-barred forms of Chiffchaff *P. collybita* and Green Warbler *P. nitidus* (nei-

ther, however, is as yellow below, or has similar bill form and colour, or bright rump). Note also that some *V. peregrina* show whitish mark on outermost tail-feathers. Active hunter of insects. Has habit of moving body from side to side while keeping head still.

Call a penetrating 'zit-zit' recalling Firecrest *Regulus ignicapillus* or 'zi' or 'zi-zi' suggesting tit *Parus*.

Habitat. Breeds in cooler northern temperate Nearctic lowlands, in grassy or boggy woodland or forest openings and clearings, mixed with dense brush and clumps of young second-growth trees; feeds in treetops in spring but needs *Sphagnum* moss for nesting; probably increased in modern times with spread of forest clearance (Pough 1949). Generally associated with deciduous forest, forest clearings, and brushy hillsides, but in Minnesota favours *Sphagnum* bogs bearing scattered spruce *Picea* or other conifers. In eastern Canada, appears commonest in low areas of streams and boggy ground, in partially cut spruce and balsam woodlands regrown with shrubby second growth (Johnsgard 1979); in Canada also in coniferous mixed and deciduous woodland, alder *Alnus* and willow *Salix* thickets and burnt forest (Godfrey 1979).

Migrants occur wherever there are trees, in gardens, orchards, cemeteries, parks, villages, and city suburbs (Forbush and May 1955). Also common in weed tangles and hedgerows (Pough 1949). Spring migrants are quite common in large trees in Louisiana cities, but in autumn they prefer willows alongside water (Lowery 1974). Wintering birds in Venezuela occur in deciduous, rain, and cloud forests and second growth up to 2200 m in mountains; forage from near ground to treetops, and in shrubs and trees at forest edge (Schauensee and Phelps 1978).

Distribution. Breeds in North America, from northern Canada south to northern USA (northern Minnesota east to north-east New York and southern Maine). Winters in Central and northern South America, from southern Mexico south to Colombia and western Venezuela.

Accidental. Iceland: specimen (Icelandic Museum of Natural History). Faeroes: 21–28 September 1984 (Boertmann *et al.* 1986). Britain: 2 different 1st-year birds, Fair Isle, 6–20 and 24 September 1975; 1st-year, Orkney, 5–7 September 1982 (Dymond *et al.* 1989).

Movements. Migratory, wintering far south of breeding range. Numbers at passage sites vary more than in other

Parulidae, reflecting population changes following outbreaks of spruce budworm *Choristoneura fumiferana*; in Ohio, maximum daily counts varied between years from less than 20 to over 1000 in autumn, and from 5 to 250 in spring (Trautman 1940; Cadman *et al.* 1987).

Migrants leave breeding areas across Canada early August to mid-September (Quilliam 1973; Sadler and Myres 1976; Tufts 1986), reaching northern USA by mid-August (Bull 1974; Bohlen 1989). Main passage through USA extends only from mid-September to mid-October, with a few stragglers into November (Imhof 1976; Potter *et al.* 1980; James and Neal 1986). Arrives in winter range late September and October, later than many Parulidae (Bent 1953; Hilty and Brown 1986; Binford 1989). Most birds move south between prairies and Appalachians, with few on east coast; only sporadic passage records west of Rockies, though regular (but rare) in California (Phillips *et al.* 1964; Garrett and Dunn 1981; Alcorn 1988). Ringing shows winter site-fidelity in Guatemala, El Salvador, and Panama, with returns up to 4 years later (Loftin 1977; Rogers *et al.* 1982a).

First spring arrivals in southern USA are early to mid-April (Oberholser 1974; Imhof 1976). Movement across Gulf of Mexico further west in spring than autumn, with more birds on western Gulf coast, and fewer in south-east states (Bent 1953; Burleigh 1958; Rappole *et al.* 1979). Birds spread north, initially west of Appalachians, to reach southern Canada and northern Rockies by early to mid-May (Palmer 1949; Bailey and Niedrach 1967; Quilliam 1973; Janssen 1987). Northern limits of breeding range (58°N in western Canada) are reached in late May and early June (Nero 1963; Todd 1963; Erskine and Davidson 1976), and stragglers are seen through most of May across eastern USA (e.g. Stewart and Robbins 1958, Mengel 1965).

Vagrant north to Hudson and Ungava Bays 56–58°N (northern Canada), late May and mid-August (Todd 1963), and to southern Greenland (Bent 1953). For vagrancy to west Palearctic, see Distribrution.   AJE

Voice. See Field Characters.

Plumages. ADULT MALE. In fresh plumage (autumn), forehead, crown, and hindneck medium bluish-grey, feather-tips washed green, partly concealing blue-grey; rarely, some rufous-chestnut mixed within grey (Raveling and Warner 1965). Remainder of upperparts bright yellowish-green, palest on rump. Supercilium narrow but distinct, white with slight yellow or pale green tinge, ending above middle of ear-coverts, bordered by long and narrow dark grey to dull black stripe from lore through eye, fading out on rear of ear-coverts. Eye-ring white, broken at front and rear. Cheek and ear-coverts white with pale green mottling, side of neck light yellowish-green. Chin and throat white with slight yellow suffusion. Side of breast light yellowish-green, shading to light ash-grey towards chest and flank; mid-chest white, chest and side of breast with pale lemon-yellow suffusion. Light ash-grey of flank tinged light green on rear; thighs grey. Belly, vent, and under tail-coverts silky white, side of belly slightly tinged

grey, mid-belly slightly yellow. Tail dark grey, both webs of central pair of feathers (t1) and outer webs of t2–t6 washed light green; inner web of t6 sometimes with large white subterminal spot, up to 7 mm long and 4 mm wide, but often absent or restricted, showing as widened white margin only (spot present in 48–78% of ♂♂, depending on population: Raveling 1965). Flight-feathers, tertials, and greater upper primary coverts greyish-black; outer webs of primaries and primary coverts narrowly and sharply fringed grey (innermost primaries green-grey or green), secondaries narrowly and sharply fringed green, whole outer web of tertials washed green; tips of flight-feathers narrowly edged white. Lesser and median upper wing-coverts dark grey with black shafts and broad green distal halves, more green-yellow on tips; greater coverts dark grey with green outer webs and narrow pale yellow-green spot on tips, latter sometimes forming faint narrow wing-bar. Under wing-coverts and axillaries white, partly washed pale yellow. *In worn plumage* (spring and early summer), forehead, crown, and hindneck pure blue-grey; green of remainder of upperparts slightly duller grass-green, slightly yellowish on rump only; supercilium white; ear-coverts and side of neck ash-grey, cheek mottled white and grey; underparts dirty white, almost without yellow, side of breast ash-grey, flank greyish-olive; green suffusion of tail partly worn off; wing-coverts have more dark grey of bases visible.

ADULT FEMALE. In fresh plumage, like adult ♂, but forehead, crown, and hindneck all-green, no blue-grey of feather-bases visible; remainder of upperparts slightly duller olive-green, less bright yellow (except on rump). Supercilium, cheek, throat, and side of belly tinged pale yellow (but markedly less so than in 1st adult ♀); side of breast and flank extensively greenish-grey; mid-belly, vent, and under tail-coverts white with faint yellow suffusion. White spot on t6 present in 33–69% of birds, depending on population (Raveling 1965). *In worn plumage*, cap and hindneck partly grey, mixed grey and green, or almost uniform green; supercilium and mid-underparts dirty white, side of breast and flank olive-grey. See also Raveling and Warner (1965).

JUVENILE. Like adult ♀, but colour of body less saturated, upperparts more greyish-olive, side of head and underparts greyish-buff or greenish-grey, a faint dark stripe through eye; tips of median and greater coverts form distinct pale yellow wing-bars; tips of tail-feathers narrower, more pointed, less truncate or broadly rounded. See also First Adult (below). FIRST ADULT MALE. In autumn like adult ♂, but cap rather more extensively green and underparts slightly more yellow, thus similar in these respects to adult ♀, but upperparts brighter and more yellowish-green. Tail-feathers juvenile, tips more pointed; flight-feathers juvenile, dark grey rather than blackish, pale fringes less sharply defined, those of inner and middle primaries and primary coverts green, not grey-white. Tips of greater upper wing-coverts narrowly fringed pale yellow, forming narrow wing-bar; sometimes a similar bar on median coverts. *In worn plumage* (spring), hardly separable from adult ♂, but traces of green often remain on cap, hindcrown, and on fringes of primaries and primary coverts. For variation and sex differences in colour of crown in spring, see Quay (1989). FIRST ADULT FEMALE. Upperparts uniform olive-green, duller than in 1st adult ♂ and without grey on cap; yellow-green restricted to rump. Dark eye-stripe more olive, less dark grey, less distinct. Supercilium, eye-ring, cheek, and underparts extensively bright sulphur-yellow; ear-coverts and side of neck yellow, mottled olive-green; side of breast and flank washed green, only mid-belly, vent, and under tail-coverts almost white. Due to almost uniform dull olive-green upperparts, largely yellow side of head and underparts, and smaller size, often appears very different from adult ♂, which shows grey and yellow-green upperparts, grey and white underparts, distinct face marks, and larger size. Tail, flight-feathers, primary coverts, and greater upper wing-coverts still juvenile; characters as in 1st adult ♂, but pale tips of greater coverts sometimes broader; median coverts either juvenile with tips off-white (soon worn off), or 1st adult with tips paler green-yellow than in adult, forming variably distinct 2nd wing-bar. Yellow of face and underparts bleaches rather rapidly; by December-January, underparts yellow-white; in spring and early summer, similar to adult ♀, with supercilium yellow-white, more distinct than in fresh plumage, underparts dirty yellowish-white, and side of breast and flank paler and less grey than in ♂ and adult ♀.

Bare parts. ADULT, FIRST ADULT, JUVENILE. Iris brown or dark brown (more reddish-brown in adult, more yellowish-brown in 1st adult: Wood and Wood 1972). Bill dark horn, dusky horn-brown, or plumbeous-black, cutting edges and basal 50-70% of lower mandible bluish horn-grey or plumbeous-grey, darkest in adult ♂ breeding; base yellowish, whitish-horn, pink, or pale flesh-grey in at least some juveniles and some birds from early autumn. Leg and foot medium grey, dark bluish-black, dull lead-grey, or plumbeous, slightly paler at side or rear of leg or on foot, in some birds with brown tinge; soles flesh, brown-grey, or grey. (Broad 1981; Meek 1984; RMNH, ZMA.)

Moults. ADULT POST-BREEDING. Complete; primaries descendent. July-August, on breeding grounds, but some may suspend flight-feather moult during migration (Pyle *et al.* 1987). Moult late July to early August (Raveling and Warner 1965). Occasionally, all moult suspended during autumn migration, with all plumage old except inner primaries or scattered feathers of body (Baird 1967). One ♂ from Wisconsin, 10 August, heavily worn, innermost primaries (p1-p2) just shed (RMNH). ADULT PRE-BREEDING. In winter quarters, January-April (Pyle *et al.* 1987) or January-March (Dwight 1900; Ridgely and Tudor 1989). Restricted to limited feathering of head and body (Pyle *et al.* 1987), mainly to head, chin, and throat (Dwight 1900), or no moult at all (Sharpe 1885). Starts mid-December to late February, completed early February to April; involves head, neck, and chin to chest, sometimes mantle, scapulars, and flank; extent variable (Raveling and Warner 1965). POST-JUVENILE. Partial: head, body, and lesser and (sometimes) median upper wing-coverts. On breeding grounds, July-August (Dwight 1900; Pyle *et al.* 1987). During migration, some birds retain a few juvenile feathers on head or neck (RMNH, ZMA). FIRST-BREEDING. As adult pre-breeding.

Measurements. ADULT, FIRST ADULT. Whole geographical range, all year; skins (RMNH, ZFMK, ZMA). Bill (S) to skull, bill (N) to distal corner of nostril; exposed culmen on average 2·5 less than bill (S).

	♂		♀	
WING	67·4 (0·91; 16)	66-69	63·0 (0·97; 14)	61-65
TAIL	43·6 (1·51; 15)	42-47	41·4 (1·80; 14)	39-45
BILL (S)	12·8 (0·66; 15)	12·0-14·0	12·8 (0·51; 13)	12·0-13·7
BILL (N)	7·6 (0·64; 15)	7·2-8·1	7·6 (0·27; 12)	7·1-7·9
TARSUS	17·0 (0·49; 16)	16·3-17·7	16·9 (0·63; 14)	16·0-17·8

Sex differences significant for wing and tail.

See also Goodpasture (1963), Raveling and Warner (1965), and Quay (1989).

Weights. Pennsylvania (USA): August-October, adult 9·6 (104) 8·0-13·6, 1st adult 9·0 (1741) 7·3-15·3; May, ♂ 10·2 (1·40; 209) 7·3-18·4, ♀ 9·8 (1·09; 164) 7·8-13·4 (Clench and Leberman 1978, which see for details). East Ship Island (off Mississippi, USA), spring: 8·2 (0·9; 60) (Kuenzi *et al.* 1991). Spring: Texas, ♂ 9·0 (7·9-10·9), ♀ 8·3 (6·5-9·5); Missouri, ♂ 9·5 (7·9-12·7), ♀ 9·4 (7·3-13·5); southern Ontario, ♂ 11·2 (9·4-12·5), ♀ 11·5 (9·4-13·5) (Quay 1989). Pelee Island (Lake Erie, Canada), May: ♂ 11·3 (3) 10·1-12·1, ♀ 11·5 (3) 10·0-13·8 (Woodford and Lovesy 1958). Kentucky (USA): September-October, ♂ 9·8 (0·99; 5) 8·2-10·8, ♀ 8·8 (1·36; 6) 7·9-11·2; May, ♂ 9·1 (Mengel 1965). Illinois (USA), averages, September: adult ♂ 10·3 (6), adult ♀ 9·6 (7), 1st adult ♂ 11·4 (2), 1st adult ♀ 10·1 (7) (Graber and Graber 1962).

Belize: autumn, 8·4 (0·9; 67) (Mills and Rogers 1990); October, ♂ 9·3 (3) 9·0-9·5, ♀ 9·2 (3) 8·7-10·0; March-April, ♂♂ 8·2, 8·5, ♀ 8·3 (Russell 1964; Roselaar 1976; ZMA). El Salvador, October: ♂ 11·5 (ZMA). Jamaica, winter: 8·5 (3) (Diamond *et al.* 1977). Mexico, April: ♂ 12·4 (RMNH). Britain, September: Fair Isle 8·6, 11·5 (Broad 1981), Orkney 13·1 (Meek 1984).

For fat-free weights, see Connell *et al.* (1960), Caldwell *et al.* (1963), and Rogers and Odum (1966).

Structure. Wing rather short, broad at base, tip bluntly pointed. 10 primaries: p7-p8 longest, p9 0-2 shorter, p6 1-2, p5 5-8, p4 8-11, p3 10-14, p2 12-15, p1 13-19 (on average, ♂ 17·2, ♀ 14·5, $n = 5$ in both). P10 reduced, narrow, pointed, a tiny white spike concealed below black reduced outermost upper primary covert: 42-48 shorter than p7-p8, 5-9 shorter than longest upper primary covert. Outer web of p6-p8 emarginated; inner web of p7-p9 with notch (sometimes faint on p7 and p9). Tip of longest tertial reaches to tip of p1-p2. Tail rather short, tip square or slightly forked; 12 feathers. Bill rather short (length of visible culmen about half head length), straight; wider and deeper at base than any *Phylloscopus* warbler, tapering more gradually to acutely pointed tip, tip of culmen less decurved. Nostrils small, oval, covered by operculum above. A few short bristles projecting forward over gape at basal side of upper mandible. Tarsus short, toes rather long and slender. Middle toe with claw 13·7 (5) 13-14·5; outer toe with claw *c.* 74% of middle with claw, inner *c.* 68%, hind *c.* 83%.

Geographical variation. Very slight, in bill length only. Northern birds (from North-west Territories and northern Alberta east to shores of James Bay) have bill slightly shorter than those occurring further south in zone from British Columbia to southern Ontario; bill of eastern populations intermediate (Raveling 1965).

Recognition. Closely resembles some *Phylloscopus* or *Hippolais* warblers, but bright yellowish-green of upperparts of ♂ *V. peregrina* not found in any of these; body heavier, bill shape different (see Structure), p10 shorter, and leg blue-grey. See also Browne (1960).

CSR

Parula americana **Northern Parula**

PLATE 2
[between pages 64 and 65]

Du. Bril-parulazanger   Fr. Paruline à collier   Ge. Meisenwaldsänger
Ru. Белоглазая парула   Sp. Reinita norteña   Sw. Messångare

Parus americanus Linnaeus, 1758

Monotypic

Field characters. 10·5–11·5 cm; wing-span 17·5–18 cm. 20% smaller than Blackpoll Warbler *Dendroica striata* with structure recalling kinglet *Regulus*. Small, short-tailed, arboreal Nearctic warbler. Remarkably pretty: bluish head and upperparts with yellowish-green patch on mantle, double white wing-bar, incomplete white eye-ring, and bright yellow underparts. Sexes dissimilar; some seasonal variation. 1st-winter separable at close range.

ADULT MALE. Moults: July–August (complete). Colourful plumage also includes black lore, intensely blue cheek contrasting sharply with throat, mottled blackish band across lower throat, adjacent mottled rufous band across breast, and often isolated rufous spots on fore-flank. Flight-feathers fringed bluish. White panels on outer tail-feathers. In fresh winter plumage, head and upperparts tinged greenish, sharp contrasts of head and breast markings reduced, and secondaries fringed greenish. Bill spiky; blackish with pale yellowish or buff-horn lower mandible. Legs bright pale brown. ADULT FEMALE. Lacks vivid contrasts of ♂, usually showing paler, only dark-streaked lore, short off-white fore-supercilium (sometimes, with broken eye-ring, forming pale spectacle), dusky eye-stripe, more diffuse border to cheek, only vestigial throat-band, and restricted breast-band. In fresh winter plumage, some indistinguishable from 1st-winter ♂. FIRST-WINTER MALE. Resembles adult, but in fresh plumage fringes of flight-feathers distinctly greenish, contrasts of face and breast markings reduced, and tail-panels restricted to subterminal spots. FIRST-WINTER FEMALE. Duller than 1st-winter ♂, with even greener fringes to flight-feathers.

Unmistakable. No other vagrant parulid is as small or as multicoloured. Tropical Parula *P. pitiayumi* (no eye-crescents or breast-bands) is very unlikely to stray; Canada Warbler *Wilsonia canadensis* much larger, with uniform blue-grey upperparts, no rufous breast-band, and entirely yellow underparts. Actions and behaviour include hovering and clinging upside-down to foliage, recalling both *Regulus* and small tit *Parus* (although movements of vagrant considered more sluggish or deliberate than this).

Apparently silent as vagrant, but commonest call in Nearctic a sharp 'chick'.

Habitat. Breeds in temperate and warmer regions of eastern Nearctic, most commonly around woodland openings by swamps, ponds, and lakes, but sometimes elsewhere where trees carry either old man's beard lichen *Usnea* in north, or its moss-like counterpart 'Spanish moss' *Til-landsia* in south; infrequent where neither available. Moisture and tall mature trees are further associated requirements; most feeding done in treetops. (Pough 1949.) In Canada, lives in both deciduous and coniferous woods, especially those which are moist and open (Godfrey 1979). Wintering birds at extreme south of migratory range in South America have been recorded in low mangroves (Schauensee and Phelps 1978).

Distribution. Breeds in North America east of Rocky Mountains, from southern Canada south to north coast of Gulf of Mexico; scarce in most areas except in north-east (Maine to Nova Scotia) and south-east (Louisiana to South Carolina). Winters in Central America from southern Mexico to Costa Rica, and from Florida and Bahamas south through West Indies to Grenada and Curaçao.

Accidental. Iceland, Britain, Ireland, France.

Movements. Migratory; a few birds winter within breeding range.

Southward movement begins August, but little migration detected until early September. Main passage south from north and east of range proceeds rapidly, peaking mid-September from Maine and Minnesota (Palmer 1949; Janssen 1987) to Florida (Stewart and Robbins 1958; Kale and Maehr 1990), where large-scale passage continues to mid-October (Bent 1953). By end of October, few remain north of winter range (Oberholser 1974; Imhof 1976). Arrivals in West Indies in August (Bond 1985; Raffaele 1989) are presumably from south-east population, as northern birds then are mostly still in breeding areas. Reaches Central America later, at end of September and in October (Monroe 1968; Binford 1989), but in Mexico there may be confusion with resident Tropical Parula *P. pitiayumi* with which it overlaps widely in winter (see Rappole *et al.* 1983). Migration is on broad front across eastern USA towards tropical America, but most birds pass through Florida to winter in West Indies (Bent 1953; Bond 1985; Pashley and Martin 1988). Movement across or around western Gulf of Mexico (Lowery 1974; Oberholser 1974) probably involves mainly birds from Louisiana and sparse populations in other western parts of range. Small numbers wander across south-west USA to California, where a few winter and, exceptionally, breed (Garrett and Dunn 1981; American Ornithologists' Union 1983; Alcorn 1988). Ringing in Puerto Rico and Jamaica shows winter site-fidelity, with returns up to 7 years later

(Diamond and Smith 1973; Faaborg and Arendt 1984).

Northward movement begins February, though many birds remain in winter quarters until April or early May (Bent 1953; Monroe 1968; Brudenell-Bruce 1975). Trans-Gulf migration, with no stopover, brings birds as far north as Louisiana by early March, when others have only reached southern Florida and southern Texas (Bent 1953; Stevenson 1957). Main passage in mid-southern states continues in March (Imhof 1976; James and Neal 1986), but further north and on east coast not until mid- or late April (Stewart and Robbins 1958; Mengel 1965; Potter *et al.* 1980; Bohlen 1989), and arrivals in northern breeding areas are from early May, continuing to late May (e.g. Palmer 1949, Speirs 1985, Janssen 1987). Rarely, occurs north-west of breeding range in Alberta and Saskatchewan (Nero and Lein 1971; Salt and Salt 1976).

Rare autumn vagrant to Atlantic seaboard of west Palearctic, where 14 records from Britain and Ireland up to 1989, especially south-west England and southern Ireland, mainly from late September to 3rd week of October, with 2 records in November (Dymond *et al.* 1989; Rogers *et al.* 1990). Vagrant also in Greenland (Salomonsen 1950-1).

AJE, PRC

Voice. See Field Characters.

Plumages. ADULT MALE. In fresh plumage, entire upperparts medium bluish-grey, each feather with narrow bright green fringe along tip, showing as green tinge on crown and nape, but hardly visible on scapulars or from back to upper tail-coverts; lower mantle contrastingly rich golden-green, showing as green triangular saddle on blue-grey remainder of upperparts. Lore plumbeous-black, area round eye plumbeous-grey, neither contrasting with medium blue-grey remainder of side of head and neck; small but contrasting white patch just above eye, similar one just below. Chin, throat, and chest bright golden-yellow, sharply demarcated from medium blue-grey lower cheek, side of neck, and side of chest; some rufous-chestnut of feather-bases usually visible on throat and chest, a little black sometimes on upper chest. Flank ash-grey (grey paler and more restricted towards rear), remainder of underparts white, tinged yellow along border with chest, rufous along border with upper flank, and pale yellow-grey on vent; thigh dark grey with yellow and white suffusion. Central pair of tail-feathers (t1) greyish-black with ill-defined blue-grey fringe along both webs; t2-t6 greyish-black to dark grey with blue-grey outer fringe and sharp white inner fringe; inner web of t4 with white blob or spot subterminally; t5 with larger white patch, inner web of t6 with square white patch 8-16 mm long, rendering terminal 25-35% of inner web white, except for *c.* 5 mm of tip. Flight-feathers, tertials, bastard wing, and greater upper primary coverts greyish-black, flight-feathers with narrow pale blue-grey outer fringes; bastard wing and primary coverts with narrow medium blue-grey fringes; tertials with broad and ill-defined blue-grey border along outer web; blue-grey of secondaries and tertials sometimes partly tinged green, but most fringes appear generally blue. Lesser and median upper wing-coverts medium blue-grey, median ones with white tip *c.* 5 mm wide; greater coverts black, tip of outer web with large white spot (forming broken wing-bar 3-5 mm wide), remainder of outer web blue-grey. Axillaries, under wing-coverts, and inner borders of flight-feathers white.

In worn plumage (spring and summer), entire upperparts, side of head and neck, and upperwing uniform slate-blue, except for golden-green lower mantle, small white spot above and below eye, and 2 white wing-bars; some sooty of feather-bases sometimes visible on forehead, and rump sometimes more ash-grey; bib on chin and upper throat golden-yellow, partly or fully suffused rufous-chestnut on throat, bordered by slate-blue or sooty bar across lower throat; feather-tips of bar partly yellow; upper chest rufous-chestnut or sooty-brown with yellow feather-tips, lower chest yellow; remainder of underparts dirty white, washed ash-grey on side of breast and upper flank, sometimes with trace of rufous where side of breast meets upper belly. ADULT FEMALE. Upperparts as adult ♂, but feather-tips more extensively green, forehead to upper mantle appearing equally mixed green and blue-grey, back to upper tail-coverts mainly blue-grey; saddle on lower mantle uniform golden-green, as adult ♂. Side of head and neck grey, slightly paler and less bluish than adult ♂, with slightly contrasting darker grey stripe through eye, bordered by short pale grey or off-white supercilium above and with similar short stripe just below eye; front part of supercilium sometimes pale yellow; eye-ring white, broken by grey in front and behind; grey of ear-coverts and of side of neck partly washed green. Underparts like adult ♂, but yellow of chin, central throat, and chest paler, and feathers without rufous tinge subterminally on upper throat and chest (sometimes a trace of sooty on lower throat or tawny on upper chest) and without blue-grey or sooty subterminally on lower throat; no rufous on side of breast or a trace only. Tail as adult ♂, but white spots on outer feathers smaller, those on t6 more rounded, 5-12 mm long. Wing as adult, but blue-grey outer fringes of secondaries, tertials, and greater upper wing-coverts partly washed green. *In worn plumage*, only traces of green feather-fringes on upperparts; grey of upperparts more ashy than ♂, less slate-blue; chin to chest paler lemon-yellow, without chestnut, slate, or sooty of adult ♂, except sometimes for hint of tawny on chest. JUVENILE. Upperparts and side of head and neck uniform slate-grey with slight olive-green tinge; chin and upper throat pale yellow; lower throat, side of breast, chest, and flank light ash-grey, remainder of underparts white; upper wing-coverts dull slate-grey, median and greater with narrow white tips. For tail, tertials, and flight-feathers, see First Adult. FIRST ADULT. Like adult, but juvenile tail, flight-feathers, tertials, greater upper primary coverts, and bastard wing retained: browner, with greener (less bluish) fringes and more worn tips than in adult at same time of year; in particular, bastard wing browner than neighbouring blackish of outer greater coverts; tips of tail-feathers more pointed, less truncate. In ♂, blue-grey of upperparts partly hidden below green feather-fringes, appearing rather similar to adult ♀ (but side of head and underparts quite different); spot on t6 6-12 mm long, as in adult ♀; chin to breast as adult ♂, but often less extensively tawny on upper and less slaty or sooty on lower throat; tawny or chestnut on upper chest extensive (unlike adult ♀); fringes of tertials and secondaries green, those of primaries and primary coverts pale grey-green or pale grey, contrasting with blue-grey of greater coverts, not as uniform blue-grey as adult ♂. In 1st adult ♀, grey of upperparts extensively washed green in fresh plumage, in particular forehead to upper mantle almost uniform green, hardly contrasting with golden-green lower mantle; in worn plumage, traces of green still present (in adult ♀, usually none by May); chin to chest pale yellow, without traces of sooty or tawny; white spot on t6 small, occasionally absent, 4-9 mm (0-12) long; white tips of median and greater coverts small, 2-3 mm, partly washed grey and green; fringes of flight-feathers and tertials green and pale green-grey, less extensively blue-grey than in adult ♀, rather similar to those of 1st adult ♂, but grey of upperparts and side of head and neck less bluish and chin to chest pale yellow without rufous.

Bare parts. ADULT, FIRST ADULT. Iris brown or dark brown. Bill dark horn-brown to black, lower mandible buff-yellow, straw-yellow, pale horn-flesh, or ivory-white. Leg flesh-brown, yellow-brown, dull horn, or olive-brown; foot paler pink-horn, yellow-horn, or buff-yellow, sole yellow. (RMNH, ZMA.) JUVENILE. No information.

Moults. ADULT POST-BREEDING. Complete; primaries descendent. On breeding grounds, July–August (Dwight 1900; Pyle *et al.* 1987). ADULT PRE-BREEDING. Partial, involving limited amount of feathering of body, mainly on head, chin, and throat; February–April (Dwight 1900), or no moult at all; in 2 of 7 examined, central pair of tail-feathers (t1) replaced (RMNH). POST-JUVENILE. Partial: head, body, and lesser, median, and greater upper wing-coverts. In summer quarters, from mid-July (Dwight 1900) or July–August (Pyle *et al.* 1987). FIRST PRE-BREEDING. Restricted to scattered feathers of head and body or no moult at all.

Measurements. ADULT, FIRST ADULT. Whole geographical range, all year (mainly eastern USA, summer, and Cuba and Jamaica, winter); skins (RMNH, ZMA). Bill (S) to skull, bill (N) to distal corner of nostril; exposed culmen on average 2·6 less than bill (S).

	♂			♀	
WING	61·4 (2·00; 22)	57–65		58·5 (1·37; 14)	56–61
TAIL	41·7 (1·46; 22)	39–45		39·8 (1·97; 14)	36–43
BILL (S)	12·7 (0·46; 22)	11·8–13·3		12·2 (0·44; 14)	11·6–12·8
BILL (N)	7·6 (0·31; 22)	7·0–8·1		7·3 (0·36; 14)	6·7–7·8
TARSUS	16·8 (0·39; 21)	16·2–17·5		16·2 (0·42; 13)	15·7–16·9

Sex differences significant. Ages combined, though retained juvenile wing and tail of 1st adult averages shorter than adult: e.g. ♂, adult wing 62·8 (1·58; 9) 60–65, 1st adult wing 60·5 (1·68; 13) 57–63; adult tail 42·7 (1·18; 9) 41–45, 1st adult tail 41·0 (1·25; 13) 39–43 (RMNH, ZMA).

Weights. East Ship Island (Mississippi, USA), spring: 7·1 (1·1; 31) (Kuenzi *et al.* 1991). Georgia and South Carolina (USA), summer: mainly July, ♂ 7·8 (8), ♀ 7·4 (6); August, ♀ 8·4 (3) (Norris and Johnston 1958). Kentucky (USA), June–July: ♂♂ 7·9, 7·9 (Mengel 1965). Pennsylvania (USA): May–June, ♂ 8·1 (0·87; 8) 7·0–9·6, ♀ 7·6 (0·61; 4) 7·1–8·5; August–September 8·0

(0·98; 10) 7·0–10·0 (Clench and Leberman 1978). November–February: ♂ 5·5 (Georgia, USA), 6·0, 6·0 (St Martin, Caribbean; lean), 8·1 (Belize), 8·1 (Curaçao, Caribbean; fat); ♀ 6·0 (Belize), ♀ 7·1 (Curaçao), unsexed 9·5 (Belize) (Russell 1964; ZMA). Jamaica, September–May: ♂ 7·0 (0·7; 7), ♀ 7·0 (0·3; 10) (Diamond *et al.* 1977).

For fat-free weight, see Rogers and Odum (1964).

Structure. Wing rather short, broad at base, tip fairly rounded. 9–10 primaries: p7–p8 longest, p9 0–1·5 shorter, p6 0·5–2, p5 3·5–6, p4 6–11, p3 8–13, p2 10–14, p1 12–16. P10 reduced, a tiny spike hidden below reduced outermost greater upper primary covert, or (in 4 of 10 birds) apparently absent; 38–46 shorter than p7–p8, 4–7 shorter than longest upper primary covert. Outer web of p6–p8 emarginated, inner web of p7–p9 with notch (often faint on p7 and p9). Tip of longest tertial reaches tip of p1 in closed wing. Tail rather short, tip square; 12 feathers. Bill rather long, straight, slender; visible culmen *c.* 67% of head length; tip of culmen and distal half of visible cutting edges slightly decurved; bill as slender as in Black-and-white Warbler *Mniotilta varia*, but relatively slightly shorter, less compressed laterally in middle; slightly less deep at base and with less straight culmen than in *Vermivora*. Some short but distinct bristles project obliquely downward from base of upper mandible. Tarsus and toes rather short and slender. Middle toe with claw 12·1 (5) 11·5–13; outer toe with claw *c.* 74% of middle with claw, inner *c.* 70%, hind *c.* 85%. Basal phalanges of front toes partly fused.

Geographical variation. Slight. Birds from north-east USA and eastern Canada average largest, those of south-central USA (Louisiana to Texas) have wing on average *c.* 4 mm shorter, those of north-central and south-east USA intermediate (Ridgway 1902), but difference too slight to warrant recognition of any race.

Forms superspecies with Socorro Parula *P. graysoni* from Socorro Island and with Tropical Parula *P. pitiayumi* which replaces *P. americana* in Central and South America south from lower Rio Grande. *P. pitiayumi* differs from *P. americana* in smaller size, more rounded wing (wing-tip formed by p6–p7, p1 7–12 shorter), absence of contrasting white spots on eye-ring, and yellow (not grey) lower cheek; also, ♂ has large, contrastingly black mask, purer green mantle, darker and bluer remainder of upperparts, entirely yolk-yellow underparts (except whitish lower flank, thigh, and under tail-coverts), and throat to chest with rufous-orange wash (no black bar across upper chest). CSR

Dendroica petechia **Yellow Warbler**

PLATE 2
[between pages 64 and 65]

DU. Gele Zanger   FR. Paruline jaune   GE. Goldwaldsänger
RU. Золотистый лесной певун   SP. Reinita amarilla   SW. Gul skogssångare

Motacilla petechia Linnaeus, 1766

Polytypic. AESTIVA GROUP. *D. p. aestiva* Gmelin, 1789, southern Canada from south-east Alberta to Nova Scotia, and north-central and north-east USA, east from Great Plains south to *c.* 33°N; *amnicola* Batchelder, 1918, Canada, north of *aestiva*, from Newfoundland west to north-east British Columbia, Yukon territory, and central Alaska. Extralimital: 4–5 races in southern Alaska, western British Columbia, western USA, and inland Mexico south to Oaxaca. PETECHIA GROUP (extralimital). Nominate *petechia* (Linnaeus, 1766), Barbados; 14–16 further races in West Indies from Florida

Keys and Bahamas south to Grenadines and islands off Venezuela (Aruba to Margarita), west to islands off Central America. ERITHACHORIDES GROUP (extralimital). About 13 races in coastal mangroves from Mexico (including southern Baja California) to north-west South America (east to Venezuela, south to northern Peru) and Galápagos.

Field characters. 11·5–13 cm; wing-span 17·5–20 cm. 15% smaller than Blackpoll Warbler *Dendroica striata*, but with typical *Dendroica* structure. Quite small, sprightly, thicket-loving Nearctic warbler, always looking remarkably yellow and with diagnostic yellow panels on tail. At close range, shows pale eye-ring, double wing-bar, and rump. Sexes rather dissimilar; no seasonal variation. 1st-winter separable.

ADULT MALE. Moults: June–August (complete). Bright greenish-yellow above, rich almost orange-yellow below. Most of head appears unmarked, but especially in fresh winter plumage, crown and rear ear-coverts washed greenish. Wings dusky, quite strongly patterned with bright yellow edges to tertials and greater and median coverts, on median coverts particularly forming obvious wing-bar. Breast and flank with distinct but soft reddish-brown lines. Bill black with grey lower mandible. Legs bright yellowish-brown. ADULT FEMALE. Noticeably duller than ♂, more olive above, particularly on crown and ear-coverts; wing markings and underparts less brilliant yellow; rufous lines on breast and flank more restricted. Conversely, tends to show more obvious yellow eye-ring. FIRST-WINTER. Resembles adult ♀, but distinguished at close range by smaller yellow panels on tail-feathers (restricted to inner edges), more olive cheek, and more obscure wing-bar. Some ♀♀ very dull, with pale buffy-yellow face and underparts, washed grey on crown and below: all ♀♀ usually lack lines on underparts. When retained from juvenile plumage, tertial-edges whitish and tips of greater and median coverts only dull yellow. Bill shows yellowish to flesh cutting edges. Legs dull brown.

Often looks all-yellow at distance, but until yellow marks on tail confirmed, subject to confusion with immature Tennessee Warbler *Vermivora peregrina* (whitish vent and under tail-coverts) and immatures of *Wilsonia* warblers (clear yellow forehead and supercilium, no wing-bars). Also needs to be separated from other potential trans-atlantic vagrants: Prothonotary Warbler *Protonotaria citrea* (bluish wings and tail, white tail-patches), Blue-winged Warbler *Vermivora pinus* (bluish wings with 2 white wing-bars), Prairie Warbler *D. discolor* (strongly marked face, white wing-bars and tail-panels, blackish flank-streaks; wags tail), and yellow form of Palm Warbler *D. palmarum* (plumage pattern like pipit *Anthus*, with dark rusty crown; wags tail). Active hunter of insects, often catching them in flight.

Call a rather full, soft 'tsep' or 'chip'.

Habitat. Widespread in Nearctic lowlands down to tropical regions. Around northern treeline, breeds in dense willow *Salix* and alder *Alnus* thickets (Gabrielson and Lincoln 1959), and in Canada in similar thickets on edges of streams, lakes, bogs, and marshes as well as in ornamental garden shrubbery (Godfrey 1979). Avoids both heavy forests and open grasslands lacking shrubs or trees, but is equally at home in both moist and dry habitats, including hedges, roadside thickets, orchards, and farms. Keeps to shrubbery, low trees, and lower parts of larger trees; rarely in tall treetops. (Forbush and May 1955; Johnsgard 1979.)

In Caribbean breeds commonly in mangroves (Lowery 1974) which are also frequented by wintering birds in South America, together with open grassy fields at sea-level and xerophytic vegetation; not commonly above 1000 m (Schauensee and Phelps 1978).

Distribution. Breeds widely over most of North and Central America, West Indies, and coastal north-west South America. Northern populations winter from southern Mexico south to northern South America; southern populations resident.

Accidental. Britain: Bardsey Island (Wales), 29 August 1964 (Evans 1965); Shetland, ♂, 3–4 November 1990 (Rogers *et al.* 1991).

Movements. Migrant (*aestiva* group, breeding from Canada to northern Mexico) to resident (*petechia* and *erithachorides* groups, breeding West Indies and mid-Mexico to north-west South America). Winter ranges of races in northern group overlap widely, in Central and South America, with each other and with southern races. (American Ornithologists' Union 1957, 1983.) Most abundant of Parulidae through much of its range, but due to early southward movement avoids autumn storms and hence only a rare vagrant to Europe (*Br. Birds* 1965, **58**, 460–1).

Eastern mid-latitude *aestiva* leaves breeding range very early, mid-July to early August, with few remaining by end of August (Palmer 1949; Stewart and Robbins 1958; Mengel 1965); presumably, July birds recorded from Texas to South Carolina are also this form (Oberholser 1974; Potter *et al.* 1980; Kale and Maehr 1990). 2 western forms (*rubiginosa*, *morcomi*) migrate on similar schedules to *aestiva* (Jewett *et al.* 1953; Phillips *et al.* 1964; Garrett and Dunn 1981; Cannings *et al.* 1987), but northern race *amnicola* moves later, mostly mid-August to mid-September (Peters and Burleigh 1951; Gabrielson and Lincoln 1959; Houston and Street 1959). Arrives southern Mexico and Central and South America from mid- or late August (Monroe 1968; Hilty and Brown 1986; Binford 1989). Although stragglers occur annually on east coast in October (Stewart and Robbins 1958; Tufts 1986), most migrants are in or beyond southernmost states of USA by end of October (Phillips *et al.* 1964; Oberholser 1974; Imhof 1976). Migration is broad-front, but eastern birds evidently bypass peninsular Florida and cross Gulf of Mexico further west, occurring east only to western Cuba; passage in Mexico and Central America is mainly in coastal lowlands, with some populations continuing east into South America (Bent 1953; Bond 1985). Winter site-fidelity shown by ringing in Belize and El Salvador (Loftin 1977).

Spring migration starts March, and earliest birds reach southern states then (Phillips *et al.* 1964; Oberholser 1974; Kale and Maehr 1990). Although a few scatter north to Nevada and Maryland by early April (Stewart and Robbins 1958; Alcorn 1988), with main arrival in Kentucky reportedly mid-April (Mengel 1965), main movement in other mid-latitude states starts only in late April (Bailey and Niedrach 1967; Bull 1974; Bohlen 1989). Soon thereafter, mass arrivals occur across southern Canada (Tufts 1986; Cannings *et al.* 1987). Far north of breeding range reoccupied only in late May or early June (Peters and Burleigh 1951; Erskine 1985; Speirs 1985). Route is reverse of autumn, with birds that winter in South America taking circuitous path north-west into Central America before turning north across Gulf (Bent 1953; Stevenson 1957).

Vagrant to Arctic Alaska and islands in Bering Sea, June–October (Kessel and Gibson 1978), and to Southampton and Baffin Islands (63–64°N) in August and October (Godfrey 1986). For vagrancy to west Palearctic, see Distribrution.   AJE

Voice. See Field Characters.

Plumages. (*D. p. aestiva*). ADULT MALE BREEDING. Small cap from forehead to centre of crown orange-yellow, feather-centres often tinged tawny-orange. Side of crown (from above eye backwards), hindcrown, and all remainder of upperparts bright green-yellow, feathers with slightly darker olive-green centres, which are usually concealed, but sometimes form faint streaking; yellow-green slightly more yellow on hindneck and rump, slightly greener on mantle and scapulars. Side of head down from supercilium bright spectrum-yellow, ear-coverts sometimes tinged or mottled green-yellow; side of neck green-yellow. Entire underparts bright spectrum-yellow; lower throat, chest, and side of breast down to upper belly and rear of flank contrastingly marked with rufous-chestnut streaks 1–2 mm wide; traces of narrow streaks sometimes on side of throat and mid-belly. Both webs of central pair of tail-feathers (t1) and outer web of others blackish-olive with green-yellow or yellow fringe on basal and middle portion; inner webs all-yellow, except for some blackish-olive along shaft at tip. Flight-feathers, inner web of tertials, greater upper primary coverts, and bastard wing greyish-black; outer webs of primaries narrowly and sharply fringed yellow to green-yellow, of secondaries and of shorter feathers of bastard wing more broadly so; outer webs and tips of primary coverts narrowly fringed green-yellow. Outer web of tertials grades from dusky olive at shaft through greenish to yellow along outer fringe. Lesser upper wing-coverts green-yellow, median blackish-olive

with *c.* 4 mm of tip yellow; greater coverts greyish-black (largely concealed), broad outer fringes yellow, narrow tips green-yellow. Under wing-coverts, axillaries, and inner borders of flight-feathers bright spectrum-yellow. ADULT FEMALE BREEDING. Like adult ♂ breeding, but cap yellow with green wash, not sharply demarcated from green-yellow of remainder of upperparts; upper cheek, ear-coverts, and side of neck tinged green; ground-colour of side of head and underparts slightly less bright yellow; rufous streaks on underparts narrower (0·5–1 mm wide), shorter, restricted to side of breast and flank, but sometimes forming narrow spotted gorget across upper chest; inner webs of tail-feathers all-yellow (t5–t6) or with broad yellow border along inner web (t2–t4) (in ♂, inner webs of t2–t6 almost entirely yellow). ADULT MALE NON-BREEDING. Intermediate between adult ♂ and adult ♀ breeding. Upperparts like adult ♀ breeding, but forehead and forecrown brighter yellow, less green-yellow, though not contrasting with remainder of upperparts; side of head and neck and all underparts bright spectrum-yellow, as adult ♂ breeding, but ear-coverts and side of neck washed green; rufous streaks on underparts almost as numerous and long as in adult ♂ breeding, but narrow (*c.* 1 mm wide); streaks more numerous and more sharply defined than in adult ♀ breeding. Wing and tail as adult ♂ breeding. ADULT FEMALE NON-BREEDING. Like adult ♀ breeding, but cap and side of head and neck green-yellow, like remainder of upperparts, with rather indistinct yellow supercilium and more distinct pure yellow eye-ring; side of breast and flank with some faint rufous streaks or none at all. JUVENILE. Upperparts and side of head and neck pale olive-brown, brown-grey, or pale grey-brown; eye-ring, chin, throat, chest, and side of breast pale grey-buff, remainder of underparts pale yellow or white, tinged buff on flank, no streaks. Upper wing-coverts dusky brown, median and greater broadly tipped yellow, greater with olive outer fringe. FIRST ADULT. Like adult, but juvenile tail, flight-feathers, greater upper primary coverts, bastard wing, and sometimes a few outer greater coverts and tertials retained. Tips of juvenile tail-feathers distinctly pointed, less broadly rounded than in adult; yellow on inner webs of tail-feathers more restricted, t2–t4 in ♂ with broad yellow border (as in adult ♀), in ♀ with narrow yellow fringe. Flight-feathers, bastard wing, and primary coverts browner and with more worn tips than those of adult at same time of year; primary coverts and bastard wing greyer-brown and with more frayed and greener fringes than neighbouring greater coverts, which are blacker and show purer yellow fringes. In non-breeding, cap and underparts of ♂ rather similar to adult ♀ non-breeding: cap mainly green-yellow, side of breast and upper flank usually with traces of faint rufous streaks; head and body of 1st adult ♀ similar to adult ♀ non-breeding, but yellow of side of head and underparts pale, white showing through on throat and belly, rufous streaks absent. In breeding plumage, head and body of 1st adult ♂ similar to adult ♂ breeding, but orange-yellow of crown sometimes less sharply defined at rear, ear-coverts tinged green, and rufous streaks on underparts sometimes slightly less extensive; head and body of 1st adult ♀ either similar to adult ♀ breeding, or yellow of side of head and underparts paler, underparts with a few traces of rufous streaks or without streaks.

Bare parts. ADULT, FIRST ADULT. Iris dark brown, dark sepia-brown, or black-brown. Bill horn-black to black, cutting edges and base of lower mandible blue-grey, but in at least some birds in 1st autumn upper mandible horn with purplish suffusion, cutting edges and lower mandible pearl-white to pale yellow, lower mandible with grey stripe at each side or greyish tip, mouth light pink. Leg and foot brown-yellow, yellowish-brown, or dark yellow-horn; in some 1st-autumn birds, medium horn with pale purple hue, toes lighter, and soles pale flesh. (Evans 1965; RMNH, ZMA.) JUVENILE. No information.

Moults. ADULT POST-BREEDING. Complete; primaries descendent. On breeding grounds, June–August, but some possibly suspend flight-feather moult over autumn migration. (Pyle *et al.* 1987). At south-west corner of James Bay (Ontario, Canada), ♂ started moult with p1 8·1 (1–19) days after young fledged, ♀ after 12·3 (9–15) days; in population, some overlap between care of fledglings and moult; p1 on average shed on 17 July (♂) or 22 July (♀), mainly 9–30 July, latest 5 August; duration of entire moult (from shedding of p1 to regrowth of s6) on average 43 days (♂) or 37 days (♀), some birds being almost flightless; moult on average completed 28 August in both sexes, but migration may start during final stages of moult (Rimmer 1988, which see for sequence of moult and many other details). ADULT PRE-BREEDING. Partial: head, body, lesser and median upper wing-coverts, and sometimes variable number of greater upper wing-coverts and tertials. In winter quarters, February–April (Dwight 1900; Pyle *et al.* 1987). POST-JUVENILE. Partial, extent as in adult pre-breeding; June–August, in breeding area (Pyle *et al.* 1987). Birds with moult completed occur from July onwards (ZMA). FIRST PRE-BREEDING. As in adult pre-breeding. Birds starting moult (chest and mantle first) occur from mid-December, birds with moult completed from mid-February; however, some ♀♀ from late March are still largely in worn non-breeding except for some feathers growing on chest and side of breast (RMNH, ZMA).

Measurements. *D. p. aestiva.* Eastern USA, summer, and Surinam, winter (latter sample may include some *amnicola*); skins (RMNH, ZMA). Bill (S) to skull, bill (N) to distal corner of nostril; exposed culmen on average 3·3 shorter than bill (S).

WING	♂	64·3 (1·37; 20) 62–67	♀	61·4 (1·74; 18)	59–65
TAIL		43·5 (1·61; 20) 41–46		42·2 (1·34; 16)	40–44
BILL (S)		13·5 (0·44; 18) 12·9–14·2		13·2 (0·40; 15)	12·5–14·0
BILL (N)		7·8 (0·28; 19) 7·4–8·4		7·6 (0·33; 15)	7·1–8·1
TARSUS		18·7 (0·62; 20) 17·7–19·5		18·5 (0·40; 15)	17·9–19·3

Sex differences significant for wing and tail. Ages combined, though retained juvenile wing and tail of 1st adult on average shorter than in adult. Thus, in ♂: wing, adult 65·4 (1·24; 6) 64–67, 1st adult 63·8 (1·14; 14) 62–66; tail, adult 44·3 (1·08; 6) 43–46, 1st adult 43·2 (1·71; 14) 41–46.

See also Raveling and Warner (1978), Wiedenfeld (1991), and (for influence of abrasion on wing length) Francis and Wood (1989).

Weights. *D. p. aestiva* and/or *amnicola.* Pennsylvania (USA): (1) adult, July–September, (2) April–June (Clench and Leberman 1978). (3) Pelee Island (Lake Erie, Canada), May (Woodward and Lovesy 1958). (4) North-west Yukon (Canada), late May and June (Irving 1960). South-west James Bay (Canada): (5) 1st half of June, (6) 2nd half of July (Rimmer 1988, which see for many other data). (7) Minnesota (USA), autumn (Raveling and Warner 1978). (8) Surinam and Trinidad, September–March (RMNH, ZMA).

(1)	♂	10·5 (1·04; 11) 9·0–12·8	♀	10·5 (1·98; 11) 8·8–16·0
(2)		9·8 (0·77; 175) 7·9–12·5		9·1 (0·72; 128) 7·4–12·0
(3)		10·7 (— ; 6) 9·9–11·8		9·4 (— ; 4) 9·3–9·5
(4)		9·3 (0·66; 22) 8·1–10·3		9·6 (0·74; 11) 8·5–10·8
(5)		9·9 (0·58; 35) —		9·1 (0·43; 16) —
(6)		9·4 (0·42; 15) —		9·6 (0·56; 29) —
(7)		10·6 (— ; 27) 9·3–12·3		10·1 (— ; 29) 9·1–11·8
(8)		9·2 (0·83; 7) 8·0–10·8		9·1 (0·91; 7) 8·1–10·5

Northern Alaska, late May: ♂♂ 8·9, 9·4 (Irving 1960). Pennsylvania, July–September: 1st adult 10·1 (1·36; 37) 8·0–13·8

(Clench and Leberman 1978). Exhausted ♂ on ship in mid-Atlantic (*c.* 38°N 46°W), August: 6·0 (ZMA). Bardsey (Wales), August: 9·0 (Evans 1965). See also Baldwin and Kendeigh (1938), Tordoff and Mengel (1956), Salt (1963), Russell (1964), Mengel (1965), Heimerdinger and Parkes (1966), and (for development of juvenile) Schrantz (1943).

Structure. Wing rather short, broad at base, tip fairly rounded. 10 primaries: p7–p8 longest, p9 0·5–1 shorter, p6 0·5–2, p5 4–6, p4 7–9, p3 9–12, p2 11–13, p1 14·2 (12) 13–17 (in extralimital races of West Indies and Central and South America, wing-tip more rounded, p6–p8 longest, p9 1–6 shorter, p5 2–4; p1 8–12, on average 10·1, $n = 10$). P10 reduced, a tiny yellow spike hidden below slightly longer black-and-green outermost greater upper primary covert; 40–46 shorter than p7–p8, 4–9 shorter than longest upper primary covert. Outer web of p6–p8 emarginated (sometimes faint on p6); inner web of p6–p9 with notch (sometimes faint or almost absent on p6). Tip of longest tertial reaches to tip of p1–p2 in closed wing. Tail rather short, tip square; 12 feathers. Bill rather short, straight, and slender; visible length of culmen *c.* 50% of head length; rather deep at base (but relatively less so than in *Vermivora*), depth at mid-nostril 3–4 mm; rather wide at base (but base less flattened than in *Wilsonia*), width at base of lower mandible 4–4·5 mm; distal half of culmen and of cutting edges of upper mandible slightly decurved (more so than in *Vermivora*). Terminal part of cutting edges of upper mandible with fine notch. Nostrils oval, partly covered by membrane above. About 6 fine short but distinct bristles projecting obliquely down from base of upper mandible. Tarsus rather long, toes rather short; slender. Middle toe with claw 13·4 (5) 13–14; outer toe with claw *c.* 76% of middle with claw, inner *c.* 71%, hind *c.* 83%. Bases of front toes partly fused.

Geographical variation. Marked and complex. Involves size (as expressed in weight, bill length, or tarsus length), relative wing length, wing shape (e.g. position of tip of p9 in relation to tips of other primaries), occurrence and extent of chestnut or tawny on head and throat, colour of upperparts (yellower or greener), and width and extent of streaks on underparts. In *aestiva* group ('Yellow Warbler') of North America south to Mexico, size smaller, wing relatively longer and more pointed (absolute wing length rather short), head of ♂ yellow or with limited tawny-orange on forehead and forecrown; in *petechia* group ('Golden Warbler') of West Indies, size larger, wing relatively shorter and more rounded, colour generally deeper yellow, head of ♂ in most races with distinct chestnut or tawny cap; in *erithachorides* group ('Mangrove Warbler') of coasts of Mexico through Central America to north-west South America, size large, wing relatively short with rounded tip, tarsus often slightly shorter than in previous group, head and throat of ♂ extensively tawny or chestnut (Ridgway 1901–11; Hellmayr 1935; Wiedenfeld 1991). Each group sometimes considered a separate species, but some races form links between groups, and others combine (e.g.) colour of one group with measurements of another (Wiedenfeld 1991); all thus united in single species. *D. p. aestiva* rather yellow-green on upperparts, bright yellow below in both sexes. *D. p. amnicola* slightly smaller and differs in darker and purer green upperparts, paler yellow underparts, and narrower and paler yellow fringes of flight-feathers; ♂ with more restricted and duller yellow on forehead and with darker chestnut streaks on underparts. Within range of *amnicola*, variation clinal, birds from Newfoundland rather near *aestiva*, those at extreme north-west of range (Mackenzie delta in northern Yukon) dullest, (grey-)green on cap and upperparts and palest yellow below (Raveling and Warner 1978, which see for details).

CSR

Dendroica pensylvanica **Chestnut-sided Warbler**

PLATE 3
[between pages 64 and 65]

DU. Roestflankzanger   FR. Paruline à flancs marron   GE. Gelbscheitel-Waldsänger
RU. Желтошапочный лесной певун   SP. Reinita de costillas Castañas   SW. Brunsidig skogssångare

Motacilla pensylvanica Linnaeus, 1766

Monotypic

Field characters. 11·5-13 cm; wing-span 19-20·5 cm. 10% smaller than Blackpoll Warbler *Dendroica striata*. Quite small, delicate, thicket-loving Nearctic warbler. Pale green upperparts with double yellow-white wing-bar; white underparts and white in outer tail. Cocks tail. Sexes dissimilar; marked seasonal variation. 1st-winter separable from breeding adult.

ADULT MALE BREEDING. Moults: June–August (complete); February–April (head, body, forewing). Crown and back greenish-yellow, streaked with blackish lines behind almost completely black nape. Supra-loral stripe white, loral patch, eye-stripe, and surround to rear ear-coverts black, joining nape; loral patch also extends down from eye, narrowing and then joining thin, then broad, chestnut stripe from breast along flank. Long white panels on 3 outer tail-feathers. Bill blackish. Legs dusky. ADULT FEMALE BREEDING. Duller than ♂, with loral area merely smudged dusky, duller and more restricted lines on back, downward facial streak less wide, and chestnut stripe on breast and flank restricted or broken up. ADULT NON-BREEDING. Appearance much changed, with brighter but rather pale green crown and upperparts, only faintly streaked when fresh, and unmarked underparts except for chestnut stripe or spots on fore-flank on ♂ and a few ♀♀. Face loses all strong markings, showing instead bright white eye-ring set off by grey lore and ear-coverts. FIRST-WINTER. Resembles non-breeding ♀ but has yellower wing-bars and smaller white panels on only 2 outer tail-feathers. ♀ never shows chestnut spots on flank. FIRST-SUMMER. As adult but duller; retained brownish wing-feathers may allow ageing.

Unmistakable, with face, breast, and flank pattern of breeding adult diagnostic, and green and white, almost ghostly, appearance of winter adult and 1st-winter equally distinctive. Beware slight chance of confusion with small, wing-barred *Phylloscopus* warblers (which always show prominent supercilium and eye-stripe and plain tail). Over distance, flight becomes slightly undulating. Spends much time perched in the open, often hunting insects on ground or flycatching in the air. Cocks tail and frequently moves with it raised and wings drooped.

Call a rich 'chip'.

Habitat. Breeds in eastern temperate Nearctic lowlands. Originally confined to limited ephemeral areas of early second growth in cleared or burnt woodland, and woodland edges created by streams, swamps, or other natural limits. Massive forest clearance, and widespread abandonment of subsequent farms and pastures, have greatly expanded suitable habitat, leading to much spread and increase. Prefers open and sunlit shrub growth, thickets, and briars, often by roads or streams, singing from tops of bushes or low trees, avoiding deep mature woods, and also farmyards, gardens, orchards, and neighbourhood of houses, although on migration it is seen on passage in city parks (Forbush and May 1955). On Great Plains, requires trees or shrubs for singing perches and shade, and moist dense low vegetation for nest-sites and foraging; these needs are met by briar thickets, forest clearings or edges, overgrown pastures, and similar open and dry habitats with varied vegetation (Johnsgard 1979).

Distribution. Breeds in North America east of Rockies, from Saskatchewan east to northern Nova Scotia, south to Nebraska, northern Ohio, and Maryland, and through Appalachians to northern Georgia and north-west South Carolina. Winters in Central America from Guatemala to central Panama.

Accidental. Britain: 1st-year, Shetland, 20 September 1985 (Rogers *et al.* 1988).

Movements. Long-distance migrant, wintering far south of breeding range. Numbers increased greatly in 1800s with forest clearance in eastern USA. Winter range much smaller than breeding range. (Bent 1953; Rappole *et al.* 1983; American Ornithologists' Union 1983.)

Autumn movement begins in August, and main migration period within breeding range is late August to mid-September (Palmer 1949; Stewart and Robbins 1958; Sprague and Weir 1984). From about mid- or late August, first arrivals appear from Great Lakes south to Mexican Gulf coast and even to Costa Rica (e.g. Imhof 1976, Janssen 1987, Stiles and Skutch 1989). Most birds leave northern USA and Canada by late September, and southern USA by end of October (e.g. Lowery 1974, Speirs 1985, Kale and Maehr 1990). Arrives Central America mostly late September to early November, later than many other Parulidae (Land 1970; Rogers *et al.* 1986a; Stiles and Skutch 1989). Movement through USA on relatively broad front, but scarce on east coast plain (Bent 1953; Stewart and Robbins 1958). Further south, data suggest birds mostly cross central Gulf of Mexico in autumn (regular east to western Florida), and western Gulf (including coast) in spring (Stevenson 1957; Rappole *et al.* 1979).

Rare in West Indies both seasons (Bond 1985). Winter ringing in Panama has shown site-fidelity, with returns up to 3 years later (Loftin 1977).

Leaves winter quarters March to early May (Land 1970; Ridgely 1981; Binford 1989). Earliest migrants reach Louisiana and Texas late March, some moving into lower Mississippi valley by early April (Bent 1953; Lowery 1974; Oberholser 1974). Main movement, mostly bypassing south-east states (Burleigh 1958; Rappole *et al.* 1979), late April to early May, with migration front advancing from southern USA to southern Canada in less than a month (Bailey 1955; Potter *et al.* 1980; James and Neal 1986; Janssen 1987). Reaches northern limits of breeding range by late May (Todd 1963; Erskine 1985; Speirs 1985).

Individuals stray west into Rocky Mountain states, and north-west to British Columbia, in spring (Bent 1953; Erskine and Davidson 1976; Cannings *et al.* 1987), but most records in south-west are in autumn (Phillips *et al.* 1964; Garrett and Dunn 1981). Vagrant north to Middleton Island (southern Alaska) and southern Greenland (Bent 1953; American Ornithologists' Union 1983). For vagrancy to west Palearctic, see Distribrution.   AJE

Voice. See Field Characters.

Plumages. ADULT MALE BREEDING. Forehead and crown bright yellow, contrasting sharply with broad black stripe running from lore, just above eye, over upper ear-coverts, and widening on side of hindneck; sometimes narrow white line between yellow and black above lore and eye. Rear of yellow crown less sharply defined; hindneck broadly streaked black, feather-fringes yellow or partly white, central hindneck sometimes with yellow spot. Upper mantle broadly streaked sooty-black and off-white (pale feather-fringes broader than on hindneck, mantle appearing paler); lower mantle, scapulars, back, and rump broadly streaked black and pale green-yellow. Upper tail-coverts black, rather narrowly fringed green-yellow (on shorter coverts) or ash-grey (on longer ones). Black of lore extends into broad black stripe along gape over lower cheek. Lower rear of eye-ring, upper cheek, lower ear-coverts (between black of gape, lower cheek, and upper ear-coverts), and side of neck greyish-white. Underparts greyish-white or white, each side of body with deep chestnut stripe, extending from lower side of neck over side of breast towards thigh, widest at side of breast, narrower towards front and rear of body, sometimes bordered by narrow black stripe at border of breast and belly. Tail black, narrow fringes along both webs of central pair (t1) and along outer webs of t2-t5 pale grey; t4 with white wedge *c.* 10 mm long on tip of inner web, t5 with larger white patch, t6 with almost entire tip of inner web white for *c.* 22-30 mm; remainder of inner borders of t2-t6 fringed white. Flight-feathers, greater upper primary coverts, and bastard wing black; primaries (except emarginated parts), primary coverts, and shorter feathers of bastard wing narrowly but sharply fringed ash-grey along outer webs, secondaries narrowly fringed bright green or pale yellow-green. Tertials black with broad pale yellow outer fringe. Greater upper wing-coverts black, outer ones with pale green fringe and yellow-white spot on tip of outer web, inner ones with broader pale yellow outer fringe and entire tip broadly white (subterminally) to pale yellow (at fringe), forming short broad bar on inner wing. Lesser and median upper wing-coverts black, lesser narrowly fringed grey

(outer) or green-yellow (inner), median broadly tipped white (subterminally) to pale yellow (on fringe), forming full broad wing-bar. Under wing-coverts, axillaries, and inner fringes of flight-feathers white, sometimes tinged pale grey or slightly yellow, shorter coverts with some black of bases exposed. *In fresh plumage* (early spring), tips of crown feathers slightly suffused olive, black of stripe over eye with narrow pale yellow feather-fringes, white of patch behind eye tinged pale yellow or cream, green-yellow fringes of upperparts fairly broad, partly concealing black of feather-centres; *in worn plumage* (about June), some black of feather-bases of cap sometimes exposed, pale fringes of upperparts, tertials, and tail partly worn off, black more prominent, whitish of underparts dirty pale greyish. ADULT FEMALE BREEDING. Intermediate in plumage between adult ♂ breeding and non-breeding plumages. Cap green-yellow with variable olive-green suffusion, not sharply demarcated from pale buff or ash-grey strip at base of upper mandible and from ash-grey hindneck, which shows green suffusion and variable black streaking. Remainder of upperparts bright green, ill-defined dark grey or sooty of feather-centres partly visible, as well as some ash-grey of feather-bases; rather narrow sharp black streaks restricted to lower mantle, inner scapulars, and upper tail-coverts. Side of head and neck mainly dull ash-grey, contrasting rather with green of cap; spot on lore and at gape dull black (not as black as adult ♂), extending into irregular and variable grey-and-black mottled stripes over eye and upper ear-coverts and on lower cheek; ear-coverts with paler cream or off-white spots. Underparts as adult ♂ breeding, but rufous-chestnut at side of body more restricted, forming broken band mainly from lower side of neck to side of breast, sometimes with a few pale rufous spots or streaks on flank, not reaching thigh. Tail as adult ♂, white patch on t6 17-27 mm long. Wing as adult ♂, but pale yellow outer fringe of tertials narrower and greener, pale tips of median and greater coverts slightly smaller (*c.* 3 mm long on median and inner greater, rather than 5-7 mm; *c.* 2 mm on outer greater, rather than 3-5 mm), fringes of lesser coverts more broadly and extensively green (less narrow and grey). ADULT MALE NON-BREEDING. Rather different from breeding plumage, showing almost no black on upperparts and side of head. Upperparts entirely bright green, ash-grey of feather-bases visible on hindneck, outer scapulars, and rump; dull black of centres of outer and lower scapulars, back, and shorter upper tail-coverts partly exposed. Longer upper tail-coverts black with broad ash-grey fringes. Side of head and neck medium ash-grey, contrasting with green of cap, paler grey towards feathering at nostril and lower cheek; narrow but conspicuous white eye-ring, ear-coverts mottled off-white. Chin to chest, side of breast, and upper side of body pale ash-grey, stripe at border of flank and belly chestnut, almost reaching thigh but not extending up to side of breast. Belly to under tail-coverts pure white. Tail and wing as adult ♂ breeding, but pale tips of median and inner greater coverts slightly smaller, more yellow. ADULT FEMALE NON-BREEDING. Like adult ♂ non-breeding, but upperparts more uniform bright green, ash-grey feather-bases not exposed unless worn; black of centres of scapulars, back feathers, and upper tail-coverts on average slightly less deep, less sharply defined, and less exposed. Side of head and neck and all underparts as adult ♂ non-breeding, but side of breast, flank, and chest partly washed green, and no chestnut at border of belly and flank or only some small, faint rufous spots. Tail and wing as adult ♀ non-breeding; wing-bars average narrower than in adult ♂ non-breeding. Often hard to distinguish from adult ♂ non-breeding, except for narrower wing-bars (not differing in colour from adult ♂, contra Pyle *et al.* 1987, bars in both sexes yellower when fresh, whiter when worn) and virtual absence of chestnut on flank. JUVENILE.

Upperparts uniform brown-olive, scapulars and upper tail-coverts with ill-defined darker centres; side of head and neck, chin to chest, and side of breast pale buff-grey or drab-grey, eye-ring off-white; flank pale buff, remainder of underparts dull white. For wing and tail, see First Adult Non-Breeding. FIRST ADULT NON-BREEDING. Like adult non-breeding, sometimes indistinguishable. Juvenile tail, flight-feathers, tertials, greater upper primary coverts, and bastard wing retained, but these little different from adult in autumn; tips of juvenile tail-feathers pointed, but in some adults also; white on t6 18–27 mm long in ♂, 16–27 mm in ♀. Upperparts almost uniform bright green, as adult ♀ non-breeding, in some birds with black feather-centres on scapulars or back (irrespective of sex); upper tail-coverts with narrower black central streak than in adult ♀ non-breeding, especially in 1st adult ♀, but some overlap; fringes of tail-coverts in some 1st adult ♀♀ largely green, not as grey as adult ♀ or 1st adult ♂. Underparts as adult ♀ non-breeding, flank, side of breast, and sometimes chest slightly washed green, vent often with some bright yellow feathers; ♂ with a few rufous feathers at border of flank and belly or with irregular rufous-chestnut stripe, similar to some adult ♀ non-breeding or sometimes more extensive; ♀ without rufous. Best ageing character is tertials: gradually narrowing towards rounded tip in 1st adult, broader and with more truncate tip in adult. FIRST ADULT BREEDING. Like adult breeding, but juvenile tail-feathers, flight-feathers, primary coverts, and bastard wing still present, browner (less black) and with more frayed tips than those of adult at same time of year, but some indistinguishable. Head and body of 1st adult breeding ♂ similar to adult ♂ breeding, but feathering at base of upper mandible whitish or pale buff, less yellow; rear of crown sometimes washed green; upper mantle sometimes streaked dark grey and off-white, dark streaks narrower and less deep black; black on longer upper tail-coverts averages narrower; black on side of head less deep, more greyish, stripes shorter and less distinct; chestnut along flank slightly less deep, less continuous; some birds hard to distinguish from adult ♀ breeding. Head and body of 1st breeding ♀ rather variable; in some birds, similar to adult ♀ breeding, cap mixed green and yellow, hindneck to rump with fairly extensive (though narrow) dark grey or black streaks, some dull black on lore and at gape, and some rufous on side of breast and flank; others almost identical to 1st adult non-breeding, upperparts bright green with limited dull black streaking on lower mantle and inner scapulars, lore, upper ear-coverts, and lower cheek ill-defined dusky grey mixed with some black, side of breast with only traces of rufous. Width of wing-bars variable, dependent on whether coverts are replaced in pre-breeding moult or not.

Bare parts. ADULT, FIRST ADULT. Iris brown or dark brown. Bill black, cutting edges and base of lower mandible black or bluish-black in spring and summer, dark blue-grey or leaden-grey in adult autumn and winter, bluish-flesh or pink-yellow in 1st adult in late summer and autumn. Leg and foot lead-colour to bluish-black, almost black in ♂ in summer, sometimes slightly greenish in ♀; fissures of joints and toes as well as soles paler dark blue-grey, lead-grey, or greyish-black. (RMNH, ZMA.) JUVENILE. No information.

Moults. ADULT POST-BREEDING. Complete; primaries descendent. On breeding grounds, June–August. ADULT PRE-BREEDING.

Partial: head, body, lesser, median, and inner greater upper wing-coverts, and variable number of tertials. In winter quarters, February–April. POST-JUVENILE. Partial: head, body, and lesser, median, and greater upper wing-coverts; on breeding grounds, late June to August. FIRST PRE-BREEDING. As adult pre-breeding, but moult in at least some ♀♀ largely suppressed, involving scattered feathering of head, mantle, chin to chest, and side of breast only. (Dwight 1900; Pyle et al. 1987; RMNH, ZMA.)

Measurements. ADULT, FIRST ADULT. Eastern USA, May–September, and Mexico to Panama, October–April; skins (RMNH, ZMA). Bill (S) to skull, bill (N) to distal corner of nostril; exposed culmen on average 3·0 less than bill (S).

	♂		♀	
WING AD	64·8 (1·51; 27)	62–68	61·5 (1·47; 17)	58–65
TAIL AD	47·3 (1·34; 21)	46–50	45·2 (1·36; 14)	43–48
BILL (S)	12·8 (0·29; 20)	12·0–13·3	12·5 (0·57; 14)	11·5–13·3
BILL (N)	7·5 (0·34; 20)	7·0–8·2	7·3 (0·24; 14)	7·0–7·8
TARSUS	17·8 (0·48; 20)	17·1–18·6	17·6 (0·60; 14)	16·5–18·4

Sex differences significant for wing and tail. Ages combined, though retained juvenile wing and tail of 1st adult on average shorter than in older birds. Thus, ♂: wing, adult 65·8 (0·79; 9) 64–67, 1st adult 64·3 (1·54; 18) 62–68; tail, adult 47·4 (1·25; 8) 46–50, 1st adult 47·2 (1·44; 13) 46–50.

Weights. Pennsylvania (USA): August–September, adult ♂ 9·7 (0·56; 19) 8·9–11·2, adult ♀ 9·3 (0·67; 16) 7·7–10·5, 1st adult 9·3 (0·71; 235) 7·8–11·6; May, ♂ 9·9 (0·95; 45) 8·1–13·1, ♀ 9·5 (0·72; 26) 7·5–10·9 (Clench and Leberman 1978). Georgia (USA), early October: ♂ 13·0 (10) 12·0–14·3, ♀ 12·2 (18) 10·2–13·9 (Johnston and Haines 1957). East Ship Island (Mississippi, USA), spring: 8·9 (1·0; 23) (Kuenzi et al. 1991). Pelee Island (Lake Erie, Canada), May: ♂ 10·0 (8) 9·1–11·1, ♀ 11·1 (6) 10·6–11·5 (Woodford and Lovesy 1958).

Mexico, ♂: April 9·5, October 9·0 (RMNH). Belize, October: ♂ 9·5 (3) 8·9–10·0, ♀ 8·9 (0·43; 5) 8·3–9·3 (Roselaar 1976; ZMA). Nicaragua, exhausted ♂, October: 6·1 (ZMA). Panama, mainly October: 8·0 (0·58; 20) 6·8–9·1 (Rogers and Odum 1966). Belize, April–May: ♂♂ 9·9, 10·5; ♀♀ 9·5, 10·0 (Russell 1964).

See also Baldwin and Kendeigh (1938), Tordoff and Mengel (1956), Norris and Johnston (1958), Rand (1961b), Graber and Graber (1962), Mengel (1965), and (for fat-free weight) Rogers and Odum (1964, 1966).

Structure. Wing rather short, broad at base, tip fairly rounded. 10 primaries: p7–p8 longest, p9 and p6 0·5–1·5 shorter, p5 4·5–7, p4 7–10, p3 9–13, p2 11–14, p1 12·5–15. P10 strongly reduced, forming a spike hidden beneath reduced outermost upper greater primary covert, which is 0–1 longer than p10; 41–46 shorter than p7–p8, 4–7 shorter than longest upper primary covert. Outer web of p6–p8 emarginated (sometimes faintly on p6 and/or p8); inner web of p7–p9 with notch (sometimes faint). Tip of longest tertial reaches tip of p1–p2. Tail rather short, tip square; 12 feathers. Bill as in Yellow Warbler D. petechia, but relatively slightly stouter at base and in middle. Tarsus rather short, toes rather long; slender. Middle toe with claw 13·9 (5) 13–14·5; outer toe with claw c. 73% of middle with claw, inner c. 68%, hind c. 77%. Remainder of structure as in D. petechia.

Geographical variation. None in plumage or size. For variation in song, see Lein (1978) and Kroodsma (1981).   CSR

Dendroica caerulescens **Black-throated Blue Warbler**

PLATE 3
[between pages 64 and 65]

Du. Blauwe Zwartkeelzanger   Fr. Paruline bleue à gorge noire   Ge. Blaurücken-Waldsänger
Ru. Синеспинный лесной певун   Sp. Bijirita sombria   Sw. Blåryggad skogssångare

Motacilla caerulescens Gmelin, 1789

Polytypic. Nominate *caerulescens* (Gmelin, 1789), southern Canada from south-east Manitoba and Ontario to Nova Scotia, and north-east USA from Great Lakes east to New England. Extralimital: *cairnsi* Coues, 1897, Appalachian mountains from West Virginia and Maryland south to north-east Georgia and north-west South Carolina.

Field characters. 12–14 cm; wing-span 17·5–19·5 cm. 15–20% smaller than Blackpoll Warbler *D. striata*, with rather more attenuated rear body and tail. Quite small, rather dark, bush-haunting Nearctic warbler; adult has diagnostic white patch at base of primaries. ♂ dark blue above, with black face, breast, and flank contrasting with otherwise white underparts; ♀ dusky-olive above and buff-yellow below, with narrow whitish supercilium and eye-ring. ♀ and immature have noticeable dark cheek; immature may lack bright primary mark. Call distinctive. Sexes dissimilar; no seasonal variation. 1st-winter difficult to separate.

ADULT MALE. Moults: July–September (complete). Crown, nape, upperparts, wings, and tail dusky-blue to blue-grey, with slightly paler-tipped, black-centred median and greater coverts and tertials, strikingly isolated white blaze on base of primaries, and blackish outer tail-feathers with white panels on inner webs decreasing in length towards centre and appearing as spots when spread. Face, breast, and broad splashed line along flank black, tinged blue on rear flank; fringe to upper breast and fore-flank white, forming 2nd isolated white mark at shoulder; rest of underparts pure white. Bill black. Legs dusky. ADULT FEMALE. Head, upperparts, wings, and tail dark dusky- to brownish-olive, with narrow pale yellow super-cilium and broken eye-ring above noticeably dark cheek; usually small dull white blaze at base of primaries, blackish covert-centres and flight-feathers, and bluish fringes and whitish panels on outer tail-feathers. Chin, throat, breast, and flank dull yellow, with dusky suffusion on side of breast and flank; belly and under tail-coverts yellowish-white. FIRST-YEAR MALE. Resembles adult, but in fresh plumage upperparts tinged greenish. Paler, greyer or greener fringes of alula and flight-feathers contrast with fully blue tips and fringes of greater coverts. FIRST-YEAR FEMALE. As adult but usually with buffier breast and duller wing-feathers. Often only vestigial or no whitish patch at base of primaries.

Adult ♂ unmistakable—as are ♀ and immature with white primary patch, but confusing in other plumages and best distinguished by noticeably dark cheek and sharp, narrow whitish supercilium and eye-crescent (this face pattern not present in any relative except Nashville Warbler *Vermivora ruficapilla*, but that smaller species lacks rear supercilium and tail-spots and has cleaner yellow underparts). Flight light and agile; fly-catches. Gait hopping. Stance level, with lengthy tail held up. Migrants often tame.

Commonest call soft, full 'tsep' or 'smack', recalling Dark-eyed Junco *Junco hyemalis*.

Habitat. Breeds in cool temperate eastern Nearctic low-lands and uplands, favouring hillsides and ridges carrying dense shady second-growth deciduous or mixed woodland, with dense undergrowth, especially laurel thickets. Often breeds near openings created by streams or roads and is active in treetops. (Forbush and May 1955.) Forages in shrub, subcanopy, and lower canopy layers up to *c.* 18 m above ground, mostly in outer branches and foliage, ♂♂ higher than ♀♀. Found in Minnesota in mixed coniferous as well as in deciduous forests, but seems to prefer latter, with undergrowth of small conifers, yews, laurel, or heather. (Johnsgard 1979.)

Distribution. Breeds in North America east of Rockies, from south-east Manitoba to Nova Scotia, south to Minnesota, Michigan, Appalachians (south to northern Georgia), and Massachusetts. Winters mainly in Bahamas and Greater Antilles, in smaller numbers also in northern Central America and northern South America.

Accidental. Iceland: ♂, Heimaey (Vestmannaeyjar), 14–19 September 1988 (Petersen 1989).

Movements. Migrant, even southernmost breeders moving nearly 1500 km to winter chiefly in West Indies. Migrates in vast numbers along Florida coasts. (Bent 1953; American Ornithologists' Union 1983.)

In autumn, migrates relatively late, few leaving breeding areas before late August (Bent 1953; Bull 1974; Janssen 1987). Peak passage in mid-latitude states mid-September to early October (Stewart and Robbins 1958; Hall 1981), and passage in Florida extends throughout October (Kale and Maehr 1990). Migrates (both seasons) chiefly in and along both sides of Appalachians, birds from west of range crossing into Florida over and through lower southern mountains; rare west of Alabama (Bent 1953; Lowery 1974; Toups and Jackson 1987). Winter site-fidelity shown by ringing in Jamaica and Haiti, with returns up to 3 years later (Diamond and Smith 1973; Woods 1975).

In spring, early birds recorded Florida in March (Kale

and Maehr 1990), but little migration before mid-April (Burleigh 1958; Imhof 1976; Potter *et al.* 1980), and last birds leave winter range early or mid-May (Bent 1953; Bond 1985; Raffaele 1989). Birds that breed in southern mountains arrive a few days earlier than migrants passing north through adjacent lowlands (Bent 1953). Earliest migrants west of Appalachians sometimes penetrate to lower Great Lakes before end of April (Quilliam 1973; Bull 1974; Hall 1983), but main arrival in breeding range is in May, reaching northern limits by end of May (Stewart and Robbins 1958; Squires 1976; Speirs 1985; Janssen 1987).

Vagrant (chiefly autumn) west to south-east British Columbia and west coast of USA, also north to Newfoundland and nearby St-Pierre-et-Miquelon (Garrett and Dunn 1981; Etcheberry 1982; American Ornithologists' Union 1983; Godfrey 1986), and to Greenland (Olsen 1991*b*); northward vagrancy seems exceptional. For vagrancy to west Palearctic, see Distribrution.   AJE

Voice. See Field Characters.

Plumages. (nominate *caerulescens*). ADULT MALE. Entire upperparts and side of neck dark slate-blue except for black feathers at base of upper mandible; centres of feathers of lower mantle and upper tail-coverts sometimes partly black. Side of head from lore, strip just above eye, and ear-coverts down to chin and throat black, extending into broad black stripe along lower side of breast and lower flank, widening at rear of flank. Patch at upper side of breast (often concealed by wing-bend) and remainder of underparts from chest to under tail-coverts white, sometimes with slight cream tinge; thigh mottled dark grey and white. Tail black, both fringes of central pair (t1) and outer fringes of t2–t5 dark slate-blue; t3 with small white subterminal spot on border of inner web (sometimes white fringe only); t4–t5 with large white subterminal blobs (larger on t5); 11–23 mm of tip of inner web of t6 white (measured along shaft), only small triangular patch on tip black. Flight-feathers, tertials, greater upper primary coverts, and bastard wing black, outer webs of primaries (except emarginated parts), primary coverts, and shorter feathers of bastard wing with narrow grey-blue fringe, secondaries and (especially) tertials with broader grey-blue fringes; bases of primaries contrastingly white, forming conspicuous patch on upperwing, white extending 9·6 (10) 6–13 mm beyond longest upper primary covert on outer web. Greater upper wing-coverts black, rather narrow fringes along outer web and tip slate-blue; median and lesser coverts slate-blue, centres of median black (mainly concealed). Under wing-coverts, axillaries, and basal inner borders of flight-feathers white, but marginal coverts black and longer primary coverts dark grey. *In fresh plumage* (autumn), feathers of mantle, scapulars, and back sometimes faintly tipped olive-green, white of underparts suffused light pink-buff (especially on vent), tips of black feathers on underparts narrowly (on throat) or more widely (on flank) fringed white. *In worn plumage* (early summer), some black of feather-bases sometimes exposed on crown, mantle, and scapulars, mid-underparts dirty white. ADULT FEMALE. Upperparts and side of neck entirely dull olive-green, occasionally with slight blue-grey cast, hindneck and rump sometimes slightly paler. Feathers at nostril and narrow but distinct supercilium and eye-ring pale cream-yellow, supercilium extending to above middle of ear-coverts, eye-ring broken in

front and behind. Lore and cheek dull grey, merging into dark olive-green on ear-coverts. Side of breast and upper flank greyish-olive with warm buff wash, remainder of underparts entirely pale buff-yellow, palest on mid-belly and under tail-coverts. Tail greyish-black, t1 and outer fringes of t2–t5 tinged bluish-grey; spots on terminal part of inner web of tail-feathers as in adult ♂, but absent on t3, reduced to fringe on t4, and smaller, more ill-defined, and tinged greyish on t5–t6, thus far less conspicuous; length of subterminal pale spot on t6 4–11 mm, but spot only slightly paler grey than remainder of inner web and hard to measure in some birds. Flight-feathers, wing-coverts, and bastard wing patterned as adult ♂, but black replaced by dark grey or greyish-black, slate-blue by olive-green; patch at base of primaries smaller, suffused isabelline, 4·6 (5) 1–10 mm long. *In worn plumage* (June–July), upperparts and upper wing-coverts duller and darker greyish-olive, less greenish, sometimes slightly bluish-grey on forehead and lesser wing-coverts; lore, cheek, and ear-coverts dark grey with faint pale spots on cheek and shorter ear-coverts; flank olive-grey, remainder of underparts pale yellow, mottled off-white and pale grey from chin to breast. JUVENILE. Upperparts and lesser and median upper wing-coverts dark olive-brown; supercilium, upper and lower part of eye-ring, and upper cheek pale buff or yellow-white, lore, front and rear of eye-ring, and stripe over lower cheek sooty-black, remainder of side of head and all side of neck dark brown; underparts ash-grey or grey-brown with olive tinge, but throat pale buff and mid-belly to under tail-coverts sulphur-yellow. Sexes differ in wing and tail: fringes of wing in ♂ partly or entirely blue-grey, in ♀ entirely olive-green; white spots on tail in ♀ inconspicuous, white patch on bases of primaries of ♀ small; see First Adult Male and First Adult Female (below). For development, see Harding (1931). FIRST ADULT MALE. Like adult ♂, but juvenile tail, flight-feathers, tertials, primary coverts, bastard wing, and sometimes a few outer greater coverts retained. Juvenile wing-feathers greyer than those of adult ♂ when fresh, less black, distinctly browner and with more frayed tips when worn; blue-grey fringes often partly tinged green, especially those of tertials, primaries, middle feather of bastard wing, and (if retained) outer greater coverts; in particular, black-brown-centred and partly green-fringed tertials and primary coverts contrast strongly with new blue-fringed greater coverts. White patch on primary-bases 7·3 (13) 5–11 mm. Tail-feathers often narrower than in adult, tips more sharply pointed; white spot on t3 usually absent, those on t4–t6 smaller and more rounded than in adult ♂, occasionally tinged grey (as in adult ♀); length of white patch on inner web of t6 5–15 mm, measured along shaft, but in some birds white does not reach shaft or is replaced by grey. Head and body as adult ♂, but fringes along feather-tips of upperparts more broadly olive-green, those of black feathers on underparts more broadly white; in worn plumage (spring), often an olive cast on hindneck and mantle (less clear slate-blue than adult ♂), flank-stripe less solid black. FIRST ADULT FEMALE. Like adult ♀, but much juvenile feathering of wing and all juvenile tail retained, as in 1st adult ♂. Difference in colour between 1st adult and juvenile feathers much less marked than in 1st adult ♂; tertials and primary coverts slightly browner and with less pure olive-green fringes than neighbouring 1st adult greater coverts, tips more frayed and less broadly rounded; patch on primary-bases 4·2 (7) 0–6 mm. Juvenile tail on average more sharply pointed than adult; subterminal pale spots on tail-feathers faint, tinged grey, usually hardly paler than remainder of inner web, rarely reaching shaft. Head, body, and 1st adult wing-coverts as in adult ♀, but without grey cast to upperparts and lesser coverts, and underparts on average slightly more yellowish when plumage fresh.

Bare parts. ADULT, FIRST ADULT. Iris brown or dark brown. Bill dark plumbeous-grey to black, darkest in breeding season; base of lower mandible tinged flesh in 1st adult in autumn. Leg and foot dull greyish-flesh or pink-grey; in ♂ in breeding season, dark plumbeous-grey or greyish-black. (RMNH, ZMA.) JUVENILE. Iris dusky blue or dark brown. Bill flesh-grey to straw, culmen and tip dark grey; traces of pale yellow gape-flanges present at fledging; mouth pink. Leg and foot pink-flesh with dark grey scutes on front of tarsus and upper surface of toes. (Harding 1931; RMNH.)

Moults. ADULT POST-BREEDING. Complete; primaries descendent. On breeding grounds, July–September. ADULT PRE-BREEDING. Restricted to limited feathering of head, or no moult at all; in winter quarters, February–April. POST-JUVENILE. Partial: head, body, lesser, median, and usually all greater upper wing-coverts; no flight-feathers, tertials, greater upper primary coverts, bastard wing, or tail. On breeding grounds, late July to early September. FIRST PRE-BREEDING. As adult pre-breeding. (Dwight 1900; Pyle *et al.* 1987; RMNH, ZMA.)

Measurements. Nominate *caerulescens*. North-east USA, May–September, and West Indies, October–April; skins (RMNH, ZMA). Bill (S) to skull, bill (N) to distal corner of nostril; exposed culmen on average 3·1 less than bill (S).

	♂		♀	
WING	66·0 (1·03; 20)	64–68	62·7 (1·17; 12)	61–65
TAIL	49·3 (1·20; 20)	47–51	48·8 (2·06; 12)	45–52
BILL (S)	13·3 (0·42; 19)	12·5–13·8	12·7 (0·30; 12)	12·2–13·2
BILL (N)	7·2 (0·37; 18)	6·5–7·9	7·1 (0·31; 12)	6·5–7·5
TARSUS	18·6 (0·55; 11)	18·0–19·5	18·5 (0·46; 11)	17·8–19·2

Sex differences significant for wing and bill (S). Ages combined, though retained juvenile wing and tail of 1st adult sometimes different from those of older birds. Thus, ♂: wing, adult 66·6 (1·18; 7) 65–68, 1st adult 65·6 (0·77; 13) 64–67; tail, adult 48·9 (1·43; 7) 47–51, 1st adult 49·5 (1·05; 13) 48–51.

Weights. Pennsylvania (USA): (1) August–October, (2) May (Clench and Leberman 1978).

(1) AD	♂	9·7 (— ; 7)	8·7–10·6	♀	10·4 (— ; 3)	9·1–12·1
1ST AD		9·8 (0·66; 35)	8·7–11·8		9·4 (0·90; 37)	8·2–12·7
(2)		9·9 (0·76; 9)	8·8–11·0		9·6 (0·84; 24)	8·1–11·7

Pelee Island (Lake Erie, Canada), May: ♀ 11·1 (3) 10·8–11·4 (Woodward and Lovesy 1958). Ohio (USA), ♂♂: 11, 12 (Stewart 1937). Kansas (USA), October, 1st adult: ♂♂ 13·8, 14·1; ♀ 11·4 (Tordoff and Mengel 1956).

On Jamaica, October–May, ♂ 9·0 (0·2; 7), ♀ 8·8 (0·4; 19); weight steady or decreasing slightly October–February, increasing again to May, when peak of 12·8 reached in ♀ (Diamond *et al.* 1977). Belize, November: ♀ 8·2 (Russell 1964). Bonaire (off Venezuela), October: ♂ 9·5 (ZMA).

For fat-free weight, see Connell *et al.* (1960).

Structure. Wing rather short, broad at base, tip fairly rounded. 10 primaries: p7–p8 longest or either one 0–1 shorter than other; p9 1–4 shorter, p6 0–2, p5 3·5–6, p4 6–10, p3 8–12, p2 10–14, p1 11–16. P10 strongly reduced, concealed below reduced outermost greater upper primary covert; 42–48 shorter than p7–p8, 4–8 shorter than longest upper primary covert. Outer web of p6–p8 emarginated; inner web of p7–p9 with notch (sometimes faint). Tip of longest tertial falls slightly short of tip of p1. Tail of average length, tip square; 12 feathers. Tarsus rather long, toes of average length; slender. Middle toe with claw 14·4 (10) 13·5–15·5; outer toe with claw *c.* 74% of middle with claw, inner *c.* 68%, hind *c.* 78%. Remainder of structure as in Yellow Warbler *D. petechia*.

Geographical variation. Slight, involving only colour; often no races recognized (Mayr and Short 1970). *D. c. cairnsi* from Appalachians differs from nominate *caerulescens* from further north in darker upperparts, ♂ having black spots or streaks, especially on mantle and scapulars, ♀ darker and duller olive above, darker olive on flank, and less yellow on belly. Grades into nominate *caerulescens* in north.

CSR

Dendroica virens (Gmelin, 1789) Black-throated Green Warbler

FR. Paruline verte à gorge noire   GE. Grünwaldsänger

A North American species, breeding from south-central British Columbia east to southern Labrador and Newfoundland, south to Minnesota, eastern Tennessee, northern Georgia, and South Carolina; winters from southern Texas and southern Florida south through Central America to Colombia, and in West Indies south to Guadeloupe and Dominica. Recorded in Germany (adult ♂, Helgoland, November 1858: Gätke 1900) and Iceland (immature ♀ found dead on ship, 19 September 1984: Petersen 1985).

Dendroica fusca Blackburnian Warbler

PLATE 3
[between pages 64 and 65]

DU. Sparrezanger   FR. Paruline à gorge orangée   GE. Fichtenwaldsänger
RU. Еловый лесной певун   SP. Silvia de Blackburn   SW. Orangestrupig skogssångare

Motacilla fusca P L Statius Müller, 1776

Monotypic

Field characters. 11–13·5 cm; wing-span 19–21 cm. 10–15% smaller than Blackpoll Warbler *D. striata*. Quite small, notably arboreal Nearctic warbler, with dark ear-coverts, pale braces on back, yellow foreparts, and mainly white outer tail-feathers. Breeding ♂ has rich orange throat and breast and white panel on mid-wing; ♀ has foreparts yellow and 2 white wing-bars; immature like dull ♀. Sexes dissimilar; some seasonal variation. Juvenile separable.

ADULT MALE BREEDING. Moults: July–August (complete); October–April (head, body, forewing). Centre of crown, supercilium, wide panel down rear ear-coverts, lower eye crescent, chin to breast, and fore-flanks orange, richest on face and throat and contrasting sharply with black crown, lore, ear-coverts, and line stretching down from behind eye. Back and rump black, with yellowish-white braces on mantle. Wings black but boldly patterned, with white mid-wing panel on inner greater coverts, tips of median coverts, and fringes and tips of outer greater coverts; white fringes to tertials. Tail black with all but outer fringes of outer feathers white, showing when spread. Underparts yellowish-white becoming pure white on rear flank and vent; lined with black spots over flanks. Bill black, with horn-grey to blue-grey base to lower mandible. Legs brown. ADULT FEMALE. Closely resembles dullest non-breeding ♂ but ear-coverts, line running down from there, and flank markings fainter, sometimes broken up on pale to rich yellow ground-colour. Dark head markings and upperparts dusky olive with blackish feather-centres and whitish braces. Wings dusky, with 2 white wing-bars only rarely joined on innermost greater coverts. For distinctions between breeding and non-breeding bird, see Plumages. ADULT MALE NON-BREEDING. Lacks vivid contrasts of breeding plumage, with mid-wing marks usually reduced to 2 white wing-bars, inviting confusion with ♀. Rest of plumage duller, with greyish-olive fringes to new black feathers and rich yellow rather than fully orange ground-colour to head and breast. FIRST-YEAR. Distinguished from adult by yellow-buff throat, indistinctly streaked upperparts, retention of dull brownish-fringed flight-feathers and bastard wing, and less white in outer tail. ♂ usually more contrasting and colourful than ♀, with more black on head and upperparts, but no clear distinction.

Adult unmistakable, with unique pattern on head, back, and foreparts, but immature troublesome, since most striking character of double white wing-bar shared by 4 known transatlantic vagrant Parulidae (including the commonest, *D. striata*) and 2 potential ones. Best distinguished from *D. striata* by much darker crown and cheek, pale braces on mantle, and fully yellow-buff fore-underparts; from Black-throated Green Warbler *D. virens* by streaked back, far less dusky wings and tail, and much yellower fore-underparts. Flight light and active though less fond of flycatching than other *Dendroica*. Hops. Stance level; silhouette rather compact.

Commonest call a rich 'chip'.

Habitat. Strictly arboreal, breeding in cooler temperate eastern Nearctic forest regions, mature stands of spruce,

hemlock, and pine being the primary breeding habitat, although deciduous and mixed second growth are also used. Likes swampy areas where old man's beard lichen *Usnea* flourishes (Johnsgard 1979). Most foraging is high in treetops (Pough 1949).

Wintering birds in South American tropics inhabit rain and cloud forests at 800–3100 m, occurring in dwarf forests at higher level and mangroves at sea-level, as well as in coffee plantations and second growth, foraging at all heights in foliage (Schauensee and Phelps 1978).

Distribution. Breeds in North America from south-central Saskatchewan east to central Quebec, south to central Minnesota, central Michigan, Appalachians south to northern Georgia, and Massachusetts. Winters in Central and South America from Guatemala south to Ecuador, central Peru, and Venezuela.

Accidental. Iceland: juvenile ♀ found exhausted on trawler, *c.* 65 km north-east of Horn (north-west Iceland), autumn 1987 (Petersen 1989). Britain: Skomer (Dyfed), 5 October 1961 (Saunders and Saunders 1992); 1st-winter ♂, Fair Isle, 7 October 1988 (Rogers *et al.* 1990).

Movements. Migrant, wintering far south of breeding range. Ecologically restricted in summer and only locally common, but in main wintering area, Colombia, occurs in various forest strata and habitats and is most abundant of Parulidae (Chipley 1980; Hilty and Brown 1986).

Autumn migration begins early August, or even late July (Bull 1974; Janssen 1987), and some migrants appear in August south to Costa Rica and Bahamas (Brudenell-Bruce 1975; Stiles and Skutch 1989), but main movement then is in north of range. Peak passage through eastern USA late August to late September (e.g. Imhof 1976, Stoddard 1978, Janssen 1987). By mid-October, many birds are in South America (Hilty and Brown 1986; Ridgely and Tudor 1989), but others straggle through eastern states throughout October. Migration mainly along or west of Appalachians, with fewer on Atlantic coast and in south-east states; most birds then cross (rather than fly round) Gulf of Mexico to Central America; spring route is further west than autumn, with yet fewer in south-east USA, and more on western Gulf coast and in Texas (Bent 1953; Burleigh 1958; Oberholser 1974; Rappole *et al.* 1979).

Spring migration begins March, earliest birds arriving north of Gulf then (Oberholser 1974; Imhof 1976). A few more move in early April (Mengel 1965; James and Neal 1986), but they penetrate little further until after mid-April. Characteristic warm air-masses moving up Mississippi valley in spring encourage earlier migration there, and arrivals occur late April north to Minnesota and Ohio, whereas migration on Atlantic coast is slower and involves fewer birds (Trautman 1940; Bailey 1955; Stewart and Robbins 1958; Janssen 1987). Peak passage late April in southern states (Imhof 1976; Potter *et al.* 1980) and early

May further north, in southern parts of breeding range (Palmer 1949; Sprague and Weir 1984; Bohlen 1989). By early May, last birds leave wintering areas (Stiles and Skutch 1989), and by late May northernmost breeding areas from Nova Scotia to Saskatchewan are occupied (Erskine 1985; Tufts 1986).

Vagrants appear (chiefly autumn) in many western states (e.g. Nevada, California) and in British Columbia (Garrett and Dunn 1981; American Ornithologists' Union 1983; Alcorn 1988.) For vagrancy to west Palearctic, see Distribrution.   AJE

Voice. See Field Characters.

Plumages. ADULT MALE BREEDING. Central forehead, broad stripe at side of forecrown, and entire hindcrown and hindneck black; contrasting orange-yellow patch on middle of forecrown, ending in front in narrow line over mid-forehead. Remainder of upperparts black with large white or yellow-white V on lower mantle and similar-coloured stripe down scapulars; some traces of off-white along sides of feathers of rump or upper tail-coverts. Long orange supercilium from side of forehead to upper side of neck. Lore and patch behind eye black, bordered behind by bright orange patch on side of neck; upper and lower part of eye-ring and small spot below eye orange, spot bordered below by narrow black line extending from gape to lower border of black patch behind eye and continued in broader black stripe over lower side of neck. Chin, throat, and chest fiery-orange; side of breast orange with bold black streaks, whiter towards axillaries. Breast and belly pale cream-yellow; flank, vent, and under tail-coverts white, flank contrastingly marked with bold black streaks *c.* 2 mm wide, longer under tail-coverts sometimes with narrower streaks. Central pair of tail-feathers (t1) black; t2-t3 black with broad white fringes along bases of both webs; t4-t5 all-white except for black shaft and sooty triangle on tip; t6 as t4-t5 but terminal half of outer web black. Flight-feathers, tertials, greater upper primary coverts, and bastard wing black; flight-feathers and tertials with narrow pale grey outer edge; bases of outer webs of outer primaries white, sometimes just visible as small patch at border of primary coverts. Outer 3 greater upper wing-coverts black with white spot on tip of outer web, inner all-white, innermost (tertial coverts) black with broad white outer fringe; median coverts white, lesser coverts black. Under wing-coverts, axillaries, and inner borders of flight-feathers white, some lesser and primary coverts mottled grey. *In worn plumage*, black of stripe at side of forecrown sometimes spotted yellow, mantle to upper tail almost uniform black, ground-colour of belly and flank dirty white. ADULT FEMALE BREEDING. Entire upperparts dark greyish-olive, each feather with black spot or short streak subterminally, showing as sharp black specks on forehead and crown, more diffuse sooty streaks on lower mantle, scapulars, and back, and bold black spots or streaks on central rump and upper tail-coverts; dark spots faint or absent on hindneck and side of rump; feather-centres on forecrown extensively yellow at base showing as yellow spot, but partly concealed by olive fringes and black specks when plumage fresh. Long distinct orange-yellow supercilium, extending into orange-yellow stripe down side of neck behind ear-coverts. Lore, upper cheek, ear-coverts, stripe from rear of cheek backwards to side of breast, and side of neck dark greyish-olive; small orange-yellow patch below eye, rear of ear-coverts almost black. Chin to chest bright orange-yellow, spotted or streaked dark

olive or sooty where bordering side of breast, sharply demarcated from pale yellow breast and belly. Flank, vent, and under tail-coverts yellow-white, flank narrowly and rather indistinctly marked with grey-olive or sooty shaft-streaks. Tail sooty-grey (less black than adult ♂), basal outer webs of outer feathers bordered white, remainder narrowly fringed olive; inner web of t3 sometimes with small white subterminal spot, inner web of t4-t6 with much white on terminal half, forming patches 12-18 mm (t4) to 22-30 mm (t6) long. Flight-feathers, tertials, greater upper primary coverts, and bastard wing sooty-grey, all (except longest feather of bastard wing) with narrow olive-grey fringe along outer web. Lesser upper wing-coverts sooty-grey, fringed dark grey-olive; median upper wing-coverts sooty-black with white or yellow-white tips 4-5 mm long, forming wing-bar; greater coverts sooty-black with narrow olive-grey fringe along outer web merging into white fringe along tip, but *c.* 4 central ones with tips to outer webs *c.* 4-8 mm wide, forming short bar or spot. Under wing-coverts and axillaries as adult ♂. *In worn plumage*, black streaks on upperparts more prominent, but olive still predominant; yellow patch on forecrown conspicuous; fringes along outer webs of flight- and tail-feathers greyish. ADULT MALE NON-BREEDING. Intermediate between adult ♂ breeding and adult ♀ breeding, but ground-colour of t1-t2, flight-feathers, and tertials much blacker than ♀. Upperparts black, each feather-tip narrowly or broadly fringed olive, but (unlike ♀) black predominant, except sometimes on hindneck; in particular, rump and upper tail-coverts almost uniform black; in birds with narrow olive fringes on crown and back, mantle and scapulars appear mainly black with contrasting yellow-white V, in birds with broader olive fringes on crown and back, mantle and scapulars more evenly streaked black and yellow-white; in all, patch on centre of forecrown distinct, uniform bright yellow or with some black specks. Pattern on side of head and neck as adult ♂ breeding, but black partly mottled olive, and orange replaced by yellow-orange. Chin to chest yellow-orange or orange-yellow, gradually merging into bright light yellow on breast and this in turn to pale yellow and white on flank, vent, and under tail-coverts; lower side of neck, side of breast, and flank with contrasting deep black streaks, rather short and 1-2 mm wide on side of breast, longer and 2-3 mm wide on flank. Tail as adult ♂ breeding, t3-t6 white except for black triangle on tip and black shaft. Wing as adult ♂ breeding, but fringes along outer webs of secondaries and p1-p8 slightly broader and more olive, those along tertials broader and whiter; central greater upper wing-coverts with some black at base, less extensively white. ADULT FEMALE NON-BREEDING. Like adult ♀ breeding, but ground-colour of side of head and neck and of chin to chest orange-yellow, buff-yellow, or deep yellow, merging rather gradually into pale yellow on breast and side of belly, black spots on crown on average smaller, no black on ear-coverts. Differs from adult ♂ non-breeding in having more olive than black on underparts, less sharply defined black streaks on mantle and scapulars, less contrasting yellow spot on forecrown, olive rather than (mainly) black ear-coverts, narrow and ill-defined olive or grey streaks on side of breast and flank, greyer t1-t2, flight-feathers, and tertials, less white on tail-feathers (t3 largely or wholly dark instead of white), and less white on central greater upper wing-coverts. JUVENILE. Upperparts dark grey-brown faintly streaked with darker brown, an indistinct and narrow paler brown patch on centre of forecrown; side of head and neck and chin to chest cream-buff or pale grey-buff, lore, upper cheek, and ear-coverts dark brown, ill-defined; remainder of underparts off-white, indistinctly mottled brown on side of breast and flank. Tail and wing as 1st adult non-breeding. FIRST ADULT NON-BREEDING AND BREEDING MALE. Rather like adult ♂, but juvenile tail, flight-feathers, greater upper primary coverts, and variable number of tertials and feathers of bastard wing retained; ground-colour greyer than adult when fresh, browner when worn, in particular primary coverts distinctly less black than neighbouring new greater upper wing-coverts; tail-feathers more pointed at tip than in adult, less truncate; white on t2–t6 less extensive than in adult ♂, base of inner webs and tips of outer webs with much black, t3 white for 8–20 mm of its length (more than in adult ♀); no white on bases of primaries. In 1st non-breeding, upperparts as in adult ♀ breeding, feather-centres of lower mantle, scapulars, mid-rump, and upper tail-coverts with more extensive and more sharply defined black than in adult ♀ non-breeding, but not as much as adult ♂ breeding (olive predominating over black, unless worn); side of head and neck and underparts as adult ♀ breeding, some black showing on ear-coverts, yellow usually rather deep orange-yellow or buff-yellow, but sometimes sulphur-yellow; remainder of underparts yellow (on breast) to yellow-white or white (on under tail-coverts), streaks on side of breast and flank sooty, sharp but rather indistinct, c. 1 mm wide (as in adult ♀ breeding, less broad and black than adult ♂ non-breeding, less olive-grey and faint than adult ♀ non-breeding); upper wing-coverts as adult ♀ breeding. 1st adult breeding like adult ♂ breeding, but orange of head and underparts on average more yellowish-orange, less flame-orange; black of hindneck to (especially) rump with traces of pale yellow fringes; black spots on side of breast narrower, less confluent; c. 5 mm of tips of median coverts white, central greater coverts either all-white or with black on bases partly visible. FIRST ADULT NON-BREEDING AND BREEDING FEMALE. Rather like adult ♀, but part of juvenile feathering retained, as in 1st adult non-breeding and breeding ♂; juvenile feathers not much different from adult ones in colour when plumage fresh, but tail-feathers and primary coverts more pointed than in adult, less rounded, tertials more rounded, tip less truncate; juvenile feathers distinctly browner and with more frayed tips in spring (in adult, more smoothly fringed greyish-black); dark on tail more extensive, t3 all-dark, white patch on t4 4–5 mm long and often not reaching shaft, occasionally absent. In 1st adult non-breeding, much like adult ♀ non-breeding, but without distinct black specks on crown and with less black on upper tail-coverts; patch on forecrown faint, yellow-olive; yellow on side of head lemon-yellow or pale buff-yellow, occasionally whitish-yellow, that on throat about similar in colour to yellow on upper belly, less orange than in some adults; yellow on chest usually brighter sulphur-yellow; streaks on side of breast and flank ill-defined, olive-grey; black of centres of median upper wing-coverts often extends into a point on white tips; 3–5 mm of tips of outer webs of central greater coverts white. In 1st adult breeding plumage, rather like adult ♀ breeding and sometimes indistinguishable; black specks on crown and black feather-centres of mantle, scapulars, and upper tail-coverts on average narrower, yellow of head and chin to chest paler, hardly orange, white tips of central greater coverts narrower.

Bare parts. ADULT, FIRST ADULT. Iris brown or dark brown. Upper mandible and tip of lower mandible dark plumbeous-grey to black, cutting edges light pinkish-horn or pale blue-grey, middle and base of lower mandible pale flesh-horn, light pink-brown, greyish-horn, or blue-grey, darkest in breeding adult. Leg and foot flesh-brown, horn-brown, or black-brown, darkest in adult ♂ breeding. (RMNH, ZMA.) JUVENILE. No information.

Moults. ADULT POST-BREEDING. Complete; primaries descendent. On breeding grounds, July–August (Dwight 1900; Pyle *et al.* 1987). ADULT PRE-BREEDING. Partial: head, body (apparently often excluding side of breast, flank, and belly to under tail-coverts), some or all tertials, and all upper wing-coverts (occasionally except a few outer greater ones). In winter quarters, starting October–November (Ridgely and Tudor 1989) or during February–April (Dwight 1900; Pyle *et al.* 1987); in March birds examined, Venezuela and Ecuador, 3 still largely in non-breeding, but new feathers growing on chin and chest, 2 others in full moult of head, body, and wing-coverts, only neck still largely old (RMNH, ZMA). POST-JUVENILE. Partial: head, body, and lesser and median upper wing-coverts. On breeding grounds, July–August (Pyle *et al.* 1987) or from early August (Dwight 1900); one from Virginia already in non-breeding 18 July (RMNH). FIRST PRE-BREEDING. As adult pre-breeding.

Measurements. ADULT, FIRST ADULT. Whole geographical range: mainly north-east USA and eastern Canada, May–September, Central America, October, and Colombia, Venezuela, and Ecuador, winter; skins (RMNH, ZMA). Bill (S) to skull, bill (N) to distal corner of nostril, exposed culmen on average 3·2 less than bill (S).

	♂		♀	
WING	69·1 (1·68; 23)	66–73	66·2 (1·73; 20)	63–70
TAIL	46·0 (1·27; 22)	44–49	45·1 (1·49; 20)	42–48
BILL (S)	13·2 (0·38; 23)	12·6–13·9	13·0 (0·42; 20)	12·2–13·7
BILL (N)	7·6 (0·32; 23)	7·0–8·1	7·5 (0·37; 20)	6·7–7·9
TARSUS	17·8 (0·56; 23)	16·8–18·8	17·6 (0·58; 20)	16·6–18·5

Sex differences significant for wing and tail. Ages combined, though retained juvenile wing and tail of 1st adult on average shorter than in older birds. Thus, in ♂: wing, adult 70·6 (1·10; 9) 69–73, 1st adult 68·2 (1·30; 14) 66–71; tail, adult 46·9 (1·14; 9) 45–49, 1st adult 45·4 (1·00; 13) 44–47.

Weights. Pelee Island (Lake Erie, Canada), May: ♂ 11·2 (4) 11·1–11·3, ♀ 10·5 (3) 10·1–11·0 (Woodford and Lovesy 1958). Ohio (USA): May, ♂ 11·1; undated, ♂ 9·9; September, ♀ 9·5 (Stewart 1937; Baldwin and Kendeigh 1938). Kentucky (USA): May–June, ♂ 9·8, ♀ 9·8; September, ♂ 11·9 (3) 10·5–14·1, ♀ 9·3 (Mengel 1965). Pennsylvania (USA): August–September, adult 10·1 (0·57; 4) 9·4–10·9, 1st adult ♂ 9·7 (0·87; 34) 8·0–12·7, 1st adult ♀ 9·0 (0·63; 26) 8·1–10·5; May–June 10·3 (0·71; 6) 9·3–11·4 (Clench and Leberman 1978). Georgia (USA), early October: ♂ 12·4 (14) 10·2–14·9, ♀ 11·7 (11) 9·6–14·9 (Johnston and Haines 1957). Illinois, September: ♂ 9·9, ♀♀ 7·6, 9·2 (Graber and Graber 1962). Belize: October, ♂ 9·7 (0·82; 4) 8·8–10·6, ♀ 9·2 (0·98; 10) 8·3–11·6 (Roselaar 1976; ZMA); March, ♂ 12·3 (Russell 1964). Birds grounded during rainstorms, October: Panama, 9·4 (Rogers 1965); on ships off Nicaragua and Colombia, 6·4, 6·7 (ZMA). Jamaica, winter: 8·8 (Diamond *et al.* 1977).

Structure. Wing relatively long, broad at base, tip pointed. 10 primaries: p8 longest, p9 0·5–2 shorter, p7 0–1, p6 2·5–3·5, p5 6·5–9·5, p4 9–13, p3 12–16, p2 13–18, p1 16–21. P10 strongly reduced, forming tiny pin hidden below reduced outermost greater upper primary covert; 44–52 shorter than p8, 6–10 shorter than longest upper primary covert, 0–1·5 shorter than outermost greater primary covert. Outer web of (p6)p7–p8 emarginated, inner web of p7–p9 with notch (sometimes faint). Tip of longest tertial reaches tip of p1–p2. Tail rather short, tip square; 12 feathers. Bill rather short; as in Yellow Warbler *D. petechia*, but more compressed laterally in middle and bristles at base of upper mandible more distinct. Tarsus and toes rather short, slender. Middle toe with claw 13·0 (6) 12·5–13·5; outer toe with claw *c.* 73% of middle with claw, inner *c.* 67%, hind *c.* 81%. Remainder of structure as in *D. petechia*. CSR

Dendroica tigrina Cape May Warbler

PLATE 4
[between pages 64 and 65]

Du. Tijgerzanger   Fr. Paruline tigrée   Ge. Tigerwaldsänger
Ru. Тигровый лесной певун   Sp. Reinita atigrada   Sw. Brunkindad skogssångare

Motacilla tigrina Gmelin, 1789

Monotypic

Field characters. 11·5–13 cm; wing-span 19–21 cm. About 10% smaller than Blackpoll Warbler *D. striata*. Quite small, conifer-loving Nearctic wood warbler, with most variable plumage of all transatlantic vagrants and thus requiring close observation. Shares pale wing-bars, yellowish rump, and pale tail-spots with several congeners, but underparts more heavily and uniformly streaked than any. Only breeding ♂ distinctive, with chestnut ear-coverts, bold white wing-panel, and heavily streaked body. ♀ and immature rather dull; combination of streaked body with (usually) pale spot behind ear-coverts provides best clue. Sexes dissimilar; some seasonal variation. Immature separable.

ADULT MALE BREEDING. Moults: July–August (complete); February–April (head, body, and fore-wing-coverts). Crown and back bright olive, obviously streaked black when worn, particularly on forehead. Face, almost-full collar, underparts, and rump bright orange-yellow, with bright chestnut ear-coverts, striking lines of black spots from lower throat to rear flanks, and white under tail-coverts. Wings olive-black, with yellowish fringes to tertials, yellowish-white margins to greater coverts, and almost wholly white median coverts, covert markings combining in bold mid-wing panel. Bill black. Legs black-horn. ADULT FEMALE. Patterned as ♂ but far less and more faintly streaked and less yellow below; shows only olive ear-coverts but these still contrast strongly with obvious yellow half-collar which reaches nape. Wing-marks do not coalesce into panel but form striking double white wing-bar. Rump distinctly greener than ♂. ADULT MALE NON-BREEDING. Resembles ♀ but distinguished by retention of bolder wing markings and yellow rump. Black on crown and chestnut on cheeks usually concealed. FIRST-WINTER MALE. Resembles non-breeding adult ♂ but browner fringes to flight-feathers may show. FIRST-WINTER FEMALE. Some resemble adult ♀ but many drabber and colder, with virtually no streaks on back, distinctly greyish tone to rear supercilium and patch behind ear-coverts, small, dull olive-yellow rump, and no more than yellowish (even dull white) ground-colour to all underparts.

Breeding ♂ unmistakable. In all other plumages, important to note combination of pale half-collar (often reduced to yellowish to pale buff spot or patch behind ear-coverts) and uniformly streaked underparts. Immature confusing, suggesting Yellow-rumped Warbler *D. coronata* (pure yellow rump and usually shoulder patch, blue fringes to tail and wing-feathers, and much browner upperparts), *D. striata* (dark rump, indistinct streaks on underparts, and bright yellowish to brown legs and feet), Palm Warbler *D. palmarum* (dark rump, yellow under tail-coverts, wags tail), and Pine Warbler *D. pinus* (dark rump, unstreaked back, only light streaks below, and much bolder double wing-bar). Flight and behaviour other *Dendroica*, but less active than most.

Call rather hard, thin 'tsip'.

Habitat. Breeds in cool temperate forested lowlands of eastern Nearctic, especially where tall spruce *Picea* and other conifers form open parklike stands, sometimes with patches of birch *Betula* (Pough 1949). Associated with fairly open stands of tall conifers or edges of coniferous forest, especially if birch or hemlock present; lives largely in upper parts of tall trees (Johnsgard 1979). In Canada, also breeds in mixed woods, and occurs on migration in various kinds of woods and thickets (Godfrey 1979). In New England found on migration in trees and shrubbery near dwellings and along village streets; sometimes also in orchards, thickets, and briar patches (Forbush and May 1955).

Distribution. Breeds in North America from southern Mackenzie east to southern Quebec, south to North Dakota, northern Wisconsin, northern New York, and Maine. Winters in West Indies, mainly in Greater Antilles, and casually in eastern Central America.

Accidental. Britain: ♂, Paisley (Strathclyde), 17 June 1977 (Byars and Galbraith 1980).

Movements. Migratory. Winter range, mainly in West Indies, much smaller than breeding range (Rappole *et al.* 1983). Often uncommon to rare, but becomes quite common locally for several years in response to outbreaks of spruce budworm *Choristoneura fumiferana* in breeding areas (Bent 1953; Cadman *et al.* 1987).

Post-breeding movement begins early August (Bull 1974; Janssen 1987), but few birds penetrate far south of summer range before September; records in Florida and Bahamas in late August (Brudenell-Bruce 1975; Kale and Maehr 1990) seem exceptional. Peak passage in southern Canada and northern USA 20 August to 20 September (Bailey 1955; Sadler and Myres 1976; Sprague and Weir 1984; Janssen 1987), extending southward through September (Stewart and Robbins 1958; Bohlen 1989), with arrivals in wintering areas usually from late September (Bent 1953; Brudenell-Bruce 1975; Raffaele 1989). Main movement between Appalachian Mountains and Mississippi River (Bent 1953), despite occasional coastal concentrations, e.g. 'thousands' at Ocean City, Maryland, 2 October 1949 (Stewart and Robbins 1958). Many turn east farther south, to pass through Florida to West Indies. Autumn migration long drawn-out, with birds frequently lingering into November in most eastern states, and occasionally into December. (Bent 1953; Potter *et al.* 1980).

Spring migration begins in March, with movement through south-east states chiefly in April (Bent 1953; Imhof 1976; Stoddard 1978), and last records in West Indies mid-May (Brudenell-Bruce 1975; Bond 1985). Main passage in May from mid-latitude states north into Canada (Stewart and Robbins 1958; Houston and Street 1959; Mengel 1965; Squires 1976), and arrives in north of range in late May (Todd 1963; Speirs 1985). Spring route is reverse of autumn, but some birds apparently fly from West Indies directly to Alabama and north-west Florida, passing by or over peninsular Florida (Bent 1953; Stevenson 1957).

Vagrant to northern Alaska (Kessel and Gibson 1978), and a few records in Newfoundland and nearby St-Pierre-et-Miquelon both seasons (Etcheberry 1982; Maunder and Montevecchi 1982; Godfrey 1986).   AJE

Voice. See Field Characters.

Plumages. ADULT MALE BREEDING. Forehead and crown black, each feather with narrow yellow-green fringe when plumage fresh, cap appearing scalloped, uniform black if worn. Sometimes a narrow rufous-chestnut stripe over middle of forehead and forecrown. Supercilium broad and distinct, deep yellow at nostril, merging gradually into rufous-chestnut above eye and above ear-coverts, bordered below by narrow but distinct dull black stripe on lore and just behind eye. Upper cheek and ear-coverts rufous-chestnut, sometimes faintly mottled deep yellow just below and behind eye, sometimes bordered below by narrow black stripe extending from gape backwards. Upper chin and broad band over lower cheek extending upwards behind ear-coverts over side of neck deep yellow, sometimes partly suffused rufous (especially on chin and at border with upper cheek), slightly mottled olive on side of neck when plumage fresh. Hindneck, mantle, scapulars, back, and shorter upper tail-coverts dark yellowish-green, each feather with contrasting black spot on centre, but these partly concealed, green prevailing unless plumage heavily worn; rump bright greenish-yellow, merging into green of back and of tail-coverts; longer upper tail-coverts black with broad olive-green fringe. Underparts deep bright yellow, merging into yellow-white on rear of flank and on vent and this in turn to white with faint yellow suffusion on feather-tips on under tail-coverts; central throat, chest, side of breast, and flank marked with sharp deep black streaks 1-2 mm wide, black on throat and central chest sometimes partly bordered tawny-rufous; dark streaks duller and wider on yellow-white lower flank, narrow but sharp on yellow of upper belly and side of belly. Thighs dark grey-brown, mixed with some off-white. Tail-feathers black, outer webs fringed yellowish-green (hardly so on outermost feathers, t5–t6); outer 4

pairs with white subterminal patch on inner web, *c.* 3–6 mm from tip; white patch 12–22 mm long on t6, gradually smaller inwards, 5–10 mm on t3, where narrow, irregular, and sometimes absent. Flight-feathers, greater upper primary coverts, and bastard wing black, flight-feathers narrowly but sharply fringed green-yellow along outer webs (widest and almost white at base of primaries, sometimes forming broken pale patch), primary coverts faintly fringed green. Tertials black, outer webs broadly bordered light olive-green when fresh, more narrowly grey-white when worn. Lesser upper wing-coverts black with olive-green fringe; median white, tips tinged yellow when plumage fresh, black of bases hidden, white forming broad conspicuous wing-bar; outer greater-coverts black, fringe along outer web pale olive-grey (at base) to off-white (on tip), inner greater black with broad white outer fringe, showing as white patch at base of wing; number of mainly white greater coverts depends on extent of pre-breeding moult, patch sometimes extensive, but occasionally absent, all coverts then black with inconspicuous olive-grey fringes. Under wing-coverts and axillaries white, partly suffused pale grey and pale yellow, those along leading edge of wing mottled dull grey. ADULT FEMALE BREEDING. Forehead and crown greenish olive-grey, each feather with small triangular speck showing on centre, much less black than adult ♂ breeding; hindneck to back as well as upper tail-coverts either dull greenish olive-grey or brighter dark yellowish-green, mainly concealed feather-centres slightly darker olive (no black), sometimes with a faintly greyer collar across hindneck or upper mantle; rump bright greenish-yellow with darker olive-green spots or streaks, merging gradually in duller green of back and tail-coverts. Supercilium narrow but distinct, deep yellow or tawny-yellow, extending from nostril to above shorter ear-coverts; lore and short line behind eye dusky grey with some fine pale yellow specks; upper cheek and ear-coverts dull olive-grey, finely spotted pale yellow, green-yellow, and (sometimes) partly tawny-rufous, olive-grey darker and more uniform on longer ear-coverts. Chin, lower cheek, and broad band down side of neck bright yellow, sometimes slightly mottled olive or tawny on cheek and usually mottled olive on side of neck, but yellow prevailing. Side of breast and upper flank streaked olive-grey and pale grey-green; throat, chest, and upper belly sulphur-yellow or pale yellow (less bright than adult ♂ breeding), sometimes almost white on throat, all boldly marked with dull black streaks and blobs (marks less sharp and deeply black than adult ♂); belly, vent, lower flank, and under tail-coverts pale cream- or yellowish-white, lower flank boldly streaked dark olive-grey; thigh dark olive-grey mixed off-white. Tail as adult ♂, but white spots on inner webs smaller and more restricted, 8–15 mm long on t6, 5–13 mm on t5, usually a narrow spot on t4, usually no spot on t3. Wing as adult ♂ breeding, but median coverts black with broad white fringe on outer web and tip (forming broken wing-bar), greater coverts black with pale olive-grey outer fringe, a few inner sometimes partly fringed white, not forming distinct spot. ADULT MALE NON-BREEDING. Upperparts as adult ♀ breeding, but cap with slightly larger and more distinct black specks, and mantle and scapulars with partly concealed black triangular feather centres (unlike ♀). Side of head and neck bright yellow with distinct dull black stripe on lore and just behind eye; upper cheek, ear-coverts, and side of neck partly speckled olive, spots sometimes forming faint malar stripe; ear-coverts usually partly mottled tawny-rufous. Underparts bright yellow (intermediate in tinge between adult ♂ breeding and adult ♀ breeding), merging into white on lower flank and vent; pronounced black streaks and elongate spots on throat, chest, side of breast, and upper belly. Wing and tail as adult ♂ breeding, but median upper wing-coverts black with broad pale yellow tips *c.* 5 mm wide, greater coverts black with green-yellow outer fringe *c.* 1 mm wide (no white). ADULT FEMALE NON-BREEDING. Like adult ♀ breeding, but cap, mantle, and scapulars greenish olive-grey, sometimes with traces of black specks on cap; sides of head and neck greenish olive-grey, with pale yellow or pale buff-yellow supercilium, some pale yellow specks below and behind eye, pale yellow lower cheek, and bright yellow feather-bases on side of neck, latter partly concealed, but some bright yellow usually just visible (the only bright yellow visible in plumage); underparts pale cream- or yellow-white, washed clearer yellow on chest and side of breast; lower throat, chest, side of breast, upper flank, and upper belly with rather faint dark olive-grey to sooty streaks or elongate spots; wing and tail as adult ♀ breeding, but median coverts more narrowly tipped green-yellow, greater coverts fringed grey-yellow or pale olive-grey. JUVENILE. Upperparts dark grey-brown, tinged olive on mantle, scapulars, and back; side of head and neck as well as underparts mouse-grey, chest, side of breast, and flank with dusky mottling or streaking, belly, vent, and under tail-coverts dull white with faint yellow tinge; upper wing-coverts dull grey-black, fringed drab-grey, tips of median and greater buff-white. FIRST ADULT NON-BREEDING. Upperparts greenish olive-grey, slightly brighter olive-green on lower mantle, back, and upper tail-coverts, more yellowish-green on rump; no dark marks. Side of head greenish olive-grey; rather ill-defined and short supercilium, fine specks below eye and on shorter ear-coverts, and lower cheek pale green-grey or greenish-yellow; some bright yellow on feather-bases on side of neck, just visible in ♂, largely concealed or even absent in ♀. Ground-colour of underparts dirty grey-white or pale buff-white, some light lemon-yellow tinge on chest (most pronounced in ♂); lower throat, chest, side of breast, and upper flank with short sharply black (♂) or less sharp but distinct dark olive-grey (♀) streaks. Tail as adult, ♂ with white patch *c.* 12–18 mm long on inner web of t6, smaller dot on t5, and narrow or irregular trace of dot on t4, ♀ with dots *c.* 5–15 mm long on t5 and t6 (often not reaching shaft, occasionally absent on t5). Wing as adult, but median upper wing-coverts dull black with yellow outer fringe and tip (♂) or with olive-green outer fringe and narrow yellow-green tip (♀), greater coverts dull black with yellow-green outer fringe, fading into small pale green-grey spot on tip. First non-breeding ♂ rather like adult ♀ non-breeding, but tail, flight-feathers, and primary-coverts of ♂ still juvenile, browner and more worn than those of adult at same time of year, tips of tail-feather more pointed, less truncate. 1st adult non-breeding ♀ duller than adult ♀ non-breeding on upperparts, less yellow on side of head and neck, and whiter below, with less pronounced streaking; flight-feathers, primary-coverts, and tail juvenile, wear and shape as in 1st adult ♂ non-breeding ♂. FIRST ADULT MALE BREEDING. Rather variable, some birds similar to adult ♂ breeding, showing black cap, much rufous-chestnut on side of head, and much white in wing, others with broad olive-green fringes to black of cap, duller olive-green mantle and scapulars, restricted tawny-rufous on ear-coverts, somewhat paler yellow chin to upper belly, and no white on greater coverts, these duller birds somewhat resembling adult ♀ breeding, but supercilium, lower cheek and side of neck extensively deep yellow, dark eye-stripe more distinct, and median coverts mainly white; in all birds, tail, flight-feathers, and primary coverts still juvenile, browner and more worn on tip than in adult, tail on average with less white than in adult ♂ breeding. FIRST ADULT FEMALE BREEDING. Variable, some as bright as adult ♀ breeding, others as dull as 1st non-breeding ♀ or intermediate between these (difference apparently mainly due to variation in extent of 1st pre-breeding moult); in all, tail, flight-feathers, and primary coverts still juvenile, brown and worn, tail with limited amount of white.

Bare parts. ADULT, FIRST ADULT. Iris brown or dark brown. Bill brown-black or plumbeous-black, cutting edges dark horn-brown or blue-grey. Leg and foot black-brown, horn-black, or plumbeous-black, soles yellowish-horn, dark horn-grey, or plumbeous-grey. (BMNH, RMNH, SMTD, ZMA.) JUVENILE. No information.

Moults. ADULT POST-BREEDING. Complete; primaries descendent. On breeding grounds, July–August. ADULT PRE-BREEDING. Partial: head, body, lesser and median upper wing-coverts, and variable number of inner greater coverts and tertials. In winter quarters, February–April. POST-JUVENILE. Partial: head, body, and lesser and median upper wing-coverts. On breeding grounds, July–August. FIRST PRE-BREEDING. Partial; as in adult pre-breeding, but often less extensive, ♀ in particular frequently retaining variable amount of 1st non-breeding on body, only head and chin to chest in partial or full breeding plumage. (Dwight 1900; Pyle *et al.* 1987; RMNH, SMTD, ZFMK, ZMA.)

Measurements. ADULT, FIRST ADULT. Whole geographical range, mainly north-east USA, May–September, migrants south-east USA, April and September–October, and Cuba, winter; skins (RMNH, SMTD, ZFMK, ZMA). Bill (S) to skull, bill (N) to distal corner of nostril; exposed culmen on average 3·3 less than bill (S).

	♂		♀	
WING	68·2 (1·41; 11)	66–71	65·6 (1·45; 9)	63–68
TAIL	45·4 (1·65; 11)	43–48	44·1 (1·19; 9)	42–46
BILL (S)	13·2 (0·39; 10)	12·7–13·7	12·6 (0·51; 9)	11·9–13·2
BILL (N)	7·8 (0·29; 10)	7·4–8·3	7·4 (0·39; 9)	6·8–7·8
TARSUS	18·0 (0·49; 11)	17·4–18·7	18·1 (0·41; 9)	17·6–18·7

Sex differences significant for wing and bill.

Ages combined, though retained juvenile wing and tail of 1st adult on average shorter than in older birds; thus, in ♂, wing, adult 69·1 (1·66; 4) 67–71, 1st adult 67·8 (1·11; 7) 66–70; tail, adult 45·9 (1·97; 4) 43–48, 1st adult 45·2 (1·55; 7) 43–48.

Weights. Pennsylvania (USA): August–October, adult ♂ 11·3 (1·76; 50) 9·3–17·3, adult ♀ 10·6 (1·20; 30) 9·5–15·3; May, ♂ 11·2 (1·06; 11) 9·9–13·1, ♀ 11·1 (1·34; 10) 9·7–14·0; September, 1st non-breeding 10·0 (0·93; 611) 8·1–16·5; October, 1st non-breeding 11·2 (2·12; 37) 8·8–17·3 (Clench and Leberman 1978). Pelee island (Lake Erie, Canada), May: ♀ 10·3 (Woodford and Lovesy 1958). Ohio (USA), August–September: average of 1st adult 10·3 (*n* = 3) (Baldwin and Kendeigh 1938). Coastal New Jersey (USA), autumn: 9·6 (1·4; 157) 7·8–15·2 (Murray and Jehl 1964). Jamaica, September–May: 9·2 (0·5; 13), maximum 10·0 (Diamond *et al.* 1977). Exhausted birds on ship off Florida, April: ♂ 7·0, ♀♀ 6·5, 6·9 (ZMA).

Structure. Wing rather short, broad at base, tip bluntly pointed. 10 primaries: p7–p8 longest, p9 0·5–1·5 shorter, p6 1–3, p5 5–9, p4 9–12, p3 11–15, p2 14–18, p1 15·5–19. P10 strongly reduced, a narrow spike hidden below reduced outermost greater upper primary covert; 44–48 shorter than p7–p9, 4–8 shorter than longest upper-coverts, 0–2 shorter than reduced outermost primary covert. Outer web of (p6–)p7–p8 emarginated, inner web of p7–p9 with notch (sometimes faint). Tip of longest tertial reaches to tip of p1–p2. Tail rather short, tip square, 12 feathers. Bill rather short; as in Yellow Warbler *D. petechia*, but base slightly finer, tip and middle slightly more compressed laterally, and tip more sharply pointed. Tarsus rather short, toes rather long, slender. Middle toe with claw 14·9 (5) 14–16; outer toe with claw *c.* 74% of middle with claw, inner *c.* 68%, hind *c.* 75%. Remainder of structure as in *D. petechia*.

Geographical variation. None.

Probably related to Magnolia Warbler *D. magnolia* and

Yellow-rumped Warbler *D. coronata* (Mengel 1964; Mayr and Short 1970).

CSR

Dendroica magnolia Magnolia Warbler

PLATE 4
[between pages 64 and 65]

Du. Magnolia-zanger   Fr. Paruline à tête cendrée   Ge. Magnolienwaldsänger
Ru. Магнолиевый лесной певун   Sp. Reinita cejiblanca   Sw. Magnoliaskogssångare

Sylvia magnolia Wilson, 1811

Monotypic

Field characters. 11–13 cm; wing-span 17–19.5 cm. About 10% smaller than Blackpoll Warbler *D. striata* but with proportionately longer tail (further emphasized by rump and tail pattern). Rather attenuated, highly decorated Nearctic wood warbler, all plumages showing broad, centrally divided white band across tail (diagnostic) and yellow throat and rump. Breeding ♂ mainly black above, with white rear supercilium below grey crown and white panel across coverts, and yellow below, with strong black streaks from breast to flanks. Winter ♂ and ♀ duller, with ♀ showing only double white wing-bar and narrower body streaks. Immature shows striking pale spectacle and greyish band across breast. Call distinctive. Sexes dissimilar; some seasonal variation. Juvenile separable.

ADULT MALE BREEDING. Moults: July–August (complete); February–April (head, body, most wing-coverts). Frontal band, lores, ear-coverts, and mantle black; pale blue-grey crown, white rear supercilium, narrow white lower eye-crescent, and brownish fringes to scapulars. Bright yellow band over lower back and upper rump. Wings greyish-black, with striking double white wing-bar coalescing with broad white fringes to inner greater coverts to form bright panel and narrow whitish fringes to tertials. Lower rump and upper tail-coverts black, fringed bluish-grey; tail black with all but central feathers broadly banded white (except on outer margins) halfway down length. Underparts yellow, with splashed lines of black spots from breast to rear flanks and white under tail-coverts. Bill black. Legs dull, rather pale brown. ADULT MALE NON-BREEDING. Appearance far less black and yellow than in spring: crown, nape, ear-coverts, and patch at shoulder bluish-grey relieved only by vestigial whitish supercilium and narrow white eye-ring. Mantle and scapulars greenish, with only a few blackish marks on edges and over rump; wing markings reduced to fully separated double white wing-bar; much weaker lines of spots on breast and fore-flanks. Bill pale horn at base. ADULT FEMALE BREEDING. Resembles breeding ♂ but face-mask less black, back green with blackish streaks, greater coverts

with less white, and finer and less black streaks on breast and flanks. ADULT FEMALE NON-BREEDING. Closely resembles winter ♂ but shows even fainter streaks on flanks and often none on breast, while upper tail-coverts fringed greenish. FIRST-WINTER. Closely resembles winter ♀, but at close range can be distinguished by retained dull brownish flight-feathers and bastard wing. Well-marked ♂ shows more black on back feathers and upper tail-coverts but not certainly distinguishable from ♀. Narrow whitish-buff fore-supercilium and similarly coloured but broader eye-ring form distinct 'spectacle'. FIRST-SUMMER. Resembles adult of either sex but still retains dull flight-feathers and bastard wing.

No other similar passerine has white band midway along tail. Flight light and dancing, with tail appearing to trail at times. Gait a hop. Stance level but occasionally allows tail to droop and spread.

Calls include rather hard, high-pitched 'dzip' or 'tlep' and distinctive disyllabic 'chip chip' or 'tizic'.

Habitat. Breeds in cool temperate eastern Nearctic, mostly in young or low conifer woods or open mixed woods and edges (Godfrey 1979). Normally in open stands of young conifers, especially spruce *Picea*, fir *Abies*, or hemlock *Tsuga*, but also in small isolated forest openings, along woodland roads, and by swamps and shallow ponds, or in scattered clumps of trees in neglected pastures. Feeds at rather low levels, staying hidden within tree. (Pough 1949.) In Great Plains, prefers spruce and fir forest with low trees and open coniferous bogs, especially hemlock and spruce, but deciduous trees are occasionally used. In taller coniferous forests occupies only edges and second growth following logging. Song-posts usually *c*. 3–14 m high. (Johnsgard 1979.) Migrants forage on trees in orchards and villages (Forbush and May 1955).

Distribution. Breeds in North America from south-west Mackenzie east to central Quebec, south to north-east Minnesota, central Michigan, Appalachians south to West

Virginia, and northern Massachusetts. Winters in eastern Mexico, Central America south to Panama, and (fewer) Greater Antilles.

Accidental. Britain: Isles of Scilly, 27–28 September 1981 (Rogers *et al.* 1982*b*).

Movements. Migratory. Winter range much smaller than breeding range (Rappole *et al.* 1983). Very common in eastern breeding areas (Palmer 1949; Erskine 1977).

Departure from breeding areas is gradual. A few birds appear in northern states in early August, but main passage there late August to September (Bull 1974; Hall 1981; Janssen 1987), with only stragglers in most breeding areas after late September (Speirs 1985; Tufts 1986). First birds reach southern states by early September (Oberholser 1974; Imhof 1976), and Central America before end of September (Land 1970; Rogers *et al.* 1986*a*; Stiles and Skutch 1989). Few remain in USA after end of October (Lowery 1974; Kale and Maehr 1990). Migration is east of Rockies, mostly along or west of Appalachians (Bent 1953); most birds then cross middle of Gulf of Mexico, with numbers much lower along Texas coast than on Veracruz coast (Mexico) (Rappole *et al.* 1979). Few appear in Florida or southern Georgia at any season (Bent 1953; Burleigh 1958), but the small numbers that winter in West Indies east to Virgin Islands (Bond 1985) must pass over south-east states. Ringing in southern Mexico, Belize, and El Salvador shows winter site-fidelity, with returns up to 3 years later (Ely *et al.* 1977; Loftin 1977; Rappole and Warner 1980).

Spring migration begins late; a few birds arrive in Texas and Louisiana in March, but most after mid-April (Lowery 1974; Oberholser 1974). Route extends further west than autumn, following coast of Mexico into Texas as well as crossing Gulf further east (Stevenson 1957; Rappole *et al.* 1979). The first thrust northward after mid-April extends to Minnesota and New York in some years (Bull 1974; Janssen 1987), perhaps due to trans-Gulf migrants overflying coastal states (Bent 1953). Thereafter, movement is more gradual, with few reaching breeding areas before early May (Quilliam 1973; Tufts 1986), and with peak movement usually in May as far south as Maryland and Arkansas (Stewart and Robbins 1958; James and Neal 1986). Northern parts of breeding range reached only in late May or early June (Peters and Burleigh 1951; Erskine 1985).

Vagrant to extreme north-east Alaska and to Bering Sea off western Alaska (Kessel and Gibson 1978); also to Greenland and nearby Davis Strait (Salomonsen 1950–1; Godfrey 1986).   AJE

Voice. See Field Characters.

Plumages. ADULT MALE BREEDING. Narrow black strip along forehead, widening on lore into broad black stripe through and just below eye over ear to side of neck. Remainder of top of head

and hindneck medium grey with slight blue tinge, separated from black eye-stripe by narrow but distinct white supercilium, extending from just above eye to upper side of neck. Small spot below eye white, together with front part of supercilium forming a distinct eye-ring, broken by black in front and behind eye. Mantle, scapulars, and upper back deep black, usually with some light green mixed in lower scapulars and back; lower back light green, mixed with some black, merging into bright yellow rump. Upper tail-coverts black, usually with traces of grey fringes. Ground-colour of underparts bright deep yellow, except for yellow-white vent, grey-white thigh and pure white under tail-coverts; lower throat, chest, side of breast, and flank contrastingly marked with deep black streaks 2–3 mm wide, coalescing on lower throat, upper chest, and side of breast, especially when plumage worn, somewhat narrower and more widely spaced towards rear of flank. Tail black, outer webs narrowly fringed grey; middle portions of inner webs of all feathers except central pair (t1) with large squarish white patch, forming distinct band of *c.* 10–13 mm wide over tail at *c.* 13–16 mm from tip; extent and shape of white on inner web of t2 similar to that of t3–t6. Flight-feathers, tertials, greater upper primary coverts, and bastard wing black (tips of primaries slightly brown in spring), narrowly but distinctly edged pale grey or grey-white on outer webs of flight-feathers (except emarginated parts of primaries) and primary coverts, more broadly so on outer webs and tips of tertials. Upper wing-coverts black, narrowly fringed pale grey, tips of median and outer greater coverts white (widest on outer webs), forming wing-bars 2–4 mm wide, outer webs and tips of inner greater coverts broadly white (or entire coverts white), forming large white patch on inner wing. Axillaries and under wing-covert white, mottled black on a variable number of small coverts along leading edge of wing. *In fresh plumage* (April), grey of crown faintly tinged green, black feathers of mantle, scapulars, and back fringed green, black spots on lower throat, upper chest, and side of belly separated by some yellow; *in worn plumage* (June–July), crown plumbeous-grey, mantle to back virtually uniform black, sharply defined from yellow of rump (but when heavily abraded, yellow sometimes largely worn off, rump appearing black with some yellow mottling); black on lower throat, upper chest, and side of breast confluent. ADULT FEMALE BREEDING. Forehead to upper mantle medium grey, slightly tinged green, less bluish than in adult ♂ breeding; lower mantle, scapulars, and back light green, brighter towards lower back (where grading into yellow of rump), centres of feathers with bold black mark, latter sometimes confluent on mantle, forming black saddle when plumage worn. Upper tail-coverts black, more broadly fringed grey than in adult ♂ breeding. Lore, upper cheek, and ear-coverts mixed grey and dull black, some yellow on ear-coverts; broad but sometimes rather ill-defined eye-ring off-white, sometimes continued into indistinct and narrow off-white supercilium, which may extend towards nostril. Side of neck dull ash-grey. Underparts bright yellow with indistinct pale grey or yellow-grey collar across lower throat; vent, thigh, and under tail-coverts white; side of breast and flank marked with grey (on side of belly) or black (on flank) streaks *c.* 1 mm wide (wider on lower flank). Tail as adult ♂, but white patch on t2 sometimes smaller, more rounded in shape. Wing as adult ♂, but white tips of median and greater coverts *c.* 2 mm wide (white sometimes largely worn off), inner greater coverts without white on outer webs or with traces only, not forming large white patch. ADULT MALE NON-BREEDING. Top and side of head and neck as adult ♀ breeding (thus, without black on forehead and without long black stripe through eye); grey of cap slightly more bluish, overlaid by some green when plumage fresh. Mantle, scapulars, and back light yellowish-green, centres of feathers boldly black (as adult ♀ breeding, but black on back of latter often less distinct, less sharply defined from somewhat duller yellow rump); upper tail-coverts black with traces of grey fringes (less grey than ♀). Underparts as adult ♀ breeding or slightly deeper yolk-yellow, a trace of a yellow-grey collar across throat; side of breast, flank, and sometimes chest boldly dappled with contrasting black spots (c. 1 mm wide on chest, if any, not confluent; 2–3 mm wide on side of breast and flank), spotting distinctly heavier and purer black than in adult ♀ breeding, but less so than in adult ♂ breeding, especially on throat and mid-chest. Tail and wing as adult ♂ breeding, white wing-bars 2–4 mm wide; outer webs of inner greater coverts fringed grey, not largely white. ADULT FEMALE NON-BREEDING. Like adult ♀ breeding, but black marks on mantle and scapulars narrow, 1–2 mm (in adult ♀ non-breeding and in part of feathering of adult ♀ breeding 3–4 mm); streaks on underparts narrow, c. 1 mm, black or dark grey, usually not extending to chest. JUVENILE. Upperparts dark sepia-brown or dark olive-brown, faintly spotted dusky grey-brown on crown, mantle, and scapulars; rump pale yellow, indistinct and poorly defined; upper tail-coverts dark sepia-brown with olive-grey fringes. Side of head and neck dull ash-grey, paler on lore and round eye. Chin and throat grey, remainder of underparts pale yellow, washed olive-brown on chest and with ill-defined dusky olive-brown streaks on side of breast and flank. Upper wing-coverts dark sepia-brown with paler olive-brown fringes; tips of median and greater coverts pale buff or yellow-buff. FIRST ADULT MALE NON-BREEDING. Closely similar to adult ♀ non-breeding, but tail, flight-feathers, and greater upper primary coverts still juvenile, slightly browner and more worn than those of adult ♀ at same time of year, tips of tail-feathers and primary coverts often (not always) more pointed than in adult; white patch on inner web of t2 round or squarish, 5–20 mm long. Forehead to upper mantle medium ash-grey with slight green tinge; lower mantle, scapulars, and back light green, some feathers with narrow and restricted amount of black on centres, largely concealed (generally, less black than in adult ♀ non-breeding); upper tail-coverts black with broad light green (shorter ones) or green-grey (longer ones) fringes. Side of head and neck light ash-grey, mottled darker grey on lore, cheek, and ear-coverts, partly mottled yellow on shorter ear-coverts, a rather broad and distinct pale buff eye-ring and (sometimes) a short and narrow supercilium of same colour. Entire underparts bright yellow, except for pale yellow-grey or pale yellow collar across throat and for white vent, thigh, and under tail-coverts; flank and (sometimes) side of breast marked with narrow elongate dull black drops, often distinct on lower flank only, on side of breast sometimes restricted to some indistinct olive streaks or olive suffusion. Upper wing-coverts black with grey fringes (latter sometimes suffused olive), tips of outer webs of median and greater coverts white for 2–3 mm, showing as narrow wing-bars. FIRST ADULT FEMALE NON-BREEDING. Like 1st non-breeding ♂, but grey of forehead to upper mantle more distinctly tinged olive; lower mantle, scapulars, and back without black on feather-centres; yellow of rump often less deep and pure, partly suffused green; black of longer upper tail-coverts less sharply demarcated from broader olive-grey fringes; streaks on underparts narrow, ill-defined, restricted to flank, distinct on lower flank only, where dark olive to olive-black (in ♂, purer black, more sharply defined); white patch on inner web of t2 rounded or elongate, 5–9 mm long; fringes of retained juvenile tertials, secondaries, and primary coverts more pale olive-grey, less pure pale grey. FIRST ADULT BREEDING. Like adult breeding, but tail, flight-feathers, and primary coverts still juvenile, as in 1st non-breeding, black less intense than in adult, white on t2 often differing in shape and extent. In ♂, on average more green

mixed into black of mantle, scapulars, and back than in adult ♂ breeding, some more olive in yellow of rump, more yellow on lower throat and upper chest, black less inclined to be confluent, number of largely white inner greater upper wing-coverts rather variable, depending on extent of 1st pre-breeding moult. In ♀, some birds have bold black centres to feathers of mantle, back, and scapulars and narrow but distinct black streaks on flank, like adult ♀ breeding, others are virtually without black and are similar to 1st non-breeding or intermediate between 1st breeding and 1st non-breeding ♀.

Bare parts. ADULT, FIRST ADULT. Iris hazel to dark brown. Bill plumbeous-grey to greyish-black, cutting edges paler blue-grey, grey-horn, or yellow-horn. Leg and foot green-brown, dark greenish-grey, brownish-black, or plumbeous-black, darkest in adult breeding, paler in 1st autumn; soles yellow, greyish-horn, or dark flesh-grey. (RMNH, ZFMK, ZMA.) JUVENILE. No information.

Moults. ADULT POST-BREEDING. Complete; primaries descendent. On breeding grounds, July–August (Dwight 1900; Pyle et al. 1987). One from 16 July, New York, not yet started (RMNH). ADULT PRE-BREEDING. Partial: head, body, lesser, median, and usually all greater upper wing-coverts, and some or all tertials. In winter quarters, February–April (Pyle et al. 1987; RMNH, ZMA). POST-JUVENILE. Partial: head, body, and lesser, median, and (sometimes) greater upper wing-coverts. On breeding grounds, July–August, but moult of greater coverts and tertials sometimes continued in winter quarters after suspension during autumn migration. FIRST PRE-BREEDING. As in adult pre-breeding, but 1–5 outer greater upper wing-coverts and some or all tertials retained, these still 1st non-breeding; moult of ♀ sometimes less extensive than in ♂, some ♀♀ retaining much feathering of 1st non-breeding on body in 1st summer. (RMNH, ZMA.)

Measurements. Whole geographical range, but mainly eastern USA and Canada, May–September; a few migrants and winter birds from Central America and north-west South America; skins (RMNH, ZMA). Bill (S) to skull, bill (N) to distal corner of nostril; exposed culmen on average 3·2 less than bill (S).

	♂		♀	
WING	61·3 (1·99; 43)	56–65	58·3 (1·36; 13)	56–61
TAIL	48·0 (1·81; 36)	44–52	45·8 (1·76; 11)	43–48
BILL (S)	12·6 (0·45; 35)	11·8–13·5	12·2 (0·38; 11)	11·7–12·8
BILL (N)	7·1 (0·28; 35)	6·6–7·7	6·9 (0·21; 11)	6·5–7·1
TARSUS	17·9 (0·41; 29)	17·3–18·8	17·7 (0·39; 11)	17·2–18·3

Sex differences significant, except for tarsus.

Ages combined, though retained juvenile wing and tail of 1st adult on average shorter than in older birds; thus, in ♂, wing, adult 62·6 (1·53; 18) 59–65, 1st adult 60·3 (1·71; 25) 56–64; tail, adult 49·1 (1·39; 16) 47–52, 1st adult 47·2 (1·71; 20) 44–51. See also Goodpasture (1963).

Weights. ADULT, FIRST ADULT. (1) Pelee island (Lake Erie, Canada), May (Woodford and Lovesy 1958). Pennsylvania (USA): (2) May, (3) August–October, adult (Clench and Leberman 1978). (4) Georgia (USA), early October (Johnston and Haines 1957).

(1)	♂	9·6 (– ; 51)	7·3–12·7	♀	9·3 (– ; 31)	7·5–11·6
(2)		8·9 (0·82; 413)	7·0–12·9		8·5 (0·81; 238)	6·6–12·6
(3)		8·7 (0·78; 115)	7·2–12·7		8·5 (0·91; 190)	6·7–12·4
(4)		9·2 (– ; 10)	7·0–10·5		9·7 (– ; 19)	7·7–11·9

East Ship island (off Mississippi, USA), spring: 7·4 (0·65; 128) (Kuenzi et al. 1991). Pennsylvania, August–October: 1st adult 8·2 (1349) 6·0–12·9 (Clench and Leberman 1978). Coastal

New Jersey (USA), autumn: 7·4 (0·7; 63) 5·8–9·3 (Murray and Jehl 1964).

Mexico: October–January, ♂ 7·5 (0·43; 9) 7·0–8·5, ♀♀ 6·8, 7·1; April, ♂ 9·4 (RMNH). Belize: autumn, 7·1 (0·8; 161) (Mills and Rogers 1990); October, ♂ 6·7, ♀ 7·5 (3) 7·1–8·0 (ZMA); February, 6·3, 6·8; April, ♂ 9·1; May, ♀ 9·0 (Russell 1964). Jamaica, September–May: 7·8 (0·9; 6) (Diamond *et al.* 1977). Exhausted birds on ship in Caribbean, October: ♂ 5·8, ♀ 5·4 (ZMA). See also Tordoff and Mengel (1956), Connell *et al.* (1960), Graber and Graber (1962), and Mengel (1965), and (for fat-free weights) Caldwell *et al.* (1963).

Structure. Wing rather short, broad at base, tip fairly rounded.

10 primaries: p7–p8 longest, p9 1–3·5 shorter, p6 0–1, p5 2·5–5, p4 5–8, p3 7–10, p2 9–12, p1 10–13. P10 strongly reduced, a tiny pin hidden below reduced outermost greater upper primary covert; 38–46 shorter than p7–p8, 5–8 shorter than longest upper primary covert, 0–2 shorter than outermost primary covert. Outer web of p6–p8 emarginated, inner web of p7–p9 with notch (often faint). Tip of longest tertial reaches to tip of secondaries. Tail of average length, tip square; 12 feathers. Bill as in Yellow Warbler *D. petechia*, but tip slightly finer. Leg rather long, toes rather short; slender. Middle toe with claw 12·5 (5) 12–13, outer toe with claw *c.* 75% of middle with claw, inner *c.* 69%, hind *c.* 80%. Remainder of structure as in *D. petechia*.   CSR

Dendroica coronata Yellow-rumped Warbler

PLATE 5
[between pages 64 and 65]

Du. Geelstuitzanger   Fr. Paruline à croupion jaune   Ge. Kronwaldsänger
Ru. Миртовый лесной певун   Sp. Reinita coronada   Sw. Gulgumpad skogssångare

Motacilla coronata Linnaeus, 1766

Polytypic. CORONATA GROUP ('Myrtle Warbler'). Nominate *coronata* (Linnaeus, 1766), Alaska, Canada south to northern British Columbia, central Alberta and Saskatchewan, and southern Manitoba, east to Labrador and Newfoundland, and north-east USA. AUDUBONT GROUP ('Audubon's Warbler') (extralimital). *D. c. auduboni* (Townsend, 1837), western North America from central British Columbia, south-west Alberta, and south-west Saskatchewan south to north-east Baja California, northern Arizona, New Mexico, and western Texas; *nigrifrons* Brewster, 1889, north-west Mexico; *goldmani* Nelson, 1897, Chiapas (south-east Mexico) and Guatemala.

Field characters. 12·5–15 cm; wing-span 21–23·5 cm. Averages slightly larger than Blackpoll Warbler *D. striata*, with rather short bill but 10–15% longer tail giving lengthy silhouette. Medium-sized, relatively robust but graceful Nearctic wood warbler. In all plumages, diagnostic combination of white eye-ring, white throat, double white wing-bar, round bright yellow rump, usually yellow blaze by shoulder, and white spots on outer tail-feathers. Fringes of wing- and tail-feathers noticeably bluish; in winter adult and immature, back always noticeably brown and well streaked. Breeding ♂ has sharply etched forepart pattern of grey, black, and white, with yellow crown-patch; breeding ♀ duller but shows similar crown-patch, usually lacking in immature. Flight undulating, with trailing tail. Call distinctive. Sexes dissimilar; quite marked seasonal variation. Immature usually separable.

ADULT MALE BREEDING. Moults: July–September (complete); February–April (head, body, some wing-coverts). Crown, nape, back, and upper tail-coverts blue-grey, strikingly marked with yellow patch on top of crown, sharply delineated yellow band over rump, and black streaks on crown, back, and upper tail-coverts. Face black, with white supra-loral mark, short rear supercilium, lower eye-crescent, and throat. Wings black, with bluish fringes; whitish margins to tertials and white tips to greater and median coverts form double wing-bar. Tail blue-black,

with white panels forming obvious spots on 3 outer pairs. Breast drenched black, creating obvious band and extending in spotted lines along flanks and contrasting with otherwise white underparts. ADULT FEMALE BREEDING. Patterned as ♂ but all dark areas browner and paler, particularly on ear-coverts and with usually incomplete breast-band and less heavy flank streaks. Shows white eye-ring rather than supercilium. ADULT NON-BREEDING. Duller than breeding bird, with fresh feather-fringes reducing streaking above and below. Plumage differences between ♂ and ♀ inconstant (see Plumages). FIRST-WINTER. Characteristically dull brown, streaked above and below, with bright yellow rump, narrow but distinct wing-bars, and whitish eye-ring; also distinguished from adult by retained juvenile flight-feathers which are browner-fringed than adult. ♀ often less streaked above and below than ♂ and may not show yellow at shoulder.

Unmistakable, with length of tail producing less compact form than most relatives, and plumage patterns complex but distinctive. Flight noticeably light and jerky or undulating; bird often appears to dance in the air when flycatching; when hovering, tail trails noticeably. Gait a hop. Stance half-upright on perch, more level on the move. Often droops or flicks wings. Aggressive towards other passerines, often charging them in flight. Hardy, surviving on berries in winter.

Calls include loud 'check' or 'chep', also described as rather hard 'chip', 'tip', 'trick', or 'tyck', and metallic 'cheep' (all harder and more metallic than other *Dendroica*); also sharp thin 'tsi' and quiet 'prit'.

Habitat. Breeds in cooler northern latitudes of Nearctic, both in lowlands and mountains up to treeline, and even in willow *Salix* scrub beyond it. Found mainly in coniferous and mixed woodland, especially if open, living often lower down than Cape May Warbler *D. tigrina* and being fond of aerial flycatching, although consuming berries and seeds in low growth. Migrants occur everywhere, especially in brushy areas, hedgerows, field borders, and weedy tangles, as well as at bird-tables. (Pough 1949; Gabrielson and Lincoln 1959.) In Great Plains, scattered evergreens, thickets near streams or lakeshores, or open plantings are more typically used than dense mature stands; breeds in pine *Pinus* as well as spruce *Picea* (Johnsgard 1979). On migration, also occurs in sheltered bushy bogs, along coasts, and even on seashores (Forbush and May 1955). Wintering birds in Louisiana occupy almost every habitat, but are conspicuously abundant in myrtle thickets, although equally at home on ground and in treetops (Lowery 1974). Breeds up to *c.* 3800 m in Rockies (Terres 1980).

Distribution. Breeds over most of North America except south-east, also in Mexico and Guatemala. Winters widely in North America, from south-west British Columbia, central Michigan, and Massachusetts southwards, through Central America south to Panama, and West Indies south to Virgin Islands.

Accidental. Iceland, Britain, Ireland.

Movements. Fully migratory (nominate *coronata* and northern *auduboni*), partially migratory, including altitudinal movements (southern *auduboni* and *nigrifrons*), or resident (*goldmani*) (Paynter 1968; American Ornithologists' Union 1983). The only short-range migrant among northern paruline warblers, and by far the most hardy, wintering locally into snow zone where adequate food (especially bayberry *Myrica pennsylvanica*) and cover coincide (e.g. Bent 1953, Tufts 1986).

Autumn migration prolonged. Post-breeding dispersal begins in early August (e.g. Sprague and Weir 1984), but directed migration begins as early as this only in north-west (Rand 1946; Gabrielson and Lincoln 1959). In south-west states, local populations and early altitudinal movements (Phillips *et al.* 1964; Garrett and Dunn 1981) confuse detection of north-south migration. Further east, early migrants reach New York, Illinois, and Texas by mid-September (Bull 1974; Oberholser 1974; Bohlen 1989), and Alabama and Bahamas soon afterwards (Brudenell-Bruce 1975; Imhof 1976). By then, most birds leave breeding areas from Alaska to Labrador (Gabrielson and Lincoln 1959; Todd 1963), but main migration con-

tinues throughout October across southern Canada (Sprague and Weir 1984; Tufts 1986; Cannings *et al.* 1987). In Bermuda, where many winter, arrives early October (Wingate 1973). Most birds reach wintering areas by mid-November, but stragglers linger and attempt to winter further north (e.g. in New Brunswick: A J Erskine) each year (Bent 1953). Ringing recoveries show passage on broad front, narrowing southwards; records of movements from Alberta to Manitoba (Sadler and Myres 1976), and from North Dakota to Arkansas and Louisiana (Bent 1953), suggest wintering east of Great Plains by many western nominate *coronata*. In Alabama, 6 recoveries of birds ringed Ontario and Illinois east to New Jersey and Maryland (Imhof 1976). Ringing in California, Alabama, and Georgia showed winter site-fidelity there, with recoveries up to 4 years later (Bent 1953; Imhof 1976).

Spring migration begins early, and birds leave Central America and West Indies during March (Ridgely 1981; Raffaele 1989). Also in March, first migrants reach coastal Washington state and British Columbia (Jewett *et al.* 1953; Butler and Campbell 1987) and upper Mississippi area (Hall 1983; Janssen 1987). Migration follows snow-line north, first arrivals reaching eastern Canada by mid-April, and penetrating far into western Canada by end of April (Rand 1946; Erskine 1964; Scotter *et al.* 1985). Peak migration occurs in April in mid-latitude states (Mengel 1965; Sutton 1967; Alcorn 1988), and in early May further north (Palmer 1949; Gabrielson and Lincoln 1959; Sprague and Weir 1984; Cannings *et al.* 1987). Most breeding areas are reached by 20 May.

Vagrant north to Alaskan and Canadian Arctic, Aleutian islands, and north-east Siberian coast; recorded both seasons in Greenland (Bent 1953; Gabrielson and Lincoln 1959; Godfrey 1986).

Rare autumn vagrant to Atlantic seaboard of west Palearctic. 17 records from Britain and Ireland up to 1989, especially south-west England and southern Ireland, mainly October, but noted once in September and November, twice in May, and a wintering individual was present in Devon, 4 January–10 February 1955 (Rogers *et al.* 1988; Dymond *et al.* 1989). The records are usually comparatively late in the season, which reflects its rather late migration in North America.   AJE, PRC

Voice. See Field Characters.

Plumages. (nominate *coronata*). ADULT MALE BREEDING. Ground-colour of upperparts light or medium grey with slight blue tinge; forehead and side of crown closely marked with fine black streaks, appearing almost uniform black when plumage worn; hindneck and upper mantle thinly streaked black, lower mantle, scapulars, back, and upper tail-coverts coarsely streaked black, black of feather-centres ending bluntly rounded towards tips; some scapulars often with olive-brown suffusion on tips. Central crown and entire rump contrastingly bright yellow, forming distinct patches. Side of head from lore and upper cheek to ear-coverts black, black of lore bordered above by narrow white line; upper rim and lower rim of eye white, forming conspicuous broken eye-ring, white above eye sometimes continued in narrow and rather faint white or pale grey supercilium. Side of neck light grey. Lower cheek, chin, and throat white, lower cheek sometimes finely speckled black, some black of featherbases sometimes visible on throat. Feathers of chest, upper side of breast, and upper side of belly black with rather narrow and even white fringes, appearing black with white scalloping when plumage fresh, but white worn off and black centres confluent on (especially) chest and upper side of belly when plumage worn. Lower side of breast with large and distinct bright yellow patch. Flank white with broad sooty-black streaks; mid-belly, vent and under tail-coverts white with variable pale cream or isabelline suffusion. Tail black, both webs of central pair (t1) and outer webs of others narrowly fringed light grey, inner webs partly edged white; inner web of t5–t6 with large white squarish subterminal patch (*c.* 20 mm long, 6–7 mm wide), t4 with similar white patch *c.* 10–12 mm long. Flight-feathers and greater upper primary coverts black, tips sometimes slightly tinged brown, outer webs and (faintly) tips narrowly and sharply fringed light grey. Bastard wing black, middle feather with narrow white outer fringe. Tertials black, outer web broadly fringed light grey or brown-grey. All upper wing-coverts black with contrasting light grey fringes; 2–4 mm of tips of median coverts and 4–8 mm of tips of outer webs of greater coverts white, forming double wing-bar. Axillaries dark grey with white borders, upper wing-coverts white, especially those along leading edge, with much black on bases. *In fresh plumage* (early spring) feather-tips of nape, lower mantle, scapulars, and rump in part slightly suffused olive-brown or olive-green, chest and breast scalloped white; *in worn plumage* (late spring and early summer) ground-colour of upperparts virtually entirely light grey, black of forehead, side of crown, rump, and side of breast very prominent. ADULT FEMALE BREEDING. Like adult ♂ breeding, but yellow patches of crown, rump, and (especially) side of breast slightly smaller, slightly paler lemon-yellow or (occasionally) greenish-yellow; black streaks on upperparts slightly narrower, forehead and side of crown rather equally streaked grey and black, black of scapulars, back, and upper tail coverts ending in blunt point rather than rounded; fringes of lower mantle and scapulars olive-brown (not grey), hindneck, upper mantle, and rump sometimes suffused olive-brown; ear-coverts tinged olive-brown; white groundcolour of underparts tinged pale cream or isabelline, especially throat less pure white; black streaks on underparts slightly narrower and less sharply defined, appearing streaked rather than scalloped when plumage fresh and mottled black rather than partly confluent black when plumage worn; white spots on t5–t6 *c.* 15–18 mm long, that on t4 *c.* 5–12 mm long, often not fully reaching border of inner web; grey fringes of tertials and upper wing-coverts partly or fully tinged olive-brown, white wing-bars narrower and less prominent. ADULT MALE NON-BREEDING. Grey of upperparts, side of head and neck, and tertials partly replaced by olive-brown (sometimes more extensively so than in adult ♀ breeding), but sides of feathers still grey, upper tail-coverts fully fringed grey, and wing and tail as adult breeding, blacker and with purer grey fringes and more prominent white marks than adult ♀ breeding. Black of forehead, side of crown, mantle, scapulars, and back rather dull, broad, ending in blunt point; yellow of crown-patch partly mottled olive-brown, but fairly distinct; yellow of rump distinct. Side of head and neck olive-brown, black mask replaced by dusky grey with some brown suffusion, a long but narrow buffish or off-white supercilium. White of underparts washed cream-isabelline; centres of chest and breast black, showing as black dots or arrowmarks; yellow on side of breast bright, but rather restricted, partly marked

black. ADULT FEMALE NON-BREEDING. Like adult ♂ non-breeding, but all grey of upperparts and tertials replaced by olive-brown, often also including part of grey on upper wing-coverts, but excluding upper tail-coverts; tinge of brown rather buffish or fuscous (in ♂, greyish-brown); yellow crown-patch more fully suffused brown; black streaks on mantle and scapulars less sharply defined; sides of head and underparts as adult ♂ non-breeding, but ground-colour slightly more buff and marks of chest and breast narrower and browner on average, slightly less sharply defined. JUVENILE. Upperparts, side of head and neck, and lesser and median upper wing-coverts buff-brown, upperparts (except hindneck) and coverts streaked dull black and dusky-grey; supercilium faint, buff; spots above and below eye white. Underparts grey-white, tinged pale yellow on belly, marked with drab-brown streaks on chest, breast, and flanks. For tail and remainder of wing, see 1st adult, below. FIRST ADULT MALE NON-BREEDING. Upperparts and side of neck drab-brown or dull olive-brown, crown with faint dark shaft-streaks, mantle and scapulars with bold dull black streaks. Feathers at centre of crown with yellow bases, yellow largely concealed or showing as a yellow suffusion only, more fully exposed when plumage worn (from late autumn onwards). Rump-patch bright yellow, narrow but conspicuous. Upper tail-coverts black, fringes along sides grey, along tips olive-brown. Side of head buff-brown, mottled grey in front of eye and on ear-coverts; lore dusky grey; eye-ring buff or off-white, broken by dusky grey in front and behind. Chin to chest, side of breast, and flanks buff (palest on throat), belly to under tail-coverts cream-white or yellow-white, lower side of breast suffused bright yellow (patch much less conspicuous than in adult); chest, side of breast, and flank narrowly streaked dull black. Tail as adult, but grey of fringes slightly tinged brown; pattern of white on t4–t6 as in adult ♀ (patch on t4 small). Flight-feathers, tertials, greater upper primary coverts, and bastard wing black-brown (browner than in freshly-moulted adult), fringes along outer webs and tips of primaries light grey with faint brown suffusion (like those of tail), those of primary coverts, secondaries, and tertials pale olive-brown (in adult ♂, fully grey; in adult ♀, partly grey). Lesser upper wing-coverts black with brown fringes, median and greater ones black with pale tawny to pink-buff tips 2–3 mm wide (mainly on outer webs), which bleach to white when plumage abraded. Upperparts less extensively grey than in adult ♂ non-breeding, yellow on crown and side of breast much less distinct, black streaks on mantle, scapulars, upper tail-coverts, and underparts distinctly narrower, tail less deep black, t4 with smaller white patch, fringes of secondaries, upper wing-coverts, and (especially) greater upper primary coverts browner; tips of tail-feathers often more pointed, less rounded. FIRST ADULT FEMALE NON-BREEDING. Like 1st non-breeding ♂ and sometimes indistinguishable; no trace of grey on upperparts; yellow of crown and side of breast concealed or virtually lacking; dark streaks on mantle, scapulars, upper tail-coverts, and underparts narrower; t4 without white patch on inner web. FIRST ADULT BREEDING. Like adult breeding, but juvenile flight-feathers, greater upper primary coverts, and sometimes bastard wing and tail retained, browner and more worn than those of adult at same time of year, t4 (if still juvenile) with less white (♂) or none at all (♀), primary coverts and often also secondaries with distinctly browner fringes than adult of same sex. Upperparts and upper wing-coverts distinctly browner than in adult (varying with sex), in part because new breeding plumage is intermediate in character between breeding and non-breeding plumage, especially tips of feathers showing brown fringes, in part because of retention of part of old non-breeding feathers (especially in ♀). Sexes separable as in adult breeding. First breeding ♀ often rather similar to 1st non-breeding ♂, but part of plumage more contrastingly worn than in latter.

Bare parts. ADULT, FIRST ADULT. Iris brown or dark brown. Bill horn-black, greyish-black, or black, cutting edges dark horn-grey to plumbeous-black. Leg and foot brown-black, plumbeous-black, or black, soles dark horn-grey, blackish-brown, or plumbeous-black. JUVENILE. Iris brown. Bill dark brown with flesh tinge. Leg and foot dark flesh-horn or dark brown. (RMNH, ZMA.)

Moults. ADULT POST-BREEDING. Complete; primaries descendent. On breeding grounds, July–September; rarely, still largely in breeding plumage during autumn migration (Howard and Dickinson Henry 1966). For sequence of moult, see Hubbard (1980). ADULT PRE-BREEDING. Partial: head and body (sometimes excluding some scapulars, some feathers of neck, part of forehead, tail-coverts, rump, or belly, especially in ♀) and variable number of lesser, median, and greater upper wing-coverts: in ♂, often all coverts except sometimes 1–3 outer median and 1–5 outer greater, in ♀ often lesser coverts only, as well as a few innermost median and greater coverts (tertial coverts) (RMNH, ZMA). See also Hubbard (1980). In winter quarters, February–April (Dwight 1900; Pyle *et al.* 1987), but 4 examined from Central America and Curaçao had just started December–January (RMNH, ZMA) and birds wintering North Carolina moulted 2nd half of March to mid-May (Yarbrough and Johnston 1965). POST-JUVENILE. Partial: head, body, and lesser and median upper wing-coverts; in breeding area, July–September (Pyle *et al.* 1987), starting *c.* 2 weeks after fledging (Grinnell 1908), when tail ¼- or ½-grown (Hubbard 1980, which see for sequence). In birds examined (RMNH, ZMA), moult completed from mid-August onwards, but sample small. FIRST PRE-BREEDING. Extent and timing variable; in winter quarters. Some start December, as adult pre-breeding, others not yet started late March. In ♂, involves head, neck, and lesser and median upper wing-coverts, but part of non-breeding feathers often retained, especially outer or lower scapulars, part of neck, forehead, side of breast, or feathering elsewhere, outer median coverts frequently retained or none replaced at all; tertial coverts (innermost greater coverts, gc7–9), often new (perhaps replaced in winter as continuation of post-juvenile moult), as well as occasionally some neighbouring greater coverts (mainly gc6 or gc5–6); in ♀, moult as in ♂ or more restricted, in some confined to some scapulars or tertial coverts only, these birds nesting in worn non-breeding plumage. (RMNH, ZMA.)

Measurements. ADULT, FIRST ADULT. Nominate *coronata*. Mainly eastern USA, spring and summer, and Mexico, autumn and winter, but some from Canada (including north-west) and from south-east USA and Curaçao, winter; skins (RMNH, ZMA). Bill (S) to skull, bill (N) to distal corner of nostril; exposed culmen on average 3·2 shorter than bill (S).

	♂		♀	
WING	74·7 (1·96; 26)	71–79	72·4 (1·77; 22)	70–76
TAIL	55·2 (1·71; 25)	51–58	54·2 (1·53; 21)	52–57
BILL (S)	12·7 (0·45; 26)	12·0–13·5	12·6 (0·39; 22)	11·8–13·1
BILL (N)	7·1 (0·22; 25)	6·8–7·6	7·0 (0·30; 22)	6·5–7·6
TARSUS	18·8 (0·52; 25)	17·8–19·7	18·7 (0·62; 21)	17·8–19·8

Sex differences significant for wing and tail.

Ages combined, though retained juvenile wing and tail of 1st adult often shorter than in older birds. Thus, wing, adult ♂ 75·3 (2·76; 7) 72–79, 1st adult ♂ 74·4 (1·59; 19) 71–77, adult ♀ 73·0 (1·68; 8) 71–76, 1st adult ♀ 72·0 (1·79; 14) 70–76; tail, adult ♂ 56·1 (1·18; 7) 54–58, 1st adult ♂ 54·9 (1·78; 18) 51–58, adult ♀ 54·2 (1·69; 8) 52–57, 1st adult ♀ 54·2 (1·49; 13) 52–56. See also

Yarbrough and Johnston (1965), Hubbard (1970), and Prescott (1981).

Weights. ADULT, FIRST ADULT. Nominate *coronata*. Pennsylvania (USA): (1) September–November. (2) April–May (Clench and Leberman 1978). (3) Pelee Island (Lake Erie, Canada), May (Woodford and Lovesy 1958). (4) Mexico, January (RMNH).

		♂		♀	
(1)	AD	13·1 (1·20; 153)	10·8–16·7	12·3 (1·07; 223)	9·9–15·3
	1ST AD	12·8 (1·19; 216)	9·7–19·7	12·0 (1·08; 345)	9·6–16·5
(2)		12·5 (1·13; 77)	10·6–18·6	11·8 (1·02; 65)	10·1–14·7
(3)		12·0 (— ; 9)	11·8–12·2	12·2 (— ; 8)	11·8–12·7
(4)		11·9 (0·63; 6)	10·9–12·7	10·8 (0·93; 7)	9·5–12·2

North-central Alaska, May: ♂ 13·8 (Irving 1960). Ohio, Kentucky, Virginia, and Indiana (USA): April, ♂ 13·2 (3) 11·9–14·1, ♀ 16·8; May, average ♀ 13·8 (n = 3); October, both sexes 12·4 (1·07; 7) 10·9–14·0 (Esten 1931; Baldwin and Kendeigh 1938; Mengel 1965; ZMA). Coastal New Jersey (USA), autumn: 11·6 (1·6; 28) 10·0–18·8 (Murray and Jehl 1964). Pennsylvania, 1st adult: September 11·3 (0·81; 46) 9·7–13·6, October 12·4 (1·22; 1299) 9·5–19·7; November 13·3 (1·20; 51) 10·9–15·7 (Clench and Lebermann 1978). Belize, November–March: ♂♂ 13·0, 13·3, ♀♀ 9·9, 10·0 (Russell 1964). Lundy (Devon, Britain), November: 11·9 (Workman 1961). For monthly averages of small samples, see Yarbrough and Johnston (1965); for daily weight changes, see King (1976c). See also Hubbard (1970) and Prescott (1981).

D. c. auduboni. For averages from various localities, see Salt (1963).

Structure. Wing rather long, broad at base, tip fairly rounded. 10 primaries: p7 longest, p8 0–1 shorter, p9 1–3, p6 0–2, p5 5–7, p4 8–13, p3 10–15, p2 12–17, p1 14–19. P10 strongly reduced; a tiny pin hidden below reduced outermost greater upper primary covert; 48–56 shorter than p7, 6–10 shorter than longest upper primary covert, 0–2 shorter than outermost primary covert. Outer web of p6–p8 emarginated, inner web of p7–p9 with (usually faint) notch. Tip of longest tertial reaches to tip of p1–p2. Tail of average length, tip square or slightly forked; 12 feathers. Bill short, similar to bill of Yellow Warbler *D. petechia*, but slightly deeper at base and in middle. Tarsus short, toes rather short; slender. Middle toe with claw 14·8 (5) 14–15·5; outer toe with claw *c.* 72% of middle with claw, inner *c.* 65%, hind *c.* 75%. Remainder of structure as in *D. petechia*.

Geographical variation. Marked, mainly involving colour. 2 groups, formerly often considered separate species: *coronata* group ('Myrtle Warbler') in north and east, *auduboni* group ('Audubon's Warbler') in west and south. Birds from *auduboni* group differ from those of *coronata* group in partly or fully yellow throat-patch (white in *coronata* group) and in having 4–5 feathers of each tail-half with subterminal white spot (in *coronata* group, 2–3 feathers); also, adult ♂ of *auduboni* group has side of head blue-grey, like upperparts (contrastingly black in *coronata* group), and lesser and median upper wing-coverts extensively white, showing as single large white patch (2 narrow white wing-bars in *coronata* group). See Hubbard (1970) for details. *Coronata* and *auduboni* groups grade into each other where ranges meet in British Columbia and Alberta (Hubbard 1969, which see for map and details of hybrid zone). Over entire range of *coronata* group, some birds show a few characters of *auduboni* group: in particular, occurrence of more extensive white than usual on wing-coverts or tail is frequent, as is presence of some yellow on throat (Hubbard 1969). Both groups are genetically close, not deserving recognition as separate species (Barrowclough 1980). Within *coronata* group, variation slight, involving size only, and no races recognized: birds from Rockies and Mackenzie (Canada) west to Alaska average larger than birds from eastern Canada and neighbouring USA, larger birds sometimes separated as *hooveri* McGregor, 1899, but difference slight, e.g. average difference in wing length only *c.* 4 mm (Ridgway 1901–11; Hubbard 1970). Within *auduboni* group, difference more marked: *nigrifrons* of north-west Mexico and *goldmani* of south-east Mexico and western Guatemala extensively black on head, neck, mantle, and breast (especially *goldmani*), *nigrifrons* has yellow crown- and throat-patches, *goldmani* yellow-and-white ones; *nigrifrons* intergrades into *auduboni* in Arizona (USA). See Hubbard (1970) for many details of variation. CSR

Dendroica palmarum (Gmelin, 1789) **Palm Warbler**

FR. Paruline à couronne rousse   GE. Palmenwaldsänger

A North American species, breeding from south-west Mackenzie east to central Ontario, south to central Alberta, north-east Minnesota, and central Michigan; wintering from south-east USA south to Belize, north-east Nicaragua, and Greater Antilles. Adult ♂ found dead on tideline, Cumbria (England), 18 May 1976 (O'Sullivan *et al.* 1977).

Dendroica striata **Blackpoll Warbler**

PLATE 5
[between pages 64 and 65]

DU. Zwartkopzanger   FR. Paruline rayée   GE. Streifenwaldsänger
RU. Пестрогрудый лесной певун   SP. Reinita listada   SW. Vitkindad skogssångare

Muscicapa striata J R Forster, 1772

Monotypic

Field characters. 12·5–14·5 cm; wing-span 20·5–23 cm. Size between Willow Warbler *Phylloscopus trochilus* and Wood Warbler *P. sibilatrix*, with proportionately slightly shorter wings but longer tail; somewhat larger than most other vagrant Parulidae but shorter than Yellow-rumped Warbler *D. coronata*. Medium-sized, quite robust but graceful, usually arboreal Nearctic wood warbler; commonest passerine vagrant from Nearctic and hence epitome of genus in west Palearctic. In all plumages, shows double white wing-bar, white tail-spots, and diagnostic yellow feet. Breeding ♂ has striking black cap, white cheeks and black malar stripe and flank spots; breeding ♀ lacks black and white contrasts, showing streaked greenish crown and ear-coverts. Immature resembles ♀ but most buffier above face and below, where less strongly streaked. Sexes dissimilar; marked seasonal variation in ♂. Immature not certainly separable from winter adult.

ADULT MALE BREEDING. Moults: December–April (head, neck, and most wing-coverts and tertials). Full head cap black, cheeks and ear-coverts white, with contrast recalling Great Tit *Parus major*. Upperparts basically olive-grey, streaked black particularly when worn. Wings dull black, with yellowish-white tips to median and greater coverts which form striking double wing-bar, narrow whitish margins to tertials and (when fresh) light green fringes to flight-feathers. Tail black, with narrow grey feather-edges; bold white panels or spots on 2–3 outer pairs visible from below, even in flight. Underparts from chin to under tail-coverts white, with striking black malar stripe further emphasizing white cheeks and breaking up into lines of heavy black streaks on side of breast and along flanks. Bill black-horn on upper mandible, usually yellow-horn on cutting edges and lower mandible. Leg and foot colour important to diagnosis: tarsus variably flesh-yellow to dusky brown but front and rear of tarsus and sides and soles of toes paler, always looking yellow, unlike any other *Dendroica*. ADULT FEMALE BREEDING. Foreparts lack strong contrasts of ♂, with crown and ear-coverts olive (as back), crown obviously streaked black but never forming cap and contrasting only with dull yellowish to buffish supercilium. Malar stripe broken up into spots, and breast and flank streaking more diffuse, contrasting less with duller, buffier ground-colour to underparts. Under tail-coverts white. ADULT NON-BREEDING, FIRST-WINTER. No reliable distinction between adult and immature. Plumage rather variable, with upperparts varying from bright yellowish-olive to greyish-green, and fore-underparts from sulphur-yellow to yellowish- or ochraceous-white. Adult (particularly ♂) generally more heavily streaked on mantle, sides of breast, and flanks than immature, with more obvious tail-spots and brighter, yellowish-green fringes to flight-feathers, these being distinctly duller and brown in retained juvenile plumage. Soles of feet pale straw, this colour often invading rest of feet and front and back of legs.

Breeding ♂ unmistakable, recalling only wholly pied Black-and-white Warbler *Mniotilta varia* but easily separated by wholly black cap. ♀ and immature have less distinctive appearance, with double white wing-bar shared by 9 other known or potential vagrant relatives. Nevertheless, in autumn, plumage ground-colour of *D. striata* greener above and yellower below than most congeners, with brown to yellow legs and uniquely yellow feet, at least on soles. In Nearctic, Bay-breasted Warbler *D. castanea* presents pitfall: plumage pattern of non-breeding adult and immature similar but, unlike *D. striata*, shows little or no supercilium, bluish fringes to flight-feathers often unstreaked, wholly pale buff underbody (frequently washed chestnut on flanks), and dusky legs and feet (see Plate 61); call a distinctive, mellow 'chip'. Another confusion species that has not yet crossed North Atlantic is Pine Warbler *D. pinus*, only differing constantly in more pronounced supercilium and eye-ring, unstreaked back, bolder wing-bars, and dusky legs and feet. Flight typical of genus, looking almost as light as *Phylloscopus*. Gait a hop. Stance level with tail often held above drooping wing-points. Characteristically feeds in outer branches and foliage of trees, picking insects from underside of leaves, but transatlantic vagrants have fed in bush and ground cover. Feeding actions rather deliberate, sometimes suggesting *Hippolais* warbler. Approachable.

Calls include loud 'smack'; also rather hard 'tsip', similar to one call of Cape May Warbler *D. tigrina*.

Habitat. Breeds in northern and north-east Nearctic to near treeline, in mountains as well as lowlands, in coniferous woods, especially spruce *Picea*, frequently stunted. Also inhabits mixed-wood edges, logged and burned areas, and alder *Alnus* thickets, favouring moist ground (Godfrey 1979). Favours woodland of stunted spruce or fir *Abies* created by frost, high winds, or bog conditions; common in treetops on migration (Pough 1949). While migrating may be found wherever trees grow, and often also along fences and stone walls in fields and pasture; in autumn even along weedy roadsides. On spring migration, common in orchards and shade trees. (Forbush and May 1955.) Winters in South America in deciduous, rain, and cloud forests up to 3000 m, and in lowlands in gallery forest, second growth, grassy fields with scattered vegetation, and coastal mangroves, foraging also on ground (Schauensee and Phelps 1978).

Distribution. Breeds in North America from north-central Alaska east to northern Labrador, south to southern Alaska and central British Columbia, and east across southern Canada to eastern New York and southern Maine. Winters mainly in South America, from Panama south to Chile and eastern Argentina.

Accidental. Iceland, Britain and Ireland (see Movements), Channel Islands, France (Ouessant).

Movements. Migrant, with longest average migration among paruline warblers, from subarctic and boreal Canada and Alaska to South America.

Autumn departure from remote parts of breeding range is seldom reported, but evidently begins in early August (Rand 1946; Gabrielson and Lincoln 1959; Todd 1963). Last records in breeding range late August to late September (Munro and Cowan 1947; Todd 1963; Sadler and Myres 1976), with major lighthouse kill of migrants in south-west Newfoundland 7–8 September 1970 (L Tuck). Peak migration through Great Lakes area begins late August (Sprague and Weir 1984; Janssen 1987), but main passage in north-east is mostly late September to mid-October (Trautman 1940; Bailey 1955; Stewart and Robbins 1958). Records from south-east states are fewer, but on similar schedules (Potter *et al.* 1980; Kale and Maehr 1990). In Bermuda, very common throughout October, with numbers diminishing to mid-November (Wingate 1973). Reaches West Indies and South America from late September, with peak movement in October (Brudenell-Bruce 1975; ffrench 1976; Schauensee and Phelps 1978). Migration almost exclusively through eastern North America, even for birds from Alaska and north-west Canada (Bent 1953). There is a long-standing dispute whether autumn route continues mainly south across ocean from New England and eastern Canada (Drury and Keith 1962; Nisbet 1970), or mainly through south-east states and Lesser Antilles (Murray 1965, 1989); both routes may be important. Chronological data do not exclude either route; very heavy birds met in New England (over twice fat-free weight: Drury and Keith 1962) indicate that some make long non-stop flights, but do not prove that all, or most, do so.

In spring, migrants leave wintering areas in March and April (ffrench 1976; Schauensee and Phelps 1978; Hilty and Brown 1986). Earliest birds appear as far north as Minnesota and New York in April (Bull 1974; Janssen 1987), with peak passage in south-east states extending from mid-April to mid-May (Lowery 1974; Imhof 1976; Kale and Maehr 1990). Although first arrivals are somewhat earlier in Mississippi valley than on east coast, peak movement begins in early or mid-May right across northern states (Stewart and Robbins 1958; Mengel 1965; Curtis 1969; Bull 1974). Rapid passage follows spring warming, bringing birds to Alaska and Mackenzie District by mid-May in some years; more often, extreme of range not reached until late May or early June (Gabrielson and Lincoln 1959; Todd 1963; Erskine and Davidson 1976; Etcheberry 1982). Main passage in southern Canada extends into early June (Sadler and Myres 1976; Speirs 1985), and in some years many birds are found outside breeding areas to mid-June (Palmer 1949; A J Erskine). Route is more westerly than in autumn, passing through southern states on broad front west to Louisiana and eastern Texas, with highest numbers in Florida; direct over-water movements between South America and New England (as in autumn) do not occur in spring (Stevenson 1957; Lowery 1974; Oberholser 1974; Rappole *et al.* 1979).

Vagrants appear frequently across western North America both seasons, especially in California (Garrett and Dunn 1981; Shuford 1981; American Ornithologists' Union 1983); presumably these belong to north-west populations. Also recorded Cornwallis Island (75°N, Arctic Canada) in May (Godfrey 1986), and in Greenland September–October (Bent 1953).

Rare autumn vagrant to Atlantic seaboard of west Palearctic. 28 records from Britain and Ireland up to 1990 (Rogers *et al.* 1991), especially south-west England and southern Ireland, mainly from late September to October (Dymond *et al.* 1989). AJE, PRC

Voice. See Field Characters.

Plumages. Adult Male Breeding. Forehead and crown black, forming contrasting black cap, reaching down to lore and middle of eye. Hindneck and side of neck narrowly streaked black-and-white. Mantle, scapulars, back, rump, and upper tail-coverts light drab-grey, mantle and scapulars broadly streaked black, back, rump, and upper tail-coverts more narrowly streaked black; black on rump and shorter upper tail-coverts sometimes less sharply defined and more concealed than on remainder of upperparts. Lower cheek to ear-coverts cream-white or off-white, partly marked with faint grey specks; black cap and white patch on side of head strongly suggestive of tit *Parus*. Contrasting black malar stripe on each side of chin, widening towards lower side of neck, merging into boldly black-and-white streaked and spotted side of breast and flank; black malar stripe narrowly mottled and streaked white when plumage fresh, hardly so when worn; bold black dots on flank gradually duller and wider apart toward rear. Thigh mixed black and drab-grey. Remainder of underparts white or off-white, slightly tinged cream or pink-buff when plumage fresh, especially on vent and rear flank. Tail black, both webs of central pair (t1) and outer web of others narrowly edged grey; tip of inner web of t4 with white fringe or shallow white spot, inner webs of t5–t6 with white wedge on tip (maximum extent of white 10–20 mm). Flight-feathers and greater upper primary coverts dull black, outer webs narrowly but distinctly fringed light green (green in closed wing rather contrasting in tinge with drab-grey and black of body). Bastard wing black, outer web of middle and (sometimes) long feather fringed white. Tertials dull black, outer webs broadly fringed pale drab-grey; almost white or pale yellow along margin. Upper wing-coverts black, fringed drab-grey; tips of median and greater coverts white or yellow-white (mainly on outer webs), forming distinct wing-bars 3–4 mm wide. Axillaries and under wing-coverts mixed pale grey and white, small coverts along leading edge of wing pale yellow with black mottling. *In fresh plumage* (March–April), black feathers of cap show narrow olive fringes, drab-grey of upperparts faintly tinged olive, white of side of head and underpart slightly tinged greyish-cream, black of chin-stripe, lower side of neck, and side of breast slightly more variegated white (but black still prominent and almost coalescent). Adult Female Breeding. Like adult ♂ breeding, but black of cap replaced by black-and-green streaking, green fringes narrowest

on forehead and side of crown, where black almost coalescing (especially when plumage worn), more evenly spaced on central crown. Remainder of upperparts as adult ♂ breeding, but grey with distinct green suffusion, especially on lower mantle, scapulars, and rump. Cream-white of side of head and neck more mottled with grey than in adult ♂ breeding, less sharply defined from cap. Ground-colour of underparts either off-white, as adult ♂ breeding, but more often with yellow tinge, especially on throat and chest, sometimes faintly elsewhere. Black stripe on side of chin and lower side of neck broken into black dots, side of breast rather evenly streaked black and off-white or pale yellow (black 1–2 mm wide, against 2–4 mm in adult ♂ breeding), flank with narrow but distinct dull black or sooty streaks, bordered drab- or buff-grey at sides (marks on flank less bold and deep black than adult ♂ breeding). Tail and wing as adult ♂ breeding, but fringes of tertials and tips of inner median and greater coverts sometimes yellowish. Adult Male Non-breeding. Upperparts bright yellowish olive-green, shading to dark ash-grey with some olive-green on feather-tips on rump and upper tail-coverts; feathers of cap, mantle, inner scapulars, back, and sometimes elsewhere with narrow black streaks on centres, partly concealed but usually showing as distinct black triangular spots or short streaks on cap and mantle. Supercilium and eye-ring yellow; lore and short streak behind eye dark green or dull black, breaking eye-ring in front and behind; upper cheek, ear-coverts and side of neck olive-green, lower cheek yellow, sometimes bordered by mottled black malar stripe below. Chest sulphur-yellow, side of breast and flank light drab-grey with yellow wash, feathers with short but sharp sooty spot or streak on shaft (sometimes on all feathers, sometimes on part of side of breast and upper chest only). Chin, throat, belly, and vent white, feather-tips washed pale yellow (not fully concealing white, especially belly prominently white), under tail-coverts pure white. Tail and wing as adult ♂ breeding, but fringes of upper wing-coverts olive-green and tips of median and greater coverts pale yellow to yellow-white; fringe of outer web of bastard wing white. Adult Female Non-breeding. Like adult ♂ non-breeding, but streaks and spots on upperparts on average narrower, rump and upper tail-coverts slightly greener, short dark streaks and spots on underparts more restricted, and yellow wash on belly on average more pronounced, sometimes extending slightly to under tail-coverts. See also wing in Measurements (below). Juvenile. Cap, mantle, scapulars, back, and sides of head light grey-brown or olive-grey with black streaks and mottlings; rump pale grey-brown or pale buff-grey, mottled or barred black; upper tail-coverts grey-brown with dusky shaft-streaks and drab fringes. Underparts dull white, partly tinged olive-yellow (except on lower belly and under tail-coverts), barred and mottled dusky-grey. Upper wing-coverts brown-black with dull olive-green fringes, median and greater tipped pale buff-yellow, becoming white with wear. First Non-breeding. Like adult non-breeding, but juvenile tail, flight-feathers, greater upper primary coverts, and bastard wing retained, tips of tail-feathers on average more sharply pointed, tip of inner web less broad and truncate (but some overlap between age groups); primaries on average more broadly fringed white along tips; primary coverts on average slightly browner and with more frayed tips; middle feather of bastard wing fringed yellow rather than white. Head, body, wing-coverts, and tertials as adult non-breeding, but ground-colour of upperparts slightly duller olive-green, less yellowish, fine black streaks restricted to mantle and scapulars (entirely absent in some ♀♀, occasionally some streaks on crown in ♂), rump more extensively green on average, less grey; underparts extensively pale yellow (more so than in adult non-breeding), under tail-coverts contrastingly white, but occasionally with pale cream-yellow tinge, side of breast and flank with ill-defined olive streaks; unlike adult non-breeding, no sharp sooty streaks on underparts, except for side of breast and flank of some ♂♂. Sexes generally indistinguishable (but see Measurements); when series of skins compared, ♂♂ show more white on average on belly than ♀♀, some ♂♂ having belly mixed with equal amounts of yellow and white, many ♀♀ fully yellow, but others of both sexes intermediate. FIRST ADULT BREEDING. Like adult breeding, but tail, flight-feathers, primary coverts, and bastard wing still juvenile; in particular, tail, primaries, and primary coverts browner and more worn on tip than in adult at same time of year; fringe of middle feather of bastard wing usually yellow; in ♂, black streaks on upperparts, side of breast, and flank on average slightly narrower, some old sooty-and-green non-breeding feathers sometimes mixed in black-and-white of hindneck or black-and-grey of remainder of upperparts; aspect often rather like adult ♀ breeding, but cap black, not streaked; in ♀, black streaks on entire head and body narrower, c. 1 mm wide (in adult ♀ breeding, c. 2 mm); upperparts brighter olive-green, less greyish; sooty-and-white streaked collar round hindneck virtually absent, and chin to belly more intensely tinged pale yellow than in adult ♀ breeding.

Bare parts. ADULT, FIRST ADULT. Iris sepia-brown, dark brown, or black-brown. Upper mandible and tip of lower mandible dark grey, dark horn-brown, brown-black, or black, cutting edges and remainder of lower mandible pale yellow, horn-yellow, horn-grey, plumbeous-grey, or yellow with blue-grey stripe on lateral lower mandible; variation apparently independent of age or sex, both adult ♂ and 1st adult ♀ may show pale yellow or dark grey lower mandible in autumn. Mouth pale flesh in 1st autumn (Evans 1970). Leg and foot flesh-yellow, light or dark horn-yellow, orange-flesh, yellow-brown, horn-brown, orange-brown, dark horn, or dusky brown; rear of tarsus, sides of toes, and soles paler pink-yellow, pale yellow, pale orange-yellow, horn-yellow, or yellow-brown; rather variable in both sexes, but on average darker in ♀ than in ♂ (Blake 1954; RMNH, ZMA). JUVENILE. Bill, leg, and foot pink-buff, bill gradually turning to dusky, leg and foot to sepia (Dwight 1900).

Moults. ADULT POST-BREEDING. Complete; primaries descendent. July–August, on summer grounds. ADULT PRE-BREEDING. Partial: head, neck, lesser, median, and greater upper wing-coverts, and tertials, but part of non-breeding feathering on body (especially neck, rump, or tail-coverts), some outer greater or median coverts (rarely, many greater), or a few tertials occasionally retained. In winter quarters; starts between mid-December and late February (mantle, cheek, chest, and side of breast first), completed March or April. POST-JUVENILE. Partial: head, body, lesser and median upper wing-coverts, and usually greater upper wing-coverts and tertials. Starts shortly after fledging, when still on or near breeding area; moult July–August (mainly 2nd half of July and 1st half of August), generally completed before autumn migration, but a few birds retain scattered juvenile (mainly on head, neck, and rear of underbody) until at least late September. FIRST PRE-BREEDING. Like adult pre-breeding, but part of non-breeding apparently more frequently retained, especially neck, back to upper tail-coverts, and (rarely) even some feathering of crown. (Dwight 1900; Preble 1908; Pyle et al. 1987; RMNH, ZMA.)

Measurements. ADULT, FIRST ADULT. Eastern USA, spring and summer, and northern South America, October–March; skins (RMNH, ZMA). Bill (S) to skull, bill (N) to distal corner of nostril; exposed culmen on average 3·4 less than bill (S).

	♂			♀		
WING	76·4 (1·91; 32)	73–79		72·3 (1·38; 30)	70–76	
TAIL	50·1 (1·53; 28)	47–53		47·9 (1·28; 27)	45–50	
BILL (S)	13·7 (0·48; 28)	12·9–14·3		13·4 (0·43; 26)	12·7–14·4	
BILL (N)	7·7 (0·31; 26)	7·2–8·3		7·6 (0·22; 26)	7·1–7·9	
TARSUS	19·1 (0·56; 28)	18·2–20·1		18·9 (0·51; 25)	18·0–19·7	

Sex differences significant for wing, tail, and bill (S). Ages combined above, though retained juvenile wing and tail of 1st adult on average shorter than in older birds. Thus, in ♂: wing, adult 77·7 (1·11; 7) 76–79, 1st adult 75·9 (2·10; 18) 73–79; tail, adult 50·4 (0·94; 7) 49–52, 1st adult 49·8 (1·89; 18) 47–53. Wing (measured with different technique), autumn, Massachusetts (USA): adult, ♂ 73·5 (2·59; 671) 66–81, ♀ 71·6 (2·32; 450) 64–79; 1st adult, ♂ 71·5 (2·57; 233) 65–78, ♀ 70·7 (2·92; 482) 65–78 (Nisbet et al. 1963).

Weights. Pennsylvania (USA): September, (1) adult, (2) 1st adult, (3) age unknown; October, (4) adult, (5) 1st adult, (6) age unknown; May–June, (7) ♂, (8) ♀ (Clench and Lebermann 1978). Massachusetts (USA), autumn: (9) adult ♂, (10) adult ♀, (11) 1st adult ♂, (12) 1st adult ♀ (Nisbet et al. 1963, which see for details and for some samples from elsewhere). New Jersey (USA), September–October: (13) data from Murray and Jehl (1964), (14) from Murray (1965), which see for daily variation.

(1)	12·2 (1·53; 61)	9·7–19·0	(8)	12·7 (1·66; 8)	11·5–16·4
(2)	11·8 (0·86; 192)	9·8–14·5	(9)	13·0 (—; 660)	9·9–21·0
(3)	12·2 (1·17; 70)	10·7–18·1	(10)	12·8 (—; 449)	9·6–21·0
(4)	15·5 (2·91; 29)	12·1–21·6	(11)	12·2 (—; 229)	9·8–19·0
(5)	13·5 (2·06; 83)	10·4–20·8	(12)	12·0 (—; 478)	9·3–21·5
(6)	15·1 (2·96; 18)	12·0–20·9	(13)	12·9 (—; 143)	8·5–22·1
(7)	13·0 (1·14; 22)	10·8–15·9	(14)	11·7 (—; 352)	9–21

North-west Yukon (Canada), late May and June: ♂ 12·4 (0·75; 6) 11·4–13·7, ♀ 12·8 (1) (Irving 1960). Range of migrants New England states (USA) 10·0–23·4 (Drury and Keith 1962, which see for details); for changes on stop-over site on spring migration Mississippi (USA), see Kuenzi et al. (1991); see also Baldwin and Kendeigh (1938).

Netherlands' Antilles (off Venezuela), October and early November: ♂ 8·4 (0·68; 11) 7·0–9·0, ♀ 8·3 (0·57; 7) 7·5–9·0. Surinam, Trinidad, and Tobago, October–February: ♂ 11·0 (1·46; 5) 8·5–12·0, ♀♀ 7·5, 13·0. (RMNH, ZMA.) Galapagos islands, May: ♂ 10·6 (Boag and Ratcliffe 1979). Bardsey (Wales), October: 11·4 (Evans 1970).

Structure. Wing rather long, broad at base, tip pointed. 10 primaries: p8 longest, p9 0–2 shorter, p7 0·5–2, p6 3–7, p5 7–13, p4 10–16, p3 13–19, p2 15–21, p1 17–24. P10 reduced, a tiny pin hidden below reduced outermost greater upper primary covert; 48–57 shorter than p8, 7–10 shorter than longest upper primary covert. Outer web of p7–p8 emarginated; inner web of p8–p9 with notch (rarely, p7 also; notch on p9 sometimes faint, especially in juvenile and 1st adult). Tip of longest tertial reaches to tip of p1–p2. Tail rather short, tip square or slightly forked; 12 feathers. Bill rather short; rather thick at base and in middle, more slender and compressed laterally at tip. Tarsus short, toes of average length; fairly slender. Middle toe with claw 14·3 (10) 13·3–15·2; outer toe with claw c. 72% of middle with claw, inner c. 68%, hind c. 76%. Remainder of structure as in Yellow Warbler D. petechia.

Recognition. First adults rather similar to 1st adult of Bay-breasted Warbler D. castanea of central and eastern Canada and north-east USA; see Howard (1968), Grant (1970), Holman (1981), Stiles and Campos (1983), and Whitney (1983).   CSR

Setophaga ruticilla **American Redstart**

PLATE 6
[between pages 64 and 65]

Du. Amerikaanse Roodstart   Fr. Paruline flamboyante   Ge. Schnäpperwaldsänger
Ru. Американская горихвостка   Sp. Candelita   Sw. Amerikansk rödstjärt

Motacilla Ruticilla Linnaeus, 1758

Monotypic

Field characters. 13–14·5 cm; wing-span 19·5–22 cm. About 10% smaller than Red-breasted Flycatcher *Ficedula parva* which it recalls in plumage pattern and behaviour as much as any close relative. Quite small but long-tailed, elegant, extremely active Nearctic wood warbler, with bright orange (♂) to yellow (♀, immature) patches at shoulder, along bases of flight-feathers, and on bases of outer tail. ♂ otherwise black, with white belly; ♀ and immature otherwise greyish-green, with white spectacle and under-parts. Flicks wings and spreads tail constantly. Sexes dissimilar; no seasonal variation. Immature separable.

ADULT MALE. Moults: July–August (complete). Head, breast, flanks, and ground-colour to upperparts, wings, and tail glossy black, with striking rufous-orange breast-sides, broad panel across bases of secondaries and primaries, under wing-coverts, and long panels on base of outer tail-feathers. Centre of lower breast, belly, and under tail-coverts mostly white. Bill and rather short legs usually black. For differences possibly related to age, see Plumages. ADULT FEMALE. Crown and face grey, with narrow white fore-supercilium and eye-ring forming indistinct spectacle; upperparts dull greyish-olive, sometimes with brown tinge, ending in dusky upper tail-coverts and tail; wings dusky. Underparts dull white. Shows yellow where ♂ is orange, but that on wing less broad (often much narrower or absent on primaries). FIRST-YEAR. Resembles adult ♀, but ♂ has yellow-orange breast-sides and can show blackish mottling on breast, back, and upper tail-coverts by summer; ♀ lacks yellow on primaries, often also on secondaries.

Unmistakable; *F. parva* lacks wing-panel and has white tail-sides. Flight light and aerobatic, allowing bird to flit like butterfly through and round foliage and to dash out from cover in accomplished flycatching. Hops. Stance level with tail often raised above body line. Restless and excitable, constantly drooping and flicking wings and raising and fanning tail. Happy with trees of medium height, their understorey and scrub, when feeding behaviour may suggest *Sylvia* warbler. Tends to feed lower in trees and in more open areas than other Nearctic warblers.

Calls include rather thin, clear 'tzit' or 'tseet' and rather hard, clicking 'tsip', also described as slightly drawn-out 'tchip'.

Habitat. Breeds in cool and warm temperate zones of Nearctic, mainly in deciduous woodlands having openings or swampy places, but also in open second growth on moist lowlands (Pough 1949). Also lives on borders of pastures, in orchards, and even among shade trees and garden shrubbery, sometimes near dwellings (Forbush and May 1955). In Great Plains region, found in moist mixed or coniferous bottomland woods, especially young or second-growth stands, and mature forest margins or openings (Johnsgard 1979).

Winters in South America in coastal mangroves, suburban areas, thorny thickets and forests to savannas, up to 3000 m, foraging actively in vegetation and at all heights (Schauensee and Phelps 1978).

Distribution. Breeds in North America from south-east Alaska east to Newfoundland, south to northern Washington and eastern Oregon and south-east to northern Colorado, north-east Texas, coastal Louisiana, and Florida. Winters in West Indies and from Mexico south to northern Peru, northern Brazil, and Guianas.

Accidental. Iceland, Britain, Ireland, France, Azores.

Movements. Migrant, breeding from Canada to south-east USA; southernmost areas only 600–800 km from wintering grounds in West Indies, although most birds move further.

Autumn migration begins early, but timing unclear, as passage masked in south by local breeders. Some birds reach Gulf coast states as early as late July (Lowery 1974; Oberholser 1974; Kale and Maehr 1990), and Central and northern South America from mid- or late August (Land 1970; Hilty and Brown 1986; Stiles and Skutch 1989). Passage chiefly late August to September in southern Canada and northern USA (Stewart and Robbins 1958; Sprague and Weir 1984; Cannings *et al.* 1987; Alcorn 1988). Peak movement September in southern states, but passage continues to end of October or beginning of November in south-east (Bent 1953; Lowery 1974; James and Neal 1986; Kale and Maehr 1990); stragglers occur throughout October all across continent, including west of coastal mountains where recorded only casually (e.g. Campbell *et al.* 1974). Migration in both seasons is on broad front, birds crossing Gulf of Mexico as well as skirting it to east and west (Bent 1953; Stevenson 1957; Peterson and Chalif 1973). In Bermuda, where wintering is regular, very common on passage in autumn, but inconspicuous in spring (Wingate 1973). Highest numbers winter in western Mexico and West Indies (Pashley and Martin 1988). Ringing data (e.g. Guatemala, Puerto Rico,

Venezuela) prove winter site-fidelity, with returns up to 7 years later (McNeil 1982; Rogers *et al.* 1982*a*; Faaborg and Arendt 1984).

Spring migration begins March, with earliest arrivals north to Arkansas and Florida (James and Neal 1986; Kale and Maehr 1990). Wintering areas are vacated gradually in April, most birds departing by early May (Land 1970; Ridgely 1981; Raffaele 1989). Progression much more uniform than in autumn, with vanguard reaching Missouri to Pennsylvania from early April, Nebraska to Massachusetts by end of April, and Idaho to Nova Scotia from early May (Bent 1953; Bailey 1955; Burleigh 1972). Despite widespread occurrence of earliest birds, peak passage is not general until May north of Gulf coast states (Stewart and Robbins 1958; Mengel 1965; Bohlen 1989), and most Canadian breeding areas are not fully occupied until last week of May (Peters and Burleigh 1951; Sadler and Myres 1976; Speirs 1985). Migration in far west is later, with little movement before May in south, and arrivals north of 55°N in early June (Jewett *et al.* 1953; Erskine and Davidson 1976; Kessel and Gibson 1978; Alcorn 1988).

Vagrant to northern Alaska and northern Canada (including Banks Island, *c.* 72°N, September), also to Greenland (Salomonsen 1950–1; American Ornithologists' Union 1983; Godfrey 1986).

Rare autumn vagrant to Atlantic seaboard of west Palearctic. In Britain and Ireland, 7 records up to 1990, mainly from south-west England and southern Ireland, October–November, but once in December (Rogers *et al.* 1988, 1991; Dymond *et al.* 1989).   AJE, PRC, DFV

Voice. See Field Characters.

Plumages. ADULT MALE. Head, neck, chin to chest, and mantle to upper tail-coverts black, slightly glossed metallic blue in some lights. Black of chest continued in broad black stripe at each side of upper belly, narrowing backwards and extending to side of lower belly and thigh; feathers of black stripe tipped white or orange when plumage fresh. Large fiery-orange or chrome-orange patch on side of breast, sharply demarcated from black of chest and side of belly, merging into paler orange on upper flank and mixed orange and white on lower flank. Narrow stripe down centre of upper belly pale orange, merging into cream-pink or white lower belly and vent. Shorter and median under tail-coverts white, median with broad dark grey stripe; longer coverts black. Central pair of tail-feathers (t1) and inner web of t2 black; remainder of tail spectrum-orange or salmon-orange, 20–23 mm (16–26) of tips contrastingly black. All upper wing-coverts, bastard wing, and tertials black; flight-feathers black or brownish-black with extensive contrasting pale orange on bases and middle portions: *c.* 25% of length of inner secondaries black on outer web, increasing to *c.* 50% of inner primaries and to *c.* 70% on 2nd outer primary (p8, ignoring reduced p10), black often slightly more extensive on inner webs; p9 usually with blackish shaft and often with dark outer web bordered by long pale orange fringe; outer web of p8 sometimes also black (in 10 birds, bases of outer primaries fully orange in 5, p9 dark in 2, p8-p9 in 3); orange on primaries extends up to 4–10 mm beyond tip of longest upper primary covert (average, 6·6 mm, $n = 10$). Under

wing-coverts and axillaries chrome-orange, merging into paler orange on broad borders along bases of inner webs of flight-feathers; marginal coverts black. 2 types of plumage in adult ♂, possibly related to age: birds described above are perhaps 3rd-year or older, while others, perhaps 2nd-years, differ as follows. Upperparts less uniform black, more grey of feather-bases visible, feathers of (especially) nape, mantle, scapulars, and rump with brown or pink-brown fringes, causing brownish tone, often contrasting with blacker cap; some white mottling on chin or throat; black stripe along upper side of belly less developed, replaced by grizzled sooty, pale orange, and white feathering; black lower border of chest sharply demarcated (not merging into black on side of belly); orange on side of breast less sharply demarcated; tail either as above or with outer web of t2 all-black; primary coverts and tips of flight-feathers distinctly browner, and often also tertials and many other upper wing-coverts; primaries with indistinct brown or orange edges along outer webs; bases of outer webs of 2–3 (0–5) outer primaries usually dark, orange on outer webs of remainder of primaries extending up to 1–7 mm beyond longest primary covert (average, 4·1 mm, $n = 15$); marginal coverts mainly orange (not black). ADULT FEMALE. Cap dark ash-grey, slightly tinged olive when fresh, gradually merging into greyish-olive-green of mantle, scapulars, back, and rump, and this in turn into dark ash-grey or greenish-grey of upper tail-coverts. Side of head and neck medium ash-grey, mottled white on short stripe above lore and on cheek; finely streaked white on ear-coverts; mottled white eye-ring. Chin and throat white, merging into ash-grey of cheek. Chest and mid-breast pale buff-yellow to light yellowish-grey; side of breast bright light yellow, forming contrasting patch. Flank light grey, belly and vent white, all with light yellow wash when plumage fresh. Under tail-coverts yellow-white or pure white, longest tipped grey. Central 2 pairs of tail-feathers (t1–t2) and 16–23 mm on tips of others sooty-black with olive-green outer fringe, remainder of tail extensively bright light yellow; occasionally, basal and middle portion of outer web of t2 yellow or inner web of t3 black. Flight-feathers dark grey or blackish-grey, outer webs fringed olive-green (except for emarginated parts of outer primaries); in about half of 10 birds examined, some light yellow visible at base of inner primaries (extending 1–5 mm beyond primary coverts); 1 bird had some yellow visible at bases of all primaries, others showed none at all; in all birds, some yellow visible at bases of secondaries (extending *c.* 3–10 mm beyond greater coverts), forming small patch on closed wing. Greater upper primary coverts and bastard wing blackish-grey, tertials and remainder of upper wing-coverts dark grey, lesser and median with broad but ill-defined olive-green tips and faint dark shafts, greater coverts and tertials extensively tinged olive-green on outer webs. Under wing-coverts and axillaries bright light yellow, similar in tone to side of breast, basal inner borders of flight-feathers pale yellow or yellow-white. Occasionally, tail-base patches tinged orange instead of light yellow but this independent of age (Spellman *et al.* 1987, which see for changes with age). JUVENILE. Upperparts and side of head and neck dark sepia; chin, throat, and chest buff-brown or grey-brown, remainder of underparts white with pale yellow tinge, some grey of feather-bases visible. Lesser upper wing-coverts sepia, median and greater dark brown with grey-buff tips, sometimes forming pale wing-bars. For tail and flight-feathers, see First Adult Male and First Adult Female (below). FIRST ADULT MALE. Rather like adult ♀. Mantle, scapulars, and back often slightly darker than adult ♀, more brownish-olive, less greenish; tail-coverts often blacker; side of head slightly darker ash-grey, short white stripe above lore to just above eye and broken white eye-ring sometimes more conspicuous. Underparts rather different: side of breast yellow-orange to pure orange, distinctly more orange than yellow under wing-coverts (in adult ♀, colour of side of breast and underwing similar, yellow); remainder of underparts as adult ♀, but slightly washed orange, not yellow. Tail as adult ♀, base bright light yellow or (occasionally) orange-yellow, t2 either all-black or with yellow base to outer web; tail-feathers narrower and with more sharply pointed tips (broader and bluntly pointed or rounded in adult ♀). Flight-feathers still juvenile; similar to adult ♀, but green fringes often less conspicuous; pattern of light yellow at bases as in adult ♀ (some yellow always present at bases of secondaries; present at bases of inner primaries in about half of birds examined; if present, extending 1–5 mm beyond primary coverts); juvenile tertials sometimes with broad pale yellow to off-white outer fringe; upper wing-coverts as adult ♀, but bases often darker, contrasting more with green-grey tips. During winter, some black feathers start to appear, first on side of head; in a few birds, these develop from August, in many others not until March–April (see Moults, and Rohwer *et al.* 1983). FIRST ADULT FEMALE. Like adult ♀, but cap and upper tail-coverts on average paler ash-grey, more distinctly washed olive; white eye-ring often more conspicuous; patch on side of breast and under wing-coverts similar or paler and less extensively yellow (side of breast not as orange as 1st adult ♂). Tail still juvenile; bases of outer feathers light yellow, of t1–t2 dark, of inner web of t3 either yellow, all-dark, or dark with elongate yellow subterminal patch; tail-feathers narrow, tips pointed (as in 1st adult ♂). Flight-feathers juvenile, browner and with more worn tips than in adult at same time of year; no yellow visible on bases of primaries; no yellow visible on bases of secondaries in about half of 20 birds examined, yellow extending 1–5(–8) mm beyond coverts in remainder. Juvenile tertials, if retained, sometimes with pale yellow or off-white fringe along tip of outer web (absent when heavily abraded).

Bare parts. ADULT, FIRST ADULT. Iris sepia-brown, dark brown, or black-brown. Bill dark horn-brown, dark grey-brown, brown-black, plumbeous-black, or black; base of lower mandible in 1st autumn flesh-grey, flesh-brown, light horn-brown, or pale grey-brown. Leg and foot dark brown-grey, dark horn-brown, brown-black, plumbeous-black, or black. (RMNH, ZMA.) JUVENILE. Bill, leg, and foot dusky pink-buff, darkening to brown-black when older (Dwight 1900).

Moults. ADULT POST-BREEDING. Complete; primaries descendent. On breeding grounds, July–August. ADULT PRE-BREEDING. In winter quarters. Restricted to scattered feathering of head and body or no moult at all; a very few ♂♂ in slight moult September–February, more (*c.* 5–17% of various samples examined) with some moult March–April; a few ♀♀ in moult throughout September–March (Rohwer *et al.* 1983). POST-JUVENILE. Partial: head, body, and lesser and median upper wing-coverts. Starts shortly after fledging, in one captive bird at age of 22 days, before juvenile flight- and tail-feathers full-grown (Petrides 1943); in heavy moult at *c.* 3 weeks old (Harrison 1984). On breeding grounds, July–August (Pyle *et al.* 1987). Single ♂ in moult 2 July (Mengel 1965). FIRST PRE-BREEDING. Partial; a continuous moult in winter quarters, with only a limited number of feathers growing at same time. In ♂, at least lore and area round eye affected; in ♀ moult very limited or nil. Of ♂♂ collected in autumn on or near breeding grounds, 12% have some black feathers; of ♂♂ collected during autumn migration, 33%; in winter quarters, 44% show some black August–December, 77% January–May; on spring migration and after arrival on breeding grounds, 96–100% of ♂♂ have at least some black; during August–February, ♂♂ show 0–40 mm² black (on head), not increasing

throughout period; some active moult March–April, after which 0–70 mm² black (on head, throat, mantle, and breast); after arrival on breeding grounds, 20–140 mm² black (occasionally, 0–185 mm²). (Rohwer *et al.* 1983.)

Measurements. ADULT, FIRST ADULT. Eastern USA, spring and summer, and Central America south to Curaçao and Trinidad, winter; skins (RMNH, ZMA). Bill (S) to skull, bill (N) to distal corner of nostril; exposed culmen on average 3·6 shorter than bill (S).

	♂		♀	
WING	64·9 (2·02; 36)	61–69	61·7 (1·53; 29)	59–65
TAIL	54·4 (1·69; 36)	51–58	53·2 (1·58; 27)	50–56
BILL (S)	12·9 (0·52; 34)	11·9–13·7	12·6 (0·45; 28)	12·0–13·4
BILL (N)	6·8 (0·36; 35)	6·2–7·4	6·7 (0·35; 28)	6·0–7·5
TARSUS	17·0 (0·51; 34)	16·1–17·8	16·8 (0·63; 27)	15·6–17·7

Sex differences significant for wing, tail, and bill (S). Ages combined above, though length of wing and tail increases with age: wing, 1st adult ♂ (juvenile wing retained) 63·2 (1·77; 13) 61–66, 2nd adult ♂ 65·5 (1·53; 13) 63–68, older ♂♂ 66·4 (1·20; 10) 64–69; 1st adult ♀ 61·2 (1·39; 17) 59–64, older ♀♀ 62·5 (1·59; 8) 60–65; tail, 1st adult ♂ (juvenile tail retained) 54·3 (1·68; 13) 52–57, 2nd adult ♂ 54·3 (1·73; 13) 51–57, older ♂♂ 54·8 (1·78; 10) 51–58.

Weights. (1) Georgia (USA), early October (Johnston and Haines 1957). Pennsylvania (USA): (2) July–October, (3) May–June (Clench and Leberman 1978, which see for details). (4) East Ship Island (Mississippi, USA), on spring migration (Kuenzi *et al.* 1991, which see for details). (5) Pelee Island (Lake Ontario, Canada), May (Woodward and Lovesy 1958).

		♂		♀	
(1)		8·7 (—; 15)	7·2–9·6	8·1 (—; 16)	6·8–9·4
(2)	AD	8·6 (0·97; 53)	7·2–12·0	8·3 (0·82; 57)	7·2–11·2
	1ST AD	8·1 (0·73; 175)	6·7–12·2	8·0 (0·63; 251)	6·5–10·3
(3)		8·4 (0·60; 89)	7·0–9·9	8·0 (0·70; 106)	6·7–10·2
(4)		7·1 (0·80; 32)	—	6·9 (0·60; 32)	—
(5)		8·2 (—; 6)	6·9–9·8	8·2 (—; 8)	7·3–9·5

Eastern USA: May–July, average 8·5 (30) (Baldwin and Kendeigh 1938; Mengel 1965; Diamond *et al.* 1977); August–September, average of ♂ 8·9 (19), ♀ 8·1 (12) (Baldwin and Kendeigh 1938; Norris and Johnston 1958; Graber and Graber 1962; Mengel 1965). Coastal New Jersey (USA), autumn: 7·9 (0·9; 358) 5·5–11·3 (Murray and Jehl 1964). See also Drury and Keith (1962).

Netherlands' Antilles (off Venezuela) and Trinidad, autumn: ♂ 6·5 (0·99; 10) 5·0–8·0, ♀ 6·6 (0·99; 7) 5·0–7·5 (RMNH, ZMA). Exhausted birds on ships, autumn: in Caribbean, ♂♂ 5·9, 5·9, ♀ 5·5 (0·34; 4) 5·2–6·0; in mid-Atlantic (*c.* 40°N 29°W) ♂ 6·6 (RMNH, ZMA). Jamaica, August–May: ♂ 7·5 (21), maximum 11·5; ♀ 7·5 (28), maximum 10·25 (Diamond *et al.* 1977, which see for monthly variations). St Martin (Caribbean), February: ♂ 5·0, ♀ 6·0 (ZMA). Central America, March–April: ♂♂ 7·1, 7·7 (Russell 1964; RMNH). Ouessant (France), October: ♀ 7·0–7·5 (Vielliard 1962).

For fat-free weights, see Caldwell *et al.* (1963) and Rogers and Odum (1964).

Structure. Wing relatively long, broad at base, tip fairly rounded. 9–10 primaries: p7–p8 longest, p9 0·5–3·5 shorter, p6 0·5–2, p5 3·5–7, p4 6–10, p3 8–12, p2 9–14, p1 11–16. P10 reduced or absent; if present (as in 8 of 10 examined), forms tiny spike hidden below reduced outermost greater primary covert; 40–49 shorter than p7–p8, 5–7 shorter than longest upper primary covert. Outer web of p6–p8 emarginated, inner web of (p7–)p8–p9 with notch. Tip of longest tertial reaches tip of longest secondaries. Tail rather long, tip square or slightly rounded; 12 feathers, t6 3-8 shorter than t1. Bill rather short, straight; broad and flattened at base (even more so than in Canada Warbler *Wilsonia canadensis*); tip of upper mandible with small hook; sides straight (not concave) as seen from above. About 8 long, distinct bristles at each side of bill-base and at nostrils. Leg rather short, toes of average length; slender. Middle toe with claw 12·7 (10) 12·0-14·0; outer toe with claw *c.* 74% of middle with claw, inner *c.* 69%, hind *c.* 76%.

Geographical variation. Slight. Birds from central and eastern USA sometimes separated from typical *ruticilla* of Canada and western USA as *tricolora* (P L Statius Müller, 1776); said to be darker above in ♀ and 1st adult plumage, and 1st adult also duller green above (Wetmore 1949), perhaps a slight difference in size (Oberholser 1938; Mengel 1965), and perhaps some difference in extent of 1st pre-breeding moult in ♂ (more extensive in east) (Rohwer *et al.* 1983); however, species treated as monotypic here, following Lowery and Monroe (1968).

Not related to Painted Redstart *Myioborus pictus* of Central America, which was formerly included in *Setophaga* (see Ficken 1965, Ficken and Ficken 1965). *S. ruticilla* is close to *Dendroica* morphologically and in behaviour (Parkes 1961; Ficken and Ficken 1965).   CSR

Seiurus aurocapillus Ovenbird

PLATE 6
[between pages 64 and 65]

Du. Ovenvogel   Fr. Paruline couronnée   Ge. Pieperwaldsänger
Ru. Золотоголовый дроздовый певун   Sp. Reinita montana   Sw. Rödkronad piplärksångare

Motacilla aurocapilla Linnaeus, 1766

Polytypic. Nominate *aurocapillus* (Linnaeus, 1766), north-east USA, and Canada from Nova Scotia and southern Quebec west to south-east Yukon and north-east British Columbia, south to western and central Alberta; *furvior* Batchelder, 1918, Newfoundland; *cinereus* Miller, 1942, south-east Alberta and south-west Saskatchewan south through Montana to Great Plains of central USA.

Field characters. 14-16 cm; wing-span 22-26 cm. Slightly longer than Northern Waterthrush *S. noveboracensis*; close in size to smaller pipits *Anthus* but dumpier, looking front-heavy. Quite large, terrestrial Nearctic wood warbler, with thrush-like plumage: crown striped orange and black, bold white eye-ring, olive upperparts, and heavily spotted and streaked underparts. Often bobs, waving raised tail. Sexes similar; no seasonal variation. Juvenile separable.

Adult. Moults: July-August (complete). Head, upperparts, wings, and tail light warm olive, with black-bordered orange crown-stripe, white eye-ring, and faintly dusky centres to tertials. Underparts white, marked with pattern recalling both thrush and pipit, with strong black malar stripe, and black spots and streaks on breast and flanks. Bill dusky, with bright flesh base. Legs bright pinkish-flesh to pale brown. First-winter. Distinguished from adult at close range by narrow rufous tips to retained juvenile tertials.

Unmistakable; lack of supercilium immediately excludes *S. noveboracensis*. Flight more reminiscent of chat (Turdidae) than smaller Parulidae, with bursts of wing-beats producing dashing progress through cover. Terrestrial, usually seen walking in dense undergrowth searching leaf litter; gait rather mincing, with tail raised above drooping wings (in manner of Hermit Thrush *Catharus guttatus*) and occasionally waved, further emphasizing frequent bobbing of rear body. Shy, hiding when disturbed.

Calls include penetrating, loud, clicking 'tzick', 'tsek', or 'tsyt', and softer 'tseet'.

Habitat. Breeds in temperate Nearctic lowlands, in woodlands, foraging on forest floor. Favours well-drained bottomland, deciduous forest, not too thick with undergrowth (Pough 1949). Sometimes found in white pine *Pinus strobus* or spruce *Picea* groves, and often on swampy ground (Forbush and May 1955). On Great Plains, also inhabits well-shaded and mature upland forest, often on north-facing slopes or in ravines (Johnsgard 1979). In Canada, favours deciduous closed-canopy woodland with good carpet of leaf litter, sometimes also occurring in open jack pine *P. banksiana* wood and more rarely in white spruce forest; also in mixed woods (Godfrey 1979).

In winter in South America, inhabits deciduous forest, forest edge, and other wooded areas near sea-level, often walking on ground among shrubbery and undergrowth (Schauensee and Phelps 1978).

Distribution. Breeds in North America from north-east British Columbia east to Newfoundland, south to eastern Colorado, south-east Oklahoma, northern Alabama, and South Carolina. Winters from northern Mexico and north coast of Gulf of Mexico to South Carolina, south through Central America and West Indies to northern Colombia and northern Venezuela.

Accidental. Britain: Shetland, 7-8 October 1973;

Devon, freshly dead, 22 October 1985; Merseyside, tideline wing found, 4 January 1969. Ireland: Lough Carra Forest (Mayo), freshly dead, 8 December 1977 (Dymond *et al.* 1989); Dursey Island (Cork) 24-25 September 1990 (Rogers *et al.* 1991).

Movements. Migrant. Breeding and winter ranges approach within 200 km in south-east USA, but most birds migrate over 1000 km, from temperate North America to Central America and West Indies.

In autumn, first arrivals south of breeding range are in early August (Bent 1953), and departure from northern breeding areas from Newfoundland west to Alberta is underway by mid-August (Peters and Burleigh 1951; Sadler and Myres 1976; Sprague and Weir 1984). Most birds have left Canada and northern states by early October (Sprague and Weir 1984; Janssen 1987), and peak passage September in mid-latitude states (Fawks and Petersen 1961; Mengel 1965), but stragglers linger into November as far north as Massachusetts and Ohio (Bent 1953). Peak migration through northern wintering areas is during October (e.g. Oberholser 1974, Rogers *et al.* 1986*a*), with first records early September in Costa Rica and Bahamas (Brudenell-Bruce 1975; Stiles and Skutch 1989) and early October in Colombia (Hilty and Brown 1986). Migration on broad front in both seasons, mainly east of Rockies (though a few appear annually on west coast), and from eastern Mexico to southern Florida (Stevenson 1957; Peterson and Chalif 1973; Fisk 1979; Garrett and Dunn 1981). Long-distance ringing recoveries exemplify broad-front migration, showing movements between Ontario and Alabama, New York and Florida, and Minnesota and southern Mexico (Bull 1974; Ely *et al.* 1977; Speirs 1985). Ringing has shown winter site-fidelity in Mexico, southern Florida, Jamaica, Puerto Rico, and elsewhere, with returns up to 5 years later (Loftin 1977; Fisk 1979; Faaborg and Arendt 1984).

Spring migration begins March, last birds leaving winter range by early May (Bent 1953; Stiles and Skutch 1989). Early arrivals penetrate north to Michigan and New York by mid-April (Bent 1953; Bull 1974). First birds arrive earlier in Mississippi valley than on Atlantic coast, but peak passage by both routes not until early May (e.g. Stewart and Robbins 1958, Mengel 1965). Mid-April records in coastal Nova Scotia correlate with overshoots by species that breed further south, e.g. Prothonotary Warbler *Protonotaria citrea* (Tufts 1986). Migrants reach breeding areas across Canada by early May (Houston and Street 1959; Speirs 1985; Tufts 1986), and northern edge of range before end of May (Todd 1963; Erskine and Davidson 1976).

Vagrant north to Alaska and southern Greenland (Bent 1953; American Ornithologists' Union 1983). For vagrancy to west Palearctic, see Distribrution. AJE

Voice. See Field Characters.

Plumages. (nominate *aurocapillus*). ADULT. Centre of forehead pale olive-brown, almost isabelline at base of culmen, merging into broad tawny-rufous central crown-stripe bordered at each side by prominent black stripe *c.* 2 mm wide extending from base of bill near nostril to side of upper hindneck. Remainder of upperparts brownish-green-olive, slightly brighter and greener on rump. Lore pale buff to off-white, sometimes with some grey mottling in front of eye, merging into brownish-green-olive of remainder of side of head and neck; patch below eye and shorter ear-coverts faintly mottled or streaked pale cream, buff, or (slightly) tawny. Broad, complete, off-white, pale cream, or pale buff eye-ring. Lower cheek white or pale cream, merging into whitish lore but contrasting sharply with olive of upper cheek and ear-coverts; bordered below by rather narrow but distinct black malar stripe which often does not quite reach base of lower mandible. Ground-colour of entire underparts white, but side of breast, lower flank, and thigh olive-brown and sometimes a faint pink-cream or buff wash on chest, upper flank, and vent; chin and throat uniform (sometimes faintly speckled grey at sides); chest, side of breast, and upper flank with contrastingly black oblong drops or short streaks, showing as heavily blotched gorget across chest; narrower black drops on upper belly and side of belly; more sooty or grey and less contrasting streaks in olive-brown of side of breast and lower flank; mid-belly, vent, and under tail-coverts uniform. Tail brownish-green-olive or brown-olive, similar to upperparts or slightly darker; terminal part of outer web of some outer feathers sometimes with white border or spot, especially in ♂; spot present in *c.* 5% of birds of all ages, up to 12 mm long, average 3·2 mm (Short and Robbins 1967). Flight-feathers and longest feather of bastard wing dark grey to greyish-black, outer web of p9 narrowly fringed pale grey, of p1-p8 pale olive (not extending to emarginated parts), of secondaries green-olive. Tertials, remainder of bastard wing, and all upper wing-coverts brownish-green-olive, like upperparts, slightly darker and greyer on centres of median coverts and on inner webs of greater coverts and tertials, slightly paler and green (not rufous) on tips of median and greater coverts. Under wing-coverts and axillaries pale grey with off-white tips, tips partly washed bright pale yellow or, occasionally, tawny-buff; inner borders of secondaries and inner primaries washed light pink-grey. *In fresh plumage* (autumn), tips of feathers of central crown washed olive, partly concealing tawny-rufous, especially on hindcrown; black of lateral crown-stripe mixed with some olive, black somewhat less sharply defined towards rear; white of lower cheek faintly mottled olive; white of throat, belly, vent, and under tail-coverts faintly washed cream. *In worn plumage*, tawny-rufous brighter, more orange-rufous, extending further down nape; black crown-stripe very prominent (less so and partly broken in some heavily worn ♀♀); olive of upperparts and upperwing slightly less green, slightly duller greyish-brown; white of underparts purer (but contaminated with dirt in some birds), black spots more sharply contrasting. Some individual variation in extent of tawny-rufous and black on hindneck, but apparently not age- or sex-related; in series of sexed birds of same date and age, olive wash on crown on average more extensive in ♀ than in ♂, but sexes generally indistinguishable. JUVENILE. Upperparts, lesser upper wing-coverts, and side of head cinnamon-brown or buff-brown; faint dark brown lateral crown-stripe; mantle and scapulars with faint darker olive-brown shaft-streaks; eye-ring pale buff, not much contrasting; lower cheek, chin, and throat pale cinnamon-buff with faint dark malar stripe; chest, side of breast and flank greyish-buff with faint darker olive-brown spots and streaks; remainder of underparts pale buff-yellow or cream-white, unmarked. For wing and tail, see 1st adult (below). For growth, see Hann (1937) and Bent (1953).

First Adult. Like adult, but juvenile tail, flight-feathers, tertials, greater upper primary and secondary coverts, and bastard wing retained; juvenile tail-feathers usually rather narrow, tips sharply pointed (in adult, broader, tip bluntly pointed or slightly rounded; intermediates occur); no white on tips of outer tail-feathers; tertials and sometimes greater coverts and central tail-feathers with rufous fringe along tip (absent in adult), usually most obvious on tips of tertials (liable to wear off, occasionally absent); tertials gradually taper towards rounded tips (in adult, broad and with squarish tips, especially 2 longest). Head and body similar to adult, but crown on average with more extensive olive or greyish-olive wash when plumage fresh (especially in ♀), tawny less conspicuous, in ♀ sometimes tawny-buff rather than tawny-rufous.

Bare parts. Adult, First Adult. Iris dark brown. Upper mandible and tip of lower mandible beige-brown, dark horn-brown, or slate-brown; cutting edges and remainder of lower mandible pale flesh-brown, pink-horn, or light beige-brown, tinged yellow towards base in autumn. Gape pink. Leg and foot pale flesh-pink to light pink-horn, sometimes tinged grey on upper surface of toes. (Robertson 1975b; RMNH, ZFMK, ZMA.) Juvenile. No information.

Moults. Adult Post-breeding. Complete; primaries descendent. On breeding grounds, July–August (Dwight 1900; Pyle et al. 1987). In birds examined, no moult up to 10 July, but sample small (RMNH, ZFMK, ZMA). Adult Pre-breeding. In winter quarters, January–March; extent limited (Pyle et al. 1987) or no moult at all. Post-juvenile. Partial: head, body, lesser and median upper wing-coverts, rarely a few tertials or some tail-feathers. On breeding grounds, starting shortly after fledging; June–August. (Dwight 1900; Pyle et al. 1987; RMNH, ZFMK, ZMA.) Moult almost completed in single bird from 6 July, Kentucky (USA) (Mengel 1965). First Pre-breeding. As adult pre-breeding.

Measurements. Nominate aurocapillus. Ontario (Canada) and north-east USA, May–September, and Central and northern South America, October–May; skins (RMNH, ZFMK, ZMA). Bill (S) to skull, bill (N) to distal corner of nostril; exposed culmen on average 3·3 less than bill (S).

	♂		♀	
WING	78·0 (1·62; 20)	75–81	74·5 (1·95; 18)	71–77
TAIL	51·6 (1·35; 20)	49–54	50·1 (2·54; 17)	46–54
BILL (S)	15·1 (0·42; 19)	14·4–15·8	14·9 (0·45; 16)	14·1–15·5
BILL (N)	8·8 (0·39; 19)	8·2–9·5	8·6 (0·36; 17)	8·2–9·2
TARSUS	22·5 (0·38; 20)	21·8–23·1	21·9 (0·66; 17)	20·8–22·9

Sex differences significant, except for bill. Ages combined above, though retained juvenile wing and tail of 1st adult on average shorter than in older birds. Thus: wing, adult ♂ 79·1 (1·93; 4) 76–81, 1st adult ♂ 77·7 (1·45; 16) 75–81, adult ♀ 74·7 (2·10; 6) 72–77, 1st adult ♀ 74·4 (1·95; 12) 71–77. See also Taylor (1972).

Weights. Sexed birds. (1) Pennsylvania (USA), May–June (Clench and Leberman 1978). (2) Eastern USA, May–September (Esten 1931; Baldwin and Kendeigh 1938; Norris and Johnson 1958; Mengel 1965; ZFMK). Kansas (USA), late September and October: (3) adult, (4) 1st adult (Tordoff and Mengel 1956). (5) Florida, September–October, adult (Taylor 1972, which see

for 1st adults). (6) Caribbean (mainly just off Belize), October (Roselaar 1976; ZMA).

	♂		♀	
(1)	♂	18·8 (0·78; 12) 17·4–20·3	♀	19·6 (1·53; 5) 18·1–21·7
(2)		19·5 (2·06; 8) 18·0–24·2		20·3 (4·11; 8) 16·3–29·6
(3)		23·2 (— ; 2) 22·5–23·8		21·4 (— ; 8) 18·3–25·7
(4)		21·9 (2·64; 14)		18·2 (— ; 6) 15·6–20·0
(5)		22·5 (2·01; 91) 17·8–27·5		21·7 (1·53; 91) 18·4–25·8
(6)		18·4 (0·70; 4) 17·4–19·1		16·9 (1·85; 5) 13·7–18·1

Illinois (USA), average September: adult ♂ 20·4 (15), 1st adult ♂ 20·6 (5); adult ♀ 19·7 (24), 1st adult ♀ 22·5 (11) (Graber and Graber 1962). Georgia (USA), early October: ♂ 22·4 (19) 20·1–24·8, ♀ 22·4 (16) 20·0–24·8 (Johnston and Haines 1957). Central America, Curaçao, and Bonaire, November–April: ♂♂ 19, 23·4; ♀ 18·7 (3) 18·0–19·1 (Russell 1964; RMNH, ZMA).

Unsexed birds. Pennsylvania: (1) adult, (2) 1st adult, (3) age unknown (Clench and Leberman 1978).

	JUL–AUG		SEP–OCT	
(1)	JUL–AUG	19·1 (14) 16·5–20·5	SEP–OCT	19·7 (34) 15·7–28·8
(2)		19·0 (59) 17·0–26·3		19·5 (209) 16·4–28·3
(3)		18·6 (13) 16·9–21·0		20·0 (66) 14·0–26·5

Belize, August–November: 16·4 (1·2; 29) (Mills and Rogers 1990). Panama, mainly October: 15·7 (1·30; 87) 12·9–19·0 (Rogers and Odum 1966). Jamaica, September–April: 19·2 (3·2; 63); lean September–October, increasing November, stable to March, fattening April (when maximum 25·5) (Diamond et al. 1977, which see for details). Louisiana (USA), spring migrants: 16·6 (2·5; 18) (Rogers and Odum 1966, which see for fat-free weights). Britain, October: 20·4, 22·0 (Robertson 1975b; Ward 1987).

Structure. Wing short, broad at base, tip bluntly pointed. 9–10 primaries: p7–p8 longest, p9 1–2 (0–3) shorter, p6 0·5–3·5, p5 5–9, p4 8–12, p3 10–14, p2 11–16, p1 13–18. P10 strongly reduced, a tiny whitish spike-like feather hidden below reduced outermost greater upper primary covert, apparently absent in 3 of 10 birds examined; 49–56 shorter than p7–p8, 7–11 shorter than longest upper primary covert, 1–2 shorter than outermost upper primary covert. Outer web of (p6–)p7–p8 emarginated; inner web of (p7–)p8–p9 with notch. Tip of longest tertial reaches to tip of p2–p3. Tail rather short, tip square; 12 feathers. Bill rather short, straight, fairly heavy at base; rather similar to Yellow Warbler Dendroica petechia, but proportionately longer and heavier, lower mandible deeper, tip of upper mandible more curved. About 4 short and rather indistinct bristles at base of upper mandible, projecting obliquely down. Leg and toes rather short compared with size of bird, fairly stout. Middle toe with claw 16·9 (12) 16·0–18·0; outer toe with claw c. 74% of middle with claw, inner c. 72%, hind c. 75%. Remainder of structure as in D. petechia.

Geographical variation. Slight, involving colour only. Difference usually visible only when series of skins examined. S. a. furvior from Newfoundland on average darker green-olive on upperparts than nominate aurocapillus from eastern USA and neighbouring Canada; dark lateral crown-stripe heavier on average, pale central crown-stripe more brownish-orange or amber-brown; spots on underparts heavier, brown of flank deeper. S. a. cinereus from Great Plains has upperparts on average paler and more greyish-olive than nominate aurocapillus (Miller 1942).

CSR

Seiurus noveboracensis **Northern Waterthrush**

PLATE 6
[between pages 64 and 65]

Du. Noordse Waterlijster   Fr. Paruline des ruisseaux   Ge. Drosselwaldsänger
Ru. Речной дроздовый певун   Sp. Reinita charquera   Sw. Nordlig piplärksångare

Motacilla noveboracensis Gmelin, 1789

Monotypic

Field characters. 12·5–15·5 cm; wing-span 21–25 cm. Size close to Meadow Pipit *Anthus pratensis* but with plumper, proportionately shorter-tailed form suggesting *Catharus* thrush. Medium-sized, plump but sleek Nearctic wood warbler adapted to ground-feeding on moist ground, where its horizontal posture, walking gait, and teetering of body and tail recall *Actitis* sandpiper. Plumage markedly pipit-like but upperparts unstreaked; long narrow yellowish supercilium distinctive. Sexes similar; no seasonal variation. Immature inseparable.

ADULT. Moults: June–August (complete). Head, upperparts, wings, and tail dark oily olive- to grey-brown, relieved by faintly yellowish or buff (wearing to off-white) supercilium (long and usually uniformly narrow, reaching nape), similarly coloured mottling on ear-coverts, and rather darker centres to tertials and primary-tips. Underparts basically very pale yellow (wearing to white), tinged distinctly lemon along upper flanks and becoming yellow-buff to pale olive-brown on sides of breast and rear flanks; copiously spotted and lined black. Pattern of marks on underparts important to diagnosis: typical bird shows spots on throat and clustered gorget of spots and streaks (but note that closely similar Louisiana Waterthrush *S. motacilla* rarely shows similar pattern of marks; see below). Bill quite strong but in proportion to head; blackish-horn with flesh base to lower mandible. Legs rather long; bright pale flesh to brown.

Genus unmistakable and separation from Ovenbird *S. aurocapillus* simple, as that species lacks supercilium and has much lighter, warmer olive upperparts and orange crown-stripe. Distinction from only slightly larger *S. motacilla* more difficult, and any vagrant waterthrush should be observed as closely as possible. Crucial to separation of the 2 species are bill size in relation to head (large in *S. motacilla*; in proportion in *S. noveboracensis*), shape of supercilium (bulging behind eye in *S. motacilla*; uniformly narrow or slightly wider before eye in *S. noveboracensis*), colour of rear supercilium (white in *S. motacilla*; uniformly yellowish or buff when fresh in *S. noveboracensis*), throat markings (diffuse in *S. motacilla*; sharp, often forming gorget in *S. noveboracensis*), and ground-colour of central flanks (strongly buff in *S. motacilla*; lemon-yellow in *S. noveboracensis*). Important also to note that *S. noveboracensis* shows quite marked plumage variation, with much-disputed pale bird in England in 1968 leading to review of identification (Wallace 1976; see also Geo-

graphical Variation); diagnosis best based on size of bill and shape and colour of supercilium. Flight dashing, suggesting darting chat (Turdidae) rather than warbler; makes low flitting or skimming landing on ground or branch. Gait a mincing, high-stepping walk; level stance strongly suggests both pipit and sandpiper, resemblance to latter particularly heightened by frequent teetering of body, wagging of tail, and bending of legs. Highly terrestrial and fond of moist ground, exploiting low mossy branches and even wading in puddles or along water margins. Shy, retreating to densest cover when disturbed. Vagrants have foraged on tide wrack.

Call distinctive: far-carrying, lengthy monosyllable with metallic, even explosive, quality, 'chwit', 'zwik', 'tsink', 'pzint', or 'peent'.

Habitat. Breeds in cooler temperate regions of Nearctic, up to northern treeline. In Alaska, confined to narrow strip of vegetation bordering streams (including mountain streams), where it moves mouse-like under debris, tangled logs, and undergrowth, often under canopy of alders *Alnus* and willows *Salix*. Sings from branches or in flight. (Gabrielson and Lincoln 1959.) Found in Canada also by ponds and lakes and in swamps, bogs, and wet parts of woodland with shallow pools of water (Godfrey 1979). In Great Plains, favours sites by standing rather than swiftly running water, especially woody borders of bogs, swamps, and second-growth swamp forests (Johnsgard 1979). Never seen far from water except on migration, when it sometimes visits gardens, trees, and shrubbery around buildings and groves of trees. Wades in shallow waters like *Actitis* sandpiper, but is also at home among treetops (Forbush and May 1955).

In winter in South America, stays always near water, by streams and lagoons or in mangroves and swamps, ascending to 2000 m (Schauensee and Phelps 1978).

Distribution. Breeds in North America from north-central Alaska east to central Labrador and Newfoundland, south to southern Alaska, central British Columbia, south-east Manitoba, northern Wisconsin, south-eastern Michigan, eastern West Virginia, and north-west New Jersey. Winters in West Indies, Central America, and South America south to northern Ecuador, north-east Peru, Venezuela, and the Guianas.

Accidental. Britain, Ireland, Channel Islands, France.

Movements. Long-distance migrant, all birds moving over 1000 km (many much further) from breeding to wintering areas.

Migration both seasons is compressed. Southward movement starts late July (Bent 1953; Janssen 1987), and in mid-August a few birds appear as far south as California, Arkansas, and Costa Rica (Garrett and Dunn 1981; James and Neal 1986; Stiles and Skutch 1989), but most are still in breeding areas then. Peak movement begins early August in Yukon Territory, mid- to late August in southern Canada (Rand 1946; Todd 1963; Sadler and Myres 1976), and main passage September across continent and south into winter range (Mengel 1965; Sutton 1967; Rogers *et al.* 1986a; Stiles and Skutch 1989). Few remain in northern states after mid-October (e.g. Bull 1974, Bohlen 1989). At Caracas (northern Venezuela), over 5 years, passage from mid-September, peaking 1st half of October; wintering birds present from mid-October (Schwartz 1964). Migration on broad front, mainly east of Rockies (American Ornithologists' Union 1983), and including passage overland through Mexico (Peterson and Chalif 1973) and through southern Florida (Bent 1953; Fisk 1979). In Bermuda, where winters regularly, common on passage in autumn, but inconspicuous in spring (Wingate 1973). Winter ringing in (e.g.) Guatemala, Panama, Venezuela, and Jamaica proves site-fidelity, with recoveries up to 4 years after ringing (Schwartz 1964; Diamond and Smith 1973; Loftin 1977; Rogers *et al.* 1982a). Very widespread in winter range, and by far the most commonly reported of Parulidae in West Indies (Pashley and Martin 1988).

In spring, recorded in south-central USA from March, but few birds appear there before mid-April (Oberholser 1974; Imhof 1976; James and Neal 1986), and most leave winter range April or early May (Schwartz 1964; Brudenell-Bruce 1975; Binford 1989). Peak passage late April in southern states (Burleigh 1958; Oberholser 1974; Imhof 1976), and in May everywhere northward. Migration in Mississippi valley averages a week earlier than at similar latitudes on east coast (Palmer 1949; Stewart and Robbins 1958; Fawks and Petersen 1961; Mengel 1965). In boreal Canada and Alaska, peak of migration is after mid-May (Gabrielson and Lincoln 1959; Houston and Street 1959; Todd 1963), and northernmost areas not occupied until early June (Nero 1967; Scotter *et al.* 1985). As in autumn, spring migration is on broad front, birds moving both around and across Gulf of Mexico (Stevenson 1957).

In late spring, occurs 200–300 km north of breeding range on Arctic coast of Alaska (Kessel and Gibson 1978), and others have reached Banks Island in Canadian Arctic (September), Greenland, and Chukotskiy peninsula (eastern Siberia) (American Ornithologists' Union 1983; Godfrey 1986; Stepanyan 1990).

Rare autumn vagrant to Atlantic seaboard of west Palearctic. 6 records from Britain and Ireland up to 1990, especially south-west England and southern Ireland, from late August to 3rd week of October (Dymond *et al.* 1989; Rogers *et al.* 1990, 1991). Only spring record is from Jersey (Channel Islands), 17 April 1977 (Long 1981).

AJE, PRC, DFV

Voice. See Field Characters.

Plumages. ADULT. Entire upperparts fuscous-brown or dark olive-brown, slightly more greenish-olive when plumage fresh (in autumn), greyer (especially on feather-tips) when worn (in spring); colour of upperparts rather similar to Bluethroat *Luscinia svecica*. Feathers of forehead and forecrown with black centres; feathers of mid-forehead and forecrown sometimes with paler olive-brown sides, forming indistinct pale patch or stripe along centre of head (exceptionally, near that of Ovenbird *S. aurocapillus* in width). Long, distinct, pale cream-buff (if fresh) to off-white (if worn) supercilium, from nostril to just behind ear-coverts; fairly even in width over entire length or slightly wider in front of eye than behind, often finely speckled dusky above eye and at rear. Narrow cream to white eye-ring, broadly broken in front of and behind eye by sooty-brown eye-stripe, extending from lore over upper ear-coverts and merging into fuscous-brown of upper side of neck. Small patch below eye pale cream-yellow or off-white, finely mottled dusky, merging into fuscous or dark olive-brown lower ear-coverts and lower side of neck, which show faint pale cream or off-white streaks. Ground-colour of underparts down from lower cheek and backwards to under tail-coverts rather variable; usually, uniform diluted pale yellow when fresh, paler yellow-white on chin and under tail-coverts, bleaching to white with yellow wash when plumage worn in spring and to off-white when heavily abraded in summer, but underparts occasionally pale buff-yellow (especially in fresh autumn birds from Newfoundland) or silky-white with faint traces of pale yellow or no yellow at all, even in early autumn (especially in population breeding at northern edge of Great Plains and Great Lakes area); upper side of breast, lower flank, and thigh washed olive-brown. Lower cheek and throat with small but distinct dark grey triangular specks; short and sometimes broken sooty-black malar stripe. Chest, side of breast, and flank with bold, long, triangular, fuscous or sooty-black streaks which become shorter and narrower on upper belly and side of belly (somewhat less bold, less black, and less rounded than in *S. aurocapillus*); mid-belly, vent, and under tail-coverts uniform, but longest tail-coverts with (usually concealed) large dark sooty-grey central mark. Tail, tertials, and all upper wing-coverts fuscous-brown, flight-feathers more blackish-brown, all closely similar to tone of upperparts; tips of 1–4 outer tail-feathers usually with white fringe or shallow white triangular spot on inner web (in 57% of birds of all ages examined May–July, thus perhaps in all older birds, as remaining 43% are probably 1-year-olds: Eaton 1957a); tertials and fringes along tips of lesser and median coverts and along outer webs of greater coverts and flight-feathers slightly more olive-brown. Under wing-coverts and axillaries grey, similar to under surface of flight-feathers; inner and shorter coverts washed olive-brown and with partly cream outer fringe, shorter coverts along leading edge of wrist mottled black and off-white. JUVENILE. Upperparts, including upper wing-coverts, dark olive-brown, feathers with grey-brown of bases partly visible and with narrow pale cinnamon or buff fringes along tips; side of head pale buff, underparts pale buff-yellow to buff-white, washed grey-brown on side of breast and flank; supercilium rather indistinct; marks on underparts as adult, but browner and less sharply defined. For tail and remainder of wing, see First Adult (below). For development and growth, see Eaton (1957b). FIRST ADULT. Like adult, but juvenile tail, flight-feathers, tertials, greater upper primary coverts, and apparently occasionally some outer greater coverts and bastard wing retained; juvenile characters of these sometimes difficult to establish, however. Juvenile tertials and sometimes outer greater coverts, primary coverts, middle feather of bastard wing, or central tail-feathers with narrow rufous fringe along tip, most obvious on tertials, but prone to wear and bleaching, generally worn off in spring, and occasionally absent even in early autumn; tips of tail-feathers generally more pointed than in adult, less broadly rounded, but some birds of any age are intermediate; tips of outer tail-feathers without white. In spring, up to about May, tips of tail-feathers and outer primaries still smooth in adult, frayed in 1st adult.

Bare parts. ADULT, FIRST ADULT. Iris sepia, brown, or black-brown. Bill dusky or greyish-black in breeding season, base of lower mandible tinged flesh or purplish; in rest of year, horn-brown to brown-black with lighter horn base to lower mandible, latter and cutting edges tinged pale greyish-flesh to pink-brown. Gape flesh-pink. Leg and foot pale flesh, light purplish-horn, greyish-flesh, dirty pink, or purplish-brown. (Harris *et al.* 1960; RMNH, ZMA.) JUVENILE. Bill flesh-grey with plumbeous culmen and tip; trace of yellow gape-flanges and some yellow at extreme base of lower mandible shortly after fledging. Leg and foot flesh. (ZMA.)

Moults. ADULT POST-BREEDING. Complete; primaries descendent. On breeding grounds, June–August. No pre-breeding moult. POST-JUVENILE. Partial: head, body, lesser, median, and many or all greater upper wing-coverts, rarely a few tail-feathers. Starts at fledging, before flight-feathers or tail full-grown; June–August, completed before start of autumn migration. (Dwight 1900; Eaton 1957a, b; Pyle *et al.* 1987; RMNH, ZMA.) In a few birds, top of head or throat appears to be newer than remainder of body in spring, perhaps as result of a partial 1st pre-breeding moult (RMNH, ZMA).

Measurements. ADULT, FIRST ADULT. Eastern Canada and north-east USA, May–September, and West Indies, Central America, and northern South America, September–May; skins (RMNH, ZMA). Bill (S) to skull, bill (N) to distal corner of nostril; exposed culmen on average 3·6 less than bill (S).

	♂		♀	
WING	77·0 (2·53; 28)	72–82	74·2 (1·60; 25)	72–78
TAIL	49·9 (2·61; 25)	46–56	48·4 (1·44; 22)	45–52
BILL (S)	15·8 (0·63; 24)	14·6–16·6	15·6 (0·63; 22)	14·3–16·5
BILL (N)	9·4 (0·42; 24)	8·6–10·2	9·5 (0·42; 22)	8·8–10·3
TARSUS	21·5 (0·74; 25)	20·4–22·8	21·4 (0·75; 21)	20·2–22·6

Sex differences significant for wing and tail. Ages combined above, though retained juvenile wing and tail of 1st adult on average shorter than in older birds. Thus, in ♂: wing, adult 79·0 (2·29; 7) 75–82, 1st adult 76·3 (2·29; 21) 72–80; tail, adult 51·2 (3·00; 7) 48–56, 1st adult 49·3 (2·32; 18) 46–54.

Weights. Sexed birds. (1) North-west Yukon (Canada), late May and June (Irving 1960). (2) Eastern USA, August–October (Tordoff and Mengel 1956; Norris and Johnston 1958; Mengel 1965). Curaçao, Trinidad, and Tobago: (3) September–November, (4) December–March (RMNH, ZMA). (5) Exhausted and just-dead birds on islands or ships in Caribbean, October (RMNH, ZMA).

	♂		♀	
(1)	16·9 (0·78; 16)	15·6–18·1	19·6 (— ; 3)	17·9–22·5
(2)	20·1 (2·86; 13)	16·9–25·6	19·7 (1·64; 5)	17·6–22·2
(3)	15·0 (1·58; 6)	12·5–17·0	14·0 (— ; 3)	12·0–16·0
(4)	16·0 (1·20; 7)	14·5–17·5	14·7 (— ; 3)	13·5–15·5
(5)	10·6 (— ; 3)	9·7–12·0	10·7 (0·58; 6)	10·0–11·5

Belize, April–May: ♀ 17·3 (3) 16·6–18·2 (Russell 1964). Freshly dead ♂ on ship in English Channel, October: 9·9 (ZMA).

Unsexed birds. Pennsylvania (USA): (6) ages combined, April–May; August–September, (7) adult, (8) 1st adult and age unknown; October, (9) adult, (10) 1st adult and age unknown (Clench and Leberman 1978). (11) Belize, late August to early November (Mills and Rogers 1990). (12) Panama, October (Rogers and Odum 1966, which see for details). (13) Louisiana (USA), spring (Rogers and Odum 1966).

(6)	17·5 (1·68; 185) 13·8–24·4	(10)	18·8 (2·51; 27) 15·3–23·1
(7)	18·5 (1·47; 58) 15·4–23·2	(11)	14·8 (1·6 ; 84) —
(8)	18·1 (1·75; 184) 13·9–23·9	(12)	14·7 (1·54; 165) 10·5–19·2
(9)	19·0 (— ; 4) 17·8–20·2	(13)	15·3 (1·78; 8) —

Jamaica, September–May: 15·8 (0·6; 10), maximum 20·8 (Diamond *et al.* 1977). See also Drury and Keith (1962). For fat-free weight, see Rogers and Odum (1966).

Structure. Wing rather short, broad at base, tip bluntly pointed. 10 primaries: p7–p8 longest, p9 0–2 shorter, p6 2–4, p5 6·5–10, p4 9–13, p3 11–15, p2 14–19, p1 16–21. P10 strongly reduced, concealed; 47–57 shorter than p7–p8, 7–12 shorter than longest upper primary covert, 1–2 shorter than reduced outermost primary covert. Outer web of (p6–)p7–p8 emarginated, inner of (p7–)p8–p9 with notch (sometimes faint). Tail rather short, tip square; 12 feathers. Bill rather long, relatively more so but with less heavy base than in *S. aurocapillus*, tip finer and more compressed laterally; tip relatively longer and base narrower than in Yellow Warbler *D. petechia* (p. 22). Bristles at base of bill short, faint. Tarsus rather short for size of bird, fairly stout; toes rather long, slender. Middle toe with claw 17·0 (15) 16·0–17·8; outer toe with claw *c.* 69% of middle with claw, inner *c.* 65%, hind *c.* 77%. Remainder of structure as in *D. petechia*.

Geographical variation. Rather slight; involves ground-colour of upperparts and underparts, and size (as expressed in length of wing, bill, and tarsus). Formerly, 3 races usually recognized: nominate *noveboracensis* from eastern Quebec, New Brunswick, and Nova Scotia (Canada) through New England states to Allegheny mountains (USA)—rather small with olive-tinged upperparts and pale yellow underparts; *notabilis* (Ridgway, 1880) from western Quebec and Great Lakes area west through Canada to Alaska, south in west to northern British Columbia and central Alberta—rather small but bill and tarsus rather long, upperparts dark and greyish-earth-brown, underparts white with little or no yellow; and *limnaeus* McCabe and Miller, 1933, in northern Rockies south from central British Columbia and south-west Alberta—still darker on upperparts, but with intermediate underparts (McCabe and Miller 1933). Later, 4th race, *uliginosus* Burleigh and Peters, 1948, was described from Newfoundland as being more olive above and deeper yellow below than others (Burleigh and Peters 1948). However, colour of upperparts mainly dependent on abrasion, and colour of underparts as well as size, though indeed showing some geographical variation, varies strongly between individuals. If underparts scored 1 (white) to 5 (yellow), underparts on average palest (score 1–2) in area from Lake Huron through southern Ontario, southern Manitoba, and Minnesota to central Alberta and Montana (Eaton 1957a; RMNH, ZMA), darkest (score 3–3·5) from New England to Newfoundland, intermediate (average score 2·3–2·6) elsewhere, including Rockies; intermediate individuals (score 2–3) form over 60% of all populations, except in Great Plains. 9 out of 12 birds from Rockies are heavily streaked below, but similar birds occur occasionally elsewhere, especially from New

England to Nova Scotia. Average difference in wing length between various populations is in order of up to 2 mm and overlap large. Bill length more variable, on average longer (in ♂, 10·3–10·6 to nostril) in southern populations (Rockies through Great Lakes area to Allegheny mountains), shorter (9·6–9·9) in north (Alaska to northern Manitoba, Quebec, and Newfoundland). (Eaton 1957a.)

Closely related to Louisiana Waterthrush *S. motacilla*, but not considered a superspecies as ranges overlap; see Mayr and Short (1970).

Recognition. Easily confused with rather similar *S. motacilla*

from eastern USA, potential vagrant to west Palearctic. *S. motacilla* differs in larger size, wing 83·2 (80–87), bill to skull 17·6 (16·6–18·2), to nostril 10·6 (9·8–11·5), tarsus 23·0 (22·0–23·7) (5 birds of each sex); broader and purer white supercilium (especially above ear-coverts); usually unspeckled chin and throat; dark spots on underparts brown-grey (less black), forming short triangles, rather than streaks; ground-colour of underparts whitish with contrasting pink-buff flank and uniform buff under tail-coverts (no such contrast in *S. noveboracensis*, and its longer tail-coverts have large dark central mark); legs on average paler pink or yellowish-flesh. See also Binford (1971) and Wallace (1976a).   CSR

Geothlypis trichas Common Yellowthroat

PLATE 7
[between pages 64 and 65]

Du. Gewone Maskerzanger   Fr. Paruline masquée   Ge. Gelbkehlchen
Ru. Желтогорлый масковый певун   Sp. Reinita Gorgigualda   Sw. Gulhake

Turdus Trichas Linnaeus, 1766

Polytypic. Nominate *trichas* (Linnaeus, 1766), central New Jersey and south-east Pennsylvania south and west through Delaware, Maryland, Virginia (except extreme south-east), upland areas of North and South Carolina and Georgia, to Arkansas and neighbouring parts of Oklahoma, Texas, and Louisiana; *brachidactylus* (Swainson, 1838), eastern USA and Canada, north of nominate *trichas*, from northern New Jersey, New York, and Connecticut to Newfoundland and southern Quebec, west to central Ontario, Minnesota, and eastern Nebraska and Kansas; *ignota* Chapman, 1890, coasts and swamps of south-east USA from coastal South Carolina to south-east Louisiana, south to Florida; *typhicola* Burleigh, 1834, south of nominate *trichas*, north of *ignota*, from extreme south-east Virginia to southern Alabama; *campicola* Behle and Aldrich, 1947, south-central Canada and northern Great Plains area of USA, west of *brachidactylus*, from western Ontario and North Dakota west to eastern British Columbia, Idaho, and Wyoming; 9–11 further races from Yukon and Alaska through western Canada and USA to Mexico.

Field characters. 11–14 cm; wing-span 16–18 cm. Close in size to Chiffchaff *Phylloscopus collybita* but with plumper form enhanced by frequent raising of tail. Quite small Nearctic wood warbler; perky but skulks in ground cover. Uniform bright yellowish-olive upperparts and yellow and buff underparts. ♂ has diagnostic black mask; ♀ has short dull supercilium and eye-ring. Behaviour recalls both Wren *Troglodytes troglodytes* and small *Sylvia* warbler. Sexes dissimilar; no seasonal variation. Immature ♂ separable.

ADULT MALE. Moults: July–August (complete). Upperparts, wings, and tail bright yellowish- to greenish-brown, looking olive in dull light or when worn. Deep frontal band and area round eye and ear-coverts black, forming striking mask and separated from rear crown by greyish-white band. Wide buff band from sides of breast to rear flanks, only slightly paler than upperparts and contrasting with vivid yellow throat and upper breast, whitish belly, and dirty yellow under tail-coverts. Bill dusky to almost black, with flesh base. Legs flesh to bright pinkish-brown. ADULT FEMALE. No mask, showing only slightly darker ear-coverts which emphasize narrow dull yellowish eye-ring and diffuse supercilium. FIRST-WINTER MALE. Moult timing as adult. Plumage worn by most vagrants. From

autumn, increasingly shows indication of mask, with greyish-black mottling from forehead to ear-coverts clearly visible at close range. FIRST-WINTER FEMALE. As adult but duller, especially on throat which can be tinged brownish or even wholly pale buff.

♂ unmistakable but ♀ can be confused with ♀ or immature of other yellow parulids, particularly Yellow Warbler *Dendroica petechia* (more uniformly coloured, with indistinct wing-bars) and Wilson's Warbler *Wilsonia pusilla* (uniformly lemon-yellow below, with similarly coloured frontal band and rear supercilium). Flight light and flitting, recalling small *Sylvia*. Hops, also shuffles and clambers in dense cover. Stance level, often with tail raised well above body line in manner of *T. troglodytes*. Secretive, feeding at ground level in dense cover.

Voice distinctive, with harsh quality recalling *T. troglodytes*, particularly when disturbed. Calls described as husky 'tchep' and 'chip', soft 'trep' and 'tep' as if clicking tongue, and quiet 'tic'.

Habitat. Breeds in temperate and subtropical regions of Nearctic in dense low cover in variety of sites, especially near water and in rank vegetation of marshes, such as cattails *Typha* and bulrushes *Scirpus*, and streamside thickets of willows *Salix*. Tangles of blackberries *Rubus* and weeds on old fields are also common breeding habitats. (Pough 1949.) Spends most time on or very close to ground or in lower branches of brambles and other bushes, often in wet meadows, near ponds and streams, or on edges of damp woods. Often in woods and orchards, but is strongly attached to dense low cover, rarely mounting to treetops. (Forbush and May 1955.) In Great Plains, lives near water or moist ground and among associated lush vegetation, including tall grasses and often shrubs and low trees. Despite preference for vicinity of water, occasionally occupies upland thickets of shrubs and small trees, poorly tended orchards, retired croplands, and weedy residential areas (Johnsgard 1979).

Distribution. Breeds over whole of North America north to south-east Alaska, southern Yukon, northern Alberta, central Ontario, central Quebec, and Newfoundland, and in Mexico south to Oaxaca and Veracruz. Winters in southern USA, West Indies, and south through Central America to Colombia and Venezuela.

Accidental. Britain: 1st-winter ♂, Lundy (Devon), 4 November 1954; ♂, Fair Isle, 7–11 June 1984; 1st-winter ♂, Isles of Scilly, 2–17 October 1984 (Dymond *et al.* 1989); 1st-winter ♂, Kent, 6 January–23 April 1989 (Rogers *et al.* 1990).

Movements. Varies from fully migratory (boreal-temperate *brachidactylus*, *campicola*, *arizela*) to partially migratory (nominate *trichas*, *typhicola*, *occidentalis*) to resident (southern forms). Winter ranges overlap widely, with up to 3 migrant and 1 resident form in some areas. In east, northern forms winter furthest south, overflying both short-range migrants and residents. (American Ornithologists' Union 1957, 1983.)

Southward migration evidently begins August, but detection difficult in this widely breeding species. Peak passage underway by late August from British Columbia to New Brunswick (Munro and Cowan 1947; Sadler and Myres 1976; Squires 1976); migration continues through September, with numbers decreasing after mid-month north of 40°N (Fawks and Petersen 1961; Mengel 1965; Burleigh 1972). Only stragglers, some of which overwinter, remain north of southernmost states after late October (Phillips *et al.* 1964; Oberholser 1974; Taylor 1976). Birds begin to appear in West Indies in early September (Brudenell-Bruce 1975; Raffaele 1989), but in Central America not until October (Land 1970; Binford 1989; Stiles and Skutch 1989). Migration on broad front, including major movements through Florida (Taylor 1976) and across Mexican Gulf (Rogers *et al.* 1986a) as well as through Mexico (Bent 1953). Winter site-fidelity shown by ringing in Haiti, Jamaica, Belize, and El Salvador, with recoveries up to 3 years later (Diamond and Smith 1973; Woods 1975; Loftin 1977).

Spring migration begins March, with arrivals from Ari-zona to South Carolina (Phillips *et al.* 1964; Lowery 1974; Potter *et al.* 1980); most birds leave Central America and West Indies April to early May (Brudenell-Bruce 1975; Stiles and Skutch 1989). In March, although a few birds occur north to Washington and Utah in west, little advance in east, where cold persists inland (Bent 1953). A few reach Maryland, Illinois, and Colorado by mid-April (Stewart and Robbins 1958; Bailey and Niedrach 1967; Bohlen 1989), but main movement from Maryland and Oklahoma north to southern Canada is early to mid-May (Munro and Cowan 1947; Stewart and Robbins 1958; Sutton 1967; Sprague and Weir 1984). Only those in extreme north-west do not arrive until early June, presumably by longer route (Rand 1946; Erskine and Davidson 1976; Scotter *et al.* 1985). Migration probably involves broad-front movement, including across Gulf of Mexico (Stevenson 1957), and concentration along Atlantic coast as in autumn (Taylor 1976).

Vagrant north of range in Alaska and Canada (Kessel and Gibson 1978; Godfrey 1986), also in Greenland (Salomonsen 1950–1). For vagrancy to west Palearctic, see Distribution. AJE

Voice. See Field Characters.

Plumages. (*G. t. brachidactylus*). ADULT MALE. Broad black mask, extending from forehead and base of bill over eye and cheek to ear-coverts and rear of lower cheek; bordered above and behind by light to medium ash-grey band *c.* 3–5 mm wide extending over forecrown and above ear-coverts down to side of neck. Remainder of upperparts including rear side of neck olive-green, tinged brown on crown, brighter olive-green on rump, and bright yellowish-green on upper tail-coverts. Chin to breast bright spectrum-yellow, feather-tips often slightly tipped orange-yellow, especially on throat and breast, merging into olive-green on side of breast. Flank and side of belly light buffish or yellowish-grey-brown, mid-belly pale yellow or off-white, vent purer white, brown of flank and white of belly often rather sharply demarcated from yellow breast; under tail-coverts contrastingly spectrum-yellow or bright buff-yellow. Tail dull greyish-green with brighter green outer webs, slightly less yellowish than upper tail-coverts but not as dull olive as mantle and scapulars. Flight-feathers, greater upper primary coverts, and longest feathers of bastard wing dark grey to greyish-black, outer webs rather narrowly fringed green; tertials and remainder of upperwing olive-green, like mantle and scapulars. Under wing-coverts and axillaries yellow, brightest on small coverts along leading edge of wing. *In fresh plumage* (autumn), upperparts slightly greener, less olive, crown more olive-brown. *In worn plumage* (from about May onwards), upperparts more greyish-olive; grey band along mask narrower; yellow of underparts slightly less bright, more greenish on breast, gradually merging into dirty white of belly; flank with more restricted pale buff-brown. ADULT FEMALE. Forehead and forecrown olive-brown, merging into brown-olive on hindcrown and this in turn to olive-green on hindneck and mantle. Side of head and neck brown-olive or greyish-olive, often slightly darker and greyer on ear-coverts, but no black mask; lore partly mottled pale yellow; narrow but conspicuous pale greyish-buff or off-white eye-ring. Chin and throat bright sulphur-yellow, sharply demarcated from olive of lower cheek, tinged buff at side of throat and on lower throat, gradually merging into pale buffish-brown on side of breast and flank and into yellow-buff on chest, and these in turn into pale yellow or off-white on mid-belly. Under tail-coverts light sulphur-yellow or buff-yellow. Thus, head and chin to breast rather different from adult ♂, remainder of body closely similar. Wing and tail as adult ♂. *In worn plumage*, forehead, crown, and side of head and neck more olive-grey, chin to chest often purer yellow, more sharply defined at lower border (unless heavily worn), but yellow of chin to chest as well as that of under tail-coverts less deep than in adult ♂ and yellow on chest does not reach as far up on side of breast and towards belly as in adult ♂. JUVENILE. Upperparts and side of head and neck uniform brown-olive, greener on upper tail-coverts; faint pale buff eye-ring; chin to chest, side of breast, and flank pale buff-olive, flank tinged cinnamon, remainder of underparts pale buff-yellow. Median and greater upper wing-coverts brown-olive, narrowly and faintly tipped pale buff-brown or cinnamon. For flight-feathers and tail, see First Adult (below). For growth, see Stewart (1952) and Bent (1953). FIRST ADULT. In early autumn, like adult ♀, but juvenile tail, flight-feathers, tertials, and greater upper primary coverts retained; tips of tail-feathers pointed (rounded in adult), tertials tapering towards rounded tips (broader and with squarish tips in adult); yellow of chin to chest replaced by yellowish-buff or light buff. During autumn and winter, 1st adult ♂ gradually develops black mask and yellow throat: at first, forehead and side of head show scattered black feathers with grey tips (unlike any ♀); later, more black visible, due partly to abrasion, partly to further moult. In autumn, a few 1st adult ♂♂ already have extensive black mask, bordered by mottled brown and ash-grey band above and behind, but eye-ring buff-white; however, most ♂♂ in 1st autumn show scattered black mottling on cheek and ear-coverts only. In spring, 1st adult ♂ has mask fully black, except sometimes for scattered brown mottling and a partly pale eye-ring; both sexes otherwise similar to adult, apart from retained juvenile feathering. By April-May, juvenile tail and tertials worn (in adult, edges smooth or slightly worn), but shape still discernible; later, tail, tertials, and primaries heavily abraded, on average more so than in adult at same time of year, but otherwise indistinguishable.

Bare parts. ADULT, FIRST ADULT. Iris brown or dark brown. Bill grey-brown to dusky brown, almost black in breeding season; basal halves of cutting edges and of lower mandible pinkish or flesh, duller flesh-grey when breeding. Leg pink-flesh, light flesh-brown, or pale pink-brown, foot similar or slightly darker flesh-brown, purple-flesh, or sepia-brown. JUVENILE. Iris brown. Bill horn-grey, cutting edges and lower mandible extensively pink-buff or flesh-coloured. Leg and foot pink-flesh, pink-buff, or flesh with shade of brown. (Bent 1953; RMNH, ZMA.)

Moults. ADULT POST-BREEDING. Complete; primaries descendent. On breeding grounds, July-August (Pyle *et al.* 1987). In Michigan, some had still not started in mid-August, 9 others in moult 9 August to early September; duration of moult *c.* 30 days (Stewart 1952). Moult not started in 6 birds examined 6-28 July, plumage worn (adults) to extremely abraded (1-year-olds). ADULT PRE-BREEDING. Partial, involving head and chin to breast or all underparts, December-March, or no moult at all. (RMNH, SMTD, ZFMK.) POST-JUVENILE. Partial or perhaps occasionally complete. On breeding grounds, July-August. Includes at least head, body, and lesser, median, and greater upper wing-coverts; in some subtropical populations, also some (perhaps all) flight-feathers and tail included (Pyle *et al.* 1987), but not in northern populations (Ewert and Lanyon 1970, contra Behle 1950 and Stewart 1952). In southern Michigan, moult starts shortly after

fledging, before tail-feathers full-grown (Stewart 1952, which see for moult sequence). FIRST PRE-BREEDING. Partial; in winter quarters, November-April. Head, chin to throat, and sometimes apparently entire underparts; sometimes face only (especially in ♀).

Measurements. ADULT, FIRST ADULT. *G. t. brachidactylus* (may include a few migrant *campicola*, which is similar in size to *brachidactylus*: Behle and Aldrich 1947). Wisconsin and Illinois east to New Brunswick, Maine, Massachusetts, and New York, May-September; skins (RMNH, SMTD, ZFMK, ZMA). Bill (S) to skull, bill (N) to distal corner of nostril; exposed culmen on average 2·8 less than bill (S).

WING	♂	57·8 (1·50; 30)	55-61	♀	54·3 (1·19; 12)	52-56
TAIL		49·8 (2·51; 30)	46-53		46·9 (2·50; 11)	44-51
BILL (S)		13·8 (0·56; 30)	13·0-14·8		13·4 (0·46; 11)	12·8-14·0
BILL (N)		8·0 (0·45; 29)	7·3-8·7		7·8 (0·47; 11)	7·3-8·4
TARSUS		20·6 (0·81; 30)	19·4-21·8		20·4 (0·66; 12)	19·5-21·3

Sex differences significant for wing, tail, and bill (S).

Nominate *trichas*. South-east Pennsylvania, Maryland, Virginia, and western North and South Carolina, May-August; skins (RMNH, ZFMK, ZMA).

WING	♂	54·2 (1·54; 11)	51-56	♀	52·7 (0·64; 4)	52-54
TAIL		47·3 (1·89; 11)	45-50		45·5 (0·82; 4)	44-47
BILL (S)		13·3 (0·53; 11)	12·5-14·0		13·2 (0·46; 4)	12·6-13·6
BILL (N)		7·7 (0·39; 11)	7·2-8·4		7·6 (0·41; 4)	7·1-8·0
TARSUS		20·4 (0·76; 11)	19·3-21·4		19·3 (0·99; 4)	17·9-20·1

Sex differences not significant. Ages combined in samples above, though retained juvenile wing and (especially) tail shorter than in older birds. Thus, in ♂ *brachidactylus*: wing, adult 57·8 (1·53; 14) 56-60, 1st adult 57·7 (1·58; 15) 55-61; tail, adult 50·4 (2·85; 14) 47-53, 1st adult 49·4 (2·16; 15) 46-52. Wing and tail of 1-year-olds often heavily worn, making trustworthy measuring difficult. For decline of wing length through wear, see Francis and Wood (1989).

Weights. *G. t. brachidactylus*. Pennsylvania (USA): (1) July-August, (2) September-October, (3) (April-)May-June (Clench and Leberman 1978). (4) North-east USA, late April to October (Stewart 1937; Baldwin and Kendeigh 1938; Mengel 1965; ZFMK, ZMA). (5) Georgia (USA), early October (Johnston and Haines 1957).

(1) AD	♂	9·8 (0·60; 66)	8·2-11·7	♀	9·6 (0·72; 74)	7·7-11·7
1ST AD		9·9 (0·71; 123)	8·3-12·5		9·3 (0·61; 83)	8·0-11·0
(2) AD		10·9 (1·27; 326)	8·6-15·5		10·2 (1·16; 246)	7·8-15·3
1ST AD		10·3 (1·17; 851)	8·0-13·7		9·9 (1·12; 607)	7·9-14·3
(3)		10·0 (0·94; 564)	7·6-13·4		9·5 (1·03; 242)	7·6-13·0
(4)		10·2 (0·98; 11)	8·6-11·6		10·2 (1·57; 9)	8·6-14·1
(5)		10·7 (—; 5)	9·1-12·5		9·9 (—; 12)	8·3-11·3

Pelee Island (Lake Erie, Canada), May: ♂ 11·0 (16) 9·2-12·8 (Woodford and Lovesy 1958). Illinois (USA), September: average 11·8 (15) (Graber and Graber 1962). Florida, winter: ♀♀ 9·2, 9·9 (ZFMK). Belize: February, ♂♂ 9·7, 10·7; April-May, ♀♀ 8·2, 9·4 (Russell 1964). Lundy (Devon, England), November: ♂ 11·7 (Whitaker 1955).

Various races. Kansas, USA (*occidentalis* or *campicola*), late September and early October: adult, ♂ 11·5 (1·18; 42) 7·9-14·1, ♀ 10·4 (0·84; 43) 8·7-12·4; 1st adult, ♂ 11·4 (0·95; 36) 9·2-13·5, ♀ 10·7 (0·95; 49) 9·2-12·9 (Tordoff and Mengel 1956). Belize: autumn, 9·7 (0·8; 32) (Mills and Rogers 1990; see Rogers and Odum 1966 for monthly averages and fat-free weight). For many winter data from Jamaica, see Diamond *et al.* (1977). For averages in western North America, see Salt (1963). For nominate *trichas* in Georgia and South Carolina (USA), see Norris and Johnston (1958).

Structure. Wing short, broad at base, tip rounded. 10 primaries: p6-p7 longest, p8 0-1 shorter, p9 1·5-4, p5 0·5-1·5, p4 2-5, p3 4·5-7, p2 6-8, p1 8·6 (10) 7-10 in *brachidactylus*, 7·5 (7) 5-9 in nominate *trichas*. P10 strongly reduced, 36-42 shorter than p6-p7, 4-7 shorter than longest upper primary covert, 0-2 shorter than reduced outermost greater upper primary covert. Outer web of p5-p8 emarginated, inner web of p6-p9 with notch. Tip of longest tertial reaches to tip of p1-p2. Tail of average length when compared with body size, long when compared with wing; tip rounded; 12 feathers, t6 4-8 shorter than t1-t3. Bill as in Yellow Warbler *Dendroica petechia*, but tip relatively longer and bill narrower at base and in middle; tip less compressed laterally than in *Mniotilta*, *Vermivora*, or *Parula*, more like *Seiurus* but less heavy at base. Bristles at base of upper mandible very short, inconspicuous. Tarsus and toes rather long, slender. Middle toe with claw 16·5 (10) 15·7-17·5; outer toe with claw *c.* 71% of middle with claw, inner *c.* 68%, hind *c.* 79%. Remainder of structure as in *D. petechia*.

Geographical variation. Marked, involving intensity of general colour, colour of border along black mask of ♂, extent of yellow on underparts, size (as expressed in wing or tarsus length and in weight), and relative length of tail and bill. Variation largely clinal, but no distinct north-south or east-west trend. Nominate *trichas* is smallest eastern race; upperparts dull olive-green, border along black mask of ♂ grey, flank buff, belly of ♂ extensively pale buff or off-white, yellow confined to chin, throat, chest, and under tail-coverts. *G. t. brachidactylus* from further north is larger (see Measurements), upperparts brighter green, flank olive-brown, remainder of underparts more extensively yellow. South of nominate *trichas*, *typhicola* and *ignota* are larger also, tail and tarsus even longer than in *brachidactylus*, wing relatively shorter and more rounded (about equal to tail length, unlike other races), upperparts duller brownish-green, flank brown-buff, remainder of underparts extensively sulphur-yellow; *typhicola* on average smaller than *ignota* (especially bill), upperparts and flank less brown. In *ignota*, 1st-autumn ♂ frequently has full black mask and ♀ rarely shows traces of mask (Taylor 1976). Further west, all races have whitish border along black mask of ♂; *campicola* otherwise like *brachidactylus*, but upperparts paler, greyish-green, yellow of underparts richer yolk-yellow, flanks paler, greyish. For other races, see Ridgway (1901-11), Hellmayr (1935), Behle (1950), and Bent (1953).

Forms species-group with a number of species with restricted range in Central America (Peninsular Yellowthroat *G. beldingi*, Yellow-crowned Yellowthroat *G. flavovelata*, and Hooded Yellowthroat *G. nelsoni*), and perhaps with similarly restricted Olive-crowned Yellowthroat *G. semiflava*, Black-polled Yellowthroat *G. speciosa*, and Chiriqui Yellowthroat *G. chiriquensis*, as well as with Bahama Yellowthroat *G. rostrata* from Bahamas, and Masked Yellowthroat *G. aequinoctialis* from northern and central South America. Closely related to and sometimes included in *Oporornis* of North America; see Bledsoe (1988) and Zink and Klicka (1990).   CSR

Wilsonia citrina Hooded Warbler

PLATE 7
[between pages 64 and 65]

Du. Monnikszanger   Fr. Paruline à capuchon   Ge. Kapuzenwaldsänger
Ru. Капюшонная вильсония   Sp. Reinita encapuchada   Sw. Svarthakad citronsångare

Muscicapa citrina Boddaert, 1783. Synonym: *Myiodioctes mitratus*.

Monotypic

Field characters. 12·5-14 cm; wing-span 20-21·5 cm. Close in size to Blackpoll Warbler *Dendroica striata* but with proportionately slightly shorter wings and noticeably longer tail; averages 10% larger than Yellow Warbler *D. petechia*. Quite large, lengthy, Nearctic wood warbler fond of moist woodland. Bright olive to greenish-brown above and yellow below; ♂ has black hood around yellow face. Often spreads tail to show white panels on outer feathers. Sexes dissimilar; no seasonal variation. Immature inseparable.

ADULT MALE. Moults: June-August (complete). Head patterned like breeding ♂ Pied Wagtail *Motacilla alba yarrellii*, with bright yellow face surrounded by black hood. Upperparts, wings, and tail bright yellowish-olive, almost unmarked except for white panels on outer tail-feathers. Underparts bright yellow. Bill black. Legs bright brown. ADULT FEMALE. Face yellow, with distinctive black loral stripe, but no hood except for hints of black mottling over eye and around ear-coverts and partial dusky breast-band. Upperparts somewhat browner than ♂; white panels on tail-feathers smaller. FIRST-YEAR. ♂ as adult, but ♀ usually lacks any black on head.

♂ unmistakable. ♀ easily separated from ♀ Wilson's Warbler *W. pusilla* by yellow face and white tail-panels. Flight rapid, recalling small *Acrocephalus* warbler. Hops. Stance level, with tail often raised and spread. In home range, prefers thickets and forest understorey.

Call a loud, somewhat metallic, musical 'chink', 'chip', or 'tsyp'.

Habitat. Breeds in temperate and subtropical zones of eastern Nearctic, mainly in lowland woods or scrub, living in dense lower layers of vegetation, but not often seen on ground. Generally favours moist mature woodland, and in north of range near streams and in ravines; in south, common in heavily forested swamps with dense tangle of shrubs (Pough 1949). In north also attracted by hillside thickets of laurel *Kazmia* (Forbush and May 1955).

Distribution. Breeds in North America from south-east Nebraska and central Iowa east to Rhode Island, south to south-east Texas, north coast of Gulf of Mexico, and northern Florida. Winters from north-east Mexico and Yucatán south to Costa Rica and Panama.

Accidental. Britain: 1st-year ♀, Isles of Scilly, 20–23 September 1970 (Edwards and Osborne 1972; C S Roselaar).

Movements. Short-distance migrant, with breeding and winter ranges separated by only 600–700 km in west (eastern Texas to northern Mexico) (American Ornithologists' Union 1983; Rappole *et al.* 1983).

Autumn migration begins early August (Oberholser 1974; Kale and Maehr 1990), although a few birds wander away from breeding areas in July (e.g. Stewart and Robbins 1958). Migrants leave northern breeding areas August to early September (Stewart and Robbins 1958; Bohlen 1989), and reverse migration then takes some north to Minnesota, Ontario, and Nova Scotia (Sprague and Weir 1984; Tufts 1986; Janssen 1987). Main passage September, and few remain north of Texas and Florida by late October (Imhof 1976; Potter *et al.* 1980; James and Neal 1986). A few reach Veracruz and Yucatán (southern Mexico) by late August, but main arrival in Central America late September to October (Monroe 1968; Land 1970; Rogers *et al.* 1986a; Ramos 1988). Movement is across western Gulf of Mexico to eastern and southern Mexico, and only a few appear in peninsular Florida and Bahamas (Bent 1953; Brudenell-Bruce 1975; Fisk 1979). Ringing in southern Mexico, Guatemala, and Belize shows winter site-fidelity, with returns up to 2 years later (Ely *et al.* 1977; Rogers *et al.* 1982a; Kricher and Davis 1986).

Spring migration starts early March (Ramos 1988), and by end of March arrivals are scattered from Georgia and Florida to Texas and Arkansas (Oberholser 1974; Stoddard 1978; James and Neal 1986; Kale and Maehr 1990)—and exceptionally (in 1950) to Illinois, Ohio, and Ontario (Bent 1953). Mississippi valley breeding areas occupied during April (Mengel 1965; Bohlen 1989), but main arrival in Maryland and New York early May (Stewart and Robbins 1958; Bull 1974). Crosses Gulf on broad front, mostly skirting southern Florida (Bent 1953; Stevenson 1957). Early migrants often overshoot or drift offshore, and in many years small numbers appear on islands around Nova Scotia, 600 km or more north-east and downwind from breeding range; many such in April or September (McLaren 1981; Tufts 1986).

Vagrants also occur widely in western USA (American Ornithologists' Union 1983), and occasionally north far beyond breeding range: one recorded at Churchill (59°N, June) and one in Newfoundland (November) (Godfrey 1986). For vagrancy to west Palearctic, see Distribrution.

AJE

Voice. See Field Characters.

Plumages. ADULT MALE. Forehead and side of head up to stripe over eye, back to ear-coverts, and down to lower cheek bright spectrum-yellow or orange-yellow, forming conspicuous mask; usually some sooty mottling just in front of eye. Black bristles at base of bill. Crown, broad band down side of neck, chin, and throat black, enclosing yellow mask and strongly contrasting with it as well as with remainder of body. Upperparts from hindneck backwards bright green, slightly more olive on hindneck, slightly more yellow-green on rump. Side of breast and flank light yellow-green, chest to under tail-coverts spectrum-yellow, brightest on mid-belly, faintly tinged olive on chest and side of belly, paler yellow on under tail-coverts. Tail dark grey, almost black on t2–t3; fringes along outer webs of t1–t4(–t5) green; terminal *c.* 20 mm of inner web of t3 contrastingly white; white on inner webs increases towards t6, which has *c.* 40 mm white. Flight-feathers, tertials, greater upper wing-coverts, and bastard wing greyish-black; outer webs of flight-feathers (except of p9 and of emarginated parts of primaries), primary coverts, and middle feather of bastard wing narrowly fringed green, tertials largely green on outer web; narrow edge along outer web of p9 off-white. Upper wing-coverts bright green, dark grey inner webs of greater coverts and centres of median sometimes partly visible. Short coverts along leading edge of wing bright yellow, shorter under wing-coverts olive or yellow, longer ones and axillaries grey with white borders; basal inner borders of flight-feathers off-white. *In fresh plumage* (autumn), tips of black feathers of head narrowly and faintly fringed yellow, mainly on hindcrown, chin, and throat. ADULT FEMALE. Basically similar to adult ♂, but black on head less extensive, though amount variable. Forehead and side of head yellow, as adult ♂, but sometimes with some dusky olive on feather-tips, especially at base of bill, above lore and eye, and at border of crown. Crown black, feather-tips narrowly fringed green on central crown, more broadly so towards rear crown, forecrown appearing almost uniform black, but black increasingly mottled green towards rear, merging into green of remainder of upperparts; black not as sharply defined at rear as in adult ♂, except sometimes when plumage heavily worn. Usually a narrow black band down front side of neck, variegated green at rear, remainder of side of neck green; black band frequently continued in narrow black collar along lower throat, which is sometimes interrupted by yellow or green in middle, but lower throat sometimes extensively mottled black and green; chin yellow, upper throat usually yellow or yellow with some green mottling; none examined had chin and throat as black as adult ♂ (but sample small). Of 198 ♀♀ (all ages) examined by Lynch *et al.* (1985), 2% had upper throat and chin with considerable amount of black, but apparently not entirely black. Tail as adult ♂, but inner web of t3 less white, often an elongated subterminal spot only; 15–25 mm of tip of inner web of t4 white, 30–35 mm of t6. Wing as adult ♂; under wing-coverts and axillaries as ♂, but often more extensively tinged yellow. Though amount of black on head variable (see Mengel 1965 and Lynch *et al.* 1985), variation is individual, not age-related once over 1 year old (Morton 1989). JUVENILE. Upperparts entirely buff- or grey-brown or grey with brown-drab feather-tips. Side of head olive-yellow. Chin, throat, chest, and side of breast light buff- or olive-brown, or mottled grey and drab, remainder of underparts yellow or white, tinged brown on flank. Upper wing-coverts grey-brown, median and greater distinctly fringed cinnamon or red-brown along tips. (Palmer 1894; Dwight 1900; Ridgway 1901–11.) For development, see Palmer (1894) and Odum (1931). FIRST ADULT MALE. Like adult ♂, but juvenile tail, flight-feathers, tertials, primary coverts, and bastard wing retained, these slightly less dark than in adult ♂, browner and with more distinctly frayed tips when worn; tips of tail-feathers and primary coverts more pointed than in adult, less smoothly rounded. Tail less extensively white, similar to adult ♀ or with t3 all-dark; white on t4 sometimes reduced to small subterminal spot. Ageing sometimes difficult; best character is difference in colour between relatively fresh 1st adult outer greater coverts and juvenile primary coverts, latter browner and with frayed and faint dirty olive fringes (brighter green on greater coverts). Head and body as adult ♂, but pale yellow fringes along tips of black feathers of head on average slightly wider, worn off less quickly. FIRST ADULT FEMALE. Like adult ♀, but head less extensively black or no black at all. Part of juvenile feathering retained, as in 1st adult ♂; tail as in 1st adult ♂. In many birds, upperparts entirely green except for orange-yellow at base of upper mandible, yellow elsewhere on forehead concealed unless plumage worn; ear-coverts spotted green, side of neck all-green. Usually a few black specks present, mainly at upper and rear border of ear-coverts and within green of forecrown; occasionally, specks join into narrow, more or less continuous line along forecrown and side of crown down front side of neck, and these birds may have forehead more fully yellow, approaching adult ♀.

Bare parts. ADULT, FIRST ADULT. Iris brown. Bill dark horn-black with slightly paler horn lower mandible; almost black during breeding season. Leg and foot light flesh-brown to purplish-flesh-brown. (Ridgway 1901–11; RMNH, ZMA.) JUVENILE, FIRST ADULT. Like adult, but base of lower mandible and leg and foot paler (Palmer 1894).

Moults. ADULT POST-BREEDING. Complete; primaries descendent. On breeding grounds, late June and July (Palmer 1894) or June–August (Pyle *et al.* 1987). No pre-breeding moult. POST-JUVENILE. Partial or, at least occasionally, complete; on breeding grounds. Starts shortly after fledging; in heavy moult late June, slower moult July (Palmer 1894, which see for moult sequence). Involves only head, body, and lesser and median upper wing-coverts, and sometimes tertials and greater upper wing-coverts in 12 birds examined (RMNH, SMTD, ZFMK, ZMA); said by Walters and Lamm (1980) and Pyle *et al.* (1987) usually to include tail and variable number of flight-feathers.

Measurements. ADULT, FIRST ADULT. Eastern USA, April–September; skins (BMNH, RMNH, SMTD, ZFMK, ZMA). Bill (S) to skull, bill (N) to distal corner of nostril; exposed culmen on average 3·5 less than bill (S).

WING   ♂ 68·3 (1·54; 27) 66–72   ♀ 64·8 (1·98; 14) 62–68

TAIL	55·4 (2·08; 19) 52–59	53·0 (1·87; 14) 50–56
BILL (S)	14·2 (0·52; 20) 13·2–15·1	13·9 (0·45; 14) 13·4–14·8
BILL (N)	7·9 (0·42; 20) 7·3–8·9	7·8 (0·38; 14) 7·3–8·4
TARSUS	20·3 (0·61; 20) 19·3–21·6	19·6 (0·54; 14) 18·7–20·4

Sex differences significant, except for bill. For birds wintering Yucatán (Mexico), see Lynch *et al.* (1985).

Weights. (1) Kentucky (USA), June and early July (Mengel 1965). Pennsylvania (USA): August–September, (2) adult, (3) 1st adult; October, (4) 1st adult; May–July, (5) adult and 2nd calendar year (Clench and Leberman 1978). (6) Georgia (USA), early October (Johnston and Haines 1957).

(1)	♂	10·7 (0·78;	4) 9·7–11·4	♀	10·0 (— ;	2) 9·9–10·1
(2)		10·4 (1·02;	6) 9·7–12·4		10·7 (0·93; 11)	9·3–12·5
(3)		9·8 (0·64; 100)	8·1–11·6		10·4 (0·67; 54)	7·4–12·0
(4)		11·6 (1·85;	6) 10·0–14·9		10·9 (0·60;	5) 10·3–11·8
(5)		10·9 (0·60; 25)	9·4–11·8		10·0 (0·70; 44)	8·6–12·2
(6)		12·4 (— ; 14)	9·5–14·3		11·9 (— ; 11)	9·8–13·4

See also Norris and Johnston (1958), Woodford and Lovesy (1958), Murrray and Jehl (1964), Lynch *et al.* (1985), and Kuenzi *et al.* (1991).

Belize: autumn 9·5 (0·9; 42) (Mills and Rogers 1990); March, ♂ 8·2 (Russell 1964). See also Rogers and Odum (1966). For fat-free weights, see Odum *et al.* (1961).

Structure. Wing rather short, broad at base, tip bluntly pointed. 10 primaries: p7–p8 longest, p9 1·5–4 shorter, p6 0–1, p5 3·5–6, p4 6–9, p3 8–10, p2 9–12, p1 11–15. P10 strongly reduced, hidden below reduced outermost upper primary covert; 43–50 shorter than p7–p8, 5–8 shorter than longest upper primary covert. Outer web of p6–p8 emarginated, inner of (p7–)p8–p9 with notch. Tip of longest tertial reaches to tip of p1–p2. Tail rather long, tip slightly rounded; 12 feathers, t6 3–4 shorter than t1. Bill closely similar to Yellow Warbler *Dendroica petechia*, rather short, straight, fairly stout; length of visible culmen *c.* 46% of head length; culmen with smooth ridge; bill relatively heavier and much less wide and flattened at base than in Canada Warbler *W. canadensis*. Differs from *Dendroica* in *c.* 8 long and distinct bristles extending obliquely forward from side of base of upper mandible and *c.* 4 shorter bristles at side of lower mandible (*W. canadensis* also has strong bristles by upper mandible but virtually none by lower mandible). Tarsus rather long, toes of average length; fairly strong. Middle toe with claw 14·9 (5) 13·8–15·8 mm; outer toe with claw *c.* 73% of middle with claw, inner *c.* 68%, hind *c.* 82%. Remainder of structure as in *D. petechia*.

CSR

Wilsonia pusilla **Wilson's Warbler**

PLATE 8
[between pages 64 and 65]

DU. Wilsons Zanger   FR. Paruline à calotte noire   GE. Mönchswaldsänger
RU. Малая вильсония   SP. Silvia de birrete del norte   SW. Svartkronad skogssångare

Muscicapa pusilla Wilson, 1811

Polytypic. Nominate *pusilla* (Wilson, 1811), extreme north-east USA and eastern and central Canada, west to Mackenzie and Alberta; *pileolata* (Pallas, 1811), Alaska and Yukon (north-west Canada) south through northern and interior British Columbia and interior western USA to Colorado and New Mexico; *chryseola* Ridgway, 1902, coasts and coastal mountain ranges of western North America, from south-west British Columbia to California.

Field characters. 11–12·5 cm; wing-span 15·5–17·5 cm.

About 10% smaller than Hooded Warbler *W. citrina* with

proportionatey shorter, finer bill; close in size to Yellow Warbler *Dendroica petechia*. Quite small, animated, willow-loving Nearctic wood warbler, with bright olive-green upperparts and lemon-yellow underparts; ♂ has black cap; no white in tail. Twitches tail. Sexes dissimilar; no seasonal variation. Immature doubtfully separable.

ADULT MALE. Moults: June–August (complete). Bright yellowish-olive-green above, with glossy black square patch from mid-crown to nape and faintly dusky markings on flight- and tail-feathers. Vivid lemon-yellow face and underparts, with faintly dusky wash on ear-coverts, sides of breast, and rear flanks. Bill blackish, with flesh to horn base to lower mandible. Legs bright brown. ADULT FEMALE. Closely resembles ♂ but crown no darker than dusky-olive, sometimes with blackish mottling on forepart. Compared to *W. citrina*, shows longer supercilium and more obvious eye-ring, due to increased contrast with more olive ear-coverts; lacks dark edges to rear crown and rear ear-coverts. FIRST-WINTER. In ♂, black crown-patch sometimes duller.

♂ unmistakable. ♀ easily distinguished from *W. citrina* and *D. petechia* by face pattern and lack of white in tail, from *D. petechia* also by lack of wing-bars. ♀ told from Tennessee Warbler *Vermivora peregrina* by dark crown, less obvious supercilium, and yellow (not white) under tail-coverts. Remarkably active and restless, with light flight and expert flycatching (during which bill snaps audibly); droops and flicks wings, and twitches tail in almost rotary action. Hops. Stance level, often holding tail above wings. Fond of damp thickets.

Calls include sharp musical 'chip', harsher 'chut', short, lisped 'tsip', loud, rather liquid 'twick' (recalling Cetti's Warbler *Cettia cetti*), and 3-syllable 'kick-kick-kick' (recalling Red-breasted Flycatcher *Ficedula parva*).

Habitat. Breeds almost throughout Nearctic climatic zones, from Arctic tundra and montane valleys and slopes in Alaska, and up to nearly 500 m in southern California. In Alaska, favours coarse grass interspersed with areas of raspberry *Rubus*, alder *Alnus*, and elder *Sambucus*, but on northern tundra may occupy farthest willow *Salix* patch, tolerating high degree of scattering of required shrub cover (Gabrielson and Lincoln 1959). In Canada, favours moist open shrubbery, including dwarf birch *Betula nana*, especially by streams, ponds, and bogs, but also in valley bottoms, on mountainsides, and in alpine meadows (Godfrey 1986).

On migration, more tolerant of drier situations (Godfrey 1986) and in New England also found seasonally in upland deciduous or coniferous woods, orchards, and even city gardens and parks. Keeps to lower levels of woodland. (Forbush and May 1955.)

Distribution. Breeds in North America from northern Alaska east to Newfoundland, south to southern California, north-central New Mexico, northern Minnesota, southern Ontario, central Maine, and Nova Scotia. Winters from north coast of Gulf of Mexico south through Central America to Panama.

Accidental. Britain: ♂, Rame Head (Cornwall), 13 October 1985 (Rogers *et al.* 1988).

Movements. Migratory, all 3 races moving at least 800 km (most birds over 2500 km) between summer and winter ranges. Winter range very limited in comparison with breeding range; highest numbers winter in Mexico. (American Ornithologists' Union 1957, 1983; Rappole *et al.* 1983; Pashley and Martin 1988.)

Autumn migration begins August, but in 1st half of month departure from western mountains predominates (Burleigh 1972; Alcorn 1988). Movement widespread from mid-August (e.g. Potter *et al.* 1980, Garrett and Dunn 1981), and most birds leave northern breeding areas then (Rand 1946; Scotter *et al.* 1985). In September, main passage moves south through USA (Jewett *et al.* 1953; Sutton 1967; Garrett and Dunn 1981; Hall 1981); most birds leave northern states by late September or early October, and by end of October only a few stragglers remain north of winter range (Bent 1953; Phillips *et al.* 1964; Burleigh 1972; Bohlen 1989). Migrants pour into Mexico and Central America from mid-September, chiefly October (Monroe 1968; Ramos 1988; Stiles and Skutch 1989). Migration on broad front in both seasons, mainly west of Appalachians, with most birds skirting Gulf of Mexico to west rather than overflying it; uncommon to rare in south-east states and West Indies (e.g. Burleigh 1958, Rappole *et al.* 1979, Bond 1985). Winter site-fidelity demonstrated by ringing (e.g. Mexico, El Salvador, Panama), with returns up to 2 years later (Ely *et al.* 1977; Loftin 1977).

In spring, birds leave winter range March to mid-May (Ramos 1988; Stiles and Skutch 1989). West-coast race *chryseola* moves early to its temperate breeding areas, with first arrivals in California and Arizona in March (Phillips *et al.* 1964; Garrett and Dunn 1981). Nominate *pusilla* and *pileolata* reach mid-latitude states in late April (Sutton 1967; Potter *et al.* 1980; Alcorn 1988). Peak movement is in May everywhere, averaging only a week later in southern Canada than in Maryland and Kentucky (Stewart and Robbins 1958; Houston and Street 1959; Mengel 1965; Squires 1976). In late May and early June, last migrants reach northern breeding areas (Peters and Burleigh 1951; Todd 1963; Erskine 1985). Some early migrants in east reach Nova Scotia mid-April, with vagrant Prothonotary Warbler *Protonotaria citrea* and Hooded Warbler *W. citrina* (Tufts 1986).

A few vagrant records July–September in arctic Canada and Alaska, on islands in Bering Sea, and in Greenland (Gabrielson and Lincoln 1959; Pedersen 1980; Godfrey 1986). For vagrancy to west Palearctic, see Distribrution.

AJE

PLATE 1. *Mniotilta varia* Black-and-white Warbler (p. 11): **1** ad ♂ breeding, **2** ad ♂ non-breeding, **3** ad ♀ non-breeding, **4** 1st ad ♂ non-breeding, **5** 1st ad ♀ non-breeding. *Vermivora chrysoptera* Golden-winged Warbler (p. 14): **6** ad ♂, **7** ad ♀. *Vermivora peregrina* Tennessee Warbler (p. 17): **8** ad ♂ summer, **9** ad ♂ winter, **10** 1st ad ♂. (NA)

PLATE 2. *Parula americana* Northern Parula (p. 20): **1** ad ♂ summer, **2** ad ♂ winter, **3** 1st ad ♂ winter, **4** 1st ad ♀ winter. *Dendroica petechia* Yellow Warbler (p. 22): **5** ad ♂ breeding, **6** ad ♂ non-breeding, **7** 1st ad ♂ non-breeding, **8** 1st ad ♀ non-breeding. (NA)

PLATE 3. *Dendroica pensylvanica* Chestnut-sided Warbler (p. 26): **1** ad ♂ breeding, **2** ad ♀ breeding, **3** ad ♂ non-breeding, **4** 1st ad ♀ non-breeding. *Dendroica caerulescens* Black-throated Blue Warbler (p. 29): **5** ad ♂, **6** 1st ad ♂, **7** 1st ad ♀. *Dendroica fusca* Blackburnian Warbler (p. 31): **8** ad ♂ breeding, **9** ad ♂ non-breeding, **10** 1st ad ♀ non-breeding. (NA)

PLATE 4. *Dendroica tigrina* Cape May Warbler (p. 35): **1** ad ♂ breeding, **2** ad ♂ non-breeding, **3** ad ♀ breeding, **4** 1st ad ♂ non-breeding, **5** 1st ad ♀ non-breeding. *Dendroica magnolia* Magnolia Warbler (p. 38): **6** ad ♂ breeding, **7** ad ♂ non-breeding, **8** ad ♀ breeding, **9** 1st ad ♀ non-breeding. (NA)

PLATE 5. *Dendroica coronata coronata* Yellow-rumped Warbler (p. 41): **1** ad ♂ breeding, **2** ad ♂ non-breeding, **3** ad ♀ breeding, **4** 1st ad ♀ non-breeding. *Dendroica striata* Blackpoll Warbler (p. 45): **5** ad ♂ breeding, **6** ad ♂ non-breeding, **7** ad ♀ breeding, **8** 1st ad ♀ non-breeding. *Dendroica castanea* Bay-breasted Warbler (for comparison): **9** 1st ad ♀ non-breeding. (NA)

PLATE 6. *Setophaga ruticilla* American Redstart (p. 49): **1** ad ♂, **2** ad ♀, **3** 1st ad ♂ autumn, **4** 1st ad ♀. *Seiurus aurocapillus aurocapillus* Ovenbird (p. 52): **5** 1st ad. *Seiurus noveboracensis noveboracensis* Northern Waterthrush (p. 55): **6** 1st ad. *Seiurus motacilla* Louisiana Waterthrush (for comparison): **7** 1st ad. (NA)

PLATE 7. *Geothlypis trichas brachidactylus* Common Yellowthroat (p. 58): **1** ad ♂, **2** ad ♀, **3** 1st ad ♂, **4** 1st ad ♀. *Wilsonia citrina* Hooded Warbler (p. 61): **5** ad ♂, **6** ad ♀, **7** 1st ad ♂ autumn, **8** 1st ad ♀. (NA)

PLATE 8. *Wilsonia pusilla pusilla* Wilson's Warbler (p. 63): **1** ad ♂, **2** 1st ad ♂ autumn, **3** 1st ad ♀ autumn. *Wilsonia canadensis* Canada Warbler (p. 64): **4** ad ♂ breeding, **5** ad ♀ breeding, **6** 1st ad ♂ non-breeding, **7** 1st ad ♀ non-breeding. (NA)

PLATE 9. *Piranga rubra rubra* Summer Tanager (p. 69): **1** ad ♂, **2** ad ♀, **3** 1st ad ♂ winter, **4** 1st ad ♀ winter. *Piranga olivacea* Scarlet Tanager (p. 72): **5** ad ♂ breeding, **6** ad ♂ non-breeding, **7** ad ♀ non-breeding, **8** 1st ad ♂ non-breeding, **9** 1st ad ♂ breeding. (CETK)

Ian Lewington

PLATE 10 (*facing*). *Montifringilla nivalis* Snow Finch (Vol. VIII, p. 386). Nominate *nivalis* : **1-3** ad ♂ summer, **4** ad ♂ winter, **5-6** ad ♀, **7** juv. *M. n. alpicola*: **8** ad ♂ summer.

Plectrophenax nivalis Snow Bunting (p. 118). Nominate *nivalis*: **9-10** ad ♂ summer, **11-12** ad ♂ winter, **13** ad ♀ summer, **14-15** ad ♀ winter, **16-17** 1st ad ♂ winter, **18** juv ♂. *P. n. insulae*: **19** ad ♂ summer, **20-21** ad ♂ winter, **22** ad ♀ winter. *P. n. vlasowae*: **23** ad ♂ summer, **24** ad ♂ winter. (IL)

Voice. See Field Characters.

Plumages. (nominate *pusilla*). ADULT MALE. Forehead and broad supercilium bright lemon-yellow, yellow of supercilium gradually merging into olive above ear-coverts. Crown black with faint blue gloss, forming sharply contrasting square-ended patch on top of head. Remainder of upperparts from hindneck to upper tail-coverts and including lesser upper wing-coverts uniform olive-green, much brighter and greener than upperparts of Willow Warbler *Phylloscopus trochilus* or Chiffchaff *P. collybita*, less olive-brown, rather similar to Wood Warbler *P. sibilatrix*, but brighter and more saturated, less pallid; hindneck and rump sometimes more yellowish-green. Lore and ring round eye lemon-yellow, feathers on lore tipped olive, sometimes forming olive patch in front of eye; upper cheek green-yellow, merging into olive-green of ear-coverts and side of neck, these not sharply demarcated from yellow of lore and throat. Entire underparts bright lemon- or sulphur-yellow, washed green on side of breast and flank, sometimes faintly mottled green on chest; yellow of vent and under tail-coverts sometimes less intense, mainly because of white feather-bases showing through. Thigh greyish-olive. Tail dark grey, outer webs of feathers fringed olive-green (except on outer pair, t6). Flight-feathers, tertials, greater upper primary coverts, and bastard wing dark grey or blackish-grey, outer webs narrowly fringed olive-green (except on emarginated parts of primaries and on longest feather of bastard wing), wider but less sharply defined on tertials. Median and greater upper wing-coverts olive-green, greyish-black centres of median and inner webs of greater sometimes partly visible. Under wing-coverts and axillaries yellow, partly washed green, longer coverts and bases of shorter coverts along leading edge of wing grey. *In fresh plumage* (August–October, April–May), yellow of forehead and of side of head slightly washed olive-green, and black feathers of crown show narrow green fringes to tips, but these do not obscure sharp outline of black cap; in worn plumage (mainly July), green of upperparts slightly duller, more olive, some dusky of feather-bases of underparts partly visible, and green fringes of tail- and flight-feathers partly or largely worn off. ADULT FEMALE. As adult ♂, but black feathers of crown broadly fringed olive-green, black cap less sharply defined; in fresh plumage, black often largely hidden, except for rather narrow black border over forecrown; in worn plumage, cap purer black on forecrown, but more sooty (less glossy) than in adult ♂, and rear crown variegated green and black, black of cap gradually merging into green of hindneck. JUVENILE. Like adult, but not black on cap; upperparts and upperwing more diluted brownish-olive, mottled darker brown, less saturated green; yellow of forehead, side of head, and underparts replaced by pale buff or pale cream-yellow, washed brown on throat, chest, side of breast and flank; median and greater upper wing-coverts narrowly tipped buff. FIRST ADULT MALE. Like adult ♂, but juvenile tail, flight-feathers, tertials, greater upper primary coverts, and bastard wing retained, tips of tail-feathers distinctly pointed (less broad and rounded than in adult), tips of tertials more rounded (less squarish); in spring, tail and flight-feathers browner and more worn than those of adult at same time of year. Remainder of plumage as adult ♂, but black feathers of cap sometimes with more pronounced green tips, especially in autumn, though generally not as much green as in adult ♀. FIRST ADULT FEMALE. Like adult ♀, but part of juvenile feathering retained, as in 1st adult ♂. Black on cap almost entirely replaced by green, 1st-autumn birds in particular showing entirely uniform olive-green upperparts, apart from rather ill-defined yellow or yellow-green forehead; a few black specks occasionally hidden on feather-centres of forecrown. In spring, more black may show on forecrown, some

birds resembling adult ♀ in this respect, others have crown still entirely olive-green.

Bare parts. ADULT, FIRST ADULT. Iris hazel, brown, or dark brown. Upper mandible and tip of lower mandible deep brown to slate-black, cutting edges and basal and middle portion of lower mandible flesh-brown, light horn-brown, or greyish-horn, paler flesh at extreme base. Leg purple-brown or horn-brown, foot light brown or flesh-brown; paler horn-yellow, flesh-horn, or horn-grey at rear of tarsus and on soles. (BMNH, RMNH, ZFMK.) JUVENILE. No information.

Moults. ADULT POST-BREEDING. Complete; primaries descendent. On breeding grounds, June–August. ADULT PRE-BREEDING. Partial: head, body, and upper wing-coverts, or head and throat only, or perhaps sometimes no moult at all. In winter quarters, February–April. POST-JUVENILE. Partial: head, body, and lesser, median, and greater upper wing-coverts; not flight-feathers, tertials, greater primary coverts, or tail. Starts immediately after fledging; in moult June–July, in breeding area. FIRST PRE-BREEDING. Like adult pre-breeding. (Dwight 1900; Pyle *et al.* 1987; RMNH, ZFMK.)

Measurements. ADULT, FIRST ADULT. Nominate *pusilla*. Eastern USA, May–September; skins (BMNH, RMNH, SMTD, ZFMK, ZMA). Bill (S) to skull, bill (N) to distal corner of nostril; exposed culmen on average 3·2 less than bill (S).

WING	♂	55·8 (1·52; 31) 53–59	♀	54·5 (1·58; 16) 52–57
TAIL		46·7 (1·59; 15) 44–49		46·1 (1·47; 14) 44–49
BILL (S)		11·8 (0·44; 15) 11·3–12·5		11·9 (0·41; 14) 11·4–12·5
BILL (N)		6·3 (0·43; 15) 5·7–7·0		6·3 (0·24; 13) 5·9–6·8
TARSUS		18·3 (0·38; 14) 17·6–18·8		18·0 (0·30; 14) 17·5–18·5

Sex differences significant for wing and bill (S).

W. p. pileolata. Alaska and western Canada, summer, and Central America, winter; skins (RMNH).

WING	♂	59·0 (1·51; 15) 57–62	♀	57·0 (0·82; 7) 56–58
TAIL		49·4 (1·82; 15) 46–53		47·8 (2·04; 7) 46–51
BILL (S)		11·9 (0·33; 15) 11·6–12·5		11·5 (0·18; 7) 11·3–11·8
BILL (N)		6·4 (0·29; 14) 5·8–6·7		6·2 (0·13; 7) 5·9–6·3
TARSUS		18·4 (0·35; 14) 17·8–18·9		18·5 (0·66; 7) 17·2–19·1

Weights. Nominate *pusilla*. Pennsylvania (USA): (1) August–October, (2) May (Clench and Leberman 1978).

(1) AD	♂	8·0 (0·74; 123) 6·4–10·5	♀	7·9 (0·83; 26) 6·5–9·5
1ST AD		8·0 (0·71; 94) 6·5–10·9		7·7 (0·76; 89) 6·0–11·9
(2)		7·6 (0·54; 297) 6·4–9·3		7·4 (0·55; 44) 6·3–8·8

Pelee Island (Lake Erie, Canada), May: 8·1 (21) 6·9–9·5 (Woodford and Lovesy 1958). Belize, winter: 6·0, 6·8 (Russell 1964).

W. p. pileolata. North-central Alaska and north-west Yukon (Canada), 20 May to 10 June: ♂ 7·6 (0·47; 11) 6·9–8·5, ♀ 8·1 (3) 7·3–8·8 (Irving 1960). Veracruz (Mexico), October–November: 7·0 (0·41; 7) 6·7–7·8 (RMNH).

Structure. Wing short, broad at base, tip bluntly pointed. 9–10 primaries: p7–p8 longest, p9 2–3·5 shorter, p6 0–1, p5 3–5·5, p4 5–8, p3 7–10, p2 8–11, p1 9–12. P10 strongly reduced, apparently absent in 2 of 10 examined; 38–44 shorter than p7–p8, 5–7 shorter than longest upper primary covert, 0–2 shorter than reduced outermost upper primary covert. Outer web of p6–p8 emarginated, inner web of (p6–)p7–p8 with notch (sometimes also a faint one on p9). Tail of average length, tip slightly rounded; 12 feathers, t6 *c.* 5 mm shorter than t1. Bill short, exposed culmen *c.* 40% of head length; shape otherwise as in Yellow Warbler *Dendroica petechia*, but tip less compressed laterally and base slightly wider, though less so than in Canada Warbler *W. canadensis*; long bristles at base of upper mandible, as in *W. canadensis* and Hooded Warbler *W. citrina*. Tarsus relatively long, toes rather short; slender. Middle toe with claw 13·4 (5) 12·8-14·3; outer toe with claw *c.* 73% of middle with claw, inner *c.* 69%, hind *c.* 76%. Remainder of structure as in *D. petechia*.

Geographical variation. Rather slight, involving colour and size. Nominate *pusilla* from central and eastern Canada and neighbouring extreme north-east USA small in size; forehead yellow, upperparts olive-green, underparts lemon- or sulphur-yellow with slight green wash. *W. p. pileolata* from Alaska, inland western Canada, and inland western USA larger, especially in wing length (see Measurements); forehead more orange-yellow, upperparts brighter yellowish-green, underparts deeper bright yellow with slight orange tinge, with restricted yellowish-green wash on side of breast and flank. *W. p. chryseola* of coastal western Canada and western USA is as small as nominate *pusilla*, but colours even brighter than *pileolata*: forehead and super-cilium orange-yellow to orange, upperparts bright green-yellow, underparts bright orange- or golden-yellow. Amount of black on cap of ♀ perhaps also differs between races, but depends on age, and much individual variation. According to Pyle *et al.* (1987), ♀♀ east of Rockies have black cap less than 8 mm long, or black entirely absent; ♀♀ west of Rockies always have black up to 12 mm long (in both, cap partly suffused green, not as uniform and shiny black as in ♂). In nominate *pusilla* examined, 1st adult ♀♀ had o, 2, 7, and 11 mm black, adult ♀♀ 10 and 13; in *pileolata*, 1st adult ♀ o (*n*=4), adult ♀ 1-3 (*n*=3); thus blacker in east than in west, but samples very small (RMNH, SMTD, ZFMK, ZMA). CSR

Wilsonia canadensis Canada Warbler

PLATE 8
[between pages 64 and 65]

Du. Canadese Zanger   Fr. Paruline du Canada   Ge. Kanadawaldsänger
Ru. Канадская вильсония   Sp. Silvia del Canadá   Sw. Kanadaskogssångare

Muscicapa canadensis Linnaeus, 1766

Monotypic

Field characters. 12·5-14 cm; wing-span 20-22 cm. Close in size to Blackpoll Warbler *Dendroica striata* but more robust, with noticeably heavier bill. Medium-sized, bold, undergrowth-haunting Nearctic wood warbler, with olive-grey to blue-grey upperparts, bright yellow under-parts and yellowish-white spectacle in all plumages. ♂ has black forecrown, lore, forecheek, and broad necklace of black spots; ♀ shows shadows of similar marks. Legs pale. Sexes dissimilar; no seasonal variation. Immature doubt-fully separable.

ADULT MALE. Moults: June–August (complete). Crown, upperparts, wings, and tail plain blue-grey; underparts bright yellow. Face and breast strikingly marked, with black forehead breaking up into crown-streaks, black lore, and forecheek emphasizing yellow fore-supercilium and yellowish-white eye-ring; black on cheek extends to side of breast and spreads out to form spotted gorget. Bill quite wide; grey to black culmen and tip and pale pinkish to yellow base. Legs yellowish-flesh. ADULT FEMALE. Duller than ♂, less blue above, and with black head markings subdued; forecrown more olive and dusky necklace far less obvious. FIRST-WINTER. Duller than adult, with greyer-brown flight-feathers; see also Plumages.

No other parulid has unmarked grey upperparts, wings, and tail. Flight and general behaviour as other *Wilsonia*.

Calls include subdued 'chip' or 'tschip'; loud 'check' recalls House Sparrow *Passer domesticus*.

Habitat. Breeds in temperate northern and eastern Nearctic deciduous and mixed forest zone, mainly in shrubby undergrowth of mature woodland, or in willows *Salix* and alders *Alnus* along streams, in swamps, and in other moist places (Godfrey 1979). In Minnesota, favours moist spruce *Picea* and birch *Betula* forest with thick brushy understorey (Johnsgard 1979). Likes cool, moist, luxuriant undergrowth of mature woodland, but also occu-pies swamp borders, streamside shrubbery, and occa-sionally young second growth in clearings (Pough 1949). Habitually forages among shrubbery, and lower limbs of trees, rarely venturing to higher treetops; often flycatches, and occasionally feeds on ground (Forbush and May 1955).

Winters in South America in rain and cloud forests and second growth, in clearings, and at forest edges up to 2100 m, foraging in both lower and upper layers (Schauensee and Phelps 1978).

Distribution. Breeds in North America, from north-central Alberta and northern Quebec south to central Minnesota, south-east New York and Maine, and along Appalachians south to Tennessee and north-west Georgia. Winters in Central America, from Belize and Honduras south to Ecuador, central Peru, and western Brazil.

Accidental. Iceland: ♂ (specimen), 29 September 1973 (ÆP).

Movements. Migratory throughout range, mostly breeding further north and wintering further south than Hooded Warbler *W. citrina*.

Southward migration commences in first half of August (Bull 1974; Tufts 1986; Janssen 1987). Most birds have left northern breeding range by early September (Houston and Street 1959; Squires 1976), and mid-latitude states by end of September (Stewart and Robbins 1958; Mengel 1965; James and Neal 1986). Final departures are later from main range, Ontario to New Brunswick, than further west (Sadler and Myres 1976; Erskine 1977). Stragglers linger to end of October near Mexican Gulf coast (Lowery 1974; Imhof 1976). First birds reach Central America by early September (Ramos 1988; Stiles and Skutch 1989), so main passage is rapid; it is also compressed, extending locally over only about 3 weeks, e.g. in Maryland (Stewart and Robbins 1958), West Virginia (Hall 1983), and Veracruz (Mexico) (Ramos 1988). Migration is mainly in and west of Appalachians, avoiding south-east states, with ongoing route through eastern Mexico and across western Gulf (Bent 1953; Rappole *et al.* 1979); scarcity in West Indies (Bond 1985) argues against major movement across Caribbean (most direct route). As most wintering is east of Andes (American Ornithologists' Union 1983), over-water (Pacific) route between southern Mexico and Ecuador (hypothesized by Rappole *et al.* 1979) seems unlikely.

Spring migration begins March (Schauensee and Phelps 1978), and a few birds reach Texas by early April (Oberholser 1974), but most move later: in Veracruz, over 2 years, 94% of passage in first 3 weeks of May (Ramos 1988). Migration mainly through Mexico, though some cross Gulf further east (Stevenson 1957; Peterson and Chalif 1973), and, as in autumn, few occur on east coast of USA south of Virginia. Main movement northward is rapid, peaking mid-May in mid-latitude states (Stewart and Robbins 1958; Mengel 1965), reaching main breeding range mid- to late May (Bailey 1955; Tufts 1986; Janssen 1987), and northern range limits by end of May or first week of June (Houston and Street 1959; Erskine 1985).

Vagrant to north Alaskan coast and Greenland (American Ornithologists' Union 1983). For vagrancy to west Palearctic, see Distribrution.   AJE

Voice. See Field Characters.

Plumages. ADULT MALE BREEDING. Narrow black forehead; crown black, feathers fringed dark grey, fringes wider towards rear; cap thus uniform black in front, heavily spotted black on mid-crown, and marked with smaller spots at rear. Remainder of upperparts, upper side of head from behind eye, and upper side of neck uniform dark bluish-grey. Short bright sulphur-yellow streak from nostril to just above front corner of eye, contrasting sharply with black forehead, with dark grey front of lore, and with black rear of lore. Broad and conspicuous white or yellow-white eye-ring, sometimes deeper yellow above eye. Feathering just in front of and below eye-ring black, extending as black stripe over upper cheek and lower side of neck to side of chest. Flank greenish-olive, under tail-coverts white (faintly yellow

when fresh), remainder of underparts from chin to vent bright sulphur-yellow, boldly marked with black rounded or triangular spots on chest; black-spotted chest forms discrete band 10-18 mm wide, becoming solidly black on side of chest; some small dusky marks on side of throat, lower throat, or side of breast. Tail greyish-black, fringes along outer webs of feathers dark blue-grey (faint or absent on outermost pair, t6). Flight-feathers, tertials, greater upper primary coverts, and bastard wing greyish-black; fringes along outer webs of primaries medium blue-grey, those along secondaries and shorter feathers of bastard wing similar but slightly darker and wider, less sharply defined, those along primary coverts dark blue-grey, narrow; outer webs of tertials dark blue-grey. Lesser, median, and greater upper wing-coverts dark blue-grey, centres of lesser and median and inner webs of greater coverts greyish-black. Under wing-coverts and axillaries white, yellowish-white, or pale grey, darker grey towards bases; axillaries and marginal coverts tipped yellow. ADULT FEMALE BREEDING. Entire upperparts dark bluish-grey, feather-centres on forehead and forecrown with narrow black arrow-marks rather variable in width, sometimes almost coalescing to form narrow stripe along side of forehead; feather-fringes from crown to mantle faintly green when plumage fresh. Yellow stripe above lore and broad eye-ring distinct, as in adult ♂, but eye-ring often more yellowish. Feathering just in front of and below eye-ring dark grey or sooty, less extensive and less black than adult ♂ breeding; side of head behind eye and side of neck bluish-grey, often with faint darker grey or sooty stripe along lower edge. Underparts as adult ♂ breeding, but side of chest with only a few large black triangular spots, and upper chest with single row of smaller black spots or none at all; remainder of chest with indistinct olive triangles, gradually more washed out towards belly. Wing and tail as adult ♂ breeding, but fringes of flight-feathers sometimes tinged green. ADULT MALE NON-BREEDING. Upperparts like adult ♀ breeding, with narrow black spots on forehead and forecrown. Side of head and neck and underparts similar to adult ♂ breeding, but black stripe from in front of and below eye-ring to lower side of neck narrower, sometimes replaced by dark grey in front of eye; black spots on chest smaller (*c.* 1·5-2 mm wide), more triangular, forming discrete band 1 cm wide (not as narrow as in adult ♀ breeding, and black spots not partly replaced by olive). ADULT FEMALE NON-BREEDING. Like adult ♀ breeding, but upperparts dark blue-grey, tinged green on forehead and slightly so from crown to back; feather-centres on forehead black, but concealed. Feathering in front of and below eye-ring dark grey, similar to or only slightly darker than side of head behind eye and side of neck, no dark stripe over upper cheek; lore grey with yellow-green suffusion, less sharply demarcated from yellow line above lore and pale yellow or yellow-white eye-ring. Spots on mid-chest olive, small (1-2 mm), and restricted in extent, grading to dull black at side of chest. JUVENILE. Upperparts drab-brown; side of head and neck and chin to chest pale buff-brown, eye-ring pale buff or off-white, remainder of underparts pale yellow, under tail-coverts white. Upper wing-coverts drab-brown, median and greater tipped buff, forming 2 distinct wing-bars. Tail as adult, but feather-tips pointed, less broadly truncate. Flight-feathers, tertials, and greater upper primary coverts as adult, but ground-colour more brown once worn, less blackish; blue-grey of fringes tinged olive-brown, especially on secondaries, tertials, and primary coverts. FIRST ADULT MALE NON-BREEDING AND BREEDING. Like adult ♂ non-breeding and breeding, but juvenile tail, flight-feathers, tertials, and primary coverts retained, slightly browner and more worn than adult at same time of year; tail-feathers more pointed; primary coverts, tertials, and secondaries slightly browner and more olive-fringed, contrasting with new 1st adult bluish-grey greater coverts. Crown and mantle slightly tinged green. 1st adult non-breeding has concealed black spots on forehead, as in adult ♀ non-breeding, and pattern on chest and side of head and neck also similar to adult ♀ non-breeding, though restricted amount of small spots on mid-chest often blacker; differs from adult ♀ non-breeding in pointed tail-feathers and less uniform wing. In 1st adult breeding, similar to adult ♂ breeding, but black feather-centres on forecrown sometimes more restricted, black stripe on upper cheek and lower side of neck on average narrower, and black-spotted band on chest narrower, c. 6–11 mm wide, often less sharply defined at rear, where black spots often grade into smaller dark or pale olive spots; tail and larger feathers of wing juvenile, blue-grey tinge on tertials and primary coverts lost due to abrasion. FIRST ADULT FEMALE NON-BREEDING AND BREEDING. Like adult ♀ non-breeding and breeding, but part of juvenile feathering retained, as in 1st adult ♂. In 1st adult ♀ non-breeding, forehead and forecrown olive-green, feathers without concealed black, remainder of upperparts blue-grey with slight green tinge; lore and feathering in front of and below eye-ring olive-grey, not sharply demarcated from yellow stripe above lore or from yellow lower cheek; triangular spots on upper chest and side of chest small and olive (not black), becoming gradually smaller and paler olive downwards. In 1st adult breeding, similar to adult ♀ breeding, but upperparts tinged green, less or no black on feather-centres of forecrown, face pattern less distinct, and black on chest mainly restricted to some spots on side; tail, tertials, and primary coverts still juvenile, more heavily worn than in adult ♀, tail-feathers more pointed (difficult to see when heavily worn), tertials and primary coverts without blue-grey. For variation in colour of forehead, crown, cheek, side of neck, and streaks on chest in both sexes at all ages, see Rappole (1983).

Bare parts. ADULT, FIRST ADULT. Iris hazel to deep brown. Upper mandible plumbeous-grey, brown, greyish- or brownish-black, or black, cutting edges and tip of lower mandible bluish-grey or horn-grey, remainder of lower mandible dull pink-white, greyish-flesh, brownish-flesh, or yellow. Leg and foot pink-flesh, yellow-flesh, buffish-pink, tan-yellow, or pale greyish-yellow. (RMNH, ZFMK, ZMA.) JUVENILE. No information.

Moults. ADULT POST-BREEDING. Complete; primaries descendent. On breeding grounds, June–August (Pyle *et al.* 1987). Moult not started in 2 ♀♀ from 10–19 July, Maine (ZFMK). ADULT PRE-BREEDING. Partial; extent variable, in some birds, all head, body, and lesser and median upper wing-coverts, in others head, neck, and underparts, or head, neck, and chin to chest. In winter quarters, February–April; some birds have moult advanced early February, others not started by mid-March. (Pyle *et al.* 1987;

RMNH, ZFMK, ZMA.) POST-JUVENILE. Partial: head, body, and lesser, median, and greater upper wing-coverts; no tertials or tail. In breeding area, June–August. FIRST PRE-BREEDING. As adult pre-breeding. (Pyle *et al.* 1987; ZMA, RMNH.)

Measurements. ADULT, FIRST ADULT. North-east USA, May–September, and Central America to Ecuador, September–April; skins (RMNH, ZFMK, ZMA). Bill (S) to skull, bill (N) to distal corner of nostril; exposed culmen on average 3·4 less than bill (S).

WING	♂	66·4 (1·62; 20)	63–69	♀ 63·6 (1·75; 9)	60–66
TAIL		53·6 (1·53; 16)	51–56	51·3 (1·82; 9)	49–54
BILL (S)		13·2 (0·59; 20)	12·3–14·3	13·2 (1·16; 9)	12·9–13·4
BILL (N)		7·3 (0·46; 16)	6·6–8·1	7·2 (0·28; 9)	6·8–7·6
TARSUS		19·2 (0·46; 16)	18·4–19·8	18·7 (0·58; 9)	17·8–19·7

Sex differences significant for wing, tail, and tarsus. Ages combined above, though retained juvenile wing and tail of 1st adult on average shorter than in older birds; thus wing of adult ♂ 67·0 (1·57; 8) 65–69, 1st adult ♂ 65·6 (1·49; 10) 63–68; adult ♀ 65·1 (0·48; 4) 64–66, 1st adult ♀ 62·4 (1·29; 5) 60–64.

Weights. Pennsylvania (USA): August–September, adult ♂ 10·8 (0·75; 20) 9·5–11·9, adult ♀ 10·4 (0·64; 31) 9·1–11·6, 1st adult 10·2 (0·83; 199) 8·4–14·7; May, ♂ 10·6 (0·80; 289) 8·7–13·5, ♀ 10·2 (0·73; 266) 8·1–12·6 (Clench and Leberman 1978, which see for details). Pelee Island (Lake Erie, Canada), May: ♂ 10·4 (20) 8·6–11·6, ♀ 10·7 (12) 9·6–12·2 (Woodford and Lovesy 1958). Maine (USA), mid-July: ♀♀ 9, 10 (ZFMK). North Carolina, mid-June: ♂ 11·4 (ZFMK). Kentucky (USA): May, ♂ 11·4; September, ♀ 12·9 (Mengel 1965). Oaxaca (Mexico), May: ♂ 10·1 (Binford 1989). Belize, April: ♂ 9·8 (Russell 1964). Panama, autumn: 6·9 (lean) (Rogers and Odum 1966). Bonaire (Netherlands Antilles), September: ♂ 6·6 (lean) (ZMA).

Structure. Wing rather short, broad at base, tip fairly rounded. 10 primaries: p7–p8 longest, p9 1–4 shorter, p6 0·5–2, p5 4–7, p4 7–9, p3 8–12, p2 9–14, p1 11–16. P10 strongly reduced, a tiny pin concealed below reduced outermost greater upper primary covert, which is c. 0·5 mm longer; 42–47 shorter than p7–p8, 4–7 shorter than longest upper primary covert. Outer web of p6–p8 emarginated, inner web of p7–p9 with notch. Tip of longest tertial reaches tip of about p1. Tail of average length, square or slightly rounded; 12 feathers, t6 3–5 shorter than t1. Bill straight, rather short, visible part of culmen about half of head length; flattened dorso-ventrally, gradually tapering to pointed tip when seen from above; tip of culmen decurved, forming small hook. 3 strong bristles projecting forward from each side of base of upper mandible. Tarsus rather long, toes of average length, slender. Middle toe with claw 14·2 (10) 13·5–15; outer toe with claw c. 74% of middle with claw, inner c. 67%, hind c. 77%.   CSR

Family THRAUPIDAE tanagers

Small to medium-sized 9-primaried oscine passerines (suborder Passeres). Some chat-like ground-feeders or specialized flycatchers but most species highly arboreal and frugivorous, diet of some also including (to greater or lesser extent) seeds, nectar, and insects; some tropical species follow ant columns. Except as vagrants, occur only in New World (from Canada to Argentina), mainly in tropics; northerly species migratory. About 261 species in c. 63 genera, of which 2 species of North American genus *Piranga* accidental in west Palearctic.

Sexes usually almost the same size. Bill variable (Storer 1969), from stout seed-eater to insect-eater type, but often of short to medium length, rather conical in shape, and somewhat hooked; upper mandible noticeably decurved with notch near tip; notch well-developed in some species but small or vestigial in most. Rictal bristles quite well developed. Wing variable, longish in some migratory species but quite short in most others; 9 primaries (p10 minute and concealed). Tail usually short or of medium length; shape variable but typically square-tipped or rounded. Leg usually quite short. Head-scratching by indirect method only, so far as known (Simmons 1957b, 1961a for *Cissopis*, *Ramphocelus*, *Thraupis*, and *Cyanerpes*). No detailed information on bathing or sunning (but see Potter and Hauser 1974 for Summer Tanager *P. rubra*). Direct (active) anting reported in at least 15 species (5 in the wild) of genera *Cissopis* (1 species), *Nesospingus* (1), *Habia* (1), *Piranga* (2, including both species accidental in west Paleartic), *Ramphocelus* (1), *Stephanophorus* (1), *Tangara* (7), and *Dacnis* (1) (Poulsen 1956; Whitaker 1957; Simmons 1961b, 1963, 1966; Potter 1970; King and Kepler 1970; K E L Simmons).

Plumages of many species vividly and boldly coloured in almost rainbow-like manner; feathers sometimes with metallic sheen or opalescence. Sexes alike or ♂ brighter than ♀; usually no marked seasonal differences.

DNA evidence indicates tanagers to be closely related to New World wood-warblers (Parulidae) and buntings (Emberizidae); see Parulidae, Fringillidae, and Emberizidae for further discussion.

Piranga rubra Summer Tanager

PLATE 9
[between pages 64 and 65]

Du. Zomertangare Fr. Tangara vermillon Ge. Sommertangare
Ru. Алая пиранга Sp. Candelo unicolor Sw. Sommartangara

Fringilla rubra Linnaeus, 1758. Synonym: *P. aestiva*.

Polytypic. Nominate *rubra* (Linnaeus, 1758), south-east USA west to Oklahoma and central Texas; *cooperi* Ridgway, 1869, south-west USA and neighbouring northern Mexico, east to western Texas and Coahuila.

Field characters. 16–17 cm; wing-span 27–30 cm. Close in size to Corn Bunting *Miliaria calandra* but with form also recalling oriole *Oriolus*, particularly in long, pointed but swollen-looking bill; averages slightly longer-billed and longer-tailed than Scarlet Tanager *P. olivacea*. Quite large, tree-haunting passerine, with long, heavy bill, bulky, peaked head, plump oval body, and lengthy wings but relatively rather short tail. Adult ♂ red except for browner wings and tail, recalling ♂ crossbill *Loxia*; adult ♀ yellowish-olive above, strongly yellow below, like brightest ♂ Greenfinch *Carduelis chloris*. Juvenile and 1st winter much as ♀ but young ♂ in 1st summer partially red. Often rather inactive. Sexes dissimilar; no seasonal variation. Immature similar to ♀.

ADULT MALE. Moults: November–March (mainly head and fore-body); mid-July to August (complete). Looks entirely dull red at distance, but at close range (particularly in fresh plumage) shows brighter, rosier tones on head, rump, and forebody, duskier wash on mantle and back, and dusky-brown inner webs to larger wing- and tail-feathers, with those of tertials and overlapping tips to longest primaries appearing dusky-black when fresh. Under wing-coverts dark red. Bill noticeably deep at base and distinctly longer than distance from frontal feathers of crown to eye; both mandibles evenly curved along distal third to form noticeable point but lacking obvious tooth; cold creamy- or dusky white to (rarely) horn-grey. Legs grey. ADULT FEMALE. Upperparts basically olive, washed bright yellow and green. Face shows faint pale yellow eye-ring and pale yellow streaks from base of bill to eye and below olive-washed ear-coverts. Wing- and tail-feathers bright olive-green, with paler yellowish-olive fringes and tips most obvious on larger wing-feathers, particularly in contrast to dark tips of primaries and inner webs of tertials. Underparts dull yellow, with orange tone from throat to rear flanks visible at close range. Under wing-coverts pale golden-yellow. FIRST-YEAR. Resembles adult ♀ but retains juvenile tail and flight-feathers. From spring, ♂ shows increasing invasion of red feathers; some ♀♀ lack orange tone on underparts.

Tanagers are essentially insectivorous and lack bustling behaviour of finches (Fringillidae). *Piranga*, and particularly *P. rubra*, best distinguished by length and shape of pale bill and by sluggish behaviour. To experienced observer, tanager form unmistakable and identification of ♂♂ of the 2 west Nearctic species in full breeding plumage is simple, as *P. rubra* lacks solidly black wings and tail of *P. olivacea*. ♀♀ and immatures far less easily distinguished, with larger green and yellow warblers (e.g. Icterine Warbler *Hippolais icterina*) presenting additional confusion species. Best separated from *P. olivacea* on (1) larger, paler, untoothed bill, (2) more marked face, (3) more intense body colours, especially below, (4) yellow under wing-coverts (white in *P. olivacea*), (5) lack of distinct contrast between body colours and those of wings and tail, and (6) sometimes multisyllabic call (see below). Flight action slower than finch, producing quite powerful but not flowing progress. Carriage usually rather level and front-heavy, with heavy bill and head contributing to somewhat neckless attitude, but may also perch upright. Hops.

Calls include disyllabic 'pi-tuck' and distinctive longer phrase 'pik-i-tuck-i-tuck' or 'chicky-tuck-tuck', with rapid, staccato utterance and descending pitch.

Habitat. Breeds in warm temperate Nearctic lowlands, especially in tall open woods with scrubby oak *Quercus* undergrowth, such as drier pine *Pinus* and hickory *Carya*, feeding and singing in treetops, but also tolerating fairly young second growth (Pough 1949; Forbush and May 1955). In Great Plains, breeds mainly in upland forest (hardwood, mixed, or coniferous), favouring lower and more open stands than those used by Scarlet Tanager *P. olivacea* (Johnsgard 1979).

Winters in tropical America in both woody and open situations including coastal mangroves, second growth, low open forest (including edges and clearings), coffee plantations, and scrubby grassland as high as 3000 m (Schauensee and Phelps 1978).

Distribution. Breeds in southern and south-central USA from California east to southern New Jersey, south to north-central Mexico. Winters from south-central Mexico south through Central America to Ecuador, northern Bolivia, and Amazonian Brazil.

Accidental. Britain: 1st-winter ♂, Bardsey Island (Wales), 11-25 September 1957 (Dymond *et al.* 1989).

Movements. Short-range migrant. Breeds entirely south of *c.* 40°N; larger western race *cooperi* winters south only to central Mexico, but nominate *rubra* south to Peru and Ecuador (Bent 1958; Rappole *et al.* 1983).

Southward movement begins late July and early August in extreme southern states (Bent 1958; Phillips *et al.* 1964; Oberholser 1974). Main migration late August into October, with most birds moving in September in northern and western areas (Stewart and Robbins 1958; Phillips *et al.* 1964; Mengel 1965), and in October further south and east (Bent 1958; Imhof 1976; Potter *et al.* 1980). Arrivals in Central American winter range are chiefly late September or early October, but they extend through November further south (Bent 1958; Schauensee and Phelps 1978; Pearson 1980; Binford 1989). Eastern birds cross Gulf of Mexico to Central America, whereas western birds move overland through Mexico (Stevenson 1957; Bent 1958). Winter site-fidelity shown by ringing in Panama and Honduras, with returns up to 2 years later (Loftin 1977).

Spring migration begins March (Bent 1958), with most birds leaving winter quarters by late April (e.g. Monroe 1968, ffrench 1976, Hilty and Brown 1986). Most birds reach south-east of breeding range late March or April; further north and west, arrivals April-May (Stewart and Robbins 1958; Ligon 1961; Phillips *et al.* 1964; Imhof 1976; Stoddard 1978). Spring route reverse of autumn (Stevenson 1957; Bent 1958).

Tends to overshoot eastern range in April, and especially in May, when recorded almost annually from New York to Nova Scotia (Bent 1958; Bull 1974; Tufts 1986); autumn vagrancy there is less frequent, but extends further, to St Pierre-et-Miquelon (off Newfoundland) (Etcheberry 1982). Also recorded north-west to Saskatchewan and Oregon (American Ornithologists' Union 1983). For vagrancy to west Palearctic, see Distribrution.

AJE

Voice. See Field Characters.

Plumages. (nominate *rubra*). ADULT MALE. Upperparts and side of neck dark ruby-red or carmine-red, darkest from mantle to back; side of head and all underparts slightly paler ruby-red or dark pink-red. Tail dark pink-red, slightly tinged grey on inner webs, brighter ruby-red on fringes. Flight-feathers blackish-grey, outer webs broadly fringed dark ruby-red, grey largely concealed when wing closed, except for tips of primaries; remainder of upperwing dark ruby-red, dark grey of inner webs and bases of greater primary and secondary coverts and of tertials largely concealed. Under wing-coverts and axillaries ruby-red or dark pink-red, longer coverts and inner fringes of flight-feathers paler, more pinkish. Breeding and non-breeding plumage similar. Bleaching and wear have limited effect, birds in fresh plumage slightly more pinkish on upperparts, worn birds more crimson. ADULT FEMALE. Upperparts and side of head and neck from eye backwards bright yellowish-olive-green or golden-green, usually brighter and more golden-yellow on forehead and on rump and upper tail-coverts. Short streak from nostril to above eye and narrow eye-ring golden-yellow, lore mottled grey and yellow, upper cheek and short ear-coverts olive-green with pale yellow shaft-streaks. Entire underparts bright golden- or orange-yellow, deeper and purer on throat, mid-belly, and under tail-coverts, slightly washed olive or buffish on side of breast, chest, flank, and thigh. Tail yellowish-green, outer webs brighter golden-green to buffish-yellow, inner webs slightly tinged grey, shafts black. Upperwing yellowish-olive-green or golden-green, often brighter greenish-yellow or golden-yellow on flight-feathers and along wing-bend; distribution of dark grey or blackish-grey as in adult ♂, thus largely concealed when wing closed, except for tips of primaries and inner webs of tertials. Under wing-coverts and axillaries golden-yellow; longest coverts and fringes of inner webs of flight-feathers pale yellow. Bleaching and wear have rather limited effect, green of upperparts, upper wing-coverts, and of fringes of flight-feathers becoming slightly more greyish-olive when worn, but marked individual variation in colour, many birds having pink, orange, or golden-buff wash, especially on cap, rump, cheek, throat, chest, and under tail-coverts, not contrasting sharply with remainder of golden-yellow of plumage. JUVENILE. Upperparts dull olive-green, tinged buff on cap, rump, and upper tail-coverts, marked all over with dark brown-grey streaks. Side of head dull olive-green with buff mottling, ear-coverts uniform olive, side of neck buff with dark grey streaks. Underparts off-white, tinged buff-yellow on side of breast, chest, and flank, orange-buff on under tail-coverts; marked all over with brown-grey streaks, narrowest on throat, mid-belly, and vent. Upper wing-coverts dull olive-grey, median with light yellow-buff fringes, greater with broad light yellow-buff to off-white tips, forming wing-bars. Remainder of wing and tail as adult ♀, fringes yellow-green in ♀, more buffish-orange in ♂. FIRST ADULT MALE. Like adult ♀, but tail, flight-feathers, greater upper primary coverts, and tertials still juvenile: tail-feathers rather narrow, tips distinctly pointed (in adult, broader, tips broadly rounded); tertial-tips rounded, fringed pale yellow or off-white, especially on tip of outer web (in adult, tips broad, truncate, pale fringe narrow and even or almost absent). In 1st autumn, colour of head and body as adult ♀, mainly golden-green above, golden-yellow below, with variable buff-orange or rusty-pink wash, especially on head, rump, and on upper or under tail-coverts. In spring, variable number of scattered contrastingly pure ruby-red feathers appear, ♂ in late spring and early summer appearing red with yellow patches or yellow with red patches (adult ♀ yellow with variable pink or orange wash; adult ♀ also has flight-feathers blacker on inner webs, less brown and worn, and tips of tail-feathers and tertials broadly rounded or truncate, scarcely abraded); some or all tail-feathers new in some ♂♂ from spring, pink-yellow to pink-red. FIRST ADULT FEMALE. Rather like adult ♀, but tail, flight-feathers, greater upper primary coverts, and tertials still juvenile, tips shaped and patterned as in 1st adult ♂; outer webs of tail- and flight-feathers greenish-yellow, less golden-buff or orange-buff than most adult ♀♀ or 1st adult ♂♂; outer webs of tertials greyer. In autumn, upperparts uniform yellowish-olive-green, slightly brighter yellow on cap and upper tail-coverts, but not as orange, pink, or buff as some adult ♀♀ or 1st adult ♂♂; underparts bright sulphur-yellow, washed olive-green on side of breast, chest, and flank, which sometimes show faint olive shaft-streaks; under tail-coverts bright yellow, without buff or orange tinge. In spring, some as in autumn, others with variable pink, orange, or golden-buff wash, like adult ♀, but flight-feathers and usually some or all tail-feathers and tertials still juvenile. Some 1st adult ♂♂ inseparable from ♀ in colour, but see Measurements.

Bare parts. ADULT, FIRST ADULT. Iris brown or dark brown. Bill pale pink-horn, yellowish-horn, or dusky horn-grey with paler cutting edges; usually pale, but occasionally dark grey and then closely similar to dark bill of Hepatic Tanager *P. flava*. Bill darker in breeding adult (both sexes), paler in immatures and in adult ♀ non-breeding, intermediate in adult ♂ non-breeding. (RMNH, ZMA.) Gape-flanges in 1st autumn ♂ yellowish (Arthur 1963). Leg and foot dull dark grey, slate-grey, bluish-slate, or purplish-grey, darkest on scutes on front of tarsus and on upper surface of toes, paler and with flesh tinge on soles (RMNH, ZFMK, ZMA). JUVENILE. Bill pinkish-tan (Pyle *et al.* 1987).

Moults. ADULT POST-BREEDING. Complete; primaries descendent. On breeding grounds, mainly August (Pyle *et al.* 1987), but sometimes from mid-July, and a single ♂ in moult on 9 September, Kentucky (Mengel 1965; Parkes 1967). ADULT PRE-BREEDING. Partial, mainly restricted to head and front part of body. In winter quarters, November–March. (Parkes 1967; Pyle *et al.* 1987.) POST-JUVENILE. Partial: head, body, and lesser, median, and greater upper wing-coverts. Mainly in breeding area, July–August, starting shortly after fledging, but scattered feathers still growing in some migrant birds examined late August to mid-October (RMNH, ZFMK, ZMA). FIRST PRE-BREEDING. Partial; as in adult pre-breeding, but often greater in extent, though strongly variable; apart from much or virtually all feath-

ering of head and body, sometimes includes variable number of tail-feathers and tertials (Parkes 1967; BMNH, RMNH, ZMA).

Measurements. Nominate *rubra*. ADULT, FIRST ADULT. Some from south-east USA, summer, but most from Belize, Costa Rica, El Salvador, Colombia, and Venezuela, on migration and in winter; skins (RMNH, ZFMK, ZMA). Bill (S) to skull, bill (N) to distal corner of nostril; exposed culmen on average 3·7 less than bill (S).

	♂		♀	
WING	96·3 (1·90; 13)	93–99	92·7 (1·80; 10)	89–95
TAIL	69·8 (1·89; 13)	65–72	67·4 (2·58; 9)	63–71
BILL (S)	21·7 (0·86; 13)	20·5–23·0	21·6 (0·76; 10)	20·9–22·7
BILL (N)	13·4 (0·56; 13)	12·5–14·3	13·3 (0·71; 10)	12·5–14·5
TARSUS	19·3 (0·64; 13)	18·4–20·2	19·7 (0·81; 10)	18·7–20·6

Sex differences significant for wing and tail. Ages combined above, though retained juvenile wing and tail of 1st adult shorter than in older birds. Thus, in ♂: wing, adult 97·2 (1·07; 8) 96–99, 1st adult 94·9 (2·19; 5) 93–98; tail, adult 70·3 (1·25; 8) 66–72, 1st adult 69·1 (2·61; 5) 65–72. See also Rea (1970).

Weights. Nominate *rubra*. ADULT, FIRST ADULT. Eastern USA, April–September: ♂ 28·9 (2·03; 7) 25·5–31·2, ♀♀ 26·9, 29·1 (Mengel 1965; Clench and Leberman 1978; see also Norris and Johnston 1958, and Murray and Jehl 1964). Louisiana (USA), spring: 28·2 (3·18; 30) (Rogers and Odum 1966). East Ship Island (Mississippi, USA), spring: 26·4 (2·37; 223) (Kuenzi *et al.* 1991, which see for weight changes during stay).

Belize: late August to early November, 28·2 (2·1; 8) (Mills and Rogers 1990); October–November, ♂♂ 28·6, 29·0, ♀♀ 29·9, 30·1; March–April, ♂ 29·3 (3) 27·6–30·9 (Russell 1964; Roselaar 1976; ZMA). Panama: October 27·3 (3·00; 22) 22·6–34·5 (Rogers and Odum 1966); March–April, in mountains, 29·3 (2·7; 10), in lowlands 27·9 (1·2; 4) (Leck 1975). On Bardsey (Wales), single ♂ 24·6 on 11 September, 36·7 on 20 September (Arthur 1963).

For fat-free weights, see Connell *et al.* (1960), Odum *et al.* (1961), and Rogers and Odum (1966).

Structure. 10 primaries: p8 longest, p9 2·0 (23) 1–3 shorter, p7 0–2 shorter, p6 2–5, p5 8–14, p4 14–19, p3 17–22, p2 19–25, p1 24·7 (23) 21–29; p9 1·2 (23) 0–4 longer than p6, once 0·5 shorter (nominate *rubra*; in *P. flava*, 0·5–3 shorter; see also Eisenman 1969). P10 strongly reduced, 63–70 shorter than p8, 9–12 shorter than longest upper primary covert. Outer web of p6–p8 emarginated, inner of p7–p9 with slight notch. Tip of longest tertial reaches tip of p2–p4. Bill strong, length of visible culmen *c.* 68% of head length; depth at base 9–10 mm, width at base 9·6–10·8 mm. Cutting edge of upper mandible straight or with slight undulation; no distinct tooth in middle. Middle toe with claw 18·1 (5) 17·5–18·8; outer and hind toe with claw both *c.* 72% of middle with claw, inner toe *c.* 67%. Remainder of structure as in Scarlet Tanager *P. olivacea*.

Geographical variation. Fairly marked, involving colour, size, and structure. *P. r. cooperi* from south-west of North America on average larger than nominate *rubra* from south-east USA, wing less pointed, bill relatively heavier, body paler in colour; average wing *c.* 6 mm longer, tail *c.* 10 mm, bill (S) *c.* 2 mm, bill (N) *c.* 1·7 mm, tarsus *c.* 2 mm; upperparts of adult ♂ dull vermilion (clearer on cap, rump, and upper tail-coverts), underparts clear light vermilion, both paler than in nominate *rubra*; upperparts of adult ♀ pale olive-grey, tinged olive-yellow on mantle and scapulars, underparts pale chrome-yellow, less yellowish-olive-green above than ♀ nominate *rubra*, purer pale yellow below (Ridgway 1869, 1901–11; Eisenman 1969; Rea 1970). For further variation, see Phillips *et al.* (1964) and Phillips (1966).

Forms species-group with Hepatic Tanager *P. flava*, occurring from south-west USA to South America, and probably represents offshoot of this species, though breeding ranges now overlap slightly (Mayr and Short 1970). *P. flava* itself may comprise 3 separate species (Schauensee 1966; Ridgely and Tudor 1989).   CSR

Piranga olivacea Scarlet Tanager

PLATE 9
[between pages 64 and 65]

Du. Zwartvleugeltangare   Fr. Tangara écarlate   Ge. Scharlachtangare
Ru. Красно-черная пиранга   Sp. Candela escarlata   Sw. Rödtangara

Tanagra olivacea Gmelin, 1789. Synonym: *P. erythromelas*.

Monotypic

Field characters. 15·5–16 cm; wing-span 27–30·5 cm. Slightly smaller than Summer Tanager *P. rubra*, with proportionately slighter and 10–15% shorter bill and 5–10% shorter tail. In adult ♂, bright red (breeding) or bright green-yellow (non-breeding) head and body contrast with wholly black wings and tail; adult ♀ and immature have greenish-olive upperparts merging with pale yellow underparts but mainly black or dusky wings and tail stand out; bright white under wing-coverts in all plumages. Sexes dissimilar; marked seasonal variation in ♂. Immature similar to ♀.

ADULT MALE BREEDING. Moults: January–April (mainly head, body, and wing-coverts). Head and body bright scarlet, contrasting vividly with black outer scapulars, wings, and tail. Under wing-coverts and bases of flight-feathers white, contrasting with body. At close range, face shows dark lore. Bill prominent, pointed, with quite deep base; upper mandible with noticeable tooth in middle; light grey. Legs dusky. ADULT MALE NON-BREEDING. Moults: July–September (complete). Unlike *P. rubra*, red of breeding plumage entirely lost, but still differs from ♀ (see below) by more uniform, greener or yellower head and body, contrasting more sharply with black wings and tail which lack all but faint olive fringes and tips. Bill less grey than in summer, mainly horn but variable (see Plumages). ADULT FEMALE. Head and upperparts mainly pale olive-green, with yellow wash most obvious above tail. Face shows mottled lore and narrow yellow eye-ring. Wings and tail basically greyish-black with larger feathers tipped and fringed olive-green, most noticeably on inner greater coverts and secondaries, and leading coverts concolorous with back. Contrast of wings and tail with body thus far more marked than in *P. rubra* but not as sharply delineated as in ♂. Underparts dull greenish-yellow, bright only on throat and rear underbody. Bill grey or dusky-horn. FIRST-AUTUMN MALE. Closely resembles adult ♀ but already shows partly to almost wholly black outer scapulars and wing-coverts, which stand out as dark patch contrasting with fading juvenile flight-feathers. If juvenile coverts retained, may show 1–2 yellowish wing-bars. FIRST-SUMMER MALE. After moult during which it appears increasingly mottled, resembles breeding adult but red plumage toned orange (rarely yellow), while black of wings and tail less uniform, retaining in particular contrast of dark wing-covert patch. Tertials pale-tipped; coverts mixed with pale feathers which rarely align to form 1 wing-bar. FIRST-YEAR FEMALE. Duller than adult, more extensively tinged olive above and below. May retain all or some juvenile wing- and tail-feathers and show 1–2 wing-bars as 1st-winter ♂ (see Plumages).

Distinction from *P. rubra* easy in ♂♂ but less so in adult and immature ♀, particularly those with dullest wings and tail. Thus important to remember that *P. olivacea* has (1) noticeably shorter, distinctly toothed and darker bill, (2) white (not yellow) under wing-coverts, (3) duller plumage overall, always lacking orange on underparts, and (4) always disyllabic call (see below). Flight, behaviour, and gait as *P. rubra*.

Call a low, toneless, hoarse or rasping disyllable, 'chip-burr', 'chip-kurr', or 'keep-back'.

Habitat. Breeds in temperate and warm temperate Nearctic in mature woodlands and groups of tall shade trees, even in suburbs (Pough 1949). Prefers white oak *Quercus alba* woods, especially in well-watered country, but will also occupy mixed woods, coppice, and orchards (Forbush and May 1955). In Canada, while preferring mature deciduous woods, is often found also in mixed and pine *Pinus* woods (Godfrey 1979).

Distribution. Breeds in North America from North Dakota and southern Manitoba east to Maine, south to western Kansas, central Arkansas, northern Georgia, central Virginia, and Maryland. Winters mainly in north-west South America, south to north-west Bolivia, rarely in Panama.

Accidental. Iceland, Britain, Ireland.

Movements. Long-distance migrant, breeding in temperate eastern North America and wintering within *c*.

10° of equator. Inconspicuous on migration, remaining in tree-tops except when driven to ground by scarcity of insects in cold weather (Bent 1958; Quilliam 1973; Bull 1974). Very inconspicuous in winter (Ridgely and Tudor 1989).

Start of autumn migration not obvious (Mengel 1965); evidently begins by late August, when arrivals first occur south of breeding range. Peak movement in northern areas ranges from mid-August to late September, and into early October in mid-latitude states (Stewart and Robbins 1958; Sprague and Weir 1984; James and Neal 1986; Janssen 1987), and birds leave south-east states by mid- or late October (Oberholser 1974; Kale and Maehr 1990). Most reach Central and South America early October to mid-November, but movement continues into December in Ecuador (Bent 1958; Pearson 1980; Ridgely 1981; Hilty and Brown 1986). Probably most birds cross central Gulf of Mexico (as in spring: Stevenson 1957); rare or uncommon on passage in Mexico, Central America, and West Indies (Peterson and Chalif 1973; Bond 1985; Stiles and Skutch 1989). Also uncommon in Colombia (Hilty and Brown 1986), so location of landfall south of Mexican Gulf not known. Non-stop flight (nearly 3000 km) to inland range in South America seems unlikely, though not impossible; more plausibly, most transients are missed in tree-tops, where they are poorly detected by sight or mist-netting.

Spring migration begins March, with departures from Ecuador and arrivals in Costa Rica, Mississippi, and southern Florida; early April records spread from Texas to Bahamas, and north to Illinois and Pennsylvania (Bent 1958). Passage in Central America continues through April, but is apparently sparse (Monroe 1968; Land 1970; Ridgely 1981). Peak passage late April in southern USA (Stevenson 1957; Oberholser 1974), and birds appear irregularly in mid-April on coasts from New York to Nova Scotia (e.g. Bull 1974, Tufts 1986), such records being associated with strong south-west/north-east airstreams. Northern breeding areas occupied early May, and main migration passes through middle and northern states in this period (Bailey 1955; Fawks and Petersen 1961; Speirs 1985; Janssen 1987).

Few birds overshoot to north; one record at James Bay in northern Ontario (Godfrey 1986), and a completely lost bird in June at Barrow (Alaska), 71°N (Gabrielson and Lincoln 1959). Occurs annually in Nova Scotia and Newfoundland in both seasons (Maunder and Montevecchi 1982; Tufts 1986), and there are scattered records across western Canada and USA (American Ornithologists' Union 1983).

Rare autumn vagrant to Atlantic seaboard of west Palearctic. In Britain and Ireland, 7 records up to 1990, especially south-west England and southern Ireland, from late September to 3rd week of October (Dymond *et al.* 1989; Rogers *et al.* 1991).   AJE, PRC

Voice. See Field Characters.

Plumages. ADULT MALE BREEDING. Head and body bright scarlet; bristles at base of upper mandible and usually some feathers just in front of eye and on lower thigh black. Entire wing and tail black; black of lesser, median, and greater upper wing-coverts, and tail deep velvety, with slight purplish gloss in some lights, black of flight-feathers, tertials, and primary coverts slightly duller, but not as grey as in 1st adult ♂ and without green fringes. Under wing-coverts, axillaries, and broad basal inner borders of flight-feathers white, marginal coverts contrastingly black. Bleaching and wear have limited effect; feathering of face sometimes slightly contaminated; some white of feather-bases may show through on underparts when plumage abraded. ADULT FEMALE BREEDING. In fresh plumage, entire upperparts and side of neck dull olive-green. Lore and front part of cheek finely mottled olive-green, pale yellow, and pale grey; narrow eye-ring pale yellow, sometimes broken in front and behind; rear part of cheek and ear-coverts uniform olive-green. Side of breast, chest, flank, and thigh yellowish-olive-green, partly suffused pale yellow on chest and lower flank, slightly tinged grey on flank; chin, throat, and belly to under tail-coverts bright light lemon-yellow. Tail greyish-black, outer webs with narrow olive-green fringe. Flight-feathers black when fresh, outer webs with narrow but sharply defined pale olive fringe; greater upper primary coverts and bastard wing greyish-black, primary coverts with faint olive outer fringe; inner web of tertials black, outer web greyish-black with rather ill-defined olive or olive-green fringe. Lesser, median, and greater upper wing-coverts olive-green, dark grey of bases of longer lesser coverts, black of bases of median ones, and black on centre and inner web of greater ones partly visible; shafts of median coverts black. Under wing-coverts, axillaries and basal inner borders of flight-feathers white; marginal coverts dark grey with olive-green fringe. *In worn plumage* (spring and early summer), tail and flight-feathers slightly greyer, less black than in fresh non-breeding plumage; fringes of flight-feathers, tertials, and upper wing-coverts olive-grey, dark grey and blackish of bases or inner webs of upper wing-coverts more exposed; upperparts slightly duller olive; pale yellow eye-ring more distinct; yellow of underparts paler, more diluted, partly tinged pale olive-grey. ADULT MALE NON-BREEDING. Easily distinguished, combining green-and-yellow head and body (as adult ♀) with uniform deep black upperwing (including flight-feathers and tertials) and tail (as adult ♂ breeding). Upperparts as adult ♀, but green often purer, less olive, sometimes tinged yellow, particularly on rump and upper tail-coverts; outer scapulars (at border of black of wing) and sometimes back variably suffused black. Side of head and neck and entire underparts as adult ♀, but lemon-yellow on average purer and brighter and side of breast, chest, and flank green-yellow, less olive; thigh partly or fully black. Tail black except for narrow white edge along inner web of outer feathers; all flight-feathers, tertials, and upper wing-coverts deep black, outer webs of tertials usually with short and narrow green fringe when plumage fresh. Underwing black and white, as in adult ♂ breeding. ADULT FEMALE NON-BREEDING. Like adult ♀ breeding, but feathering of wing and tail fresh. JUVENILE. Upperparts and side of head and neck olive-green, faintly mottled dusky. Underparts white with pale yellow tinge, brighter yellow on mid-belly and under tail-coverts; side of breast, chest, and flank broadly streaked greyish olive-brown. Upper wing-coverts dark grey with narrow olive-green fringes, median and greater with narrow pale yellow spot on tip, forming narrow broken wingbars. Flight-feathers, tertials, tail, primary coverts, and bastard wing as in fresh adult ♀ breeding, but ground-colour greyer, less blackish, fringe along tip of outer web of primaries white, tip not uniform black; tips of tail-feathers often more rounded, less pointed; tips of tertials rounded, less truncate; green fringe along outer web of tertials broader and more distinct, ending in distinctly pale yellow or white fringe 1–2 mm wide on tips (no distinct pale tip in adult ♀). Sexes similar, but separation possible as soon as 1st adult upper wing-coverts start to grow: these uniform black or black with green fringe in ♂, dark grey with green fringe in ♀. FIRST ADULT NON-BREEDING MALE. Head and body as adult ♀ breeding; tail, flight-feathers, tertials, and primary coverts still juvenile, and thus rather like adult ♀ (for differences, see Juvenile, above); differs from both adult ♀ and juvenile in mainly black upper wing-coverts (traces of green fringes usually present), showing as strongly contrasting patch; these coverts similar to those of adult ♂ non-breeding, contrasting strongly with green body (as in adult), but (unlike adult) also with mainly green flight-feathers and tertials; during autumn migration, usually has all lesser, median, and greater upper wing-coverts black, but a few birds have only a variable number of lesser and median coverts black, with remaining coverts still juvenile. FIRST ADULT BREEDING MALE. Head and body scarlet, as adult ♂ breeding, but general tone often slightly more orange-scarlet or flame-scarlet, rarely cadmium-yellow with variable orange tinge. Tail and tertials either all or partly juvenile, as 1st non-breeding ♂, or all or partly new, all-black (unlike 1st non-breeding ♂); tertials with short white or pink fringe along tip (unlike adult ♂ breeding); flight-feathers, greater upper primary coverts, and bastard wing still juvenile, greyish- or brownish-black with traces of olive or pale grey fringes, contrasting with deep black of remainder of upperwing; underwing as adult ♂ breeding. Occasionally, some red mixed into white of under wing-coverts, or some red, orange, yellow, or white mixed into black of median or greater upper wing-coverts, rarely forming wing-bar (Dwight 1900; Ridgway 1901–11; Parkes 1988b; Raymond 1988; ZMA). FIRST ADULT NON-BREEDING AND BREEDING FEMALE. Like adult ♀, but upperparts on average duller, more olive, yellow of underparts less bright, tinged olive or green. In non-breeding, juvenile tail, flight-feathers, tertials, greater upper primary coverts, bastard wing, and occasionally some outer greater upper wing-coverts retained; tail-feathers more pointed than in adult, tertials more rounded at tip and with partly white fringe (see Juvenile). In breeding, tail and tertials either all or partly new (similar to those of adult ♀, tertials contrasting in wear with inner secondaries) or all or partly juvenile, as in 1st adult non-breeding; flight-feathers, greater upper primary coverts, and bastard wing juvenile, greyer or browner and with distinctly more frayed tips than in adult ♀ at same time of year, but some birds indistinguishable, especially once nesting started.

Bare parts. ADULT, FIRST ADULT. Iris brown or dark brown. Bill of breeding ♂ and some breeding ♀♀ blue-grey or steel-grey, in others greenish-horn, dull olive-grey, or horn-grey with flesh-pink to light horn lower mandible. Leg and foot lead-grey, darkest on scutes on front of tarsus and on upper surface of toes. (RMNH, ZMA.) JUVENILE. No information.

Moults. ADULT POST-BREEDING. Complete; primaries descendent. In breeding area, July–September (Pyle *et al.* 1987). Single ♂ from early September had primary moult score 20 (inner 3 primaries new, p4–p5 growing); tail, upper wing-coverts, and tertials new or in last stage of growth, head and body mainly old with many new feathers on mantle, side of breast, and chin, and a few on crown, back, and chest (RMNH). ADULT PRE-BREEDING. Partial: head, body, lesser, median, and greater upper wing-coverts, sometimes variable number of tertials and tail-feathers.

In winter quarters, January–April. POST-JUVENILE. Partial: head, body, lesser and median upper wing-coverts (occasionally a few retained), and variable number of greater coverts; of greater coverts, some birds replace innermost (tertial coverts) only, others replace most or all. On breeding grounds, July–September; some still fully juvenile by early September, others in 1st adult non-breeding plumage from mid-August onwards. FIRST PRE-BREEDING. Partial; in winter quarters, January–May. Involves head, body, lesser and median upper wing-coverts, all greater upper wing-coverts or (probably as continuation of post-juvenile) outer ones only (these in spring then blacker and newer than older inner ones replaced earlier in post-juvenile), variable number of tail-feathers and tertials (in ♂, often all), occasionally innermost secondary (s6), and a few feathers (rarely all) of bastard wing. At arrival on breeding grounds, ♂ retains juvenile flight-feathers, greater upper primary coverts, and variable number of feathers of bastard wing only, and frequently a few innermost or outermost greater upper wing-coverts and occasionally some lesser and median coverts remain of 1st adult non-breeding.

Measurements. ADULT, FIRST ADULT. Mainly eastern USA, May–September; some from Belize, Curaçao, Bonaire, and Peru, on migration and in winter; skins (RMNH, ZFMK, ZMA). Bill (S) to skull, bill (N) to distal corner of nostril; exposed culmen on average 3·7 less than bill (S).

WING	♂ 97·2 (2·35; 29) 92–102	♀ 94·3 (1·82; 18) 91–98
TAIL	65·3 (2·27; 24) 61–68	64·0 (1·60; 16) 61–67
BILL (S)	18·9 (0·72; 24) 17·8–20·2	19·0 (0·69; 15) 18·0–20·0
BILL (N)	11·0 (0·48; 20) 10·2–12·0	11·4 (0·57; 13) 10·5–12·2
TARSUS	19·6 (0·44; 20) 18·8–20·2	19·4 (0·73; 13) 18·6–20·4

Sex differences significant for wing. Ages combined above, though juvenile wing on average shorter than that of older birds and tail perhaps longer. Thus in ♂: wing, adult 98·8 (2·09; 9) 96–102, 1st adult 96·5 (2·13; 18) 92–100; tail, adult 65·1 (2·50; 8) 62–68, 1st breeding 64·2 (2·53; 7) 61–67, 1st non-breeding (tail juvenile) 66·3 (1·54; 9) 63–68 (averages in ♀ 94·9, 93·9, 63·6, 64·2, and 64·6 respectively).

Weights. Pennsylvania (USA): (1) adult and 1st breeding, May–June, (2) adult, July–September; (3) juvenile and 1st non-breeding, July–August: (4) 1st non-breeding, September–October (Clench and Leberman 1978).

(1)	♂ 28·3 (1·77; 58) 24·4–34·3	♀ 28·3 (2·78; 49) 21·8–35·2
(2)	29·9 (2·32; 25) 26·3–34·6	28·0 (2·18; 19) 21·5–30·5
(3)	28·3 (1·54; 42) 24·3–31·4	28·0 (1·86; 35) 23·4–30·8
(4)	29·5 (1·86; 140) 24·2–36·0	28·5 (1·83; 102) 24·5–33·5

East Ship island (Mississippi, USA), spring: 26·0 (2·94; 191) (Kuenzi *et al.* 1991). Ohio (USA), May: ♂ 24·0 (Baldwin and Kendeigh 1938). Kentucky (USA), June: ♀ 31·5 (Mengel 1965). Maine (USA), July–September: ♂ 36·0, ♀ 27·3 (3) 25–30 (ZFMK). Coastal New Jersey (USA), autumn: 29·5 (4·0; 18) 24·5–42·5 (Murray and Jehl 1964). Illinois (USA), September: ♂ 32·0 (3), ♀ 32·8 (1) (Graber and Graber 1952). Georgia (USA), early October, adult: ♂♂ 30·5, 31·6, ♂ 36·9 (25) 31·9–44·9 (Johnston and Haines 1957).

Belize, October: ♂ 31·2 (1·93; 5) 28·9–33·9, ♀ 30·2 (3·01; 4) 27·2–34·3 (Roselaar 1976; ZMA); April, ♂ 37·8 (Russell 1964). Peru, December: ♂ 19·6 (ZFMK). Curaçao, February: ♂ 17·9 (ZMA). Bonaire, April: ♂ 18·7 (ZMA).

For fat-free weights, see Connell *et al.* (1960), Odum *et al.* (1961), and Caldwell *et al.* (1963).

Structure. Wing fairly long, broad at base, tip pointed. 10 primaries: p8 longest, p9 1–2·5 shorter, p7 (0–)1–2 shorter, p6 4–7, p5 10–15, p4 15–21, p3 19–24, p2 22–28, p1 26·8 (15) 24–

30. P10 greatly reduced, a tiny spike-like feather hidden below reduced outermost greater upper primary covert; 65–73 shorter than p8, 10–13 shorter than longest upper primary covert, 0–2 shorter than outer upper primary covert. Outer web of (p6–)p7–p8 emarginated, inner of p8 (p7–p9) with faint notch. Tip of longest tertial reaches tip of p1–p3. Tail rather short, tip straight or slightly forked; 12 feathers. Bill stout, rather short, length of visible culmen *c.* 60% of head length; bulbous at base, where *c.* 7·7–9·8 mm deep and *c.* 8·7–9·8 mm wide; culmen distinctly, and cutting edges slightly, decurved; cutting edge of upper mandible with horny tooth in middle and notch just behind (tooth sometimes sharp, more often blunt; occasionally tooth and notch barely indicated). Nostril rather small, rounded, bordered behind by short plush-like feathering. About 4–5 rather short bristles on each side of base of upper mandible; some shorter ones higher at base of upper mandible and at side of lower mandible. Tarsus and toes short, rather thick. Middle toe with claw 17·5 (10) 17·0–18·0; outer toe with claw *c.* 75% of middle with claw, inner *c.* 69%, hind *c.* 78%.

Geographical variation. None.

Forms superspecies with Western Tanager *P. ludoviciana* from western North America (Mayr and Short 1970). Entirely allopatric, but some possible hybrids reported (Tordoff 1950; Mengel 1963).   CSR

Family EMBERIZIDAE buntings and allies

Small to medium-sized, thick-billed 9-primaried oscine passerines (suborder Passeres). Occur in both New World and Old, with main diversity in former. Like finches (Fringillidae), emberizids are specialist seed-eaters but more conservative in choice of seeds (see below) and diet more varied, being often supplemented by insects and other small invertebrates and by fruit (berries). Bill typically strong, hard, and deep; as in Fringillidae, structurally designed internally for shelling seeds (with aid of tongue and strong jaw muscles), but emberizids adaptively closer to Fringillinae (chaffinches) than to Carduelinae (typical finches). Unlike finches, have distinct preference for monocotyledonous seeds (grasses) which they shell by crushing them in bill (see Ziswiler 1965 for trials on captive birds and many other details; also Ziswiler 1979).

Following Voous (1977), Emberizidae here treated as a separate family from other 9-primaried and often finch-like birds, Fringillidae, Parulidae (New World wood-warblers), Thraupidae (tanagers, etc.), and Icteridae (New World blackbirds, etc.). Often classified in 2 subfamilies, Emberizinae and Cardinalinae, though DNA evidence (Sibley and Ahlquist 1990) now suggests that affinities of Cardinalinae are closer to tanagers and New World orioles than to Emberizinae. See further under Passeridae and Fringillidae.

NB. Thick, seed-eater bills have apparently evolved independently in several groups of 9-primaried oscines. Externally, for instance, massive bills of some Carduelinae (e.g. grosbeaks *Coccothraustes*, *Hesperiphona*, and *Eophona*) are closely similar to those of some Cardinalinae (e.g. grosbeaks *Caryothraustes*, *Pitylus*, and *Guiraca*), some Emberizinae (e.g. ground-finches *Geospiza*, seed-finches *Oryzoborus*), and some Icteridae (e.g. Bobolink *Dolichonyx oryzivorus*). Internal morphology of bill, jaw musculature, and alimentary tracts quite different however, linked with marked differences in method of crushing seeds (as indicated above and elsewhere in this volume). Superficially similar large bills also found in members of less closely related groups such as Estrildidae (e.g. Java Sparrow *Padda oryzivora*) and Ploceidae (e.g. White-billed Buffalo-weaver *Bubalornis albirostris*) and here too internal morphology different. See Ziswiler (1965, 1967b, c) for full details.

Subfamily EMBERIZINAE buntings, New World sparrows, and allies

Small and medium-small emberizids variously known as buntings, longspurs, sparrows, juncos, towhees, etc.; occur in many open habitats, including savanna, steppe, alpine tundra, scrub, and desert, but mostly in parkland, fields, hedgerows, cultivation, etc. (especially in northern hemisphere). Often largely terrestrial, picking up seeds (etc.) from ground, though perching freely just above it. As in fringilline finches, insects and other small invertebrates eaten in summer, forming sole food of nestlings; caught by aerial flycatching at times. Found in New World and Old but main radiation in former, with 58 species of 17 genera in North and Central America, including 2 species also found in northern Palearctic (see below), and *c*. 176 species of *c*. 45 genera in Central and South America from where 14 species of 4 genera (*Geospiza*, *Camarhynchus*, *Certhidea*, *Pinaroloxias*) on Galapagos Islands (Darwin's finches) and 3 species of 2 genera (*Rowettia*, *Nesospiza*) on islands of Tristan da Cunha group evolved. Only *c*. 41 species—all buntings in large genus *Emberiza* and 3 other monotypic genera (see below)—restricted to Old World; of these, only 3 species (all in *Emberiza*) strictly Afrotropical. Emberizids sedentary or, in case of more northerly species, migratory. Some 290 species; 29 in west Palearctic—19 breeding, 10 accidental (7 from North America).

About 72 genera. New World sparrows of North and Central America include (e.g.) (1) *Pipilo* (towhees), 7 species: (2) *Aimophila*, 13 species—of which one species reaches South America; (3) *Spizella*, 7 species; (4) *Ammodramus*, 6 species; (5) *Zonotrichia*, 5 species—including one in South America; (6) *Melospiza*, 3 species; (7) *Passerella* (Fox Sparrow *P. iliaca*), monotypic; (8) *Junco* (juncos), 4 species. True buntings (to which rest of this account now largely confined) comprise following genera: (1) *Calcarius* (longspurs), 4 species—North America, 1 species (Lapland Bunting *C. lapponicus*) also northern Eurasia (circumpolar); (2) *Plectrophenax* (Snow Bunting *P. nivalis*), monotypic—much as last; (3) *Emberiza* (typical buntings), 38 species—Europe and especially Asia, extending into Africa; (4) *Miliaria* (Corn Bunting *M. calandra*), monotypic—Eurasia and North Africa; (5) *Melophus* (Crested Bunting *M. lathami*), monotypic—southern Asia; (6) *Latoucheornis* (Fokien Bunting *L. siemsseni*), monotypic—central China.

In buntings (with exception of *M. calandra*), sexes almost the same size. Bill typically short, conical, and pointed—with cutting edges incurved. *M. calandra* has bony hump in roof of mouth against which seeds are crushed; this feature also found in many other buntings, but usually less well developed and absent in some (e.g. *P. nivalis*). Nostrils sited near base of bill; open, oval in shape, and largely hidden by feathers. Rictal and nasal bristles present but minute. Food brought to young in bill (not regurgitated as in cardueline finches). Wing rather long and bluntly pointed in migratory species, shorter and more rounded in sedentary ones; 9 primaries (p10 rudimentary and hidden). Flight typically undulating and finch-like; long-flights occur in a few species but not a typical feature of group as a whole. Tail rather short to long; slightly rounded, straight, or shallowly forked; 12 feathers. Leg rather short or of medium length with strong toes and claws, hind claw long and straight in *Calcarius*; tarsus scutellate, booted at side, forming sharp ridge at rear. Foot not used for holding food or for uncovering it on ground when feeding; special 'double-scratch' action

of New World sparrows lacking (Harrison 1967). Gait a hop, walk, or run. Head-scratching by indirect method (see Simmons 1961*a* for *Emberiza*). Bathe only in typical passerine stand-in manner so far as known; apparent communal bathing in snow reported from *P. nivalis* (Riddoch 1986). Sunning records few: lateral sunning reported in the wild from Yellowhammer *E. citrinella* (see Simmons 1986), also from a number of New World sparrows (Hauser 1957; Potter and Hauser 1974); wings-down or partial spreadeagle sunning observed in *C. lapponicus* (E K Dunn). With possible exception of *E. citrinella*, no records of anting for any bunting so far as known (Simmons 1957*a*), but active (direct) anting reported in a number of New World sparrows of genera *Pipilo*, *Junco*, and *Zonotrichia* (Whitaker 1957; Simmons 1961*b*; Potter 1970; Potter and Hauser 1974).

Buntings in general not so vocal as cardueline finches, but have equally complex repertoire made up of simpler, less twittering, but quite loud calls. Song of ♂ of typical advertising type: varies from relatively short and simple to loud and musical. Although usually gregarious outside breeding season, feeding in small parties or larger flocks and roosting communally, buntings typically solitary and territorial when nesting. Monogamous mating system the general rule but polygyny occurs in some populations of *M. calandra* and Reed Bunting *E. schoeniclus*. Like fringillids, buntings are non-contact birds, avoiding physical contact with conspecifics even when roosting and never allopreening; courtship-feeding, however, occurs in a few species (e.g. *P. nivalis*) though poorly developed in group as a whole and probably often facultative. Nest a cup, usually placed on or near ground; built by ♀ only, with ♂ in close attendance when she collects material. Incubation by ♀ only (fed by ♂). Young fed by both sexes, exceptionally by ♀ only. Young typically leave nest at early age (from as little as 8 days) while still quite incapable of flight. Distraction-lure display recorded from breeding adults of a number of species.

Except in a few species (e.g. *M. calandra*), sexes of buntings differ in plumage, ♂ the brighter bird to greater or lesser extent. Plumages of emberizines as a whole are varied, but many species are streaked with contrasting colours mostly on head, throat, and chest, forming bold pattern; many species (especially in Palearctic) have white spots on outer tail-feathers. Juvenile like adult ♀. Nestling covered with plentiful down; no spots on tongue. Adult post-breeding moult complete, starting shortly after fledging of last brood of young. Partial pre-breeding moult also in some species, especially in long-distance migrants; restricted to head and body. Post-juvenile moult complete in *M. calandra* but partial in other species, restricted to head, body, and wing-coverts; starts *c*. 1–2 months after fledging.

Pipilo erythrophthalmus **Rufous-sided Towhee**

PLATE 12
[between pages 256 and 257]

Du. Roodflanktowie   Fr. Tohi à flancs roux   Ge. Rötelgrundammer
Ru. Красноглазый тауи   Sp. Chingolo punteado   Sw. Brunsidad busksparv

Fringilla erythrophthalma Linnaeus, 1758

Polytypic. *ERYTHROPHTHALMUS* GROUP. Nominate *erythrophthalmus* (Linnaeus, 1758), north-east USA and neighbouring eastern Canada, west to southern Manitoba (Canada), Minnesota, Iowa, eastern Kansas, and north-east Oklahoma (USA), south to northern Arkansas, Tennessee, southern Appalachians, and Virginia. Extralimital: *canaster* Howell, 1913, south of nominate *erythrophthalmus*, in narrow zone from Louisiana through Alabama and Georgia to North Carolina; *rileyi* Koelz, 1939, south of *canaster*, from south-east Alabama and central Georgia to south-east North Carolina, south to northern Florida; *alleni* Coues, 1871, southern Florida. *MACULATUS* GROUP (extralimital). 19-20 races in western North America (west of nominate *erythrophthalmus*), south through Mexico to Guatemala.

Field characters. 17-18 cm; wing-span 25-30 cm. Close in size to Corn Bunting *Miliaria calandra* but with shorter, more rounded wings, proportionately longer and more rounded tail, and longer legs. Rather large, robust bird of dense cover, with striking tri-coloured plumage in both sexes and ample tail, often cocked. Adult ♂ has black hood, back, wings, and tail, contrasting with rufous flanks and vent and white belly, primary-patch, and outer tail; adult ♀ brown where ♂ black. Sexes dissimilar; no seasonal variation. Immature separable at close range. 2 races reported in west Palearctic, but only nominate *erythropthalmus* of north-east USA (described here) accepted as certain vagrant (see also Geographical Variation).

ADULT MALE. Moults: July-September (complete). Head, breast, upperparts, wings, and tail black, with white bases to primaries, buff-white distal margins to tertials and tips to secondaries, and white margins and tips to 3 outermost tail-feathers, striking in flight. Broad band from side of breast to rear flank rusty orange-brown, contrasting with black wings and white belly but not with slightly paler rufous-buff to yellowish-buff under tail-coverts. Bill stout and conical but slightly decurved; blue-black. Eye red. Legs bright flesh to purplish-brown. ADULT FEMALE. Warm umber-brown where ♂ black; primary-patch narrower and less pure white. FIRST-WINTER. Resembles adult, best distinguished by retained brownish juvenile flight-feathers. ♂ also aged by retained brown juvenile outer greater coverts. Eye brown, not reddening until 1st autumn.

Unmistakable, with bustling, noisy progress as distinctive as vivid plumage contrasts. Flight low and rushed, with beats of round wings producing fluttering sounds, and tail pumped and often spread in turns and landings. Gait a vigorous hop, becoming double-footed backwards kick when searching through leaf litter and making audible scratching noise. Active and restless, with presence in dense cover further signalled by frequent cocking and flirting of tail and flicking of wings. Usually solitary after breeding.

Calls include loud 'towhee', also rendered as slurred 'chewink'; slightly metallic 'tyst' and soft 'heu'.

Habitat. Breeds in temperate Nearctic lowlands in scrub, and any sort of dense low woody vegetation, including abandoned or poorly kept pastureland, isolated forest openings, field edges, parks, and roadsides, where ground is accessible for foraging (Pough 1949). In west, favours bushy hillsides, grown-over gullies, and all kinds of scrub at low to moderate altitudes (Larrison and Sonnenberg 1968). Not commonly found in tall mature trees, wetlands, or arid places and normally keeps close to cover. Will occupy city parks or suburbs with trees and tall shrubbery (Johnsgard 1979). In Canada, also favours willow *Salix* and alder *Alnus* patches and shrub cover along streams, or on mountainsides and in valleys, especially where leaf litter plentiful (Godfrey 1979).

Distribution. Breeds in North America from southern British Columbia east to Maine, south throughout USA to highlands of Mexico and central Guatemala. Winters within breeding range, vacating extreme north.

Accidental. Britain: Lundy Island, Devon, 7 June 1966 (Waller 1970).

Movements. Partial migrant. Southern races mainly resident; northern races (except Pacific coast *oregonus*, largely sedentary) withdraw southward from part (*curtatus*, nominate *erythrophthalmus*) or all (*arcticus*) of their ranges in winter (American Ornithologists' Union 1957; Bent 1968). Most migration is into ranges of more southern forms, and is largely undetected except from specimens.

Movements in August bring birds to east coast from New York to Nova Scotia (Bailey 1955; Bull 1974; Tufts 1986), but southward migration begins only in September (Palmer 1949; Burleigh 1972; Janssen 1987). Most birds have left British Columbia interior and Canadian prairies by end of September, when first migrants appear in Texas and Arizona and also along Atlantic coast (Phillips *et al.* 1964; Bent 1968; Oberholser 1974; Cannings *et al.* 1987). Recorded autumn movements south of 35°N are chiefly in October, as by November migrants are mingled with southern residents (Bent 1968). Ringing data show broad-front movement, with recoveries (e.g.) from New

York to south-east states and from Michigan, Maryland and Tennessee to Alabama (Bull 1974; Imhof 1976).

Many spring reports represent start of singing rather than movement; most 'first records' in March in east are within winter range, but March birds in western inter-mountain regions from California to British Columbia are probably early migrants (Grinnell and Miller 1944; Jewett *et al.* 1953; Cannings *et al.* 1987). Main movement continues to mid-April in west, and extends from 2nd half of April to early May across north-east (Bent 1968; Quilliam 1973; Bull 1974; Janssen 1987). Arrivals continue to mid-May only near northern range limits (Palmer 1949; Nero and Lein 1971; Sprague and Weir 1984). Migrants regularly overshoot known range in eastern Canada (Squires 1976; Speirs 1985; Tufts 1986; Erskine 1992).

Vagrancy uncommon, reaching north only to Fort Severn (55°N, Ontario), Moisie River (Quebec), and Newfoundland (Godfrey 1986).   AJE

Voice. See Field Characters.

Plumages. (nominate *erythrophthalmus*). ADULT MALE. Entire upperparts, side of head and neck, and chin to chest black, often with faint gloss on head to scapulars, duller and slightly tinged grey from chin to chest; tips of feathers of rump and upper tail-coverts with traces of rufous fringes. Side of breast and flank deep cinnamon-rufous; rear of flank and all under tail-coverts buff-cinnamon; thigh mixed black and pale cream-buff or white (darkest on rear); remainder of underparts contrastingly white. Tail black, 3rd pair (t3) sometimes with small white spot or fringe on tip, tip of inner web of t4 white for 15-20 mm, tip of inner web of t5 for 25-35 mm, of t6 for 35-45 mm; some traces of white on tip of outer web of t4-t5, outer web of t6 white. Flight-feathers dull black, inner webs and tips greyish-black; bases of outer webs of middle and outer primaries (about p3 to p8) white, showing as white patch which extends *c.* 5-10 mm beyond tips of greater upper primary coverts; outer webs of p5-p8 with broad white border just below emarginated part, connected through narrower white outer border with white at base; base of outer web of p9 fringed white; basal borders of inner webs of flight-feathers greyish-pink or isabelline-white. Tertials black, shortest feather with pink-cream or white outer web, middle feather with broad pink-buff to off-white outer border, longest feather with narrow buff or rufous fringe near tip of outer web. Upper wing-coverts and bastard wing black, deepest and sometimes slightly glossy on lesser, slightly duller on greater upper primary coverts and on longest feather of bastard wing. Under wing-coverts and axillaries white, longer primary coverts mottled dull black and fringed pale buff, marginal primary coverts white. *In fresh plumage* (autumn), some traces of rufous fringes along tips of feathers of upperparts (most pronounced from back to upper tail-coverts, but even here fringes narrow and indistinct); white on outer webs of tertials tinged cream, partly fringed rufous; *in worn plumage* (May–June), pale fringes of tertials bleached to white, that on longest sometimes fully worn off; narrow white borders along middle portions of outer primaries worn off, white patch at primary base not connected with white patch formed by fringes near emarginations; some grey of feather-bases sometimes visible on throat; rufous of flanks sometimes slightly paler, thigh blacker. ADULT FEMALE. Like adult ♂, but black of head, body, central tail-feathers, and wing replaced by brown. Entire upperparts, side of head and

neck, and chin to chest dark cinnamon-brown, slightly tinged dark rufous when plumage fresh, tinged russet on cap to mantle and on tail-coverts, more fawn or olive-drab on back and rump, especially when worn, more greyish-cinnamon on chin to chest if worn. Remainder of underparts deep cinnamon-rufous (on flank), buff-cinnamon (at rear), and white (in middle), as in adult ♂, but black of thigh replaced by olive-brown. Tail as adult ♂, but both webs of t1 and outer webs of others russet-brown (if fresh) to olive-brown (if worn), black of inner webs of t2-t5 slightly duller and more olive, except at border with white; white tips to inner webs of t4-t6 as in adult ♂, but white slightly less extensive (*c.* 30-35 mm on t6, 10-15 mm on t4), outer web of t6 white with partly olive-brown tip. Pattern of white on flight-feathers and tertials as adult ♂, but black on tertials and on outer webs of flight-feathers replaced by russet-brown (if fresh) or dark olive-brown (if worn); as in adult ♂, pale fringes of tertials tinged pink-buff to rufous when fresh, white when worn. Upper wing-coverts, including greater upper primary coverts, dark olive-brown to black-brown, fringes slightly brighter russet-brown or medium olive-brown. Underwing as adult ♂. JUVENILE. Upperparts, side of head, and lesser upper wing-coverts olive-brown or russet-brown, marked with broad ill-defined dark brown to dull black streaks on mantle and scapulars and narrower ones on cap. Cheek, side of neck, and underparts dull white, tinged buff; side of head marked with fuscous-brown stripe through eye and malar stripe; chest, flank, and under tail-coverts tinged cinnamon, throat, chest, and side of breast streaked dark grey or dull black. Median and greater upper wing-coverts olive-brown or russet-brown with dull black centres and buff-white tips. Sexes differ in colour of flight-feathers and tail, as in adult. See also 1st adult, below. FIRST ADULT MALE. Like adult ♂, but black of head and body on average slightly duller, more sooty on upperparts, more greyish-black on throat and chest, throat with some grey or off-white of feather-bases visible when plumage worn; many tips of feathers of upperparts, upper wing-coverts, and tail fringed rufous or brown when plumage fresh. Juvenile flight-feathers, greater upper primary coverts, tail, and usually tertials, greater upper wing-coverts, and feathers of bastard wing retained; primaries and primary coverts less deep black than in adult, more greyish-black when fresh, more sepia-brown when worn; pale outer fringes of tertials sometimes narrower and more extensively tinged rufous; tail as adult, but white partly washed buff or pale cream, outer web of t6 less pure white, partly tinged dusky along shaft and on tip; white on juvenile primaries on average less extensive, white of rather small patch on primary-bases sometimes not connected with white near emargination, or connection faint (especially when worn); white of tertials and of middle portions and tips of primaries sometimes worn off when heavily abraded. FIRST ADULT FEMALE. Like adult ♀, but part of juvenile feathering retained, as in 1st adult ♂. Differences between, e.g., paler brown and more frayed juvenile greater upper primary coverts of 1st adult and darker brown smoothly-edged primary coverts of adult often difficult to see. Tips of juvenile tail-feathers slightly more pointed than in adult, less rounded or truncate; white on tips of t4-t6 partly suffused cream or pale buff, borders of white less regular, and white less extensive, 8-15 mm long on t4, 15-25 on t5, 20-35 on t6 (sample small, however); outer web of t6 dark except for white or cream outer margin (in adult, mainly white). Patch at base of juvenile primaries on average smaller, 1-8 mm long. For ageing, see also Davis (1957).

Bare parts. ADULT, FIRST ADULT. Iris of nominate *erythrophthalmus* bright deep red, carmine, rose-red, or (occasionally) brownish-red in ♂, dark red-brown, bright brick-red, or dark deep red in ♀; in *canaster*, mainly red, sometimes orange or yellow; in *rileyi*, mainly yellow, sometimes red or orange; in *alleni*, brownish-yellow to yellow-white; in *maculatus* group, mainly red. Bill of ♂ greyish-black or steel-black, base of lower mandible paler in winter; bill of ♀ blue-grey to plumbeous-black on culmen, light yellow-horn on cutting edges, and greyish-horn on lower mandible. Leg and foot pink-flesh, purplish-flesh, purplish-brown, or dull flesh-brown, darkest on toes. JUVENILE, FIRST ADULT. At fledging, iris of nominate *erythrophthalmus* and of *maculatus* group mud-brown, sepia-brown, grey-brown, or dark brown; in autumn, an orange or reddish inner ring develops, gradually becoming larger to include whole iris by winter or spring, when iris uniform pale or bright orange, reddish-orange, orange-red, or pale red; by late May, iris as in adult. In *alleni*, iris pale neutral grey at fledging, bill shading gradually to slate-blue (in ♂) or greyish-horn (in ♀) during autumn, and leg and foot to dusky sepia-brown; similar to adult in spring. (Dwight 1900; Ridgway 1901–11; Dickinson 1952; Nichols 1953; Davis 1957; Wood and Wood 1972; RMNH, ZMA.)

Moults. ADULT POST-BREEDING. Complete; primaries descendent. Starts late June to late August, completed August or early September (Dwight 1900; Bent 1968; RMNH). ADULT PRE-BREEDING. Restricted to a few feathers on chin or elsewhere or no moult at all (Dwight 1900). POST-JUVENILE. Partial: head, neck, lesser and median upper wing-coverts, sometimes some or all greater upper wing-coverts, tertials, and feathers of bastard wing; in some races also tail. Starts late June to late August, completed late July to September; much feathering of face still growing in one from mid-October. (Dwight 1900; Sutton 1935; Davis 1957; Bent 1968; Pyle *et al.* 1987; RMNH.)

Measurements. ADULT, FIRST ADULT. Nominate *erythrophthalmus*. North-east USA, whole year; skins (RMNH, ZMA). Bill (S) to skull, bill (N) to distal corner of nostril; exposed culmen on average 3·7 less than bill (S).

WING	♂	88·7 (3·10; 17)	85–95	♀ 84·4 (1·30; 8)	82–87
TAIL		90·9 (2·83; 17)	86–96	86·2 (2·31; 8)	81–89
BILL (S)		17·1 (0·79; 17)	15·8–18·0	16·6 (0·43; 8)	16·1–17·2
BILL (N)		9·7 (0·60; 17)	8·7–10·5	9·5 (0·38; 8)	8·9–10·1
TARSUS		28·5 (0·62; 17)	27·6–29·5	27·9 (0·59; 8)	26·9–28·8

Sex differences significant, except for bill.
See also Dickinson (1952).

Weights. Nominate *erythrophthalmus*. Pennsylvania (USA): adult and 1-year-old, (1) March–June, (2) July–October; juvenile and 1st adult, (3) July–August, (4) September–November (Clench and Leberman 1978).

(1)	♂	41·1 (2·43; 142)	32·1–48·2	♀ 38·5 (3·61; 62)	32·1–52·3
(2)		43·0 (2·77; 61)	37·8–50·0	39·7 (2·66; 61)	33·9–44·7
(3)		38·9 (2·20; 44)	34·8–44·5	36·7 (1·84; 21)	33·9–41·2
(4)		42·9 (3·06; 167)	29·3–51·5	39·3 (2·49; 144)	30·8–47·6

Coastal New Jersey (USA), autumn: 39·1 (121) 31·4–47·5 (Murray and Jehl 1964). Averages Ohio (USA): July–November, adult 40·5 (72), 1st adult 38·3 (69) (Baldwin and Kendeigh 1938, which see for average per sex per month). Lundy (Devon, England), June: ♀ 40 (Waller 1970).

For nominate *erythrophthalmus* and/or other races, see also Grinnell *et al.* (1930), Linsdale and Sumner (1937), Becker and Stack (1944), Norris and Johnston (1958), and Rand (1961a). A single ♀ of *maculatus* group (probably race *arcticus*) in Britain between 42·6 and 53·0 on various dates in autumn and winter (Cudworth 1979).

Structure. Wing short, broad at base, tip rounded. 10 primaries: p6 longest, p5 and p7 0–2 shorter, p8 2·5–5, p9 11–14, p4 1·5–4, p3 4–7, p2 7–11, p1 8–13; tip of p9 about equal to tip of p1 or slightly shorter. P10 strongly reduced, a tiny feather hidden below reduced outermost greater upper primary covert; 57–68 shorter than p6, 9–12 shorter than longest upper primary covert. Outer web of p4–p8 emarginated, inner web of p6–p9 with faint notch. Tail long, tip slightly rounded; 12 feathers, t6 7–13 shorter than t2–t3, t1 often slightly shorter than t2–t3. Bill short, length about half of head-length; stout, *c.* 9–10·5 mm deep and *c.* 8–9·5 mm wide at base; culmen bluntly ridged, slightly decurved towards tip, often ending in a small hook; cutting edges strongly inflected downwards on basal half. Nostril rather small, oval, partly covered by narrow membrane above. Numerous long fine bristles along base of upper mandible and on chin, some shorter ones at side of lower mandible. Leg and foot rather long, strong. Middle toe with claw 23·5 (10) 21·5–25·5 mm; outer toe with claw *c.* 79% of middle with claw, inner *c.* 76%, hind *c.* 80%.

Geographical variation. Marked, involving body size, bill size, amount of white in wing and tail, colour of eye, occurrence of white spots or streaks on upperparts and upper wing-coverts, and in general colour. 2 main groups: (1) *erythrophthalmus* group in east, with uniform upperparts, broad white borders along outer primaries (forming patch), and pronounced sexual dimorphism (black of ♂ replaced by brown in ♀); (2) *maculatus* group in west, south to Guatemala, which shows white spots or streaks on mantle, scapulars, and upper wing-coverts in all plumages, reduced white fringes along primaries, and in which colour of ♀ is dark grey, more closely similar to black of ♂. Within *erythrophthalmus* group, nominate *erythrophthalmus* of north-east USA and neighbouring Canada is large, with relatively small bill, red iris in adult, and much white in tail; *alleni* from southern Florida is small, bill of medium size, iris whitish in adult, and tail with very little white; intervening races *canaster* and *rileyi* from south-east USA (north of *alleni*) are intermediate in amount of white in tail and variable in iris colour; general size large in *canaster*, medium in *rileyi*; bill in both relatively large; see Dickinson (1952) for details. For races of *maculatus* group, see Hellmayr (1938) and Sibley (1950). For intergradation zone between nominate *erythrophthalmus* group and *maculatus* group in central North and South Dakota and central Nebraska, see Sibley and West (1959).

Forms superspecies with Collared Towhee *P. ocai* from Mexico and Socorro Towhee *P. socorroensis* of Socorro Island (off Baja California, Mexico); ranges of *P. erythrophthalmus* and *P. ocai* overlap locally in southern Mexico, resulting in extensive hybridization in some areas of overlap, but none in others (Sibley 1950, 1954; Sibley and West 1958). CSR

Spizella pusilla (Wilson, 1810) **Field Sparrow**

FR. Bruant des champs   GE. Klapperammer

A North American species breeding from Montana and Minnesota east to southern Quebec and Maine, south to south-central and south-east Mexico; northern populations winter from eastern Kansas east to Maryland and southern Massachusetts, south to north-east Mexico and in USA to coast of Gulf of Mexico and central Florida; southern populations resident. 5 came aboard eastbound ship off North America in 1962, 1 staying until at least 15°W on 12 October (Durand 1963).

Chondestes grammacus (Say, 1823) **Lark Sparrow**

FR. Bruant à joues marron   GE. Rainammer

A western and central North American species, breeding very locally and irregularly east of Mississippi valley, north to southern British Columbia, southern Manitoba, and southern Ontario, south in the west to northern Mexico. Winters in southern USA and Mexico. Recorded in eastern England: Landguard Point (Suffolk), 30 June to 4 July 1981 (Rogers *et al.* 1982), originally considered ship-assisted but recently (1993) accepted as of natural occurrence; Waxham (Norfolk), 15–17 May 1991 (*Birding World* 1991, **4**, 156–9), under review.

Ammodramus sandwichensis **Savannah Sparrow**

PLATE 12
[between pages 256 and 257]

DU. Savannahgors   FR. Bruant des prés   GE. Grasammer
RU. Саванная овсянка   SP. Chingolo sabanero   SW. Gulbrynad grässparv

Emberiza Sandwichensis Gmelin, 1789. Synonym: *Passerculus sandwichensis*.

Polytypic. *A. s. princeps* Maynard, 1872 ('Ipswich Sparrow'), Sable Island (off Nova Scotia); *labradorius* Howe, 1901, Newfoundland, Labrador, and Quebec; *savanna* (Wilson, 1811), Nova Scotia and New Brunswick; *mediogriseus* Aldrich, 1940, north-east USA from Great Lakes area (except in west) to New England states, including neighbouring southern Ontario and south-east Quebec, east to Gaspé peninsula; *oblitus* Peters and Griscom, 1938, from Keewatin, Manitoba (except south-west), and western Ontario south to Minnesota, Wisconsin, and Michigan, and perhaps this race further south in USA; *nevadensis* Grinnell, 1910, eastern British Columbia, Alberta, Saskatchewan, and south-west Manitoba, south through Great Plains of USA; *anthinus* (Bonaparte, 1853), northern Canada and Alaska (except part of Pacific coast), west of *oblitus*, north of *nevadensis*, and eastern Chukotskiy peninsula (north-east Siberia); *c.* 14 races along Pacific coast of Alaska and Canada south through western USA to Central America.

Field characters. 14–16 cm; wing-span 22·5–25 cm. Close in size to Reed Bunting *Emberiza schoeniclus* but with proportionately longer, more pointed bill and less full and noticeably shorter, notched tail, with pointed feathers; relatively much shorter-tailed than Song Sparrow *Melospiza melodia*, with east Nearctic race *savanna* smaller than that species but Sable Island race *princeps* as large. Among vagrant Nearctic sparrows, relatively small to medium-sized and tubby; a uniformly streaked, open-country species. Shows yellow fore-supercilium, narrow pale central crown-stripe, and buffish-white double wing-bar. *A. s. princeps* noticeably pallid, with cryptic sandy appearance. Sexes similar; no seasonal variation. Immature

not certainly separable. 2 races identified in west Palearctic and at least 2 others could occur.

ADULT. Moults: July–September (complete), February–April (mostly head). Most uniformly streaked of vagrant Nearctic sparrows, with plumage recalling ♀ or immature of smaller *Emberiza* though more completely streaked below. Tone and clarity of features varies considerably, particularly in races that have reached west Palearctic, but following are constant: (a) narrow pale cream central crown-stripe contrasting with very dark brown lateral stripes, (b) yellow fore-supercilium, off-white to greyish rear supercilium (palest in non-breeding *princeps*), and bright pale yellow eye-ring, (c) dark upper and lower border to ear-coverts, (d) fully streaked back, (e) buff-white tips to median and greater coverts, forming double wing-bar (usually indistinct in *savanna*), (f) striking white submoustachial stripe contrasting with long dark malar stripe and (g) fully streaked or spotted breast and flanks (marks narrow in *princeps*, much broader in *savanna*), (h) lack of prominent rufous areas on wing (except on fringes of tertials, secondaries, and greater coverts in *savanna*), (i) rather white ground to underparts, and (j) pale grey outer tail-feathers. In general appearance, rather small *savanna* noticeably darker than plump *princeps*, as ground of face, back, wings, and upper flanks distinctly browner and streaks heavier. *A. s. princeps* further distinguished by less obvious lateral crown-stripes and almost isabelline-grey ground to central mantle and virtually unstreaked rump. Bill long; greyish- to pinkish-horn, with dark culmen and paler lower mandible. Legs pale pink to pale yellow-brown.

In Nearctic, *A. sandwichensis* shares streaked plumage with 12 other relatives, but confusion among transatlantic vagrants only likely with *M. melodia*, which is similar in size to *princeps* but differs from both races in proportionally much longer and slightly rounded tail, greyish supercilium, chestnut streaks on head and back, large central dark spot on breast, more chestnut in wings and along tail-base, and pale brown (less pinkish) legs. Flight swift but rather erratic, even zigzagging over short distance, with bird soon landing with depressed tail; over longer distance, develops undulations even at low level. Under repeated pressure, flies far off. Gait of *savanna* primarily a hop, sometimes runs, rarely walks; of *princeps* primarily a walk, with head and shoulders moving in dove-like manner, but often runs, with head held low parallel with back and tail, and sometimes hops. Both races recall small bunting or Dunnock *Prunella modularis* when foraging on ground. *A. s. savanna* perches typically on small bushes, posts, or low wire fences, but has been seen in trees; *princeps* is much more strictly terrestrial, walking along open beaches and in beach grass, perching only on hummocks or bushes. *A. s. princeps* usually solitary but on occasion joins other terrestrial passerines, particularly Shore Lark *Eremophila alpestris* and Snow Bunting *Plectrophenax nivalis*.

Calls include high, sharp, dry 'tsip' and sharp 'chirp' (in alarm) from *princeps*; rather hard 'tsep' and thin 'tsi' not attributed to race.

Habitat. Breeds in Nearctic, from arctic tundra through boreal and temperate zones to subtropics and tropics, in various types of open unwooded habitat, especially with short herbage, either moist or dry, lowland (including coastal) or locally montane, as on higher mountain ranges of Alaska, where it also frequents shoreline driftwood and debris (Gabrielson and Lincoln 1959). Freely uses low or even more elevated perches, but is predominantly a ground bird. Although fond of sand-dunes and beaches, especially near salt-marshes, prefers walking or running on grassy surfaces, such as meadows or hayfields; also frequents taller rank grasslands. On migration, found in upland pastures, weedy fields, orchards, and gardens (Pough 1949; Forbush and May 1955). Sometimes found in bogs, fens, and marshes, but not characteristically a wetland bird. *A. s. princeps* chiefly frequents coastal sand-dunes, both summer and winter.

Distribution. Breeds in North America, from western and northern Alaska east to northern Labrador and Newfoundland, south to central California, northern New Mexico, Nebraska, Kentucky, western Virginia, Maryland, and northern New Jersey; in highlands of Mexico, from Chihuahua and Coahuila south to Guerrero and Puebla, and of south-west Guatemala; in eastern part of Chukotskiy peninsula (north-east Siberia). Winters along Atlantic coast from Nova Scotia to Florida, inland from southern British Columbia, New Mexico, Tennessee, and Massachusetts south throughout most of Mexico to Guatemala, Belize, and northern Honduras.

Accidental. Britain: Portland, Dorset, 11–16 April 1982 (*princeps*) (Rogers *et al.* 1985); 1st-winter, Fair Isle, 30 September–1 October 1987 (perhaps *labradorius*) (Rogers *et al.* 1988).

Movements. Great variation, from long-distance migrants (northern inland-breeding races) through short-distance and partial migrants (northern coastal and mid-latitude races) to altitudinal migrants (breeding in southern alpine areas) and residents (southern coastal races) (American Ornithologists' Union 1957, 1983; Bent 1968). Frequents only open grass/herb layer habitat at all seasons, so birds retreating from winter snow (covering plant seeds on which they feed) may move considerable distance before finding suitable habitat; wintering is south of or below snow-line. Highest numbers winter in southern states (west to Texas) and southern California. (Rising 1988.)

Post-breeding movements start in August in west, including arrivals in Texas and Arizona, altitudinal movements in Washington and Alberta, and beginning of exodus from Yukon Territory (Rand 1946; Jewett *et al.* 1953; Phillips *et al.* 1964; Oberholser 1974; Sadler and Myres 1976). Far northern areas (both east and west) are vacated by mid- or late September (Todd 1963; Scotter *et al.* 1985), when migration becomes general across southern Canada (Munro and Cowan 1947; Knapton 1979; L M Tuck). Main passage in east continues through October, with peak period becoming later southward (Bent 1968; Potter *et al.* 1980; Janssen 1987). Present in southern states mostly from late September (Lowery 1974; Oberholser 1974; Kale and Maehr 1990); longest-distance migrants recorded in West Indies and Mexico in October, chiefly from November (Bond 1985; Binford 1989). Northern coastal populations (*princeps* in east, nominate *sand-*

wichensis in west) winter chiefly in coastal areas further south, and migrate later than northern inland populations (Gabrielson and Lincoln 1959; Tufts 1986). In California, 3 largely sedentary races are joined in winter by migrants from north (2 or 3 races), north-east (*nevadensis*), and even south-east (*rostrata*) (Grinnell and Miller 1944). Situation in north-west Mexico similarly complex, but not yet documented. Migration (both seasons) is on broad front east of Rockies, but tends to follow mountain meadows, open valley-bottoms, or sea-coasts, further west (Jewett *et al.* 1953; Bent 1968). Earlier suggestion that Aleutian birds make a long flight across Gulf of Alaska is unsupported by more recent data (Bent 1968). Ringing in South Carolina and Alabama (and elsewhere) has confirmed winter site-fidelity, with returns up to 3 years later (Bent 1968; Imhof 1976).

In spring, *princeps* migrates March to early April, but little movement of other races in March. Full tide of migration moves north across continent in April, and first arrivals reach New Brunswick, Manitoba, and far north-west. (Munro and Cowan 1947; Bent 1968; Squires 1976; Scotter *et al.* 1985.) Dates of last stragglers range from late April in tropics (Brudenell-Bruce 1975; Binford 1989) to mid- or late May across USA (Bent 1968). Only subarctic regions of north-central Canada receive most of their breeding birds after mid-May (Cooke *et al.* 1975; Todd 1963).

Vagrant to Canadian Arctic (Godfrey 1986), and one record from Hawaii (Pratt *et al.* 1987). Records in Wrangel Island and southern Ussuriland (eastern Siberia) (Stepanyan 1990) and in Japan (including flock of up to 6, January–February 1988) (Brazil 1991) perhaps involve birds breeding in Chukotskiy peninsula.   AJE, DFV

Voice. See Field Characters.

Plumages. ADULT (*A. s. princeps*). In fresh plumage (autumn), top of head greyish sandy-buff, marked with rather narrow and ill-defined short dark sepia or sooty streaks; a narrow almost uniform cream or buff-white stripe runs from mid-forehead to mid-crown. Hindneck light ash-grey with faint buff-brown streaks. Mantle, scapulars, back, rump, and upper tail-coverts buff-brown, each feather with ill-defined sooty streak on centre and with light brown-grey fringe; fringes tinged sandy pink-cream when freshly moulted (late summer), but bleached to grey by October–November, last so on rump. Supercilium fairly broad and distinct from base of upper mandible to above eye, pale yellow to yellow-white, gradually fading out behind eye: tinged off-white above eye, merging into dull grey-brown with dusky specks towards rear. Eye-ring pale buff, broken in front and behind. Lore and patch just below eye finely mottled grey, cream, and off-white, bordered at rear by warm pink-buff on ear-coverts; upper border of ear-coverts rather faintly darker buff-brown, lower border distinctly black-brown, forming stripe running backwards from gape. A broad and distinct pale cream-buff stripe over lower cheek, widening towards rear, bordered below by mottled black-brown malar stripe. Side of neck behind ear-coverts cream or off-white with fine dark brown streaks, merging into brown-grey with more obsolete dusky streaks at rear.

Ground-colour of underparts white with cream-buff wash, palest on throat and mid-belly, more buffy on rear of flank and on under tail-coverts; lower side of throat, side of breast, all chest, and flank with rather narrow and diffuse buff-brown streaks, latter blackish at shaft. Tail dark sepia-brown, central pair (t1) tinged buff-brown on both webs, t2–t6 with narrow pale pink-buff outer fringe. Flight-feathers and greater upper primary coverts dark sepia-brown, outer webs of primaries and primary coverts narrowly fringed pale pink-buff, of secondaries more broadly warm pink-cinnamon. Lesser upper wing-coverts buff-brown; median and greater coverts and tertials sepia-black with broad light pink-cinnamon borders. Under wing-coverts and axillaries pale sandy pink-cream, longer coverts tinged grey. *In worn plumage* (spring and early summer), ground-colour of upperparts bleached to pale sandy-grey, narrow dark streaks more contrasting; ground-colour of side of head and neck and of underparts white, stripes on side of head and streaks on chest more sharply-defined; t1 and pale fringes on t2–t6 and primaries bleached to off-white, fringes partly or almost completely worn off; pink-cinnamon of upper wing-coverts, tertials, and outer webs of secondaries bleached to greyish-buff or grey-white, especially on tips of greater coverts and on exposed parts of tertials. JUVENILE (*A. s. savanna*). Cap streaked black, buff-brown, and buff, with buff-yellow median crown-stripe (occasionally faint); nape similar, but black streaks reduced. Mantle and scapulars buff-brown and buff-yellow, heavily streaked black; rump buff-brown to buff with narrower black marks. Upper tail-coverts dark brown, broadly fringed buff-brown. Supercilium off-white, partly speckled dusky, tinged yellow above lore, bordered below by dull black eye-stripe. Eye-ring white or buff-white. Ear-coverts buff, sandy-grey, or olive-brown, partly bordered black; bordered by buff-yellow patch below. Side of neck white with black streaks. Underparts buff-yellow (darkest on chest; palest, almost white, on chin and from belly to under tail-coverts), marked with black or dark-brown streaks on lower cheek, chest, side of breast, and flank. Tertials black, outer webs broadly fringed rufous-brown, shortest edged buff-white. Upper wing-coverts black, lesser and median edged buff-white, greater fringed rufous along outer web and buff-white on tip, pale wing-bars more distinct than in adult. Some juveniles much more buff-yellow than others. (Sutton 1935; Bent 1968.) FIRST ADULT. Like adult, but juvenile tail, flight-feathers, and greater upper primary coverts retained. Tips of juvenile tail-feathers more pointed than in adult, less square or truncate (for shape, see Pyle *et al.* 1987); tips of tail-feathers, outer primaries, and primary coverts more worn than in adult at same time of year, difference especially marked in spring.

Bare parts. ADULT, FIRST ADULT. Iris olive-brown to dark brown. Bill light pink-horn, flesh-pink, pinkish-yellow-horn, or orange-pink, culmen and tip of lower mandible slate-black, steel-grey, or slate-grey. Leg and foot bright flesh-pink, orange-pink, dull flesh, or light yellow-brown. Bill, leg, and foot of same dark western races darker than in paler races of eastern North America. JUVENILE. Bill flesh-pink-grey with yellow cutting edges. Leg and foot pale flesh-pink. (Ridgway 1901–11; Sutton 1935; Peters and Griscom 1938; RMNH.)

Moults. ADULT POST-BREEDING. Complete; primaries descendent. In breeding area, July–September; many in full moult late July. ADULT PRE-BREEDING. Partial, extent limited (especially in ♀), involving head, throat, mantle, part of chest, and a few tertials or some feathers elsewhere on body, February–April, or no moult at all. POST-JUVENILE. Partial: head, body, and lesser, median, and greater upper wing-coverts. Starts from late June or early

July; in full moult July; moult completed early August to early September. In some birds, moult of greater coverts and tertials not until winter. FIRST PRE-BREEDING. As adult pre-breeding; extent highly variable. In some birds, involves all head, body, lesser, median, and greater upper wing-coverts, tertials, and (occasionally) t1; in others, head and forepart of body only, or moult restricted to top and side of head and part of mantle and scapulars; in these latter birds, greater upper wing-coverts and tertials sometimes still in relatively fresh 1st non-breeding. (Dwight 1900; Sutton 1935, Pyle *et al.* 1987; RMNH, ZMA.)

Measurements. ADULT, FIRST ADULT. *A. s. princeps.* Sable Island (off Nova Scotia), late May (Broyd 1985).

WING	♂	77·8 (23)	73·3-83·4	♀	73·1 (15) 68·7-76·9
TAIL		59·3 (23)	56·1-63·0		56·0 (13) 51·0-58·6
TARSUS		22·6 (21)	21·7-23·5		21·9 (12) 21·4-22·9

Single ♀ from Massachusetts (USA), November: wing 76, tail 55, bill (S) 13·7, bill (N) 8·6, tarsus 22·2 (RMNH).

A. s. savanna. Northern Nova Scotia, late May and June (Broyd 1985).

WING	♂	70·0 (42)	65·8-74·8	♀	66·0 (36) 62·4-69·7
TAIL		53·3 (40)	49·1-58·6		49·9 (32) 46·2-52·8
TARSUS		20·8 (33)	18·7-22·0		20·2 (31) 18·2-21·5

A. s. savanna, mediogriseus, and *oblitus,* combined. Eastern Canada, summer, and eastern USA, whole year; skins (RMNH, ZMA). Bill (S) to skull, bill (N) to distal corner of nostril; exposed culmen on average *c.* 2·2 less than bill (S).

WING	♂	72·0 (1·62; 10)	70-74	♀	68·6 (0·79; 7)	67-70
TAIL		49·4 (2·65; 10)	45-53		45·8 (1·58; 7)	44-48
BILL (S)		12·5 (0·46; 10)	11·9-13·2		12·5 (0·39; 7)	11·9-13·1
BILL (N)		7·6 (0·41; 10)	7·0-8·1		7·6 (0·35; 7)	6·9-7·9
TARSUS		21·0 (0·73; 10)	20·2-22·0		20·4 (0·40; 7)	19·8-20·8

Sex differences significant for wing and tail.

See also Peters and Griscom (1938) and Aldrich (1940). For sexual dimorphism in measurements throughout geographical range, see Rising (1987).

Weights. ADULT, FIRST ADULT. *A. s. princeps.* Sable Island: late May, ♂ 27·9 (25) 21·0-32·5, ♀ 26·1 (16) 22·4-32·5 (Broyd 1985), breeding adult 25·6 (5·7; 108) (Stobo and McLaren 1975, which see for details). Single bird Dorset (England), April: 26·7-31·0 (Broyd 1985). For growth of nestlings, see Ross (1980).

Other northern and eastern races (all of about similar size). Sexed birds. *A. s. savanna,* Nova Scotia, (1) late May, (2) late June; (3) *mediogriseus,* extreme southern Ontario (Canada), late May and early June (Broyd 1985); (4) *anthinus,* central-north Alaska, May-July (Irving 1960). (5) Kentucky (USA), April-May (Mengel 1965).

(1)	♂	21·0 (12)	18·5-23·2	♀	19·1 (15) 17·5-20·6
(2)		20·0 (31)	17·4-22·5		18·9 (22) 16·0-22·5
(3)		19·8 (41)	16·9-23·6		18·9 (14) 16·5-22·2
(4)		17·8 (9)	16·1-19·8		18·4 (2) 14·9-21·8
(5)		20·4 (6)	18·9-21·6		17·8 (2) 17·6-18·0

Sexes combined. Pennsylvania (USA): (6) August-October, (7) March-May (Clench and Leberman 1978). Coastal New Jersey (USA): (8) autumn (Murray and Jehl 1964).

(6)	17·4 (2·07; 29) 12·4-22·3	(8) 17·4 (— ; 16) 14·8-19·9
(7)	18·3 (1·47; 31) 15·4-23·0	

See also Grinnell *et al.* (1930), Baldwin and Kendeigh (1938), Tordoff and Mengel (1956), Norris and Hight (1957), Drury

and Keith (1962), and (for fat-free weights) Connell *et al.* (1960) and Odum *et al.* (1961).

Structure. Wing rather short, broad at base, tip rounded. 10 primaries: p7-p8 longest, p6 and p9 0·5-1·5 shorter, p5 2-4, p4 6-9, p3 9-12, p2 11-14, p1 12-16 (*savanna* and *mediogriseus*). Outer webs of p6-p8 emarginated, inner of (p7-)p8-p9 with notch. Tip of longest tertial reaches tip of p4-p5. Tail short, tip slightly forked; 12 feathers, t1 2-5 shorter than t5, t6 1-2 shorter. Bill short, fairly slender, conical; *c.* 5·5-6·5 mm deep and *c.* 5·7-6·3 wide at feathering near base in *savanna* and *mediogriseus,* *c.* 6·5 mm wide and deep in *princeps.* Base of culmen slightly elevated, tip slightly decurved; bill-tip sharp. Cutting edges straight, but sharply kinked downward below nostril, where edge of lower mandible forms blunt tooth. Nostril small, oval, partly covered by operculum above, partly covered by feathering of lore at rear. Some short and fine hair-like bristles at base of mandibles. Tarsus short, toes rather short, slender. Middle toe with claw 18·4 (10) 16·5-20·0 mm in *savanna* and *mediogriseus,* 20·3 (3) 19·5-21·0 in *princeps;* inner toe with claw *c.* 70% of middle with claw, outer *c.* 72%, hind *c.* 29%.

Geographical variation. Marked; mainly involves size (wing, tail, and tarsus length, or weight), relative length and depth of bill, ground-colour of head and body, and coarseness and quantity of dark streaking on body. Apart from large pale *princeps* of isolated Sable Island (off Nova Scotia), most other northern and eastern races closely similar in size and rather similar in colour; in non-breeding plumage, only extremes identifiable (Mengel 1965; Broyd 1985; Rising 1987; but see Norris and Hight 1957); races along Pacific seaboard often more distinct in colour or size, thick-billed birds from western Mexico and San Benito Island being particularly aberrant; these and some other races (including *princeps*) may each form separate species (Zink *et al.* 1991*b*). For western races, see Ridgway (1901-11), Hellmayr (1938), Peters and Griscom (1938), Aldrich (1940), Godfrey (1986), and Rising (1987). Eastern races darker and smaller than *princeps* described in Plumages: broad black-and-brown stripe at side of crown, broader black streaks on mantle, scapulars, upper tail-coverts, throat, chest, side of breast, and flank (especially in *labradorius* and *oblitus*; *savanna, mediogriseus,* and *nevadensis* intermediate between these and *princeps*), browner ground-colour of upperparts and side of head and neck, and warmer cinnamon to rufous tinge on greater upper wing-coverts, tertials, and outer fringes of secondaries. Upperparts of *savanna* medium brown, feather-fringes light brown; of *mediogriseus* greyish-brown, fringes light grey-brown or brown-grey; of *nevadensis* pale grey- or pink-brown, fringes ash-grey; of *labradorius* heavily marked rich brown and black; of *oblitus* heavily marked light grey and black. Side of head of *savanna* pale, often tinged buff, lore usually yellow; of *mediogriseus* grey-brown, sometimes tinged ochre, lore yellow or whitish; of *nevadensis* greyish, lore with limited cream or pale yellow; of *labradorius* dark, predominantly buff and brown, lore bright yellow; of *oblitus* dark, but without buff and brown, lore bright yellow. Greater coverts and fringes of tertials and secondaries medium cinnamon-brown in *savanna* (but much subject to bleaching in this and other races); of *mediogriseus* medium grey-brown; of *nevadensis* pale brown-grey; of *labradorius* dark rufous-brown; of *oblitus* rather pale cinnamon-brown or grey-brown. (Aldrich 1940; Norris and Hight 1957.) CSR

Passerella iliaca Fox Sparrow

PLATE 12
[between pages 256 and 257]

Du. Roodstaartgors   Fr. Bruant fauve   Ge. Fuchsammer
Ru. Пестрогрудая овсянка   Sp. Chingolo zorruno   Sw. Rävsparv

Fringilla iliaca Merrem, 1786. Synonym: *Zonotrichia iliaca*.

Polytypic. Nominate *iliaca* (Merrem, 1786), eastern Canada west to Ontario. Extralimital: *zaboria* Oberholser, 1946, western Canada and Alaska, north and east of coastal ranges and Rockies, east to Manitoba; *c.* 14–16 races from coasts and islands of southern Alaska through Rockies and Pacific coast of USA to Colorado and southern California.

Field characters. 17–19 cm; wing-span 26–28 cm. Close in size to Corn Bunting *Miliaria calandra* but with proportionally smaller, conical bill and rather longer tail. Large robust thicket-loving Nearctic sparrow; habitually scratches through leaf litter. Within heavily-streaked appearance, rusty or fox-red tones on head, wing, rump and tail catch eye. Flight bunting-like, trailing bright tail. Sexes similar; no seasonal variation. Juvenile inseparable. Nominate *iliaca* described here.

ADULT. Moults: July–September (complete). Ground of head, back, breast-sides and upper flanks light grey, of wings, lower rump and tail bright fox-red and of underparts white. Most striking character, particularly when flying away, is area of intense rufous on upper tail-coverts and tail. At closer range, shows rufous streaks on crown, rufous cheek and broad malar stripe, rufous and brown streaks on back, breast, and along flanks (often triangular in shape on flanks), isolated blackish and whitish terminal tips to median and lesser wing-coverts, and sharp almost black streaks on sides of belly. At some angles, narrow, pale almost white eye-ring and variable but usually deep grey supercilium, nape, and half-collar stand out. Head on, breast-streaks merge into loose patch. Bill yellowish-horn, with dusky culmen. Legs pale clay to flesh.

Unmistakable, being larger than House Sparrow *Passer domesticus* and all Nearctic relatives; also oddly coloured, recalling Dunnock *Prunella modularis* until rump and tail show. Flight strong and fast but with somewhat fluttering action and broad-rumped and broad-tailed silhouette; usually content to move away at low height but under pressure escapes into thicket or up to tree. Gait a hop; scratches vigorously in debris of undergrowth, using double-footed kick. Stance quite erect, trailing tail. Flicks wing and tail.

Commonest calls 'click' and 'chip'.

Habitat. Breeds in northern Nearctic lowlands and in some mountains further south, from Arctic to California. Lives in dense woodland, either coniferous or deciduous, favouring streamside growths of willow *Salix* and alder *Alnus*, regrowth on burnt patches of forest, stunted conifers on coast, and woodland thickets and edges. Forages largely on ground amid leaf litter, rarely showing itself in the open and avoiding human neighbourhood. Will sing from prominent perch but tends to avoid tall trees. (Pough 1949; Gabrielson and Lincoln 1959; Godfrey 1979.) In winter, may emerge into more open places and is at home in heavy snow. In Pacific coastal region, where it ascends well above 1500 m, it requires heavy shade, moist leaf litter-covered ground, and solitude from human activities (Larrison and Sonnenberg 1968).

Distribution. Breeds in North America, from western and northern Alaska east to northern Labrador, south to north-west Washington, in western mountains to southern California, central Utah, and central Colorado, and east of Rockies to central Alberta, central Ontario, southern Quebec, Nova Scotia, and Newfoundland. Winters from southern Alaska south through Pacific states to Baja California, and from New Mexico, Kansas, southern Wisconsin, southern Ontario, Nova Scotia, and southern Newfoundland south to southern Texas, coast of Gulf of Mexico, and Florida.

Accidental. Iceland: Borg, 5 November 1944 (Ólafsson in press). Ireland: Copeland Island (Down), 3–4 June 1961 (Dymond *et al.* 1989). Germany: Mellum, 13 May 1949 (Niethammer *et al.* 1964); Scharhörn, 24 April 1977 (Schmid 1979). Italy: Liguria 1936 (Brichetti and Massa 1984).

Movements. Migrant, varying greatly in status. Boreal races (nominate *iliaca*, *zaboria*) fully migratory; north-west races (*unalaschcensis* group) show 'leapfrog migration' down west coast, northernmost races wintering furthest south; western mountain races (*schistacea* group) vary from fully migratory to nearly sedentary with minor altitudinal movements (American Ornithologists' Union 1957; Bent 1968). Boreal races occupy two-thirds of total breeding area, and most migration data refer to them; some members of all other forms winter in California, where only specimens provide unambiguous data (Grinnell and Miller 1944; Bent 1968).

Autumn migration begins mostly in September, continuing to November with stragglers in December (Bent 1968). In centre and east of range, some migrants reach southern Alberta, Minnesota, and New York by early October (Bull 1974; Sadler and Myres 1976; Janssen 1987), but others remain in breeding areas well into October (e.g. Peters and Burleigh 1951, Speirs 1985). Main broad-front movement southward late October to November (Stewart and Robbins 1958; James and Neal 1986;

Bohlen 1989), by which time all parts of eastern winter range are occupied (Oberholser 1974; Imhof 1976; Potter *et al.* 1980). In west, patterns are less evident; movement is under way in August in Alaska, and by late September in southern British Columbia and Washington (Jewett *et al.* 1953; Gabrielson and Lincoln 1959; Cannings *et al.* 1987); in southern California, most local birds have left by mid-September, when immigrants begin to arrive (Garrett and Dunn 1981). North-west races move to California along narrow coastal flyway; inland, birds from British Columbia mountains move at high elevations without descending to valleys (Munro and Cowan 1947; Bent 1968).

In spring, birds depart from south-west wintering areas March–April (Phillips *et al.* 1964; Garrett and Dunn 1981), and reach northern British Columbia and southern Alaska late April or early May (Munro and Cowan 1947; Gabrielson and Lincoln 1959). Further east, birds begin migration late February or early March, moving north gradually at first, with peak passage in northern states and southern Canada in April (Palmer 1949; Bent 1968; Sadler and Myres 1976; Janssen 1987); reaches northern breeding areas in Labrador and Yukon from early May (Rand 1946; Todd 1963). Records and ringing recoveries show marked passage along Atlantic coast, presumably chiefly to Newfoundland, where many breed (Bent 1968).

Vagrant to Alaskan and Canadian Arctic, and to Greenland (American Ornithologists' Union 1983); *unalaschcensis* recorded in Chukotskiy peninsula and Komandorskie islands (USSR) (Stepanyan 1990), and twice in Japan (Brazil 1991). AJE

Voice. See Field Characters.

Plumages (nominate *iliaca*). ADULT. In fresh plumage (September to about April), cap, hindneck, upper side of neck, and upper mantle medium grey, each feather with broad deep rufous-chestnut tip, latter partly or largely concealing grey, usually except for almost uniform grey stripe above eye. Lower mantle and scapulars deep rufous-chestnut, sides of feathers fringed light olive-brown or buff-brown on tip and grey on base; outer webs of outer scapulars deep rufous-cinnamon. Back and rump light olive-brown or buff-brown, sometimes partly mottled rufous-chestnut, contrasting sharply in colour with bright cinnamon-rufous or fox-red upper tail-coverts; tips of tail-coverts narrowly fringed pale grey-buff when quite fresh. Lore and eye-ring mottled off-white, dull grey, and (sometimes) rufous, merging into finely streaked off-white and rich rufous-chestnut upper cheek and shorter ear-coverts; sometimes a short uniform pale buff or off-white stripe above lore. Lower cheek and longer ear-coverts rich rufous-chestnut, some pale olive or grey of feather-bases often visible on central ear-coverts, chestnut of lower cheek broken into 2 stripes (backwards from gape and backwards from lower corner of lower mandible) by short off-white stripe backwards from side of lower mandible, which is usually partly speckled chestnut. Lower side of neck rufous-chestnut, white of feather-bases partly visible. Ground-colour of underparts pale cream to white, tinged cream-buff on lower flank, thigh, and under tail-coverts; chin and throat either uniform or with small rich rufous-chestnut specks (sometimes joining to form narrow band across upper throat); chest, side of breast, and

flank with coarse broad-triangular rufous chestnut dots, which sometimes form a coalescent band across upper chest and side of breast and often form a rufous or fuscous patch on middle of lower chest; chestnut marks on flanks more elongate; upper belly and side of belly with smaller fuscous-brown or dull black triangles. Tail deep rufous-cinnamon, tinged brown or sooty on inner webs. Flight-feathers, greater upper primary coverts, and bastard wing greyish-black, outer web with rather narrow and ill-defined rufous-cinnamon fringe (faint on primary coverts, absent from longest feather of bastard wing), but fringe of p9 and emarginated parts of p5–p8 pale cinnamon to isabelline-white. Lesser upper wing-coverts grey or olive-grey with variable amount of rufous suffusion; median and greater upper wing-coverts and tertials greyish-black with broad rufous-chestnut outer fringe, latter largely concealing black except on tertials; median and greater coverts with small pale buff to off-white spot on tip of outer web, forming narrow broken wing-bar. Under wing-coverts and axillaries pale grey to off-white, partly variegated dull grey. *In worn plumage* (about May–July), rufous-chestnut feather-tips from cap to upper mantle partly or fully worn off, plumage here mainly grey with traces of chestnut spots (especially on crown) or with brown suffusion; lower mantle and scapulars olive-brown to greyish with deep rufous-chestnut streaks; white stripe backwards from lower mandible and pale olive-grey centre to ear-coverts often more distinct and usually a prominent white bar at lower rear of ear-coverts; rufous triangular spots on underparts narrower, smaller, tending less to coalesce; ground-colour of lower flank and under tail-coverts cream-white; spots on tips of median and greater upper wing-coverts whiter, but sometimes largely worn off. Much individual variation in amount of chestnut and grey on cap and neck, in part due to individual differences in abrasion, in part to possible geographic origin: spring migrants from Nova Scotia have extensive chestnut on cap and mantle and deep rufous on tail-coverts and tail, those of Massachusetts slightly less so, those of south-east Ontario and Michigan mainly grey on cap and mantle and paler cinnamon-rufous on tail-coverts and tail; those of Wisconsin greyer still, but latter perhaps attributable to western race *zaboria* or intergrades with this race. JUVENILE. Rather like adult, differing mainly in looser and softer texture of feathers of body, especially of vent and under tail-coverts; upperparts on average duller, dark feather-centres less sharply demarcated from pale fringes; underparts washed buff, white ground-colour less pure; streaks on chest narrower, less blotched; top and side of head and neck uniform chestnut, but lore and eye-ring buff-white and small white bar behind ear-coverts; mantle, scapulars, and back marked with ill-defined dark chestnut and rusty-buff streaks. For wing and tail, see First Adult. FIRST ADULT. Like adult, but juvenile tail, flight-feathers, and greater upper primary coverts retained. Juvenile p9 and tail-feathers on average narrower than in adult, tips more pointed, less truncate than in adult (see Pyle *et al.* 1987 for shape), but some birds intermediate in shape and these difficult to classify; tips of juvenile outer primaries, greater upper primary coverts, and (especially) tail-feathers more worn than those of adult at same time of year, especially in spring and summer.

Bare parts. ADULT, FIRST ADULT. Iris hazel, light brown, or dark brown. Bill greyish-brown, dusky brown, dark plumbeous-grey, or greyish-black, basal half of lower mandible yellow, orange-flesh, yellow-orange, bright orange, or pinkish-grey. Leg and foot dull flesh-grey, brownish-pink, light brown, or purplish brown-grey. (Wilde 1962; RMNH, ZMA.) JUVENILE. No information.

Moults. ADULT POST-BREEDING. Complete; primaries descendent. On breeding grounds, July–September (Pyle *et al.* 1987). ADULT PRE-BREEDING. Restricted to some feathers on chin (Dwight 1900), March–April (Pyle *et al.* 1987), or no moult at all (Swarth 1920). POST-JUVENILE. Partial: head, body, and lesser and (usually) median upper wing-coverts, in breeding area, July–September; sometimes also greater upper wing-coverts and tertials, either all or partly in breeding area or in winter quarters (Pyle *et al.* 1987; RMNH, ZMA).

Measurements. ADULT, FIRST ADULT. Nominate *iliaca*. Eastern Canada, summer, and eastern USA, September–May; skins (RMNH, ZFMK, ZMA). Bill (S) to skull, bill (N) to distal corner of nostril; exposed culmen on average 4·2 less than bill (S).

WING	♂	90·2 (1·45; 14)	87–92	♀ 85·6 (2·69; 7)	81–89
TAIL		70·6 (2·08; 14)	67–74	66·2 (3·11; 7)	62–70
BILL (S)		15·3 (0·50; 14)	14·7–16·1	15·1 (0·86; 7)	14·2–16·5
BILL (N)		8·8 (0·41; 14)	8·1–9·4	8·5 (0·35; 7)	8·1–9·0
TARSUS		25·1 (0·55; 14)	24·3–25·9	24·8 (0·62; 7)	24·0–25·7

Sex differences significant for wing and tail.
    See also Swarth (1920) and Linsdale (1928).

Weights. ADULT, FIRST ADULT. Nominate *iliaca*. Pennsylvania (USA): October–November, (1) adult, (2) 1st adult, (3) age unknown; (4) March–May (Clench and Leberman 1978, which see for details).

(1)	37·0 (3·40; 84) 27·2–49·0	(3)	38·4 (4·00; 43) 42·0–46·2
(2)	36·2 (2·80; 106) 29·6–43·9	(4)	36·9 (2·72; 212) 29·4–47·0

Indiana (USA): October 33·6 (Esten 1931). Ohio (USA): April 36·6 (*n*=9), October–November 36·3 (*n*=9) (Baldwin and Kendeigh 1938). Kentucky (USA), November–January: ♂♂ 38·3, 45·6; ♀ 39·5 (3) 37·0–43·2 (Mengel 1965). Ireland, June: 40·3 (Wilde 1962). See also Weise (1962).
    P. i. zaboria. North-central Alaska, May–August, 37·4 (4) 34·3–42·7, ♀ 32·7 (3) 30·2–34·0; north-west Yukon (Canada), May–June, ♂ 35·9 (*n*=25), ♀ 34·3 (*n*=3) (Irving 1960).
    Other races. All races combined: ♂ 32·2 (362), ♀ 31·5 (239) (Linsdale 1928, which see for details of many races). See also Grinnell *et al.* (1930) and Linsdale and Sumner (1934).

Structure (nominate *iliaca*). Wing fairly long, broad at base, tip bluntly pointed. 10 primaries: p7 longest, p8 and p6 0–2 shorter, p9 3·5–7·5, p5 2–5, p4 6–11, p3 10–15, p2 12–18, p1 14–20; p10 strongly reduced, a tiny pin hidden beneath reduced outermost greater upper primary covert, 54–65 shorter than p7, 9–14 shorter than longest upper primary covert. Outer web of p5–p8

emarginated, inner web of (p7-)p8–p9 with notch. Tip of longest tertial falls slightly short of tip of p1 or reaches up to tip of p2. Tail long, tip square or slightly notched; 12 feathers, t3–t5 longest, t1 and t6 0–3 mm shorter. Bill short, conical, slightly shorter than half of head-length; rather slender at base (depth 7·6–9·2, width 7·5–9·0 mm), sharply pointed; culmen, cutting edges, and gonys almost straight. Nostril small, rounded or triangular, covered by short tuft of feathers projecting from base of upper mandible; some short fine bristles at base of upper mandible projecting over gape. Tarsus and toes strong, rather long, claws markedly long and strong. Middle toe with claw 21·8 (6) 20–24 mm; outer toe with claw *c.* 85% of middle with claw, inner *c.* 84%, hind *c.* 82%.

Geographical variation. Marked, both in colour and size. About 18 races recognized (Swarth 1920; Paynter 1970), but should probably be fewer, as some races in south apparently not valid (Phillips *et al.* 1964). In general, the 3 northern and eastern races (including nominate *iliaca* and *zaboria*, which are the most widespread) have mantle and scapulars reddish to grey, spotted fox-red; tail and spots on underparts fox-red; wing-tip rather bluntly pointed; tail shorter than wing; bill of medium size; *zaboria* on average greyer than nominate *iliaca* (less rufous), especially on head, spots on underparts more sooty, less rufous, and size somewhat larger, but with much individual variation (Ridgway 1901–11; see also Plumages). The 6 races of Pacific coast from Alaska to north-west Washington have mantle and scapulars rather uniform fuscous-brown or dull sooty-brown (brown predominant, only a small amount of grey), spots on underparts fuscous-brown or olive-brown (gradually darker from west to east); tail sooty-brown; wing-tip short, rounded; tail shorter than wing; bill of medium size (relatively smaller eastward). Remaining races, occurring inland from British Columbia to California and in coastal ranges of south-west USA, have rather pale and uniform grey head, mantle, scapulars, and rump (grey predominating, especially in south), grey contrasting with dull red-brown of wing and tail; spots on underparts dull; tail equal in length to or longer than short and rounded wing; bill large and swollen (especially in west and south). For further details, see Ridgway (1901–11), Swarth (1920), Linsdale (1928), Hellmayr (1938), Phillips *et al.* (1964), and Bent (1968).
    Monotypic genus *Passerella* sometimes included in *Zonotrichia* (Paynter 1964), and *Passerella*, *Zonotrichia*, and *Melospiza* sometimes in *Junco* (Short and Simon 1965), but *iliaca* is a highly distinctive form, differing from *Zonotrichia*, *Melospiza*, and *Junco* in many details, and therefore here retained in separate genus, following Zink (1982) and Parkes (1990), contra Voous (1977).   CSR

Melospiza melodia **Song Sparrow**

PLATE 12
[between pages 256 and 257]

DU. Zanggors   FR. Bruant chanteur   GE. Singammer
RU. Певчая овсянка   SP. Chingolo melodioso   Sw. Sångsparv

Fringilla melodia Wilson, 1810. Synonym: *Zonotrichia melodia*, *Passerella melodia*.

Polytypic. Nominate *melodia* (Wilson, 1810) eastern Canada and neighbouring north-east USA, west to eastern Ontario, south through New England, eastern New York, and eastern Pennsylvania to central Virginia. Extralimital: *atlantica* Todd, 1924, coastal strip of eastern USA from Long Island to Virginia; *euphonia* Wetmore, 1936, Appalachians (eastern USA), south from western New York, west of nominate *melodia*; *juddi* Bishop, 1896, central Canada and north-central

USA, west from central Ontario in Canada and from western foot of Appalachians west to eastern foot of Rockies; *c.* 35 further races in western North America, west to Aleutian Islands and south to western Mexico.

Field characters. 15–16·5 cm; wing-span 20–21·5 cm. Close in size to Reed Bunting *Emberiza schoeniclus* but shape and stance also reminiscent of Dunnock *Prunella modularis*; averages distinctly larger than continental Nearctic races of Savannah Sparrow *Ammodramus sandwichensis* and has noticeably longer rounded tail. Medium-sized but quite long Nearctic sparrow; epitome of streaked members of that group. Plumage sharply streaked, with lined head, rather pale nape, and black spot in centre of chest of adult; no yellow on face. Pumps tail in flight. Call distinctive. Sexes similar; no seasonal variation. Immature not separable. Eastern nominate race *melodia* described here.

ADULT. Moults: August–September (complete). Supercilium, cheeks, and hindneck pale grey; head further marked by narrow, inconspicuous buff-grey central crown-stripe, broad rufous-brown lateral crown-stripe, rufous-brown rear eye-stripe and broken edge of ear-coverts, and fainter rufous streaks on shawl. Nape olive-grey, faintly streaked. Back greyish-brown, sharply streaked brown-black on mantle; rump more olive, becoming almost rufous near dark-centred tail-coverts. Wings rufous-brown, with median and greater coverts blackish, tipped buff (but hardly showing wing-bars). Tail distinctly rufous on sides of base but dusky-brown on end. Underparts off-white, with pale submoustachial stripe contrasting with rufous-black malar stripe which ends in broken patch; breast and flanks heavily streaked rufous-black to brown-black, marks usually coalescing in centre of breast to form large spot. Bill greyish-horn with darker tip and upper mandible. Legs pale bright brownish-flesh. FIRST-YEAR. As adult.

With size and appearance midway between Fox Sparrow *Passerella iliaca* and *A. sandwichensis savanna* and not unlike *E. schoeniclus*, *M. melodia* not easy to identify. Nevertheless, *E. schoeniclus* shows white outer tail-feathers freely, and larger *P. iliaca* has fully chestnut rump and tail. Separation from *A. sandwichensis* requires close observation of tail shape (noticeably short and notched in *A. sandwichensis*), supercilium (not grey, but wholly or partly yellow or white in *A. sandwichensis*) and wings (far less strongly rufous in *A. sandwichensis*). Flight light and fast but made to look laboured and awkward by pumping of relatively long, rounded tail. Gait a hop. Stance rather level; often assumes slightly hunched posture recalling *P. modularis*. Perches on top or side of bushes, but tends to skulk in cover when disturbed. Will feed on open ground.

Calls include characteristic, slightly harsh 'chirup' or 'chepp', recalling House Sparrow *Passer domesticus* but croakier; also fine, high-pitched 'tsii'.

Habitat. Breeds widely in temperate and adjoining climatic zones of Nearctic, mainly in lowland or upland, but locally to 1500 m or higher. A typical edge species, inhabiting thickets of shrubs and trees among grassland, brushy margins or openings of forest, brushy edges of ponds or lakes, shrub swamps, shelterbelts, farmsteads, and sometimes parks or suburbs (Johnsgard 1979). Commonly occupies cleared land or abandoned farmland, between establishment of first few shrubs and formation of closed canopy. Also lives in wild alder *Alnus* swamps, by shrub-fringed woodland lakes and around houses. (Pough 1949.) Likes well-watered fertile country where thickets abound, with great fondness for water (Forbush and May 1955). In Canada, also not a bird of closed-canopy mature forest, living in farmland thickets and bushy places along margins of woods, ponds, lakes, and streams, as well as in hedgerows and about buildings (Godfrey 1979). On Pacific coast, prefers low dense brush and moist ground, also often occupying salt-marshes and driftwood along upper ocean beaches (Larrison and Sonnenberg 1968). In Alaska, however, infrequently found in typical land habitats outlined above, mainly occupying rocky marine beaches, and even docks or wharves (Gabrielson and Lincoln 1959).

Distribution. Breeds in North America, from southern Alaska (including Aleutian Islands) east to Newfoundland, south to Baja California, highlands of Mexico south to Puebla, northern New Mexico, north-central Arkansas, northern Georgia, and coastal South Carolina. Winters from southern Alaska (Aleutian Islands) and coastal and southern British Columbia east through northern USA to south-east Canada, south throughout rest of breeding range and to southern Texas, Gulf of Mexico coast, and southern Florida.

Accidental. Britain: see Movements. Norway: ♂, Østfold, 11 May 1975 (Ree 1976).

Movements. Status varies, showing little correlation with racial subdivision. Only *inexpectata* from northern Rockies withdraws entirely from breeding area, but *juddi* and nominate *melodia*, the other northern inland races, are also mainly migratory. In general, populations of northern coasts and of mid-latitudes inland are partly migratory, with some resident races in Aleutian Islands (Alaska); southern populations, especially in south-west states and Mexico, are sedentary. (Grinnell and Miller 1944; American Ornithologists' Union 1957, 1983; Bent 1968; Godfrey 1986.) Studies have shown that in same populations some birds migratory, others sedentary. Individuals reported in winter virtually throughout breeding range. (Bent 1968.)

Leaves breeding grounds late, and returns early. In

autumn, a few birds are detected away from breeding areas in early September, mostly in northern states, and soon after mid-September migration is under way throughout range. Main passage late September to October almost everywhere, with peak numbers in October. (Bent 1968.) Most birds have left northern inland areas by early October, but some linger and attempt to winter (Salt and Salt 1976; Speirs 1985; Janssen 1987). In areas with less harsh winters, migration is more prolonged, extending regularly into early November (e.g. Bailey 1955, Tufts 1986, Bohlen 1989). In west, wintering is widespread in valleys north to interior British Columbia, and, except where flocks are seen moving, specimens are often the only evidence of migration of more northern races; there too, most passage late September to October (Munro and Cowan 1947; Cannings *et al.* 1987). Movements (often short) occur on broad front east of Rockies; routes of western races tend to follow coast, or valleys in mountainous terrain (Bent 1968). Birds ringed New York recovered late autumn or winter from Mississippi east to South Carolina, and in spring or summer in Nova Scotia and eastern Quebec (Bull 1974); winter recoveries in Alabama include birds ringed Ohio east to Atlantic coast (Imhof 1976).

Spring migration begins early March in many areas; reports of migrants in late February frequently reflect resumption of singing by birds that have wintered unseen. Wintering areas south of breeding range are usually vacated by late April or early May (Phillips *et al.* 1964; Sutton 1967; Kale and Maehr 1990). First arrivals in Great Lakes area are before mid-March, and main passage reaches north-east states and New Brunswick by late March (Nice 1937; Quilliam 1973; Bull 1974; Squires 1976). Northward migration in west is rapid, some birds passing beyond 60°N before mid-April, though some breeding areas in far north-west are not occupied until early May (Munro and Cowan 1947; Nero 1967; Bent 1968; Scotter *et al.* 1985). Northern breeding areas in east, where snowfall is greater, become accessible later, and birds reach northern Ontario, central Quebec, and Newfoundland during May (Peters and Burleigh 1951; Todd 1963).

No reports of vagrancy north to Arctic North America. Rare spring vagrant to west Palearctic, including 6 records from Britain up to 1990, mid-April to early June (Dymond *et al.* 1989; Rogers *et al.* 1990, 1991).   AJE, PRC

Voice. See Field Characters.

Plumages. (nominate *melodia*). ADULT. In fresh plumage (autumn and winter), forehead and crown dark rufous-cinnamon with narrow black streaks; narrow light olive- or brown-grey stripe on middle of forehead, widening on mid-crown, dividing rufous-cinnamon of cap in 2 broad stripes. Supercilium broad and contrasting, widest above eye; mainly light ash-grey, but tinged buff where reaching base of upper mandible and there gradually merging in buff-and-grey mottled lore and short streak below eye. Eye-ring pale buff, inconspicuous, broken in front

and behind. Dark rufous-cinnamon stripe from just below and behind eye backwards over upper ear-coverts, bordered below by buff, olive-buff, or buff-grey patch on middle of ear-coverts (partly or fully bordered by rufous or sooty bar at rear), and this in turn by dark rufous-cinnamon stripe extending from gape backwards over lower ear-coverts. A contrasting cream or off-white stripe extends backwards from lower mandible over lower cheek, widening towards rear; bordered below by mixed black and rufous-brown malar stripe, widening into patch on side of lower throat. Hindneck and side of neck olive-grey with narrow and ill-defined rufous streaking. Feathers of mantle and scapulars deep rufous-cinnamon, each streaked black on centre and fringed light olive-grey on side. Back, rump, and upper tail-coverts olive-grey or buff-olive, tail-coverts and sometimes feathers elsewhere with rufous centre and black shaft-streak. Chin and throat white with buff tinge, sometimes partly marked with indistinct grey or rufous triangular spots. Chest, side of breast, and flank pale cream-buff to cream-white with contrasting black elongate-triangular streaks, each of which shows rufous at border of black; lower flanks with similar but longer streaks; short streaks on lower mid-chest often confluent, forming distinct patch. Belly and vent cream-white; under tail-coverts pale cream-buff with small rufous-and-black spots on centres (partly concealed). Tail rufous-brown, slightly brighter rufous towards outer borders of feathers; central pair (t1) sooty along shaft, t2-t6 tinged dark grey on inner webs. Flight-feathers, greater upper primary coverts, and bastard wing greyish-black, outer webs of primaries, primary coverts, and bastard wing fringed buff-brown, fringe almost white on outer web of p9, at emarginations of p6-p9, and on longest feather of bastard wing, warmer rufous-cinnamon on outer webs of secondaries. Lesser upper wing-coverts olive-grey with rufous centres and tips; median dusky cinnamon-brown with sooty centre and pale olive-grey fringe; greater coverts and tertials deep rufous-cinnamon with broad black drop subterminal on centre, pale rufous-cinnamon sides, and narrow pale cream-buff fringe along tip. Under wing-coverts and axillaries pale greyish- or creamy-white, longer coverts grey. *In worn plumage* (spring and early summer), central crown-stripe, fringes of mantle and scapulars, and rump paler and greyer; black streaks of mantle and scapulars shorter, more sharply-defined; dark stripes on side of head more contrasting with head ground-colour; underparts whiter, pale cream-buff more restricted, dark spots and streaks smaller and more contrasting; tail and fringes of flight-feathers, tertials, and greater upper wing-coverts bleach to buff-brown or greyish-buff, less warm rufous. JUVENILE. Head marked with distinct fuscous-brown stripes at each side of crown, over upper and lower ear-coverts, and below lower cheek, separated by faint fuscous-mottled pale buff stripe over mid-crown, distinct broad pale buff supercilium, warmer buff stripe over upper cheek, and paler buff stripe over lower cheek. Hindneck, upper mantle, and back to upper tail-coverts olive-brown with dusky grey mottling (blacker on tail-coverts); lower mantle and scapulars black with broad rufous (on outer webs) or buff (on inner webs) fringes. Underparts cream-yellow to grey-white, marked with short and narrow triangular sooty spots on side of throat, chest, side of breast, and flank, forming streaks. Lesser upper wing-coverts greyish-olive-brown with black centres; median black, narrowly tipped cream-buff or white; greater coverts and tertials rufous-cinnamon with black central mark and narrow cream-buff to white tip. ♂ on average more heavily streaked on chest than ♀; occasionally, a large dark solid patch on mid-chest (Sutton 1935). Texture of plumage looser than in adult. For tail and remainder of wing, see 1st adult, below. For development, see Nice (1943). For identification of juvenile of this and some related species, see Rimmer (1986). FIRST ADULT.

Like adult, but juvenile greater upper primary coverts and some or all flight-feathers retained, these more pointed and more frayed on tip than in adult at same time of year; in birds with part of flight-feathers new (usually outer primaries or inner secondaries), new feathers often show a slight contrast in colour and wear with neighbouring older juvenile ones (unlike adult). Many birds apparently indistinguishable from adult, especially ♀♀.

Bare parts. ADULT, FIRST ADULT. Iris brown or dark brown. Upper mandible and tip of lower mandible dark grey, dark slate, or black, cutting edge and remainder of lower mandible dusky horn-grey in winter, pale grey or steel-grey in summer, sometimes with some pink tinge at base. Leg and foot pink-flesh, tinged brown on front of tarsus and on upper surface of toes. JUVENILE. Iris brown. Bill pink-flesh to dull flesh-grey, cutting edges paler yellowish-horn-grey. Leg and foot pink-flesh or pale greyish-flesh. (Dwight 1900; Sutton 1935; Davis and Dennis 1959; Spence and Cudworth 1966; Wright 1972; RMNH.)

Moults. ADULT POST-BREEDING. Complete; primaries descendent. On breeding grounds, mid-August to late September (eastern races: Dwight 1900) or July–October (whole geographical range: Pyle *et al.* 1987). On Vancouver Island (western Canada), primary moult starts with loss of p1 1–20 July (score 1), completed with regrowth of p9 (score 45) early to late August, occasionally up to 10 September; secondaries start with primary moult score 20–25, completed with outer primaries or a few days later; tertials start at score 5–20, completed at score 30–40; tail starts at score 5–20, completed at 35–45; sexes start at same time, when young independent in early breeders, when fledglings fed in late breeders (Dhondt and Smith 1980). No pre-breeding moult. POST-JUVENILE. Partial or almost complete. Starts between early July and late September, completed September to early November; individual duration *c.* 2 months. Includes head, body, and lesser, median, and greater upper wing-coverts, often bastard wing and tertials (sometimes excluding longest feather), usually all tail, and (especially in birds from early broods) 5–6 outer primaries; rarely, some secondaries; at least greater upper primary coverts, inner primaries, and most secondaries retained. (Dwight 1900; RMNH, ZMA.)

Measurements. ADULT, FIRST ADULT. Eastern Canada and USA (east from Ontario and Great Lakes), comprising nominate *melodia* and *euphonia* which are basically similar in measurements; all year, skins (RMNH, ZMA). Bill (S) to skull, bill (N) to distal corner of nostril; exposed culmen on average 3·2 less than bill (S).

	♂		♀	
WING	67·6 (1·82; 13)	64–70	64·4 (1·46; 8)	62–67
TAIL	63·5 (3·13; 13)	59–69	61·4 (3·05; 7)	57–66
BILL (S)	13·8 (0·36; 13)	13·1–14·4	13·8 (0·19; 8)	13·5–14·0
BILL (N)	8·3 (0·37; 13)	7·8–8·8	8·2 (0·20; 8)	8·0–8·5
TARSUS	21·8 (0·60; 12)	20·8–22·5	21·3 (0·52; 8)	20·4–22·1

Sex differences significant for wing.

Wing of ♂ mainly 64–69, of ♀ mainly 60–64 (Nice 1937). For data from many populations across whole of North America, see Aldrich (1984). For heritability of some measurements, see Smith and Zach (1979). Bill in summer slightly longer than in rest of year (Davis 1954).

Weights. ADULT, FIRST ADULT. Nominate *melodia* and *euphonia*, combined. Pennsylvania (USA) (Clench and Leberman 1978):

DEC–FEB	23·8 (35) 20·2–28·1	JUN–AUG	19·4 (464) 11·9–25·2
MAR–MAY	21·2 (2045) 11·4–29·1	SEP–NOV	20·6 (3052) 11·7–28·0

In these samples, all ♂♂ 21·0 (259) 18·2–29·8, all ♀♀ 20·4

(187) 11·9–26·1, all unsexed birds 20·7 (5150) 11·4–29·1 (Clench and Leberman 1978, which see for many details).

Ohio (USA), average and sample size only (Baldwin and Kendeigh 1938, which see for details).

	AD ♂	AD ♀	BOTH SEXES AND ALL AGES
APR–MAY	21·2 (288)	20·8 (118)	21·3 (846)
JUN–JUL	20·7 (171)	19·9 (115)	19·3 (1835)
AUG–SEP	21·8 (27)	20·3 (42)	20·5 (1202)
OCT–MAR	—	—	22·0 (106)

Ohio: all year, ♂ 23·0 (463) 18·8–30·0, ♀ 21·4 (283) 17·0–26·0 (Nice 1937, which see for details); ♂ only, October–November 21·8 (51) 18·9–24·3, December–January 24·4 (55) 21·7–30·0, February–March 23·7 (152) 20·1–28·4 (Nice 1946, which see for differences between local birds and visitors). Coastal New Jersey (USA), autumn: 20·3 (2·27; 115) 16·4–24·9 (Murray and Jehl 1964). Massachusetts (USA), April: 21·9 (1·78; 105) 17·4–26·4 (Helms 1959, which see for details). Britain, late April and May: 2 ♂♂, 22·9–24·5 and 24; a possible ♀, 18·6–19·2 (Davis and Dennis 1959; Spence and Cudworth 1966; Wright 1972). For USA, see also Stewart (1937), Nice (1938), Drury and Keith (1962), and Mengel (1965).

Other races. See Grinnell *et al.* (1930), Hoffman (1930), Smith and Zach (1979), and Rimmer (1986).

Structure. Wing short, broad at base, tip rounded. 9–10 primaries: p6–p7 longest, p8 1–3 shorter, p9 6–10, p5 0–2·5, p4 2–4, p3 4·5–7·5, p2 6–9, p1 (6–)7–12; tip of p9 between tip of p1 and p3. P10 strongly reduced, a tiny pin hidden below reduced outermost greater upper primary covert, or apparently entirely absent; if present, 43–50 shorter than p6–p7, 6–10 shorter than longest upper primary covert. Outer web of p5–p8 emarginated; inner web of (p5–)p6–p9 with notch (sometimes indistinct). Tip of longest tertial reaches to about tip of p1. Tail rather long, tip rounded; 12 rather soft and narrow feathers with rounded tips, t6 6–12 mm shorter than t1. Bill short, conical, rather heavy at base in races of eastern North America (depth at base 7–8 mm, width 6·5–7·5 mm), relatively more slender in many western races. Culmen slightly decurved, gonys straight or slightly convex; cutting edges straight, except for marked downward kink at extreme base. Nostril small, almost round, partly covered by short bristly feathers extending from lore. Some short and fine bristles project obliquely downward from lateral base of upper mandible over gape. Tarsus and toes rather long, strong. Middle toe with claw 19·0 (10) 18–20 mm; outer toe with claw *c.* 73% of middle with claw, inner *c.* 71%, hind *c.* 82%.

Geographical variation. Marked, but largely clinal. Involves mainly size (wing, tail, tarsus, or bill length), relative bill depth, ground-colour of upperparts and underparts, and width, extent, and contrast of black streaking on upperparts and underparts. Eastern and central races (nominate *melodia*, *atlantica*, *euphonia*, and *juddi*) of intermediate size with relatively heavy bill. Size of races from Rockies (from inland British Columbia south to California) about equal to eastern ones, but bill more slender. Races from Pacific coast (British Columbia to Oregon) and of Mexico larger and bill even more slender. Most races from coastal California, including offshore islands, markedly small and with slender bills. Those of Aleutian Islands and southern Alaska are giants, with long and slender bills. Variation in colour more difficult to assess due to marked influence of bleaching and wear on ground-colour and contrast of dark streaks, birds in fresh plumage markedly different from those of same race in worn plumage. In spring, *atlantica* much greyer above than nominate *melodia*, black streaks more distinct, red-brown of feather-centres reduced in extent; *euphonia* has upperparts greyer and darker than nominate *melodia*, side of head greyer, less buff and brown, tail on average darker (Wetmore 1936); *juddi* has contrast and extent of black of streaks as in *atlantica*, but ground-colour of upperparts much paler grey, red-brown feather-centres paler and more restricted; dark streaks on underparts restricted in extent, more sharply-defined, ground-colour more clearly white. In western races, colour generally much darker than in eastern ones, but much variation according to race. For survey of races, see Ridgway (1901–11), Wetmore (1936), Hellmayr (1938), and Bent (1968). For trends of variation in size, see Aldrich (1984).

For retention of separate genus *Melospiza* for *melodia* and related species rather than merging it with *Passerella* (as advocated by Linsdale 1928 and Mayr and Short 1970) or with *Zonotrichia* (as done by Paynter 1964, 1970), see Zink (1982) and American Ornithologists' Union (1983).   CSR

Zonotrichia georgiana (Latham, 1790) Swamp Sparrow

Fr. Bruant des marais   Ge. Sumpfammer

A North American species, breeding east of Rocky Mountains from northern Canada south to Nebraska, Illinois, and Maryland, wintering in southern parts of breeding area and south to central Mexico. Up to 7 present on eastbound ship in North Atlantic in October 1962, at least 1 staying until *c.* 30°W on 11 October (Durand 1963).

Zonotrichia leucophrys White-crowned Sparrow

PLATE 13
[between pages 256 and 257]

Du. Witkruingors   Fr. Bruant à couronne blanche   Ge. Dachsammer
Ru. Белоголовая зонотрихия   Sp. Chingolo piquiblanco   Sw. Gråstrupig sparv

Emberiza leucophrys J R Forster, 1772

Polytypic. Nominate *leucophrys* (J R Forster, 1772) (synonym: *nigrilora*), eastern Canada from Newfoundland and Labrador west to north-central Ontario, where it grades into *gambelii* in Fort Severn area (southern shore of Hudson Bay). Extralimital: *gambelii* (Nuttall, 1840), central and western Canada (west from north-west Ontario) and Alaska, south in Rockies to *c.* 52°N; *oriantha* Oberholser, 1932, inland western USA south to east-central California, Arizona, and north-west New Mexico, north in Canada to south-east British Columbia, southern Alberta, and south-west Saskatchewan, grading into *gambelii* in Rockies; *pugetensis* Grinnell, 1928, coastal areas of western North America from south-west British Columbia to north-west California; *nuttalli* Ridgway, 1899, coastal strip of central California.

Field characters. 16–18·5 cm; wing-span 23–25·5 cm. Similar in size to Reed Bunting *Emberiza schoeniclus*; up to 10% longer than White-throated Sparrow *Z. albicollis* with longer neck and often erect posture. Quite large, relatively elegant Nearctic sparrow, with bunting-like form and stance. Plumage mainly unstreaked pearly-grey below and heavily streaked rufous above; adult has black and white crown-stripes, whitish throat, and double white wing-bar. Sexes similar; no seasonal variation. Immature separable. Eastern race, nominate *leucophrys*, described here.

ADULT. Moults: June–August (complete), November–April (mainly head and body). Narrow frontal band, lateral crown-stripe, lore near eye, and narrow rear eye-stripe black, contrasting with pure white central crown-stripe and white rear supercilium and pearl-grey face, nape, breast, and fore-underparts. Back grey to pale brown, with dark brown streaks; rump and rear flanks tawny. Wings warm buff-brown, brightest on coverts and edges of tertials; median and greater coverts have blackish centres and white tips forming double wing-bar; tertials show similar terminal marks. Tail dark brown. Rear belly, vent, and under tail-coverts buff-white. Bill pink-horn, with dark greyish upper mandible. Legs bright flesh or brown. FIRST-WINTER. Differs distinctly from adult in grey-buff central crown-stripe, duller dark brown head-stripes, duller cream throat, partly brown ear-coverts, and buff-brown (less grey) nape and mantle. Pearl-grey of underparts restricted to collar and breast.

In brief view, adult and immature easily confused with *Z. albicollis*, separation requiring clear view of head markings and (with immature) underparts: *Z. albicollis* always has sharp-etched white throat and at least yellowish fore-supercilium in adult, streaked chest and flanks in winter (particularly 1st winter), and wholly dark bill at all ages. Pine Bunting *E. leucocephalos* is another potential confusion species but differs in strongly rufous rump and white outer tail-feathers. Flight and behaviour recall *E. schoeniclus* and other buntings. Usually more extended neck and more erect posture give quite different posture from *Z. albicollis*. Feeds in open near dense cover, being less secretive than *Z. albicollis*.

Calls include rather sharp, metallic 'pzit', suggesting alarm-call of Pied Flycatcher *Ficedula hypoleuca*; also a thin, high 'tssiip' or 'seeet'.

Habitat. Breeds extensively in Nearctic from northern Alaska and Canada south through temperate zone. In high latitudes inhabits patches of dwarf birch *Betula* and dwarf willow *Salix* on tundra edge or along streams and depressions, and areas of recumbent and wind-swept stunted spruce *Picea* on exposed coasts. Throughout range, primarily a bird of woody shrubbery and thickets in more open situations; it also occupies bushy edges of woodlands, openings, old burns, and mountainside shrubbery. (Godfrey 1979.) In Alaska, inhabits small shoreline patches of alder *Alnus* and willows, ranging up to limit of trees; during migration abundant along roadsides (Gabrielson and Lincoln 1959). In New England, spring migrants frequent cultivated fields, pastures, roadsides, and bordering thickets; in autumn, feeds wherever weed seeds are abundant, in cornfields, potato fields, or by roadsides, preferring to be near cover (Forbush and May 1955). Sings usually from high perch on bush or tree and feeds mostly on ground (Pough 1949). On American west coast, particularly abundant near salt water, especially in brush or small tree growth; also favours brush and willow thickets by meadows, streams, and lakes, generally requiring nearby water, damp grass, and bushy retreats. Often occurs also in brushy places in gardens and parks; ranges from low to high elevations, including alpine ridges (Larrison and Sonnenberg 1968).

Distribution. Breeds in North America, from western and northern Alaska east to central Keewatin, south in west coastal areas and mountains to southern California, central Arizona, and northern New Mexico, and from northern Saskatchewan and northern Manitoba east across northern Ontario to Labrador, northern Newfoundland, and central Quebec. Winters mainly from southern British Columbia, Idaho, and Wyoming south to northern Mexico, southern Texas, and coast of Gulf of Mexico east to north-west Florida, less regularly in eastern coastal areas of USA from Massachusetts south to Florida, Bahamas, and Greater Antilles.

Accidental. Iceland: Heimaey, 4–6 October 1978 (Ólafsson in press). Britain: Fair Isle, 15–16 May 1977; Humberside, 22 May 1977 (Rogers *et al.* 1979). France: Barfleur (Manche), 25 August 1965 (Dubois and Yésou 1986). Netherlands: December 1981 to February 1982 (Berg 1989).

Movements. Status varies: subarctic races *gambelii* (in west) and nominate *leucophrys* (in east) are long-distance migrants, but a few southern *gambelii* winter within breeding range; western subalpine *oriantha* vacates breeding area, but its movements are often short; of west coast races, *nuttalli* is sedentary, but more northern *pugetensis* is migratory (short-distance) although a few winter in breeding areas (American Ornithologists' Union 1957, 1983; Bent 1968; Cannings *et al.* 1987). For review, see Cortopassi and Mewaldt (1965). Rest of account refers to nominate *leucophrys* and *gambelii*.

Southward migration begins late July in Alaska, and most birds leave north of range by late August, with arrival of first snows (Rand 1944, 1946; Bent 1968; Scotter *et al.* 1985). Movement is gradual at first, with main passage through southern Canada and northern states mid- or late September to mid-October (Bull 1974; Sadler and Myres 1976; Squires 1976; Janssen 1987). Reaches southern California from mid-September (Garrett and Dunn 1981; Morton and Pereyra 1987), but arrivals later further east: from mid-October in Kentucky (Mengel 1965) and late October in Texas (Oberholser 1974). Ringing data show that nominate *leucophrys*, from breeding areas as much as 3000 km apart (east to west), head between south and south-west and converge to winter only a few hundred km apart, chiefly central Texas to western Kentucky; winter range of *gambelii* overlaps with nominate *leucophrys* but is chiefly from western Texas westward; western populations of *gambelii* migrate through inland valleys and winter further inland than south-west races. Birds apparently do not follow identical route in successive years; no returns to passage sites recorded, despite intensive ringing. Winter site-fidelity (including south-west races) shown from many areas, with returns up to 7 years after ringing. (Cortopassi and Mewaldt 1965; Bent 1968.)

Spring migration is late and rapid. Main movement begins mid-April right across continent, peaking late April to mid-May (Erskine 1964; Cortopassi and Mewaldt 1965; Garrett and Dunn 1981; Bohlen 1989). In Okanagan valley (southern British Columbia), one of most abundant migrants, passing through 'in spectacular waves' (Cannings *et al.* 1987). Some birds reach subarctic nesting areas before mid-May (Gabrielson and Lincoln 1959; Todd 1963; DeWolfe *et al.* 1973). Snow-melt is later in central Ungava (northern Quebec), where arrival is in late May (Todd 1963; Bent 1968).

Frequent vagrant north of breeding range (mainly spring), to Pribilof islands (Bering Strait), islands in Canadian Arctic, 72–76°N, even Fletcher's Ice Island (82°N), and to Greenland. (Bent 1968; American Ornithologists' Union 1983; Godfrey 1986.) Also several records on Wrangel Island (north-east Siberia) summer and autumn (Stepanyan 1990), and in Japan October–April (Brazil 1991). AJE, DFV

Voice. See Field Characters.

Plumages. (nominate *leucophrys*). ADULT. Forehead and broad stripe on each side of crown black, stripe ending blunt- or round-ended. Broad stripe over centre of crown white, slightly tinged grey when plumage fresh. Upper part of lore to front corner of eye black. A distinct white supercilium, extending from above front of eye to side of upper nape, connected with white of centre of crown by white on upper nape; rear of supercilium and upper nape tinged light grey to varying extent when plumage fresh. Eye-ring white, narrowly broken by black in front and behind; a narrow black stripe behind eye, extending to and widening at side of lower nape. Lower part of lore and side of head and neck below white of eye-ring and black eye-stripe light ash-grey, hindneck similar but grey slightly darker and slightly suffused brown. Mantle and scapulars light ash-grey on centre, more brown-grey towards side of body, partly tinged dull chestnut-brown on shorter outer scapulars, all feathers with broad fuscous-brown to dull chestnut-brown central streak. Back, rump, and upper tail-coverts pale buff-brown or tawny-olive, tips of tail-coverts narrowly fringed off-white when plumage quite fresh. Chin greyish-white, gradually merging into light ash-grey of cheeks, chest, side of breast, and upper flank and this in turn into cream-white or pure white of belly and vent and into warm buff or light buff-brown of lower flank, thigh, and under tail-coverts. Tail greyish-black, tinged sepia on central pair (t1), narrowly fringed buff-brown along outer webs, faintly grey-white along inner webs and tips. Flight-feathers, greater upper primary coverts, and bastard wing greyish-black, outer webs of primaries narrowly fringed buff-brown (faint and almost white along p9 and emarginated parts of p6–p8), outer webs of secondaries more broadly fringed with dull cinnamon, outer webs and tips of primary coverts faintly fringed cinnamon. Lesser upper wing-coverts dull brown-grey with slightly paler and greyer fringes; median and most outer greater coverts dull black with narrow buff-brown fringe and boldly contrasting pale cream-buff to pure white spot on tip of outer web, 3–5 mm long, forming distinct broken wing-bars; innermost greater coverts (tertial coverts) and tertials dull black or brown-black, outer web broadly fringed rich rufous-chestnut at base, chestnut gradually merging into narrower cream-white or pure white fringe along terminal part of outer web and along tip. Under wing-coverts and axillaries dull grey with grey-white tips, base of axillaries and shorter coverts below leading edge of wing sometimes partly suffused pale yellow. *In fresh plumage* (September to about April), white of crown-stripe, supercilium, and nape washed light grey, less pure, and buff of lower scapulars and back to upper tail-coverts sometimes slightly tinged yellow-olive; *in worn plumage* (about May–July), white of head-stripes purer, but grey of hindneck, side of head and neck, and underparts duller, less pure, chin and belly dull grey-white, not contrasting, lower flank with more restricted greyish-buff, cinnamon and chestnut of secondaries and tertials paler, white tips of median and greater coverts and fringes along tips of tertials partly worn off, sometimes hard to see. Sexes similar in nominate *leucophrys*, but ♀ *gambelii* averages more olive-grey on mid-crown stripe than ♂, and browner lateral crown-stripes, making head pattern less contrasting; however, much overlap in colour and contrast with ♂ (Fugle and Rothstein 1985). For sexing with help of wing-length, see Mewaldt and King (1986). JUVENILE. Lore and broad lateral crown-stripe dull rufous-brown or fuscous-brown; mid-crown stripe and supercilium pale buff, crown-stripe with dusky spots, supercilium not extending to front of eye. Narrow broken eye-ring pale buff; an ill-defined fuscous-brown stripe through eye, a short dull black malar stripe. Hindneck, side of neck, cheek, chin and throat buff with narrow dusky streaks. Mantle to upper tail-coverts cinnamon-brown or buff-brown, broadly streaked

black on mantle and scapulars, more faintly mottled sooty-grey on back, rump, and upper tail-coverts. Chest and side of breast rather heavily streaked dull black and buff, remainder of underparts buff, narrowly streaked black on flank, mottled grey on thigh. Lesser and median upper wing-coverts black with broad buff fringe (black extending into point at tip of each covert); greater coverts and tertials dull black or black-brown with ill-defined cinnamon-brown outer fringe and more contrasting tawny-cinnamon fringe along tip. Flight-feathers, greater upper primary coverts, bastard wing, and tail as in adult, but see 1st adult, below. In worn plumage, buff colours in part bleach to off-white or cream, especially of pale stripes on head and of tips of median and greater coverts, latter then forming more distinctly contrasting wing-bars. For development, see Banks (1959) and Morton *et al.* (1972). FIRST ADULT NON-BREEDING. Stripe over mid-crown greyish-cinnamon, bordered at each side by broad deep rufous-chestnut or dark chestnut-brown lateral crown-stripe; crown-stripes well-defined, but colours less contrasting than in adult. Lore grey-brown or dark brown, extending above eye into narrow pink-buff or cream supercilium, which is ill-defined at rear. Upper cheek and ear-coverts buff-brown or tan, somewhat contrasting with darker brown lore, faintly speckled cream just in front of and below eye, faintly streaked cream on ear-coverts; upper border of ear-coverts dark brown or fuscous, forming rather ill-defined stripe behind eye. Hindneck and upper mantle tan or buff-brown, faintly marked with dull olive-grey, tinged paler and greyer towards side of neck. Remainder of upperparts as in adult, but fringes of lower mantle and scapulars warmer buff, less pale and grey. Side of neck and entire underparts as adult, mainly grey with paler grey or whitish chin, belly, and vent and buff-washed flank and under tail-coverts, but grey slightly tinged buff or brown and often a faint sooty malar stripe. Tail, flight-feathers, greater upper primary coverts, tertials, and bastard wing as adult, but tips of tail-feathers and primary coverts often more pointed than in adult, less truncate or rounded; tips of tail, outer primaries, and primary coverts often distinctly more worn than in adult at same time of year, especially in spring and early summer. FIRST ADULT BREEDING. Like adult, but flight-feathers, primary coverts, and sometimes all or part of tail still juvenile, as in 1st non-breeding. New feathers on head and body as adult, streaks on head black and white, but *nuttalli* in particular tends to retain part of 1st non-breeding feathers on (especially) hind-crown and nape, these mottled black, grey, brown, and buff, variation depending on extent of pre-breeding moult. For relation of crown colour to dominance in flock or breeding success, see Ralph and Pearson (1971), Parsons and Baptista (1980), and Fugle *et al.* (1984).

Bare parts. ADULT, FIRST ADULT. Iris brown. Culmen dark grey-horn to grey-black, colour of remainder of upper mandible and of entire lower mandible dependent on race: dark pinkish in nominate *leucophrys* and *oriantha*, orange, saffron-yellow, pink-brown, or flesh in *gambelii*, yellow to orange-yellow in *pugetensis* and *nuttalli*. Leg and foot light horn-brown, flesh-brown, or purple-brown, paler flesh-pink on rear of tarsus and soles. JUVENILE. Iris dark brown. Bill flesh-pink, pink-yellow, or corn-yellow (depending on race), culmen brown to dark grey. Leg and foot flesh-pink or lilac, toes tinged grey. (Bent 1968; Godfrey 1986; Pyle *et al.* 1987; RMNH.)

Moults. ADULT POST-BREEDING. Complete; primaries descendent. In breeding area (Pyle *et al.* 1987). In *oriantha* (California), moult starts with shedding of p1 in late July or early August (day 1); p9 (outer functional primary) shed on about day 24, regrown on about day 46; tertials start with shedding of s8 on about day 10, all regrown by about day 37(-50); secondaries start with shedding of s1 about day 20, last feathers (s4-s5) regrown about day 51(-61); tail starts with t1 about day 4, moult centrifugal, t6 shed 1-6 days after t1, all feathers growing simultaneously during days 10-35; moult of body spans complete period of moult of flight-feathers, and often the last to be completed; total duration of moult $48 \cdot 8 \pm 5 \cdot 2$ days, similar for both sexes, though ♂ starts 0-5 days before ♀ (Morton and Welton 1973). For many details of moult of *oriantha* in California, see also Morton and Morton (1990). In Oregon, duration of moult in *oriantha* $60 \cdot 7 \pm 2 \cdot 7$ days in ♂, $52 \cdot 8 \pm 3 \cdot 3$ days in ♀, ♂ starting on average 10 July, ♀ 15 July, range of starting dates *c.* 25 June to 30 July (King and Mewaldt 1987). In *gambelii* (Alaska), p1 shed *c.* 10 July, p9 *c.* 6 August; moult of primaries (including regrowth of outer feathers) lasts average $37 \pm 3 \cdot 4$ days, secondaries $25 \cdot 7$ days, tail $25 \cdot 1$ days (Morton *et al.* 1969; Farner *et al.* 1980; but see Chilgren 1978); total duration of moult (including body) $49 \cdot 2 \pm 5 \cdot 2$ days (Morton *et al.* 1969; Morton and Welton 1973). In northern populations of *pugetensis*, moult starts mid- or late June (Mewaldt *et al.* 1968) or 19 July to 2 August, duration of entire moult 47 days (Mewaldt and King 1978); in *nuttalli*, starts mainly early June (Mewaldt *et al.* 1968), 6-29 June (Mewaldt and King 1977), or 3-22 July, duration of moult 83 days; ♂ starts 3-13 days before ♀ (Mewaldt and King 1978). For influence of day length on moult, see Wolfson (1954*b*), Chilgren (1978), and Farner *et al.* (1980); for influence of food, see Murphy and King (1984) and Murphy *et al.* (1988). ADULT PRE-BREEDING. Partial, extent depending on race: more limited in *nuttalli*, less so in nominate *leucophrys*, *oriantha*, and *pugetensis*, most extensive in *gambelii*. Usually involves head and much of body, but in *gambelii* often also tertials (sequence s8-s9-s7), most wing-coverts (except primary and carpal coverts, longer under wing-coverts, and sometimes a few outer greater coverts), and t1; in *nuttalli*, largely restricted to cap and throat. Moult of crown starts early November to mid-March, completed by April; moult generally mainly late February to April, but January-April in *nuttalli* and mainly mid-February to late March in *pugetensis*. (Grinnell 1928; Blanchard 1941; Michener and Michener 1943; Blanchard and Eriksson 1949; Banks 1964; Mewaldt *et al.* 1968; Ralph and Pearson 1971). POST-JUVENILE. Partial: head, body, lesser, median, and greater upper wing-coverts, occasionally some or all tertials or central pair of tail-feathers; in breeding area; July-October, mainly August. (Dwight 1900; Bent 1968; Pyle *et al.* 1987, RMNH, ZMA.) Starts at age of $34 \cdot 3 \pm 5 \cdot 2$ days (Morton *et al.* 1972), *c.* 20 days after completion of growth of juvenile feathering (Mewaldt and King 1977). In *gambelii*, Alaska, duration in individual birds 33-34 days (DeWolfe 1967; Morton *et al.* 1969), starting early July to late August, completed from early August onwards (DeWolfe 1967), or mainly between 5 July and 20 August, occasionally starting late June or completing up to late August (Morton *et al.* 1969). In *oriantha* from Oregon, duration in individuals $58 \cdot 0 \pm 3 \cdot 4$ days, starting 2 July (15 June to 25 July) (King and Mewaldt 1987). In *oriantha* from California, duration in individuals $32 \cdot 4 \pm 4 \cdot 6$ days, population as a whole 20-30 July to 15-25 September (Morton *et al.* 1972); in *nuttalli* (California), duration in individual birds 60 (20) 55-65 days, late May to mid-October; in *pugetensis*, starts from early June (Mewaldt and King 1977).

Measurements. ADULT, FIRST ADULT. Nominate *leucophrys*. Eastern Canada, summer, and eastern USA, winter; skins (BMNH, RMNH, ZMA). Bill (S) to skull, bill (N) to distal corner of nostril; exposed culmen on average *c.* $3 \cdot 2$ shorter than bill (S).

WING	♂	$81 \cdot 8$ ($2 \cdot 18$; 24)	78-86	♀ $77 \cdot 1$ ($2 \cdot 68$; 12) 73-80

TAIL	$70 \cdot 5$ ($3 \cdot 10$; 11) 66-75	$67 \cdot 8$ ($2 \cdot 75$; 12) 64-72
BILL (S)	$13 \cdot 6$ ($0 \cdot 45$; 10) $13 \cdot 0$-$14 \cdot 4$	$13 \cdot 6$ ($0 \cdot 50$; 12) $12 \cdot 8$-$14 \cdot 4$
BILL (N)	$8 \cdot 2$ ($0 \cdot 62$; 10) $7 \cdot 6$-$9 \cdot 2$	$8 \cdot 0$ ($0 \cdot 32$; 12) $7 \cdot 5$-$8 \cdot 6$
TARSUS	$23 \cdot 7$ ($0 \cdot 56$; 11) $22 \cdot 7$-$24 \cdot 6$	$23 \cdot 0$ ($0 \cdot 83$; 12) $21 \cdot 4$-$24 \cdot 3$

Sex differences significant, except bill. For influence of sex and age on measurements, see Fugle and Rothstein (1985). Bill in summer slightly longer than during rest of year (Banks 1964; Morton and Morton 1987).

Weights. ADULT, FIRST ADULT. Nominate *leucophrys*. Eastern Canada, summer: ♂ $29 \cdot 7$ (16) $24 \cdot 8$-$32 \cdot 5$ (Banks 1964). Pennsylvania (USA): May $30 \cdot 0$ ($3 \cdot 22$; 118) $24 \cdot 6$-$38 \cdot 5$; September-November $27 \cdot 7$ ($2 \cdot 72$; 113) $18 \cdot 9$-$33 \cdot 3$ (Clench and Leberman 1978). Ohio (USA): May $32 \cdot 8$ ($n = 151$); September-November, adult $29 \cdot 8$ ($n = 132$), 1st adult $28 \cdot 5$ ($n = 96$) (Baldwin and Kendeigh 1938; see also Stewart 1937). Ohio (USA): autumn and spring, ♂ $30 \cdot 7$ (8) $27 \cdot 4$-$32 \cdot 8$, ♀ $27 \cdot 1$ (6) $25 \cdot 1$-$28 \cdot 1$; winter, ♂ $35 \cdot 0$ (8) $30 \cdot 5$-$36 \cdot 8$ (Nice 1938). Michigan (USA): $32 \cdot 1$ ($2 \cdot 1$; 48) $22 \cdot 9$-$38 \cdot 7$ (Becker and Stack 1944). See also Mengel (1965).

Z. l. gambelii. North-central Alaska, May-August: ♂ $25 \cdot 6$ (29) $22 \cdot 2$-$29 \cdot 5$, ♀ $23 \cdot 3$ (9) $21 \cdot 7$-$27 \cdot 4$ (Irving 1960). Fairbanks (Alaska), August: adult ♂ $26 \cdot 9$ ($1 \cdot 70$; 16) $24 \cdot 0$-$29 \cdot 9$, 1st adult $26 \cdot 6$ ($1 \cdot 12$; 21) $24 \cdot 6$-$29 \cdot 8$ (DeWolfe 1967). North-west USA: January-February $27 \cdot 4$ ($0 \cdot 80$; 18); March to mid-April (in pre-breeding moult) $28 \cdot 1$ ($1 \cdot 97$; 58); late April and early May (during pre-migration fattening) $29 \cdot 7$ ($3 \cdot 10$; 38) (King and Farner 1959, which see for lean weights and other details). California (USA), January-March: ♂ $27 \cdot 7$ ($1 \cdot 96$; 92) $22 \cdot 5$-$31 \cdot 5$, ♀ $25 \cdot 3$ ($2 \cdot 00$; 73) 21-31 (Fugle and Rothstein 1985). See also McGreal and Farner (1956), King (1961), King *et al.* (1963, 1965), King and Farner (1966), Chilgren (1977), King and Mewaldt (1987), Schwabl *et al.* (1988), and Schwabl and Farner (1989).

Other races. See Grinnell *et al.* (1930), Blanchard (1941), Banks (1964), Mewaldt *et al.* (1968), Morton *et al.* (1972, 1973), Mewaldt and King (1977), and Morton and Morton (1990).

Structure. Wing rather long, broad at base, tip bluntly pointed, 10 primaries: p7 longest, p8 and p6 0-$0 \cdot 5$ (-$1 \cdot 5$) shorter, p9 3-6, p5 $2 \cdot 5$-4, p4 6-8, p3 8-11, p2 10-14, p1 13-16; p10 strongly reduced, a tiny pin hidden beneath reduced outermost upper primary covert, 52-60 shorter than p7, 8-13 shorter than longest upper primary covert. Outer web of (p5-)p6-p8 emarginated; inner web of (p7-)p8-p9 with notch (sometimes faint). Tip of longest tertial reaches to tip of p1-p3. Tail rather long, tip slightly rounded; 12 feathers, t3-t4 longest, t1 0-4 shorter, t6 2-5 shorter. Bill short, conical, about equal to half of headlength, fairly slender at base (depth $6 \cdot 5$-$7 \cdot 2$ mm, width $6 \cdot 4$-$7 \cdot 0$ mm), tip sharply pointed; culmen slightly decurved, cutting edges and gonys virtually straight, extreme base of cutting edges kinked sharply downward, forming small tooth at kink, but base largely hidden by feathering and a few fine short bristles when bill closed. Tarsus rather long, toes and claws short, all fairly strong. Middle toe with claw $20 \cdot 0$ (5) $19 \cdot 0$-$20 \cdot 5$ mm; outer toe with claw *c.* 76% of middle with claw, inner *c.* 68%, hind *c.* 78%.

Geographical variation. Marked, differing in colour of lore (pale supercilium extending to nostril in races with white or buff lore, supercilium extending from above eye backwards in races with black or dark brown lore), colour of bill (yellow or pink), colour of upperparts (either mainly brownish, or with rufous feather-centres, greyish fringes, and grey rump), colour of bend of wing (either bright yellow, or white to yellow-white), and size (wing length). In nominate *leucophrys* from eastern Canada and in *oriantha* from inland western USA, lore black (adult) or dark brown (1st adult non-breeding), bill pinkish or reddish, upperparts mixed rufous and grey (paler and greyer in *oriantha* than in nominate *leucophrys*, and breast of *oriantha* markedly paler grey: Godfrey 1965), and bend of wing whitish. In *gambelii* from western Canada and Alaska, bill, upperparts, and bend of wing as in nominate *leucophrys*, but lore white (adult) or buff (1st adult non-breeding) and supercilium long. For populations intermediate between nominate *leucophrys* and *gambelii*, see Banks (1964). In *pugetensis* and *nuttalli* from south-west British Columbia (Canada) and coastal ranges of western USA, lore usually white or pale grey, as in *gambelii*, but upperparts (including rump) mainly dark olive-brown, streaked black on lower mandible and scapulars, bill yellowish, and bend of wing bright yellow; also, breast dark brown or brown-grey (pale brown or grey in *gambelii*) and underwing washed yellow (silver-grey in *gambelii*); both races differ mainly in ratios of lean weight to wing, bill to wing, or bill to tail; also in wing shape, length of hind claw, timing and extent of moults, and (slightly) colour (Hellmayr 1938; Banks 1964; Mewaldt 1977). Nominate *leucophrys* and *gambelii* large (wing ♂ 73-84, ♀ 68-80), *oriantha* rather large (wing ♂ 72-81, ♀ 66-76), *pugetensis* and *nuttalli* small (wing ♂ 67-74, ♀ 63-72) (Pyle *et al.* 1987). For variation in song, see Baker (1975) and Baptista (1977). For details on morphology of races, see Swenk (1930), Blanchard (1941), Banks (1964), Godfrey (1965), Bent (1968), and Mewaldt (1977).

Closely related to Golden-crowned Sparrow *Z. atricapilla* of north-west North America (Zink 1982; Zink *et al.* 1990).   CSR

Zonotrichia albicollis White-throated Sparrow

PLATE 13
[between pages 256 and 257]

Du. Witkeelgors   Fr. Bruant à gorge blanche   Ge. Weisskehlammer
Ru. Белошейная зонотрихия   Sp. Chingolo gorgiblanco   Sw. Vitstrupig sparv

Fringilla albicollis Gmelin, 1789

Monotypic

Field characters. 15·5-18 cm; wing-span 22-25 cm. Similar in size to Reed Bunting *Emberiza schoeniclus*; up to 10% shorter than White-crowned Sparrow *Z. leucophrys*, with neckless and less upright form. Quite large, long, and bunting-like Nearctic sparrow, with rather secretive behaviour. Dimorphic, with white or tan head-stripes; unstreaked below, with rest of plumage dominated by diagnostic black-edged white throat and yellow fore-supercilium. Dark ridge to bill. Sexes similar; some seasonal variation. Immature separable.

ADULT BREEDING. Moults: July–September (complete), March–April (front of body). Plumage pattern like *Z. leucophrys* but differs in (a) lack of black frontal band, with white, grey, or cream-buff central crown-stripe reaching bill, (b) full-length supercilium, always bright yellow to dirty yellowish-buff in front of eye, (c) white throat, usually with narrow black edge, (d) more rufous-brown mantle with more continuous black streaks, (e) faint rufous or blackish streaks and spots on collar and sides of breast, and (f) dull bluish-grey bill, with dusky-brown to black culmen. In white-striped morph, central crown-stripe essentially white and lateral stripes and narrow rear of eye-stripe black; in tan-striped morph, central crown-stripe greyish-buff and lateral stripes and rear of eye-stripe rufous-brown, flecked black. Eye red-brown. Legs flesh to brown. ADULT NON-BREEDING. Head pattern of white-striped morph much less contrasting, more buff and dark brown. Streaks on sides of breast stronger. FIRST-WINTER. Almost all resemble tan-striped winter adult but distinguished by lack of clear grey on face and breast, these areas being mottled or more strongly but still diffusely streaked. Throat dull white, with broken malar stripe joining breast streaks. On some, diffuse streaks continue along flanks, becoming indistinct at vent. Eye brown.

Commonest vagrant Nearctic sparrow in west Palearctic, but difficult to separate in glimpse from *Z. leucophrys*; see that species. Paler immature may also be confused with Song Sparrow *Melospiza melodia* but is larger, has dark-striped head with buff (not grey) supercilium, and much brighter wing-bars. Flight active and fast, with flirting tail; usually at low level. Escapes into thickets or undergrowth. Gait a hop, but scratches leaf-litter with double-footed kick. Stance and posture recall *Passer* sparrow, with compact, neckless form. Secretive, usually on ground within cover.

Commonest call a loud 'chink'; also a thin, high, drawn-out 'tseet' or 'tseep', and loud, rather metallic 'chink'.

Habitat. Ranges across cool boreal and temperate forested regions of northern Nearctic. In Canada, mostly coniferous or mixed forest, especially in clearings cluttered with slashing, burntwoods, and open young woodlands and thickets (Godfrey 1979). Further south in central USA, favours various semi-open wooded habitats such as coniferous forest with well-developed woody undergrowth, groves of aspen *Populus* with shrubby understorey, marshes bordered with willow *Salix*, and sometimes conifer plantations, but typically nests on ground, although singing posts are usually *c.* 6-12 m up in spruce *Abies* (Johnsgard 1979). In eastern North America, feeds chiefly on ground, often among dead leaves; wintering flocks found wherever there are thickets, brushy field borders, and weed tangles (Pough 1949). Flocks wintering in Louisiana frequent shrubbery or other woody vegetation and do not like to venture far into open (Lowery 1974). On spring migration through New England, seems to prefer moist thickets, but in autumn many visit weedy gardens and cornfields, usually remaining on or near ground, rarely perching high in trees; also favours bush-bordered roads and edges of pinewoods (Forbush and May 1955).

Distribution. Breeds in North America, from south-east Yukon east to Newfoundland, south to central British Columbia, central Saskatchewan, North Dakota, northern Wisconsin, northern Ohio, northern Pennsylvania, and northern New Jersey. Winters from south-east Iowa, northern Ohio, Pennsylvania, and Massachusetts south to California, northern Mexico, southern Texas, and Gulf of Mexico coast east to southern Florida.

Accidental. Iceland, Britain, Ireland, Netherlands, Denmark, Sweden, Finland, Gibraltar.

Movements. Migratory; insignificant proportion of population winters in south-east part of breeding range (Bent 1968; American Ornithologists' Union 1983).

Autumn migration begins in August, with passage reported late August to September in Alberta (Sadler and Myres 1976; Salt and Salt 1976), and first records south of breeding range in early September (Bent 1968). Main passage through northern and mid-latitude states late September to October (Mengel 1965; Bull 1974; Janssen 1987), with most birds leaving Canada by end of October (Peters and Burleigh 1951; Sadler and Myres 1976; Speirs 1985). Reaches southern states mostly from early October, extreme south-east (Florida) not until late October; arrivals continue into November (Lowery 1974; Oberholser 1974; Kale and Maehr 1990). Individuals often winter west and north of main winter range, especially where artificial food available (e.g. Peters and Burleigh 1951, Janssen 1987). Migration is on broad front across North America east of Rockies, but numbers are much lower west of 95°W (Erskine 1977; American Ornithologists' Union 1983). Ringing data show mostly south-west heading from east of Great Lakes, and SSE or south-east heading from western Canada (reverse in spring), with most recoveries from Carolinas west to Arkansas (Houston and Street 1959; Bull 1974; Imhof 1976; James and Neal 1986). Ringing in Georgia and Arkansas also shows winter site-fidelity, with returns up to 6 years later (Stoddard 1978; James and Neal 1986).

Northward migration only becomes evident about mid-April, and most birds leave wintering areas by end of April (Stewart and Robbins 1958; Oberholser 1974; Bohlen 1989; Kale and Maehr 1990); earlier reports (from mid-March) partly involve birds wintering locally. Movement is concentrated, with hundreds or thousands following close behind first arrivals, and by 1 May birds are pouring into eastern Canada (Bailey 1955; Sprague and Weir 1984), though few birds reach north-east breeding areas until after mid-May (Todd 1963). West of Great Lakes, smaller movement is less obvious, with first arrivals in late April and peak movement in mid-May (e.g. Houston and Street 1959; Sadler and Myres 1976).

Some birds overshoot breeding range in spring, giving records in arctic Alaska, and in south Baffin and Coats islands (Canada), but most records in west and south Alaska are in autumn (Kessel and Gibson 1978; Godfrey 1986).

Rare vagrant to west Palearctic, especially in spring. 17 records from Britain and Ireland up to 1990, of which 12 in April–June and 5 in October–December (Dymond *et al.* 1989; Rogers *et al.* 1990, 1991). Bird recorded 1 December 1984 remained until at least 7 April 1985 (Rogers *et al.* 1986*b*).
AJE, DFV

Voice. See Field Characters.

Plumages. ADULT MALE BREEDING. 2 morphs, one characterized by black lateral crown-stripes and white mid-crown stripe and supercilium, other by mottled black-and-chestnut lateral crown-stripes and less contrasting tan or tawny-buff mid-crown stripe and supercilium (Lowther 1961); intermediates between both morphs occur, and occasional birds combine mainly black with fully tan striping or largely chestnut lateral stripe with contrasting pale cream-grey mid-stripe and supercilium. WHITE-STRIPED MORPH. Stripe from middle of forehead over mid-crown to central hindneck white, feathers slightly tipped olive-grey or buff when plumage fresh, rear partly streaked black or (when pre-breeding moult incomplete and non-breeding plumage partly retained) heavily suffused buff or tan. Broad black lateral crown-stripe, extending from nostril to upper side of mantle, some feathers at rear of stripe fringed rufous-chestnut when plumage fresh, or (when non-breeding plumage partly retained), rear extensively mottled rufous-chestnut and black. Front part of supercilium contrastingly bright yellow (extending from side of upper mandible to above eye), rear part (from above eye to rear side of neck) contrastingly white, faintly tinged cream when fresh, variably tinged buff or tan at upper side of neck. Lore and patch below eye medium ash-grey with fine off-white mottling, merging below into narrow solid black line extending from gape to below ear-coverts; sometimes a narrow black line between yellow of supercilium and grey of lore. Eye-ring narrow, mottled buff and grey. Ear-coverts medium ash-grey, separated from white of supercilium by distinct black stripe behind eye. Side of neck variegated black, grey, and rufous-chestnut. Mantle and scapulars closely marked with black and rufous-chestnut streaks, some sides of feathers partly paler pink-buff or grey-buff. Rump and upper tail-coverts uniform greyish olive-brown or drab-grey, back similar but partly streaked chestnut and black. Chin, throat, and lower cheek pale cream or white, sometimes broken at each side by narrow black malar stripe, sharply demarcated from medium ash-grey upper cheek, lower side of neck, and upper chest, from which it is sometimes separated by narrow black border. Lower chest, side of breast, and upper flank variable, either almost uniform medium ash-grey with traces of dull black bars when fully replaced in pre-breeding moult, or more brownish-grey with more numerous traces of pale streaks and blackish specks or bars when plumage still in non-breeding. Lower flank, thigh, and under tail-coverts cream-buff with darker olive-brown streaks; belly and vent white, tinged cream when plumage fresh. Tail greyish-black, tinged drab-grey on central pair (t1) and outer webs of others, fringes narrowly pink-cinnamon. Flight-feathers, greater upper primary coverts, and bastard wing greyish-black, all except longest feather of bastard wing with narrow rufous-cinnamon fringe along outer web (brightest and widest on secondaries), but feathers soon bleach to dark brown and fringes to buff or (on exposed parts of primaries) almost off-white. Lesser upper wing-coverts rufous-brown with sooty-grey centre and pale olive-grey tip; median and greater coverts and tertials black with rufous-cinnamon fringes along sides and pale pink-cinnamon tips, these soon bleaching to pale cinnamon or buff on tertials and on exposed sides of median and greater coverts and to contrasting white on tips of median and greater coverts, forming broken wing-bars. Under wing-coverts and axillaries white with cream-buff suffusion, shorter coverts along leading edge of wing bright pale yellow mottled dark grey. TAN-STRIPED MORPH. Mid-crown stripe tan or greyish-buff, often narrower and always less sharply defined than white stripe of white-striped morph. Lateral crown-stripe rufous-chestnut (when fresh) or rufous-cinnamon (when worn), streaked or spotted black, especially on forehead, where sometimes almost fully black. Front part of supercilium yellow (often less bright and more restricted than in white-striped morph, sometimes tinged buff), rear part cream, buff, or tan, deepest above ear-coverts, from which it is separated by narrow black-brown stripe. Side of head below eye and side of neck as in white-striped morph, but grey usually tinged brown. Remainder of upperparts as in white-striped morph. White throat-patch variable, either fully white, occasionally partly or fully outlined by black (as in white-striped morph), or grey-white and rather gradually merging into brown-grey of cheek, lower side of neck, and upper chest; patch either uniform, or indistinctly streaked grey and grey-white, or marked with single, forked, or double black malar stripes. Chest, side of breast, and upper flank brownish- or buffish-grey, variably marked with short dull black streaks or triangular spots bordered by buff (these usually most conspicuous at border of white of belly); amount of streaks and spots depends on extent of pre-breeding moult (newer feathers of breeding plumage on upper chest greyer and more uniform than browner and more streaked ones on lower chest; some birds have chest virtually all new and grey, others mainly old and brownish-grey) and probably on age, birds showing extensive streaking all over chest, side of breast, and flank (resembling juvenile) likely to be in 1st non-breeding plumage, or in 1st breeding with much of non-breeding retained. Streaks on lower mid-chest sometimes form a dark patch. Wing and tail as in white-throated morph. Sexes similar, except for wing length (Brewer 1990; Piper and Wiley 1991); however, ♀ more often tan-striped, ♂ more often white-striped; with help of wing length and colour morph (with separate scoring of colour of mid-crown stripe, lateral crown-stripe, tinge of yellow on supercilium, occurrence and extent of dusky malar stripes, and colour and contrast of throat patch) sexing can be attempted by discriminant function analysis (Rising and Shields 1980) or multiple logistic regression analysis (Schlinger and Adler 1990). ADULT NON-BREEDING. Like adult breeding, but variation in head, throat, and chest colour less clearly bimodal (2 distinct morphs not apparent), more gradual and continuous, with many birds showing intermediate patterns (Vardy 1971; Atkinson and Ralph 1980); this mainly because white mid-crown stripe and supercilium of white-striped morph partly washed buff on feather-tips, and black lateral stripes fringed brown (Piper and Wiley 1989).

No sexing possible, except by wing length. JUVENILE. Median crown-stripe and entire supercilium olive-grey-buff, lateral crown-stripes dark chestnut-brown or fuscous-brown. Remainder of upperparts chestnut-brown, streaked black, fringes of feathers of mantle and scapulars buff. Underparts dull white, washed buff on throat, chest, side of breast, and flank, finely spotted brown on chin, uniform dull white on belly and vent, heavily streaked brown on remainder. (Dwight 1900.) For wing and tail, see 1st adult. FIRST ADULT NON-BREEDING AND BREEDING. Like adult, but juvenile flight-feathers, primary coverts, and tail retained; tips of primary coverts and tail more pointed than in adult, less bluntly rounded or truncate, but many birds intermediate in shape and hard to age; tips also more worn than in adult at same time of year. In 1st adult non-breeding, morphs as in adult non-breeding, not as distinct as in adult breeding, variation in colour of stripes of head and of throat to chest seemingly continuous; see Piper and Wiley (1989). However, only 5% of 209 birds in 1st adult non-breeding plumage examined by J K Lowther and J B Falls (in Bent 1968) had stripes of head black and white, remainder olive-grey or tan and brown. In 1st breeding and subsequent plumages, relative proportion of white-striped may increase (Vardy 1971), though this is contradicted by supposition that morph is fixed genetically (Thorneycroft 1966, 1975). For characteristics of various plumage types, see Watt (1986), Piper and Wiley (1989), Brewer (1990), and Schlinger and Adler (1990). For behavioural differences between morphs, see Ficken et al. (1978) and Watt et al. (1984); for differences in nesting habitat, see Knapton and Falls (1982).

Bare parts. ADULT, FIRST ADULT. Iris of adult reddish-brown, of 1st adult in autumn grey-brown, gradually grading through brown into reddish-brown in spring. Bill brown-olive to dark greyish-horn, almost black on tip of culmen; cutting edges paler, more yellowish-horn. Leg and foot light brown, buff-brown, dull flesh-brown, or light purplish-brown, paler flesh on rear of tarsus and on soles; in 1st adult in autumn sometimes pink or flesh-grey with brown-grey toes. JUVENILE. Iris grey-green. Bill slaty-brown, leg and foot pinkish-buff, gradually becoming darker with age. (Dwight 1900; Yunick 1977b; RMNH, ZMA.)

Moults. ADULT POST-BREEDING. Complete; primaries descendent. In breeding area, July–September (Pyle et al. 1987), mainly from early August (Bent 1968); moult not started in 5 birds from Maine (USA) in mid-July (ZFMK). In captivity, moult of ♂ early July to late September, peaking mid-July to late August; in ♀ late June or early July to late September or early October, peaking late July to late August; no difference between morphs (Kuenzel and Helms 1974). ADULT PRE-BREEDING. Partial: head, throat, chest, flank, usually all or part of mantle and scapulars, and occasionally central pair of tail-feathers or some tertials. Starts mid-March to early April, completed by late April or early May (Vardy 1971; Piper and Wiley 1989; RMNH, ZMA); in Georgia (USA), starts mid-March to mid-April, many in heavy moult in 1st half of April, but a few not started in early May and some still moulting on arrival in breeding area in May (Odum 1949; RMNH, ZMA). In captivity, limited moult between late February or early March and mid- or late April (Kuenzel and Helms 1974). POST-JUVENILE. Partial: head, body, lesser, usually median coverts and variable number of inner or all greater upper wing-coverts, especially innermost (tertial coverts) and some or all tertials, occasionally central pair of tail-feathers (t1). Some birds start late July, all in moult by about mid-August; generally completed by late August or early September. FIRST PRE-BREEDING. Like adult pre-breeding, but often more restricted in extent, usually involving only head (except nape), upper chest, and scattered feathers of mantle, side of breast, or inner scapulars. (Bent 1968; Pyle *et al.* 1987; RMNH, ZMA.)

Measurements. ADULT, FIRST ADULT. Eastern Canada and eastern USA, all year; skins (RMNH, ZFMK, ZMA). Bill (S) to skull, bill (N) to distal corner of nostril; exposed culmen on average 3·6 less than bill (S).

WING	♂	76·8 (1·62; 16) 74–80	♀	72·4 (1·46; 14) 70–75
TAIL		72·1 (2·32; 15) 66–76		68·9 (2·69; 14) 64–73
BILL (S)		14·4 (0·40; 16) 13·5–15·0		13·8 (0·58; 14) 13·1–14·9
BILL (N)		8·5 (0·31; 16) 8·1–9·1		8·0 (0·23; 14) 7·7–8·5
TARSUS		24·0 (0·77; 16) 22·6–25·3		23·7 (0·75; 14) 22·6–25·2

Sex differences significant for wing and tail.

See also Odum (1949), Atkinson and Ralph (1980), Watt (1986), Schlinger and Adler (1990), and Piper and Wiley (1991).

Weights. ADULT, FIRST ADULT. Ohio (USA): (1) autumn, (2) spring (Nice 1938). Georgia (USA): (3) January to mid-February, (4) mid-April to mid-May (Odum 1949, which see for details).

(1)	♂	26·5 (— ; 50) 23·2–30·8	♀	24·7 (— ; 75) 21·0–29·2
(2)		29·4 (— ; 17) 26·1–35·0		26·4 (— ; 18) 22·5–32·5
(3)		29·2 (1·93; 36) —		27·5 (1·66; 21) —
(4)		30·8 (1·77; 26) —		27·0 (2·10; 34) —

Ohio: (5) April–May, (6) September–November, (7) undated (Stewart 1937; Baldwin and Kendeigh 1938). (8) Michigan (USA), undated (Becker and Stack 1944). Pennsylvania (USA): September–December, (9) adult, (10) 1st adult; ages combined, (11) January–March, (12) April–May (Clench and Leberman 1978). (13) Coastal New Jersey (USA), autumn (Murray and Jehl 1964).

(5)	29·3 (— ; 106)	—	(10) 25·7 (2·27; 1617) 19·3–37·1
(6)	26·0 (— ; 875)	—	(11) 27·3 (3·35; 46) 20·7–34·3
(7)	24·8 (— ; 26) 20·8–29·5		(12) 26·9 (2·58; 399) 21·2–35·4
(8)	27·2 (1·9 ; 375) 19·1–34·5		(13) 23·7 (2·1 ; 341) 19·0–33·7
(9)	25·7 (2·13; 1150) 19·0–33·1		

Range of migrants New England coast 17·1–37·3 (Drury and Keith 1962). For fat-free weight, see Connell *et al.* (1960) and Odum *et al.* (1961). For variation in weight and fat content with season or month, see Odum and Perkinson (1951), Wolfson (1954*a, c*), and Weise (1962); see also Odum (1958), Caldwell *et al.* (1963), and Prescott (1986). For variation in daily weight cycle, see Kontogiannis (1967). For variation throughout year for each sex and both morphs, see Kuenzel and Helms (1974); average of tan-striped morph always 0·5 g below that of white-striped morph, except in ♀ in winter.

Structure. Wing rather short, broad at base, tip fairly rounded. 10 primaries: p6 longest, p7 equal or (occasionally) slightly shorter, p8 1–2(–3) shorter than p6, p9 7–11, p5 0–1, p4 3–5, p3 5–10, p2 7–12, p1 9–14, p10 strongly reduced, 50–58 shorter than p6, 8–10 shorter than longest upper primary covert. Outer web of p5–p8 emarginated (rarely, p4 slightly also), inner web of (p6–)p7–p9 with notch. Tip of longest tertial reaches to about tips of secondaries. Bill short, conical, as in White-crowned Sparrow *Z. leucophrys*; depth at base of feathering 6·8–8·0 mm, width 6·6–7·5 mm. Middle toe with claw 20·5 (10) 19–22; outer toe with claw *c.* 73% of middle with claw, inner *c.* 70%, hind *c.* 77%. Remainder of structure as in *Z. leucophrys*.

Geographical variation. None. For relationships, see Zink (1982) and Zink *et al.* (1991*a*).

CSR

Junco hyemalis **Dark-eyed Junco**

PLATE 13
[between pages 256 and 257]

DU. Grijze junco   FR. Junco ardoisé   GE. Junko
RU. Серый юнко   SP. Chingolo pizarroso   SW. Snöfågel

Fringilla hyemalis Linnaeus, 1758

Polytypic. HYEMALIS GROUP (Slate-coloured Junco). Nominate *hyemalis* (Linnaeus, 1758), north-east USA, Canada (except south-west), and Alaska; *carolinensis* Brewster, 1886, Appalachians (eastern USA) north to north-east West Virginia and western Maryland, grading into nominate *hyemalis* in northern and western Pennsylvania (extralimital); *cismontanus* Dwight, 1918, north Canadian Rockies from south-central Yukon to central British Columbia and west-central Alberta. Extralimital: AIKENI GROUP (White-winged Junco), 1 race in Great Plains of north-central USA; OREGANUS GROUP (Oregon Junco), 8 races along west coast of North America from British Columbia (Canada) to Baja California (Mexico), and in Rockies from southern British Columbia and southern Alberta south to central Idaho and north-west Wyoming; INSULARIS GROUP (Guadelupe Junco), 1 race on Guadelupe island (off Baja California); CANICEPS GROUP (Grey-headed Junco), 3 races in inland south-west USA, south and east of *oreganus* group.

Field characters. 13·5–15 cm; wing-span 23–25 cm. Close in size to Tree Sparrow *Passer montanus* but with bunting-like bill and structure, including rather long tail; 30% smaller and more lightly-built than Rufous-sided Towhee *Pipilo erythrophthalmus*. Medium-sized, perky, rather tame Nearctic sparrow, differing from other vagrant Emberizidae in pink bill, and uniformly dusky-grey to brown hood, upperparts, and upper flanks contrasting with pure white underbody and tail sides. Flicks tail. Call distinctive. Sexes similar; no seasonal variation. Juvenile separable. Nominate *hyemalis* described here.

ADULT MALE. Moults: July–October (complete), February–April (head). Dark plumage plain dusky-grey to slate-grey, showing no obvious characters other than white underbody and white outer tail-feathers which provide noticeably straight panels on edges of tail. Bill short and conical; pale pink with fine dusky tip; eye dark red-brown. Legs pale pink to orange-brown. ADULT FEMALE. Paler than ♂, with brownish wash to dark plumage. FIRST-WINTER MALE. Resembles adult ♂ but usually retains juvenile tertials with browner fringes and faint whitish tips, on some also similarly marked outer greater coverts. Eye greyish to dark brown. FIRST-WINTER FEMALE. Resembles adult ♀ but shows similar contrasts on tertials and greater coverts to 1st-winter ♂, while rear flanks noticeably brown and back paler brown.

Unmistakable. Flight light and fluttering, with quick wing-beats; tail noticeably straight in silhouette. Gait a hop. Stance level or half-upright, with tail carried up and often flicked. Feeds on ground, usually in or near cover.

Calls include characteristic smacking, or clicking sound, run together into squeaky twitter, metallic 'clink', and slightly liquid 'chek' in alarm.

Habitat. Breeds in northern Nearctic, in Canada in coniferous and mixed woodland (especially openings and edges), on burntlands, and occasionally in gardens (Godfrey 1979). In Alaska, where it also breeds up to treeline north of Arctic Circle, favours spruce *Picea* forests in summer but after young fledge they shift to trails, roadsides, glades, and weed patches, usually travelling by short flights (Gabrielson and Lincoln 1959). A few breed south of conifer zone in mainly deciduous woodland, where felled areas with slash piles often attract them. Many nests are along old roads and on edges of clearings; song perches are on forest treetops or dead timber. In winter, frequents woodland and fields, but most occur along hedgerows and brushy field borders; feeds then on ground. (Pough 1949.) In New Hampshire (USA), nests above treeline, usually in cool damp woods, but sometimes in open blueberry pasture, lumber clearings, or deserted logging camps; in winter shifts to neglected bush-covered fields and to weed-grown gardens, and in cold weather often comes to bird-tables (Forbush and May 1955). In South Dakota, breeding habitat includes groves of aspen *Populus* and deciduous woods in hollows and canyons (Johnsgard 1979).

Distribution. Breeds in North America, from western and northern Alaska east to Labrador and Newfoundland, south to northern Baja California, southern New Mexico, north-west Nebraska, east-central Minnesota, south-east Wisconsin, central Michigan, Appalachians south to northern Georgia, and south-east New York. Winters from southern Alaska and southern Canada east to Newfoundland, south to northern Mexico and coast of Gulf of Mexico east to Florida.

Accidental. Iceland, Britain, Ireland, Netherlands, Norway, Poland, Gibraltar.

Movements. Status varies. Boreal race (nominate *hyemalis*) and 3 northern montane races (*cismontanus, montanus, mearnsi*) largely migratory, although a few winter within southern breeding range. Other races partially migratory to sedentary according to latitude; some make altitudinal movements. (Grinnell and Miller 1944; American Ornithologists' Union 1957, 1983; Bent 1968.)

A few birds appear south of breeding areas in August (Burleigh 1972; Sprague and Weir 1984; Janssen 1987), but these reflect mostly local movements, as migration is not general and widespread until September, when snow first comes to northern and mountain areas. Birds start to leave Yukon and Mackenzie District in early September, and altitudinal movements from western mountains also begin then (Rand 1946; Bent 1968; Scotter *et al.* 1985). By late September, migration front extends south to 40°N, from Rockies to Atlantic, though most birds are still in Canada then (Bailey 1955; Bailey and Niedrach 1967; Bohlen 1989). Main passage almost everywhere October to early November (e.g. Palmer 1949, Sadler and Myres 1976, Garrett and Dunn 1981, Hall 1983). Most wintering areas are occupied by November, although many 'half-hardy' flocks and individuals linger north of snow line, some wintering successfully (Phillips *et al.* 1964; Bent 1968; Stoddard 1978). Migration (both seasons) on broad front east of Rockies, with channelling through valleys in west, but movement along coasts is not typical (Bent 1968). Winter recoveries in Alabama include birds ringed Illinois east to Atlantic states (Imhof 1976); in Arkansas, birds ringed north-west to Saskatchewan (more than 2000 km) and east to Indiana (James and Neal 1986). Ringing in Ontario, New Mexico, and other areas, shows winter site-fidelity, with returns up to 4 years later (Ligon 1961; Speirs 1985).

In spring, pattern of 'earliest arrivals' is confused by birds wintering north of main range. Some records in early March, occasionally even late February, in most northern states, may be early migrants. By late March, migration is under way across continent, and some birds are ascending into still-snowy highlands (Bent 1968; Sprague and Weir 1984; Tufts 1986; Janssen 1987). Main passage in April, when most winter areas are vacated and first arrivals in north-west extend beyond 60°N (Rand 1946; Gabrielson and Lincoln 1959; Bent 1968). Arrivals in Ungava (northern Quebec) and Labrador are in mid- or late May, when migration further west has ended (Todd 1963; Knapton 1979; Speirs 1985).

Frequent vagrant to Canadian and Alaskan Arctic; also recorded north-west to Pribilof islands, and (both seasons) eastern Siberia (Chukotskiy peninsula and Wrangel Island) (Gabrielson and Lincoln 1959; American Ornithologists' Union 1983; Godfrey 1986; Stepanyan 1990).

Rare vagrant to west Palearctic, especially in spring. 15 records from Britain and Ireland up to 1990, of which 12 in April-May and 3 in December-January; 2 birds present December-March (Dymond *et al.* 1989; Rogers *et al.*

1991). Individual recorded at Gibraltar 18–25 May 1986 coincided there with arrival of White-throated Sparrow *Zonotrichia albicollis* (Holliday 1990). AJE, PRC

Voice. See Field Characters.

Plumages. (nominate *hyemalis*). ADULT. Entire upperparts, side of head and neck, and chin to chest and side of breast as well as lower flank dark sooty-grey or dark slate-grey, darkest (slate-black) on cap, slightly paler slate-grey on rump and chest; lower mantle and scapulars with a few faint traces of olive-brown on feather-tips. Thigh mixed slate-grey and white, remainder of underparts white, sometimes tinged pale cream-pink, sharply contrasting with slate-grey of chest. Central 3 pairs of tail-feathers (t1–t3) black, outer webs fringed slate-grey, inner webs of t1 washed grey; t4 variable, black with long white wedge (up to 40 mm long and 6 mm wide) on terminal part of inner web, or with elongated white streak subterminally along shaft of inner web (10–20 mm long, 2–4 mm wide); t5 white, sometimes with trace of black on tip and extreme base, t6 white (RMNH, ZMA); 95–100% of surface of t6 white, *c.* 90% (60–100%) of t5, *c.* 25% (0–90%) of t4, 0–5% of t3 (Yunick 1972). Flight-feathers and greater upper primary coverts sooty black, outer webs narrowly fringed slate-grey or (on basal and middle portions of outer primaries) light grey. Tertials black, outer webs broadly bordered slate-grey (grey sometimes with traces of sooty bars). Upper wing-coverts dark slate-grey, median with sooty-black centre, greater with black inner web, terminal part of slate-grey fringe along outer web usually paler grey or almost white; rarely, white forming narrow wing-bar (Killpack 1986). Axillaries and under wing-coverts slate-grey with white tips, but longer under secondary coverts white; basal inner borders of flight-feathers shiny pale grey. *In fresh plumage* (autumn), olive tinge of grey feather-tips on upperparts more pronounced, especially from nape to back (but slate predominant), throat and chest with traces of narrow pale grey or whitish feather-fringes, tip of outer web of secondaries and greater upper wing-coverts with small pale grey to off-white spot. *In worn plumage* (May–June), upper-parts and chin to chest uniform deep slate-grey, virtually black on forehead and lore; slate on flank and thigh more extensive; grey fringes and small off-white tips of tertials and greater coverts worn off, tertials and greater coverts appearing blacker. ADULT FEMALE. Like adult ♂, but slate of head, body, and underwing distinctly paler, medium grey; tips of grey feathers on head and body washed olive-brown, less contrastingly fringed, cap showing about equal amounts of grey and olive-brown in fresh plumage, lower mantle and scapulars mainly olive-brown, rump to upper tail-coverts, side of head and neck, and chin to chest mainly grey. Tail as adult ♂, but t4 black with elongated white subterminal mark, t5 with dark wedge at base of inner web and often some dusky marks on tip of outer web (RMNH, ZMA); (60–)90–100% of surface of t6 white, *c.* 80% (25–98%) of t5, *c.* 10% (0–10%) of t4, none on t3 (Yunick 1972). Wing as adult ♂, but grey of lesser, median, and greater upper wing-coverts paler, tips often partly suffused olive-brown, especially on inner coverts, grey of fringes of secondaries suffused buff, broad fringes along outer webs of tertials light olive-brown to greyish-buff. *In worn plumage*, olive-brown of feather-tips largely worn off, except for traces on lower mantle and scapulars, but grey of head and body distinctly paler medium or dark neutral grey than in adult ♂, less dark slaty, usually with slight olive-brown tinge; pale brown or buffish fringes of inner coverts and tertials partly worn off. JUVENILE. Upperparts drab-grey or grey-brown, rather broadly streaked black; chin, throat, side of breast, chest and flank dull buff or buff-grey, spotted and streaked dusky grey (except on chin), remainder of underparts white, spotted dusky grey on upper belly. Upper wing-coverts drab-grey with darker centres and paler brown-grey fringes, tip of median and greater coverts with indistinct off-white spot. FIRST ADULT MALE. Like adult ♂, but tips of feathers of upperparts slightly more extensively fringed olive-brown than in adult ♂ at same time of year, side of head and neck and chin to chest narrowly but extensively fringed pale buff to off-white, flank partly washed pale buff; if worn, grey dark and slaty, as in adult ♂, but with fairly distinct olive-brown wash; amount of brown suffusion rather as in adult ♀ at same time of year, but grey darker, more slaty. Tail still juvenile; as adult ♀, but tips of tail-feathers often somewhat more pointed and more frayed than in adult at same time of year; *c.* 90–100% of surface of t6 white, *c.* 85% (65–100%) of t5, *c.* 10% (0–45%) of t4, none on t3 (Yunick 1972). Flight-feathers, greater upper primary coverts, and tertials still juvenile; broad borders along outer webs of tertials pale cinnamon-buff or light buff, contrasting markedly with slate-grey of greater coverts (unlike adult ♂), fringes of secondaries and primary coverts often slightly tinged pale brown, less pure grey than in adult ♂. FIRST ADULT FEMALE. Like adult ♀, but medium grey of upperparts completely concealed below broad and ill-defined olive-brown or buff-brown feather-tips when plumage fresh, except for rump; side of head and neck, chin to chest, side of breast, flank, and thigh mixed with about equal amounts of medium grey and buff-brown; if worn, grey of body retains brown cast. Tail juvenile, t4 wholly black or with a narrow elongated white patch on inner web at most; tips of feathers more pointed and more frayed than in adult (RMNH, ZMA); *c.* 90% (60–100%) of surface of t6 white, *c.* 70% (30–98%) of t5; *c.* 10% (0–50%) of t4, none on t3 (Yunick 1972). Flight-feathers, greater upper primary coverts, and tertials juvenile, similar in colour to those of adult ♀; fringes of greater coverts usually browner than in adult ♀. For ageing and sexing by use of combination of wing-length and amount of brown on head or body, see Blake (1962, 1964, 1967), Grant and Quay (1970), and Balph (1975); see also Dow (1966), Yunick (1981, 1984), and Pyle *et al.* (1987).

Bare parts. ADULT, FIRST ADULT. Iris dark reddish-brown to claret-purple; in 1st adult, grey-brown or brown up to March. Bill light flesh-pink or lilac-white; in winter, with small plumbeous-grey tip; upper mandible dark in race *dorsalis* of *caniceps* group. Tarsus light brown, flesh-brown, pinkish-horn-grey or dull flesh-grey, toes dull grey-horn, slate-grey, or dark grey. (Dwight 1900; Ridgway 1901–11; Yunick 1977*b*; Pyle *et al.* 1987; RMNH, ZMA.) JUVENILE. Bill dusky pink-buff becoming flesh when older; leg and foot pink-buff, dusky grey in older young (Dwight 1900; Bent 1968).

Moults. ADULT POST-BREEDING. Complete; primaries descendent; exceptionally, a few secondaries retained (Yunick 1976). On breeding grounds, July–October. In New York State (USA), starts from late July, completed about late September (Yunick 1976). In Virginia (USA), mainly from August (Nolan *et al.* 1992). Some moult on head frequent mid-November to December, Maryland (USA) (Brackbill 1977). For moult under various conditions in captivity, see Ketterson and Nolan (1983) and Nolan and Ketterson (1990): in captive groups held under natural conditions originating from Ontario (Canada) and Indiana (USA), moult in both well under way by mid-August, completed mid- or late October. For influence of breeding cycle and gonadal regression on onset of moult, see Nolan *et al.* (1992). ADULT PRE-BREEDING. Partial; extent limited. February–April. Of 57 birds examined, Maryland, 56% had some moult on head or throat late February to mid-April, once also late January (Brackbill 1977). POST-JUVENILE. Partial: head, body, lesser and median upper wing-coverts, occasionally some or all tertials or feathers of bastard wing, and variable number of greater coverts (or none); on breeding grounds, July-October (Pyle *et al.* 1987; RMNH, ZMA). In Maryland, generally completed by late September, occasionally up to mid-October (Brackbill 1977; Pyle *et al.* 1987; RMNH, ZMA); some outer greater coverts sometimes replaced in winter, these then much newer than neighbouring inner coverts replaced in autumn (RMNH, ZMA).

Measurements. ADULT, FIRST ADULT. Nominate *hyemalis*. North-east USA and eastern Canada, whole year; skins (RMNH, ZMA). Bill (S) to skull, bill (N) to distal corner of nostril; exposed culmen on average 3·2 less than bill (S).

WING	♂	80·8 (1·16; 21)	78–83	♀ 75·0 (2·91; 10)	70–80
TAIL		64·6 (1·69; 21)	61–68	61·5 (2·46; 10)	58–65
BILL (S)		12·9 (0·40; 19)	12·1–13·6	12·5 (0·62; 10)	11·5–13·4
BILL (N)		7·9 (0·37; 20)	7·2–8·8	7·7 (0·32; 10)	7·3–8·3
TARSUS		21·4 (0·50; 21)	20·5–22·3	21·7 (0·72; 10)	20·2–22·6

Sex differences significant for wing and tail. No difference in average wing length between adult and 1st adult (Grant and Quay 1970, which see for winter data from North Carolina). Bill in summer longer than rest of year, depending on food (Davis 1954).

Weights. ADULT, FIRST ADULT. Nominate *hyemalis*. Ohio (USA) (Nice 1938):

autumn	♂	19·7 (55)	17·7–22·0	♀ 18·9 (20)	16·3–20·5
winter		20·8 (48)	17·4–25·2	20·1 (14)	18·1–23·1
spring		20·6 (16)	18·3–25·2	18·6 (1)	—

(1) Michigan (USA), winter (Becker and Stack 1944). (2) New Jersey (USA), winter (Bender 1949). (3) Massachusetts (USA), winter (Helms and Drury 1960, which see for data per half month mid-November to mid-April and for influence of time of day). (4) Coastal New Jersey, autumn (Murray and Jehl 1964). Pennsylvania (USA): (5) December-February, (6) March, (7) April, (8) May, (9) September-October, (10) November (Clench and Leberman 1978, which see for details).

(1)	19·5 (597)	14·7–27·6	(6)	20·3 (2061)	13·5–25·4
(2)	21·2 (169)	18·2–24·9	(7)	19·7 (2737)	14·2–26·7
(3)	21·2 (856)	16·2–26·6	(8)	19·9 (10)	17·9–22·2
(4)	17·5 (75)	15·0–20·4	(9)	18·8 (1498)	12·9–23·4
(5)	21·7 (178)	17·9–26·2	(10)	19·5 (1231)	13·9–25·8

In Pennsylvania sample above, all ♂♂ 20·3 (3020) 14·3–26·7, all ♀♀ 18·7 (1570) 14·3–25·1, all of unknown sex 19·5 (3125) 12·9–26·2. Ohio: averages, October 18·9 (330), November-February 20·4 (110), March-April 21·0 (285) (Baldwin and Kendeigh 1938); undated 20·9 (171) 14·8–25·1 (Stewart 1937). For table of weights in morning and afternoon per half month mid-October to mid-April Delaware (USA), see Knowles (1972). For winter weight and fat content at various latitudes in USA, see Nolan and Ketterson (1983): weight about similar on southern sites in Tennessee, South Carolina, and Alabama, *c.* 3 g higher for each sex- and age-group further north in Michigan, intermediate in Indiana. For relationship between weight and wing length, see Prescott (1978). For weight and fat content in winter, see Helms *et al.* (1967); for weight and fat in spring under natural and unnatural captive conditions, see Wolfson (1942, 1945), Johnston (1962), Weise (1962), Ketterson and Nolan (1983), and Nolan and Ketterson (1990). For weight loss during night, see Blake (1956) and Ketterson and Nolan (1978). For lean weights and fat contents, see Johnston (1962), Farrar (1966), Helms and Drury (1960), and Nolan and Ketterson (1983); for weight and wing-loading, see Chandler and Mulvihill (1992).

Structure. Wing rather short, broad at base, tip bluntly pointed. 10 primaries: p7 longest, p8 0–2 shorter, p9 5–8, p6 0–1, p5 1–3·5, p4 5–9, p3 10–13, p2 11–15, p1 13–18; p10 strongly reduced, hidden below reduced outermost upper primary covert which is 0–2 longer, p10 50–60 shorter than p7, 6–11 shorter than longest upper primary covert. For relation of wing-shape with sex, age, and race, see Chandler and Mulvihill (1988, 1990) and Mulvihill and Chandler (1990, 1991). Outer web of p5–p8 emarginated, inner web of p6–p9 notched (often faint, especially in juvenile). Tip of longest tertial reaches to about tip of p1. Tail rather long, tip square; 12 feathers, t1 and t6 often slightly shorter than t2–t5. Bill short, conical, slightly shorter than half of head-length; fairly slender at base (depth 5·8–6·4 mm, width 5·6–6·3 mm), tip sharply pointed; culmen, cutting edges, and gonys virtually straight, except for hidden kink at extreme base of cutting edges, as in *Zonotrichia*. Nostril small, rounded, covered by short tuft of feathers projecting from base of upper mandible; some short bristle-like feathers project over side of gape. Tarsus and toes fairly long and slender, claws short and slender. Middle toe with claw 18·5 (10) 17–20; outer toe with claw *c.* 77% of middle with claw, inner *c.* 68%, hind *c.* 83%.

Geographical variation. Marked, especially in colour, less so in size. 5 groups recognized, each sometimes formerly considered separate species; (1) *hyemalis* group (Slate-coloured Junco) in east and north, (2) *aikeni* group (White-winged Junco) in north-western Great Plains, (3) *oreganus* group (Oregon Junco) along Pacific coast and in interior north-west USA and south-west Canada, (4) *caniceps* group (Grey-headed Junco) in interior of south-west USA, and (5) *insularis* group on Guadelupe island off Baja California (Mexico). Single race *insularis* of *insularis* group is isolated (though probably related to *oreganus* group), but all others intergrade locally to varying degree (American Ornithologists' Union 1983); 2 more groups sometimes split off, *mearnsi* group (Pink-sided Junco), with one race (*mearnsi*) in north-west interior USA and Cypress Hills of southern Canada (here included in *oreganus* group, but may be found to belong to *caniceps* group), and *dorsalis* group (Red-backed Junco), with one race (*dorsalis*) in south-west interior USA (here included with *caniceps* and *mutabilis* in *caniceps* group, following American Ornithologists' Union 1983, though differing rather from those races in relative proportions of tail and bill and in bare part colours).

Hyemalis group comprises 3 races: (1) nominate *hyemalis* through entire northern part of species' range from Alaska to Newfoundland and north-east USA, characterized by mainly slate-grey to black colour, variably tinged brown (depending on age, sex, and season), apart from white belly to under tail-coverts and white on outer tail-feathers; (2) *carolinensis* in Appalachians of eastern USA, similar to nominate *hyemalis*, but slightly larger and head of ♂ more uniform and paler slate, less dark and less tinged brown than nominate *hyemalis*, grey of ♀ less mixed buff and brown, both often with dark spot on base of upper mandible (Miller 1941; Mulvihill and Chandler 1991); (3) *cismontanus* in Canadian Rockies, which shows characters of stabilized hybrid population between nominate *hyemalis* and *oreganus* group, but nearer to nominate *hyemalis* (Miller 1941). Single race *aikeni* of *aikeni* group is similar to nominate *hyemalis*, but size much larger and with large white tips on median and greater upper wing-coverts (traces of this rarely in some nominate *hyemalis* also), white fringes along tertials, and more white in tail. Races of *oreganus* group have black or grey hood, extending to chest and forming rounded bib contrasting with white belly; black or grey of chest does not reach side of breast and flank, which are rufous, cinnamon, or yellow-buff (depending on race); black or grey of neck sharply demarcated from mantle, upperparts being various shades of brown (depending on race). In *caniceps* group, head and body grey, apart from black lore, rufous saddle on mantle and scapulars, and white belly to under tail-coverts. *J. h. insularis* is dark brown-grey on head and upperparts, light grey from chin to chest, and vinous-cinnamon on flank; differs from races of *oreganus* group in relatively long and heavy bill, strong foot, and short wing and tail. For survey of races, see Ridgway (1901–11), Hellmayr (1938), Miller (1941), and Bent (1968). For relationships, see Miller (1941) and Zink (1982). CSR

Calcarius lapponicus **Lapland Bunting**

PLATE 14
[between pages 256 and 257]

Du. IJsgors   Fr. Bruant lapon   Ge. Spornammer
Ru. Лапландский подорожник   Sp. Escribano lapón   Sw. Lappsparv   N. Am. Lapland Longspur

Fringilla Lapponica Linnaeus, 1758

Polytypic. Nominate *lapponicus* (Linnaeus, 1758), arctic Eurasia, east to Kolyma delta; *subcalcaratus* Brehm, 1826, Greenland and northern Canada, west to mouth of Mackenzie, wintering partly in Europe. Extralimital: *alascensis* Ridgway, 1898, eastern Chukotskiy peninsula, Alaska, and islands of Bering Sea (except Komandorskiye Islands), east to mouth of Mackenzie river; *coloratus* Ridgway, 1898, Komandorskiye Islands; *kamtschaticus* Portenko, 1937, Kamchatka and shores of northern Sea of Okhotsk to Koryakland.

Field characters. 15–16 cm; wing-span 25·5–28 cm. Slightly smaller than Snow Bunting *Plectrophenax nivalis* but with similar form, differing from typical *Emberiza* bunting in stubbier bill, larger head, bulkier build, noticeably oval, pointed wings, proportionately shorter forked tail, and long hind claw. Robust, slightly squat bunting, with flickering flight and plumage pattern like lark (Alaudidae); often runs. In all adult plumages, shows variably rufous nape; breeding ♂ also has striking black head with white zigzag line from eye to nape and down neck. Juvenile has bright reddish greater coverts, inviting confusion with Reed Bunting *E. schoeniclus* but distinguished by pale bill, more open face pattern, and characteristic mottling of central breast and foreflanks, also shown by ♀ and 1st-winter ♂. Underparts of adult noticeably white. Voice distinctive. Sexes dissimilar; marked seasonal variation in ♂. Juvenile separable.

ADULT MALE BREEDING. Moults: June–August (complete); April (face), strong contrasts of breeding plumage resulting mainly from wear. Foreparts dominated by black head and breast and patches on foreflanks, with glowing chestnut nape and sandy-white supercilium, continuing as bold white zigzag line which turns down and forward along rear ear-coverts and down sides of and round breast, thus separating deep bib from foreflank marks. Upperparts broadly streaked black on chestnut, with narrow whitish and sandy fringes which often form obvious pair of pale braces on mantle; back darker than any west Palearctic *Emberiza*. Wings show noticeably black feather-centres to coverts and tertials, and chestnut edges and whitish tips to median and greater coverts forming bright panel on median and (when fresh) quite distinct wing-bar on greater. Long primaries and rather short secondaries black-brown, with paler whitish or buff fringes and tips obvious when fresh. Underwing mainly white; like belly, catches eye when lit in flight. Tail black-brown, inner feathers margined buff but outermost broadly fringed buffish-white, forming pale edges which can be noticeable, but far less obvious than on *P. nivalis*, *Emberiza* or lark with white outer tail-feathers. Underparts from breast to under tail-coverts almost pure white, often appearing unmarked but with long black streaks on upper rear flanks sometimes visible. Bill bright yellow, with dusky tip; legs dark brown. Hind claw rather straight and noticeably long. ADULT MALE NON-BREEDING. Underlying plumage pattern identical to breeding ♂ but strong contrasts mostly obscured by long tips to fresh feathers of head, foreparts, and upperparts, which are coloured buff and brown on head and breast and sandy- to whitish-buff on back and larger inner wing-feathers. In spite of pale tipping, patches of black usually visible on crown, rear ear-coverts, moustachial area, and lower throat, and rufous nape also discernible, while sandy lore and supercilium (creating pale face), pale buff patch in front of nape, and buffish-white chin and throat form characters at least as striking as in spring. Fresh pale buff tips to coverts and margins to tertials form brighter wing markings. Underparts less pure white, with noticeable buff tinge at close range and heavy black and brown streaks on flanks. Dark eye prominent. ADULT FEMALE. Resembles ♂ in partly worn plumage, being heavily streaked black above, but never having such boldly contrasting colours and lacking unspotted rufous nape. Easily distinguished from breeding ♂ by brown and black mottling on crown and ear-coverts, more or less white chin, throat, and submoustachial stripe, and erratically placed black and brown streaks and patches on breast and all along flanks. After autumn moult, usually shows striking pale buff central crown-stripe, lacking in adult ♂ but present in immature. JUVENILE. Far less distinctive than adult, recalling young *Emberiza*, particularly *E. schoeniclus*. Head, upperparts, and wings far less black than adult, with noticeably yellow-buff and rufous ground and much narrower and less continuous dark streaks; underparts noticeably tinged yellowish-buff. Best clues are broad sandy central crown-stripe (rarely shown by *E. schoeniclus*), dark spots at end of eye-stripe and along lower edge of cheeks, white spot on rear ear-coverts, almost white double wing-bar emphasizing noticeably rusty panel on greater coverts, and erratically placed black or brown streaks on malar area, breast, and flanks; also structure (see above) and voice (see below). Bill flesh-yellow; legs flesh to bright brown. FIRST-YEAR. Following moult of all juvenile plumage except for flight-feathers, greater coverts, and tail, assumes more adult-like plumage, particularly when worn. ♂ distinguished from ♀ by more extensive black patches on sides of crown, ear-coverts, and bib (often boldly barred) and cleaner, less streaked rufous shawl. Some ♀♀ indistinguishable from juvenile except by clearer white throat and lower underparts.

Breeding ♂ unmistakable if seen well, but in brief glimpse or when silent can be confused with *E. schoeniclus* (less robust, with proportionately shorter wings and longer tail; lacks pale supercilium but has white submoustachial stripe, whitish-grey nuchal band, chestnut lesser wing-coverts, and often-flicked bright white outer tail-feathers) and Rustic Bunting *E. rustica* (noticeably smaller and slighter, with silky-white underparts splashed chestnut across breast and along flanks). Until general character and voice learnt, winter ♂, ♀, and immature far less distinctive; can be confused with *Alauda* and *Calandrella* larks, ♀ and young sparrows *Passer*, several *Emberiza*, *P. nivalis*, and all but 2 of the Nearctic vagrants in Emberizinae. Most liable to cause confusion are House Sparrow *P. domesticus*, Pine Bunting *E. leucocephalos*, *E. schoeniclus*, *P. nivalis* (particularly juvenile and 1st-winter ♀ with little white in wing), and Corn Bunting *Miliaria calandra*; thus note especially structure (particularly extension of folded primaries which equals length of tertials, matched only by *P. nivalis*), manner of flight (see below), habit of running (matched only by larks and *P. nivalis*), and voice (see below). At close range, close inspection of nape for sign of rufous feathering usually provides diagnosis. Flight recalls both *P. nivalis* and even more frequent companion Skylark *A. arvensis*, with closely similar action to both but not as deeply undulating as *P. nivalis* and not hovering before landing as *A. arvensis*; lands with sudden collapse at end of skimming glide. Flight silhouette somewhat less bulky than *P. nivalis*, again so closely recalling *A. arvensis* as to be indistinguishable from that species within mixed flock but usually showing longer, more oval outer wings and shorter, noticeably forked tail. Gait includes characteristic rapid run, loping walk, and hop. Will perch on low plants and rarely trees. Stance usually level on ground or perch, but raises head and occasionally body in alarm. Under pressure, often runs off but less regularly than *P. nivalis*. Flocks usually shy, particularly when with larks, but some birds allow close approach. Gregarious, forming flocks of scores on migration and at favoured wintering sites. Regularly mixes with other terrestrial passerines, often migrating and wintering alongside *A. arvensis*.

Song usually given in circular display-flight: a short but lively and musical repeated phrase, suggesting *A. arvensis*, 'teetooree-teetooree-trree-oo' or 'kretle-krIEEE-trr-kritle-kretle-tru'. Calls on breeding grounds, very musical, piping 'teeleu' clearly enunciated in alarm, distinguished from rather metallic 'teeuu' or 'TEElu' and quiet tuneless but hard, clipped 'tututuee' or 'ticky-tick'. On migration and in winter, last call becomes diagnostic dry, slightly rattled 'trill', usually ending in more melodious soft fluted 'tick-tick-tick-teu', 'ticky-tick-teu', or 'prrrt. . . chu' (in 1st part of latter closely resembling flight-call of Lesser Short-toed Lark *C. rufescens*). Other calls include variant of fluted note, 'teu' (when muffled, like piped call above, indistinguishable from distant *P. nivalis*), sharp 'zit', and explosive squeaky 'peet-teu' (both recalling Linnet *Carduelis cannabina* but shriller), and harsh 'jeeb' (given by night migrant).

Habitat. Extends across arctic tundra and boreal region of west Palearctic, between July isotherms of 2°C and 14-15°C. In contrast to equally arctic Snow Bunting *Plectrophenax nivalis*, avoids rocky and precipitous or bare terrain, favouring low shrubby tundra and damp hummocky moss-tundra with dwarf birch *Betula*, willow *Salix*, and heath plants (Voous 1960b). Avoids shrub thickets and all kinds of forest, occupying large expanses of open grassy or marshy lowland tundra (Dementiev and Gladkov 1954). In southern Greenland, prefers valley bottoms with peat bogs and plenty of pools, normally below c. 150 m, but sometimes up to nearly 500 m (Nicholson 1930). In low-arctic Greenland, generally occurs more commonly in interior than on coast, frequenting the arctic heath, a low carpet of close vegetation dominated by crowberry *Empetrum nigrum* with *Betula nana* and other shrubby plants, covering level or rolling, often wet and hummocky ground. Sometimes nests under rather high willow scrub and in birches, but prefers wet swampy ground, often near lakes. Stones as well as hummocks and twigs sometimes used as song perches. Avoids human settlements in breeding season, but flocks visit them on autumn passage (Salomonsen 1950-1).

In Scandinavia, breeds on open fells and mosses above or beyond treeline, although sometimes within some forest-encircled swamp. In addition to terrain similar to that described for Greenland, occupies grassland and even hay meadows (Bannerman 1953a).

In Canada, breeding habitats are similar; on migration and in winter frequents weedy and grassy fields, grain stubble, airfields, and shores, being essentially a ground

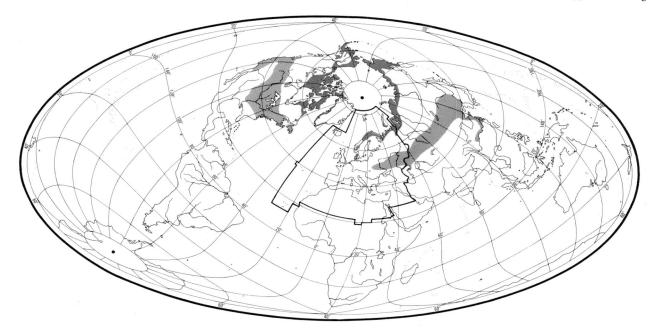

bird (Godfrey 1979). Generally winters on flat and open grassy areas bordering coasts and estuaries, grass moorland, grass steppes, and bare open cultivated areas (Harrison 1982), in Britain favouring rough grassland or stubble at no great distance from sea (Bannerman 1953*a*).

Distribution. Contraction of southern limit to north in Finland, but evidence of expansion further east (Kola peninsula).

FINLAND. Range has contracted; formerly extended to northern end of Gulf of Bothnia (Koskimies 1989). RUSSIA. Kola peninsula: expanding, from former range in tundra and forest tundra, into adjacent boreal taiga (Mikhaylov and Fil'chagov 1984).

Accidental. Iceland, Faeroes, Spain, Austria, Switzerland, Czechoslovakia, Yugoslavia, Bulgaria, Malta.

Population. No evidence of long-term changes.

SWEDEN. Estimated 300 000 pairs (Ulfstrand and Högstedt 1976). FINLAND. Marked annual fluctuations, but no known long-term trends. Estimated 100 000–200 000 pairs (Koskimies 1989; OH).

Survival. Oldest ringed bird (Alaska) 6 years (Custer and Pitelka 1977).

Movements. Migratory. European birds head between south-west and (chiefly) south-east, Asian and North American birds chiefly south, and Greenland birds both south-west (to North America) and south-east (to north-west Europe).

European birds winter mainly in south European Russia and Ukraine, south of *c*. 50–55°N (Dementiev and Glad-kov 1954), but no detailed information from this area. Most data relate to western and central Europe. Passage records are far more numerous than winter records, but fluctuate markedly from year to year. Unusually large numbers in 1950 and 1956 thought to be chiefly of Scandinavian origin, but in 1953 of Greenland origin. Reports of passage or wintering have increased considerably since 1950s, due at least in part to greater observer coverage; not known whether larger numbers or change in migratory pattern are also involved. (Jacobsen 1963, which see for discussion.)

Winter distribution in Britain confined almost entirely to east coast from Kent to Firth of Forth, especially at estuaries and at various sites in Norfolk, with few inland. Number difficult to estimate, perhaps *c*. 200–500 birds, but many more in peak years. Many birds appear to move on in late winter. (Lack 1986.) Some are from Scandinavia, others from Greenland (see below). Passage records are chiefly in north and north-west, with some in south-west. Since 1950s (and probably before) regular on passage on north and north-west coast of Ireland; parties of *c*. 11–20 recorded wintering in south in some years (Hutchinson 1989). In Scotland, regular on passage both seasons in Shetland, and fairly regular in autumn in Outer Hebrides and on east coast; occasional winter records (annual recently in Musselburgh area, Firth of Forth) (Thom 1986). Recent annual farmland records from central England suggest overland autumn passage with Skylark *Alauda arvensis* much overlooked (D I M Wallace). No ringing recoveries involving movements to or from Britain.

In Sweden, a few winter annually in south. Only small numbers migrate via Ottenby in south-east (Edelstam

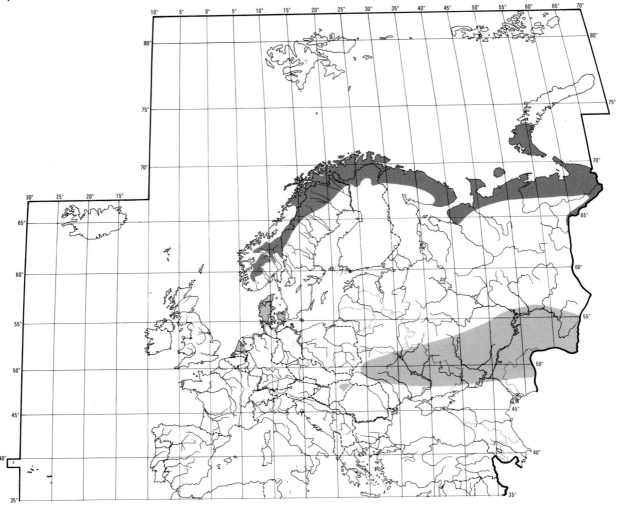

1972; Enquist and Pettersson 1986) or Falsterbo (southern Skåne) in extreme south-west, but considerable passage in Halland further north on south-west coast; perhaps birds then cross to Denmark from northern Skåne (Götmark *et al.* 1979; Wirdheim and Carlén 1986). Most birds fly south-east from northern Scandinavia; usually inconspicuous in autumn (though strong passage reported 1990: *Br. Birds* 1991, **84**, 236), but marked passage in spring through Finland and across Kvarken (*c.* 64°N, Gulf of Bothnia), with flocks up to 500 observed in Västerbotten (north-east Sweden) (Jacobsen 1963; SOF 1990). In Denmark, regular in small numbers on passage and (fewer) in winter; highest numbers seen in north-west (Hanstholm) in autumn, but in north-east (Skagen) in spring (Jacobsen 1963; Møller 1978*a*). Occurs only on passage on Helgoland (north-west Germany), and no evidence of long-term change in frequency. Regular and fairly common in autumn, and annual but uncommon in spring (only 10% of 521 birds, 1976–87). (Vauk 1972; Dierschke 1989.) Also regular on west coast of Schleswig-Holstein (northern

Germany) on passage, and a few winter in some years (Busche 1980). Not rare in Netherlands in winter, occurring chiefly on coast but also inland (except south-east); passage (in higher numbers) shows marked annual fluctuations; at Castricum on west coast, fewer than 10 ringed in some years, over 100 in other years (Yésou 1983; SOVON 1987; Schekkerman 1989). On Belgian coast, several hundreds winter in some years, but none in other years, and perhaps birds then winter further north-east; this supported by recovery at Wilhelmshaven (north-west Germany), November 1961, of bird ringed 410 km south-west at Zeebrugge (Belgium), November 1960. Rare and irregular inland in Belgium. Several ringing recoveries show Scandinavian provenance, e.g. bird ringed as nestling at 66°02′N in northern Sweden, June 1963, recovered 1790 km SSW at Knokke on Belgian coast, January 1966. (Lippens and Wille 1972.) In France, increase in birdwatching since late 1950s has revealed regular passage and wintering in small but varying numbers; data from 19th century suggest little recent change in status; possibly only 100-150 birds winter (with increase in colder years), but further studies may show wintering on larger scale, in view of passage numbers and extensive suitable habitat. Most records are from northern coastal areas, with a few along west coast (as far as extreme south). Rarely observed inland; 5 birds present 12 November 1986 to 4 February 1987 provide 1st record of overwintering far from coasts. In north, mostly reported (both passage and winter) from various sites in extreme north-east (especially Grand-Fort-Philippe) and in north and south Finisterre (Bretagne), also in bay of Mont-Saint-Michel; birds wintering on north Finisterre coast disappear at end of December, perhaps to go further inland. (Yésou 1983; Yeatman-Berthelot 1991.) Only 2 confirmed records in Spain despite widespread trapping of finches, so presumably few occur; one recorded from extreme north-west, one from Columbretes islands off east coast (Zink 1985).

In Mecklenburg (north-east Germany), regular in and near coast, and occasional inland; since 1950s, 20-150 or more each winter (Klafs and Stübs 1987). Also reported immediately south in Brandenburg since 1959 (no previous records); in Berlin area, 1959-76, 30 records involving 38 individuals (Dittberner *et al.* 1969; Bruch *et al.* 1978). Rarely recorded in southern Germany or Austria (Eifler and Blümel 1983; Zink 1985; Wüst 1986), and only 7 records from Switzerland (Winkler 1984; *Br. Birds* 1988, 81, 338). In Poland, scarce but probably regular autumn and winter visitor, mostly along coast (Tomiałojć 1990). Schüz (1941) reported autumn passage on Kaliningrad coast (western Russia), but apparently irregular there. Only sporadic records in Baltic states (Transehe 1965; Vīksne 1983), and reported only in fairly small numbers in St Petersburg region, even in well-studied areas (Noskov *et al.* 1981; Mal'chevski and Pukinski 1983). Irregular in Moscow region (Ptushenko and Inozemtsev 1968), and exceptional in Belorussiya (Fedyushin and Dolbik 1967). Winters regularly far south of observed passage, however, so perhaps birds overfly at high altitude. In Italy, records are chiefly in north (Lombardia, Veneto, Friuli), but apparently regular as far south as Lazio, *c.* 42°N (Brichetti 1976; Carlo 1991). One record from Malta (Sultana and Gauci 1982). Higher winter numbers are reported from Hungary; 1st record 10 January 1960, involving 50 birds. Since 1971, recorded annually in Hortobágy National Park in north-east and adjoining areas, with increase in 1980s; occurs only in small flocks in some years, but in flocks of 100 (up to 480) in other years; in severe cold or heavy snowfall, regularly moves on to unknown site in midwinter, returning after a few weeks. (Szabó 1976; Kovács 1981; Haraszthy 1988.) Apparently rare in Balkans, though more frequent since 1960s (Matvejev 1976). In south European Russia and Ukraine, however, winters in considerable numbers; rare in extreme south, in Crimea and northern Caucasus (Dementiev and Gladkov 1954). Winters fairly regularly in southern part of Volga-Kama region (Popov 1978).

Winter populations in north-west Europe include birds from Greenland as well as Scandinavia (but no direct evidence from ringing). Small numbers occur regularly in western and southern Iceland; 42 records 1979-88; biometric data show closer similarity to Greenland than Scandinavian birds, though these may also occur (Pétursson *et al.* 1991; Fox *et al.* 1992). In Britain, arrivals on north and north-west coasts often coincide with Greenland races of other species; largest influxes occur in anticyclonic conditions with westerly airstream; markedly earlier arrival in north-west than in east also supports Greenland provenance. Birds occurring irregularly in southern Norway may have same origin; also, small numbers reaching Fair Isle (Scotland) from south-east may include birds which had previously drifted to Norway from west. (Williamson 1953; Williamson and Davis 1956.) Similarly, some early arrivals elsewhere in north-west Europe may be Greenland birds; in France, reaches north-west before north-east (Yésou 1983); in Netherlands, wing measurements average longer at start of passage, showing no significant difference from Greenland birds; in midwinter, wing measurements in Netherlands and also in Wash (eastern England) are significantly shorter than Greenland birds, but show no significant difference from Scandinavian birds (Francis *et al.* 1991).

Autumn migration begins August. Most birds leave Greenland mid-August to end of September (Salomonsen 1950-1), and recorded chiefly mid-September to mid-October in Iceland (Pétursson *et al.* 1991). Vacates Sweden late August to October, with mid-September peak in south-west (Jacobsen 1963; Götmark *et al.* 1979; SOF 1990). Reported times of passage vary greatly; average earliest record 27 August over 17 years on south-west Finnish islands (Hildén 1974*b*), 20 September over 11 years on Helgoland (Dierschke 1989), 14 October over 10 years in northern Denmark (Møller 1978*a*). At Revtangen (south-west Norway), 1949-63, recorded 5 September to 8 November, chiefly 12 September to 16 October (Haftorn 1971). At Castricum (western Netherlands), 1965-86, early birds recorded late August or early September, but main passage from 2nd half of September, peaking late October; adults appear to pass later than juveniles, but very few adults caught; no significant difference in median date of ♂♂ and ♀♀ (Schekkerman 1989). In France, earliest records in north-west at end of August or beginning of September, with gradual increase to October and November; recorded only from mid-September at Cap Gris-Nez in north-east (Yésou 1983). First arrivals in Scotland and Ireland in late August also; long-term data at Fair Isle (Scotland) show earliest bird 23 August and build-up from early September, peaking 2nd half of September and gradually diminishing in October (Hutchinson 1989; Dymond 1991). Arrives in eastern Britain from 2nd week September, with numbers increasing October (Davis 1954; Williamson and Davis 1956); in Norfolk, 1986-90, first arrivals 8-14 September (*Norfolk Bird Reports*). In

Mecklenburg, recorded from mid-September, with most birds early or mid-October (Klafs and Stübs 1987). Reaches Hungary from mid-September, with numbers building up October–November (Kovács 1981; Haraszthy 1988).

Spring migration (February–)March–May. Leaves Hungary mostly in March; occasionally recorded early April, and late flock reported 1 May 1979 (Kovács 1981; Haraszthy 1988). Movement in Mecklenburg February to mid-April (Klafs and Stübs 1987), and in Lake Ladoga region (western Russia) mid- or end of April to mid-May (Noskov et al. 1981). In France, some evidence of small movement mid-February to mid-March (Yésou 1983). In eastern England, reported until mid-May in some years, exceptionally later (*Norfolk Bird Reports*). Main passage April in north-west Germany and Denmark, with latest reports early May (Møller 1978a; Dierschke 1989). In Helsinki area (Finland), average earliest record 6 April (22 March to 21 April) over 20 years (Tiainen 1979). Reaches breeding grounds mostly in May (Dementiev and Gladkov 1954; Semenov-Tyan-Shanski and Gilyazov 1991); in Finnmark (northern Norway), average first arrival 20 May (7–27 May) over 12 years (Haftorn 1971). Return passage towards Greenland inconspicuous; small numbers recorded on Fair Isle, mostly late April to early May, in Ireland March–April, and in Iceland chiefly in May (Hutchinson 1989; Dymond 1991; Pétursson et al. 1991). Arrives in Greenland May to early June (Salomonsen 1950–51). Vagrant to Franz Josef Land, but not reported from Spitsbergen (Løvenskiold 1963).

Birds breeding in west Greenland winter in North America (only 1 ringing recovery, at 52°N in Canada); study shows that birds can accumulate sufficient fat for direct flight to North American continent, rather than making an initial shorter sea-crossing to Baffin Island which would result in longer route. 2 recoveries, in Quebec (Canada) and Minnesota (USA), of birds ringed Angmagssalik (65°36′N, south-east Greenland), show that at least some birds breeding on east coast also migrate south-west; perhaps they move gradually towards southern tip, where birds are known to congregate in autumn. Greenland birds wintering in north-west Europe may originate further north on east coast (notably Scoresby Sund); from there, migration south-east to Iceland and Britain would involve markedly shorter route than to North America. (Francis et al. 1991; Fox et al. 1992.) Exceptional influx into north-west Europe in 1953 coincided with unusual winter abundance in USA (Williamson and Davis 1956, which see for analysis).

Population breeding in North America far more conspicuous on passage and in winter than European birds, due at least in part to greater abundance. Migrates through most parts of southern Canada; especially common in prairie provinces but rare in Newfoundland. Winters in southern Canada in small numbers, also in northern and central USA; overall distribution roughly coincides with regions where winter wheat and oats are sown; makes considerable wandering movements in winter, seeking localities with good food supply, so disjunct populations frequently occur. (Godfrey 1986; Root 1988.) In autumn, leaves breeding range mostly late August to September (Gabrielson and Lincoln 1959; Bent 1968), and reaches south of winter range in 1st half of November (James and Neal 1986). Spring migration begins February, and main movement through northern USA late March to early May (Bull 1974; James and Neal 1986; Janssen 1987). Reoccupies breeding areas mid-May to mid-June; movement along west coast is earlier, with arrivals in south-west Alaska from late April; tends to have marked fat reserves on arrival (Bent 1968).

In Asia, winters from southern Siberia and Mongolia east to Korea and Japan. Local numbers vary greatly according to severity of winter (Dementiev and Gladkov 1954). Apparently some winter south of west Siberia: regular on passage in Kazakhstan and sometimes recorded in winter (not in north or centre), probably only in severest winters (Korelov et al. 1974). Common on passage on north-east shore of Lake Baykal (Ananin and Fedorov 1988), and in winter in eastern Mongolia (Piechocki and Bolod 1972). In Ussuriland, winters in small numbers on coastal plain (Panov 1973a). In China, winters south to northern Chekiang, and in Yangtze valley west to Hupeh and Szechwan (Schauensee 1984). Infrequent and sporadic in Japan; records mostly on east coast of Hokkaido, but probably regular as far south as Hegura island (37°52′N) off west Honshu (Brazil 1991).

Leaves Asian breeding grounds mainly late August to September (Dementiev and Gladkov 1954; Kishchinski 1980; Danilov et al. 1984); most have left Wrangel island in extreme north-east by end of August (Stishov et al. 1991). Most passage records September–October (Kozlova 1933; Johansen 1944; Dementiev and Gladkov 1954; Vorobiev 1963). Reaches Ussuriland from late October (Panov 1973a); in Beidaihe area (north-east China), earliest record 13 October over 5 years (Williams et al. 1992). Spring migration prolonged, March–June according to latitude. Present in north-east China until mid-March (Hemmingsen 1951). Passage March–April in Ussuriland (Panov 1973a), and average earliest record on Kamchatka peninsula 10 April over 7 years; arrivals continue to end of May (Kishchinski 1980). Present in Mongolia until April (Piechocki and Bolod 1972). In Kurgal'dzhin (northern Kazakhstan), very rare in autumn and only 1 winter record, but regularly stops over in 1st half of March in numbers varying from thousands to few; birds continue northward in early April (Kovshar' 1985). In 1976 (but not in other years), common on passage late March to early April in valley of Irtysh (western Altai) (Berezovikov 1983), and in May further north at Chany lakes (Yurlov et al. 1977). Main movement through western and central Siberia April–May (Johansen 1944). Reaches breeding grounds in Yakutiya in 2nd half of May

(Vorobiev 1963), and Yamal peninsula and Wrangel island late May to early June (Danilov *et al.* 1984; Stishov *et al.* 1991).

DFV

Food. Invertebrates (especially flies Diptera) in peak breeding season, otherwise seeds of grasses and low herbs. On breeding grounds, forages busily on ground, running from tussock to tussock picking invertebrates from surface of vegetation, very rarely in bare places; sometimes in shrubs, even 2–3 m up in tree, e.g. birch *Betula*, taking insects. Very often recorded plucking invertebrates from Rosaceae flowers (e.g. *Rubus*, *Dryas*). Jumps up to snatch flying insects and has been observed hawking *c*. 75 cm above ground. (Nicholson 1930; Rowell 1957; Custer and Pitelka 1978; Stishov *et al.* 1991; D T Holyoak.) In northern Canada, recorded feeding on blow-flies (Calliphoridae) at animal corpses (D T Holyoak). In winter, Kaliningrad region (western Russia), fed on ground in rough grassland and harvested root-crop fields, but particularly in stubble and on grassy rutted tracks (Schüz 1941); in Hungary, on pasture near wetlands, preferring places with tussocks of dead and trampled grass and low herbs, as well as in various kinds of stubble (Szabó 1976; Kovács 1981). In Britain, autumn and early winter, frequents coastal salt-marsh and adjacent rough pasture or stubble, and also forages along tideline (Lack 1986). See Yésou (1983) for France, and SOVON (1987) for Netherlands. In south-west Greenland, summer, picked up 72 invertebrates in 6·5 min, making 32 feeding excursions per hr (Nicholson 1930).

Diet in west Palearctic includes the following. Invertebrates: larval Lepidoptera, adult and larval flies (Diptera: Tipulidae, Culicidae, Chironomidae, Bibionidae, Muscidae), Hymenoptera (larval Tenthredinidae), beetles (Coleoptera: Staphylinidae), spiders (Araneae), earthworms (Lumbricidae). Plants: seeds of knotgrass, etc. *Polygonum*, dock, etc. *Rumex*, chickweed *Stellaria*, shepherd's purse *Capsella*, medick *Medicago*, eryngo *Eryngium*, crowberry *Empetrum*, plantain *Plantago*, mint *Mentha*, mayweed *Matricaria*, hawkbit *Leontodon*, grasses (Gramineae, including oats *Avena*), sedges (Cyperaceae), rushes (Juncaceae). (Mikheev 1939; Schüz 1941; Rowell 1957; Sterbetz 1967; Szabó 1976; Cumming 1979; Kovács 1981.)

Apparently no detailed study of diet from west Palearctic. In Barrow area (Alaska), 235 stomachs and gullets contained 8849 items as follows: in late May, 76% of diet by dry weight (corrected) seeds, mainly *Luzula* (Juncaceae) and *Cerastium* (Caryophyllaceae), 18% adult beetles and Hemiptera, and 6% larval Diptera, mostly Chironomidae; in mid-June, 86% larval Diptera (63% crane-flies Tipulidae), 10% adult beetles and Hemiptera, 2% larval Tenthredinidae, and 2% spiders; in mid-July, 78% adult Diptera (67% crane-flies), 7% larval Tenthredinidae, 6% adult beetles and Hemiptera, 4% seeds, 3% larval crane-flies, and 2% spiders; in mid-August,

27% larval Tenthredinidae, 26% larval beetles, 25% seeds, 8% adult beetles and Hemiptera, 7% adult Diptera, 4% spiders, and 3% other adult Hymenoptera. Springtails (Collembola) taken in some quantities at beginning of June (*c*. 15% of items by number) but hardly show up in analysis by weight; in early June, Chironomidae larvae 34% by number. Only seeds available at start of breeding season since snow still on ground and invertebrates inactive; Chironomidae larvae more readily obtained early on than those of crane-fly, while large Tenthredinidae larvae on prostrate willow *Salix* are easily captured late in season. (Custer and Pitelka 1978, which see for dietary overlap and competition with 4 sandpiper *Calidris* species.) Similarly in north European Russia, feeds mainly on seeds of cotton-grass *Eriophorum* on arrival, proportion of invertebrates in diet increasing gradually to 100% in July (Mikheev 1939). In Swedish Lapland, eats crowberry seeds before Diptera emerge (Rowell 1957). On Yamal peninsula (northern Russia), of 7 stomachs, May–June, 3 contained only seeds, 3 contained seeds, crane-flies, and beetles (Staphylinidae), and 1 only Staphylinidae and Psyllidae (Hemiptera); 2 stomachs, late June and July, held beetles, Diptera, and Psyllidae (Danilov *et al.* 1984). On Wrangel Island (north-east Russia), beetles (Carabidae) common in diet (Portenko 1973). For 6 stomachs from Manitoba (Canada), see Grinnell (1944). In western Greenland, huge numbers of caterpillars of *Eurois occulta* (Noctuidae) eaten when abundant, but some years completely absent (Fox *et al.* 1987). In winter quarters, feeds on seeds of low Polygonaceae, Caryophyllaceae, Cruciferae, Labiatae, Compositae, and similar agricultural weeds, also on grasses and rushes (Schüz 1941; Sterbetz 1967; Szabó 1976; Kovács 1981).

In Barrow area, needs to consume 10 000–14 000 items per day at end of May (mostly seeds) and up to 17 000 in August (including many seeds); in June–July, needs only 2000–4000 invertebrates per day; in May, needs intake of 18 items per min of foraging, in August 10–20 per min, but in June–July only 4–7 invertebrates per min. Over summer, average daily energy requirement of ♂ was 188 kJ, of ♀ 169 kJ. (Custer *et al.* 1986.)

Young in Swedish Lapland fed almost wholly adult and larval Diptera (especially mosquitoes Culicidae and midges Chironomidae), small beetles such as Staphylinidae, and earthworms (Rowell 1957). Food of young collected 20–30(–60) m from nest (Grote 1943a). In Scotland, adult crane-flies brought by adults to nest (Cumming 1979), and in north European Russia nestlings given Diptera (crane-flies, Bibionidae, Muscidae), Tenthredinidae, and spiders (Mikheev 1939). On Yamal peninsula, 93 collar-samples contained 578 items, of which 76% Diptera, 14% sawfly larvae (Hymenoptera: Symphyta, probably Tenthredinidae), 6% adult Lepidoptera, 3% spiders, and 1 snail (Pulmonata) (Danilov *et al.* 1984). In western Greenland, in lowland (and upland), total of 120 items were 12% (79%) by number adult Diptera, 77% (10%) larvae, 12%

(8%) spiders, 0% (3%) seeds; diet according to abundance at each site (Madsen 1982). In same study area, piles of *Eurois occulta* caterpillars left at nests in years of abundance (Fox *et al.* 1987). In Alaska, in Arctic (and subarctic), total of 2767 items from collar-samples were 44% (15%) by weight larval and pupal crane-flies, 25% (78%) adults, 15% (0·1%) larval Tenthredinidae, 10% (2%) other Diptera, and 5% (3%) spiders; variation suggests difference in preferred feeding habitat and change in timing of nesting with respect to emergence of adult crane-flies (Seastedt 1980). For diet of young in northern Canada, see Hussell (1972).

In Barrow area, young 10-12 days old had average daily energy intake of 117 kJ, and 1800 kJ from hatching to independence; parents brought maximum 1200 items per nestling per day (Custer *et al.* 1986). At one nest in Swedish Lapland, each nestling received *c.* 80 meals per day (Rowell 1957).   BH

Social pattern and behaviour. Important studies of nominate *lapponicus* in Swedish Lapland (Rowell 1957; Gierow and Gierow 1991) and Malozemel'skaya tundra, European Russia (Mikheev 1939); also on extralimital Bylot Island (Northwest Territories, Canada) (Drury 1961). Valuable studies on *alascensis* in Barrow (Alaska) by Custer and Pitelka (1957), Seastedt and MacLean (1979), Tryon and MacLean (1980), and McLaughlin and Montgomerie (1985, 1989a, b). For useful review of pioneering studies in Russia, see Grote (1943a). No obvious differences in behaviour between races. For comparison of (some) behaviour of captive birds with other Emberizidae, see Andrew (1957a).
    1. Gregarious at all times, much less so during breeding season, but even then small flocks of unpaired ♂♂ may occur (see Flock Behaviour, below). In immediate post-breeding period, juveniles form small nomadic flocks (Pleske 1928; Nicholson 1930). Autumn migration occurs in flocks, with large numbers reported in some places, e.g. in extralimital Lena valley (Russia), September, flocks of hundreds and sometimes thousands (Vorobiev 1963). On wintering grounds, typically in small flocks; in Kaliningrad region (western Russia), winter flocks strongly attached to highly circumscribed areas, instantly returning to them if flushed, though nomadism to some extent induced by regular association with more mobile Skylarks *Alauda arvensis* (Schüz 1941). In Hungary, mostly in conspecific flocks, but sometimes with Snow Bunting *Plectrophenax nivalis*, Reed Bunting *Emberiza schoeniclus*, Twite *Carduelis flavirostris*, and also reported seen with Fieldfares *Turdus pilaris* (Szabó 1976). Flock associates in Sarthe (France) additionally included Meadow Pipit *Anthus pratensis* and various finches (Fringillidae) (Lapous 1988). Spring passage in Denmark usually consisted of single birds or small flocks of up to 5-6, often in association with *A. arvensis*, also occasionally *P. nivalis*, Shore Lark *Eremophila alpestris*, *A. pratensis*, and *C. flavirostris* (Jacobsen 1963). At beginning of June in east Finnmark (Norway) small parties fed with Bluethroats *Luscinia svecica* and Red-throated Pipits *A. cervinus* (Blair 1936). In extralimital Russia, spring migrant flocks from less than 10 birds to hundreds (Berezovikov 1983; Morozov 1984). First arrivals (♂♂) on breeding grounds (Malozemel'skaya tundra) in flocks of 10-15 but main influx thereafter in flocks of up to 300-400, exceptionally 1000; initially ♀♀ kept mostly apart from ♂♂; flocks initially associated with *P. nivalis* and *E. alpestris*, but break up and disperse onto tundra as snow melts

(Mikheev 1939). Similar pattern reported in Northwest Territories, with birds readily reverting to flocks during early occupation, according to weather conditions (Van Tyne and Drury 1959; Drury 1961). BONDS. Mainly monogamous but to some extent ♂♂ also polygamous and promiscuous (Lyon and Montgomerie 1987). In study in Hardangervidda (Norway) 3 out of 18 ♂♂ were bigamous (Bjørnsen 1988). In *alascensis*, low incidence of polygamy; arises when ♂, after mate starts incubating, courts unattached ♀♀ or ♀♀ with incomplete clutches (Custer and Pitelka 1977; Seastedt and MacLean 1979). In Scotland, where breeding occurred in 1977, ♂ was apparently paired with 2 ♀♀, nests *c.* 40 m apart (Cumming 1979). Some relationships between neighbours are complex, perhaps involving shared paternity; e.g. Tryon and MacLean (1980) describe case in which unpaired ♂ drove pair from their territory and took over ♀, and nesting followed; thereafter, annexing ♂ was seen increasingly less and did not contribute to feeding young; however, a new neighbouring ♂ began to share feeding the young (suggesting that possibly he also had copulated with ♀). No information on duration of pair-bond. ♀ builds nest, incubates, and broods; according to Blair (1936) both sexes incubate and brood, but not infrequent occurrence of ♂-characteristics in ♀-plumage (Drury 1961; Bjørnsen 1988) might have given impression that ♂♂ were involved. Both sexes feed young about equally (Drury 1961; see also Relations within Family Group, below). Brood-division occurs as soon as young leave nest; young start self-feeding 9-10 days after leaving nest, and are independent *c.* 5 days later (i.e. at *c.* 23 days old) (McLaughlin and Montgomerie 1985, 1989a). Statement by Rowell (1957) that young are self-supporting after 4 days out of nest is presumably an underestimate. Age of first breeding 1 year (Gierow and Gierow 1991); in *alascensis*, probably 1 year for most ♀♀, but proportion of 1-year-old ♂♂ were non-breeders (Custer and Pitelka 1977); see also Flock Behaviour (below). No such surplus of ♀♀ ever recorded (Lyon and Montgomerie 1987). BREEDING DISPERSION. Territorial and in neighbourhood groups, probably due to clustering in favourable habitat (e.g. Danilov *et al.* 1984). No good studies of territory size in west Palearctic, but in Northwest Territories, where density high, territories typically 0·2 ha or less, as little as 0·1 ha in some cases (Drury 1961). In Barrow, average 1·76 ha (1·09-2·83, *n* = 20); the better the habitat (and therefore food supply) the smaller the territory (Seastedt and MacLean 1979). In Chukotka (north-east Siberia) 100-200 m between nests, minimum 60-70 (Kishchinski 1980; see also Kishchinski *et al.* 1983). Territory size difficult to establish because little overt defence seen at low densities (Wynne-Edwards 1952) and sometimes even when density not so low (Rowell 1957). Territories not contiguous (Seastedt and MacLean 1979) or overlapping, though feeding areas (which typically extend beyond territory: see below) of neighbours often overlap (Danilov *et al.* 1984) and concentrated food sources may attract birds from various territories (Gierow and Gierow 1991). Territory thus serves for variable proportion of feeding, but courtship and nesting typically within territory (although see subsection 5 of Heterosexual Behaviour, below). Seastedt and MacLean (1979) reported that over 90% of foraging by ♂ and ♀ occurred on territory early in season (period of active defence); less than 90% later in season during chick-feeding, at which time less regard paid to territorial boundaries (Seastedt and Maclean 1979; Bjørnsen 1988; Tryon and MacLean 1980). ♂'s foraging range markedly larger than ♀'s, extending to 8·2 ha compared with territory of 3·4 ha (Tryon and MacLean 1980). Density highly variable, e.g. from 17-50 pairs per km² (Wynne-Edwards 1952) to up to 200 pairs per km² in localized optimal patches (Custer and Pitelka 1977); even 400 pairs per km² estimated in one valley in north-east Siberia (Portenko 1973). Apart from strong habitat effects on density (see below for details), also considerable variation between years (Gierow and Gierow 1991); e.g. in Barrow, density declined from average 75 pairs per km² (50–95) in 1951–69 to 25 in 1972, due probably to scarcity of lemmings (Microtinae) after 1969 causing predation to be redirected at *C. lapponicus* (Custer and Pitelka; see also Seastedt and MacLean 1979). For density in relation to proximity to Peregrine Falcon *Falco peregrinus* eyries in west Greenland see Meese and Fuller (1989). Densities in west Palearctic as follows: in Hardangervidda plateau, over 2 years, 24–6 territories per km² (Bjørnsen 1988); in Ammarnäs (Swedish Lapland), 1964–83, 9·65 (2–28) territories per km² in alpine habitat, 23·4 (15–32) in lowland (Svensson *et al.* 1984); for higher densities in same region, 1984–9, see Gierow and Gierow (1991). In west Greenland, 42 and 128 pairs per km² on south-facing lowland willow *Salix glauca* in 2 years (Fox *et al.* 1987); in north-west Greenland, 37 pairs per km² (Joensen and Preuss 1972). In Northwest Territories, up to 25 pairs per km² (Lyon and Montgomerie 1987). Density typically highest in lowland hummocky tundra, e.g. 74–94 birds per km² over 3 years in Chukotka, compared with 29–36 for montane tundra and 24–33 for mossy grassy bogs (Tomkovich and Sorokin 1983). In Malozemel'skaya tundra, up to 100 nests per km² in hummocky tundra, as long as no competing *A. cervinus* present (Mikheev 1939). On Wrangel Island (north-east Russia), 2–5(–8) birds per km² on continuous moss cover along rivers, 20–30 in tundra without tussocks, 70–80 in sedge/hummocky tundra (Stishov *et al.* 1991). See also Danilov *et al.* (1984) for Yamal peninsula (northern Russia). In Norway, 62% of territorial ♂♂ ringed in 1 year were present the next, demonstrating marked site-fidelity; ♂♂ that bred in the area the previous year occupied territories earlier than newcomers, and older ♂♂ also claimed higher-quality territories (Bjørnsen 1988; see also Antagonistic Behaviour, below). In Ammarnäs, birds that returned to same area for several years did not shift breeding site more than 100–150 m (Gierow and Gierow 1991). Custer and Pitelka (1977) reported return rate of 11·5% for ♂♂ in Barrow during 7-year study, and Drury (1961) found nests at same site in successive years had characteristics suggesting building by same ♀ in some cases. However, lower return rates than these recorded in Russia (Danilov *et al.* 1984; Dobrynina 1986; Stepanitskaya 1987). Virtually no information on fidelity to natal site of 1st-time breeders, but Tomkovich and Sorokin (1983) recorded 1-year old ♀ breeding not far from ringing site in Chukotka. ROOSTING. Little information. During incubation and brooding, ♀ roosts on nest for 2–5 hrs around midnight (Grote 1943a; Rowell 1957; Hussell 1972); during this time, ♂ also rests, usually in thick cover near nest (Rowell 1957), although hiatus in his singing may be much shorter (see Song-display, below). Several pairs found roosting communally in tussocks in cold weather, Baffin Island (Watson 1957a). In Hungary, winter, birds seen sleeping and preening in deep tyre-ruts during midday sun (Szabó 1976). See Flock Behaviour (below) for resting in trees on passage.

2. Readily allows man to approach to within a few metres throughout the year (e.g. Nicholson 1930, Lapous 1988). However, has a variety of tactics for evading detection or distracting observer, and is unusual in performing mobile distraction-lure display of disablement type (see details, below; also Parental Anti-predator Strategies) at apparently any time of year. In Kaliningrad region, outside breeding season, flock-members typically press close to ground and are reluctant to flush even when approached to within *c*. 1·5 m (thus very hard to detect at any distance). Sometimes, rather than flying away, birds spring up briefly into the air and down again, then run off, crouching as if

injured. If cautiously driven, some will move just ahead of observer, and one was followed thus for 45 min, making only 2–3 very short flights. (Schüz 1941.) Same distraction-display described by Szabó (1976) in Hungary and by Lapous (1988) in France, November, involved the following: moving in hunched posture with spread or drooped wings and fanned tail, performer appearing to slide along or drag itself; in French observation, bird took off after several metres, landed, and gave an alarm-call (see 5a in Voice) before allowing itself to be approached again. FLOCK BEHAVIOUR. In winter, marked site-fidelity at least in short term (see introduction to part 1, above) linked with reluctance of flocks to fly, even when flock-associates flew away; *C. lapponicus* flies up, circles, and returns precisely to place of departure (Schüz 1941; Szabó 1976). On Putorana plateau (north-central Siberia) flocks of 3–20 (both sexes), migrating in early June, frequently came down to feed and rest, often in trees; quite a lot of song was heard (see Song-display, below) but no Song-flights (Morozov 1984). In Hardangervidda, large numbers of unpaired ♂♂ formed parties of up to 4 which intruded on territories to solicit unpaired ♀♀ (Grampian and Tay Ringing Group 1981; see also subsection 4 of Heterosexual Behaviour, below). SONG-DISPLAY. Serves for mate-attraction and territorial defence (see Antagonistic Behaviour, below). In territory, boulder, bush, or hummock near nest serves as vantage point from which ♂ sings or launches Song-flights. According to Salomonsen (1950–1), each ♂ had single song-perch in territory, but 2 or more used in territories studied by Drury (1961). When singing (see 1 in Voice) from perch, ♂ lowers primaries, raises bill obliquely, and slightly ruffles belly, flanks, and sometimes crown (Drury 1961). Song-flight rather like pipit *Anthus*; following description combined from several sources: ♂ ascends suddenly, and typically silently (but see below) on steep diagonal; after reaching *c*. 6–15(–20) m, starts singing, swings from side to side, and turns to commence slow, widely spiral, gliding descent (Fig A) with outspread wings and tail fanned upwards

A

(Hortling and Baker 1932; Witherby *et al.* 1938; Grote 1943a; Portenko 1973); tends to land where he took off (Portenko 1973) or, just before landing, glides to next eminence (Grote 1943a). Especially in song-duel, and when suddenly responding to conspecific intruder, ♂ often sang on ascent into Song-flight, and in these cases wingbeats were rapid and shallow. In all Song-flights, trajectory in descent was usually a semi-circle, and as often on 'set' as on quivering wings; when wind blew 10–14 knots, ♂ in Song-flight hung suspended, occasionally beating partly folded wings, tail fanned wide, and sang 5–6 songs per flight. (Drury 1961.) Birds may sing from perches on spring migration (see Roosting, above) but Song-flights only begin with territorial establishment. Resident may be silent for hours and occasionally 'explode' into song, e.g. when ♀ visits territory (Bjørnsen 1988).

Song largely from ground in nest-area during building and laying (♂ then mate-guarding) but after clutch complete nearly all singing was in flight and over whole territory (Drury 1961). Singing typically declines rapidly after ♀ begins incubating (so song-period per ♂ not more than *c.* 10 days) but may continue sporadically through incubation and for 1–2 days after hatching (Drury 1961; Gierow and Gierow 1991). However, strong resurgence if nest predated, and ♂♂ (perhaps would-be polygynists) also heard singing late in breeding season at times when territorial boundaries have usually dissolved (Bjørnsen 1988; see Bonds, above). Grote (1943*a*) reports that song-period in Siberia lasts *c.* 3 weeks and that song late in season was confined to the ground (see Heterosexual Behaviour, below). On Bylot Island, no song heard after 10 July (Drury 1961). In Kaliningrad region, on autumn passage, Song-flight seen 26 September, song from the ground on 1 October (Schüz 1941). On Bylot Island, Song-flights commonest 06.00–10.00 hrs and 17.00–22.00 hrs (Drury 1961). In arctic Finland, song heard almost all day and night, with lull confined to period immediately after midnight (Ruthke 1939*a*). However in west Greenland, song-output, which had declined markedly by end of June, was almost restricted to widespread 'dawn chorus' of counter-singing ♂♂ from *c.* midnight to 01.00 hrs (Nicholson 1930). ANTAGONISTIC BEHAVIOUR. (1) General. ♂♂ aggressive during territorial establishment, with direct confrontations mostly evident early in season and where density is relatively high. In Hardangervidda, adult ♂♂ were present, foraging amicably, in the study area 7–10 days before territories were established; ♂♂ started singing (see Song-display, above) during a 4-day period and vigorously defended their territories against rivals (Bjørnsen 1988). In same area, 1981, ♂♂ seen defending territories against intruding unpaired ♀♀ throughout breeding cycle (Grampian and Tay Ringing Group 1981). In Yamal peninsula, residents showed aggression until middle of chick-rearing period but not thereafter; ♀♀ did not participate in territorial disputes (Danilov *et al.* 1984). In Northwest Territories, minority of ♂♂ which failed to obtain mates continued defending territories well past the latest date of clutch initiation (Lyon and Montgomerie 1987); on Bylot Island, no disputes seen after 1 July (Drury 1961). (2) Threat and fighting. In study on Bylot Island, territorial disputes consisted of vigorous answering Song-flights (see Song-display, above) or pursuits which ended in short fast dash at rival, often followed by Song-flight or rapid return to territory. Threat also includes dive-attacks and Forward-threat display which may be static or mobile (Fig B) in which ♂ runs low towards rival with head

B

thrust forward and slightly up (exposing markings), plumage sleeked (or ruffled only on rump), and giving Threat-calls (see 4b in Voice); victim, attacked when side-on to attacker, flew away with shallow quivering wingbeats ('moth flight'). (Drury 1961.) When stationary, Forward-threat display includes bill-thrusting at rival (Andrew 1957*a*; Drury 1961). In forward advance elicited on stuffed ♂, aggressors flicked tail and wing-tips, and also directly attacked specimen (Danilov *et al.* 1984). Drury (1961) did not record any physical fighting in the field, but Mikheev (1939) describes how territorial dispersion was

achieved by violent fighting both on the ground and in the air, involving pecking and pulling out feathers; apart from conspecifics, *A. cervinus* was also occasionally expelled from territories. Nicholson (1930) reported skirmishing with Redpolls *Carduelis flammea* and *P. nivalis*. HETEROSEXUAL BEHAVIOUR. (1) General. In North American populations, ♂♂ usually precede ♀♀ to breeding grounds by 2–3 days (Hussell 1972), and in west Greenland by periods ranging from 2–3 days to 2 weeks (Madsen 1982; Fox *et al.* 1987). In Malozemel'skaya tundra, in one breeding season, mass arrival of ♂♂ preceded that of ♀♀ by 9 days; after initial separation of the sexes for *c.* 1 week thereafter, pair-formation took place as snow receded from tundra (Mikheev 1939); not known if this pattern typical of European populations or a response to late springs (Gierow and Gierow 1991). However, onset and duration of pair-formation do vary with geographical situation and weather, e.g. pairing starts at end of May in Malozemel'skaya tundra, but 1–2 weeks later in higher latitudes, and in far north of Russia birds sometimes arrive already paired (Grote 1943*a*). Nominate *lapponicus* in North America forms pairs only after arrival on breeding grounds (Drury 1961; Hussell 1972); in Alaska, process hastened by ♂♂ establishing territories while ground almost completely snow-covered (Tryon and MacLean 1980). (2) Pair-bonding behaviour. Account based largely on Drury (1961). Prospecting ♀♀ visit territories of ♂♂. First interactions are typically aerial chases, preceding any ground-display. Such chases are characteristically longer, slower, and less zigzagging than territorial disputes. If ♂ caught up with ♀, there was a burst of rapid zigzagging, but if he fell far behind, ♀ slowed down till he overtook her. Example of ground-display as follows: ♂ ran, singing the while, in front of or alongside ♀, performing Advertising-display (Fig C): stood with forebody raised, breast

C

puffed out, head and bill held high, wings half-spread, drooped and quivering, tail spread and lowered; ♀ ran slowly ahead of ♂ and performed Inciting-display (Fig D): with head partly lowered, wings partly spread, tail raised somewhat, and giving Inciting-calls (see 3a in Voice); displaying ♀ and ♂ often pecked stiffly at ground, ♂'s chestnut nape emphasizing his stiff bow;

D

this presumably the same as repeated bill-lowering described by Andrew (1957a). ♀ ran around and ahead of ♂, then flew; if he did not follow, she returned and incited again till he did. During chases, she landed several times, ran along the ground, then flew again, either fast and darting, or slowly with quivering wings. Song-display of ♂♂ chiefly confined to the ground during such interactions. Once paired, ♂ followed ♀ closely during nest-building and laying, but mate-guarding gave way to Song-flights after clutch completed. (3) Courtship-feeding. Does not normally occur, and none recorded in some studies (Grote 1943a; Drury 1961; Hussell 1972). In 5-year study by Lyon and Montgomerie (1987) low frequency of courtship-feeding during incubation was performed by most ♂♂ in one year but in none of the other 4 years; food supply seemed to be poorest in the year Courtship-feeding occurred, suggesting that ♀♀ perhaps only beg when hard-pressed. In Scotland, ♀ performed possible Begging-display (Fig E): approached ♂ and crouched, gaping

E

and wing-quivering; despite soliciting thus several times, ♂ did not respond (Cumming 1979). (4) Mating. The only detailed description of sequence is by Drury (1961): ♂ adopted Advertising-posture (see Advertising-display (above), singing with billful of vegetation, and then, still singing, flew up almost vertically for *c.* 6 m, then down again; on landing, collected some more material, then pointed bill up and fluttered wings without taking off (Grote 1943a reports both ♂ and ♀ fluttering repeatedly upwards together). After several apparent attempts to take off, ♂ collected yet more material while ♀ crept low and hesitantly through the grass, tail slightly raised, wings occasionally spread and quivering. While approaching ♀, ♂ performed intense Advertising-display, bill pointing straight up (after he dropped load of material) and dragging his wings as he neared ♀. ♂ now lowered his head, ruffled plumage on back and rump, and widely spread his tail (Fig F: Fluffed-display, Andrew 1957a) accom-

F

panied by a chattering sound. ♂ then copulated in response to ♀'s Soliciting-display (Fig G, left): ♀ crouched with head low, tail spread, partly-spread wings quivering, and calling (see 3b in Voice). After dismounting, ♂ walked in front of ♀ in Fluffed-display, while she stood up, raised her bill and tail (Fig G, right) and gave a chattering call. ♀ often performed apparent Soliciting-display on leaving nest after bout of building. (Drury 1961.) Extra-pair copulations probably common (see Bonds and Flock Behaviour, above) but no details. (5) Nest-site selection. Apparently by ♀; occasionally she chooses site outside ♂'s territory, forcing him to extend territory to encompass it (Seastedt

G

and MacLean 1979). (6) Behaviour at nest. Nest built by ♀; ♂ only carries nest-material for display (Gierow and Gierow 1991; see Mating, above). During breaks from incubation in early morning, ♀ feeds, preens, and once seen to bathe; at this time ♂ feeds near ♀, keeping in touch with regular Contact-alarm calls (Rowell 1957; see 2 in Voice). RELATIONS WITHIN FAMILY GROUP. ♀ removes eggshells immediately after hatching and carries them well away from nest (Rowell 1957). ♀ broods young at night for most of the nestling period (Hussell 1972; see also Roosting, above). During day, only the heaviest rain induced ♀ to stop feeding and brood the (presumably well-grown) young. ♂ never seen to brood; he provided most food for young in first few days after hatching, ♀ taking equal share later on. ♂ typically flew to song-perch and adopted Advertising-posture after feeding young. (Rowell 1957.) Nestlings mostly silent; gape at any disturbance on 2nd or 3rd day; eyes open 3–5 days after hatching, and thereafter gaping is orientated at arriving parent (Rowell 1957; Drury 1961). Both parents seen carrying off faecal sacs (Nicholson 1930), throughout nestling period and even initially after leaving nest (Rowell 1957). Young leave nest 3–5 days or more before able to fly, even when nest not disturbed (Wynne-Edwards 1952). From the few hours before first young leave, ♀ periodically probes bottom of nest and carries off lining material (Rowell 1957). When young left nest asynchronously, ♂ fed those outside nest while ♀ looked after those remaining in nest (Nicholson 1930). For subsequent brood-division, see Bonds (above). For 8–10 days after departure from nest, parent made repeated foraging trips to feed one chick (stationary in territory) before moving on to feed the next in similar fashion. During the last 5–7 days before independence (see Bonds, above) young are quite mobile, moving onto neighbouring territories and following foraging parents; those tended by ♀ moved furthest from nest. (McLaughlin and Montgomerie 1985, 1989 a, b, which see for interpretation of brood-dispersal strategy.) ANTI-PREDATOR RESPONSES OF YOUNG. For 1–2 days before leaving nest, young crouch when disturbed (Drury 1961) or, presumably if more alarmed, may leave nest at 8 days compared with usual 10–11 (Gierow and Gierow 1991). However, in some studies, 8 days falls within normal range of leaving nest (see Breeding). After leaving nest, young hide in tussocks (etc.) where very difficult to flush (Grote 1943a). PARENTAL ANTI-PREDATOR STRATEGIES. (1) Passive measures. Some ♀♀ much more confiding at nest than others (Rowell 1957; Drury 1961). Some are tight sitters, especially around hatching (Hortling and Baker 1932; Rowell 1957), sometimes allowing themselves to be almost touched on nest (Grote 1943a; Drury 1961). Most ♀♀, however, leave nest when human intrudes, at approach-distances from less than 1 m to 50 m, often 20–30 m (Congreve 1936; Grinnell 1944; Drury 1961; Madsen 1981, 1982). If given sufficient warning by mate, ♀ creeps furtively some distance from nest before flying (Congreve 1936; Rowell 1957; see below for flight-style); may likewise return covertly to nest (Grinnell 1944). (2) Active measures: against birds. Apart from alarm-call (5a in Voice) given when Peregrine *Falco peregrinus* flew over (Drury 1961), no active measures described. (3) Active measures: against man (no information for other animals). Intruder in vicinity of nest typically elicits alarm-calls from guarding ♂, becoming more agitated if ♀ is on nest, and if nest is approached (Drury 1961). ♂ then flies low and hesitantly back and forth, around, and sometimes up to feet of intruder. Not uncommonly escorts intruder out of territory (Nicholson 1930; Grote 1943a; Wynne-Edwards 1952; Rowell 1957). ♀ flushed suddenly from nest flies off hurriedly, sometimes in laboured flight (quivering, hesitant), calling continuously; also reported flying within 1 m of intruder's head, often stimulating ♂ to mob likewise (Rowell 1957; Drury 1961). Occasionally, especially around hatching (Rowell 1957), ♀ also performs distraction-display of disablement type on ground, apparently like that sometimes performed outside breeding season (see introduction to part 2, above): hops and flutters over ground with neck extended, head lowered, tail fanned; at low intensity, one or both wings partially raised, at higher intensity wings variously spread, dragged, or jerked (Nicholson 1930; Williams 1941; Rowell 1957; Drury 1961). In this manner, sometimes with open bill, ♀ circled intruder by nest, or moved away from him when she was approached (Rowell 1957). ♀ normally silent, but gave 'si-' calls in intense display (Rowell 1957: see 5b in Voice). Extreme wing-drooping (Fig H) by perched ♂ suddenly alarmed

H

by observer in hide may have been mild distraction-display (Vaughan 1979) or perhaps mild Advertising-display (see Fig C).

(Figs by D Nurney: A–D and F–G based on drawings in Drury 1961; E from drawing in Cumming 1979; H from photograph in Vaughan 1979.)   EKD

Voice. Freely used throughout year. Rather variable according to context and individual, and also readily heard (and described) differently by different human observers. Further information needed on functional significance of diverse Contact-alarm calls. Following scheme therefore provisional, and perhaps conservative in identifying a few basic call-types, with possibility not being discounted that this masks a genuinely wider repertoire. See especially Rowell (1957: Swedish Lapland) and Drury (1961: Bylot Island, Northwest Territories, Canada) for various other sounds which may add to those listed below but which, on present evidence, have not been given separate status. No details of racial or other geographical variation, if any. For further sonagrams, see Bergmann and Helb (1982).

CALLS OF ADULT. (1) Song of ♂. Short, jingling phrases with hard timbre, repeated with little variation (L Svensson), though recordings indicate units may be richly varied, including vibrant sounds and diads (W T C Seale): 'kretle-KRLEE-trr kritle-kretle-trü', 2nd unit high-pitched, stressed, and drawn-out (Bruun *et al.* 1986; L Svensson). Short lively phrase suggesting start of song of Skylark *Alauda arvensis* but of harder quality, e.g. 'teeTOOree-teeTOOree-trreeoo' (Witherby *et al.* 1938). Also resembles Dunnock *Prunella modularis* (Bergmann and Helb 1982), though this comparison apparently only apt for Song-flight, in which delivery is typically more prolonged (L Svensson) or faster (E K Dunn) than when perched. Each phrase lasts 1–4 s, usually *c.* 1·5–2 s, clearly separated from next by 4–5 s; in Song-flight, longer phrases are interspersed among typical shorter ones, with result that average phrase (*c.* 2–3·3 s) is longer in Song-flight than when perched (*c.* 2–2·5 s) (Bjørnsen 1988; L Svensson). Phrases of a particular individual are quite similar, but there are differences between individuals (Bergmann and Helb 1982). Song given from perch (e.g. Figs I–II, representing 2 different ♂♂) starts very quietly with marked crescendo over first few units (quietest sounds have been accentuated on sonagram); phrases of individual exemplified by Fig I often end in vibrant unit;

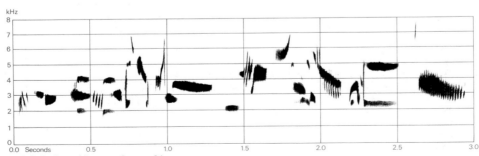

I   J G Corbett   Norway   June 1986

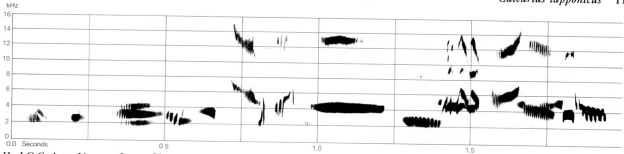

II J G Corbett Norway June 1986

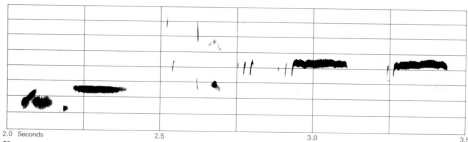

II *cont.*

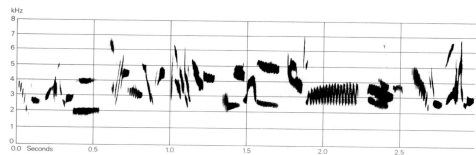

III P A D Hollom Finland June 1989

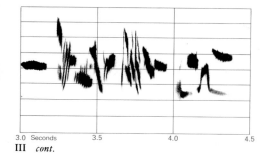

III *cont.*

each burst of rather scratchy song depicted (half-speed) in Fig II ends in series of high, squeaky notes (W T C Seale). Song of *C. lapponicus* also sometimes ends with a quite long unit which first rises then falls in pitch (Bergmann and Helb 1982). Example (Fig III; quietest sounds at start have been accentuated) of song given in Song-

flight shows diads and repetition of considerable part (sub-phrase) of phrase after short (unrepeated) section in middle (W T C Seale). Sounds given during descent of Song-flight are rendered almost identically by two sources: 'tshevi-' (Hortling and Baker 1932) and accelerando 'tschive-' units (Ruthke 1939). (2) Contact-alarm calls. (2a) Commonest flight-call throughout the year (but also given on ground) a hard dry rippling 'prrr(r)t', alternating to varying extent with call 2b (Bruun *et al.* 1986; Jonsson 1992). Also described as a delicate rapid 'pititi', similar to Snow Bunting *Plectrophenax nivalis*, reduced to one unit (i.e. 'pi' or 'ti') as flight-intention signal (Bergmann and Helb 1982); hard tuneless 'tutuTUCC' or 'ticky-tick' (Witherby *et al.* 1938). Fig IV (from migrating birds) rendered 'jip-ip-ip'; Fig V shows (from alarmed ♀) 'tik-ik-ik' followed by ringing sound (rather like rapidly damped bell), and finally a descending portamento 'siu' (apparently call 2b) (W T C Seale). Timbre and delivery

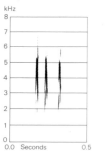

IV P J Sellar Greenland August 1979

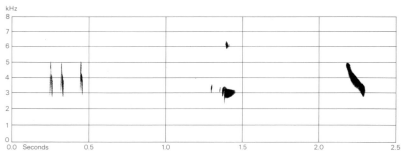

V B N Veprintsev Russia May 1980

vary with degree of excitement: at close quarters a soft 'trrr' rather like watch being wound; sometimes in louder extended bursts when alarm causes flock to fly up suddenly, e.g. 'trrrrtrtr tr tr' (Schüz 1941). For other renderings see especially Szabó (1976). (2b) Nasal 'tew' (Jonsson 1992) or 'tschü' (Szabó 1976) or brief 'chu' which may at times be very like 'piU' of *P. nivalis* (Bruun *et al.* 1986); recordings include 'tiu chiu' (Fig VI) and 'chiu chioo' (Fig VII: W T C Seale). Main call also rendered

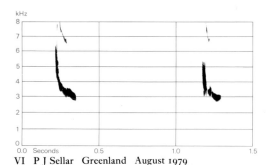

VI P J Sellar Greenland August 1979

VII P J Sellar Greenland August 1979

as simple whistle, 'dö', 'tö', 'tjö', etc., variable in pitch and duration (sometimes very short, sometimes drawn out), very similar to *P. nivalis*; used widely in pair-contact and between flock-members, e.g. from birds on ground to others flying overhead (Schüz 1941). Quiet introverted 'dög' or 'dät', said to be like Reed Bunting *Emberiza schoeniclus*, heard in winter (Kroyman 1967), presumably the same. In

flight, usually preceded by call 2a (Jonsson 1992) and may grade into call 5a during disturbance. (2c) Sharp 'zit' like Linnet *Carduelis cannabina* (Witherby *et al.* 1938), 'zip zip' (Schüz 1941), 'chip chip' (Drury 1961), harsh 'jeeb', including from nocturnal migrants (Bruun *et al.* 1986). (3) Courtship-calls. (3a) Inciting-call of ♀. 'Zeep zeep' given by ♀ to attract ♂ (Inciting-display) during courtship (Drury 1961). (3b) Soliciting-call of ♀. Listed by Drury (1961) as 'begging note' from ♀ inviting copulation, but no detailed description. (3c) 'Chattering' (not further described) heard from ♂ performing Fluffed-display to ♀, and from ♀ immediately after copulation (Drury 1961); perhaps only a rapid series of call 2a. (4) Threat-calls. (4a) Vibrant 'chreep', like food-call of young (see below), from ♂ in stationary Forward-threat display (Drury 1961). (4b) 'Sput dirrrr' from ♂ running at rival in Forward-threat display (Drury 1961). 4a–b are perhaps only intense Contact-alarm calls. (5) Alarm- and warning-calls. (5a) Main alarm-call throughout the year a melodious but plaintive, piping call of 2 separate units or often disyllabic, e.g. 'TEE-hü' (Bruun *et al.* 1986) or 'piü' or 'tiü' (Schüz 1941). Wheezy 'dyew' or 'dzeeu' from ♂ when threatened in territory by human intruder, also when Peregrine *Falco peregrinus* flew over (Drury 1961); outside breeding season, 'tiiouu' heard in pause during distraction-display, thus much longer than call 2b (Lapous 1988). Fig VIII

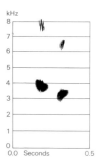

VIII B N Veprintsev and V V Leonovich Russia June 1977

shows vibrant 'ji-ju' from alarming ♂ (same also heard from ♀); Fig IX shows commmon alternation of disyllabic (or in this case 2 separate units) call with monosyllable, i.e. 'tri loo' followed by lower-pitched, mellower, plaintive

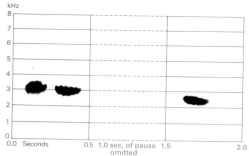

IX   P A D Hollom   Finland   June 1989

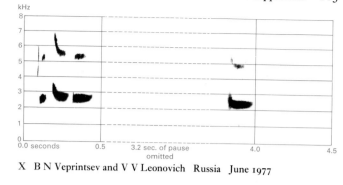

X   B N Veprintsev and V V Leonovich   Russia   June 1977

'hoo' reminiscent of Bullfinch *Pyrrhula pyrrhula* (W T C Seale). In west Greenland, this combination heard as a loud 'tew' followed by an almost inward 'inh' or 'pin', thus (e.g.) 'TEW-pin TEW TEW-pin' (E K Dunn). Jonsson (1992) also describes a regular slow alternation of 2 call-types: ringing 'dyuee' (softer than 2b) followed after *c.* 2 s by 'triü'. Similar sequence shown in Fig X: bell-like 'twa-dil-oo' followed by more vibrant 'droo' (W T C Seale). Variability not unexpected since a medley of related calls (evidently incorporating 2b and 5a) may be given when disturbed in territory, e.g. 'tjüb', 'tije', 'drü', and 'tjü' (Bergmann and Helb 1982). (5b) 'Si-si-si-si' heard during high-intensity distraction-display in breeding season (Rowell 1957). Rapid 'pitze-pitze', sometimes heard from ♀♀ watching intruder from safe distance after flushing from nest (Drury 1961), perhaps the same. On Baffin Island (Northwest Territories), low screeching sound heard during distraction-display (Watson 1957*a*).

CALLS OF YOUNG. Thin piping note of uniform pitch given as food-call in nest, and also when handled (Rowell 1957). After leaving nest, food-call a quavering 'cheep'; loud 'pseep' calls given by well-grown young when handled or when stumbling in alarm from nest (Drury 1961). Recording of 2-week-old young includes brief 'chi' and longer grating or rasping 'chzree' (W T C Seale). Free-flying juveniles give a rattling churring like finch (Fringillidae) (Nicholson 1930), similar to adult call 2a. Juveniles also heard giving apparently calls 2b-c (D T Holyoak).   EKD

Breeding. SEASON. In general, probably timed so that young leave nest during peak abundance of adult flies (Diptera). Scotland: eggs laid in 2 nests *c.* 20 June (Cumming 1979). Southern Norway: nest-building and fledged young recorded 2nd half of June (Grampian and Tay Ringing Group 1981). Swedish Lapland: eggs laid mid-June (end of May to mid-July); laying highly synchronized within population (Rowell 1957; Gierow and Gierow 1991). Northern Russia: eggs laid 1st half of June, late June in cold springs; young from replacement clutches can leave nest as late as mid-August (Mikheev 1939; Danilov *et al.* 1984; Stishov *et al.* 1991). For Alaska, see Custer

and Pitelka (1977); for western Greenland, see Madsen (1982) and Fox *et al.* (1987). SITE. On ground, commonly in slight depression or sheltered position in lee of hummock or tussock; in dry spot but often near water, very frequently protected by overhanging twigs of birch *Betula*, willow *Salix*, heather, or similar shrub; never in bare or rocky areas (Nicholson 1930; Rowell 1957; Portenko 1973; Danilov *et al.* 1984; Gierow and Gierow 1991). Nest: tightly built structure of dry grass, sedge, rootlets, leaves, lichen, etc., lined with soft grasses, plant down, hair, and many feathers, particularly of grouse *Lagopus*; sometimes no lining (Wynne-Edwards 1952; Rowell 1957; Portenko 1973; Gierow and Gierow 1991). Outer diameter 7–13 cm, inner diameter 5–7 cm, depth of cup 4–6 cm (Mikheev 1939; Grinnell 1944). Building: by ♀ only, accompanied by ♂; material gathered in immediate surroundings of nest (Drury 1961, which see for details of technique; Makatsch 1976; Gierow and Gierow 1991). EGGS. See Plate 33. Sub-elliptical, smooth and slightly glossy; very variable, pale greenish, greyish, or buffish, usually very heavily marked with olive-brown, rusty, or purplish-black blotches, spots, and scrawls (Harrison 1975; Makatsch 1976). Nominate *lapponicus*: 20.8×15.0 mm ($18.0–23.9 \times 14.0–17.3$), $n = 600$; calculated weight 2.46 g (Schönwetter 1984). Clutch: 5–6 (3–7), very rarely 8. In Swedish Lapland, of 185 clutches: 3 eggs, 3%; 4, 9%; 5, 40%; 6, 43%; 7, 5%; 8, 0.5%; average 5.39 (4.94–5.80 over 6 years); before 13 June 5.59 ($n = 78$), 14–20 June 5.02 ($n = 50$) (Gierow and Gierow 1991, which see for effect of age of ♂ and ♀ on clutch size); Rowell (1957) found 5.14 ($n = 14$). Of 17 clutches in southern Norway, 1 of 4 eggs, 1 of 5, 10 of 6, 5 of 7; average 6.12 (Grampian and Tay Ringing Group 1981). Average in Yamal peninsula (northern Russia), 5.16, $n = 119$ (Danilov *et al.* 1984). For Alaska, see Custer and Pitelka (1977); for northern Canada, see Wynne-Edwards (1952), Hussell (1972), and Lyon and Montgomerie (1987); for western Greenland, see Fox *et al.* (1987). In Alaska, replacements laid by 4 of 26 ♀♀ which lost clutches during laying, 2 of 4 which lost clutches during 1st week of incubation, but none of 18 which lost clutches later than this (Custer and Pitelka 1977). Eggs laid daily. One brood (Rowell 1957;

Hussell 1972; Grampian and Tay Ringing Group 1981; Madsen 1982); very rarely, 2 broods (Custer and Pitelka 1977). INCUBATION. 11-13 days; in Swedish Lapland, average 12·4 days ($n = 24$) from laying to hatching of last egg (Gierow and Gierow 1991); 11·5 days ($n = 17$) given for western Greenland (Madsen 1982); by ♀ only (Mikheev 1939; Danilov *et al.* 1984; Gierow and Gierow 1991). Starts with 3rd or 4th egg, hatching over 1-2(-3) days (Drury 1961; Danilov *et al.* 1984; Gierow and Gierow 1991), but apparently from 1st egg, hatching over 4 days according to Wynne-Edwards (1952) and Rowell (1957). At one nest, ♀ on eggs for 3-4 stints of *c.* 15 min 3 days before clutch completion, and sat all day when last egg laid (Drury 1961); for stints, see also Grinnell (1944) and Lyon and Montgomerie (1987). YOUNG. Fed and cared for by both parents about equally, though ♂ often observed doing more; brooded only by ♀ (Mikheev 1939; Grinnell 1944; Rowell 1957; Drury 1961). FLEDGING TO MATURITY. Fledging period 9-10 days (8-11), young leaving nest *c.* 2-5 days before able to fly; pick up insects immediately on leaving nest and independent *c.* 7-10 days later (Mikheev 1939; Wynne-Edwards 1952; Madsen 1982; Gierow and Gierow 1991). Fledging period given as 11-12(-14) days by Danilov *et al.* (1984). For development of young, see Grinnell (1944), Maher (1964), Hussell (1972), and Madsen (1982). Age of first breeding 1 year (Custer and Pitelka 1977; Gierow and Gierow 1991). BREEDING SUCCESS. In Swedish Lapland, of 583 eggs in 97 nests, 92% hatched and 87% produced young 6-8 days old; in 5 of 6 years, average 14% of nests lost to predators, but in one year 88% of 26 were lost and possibly no young at all reached fledging; most likely predators were Raven *Corvus corax* and fox *Vulpes* (Gierow and Gierow 1991). In southern Norway, average brood size 5·5 ($n = 8$), 90% of clutch size (Grampian and Tay Ringing Group 1981). In northern Russia, many nests lost early in season to floods during snow- and ice-melt and to sudden cold spells; eggs and young taken by Arctic Skua *Stercorarius parasiticus* and Long-tailed Skua *S. longicaudus* (Grote 1943a). In Alaska, over 7 years, of 1244 eggs, 51·6% hatched (35·6-88·2%) and 36·8% produced fledged young (24·2-75·5%), giving 2·23 fledged young per ♀ (1·43-4·27); 52·6% of 95 nests found during laying suffered predation; of egg losses, 73% due to predators (especially *S. parasiticus* and Pomarine Skua *S. pomarinus*), 17% failed to hatch, and 7% abandoned; of nestling losses, 72% predated, 12% abandoned, 12% starved, 2% due to weather (Custer and Pitelka 1977). Main predators in western Greenland were *C. corax* and Arctic fox *Alopex*; all predation apparently on eggs, perhaps because ♀ leaves nest if threatened (Madsen 1982; Fox *et al.* 1987; see these sources for details). For northern Canada, see Wynne-Edwards (1952), Drury (1961), and Lyon and Montgomerie (1987). BH

Plumages. (nominate *lapponicus*). ADULT MALE BREEDING. Forehead, crown, lore, cheek, chin, throat, and chest completely black; when fresh (April-May), crown and nape with yellow edges. Very broad pale yellowish supercilium, slightly narrower above eye, narrow or absent in front of eye, extending over ear-coverts into bar on side of neck, connected with white sides of chest and with narrow yellowish band across nape. Neck rufous, in early breeding season sometimes with some narrow pale feather-edges. Remainder of plumage as in adult male non-breeding (see below), but only a trace left of light feather tips of mantle, rump, scapulars, and wing-coverts, or these completely worn off, feathers showing black centres and narrow rufous sides. Flank, thigh, and (sometimes) side and upper belly marked with black spots and streaks, or (later in season) flank solid black with narrow white marks. ADULT MALE NON-BREEDING. In fresh plumage (August-December), forehead and crown black, feathers with rusty- to yellowish-brown fringes, black extending in wedge to tip; fringes broader on central crown, forming inconspicuous crown-stripe. Feathers of lore, supercilium, and ear-coverts rusty-brown to yellowish with black bases. Narrow black eye-stripe, partly obscured by yellowish-straw feather-tips; stripe widening at end of upper ear-coverts, forming black triangular spot. Eye-ring yellowish-white. Ill-defined black moustachial stripe bordering ear-coverts and connected to eye-stripe and black of crown. White to off-white malar stripe, connected to off-white patch on upper side of breast. Broad yellowish or rusty submoustachial stripe. Broad rufous band across nape, feathers tipped yellow or (occasionally) black. Feathers of mantle, scapulars, and back tricoloured, with broad black centre bordered by narrow rufous, rusty, or chestnut fringe and with broad pale yellow or rufous edge. Rump and upper tail-coverts similar to mantle, but rufous edges broader. Chin and throat off-white to buffish. Centres of feathers on lower throat and side of throat black, concealed by broad white or buffish tips. Broad black patch across chest, bordered by whitish on throat, side of breast, and belly, black largely concealed by buffish or whitish tips. Belly, vent, and under tail-coverts white to buffish-white; longest under tail-coverts occasionally with inconspicuous brownish-black marks. Feathers of flank and thigh with dark centres, showing as streaks or spots, especially on upper flank, black sometimes bordered with narrow rufous fringe. Tail feathers black with narrow yellowish-buff or rusty fringes; t5 usually with narrow white wedge on tip of inner web, covering 30-50% of length, but amount of white variable and t5 occasionally all-black; t6 largely white, black restricted to basal half of inner web and stripe along shaft, latter widening at tip of feather; white on tail sometimes sullied buff or grey-buff. Flight-feathers, primary coverts, and bastard wing grey-black, outer webs with narrow light brown or yellowish fringes, broader on secondaries; p1-p6 with broad off-white tip, occasionally extending as narrow fringe onto inner web. Tertials grey-black, outer web and tip of inner web with broad chestnut fringe and pale edge, latter extending onto inner web. Greater upper wing-coverts grey-black, outer web with broad chestnut or rusty fringe and white tips forming narrow white wing-bar; median coverts similar to greater, but brown fringes often slightly paler and narrower, sometimes absent, white tips showing as white wing-bar; lesser coverts grey-black with rusty to yellowish fringes. Under wing-coverts and axillaries white with dark centres. In worn plumage, crown, eye-stripe, malar stripe, throat, and chest gradually become blacker and band across neck becomes more obviously rufous. ADULT FEMALE BREEDING (March-August). Like adult ♂ breeding, but forehead, crown, and nape black with yellowish edges on feathers. Crown becomes darker through wear, leaving narrow ill-defined crown-stripe. Lore yellowish. Broad yellowish supercilium, narrowly and indistinctly extending across nape.

Narrow inconspicuous black eye-stripe, widening on ear-coverts. Border of ear-coverts black-brown, joining malar stripe later in season; centre of ear-coverts rusty to yellowish, forming pale spot. Band across nape yellowish-rufous, mixed with black spots. Feathers of chin and throat with black base, concealed by yellowish-white distal half. Later in season, chin and throat brown. Inconspicuous black band across chest, becoming more obvious through wear later in season. Side of breast with black-brown spots and streaks; black centres bordered by dark straw or rusty edges and edged and tipped yellowish-white; feathers of flank and thigh similar, but with long dark feather-centres. Remainder of plumage as adult ♂ non-breeding, but only a trace left of pale fringes of mantle, rump, scapulars, and wing-coverts; or these completely worn off, leaving black centres and narrow rufous edges. ADULT FEMALE NON-BREEDING. In fresh plumage (August–December), like fresh adult ♂, but feathers on crown have narrower black centres, showing as streaks, especially in centre of crown, and broader yellowish or rusty edges; black centres not as glossy as adult ♂. Black feather-centres of lore, supercilium, and chin much narrower or virtually absent. Head pattern as ♂, but dark feathers more brownish. Nape rufous as in ♂, but much less conspicuously so, feathers with black streaks along shafts and with broad yellowish fringes. NESTLING. Down long and plentiful; dark brown to light tan, shading to yellow and grey for distal third (Maher 1964; Harrison 1975). JUVENILE. Upperparts broadly streaked black, slightly marked orange-brown on crown and scapulars, all feathers edged buffish-yellow. Side of head as upperparts, but ear-coverts browner and with white spot at rear. Chin and upper throat whitish-buff with brown-black spots; lower throat, chest, and flank buffish-yellow, narrowly streaked brownish-black; remainder of underparts whitish-buff. Tail, wing, and greater upper wing-coverts as adult; median coverts with white tips, divided by central black streak; lesser coverts fringed white. (Witherby *et al.* 1938). FIRST ADULT MALE NON-BREEDING. Like adult ♂ non-breeding; black spots on rufous nape more common than in adult ♂. Tail-feathers narrower, tips more brownish-black. FIRST ADULT MALE BREEDING. Like adult ♂ breeding, but crown and throat not as uniformly black in early breeding season, and rufous nape has black spots on distal part of feathers; primaries browner and more abraded. FIRST ADULT FEMALE NON-BREEDING. Like adult ♀ non-breeding, but less black on feathers of side of head and malar stripe. Feathers on lower throat and chest streaked black and tawny, without large black centres (Witherby *et al.* 1938). Tail as 1st adult ♂ non-breeding. FIRST ADULT FEMALE BREEDING. Like adult ♀ breeding, but rufous of nape sometimes virtually absent. Tail-feathers narrower and with more pointed tips than in adult.

Bare parts. ADULT, FIRST ADULT. Iris brown. In summer, bill pale yellow to orange with small black tip (Bjørnsen 1988); in autumn and winter, light brown or flesh-grey with dark grey to black tip. Leg and foot brown to black. NESTLING. Inside of bill red, gape-flanges yellow-white (Harrison 1975). JUVENILE. Bill light flesh-horn with slate-grey culmen. Leg and foot flesh-pink or dull flesh-brown (Delin and Svensson 1988).

Moults. ADULT POST-BREEDING. Complete; primaries descendent. In eastern Greenland, primary moult mid-June to late August; duration 34–65 days, average for ♂ *c.* 38 days, for ♀ *c.* 30 days; secondaries start at primary moult score *c.* 23 and finish after primaries; tertials start at primary moult score *c.* 17. Tail-feathers moulted centrifugally, during primary moult, but occasionally all feathers shed simultaneously. Moult of head and body starts after onset of primary moult; moult of upper and under wing-coverts starts also after onset of primary moult and finishes earlier. (Ginn and Melville 1983; Fox *et al.* 1987; Francis *et al.* 1991.) Moult completed on arrival in western Europe, August (ZMA, RMNH). ADULT PRE-BREEDING. Perhaps partial: chin and upper throat. Colours of rest of head and of throat-patch apparently obtained by abrasion rather than by moult. POST-JUVENILE. Partial: body and lesser and median upper-wing coverts. Starts early July, *c.* 3 weeks after fledging. (Bent 1968; Francis *et al.* 1991.) Many juveniles in western Greenland still in active body moult mid-September (Fox *et al.* 1987), but moult completed in birds arriving in western Europe, August (ZMA, RMNH).

Measurements. Nominate *lapponicus*. Scandinavia (summer) and western Europe (autumn to spring); latter perhaps including migrant *subcalcaratus*; ages combined; skins (RMNH, ZMA). Bill (S) to skull, bill (N) to distal corner of nostril; exposed culmen on average 3·2 mm less than bill (S).

	♂		♀	
WING	96·1 (1·55; 18)	93–99	89·3 (2·18; 16)	85–92
TAIL	59·9 (3·00; 22)	55–65	56·0 (3·02; 15)	50–60
BILL (S)	13·0 (0·72; 16)	12·1–15·0	12·3 (0·79; 13)	11·0–13·6
BILL (N)	7·7 (0·70; 17)	6·8–9·1	7·0 (0·51; 14)	6·4–8·4
TARSUS	20·8 (0·71; 20)	19·5–22·2	20·9 (0·89; 10)	19·5–22·5

Sex differences significant, except for tarsus.

Nominate *lapponicus*. Wing. Live birds, Norway: ♂ 93 (2·7; 29) 87–98 (Bjørnsen 1988). Scandinavia, skins: ♂ 92 (10) 90–96 (Salomonsen 1931*b*). Yamal peninsula: ♂ 91·3 (10) 88–95, ♀ 85·6 (5) 84–88 (Danilov *et al.* 1984).

C. l. subcalcaratus. Western Greenland, live birds: (1) adult, (2) juvenile (Francis *et al.* 1991).

		♂		♀	
WING	(1)	97·6 (2·20; 16)	—	91·5 (1·75; 32)	—
	(2)	98·3 (1·77; 185)	—	92·9 (1·95; 225)	—
TAIL	(1)	59·7 (4·04; 13)	—	54·3 (2·89; 24)	—
	(2)	60·5 (2·22; 7)	—	55·8 (1·27; 8)	—

Western Greenland, skins: wing ♂ 97 (12) 95–101 (Salomonsen 1931*b*).

Race unknown. Netherlands, live birds: wing, ♂ 95·9 (2·55; 64) 89–102, ♀ 90·4 (2·10; 35) 86–94 (Schekkerman 1989).

Wing and bill of (1) nominate *lapponicus*, Kola peninsula to Kolyma delta; (2) *alascensis*, Alaska, Aleutian Islands, Wrangel, and Pribilof Islands; (3) *subcalcaratus*, Canada and western Greenland; (4) *kamtschaticus*, northern Sea of Okhotsk, Kamchatka, and Anadyr; (5) *coloratus*, Komandorskiye Islands (Portenko 1973).

	♂	♀
WING (1)	91·7 (147)86·4–95·8	85·7 (89)79·8–89·5
(2)	93·3 (39)89·1–98·2	87·3 (13)84·0–91·5
(3)	92·3 (4)91·5–93·9	87·1 (3)82·8–89·4
(4)	91·4 (43)86·5–95·9	85·6 (10)83·0–88·7
(5)	97·4 (16)93·4–99·7	90·2 (9)88·2–93·7
BILL (1)	8·0 (145) 6·9–9·2	7·7 (90) 6·8–8·6
(2)	8·7 (40) 7·3–9·7	8·2 (13) 7·1–9·6
(3)	9·0 (11) 8·6–9·4	8·8 (3) 8·1–9·3
(4)	8·5 (40) 7·6–9·8	8·4 (9) 7·9–9·1
(5)	9·1 (17) 8·5–10·3	8·6 (7) 8·0–9·2

Weights. Nominate *lapponicus*. (1) Yamal peninsula, May–July (Danilov *et al.* 1984); (2) Norway, May–June (Bjørnsen 1988); (3) Kazakhstan, winter (Korelov *et al.* 1974); (4) northern Russia (Dementiev and Gladkov 1954). (5) Race unknown, Netherlands, September–December (Schekkerman 1989).

	♂		♀	
(1)	24·4 (— ; 11) 18·7–27·3		22·9 (— ; 8) 20·7–24·0	
(2)	25·2 (1·42; 29) 22·0–27·8		— (— ; —)	—
(3)	22·9 (— ; 4) 20–26		— (— ; —)	—
(4)	25·3 (— ; 32) 23–28		25·6 (— ; 27) 20·3–30·5	
(5)	24·8 (1·71; 35) —		22·5 (1·95; 17) —	

C. l. alascensis. Central and northern Alaska: May–July, adult ♂ 28·6 (27) 24·6–33·1, adult ♀ 25·4 (22) 22·3–30·6; July–September, immature ♂ 23·0 (8) 22·3–24·7, immature ♀♀ 24·4, 27·4. North-west Yukon (Canada): May, adult ♂ 26·2 (2·27; 5) 23·0–28·4, adult ♀ 24·3. (Irving 1960.) Chukotskiy peninsula (eastern Siberia): ♂ 26·2 (3·55; 9) 20·3–33·7 (Portenko 1973).

C. l. kamtschaticus. Eastern Siberia: ♂ 31·2 (11) 25·2–36·2, ♀♀ 21, 22 (Dementiev and Gladkov 1954).

C. l. subcalcaratus. For seasonal and diurnal variation in body weight, see Fox *et al.* (1992). For growth of young, see Fox *et al.* (1987). For details on relationship between moult score and weight, see Fox *et al.* (1992).

Structure. Wing rather long, broad at base, pointed at tip. 10 primaries: p9 longest, p8 0–1 shorter, p7 0–3, p6 6–11, p5 12–19, p4 18–24, p3 22–27, p2 25–31, p1 29–34; p10 strongly reduced, 61–70 shorter than p9, 10–14 shorter than longest upper primary covert. Outer web of p7–p8 clearly emarginated, p6 occasionally slightly emarginated; inner web of p9 with inconspicuous notch. Secondaries broad with notched tips. Tip of longest tertial reaches tip of (p1–)p3–p4. Tail rather short, tip slightly forked; 12 feathers, t5 longest, t1 2–6 shorter. Bill conical; similar to that of Yellowhammer *Emberiza citrinella*, but relatively longer and blunter, culmen and gonys slightly convex. Middle toe with claw 18·4 (12) 17–20 mm; outer and inner toe with claw both *c.* 71% of middle with claw, hind *c.* 102%. Hind claw 11·1 (19) 6–15; long, slightly curved, lark-like.

Geographical variation. Rather slight, mainly involving colour of upperparts and size. *C. l. alascensis* from eastern Chukotskiy peninsula, islands of Bering Sea (except Komandorskiye Islands), and Alaska, east to mouth of Mackenzie river, is pale on upperparts, especially in winter plumage; in summer, upperparts lightly streaked buffish-grey-brown, virtually without rusty tinge, and with narrow black feather-centres; wing and bill rather large. *C. l. coloratus* from Komandorskiye Islands is largest race; upperparts including nape darker and more rufous; edges of flight-feathers rusty-brown, not ochre-rufous (Witherby *et al.* 1938; Dementiev and Gladkov 1954; RMNH). *C. l. kamtschaticus* from Kamchatka and shores of northern Sea of Okhotsk to Koryakland is similar in colour to *coloratus*, but smaller in size (Portenko 1973). *C. l. subcalcaratus* from Greenland and Canada has bill slightly longer than nominate *lapponicus*, heavier and deeper at base, wing on average longer; in summer, nape and feather-edges of mantle paler (Salomonsen 1931b); in winter, differences doubtful (ZMA, RMNH).   GOK

Plectrophenax nivalis Snow Bunting

PLATE 10
[between pages 64 and 65]

Du. Sneeuwgors   Fr. Bruant des neiges   Ge. Schneeammer
Ru. Пуночка   Sp. Escribano nival   Sw. Snösparv

Emberiza nivalis Linnaeus, 1758

Polytypic. Nominate *nivalis* (Linnaeus, 1758), Arctic America from west coast of Alaska and Alaska peninsula east to Greenland, also Svalbard, Fenno-Scandia, Kola peninsula, and Faeroes; *vlasowae* Portenko, 1937 (synonym: *pallidior* Salomonsen, 1947), Pechora area (north-east European Russia) east through northern Siberia to Chukotskiy peninsula and Anadyrland, including offshore islands, grading into nominate *nivalis* from Kanin peninsula to Pechora area and perhaps Novaya Zemlya; *insulae* Salomonsen, 1931, Iceland and (in part mixed with nominate *nivalis*) Scotland. Extralimital: *townsendi* Ridgway, 1887, west coast of Bering Sea (south of Anadyrland), Komandorskiye and Pribilof Islands, and western Aleutian chain from Attu to Adak, grading into nominate *nivalis* in eastern Aleutians; *hyperboreus* Ridgway, 1884, Hall and St Matthew Islands (Bering Sea).

Field characters. 16–17 cm; wing-span 32–38 cm. Close in size to Corn Bunting *Miliaria calandra*, with even longer and more pointed wings but less chesty body; slightly larger than Lapland Bunting *Calcarius lapponicus* but with similar build, forming with it pair of highly terrestrial passerines. 2nd largest bunting of west Palearctic, with deep, stubby bill, rather round but usually flat-crowned head, rather long, quite deep body, proportionately long pointed wings, and noticeably forked tail; distinctive robust, low-slung silhouette on ground. Plumage predominantly white, strikingly pied on back and wings in breeding adult and softly variegated warm buff on head, back, and chest of winter adult and immature. In flight, shows bold white panels on wings and along sides of tail though former much restricted in juvenile and 1st-winter ♀. Flight powerful but action cum wing-pattern create partly illusory flickering or drifting progress. Runs freely. Voice distinctive. Sexes dissimilar; marked seasonal variation. Juvenile separable. 3 races in west Palearctic, separable on upperpart pattern. Only nominate *nivalis* described in detail here (see also Plumages and Geographical Variation).

ADULT MALE SUMMER. Moults: July–September (complete). Immaculate appearance entirely result of wear. White of head, neck, underbody, rump (usually), most of wings, and sides to tail contrasts dramatically with black of back, scapulars, tertials, bastard wing, outer primaries, upper tail-coverts, and tail-centre. When not fully worn, may show warm buff tinge, particularly on crown and sides of breast and usually retains vestiges of tawny- to whitish-buff fringes and tips on outer mantle, scapulars, and tertials. Black bill and dark brown eye stand out. Legs black. ADULT MALE WINTER. Underlying plumage pattern identical to summer ♂ but rather less pied, with long tawny-buff crown, buff tints to lores and ear-coverts and from sides of breast along flanks, obvious pale tawny- to whitish-buff fringes to back, scapulars, and tertials long enough when fresh almost to conceal black bases and variable buff, black, and white mottling on rump. Note that clarity of rump pattern varies not only individually but also from race to race (see Plumages and Geographical Variation). In spite of quite marked change in body plumage, ♂ retains strikingly white-based and black-ended wings ('dipped in ink' appearance). Bill bright straw with dusky edges and tip. Legs remain black. ADULT FEMALE. Plumage pattern similar to ♂, but even when breeding never attains such a pied appearance, usually retaining buff or black streaks on crown, ear-coverts, and nape, buff or black patch on sides of breast, grey or brown cast to mantle and fringes to scapulars and tertials, brown and black mottled rump, and duller, less contrasting wings and tail pattern. At close range, wing shows white-tipped, brown-black lesser and median coverts, black primary coverts (white with black tips in ♂), blackish marks on ends of secondaries, and more black on innermost primaries. In fresh plumage, further distinguished from ♂ by tawny-buff fore-face, ear-coverts, and partial breast-band. Bill dusky when breeding, otherwise as winter ♂. JUVENILE. Lacks mainly white appearance of adult, with ground-colour of plumage mainly ashy- to buffish-grey, mottled and streaked black on upperparts, more softly so on underparts, and unmarked only on centre of underbody. Wings much less distinctly patterned even than ♀, with all or most greater coverts brown-black, tipped white, and white bases to secondaries restricted, even clouded; often no white patch obvious in the field. Bill brown. FIRST-YEAR. Following moult of all juvenile plumage except for flight-feathers, greater coverts, and tail, assumes more adult-like appearance, particularly when worn. However, both sexes retain buffier head, buffier and browner upperparts, darker rump, and less vivid wing pattern. Bird showing white bases to primaries, much white on secondaries and distinct linear contrast between pale buff or white shawl and blacker mantle is ♂. Rump of ♂ never pure white as in some adults.

Adult unmistakable in west Palearctic, there being no known overlap with usually montane Snowfinch *Montifringilla nivalis* (slightly larger, longer, and more upstanding, with grey head, brown upperparts, black bib in summer, and always fully white inner wings). Immature, particularly ♀♀ with least marked wings, subject to confusion with *C. lapponicus* which shares both breeding and wintering habitats, and much behaviour, and has 2 similar calls. Mistake unlikely with bird at close range, however, as *P. nivalis* always shows softly marked head and lacks (a) any discrete dark patches or streaks on underparts and (b) strong rufous on nape and wing-coverts; with birds in flight, frequent utterance by *P. nivalis* of diagnostic lilting ripple allows instant identification. Flight swift, with strong beats of long wings (in bursts) alternating with closed-wing attitudes and, when landing, being varied by side slips and lengthy skims; markedly undulating over distance. Flocks on migration and in winter quarters often ascend to considerable height to tumble abruptly to ground level and suddenly land. Disruptive plumage pattern of adult conveys added lightness to flying bird so that flocks show remarkable resemblance to squall of snow. Gait essentially a light, quick run, with curious rather clockwork action exaggerated by frequent hiding of legs under fluffed-out body; also hops and clambers, particularly in breeding habitat. Stance usually noticeably level, with head and neck raised in alarm but body rarely lifted on legs. Notably terrestrial, usually preferring in winter to go straight to ground; when breeding, however, perches on prominences, particularly rocks, but even trees on occasion. Under pressure, tends to run off before taking flight; also displays unease by flicking wing-tips and tail. Usually allows relatively close approach. Markedly gregarious on migration and in winter quarters, regularly forming flocks of hundreds. Rarely mixes with other species, and more likely to do so with Shore Lark *Eremophila alpestris* than *C. lapponicus*.

Song musical, somewhat lilting and loud but curiously ventriloquial phrase, with fluted di- and trisyllabic notes that lack jingling quality of *C. lapponicus*, 'turee-turee-turee-turiwee' or 'sweeto-swevee-weetuta-swee'. One flight-call diagnostic: soft, charming, musical, and distinctly rippled twitter, 'tirrirriripp', 'dirrirrt', or 'tir-rirrillit'. Other calls: short, soft, plaintive, musical 'tuu' or 'piu', louder, more whistled or ringing, and less formed than similar note of *C. lapponicus*, and usually commonest call of solitary bird; longer, rippled, but stonier 'stirrrp', lacking ticking, rattling quality of *C. lapponicus*; loud, high-pitched 'tweet' in excitement and chorus of more rasping notes from flock, recalling Sand Martin *Riparia riparia*.

Habitat. Extends across arctic and higher boreal zones, beyond or above treeline, between July isotherms of 2°C and 14-15°C. Breeds in usually treeless, uncultivated, barren, rocky terrain, often near snow and ice, and even on nunataks deep within icecap. Often on seacliffs, including those with mass seabird colonies, but also on rocky terrain from lowland to plateau, bearing low and often sparse plant cover. In parts of range adapts to human settlements (Voous 1960). Generally avoids forest but in some regions perches freely on trees; in winter, especially after snowfall, will perch on walls, railings, roofs, and overhead wires (Bannerman 1953a).

In Russia, breeds on high rocky tundra, broken by valleys, ravines, precipices, and cliffs on river banks and sea-coasts. Extralimitally in Kamchatka, found on old lava streams on volcano slopes at up to 2300 m. Favours proximity of flowing or standing water, and patchy moss or herbaceous cover. (Dementiev and Gladkov 1954.) In Britain, breeds exclusively on scree and boulder fields on high tops (Buckland *et al.* 1990); shows strong preference for rocks and boulders, but shifts to other habitats after rearing young (Fuller 1982). Data for well-studied Greenland population show that birds there frequent broken terrain with rocky and stony ground, and very open and low vegetation, thus barren mountain slopes, blockfields, and screes; ranges further north than any other passerine, and occurs at greater altitude, above 1000 m, also differing in preferring coastal regions to interior, and in showing adaptation to human settlements almost as pronounced as House Sparrow *Passer domesticus* in temperate zones (Salomonsen 1950-1).

Migrating and wintering birds in USSR resort to open countryside, roads, threshing floors, and outskirts of settlements, and to forest edges (Dementiev and Gladkov 1954; Flint *et al.* 1984). In Belgium, winters in littoral zone along beaches and treeless tracts (Lippens and Wille 1972). In northern Britain, some try to winter on upland moors, unless displaced by snowfall, while others favour stubble and turnip fields, remainder choosing marram grass *Ammophila* behind sandy beaches (Buckland *et al.* 1990). In England, favours shingle beaches and salt-marshes as well as sand-dunes and neighbouring stubble fields (Bannerman 1953*a*).

Distribution. No changes reported.

FAEROES. Irregular breeder (Bloch and Sørensen 1984).

Accidental. Bulgaria, Malta, Algeria, Morocco, Madeira (possibly regular), Canary Islands.

Population. Marked fluctuations, but no long-term trends reported.

FAEROES. Irregular, in small numbers; 0-10 pairs (Bloch and Sørensen 1984). BRITAIN. Population very small and fluctuating. Main population in central Highlands 2-5 pairs 1976-83, up to 5 pairs elsewhere in Scotland. (Thom 1986.) Probably increased recently; up to 50 pairs may have bred in late 1980s (Watson and Smith 1991). SWEDEN. Estimated 60 000 pairs (Ulfstrand and Högstedt 1976). FINLAND. About 10 000 pairs (Koskimies 1989).

Survival. Scotland: adult mortality 37% (R D Smith). Oldest ringed bird 4 years 1 month (Hickling 1983).

Movements. Partially migratory to migratory, many birds wintering far south of circumpolar breeding range; northernmost areas are vacated. Migration nocturnal, also diurnal (Salomonsen 1950-1; Bent 1968). See Banks *et al.* (1991*a*) for discussion of origin of birds wintering in western Europe. For review, see Zink (1985).

In Europe, winters mostly in coastal areas and on inland plains. Numbers vary greatly from year to year, and also fluctuate over long periods (Zink 1985). In Sweden, winters in south, and small numbers remain in breeding areas in mountains in some years (SOF 1990). Winters on south and west coasts of Norway, north mostly to 65-66°N (Haftorn 1971); in Finland, large flocks (up to 100-200) move south along west coast in early winter, and small numbers occur midwinter in south-west (Haila *et al.* 1986). On Kola peninsula (north-west Russia) most birds depart, but a few are seen in towns in some years (Semenov-Tyan-Shanski and Gilyazov 1991). Present in Iceland all year, by far the commonest wintering passerine (Æ Petersen); previously thought sedentary, but considerable

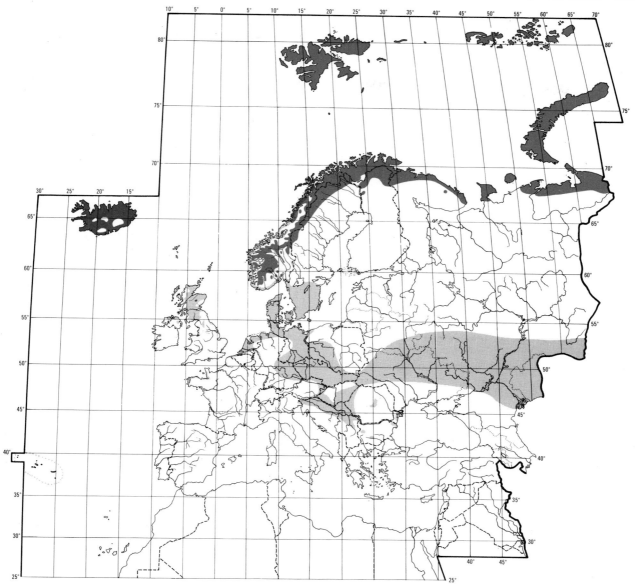

numbers migrate (see below). Some Greenland birds may winter in Iceland, but no proof. On Faeroes, very scarce in summer, but rather common on passage and in winter (Bloch and Sørensen 1984). Perhaps 10 000–15 000 winter in Britain and Ireland, but numbers vary greatly; most are in Scotland (including islands), and on east coast of England south to Kent (Lack 1986). Some Scottish breeding birds present all year; in study area in Cairngorms, 1988–91, 17–35% of local birds recorded during November–April (R D Smith). Distribution chiefly coastal in Ireland, but possibly widespread in small numbers in northern mountains (Hutchinson 1989). In France, annual in small numbers, mostly on north coast, with a few on west coast (south to Bassin d'Arcachon) and Medi-

terranean coast (chiefly Camargue); occasional records inland (Yeatman-Berthelot 1991). Regular in coastal Spain, at least in north-west (Cantabrias, Galicias, and Asturias), and probably in Catalonia in north-east; also recorded south to Cadiz, and from Balearic Islands (Souza 1991). Many winter in Denmark, north-west Germany, and Netherlands, especially in coastal areas, with fewer in Belgium (Lippens and Wille 1972; Møller 1978a; Busche 1980; SOVON 1987). In north-east Germany, many winter on coast and on Oder river plain, with fewer further south and west (Dittberner *et al.* 1969; Erdmann 1985); fairly regular in Rheinland (western Germany) (Mildenberger 1984); in Bodensee area (southern Germany), annual in autumn and occasional in winter (Jacoby *et al.* 1970). Very few midwinter records from Switzerland (Winkler 1984). Regular in north-east Austria (apparently scarce in other areas) (Straka 1991). In Italy, probably winters each year in small numbers in north and on Adriatic coast (Pandolfi 1987). Regular on passage in Poland and Baltic states, and small numbers winter (very few in Estonia) (Vīksne 1983; Rootsmäe 1990; Tomiałojć 1990). In St Petersburg region, far more common on passage than Lapland Bunting *Calcarius lapponicus*, and winters in some years (Mal'chevski and Pukinski 1983). Winters in large numbers on east Hungarian plains (Kovács 1981), and in adjoining areas of north-west Rumania, where not recorded east of 24°E (Kováts 1973). A few records in recent decades from Bulgaria (T Michev); rare in Yugoslavia (Matvejev and Vasić 1973). In former European USSR, winters mainly between *c.* 52°–55°N, fewer and less regular further north and south (Dementiev and Gladkov 1954). Vagrant to Malta, Sicily, Portugal, and Gibraltar (Nethersole-Thompson 1966; Sultana and Gauci 1982; Iapichino and Massa 1989; Finlayson 1992), and also to Turkey; 8 birds observed 2 February 1992 in Kizilirmak delta on Black Sea coast were 1st report since at least 1966 (Dijksen and Klemann 1992). Not reported elsewhere in Levant or Middle East. In North Africa, vagrants reported only in northern Morocco and northern Algeria (Pineau and Giraud-Audine 1979; Zink 1985). Recorded on Atlantic islands south to Tenerife (Canary Islands); regular in small numbers on Azores, where recorded from all islands; annual on São Miguel since 1978, with flocks up to at least 50 birds (more usually *c.* 10) (Bannerman 1963; Bannerman and Bannerman 1965; G Le Grand).

Greenland is mostly vacated, but small minority winters in south (Salomonsen 1950–1). Island populations of Franz Josef Land, Spitsbergen and Wrangel Island wholly migratory (Parovshchikov 1962; Løvenskiold 1963; Stishov *et al.* 1991).

In Asia, winters in steppes of western Siberia (where very numerous: Johansen 1944) and central Siberia, in Transbaykalia, and east to east coast (Dementiev and Gladkov 1954). In Kazakhstan, locally common in north, rare south of 48°N (Korelov *et al.* 1974). Regular along Lena river at Yakutsk in north-east Russia (E R Potapov); further north-east, common in southern Koryak highlands (Kishchinski 1980), and very common on Kamchatka peninsula (Lobkov 1986). Occasional winter visitor to northeast and north-west China (Schauensee 1984), and uncommon and local along coasts of northern Japan (Brazil 1991). Several records of Siberian race *vlasowae* reaching western Europe (Banks *et al.* 1991*a*).

P. n. townsendi present all year in island breeding areas (Pribilof, Komandorskiye and western Aleutian Islands), but records from Sakhalin island and mainland Alaska show that some migrate (Gizenko 1955; Gabrielson and Lincoln 1959).

North American birds winter from southern Canada south to central USA, also in west and north-west Canada, overlapping with breeding range in Alaska. Vagrant as far south as Florida and Texas. (American Ornithologists' Union 1983.) In Alaska, winters in central and southern areas, locally in considerable numbers (Gabrielson and Lincoln 1959); perhaps present throughout year in low arctic regions of Canada (Snyder 1957). 2 recoveries in Yakutiya (north-east Russia) of birds ringed Barrow (northern Alaska) (Nethersole-Thompson 1966) show that at least some individuals migrate west or south-west.

Data (including ringing recoveries) from field studies in Iceland, northern Scotland, and Netherlands show that Icelandic race *insulae* winters in Britain and Netherlands, and this supported by museum specimens (Banks *et al.* 1991*a*; Jukema and Fokkema 1992); some specimens from France, southern Scandinavia and Azores are also referable to this race; *insulae* also breeds on Faeroes, but population there too small to account for wintering numbers in western Europe (Banks *et al.* 1991*a*). Winter quarters of Scandinavian birds (nominate *nivalis*) largely unknown, but are probably chiefly in southern Scandinavia, southern shores of Baltic, and countries bordering North Sea (Banks *et al.* 1991*a*). Winter population in northern Scotland *c.* 70–85% *insulae*, and 15–30% nominate *nivalis* (Banks *et al.* 1991*a*); in Netherlands, *c.* 63–64% *insulae*, and 36–37% nominate *nivalis* (Jukema and Fokkema 1992). Scandinavian origin thought probable by Banks *et al.* (1991*a*) for most nominate *nivalis* wintering in Britain, though south-east Greenland origin suggested by Jukema and Fokkema (1992) for Netherlands. Study in northern Scotland showed that nominate *nivalis* gained less weight on average before migrating than Icelandic birds, suggesting preparation for shorter journey (i.e. to Scandinavia), but sample small. Some ringing recoveries involving Britain indicate movements to or from Greenland, however, and passage records on Surtsey off southwest Iceland suggest movement between southern Greenland and north-west Europe. (Banks *et al.* 1991*a*.) Bird ringed on Fair Isle (Scotland), April 1959, and recovered in Newfoundland (Canada), May 1960, had probably originated in Greenland, and wintered on different sides of Atlantic in successive years (Spencer 1961). Recovery data (irrespective of race) also show passage south-east along east coast of Britain and beyond; exceptional recovery in northern Italy, 18 November 1962, of bird ringed in Suffolk (eastern England), 4 November 1962 (Spencer 1963; Banks *et al.* 1991*a*). Up to 1991, ringing programmes in Britain and Iceland have resulted in 37 recoveries showing movements between these countries; other recoveries involving movements to or from Britain as follows: Low Countries (9), Scandinavia (2), north-east Germany (1), Italy (1: see above), North Sea (1), Faeroes (1), Greenland (1), and Canada (1) (Mead and Clark 1993).

Ringing data show that birds from north-east Greenland migrate ESE, and Spitsbergen birds south-east, to northern Russia, and continue to winter quarters in Russian steppes further south. Most recoveries are of birds on spring passage in coastal areas between Arkhangel'sk region and northern Urals, with a few also from north-west Scandinavian coast, and an autumn recovery from Shetland (Scotland); indication of actual wintering grounds given by recovery midwinter at Kuybyshev (53°10′N 50°10′E) on middle Volga. Also bird ringed on Novaya Zemlya recovered south in Komi region. (Salomonsen 1959; Zink 1985.) Biometric data suggest that few north-east Greenland birds winter in western Europe (Banks *et al.* 1991*a*). Ringing data and observations show that west Greenland populations migrate across Davis Strait and along east coast of Labrador, then west to main winter quarters in southern Canada and northern USA; one recovery as far west as 96°13′W in Minnesota; also several recoveries in western Greenland of birds ringed in Michigan (Salomonsen 1950–1; Nethersole-Thompson 1966).

Autumn movement prolonged, September–December, with most passage records October–November. Leaves breeding grounds September–October, e.g. Spitsbergen (Løvenskiold 1963), Kola peninsula (Semenov-Tyan-Shanski and Gilyazov 1991), and Yamal peninsula (Danilov *et al.* 1984). Departure from Greenland begins early September in north, mostly end of September to mid-October in south, with small numbers to early or mid-November (Salomonsen 1950–1); on Wrangel Island, inland areas almost entirely vacated by early September, and most leave coastal areas by end of September (Stishov *et al.* 1991). In Canadian Arctic, the last land bird to depart; local movements evident by end of August, southward movement mostly September, with only small numbers in October (Snyder 1957; Bent 1968). In north-east Scotland, autumn arrival protracted, chiefly October–November, with early birds in September in most years. Winter site-fidelity shown by retraps in successive years. (Banks *et al.* 1991*a, b*.) On Fair Isle (Scotland), passage shows 2 clear peaks, early October (sometimes mid-September) and 1st half of November (Dymond 1991). Main arrival in south of European range from November (Jacoby *et al.* 1970; Kováts 1973; Olioso 1973; Kovács 1981; Pandolfi 1987.) Present in Azores from October (G Le Grand). Data elsewhere show similar pattern. Arrives in Volga-Kama region (east European Russia) October–November (Popov 1978); in west Siberia, sometimes reaches Tomsk at end of September, and present in Altai from early or mid-November (Johansen 1944). Recorded in Japan mostly from December (Brazil 1991). Reaches wintering grounds in southern Canada and northern USA mainly from October (Bent 1968; Squires 1976; Janssen 1987).

Spring movement northward begins early or mid-February. Leaves southern France February–March (Olioso 1973); latest record 28 February in Rumania (Kováts 1973), and rare by March in Hungary (Kovács 1981). Passage peaks end of February to early or mid-March in Denmark, north-east Germany, and Poland (Dittberner *et al.* 1969; Møller 1978*a*; Górski 1982), and last third of March, continuing to last third of April, in Estonia (Rootsmäe 1990) and St Petersburg region (Mal'chevski and Pukinski 1983). First records in southern Finland usually mid-March (Tiainen 1979). Reaches southern Norway mid- or late March to April, and northern Norway at beginning of May (Haftorn 1971). In north-east Scotland, spring departure rapid; most birds leave in March, a few still present in 1st half of April; ♂♂ depart *c.* 9 days before ♀♀ on average (Banks *et al.* 1991*a*); *insulae* leave with sufficient fat reserves for single non-stop flight to Iceland (Banks *et al.* 1989). In Asia, main movement north March–April (Dementiev and Gladkov 1954). Leaves southern part of winter range in North America from mid-February to early April (Bent 1968; Janssen 1987). On breeding grounds, a harbinger of spring. In Greenland, arrives with earliest seabirds; first arrivals in south usually late March or early April, in north-east in 1st half of April; reaches north coast after mid-April or at beginning of May; migration continues throughout May, sometimes into early June. ♂♂ arrive 3–4 weeks before ♀♀. (Salomonsen 1950–1.) In Arctic Canada, the first land bird to arrive, by end of March locally in southern areas, mostly from April, continuing as late as June in some years (Snyder 1957). In Asia, arrives in southern breeding areas in March, e.g. Yamal peninsula (Danilov *et al.* 1984), Yakutiya (Vorobiev 1963), not until April on Wrangel Island (Stishov *et al.* 1991), also Spitsbergen (Løvenskiold 1963). Vagrants recorded north of breeding areas at 86°59′N and 87°05′N, even 'quite close to North Pole' (Nethersole-Thompson 1966).

Evidence suggests ♀♀ more migratory than ♂♂. Birds remaining to winter in Greenland are chiefly ♂♂, and winter populations in western Europe predominantly ♀♀ (Banks *et al.* 1991*a*). In Cairngorms, almost twice as many locally breeding ♂♂ as ♀♀ were seen November–April, and most observations of ♀♀ were March–April; probably, some ♂♂ visit breeding grounds throughout year when conditions allow (R D Smith). In Rumania, arrival of ♀♀ peaks in December, but ♂♂ not until January (Kováts 1973), suggesting they stayed further north initially. Also, data from 55–57°N in west Siberia suggest that ♀♀ winter further south than ♂♂; ♀♀ were collected only in passage periods, but ♂♂ both on passage and in winter (Larionov 1927). Probably ♂♂ are better able to withstand conditions further north, and numbers of ♂♂ migrating vary depending on food availability in north of winter range (Rae and Marquiss 1989). At wintering sites in north-east Scotland, proportion of adult ♂♂ and ♀♀ increases with altitude, and proportion of 1st year ♀♀ declines; higher-altitude wintering sites perhaps preferred because of food availability (large areas of suitable vegetation, when snow-free, mostly unexploited by other species), reduced risk of predation, and similarity of habitat to breeding grounds. Latitudinal segregation may occur for similar reasons. (R D Smith.) See Zink (1985) and Banks *et al.* (1991*a*) for short-distance movements within same winter in north-west Europe, and individuals wintering at different sites in different years.   DFV

Food. Mainly seeds, with addition of insects in breeding season; young given only invertebrates (Tinbergen 1939; Salomonsen 1950-1; Løvenskiold 1963; Portenko 1973; Semenov-Tyan-Shanski and Gilyazov 1991). Feeds almost wholly on ground, sometimes perching on grasses or herbs to reach seeds (Tinbergen 1939; Nethersole-Thompson 1966; Straka 1991). In Cairngorms (Scotland), spring and summer, snow patches very important foraging areas because of large numbers of insects carried up by air currents and left immobilized on snow (Nethersole-Thompson 1976). In winter quarters or on passage, feeds on arable land, especially winter crops, stubble, ploughed fields, etc.; pasture apparently often avoided; generally far from hedges, trees, and buildings; in eastern Germany, also in fields where dung spread, and on rough grassy edges and other weedy places (Rinnhofer 1976; Eifler and Blümel 1983; Oeser 1984); in Brandenburg, 26% of 118 foraging observations were on wasteland and rubbish tips (Dittberner *et al.* 1969). In eastern Austria, 93% of 594 observations were on winter crops, bare earth, and riverside gravel (Straka 1991, which see for flock feeding technique). However, will feed at grain silos and dung heaps in eastern Germany (Eifler and Blümel 1983), and in northern Scotland descends to farmyards and animal feeding places. In Iceland, enters towns in winter (Nethersole-Thompson 1966). In (e.g.) Britain, also on moorland and grassy uplands (Armitage 1932, 1933; Philipson 1939; Banks *et al.* 1991*b*). For winter foraging sites in Hungary, see Sterbetz (1971) and Kovács (1981). At this time of year commonly feeds on coastal saltmarshes, dunes, and similar areas of low vegetation (Lack 1986; SOVON 1987; Yeatman-Berthelot 1991), and on shore often at seaweed on tideline for invertebrates, especially crustaceans such as sandhoppers *Talitrus* and sea slaters *Ligia oceanica*, which are eagerly taken even when up to 2·5 cm long (King and King 1968; Hopkins 1985; Lack 1986; Banks *et al.* 1991*b*). Stretches up from ground to reach seed-heads (Tinbergen 1939), or, when feeding on rush *Juncus*, either pulls head apart, jumps against base of stem with body to scatter seeds, or bends it over by perching on it (Philipson 1939; Henty 1979, which see also for flock foraging technique). On breeding grounds, commonly makes sallies after flying insects, particularly crane-flies (Tipulidae) and Lepidoptera (Nicholson 1930; Wynne-Edwards 1952; Nethersole-Thompson 1966). In Scotland, forages around ski stations and other tourist resorts in mountains (Watson and Smith 1991); entered tent to take scraps in Greenland (Nicholson 1930); in Norway, pecks at meat and fish hung up for drying (Nethersole-Thompson 1966), and on Kola peninsula (north European Russia), seen feeding on roads and extracting seeds from horse-droppings (Semenov-Tyan-Shanski and Gilyazov 1991).

Diet in west Palearctic includes the following. Invertebrates: bristle-tails (Thysanura), stoneflies (Plecoptera), grasshoppers (Orthoptera: Acrididae), earwigs (Dermaptera: Forficulidae), bugs (Hemiptera: Auchenorrhyncha), adult and larval Lepidoptera (Noctuidae), caddis flies (Trichoptera: Limnephilidae), adult and larval flies (Diptera: Tipulidae, Chironomidae, Cecidomyidae, Dolichopodidae, Coelopidae, Muscidae), Hymenoptera (Tenthredinidae, ants Formicidae), beetles (Coleoptera: Staphylinidae, Elateridae, Chrysomelidae, Curculionidae), spiders (Araneae), harvestmen (Opiliones), mites (Acari), crustaceans (sandhoppers Talitricidae, sea slaters Ligiidae), worms (Oligochaeta: Lumbriculidae). Plants: seeds, berries, shoots, etc., of spruce *Picea*, birch *Betula*, dock *Rumex*, knotgrass *Polygonum*, goosefoot *Chenopodium*, orache *Atriplex*, *Camphorosma*, amaranth *Amaranthus*, *Celosia*, mouse-ear *Cerastium*, buttercup *Ranunculus*, poppy *Papaver*, scurvy-grass *Cochlearia*, cloudberry *Rubus*, rose *Rosa*, rowan, etc. *Sorbus*, saxifrage *Saxifraga*, clover *Trifolium*, heather *Calluna*, crowberry *Empetrum*, Boraginaceae, self-heal *Prunella*, plantain *Plantago*, nightshade *Solanum*, bur-marigold *Bidens*, mugwort *Artemisia*, duckweed *Lemna*, grasses (including barley *Hordeum*, oats *Avena*, wheat *Triticum*, *Festuca*, *Lolium*, *Poa*, *Trisetum*, *Aira*, *Arctagrostis*, *Ammophila*, *Phleum*, *Molinia*, *Echinochloa*, *Setaria*), sedges (including *Carex*, *Scirpus*), rushes (including *Juncus*, *Luzula*), moss (Musci), lichen *Cladonia*. (Rey 1907*a*; Witherby *et al.* 1938; Turček 1961; Løvenskiold 1963; Nethersole-Thompson 1966; King and King 1968; Sterbetz 1971; Kováts 1973; Balfour 1976; Rinnhofer 1976; Popov 1978; Eifler and Blümel 1983; Hopkins 1985; Lack 1986; Semenov-Tyan-Shanski and Gilyazov 1991; Straka 1991.)

In Transylvania (Rumania), winter, 11 stomachs contained 1322 items of plant material, of which 70% (by number) seeds of *Polygonum*, 13% *Celosia*, 7% *Setaria*, 7% wheat, 2% *Echinochloa*, and 1% self-heal; 3 stomachs contained some incidental insect material, mainly grasshoppers, Hemiptera, ants, and beetles (Kováts 1973). On Helgoland (north-west Germany), winter, of 7 stomachs, 5 contained seeds of orache and *Polygonum*; other material was grass seeds, pieces of Rosaceae berries, and a few insect fragments (Rey 1907*a*). In Spitsbergen, June–July, 17 stomachs held only vegetable matter; on arrival in April, probably has to depend on previous year's seeds in snow since invertebrates unavailable; seeds main part of diet again in autumn (Løvenskiold 1963). In western Greenland, summer, 15 stomachs all contained plant material and 14 animal material; Diptera (Tipulidae, Chironomidae, Mycetophilidae) were in 10 stomachs, beetles (Byrrhidae, Curculionidae) in 9, Hymenoptera (Ichneumonidae, Chalcidoidea) in 5, and Hemiptera in 4; other invertebrates included adult Lepidoptera, spiders, and snails (Pulmonata); seeds of crowberry were in 9 stomachs (one held 51 seeds), *Carex* in 7, Polygonaceae in 6, flowers and seeds of bilberry *Vaccinium* in 3, as well as seeds and spores of Caryophyllaceae, Cruciferae, *Juncus*, and moss (Longstaff 1932). Probably faces shortage of food in Greenland when first on breeding grounds since only seeds available in snow; birds forage around Inuit settlements and turn over lichen on rocks for small items, possibly springtails (Collembola) or mites (Tinbergen 1939; Salomonsen 1950–1). On Wrangel Island and Chukotskiy peninsula (north-east Russia), one spring stomach contained 5 Diptera and other larvae and 28 small seeds; July–August stomachs held caterpillars, spiders, and many seeds; September–October, caterpillars, caddis fly (Trichoptera) larvae, and seeds; late-autumn stomachs contained only seeds (Portenko 1973). Caddis flies taken in large numbers in summer in Iceland (Timmermann 1938); in Cairngorms, takes mostly adult crane-flies (Nethersole-Thompson 1976). In Pennines (north-central England) in winter, preferred food was *Molinia* grass infested with gall-midge larvae (Cecidomyidae) in stems (Armitage 1932, 1933); also very fond of seeds of *Juncus squarrosus* (Philipson 1939). For winter diet in Hungary, see Kovács (1981); for Volga-Kama region (east European Russia), see Popov (1978). In Manitoba (Canada), on spring migration feeds on leaves of salt-marsh grass *Puccinellia* (Bazely 1987, which see for discussion). For digestion rates and identification problems of stomach contents, see Custer and Pitelka (1975).

Diet of young wholly invertebrates. On Devon Island (northern Canada), collar-samples from 11 nests contained 2190 items, of which 96% (by number) adult Diptera (82% midges Chironomidae, 7% Tipulidae, 4% Empididae, 3% Muscidae), 2% spiders, 1% Hymenoptera, 1% larvae and pupae, 0.5% adult Lepidoptera, and 1% plant material. Tipulidae declined from 22% in mid-July to 4% in late July, and midges rose from 56% to 87%. (Hussell 1972.) In north-east Greenland, 18 collar-samples contained 193 items of animal prey, of which 57% (by number) spiders, 19% Lepidoptera (17% caterpillars), 16% Tipulidae (9% larvae and pupae), and 4% mosquitoes (Culicidae) (Asbirk and Franzmann 1978). In other Greenland studies, only Diptera, Lepidoptera, and spiders seen to be fed to young; on rainy days in July, nestlings given small Lepidoptera gathered in shrubs, while in sunny weather young mosquitoes were picked from stones by pools; adults often foraged far from nest (Longstaff 1932; Tinbergen 1939). In Cairngorms, young received mainly adult crane-flies brought bundled in bill, sometimes from 500 m away; before crane-flies emerge, young may go short of food in cold wet weather (Nethersole-Thompson 1966, 1976). See also Kareila (1958) and Lyon and Montgomerie (1987) for adult foraging distances. In northern Kamchatka (north-east Russia), nestlings fed adult and larval Lepidoptera, adult crane-flies, spiders, and pieces of fungi (Kishchinski 1980). BH

Social pattern and behaviour. Important pioneering study by Tinbergen (1939) in Angmagssalik region (eastern Greenland). Long-term study by Nethersole-Thompson (1966) in Cairngorms (Scotland) revealed major differences in dispersion and behaviour, associated particularly with status as small, sparse population breeding at fringe of range. Account includes recent information for Cairngorms supplied by A Watson, also by R D Smith from intensive study of marked birds, 1987–92. Account also includes reference to significant extralimital studies on nominate *nivalis*, notably by Sutton and Parmelee (1954), Watson (1957a), Lyon and Montgomerie (1985, 1987), and Lyon *et al.* (1987) in Northwest Territories (Canada).

1. Usually gregarious outside breeding season, but quite often encountered singly (A Watson). Post-breeding nomadic flocks of juveniles occur prior to migration (Nicholson 1930; Tinbergen 1939), e.g. in Ellesmere Island (Northwest Territories) flocks of up to 40 (Knights and Walker 1989) and in Greenland flocks exceeding 100 (Salomonsen 1950–1). Migrates (both autumn and spring) and winters in groups varying from small to very large (Dementiev and Gladkov 1954), e.g. in Oberlausitz (eastern Germany) wintering flocks mostly 2–50, sometimes more than 100 (Eifler and Blümel 1983); on Polish Baltic coast, typically 2–5 per flock but sometimes up to several hundreds (Górski 1982). For Hungary see Kovács (1981), for Austria see Straka (1991). In Caithness (Scotland), flocks usually 10–300, maximum 2000 (Banks *et al.* 1991b, which see for seasonal variation). Other exceptional flocks as follows: 1000 in Co. Derry (Northern Ireland) (Fitzharris and Grace 1986); more than 1000 in Glen Esk (Scotland) (Nethersole-Thompson 1966); up to 1500 in Orkney (Booth *et al.* 1984); 10 000 in April in Hailuoto, Finland (Markkola and Vierikko 1983). In northern Scotland, some winter flocks moderately site-faithful from year to year, and relatively sedentary during winter (Marshall and Rae 1981; Banks *et al.* 1991b); other areas have substantial population in some years, none in others (A Watson). Population in Caithness appears to consist of one or more flocks within fairly extensive area (Banks *et al.* 1991b). In Cairngorms, where local breeding population overwinters to variable extent, some birds seen on 50 or more days each winter (R D Smith). ♀♀ generally outnumber ♂♂ in flocks wintering in western Europe (Banks *et al.* 1991a, b; see also Movements). Degree of association with other species varies. In Caithness, flocks usually pure *P. nivalis*, typically keeping apart from (e.g.) Skylarks *Alauda arvensis*, Reed Buntings *Emberiza schoeniclus* and finches (Fringillidae) (Banks *et al.* 1991b). In France, flock of *P. nivalis* did not mix with other seed-eaters in vicinity until it joined large flock of Linnets *Carduelis cannabina* at end of January (Olioso 1972). In Hungary, seen with Redpolls *C. flammea* and especially Shore Larks *Eremophila alpestris*; in Hortobágy, mostly with Twite *C. flavirostris*, far less with Lapland Bunting *Calcarius lapponicus*, Fieldfares *Turdus pilaris* and *A. arvensis* (Kovács 1981). In Oberlausitz, flock of at least 250 *P. nivalis* with c. 1000 Yellowhammers *E. citrinella*, several hundred *A. arvensis*, several *C. flavirostris*, *T. pilaris*, and *E. alpestris* (Creutz 1988). On return to breeding grounds, flock structure maintained until receding snow cover favours dispersal onto territories; flocks initially all ♂♂, but, with arrival of ♀♀ (see Heterosexual Behaviour, below), contain both sexes (Tinbergen 1939). After break-up, flock may re-form if bad weather returns (Manniche 1910; Stishov *et al.* 1991); Tinbergen (1939) found unpaired birds, but not pairs, resorting to flocks in bad weather. In Cairngorms, some ♂♂ probably visit breeding areas throughout winter when conditions allow, and are regularly found on future territories from March; in April–May, pair-bonds temporarily severed in favour of flocking if bad weather intervenes (R D Smith). BONDS. Mating system typically

monogamous (almost invariably so at high latitudes: see below), exceptionally bigamous, i.e. ♂ with 2 ♀♀. In Angmagssalik, up to 5% of ♂♂ bigamous in some years, achieved as follows: ♂, during incubation of his 1st mate, mated (and raised brood) with 2nd ♀ which had just fledged her 1st (very early) brood; in another case, ♀ copulated with neighbouring ♂ while she was looking after part of fledged brood (see below for details of brood-division) which she then abandoned to die, while her 1st mate continued feeding his part of brood (Tinbergen 1939). In Cairngorms, promiscuity and polygamy recorded. Unpaired ♂♂ tend to stay on territory throughout breeding season (R D Smith), but in earlier study, when density lower, unpaired ♂♂ apparently prospected widely for copulations in other territories; sometimes succeeded in pairing with resident ♀♀ which left 1st fledged broods in mate's care; at least 1 ♀ produced 2 clutches with 2 different ♂♂ in same year (Nethersole-Thompson 1966); such re-pairing very rare, but another case recorded in 1992: marked ♀ produced clutches for 2 different ♂♂ in different territories 2 km apart (1st ♂ still alive when 2nd clutch laid). Also in Cairngorms, bigamy (and once ♂ with 3 ♀♀) occurs in ♂♂ at least 2 years old (none recorded in 1-year-olds); movements of 2-5 km by paired ♂♂ not uncommon (usually when own mate incubating), often to territory where ♀ was in fertile period. (R D Smith.) In study by Nethersole-Thompson (1966, 1976), one ♂ had 2 mates in nests c. 140 m apart; clutches were completed 4 days apart; 2nd ♀ lost her clutch and re-laid; meanwhile ♂ was helping to feed 1st ♀'s brood; after both broods fledged, ♂ mated with a 3rd ♀ c. 800 m away; her previous ♂ fed and reared 2 fledglings unaided. Another ♂ had 2 ♀♀ (nests c. 250 m apart), one of which later paired with another ♂ and laid 2nd clutch. Bigamy thought to be virtually impossible in highest latitudes where degree of nesting synchrony would force bigamous ♂♂ to feed (unsuccessfully) 2 broods at once; thus experimental removal of ♂♂ at Sarcpa Lake (Northwest Territories), where *P. nivalis* always monogamous, depressed breeding success and indicated importance of ♂-role in provisioning young, at least in years of scarce food (Lyon *et al.* 1987). No evidence that pair-bond maintained outside breeding season, but several cases of bond renewed in 2 (once 3) successive breeding seasons (R D Smith). Some birds arrive on breeding grounds already paired (see Heterosexual Behaviour, below). ♀ builds nest, incubates, and broods (Manniche 1910; Sutton and Parmelee 1954; Nethersole-Thompson 1966). ♂ occasionally recorded carrying nest-material (Tinbergen 1939; Watson 1957a; Nethersole-Thompson 1966, 1976) but never to future nest-site and perhaps only for display purposes (R D Smith). ♂ regularly feeds ♀ throughout incubation but only rarely before incubation (see Heterosexual Behaviour, below, for details). Both sexes feed young till independence (Tinbergen 1939; Sutton and Parmelee 1954; Nethersole-Thompson 1966). Brood-division typical (Tinbergen 1939; Kareila 1958), often starting with asynchronous fledging such that ♂ cares for older young out of nest, leaving ♀ to tend those still in nest (Sutton and Parmelee 1954; Watson 1957a; Nethersole-Thompson 1966). Young begin self-feeding 3-4 days after leaving nest, and independent c. 8-12 days after leaving nest (Tinbergen 1939; Salomonsen 1950-1). Age of first breeding by both sexes typically 1 year in Cairngorms (Smith 1991; R D Smith). BREEDING DISPERSION. Solitary and territorial. Marked variation in territory size with region and habitat. In Angmagssalik, before ♀♀ arrived, ♂♂ established territories initially c. 600 m across, diminishing to c. 50-100 m across during May with competition from new settlers (Tinbergen 1939). At Ellesmere Island, later-arriving ♂♂ (up to 1 month after 1st ♂♂) tended to have smaller territories and were often limited in choice of nest-site (Knights and Walker 1989). In optimal habitat, territories commonly small, leading to aggregation: e.g. 'almost in colonies' in old packing cases in Longyearbyen, Spitsbergen (Nethersole-Thompson 1966); in Iceland 4-5 pairs in sheltered hollow within radius of c. '300 yards' (Yeates 1951); in Alaska often 1 pair per 0·4 ha, nests sometimes only c. 25 m apart (Nethersole-Thompson 1966). See below for high density on Baffin Island (Northwest Territories). In Cairngorms, some ♂♂ flew 500 m to engage in song-duels with neighbours; smallest territory ('song range') 3 ha, largest c. 40 ha, on average larger in years when breeding population smaller (Milsom and Watson 1984; A Watson). From 1988-91, median nearest-neighbour distance decreased from c. 580 m to 330 m in one area, and from 700 m to 400 m in another; repeat nests and 2nd broods were commonly less than 100 m from 1st nests (2nd clutch never in same nest as 1st) (R D Smith). See Bonds (above) for distance between nests of bigamous ♂♂ in Cairngorms. Territory serves for courtship, nesting, and feeding; even where nests not aggregated, and when territories diffuse and large at start of breeding season, feeding typically extends to neutral ground outside territory, up to 260-640 m from nest in different studies (Kareila 1958; Nethersole-Thompson 1966, which see for references; Lyon and Montgomerie 1987). Densities as follows: in Norway: 28 territories per km² (19-42) in Nedal (Moksnes 1973); average 4 (maximum 7) in Finse (Lien *et al.* 1970). In Finnish Lapland: c. 6·7 pairs per km² (Kareila 1958); 8 pairs per km² (Silvola 1966). In Vaygach and Yugorskiy peninsula (north European Russia), 20-120 pairs per km² in gravelly tundra, 40-70 in shoreline storm debris (trees, etc.), typically 200-250 (10-500) on mixture of scree, cliff, and rocks (Uspenski 1959); these figures, originally in pairs per ha, are so high as to suggest they are based on areas strictly circumscribed by neighbourhood groups. Extralimital densities (Greenland and Northwest Territories): 8-13 pairs per km² (Joensen and Preuss 1972; Freedman and Svoboda 1982; Montgomerie *et al.* 1983; Lyon and Montgomerie 1987). On Baffin Island, 38 pairs and 10 unpaired ♂♂ per square mile (i.e. c. 15 pairs and 4 unpaired ♂♂ per km²), but much of habitat unsuitable; in 16 ha of good habitat, 12 pairs and 4 unpaired ♂♂ (i.e. 75 pairs and 25 unpaired ♂♂ per km²); almost all birds on less favourable habitats were unpaired ♂♂ (Watson 1957a, 1963). In Chukotskiy peninsula (north-east Siberia) distance between pairs 20-500 m (Portenko 1973). For density in relation to proximity to Peregrine Falcon *Falco peregrinus* in western Greenland, see Meese and Fuller (1989). In Cairngorms, some ♂♂ return to same or nearby territory from year to year; fidelity for 2-3 years not unusual; in 2 cases same territory held for 4 years (♀) and 5 years (♂); however, large displacements (2 km or more) also sometimes occur, once c. 25 km; site-fidelity very rarely extends to nest-site: of c. 150 nest-sites found in Cairngorms in 6 years, only 2 were used twice (consecutive summers); natal site-fidelity low: of 270 ringed young, 11·5% returned to breed in study area in 1st year but seldom close to where reared (R D Smith). ROOSTING. Best account of seasonal change in roosting sites from study on Ellesmere Island: on arrival, late April, singletons or groups of 24 roosted (not close together) in timber piles or niches in sandstone outcrops; in May both sexes roosted in eroded snowbanks; in late June ♂♂ roosted in territory while ♀♀ on nest; in August some groups dispersed among rocks on steep banks while other groups (both sexes) roosted among boulders in nearby streambeds and under large snowbanks; in September flocks of up to 100, sometimes including *C. lapponicus*, roosted in open tundra (Parmelee and Macdonald 1960). In Cairngorms, before nesting, both sexes roosted in scree in territory; during laying and incubation ♂ roosted in crevice not far from nest (Nethersole-Thompson 1966). In Angmagssalik, before ♀♀ arrived, ♂♂ used same hole in territory for several nights but occasionally moved to new site (Tinbergen 1939). In north-east Greenland, post-breeding groups found crouching together in hollows in rock walls inaccessible to ground predators (Manniche 1910). In Spitsbergen, June–July, communal roosting on ledges in disused quarry and huts (including shelves inside: Løvenskiold 1963), and under pieces of wood; roost-sites were used regularly though not necessarily by same birds; number of birds in any roost varied greatly from day to day, partly due to disturbance (see Flock behaviour, below), partly with wind direction (birds preferring to roost in lee). Only adult ♂♂ used roost in June; from late July to early August (i.e. after fledging) total 358 birds (73% adults, of which 82% ♂♂) seen in roosts. (Swann 1974.) Roost-sites and strategy (single or communal) in winter equally varied, including: on Pennines (England), solitary in peat hags under tufts of *Molinia*, communally in hill-top quarry or sheltered valley (Armitage 1933); in Norway 'migratory' flocks often roosted in trees, and in Invernessshire (Scotland) flocks occasionally roosted in rowan *Sorbus* (Nethersole-Thompson 1966). In France, winter, seen roosting on ground among grass and stones (Olioso 1972). Even when conditions not extreme (A Watson), will burrow into snow for overnight roosting and daytime loafing, e.g. in Massachusetts (USA) birds made individual holes in snowdrifts and huddled there for most of day, occasionally emerging to feed (Bagg 1943). For review of roosting in snow see Thiede (1982, 1989). For apparent snow-bathing see Riddoch (1986). Circadian activity rhythm on breeding grounds in high latitudes well studied, e.g. Haarhaus (1968), Krull *et al.* (1985). In Angmagssalik, end of March, birds awoke at *c.* 03.30 hrs and progressively earlier during April, stabilizing at *c.* 01.00 hrs by beginning of May, after which no earlier waking; sleeping time therefore *c.* 2–3 hrs (Tinbergen 1939). Franz (1949) also found that quiescent period tended to remain fixed in spite of increasing daylength. In north-east Greenland, around hatching, ♂ and ♀ (on nest) showed minimum activity around midnight; ♂ slept from *c.* 20.35 to *c.* 02.10 hrs (Asbirk and Franzmann 1979). In Spitsbergen, from June onwards, birds roosted communally (see above) from exceptionally early in day: usually began arriving in roosts 14.00–15.00 hrs (later on warm days), peak numbers 18.00–22.00 hrs; morning dispersal more sudden, usually 23.00–01.00 hrs, earlier in June when only adult ♂♂ used roost (Swann 1974).

2. Relatively approachable in breeding season, especially where nesting around human habitation (Salomonsen 1950–1). Winter and spring flocks much less approachable, typically flushing at least disturbance (Tinbergen 1939; Henty 1979; Creutz 1988). In Sachsen (eastern Germany), larger winter flocks (more than 50 birds) allowed approach by man to 12–22 m (Oeser 1984); see Flock Behaviour (below) for response to infringement of this distance. On South Uist, flying flock thwarted Sparrowhawk *Accipiter nisus* by forming tight ball (Meinertzhagen 1959). Ground-hugging flight in vicinity of *F. peregrinus* eyrie adaptive for evading capture (Meese and Fuller 1989). When uneasy, flicks tail and wings upwards (Witherby *et al.* 1938; Tinbergen 1939). FLOCK BEHAVIOUR. For flock contact-calls see 2 in Voice. Winter flocks characteristically restless and volatile, frequently flushing (low and fast) and sweeping over ground repeatedly before landing again; foraging flock moves rapidly over ground in constant direction, interspersed with short flights, birds at rear constantly leap-frogging rest in roller-feeding fashion (Henty 1979). In Sachsen, winter, when man approached flock too closely (see above), peripheral birds ran into middle of flock or flew over it to land on far side; flock usually very compact in flight (see above for anti-predator benefit); flock very ground-hugging and never seen to settle on trees or man-made structures (Oeser 1984) but in north-east

Scotland often perch by day on fences, buildings, wires (etc.) and occasionally hardwood trees (A Watson); see also Witherby *et al.* (1938) and Roosting (above) for perching on trees, etc. In Spitsbergen, roosting birds (see Roosting, above) were easily disturbed and would then quickly depart to neighbouring roosts (Swann 1974). SONG-DISPLAY. Serves for mate-attraction and territorial defence (see Antagonistic Behaviour, below). ♂ sings (see 1 in Voice) mostly from ground (using high boulders, etc., in territory as preferred vantage points), occasionally in Song-flight shortly before nesting. When singing from perch, ♂ tail-flicks between bouts. (Salomonsen 1950–1; Nethersole-Thompson 1966; A Watson.) In Song-flight, ♂ rises steeply and silently with rapid wing-beats for a few metres (often up to 10–15 m); at peak of ascent (and not until then) starts singing with outspread, slightly trembling wings, and glides down (in direction of rival if Song-flight provoked by him) to vantage point; after landing, sometimes continues singing and may keep wings raised before closing them and crouching forwards. (Tinbergen 1939; Salomonsen 1950–1; Nethersole-Thompson 1966.) Also seen singing during normal flight (starting a few seconds before landing and completing song after landing) and while feeding young (Salomonsen 1950–1). Singing starts on spring migration (Rüppell 1970), and Song-flights seen during pair-formation at staging post (Watson 1957*a*). In Angmagssalik, quiet song was one of first signs of break-up of flocks and subsequent territorial establishment. On territory, ♂♂ spent most of morning singing, at first only quietly at long intervals during foraging (would stop feeding, jump onto any stone, give a few phrases, then resume feeding); later, song-bouts more frequent and longer (including for more than 1 hr soon after awaking) from favoured song-posts. ♂ stopped singing as soon as ♀ arrived on territory, but resumed if she left or temporarily dropped out of ♂'s sight; also resurgence when ♀ laying and even more so during incubation. (Tinbergen 1939.) In Cairngorms, ♂♂ heard giving low-intensity song from mid-March (when still in winter plumage) on territory before pairing and breeding there (R D Smith, A Watson); some ♂♂ continue singing during courtship, nest-prospecting and building, also when visiting different areas before their mates have begun site-prospecting; in some cases, resurgence of song occurs when ♀ starts incubating (presumably ♂ trying to attract new mate), and song-output of such birds difficult to distinguish from unpaired birds (Nethersole-Thompson 1966). Main song-period in Cairngorms mid-April to mid-July, some ♂♂ continuing at reduced rate till end of July (R D Smith, A Watson); same rather sudden cessation of song (latest 20 July) in Iceland (Nethersole-Thompson 1966, which see for extralimital song-periods). Singing also continues till 3rd week of July in Greenland (Nicholson 1930; Salomonsen 1950–1). In breeding season, song begins on awakening (see above); lull in middle of day, especially 11.00–14.00 hrs (Portenko 1973). On Baffin Island, some ♂♂ in denser populations sang freely all night in June, with peak at *c.* 03.00–04.00 hrs on 3 mornings; sang less in sparser populations (Watson 1957*a*). ANTAGONISTIC BEHAVIOUR. (1) General. On breeding grounds, growing ♂-♂ aggression in pre-breeding flocks signals imminent dispersal onto territories. Territorial establishment associated with prolonged fighting (see below); less aggression during pair-formation, but resurgence during incubation. During chick-rearing, ♂♂ tolerate one another on neutral ground but continue to defend own territories. (Tinbergen 1939.) However, in Finnish Lapland, aggression arose later in summer (post-fledging) as soil became drier and feeding areas more restricted (Kareila 1958). (2) Threat and fighting. Following account based mainly on observations by Tinbergen (1939). Birds challenge conspecifics with Forward-threat display (Fig A): advance on rival with head

A

lowered and retracted, bill pointing at opponent, also sometimes wing-fluttering and giving Threat-calls (see 3 in Voice). ♂♂ behaving thus at Angmagssalik did not stay in flock more than 2 days. Once on territory, resident ♂ immediately flew to one of his song-posts on hearing strange conspecific. Intruder confronted with Forward-threat usually flew on but, if it landed, resident flew towards it and performed Song-flight (see Song-display, above), sometimes developing into song-duel and aerial fight in which combatants use feet and bill, accompanied by Anger-calls (see 4 in Voice), sometimes falling interlocked and rolling down snow-slopes. On separating, resident often resumed chase, but roles of pursuer and pursued reversed when territorial boundary crossed, often leading to extended flights to and fro. Rivals often interrupted expelling-flight to land, face each other, walk up and down, performing Forward-threat and mock-feeding. Sometimes intruder remained unnoticed for some time, crouching submissively when resident performed Song-flight. (Tinbergen 1939.) In Cairngorms, lack of rigid territorial boundaries and little defence, though resident will expel any ♂ intruding on song-post or near nest (Nethersole-Thompson 1966). Generally, aggression is ♂ against ♂ and ♀ against ♀ during territorial intrusions and border disputes: e.g. if ♀ from territory A trespasses on territory B and B-♀ challenges her, then A-♂ (after some hesitation) usually joins in and engages B-♂ (Tinbergen 1939). Sutton and Parmelee (1954) describe 2 ♀♀ fighting almost to the death. Newly established ♂ may initially respond aggressively to ♀ flying over his territory, switching to courtship if she lands (Tinbergen 1939); for mate-guarding, see Heterosexual Behaviour, below. Tinbergen (1939) found no significant aggression towards other species. Sutton and Parmelee (1945) recorded ♂ (with laying ♀) repeatedly attacking Wheatears *Oenanthe oenanthe* with nest nearby; in Cairngorms, *O. oenanthe* were also evicted but Meadow Pipits *Anthus pratensis* were ignored (Nethersole-Thompson 1966), though interactions are not rare when all 3 species are in flocks (A Watson). HETEROSEXUAL BEHAVIOUR. (1) General. ♂♂ typically precede ♀♀ to breeding grounds by 3–4 weeks in Greenland (Salomonsen 1950–1). In Scotland, no heterosexual behaviour ever seen in wintering areas (i.e. at altitudes below 900 m), even in March–April (R D Smith). However, some birds arrive on breeding grounds having already paired *en route* (Watson 1957a; Parmelee and Macdonald 1960; Nethersole-Thompson 1966). Otherwise, ♂♂ establish territories and try to attract mates there (e.g. Tinbergen 1939). (2) Pair-bonding behaviour. Until paired (and sometimes even after: see Bonds, above) resident ♂ performs Advertising-display ('mannequin display') to every ♀ who lands in his territory: in upright posture, with wings spread back and down, and tail fanned and lowered, ♂ scuttles a short distance from ♀ (thus displaying bold piebald rear view to her), then shuts his wings and tail and runs back to her before turning and displaying anew; ♂ sometimes displays thus while crawling down boulder on which ♀ perched. (Nethersole-Thompson 1966.) In Advertising-display (Fig B) prior to copulation (see subsection 4, below) ♂ danced around ♀ with his wings raised (sometimes waved or fluttered), head raised, singing continuously; sometimes tilted body so that wing-tip touched ground, and proceeded straight from display to mounting (E K Dunn). ♂ supplements Advertising-display variously with Inciting-call (see 5a in Voice),

B

panting, mock-feeding, and sexual chasing (sometimes high into air); see 5 in Voice for associated calls. ♀ shows no obvious reaction to Advertising-display. (Tinbergen 1939.) ♂ seen raising wings high over back (Sutton and Parmlee 1954); both sexes reported to do this by Witherby *et al.* (1938). If ♀ remains on territory, ♂'s aggression to her gradually wanes and they keep close company (i.e. mate-guarding), feeding together, communicating regularly with Ripple-calls (see 2a in Voice), and soon attempting copulation (Tinbergen 1939). If ♂ loses sight of ♀ he flutters up, wings vibrating, then, when 3–7 m up, rocks from side to side, hanging in wind and drifting in semicircles, all the while looking down; on landing, sings loudly (Nethersole-Thompson 1966). (3) Courtship-feeding. Plays little part in courtship, but ♂ seen feeding ♀ from day before she laid 1st egg. ♀ solicits food with Begging-display: raises and opens bill, giving Begging-calls (see 5c in Voice) and wing-shivers. Incubating and brooding ♀ is fed on nest, or flies off to meet mate with Begging-display (Sutton and Parmelee 1954; Watson 1957a). ♀ also seen begging, sometimes successfully, when pair at nest together to feed young (Nethersole-Thompson 1966: Fig C). In Cairngorms, *c.* 50% of ♂♂ feed their mates, but mostly

C

sporadically and effecting little nutritional benefit; rate of Courtship-feeding seems to depend more on particular ♂ (e.g. some ♂♂ consistently perform in different years) than on his mate or her circumstances (e.g. no apparent increase in bad weather or in presumably more arduous incubation of 2nd clutches) (R D Smith). However, Courtship-feeding apparently more prevalent at higher latitudes where low temperatures may favour more continuous incubation (Nethersole-Thompson 1966). Thus in Northwest Territories, Courtship-feeding regular throughout incubation (Sutton and Parmelee 1954; Watson 1957a; Hussell 1972), but rare before incubation (Lyon 1984); ♂-removal experiments demonstrated that Courtship-feeding increased hatching success and shortened incubation periods (Lyon and Montgomerie 1985). (4) Mating. Occurs on ground in territory, including on snow (Løvenskiold 1963). ♂ typically initiates, though ♀ solicits when receptive. ♂ walks ('in a slightly unusual way... difficult to describe') or flies to ♀. When near her he flies and tries to mount but early in season ♀ resists, flying away and provoking wild erratic chase which typically results in ♂ giving up and landing to perform excited Advertising-display etc. (see subsection 2, above). Such attempted copulation and associated display may continue for weeks before ♀ receptive. (Tinbergen 1939.) She then performs Soliciting-display: crouches horizontally, points bill up, lifts tail, and wing-shivers, whereupon ♂ mounts (Tinbergen 1939; Nethersole-Thompson 1966). ♂ seen seizing ♀'s tail before mounting (Nethersole-Thompson 1966). ♀ may solicit when ♂ shows no readiness to mate and refuses her; after unsuccessful copulation, ♀ may resume soliciting while ♂ may perform apparent low-intensity Advertising-display: stands erect with tail lowered and somewhat spread but wings folded (Fig D). In Greenland, start of suc-

D

cessful copulation coincides with onset of nest-building, 8–13 days before 1st egg laid (no copulation seen thereafter). (Tinbergen 1939.) In Cairngorms, ♀♀ seen soliciting successfully while site-prospecting (for 1st or replacement nests) and mating continued on days of laying, including once shortly after last egg laid (following which ♀ bathed in puddle while ♂ crouched flat on rock just above her) (Nethersole-Thompson 1966). Also in Cairngorms, fertile ♀♀ were almost always accompanied and defended by their mates (mate-guarding), preventing extra-pair copulations; however, opportunistic rape sometimes suspected when (e.g.) ♀ left nest to feed on her own (R D Smith). In Greenland, copulation most prevalent 02.00–06.00 hrs (Tinbergen 1939); in Cairngorms, seen 04.00–20.30 hrs (Nethersole-Thompson 1966). (5) Nest-site selection. Pair prospect together, periodically interrupting feeding to examine crevices (etc.), sometimes entering several in rapid succession; typically both enter, but sometimes ♂ or ♀ alone. Flit in and out of crevices with much nervous pecking at lichen by ♀. She appears to make final choice, and building may begin same day. Several nests often begun by ♀ before choosing one. (Tinbergen 1939; Nethersole-Thompson and Nethersole-Thompson 1943b; Nethersole-Thompson 1966, which see for exceptional case of ♂ choosing site which ♀ rejected.) In 1 out of 4 cases, ♀ built nest outside territory, ♂ extending territory accordingly (Tinbergen 1939). (6) Behaviour at nest. Building by ♀ (see Bonds, above, for apparent exceptions). ♀ continued lining nest for 2–3 days after 1st egg laid (Tinbergen 1939). During laying, nest-sites may be seldom visited and much time spent feeding on neutral ground (Nethersole-Thompson 1966), though Tinbergen (1939) found almost all feeding within territory during laying. For ♂ feeding ♀ during incubation see subsection 3 (above). In Cairngorms, ♂ seldom visits incubating ♀ except sometimes to feed her or to escort her to feeding grounds (♂ calls ♀ off nest) or back again. ♀ sometimes leaves nest unaccompanied but typically for shorter spells. (Nethersole-Thompson 1966.) RELATIONS WITHIN FAMILY GROUP. Both sexes feed young though contributions vary (Nethersole-Thompson 1966, which see for details). ♀ apparently brings most food during first few days

after hatching (Manniche 1910; Tinbergen 1939) though this perhaps related to ♂'s greater wariness at nest under observation (Nicholson 1930). Initially young gape vertically and silently; food-calls (see Voice) heard from day 3; eyes begin to open *c.* day 5, after which begging oriented towards parent (Rüppell 1970). After 5 days, ♀ seldom broods except at night (Nethersole-Thompson 1966). On Baffin Island, Watson (1957a) found ♀ brooding young at night up to 12 days old. In first few days after hatching, faecal sacs eaten by ♀, possibly also by ♂; later, both parents carry and drop them well away from nest; *c.* 1 day before fledging, faeces are left to accumulate in nest or at nest-entrance (Tinbergen 1939; Nethersole-Thompson 1966). From 7–8 days old, young make frequent short excursions to nest-entrance; sometimes older young stay there till all of brood are ready to leave together, or ♂ (parent) may lead off older young while ♀ tends those still in nest (Watson 1957a; see Bonds, above, for brood-division). For change in calls of young at fledging see Voice. On first leaving nest, young mobile on ground but barely able to fly; stay near nest-entrance for a few hours, then disperse, each brood-member adopting separate refuge and attracting parent with locatory calls (see Voice). Dispersing young may eventually enter neighbouring territories, apparently with impunity. See Bonds (above) for timing of independence, after which young gather into roaming flocks. (Tinbergen 1939.) ANTI-PREDATOR RESPONSES OF YOUNG. Fledged young flee into holes (etc.) to escape from Glaucous Gulls *Larus hyperboreus*, but evade foxes *Alopex* by direct flight (Rüppell 1970). PARENTAL ANTI-PREDATOR STRATEGIES. (1) Passive measures. Variability in tightness of sitting found by Nethersole-Thompson (1966) perhaps related partly to stage of incubation: ♀ easily flushed just after laying, but sat more tightly as incubation proceeded (Balfour 1976). However, R D Smith found attentiveness to clutch more dependent on individual ♀ than on stage of incubation. One ♀ became tame enough to approach observer and take crane-flies (Tipulidae) from his hand (Sutton 1932). Typical reaction to Merlin *Falco columbarius* or *F. peregrinus* overhead is to stay on ground and give 'weee' call (Tinbergen 1939: see 6a in Voice). (2) Active measures: against birds. One record of mobbing *F. peregrinus* (Koes 1989). (3) Active measures: against man and other animals. When man, dog, or fox approached nest or fledgling too closely, parents gave 'weee' call and song fragments (see 6 in Voice for other calls given in this context), and often fluttered in circles round head of intruder (Tinbergen 1939). In Spitsbergen, report of pair at nest attacking man (who was thus able to catch ♀ in his cap); ♂ will escort fox for a long way, calling constantly (Løvenskiold 1963). In Angmagssalik, no distraction-display seen when alarmed by man, but some ♀♀ in Cairngorms perform mobile distraction-display of disablement-and/or 'rodent-run' type: one ♀ seen spreading tail and fluttering from nest; others run off, head tucked in and plumage sleeked, 'looking remarkably like large mice'; scuttle thus under and over rocks, then double back. Rodent-runs frequently seen performed by ♀♀ in Iceland but never by ♂♂ (Nethersole-Thompson 1966). Mobile distraction-display implicit in description from Norway: flushed ♀ did not fly away but ran ahead, trying to lead intruder from nest, returning only once danger passed (Balfour 1976).

(Figs by D Quinn: A and D from drawings, C from photograph, in Tinbergen 1939; B from *Realms of the Russian Bear*, BBC film 1992.)   EKD

Voice. Wide repertoire, mostly confined to breeding season, but some calls (2a–b, 4) also used in winter. On breeding grounds, sometimes gives medley of different calls in short bout (Bergmann and Helb 1982, and from recordings). Voice well studied, notably in Cairngorms (Scotland) by Nethersole-Thompson (1966) and in Greenland by Tinbergen (1939) and Salomonsen (1950-1). Account includes renderings by E K Dunn from north-west Greenland. For additional sonagrams see Bergmann and Helb (1982).

CALLS OF ADULT. (1) Song of ♂. Brief musical warble, quite variable in quality and at times reminiscent of Dunnock *Prunella modularis* or Black Redstart *Phoenicurus ochruros*, though richer and less monotonous (Nicholson 1930); also somewhat reminiscent of Blue Rock Thrush *Monticola solitarius* (Bergmann and Helb 1982), and clear timbre suggests Rustic Bunting *E. rustica* (Bruun *et al.* 1986). Described as short rippling warble of distinct structure but rather varied, consisting of 9-14 units; sometimes repeated at intervals of 5-10 s, at other times (as in Songflight) delivered 3-4 times without pause; examples are 'ditrée-ditréedipitree-ditrée-divée' and 'vee-divéedi-divéedi-divée' (Salomonsen 1950-1, which see for renderings of other songs). Another song in north-east Greenland described as 'tzwee-tzwee-chu-we-tu-(wee)' (E K Dunn). Analyses by J Hall-Craggs of 11 recordings by 7

recordists in 5 countries as follows: phrases are short, average 2·1 s (1·25-2·7, $n = 14$). Unit/note delivery is slow (*c.* 4-8 per s) but some auxiliary notes not heard as discrete can be seen in all figures. Frequency range in Iceland and Greenland 2·0-7·5 kHz, while upper limit in Finland, Norway and Russia rarely exceeds 6·5 kHz. Notes of fairly steady pitch mingle with short ascending and long descending portamento tones, some with 'buzzing' quality due to frequency modulation or amplitude modulation or both. Temporal organisation within phrase is unusually regular with notes grouped into subphrases (A, B, etc.). Some of these bars (of music) are of almost constant duration (e.g. 0·6-0·62 s in Figs I-II, sung by same bird). Note that Fig II begins with 3rd note of Fig I and ends with 1st and 2nd notes of Fig I. This gives a more freely flowing Fig II. Bird then repeats both forms of phrases. Another unusual feature is that most songs comprise 3 subphrases where 3rd same as 1st (i.e. symmetrical ABA, BCB or ternary form). Although binary form (AB) is known in birds, ternary is rare. Of 14 phrases analysed, 7 (made up of 3 each from Iceland and Greenland, and 1 from Finland) are clearly (and audibly) ternary (Figs I-

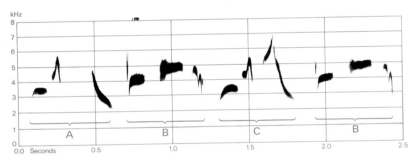

I P J Sellar Iceland June 1967

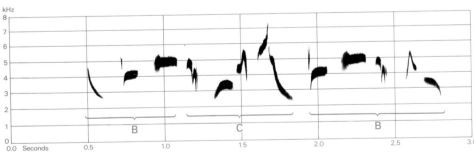

II P J Sellar Iceland
June 1967

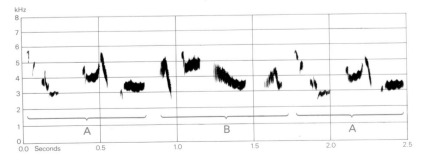

III P A D Hollom Iceland May 1991

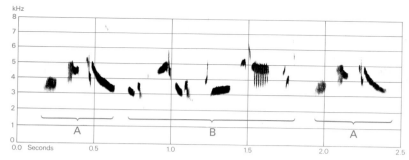

IV Roché (1970) Finland June 1968

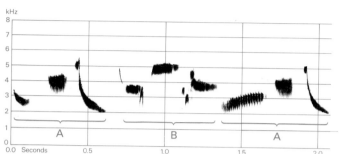

V P J Sellar Greenland July 1969

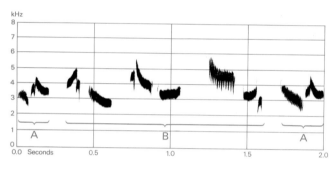

VI P J Sellar Greenland July 1969

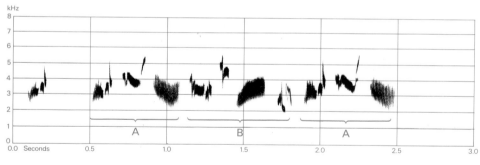

VII P J Sellar Greenland
July 1969

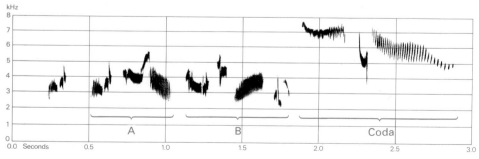

VIII P J Sellar Greenland
July 1969

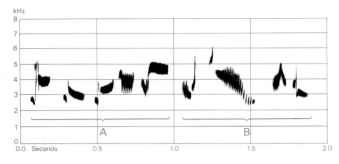

IX   B N Veprintsev and V V Leonovich   Russia   June 1977

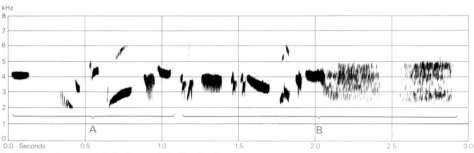

X   B N Veprintsev and V V Leonovich   Russia   June 1977

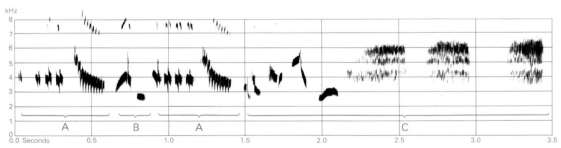

XI   S Palmér/Sveriges Radio (1972–80)   Norway   June 1963

VII). Fig VIII shows quiet, high frequency coda sung as an alternative to 2nd A-subphrase in Fig VII (same bird). The 2 Russian recordings (Figs IX–X) are binary in form; in Fig IX, moderately fast delivery of A contrasts with rather slow delivery of B, while in Fig X reverse is true, spirited short motifs in B following quite leisurely melody in A. Single Norwegian recording (Fig XI) is example of compound binary: ABAC where 2 A-subphrases are almost identical, but B is brief and C more than twice duration of B. Only Russian and Norwegian recordings have rasp- or skirr-like sounds appended (see Figs X–XI and call 4, below). (J Hall-Craggs.) Although each ♂ has several different phrase-types (Borror 1961; Bergmann and Helb 1982) local dialects are striking (Tinbergen 1939), so much so that Inuit were known to use *P. nivalis* dialects for route-finding in dense fog (Nethersole-Thompson 1966). At start of breeding season (birds still in flocks) song quiet and restrained; fragmentary song given when alarming at predator near young (Tinbergen 1939). (2)

Contact-alarm calls. (2a) Ripple-call. Melodious (rippling) 'pirrr' or 'pirrr-rit' (Salomonsen 1950–1), 'tir-r-ip' (Nethersole-Thompson 1966) or trembling 'pjrrr' (Tinbergen 1939). Pleasant bubbling sound (E K Dunn); Fig XII shows 3 ripples after introductory unit; middle units in each of Figs XIII–XIV show Ripple-call within series of different calls. Likened variously to call of Crested Tit *Parus cristatus* (Rosenberg 1953). Along with call 2b (below), Ripple-call is most commonly heard, used all year for contact between flock-members (commonly heard when taking off or in full flight), and between pair-members (Tinbergen 1939; Salomonsen 1950–1; Nethersole-Thompson 1966). Equivalent call of Lapland Bunting *Calcarius lapponicus* (see call 2a of that species) sounds harder (Kroymann 1967). (2b) Peee-call. Long mono-syllabic 'peee' (Tinbergen 1939), also rendered soft 'chee' or 'cheep' (Salomonsen 1950–1), soft 'twee' or 'tweet' (Nethersole-Thompson 1966). Same or similar sound, but not drawn-out, often immediately precedes call 6b (below),

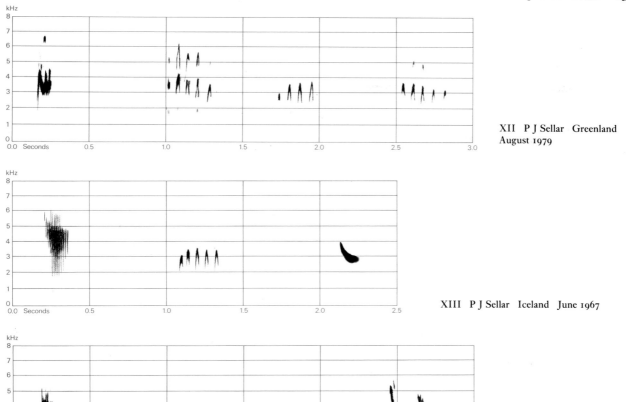

XII P J Sellar Greenland
August 1979

XIII P J Sellar Iceland June 1967

XIV P J Sellar Greenland
August 1979

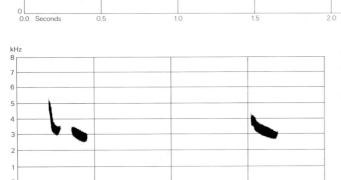

XV P J Sellar Greenland July 1969

e.g. 1st syllable ('pee-') of 'pee-tu' in Fig XV. Peee-call widely used for contact all year, often heard in flight or when about to fly, also sometimes between song-phrases (Tinbergen 1939; Nethersole-Thompson 1966). (3) Threat-call. High-pitched disyllabic 'pee-ee' (Nethersole-Thompson 1966: 'territory establishment call'), or (disyllabic) 'pEEE', roughly resembling 'sawing' song of Coal Tit *Parus ater*, and used by resident ♂

directing threat-display at rivals, including sometimes mate early in pair-formation (Tinbergen 1939). (4) Anger-call. Shrill trembling 'cherr' heard during fights with rivals (Tinbergen 1939); harsh 'skirr', also heard sometimes from ♂ when man approaches nest (Nethersole-Thompson 1966), and from 2 birds chasing in winter flock (Olioso 1972). This probably the call heard as drawn-out penetrating 'tsrr' from flocks landing in winter (Bergmann and Helb 1982). Rasping sounds at end of some songs (Figs X–XI) apparently of this type. Harsh 'chree' (Salomonsen 1950-1) perhaps also this call. (5) Courtship-calls. (5a) Inciting-call of ♂. Long high-pitched sound like call of swift *Apus*, but much softer, often given 2–3 times in rapid succession and accompanied by wing-trembling and panting; sometimes intersperses full song, also accompanies Advertising-display, e.g. after sexual chase (Tinbergen 1939: 'swift call'). Rendered 'sis-sis-sis' or harsh 'ziz-ziz-ziz', reminiscent of fledgling food-calls (see below), indicating ♂'s willingness to copulate with ♀ (Tinbergen 1939; Nethersole-Thompson 1966). (5b) Rapid series of shrill squeaking sounds reminiscent of Ptarmigan *Lagopus mutus* chick, different from call 5a, heard from ♂ flapping his wings rhythmically (Nethersole-Thompson 1966). (5c) Begging-call of ♀. Repeated 'skirr' very like call 4, given when fed by ♂ on or off nest (Nethersole-Thompson 1966). Also described as growling 'churr' very like food-call of well-grown young (Sutton and Parmelee 1954). (5d) Rather shrill 'see-sis-ip' from sexually receptive ♀ (in presence of mate) when approached by strange ♂ (Nethersole-Thompson 1966). (6) Alarm- and warning-calls. (6a) Monosyllabic soft 'weee' given on detecting approach of raptor, man, dog, or Arctic fox *Alopex lagopus* (Tinbergen 1939); soft low 'seep' or 'weep', indicating strong fear (Nethersole-Thompson 1966). (6b) Soft rather mournful 'tew' like Siskin *Carduelis spinus* given in mild alarm or anxiety, e.g. when separated from mate or flock (Nethersole-Thompson 1966). Sharp, slightly descending, fluty 'tiu' (e.g. last unit in Fig XIII, 2nd unit in Fig XV), heard from both sexes when young were approached (E K Dunn). Often preceded by call resembling 2b, thus 'twee-chew' (Nethersole-Thompson 1966), clear-toned 'didü' (Bergmann and Helb 1982), 'pi-tu' (Fig XV, 1st double unit); 'siu cheeulie' (complex unit at end of Fig XIV) perhaps related (J Hall-Craggs). Strident 'chi-chew chi-chu-t-t' (stuttering end), also 'chi-chu-chi...', etc., when observer approached fledglings (E K Dunn). (6c) A 'chis-ick' in mild alarm often given by unpaired ♂♂ on breeding grounds (Nethersole-Thompson 1966); incisive noisy 'chi-tik' (Fig XVI) and 'chut-ut' (Fig XVII, 1st double unit) (J Hall-Craggs). When observer approached juvenile flock, nearby ♂ first gave call 6b, then, on closer approach, staccato 'tre-trip' (E K Dunn). Recordings include various 'chizz'-type calls which may be related,

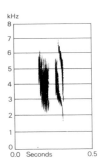

XVI P J Sellar Greenland July 1969

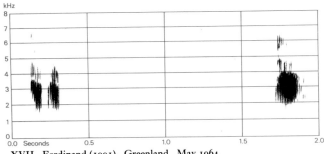

XVII Ferdinand (1991) Greenland May 1964

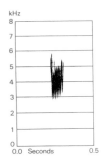

XVIII P J Sellar Greenland August 1979

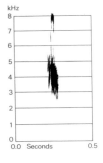

XIX P J Sellar Greenland July 1969

e.g. descending 'chuz' (Fig XIV, 1st unit), ascending 'chiz' (Fig XVIII); some are more rasping, e.g. Fig XIII (1st unit), Fig XVII (2nd unit) (J Hall-Craggs). (6d) Sharp 'ting' given by ♂ to warn mate under rock of human presence (Nethersole-Thompson 1966). (7) Other calls. A 'chüp' (Salomonsen 1950-1). Recordings of various chirps (recalling sparrow *Passer*) on breeding grounds may include this call, e.g. Fig XIX (from ♀), and 1st unit ('whik' or 'kwik') in Fig XII (J Hall-Craggs).

CALLS OF YOUNG. First food-call, audible at day 3-4, a soft very high-pitched 'peep' (Nethersole-Thompson 1966). As young grow, this develops into loud metallic chittering (Nicholson 1930) rendered 'chirr' or 'tchirrr' (E K Dunn); later audible at considerable distance, especially when it becomes continuous monotonous screaming as young are fed on nest; fledglings give metallic hissing 'ziz-ziz-ziz' as they receive food (Nethersole-Thompson 1966), also heard as excited 'tchrz' with wing-fluttering (E K Dunn). From day of fledging, food-calls are supplemented by locatory call described as short, shrill, far-carrying, monosyllabic squeak, repeated with gaps of c. 20-40 s (Tinbergen 1939), e.g. sharp 'pee(t)' (E K Dunn). However, this call also repeated at much faster rate, as following renderings indicate: 'pitt-pitt' (Salomonsen 1950-1), 'tyip-tyip-tyip' or 'zip-zip-zip' (Nethersole-Thompson 1966). Adult Ripple-call (2a, above) first heard from young 9 days after leaving nest (Tinbergen 1939), and other 'social and flock' calls soon follow (Nethersole-Thompson 1966). EKD

Breeding. SEASON. Cairngorms (Scotland): eggs laid first 3 weeks of June (2nd half of May to 2nd half of July)

(Nethersole-Thompson 1966, 1976; R D Smith). Spitsbergen: eggs laid 2nd half of June (late May to July) (Løvenskiold 1963). Kola peninsula (north-west Russia): laying starts 1st half of June (Semenov-Tyan-Shanski and Gilyazov 1991). Iceland: eggs laid from about mid-May into July (Timmermann 1938-49). For Greenland, see Nicholson (1930), Tinbergen (1939), Salomonsen (1950-1), and Asbirk and Franzmann (1978); for Canada, see (e.g.) Sutton and Parmelee (1954), Hussell (1972), and Knights and Walker (1989); for Wrangel Island (northeast Russia), see Portenko (1973). SITE. In cleft or crevice in scree, between rocks, in rock-face (including sea cliffs among auks Alcidae), under boulder on grass, etc; more rarely, under turf or in hole in (e.g.) river-bank; nest up to *c.* 1 m inside cavity and entrance inaccessible to, or too narrow for, predators; seldom far from vegetation (Timmermann 1938-49; Sutton and Parmelee 1954; Nethersole-Thompson 1966; Haftorn 1971; Balfour 1976). Sometimes, though rarely, open on ledge, etc. (Watson 1957a; Løvenskiold 1963). Commonly in Arctic settlements inside buildings, under roofs, in nest-boxes, native cairns, graves, etc., and recorded inside any suitable object, (e.g.) tin can, box, human corpse (Salomonsen 1950-1; Løvenskiold 1963; Vaughan 1992). On Wrangel Island and Chukotskiy peninsula, in driftwood on shore (Portenko 1973). Nest: foundation of grass, stalks, leaf stems, moss, and lichen, lined with fine grass, hair, and many feathers (Tinbergen 1939; Salomonsen 1950-1; Wynne-Edwards 1952; Løvenskiold 1963; Nethersole-Thompson 1966, which see for many details and comparisons). Size depends on cavity: outer diameter 10-20 cm, inner diameter 6·5-11 cm, depth of cup 1-6 cm (Timmermann 1938-49; Makatsch 1976). Building: by ♀ only, accompanied by ♂; ♀ probably excavates scrape where necessary; takes from less than 1 day to 4 days in northern Canada; may start 2-3 nests before choosing one; feathers can still be added to lining 3 days after 1st egg laid; occasionally, old nest re-lined (Tinbergen 1939; Sutton and Parmelee 1954; Watson 1957a). In Cairngorms, 7-11 days (n = 5) recorded between start of building and 1st egg; sometimes pulls grass from ground and will fly up to 700 m from nest to gather feathers; replacement nests built more quickly (Nethersole-Thompson 1966, which see for details of bouts and technique). EGGS. See Plate 33. Variable in size and markings; short to long sub-elliptical, smooth and slightly glossy; pale blue or greenish-blue, occasionally buffish, fairly evenly covered with reddish-brown to purplish-black spots and blotches, though can be concentrated at broad end; some have scrawls and violet-grey undermarkings (Timmermann 1938-49; Harrison 1975; Makatsch 1976). Nominate *nivalis*: 22·8 × 16·4 mm (20·6-24·8 × 15·1-17·7), n = 36; calculated weight 3·16 g (Makatsch 1976); 85 eggs from Scotland were 21·5 × 16·3 mm (19·5-25·2 × 15·3-17·6) (Nethersole-Thompson 1966, which see for review). *P. n. insulae*: 21·6 × 15·9 mm (18·7-24·8 × 14·6-16·9), n = 133; calculated weight 2·87 g

(Schönwetter 1984). *P. n. vlasowae*: 23·3 × 16·9 mm (21·2-24·8 × 16·0-17·0), n = 26 (Makatsch 1976). Clutch: 4-6 (3-8). In Cairngorms, of 51 clutches: 2 eggs, 2%; 3, 6%; 4, 18%; 5, 39%; 6, 35%; average 5·0; average late May to mid-June 4·6 (n = 18), mid-June to early July 5·6 (n = 15), early to mid-July 4·8 (n = 7); 1st clutches 4·8 (n = 33), replacements 5·6 (n = 15). Replacement nest built and eggs laid 4-6 days after loss of 1st clutch. (Nethersole-Thompson 1966, which see for comparison with other studies.) In another study, 7-10 days recorded between loss of clutch and start of replacement; no change in size over season and no variation with age of ♀; in 2 years, 48% of 25 pairs reared 2 broods; older ♀♀ more likely to have 2 broods than 1-year-olds (Smith 1991; R D Smith). See also Milsom and Watson (1984). On Spitsbergen, of 42 clutches: 3 eggs, 14%; 4, 14%; 5, 29%; 6, 36%; 7, 5%; 8, 2%; average 5·1 (Løvenskiold 1963); see also Balfour (1976). In Norway, of 39 clutches: 4 eggs, 3%; 5, 44%; 6, 39%; 7, 15%; average 5·6 (Haftorn 1971). Average in Greenland 5·18, n = 22 (Nicholson 1930; Asbirk and Franzmann 1978), on Baffin Island (northern Canada) 5·48, n = 25 (Sutton and Parmelee 1954; Watson 1957a), and elsewhere in Arctic Canada 5·47, n = 90 (Hussell 1972, which see for discussion of variation; Knights and Walker 1989). Eggs laid daily in early morning (Tinbergen 1939; Sutton and Parmelee 1954; Watson 1957a; Nethersole-Thompson 1966). INCUBATION. 12-13 days; by ♀ only (Sutton and Parmelee 1954; Watson 1957a; Nethersole-Thompson 1966). In Cairngorms, of 22 ♀♀, 16 started incubation proper before clutch complete (only 1 with 1st egg), 5 with last egg, and 1 after completion (Nethersole-Thompson 1966). According to Tinbergen (1939), starts 1-3 days after completion, but on Baffin Island usually with 1st egg and hatching over 2-4 days (Sutton and Parmelee 1954; Watson 1957a; Hussell and Holroyd 1974). For discussion of hatching asynchrony, see Hussell (1985). In Cairngorms, 3 ♀♀ had average stints on eggs of 16·5-68 min (6-91), and breaks of *c.* 12 min (6·5-22) (Nethersole-Thompson 1966); in northern Canada, average time off eggs 15·9 min per hr (5-30, n = 17) (Lyon and Montgomerie 1987); in north-east Greenland, 8·4 min (Asbirk and Franzmann 1978). YOUNG. Fed and cared for by both parents, routine varying considerably between nests (Nicholson 1930; Tinbergen 1939; Sutton and Parmelee 1954; Nethersole-Thompson 1966). FLEDGING TO MATURITY. Fledging period *c.* 12-14 days (10-17); older nestlings usually remain in cavity after leaving nest and all young leave cavity together, though commonly still unable to fly (Salomonsen 1950-1; Sutton and Parmelee 1954; Watson 1957a). At Barrow (Alaska), left nest at average age 10·5 days (9-13, n = 23) and cavity at 13·1 days (12-15, n = 43) (Maher 1964, which see for development of young and comparison with other seed-eaters). May even leave cavity at 8 days (Nethersole-Thompson 1966). Fed by parents for 8-12 days after fledging, though can catch flies 3-5 days after leaving cavity (Tinbergen 1939; Salomonsen 1950–1; Nethersole-Thompson 1966). For development of young, see also Hussell (1972). Age of first breeding 1 year (Smith 1991; R D Smith). BREEDING SUCCESS. On Ellesmere Island (northern Canada), of 131 eggs in 21 nests, 80·9% hatched and 74·8% produced fledged young (4·7 per nest overall); predation levels very low because of inaccessibility of nest cavity (Knights and Walker 1989). Similarly on Baffin Island, where of 72 eggs in 14 nests, 88·9% hatched and 80·6% produced fledged young, and no predation recorded (Sutton and Parmelee 1954); in another study, 84% of 44 eggs hatched and 61% produced young which left vicinity of nest (Watson 1957a). Elsewhere in northern Canada, only 28% of 61 nests suffered any predation (Lyon and Montgomerie 1987); see also Lyon and Montgomerie (1985) for success of manipulated broods. However, on Devon Island, many nests predated, young fledging from only 39% of 36 nests, although 91% of 253 eggs hatched; main predators were Arctic fox *Alopex lagopus* and stoat *Mustela erminea* (Hussell 1972; Hussell and Holroyd 1974). In north-east Greenland, average brood size (4·4, $n = 8$) was 83·0% of clutch, and number of fledged young (2·2, $n = 17$) was 41·5% (Asbirk and Franzmann 1978). In Cairngorms, 14 nests contained at least 64 eggs, of which 73% hatched and 55% produced fledged young, giving at least 2·5 per nest overall; predation hardly recorded but young killed by rain and sleet after leaving nest cavity (Nethersole-Thompson 1966). In another study, 25 pairs reared at least 28 broods of average 4·0 to fledging; 92% of 115 ♀♀ reared fledged young; at least 16% of 136 nests with eggs failed completely, and 13% of eggs in 118 nests infertile or addled. Most failures due to starvation of young, perhaps because of poorer food supply late in season; predation almost non-existent. Younger ♀♀ had lower hatching success, as did nests closer to paths; in late nests, brood size lower and rate of failure higher; however, disturbance by walkers (etc.) probably does not affect breeding success. Although clutch size smaller than further north, greater success and longer season can mean higher productivity in Scottish population. (Smith 1991; R D Smith.)   BH

Plumages. (nominate *nivalis*). ADULT MALE. In fresh plumage (August–February), small nasal bristles black; feathers of forehead and crown white with narrow grey bases; feather-tips of forehead and forecrown black, greyish-black, or orange, bordered subterminally by narrow orange-brown band; feathers of central crown to nape white with broad orange-brown tips; latter sometimes very narrow or absent on hindcrown. Supercilium broad, white, often partly obscured by narrow orange tips of feathers; front part often slightly or deep orange. Lore and eye-ring orange or white. Feathers of neck like hindcrown, but orange-brown tips narrower, giving impression of paler orange collar. Feathers of ear-coverts white, tips orange-brown or dark brown. Mantle, back, and upper rump black, feathers with broad yellowish or orange-brown fringes, latter occasionally with white subterminal band; fringes of back and upper rump broader; black feather-centres extend into point at shaft. Scapulars like feathers of mantle, but fringes usually more rufous-brown, black centres

round-ended or with blunt tip. Feathers of lower rump and shorter upper tail-coverts white, tipped broadly orange-brown or rusty-brown, occasionally a few feathers with narrow black centres; rarely, many feathers of rump black. Longest upper tail-coverts black with broad orange-brown or whitish-orange fringes. Underparts white, but feathers of chin, centre and side of chest, and flank with orange-brown tips, often forming gorget on upper chest. Tips of tail-feathers broad and rounded; central pairs (t1–t3) black with broad white fringes, black extending into blunt point in white of tip; outer pair (t6) white, *c*. 25% of tip of feather brownish-black, mainly on outer web; t5 similar to t6, but brownish-black sometimes narrower, extending along shaft towards base of feather and outer fringe buff or buff-white; t4 similar to t5, but tip of feather brownish-black on inner web also, or whole feather brownish-black with narrow white streak along shaft in centre of feather. Outer primaries (p3–p10) black, bases with increasing white towards innermost primary, mostly concealed by upper primary coverts on outermost primaries; p2 white, tip black, p1 fully white or with small subterminal black spot, or with black extending into narrow wedge or streak towards base for about half feather length; tips and fringes of p1–p7 off-white, fringes of p8–p9 off-white, tips sometimes inconspicuously grey. Secondaries white, sometimes with some grey or black on outer webs of s1–s3. Tertials black with broad rusty-brown, orange-brown, or earth-brown fringes and tips; longest tertial bordered white, basal two-thirds of outer web and basal half of inner web white, but white concealed by other tertials. Upper primary coverts white, outer ones sometimes with narrow black subterminal patch; black of tips sometimes extends along shaft towards base of feather; when black on tips large, some black also found on tips of shorter upper primary coverts. Greater upper wing-coverts white, but innermost black with broad white tips; occasionally, outermost greater coverts with some black or grey on outer webs. Median and lesser upper wing-coverts white. Bastard wing black, usually with narrow white fringe. Undersurface of primaries black, sharply demarcated from white of bases. Under wing-coverts and axillaries white; marginal coverts black with white fringes. During late winter and spring, fringes of feathers bleach to pale yellow or white. *In worn plumage* (from about May), brown, orange, buff, or white fringes completely worn off, feathers of mantle, scapulars, back, tips of primaries, tertials, bastard wing, and occasionally some feathers of central rump black, remainder of plumage white. ADULT FEMALE. In fresh plumage (August–February), feathers of forehead, crown, and nape grey, tipped dark grey-brown or orange-brown, often with narrow white subterminal mark; rarely, feathers of forehead and centre of crown with little grey at base and with broad white subterminal bar, resembling adult ♂. Feathers of neck similar to crown, but tips pale orange-buff, white subterminal bar often broader than on crown, forming inconspicuous pale collar. Supercilium white, sometimes with orange-brown mottling. Lore white or orange-brown, ear-coverts orange-brown. Feathers of mantle grey with pale orange-brown edge and often with yellowish fringe; grey centre of feathers extends into sharp point at shaft. Scapulars similar to feathers of mantle, but edges darker brown. Feathers of rump and shorter upper tail-coverts orange-brown with yellow-buff or whitish edges, centre of feathers broadly or narrowly grey, rarely completely black; feathers on side of rump white, sometimes with orange-brown tips; longer upper tail-coverts orange-brown, white, or white with orange-brown tips, partly with conspicuous grey centres. Underparts white, but side of breast orange-brown, often extending into gorget across upper chest; flank sometimes washed orange-brown, feathers occasionally with faint dark shaft-streaks. Tail as adult ♂. Primaries brown-grey with off-white fringes; off-white tips gradually wider from p7 inwards, as in adult ♂; bases of p1-p3 white, partly concealed by upper primary coverts. Secondaries white, s1-s3 with brown-grey subterminal spot, mainly on outer web. Tertials as fresh-plumaged adult ♂. Upper primary coverts and bastard wing grey-brown with narrow white edges, longest upper primary coverts white on bases. Greater upper wing-coverts grey, outer feathers broadly fringed white, mixed with some brown or orange-brown, inner ones fringed brown or orange-brown; median coverts brownish-black, brown, or buff-brown, tipped white or brown-white; lesser upper wing-coverts pale grey with broad white fringes. Undersurface of primaries grey, poorly demarcated from white bases; marginal under wing-coverts grey with broad white fringes, remainder of underwing white. *In worn plumage* (April-July), central crown, mantle, and scapulars black with many traces of white fringes; ear-coverts, hindneck, and side of neck mottled grey and white, head and neck less pure white than in adult ♂. For sexing, see also Rae and Marquiss (1989) and Svensson (1992). NESTLING. Down long and plentiful, dark grey, on upperparts only (Harrison 1975). JUVENILE. Upperparts buffish-grey or brownish-grey, tinged rufous, sparsely streaked black on head and neck, strongly streaked black on mantle and scapulars. Rump and upper tail-coverts rufous-brown with narrow black shaft-streaks. Chin, throat, chest, and flank light brown-grey with orange tinge, streaked black. Remainder of underparts off-white, with sparse greyish or brownish-black shaft-streaks. Tail and wing as 1st adult; median upper wing-coverts greyish-black, tipped white, lesser coverts greyish-black, edged greyish-white. (Witherby *et al.* 1938.) FIRST ADULT. Like adult, but juvenile flight-feathers, tail, greater upper primary coverts, and most or all greater upper wing-coverts retained. Tail often narrower and more pointed at tip than in adult, less broad and rounded, black on centre of feathers more sharply pointed (Svensson 1992), but difference sometimes slight. Secondaries and primary coverts on average more extensively dark on tip than in adult, but differences bridged by large individual and geographical variation in extent of black at all ages (Natorp 1931; Salomonsen 1931a, 1947a; Smith 1992); amount of black decreases with age in individual ♂♂ (also in older birds), but more or less constant in individual ♀♀ (Fokkema *et al.* 1978; J Jukema), and ageing on amount of black therefore doubtful; pattern and wear of tertials individually variable, no constant difference between age groups or races (contra Smith 1992).

Bare parts. ADULT, FIRST ADULT. Iris dark brown. Bill plumbeous-grey to black in summer, changing to pale yellow or orange-yellow with tiny dark tip during post-breeding moult, darkening again during March-May (Stresemann and Stresemann 1970; ZMA). Leg and foot dark grey or black. NESTLING, JUVENILE. Inside of bill deep red, gape-flanges yellow. When older, bill light brown with dark brown tip. Leg and foot light brown. (Witherby *et al.* 1938; Harrison 1975.)

Moults. ADULT POST-BREEDING. Complete, primaries descendent. In Greenland and Labrador, starts with p1 *c.* 10 July to *c.* 10 August, completed with regrowth of p9-p10 *c.* 15 August to *c.* 15 September, duration 36-37 days. Tail-feathers shed rapidly in centrifugal sequence, feathers growing simultaneously; starts at primary moult scores 10-20, completed with regrowth of p9-p10 (score 45-50). Secondary moult ascendent, s1 shed shortly after p6, s6 after p8; tertials moult in sequence s8-s9-s7, s8 shed shortly after p1, s9 after p3, s7 with p6-p7. Body starts with underparts at shedding of p3, head when p6 shed, all completed with regrowth of p9. On Iceland, moult approximately 20 July

to late August, but single ♀ with dependent young not yet started 22 August. (Stresemann and Stresemann 1970.) In eastern Greenland, some ♂♂ start when young 5-10 days old, ♀♀ generally not until young independent. Starts mid-July to early August, completed mid-August to early September; duration of primary moult *c.* 28 days, up to 6 feathers growing at same time in each wing, rendering birds almost flightless; duration of secondary moult *c.* 15 days, tertials *c.* 17 days, tail *c.* 23 days. (Manniche 1910; Green and Summers 1975; Asbirk and Franzmann 1978.) In Scotland, moult starts late July or early August (Ginn and Melville 1983). ADULT PRE-BREEDING. Verification needed for partial moult of head in spring, as recorded by Salomonsen (1949). No moult found in birds examined by Svensson (1992) and C S Roselaar. POST-JUVENILE. Partial: head, body, and lesser, median, and sometimes inner greater upper wing-coverts and tertials. Starts *c.* 3 weeks after fledging; 31 birds from eastern Greenland in heavy moult 12-15 August (Asbirk and Franzmann 1978); completed late August in Taymyr (Siberia) (Johansen 1944). 2 ♀♀ from Spitsbergen, 4 and 23 September, still in body moult (ZMA).

Measurements. ADULT, FIRST ADULT. Nominate *nivalis*. Greenland, Spitsbergen, and western Europe, all year; skins (ZMA). Bill (S) to skull, bill (N) to distal corner of nostril; exposed culmen on average 4·3 less than bill (S).

	♂			♀		
WING	110·6 (3·33; 24)	104-118		104·2 (1·90; 19)	100-107	
TAIL	63·8 (2·56; 24)	59-69		58·8 (2·39; 18)	56-63	
BILL (S)	14·5 (0·75; 22)	12·4-15·8		13·8 (0·79; 16)	11·9-14·9	
BILL (N)	8·1 (0·46; 24)	7·1-8·7		7·7 (0·30; 17)	7·2-8·3	
TARSUS	21·1 (0·88; 16)	20·0-22·7		20·9 (0·84; 14)	19·1-22·3	

Sex differences significant, except for tarsus.

P. n. insulae. Iceland and western Europe, all year; skins (ZMA).

	♂			♀		
WING	109·4 (1·87; 17)	107-114		103·3 (2·76; 26)	96-108	
TAIL	63·5 (2·63; 16)	58-69		56·8 (1·62; 24)	54-60	
BILL (S)	14·7 (0·39; 12)	13·9-15·2		14·3 (0·86; 26)	12·7-15·6	
BILL (N)	8·0 (0·24; 13)	7·8-8·5		7·9 (0·40; 26)	7·2-8·7	
TARSUS	21·3 (0·45; 10)	20·8-22·2		21·0 (0·67; 22)	19·6-22·0	

Sex differences significant for wing and tail.

Wing. Nominate *nivalis*. (1) Netherlands, winter (Jukema and Fokkema 1992). *P. n. insulae*. (2) Netherlands, winter (Jukema and Fokkema 1992). Nominate *nivalis* and *insulae* combined. (3) Scotland, winter (Banks *et al.* 1989). (4) Scotland, winter (Rae and Marquiss 1989). (5) England, winter (Smith 1992). *P. n. vlasowae*. (6) Northern Russia (Portenko 1973). (7) Anadyrland (Démentieff 1935). *P. n. townsendi*. (8) Komandorskiye Islands (Bering Sea), autumn (Portenko 1973).

	♂			♀		
(1)	110·1 (2·65; 48)	—		103·6 (2·91; 23)	—	
(2)	108·7 (2·43; 20)			102·3 (2·39; 25)	—	
(3)	110·8 (2·45; 175)	104-117		105·0 (2·40; 201)	98-110	
(4)	111·7 (1·90; 41)	107-115		105·8 (2·10; 148)	101-110	
(5)	112·4 (2·39; 678)	107-119		106·1 (2·28; 1395)	99-113	
(6)	110·8 (0·89; 27)	106-115		103·5 (0·72; 19)	100-107	
(7)	110·1 (1·69; 7)	108-112		102·7 (—; 3)	102-104	
(8)	117·2 (1·08; 8)	111-122		108·0 (0·44; 5)	106-110	

Wing. Lapland to Novaya Zemlya (nominate *nivalis*), ♂ 102-113, ♀ 100-109; Ob' to Taymyr (*vlasowae*), ♂ 103-114, ♀ 99-107; Indigirka to Anadyr' (*vlasowae*), ♂ 102-114, ♀ 102-110 (Démentieff 1935); Wrangel Island (north-east Siberia, *vlasowae*), ♂ 113·8 (2·14; 4) 112-117; Komandorskiye Islands (*townsendi*), ♂ 118·1 (1·55; 8) 115-120; Pribilof Islands (Bering Sea, *townsendi*), ♂ 118·1 (2·10; 4) 115-119·5; western Aleutian Islands (Bering sea, *townsendi*), ♂ 115·3 (2·49; 5) 113-119·5 (Vaurie 1956e).

For growth of young, see Maher (1964); for decrease in wing

length due to wear of primaries, see Portenko (1973); for seasonal variation in bill length, see Jukema (1992b). See also Salomonsen (1931a).

Weights. ADULT, FIRST ADULT. Nominate *nivalis*. (1) North-west Yukon (Canada) and Spitsbergen, April–May (Irving 1960; Korte 1972). *P. n. vlasowae*. (2) Russia (Dementiev and Gladkov 1954). (3) North-east Siberia (Portenko 1973). Nominate *nivalis* and *insulae*, combined. Scotland, winter: (4) (Banks *et al.* 1989), (5) (Rae and Marquiss 1989). Netherlands, winter: (6) condition fair to good, (7) exhausted birds (ZMA). Race unknown. (8) Germany, winter (Niethammer 1937).

(1)	♂	38·7 (3·52; 20)	34–45	♀	36·0 (2·16; 4)	34–39
(2)		39·8 (— ; 41)	33–50		34·6 (— ; 41)	28–42
(3)		44·5 (— ; 77)	34–53		42·6 (— ; 2)	41–44
(4)		34·0 (2·38; 522)	—		31·4 (1·95; 467)	—
(5)		34·4 (1·73; 41)	31–38		31·1 (1·68; 138)	28–38
(6)		34·9 (4·81; 12)	28–44		32·1 (5·61; 29)	26–43
(7)		21·2 (1·61; 3)	20–23		20·6 (2·33; 8)	18–23
(8)		34·0 (— ; 20)	28–42		30·9 (— ; 21)	27–38

Spitsbergen (nominate *nivalis*): May, ♀ 52; September, ♀ 36 (Korte 1972). Iceland (*insulae*): January, ♂ 39; May–June, ♂ 34·0 (3) 33–35, ♀ 33·4 (3) 31–35 (Timmermann 1938–49; ZMA). For daily variation in winter and pre-migration fattening in spring, see Banks *et al.* (1989). For daily variation in ♂, see Bentz (1990). See also Stewart (1937) and Johnston (1963).

NESTLING, JUVENILE. For growth, see Maher (1964).

Structure. Wing rather long, broad at base, tip pointed. 10 primaries: p9 longest, p8 0–1 shorter, p7 3–4, p6 9–15, p5 19–24, p4 27–33, p3 32–37, p2 36–42, p1 40–45; p10 strongly reduced, 75–80 shorter than p9, 13–19 shorter than longest upper primary covert. Outer web of p7–p8 clearly emarginated. Tip of longest tertial reaches tip of p2–p5. Tail rather short, tip slightly forked; 12 feathers, t4(–t5) longest, t1 3–5 shorter. Bill short, conical, culmen straight or slightly decurved, gonys slightly convex, depth at base in *insulae* and nominate *nivalis* 6·0 (50) 5·3–6·9, width 6·3 (50) 5·6–7·0; similar to bill of Yellowhammer *Emberiza citrinella*, but cutting edges only slightly toothed at base. Middle toe with claw 19·2 (20) 17·2–21·4; outer toe with claw *c.* 73% of middle with claw, inner *c.* 72%, hind *c.* 85%; hind claw 9·4 (19) 8·0–11·6.

Geographical variation. Based mainly on Salomonsen (1931a, 1947a, b) and Vaurie (1956e, 1959). Involves colour of fringes of upperparts in fresh plumage, amount of black on rump, flight-feathers, and upper wing-coverts, and size (wing or tail). *P. n. insulae* from Iceland is darkest race; black on wing and tail more extensive (see Banks *et al.* 1991, Jukema and Fokkema

1992, and Svensson 1992 for identification on extent of black); feather-tips of upperparts of fresh-plumaged ♂ darker and more extensively deep rusty-brown or earth-brown than those of ♂ nominate *nivalis* (in both races, gradually becoming paler with abrasion, and completely worn off when breeding), band across chest darker, more distinct; upper rump, centre of lower rump, and shorter upper tail-coverts black, concealed by rusty-brown when plumage fresh, but exposed in breeding plumage (lower rump and shorter upper tail-coverts of ♂ nominate *nivalis* largely white when worn); feather-fringes of fresh ♀ dark earth-brown or chestnut-brown, back, rump, upper tail-coverts, and upper wing-coverts extensively black on centres, much darker than in ♀ nominate *nivalis* when plumage worn; ♀ in worn plumage mottled and streaked grey-brown and sooty on upperparts and from side of head to chest, brown of cap and rump tinged rufous, lower mantle and scapulars more extensively black; nape, supercilium, rump, side of neck, and side of chest not as white as in worn ♀ nominate *nivalis*. *P. n. vlasowae* from Pechora basin (north European Russia) east to Wrangel Island and Anadyrland (eastern Siberia) paler than nominate *nivalis*; fresh fringes of upperparts pale yellow-brown to cream-white, less tawny-buff, much white showing through on rump of ♂ when plumage fresh, rusty on side of breast restricted and pale, no black on lower back, rump, and most upper tail-coverts, here all-white when worn, except for longest tail-coverts; on average, less white in wing and tail than in nominate *nivalis*. Birds occurring White Sea to Pechora are intermediate between nominate *nivalis* and *vlasowae*; also, part of population of nominate *nivalis* breeding Ellesmere Island (Canada), north-east Greenland, and Spitsbergen is similar to *vlasowae*, and the few *vlasowae*-like birds recorded Norway, Denmark, Netherlands, and perhaps Britain (Fjeldså 1976; Banks *et al.* 1991; British Ornithologists' Union 1992; Jukema 1992a) are likely to originate from high-arctic Atlantic localities rather than from Siberia, though this not likely for the more numerous *vlasowae*-like birds occurring Hungary and Rumania (Horváth and Hüttler 1963; Kováts 1973). *P. n. townsendi* from islands in Bering Sea similar to *vlasowae*, but larger, especially on Komandorskiye and Pribilof Islands, slightly smaller on coast of western Bering Sea and on Aleutian Islands, on latter grading into nominate *nivalis* in east, boundary of races hard to define (Hellmayr 1938); some birds from eastern Siberia and Komandorskiye, Pribilof, and St Lawrence Islands are very white, rather like palest race *hyperboreus* of Hall and St Matthew Islands (American Ornithologists' Union 1983; Pyle *et al.* 1987), which is slightly larger than nominate *nivalis* (particularly bill) but which has upperparts fully white in ♂ and black on wing and tail strongly reduced. *P. n. hyperboreus* sometimes considered a separate species, McKay's Bunting *P. hyperboreus* (see American Ornithologists' Union 1983). GOK

Emberiza spodocephala Black-faced Bunting

PLATES 15 and 16 (flight)
[between pages 256 and 257]

DU. Maskergors   FR. Bruant masqué   GE. Maskenammer
RU. Седоголовая овсянка   SP. Escribano enmascarado   SW. Gråhuvad sparv

Emberiza spodocephala Pallas, 1776

Polytypic. Nominate *spodocephala* (Pallas, 1776), from upper Ob' valley east to Sea of Okhotsk and Sea of Japan, south to Manchuria and North Korea; *sordida* Blyth, 1844, central and eastern China; *personata* Temminck, 1835, Kuril Islands, Sakhalin, and Japan.

Field characters. 13·5–15 cm; wing-span 20–23 cm. Slightly smaller and proportionately a little shorter tailed than north-western race of Reed Bunting *E. schoeniclus schoeniclus* but noticeably larger than Pallas's Reed Bunting *E. pallasi*. Rather small, quite slim bunting, with structure intermediate between *E. schoeniclus* and *E. pallasi* but having long, stout, conical bill. Adult ♂ has diagnostic black lores, greyish hood, and pale yellow underparts. Basic plumage colours and pattern of ♀ and immature recall *E. schoeniclus*, but, with face and sides of neck clouded grey, also suggest Dunnock *Prunella modularis*. Sexes dissimilar; marked seasonal variation in ♂. 1st-year birds difficult to separate from ♀. Siberian race, nominate *spodocephala*, described here.

ADULT MALE BREEDING. Moults: September (complete); November–March (mainly face and parts of forebody). Head, shawl, and deep breast olive-grey, looking dark in shadow; relieved only by bright pinkish to yellowish base to grey-black bill and black surround to bill (comprising lore, lower forehead, and point of chin and nearby speckles), producing depressed expression. Rest of upperparts and wing basically rufous-brown and underparts pale yellow to almost white on under tail-coverts. Mantle evenly streaked black (lacking paler stripes since feather margins uniformly coloured); lesser coverts grey, median coverts noticeably black-based and yellowish-buff-tipped (showing as black band and narrow pale wing-bar), greater coverts black-centred, rufous-fringed and pale yellowish-buff-tipped (presenting narrow pale buff lower wing-bar), and folded inner flight-feathers showing deep rufous panel. Rear flanks lightly streaked dark brown. Rump dull brown to greyish-green, mottled paler but not streaked. Tail dark brown, with full white panels on 2 outer pairs of feathers showing in flight. Legs brown-flesh. ADULT MALE NON-BREEDING. Lacks essentially tricoloured appearance of breeding bird, due to break-up of head pattern by distinct dark black-brown spots and streaks on crown and nape, dark speckling of cheeks and clear indications of pale submoustachial and dark malar stripes and pale yellowish-white chin and throat. Some have marks sufficiently strong to resemble ♀, showing face pattern recalling *E. schoeniclus*. Still shows fully grey 'scarf' round neck but sides of breast and flanks more heavily streaked on more rufous-buff ground. Bill-base always pinkish. ADULT FEMALE. When breeding, a few are like duller ♂♂, showing dusky, little-marked face and greyish, virtually unstreaked breast, but most lack that distinctive tricoloured appearance, having instead plumage that lacks any striking characters except for strong off-white submoustachial stripe between grey-brown ear-coverts and dark malar stripe. Suggests particularly duller, greyer *E. schoeniclus*. Diagnosis therefore requires lengthy, close observation, with concentration on (a) grey to olive-brown lesser coverts (eliminating *E. schoeniclus*), (b) grey-brown crown, (c) rather uniform pale brown to grey central ear-coverts (mottled rufous-brown in *E. schoeniclus*), (d) pat-tern of median and greater coverts (wing-bars insignificant in *E. schoeniclus*), and (e) inconspicuous supercilium (unlike *E. schoeniclus*). Rufous fringes to tertials and secondaries form bright wing-panel but this matched by many *E. schoeniclus* and *E. pallasi*; marked indentation of rusty outer fringes into black centres of tertials said to produce characteristic line of colour (Bradshaw 1992), but difficult to see, even in photographs. Ground-colour of underparts variable, dull white to cream-yellow: grey to buffy wash across breast, often sharply delimited on lower edge; streaking variable, usually much less heavy on lower breast than in *E. schoeniclus* but always distinct on flanks. Bill as ♂, with pale lower mandible. FIRST-WINTER MALE. Some closely resemble adult ♂, others show at least blackening of fore-face by late December. FIRST-SUMMER MALE. Normally resembles adult; may only be separable by retained worn tail-feathers but a few retain incomplete face markings.

Breeding ♂ unmistakable but ♀ and immature lack distinctive features; ♀♀ of south Siberian population of nominate *spodocephala* have characteristic dull, rather cold and greyish, copiously streaked appearance unlike any other sympatric bunting (D I M Wallace), but this difficult to convey other than by likening to *P. modularis*. Differentiation from *E. schoeniclus* and *E. pallasi* covered above; confusion with other *Emberiza* far less likely, common buntings of western Europe all being noticeably larger and other smaller vagrant relatives more boldly marked on head. Of latter, 1st-year Chestnut Bunting *E. rutila* most recalls *E. spodocephala* but lacks white outer tail-feathers and has usually unstreaked rufous rump and pale eye-ring (Bradshaw 1992). Flight and gait apparently much as other small *Emberiza*. Migrants in southern Siberia skulking, creeping about under bushes and thereby again recalling *P. modularis*. Frequently flicks tail open, like *E. schoeniclus* and *E. pallasi*, and may raise crown feathers in alarm. Sociable but rarely forming large parties, even on migration.

Commonest call quiet but sharp 'tzit', or quiet, slightly sibilant 'tsick' or 'tick', slightly thinner in tone than monosyllables of other buntings and often repeated.

Habitat. Breeding in east Palearctic in tall dense grass and shrubs, especially in river valley floodlands, in moist coniferous taiga forests and occasionally in mountain forest, which may be broadleaf, up to 600 m in Altai and 1500 m in Japan. Uses top or middle of a bush or side branch of low tree as song-post (Dementiev and Gladkov 1954). In Indian winter quarters, feeds on ground in rice stubbles or on moist edges of pools, usually resorting to cover near water (Ali and Ripley 1974).

Distribution. Breeds in central and eastern Asia, from western Altai mountains east to Sea of Okhotsk, Sakhalin, southern Kuril Islands, and Japan, south to northern Tibet, south-west China, eastern China, and central Honshu. Winters in Korea, eastern and southern China, eastern Nepal east to northern Indo-China, and central and southern Japan.

Accidental. Netherlands: 1st-winter ♂, Westenschouwen, 16 November 1986. Germany: Helgoland, 5 November 1910 and 23–26 May 1980. Finland: ♂, Dragsfjärd, 2 November 1981. (Alström and Colston 1991.)

Movements. Chiefly migratory. Northern race, nominate *spodocephala*, migrates through Mongolia, south-east Russia, north-east China and Korea to winter in southern Korea, eastern and southern China from Hopeh (few) south to extreme south (including Hainan), west to Kwangsi and Hunan, also in Taiwan (Vaurie 1959; Gore and Won 1971; Chang 1980; Schauensee 1984). Southern race *sordida* disperses widely between south-west and east, to winter from Bangladesh and eastern Nepal east through northern Burma to extreme north of Thailand (rare), northern Laos and northern Vietnam (Ali and Ripley 1974; King *et al.* 1975; Smythies 1986; Lekagul and Round 1991); also in Taiwan (Chang 1980), and in eastern and southern China (including Hainan), north mostly to *c.* 26°N, but to *c.* 32°N in east. Winter and summer ranges overlap slightly in southern China. (Schauensee 1984.) Thus winter ranges of nominate *spodocephala* and *sordida* overlap in China. Island race *personata* migratory in north of range, partially migratory in south. Winters in central and southern Japan; common in foothills and lowlands from northern Honshu south to Nansei islands; rare in Hokkaido (Brazil 1991); also winters on Ryukyu islands (Vaurie 1959), and reaches east coast of China (Schauensee 1984) and occasionally South Korea (Gore and Won 1971).

4 long-distance ringing recoveries: 2 birds ringed Korea, October 1964, recovered respectively 1040 km NNE in Ussuriland, October 1968, and 1760 km SSW in Taiwan, March 1966 (McClure 1974); bird ringed on Sakhalin island, September 1974, recovered 3 months later in southern Kyushu (Japan), and bird ringed Honshu (Japan), October 1967, recovered on Sakhalin, September 1970 (Ostapenko 1981).

In autumn, nominate *spodocephala* leaves central Siberia chiefly in September (Reymers 1966); further east also, in southern Ussuriland, departs mainly September, though sometimes considerable numbers still present in early October, and last stragglers mid-October (Panov 1973*a*; Polivanova and Polivanov 1977). Main passage September through Manchuria (north-east China) and Mongolia (Piechocki 1959; Mey 1988), September–October in Korea (where abundant both seasons) (Gore and Won 1971). Not known if ongoing route for birds passing through Korea to Taiwan (see ringing recovery above) is via southern Japanese islands or Chinese coast (McClure 1974). No information on autumn departure or spring return of *sordida*. Birds (nominate *spodocephala* or *sordida*) present on Taiwan October–May (Chang 1980); large numbers

winter; individuals ringed in roosting flocks made local movements up to 173 km (McClure 1974). In Hong Kong, the commonest wintering bunting, mainly early November to end of April (extremes 25 September to 19 May) (Chalmers 1986). In north-east India, recorded 24 October to 7 April; passage reported 19 December at *c.* 2700 m in Sikkim (Ali and Ripley 1974).

Spring migration prolonged. Begins March, with heavy passage in Hong Kong (Chalmers *et al.* 1991). In Hopeh (north-east China), passage throughout April and May, mainly end of April to 3rd week of May (Williams 1986); also April–May in Korea (Gore and Won 1971). Rare records of nominate *spodocephala* in Japan are chiefly in spring (Brazil 1991). In Ussuriland, vanguard from early to mid-April; main arrival and passage very concentrated and punctual, beginning regularly 19–23 April over 11 years; movement continues to end of April (Polivanova and Polivanov 1977). Further west, arrives later; in Chita region (south-east of Lake Baykal), recorded from 11 May in one year (Leontiev and Pavlov 1963); in Mongolia, movement 2nd half of May continuing into June (Piechocki and Bolod 1972; Mauersberger *et al.* 1982). Arrivals and passage mid-May to mid-June in southern taiga of central Siberia (Reymers 1966), and not present in Altai in extreme west of range until June (Johansen 1944; Ravkin 1973).

E. s. personata apparently migrates later in autumn, leaving Sakhalin and Kuril islands mainly in October (Gizenko 1955; Nechaev 1969). In Japan, departure from highland breeding grounds, and passage from further north, September–November. Return movement is from March; birds present on Hokkaido breeding grounds from mid-April. (Brazil 1991.) Arrival on Kuril Islands reported from end of April to beginning of June (Nechaev 1969; Il'yashenko *et al.* 1988). Reaches southern Sakhalin end of April or beginning of May, and northern Sakhalin 1½ or 2 weeks later (Gizenko 1955).

'Record' 1894 in Philippines was misidentified Yellow-breasted Bunting *E. aureola* (Dickinson *et al.* 1991). For vagrancy to west Palearctic, see Distribution.　DFV

Voice. See Field Characters.

Plumages. (nominate *spodocephala*). ADULT MALE BREEDING (January–August). Nasal bristles small, black. Feathers of head, neck, upper mantle, lower throat, and chest uniform greyish-green, feathers on throat, lore, and round base of bill black. Feathers of crown (especially at side), longest ear-coverts, chin, and lower throat sometimes with black tips; black spots on side of throat may form inconspicuous malar stripe. Eye-ring greyish-green with yellowish tinge. Feathers of central and lower mantle and scapulars black with narrow pale chestnut edge, broadly fringed buff; feathers on upper mantle intermediate in pattern and colour between neck and central mantle. Rump and upper tail-coverts brown, greyish-brown, or grey-green with slight olive tinge, feathers with pale edges; longest upper tail-coverts earth-brown. Belly and under tail-coverts pale yellowish-white in west of range, rich yellow in east; side of breast with buff hue, flank with a few long dark brown or grey-brown streaks. Central pair of tail-feathers (t1) earth-brown; t2–t4 dark brown, t4 usually with small white spot on tip, t5 blackish-brown with white wedge on inner web, covering about half of distal part; t6 mainly white with blackish-brown wedge at base of inner web and grey-brown spot on tip of outer web. Flight-feathers, upper primary coverts, and bastard wing grey-brown; outer edge of p9 and of distal part of p6–p8 white, that of basal part of p5–p8 yellowish-green; outer edges of remainder of primaries and of all secondaries rusty-brown; primary coverts and bastard wing with inconspicuous brown or green edges. Tertials black or grey-black, broadly fringed light rusty-brown on outer web, narrower on tip. Greater upper wing-coverts black, outer webs fringed pale brown with whitish or buff tips; median coverts like greater, but brown less extensive and with brown or buffish-white tips, latter extending to inner web; lesser coverts grey with broad greyish-green edges. Greater under wing-coverts white, median and lesser under wing-coverts and axillaries lemon-yellow. ADULT FEMALE BREEDING (January–August). Forehead, crown, nape, and neck greyish-green or greenish-olive, tips of feathers with brown shaft-streaks, most conspicuous on side of crown, forming faint lateral crown-stripe. Eye-ring yellowish-white. Supercilium pale yellow or buffish in front of eye, yellowish-white above and behind eye, usually rather indistinct, rarely broad and reaching beyond ear-coverts. Eye-stripe brown. Lore pale yellow or buffish; ear-coverts pale-brown, brown-grey, or greyish-green with yellowish tinge; small pale yellowish or pale brown spot on rear of ear-coverts; inconspicuous brown malar stripe, not reaching base of lower mandible. Moustachial stripe pale brown, rather indistinct or (rarely) almost absent; sub-moustachial stripe broad, off-white or yellowish. Rump uniform pale brown or earth-brown, sometimes with rusty-brown tinge, rarely with slight greenish tinge, not contrasting with mantle. Mantle, scapulars, upper tail-coverts, belly, under tail-coverts, tail, and wing like adult ♂ breeding. Chin greyish-yellow, upper chest spotted brown, forming faint gorget. Feathers on side of breast with brown shaft-streaks or slightly darker ill-defined spots. Flank sometimes more strongly streaked than in adult ♂ breeding. ADULT MALE NON-BREEDING (October–January). Like adult ♀ breeding, but feathers of crown and nape with black distal spot, each bordered rusty-brown, or feathers tipped rusty-brown. Ear-coverts and lore perhaps slightly darker than in adult ♀ breeding. Feathers of neck plain greyish-green or similar to feathers of crown. Chin and throat yellowish-white, faintly tipped greyish or brownish; feathers of lower throat and upper chest broadly edged and tipped greyish-green, forming ill-defined chest-band. Dark streaks on flanks obscured by greyish-white feather-edges. ADULT FEMALE NON-BREEDING (October–January). Like adult ♀ breeding, but feathers of crown edged brownish, especially on lateral crown-stripe. JUVENILE. Rather like ♀, but upperparts strongly rusty-brown or red-brown; cap brown, closely mottled and streaked dark brown on forehead and on each side of crown, mid-crown paler; supercilium rusty-brown, more distinct than in adult ♀, ear-coverts streaked. Underparts yellowish, throat and chest marked with black-brown arrow-marks, side of breast and flank tinged brown. (Hartert 1903–10.) FIRST ADULT NON-BREEDING. Like adult non-breeding, but part of juvenile feathers retained, contrasting in shape and abrasion with fresh neighbouring feathers; many birds have outer primaries new, contrasting somewhat with retained juvenile inner ones (all uniform in adult); in those with all primaries new, at least greater primary coverts and middle secondaries still old (in particular, tips of primary coverts distinctly more worn than those of outer primaries); in those with all flight-feathers old, part of juvenile tail and occasionally tertials or outer

greater coverts retained; when old, t4 without white on tip, in contrast to adult. For plumage details, see also Bradshaw (1992). FIRST ADULT BREEDING. Like adult breeding, but part of juvenile feathering retained, as in 1st adult non-breeding. See also Svensson (1992) for sexing and ageing.

Bare parts. ADULT, FIRST ADULT. Iris reddish-brown; in 1st autumn, grey-brown. Upper mandible black or grey-black with flesh or whitish-horn cutting edges, widening towards base of bill; lower mandible light horn, pink, or flesh with small dark grey tip. Leg and foot pale pink-brown or pinkish-flesh with paler soles; sometimes with some brown spotting. (Ali and Ripley 1974; Ree and Berg 1987; Bradshaw 1992; Svensson 1992.) JUVENILE. No information.

Moults. ADULT POST-BREEDING. Complete; primaries descendent. In Manchuria, 6 adults heavily worn but not yet moulting mid-July to early August, another in wing moult early September (Piechocki 1959). Moult completed early September (Dementiev and Gladkov 1954). ADULT PRE-BREEDING. Partial: feathering at bill-base, lore, supercilium, part of ear-coverts, chin, throat, and chest; moult in ♀ probably more restricted than in ♂ or entirely absent. Some start from late November, generally completed by March. (Bradshaw 1992; RMNH.) POST-JUVENILE. Partial or perhaps occasionally complete. Starts late July to late September, completed early September to early October, sometimes mid-October. Involves head, neck, lesser, median, and often all greater upper wing-coverts, usually all tertials, usually t1, and frequently all tail-feathers. In Japan, many birds replace variable number of primaries, starting descendently from inner or central outwards; of 481 birds, 5% retained all juvenile primaries, 5% replaced all 9 functional primaries (but apparently not all secondaries); 2% replaced outer 6 primaries, 40% outer 5, 28% outer 4, 4–9% outer 1–3 (Ishimoto 1992). FIRST PRE-BREEDING. Like adult pre-breeding, but sometimes less extensive, some ♂♂ retaining much non-breeding plumage in summer.

Measurements. ADULT, FIRST ADULT. Nominate *spodocephala*. Whole geographical range, all year; skins (RMNH, ZMA). Bill (S) to skull, bill (N) to distal corner of nostril; exposed culmen on average 3·5 less than bill (S).

	♂		♀	
WING	72·3 (1·43; 27)	70–75	68·3 (1·97; 19)	66–72
TAIL	57·8 (1·81; 25)	55–61	56·1 (2·01; 16)	54–60
BILL (S)	14·0 (0·65; 23)	13·0–15·0	13·9 (0·69; 17)	12·4–15·1
BILL (N)	7·9 (0·39; 23)	7·3–8·5	7·8 (0·75; 16)	7·1–9·1
TARSUS	19·4 (0·80; 18)	18·9–20·1	19·1 (0·91; 12)	18·3–20·9

Sex differences significant for wing and tail.

Wing. Nominate *spodocephala*. (1) Mongolia and western Manchuria (Piechocki 1959; Piechocki and Bolod 1972). (2) Central and eastern Manchuria ('*extremi-orientis*') (Meise 1934a; Piechocki 1959). *E. s. sordida*. (3) Kansu (central China) (Vaurie 1972; see also Stresemann *et al.* 1937). *E. s. personata*. Live birds, Japan: (4) adult, (5) 1st adult (Dornberger 1983); (6) adult, (7) 1st adult (Ishimoto 1992, which see for other measurements).

	♂			♀		
(1)		71·0 (— ; 11)	69–77		68·1 (— ; 4)	66–72
(2)		70·0 (— ; 37)	66–73		67·3 (— ; 10)	65–71
(3)		74·5 (— ; 11)	70–78		72·0 (— ; 2)	71–73
(4)		74·2 (1·09; 33)	71·5–76·0		70·0 (1·30; 12)	68·0–72·5
(5)		72·8 (1·45; 41)	68·5–75·5		69·2 (1·26; 33)	66·5–71·0
(6)		71·5 (1·19; 11)	68·2–74·4		67·7 (1·32; 49)	65·0–71·4
(7)		69·9 (1·31; 299)	65·1–73·4		66·6 (1·48; 236)	62·5–72·3

Weights. Nominate *spodocephala*. (1) Manchuria (China), mid-July to early September (Piechocki 1959). *E. s. personata*. Japan, late September to early November: (2) adult, (3) first adult (Dornberger 1983; Ishimoto 1992).

(1)	♂	18·2 (— ; 16)	15–23	♀ 16·7 (— ; 11)	15–18
(2)		20·5 (1·42; 44)	16·2–23·9	19·7 (0·91; 15)	17·9–22·0
(3)		20·2 (1·62; 63)	14·7–26·9	19·1 (2·30; 58)	17·7–21·4

Nominate *spodocephala*. Mongolia, May: ♂ 19 (Piechocki and Bolod 1972). Amur region, eastern Siberia: ♂ 17–18, ♀ 15·4–17·1 (3) (Dementiev and Gladkov 1954). Taiwan, January–February: ♂ 18·5 (2·78; 5) 16·0–22·5 (RMNH).

E. s. sordida. Kansu (China), September: ♂ 19·9 (3) 19·2–20·5 (Stresemann *et al.* 1937).

E. s. personata. Sakhalin (eastern Siberia): ♂ 21·6 (5) 19–23 (Dementiev and Gladkov 1954).

Structure. Wing rather short, broad at base, tip bluntly pointed. 10 primaries: p7 longest, p8 0–1 shorter, p9 2–4, p6 0–2, p5 2–6, p4 7–8, p3 9–11, p2 11–14, p1 13–17; p10 strongly reduced, 47–52 shorter than p7, 7–12 shorter than longest upper primary covert. Outer web of p5–p8 emarginated. Tip of longest tertial reaches tip of p1–p5. Tail of moderate length, tip slightly forked; 12 feathers, t3(–t4) longest, t1 2–5 shorter. Bill conical, 5·8 (10) 4·9–6·3 mm deep at base, 5·7 (10) 4·8–6·3 wide (nominate *spodocephala*). Middle toe with claw 22·2 (9) 20·0–24·4; outer toe with claw *c.* 68% of middle with claw, inner *c.* 67%, hind *c.* 69%; hind claw 7·8 (18) 7·0–9·0. Remainder of structure as in Yellowhammer *E. citrinella*.

Geographical variation. Involves wing length, depth and width of bill, colour of face-mask and throat in breeding plumage, and amount of white in tail (Vaurie 1956*e*). 3 races recognized by Vaurie (1959). In nominate *spodocephala*, saturation of yellow on underparts and of grey-green on head varies clinally from Ob' river in west to Sea of Okhotsk in east: westernmost populations (west of Angara) sometimes separated from nominate *spodocephala* from Transbaykalia and Yakutiya as *oligoxantha* Meise, 1932, with colour generally lighter, little grey-green on head, and little yellow on underparts, and with faint streaks on flanks; easternmost populations, from Great Khingan through Manchuria and Amurland to Sea of Japan, sometimes separated as *extremi-orientis* Shulpin, 1928, being darker, with much grey-green on head and rich yellow underparts. *E. s. sordida* (synonym: *melanops* Blyth, 1845) from central and eastern China is a rather dark and bright race; yellow on underparts intense, streaks on flank brighter and stronger, and black face-mask more extensive than in nominate *spodocephala*, bill larger, wing slightly longer. *E. s. personata* from Kuril Islands, Sakhalin, and Japan brightest and darkest of all races, with strong greenish tinge on head and neck; lore and chin black, throat, chest, and belly bright yellow (no grey-green on central throat), streaks on flank poorly developed, and outer tail-feathers with less white; ♀ without black on lore and chin. Similar in size to *sordida*, but bill depth at feathering 6·2 (11) 5·2–6·5, width 6·2 (11) 5·6–6·7.

Recognition. See Bradshaw (1992) and Svensson (1992).

GOK

Emberiza leucocephalos **Pine Bunting**

PLATES 11 and 16 (flight)
[between pages 256 and 257]

Du. Witkopgors   FR. Bruant à calotte blanche   GE. Fichtenammer
Ru. Белошапочная овсянка   Sp. Escribano cabeciblanco   Sw. Tallsparv

Emberiza leucocephalos S G Gmelin, 1771

Polytypic. Nominate *leucocephalos* S G Gmelin, 1771, east European Russia and Siberia from Perm' east to *c.* 67°E. Extralimital: *fronto* Stresemann, 1930, China from Bogdo-Ula mountains (Sinkiang) to north-east Tsinghai and Kansu.

Field characters. 16·5 cm; wing-span 25–30 cm. Slightly larger than Yellowhammer *E. citrinella*, with tail usually 5–10% longer; slightly larger and more robust than Rock Bunting *E. cia*. Eastern counterpart and close relative of *E. citrinella*. Unlike *E. citrinella*, adult plumage of ♂ and ♀ strikingly different but both show white ground-colour to underparts, long rufous rump, and bright white outer tail-feathers. Breeding ♂ has striking white central crown and cheeks contrasting with bold black and chestnut stripes on face and chestnut throat; white underparts, interrupted by chestnut-spotted chest-band and flanks; similar but much duller in fresh plumage. ♀ duller and patterned more like *E. citrinella* but no trace of yellow, with dull white ground-colour to plumage most obvious in pale head and throat markings, tips to median coverts, and belly. Sexes dissimilar; marked seasonal variation in ♂. Juvenile separable. Hybridizes freely with *E. citrinella*

erythrogenys in western Siberia, with vagrant offspring reaching Britain.

ADULT MALE. Moults: August–September (complete); February–March (part of face). Fresh plumage has noticeable pale tips and fringes initially obscuring head and forepart pattern; strong contrasts of breeding bird's plumage are due to wear. Head shows remarkably painted appearance: long white crown, flecked brown and black; black forehead and circlet round crown; broad reddish-chestnut stripe through eye, with dull white supra-loral streak and mark under eye only visible at close range; white ear-coverts, edged black at rear and forming bold patch vividly contrasting with eye-stripe; reddish-chestnut bib and surround to cheeks. Malar area shows black mottling; variably whitish patch may show between face and uniform greyish-brown rear neck. Mantle and scapulars greyish- to buff-brown, with black streaks in similar pattern to *E. citrinella*, but chestnut rump shows many buff tips, often looking paler than in *E. citrinella*. Wings show similar basic colours to back, but fringes and tips of feathers have characteristic pale appearance, with greyish-brown lesser coverts, whitish fringes to tips of median coverts, pale buff fringes to chestnut-tipped greater coverts, and (particularly) white outer fringes to outer 3–4 primaries all creating stronger markings than in *E. citrinella*. When fully worn and bleached, pale covert-tips form even more obvious bars. Tail brown-black, with whitish edges to outermost feathers as well as bright white ends on 2 outer pairs. Underparts pure white, with partly black but mostly chestnut mottling and spots forming obvious chest-band and flank marks; effect of pattern is to give bird white collar below bib and strikingly pale belly and vent below dark chest and flanks. Underwing shows cream or white axillaries, contrasting with greyish undersurface of flight-feathers. Bill brown, with dark culmen and lower mandible; legs yellowish-brown. A few aberrant birds have throat white. ADULT FEMALE. Much duller than ♂ but basic head pattern of worn bird similar and still distinctive, though differing in pale buff lore, throat, and supercilium, last only mottled chestnut. More subtly variegated than *E. citrinella*, with white in centre of crown (occasionally showing when feathers raised), dusky lateral crown-stripes, rear eye-stripe, and lower border to obvious buff-white cheeks, sharp dusky malar streaks, (on some) almost white patch between throat and breast (echoing ♂'s half-collar), and pale greyish-buff nape. At close range, typical bird shows sharper, duller streaks on lateral crown-stripes and darker malar and chest streaking than *E. citrinella*. Rest of upperparts, wings, and tail as ♂ but underparts duller white, with brown streaks on breast typically almost black in centre, unlike *E. citrinella*. JUVEN-ILE. Closely resembles fresh-plumaged ♀, showing little indication of adult head pattern; rump duller and streaked black, upperparts tawnier, breast and flanks streaked more heavily with black, and ground-colour of underparts buffish-white.

Certain identification bedevilled by (1) general similarity to *E. citrinella*, (2) existence of dilute morphs of *E. citrinella*, (3) hybridization and intergradation of all characters with *E. citrinella* (U Olsson), and (4) convergent appearance of Rustic Bunting *E. rustica*, Meadow Bunting *E. cioides*, and *E. cia*, though simultaneous occurrence of *E. cia* with *E. leucocephalos* unlikely. Adult ♂ unmistakable once precise head pattern confirmed but certain separation of ♀ and immature requires close, detailed observation. *E. rustica* shares chestnut-marked silky-white underparts with *E. leucocephalos* but confusion unlikely to persist since *E. rustica* over 10% smaller, with proportionately shorter tail, different head and throat pattern, and distinctive call (see that species). Separation from *E. cia* best based on underpart tone: almost uniform buff to pale rufous from throat to belly on *E. cia*, always whitish on throat and white on belly in *E. leucocephalos*. *E. cia*

also shows unstreaked throat, dark eye-stripe, more rufous back and (particularly) rump, and narrow tail. For *E. cioides*, see Plates 15–16. Elimination of dilute *E. citrinella* and hybrids highly problematic, with identification failures persisting in England even with photographed adults (Lansdown and Charlton 1990) and apparent immature *E. leucocephalos* showing at close range tell-tale traces of yellow on belly, as in case of 2 October vagrants (D I M Wallace). Best chance of certain distinction is with adult ♀ in worn plumage, showing diagnostic whitish flecks in mid-crown, chestnut-tinged supercilium, grey-white rear ear-coverts and throat patch, pale chestnut spots on sides of breast and along flanks, and generally paler, buffer appearance, even on rump. Separation most difficult with darker, more streaked immature; crucial to confirm pale whitish tips to median coverts and lack of any yellow below; if visible, lesser coverts show as rather greyish-brown patch, paler than on *E. citrinella*. Important also to note that (1) outer fringes of all flight-feathers and greater coverts are paler in *E. leucocephalos* than *E. citrinella*, those on outer primaries particularly being white, not greenish-yellow, and (2) axillaries of *E. leucocephalos* are white, not yellow. Flight, gait, stance, and behaviour as *E. citrinella*, with no differences in actions seen between birds of the 2 species, apparently paired, in south-central Siberia (D I M Wallace). Strongly migratory, unlike *E. citrinella*; increasingly occurring as vagrant to west Palearctic, in seasonal pattern similar to Little Bunting *E. pusilla* and freely associating with other *Emberiza* and finches (Fringillidae).

Voice similar to *E. citrinella* (see Voice), but song tends to have slightly fewer introductory 'tzi' units or motifs before terminal 'teeee'; call of wintering bird in Israel reported as sharper.

Habitat. Breeds in Asia, overlapping in range and partly in ecological niche with Yellowhammer *E. citrinella*. Lacks subarctic element to match that of *E. citrinella*, but is predominantly boreal and cool temperate in breeding distribution, and is accordingly a much more pronounced migrant (Harrison 1982). Winters commonly in flocks on foothills and plains of India and Pakistan, up to 1500 m, occasionally to nearly 2700 m; here it feeds on ground, perching in trees, on bush-covered grassy slopes, and on stubble and fallow fields (Ali and Ripley 1974).

Within breeding range, avoids treeless steppe or grassland, nesting in margins of sparse forest in wet valleys, depressions, floodplains, and even ravines overgrown with shrubs and trees (Shkatulova 1962). Prefers thin woods, forest clearings, or glades, favouring overgrown bushes, but sings also from treetops (Flint *et al.* 1984). Sings incessantly from tops of birches *Betula* (Dresser 1871–81). Common up to around 1300 m, and occasionally up to 2000 m in USSR. Favours well-lit forests of conifers, or in some regions birches and other deciduous trees, but avoids riverain deciduous woods, as well as mountain

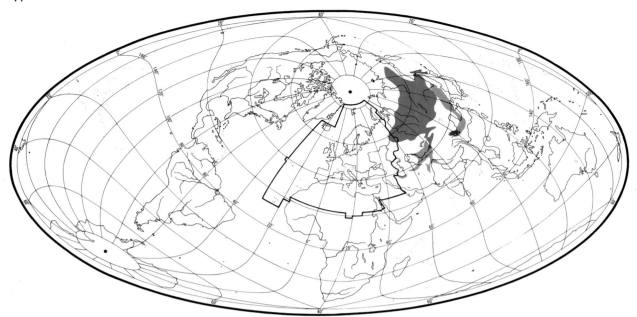

taiga. Will tolerate steppes if grassy, with clumps of trees. (Dementiev and Gladkov 1954.) In Mongolia, occurs on roads and tracks on mountain slopes where no thick forest; in winter, sometimes feeds in yards in villages (Kozlova 1933). Commonly near rivers or other water (Witherby *et al.* 1938).

Distribution and population. No changes reported.

Accidental. Iceland, Britain, France, Belgium, Netherlands, Denmark, Norway, Sweden, Finland, Germany, Switzerland, Austria, Czechoslovakia, Hungary, Yugoslavia, Greece, Bulgaria, Spain, Italy, Malta.

Movements. Migratory, birds moving chiefly south to winter in southern and central Asia. Winter range overlaps slightly with breeding range. In zone of sympatry with Yellowhammer *E. citrinella* (western Siberia), more migratory than that species and makes longer movements.

In winter, fairly common in foothills and at lower levels in south and south-east Kazakhstan, but numbers vary greatly (Korelov *et al.* 1974); regular in southern and eastern Turkmeniya (Murgab basin and Kopet-Dag) (Rustamov 1958), and also winters in south-east Uzbekistan (Stepanyan 1970) and Tadzhikistan; in periods of heavy snowfall birds descend to valleys, moving up to 20–30 km (Abdusalyamov 1977). Fairly widespread in Afghanistan, chiefly in east and north-west (S C Madge). In Iran, recorded in north and south-west; probably more widespread, but much overlooked (D A Scott); occasionally recorded from Iraq (Vaurie 1959). Locally common winter visitor to Pakistan and north-west India, from North West Frontier Province south to Quetta and east through Punjab plains, Gilgit, Kashmir, and Himalayan

foothills to west-central Nepal, *c.* 84°E (Ali and Ripley 1974). Regularly winters in Mongolia, though many local birds depart (Kozlova 1933; Piechocki *et al.* 1982; Mey 1988); winter population presumably partly involves birds from further north. In central Siberia, some birds remain in Angara river area at *c.* 57°30′N in mild winters (Reymers 1966). In China, breeds in north, and winters in north-east, west, and central areas (Schauensee 1984); winter visitor to Beijing area (J Palfery). Rare on passage and in winter in Korea (Gore and Won 1971). Some birds on northern Sakhalin island move south to southern Sakhalin (where few breed), and in harsh winters migrate further south (Gizenko 1955). In Japan, previously occasional winter visitor, but now annual in very small numbers, almost exclusively to Hegura island off western Honshu, with a few records from Hokkaido and northern Honshu (Brazil 1991). In Israel, birds of unknown provenance winter regularly in small numbers on Mount Hermon (mostly at 1300–1700 m), in eastern Galilee and Jerusalem hills; first recorded 1971 (Shirihai in press). Apparently also reaches north-east Italy regularly (see below).

Autumn migration August–November; in sympatric zone, begins earlier than in *E. citrinella* (Panov 1973*b*). In Yakutiya in north-east of range, movement throughout September (Vorobiev 1963). In western Siberia, dispersal begins July, with movement to larger river valleys; migration begins 2nd half of August, and most birds leave north-east Altai then, with diminishing numbers in 1st half of September when main migration of *E. citrinella* begins (Panov 1973*b*; Ravkin 1973, 1984). In Kazakhstan, migration is chiefly in east, mid-September to October; birds usually reach Talasskiy Alatau several days before *E. citrinella* (Korelov *et al.* 1974). Large numbers migrate

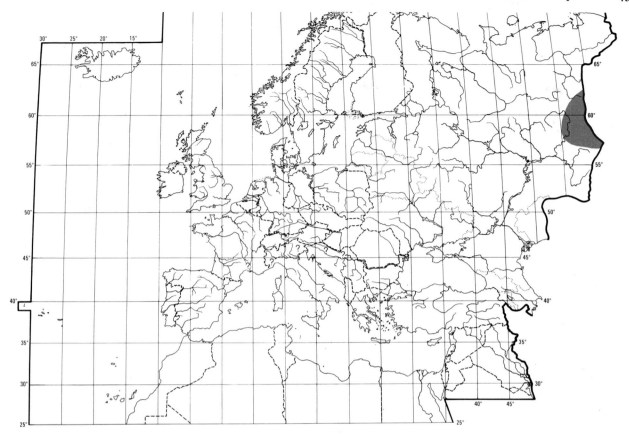

through Chokpak pass (western Tien Shan); birds appear in last third of September, with main passage 11–31 October; adults start arriving slightly before juveniles, but median date similar; ♂♂ start arriving slightly before ♀♀, but their median date is slightly later (Gavrilov and Gistsov 1985). Main arrival in Tadzhikistan mid-November (Abdusalyamov 1977), and present in Afghanistan, India, and Pakistan from November (Ali and Ripley 1974; S C Madge). In upper Indus valley (Ladakh, north-west Himalayas), in one year, earliest bird 31 October, with slight increase to 9 birds on 19 November, when trapping ceased (Delany *et al.* 1982). Passage in Mongolia reported September to early October (Piechocki and Bolod 1972; Mauersberger *et al.* 1982). Occasionally recorded in Ussuriland (far eastern Russia) October–November (Panov 1973*a*), and present in Japan from early October (Brazil 1991), in Hopeh (north-east China) from October (Hemmingsen and Guildal 1968). Winter visitors reach Israel gradually from early November, chiefly mid-November to mid-December; arrive a few days later than *E. citrinella* (Shirihai in press).

Spring migration early and rapid, chiefly March–April. In Israel, birds depart more or less simultaneously in 2nd and 3rd weeks of March (Shirihai in press). Recorded in Iran to late March (D A Scott), in Afghanistan until April (S C Madge), and leaves India and north-east China March–April (Hemmingsen and Guildal 1968; Ali and Ripley 1974). Main departure from Tadzhikistan in 2nd half of March, though sometimes begins February in Gissar valley (Ivanov 1969; Abdusalyamov 1977). In Kazakhstan, passage is on broader front spring than autumn; fairly large numbers migrate through foothills of Tien Shan and Dzhungarskiy Alatau in east, and small numbers through Chu and Syr-Dar'ya valleys and central Kazakhstan; movement in southern Kazakhstan mid-February or early March to early April, in northern Kazakhstan late March to April (Korelov *et al.* 1974). In Chokpak pass, migration is chiefly along wooded mountain slopes, with fewer in open areas; average earliest record over 8 years 4 March (1–9 March), with main passage 6–25 March and last records in first 2 weeks of April; ♂♂ migrate a few days earlier than ♀♀ (Gavrilov and Gistsov 1985). Marked passage at Ulan-Bator (Mongolia) at end of March (Baumgart 1978), and recorded in Ussuriland March–April (Panov 1973*a*). Date of arrival on breeding grounds varies from year to year, dependent on weather and snow cover; coincides with thaw on south-facing slopes. Reaches southern Siberia end of March or early April (in some localities slightly later, in others slightly earlier, than *E. citrinella*) (Shkatulova 1962; Reymers 1966; Panov

1973*b*); present in Yakutiya from 3rd week of April (Vorobiev 1963).

Widespread vagrant west of normal range, south-west to Iberia and north-west to Iceland (see also Distribution). Occurs mostly in autumn, and sightings in winter and spring probably chiefly involve birds which arrived in previous autumn (Niethammer and Thiede 1962). Regular in north-east Italy (Moltoni and Brichetti 1978; Brichetti and Massa 1984), and probably overlooked there (Moltoni 1951), so perhaps status is of winter visitor rather than vagrant. Of 124 published records up to 1960 in Europe, 105 (*c.* 85%) in Italy, of which only 1 in south; old records from southern Poland and Vienna area (Austria) suggest birds may move west along northern edge of Carpathians and then south along March valley to reach northern Italy; ♂♂ and ♀♀ occur there in similar numbers (Niethammer and Thiede 1962). Elsewhere in Europe, highest numbers in Britain, Low Countries, France and Yugoslavia (Alström and Colston 1991). In Britain, 19 records up to 1990, of which 17 since 1967, chiefly in Orkney and Shetland (11), with 4 in east (Humberside to Sutherland), and 4 in south (Surrey, Dorset, Isles of Scilly); most are in October–November (13), others in August (1), January (2), and April (3) (Dymond *et al.* 1989; Rogers *et al.* 1991). Hybrid with *E. citrinella* recorded Suffolk, April 1982 (Lansdown and Charlton 1990). In Netherlands, 20 records since 1900, of which 18 since 1960, all in October–November (Berg *et al.* 1992). In France, 12 records since 1900, mid-September to February, chiefly November–January; 8 in 1958-64, and only 1 subsequently; 9 were from Bouches-du-Rhône in extreme south (presumably arriving via northern Italy), and lack of more recent records there may be due to cessation of winter trapping (Dubois and Yésou 1992). In Fenno-Scandia, recorded only in recent decades, in Sweden from 1959, Norway 1966, and Finland 1968 (Haftorn 1971; Lewington *et al.* 1991). First Swiss record was in 1991 (*Br. Birds* 1992, **85**, 15).

North-central Chinese race *fronto* apparently resident (Vaurie 1959; Schauensee 1984).   DFV

Food. Almost all information extralimital. Diet seeds and other plant material; insects in breeding season. Forages primarily on ground and in low bushes; on breeding grounds, feeds at forest edge or in large clearings; in winter quarters, where specializes on cereal grains, searches for food in flocks, often with other seed-eaters, on arable fields (bare soil or stubble), waste ground, in orchards, villages, parks, by roads and tracks, etc.; often near water and swampy places. In north-east China, recorded in winter foraging in poplar *Populus* and willow *Salix* trees. (Briggs and Osmaston 1928; Dementiev and Gladkov 1954; Niethammer and Thiede 1962; Cheng 1964; Reymers 1966; Roberts 1992; Shirihai in press; J Palfery.)

Diet includes the following. Invertebrates: adult damsel flies and dragonflies (Odonata), adult grasshoppers, etc. (Orthoptera), bugs (Hemiptera: Cicadidae), larval Lep-

idoptera, adult flies (Diptera), adult Hymenoptera, adult and larval beetles (Coleoptera: Curculionidae), spiders (Araneae), snails (Pulmonata). Plants: seeds and other parts of Polygonaceae, rose *Rosa*, Leguminosae, grasses (Gramineae, including *Elymus*, oats *Avena*, wheat *Triticum*, barley *Hordeum*, rye *Secale*, millet *Panicum*, rice *Oryza*). (Dementiev and Gladkov 1954; Piechocki 1959; Sollenberg 1959; Yanushevich *et al.* 1960; Pek and Fedyanina 1961; Cheng 1964; Salikhbaev and Bogdanov 1967; Ali and Ripley 1974; Tucker and Tucker 1978; Germogenov 1982.)

In Kirgiziya, autumn–spring, 72% of 46 stomachs contained grains of wheat, oats, and barley, 26% seeds of herbs, mainly Polygonaceae and Leguminosae, and 2% seeds of rose. Stomachs in late spring and summer held adult and larval insects and snails. (Yanushevich *et al.* 1960.) In another study in same area, almost half of winter diet wheat grains and remainder other cereals, including rye, and wild grasses; spring stomachs contained Orthoptera, Hemiptera, and beetles (Pek and Fedyanina 1961). In Uzbekistan, November–March, of 14 stomachs, 7 contained wheat, 3 millet, and 1 seeds of herbs; 13 stomachs, autumn and spring, each held 2-50 seeds, and 9 of them also contained insects (Salikhbaev and Bogdanov 1967). In Chita region (south-east of Lake Baykal, Russia), April–July, 3 stomachs contained oats and wheat, caterpillars, and adult insects (Dementiev and Gladkov 1954). In northern China, late summer, many stomachs contained small Orthoptera (Piechocki 1959); 75% of food over year was seeds of grasses and cereals; in summer, takes insects and spiders (Cheng 1964). In winter quarters in Pakistan and northern India very fond of rice grains (Ali and Ripley 1974).

Young fed mostly insects and some seeds. In Lena valley (eastern Russia), unknown number of collar-samples contained 106 insects, of which 48% by number adult Orthoptera, 16% adult Hymenoptera, 14% caterpillars, 12% adult Odonata, 7% larval beetles, 3% adult Diptera (Germogenov 1982). In north-east China, young given Orthoptera, Hemiptera, Diptera, and adult and larval beetles (Cheng 1964). In Kirgiziya, small Orthoptera fed to nestlings (Yanushevich *et al.* 1960). Cicadas (Cicadidae), as well as cereals, also recorded in diet of young (Witherby *et al.* 1938; Sollenberg 1959).   BH

Social pattern and behaviour. Limited information, though relationship with Yellowhammer *E. citrinella*, with which it hybridizes regularly (see Bonds, below), well studied.

1. Gregarious outside breeding season. Small flocks from late July–August (Dementiev and Gladkov 1954). Migrates usually in small parties (Reymers 1966). In Israel, first arrivals in autumn are singletons or small groups which associate in larger flocks (up to 40 together) on wintering sites (Shirihai in press). Postbreeding flocks typically mix with other ground-feeding granivores, notably *E. citrinella* (Dementiev and Gladkov 1954; Kovshar' 1966; Shirihai in press). In Pakistan, associates variously with Black-throated Accentor *Prunella atrogularis*, Rock Sparrow *Petronia petronia*, Linnet *Carduelis cannabina*, Red-

mantled Rosefinch *Carpodacus rhodochlamys*, Rock Bunting *E. cia*, and White-capped Bunting *E. stewarti* (Roberts 1992). In Pamir-Alay (south-central Asia) usually associated with Bramblings *Fringilla montifringilla* (Ivanov 1969). In Beijing (China), flocks of up to 50, sometimes including *F. montifringilla* and Rustic Buntings *E. rustica* (J Palfery). Bonds. No evidence for other than monogamous mating system. No information on duration of pair-bond. ♀ alone builds nest (probably) and alone incubates (Dementiev and Gladkov 1954). Shkatulova (1962) presents contradictory information on role of sexes in brooding and feeding young, but apparently only ♀ broods and one or (more likely) both sexes feed young. No information on age at independence or at first breeding. Hybridization with *E. citrinella* common in sympatric zone, and widely studied (Radzhabli *et al.* 1970; Panov 1973*b*, 1989; Khakhlov 1991). F1-hybrids are fertile, leading to varying degrees of departure from pure *E. leucocephalos* or *E. citrinella*. In sympatric zone, 2·5% of total population are F1-hybrids; *c.* 15% of all *E. leucocephalos* and *c.* 20% of *E. citrinella* show intermediate features indicative of varying degrees of mixed heredity (Panov 1973*b*, 1989, which see for discussion of isolating mechanisms; see also 1 in Voice). Breeding Dispersion. Solitary and territorial. In southern taiga (central Siberia), in pine *Pinus* forest not far from fields and meadows, minimum territory 0·25-0·5 ha, more often 1 pair per 4-6 ha in burned pine forest, not more than 1 pair per 9-10 ha in wet (marshy) burned forest (Reymers 1966). Densities found by Ravkin (1984) in western Siberia support preference for more open (i.e. burned or logged) dry pine forest: 50 birds per km² in logged-out pine forest retaining undergrowth; 18 in not very dense pure pine forest, 2-6 in typical (denser) pine forest, 1-2 in bogs. In north-east Altai, 5-7 birds per km² in mixed pine and birch *Betula*, up to 10 in montane fields (Ravkin 1973). In review by Panov (1973*b*) highest density recorded anywhere was 93 birds per km² in mixture of valley forest, meadows and willow *Salix* thickets of Irtysh river, compared with 19 in pure valley forest (i.e. more densely wooded) in same region; in Tomsk region, 28 birds per km² in fields, 3 in mixed forest and bogs, 2 in valley forest (Panov 1973*b*, which see for other densities and comparison with sympatric *E. citrinella*). Roosting. During breeding, ♀ roosts on nest, ♂ in tree nearby (Shkatulova 1962). In winter, Israel, roosts communally among pines and cypresses *Cupressus* (Paz 1987), mixed with *E. citrinella* (Shirihai in press); in January, *E. leucocephalos* and *E. citrinella* once found roosting together (numbers unknown) in same tree as 4 Goshawks *Accipiter gentilis* (Anon 1986).

2. Little information on approachability. In Beijing, winter, typically wary when disturbed, flying into tree-tops or well away from immediate area (J Palfery). April vagrant to Dorset (southern England) tail-flicked constantly and called softly from perch on fence (Walbridge 1978). Somewhat more lively and agile than *E. citrinella* (Johansen 1944). Flock Behaviour. No information. See introduction to part 1 (above) for flock associates. Song-display. Very like *E. citrinella* (see 1 in Voice). Readily sings from top or side of shrub or tree. In central Siberia, ♂♂ start singing from arrival on breeding grounds, i.e. from 1st half of April, later further north; song begins to decline by beginning of May but resurgence at end of May, presumably associated with 2nd clutches (Dementiev and Gladkov 1954; Reymers 1966). In Baraba (western Siberia), intense singing heard in 1st 10 days of June (Pukinski 1969). End of song-period in central-southern Siberia typically mid-July (Dementiev and Gladkov 1954; Shkatulova 1962). Sings more or less throughout day, slightly more in morning (Podarueva 1979). Antagonistic Behaviour. At start of breeding season, ♂♂ sing energetically and fight with rivals; pursue one another in the air, on ground,

and in dense shrubs (Dementiev and Gladkov 1954). Territorial disputes with neighbours are rare, and only occur at high density and early in season. Threat-postures of ♂♂ are not very obvious, but sometimes include Advertising-display (see Heterosexual Behaviour, below). During pair-formation and nest-building, ♀ very aggressive; if ♂ approaches, she performs Forward-threat posture; in mild display, feathers are ruffled on crown and back (Fig A); in intense display, ♀ crouches lower, sleeks plumage on head, and gapes (Fig B); tries to peck ♂ and later chases him

(Panov 1973*b*). In study of captive ♂ *E. leucocephalos* and ♀ *E. citrinella*, after ♂ began singing (start of May), he was often involved in quite serious fights through aviary wire with ♂♂ *E. citrinella* attracted to ♀ (Löhrl 1967). Heterosexual Behaviour. (1) General. Most ♂♂ arrive on breeding grounds somewhat earlier than ♀♀; pairs form during spring migration (Dementiev and Gladkov 1954), as well as on breeding grounds. (2) Pair-bonding behaviour. Courtship-display of ♂ similar to *E. citrinella* (Panov 1973*b*, on which following description based): ♂ first performs Advertising-display (Fig C, left) in which plumage markedly ruffled on head and throat, and tail flicked and flirted, while drooped wings are gently flicked outwards (occasionally also quivered, though this more common in *E. citrinella*); Advertising-display interspersed variously with bill-wiping, collection in bill of nest-material (for display purposes only, Stem-display: Fig C, right), and whirring flight (noisy wingbeats) after which ♂ always landed with wings briefly raised (Fig D);

D

wing-raising indicates high motivation (perhaps also element of threat), and seen several times in captivity, but no observations in the wild; may be followed by horizontal posture in which plumage markedly ruffled and bird turns head (Fig E). For

E

aggressive responses by ♀ to ♂-display, see Antagonistic Behaviour (above). (3) Courtship-feeding. None. (4) Mating. In most stereotyped posture of ♂ ('pre-copulation posture': Panov 1973b) body feathers are raised except on head which is sleeked (Fig F). ♀ performs Soliciting-display with body horizontal, wing-quivering, head forward or raised (Panov 1973b: Fig G). (5)

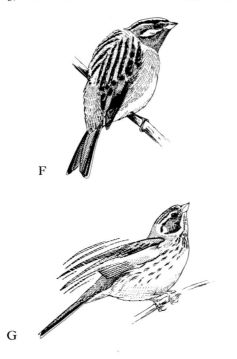

F

G

Nest-site selection and behaviour at nest. No information. RELATIONS WITHIN FAMILY GROUP. Little known. Eyes of young begin to open at 4 days, fully open 6–7 days. In last 2 days

before leaving nest, young are not brooded at night, and only slightly by day. After fledging, do not stay long in breeding territory. (Shkatulova 1962.) ANTI-PREDATOR RESPONSES OF YOUNG. Departure from nest at 9 days (Shkatulova 1962) suggests premature fledging (see Breeding). No further information. PARENTAL ANTI-PREDATOR STRATEGIES. (1) Passive measures. No information. (2) Active measures: against man. In Chita region (south-east of Lake Baykal, Russia), when nest examined, ♂ always tried to lead observers away with mobile distraction-lure display of disablement type: hopped along, trailing one wing, giving no sound. When this was of no avail, ♂ began to fly, calling loudly, and hopping 0·5–1 m from nest; only became passive when observers left nest. (Shkatulova 1962.) Distraction-display also reported in Altai when pair had young in nest (Dementiev and Gladkov 1954). No further information.

(Figs by D Nurney from drawings in Panov 1973b.)   EKD

Voice. Freely used throughout year. Song well studied (because of similarity to Yellowhammer *E. citrinella*, with which *E. leucocephalos* regularly hybridizes). Calls similar to those of *E. citrinella* (L Svensson). Main study (whole repertoire) by Panov (1973b, 1989; see Panov 1973b for detailed quantitative comparison with *E. citrinella*); also quantitative comparison of song by Wallschläger (1983). See these sources for additional sonagrams. Audible wing noise heard during flights which intersperse courtship-display (Panov 1973b). Account includes notes by J Palfery on wintering birds in Beijing (China).

CALLS OF ADULTS. (1) Song. Very like both *E. citrinella* and *E. leucocephalos* × *E. citrinella* hybrids (e.g. Löhrl 1967), and doubtfully separable from them (L Svensson). Similarly consists of rapid introductory series (with marked crescendo) of high-pitched tinkling buzzy units or motifs (e.g. 'tzee'), typically followed by 1–2 units of differing structure which may be considerably drawn out (W T C Seale), e.g. whole song 'sri-sri-sri-sri-sri-zyyh' (Sollenberg 1959). Panov (1973b) compared 79 phrases by 7 ♂♂ of *E. leucocephalos* with 47 phrases by 4 ♂♂ of *E. citrinella*; only major difference was that on average there were fewer introductory ('tzee') units/motifs in song of *E. leucocephalos* (7·9) than in *E. citrinella* (10·9). This difference confirmed by Wallschläger (1983), which see for further discussion, including evaluation of other differences found by Panov (1973b). Another possible difference is that in none of available recordings of *E. leucocephalos* do units in main/initial part of song have strong starting transients, whereas these are common in *E. citrinella* recordings (W T C Seale). Some songs of the 2 species indistinguishable to human ear (Löhrl 1967; Boswall 1970; Panov 1973b) but individual variation within each species is considerable: during short period, ♂ *E. leucocephalos* may use 3–4 song-types, giving 10–15 of one type before switching to another (Panov 1973b, 1989). If enough song is compared, slight structural differences between the species emerge, some apparent only from sonagrams, others audible. In study by Wallschläger (1983) based on 38 phrases from 2 ♂♂ in Irkutsk (central Siberia) and other recordings, *E. leucocephalos* sings at least 3 different song-types; for 1st apparently unusual type (not further discussed here) see Löhrl (1967); 2 other types (found also in *E. citrinella*) are defined by differences between end of phrase; in some songs (type-A), ending is simple extended 'ty' (pronounced 'teee'); in others (type-B) 'ty' is preceded by short, quiet, high-pitched (*c.* 8 kHz)

'zi', thus 'zity'; of 2 ♂♂ from Irkutsk, one sang both types, other only 'ty' type. Recording by K Mild of songs from Irkutsk bird similarly includes type-A (Fig I: 'dzu dzu. . . dzeeeee') and type-B (Fig II: 'd-zi d-zi. . . see dzeeeee', and Fig IV: 'dzu dzu. . . see dzeeeee'); these type-B songs (especially Fig IV) look superficially similar to Fig I

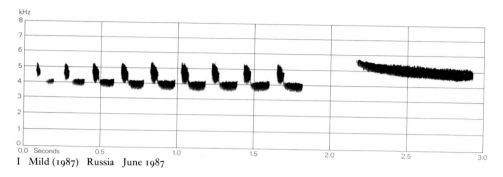

I   Mild (1987)   Russia   June 1987

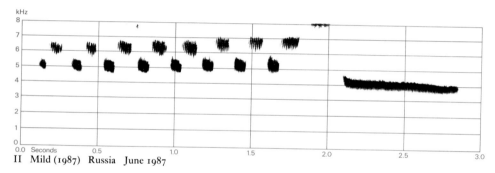

II   Mild (1987)   Russia   June 1987

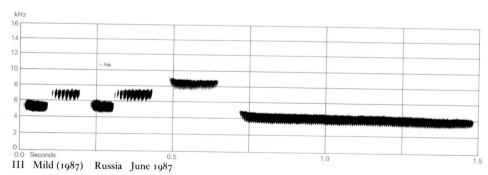

III   Mild (1987)   Russia   June 1987

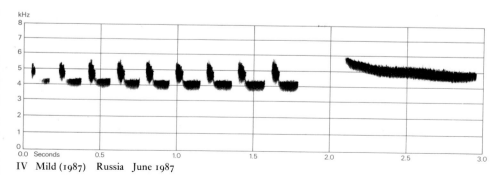

IV   Mild (1987)   Russia   June 1987

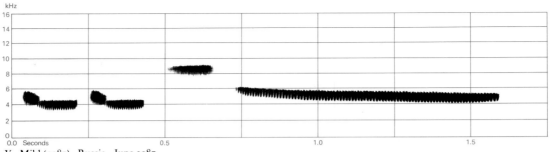

V   Mild (1987)   Russia   June 1987

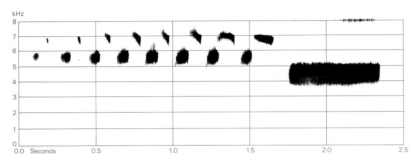

VI   B N Veprintsev and V V Leonovich
Russia   June 1975

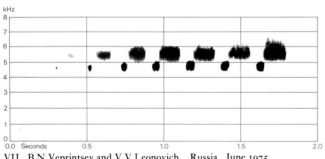

VII   B N Veprintsev and V V Leonovich   Russia   June 1975

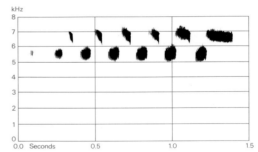

VIII   B N Veprintsev and V V Leonovich   Russia   June 1987

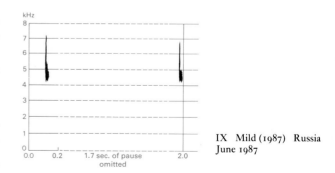

IX   Mild (1987)   Russia
June 1987

because penultimate high-pitched unit runs off top of scale, but these become visible in half-speed versions of song-endings (Figs III, V). Fig VI is effectively hybrid in structure between type-A and type-B in that there is no penultimate unit of particularly high pitch, yet unit which occupies this position differs somewhat in structure from those preceding it; final unit in Fig VI is particularly nasal 'dzzzeee'. Fig VII shows 3rd song-type (C), consisting only of compound 'tzee' units and lacking any distinctive ending; Fig VIII a variant of this in which, however, token ending is provided by slight lengthening (etc.) of final unit. (W T C Seale.) See Calls of Young (below) for juvenile song. (2) Contact-alarm calls. (2a) A 'tsik' (Panov 1973b) given both when perched and in flight, commonly when first flushed. Also rendered a metallic 'tick' (Walbridge 1978) or 'tic' (Fig IX). In Pakistan, thin 'twik' easily distinguished from more metallic, 'less rounded' call of Rock Bunting E. cia (Roberts 1992). Variant is quiet muffled 'tsik' given by ♂ E. leucocephalos during

Advertising-display (Panov 1973b, 1989). (2b) Short hoarse whistle rendered 'tseu' is used as often as call 2a for contact, also when alarmed (Panov 1973b, 1989). In recordings (Fig X, also 1st unit of XI), buzzy descending 'zeeu' (W T C Seale). This apparently the call heard from Swedish vagrant between song-phrases (Sollenberg 1959). (2c) 'Trilling'-call. Similar to E. citrinella (presumably

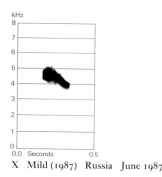

X   Mild (1987) Russia June 1987

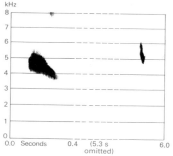

XI   B N Veprintsev and V V Leonovich
Russia June 1975

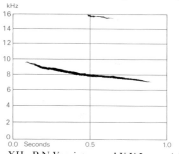

XII   B N Veprintsev and V V Leonovich
Russia June 1975

flight-call of that species) but distinguishable by usually being shorter; longer calls indicate greater excitement. Given by both sexes in all sorts of social interactions, notably during pair-formation. (Panov 1973*b*, 1989.) A 'tititic' (like rapid series of call 2a) or 'churtitik' (J Palfery) is presumably this call. (3) Soliciting-call of ♂. Series of quiet squeaking 'peeping' sounds given by ♂ just before copulation (Panov 1973*b*, 1989). (4) Threat-call. Squeaky 'trill' comprising long series of short units; same call in *E. citrinella* slightly higher pitched, more incisive, and drier (Panov 1973*b*, 1989). (5) Alarm-calls. See call 2a. Also gives high-pitched descending (from *c.* 10 kHz to *c.* 7 kHz) drawn-out whistle 'see'; no details of context (Panov 1973*b*, 1989, which see for sonagram) but perhaps homologous to 'zieh' given by *E. citrinella* when raptor overhead (Bergmann and Helb 1982). Following calls perhaps in this category: plaintive 'seeeoo' or 'seeeuw'; also 'tit seeoo' (apparently combination with call 2a) and reedier 'swee-ee' or 'sree' (J Palfery); in recording (Fig XII, half speed, narrow band), high-pitched, penetrating, descending portamento 'seeeoo' (W T C Seale). (6) Other calls. (6a) In flight, 'swit' or sometimes 'switit' like wagtail *Motacilla* (J Palfery). Soft 'tillip', given in flight by April vagrant to southern England (Walbridge 1978) perhaps same or related. (6b) In recording (Fig XI, 2nd unit), very brief, very quiet 'jp', not easily reconciled with other calls listed above (W T C Seale).

CALLS OF YOUNG. No details of food-calls. Song of 1st-year much more variable and complex than adult, with successive units markedly different from one another (Panov 1973*b*, which see for sonagram).   EKD

Breeding. All information extralimital. SEASON. Western Siberia: eggs laid from beginning of May; exceptional clutches at end of June are perhaps 2nd broods (Johansen 1944; Pukinski 1969). Kazakhstan: eggs laid from early May; nest-building, eggs, and fledged young all seen mid-June, some young still in nest end of July (Korelov *et al.* 1974). Central Siberia: eggs laid about mid-May; 2nd broods fledge towards late July (Reymers 1966). Eastern Siberia: eggs laid from late April (Panov 1973*a*, which see for review). Chita region (south-east of Lake Baykal,

Russia): eggs laid mid-May to early July (Shkatulova 1962). See also Vorobiev (1963) and Ravkin (1973). SITE. In depression on ground, under bush, grass tussock, fallen branch or tree, etc. (Dementiev and Gladkov 1954; Shkatulova 1962; Cheng 1964; Korelov *et al.* 1974). Nest: very like that of Yellowhammer *E. citrinella*; bulky foundation of tightly woven stalks, rootlets, and dry grass, lined with soft grasses and very often with horsehair (Shkatulova 1962; Vorobiev 1963; Korelov *et al.* 1974; Makatsch 1976). 10 nests in Chita region had average outer diameter 10·5 cm (7·0–13·0), inner diameter 6·2 cm (5·6–7·0), depth of cup 4·7 cm (3·2–6·0) (Shkatulova 1962). Overall height of one nest 5 cm (Taczanowski 1873). See also Cheng (1964). Building: by ♀ only (Cheng 1964; Korelov *et al.* 1974). EGGS. See Plate 33. Sub-elliptical, smooth and slightly glossy; very like *E. citrinella*; very pale whitish blue/green to pinkish or grey with faint purplish and lavender-grey spots and blotches plus a few brownish-black hairstreaks (Witherby *et al.* 1938; Harrison 1975; Makatsch 1976). Nominate *leucocephalos*: 21·5 × 16·2 mm (19·0–23·5 × 14·2–17·3), *n* = 79; calculated weight 2·96g (Schönwetter 1984). 15 eggs from Chita region had average dimensions 23·7 × 17·6 mm (Shkatulova 1962). Clutch: 4–5 (3–6); in Kazakhstan, mostly 4–5 (Korelov *et al.* 1974); in Chita region, of unknown number of clutches: 3 eggs, 10%; 4, 10%; 5, 40%; 6, 40% (Shkatulova 1962); in western Siberia, 5 is commonest size (Johansen 1944). Some pairs have 2 broods in Chita region and in central Siberia (Shkatulova 1962; Reymers 1966), and probably in other parts of range, especially in south (Johansen 1944; Dementiev and Gladkov 1954; Ravkin 1973; Korelov *et al.* 1974). Record of ♀ starting 2nd nest once young had fledged (Taczanowski 1873). Eggs laid daily in early morning (Dementiev and Gladkov 1954; Shkatulova 1962; Makatsch 1976). INCUBATION. 13 days, by ♀ only (Shkatulova 1962; Cheng 1964; Korelov *et al.* 1974; Makatsch 1976). At one nest, ♀ sat for 20–22 hrs per day in stints of 46–52 min (Shkatulova 1962). YOUNG. Fed and cared for by both parents; at one nest in Kirgiziya, ♀ seemed to do most feeding (Yanushevich *et al.* 1960; Shkatulova 1962). Said to be brooded by ♀ alone (Shkatulova 1962). FLEDGING TO MATURITY. Fledging period 9–10(–14) days; often leaves nest before able to fly, which occurs 1–3 days later (Johansen 1944; Yanushevich *et al.* 1960; Shkatulova 1962; Cheng 1964; Korelov *et al.* 1974). For development of young, see Shkatulova (1962). BREEDING SUCCESS. No information. BH

Plumages. (nominate *leucocephalos*). ADULT MALE. *In fresh plumage* (September–January), feathers of forehead and sides of head sepia-black with drab-brown sides, crown largely white but this much obscured by black and dark drab feather-tips. Nape dark drab with black streaks, feathers on lower nape with white or grey tips, showing as a faint broken pale collar. Eye-ring narrow, mottled buff and rufous-chestnut, sometimes a faint buff-brown supercilium. Lore brick-red, extending into broad brick-red stripe over eye and over upper ear-coverts to side of nape, mottled light drab at rear. Stripe behind eye dull black, mottled brown. Upper cheek (below eye) and lower ear-coverts contrastingly whitish, tinged buff at rear, bordered below at rear end by short dull black stripe. Lower side of neck chestnut, or brick-red, bordered behind by mixed black, white, and pale buff bar. Upper mantle light drab changing gradually into rufous-cinnamon at rear, feathers of central mantle to back with black-brown to black centres and rufous-cinnamon to pale buff fringes, scapulars with narrow black-brown centres and broad bright rufous-cinnamon fringes. Feathers of rump bright rufous-cinnamon with pale buff fringes; upper tail-coverts dark rufous-cinnamon, darker than rump, with pale grey or buff tips and sometimes with contrasting sooty shafts. Chin and throat brick-red, mottled with black-brown in malar region, feathers tipped pale buff. Broad and conspicuous white or pale cream collar across upper chest, bordered below by bar of bold black spots and this in turn by broad but ill-defined rufous-and-buff mixed bar on lower chest. Ground colour of remainder of underparts cream-white to white; flank streaked with black-brown, rufous, and pale buff, streaks sharpest on lower flank; longer under tail-coverts usually with sharply contrasting narrow brown shafts. Tail-feathers mainly brown-black; fringes along both webs of central pair (t1) and along outer webs of t2–t4 pale rufous, small fringe along tips off-white; inner web of t5–t6 with long white wedge, occupying well over half of t6 and slightly under half of t5; outer edge of t5–t6 white. Flight-feathers greyish-black with sharp but narrow rufous-cinnamon fringe along entire outer web; fringes lighter on primaries, white on outermost 3–4. Primary coverts greyish-black with narrow and clear-cut cinnamon fringes. Tertials and greater upper wing-coverts black or black-brown, distal half of outer web and whole tip broadly fringed rufous-cinnamon, tertials sometimes with narrow pale buff border, greater coverts usually with ill-defined pale buff or off-white tip. Median upper wing-coverts with broad dark brown centre and broad cinnamon-rufous tip, latter fringed paler buff or almost white forming indistinct wing-bar. Lesser upper wing-coverts dark grey-brown with broad sandy-buff tips, some larger ones sometimes partly suffused cinnamon. Under wing-coverts and axillaries white or light grey, greater under primary coverts with dark grey mottling. *In worn plumage* (about February–May), forehead, side of crown, and upper hindneck black, remainder of crown and nape pure white. Lore and broad stripe through eye uniform brick-red, contrasting with white upper cheek and ear-coverts, latter bordered with black on both upper and lower margins; rear side of neck white or light ash-grey, sometimes a narrow light grey stripe between black of side of crown and brick-red of eye-stripe. Mantle and back more richly rufous-cinnamon, less streaked with buff and black; rump and upper tail-coverts deep rufous-cinnamon, slightly mottled

grey or pale buff on feather-tips. Chin, throat, and lower side of neck uniform brick-red, contrasting sharply with white of lower cheek and upper chest. A more or less well-defined tawny to rufous-cinnamon band separates white of upper chest from white of belly. Flank whitish with bright rufous-cinnamon streaks. Vent and under tail-coverts white or slightly buff, coverts with sharp and narrow dark shafts. Buff fringes of tail, tertials, and greater upper wing-coverts bleached, partly or fully worn off, tips of median coverts partly off-white. ADULT FEMALE. Forehead dark brown with cinnamon mottling; centre of crown streaked pale buff, white, and black-brown (white largely concealed, extending over less than half length of feather), side of crown closely marked with black-brown to black streaks; nape pale buff, more or less heavily streaked with black. Lore pale buff, occasionally mottled brick-red; eye-ring cream-white; streak over upper ear-coverts light drab-brown or buffish-grey, faintly streaked dark grey; upper cheek and lower ear-coverts sepia-brown or sooty, lower ear-coverts with paler grey-buff centre; patch on lower side of neck contrastingly cream-white. Remainder of upperparts similar to those of adult ♂. Chin and throat light cream-buff closely streaked dusky grey or sooty at side, forming broken malar stripe, speckled dusky on remainder; in some birds, throat mottled with some rufous-cinnamon and in these sometimes an ill-defined off-white bar across upper chest. Ground-colour of chest, side of breast, and flank pale cream-buff or pink-cream, marked with ill-defined pale rufous-cinnamon streaks; upper border of chest (below white, if any) blotched sooty grey, streaks on flanks partly black along shaft. Remainder of underparts white or cream-white, vent and under tail-coverts with narrow black-brown shaft-streaks. Tail-feathers as in adult ♂, but fringes on t1 buff rather than rufous. Wing as in adult ♂, but tips of greater and median upper wing-coverts sometimes more contrasting pale buff to cream-white, showing as broken double wing bar. NESTLING. Down smoke-grey; on head, back, scapulars, wing, thigh, and belly (Korelov *et al.* 1974). JUVENILE. Very similar to adult ♀, but pale bases to feathers of central crown faint. Rump dull cinnamon, streaked black; rest of upperparts duller and more tawny than in adult ♀. Throat whitish with black streaks. Chest and flank tawny-buff, streaked black-brown; under tail-coverts pale buff with fine black streaks (Witherby *et al.* 1938). FIRST ADULT. Like adult, but juvenile tail, flight-feathers, and greater upper primary coverts retained; tips of tail-feathers and primary coverts sharper than those of adult (less truncate or broadly rounded); tips of primaries, primary coverts, and tail-feathers more heavily worn than in adult at same time of year, but difference often hard to see.

Bare parts. ADULT, FIRST ADULT. Iris dark brown. Upper mandible slate-grey, dark brown or brown-black, cutting edges and lower mandible pale leaden-blue or pale horn with grey tinge. Leg and foot brownish-flesh, pale flesh-brown, or yellow-brown. (Richmond 1895a; Witherby *et al.* 1938.) NESTLING, JUVENILE. No information.

Moults. ADULT POST-BREEDING. Complete; primaries descendent. In eastern Tien Shan (north-west China), ♂ in full moult on 30 August, ♀ on 9 September (Hellmayr 1929). In lowlands of Manchuria, moult completed in single ♂ from mid-August, but in mountain areas 5 birds still moulting in early September (Piechocki 1959). POST-JUVENILE. Partial: head, body, lesser, median, and greater upper wing-coverts, and tertials; only body feathers, flight- and tail-feathers, and greater upper primary coverts retained until summer of 2nd calendar year. Completed on start of autumn migration.

Measurements. ADULT, FIRST ADULT. Nominate *leucocephalos*. Whole geographical range, all year; skins (RMNH, ZMA; wing includes data from Piechocki and Bolod 1972). Bill (S) to skull, bill (N) to distal corner of nostril; exposed culmen on average 3·1 less than bill (S).

WING	♂	94·0 (2·82; 39) 88–100	♀	88·9 (2·76; 17) 84–95
TAIL		75·0 (3·45; 27) 68–83		69·1 (3·75; 10) 64–75
BILL (S)		11·5 (0·97; 27) 13·2–15·4		14·2 (0·84; 10) 13·3–15·5
BILL (N)		8·0 (0·24; 27) 7·4–8·5		8·2 (0·30; 10) 7·7–8·5
TARSUS		19·5 (0·79; 27) 17·6–20·8		19·5 (0·34; 10) 18·8–19·8

Sex differences significant for wing and tail.

South-west Siberia: wing, ♂ 91·4 (2·30; 7) 89–95, ♀ 84·3 (1·49; 4) 83–86 (Havlín and Jurlov 1977). Kazakhstan: wing, ♂ 90–97 (51), ♀ 84–91 (19) (Korelov *et al.* 1974).

E. l. fronto. China: wing, ♂ 96·9 (12) (93–)95–100·5, ♀ 90·6 (5) 88–92 (Stresemann *et al.* 1937); ♂ 98·3 (16) 95–100, ♀ 92 (4) 91–94 (Vaurie 1972).

Weights. ADULT, FIRST ADULT. Nominate *leucocephalos*. (1) Mongolia, April–August (Piechocki and Bolod 1972). (2) South-west Siberia, June–September (Havlín and Jurlov 1977). (3) Manchuria, August–September (Piechocki 1959). (4) Kirgiziya, all year (Yanushevich *et al.* 1960). (5) Kazakhstan, all year (Korelov *et al.* 1974). (6) Netherlands, Belgium, and France, October–February (Blondel 1963; Cox 1981; ZMA).

(1)	♂	30·0 (2·27; 12) 26·0–35·0	♀	27·1 (0·83; 7) 26·0–28·0
(2)		28·4 (2·67; 7) 26·0–33·5		24·6 (0·84; 4) 24·0–25·8
(3)		30·0 (— ; 4) 28·0–32·0		26·0 (— ; 1) —
(4)		— (— ; 14) 27·7–35·0		— (— ; 8) 25·0–34·0
(5)		29·4 (— ; 32) 25·0–36·5		29·2 (— ; 16) 25·0–36·0
(6)		25·1 (6·04; 4) 19·8–30·7		28·0 (3·31; 3) 23·5–31·4

Kuril Islands, January: ♀ 32·4 (Nechaev 1969).

E. l. fronto. China: March, ♂ 31·7; June, ♂ 24·8, ♀ 25·5 (Stresemann *et al.* 1937).

Structure. 10 primaries: p7 and p8 longest or either one 0–1 shorter than other; p9 0–2 shorter, p6 0–3, p5 6–11, p4 14–18, p3 18–22, p2 21–24, p1 22–26; p10 strongly reduced, narrow, 11·9 (10) 9–15 shorter than longest upper primary covert. Depth of bill at base 6·5 (10) 6·0–6·8, width 7·3 (10) 6·6–7·8. Middle toe with claw 18·7 (8) 16·7–20·5; outer toe with claw *c.* 69% of middle with claw, inner *c.* 63%, hind *c.* 82%. Remainder of structure as in Yellowhammer *E. citrinella*.

Geographical variation. *E. l. fronto* of north-east Tsinghai and Kansu (China) very similar to nominate *leucocephalos*, but black bands on crown of ♂ broader, that on forehead 6·5–9 mm wide (in nominate *leucocephalos*, 4–6 mm), and chestnut sometimes darker (Stresemann *et al.* 1937; Vaurie 1959). Wing of *fronto* averages slightly longer.

Closely related to *E. citrinella*, differing mainly in replacement of yellow or yellowish colour of plumage by white or whitish; some authors (e.g. Johansen 1944) therefore consider them conspecific. Furthermore, many hybrids known from overlap area in western Siberia (Johansen 1944; Panov 1973b, 1989), showing varying mixtures of white and yellow in plumage. Also, birds of westernmost populations of *E. leucocephalos* often show yellowish tinge (especially on head), while easternmost *E. citrinella* are often very pale with well-marked chestnut band across lower chest (Johansen 1944). Recognition of 2 separate species supported by observation near Irkutsk (eastern Siberia) of ♂ *E. citrinella* and ♂ *E. leucocephalos* singing together in same tree without any social interaction, and no obvious hybrids noted in that particular region (Mauersberger 1971), though D I M Wallace recorded apparent mixed pair there.

N.B. Name *E. leucocephalos* conforms with International Code of Zoological Nomenclature and is thus correct (i.e. not *E. leucocephala*). MP

Emberiza citrinella Yellowhammer

PLATES 11 and 16 (flight)
[between pages 256 and 257]

DU. Geelgors   FR. Bruant jaune   GE. Goldammer
RU. Обыкновенная овсянка   SP. Escribano cerillo   SW. Gulsparv

Emberiza Citrinella Linnaeus, 1758

Polytypic. Nominate *citrinella* Linnaeus, 1758, western Europe from Norway, south-west England, and northern Spain east to extreme north-west European Russia, Poland, western Czechoslovakia, western Hungary (west of Tisza river), and western Yugoslavia, grading into *caliginosa* in western half of England and into *erythrogenys* in west European Russia, Baltic states, western Belorussiya, western Ukraine, Slovakia, eastern Hungary, western and central Rumania, eastern and southern Yugoslavia, and Greece; *caliginosa* Clancey, 1940, Scotland, Ireland, Isle of Man, and Wales; *erythrogenys* C L Brehm, 1855, eastern Europe and Asia, east from *c.* 40°E in north and from *c.* 25°E in south.

Field characters. 16–16·5 cm; wing-span 23–29·5 cm. About 10% longer than Chaffinch *Fringilla coelebs*, with distinctly longer and more forked tail; slightly longer and noticeably less compact than Cirl Bunting *E. cirlus*. Rather large bunting, with noticeably attenuated rear body and tail. Matched closely in size and form only by Pine Bunting *E. leucocephalos*. Adult plumage features basically lemon-yellow head (little marked and brilliant in ♂), streaked warm brown upperparts, long rufous-chestnut rump, yellow underparts with streaked chest, and bright white outer tail-feathers. Immature far less distinctive, with less yellow and more obvious streaks on underparts.

Song distinctive. Sexes dissimilar; little seasonal variation. Juvenile separable. 3 races in west Palearctic, 2 western forms intergrading in central England and more distinctive eastern one frequently hybridizing with *E. leucocephalos* in western Siberia.

ADULT MALE. (1) European nominate race, *citrinella*. Moults: August–October (complete); brightness of breeding plumage produced by wear. Ground-colour of plumage lemon-yellow on head, throat, upper breast, and whole underbody, chestnut on sides of breast and flanks, buff-brown on upperparts, rufous-chestnut on rump, and brown-black on flight-feathers and tail; at any distance, most obvious features are bright head and clouded breast. From behind, particularly when wings drooped, long rufous rump noticeable. At close range, plumage shows more intricate pattern including (a) brownish-green circlet on crown, rear eye-stripe, and border to rear and lower ear-coverts (all these marks rather variable), (b) light and dark furrows down sides of breast, merging with olive-green band across upper breast and round shoulders, (c) dark brown streaks along flanks, (d) bold blackish streaks on back, (e) faint greyish feather-tips on rump (when fresh), (f) yellowish-buff margins to wing-coverts and tail-feathers, just pale enough on tips of median and greater coverts to form 2 indistinct wing-bars, and (g) bold white panels on 2 outermost tail-feathers, forming strikingly pale edges to long, noticeably forked tail. Underwing yellow; under tail-coverts yellow, heavily streaked chestnut-brown. Plumage variations include more or less sparse olive streaks on chin and throat (in fresh plumage), chestnut on throat (rare), and, in worn plumage, greater or lesser divide of breast markings by yellow of belly and loss of greenish-yellow fringes to flight-feathers. Bill quite strong and conical, with dusky culmen on bluish-horn upper mandible and paler lead-blue lower mandible. Legs pale flesh-brown. ADULT FEMALE. Plumage even more variable than ♂ with essentially similar pattern of marks but rather duller ground-colours above and paler yellow below. Differs most distinctively in (a) much broader olive circlet on crown, ear-covert markings, malar stripe, and throat streaks, (b) more isolated pale yellow marks in centre of crown, around eye, on cheeks, and on sub-moustachial stripe which broadens under ear-coverts to form noticeable pale patch, (c) olive nape and band round lower neck and shoulders, breaking up on sides of breast, and (d) more streaked, less furrowed sides of breast, flanks, and under tail-coverts, with streaks much darker, more black than chestnut. With wear, becomes brighter and more noticeably streaked particularly below. Separation of brightest ♀ from duller 1st-year ♂ impractical. JUVENILE. Plumage again variable but generally more intensely streaked on even paler, duller ground-colours than ♀. Best distinguished by pale yellowish-buff tips to median and greater coverts forming bright double wing-bar, and usually paler legs; may also show trace of yellow gape. Unusually for bunting, brightest ♂ distinguishable from ♀ by

more distinctly yellowish tone on crown, throat, and belly. Some ♀♀ drab, even washed out in appearance, inviting confusion with other buntings but lacking obvious supercilium. FIRST-YEAR MALE. Moults all but secondaries and outer wing feathers July–October, then resembling adult ♀ except for brighter yellow and chestnut tracts and reduced overlay of marks on head. FIRST-YEAR FEMALE. Duller than adult ♀, still lacking olive on crown and nape and showing paler, distinctly buff-toned brown-streaked breast and flanks and weaker yellow on rest of underparts. (2) East European and Russian race, *erythrogenys*. ♂ paler, sandier, and less streaked above than nominate *citrinella*, with greyish rather than pale olive nape, shoulders, and vestigial breast-band, darker more chestnut breast and flanks (and sometimes throat), and richer, less lemon-toned yellow underparts.

Commonest, most widespread bunting of west Palearctic. ♂ unmistakable. Adult and (particularly) juvenile ♀♀ constitute pitfalls for observers unaware of their marked plumage variation and eager to identify other congeners, particularly *E. cirlus* and closely related *E. leucocephalos*; see those species for diagnosis, best based for *E. cirlus* on structure and voice and for *E. leucocephalos* on lack of yellow in ground-colour of plumage. Problems posed by hybrids with *E. leucocephalos* are daunting; see that species. Flight recalls *F. coelebs* but action rather stronger, with often quicker bursts of wing-beats on take-off giving marked acceleration and rather shorter wing closures producing at times remarkably direct progress and always less marked and less regular undulations. Landing accomplished with apparently untidy collapse due to length of tail which produces characteristically more attenuated silhouette in full flight; tail can also look bulbous at tip. Escape-flight often markedly long and steep, taking bird straight to top of hedge or tree. Gait a hop. Stance variable; often markedly upright on perch, with long forked tail below level of body, but noticeably horizontal on ground. Raises head in alarm but usually looks rather neckless (especially pale-headed ♂). Flicks tail gently, particularly when uneasy. ♂♂ easy to spot, with heads shining at great distance, but ♀ and juvenile comparatively unobtrusive; difficult to approach at times. Gregarious outside breeding season, mixing freely with other ground-foraging passerines. Secretive as migrant, being rarely observed on diurnal passage and not associating with allies.

Typical song one of most memorable passerine sounds, but very similar to *E. leucocephalos*, with which confusion likely: a quite long, rhythmic, slightly jingled repetition of somewhat sibilant, high-pitched unit or motif followed immediately by more plaintive higher-pitched note then longer monosyllable—'zeen-zeen-zeen-zeen-zeen-zeen-zeese'; common variants lack one or both terminal notes; when both omitted much more likely to be confused with flat, rhythmic phrases of other passerines, e.g. Cirl Bunting *E. cirlus*, Lesser Whitethroat *Sylvia curruca*, River Warbler *Locustella fluviatilis*. Commonest call a clipped,

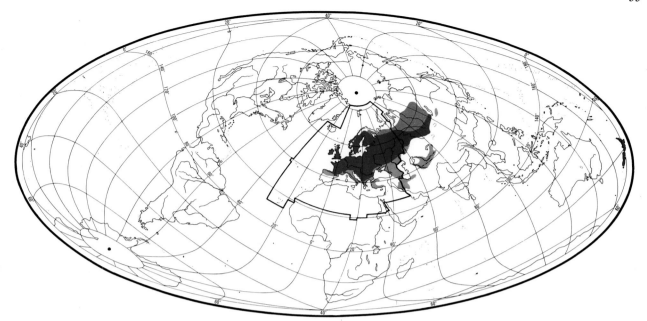

rather metallic monosyllable, 'zit'. Alarm-call a high-pitched 'see' or 'eee'. In flight, 'tit' or 'zik', and in flocks often 'tsirrup' or 'trrp'.

Habitat. Breeds across temperate and boreal zones of west Palearctic, within July isotherms 12–23°C, mainly in open lowlands or hilly country, in both continental and oceanic climates. Prefers dry sunny habitats with fairly rich and varied vegetation, avoiding dense forest, undrained wetlands, towns, or busy inhabited areas. Probably originally based on edges of open areas of forest (including coniferous taiga as well as broad-leaved woods) and fringing scrub of gorse *Ulex*, broom *Cytisus*, hawthorn *Crataegus*, and juniper *Juniperus*, together with northern birch zone. Has profited by farming to extend widely across cultivated land with hedges, plantations, and paths or highways flanked by trees and bushes, but stops short of gardens, cemeteries, and ornamental parks (Voous 1960*b*). Will occupy sea-cliffs carrying shrubby growth with patches of bracken *Pteridium* and heath plants, and also coastal sand-dunes, grassland in course of being overgrown with scrub, young tree plantations, and road or railway embankments. Main requirements seem to be juxtaposition of low woody vegetation, taller song-posts (including human artefacts), open ground for foraging without impediment, and freedom from undue disturbance (Sharrock 1976; Fuller 1982). In winter, ground-feeding on stubble or turnip fields is important (Buckland *et al.* 1990).

In Switzerland, inhabits mainly transitional habitats between open areas and woodland, preferring traditional agricultural landscapes to modern monocultures, and being most abundant at 600–900 m. Where cereal cultivation goes higher, however, ascends to 1500 m, and in south even to 1800 m or exceptionally higher, but only where human land use favourable. Snowfall leads it into settlements. (Glutz von Blotzheim 1962.) In USSR, mainly below 500 m but in Caucasus ascends to *c.* 2000 m in subalpine zone. Spreads through northern forest tundra into true tundra in limited numbers, but prefers to nest in sparse forest tracts, margins, clearings, burnt patches, and felled areas. Inhabits both coniferous and deciduous forest, in Ukraine spreading into steppe oakwood, edges of floodland wood, and orchards; avoids dense stands and tall trees (Dementiev and Gladkov 1954).

As a ground feeder with strong attachment to bush and scrub cover and commanding song-posts, mixed habitat requirements have created uncertainty whether it should be classed as a forest species spreading over regions converted to farming or as a heath species also occupying forest-edge habitats. Dislike of both closed and wholly open habitats is well served by modern land use (Yapp 1962). Shows little attachment to water, or to most human artefacts. Unlike Cirl Bunting *E. cirlus*, appears robustly indifferent to weather unless exceptionally rough or cold.

Distribution. No major changes reported. (Introduced in New Zealand.)

IRELAND. Apparently some range contraction in last 50 years (Hutchinson 1989). NETHERLANDS. Retreating from many former breeding areas in west and north (SOVON 1987).

Accidental. Iceland, Malta, Jordan.

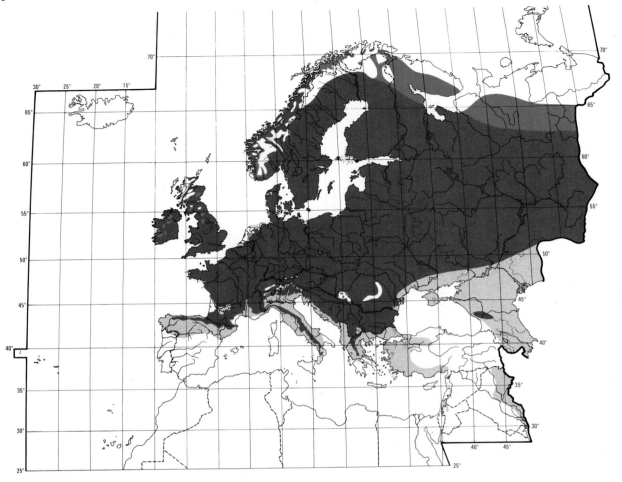

Population. Has declined in several countries, and increased in Finland; all changes apparently due to changing land use and farming practices.

BRITAIN. No trends established; about 1·5 million pairs (Hudson and Marchant 1984; Marchant *et al*. 1990). IRELAND. Some local decrease in last 50 years (Hutchinson 1989). FRANCE. More than 1 million pairs (Yeatman 1976). BELGIUM. Estimated 48 000 pairs; declining, due to changes in agricultural practice and urbanization (Devillers *et al*. 1988). NETHERLANDS. Estimated 25 000–30 000 pairs (SOVON 1987). GERMANY. Estimated 1·7 million pairs (Rheinwald 1992). East Germany: 280 000 ± 150 000 pairs (Nicolai 1993); marked decline since *c*. 1970 (SS). SWEDEN. Estimated 1 million pairs (Ulfstrand and Högstedt 1976). Decreased greatly (locally by *c*. 95%) in 1950s due to pesticides; recovered to former levels within a decade after banning of most destructive chemicals (LR). FINLAND. Long-term increase (Järvinen and Väisänen 1978). Increase due to fragmentation of forests and expanding cultivation; perhaps 1–2 million pairs (Koskimies 1989). SWITZERLAND. Decreased in 1950s, but no exact data (RW). LATVIA. Marked decrease from 1950s (Vīksne 1983).

Survival. Britain: annual mortality of adult ♂♂ 42 ± SE6%, ♀♀ 52 ± SE9% (Dobson 1987); sexes combined 47 ± SE2·7% (Prŷs-Jones 1977). Finland: annual mortality 46 ± SE9·4% (Haukioja 1969). Switzerland: 46 ± SE6% (Prŷs-Jones 1977). Oldest ringed bird 9 years 6 months (Hickling 1983).

Movements. Sedentary to migratory, with most populations partial migrants; also dispersive. Vacates entirely only extreme north of range, and winters chiefly within breeding range, especially in milder years. European migrants head chiefly south-west, usually moving only short or medium distances (up to *c*. 500 km in northern Europe, and *c*. 250 km in central Europe), so birds wintering south of range in Mediterranean region are mostly from central or southern Europe. No evidence that juveniles migrate more than adults. Account based chiefly on Zink (1985).

Sedentary in Britain; winter distribution similar to summer, except for tendency to withdraw from uplands; in Ireland, difference between winter and summer distribution more marked. Ringing in Britain shows that *c.* 70% of adults winter within 5 km of breeding territories. (Lack 1986.) Of 400 recoveries of British-ringed birds, only 7 at 50 km or more; most of these longer movements involve East Anglian birds moving to coast. No evidence of difference between sexes either in distance moved or in proportion making long movements. (Prŷs-Jones 1977.) Passage birds, probably from Scandinavia, appear in small numbers on east coast of Britain in autumn north to Shetland, and occasionally winter, at least in extreme north (where none breed); bird ringed Norway recovered Kent; 3 British-ringed birds recovered in winter in Netherlands (2) and France (1) were probably also of foreign origin (Lack 1986; Thom 1986; Mead and Clark 1987, 1991).

Sedentary also in France, but winter distribution spreads further south than summer, reaching Mediterranean. Recoveries showing long movements are in southern France, in triangle Poitiers–Tarbes–Nice, and relate to birds ringed in Low Countries, western Germany, Switzerland, and north-east France. (Yeatman-Berthelot 1991.) Some birds overwinter in Scandinavia, occasionally even in extreme north, and numbers migrating vary greatly from year to year (e.g. at Falsterbo in southern Sweden, 18 309 in 1959, but 52 in 1960: Ulfstrand *et al.* 1974); local ringing shows that some remain in breeding areas throughout year. Regular in Norway as far north as Nordland, and fairly regular north to Bodø (67°20′N). Most Scandinavian and Danish migrants winter south only to Low Countries and north-west Germany; exceptional recovery in south-west France, October, of bird ringed Denmark, April. (Zink 1985; SOF 1990.) Considerable part of Finnish population remains for winter, especially in rural areas (Järvi and Marjakangas 1985); limited evidence that Finnish ♀♀ may disperse to greater extent than ♂♂ (Prŷs-Jones 1977). Many Finnish birds winter in Sweden (Hjort and Lindholm 1978), and migrants from Finland, Baltic states, western Russia and Ukraine winter south or south-west to northern Italy; also bird ringed at *c.* 46°N 24°E in Rumania, winter, recovered as far north-east as *c.* 56°N 55°E in Russia, autumn (Zink 1985).

Central European birds also partially migratory; unusually high number from wide area migrate in some years, e.g. 1959, 1964 (Zink 1985). In Belgium, *c.* 75% sedentary; few local birds move long distances, but birds ringed as nestlings recovered as far south as southern France and Mallorca (Balearic Islands), and birds ringed on passage or in breeding season have reached Portugal, Morocco (1650 km), and Algeria. Birds wintering or on passage in Belgium are chiefly from Germany and Netherlands; bird from as far as 680 km ESE in Czechoslovakia also recorded. (Lippens and Wille 1972; Zink 1985.) In Germany, of 20 recoveries of birds ringed as nestlings, 17 within 11 km of ringing site, 3 at 210–890 km south-west or SSW; of 133 recoveries of adults (ringed at all seasons), 86 within

10 km, 32 at 12–230 km within Germany, and 15 elsewhere, in France (10, up to 1850 km south-west), Italy (4, up to 450 km south), and Finland (1, at 1730 km NNE) (Dornberger 1977). In Switzerland, most local birds sedentary, and few occur on passage; winters in all low-lying areas, also in small numbers in mountain villages, up to 1500 m in Valais in south (Schifferli *et al.* 1982; Winkler 1984).

In Spain, winters far south of breeding range, and many immigrants reach northern areas, especially in cold years (Muntaner *et al.* 1983; Zink 1985; Noval 1986). Irregular in Gibraltar (Finlayson 1992), only 13 records October–December in Malta (Sultana and Gauci 1982), and at least 6 in Sicily (Iapichino and Massa 1989). More widespread winter than summer in Greece (G I Handrinos). In Turkey (perhaps resident in north-west), rather local and uncommon in passage periods, wintering in western two-thirds (Martins 1989). Scarce winter visitor to Cyprus (Flint and Stewart 1992), and to northern highlands of Jordan (I J Andrews). Rare until late 1960s in Israel, but now fairly common in mountains due to extensive planting of fruit orchards and pine woods (Shirihai in press). Single March record in United Arab Emirates is only report from Arabia (F E Warr). Winter visitor to Iraq, mostly to north (Allouse 1953; Moore and Boswell 1957), and common in northern and western Iran, especially in south Caspian region and Tehran area (D A Scott). Records (few) in north-west Africa include recoveries of birds ringed Belgium (2) and Hungary (1) (Heim de Balsac and Mayaud 1962; Zink 1985). Single February record from Egypt (Goodman and Meininger 1989).

Ringing data show that individuals winter in widely differing areas in different years: e.g. bird ringed winter in northern France recovered in south-west France in later winter, one ringed eastern France was recovered in northern Italy, and 2 ringed Germany recovered in Spain. Hard weather movements occur midwinter, e.g. in France and southern Germany; exceptional recovery in north-east Spain, 21 February 1933, of bird ringed Poland 3 weeks earlier. (Zink 1985; Yeatman-Berthelot 1991.)

Autumn movement September–November(–December), peaking early October at Lista in southern Norway, mid-October in south-east Sweden and northern Denmark, and mid- to end October in Switzerland (Haftorn 1971; Møller 1978a; Winkler 1984; Enquist and Pettersson 1986). Most records on Fair Isle mid-October to mid-November (Dymond 1991). In south of winter range, recorded chiefly December–February in Camargue (southern France) and Cyprus (Blondel and Isenmann 1981; Flint and Stewart 1992), November–February in Jordan and Israel (Shirihai in press; I J Andrews), and end of October to 3rd week of March in Iran (D A Scott).

Spring movement February–May, but mostly March–April (Winkler 1984; Zink 1985; SOF 1990). On Fair Isle, records mainly in 1st half of April (Dymond 1991), and reaches Norway from beginning of April (Zink 1985).

Peaks end of March in northern Denmark (Møller 1978a), mid-April at Ottenby in south-east Sweden (Enquist and Pettersson 1986), and reaches extreme north of Scandinavia late April to May (Wallgren 1954).

In east of range, dispersive and migratory, wintering south chiefly to southern Kazakhstan (rare in Turkmeniya and Tadzhikistan) and northern Mongolia; vagrant to northern China and Japan (Johansen 1944; Rustamov 1958; Vaurie 1959; Abdusalyamov 1977). Some birds remain in western Siberia, but many depart (Johansen 1944); movement begins 1st half of September, peaking in 2nd half (Panov 1973b), and continues locally to end of October (Gyngazov and Milovidov 1977). In Kazakhstan (breeds in north), occurs almost throughout on passage and in winter, in strongly fluctuating numbers; most numerous in east—in Zaysan depression on passage, and in foothills of Dzhungarskiy Alatau and Talasskiy Alatau in winter (Korelov et al. 1974). In Chokpak pass (western Tien Shan), average earliest record over 13 years 3 October (20 September to 13 October), and main passage in 2nd half of October (Gavrilov and Gistsov 1985). Spring movement mostly March–April. Begins end of February in Talasskiy Alatau, continuing throughout March (Kovshar' 1966). In Chokpak pass, over 7 years, average earliest bird 7 March and latest 24 April, with peak 16–25 March, ♂♂ a few days before ♀♀ (Gavrilov and Gistsov 1985). Returns to breeding grounds in Ural valley late March to mid-April (Levin and Gubin 1985), and to western Siberia from end of March (Panov 1973b; Gyngazov and Milovidov 1977). For comparison of timing with Pine Bunting E. leucocephalos in zone of overlap, see that species.   DFV

Food. Seeds, chiefly of grasses (Gramineae); invertebrates in breeding season and casually throughout remainder of year. Feeds almost wholly on ground; in spring, forages near nest-sites, in woodland clearings and borders, by hedges and tracks, in newly-sown fields, etc.; in summer and autumn, on pasture and arable land, waste ground, stubble, and other harvested fields; in winter, also in agricultural areas, but in severe weather, particularly snow, comes to settlements, farmyards, animal feed, etc., though not often to gardens. Outside breeding season, feeds in flocks, often with other seed-eaters. (Eber 1956; Inozemtsev 1962; Hasse 1963; Prŷs-Jones 1977; Eifler and Blümel 1983; Härdi 1989.) In Schleswig-Holstein (northern Germany), of 4528 foraging birds, 55% on open agricultural land (mostly early spring and autumn to winter), 31% in farmyards and at stored grain (winter), 10% in woodland clearings (all year), and 4% in hedges (spring and summer) (Eber 1956). In Switzerland, winter, of 12 961 observations, 19% on stubble, 18% animal feeding places, 13% unharvested cereal fields, 9% pasture, 9% orchards, 7% woodland, and 5% waste ground (Härdi 1989, which see for details). In eastern England, 62% of 116 feeding observations on ground, 28% at less than c.

45 cm above ground, 10% at c. 45–90 cm (Kear 1962); in Oxford area, southern England, 90% of all observations over year were on ground, almost all of remainder on cereal in late summer, and 85% of 250 pecks in short grass were downwards towards soil (Prŷs-Jones 1977, which see for comparison with Reed Bunting E. schoeniclus). Will stand on short grass or cereal stem to reach seed-head and sometimes seen clinging to tall herbs, especially in snow (Kear 1962; Hasse 1963; Prŷs-Jones 1977; Härdi 1989). Very infrequently in trees, but has been recorded foraging for insects high up in pines Pinus, moving nimbly around like tit Parus in outer twigs, even hanging head-down or briefly hovering (Hasse 1963); also noted taking buds of beech Fagus in winter (Härdi 1989). Flying insects taken in brief pursuit flights or by jumping up from ground (Eber 1956; Hasse 1963). Grass stems pushed gradually through bill from side and seeds extracted from head with help of tongue (Eber 1956). In contrast to de-husking method employed by finches (Fringillidae), using sharp mandible edges to cut through husk, E. citrinella manoeuvres seeds to middle of bill with tongue (positions long seeds across lower mandible with ends often protruding at each side) and crushes them open by rapid up-and-down movement, so that husk is split against projection in upper mandible and between blunt mandible edges; large seeds held towards base of bill, small ones towards tip (Ziswiler 1965, which see for comparison of technique, bill morphology, etc. with other seed-eaters). See also Eber (1956) and Prŷs-Jones (1977).

Diet in west Palearctic includes the following. Invertebrates: springtails (Collembola), mayflies (Ephemeroptera), grasshoppers, etc. (Orthoptera: Gryllidae, Tettigoniidae, Acrididae), cockroaches (Dictyoptera: Blattidae), earwigs (Dermaptera: Forficulidae), bugs (Hemiptera: Pentatomidae, Coreidae, Berytinidae, Tingidae, Aphidoidea, Cicadidae), lacewings (Neuroptera: Chrysopidae), adult and larval Lepidoptera (Nymphalidae, Pieridae, Tortricidae, Noctuidae, Lymantriidae, Sphingidae, Geometridae), caddis flies (Trichoptera), adult and larval flies (Diptera: Tipulidae, Culicidae, Chironomidae, Stratiomyidae, Rhagionidae, Therevidae, Asilidae, Syrphidae, Platystomidae, Trypetidae, Tachinidae, Muscidae), Hymenoptera (Tenthredinidae, Cephidae, Cynipidae, Ichneumonidae, Braconidae, ants Formicidae, bees Apoidea), adult and larval beetles (Coleoptera: Carabidae, Silphidae, Staphylinidae, Scarabaeidae, Elateridae, Cantharidae, Ptinidae, Tenebrionidae, Cerambycidae, Chrysomelidae, Bruchidae, Curculionidae), spiders (Araneae), woodlice (Isopoda: Armadillidiidae), millipedes (Diplopoda), earthworms (Lumbricidae), snails (Pulmonata: Arionidae). Plants: seeds and other parts of spruce Picea, pine Pinus, beech Fagus, grape Vitis, mistletoe Viscum, nettle Urtica, dock Rumex, knotgrass Polygonum, goosefoot Chenopodium, chickweed Stellaria, mouse-ear Cerastium, bramble Rubus, vetch Vicia, clover Trifolium, crowberry Empetrum, forget-me-not

Myosotis, Labiatae, fleabane *Coryza*, dandelion *Tarax-acum*, knapweed *Centaurea*, sow-thistle *Sonchus*, yarrow *Achillea*, sunflower *Helianthus*, plantain *Plantago*, grasses (Gramineae, including especially all types of cereals, *Festuca*, *Lolium*, *Poa*). (Collinge 1924–7; Tarashchuk 1953; Eber 1956; Bösenberg 1958; Khokhlova 1960; Turček 1961; Glutz von Blotzheim 1962; Inozemtsev 1962; Hasse 1963; Korenberg *et al.* 1972; Prŷs-Jones 1977; Soper 1986; Härdi 1989.)

Some plant families (e.g. Cruciferae) completely ignored in wild although among commonest in habitat; seemingly avoids oily seeds, preferring those rich in starch (Eber 1956). In tests, only monocotyledonous seeds (grasses in this case) accepted, and took seeds of herbs only when very hungry (Ziswiler 1965). In Moscow region (Russia), over year, 34 stomachs contained 478 items, of which 39·3% by number adult beetles (33% Curculionidae), 1% caterpillars, 1% Tipulidae, 12·3% wheat *Triticum*, 8·8% oats *Avena*, 8·7% pine, 4·6% spruce, 21·1% other seeds. Beetles rose from 0 at beginning of March to 70% in June, declining to 0 in September. (Inozemtsev 1962, which see for variation in diet with feeding habitat.) For 100 stomachs from European Russia, see Korenberg *et al.* (1972). In Ukraine, over year, 39 stomachs contained 463 items, of which 19·4% by number beetles (15% Curculionidae), 12·5% Lepidoptera (8% caterpillars), 4·8% Orthoptera (4% Acrididae), 3·0% Hemiptera, 2·4% Hymenoptera, 42·8% grass seeds, 10·2% wheat, 2·2% barley *Hordeum* (Tarashchuk 1953); *c.* 260 insects in 49 stomachs in spring were 42% by number caterpillars, 40% beetles (mostly Curculionidae), and 14% Hymenoptera; 32 autumn stomachs held 126 insects, of which 64% Acrididae, 30% beetles (almost all Staphylinidae); stomachs contained more seeds in autumn than in spring (Khokhlova 1960). In England, 58 stomachs obtained over the year held 71% by volume plant material, including 58% seeds of weeds and wild fruits and 10% cereals (36% in September); animal prey included millipedes, earthworms, and snails (Collinge 1924–7). In Oxford area, December, animal material present in 13% of stomachs and gullets (*n* = 8) but comprised less than 1% by corrected volume; April–June in 100% (*n* = 12), almost 80% by volume May–June; seeds present in 100% every month except May–June when in 80%. In October, 96% (by number) of seeds eaten were Gramineae (including cereals) (*n* = 228), November–January 100% (*n* = 6522), February 97% (*n* = 3339), March 99% (*n* = 393), April 84% (*n* = 93); by weight however, Gramineae virtually 100% of all seeds October–March and 99% in April; other seeds eaten were Urticaceae, Polygonaceae, Chenopodiaceae, and Leguminosae. Cereals 51–99% by weight of all plant material over year, remainder almost wholly seeds of grasses *Festuca* and *Lolium*, which are also large; proportion of cereals lowest in January (10% by number, *n* = 4722, 60% by weight) and February (7% and 51%); consumption rises in March at sowing. Animal prey in October mainly

beetles, and in January–February springtails (82% of all insects taken), though this due mainly to specialization by a few birds. (Prŷs-Jones 1977, which see for comparison with *E. schoeniclus* and Corn Bunting *Miliaria calandra*.) In Schleswig-Holstein, 11% of 4106 feeding observations over year were invertebrates (1% February, *n* = 1350; 53% June–August, *n* = 247); May–August fed on seeds of *Lolium* and *Poa* and sometimes milky seeds of dandelion; September–April cereals, some seeds of mouse-ear; over year, cereals (especially oats) were 73% of plant material (Eber 1956). For gizzard and gullet contents, and energy requirements when roosting (Oxford area), see Evans (1969*b*) and Prŷs-Jones (1977).

Young fed mainly invertebrates. In eastern Germany, 124 collar-samples contained 986 items, of which 29·3% by number Lepidoptera (9% adults; 15% Noctuidae, 4% Nymphalidae, 3% Geometridae), 22·0% beetles (15% adults; 9% Chrysomelidae, 5% Curculionidae, 4% Elateridae), 10·8% Diptera (8% adults), 6·0% earwigs, 5·2% Hymenoptera (5% Tenthredinidae), 5·2% spiders, 3·6% Orthoptera (3% Acrididae), 2% woodlice, 1% snails, 6·0% wheat seeds, 4·8% barley, 2% sunflower, 1% oats. Adult Lepidoptera very often given to young with wings attached, and *c.* 30% of caterpillars hairy; spiders apparently avoided in relation to abundance and *E. citrinella* possibly specializes in slow-moving invertebrates; little change in composition of diet over nestling period. (Bösenberg 1958.) No plant material recorded in nestling diet in other studies: 17 stomachs, England, held only insects (Collinge 1924–7); in Schleswig-Holstein, young fed Orthoptera, adult and larval Lepidoptera, Diptera, and snails (Eber 1956); 10 collar-samples from Ural valley (north-west Kazakhstan) contained Odonata, adult and larval Lepidoptera, and beetles (Levin and Gubin 1985); see also review by Inozemtsev (1962).   BH

Social pattern and behaviour. Substantial studies by Diesselhorst (1949, 1950) and Wallgren (1956), and monograph by Hasse (1963); also historically important studies on territorial behaviour by Howard (1929, 1930). Most detailed study of displays by Andrew (1957*a*), on which relevant sections of following account largely based. No major recent studies.

1. Usually in flocks outside breeding season, but these usually loosely knit, primarily associations at good feeding sites, often with other species; during migration, flocks more closely integrated (Wallgren 1956; Hasse 1963); very variable in size, one of *c.* 1000 birds (Creutz 1988) perhaps largest recorded. Flocks begin to form in late summer; small parties from mid- or late July, at first mainly young birds, adults gradually joining in. Flocks begin to break up in early spring, in continental areas a few days after temperatures rise above 0°C. At first, ♂♂ leave flocks in morning and evening, visiting future territories, some returning to flocks in middle of day. ♀♀ leave flocks a little later than ♂♂. Whole process, from winter flocking to establishment of breeding territories, takes *c.* 2 months; interrupted by cold spells, when birds revert to flocks. (Hasse 1963.) BONDS. Monogamous mating system. 2 cases of polygyny recorded (Diesselhorst 1950). ♀ alone builds nest, and incubates. ♀ alone normally broods young (but see Relations within Family Group,

below). Both sexes feed young, and care for fledged young (Hasse 1963). Age of first breeding 1 year (Paevski 1985). BREEDING DISPERSION. Solitary and territorial. Territories often more or less linear, e.g. along hedges or borders between woodland/scrub and open land, and area difficult to assess, as borders tend to be 'fluid', and some overlap may occur (Diesselhorst 1949; Wallgren 1956). In farmland in Cambridgeshire (England), average length of territories along hedges 60 m; trespassers ignored if more than 10–15 m out into adjacent fields, thus average defended area *c.* 2000 m² (Andrew 1956c). Territories in transition zone between woodland/bush and open fields, Germany, from *c.* 2500 m² to at least 20 000 m²; average size in good habitat *c.* 3000 m² (Diesselhorst 1949). 4 territories in similar transition zone, southern Finland, 4400, 5700, 7000, and 7500 m² (Wallgren 1956). 8 adjacent territories on bushy commonland, England, occupied *c.* 6 ha, i.e. 7500 m² per territory (Howard 1930). Territory serves for pair-formation and nesting, but has little importance as source of food, most of which is obtained on neutral ground beyond territory borders (Wallgren 1956; Hasse 1963). Breeding densities vary widely, depending on amount of undefended ground away from woodland edge, bushy areas, or hedges which form core of territories. In farmland with hedges, eastern England, average density over 8 years 21·8 pairs per km² (Benson and Williamson 1972). In farmland on Damnica plateau (Poland), densities in 2 areas 3·14 and 2·64 pairs per km² (Górski 1988). In open habitats, Sweden, typically *c.* 15 pairs per km² (Ulfstrand and Högstedt 1976); in sea-shore meadows, Öland (Sweden), average 6·2 territories per km² (maximum 30) (Ålind 1991). Densities in wooded country low, presence of territories dependent on open areas; thus recorded at density of 1 pair per km² in only one of several plots in mixed ash *Fraxinus* and alder *Alnus* forest, Białowieża (Poland), and not at all in other forest types (Tomiałojć *et al.* 1984). In deciduous forest, Öland, with 11% open meadows, 28 pairs per km² (Fritz 1989). In broad-leaved woodland, southern Finland, 11–20 pairs per km², territories being in open areas; 4 pairs per km² in cut-over spruce *Picea* woodland (Palmgren 1930). In oak *Quercus robur* forests in Bourgogne (France), 2 pairs per km² in coppice-with-standards, 3–56 in shelter-woodland (Ferry and Frochot 1970). ROOSTING. Outside breeding season, sites very varied; usually in low, thick vegetation, commonly in company with other species, but various other sites recorded. In southern Finland, roosts mainly in thick stands of young spruce, often less than 1 m above ground; alder and other bushes used when in leaf, and occasionally roosts on ground. When not cold, may roost in hedges far from woods; in winter, always in woods. (Wallgren 1956, which see for times of going to roost and leaving roost throughout year.) In Germany, roosting recorded in thick bushes, but mainly in reedbeds, reeds over water being avoided (Dorsch 1970); but in England one case recorded of roosting in area of reeds where water was deepest (Frost 1979). Roosting in reedbeds also recorded in France, also in old hops *Humulus* and brambles *Rubus*, 30–80 cm above ground (Labitte 1937). In open agricultural country may roost in crevices and cracks in barns (Stiefel 1976). In very cold weather may choose roost-sites with protection afforded by snow. Thus in southern England found roosting with Skylarks *Alauda arvensis* at bottom of slanting snow tunnels at base of grass tussocks in field, but not certain if birds had tunnelled into snow, or were using established roost-sites which were snowed up and from which tunnels were made as birds emerged (Gladwin 1985). In Finland at 64°N, birds sheltered from cold during day in crevices, under overhangs (etc.), and during blizzard in small hollows in snow, with backs to wind; at night apparently roosted in crevice between house wall and snow piled against it (Järvi and Marjakangas 1985). Roosting communal; generally 4–10 birds

together (Labitte 1937); at least 8 (4 pairs), early April, consisting of territory-owners and neighbouring pairs, who were driven away at dawn (Wallgren 1956). In breeding season, pairs begin to roost in territory, ♂ and ♀ at first together, then ♀ on nest (in one case, ♂ *c.* 25 m away from nest, 2·5 m up in small spruce, near trunk) (Hasse 1963). See also Relations within Family Group (below). Young birds, while still not flying strongly, roost on ground near where last fed, hiding in low vegetation (Hasse 1963).

2. Not shy, but wary when in flocks, reacting to alarm before other species (Wallgren 1956). Generally indifferent to presence of other species which pose no threat, but may behave aggressively to food competitors. For mobbing of supposed nest-predator, see Parental Anti-predator Strategies (below). Violent attack recorded on shrew *Sorex* at feeding place, early spring; reason not clear (Marjakangas 1983). FLOCK BEHAVIOUR. Not closely coordinated. Feeding flocks form at restricted food sources, individuals coming and going independently. Little aggression, but occasional conflicts over favoured food items; generally indifferent to other species. In captive flock in aviary, dominance order established between individuals; ♂♂ more aggressive to other ♂♂ than to ♀♀, old ♀♀ equally aggressive to ♂♂ and ♀♀. Aggression resulted from shortage of some resource such as food or water. Dominance order not observed in free-living flocks, but not studied in detail. In late winter, ♂♂ may be aggressive to one another in flocks (early appearance of territorial behaviour), and may sing; such song rare in Germany (Hasse 1963), apparently commoner in Finland (Wallgren 1956); see below. For seasonal changes in flock composition, see 1 above. Flocks formed by migrants more coordinated than feeding flocks, and tend to be larger (Wallgren 1956); call given on taking off, and flight-call, presumably serve to keep contact between flock members and coordinate flock movements. SONG-DISPLAY. Song (see 1 in Voice) usually given from tree or bush; occasionally from clod of earth (etc.) on ground (Hasse 1963), rarely from building (Witherby *et al.* 1938; Scherner 1972c). Apart from head being thrown back (Fig A), no special display asso-

A

ciated with song, but characteristic behaviour of singing ♂♂, southern Finland, may have display function: from end of April, immediately after leaving roost, sang intensely for 15–20 min (see below) in low bushes, on tree-stumps, stones, etc., flying round, mainly on open ground, in territory or its immediate vicinity; first seen 21 April, occurred when ♀ was building or incubating, then stopped, and began again when young of 1st brood independent. (Wallgren 1956.) Song-period: from taking up of territories in early spring to July–August or later. Beginning depends on weather: in Germany, a few days after temperature rises above 0°C (Hasse 1963); in eastern England, from early February (Andrew 1956c); in Denmark, from end of February (Møller 1988); in southern Finland, 10 March, 27 March, 5 April, and 20 March in 4 successive years (Wallgren 1956). Ending of song also geographically variable: in Denmark early September (Møller 1988); in southern Finland, usually July (last heard 21 July), with occasional song in autumn (Wallgren 1956); in Germany, may continue into September or early October, with exceptional record (Luxembourg) 18 October (Hasse 1963). In northern England, pronounced resurgence of singing in August (in one case, 3482 songs given by one ♂ in a day), contrasting with arctic Norway, where song ended in July, with no late resurgence (Rollin 1958). Diurnal pattern of song, southern Finland, shows marked peak immediately after leaving roost, with intensive song for 15–20 min (see above), followed by period of foraging; song then continued until 10.30–11.00 hrs, then sporadic until 16.30 or 17.00 hrs, when increased to intensity similar to morning song, ending *c.* 15 min before going to roost (Wallgren 1956). Quantitative assessment of diurnal rhythm of song in Denmark (Møller 1988) showed similar pattern, with most song around sunrise, less song around sunset, and least song at midday. Pattern also similar near Moscow, with peaks 03.00–04.00 hrs and 19.00–20.00 hrs; abrupt decline after 20.00 hrs, and ceased *c.* 1 hr after sunset (Denisova and Gomolitskaya 1967). Detailed study in northern England showed that song began earlier relative to sunrise as season progressed: at sunrise in March, but when sun more than 6° below horizon in June–July (Rollin 1958). When mate incubating, ♂ has peaks of song at midday as well as morning and evening; does not usually sing while ♀ off nest, but begins when she returns to nest; usually sings 20–30 m from nest (Wallgren 1956). In contrast to Germany, where song strictly associated with territory (Steinfatt 1940; Diesselhorst 1949), in southern Finland song regularly given by birds in flocks, both migrating and overwintering. Birds in migrating flocks often sing late April (once heard also in early April); wintering birds in flocks sing in snowy weather if sunny and fairly warm, heard 7 March to 16 April. (Wallgren 1956.) Main function of song, as indicated by detailed analysis of its spatial and temporal distribution, is to reduce intrusion by other ♂♂ during ♀'s fertile period (Møller 1988); but advertising for mate clearly also important, as unpaired ♂♂ continue to sing intensively until late May, when song of nesting ♂♂ has declined (Wallgren 1956). ANTAGONISTIC BEHAVIOUR. Main aggressive display is Head-forward threat-posture (Fig B). Legs flexed,

B

body horizontal, head in line with body, bill opened wide; head feathers may be somewhat sleeked. Head withdrawn, then thrust forward (sometimes repeatedly) at opponent. Display may be accompanied by calls (see 5 in Voice). Mutually threatening birds face each other thus, often simultaneously raising and lowering heads. Threatening may pass insensibly into attack or retreat. At high intensity, tendency to bring open bill as close to opponent as possible may abolish or reduce other components of display, e.g. body may be raised to bring bill opposite a

bird perched higher, or threatening bird may hover. In conflicts between ♂♂ over territory, birds usually silent, and usually do not gape. ♂ often bows repeatedly when near rival; in common variant, head and forepart of body repeatedly lowered, with bill horizontal. Displaying birds may pivot: body swung round laterally, turning about hips or about feet; bill often lowered. At high intensity, pivots pass into leaping round, first to face one way then the other; this sometimes followed by brief retreats and advances. When near rival, ♂ often trails wings (usually low-intensity sexual response; see below). Other behaviour associated with conflicts over territory includes slow chases; birds fly with slow wing-beats of normal amplitude, sometimes separated by very brief glides; tail held spread, both in flight and in brief perching between flights (Fig C). ♂ being chased often

C

gives alarm 'see' (see 3 in Voice); sometimes crouches motionless, glancing around. (Andrew 1957a.) HETEROSEXUAL BEHAVIOUR. (1) Pair-bonding behaviour. Pair-formation takes place in ♂'s territory, beginning soon after break-up of winter flocks. ♀♀ seek out territorial ♂♂. Whole process, from first meeting to establishment of pair, may take several weeks. Characteristic mutual display ('Symbolpicken': Diesselhorst 1949, 1950) associated with pair-formation at beginning of breeding season (later, pair-formation follows rather different course—see below). When ♀ appears, ♂ ceases to sing and flies to ground. With plumage slightly ruffled, wings slightly drooped, and tail a little raised, ♂ and ♀ pick up and drop small particles such as pebbles, short bits of stalk (Fig D), pieces of leaf, or other plant

D

material. ♂ often hops about in front of ♀, but they may be close together or some metres apart. (Diesselhorst 1950, Hasse 1963.) Birds may repeatedly lower bill without touching ground; ♂ may trail wings and slightly ruffle rump feathers (Andrew 1957a). Pair-formation more rapid later in season (e.g. after death of bird or territorial changes), accompanied by 2 distinct displays: (a) with plumage ruffled, ♂ flies and lands in front of ♀, then makes short runs to and fro, finally turning sharply away from her; (b) ♂ flies some metres away from ♀, picks one (Fig D) or several long stalks (much as ♀ when collecting nest-material) and runs about for a few seconds with them before letting them fall (Diesselhorst 1950); same as Stem-display in other *Emberiza*. From time of pair-formation to start of breeding, pair show no marked sexual activity (Andrew 1957a). ♂ associates especially closely with ♀ during her fertile period; presumably mate-guarding (Møller 1988). Open spaces on ground chosen for courtship; running displays (see below) clearly related to this. ♂'s courtship behaviour also comprises following components. (a) Fluffed-run. With legs often markedly bent, bill occasionally lowered, usually only rump feathers ruffled, tail occasionally spread, and wings usually drooped (occasionally trailed or raised), ♂ makes series of slow runs, usually away from ♀; legs less bent between runs; this display often followed by nest-material carrying (see above). (b) Bill-raised run. Bill markedly raised (pointed obliquely up); body varies from almost erect to rather lowered; rump feathers sometimes ruffled; wings raised with little extension until above horizontal, occasionally fully extended vertically up; occasionally quivered, but only while actually being raised. In low-intensity displays, bursts of wing-quivering occasionally repeated. Tail widely spread, sometimes lowered. In this posture, ♂ makes series of runs, often towards ♀; in pauses between runs, display components usually decrease. All body feathers may be somewhat ruffled (probably fear reaction) if ♂ approaches very near to ♀. At the end of Bill-raised run, ♂ often runs to behind ♀, raising bill and wings even further just before leaping onto her back to attempt copulation. (Andrew 1957a.) ♂ in intense bill-raised display gave quiet churring call (Salmon 1948). (2) Courtship-feeding. Does not occur. (3) Mating. Takes place either in trees and bushes or on ground. If both partners ready to mate, mating not preceded by any special display by ♂. ♀ performs Soliciting-display (Fig E): crouches with legs flexed, body more or less

E

horizontal, bill and tail markedly raised; wings raised up to 20° above horizontal and rapidly quivered, tail rarely spread. At high intensity, wings vibrate so strongly that bird is lifted from ground. Soliciting ♀ gives rapid, soft 'tititi. . .' calls (see 6 in Voice). ♂ usually flies straight to ♀, then hovers over her, or lands directly on her back, where he stands with beating wings, rather erect body, and somewhat bent legs, looking down. Immediately, or after a few seconds, ♂ lowers rear body and brings tail down to one side of ♀'s; wings may cease to beat as cloacas come into contact. Tail held down usually for c. 5 s; then ♂ flies off for some distance. (Andrew 1957a.) If ♀ less ready, mating or attempted mating may be preceded by ♂'s courtship display. ♀ may avoid ♂, or may actively attack, both birds rising into air (Andrew 1957a; Schneider 1964; Hasse 1965). Reversed sexual behaviour not unusual. ♂ may land on ♀'s back to mate, then fly off and land beside ♀, and assume posture similar to soliciting ♀ except that tail not raised. ♂ may rise from this posture and mount ♀ again, and procedure may be repeated several times. When ♂ 'solicits', ♀ usually ceases to solicit and flies onto ♂'s back, but slips off again almost immediately without bringing tail down. (Andrew 1957a.) Other accounts of mating describe what are apparently variations or incomplete sequences of behaviours

summarized above (Howard 1929; Witherby et al. 1938; Cawkell 1947; Yeo 1947). (4) Nest-site selection and behaviour at nest. ♀ apparently chooses nest-site, but this not studied in detail. Accounts of nest-building vary. ♂ not seen to accompany nest-building ♀ (Geyr von Schweppenburg 1942b; Wallgren 1956); described as following nest-building ♀ closely (Howard 1929). ♀ alone incubates. During incubation, ♂ usually calls ♀ off nest; ♀ answers, and both birds go off and forage together. Less often, ♀ leaves nest without contact with ♂, and forages alone; occasionally ♀ calls first, and ♂ answers. ♀ usually returns to nest accompanied by ♂, rarely alone. (Wallgren 1956.) RELATIONS WITHIN FAMILY GROUP. Usually ♀ alone broods young; roosts at night on nest until young 6–7(–11) days old (Wallgren 1956; Hasse 1963). Occasional brooding by ♂ recorded, and in one case, when ♀ disappeared, ♂ regularly brooded nestlings both day and night (Hasse 1961). Both parents feed young and remove faeces; faeces swallowed up to 3rd–4th day after hatching, then carried away and dropped. ♂ may either feed young directly, or, if ♀ is at nest, pass food to ♀ who then feeds young; this also usual if ♂ and ♀ go to nest together. Both parents feed young after fledging, for 8–14 days (Steinfatt 1940), 13 days (one case only) (Wallgren 1956). ANTI-PREDATOR RESPONSES OF YOUNG. No details. PARENTAL ANTI-PREDATOR STRATEGIES. ♂ gives very high-pitched call, scarcely audible to human, with bill wide open (see 3b in Voice) when ♀ on nest and intruder nearby (Howard 1929). ♀♀ frightened from nest gave distraction-displays in about one-seventh of cases recorded; flew off rather slowly with tail spread, usually landing nearby with tail still slightly spread before flying away. In other cases, ♀♀ flew swiftly away as soon as put off nest. (Andrew 1956b.) Crouch (1948) recorded distraction-display on ground, with one wing trailing. ♂ and ♀ gave intense threat-displays (♀ less intensely than ♂) to dummy Woodchat Shrike Lanius senator placed near nest with young. Plumage very ruffled so that body appeared like ball, wings and tail a little raised; tail fully spread so that white outer feathers very conspicuous, and held obliquely down. ♂ approached dummy with small hops. (Dancker 1956.) At nest with small nestlings overrun by ants Lasius niger, ♂ ate hundreds of ants (but did not attempt to feed them to young) until they disappeared, thus almost certainly saving nestlings (Wallgren 1956).

(Figs by D Nurney: A from photograph in Ornis 1991, 5, 8; B, D, and E from drawings in Panov 1973b; C from photograph in Vogeljaar 1991, 39, (1), 11.)   DWS

Voice. Well studied. For discussion of function of song, see Møller (1988); for further data on song dialects, see Hiett and Catchpole (1982), Hansen (1985), Bergmann and Helb (1987), and Glaubrecht (1989). Song and calls very similar to Pine Bunting E. leucocephalos; see that species. Account includes unpublished data from P Hansen.

CALLS OF ADULTS. (1) Song. Series of reiterated units (complex units may be considered as short motifs; series can take form of rapid trill) (part A) followed by 1–2 drawn-out units (part B); very characteristic (apart from similarity to E. leucocephalos) and variously transcribed in different languages, in English as 'little-bit-of-bread-and-no-cheese'; these transcriptions not however indicating uniformity of units in part A or variations in part B. In incomplete songs, which are common, part B may be partly or entirely omitted; in latter case, song easily confused with Cirl Bunting Emberiza cirlus. In part A, suc-

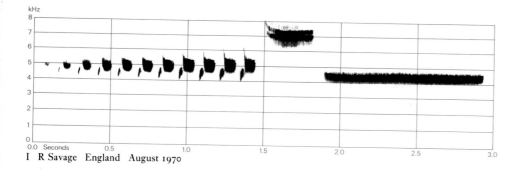

I R Savage England August 1970

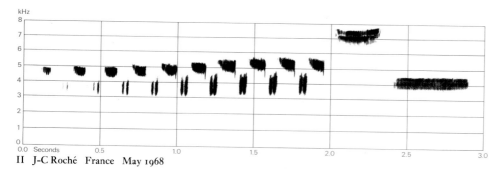

II J-C Roché France May 1968

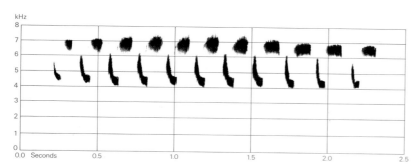

III J-C Roché France May 1968

cessive units or motifs typically increase in volume, and pitch may rise or (less usually) fall slightly; each unit typically composed of 2 sub-units of different frequencies, and same also true of units within a motif, giving characteristic 'chinking', but sometimes more buzzing, quality, and can then resemble song of River Warbler *Locustella fluviatilis* (L Svensson). In geographically most widespread dialect (see below), part B consists of shorter, very high-pitched unit (not always audible to older people) followed by longer, lower pitched unit (Figs I–II). Incomplete song-phrases (Fig III) normally lack final units in order from last one (P Hansen). In dialect with more restricted distribution, in northern Europe, part B consists of 2 units of more equal length, 1st lower pitched (about same frequency as part A) and 2nd higher pitched and with marked descent in frequency (Fig IV), or 2 lower-pitched units of nearly the same frequency (Fig V: Hansen 1985). During bout of singing, song-phrases may begin in incomplete form (part A only, short at first, then length-

ening, part B then added), and occasional incomplete songs may also occur later in sequence; average interval between songs 7·01 s ($n = 28$) (Helb 1985, which see for detailed analysis of development of songs in a sequence). Differences in dialect affect only part B of song. These 2 main dialects usually referred to in German literature as 'zity' and 'tysieh' (or 'tysiih') respectively, in English perhaps better as 'zi-teee' and 'tee-sii'. Individual ♂♂ have repertoires of 1–3 (mostly 2) song-types (Hiett and Catchpole 1982), 1–4 (mostly 2–3) (Hansen 1984), or at least 2 (Helb 1985); differ in both part A and part B (Hiett and Catchpole 1982) or only in part A (Hansen 1984; Helb 1985; Bergmann and Helb 1987). Population contains *c.* 50 different part A units. Individual song repertoires remain stable for life. (P Hansen.) In normal singing, bird sings a number of song-phrases of one type, then switches to another type, and so on (Hiett and Catchpole 1982); a change of song-type during bout of singing may also be stimulated by presence of intruding ♂, or sight of strange

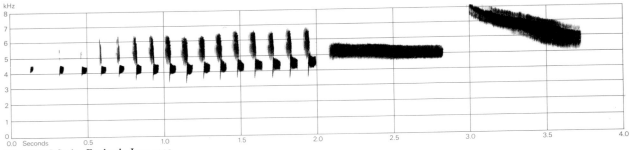

IV W T C Seale England June 1990

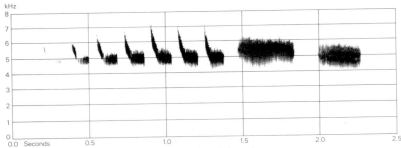

V S Palmér/Sveriges Radio (1972–80) Sweden July 1965

VI P A D Hollom England July 1987

♀ or singer's mate (Hasse 1963). Neighbouring ♂♂ with shared song-types usually avoid singing them simultaneously (Hansen 1981b, which see for discussion of coordinated singing and song matching). 2 main dialects so far known, with roughly western and eastern distributions (song not yet studied in detail in all parts of range), the 'tysieh' (or 'tee-sii') dialect occurring in Baltic area, Finland, eastern Europe, Denmark, and northern Germany (Schleswig-Holstein), the 'zity' (or 'zi-teee') dialect in rest of Germany, Austria, France, Netherlands, Norway, and Britain. In Sweden, mixture of dialects found (P Hansen, L Svensson). Further study needed in western Europe, as individuals singing 'tee-sii' songs have been recorded France (J-C Roché) and England (W T C Seale). No studies reported from south-west Europe. Where the two dialects abut, in Baltic area and Danish islands, situation is complex, with mosaic distribution and evidence of recent change apparently caused, at least in some cases, by immigration of birds of one dialect into area of other (Hansen 1985; Glaubrecht 1989, which see for further details, including development of sub-dialects). On island of Bornholm, only 'zity' dialect recorded in 1930s, whereas birds singing both dialects present by 1979–80 (Conrads 1984), but evidence for change inconclusive (P Hansen). Some mixed singers (birds having songs of both dialects in repertoire) occur in areas of overlap (Møller 1982b; Kaiser 1983). In Jylland, where dialects overlap, birds with different dialects also show ecological differences (Møller 1982b). (2) Subsong. Twittering subsong given in autumn, by adults of both sexes

kHz

VII W T C Seale England June 1985

and by young birds (Naumann 1900); see Calls of Young (below). (3) Contact-alarm and alarm-calls. Main study by Andrew (1957b), on which following account largely based. (3a) Commonest call, used for contact throughout year, perched or in flight, and, with modifications, also in other contexts, 'zit' or 'tzit', also rendered as 'styff' (Rosenberg 1953), 'cheep' (Andrew 1957b), 'tipp', 'tjipp', 'dsipp' (Hasse 1963), and many other transcriptions (Fig VI, 2nd unit; Fig VII, 1st unit). Duration c. 0·02–0·07 s, frequency range 3·3–6·4 kHz, mean frequency often descending sharply. This call not rapidly reiterated, but usually delivered at measured rate and at more or less regular intervals; birds calling on ground look round continually while giving it, and often fly to passing conspecifics. Often alternates with 'see' calls (see below) without perceptible cause; groups of 1–3 calls sometimes alternate. (3b) Usual alarm call higher-pitched 'see'; lasts

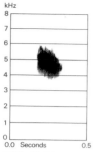

VIII V C Lewis England
July 1978

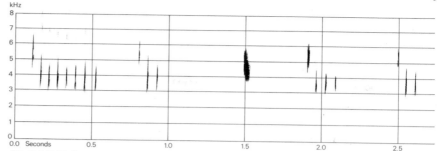

IX P A D Hollom England December 1975

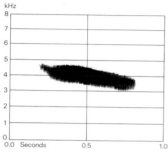

X W T C Seale England June 1985

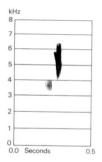

XI W T C Seale England June 1985

vibrant, or less loud, higher-pitched 'chu' or 'chee'. High-pitched 'eee' call, given during fighting when fear component is strong, may be homologous with 'see' call (see 3b, above). At maximum urgency, these calls may all be rapidly repeated as chatters. (Andrew 1957b.) (6) Soliciting-call. ♀ soliciting mating gives 'tititi...', resembling series of rapidly repeated flight-calls (Andrew 1957b). (7) Other calls. (7a) Soft 'chup', given by ♀ as she put food into nestling's gape, or when nestlings would not gape despite proffered food (Andrew 1957b). (7b) A 'zieh' (drawn-out version of alarm 'see'; call 3b) given as warning for flying predator (Fig X: Bergmann and Helb 1982, see also for sonagram). Call shown in Fig X was preceded c. 1·4 s before by abrupt 'z-lit' (Fig XI), perhaps anxiety call (W T C Seale).

CALLS OF YOUNG. Nestlings give very soft 'zee' (Andrew 1957b), 'ssississi' (Hasse 1963), or 'tsi' calls (Bergmann and Helb 1982, which see for sonagram) from 4th day; with increased excitement, rapidly repeated in bursts. Single calls last c. 0·05–0·13 s; frequency range usually 6–7 kHz. Fledglings give similar 'zee' in alternation with 'tip' (see below) as parent approaches, then burst of 'zee' calls as they beg; also give 'zee' calls as they fly after parent. At about time young leave nest, distinct 'tip' calls begin to be given, closely resembling adult's flight-call; duration c. 0·03 s, frequency range 4–7 kHz. This call often separated by fairly long intervals when fledglings perched alone; becomes more frequent, often in bursts, when parent seen approaching. Older fledglings give 'tip' calls in flight, which appear to develop into adult flight-call. (Andrew 1957b.) When at least 10–11 weeks old, juveniles give 'juvenile subsong': series of sounds with wide frequency range, lacking rhythm and structure of adult song, to which it bears no resemblance. In late summer, all juvenile subsong is from non-territorial ♂♂ of the year; highly variable between individuals, and continues into autumn, then ceases from about November until middle of February, after which young ♂♂ begin again with subsong, which quickly develops into full song. (Diesselhorst 1971b.) DWS

c. 0·11–0·17 s, mean frequency usually descending slightly (Fig VI, 1st unit; Fig VII, 2nd unit; Fig VIII). Expresses strong fear; elicited by humans near nest or fledglings, also occurs during conflicts over territory. (Andrew 1957b.) (4) Flight-call. Single or multiple 'tit' units (Fig IX: Andrew 1957b). Given singly when perched, if there is tendency to fly, usually accompanied by tail-flicks. Rapid bursts of 4–5 'tit' units often given as bird takes off, and bursts of (1–)4–5 units given at intervals in flight; sonagram starts with a burst of 8 such units. When given by perched birds, expressing fear (tendency to fly), often alternates with 'cheep' calls (call 3a) (Fig IX), singly or in groups of 2 or more. (Andrew 1957b.) Bergmann and Helb (1982, which see for sonagram) described 'trilling' 'tirr' as flight-call. (5) Calls associated with aggression. Head-forward aggressive display (see Social Pattern and Behaviour) usually accompanied by 'chaa' call, harsh and

Breeding. SEASON. Britain: eggs laid beginning of April to beginning of September; of 388 clutches, 49% laid 6

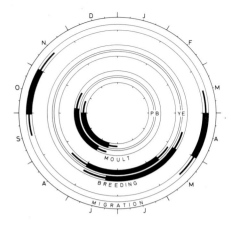

May to 2 June; 71% May-June (Parkhurst and Lack 1946, which see for regional variation); of 680 clutches, 75% laid 22 April to 16 June; 53% 6 May to 2 June (Peakall 1960, which see for variation). Finland: see diagram; in south, 42% of clutches started before mid-May, 6% in April; further north only 19% started before mid-May and none in April (n = 105); clutches occasionally found August (Haartman 1969). France: eggs laid from end of April to July, rarely into September; average date of first egg over 21 years 3 May; 65% of 105 clutches laid mid-May to mid-June (Labitte 1954, which see for variation). Italy: egg-laying from end of April to July (Brichetti 1976). Czechoslovakia: eggs laid early April to end of August, mainly end of April to end of June, peak late April to late May (Hudec 1983). Belorussiya: in Minsk, complete clutches early to mid-May; 2nd clutches early to mid-June. Ukraine: in Khar'kov region, eggs laid from early May, 2nd clutches started early June. (Dementiev and Gladkov 1954.) SITE. Nearly always on or very close to ground, well hidden among grass or herbage (Dementiev and Gladkov 1954; Haartman 1969; Brichetti 1976). Typically against bank or base of hedge, small tree, bush, or well inside bramble *Rubus* (Haartman 1969; Campbell and Ferguson-Lees 1976). Of 603 nests in Britain, 31% on ground (12% in grass and 8% on bank), 57% in bush or tree (17% in hawthorn *Crataegus* and blackthorn *Prunus*, 15% in bramble, 10% in gorse *Ulex* and broom *Cytisus*, 3% in tree), and 12% in small plants; average height 46·2 cm (0-182) (Peakall 1960). See also Parkhurst and Lack (1946), where 12% of 149 nests on ground, 64% less than 0·5 m above ground, and 0·7% above 1·25 m. In Rheinland (western Germany), 25% of 277 nests on ground (mostly in grass or cereals), 23% in conifers (mostly spruce *Picea*), 17% in broadleaved trees, 17% in thorn bushes (mostly bramble), and 17% in herb layer (mostly nettles *Urtica*); 76% of nests less than 1 m above ground (Mildenberger 1984); for Oberlausitz (eastern Germany), see Eifler and Blümel (1983). In parts of Scandinavia and USSR, all nests on ground, often in heather and blueberry *Vaccinium* (Hasse 1963). Ground nests more common in April than

later in season (Peakall 1960; Glutz von Blotzheim 1962; see also Parkhurst and Lack 1946). Nest: dry grass, plant stems, straw, leaves, and some moss lined with rootlets, fine grass, and horsehair (Jourdain 1906; Haartman 1969; Campbell and Ferguson-Lees 1972; Harrison 1975) although according to Hasse (1963) and Makatsch (1976) only rarely with hair. In USSR, horsehair apparently much more abundant in nests in wet years (Dementiev and Gladkov 1954). Nests in bushes larger and bulkier than those on ground (Jourdain 1906; Makatsch 1976). Outer diameter 11·5-13·0 cm, inner diameter 5·5-7·0 cm, overall height 5·5-8·0 cm, depth of cup 4·0-4·5 cm (Dementiev and Gladkov 1954; Makatsch 1976); outer diameter up to 28 cm recorded (Hasse 1963). Building: by ♀ only, mostly in early morning and evening, sometimes accompanied by ♂; material gathered from immediate vicinity of nest, no more than c. 40 m distant; takes 4-5(-8) days (Wallgren 1956; Hasse 1963; Harrison 1975; Makatsch 1976). EGGS. See Plate 33. Sub-elliptical to short sub-elliptical; smooth and slightly glossy (Harrison 1975), or non-glossy (Makatsch 1976). Highly variable in colour; white, tinted bluish, greyish, or purplish, usually with faint, fine spotting or thin scribbles in pale violet-grey or reddish-purple, and with sparse, bold, irregular scrawls, small blotches, and fine hairlines in black or purplish-brown (Harrison 1975; Makatsch 1976); hair-streaks may be distributed uniformly over whole surface or may form ring around broad end; some eggs warm red-brown with a few blackish hairlines, others unmarked white or pale blue (Witherby *et al.* 1938; Dementiev and Gladkov 1954; Harrison 1975; Makatsch 1976). Nominate *citrinella*: 21·7 × 16·4 mm (19·0-24·8 × 15·0-18·2), n = 320; calculated weight 3·0 g. *E. c. caliginosa*: 21·9 × 16·2 mm (19·7-25·9 × 15·0-17·5), n = 100. (Makatsch 1976.) *E. c. erythrogenys*: 21·0 × 16·3 mm (19·8-22·8 × 15·0-17·2), n = 54; calculated weight 2·86 g (Schönwetter 1984). Clutch: 3-5 (2-6). In Finland, of 126 clutches: 3 eggs, 7·1%; 4, 39·7%; 5, 50·0%; 6, 3·2%; average 4·49 (Haartman 1969). In Czechoslovakia, of 259 clutches: 2 eggs, 2·3%; 3, 20·5%; 4, 57·1%; 5, 20·1%; average 3·95 (Hudec 1983). In Britain, average 3·29 (n = 809); increases from early spring until June and then declines, with 3·28 in May (n = 451), 3·35 June (n = 225), 3·20 July-August (n = 98); maximum 27 May to 9 June, 3·6 (n = 173) (Parkhurst and Lack 1946); see also Peakall (1960). Similar pattern of seasonal variation in Switzerland and Germany (Glutz von Blotzheim 1962; Mildenberger 1984). 2(-3) broods over west and south of range (Harrison 1975; Brichetti 1976; Makatsch 1976; Mildenberger 1984); 2 broods in USSR (Dementiev and Gladkov 1954) and Finland (Wallgren 1956). Eggs laid daily, in early morning (Dementiev and Gladkov 1954; Hasse 1963; Harrison 1975). INCUBATION. 12-14 days (11-19) (Dementiev and Gladkov 1954; Peakall 1960; Hasse 1963); average in Britain 13 days (n = 36) (Peakall 1960). Generally by ♀ only (Wallgren 1956; Makatsch 1976), but see Witherby *et al.* (1938), Dementiev and Gladkov (1954), Hasse (1963), and Haartman (1969) for reports of ♂ occasionally taking over for short periods. Starts with last egg so that hatching takes place over a few hours (Glutz von Blotzheim 1962; Hasse 1963; Makatsch 1976). At 2 nests, in 60 hrs observation, ♀♀ sat on eggs for *c.* 77–88% of time, apparently irrespective of incubation stage (Wallgren 1956). YOUNG. Fed and cared for by both parents (Witherby *et al.* 1938; Wallgren 1956; Hasse 1963; Haartman 1969). If ♀ starts 2nd nest, ♂ feeds 1st brood alone for 1–2 weeks (Glutz von Blotzheim 1962). Usually only ♀ broods: in warm weather, all day for only *c.* 2 days, but in cool weather brooding may continue until fledging; ♂ which had lost mate brooded normally (Hasse 1963, which see for development of young). FLEDGING TO MATURITY. Fledging period 11–13 days (9–18); young may leave nest before fully fledged (Dementiev and Gladkov 1954; Peakall 1960; Haartman 1969; Brichetti 1976). In 26 nests, average fledging time 12·4 days (Peakall 1960). Young independent 8–14 days after fledging (Wallgren 1956; Hasse 1963). First breeding at 1 year old (Paevski 1985). BREEDING SUCCESS. In Britain, of 244 eggs in completed clutches, 58% hatched; hatching success increased through season from 32% (n = 53) during 2nd half of May to 73% (n = 330) late July and August; 31% of 70 nests with full clutches failed completely (Parkhurst and Lack 1946); 70·7% of 135 clutches produced nestlings and 44·6% fledged young (Peakall 1960). In Oberlausitz, of 34 clutches, 27% destroyed as eggs, 21% as nestlings, so 52% produced fledged young; *c.* 3 fledged young per successful pair (Eifler and Blümel 1983). Nests frequently destroyed by small rodents, crows *Corvus*, Jay *Garrulus glandarius*, and Magpie *Pica pica* (Labitte 1954; Hasse 1963, which see for review of predators). MK, BH

Plumages. (nominate *citrinella*). ADULT MALE. In fresh plumage (October–January), tips of feathers of forehead and crown dark olive-brown to greenish-olive with black shaft-streak, bases extensively bright yellow. Supercilium bright yellow, partly mottled greenish-olive from above the eye to above rear of greenish-olive upper ear-coverts. Lore mottled grey and yellow; eye-ring narrow, yellow. Stripe from gape to middle of ear-coverts yellow, partly mottled greenish-olive, bordered by narrow dusky line below, latter extending into broad olive-black crescent on lower ear-coverts. Lower cheek and side of neck yellow, marked with narrow or wide dusky olive or chestnut malar stripe. Nape yellow with olive specks, marked olive-black at each side. Upper mantle lime-green, merging into rufous-cinnamon on lower mantle and scapulars; centres of feathers streaked black, sides fringed yellow or lime-green. Back, rump, and upper tail-coverts bright rufous-chestnut, feathers with yellowish or pale buff tips. Chin green-yellow; throat sulphur-yellow to bright yellow, marked with olive dots; band across upper chest greenish-olive, sometimes mixed with some yellow, usually bordered above by spotted dusky olive bar, which is sometimes speckled black; band across lower chest, side of breast, and upper flank rufous-cinnamon with ill-defined olive and yellow streaks. Lower flank streaked rufous-cinnamon, dull black, and pale yellow; belly, vent, and under tail-coverts bright sulphur-yellow, mottled grey or rufous on thigh and marked with

ill-defined grey or rufous streaks on coverts. Tail-feathers dark sepia-brown with narrow whitish fringes on inner webs and green-yellow fringes on outer webs; tip of inner web of outer two pairs (t5 and t6) each with a long white wedge, extending over about half of length of t6 and about a quarter of t5. Flight-feathers, greater upper primary coverts, and bastard wing greyish-black, outer web of most of them with narrow and clear-cut yellow or green-yellow fringe, but outer web of secondaries more broadly fringe buff-yellow (outer feathers) or rufous-yellow (inner ones). Greater and median upper wing-coverts and tertials greyish-black or dull black, tips of median and outer webs and tips of greater coverts and tertials broadly fringed rufous-cinnamon, fringes at bases partly yellow. Lesser upper wing-coverts olive-yellow, dull grey centres largely concealed. Under wing-coverts and axillaries pale yellow, longer with pale grey centres, shorter with dark grey centres. *In worn plumage* (about February–May), head and chin to throat yellow, usually except for remnants of olive-and-black feather-tips on forehead, side of crown, rear crown, side of nape, and shorter upper as well as longer lower ear-coverts; malar stripe either prominent or virtually absent, olive or (occasionally) chestnut. Band from lower side of neck extending over upper chest more or less uniform greenish-olive, but sometimes bleached and abraded to green-yellow; band across lower chest and side of breast more clearly rufous-cinnamon, though usually mixed yellow, and rufous-cinnamon sometimes confined to rather narrow streaks. Flanks more sharply streaked rufous and black on clearer light yellow ground. Back, rump, and upper tail-coverts virtually uniform rufous-chestnut or cinnamon. Pale fringes of tail-feathers bleached to off-white, sometimes largely worn off; fringes of flight-feathers, primary coverts, and bastard wing bleached to pale green-yellow or grey-yellow; cinnamon borders of tertials and greater coverts partly or fully bleached to pale grey. ADULT FEMALE. Like adult ♂, but forehead and crown rather evenly streaked buff-brown and black, hardly any yellow on feather-bases visible: generally, less than half of each of the crown-feathers yellow, and this largely concealed (in ♂, over half of each feather yellow, more exposed) (Svensson 1992). Side of head and neck olive-brown or grey-olive, pure yellow restricted to patch in front of eye and to some mottling on middle of ear-coverts, on upper cheek, and on upper side of neck; usually no clear yellow supercilium and no clear yellow streaks behind eye or on cheeks. Lower cheek yellow, marked with indistinct broken dusky olive malar stripe. Upperparts as in adult ♂; underparts as adult ♂, but ground-colour often slightly paler yellow, side of breast and entire chest marked with ill-defined greyish-olive or olive-brown streaks, upper chest less uniform greenish-olive than adult ♂, lower chest virtually or fully without rufous; streaks of flanks and under tail-coverts on average narrower, sharper, and less rufous, upper belly and side of belly marked with narrow dusky shaft-streaks. Tail and wing as adult ♂, but broad borders of greater upper wing-coverts and tertials sometimes more pink-cinnamon, less deep rufous. *In worn plumage*, head and chin to throat more extensively yellow, but heavily mottled olive, much less yellow than in worn adult ♂, head pattern less distinct; chest more clearly streaked olive (without rufous of adult ♂). NESTLING. Down smoke-grey, plentiful and up to *c.* 1 cm long; on head, shoulder, and back (Heinroth and Heinroth 1924–6; Witherby *et al.* 1938; Hasse 1963). For development, see Hasse (1963). JUVENILE. Forehead and crown evenly marked with fine dull black and buff-brown to pink-buff streaks. Supercilium indistinct, pale buff or pink-buff, speckled dull black. Eye-ring narrow, pale buff, fairly distinct. Lore and patch below eye mottled pale buff and dusky grey. Upper cheek and ear-coverts sepia-brown or dark olive-brown, often somewhat paler olive-brown on middle of ear-coverts and often bordered above by short dark brown stripe behind eye and below by dull black stripe backwards from gape. Remainder of upperparts like adult, but black streaks on mantle and scapulars broader, back, rump, and upper tail-coverts mottled black, and ground-colour of back and rump hardly more rufous than mantle and scapulars, less chestnut than in adult. Some variation in ground-colour of upperparts and side of head and neck, some birds more buffish (as described above), others more yellowish, and brightest yellow birds of latter type somewhat resemble adult ♀, but dark streaks on upperparts broader, back and rump paler, more golden-buff, and entire feathering shorter and looser. Ground-colour of underparts light yellow to yellowish-white, sometimes washed pink-buff on side of breast, chest, and flank; chin to chest and flank heavily marked with sharp dull black streaks. Tail, wing, and tertials as adult, but see 1st adult, below. FIRST ADULT. Like adult, but juvenile tail, flight-feathers, and greater upper primary coverts retained. Tips of tail-feathers, primaries, and primary coverts more worn than in adult at same time of year; tips of tail-feathers often distinctly pointed, usually less broadly rounded or truncate than in adult. See Svensson (1992) for shape of tail-feathers and abrasion. Head and body as in adult, but on average less yellow on feathers of forehead and crown, some ♂♂ with only half of feather yellow or slightly less, some ♀♀ virtually without yellow or whitish at feather-centre or -base. 1st adult ♂ often closely similar to adult ♀, apart from tail; 1st adult ♀ usually much paler yellow below than adult ♀ and 1st adult ♂, more clearly streaked below, and with more diffuse white on tail.

Bare parts. ADULT, FIRST ADULT. Iris dark brown. Bill blue-grey, darkest on culmen and gonys, paler on cutting edges. Leg and foot light brown, pale flesh-brown, or pink-flesh, tinged grey on toes. (Hartert 1903–10; Witherby et al. 1938; RMNH, ZMA.) NESTLING. Bare skin, including leg, pink-flesh or reddish-flesh. Upper mandible brownish, lower grey-yellow; mouth flesh-red, gape-flanges yellow (Heinroth and Heinroth 1924–6; Hasse 1963). JUVENILE. Bill pale horn-grey, paler and less bluish than in adult. Leg and foot light pink-flesh. (Heinroth and Heinroth 1924–6.)

Moults. ADULT POST-BREEDING. Complete; primaries descendent. Exceptionally, primary moult suspended (Herroelen 1980). In Britain, starts with p1 between mid-July and end of August, occasionally from early July or as late as mid-September. Total duration of primary moult c. 54 days; moult completed early September to mid-October, occasionally mid-August to early November (Ginn and Melville 1983). Individual duration of moult has also been estimated at over 80 days, with onset of moult in population spread over c. 2 months; of 64 birds in moult, mean 3·8 primaries growing at same time (Newton 1968). In north-west Russia, starts with p1 1 July to 10 August (mainly 15 July), completed with p9–p10 10 September to 31 October (mainly 25 September); ♂ starts c. 2 weeks before ♀ (Rymkevich 1990). POST-JUVENILE. Partial: head, body, lesser, median, and usually all greater upper wing-coverts and tertials, and often central pair of tail-feathers. Starts soon after fledging, completed August–October. (RMNH, ZMA.)

Measurements. ADULT, FIRST ADULT. Nominate citrinella. From south-east England and Fenno-Scandia south through western and central Europe to Spain and Italy, all year, skins (RMNH, ZMA). Bill (S) to skull, bill (N) to distal corner of nostril; exposed culmen on average 3·0 shorter than bill (S).

WING	♂	88·3 (3·19; 162)	80–96	♀ 83·6 (2·19; 74)	79–90
TAIL		70·2 (3·38; 83)	62–79	66·9 (3·09; 50)	60–76
BILL (S)		14·2 (0·88; 83)	12·4–16·0	13·8 (0·90; 49)	11·9–15·5
BILL (N)		8·0 (0·45; 83)	6·8–9·4	8·0 (0·39; 49)	7·1–8·9
TARSUS		19·4 (0·88; 78)	17·0–21·3	19·2 (0·83; 46)	17·4–21·8

Sex differences significant for wing and tail. Retained juvenile wing and tail of 1st adult shorter than in older birds; thus, wing, adult ♂ 91·8 (1·64; 13) 89–95, 1st adult ♂ 90·0 (2·71; 29) 86–96, adult ♀ 86·7 (1·89; 10) 84–90, 1st adult ♀ 84·4 (1·87; 8) 81–87 (ZMA).

E. c. caliginosa, Ireland, Scotland, and western England, all year; skins (RMNH, ZMA).

WING	♂	86·4 (3·35; 10)	80–91	♀ 79·0 (1·41; 3)	77–80
TAIL		67·9 (2·51; 10)	63–72	61·0 (0·82; 3)	60–62
BILL (S)		13·8 (0·82; 10)	11·9–14·6	14·0 (0·66; 3)	13·4–14·7
BILL (N)		7·9 (0·30; 10)	7·4–8·4	7·6 (0·86; 3)	6·6–8·7
TARSUS		19·4 (0·87; 9)	17·9–20·7	18·7 (0·67; 3)	18·0–19·6

Sex differences significant for wing and tail.

E. c. erythrogenys. Rumania, European Russia, and central Asia, all year; skins (RMNH, ZMA).

WING	♂	89·9 (2·81; 13)	85–96	♀ 86·3 (2·68; 9)	83–91
TAIL		71·0 (2·86; 13)	65–76	67·3 (3·10; 7)	64–74
BILL (S)		14·2 (0·75; 13)	12·5–15·4	13·8 (0·50; 7)	13·1–14·7
BILL (N)		8·1 (0·48; 13)	7·6–9·4	7·9 (0·43; 7)	7·0–8·4
TARSUS		19·6 (0·61; 13)	18·3–20·5	19·4 (0·70; 6)	18·4–20·6

Sex differences significant for wing and tail.

Wing. *E. c. caliginosa*. (1) Britain (except central, south, and east England) (Harrison 1955; Niethammer 1971; ZMA). Nominate *citrinella*. (2) South-east England (Harrison 1955; Niethammer 1971; ZMA; see also Eck 1985). (3) South-east France, winter (G Olioso). (4) Central and north-east France and German Rhine valley (Stresemann 1920; Eck 1985). (5) Netherlands, May–July (ZMA). (6) Netherlands, September–April (ZMA). (7) Southern Germany (Dornberger 1978). (8) Germany (Niethammer 1971). (9) Eastern Germany (Sachsen, Jerichow, and Lausitz), whole year (Eck 1985, 1990). (10) Norway (Haftorn 1971). (11) Sweden (Eck 1985; ZMA). (12) Scandinavia (Harrison 1955). (13) Croatia (Simeonov and Doïchev 1973). Intermediates between nominate *citrinella* and *erythrogenys*. (14) Southern Yugoslavia (Stresemann 1920). (15) South-east Lithuania and north-west Belorussiya (Stresemann 1920; Eck 1985). (16) Central and southern Belorussiya (Alex 1985). (17) North-east Belorussiya and western European Russia (Smolensk and Novgorod) (Alex 1985; Eck 1985). *E. c. erythrogenys*. (18) Rumania (ZMA). (19) Bulgaria (Simeonov and Doïchev 1973). (20) Central European Russia (Moscow to Saratov) (Stresemann 1920; Eck 1985). (21) Western and central Asia (Stresemann 1920; Eck 1985; ZMA).

(1)	♂	86·1 (2·37; 74)	81–92	♀ 81·3 (2·13; 18)	76–84
(2)		86·1 (— ; 101)	82–93	83·8 (2·32; 28)	77–87
(3)		89·8 (3·03; 27)	83–94	85·9 (1·70; 26)	81–88
(4)		90·2 (— ; 37)	85–99	88·0 (5·10; 5)	85–97
(5)		89·9 (1·95; 9)	87–94	86·9 (2·33; 6)	83–90
(6)		90·6 (2·53; 14)	86–95	83·2 (1·55; 4)	81–85
(7)		91·7 (1·81; 34)	88–97	— (— ; —)	—
(8)		89·7 (2·32; 90)	85–95	84·1 (2·21; 25)	83–89
(9)		90·7 (2·24; 152)	85–96	85·7 (2·19; 74)	81–90
(10)		— (— ; 21)	87–97	— (— ; 12)	82–92
(11)		91·1 (2·54; 18)	86–97	86·1 (1·06; 7)	85–88
(12)		90·8 (2·39; 29)	85–95	— (— ; —)	—
(13)		90·2 (2·51; 52)	—	83·3 (2·21; 14)	—
(14)		90·7 (2·02; 36)	86–96	84·8 (1·33; 16)	83–87
(15)		89·5 (2·87; 65)	82–95	85·6 (2·01; 9)	83–89
(16)		88·1 (2·59; 36)	83–93	84·1 (2·11; 25)	81–89
(17)		88·6 (1·84; 27)	86–92	84·8 (1·59; 26)	81–87

(18)		92·6 (2·12; 8)	88–95		87·7 (1·92; 5)	85–91
(19)		90·1 (2·89; 140)	—		86·3 (2·27; 72)	
(20)		92·1 (2·41; 17)	87–98		86·6 (1·29; 9)	85–90
(21)		91·7 (1·40; 9)	89–94		86·7 (2·35; 9)	84–90

For data from Carpathian basin (Croatia to central Rumania), with wing measured without full stretching, see Horváth and Keve (1956).

Weights. ADULT, FIRST ADULT. Nominate *citrinella*. (1) Netherlands, Germany, France, Sweden, and Switzerland, all year (Krohn 1915; Bacmeister and Kleinschmidt 1920; Weigold 1926; RMNH, ZMA). (2) South-east France, winter (G Olioso). (3) Norway, October–May (Haftorn 1971). (4) Sweden, all year (Zedlitz 1926). (5) Czechoslovakia, all year (Havlín and Havlínová 1974). (6) Germany, all year; (7) New Zealand (introduced birds of English origin), all year (Niethammer 1971). South-east Germany: (8) September–February, (9) March–July (Eck 1985). (10) Southern Germany, winter (Dornberger 1978).

(1)	♂	26·7 (4·94; 34)	14–36	♀	26·8 (5·22; 15)	18–35
(2)		28·9 (3·19; 27)	25–38		27·4 (1·98; 26)	23–30
(3)		31·5 (— ; 8)	29–36		31·4 (— ; 6)	28–35
(4)		32·8 (1·04; 11)	31–35		30·5 (1·04; 5)	29–32
(5)		29·9 (2·04; 57)	25–34		29·5 (2·37; 38)	26–36
(6)		28·4 (— ; 43)	25–32		27·7 (— ; 14)	14–31
(7)		25·7 (— ; 10)	23–30		25·2 (— ; 5)	24–28
(8)		31·0 (— ; 14)	26–34		30·8 (— ; 19)	28–36
(9)		29·8 (— ; 64)	26–35		29·1 (— ; 20)	26–36
(10)		29·8 (1·83; 35)	25–33		— (— ; —)	—

Baltic states, averages autumn: ♂ 28·4, ♀ 27·5 (total n = 427) (Dol'nik and Blyumental 1967). Poland, all year: sexes combined, 28·2 (62) (Busse 1970).

E. c. caliginosa. Skokholm Island (Wales): 24·4 (3) 23·1–26·4 (Browne and Browne 1956). Frost victims England: 19·0 (3) 18–21 (MacDonald 1962, 1963).

E. c. erythrogenys. Kazakhstan: ♂ 31·0 (39) 27–36, ♀ 30·6 (11) 27·5–34 (Korelov *et al.* 1974). Kirgiziya: ♂ 29·5–32·0 (3), ♀ 29·0–35·5 (6) (Yanushevich *et al.* 1960). Northern Iran, March: ♀♀ 27·0, 29·5 (Schüz 1959). Turkey, December: ♀ 29·0 (Kumerloeve 1970a).

JUVENILE. Nominate *citrinella*. Czechoslovakia, June–August: 29·7 (2·28; 12) 27–34 (Havlín and Havlínová 1974).

Structure. Wing rather short, broad at base, tip bluntly pointed. 10 primaries: p7 and p8 longest or either one 0–1 shorter than other, p9 0–4 shorter, p6 0–2, p5 4–8, p4 12–17, p3 16–20, p2 18–22, p1 19–23; p10 strongly reduced, narrow, 11·9 (10) 10–13 shorter than longest upper primary covert. Outer web of p6–p8 emarginated, inner web of p6–p9 with (very) faint notch. Tip of longest tertial reaches to about tip of p4. Tail rather long, tip forked; 12 feathers, t1 5–10 shorter than t4–t5, t6 often slightly shorter than t4–t5. Bill fairly short, about half of head length, rather deep but narrow at base (depth about 60% of length of visible part of culmen), tip laterally compressed. Culmen and cutting edges slightly decurved; upper mandible flattened dorso-ventrally, with distinct 'tooth' in middle, halfway along cutting edges; lower mandible deep at base, with distinct kink upward at gonys fusion, cutting edges strongly kinked downward at extreme base, kink with blunt 'tooth'. Nostril small, oval, partly covered at base by feathers of lore; some short soft bristles project obliquely downwards from each side of base of upper mandible. Tarsus and toes fairly short and slender. Middle toe with claw 19·8 (7) 18·5–21·6; outer toe with claw *c.* 75% of middle with claw, inner *c.* 70%, hind *c.* 78%.

Geographical variation. Rather slight, mainly involving colour. No differences in size except for birds of Britain, which are rather small (see Measurements). Nominate *citrinella* of Scandinavia and western and central Europe is generally rather dark and green in colour, with restricted amount of rufous streaking on chest and side of body. *E. c. erythrogenys* in east of range generally paler, more sandy-brown above, and with extensive deep rufous wash on chest, side of breast, and flank. These races connected by broad zone in which highly variable populations occur, running through west European Russia, Baltic states, western Belorussiya and Ukraine, western and central Rumania, and eastern Hungary (east of Tisza river) to eastern and southern Yugoslavia; see Stresemann (1920), Harrison (1954, 1955), Horváth and Keve (1956), Vaurie (1956e, 1959), Alex (1985), and Eck (1985). In both west and east, many populations can be recognized which differ slightly in colour when series are compared, each of which is sometimes considered a separate race, but none recognized here except for *caliginosa* from Scotland, Ireland, northern England, and Wales, which forms end of cline of increasing colour saturation running from central Europe to north-west. Mantle and scapulars of *caliginosa* warm rufous-brown, with limited yellow-olive fringing along sides of feathers, limited olive in hindneck and across upper chest, faint greenish tinge to yellow of head and underparts, and extensive but dull rufous on chest and side of body; in fresh plumage, top and side of head rather dark brown-green. (Clancey 1940; Harrison 1954, 1955; ZMA.) Typical nominate *citrinella* from Scandinavia and Denmark has mantle and scapulars rufous-brown but slightly paler than in *caliginosa*, yellow-olive fringes slightly broader, dark central streaks slightly narrower, olive of hindneck and of chest-band fairly distinct, yellow of head and underparts pure and bright, rufous on underparts rather restricted, less continuous than in *caliginosa*. Populations occurring south-east England, Netherlands, Belgium, western Germany, and northern and western France sometimes separated as *nebulosa* Gengler, 1920, but here included in nominate *citrinella*; mantle and scapulars of these populations lighter greenish-olive-brown than in Scandinavia, olive band across chest distinct (more so than in Scandinavia), yellow rather dull and pale; rufous on underparts restricted; birds from western half of southern England intermediate between 'nebulosa' and *caliginosa*. Throughout much of central Europe, from central Germany and Poland south to northern Spain, Italy, and northern Yugoslavia, birds occur which show greenish-olive-brown upperparts, like 'nebulosa', but with greener olive band on hindneck, brighter pure yellow head and underparts, and with limited amount of olive on upper chest (clear bar of dark olive dots when plumage fresh, not as full as in 'nebulosa', almost no olive on mid-chest when worn); these birds sometimes separated as *sylvestris* Brehm, 1831, but included here in nominate *citrinella*. In *erythrogenys*, mantle and scapulars rather pale and with narrow dark shaft-streaks, as in 'sylvestris', but general colour here pinkish-cinnamon, light buff-brown, or sandy-grey, depending on population, not greenish-olive; yellow of head and underparts pure but often less deep (especially in ♀); olive on chest often reduced (depending on population), rufous on chest and side of body extensive, often continuous, as in *caliginosa*, but often brighter, less dull. Amount of white in tail somewhat variable, but depends mainly on age and sex, scarcely on locality (C S Roselaar; see also Alex 1985 and Eck 1985). Birds with chestnut spotted or full malar stripe occur frequently in various populations, without clear trend; birds with patches of rufous elsewhere on head, throat, or upper chest occur mainly in east European Russia and Asia, probably due to introgression of characters of Pine Bunting *E. leucocephalos*.

MP

Emberiza cirlus Cirl Bunting

Du. Cirlgors   Fr. Bruant zizi   Ge. Zaunammer
Ru. Огородная овсянка   Sp. Escribano soteño   Sw. Häcksparv

Emberiza Cirlus Linnaeus, 1766

Monotypic

Field characters. 15·5 cm; wing-span 22–25·5 cm. Slightly smaller than Yellowhammer *E. citrinella*, with 20% proportionately shorter bill and 5–10% shorter, rather more rounded wings, but almost as long tail. Medium-sized bunting, recalling *E. citrinella* but with rather smaller, slighter, and more compact form most obvious in often flatter-headed and more hunched appearance. Plumage pattern also recalls *E. citrinella* but shows at all times dull greyish-olive to greyish-brown rump. Adult ♂ colourful, but much less yellow than *E. citrinella*, with striking grey-olive and black crown, eye-stripe, and bib, and russet-sided olive chest. ♀ much less easy to distinguish from *E. citrinella*, but ground-colour of plumage more buff than yellow with more linear face pattern and finer streaks below. Unobtrusive, spending much time on ground close to cover. Typical song and commonest calls differ from *E. citrinella*. Sexes dissimilar; little seasonal variation. Juvenile separable at close range.

ADULT MALE. Moults: July–October (complete); March–April (part of face). Forehead, crown, and nape grey to olive-green, noticeably streaked blackish except on nape; eye-stripe greenish-black, bib black, with pale fringes; all join to isolate long buffish or clear yellow supercilium and stripe from base of bill to rear of cheek and thus form striking linear pattern. Although often traces of yellow on middle and rear of cheeks below lower stripe, such marks insufficient to create deep panel (contra illustrations in, e.g., Peterson *et al.* 1983, Heinzel *et al.* 1972). Below bib and reaching sides of neck, striking but usually compressed yellow collar contrasts with olive breast-band and chestnut and buff patches on sides of chest. Mantle and scapulars chestnut, feathers with yellowish to greyish-buff fringes and terminal black streaks (latter showing particularly on mantle, leaving almost unmarked scapulars to form brighter rufous panels above wings); rump yellowish-grey to olive, only indistinctly streaked, but longest upper tail-coverts tinged rufous-brown, fully streaked black. Wings mainly rufous-brown, with broad chestnut edges to tertials and inner greater coverts; no obvious wing markings, but pale greyish tips to median coverts and particularly greyish- to olive-green lesser coverts may show. Tail dark brown, with paler greenish fringes and white outer web and large wedge on feather and small white wedge on next outermost showing well in flight (but white forms narrower edge than in *E. citrinella* which has outer webs of both these feathers fully white). Underparts below chest buffish-yellow, cleanest on belly

and rather finely streaked dark brown on flanks; visible under tail-coverts pure yellow or white. Bill distinctly bicoloured, with blackish-horn culmen and pale bluish lower mandible. Legs brownish-flesh, feet often more reddish. With wear, bird becomes more immaculate, showing greyer, more black-streaked crown, richer chestnut on back, blacker bib, and purer green and chestnut on chest. A few lack obvious yellow ground-colour; not known if these are immatures. ADULT FEMALE. Descriptions vary, mostly in strength of buff to dull yellow ground-colour to head and underparts but also in definition of head markings. All birds lack relative brightness of most ♀ *E. citrinella* and their strongly striated backs and underparts. Head basically pale yellowish-buff to yellowish-brown; far less distinctly patterned than in ♂ but more so than in ♀ *E. citrinella*, with (a) grey-washed and finely black-streaked crown, said to lack obvious pale crown-stripe of *E. citrinella* (Harris *et al.* 1989), but paler crown-centre clearly visible in photograph (Delin and Svensson 1988) and vagrant (D I M Wallace), (b) dull finely streaked fore-supercilium, usually strongly yellow in *E. citrinella*, (c) grey to olive-brown stripe behind eye, emphasized by fine dark streaks, (d) narrow to quite broad grey moustachial stripe reaching end of ear-coverts and, with eye-stripe, isolating pale yellowish central cheeks and ear-coverts which sometimes form even paler rear spot (all dark stripes rather more distinct than in *E. citrinella* though lacking striking linear pattern of ♂), (e) narrow, dark malar streak, unlike variable but often broader stripe of *E. citrinella*. Mantle greyish-buff to greyish-brown; scapulars buff-brown to dull chestnut; both tracts streaked black-brown. Important to note that scapulars, particularly when worn, form noticeably warm panels above wings, with their chestnut tone matched only on inner greater coverts and tertial edges; such contrast scarcely visible on ♀ *E. citrinella*. Rump dull, quite cold greyish- to olive-brown, variably streaked but ending in more rufous upper tail-coverts, as ♂. Wings patterned as ♂ but duller, with fringes on greater coverts and tertials less rufous. Tail as ♂ but with less white on outer feathers. Breast suffused grey-buff, with pale chestnut patch on side of chest, strongly but finely streaked overall with black-brown pencil marks and these continuing along pale, dull yellow flank and onto under tail-coverts; centre of belly unmarked yellow. Important to note that tone of yellow never matches characteristic lemon shade of *E. citrinella*, being duller or warmer. Bare parts as ♂. With wear, crown, breast, and rump become greyer, eye-stripe brighter, mantle and chest-patches more rufous, and underparts generally more streaked. JUVENILE. Resembles ♀ but ground-colour of plumage rather cleaner and less buff while all streaks more pronounced and noticeably thicker on throat and breast. Distinction from *E. citrinella* apparently unstudied but lesser coverts dark greyish-olive, not brown as *E. citrinella*, and tips of larger coverts duller. FIRST-YEAR. Both sexes resemble adult; may show more streaks on nape and rump.

Due to its scarcity in north-west Europe, field identification relatively little studied and subject to somewhat conflicting comments. Adult ♂ unmistakable, being separated from all other yellow-bellied *Emberiza* by its strongly linear face pattern and pale, dull rump. Immature ♂ and ♀ at all ages far less distinctive, easily confused with dull *E. citrinella* and ♀♀ and immatures of Pine Bunting *E. leucocephala* and Rock Bunting *E. cia* as all these may hide their rufous rumps and share general buffy look of many *E. cirlus*: see those species. Beware in particular similarity of one short call in *E. cirlus* and *E. cia*. Confusion not likely with Yellow-breasted Bunting *E. aureola* (distinctly smaller and shorter-tailed, with cleaner plumage tones, bold, unstreaked supercilium, paler stripes on mantle, white bar across tips of median coverts, and more restricted streaks on breast and flanks). Structure and behaviour variably described, and following notes stem from review of identification literature (from 1940) and recent English study (A Evans, RSPB). Somewhat slighter or slimmer than *E. citrinella*, with (1) slightly drooping appearance to bill (extension of culmen tip beyond lower mandible visible in photographs), (2) usually flatter head shape (enhanced by plumage pattern), but often slightly crested or peaked crown (particularly in excited ♂♂), (3) apparently short neck or head tucked into shoulders (as if half-shrugged, especially in winter), even back outline and hence more compact head and body contributing to hunched appearance on perch or crouched posture on ground, and (4) less ample but scarcely less long tail (often carried up, not trailing on ground). Compared to *E. citrinella*, appears less robust in flight; surprisingly, tail does look noticeably less long in silhouette and also lacks terminal width of *E. citrinella*. Measured differences in wing shape not yet recognized in field. Descriptions of flight action also vary. In tight spaces, appears lighter, less untidy on the wing than *E. citrinella*, dropping quickly from perch to ground or flying horizontally into cover; over longer distances flight may appear more dipping and volatile but typically (in English stronghold or as vagrant among *E. citrinella*) it shows only shallow undulations, lacking thrust of retreating *E. citrinella* and looking weak (A Evans, R A Hume). Gait a hop, at times slow and apparently shuffling like Dunnock *Prunella modularis*, at others fast and sprightly. Vagrant within flock of *E. citrinella* moved and fed more quickly (D I M Wallace). Stance less upright and attenuated than *E. citrinella*, with hunched form and particular habit of sitting down on one

or both tarsi, depending on angle of perch; often moves or sits with wing-tips drooped to greater extent than *E. citrinella*, fully exposing rump, but not observed to flick wings or tail or bob head (A Evans). Secretive in tree and bush cover, even skulking among ground plants, though ♂♂ perch openly when singing. Mainly sedentary but in England forms winter flocks of up to 80 birds which prefer stubbles and weedy fields, usually staying within 30 m of hedges, unlike *E. citrinella* (A Evans, RSPB). Normally keeps own company but vagrants outside breeding range join mixed finch and bunting flocks.

Voice distinctive. Typical song a brief, rapid, rattling or slightly harsh, rolled trill or tremolo, much recalling terminal rattle of Lesser Whitethroat *Sylvia curruca* but not carrying as far: 'zezezeze...'; quieter, abbreviated form given in winter (for variant songs, see Voice). Commonest call, used in contact by birds on ground, by paired ♂ and ♀, and occasionally in flight, a single short, rather thin and quiet monosyllable, 'sit', 'zit', 'tzip', 'sip', or, with melancholy timbre, 'tzepe' (A Evans); noticeably less metallic than common call of *E. citrinella*, recalling Song Thrush *Turdus philomelos* and (though less shrill) *E. cia*. Call of North African birds apparently disyllabic, 'zib-zib'. Escaping flock produces chattering chorus (A Evans). Rare flight-call runs the above units together, producing characteristic sibilant 'sissi-sissi-sip'. Agitation-calls include repeated loud 'tzip', accelerated into descending trill (from ♂♂), and single, loud, shrill 'tzeeep' (parents to young) (A Evans).

Habitat. Breeds within Mediterranean and adjoining oceanic temperate zones of south-west Palearctic; further limited by highly selective climatic, topographical, and ecological requirements. Except in England, is bounded by 20°C July isotherm (Voous 1960*b*). Extends to 17°C isotherm in southern Britain but perhaps more importantly limited as British resident to areas with mean January temperature above 6°C, and either at least 1500 hrs of sunshine or less than 105 cm of rainfall per year (Yapp 1962); other limiting factors are wind, night-frosts, altitude, slope, aspect, and exposure. In addition to such combination of sunshine, low rainfall, and mild winters, has equally exacting ecological requirements. These take somewhat different forms in north and south of range, although preference for benign, often sloping, and sunny terrain is general. In Britain, in contrast to Yellowhammer *E. citrinella*, avoids extensive open farmland, being confined mainly to small fields with plenty of hedgerow growth and tall trees, elms *Ulmus* being favoured before their widespread demise through Dutch elm disease. While not depending on access to water, favoured sites are often by or near coasts or tidal inlets, or in valleys of rivers and streams. Where such sites have been taken by human settlements their fringes are occupied, including large gardens and orchards, although the bird's love of cover and somewhat secretive habits prevent it becoming at all a familiar or tame neighbour. Peripherally, may occupy other sites such as rubbish tips, gravel pits, and heaths with scattered clumps of birch *Betula* and gorse *Ulex* (Sharrock 1976). Despite retiring nature, ♂ requires high and prominent song-posts, used most of year; can include not only outer branches of tall trees but human artefacts such as overhead cables. In winter, resorts to stubble but infrequently to stackyards, often showing preference for quite marshy ground (Bannerman 1953a). Declining British population depends on weedy stubbles, preferably bordered by large mature hedges (Evans 1992).

Near northern limits on European continent, where numbers and range have also recently declined, habitats include walls and ruins, and locally pinewoods on dry ground (Lippens and Wille 1972). In south-west Germany, favours eastern slopes even when stony and poorly vegetated, but further north in Rheinland prefers orchards and graveyards on gentle slopes (Niethammer 1937).

Study in Pfalz (Germany) showed restriction of occupancy to narrow linear belt *c.* 60 km long and only a few hundred metres wide, mostly between 200-300 m altitude with slope of 20-35° and mostly ESE-SSE aspect. Cultivation of vines accounted for 70-80% of most territories, with minimum of *c.* 10%. Favoured bramble *Rubus* and ivy *Hedera* on low walls, and also hawthorn *Crataegus* and blackthorn *Prunus*; avoided cleared areas and monocultures without trees. Microclimates were significant within general regime of long warm summer and, especially, short mild snow-free winter. Accordingly categorized as thermo-helio-xerophile species, subject even in optimum habitat to considerable fluctuations and withdrawals from occupancy. (Groh 1975.) In Switzerland, situation transitional between northern and southern patterns, most breeders below 800 m, although some slightly above 1400 m. Favours sunny, dry benign slopes with hedges or thick bushes, but also eroded, sparsely-vegetated terrain. Occupies vineyards, terraced cultivation, verges of railways and motorways and natural gulleys or screes overgrown with thickets. Appears to avoid neighbourhood of other breeding *Emberiza*. (Glutz von Blotzheim 1962; Lüps *et al.* 1978.)

Study of wintering population near Basel showed need for 2 fundamental habitat components: dense thicket with branches reaching to ground, and scattered taller trees for resting and preening immediately adjoining foraging areas, forming compact mosaic of open dry sparsely vegetated or fallow ground mixed with low growth up to 20 cm tall and many stems lying broken on ground; such habitats strongly preferred (Salathé 1979).

In Spain, breeding occurs in glades and open patches of cultivation in cork oak *Quercus suber* woods near Gibraltar, but also at *c.* 1000 m in mountains, while in North Africa breeding occurs up to *c.* 1500 m in Atlas, as well as in lower valleys and on slopes covered with scrub and myrtle *Myrtus* on outskirts of cork oak forest (Bannerman

1953a); also recorded in clearings in palm *Phoenix* groves (Etchécopar and Hüe 1967). Evidently exceptionally responsive to wide diversity of environmental factors, and to subtle changes in them. Rejects as intolerable most habitats on offer.

Distribution. Range has contracted in north-west, probably due to climatic change, but has spread north in parts of south-central Europe and Balkans.

BRITAIN. Marked range contraction during 20th century, probably due to climatic and agricultural changes, to southern and south-western areas with warm summers and mild winters (Sharrock 1976); contraction continued throughout 1980s (Evans 1992). FRANCE. Some recent range contraction in north (Yeatman 1976). BELGIUM. Former breeder, confined to warmest areas. Nearly extinct at beginning of 1950s; singing ♂♂ present in small numbers up to 1984. (Devillers *et al.* 1988.) GERMANY. Formerly confined to south-west, near border with France. Breeding first recorded in Bayern 1984 (Thiede 1987). AUSTRIA. Breeding suspected in earlier years in southern Steiermark, confirmed in 1989; may also breed in Inntal west of Innsbruck (H-MB). HUNGARY. First recorded breeding 1975 (G Magyar). YUGOSLAVIA. Spreading slowly northwards inland (Ham and Šoti 1986; VV). RUMANIA. Has spread north and north-west from Danube defile (Talpeanu and Paspaleva 1979).

Accidental. Netherlands, Denmark, Poland, Czechoslovakia, USSR, Malta, Egypt, Canary Islands.

Population. Has declined markedly in Britain in last 60 years, become extinct in Belgium, and probably declined in northern France (see Distribution). Presumably increasing in parts of south-east Europe (see above), but no exact data available.

BRITAIN. Marked long-term decline (see Distribution). Most numerous and widespread in 1930s, after which steady decline. Probably 252-319 pairs 1968-72, 210-240 1973-6, maximum of 167 in 1982, 118 in 1989 (Sitters 1982, 1985; Marchant *et al.* 1990; Evans 1992). FRANCE. 100 000 to 1 million pairs (Yeatman 1976). GERMANY. Estimated 250 pairs (Rheinwald 1992). Decreased in southern Baden (western foothills of Schwarzwald) from 12 to 6 territories after hard winters of 1984-5 and 1985-6 (Federschmidt 1988). AUSTRIA. Fewer than 5 pairs (H-MB). HUNGARY. 2 pairs 1975; since 1981, 5-12 pairs (G Magyar).

Survival. Oldest ringed bird 6 years 1 month (Dejonghe and Czajkowski 1983).

Movements. Most populations essentially sedentary, but many leave colder parts of range in continental Europe in winter. Review by Zink (1985).

Essentially resident in England, but tends to wander in autumn and winter, occasionally occurring then away from breeding areas, and distribution of winter records implies

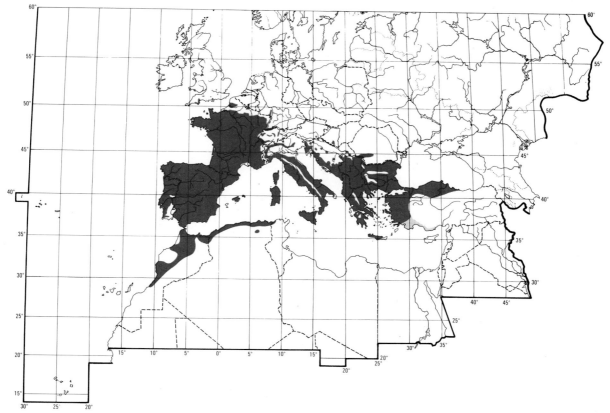

normal movements within 100 km of breeding areas (Witherby *et al.* 1938; Lack 1986). Ringing studies in Devon show most birds wintering within a few km of nest-sites, some within few hundred metres, but 2 birds moved 40 km south (Smith *et al.* 1992). Exceptional record of bird ringed as juvenile in Sussex in July and recovered on Isle of May (Scotland) in June of following year (Eggeling 1960).

Mainly sedentary in much of France, but local populations largely vacate north-east France, Alps and foothills, Landes, and basin of lower Garonne in winter (Yeatman-Berthelot 1991). Some parts of Switzerland also vacated in winter; at Col de Bretolet passage observed October–November and late February to mid-April, with over 100 recorded 1–2 November 1963 (Dorka 1966; Winkler 1984). In 4 years, latest records at breeding area near Thun (Switzerland) between 26 October and 23 November (Blaser 1974). Timing of occupation of wintering areas in Switzerland apparently variable: December–February at several, but from end of October (once 18 September) at another (Jacoby *et al.* 1970; Link and Ritter 1973*b*; Sermet 1973; Salathé 1979). In Pfalz (south-west Germany), only minority remain in breeding areas in winter, mainly ♂♂ of breeding pairs, in variable numbers from year to year (Groh 1975; see also Social Pattern and Behaviour).

13 ringing recoveries from continental Europe show long-distance movements: 11 involved wintering areas between west and SSW of ringing localities, 2 moved to wintering areas to south-east. Longest movements were from central France to north-east Spain, from Belgium to south-west France (725 km to south-west), from Italy to France (519 km WSW), and from north-west to southern Italy (650 km SSE). Recoveries of birds ringed in winter in south-west France at 160 km NNW and 255 km NNE in following winter suggest year-to-year variation in extent of movement. (Zink 1985.)

In Sardinia and Crete, probably makes altitudinal movements from montane to coastal regions in winter (Stresemann 1943*c*, 1956; Altner and Reger 1959; Mocci Demartis 1973; Zink 1985). Irregular winter visitor to Gibraltar, October–March, and perhaps also passage migrant (Cortés *et al.* 1980). Vagrant in Malta, where recorded at least 9 times, November–December (Sultana and Gauci 1982). Scarce and irregular passage migrant in western Sicily and on Ustica and Pantelleria islands, November–December and February–April (Iapichino and Massa 1989). In north-west Africa, local and rather erratic southward movements reported (Etchécopar and Hüe 1967); occurs more widely in winter, reaching northern edge of Sahara; no indication of immigration from Europe (but see above for evidence of passage at Gibraltar, Sicily, and elsewhere in Mediterranean region) (Heim de Balsac and Mayaud 1962). In Turkey mainly resident, wandering a little in winter (Hüe and Etchécopar 1970; Hollom *et al.* 1988). DTH

Food. Seeds, mostly of grass or cereals (Gramineae); invertebrates in breeding season. Feeds almost wholly on ground, sometimes on stems of grasses or low herbs, most commonly on trampled or grazed grass in fields, by tracks, at vineyard edges, and similar weedy places, rarely on bare soil; in winter, very often on rough pasture and stubble (in Devon, south-west England, usually barley *Hordeum*), perhaps taking seeds of weeds and grass rather than cereal grains. Structure of foraging sites more important than vegetation type; rarely more than 30 m from cover of bushes or hedge, in grass (etc.) no taller than *c.* 20 cm for good visibility. (Groh 1975; Salathé 1979; Hölzinger 1987; Cole 1990; Smith *et al.* 1992; A Evans.) In Devon, searches for invertebrates on rough pasture, embankments of sunken lanes, herb-rich meadows, in traditional orchards, and in vegetation among coastal rocks (Sitters 1991, which see for discussion of foraging strategy). Beetles (Coleoptera) and grasshoppers, etc. (Orthoptera) captured in short pursuit flight from ground; removes wings and legs of larger insects before eating (Groh 1975). Increase in winter cereal means less stubble, and loss of this as winter feeding site is probably important factor in population decline (Sitters 1991; Smith *et al.* 1992; A Evans). Winter feeding flocks very often monospecific, only rarely feeding with Yellowhammer *E. citrinella* and finches (Fringillidae); probably takes smaller seeds and less hard cereal grain then *E. citrinella* (Link and Ritter 1973b; Salathé 1979; Cole 1990; A Evans); commonly comes to scattered seed in farmyards and (apparently more readily than *E. citrinella*) at bird-tables (Glutz von Blotzheim 1962; Groh 1975; Smith *et al.* 1992); in Devon, 85% of winter observations in stubble, 15% at bird-tables (A Evans). In standing cereal, usually feeds in flattened patches or wheel-tracks, or at edges (Cole 1990; Sitters 1991). Most common feeding method is picking seeds from ground, but also climbs up stems to bend them over, stretches up to reach seed-head and pull it down if less then *c.* 20 cm high, or, less usually, clings to neighbouring stout stalk or wire and reaches over to get at head; in snow, feeds on tall stems or snow-free patches under bushes (Groh 1975; Salathé 1979).

Diet in west Palearctic includes the following. Invertebrates: mayflies (Ephemeroptera), grasshoppers, etc. (Orthoptera: Tettigoniidae, Acrididae), earwigs (Dermaptera), bugs (Hemiptera: aphids Aphidoidea), lacewings (Neuroptera: Chrysopidae), adult and larval Lepidoptera (Tortricidae, Noctuidae), flies (Diptera: Tipulidae), Hymenoptera (sawflies Symphyta, ants Formicidae, wasps Vespidae), beetles (Coleoptera: Carabidae, Staphylinidae, Curculionidae), spiders (Araneae), earthworms (Lumbricidae), snails (Pulmonata). Plants: seeds and other parts of olive *Olea*, nettle *Urtica*, knotgrass *Polygonum*, bindweed *Fallopia*, chickweed *Stellaria*, bittersweet *Solanum*, groundsel *Senecio*, grasses (Gramineae, including barley *Hordeum*, oats *Avena*, wheat *Triticum*, *Sorghum*, *Elymus*, *Poa*, *Lolium*, *Setaria*, *Digitaria*), moss (Musci). (Witherby *et al.* 1938; Tutman 1950, 1969; Kovačević and Danon 1952, 1959; Brighouse 1954; Groh 1975; Salathé 1979; Cole 1990; Sitters 1991; A and J Evans.)

Apparently no detailed study of adult diet. Probably takes only plant material outside breeding season, and invertebrates according to availability; preferred seeds seem to be couch-grass *Elymus* (probably main winter food in Pfälzerwald, south-west Germany), rye-grass *Lolium*, meadow-grass *Poa*, and soft cereal grains (Groh 1975; Salathé 1979; Cole 1990).

Young given mostly invertebrates, but also seeds. In Devon, of 174 adult visits (where items identified) to 2 nests, 34% with adult Lepidoptera, 20% Tettigoniidae, 18% larvae, 17% cereal grain, and 10% Acrididae; at a 3rd nest, 51% of 1108 visits were with barley and wheat grains, 25% invertebrates, and 24% unidentified; at 4th, of 41 invertebrates, 18 were Lepidoptera (16 caterpillars), 9 other larvae, 8 spiders, 3 Acrididae, 2 Diptera, and 1 Tettigoniidae. Of 7 faecal sacs from young, all 7 contained Diptera, 5 caterpillars, 5 Staphylinidae, 3 sawfly larvae, 3 Curculionidae; 6 contained seeds of barley, 5 nettle, and 4 *Poa*. At one nest, ♀ brought significantly more Lepidoptera, and ♂ more Diptera and spiders; more larval than adult Lepidoptera brought to earlier nests, vice versa to later ones, perhaps reflecting abundance. Also, only invertebrates recorded at early nests, grain being taken as it ripens towards end of July, and adults collect grain even when invertebrates easily obtainable, so cereals apparently important component of diet for later broods, especially in rain. Most food gathered 20-100 m from nest, though one pair flew 250 m to barley field. (Sitters 1991, which see for many details.) At one nest in Devon, *c.* 75% of food brought was adult Lepidoptera, remainder Orthoptera, beetles, Tipulidae, and grass seeds in some quantity (Brighouse 1954). In Pfälzerwald, insects and larvae main food, especially caterpillars of oak-roller moth *Tortrix viridana*; also adult Lepidoptera, Orthoptera, mayflies, aphids, beetles, and earthworms. Insects brought in gullet or bill, singly or bundled, and large ones are chewed before being fed to young. Food collected mainly in vicinity of nest, but sometimes from 200 m away (Groh 1975). In Switzerland, Orthoptera very important food of nestlings (Melcher 1951; Fuchs 1964). BH

Social pattern and behaviour. Well known. 10-year study of marked birds in Pfälzerwald (south-west Germany) by Groh (1975). Other studies on winter flock in Switzerland by Salathé (1979) and in Devon (south-west England) (Sitters 1991; A Evans, J Evans).

1. Outside breeding season some adult ♂♂ remain on territory, often accompanied by mate; otherwise occurs mainly in small parties and flocks. Birds remaining on territory only associate

with flocking conspecifics for at most a few hours or days. In south-west Germany, in hard winters most leave territories, but in mild winters most may remain: e.g. in mild winter 1974–5, when flocks did not form, of 23 territories investigated, pairs were found on 14, ♂♂ alone on 3. (Groh 1975.) Largest flocks recorded: 83, England (D Buckingham, A Evans); *c.* 44, Switzerland (Link and Ritter 1973*b*; Winkler 1984); 17 (usually fewer than 10), south-west Germany (Groh 1975). Pairs may remain together in flocks, which thus consist of 1st-years accompanied by variable proportion of adults (paired and unpaired). Adults in flocks are mainly from nearby breeding areas. (Groh 1975.) Winter densities in Strait of Gibraltar area (Spain) averaged 6·1 birds per km² in shrublands, 15·7 in forests of cork oak *Quercus suber* (Arroya and Tellería 1984). Winter flocks in Devon are typically single-species, and only occasionally include (e.g.) Linnet *Carduelis cannabina*, Chaffinch *Fringilla coelebs*, or Yellowhammer *E. citrinella* (A Evans). Similarly in Switzerland, flock rarely included other species (Salathé 1979). However, in some regions often flocks with *E. citrinella*, other *Emberiza*, and finches (Fringillidae) (e.g. Witherby *et al.* 1938, Winkler 1984, Lack 1986, Delin and Svensson 1988). In south-west Germany, often flocks with other birds in winter, mainly Tree Sparrow *Passer montanus*, Greenfinch *Carduelis chloris*, and *F. coelebs*, occasionally *E. citrinella* or Rock Bunting *E. cia* (Groh 1975). BONDS. Generally monogamous; 2 instances of ♂ with 2 ♀♀ both resulted from death of ♂ in adjacent territory (A Evans). Pair stays together in subsequent years, if both partners survive (Groh 1975; A Evans). No record of change of partner or of pair moving from territory; with 13 ringed pairs, 7 were together for single season, 4 for 2 years and 2 for 3 years, mortality probably accounting for changes of mate. (Groh 1975.) Polygamy not recorded in study by Groh (1975), but reported by Magnenat (1969). Rarely hybridizes with *E. citrinella* (Groh 1975; Brandner 1991) and *E. cia* (Groh 1975); see also Breeding Dispersion (below). Age of first breeding often 1 year (Groh 1975). Nest-building almost entirely or entirely by ♀ (see Heterosexual Behaviour, below). Incubation by ♀ only, occasionally fed on nest by ♂; young fed by both parents (see Relations within Family Group, below). BREEDING DISPERSION. Strongly territorial; ♂♂ vigorously exclude other ♂♂ from within *c.* 150 m of nest. Territories can be clumped (e.g. Sharrock 1976), with less than 200 m between adjacent nests, giving rise to erroneous notion that species is semi-colonial (A Evans). In south-west Germany, territory size 0·48–1·83 ha (mean 0·94, *n* = 22). Isolated territories were almost always larger than those grouped together. Sporadic feeding occurs outside territory, but pair usually remains within radius of *c.* 200 m. (Groh 1975.) See Groh (1975) for characteristics of features delineating territorial boundaries. In south-west England, breeding ♂♂ from 2 adjacent territories frequently trespassed on territory of 3rd ♂ and were pursued to territory boundaries; territories perhaps separated by strips of undefended land (Gait 1947). In Devon, some birds feeding nestlings regularly foraged 150 m from nest, exceptionally more than 200 m (Sitters 1991). Individual pairs are prone to move about from year to year (Sitters 1982). Replacement nest of one pair was *c.* 150 m from original nest-site (Sitters 1991), but nests used for 1st, 2nd, and 3rd broods of Swiss pair all within 12 m of each other (Fuchs 1964). Nests only used once, although same pair will sometimes nest in same bush in subsequent years (A Evans). In south-west Germany, territories often occupied all year, but no song or territory defence in cold winter months. Active territory defence begins in mild weather in January and may become intense in mild weather in February; *c.* 90% of territories occupied by end of March. Unpaired ♂♂ mostly have poor territories; may occupy small areas near established territ-

ory, where they sing and perch conspicuously, often coming to be tolerated there. Older ♂♂ occupy best territories and defend core areas intensively. Territory boundaries established mainly with song, backed up by threat and fighting (see Antagonistic Behaviour, below). Some parts of boundary between adjoining territories appear to be flexible but others defended rigidly. (Groh 1975.) In Devon, concurrent nests of *E. cirlus* and *E. citrinella* once found within 3 m of each other and within 50 m on several occasions, without evident interspecific conflict (A Evans). Occupied nests of these species similarly found 3·33 m apart in Switzerland (Blaser 1973). Nests of *E. cirlus* also reported very close to concurrently occupied nests of *E. cia*: 2 m and 9 m apart (Heseler 1966) and 10 m apart (Groh 1975). Few data on overall breeding densities (pairs per km²): in Bulgaria, range from 2 in Sofia city park (Iankov 1983) to 8 in vegetation dominated by *Paliurus spina-christi* and 40 where *Juniperus oxycedrus* dominant (Petrov 1982); in maquis in Morocco, 3·8–25·5, mean 16·0 (Thévenot 1982). German study showed considerable fidelity of ♂♂ to territory from year to year, although some moved: e.g. of 102 ringed adult ♂♂, 48 were found next year, 32 in same territory, 16 elsewhere (9 more than 1 km away, 4 at 1–5 km, others up to 14 km); in 3rd year after ringing, of 8 ♂♂ located, 5 still in same territory, 3 elsewhere (600 m, 3 km, and 20 km away). (Groh 1975.) Little evidence of natal site-fidelity, but ♀ ringed as nestling found breeding 400 m away the following year (Groh 1975). ROOSTING. In thick vegetation, 1 m or more above ground; in winter also in leafless bushes. Sleeps singly, in pairs and (in winter) in groups but individuals remain well separated on sleeping perches. (Groh 1975.) In continental Europe, generally roosts in bushes and scrub, in England probably also in hedgerows, but information lacking (Witherby *et al.* 1938).

2. In England, not shy when feeding on ground, typically allowing approach to within *c.* 15 m (A Evans). Similarly, in south-west Germany flies at *c.* 15 m, sometimes 8–10 m (Groh 1975). In Hampshire (England), family parties reported to approach to within 2–3 m of observer sitting motionless, as if inquisitive (Christie 1983). Apparently much shyer in winter flocks in Switzerland, flying at 25 m or more (Melcher 1951; Salathé 1979). Often freezes when disturbed on ground, before taking flight (A Evans), like some other *Emberiza* (Andrew 1956*b*). When anxious, spreads tail with flicking action that exposes white outer feathers, although this movement less evident than in *E. cia*. Similar tail-spreading seen in other contexts such as encounters between rival ♂♂, when it perhaps signals threat to approach. (Groh 1975.) Raising of crest feathers seen in many contexts (see also Song-display, below), including when seeking food, when giving Contact-call (see 2 in Voice) to warn young of potential predator, when in company with conspecifics and other birds, and once when lizard *Lacerta muralis* encountered (Groh 1975). Gives Warning-call (see 3 in Voice) on approach of raptor (A Evans). FLOCK BEHAVIOUR. Study of winter flock in Switzerland showed loose cohesion of flock when disturbed, so that it frequently split into smaller groups. Individual distance, usually of 0·3–0·4 m, was maintained when perching in bushes and feeding on ground. White outer tail-feathers may facilitate cohesion of flocks. After disturbance, flock moved progressively lower in cover before flying down to feed on ground adjacent to edge of cover. Retreat when disturbed also occurs in stereotyped manner. (Salathé 1979.) Feeding flocks in Devon rarely venture more than 30 m from hedgerows and when disturbed fly low back to cover, giving Contact-call; remain in hedge until danger passed and then drop singly to ground at base of hedge before working out towards middle of field (A Evans). SONG-DISPLAY. Song (see 1 in Voice) typically given from high in tall tree, but also from telegraph wire, bush, wall,

etc.; exceptionally on ground (e.g. Witherby *et al.* 1938, Groh 1975). Almost any raised perch may be used (A Evans). Sometimes also sings from inside thorny bushes (Christie 1983). ♂ of Swiss pair used several song-posts, some at 85–100 m from nest (Fuchs 1964). Rarely sings in flight; seen only 6 times in German study; rarely sings with food or other material in bill (Groh 1975). Main song-period in Devon late March to mid-August, but throughout range some song may occur in any month in fine weather (Witherby *et al.* 1938; Labitte 1955a; Lack 1986; A Evans). In April ♂♂ sing almost constantly during daylight (A Evans). During breeding season song begins soon after dawn; most persistent in morning and early afternoon. Little song used for 4–6 weeks after breeding season, when birds moulting. From mid-September to end of October, singing becomes commoner. (Groh 1975.) ♂ usually sings briefly before entering and after leaving roost, not infrequently doing this even in winter (Groh 1975). Song functions mainly in advertisement and defence of territory (see Antagonistic Behaviour, below), but ♂♂ in flocks occasionally give brief song (Groh 1975; Kreutzer 1985). Singing accompanied by conspicuous visual display that emphasizes head and throat pattern: with each song ♂ throws head right back, sometimes to vertical or even beyond, bill wide open, and whole body, lowered wings, and lowered tail vibrate; crown feathers often raised so as to show small crest (Groh 1975; A Evans). ANTAGONISTIC BEHAVIOUR. (1) General. Overt aggression noted mainly from ♂♂ in territory-defence, which involves song, threat-display, pursuit-flights, and fights (Groh 1975). In southwest Germany, seasonal variation in vigour of territory defence indicated by extent to which ♂♂ attracted by playback of song: mainly February–September (strongly from end March to beginning of June), but only a few approach in October–January. Both paired and unpaired ♂♂ respond; ♀♀ never react strongly. (Groh 1975.) Disputes between territorial ♂♂ reached peak in March–April. ♀♀ only rarely involved in expulsion of territorial intruders but they were twice seen to become involved while ♂♂ fighting. ♂♂ mostly unaggressive when in flocks, but sometimes Threat-call (see 4 in Voice) and Head-forward threat-display (see below) directed at other ♂♂ (Groh 1975). Fear responses apparently much like those of other *Emberiza* (Andrew 1956b) but not studied in detail. Crest is raised both in alarm and in aggression (A Evans), as in some other *Emberiza* (Andrew 1956b). (2) Threat and fighting. ♂ threatens other ♂♂ intruding on territory with Head-forward threat-display (Groh 1975): bird crouches low, holding wings out to side, slightly raising head and sometimes also tail, most often with bill agape, sometimes showing intention movements of approach, or raising head feathers to form small crest. This display used in all months of year, especially by ♂♂ but also ♀♀; very frequent in territorial threats in spring and summer and similarly frequent from ♂♂ responding to song playback of another ♂, before flying at tape-recorder (Groh 1975; A Evans). Same display used against conspecifics in winter in flocks and at roosts and sometimes also against other species of small bird in threats over singing perch, feeding place, or drinking place (Groh 1975). Threat-call commonly given during threat-display and before or during aggressive pursuit-flights. (Groh 1975; A Evans.) Territory defence by ♂ also involves pursuit-flights and sometimes fighting breast-to-breast in air. ♂♂ excited by territorial threat may make zigzag flights with short straight stretches. Fights occur mainly on borders of territories. ♂♂ sing during pauses between fights but seldom during fights. After fighting, ♂♂ often retreat further into territories, singing competitively. Pursuit-flights and fights with *E. citrinella* recorded occasionally in breeding season in Germany. (Groh 1975.) 2-year study in Devon gave no evidence of aggressive interactions with, or exclusion by, *E. citrinella* (A Evans),

although literature has suggestions (e.g. Sharrock 1976) that they compete in England. ♂ chased away singing Ortolan Bunting *E. hortulana* with pursuit-flight. Similar aggression against *E. cia* also recorded. (Groh 1975.) HETEROSEXUAL BEHAVIOUR. (1) General. Described as very undemonstrative (Witherby *et al.* 1938). Few data available on pair-formation. Established pairs remain together all year. Pairs may arrive at territory together early in year, perhaps implying that pair-formation occurs in winter flocks before territory established. Unpaired ♂♂ often occupy poor territories, where they sing and perch conspicuously. Very few unpaired ♀♀ evident in early spring, but these sometimes approach unpaired ♂♂ on their territories and this can result in pair-formation (see below). (Groh 1975.) (2) Pair-bonding behaviour. Apparently no elaborate courtship-display (Witherby *et al.* 1938; Groh 1975). In German study, Groh (1975) twice saw first meeting of ♀ arriving on territory of unpaired ♂. Both gave Contact-calls and picked here and there on ground without taking food. Thereafter ♂ followed ♀ closely and pair remained together, keeping close company. Generally, pair on territory appear inseparable: whenever ♀ flies, ♂ follows closely. (Groh 1975.) (3) Courtship-feeding. Not reported during pair-formation, although ♂ may feed incubating or brooding ♀ (see below). (4) Mating. Seen 4 times (March–April) by Groh (1975): ♀ crouched on perch, fluttering wings like juvenile begging for food; ♂ flew to her, mounted while ♀'s tail held 45° above horizontal, remaining on her back for 3 s; ♂ then flew away and ♀ ruffled plumage; no calls heard. Older report of single observation also noted mating preceded by ♀ quivering half-open wings and flattening out body and head (Witherby *et al.* 1938). (5) Nest-site selection and behaviour at nest. Site apparently chosen by pair together, as pairs seen inspecting possible sites up to a month before building (Groh 1975). Stem-display of ♂ occurs in breeding season, consisting of carrying of possible nest-material (leaves, moss, etc.). 33 observations were mostly in April (extreme dates 16 March to 26 July), display lasting 5–15(–70) s; mate present on 24 of 33 occasions. In view of frequent absence of ♀, display may function to synchronize efforts of partners at beginning of nest-building, rather than to stimulate onset of nest-building by ♀ (as in *E. hortulana*: Conrads 1969). Misinterpreted observations of Stem-display by ♂ may account for assumption in literature (e.g. Niethammer 1937, Mildenberger 1940) that both sexes build nest. (Groh 1975.) Observations of ♂ adding material to nest are few, and each involved only a few pieces of material being added (Venables 1940; Beneden 1946; Groh 1975), so little doubt exists that ♂ plays at most very minor role in nest construction. However, ♂ often accompanies ♀ collecting material, without playing any active part (Böhr 1962; Groh 1975). For details of collection of material and building, see Groh (1975). Incubation by ♀ alone, occasionally fed on nest by ♂ (Witherby *et al.* 1938; Sitters 1991; A Evans); one report of ♂ incubating (Truscott 1944). RELATIONS WITHIN FAMILY GROUP. Young fed by both sexes (e.g. Richmond 1963, Groh 1975, Sitters 1991, A Evans). Reports that young fed chiefly by ♀ and only exceptionally by ♂ (Witherby *et al.* 1938; Harrison 1975) apparently incorrect: ♂ made more feeding visits than ♀ at several nests, fewer than ♀ at others, but was involved to substantial extent at all of 10 nests studied by Sitters (1991). Young are brooded by ♀ to gradually decreasing extent during 1st week, especially during mornings. Older young, even those within a few days of fledging, are brooded (or at least sheltered) during rain. When wet weather necessitates more brooding by ♀, nestlings receive less food in total. Some of food brought to nest by ♂ is passed to ♀ when she is brooding; this food mainly delivered to nestlings but ♀ swallowed food at least once. ♀ removes faecal sacs while nestlings small. (Sitters 1991.)

At nest in Devon, ♂ fed ♀ before dividing food between small young (Brighouse 1954). See Groh (1975) for details of development of young; eyes open on 5th day. Young fed by parents for 2–3 weeks after fledging, although ♀ may start to build a new nest after 1 week or even earlier. Fledged young often remain close to nest, particularly in first few days, but they can move considerable distances, up to 100 m in several cases. (A Evans.) Groh (1975) recorded that fledglings beg for food with fluttering wings in usual manner for small passerines; begin to find own food on 18th day and cease giving food-call after 20th day; 16-day-old young recorded 140 m from nest-site, 18-day-old at 220 m. ANTI-PREDATOR RESPONSES OF YOUNG. Well-feathered nestlings crouched motionless in nest as observer approached, but flew from it when hand extended close to them (Jouard 1934). PARENTAL ANTI-PREDATOR STRATEGIES. (1) Passive measures. When parent returning to nest with food sees potential predator in vicinity it will perch until danger past, sometimes for 10–15 min, occasionally dropping or eating food it has brought (Sitters 1991; A Evans). (2) Active measures: against man. Readily deserts nests with eggs following disturbance (Jouard 1934). When disturbed by observer, ♂ of pair feeding young much bolder than ♀, approaching nest in open, whereas ♀ kept under cover (Richmond 1963). Apparent distraction-display (injury-feigning) reported by several observers, from both sexes by one observer, but this display apparently unusual (Witherby *et al.* 1938). Display in which wings were drooped by bird disturbed from nest is described by Zumstein (1927). Contact-call and apparently also Threat-call used to warn young of approach of man (Groh 1975). ♂ gave threat-display (see Antagonistic Behaviour) and Threat-call when nestlings handled (A Evans). Paillerets (1934) described 2 instances in which small nestlings were apparently carried from nests by parents following disturbance at nest-site: in 1st case, single nestling was re-found *c.* 100 m away; in 2nd case, 4 nestlings re-found widely separated *c.* 65 m away; with both broods, 'removed' young were in cover and parents continued to feed them. (3) Active measures: against other animals. Contact-call used to warn young of approach of potential predators, including cat and Jay *Garrulus glandarius* (Groh 1975). Parents warn fledglings of approaching raptor with Warning-call, which causes young immediately to stop begging and freeze (A Evans). ♂ gave threat-display (see Antagonistic Behaviour) and Threat-call while hopping/flying down a lane in apparent attempt to distract squirrel *Sciurus* away from newly fledged brood (A Evans).   DTH

Voice. Frequently used. This account based mainly on details of calls in Groh (1975) and analytical and experimental investigations of song by Kreutzer (1979, 1983, 1985, 1987, 1990) and Kreutzer and Güttinger (1991). Numerous other studies of song, e.g. Stadler (1924), Dobbrick (1933), Mayaud (1941), Haas (1943), Labitte (1955*a*), Voigt (1961), Groh (1975). For additional sonagrams, see Groh (1975), Bergmann and Helb (1982), and Kreutzer (1985, 1990).

CALLS OF ADULTS. (1) Song. (1a) Full song of ♂. Quick repetition of same unit or motif, typically 10–20 times, suggestive of song of Lesser Whitethroat *Sylvia curruca* (e.g. Witherby *et al.* 1938, Jonsson 1992) or Arctic Warbler *Phylloscopus borealis* (Delin and Svensson 1988). Several observers report more than one song-type given by individual ♂♂ (e.g. Chappuis 1976, Vinicombe 1988, Pinder 1991). Several different songs evident from record-

ings (P J Sellar), e.g. Figs I–V; Fig V perhaps shows mimicry of Yellowhammer *E. citrinella*. 2180 songs of 89 individuals from several countries analysed by Kreutzer (1985, 1990) showed duration of song $1.26 \pm SD0.22$ s, number of units $17.34 \pm SD5.70$; individual units have duration $43 \pm SD15.5$ ms, silences between units $31.5 \pm SD9.3$ ms; maximum frequency $5.59 \pm SD0.28$ kHz, minimum $3.21 \pm SD0.35$ kHz, range $2.37 \pm SD0.62$ kHz.

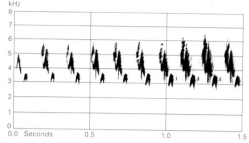

I P A D Hollom Spain April 1990

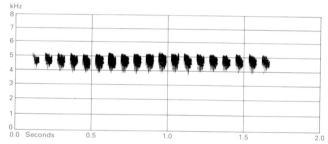

II V C Lewis England May 1977

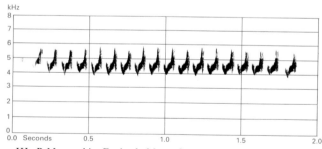

III R Margoschis England May 1982

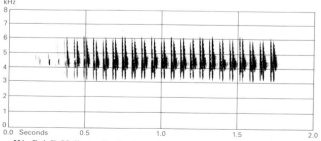

IV P A D Hollom England August 1979

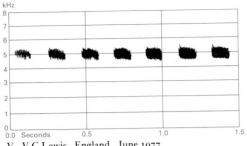

V   V C Lewis   England   June 1977

Individuals have 1–6 (mean 2·4 ± SD1·0) different songs. Differences between song-types consist mainly of differences in structure of units; a finite total of 61 song-types was found, which thus corresponds to 61 different units. Only 5 types of sub-unit are used, in various combinations, to form all of these units (usually 2–3 sub-units per unit, range 1–7); some combinations of sub-units are used frequently, some rarely, others not at all. Song is learned. For details, see Kreutzer (1985, 1990). Chappuis (1976) reported regional differences in songs, but fuller studies show that range of song-types occurring within populations is as large as occurs between different populations, so no true dialects. Range of song-types used is apparently the same throughout the species' geographical range, and ♂♂ recognize recorded songs from all parts of it. Individual ♂♂ show significant tendency to use song-types different from those of neighbouring ♂♂. Mimicry of first part of song of *E. citrinella* occurs. (Kreutzer 1983, 1985, 1990.) For detailed study of song recognition using experimentally synthesized songs, see Kreutzer (1983, 1985, 1990) and Kreutzer and Güttinger (1991). (1b) Occasional song by ♀ 'well authenticated' (Witherby *et al.* 1938), but not found by Groh (1975) or Kreutzer (1985, 1990). (1c) Subsong sometimes given in winter consists of solitary phrase of 4–5 units of full song, given quietly (A Evans). For information on incomplete songs of young birds and on song 'maturation', see Groh (1975) and Kreutzer (1985). (2) Contact-call. Single, short, thin, quiet 'tzip' or melancholy 'tzepe', less metallic-sounding than call of *E. citrinella* (A Evans). Rendered 'zree' or 'zi' (Fig VI) and thin 'zee' (Fig VII); also 'tit', given singly

or repeated (Andrew 1957); 'tsi' (Bergmann and Helb 1982); rather thin 'zit' or 'sip', sometimes run together in flight to produce characteristic sibilant 'sissi-sissi-sip' (Witherby *et al.* 1938); thin, sibilant, but quite audible 'ssi' with no final consonant (Christie 1983); 'zit' like Song Thrush *Turdus philomelos* (Delin and Svensson 1988); 'dsib', 'sib', 'zip', or 'schieb' ('Stimmfühlungslaut' of Groh 1975). Commonest call, given throughout year, but especially in breeding season. Used in many contexts: as contact-call between pair, by flocking birds, in fights with rivals, to warn young of predators, when seeking roost-site, by ♀ incubating, or by ♀ approaching nest (Groh 1975; A Evans). Ventriloquial; can be repeated, or elicit a reply. For example, before flock takes off in response to approaching predator, call is taken up by increasing numbers of birds and repeated more and more frequently, resulting in 'chattering' effect which reaches peak as flock takes off. (A Evans.) Used by young from day 16 (Groh 1975). Groh (1975) distinguished a quick, often thin, and rather loud 'si', 'sisisi', or 'dsibdsibdsib', given by both sexes in flight, as greeting-call, and in encounters with rivals, but may be better regarded as variant of 'dsib' call since it does not differ clearly in sound or context. (3) Warning-call. Longer and more drawn-out than call 2: 'tzeeep'; given between members of pair on approach of mild threat (A Evans); long, loud, descending 'zeee' (Fig VIII). Apparently same call described by Richmond

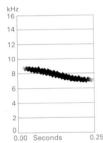

VIII   P A D Hollom   England   August 1979

(1963) as plaintive 'tsee' given by both parents when anxious near nest, reminiscent of call of Reed Bunting *E. schoeniclus*. Rendered 'siu' by Delin and Svensson (1988); 'zieh' (Bergmann and Helb 1982); 'zieh', not varying much, similar to call of *E. schoeniclus* and recalling warning call of *Turdus philomelos* (Groh 1975). A whiny, squeaky, indrawn 'week' or 'weet' given near nest (Witherby *et al.* 1938) may also be the same. Groh (1975) described this as an excitement- and alarm-call, given by both sexes; commonest in breeding season (not given November–January, not often in February); used in varied contexts including warning of potential predators (men, cats, dogs, Sparrowhawk *Accipiter nisus*). Variant is single, louder, more urgent-sounding, shrill 'tzeeep', given as raptor alarm-call, particularly by parents to recently fledged young (A Evans). (4) Threat-calls. Groh (1975) described 3 different calls used mainly in threat: quick 'dsibdsib-

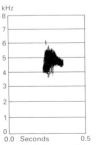

VI   D J Sutton   France   August 1975

VII   V C Lewis   England   June 1977

zizirrrdsibdsibdrrr' with many variants (with smacking and buzzing timbre), 'trrrp' or 'trr' (short, dry, buzzing, fairly loud), and 'rrrrrrrrr' or 'drrrrrrr' (quick, abrupt, harsh 'trill' recalling Wren *Troglodytes troglodytes*, mostly loud); in order listed, appear to indicate increasing anger. Given by both sexes, but especially by ♂♂ in territorial encounters, 'rrrrrrrrr' often being used in territorial fights. Lower-intensity versions used to warn young of presence of man. (Groh 1975.) Recording is of rattling tremolo (Fig IX) given in bursts. Other apparent renderings of these

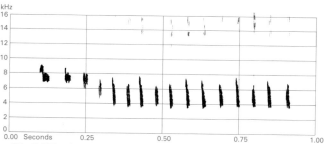

IX V C Lewis England June 1977

calls: loud, repeated 'tzip', accelerating into descending 'trill': 'tzip tzip tzip tzip-tzip-tziptziptzipppp', given by ♂ before or during antagonistic flight directed at another ♂ or (occasionally) predator (A Evans); lisping 'tsip-tsip-tsipipip', at times slurred off into a 'trill', given near nest with young (Richmond 1963); very fast series of clicks, 'zir' r' r' r' r', like crackling electricity (Delin and Svensson 1988); rattling 'tchrrr' or 'trrr' given chiefly by ♂ as alarm-call at nest (Witherby *et al.* 1938). (5) Fear-call. Gurgling, throaty sound heard 4 times from ♂♂ being handled (Groh 1975). (6) Other calls. (6a) Quick and rather loud 'dididi' or 'tititi' heard several times, given by ♀ to young when observer visited nests (Groh 1975). Rather forcible 'chit' given rarely when alarmed at nest (Witherby *et al.* 1938) may be the same. (6b) 'dsick' or 'slick', resembling short call of *E. citrinella*, heard from birds in flight and in territorial fights (Groh 1975). (6c) A low 'djä-djä' heard 3 times during encounters between 2 pairs (Groh 1975). (6d) Recording contains deep, abrupt 'chut' (Fig X: P J Sellar).

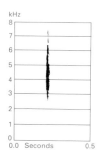

X V C Lewis England June 1977

CALLS OF YOUNG. From day 6, 'dsi-dsi', 'si', or 'sib', mostly disyllabic (seldom 1–3 units); quiet at first, louder from day 8, and loud and repeated ('dsidsidsidsi') when young fed (Groh 1975). Recordings contain fairly regular (Fig XI), and paired (Fig XII), call repetition, described

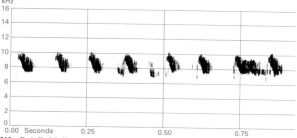

XI P A D Hollom England August 1977

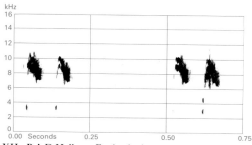

XII P A D Hollom England August 1977

as frenzy of tittering cries when parents arrived with food (Richmond 1963). Food-calls of young are loud, audible 70 m away (Sitters 1991; A Evans). DTH

Breeding. SEASON. Extended throughout range. Devon (south-west England): 1st clutches started 1st week of May, exceptionally late April; latest clutches started towards end of August (A and J Evans); for discussion, see Simson (1958). North-central France: over 10 years, mean start of laying 23 April; eggs recorded beginning of

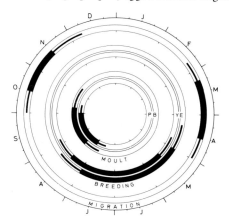

September (Labitte 1955a). Pfälzerwald and Rheinland (western Germany): see diagram; eggs laid May (early April to late August) (Groh 1975; Mildenberger 1984). For Hungary, see Waliczky et al. (1983). Sicily: once bred November (Iapichino and Massa 1989). Valencia (eastern Spain): mean laying dates for 1st and 2nd clutches 19 April and 25 May (end of March to mid-June) (Barba and López 1990). North-west Africa: beginning of April to June (Heim de Balsac and Mayaud 1962). SITE. Low down and well hidden in dense tree, shrub, hedge, or creeper; often on wall behind vegetation; rather uncommonly on ground (Groh 1975; Makatsch 1976; Federschmidt 1988; Cole 1990); in Sicily, also in rock crevice (Iapichino and Massa 1989). In Pfälzerwald, of 73 nests, 11% on ground, 12% below 0·5 m, 36% 0·5-1 m, 33% 1-2 m, 7% 2-3·2 m; 52% in bush or tree, 37% on vineyard wall; principal covering or supporting plants were ivy Hedera, bramble Rubus, grass (Gramineae), hawthorn Crataegus, and blackthorn Prunus; probably used according to abundance (Groh 1975). Of 53 nests in Rheinland, 26% in conifers, 26% thorny shrubs, 23% broad-leaved trees, 17% on ground, 8% on herbs (Mildenberger 1984). For tree species in Switzerland, see Glutz von Blotzheim (1962). In Devon, none of 97 nests were on ground; many in gorse Ulex and hazel Corylus (A and J Evans). Recorded at 6 m in elm Ulmus (Makatsch 1976). For positions on conifer branches, see Fuchs (1964). High proportion of sites on south-facing slopes (Groh 1975; Cole 1990). Nest: rather bulky and untidy; foundation of rough stalks, roots, grass, leaves, and moss (which is sometimes main material), lined with fine stems and much hair but not feathers (Haas 1943; Groh 1975; Makatsch 1976); in Pfälzerwald, 20 nests had average outer diameter 12·3 cm (9·9-17·5), inner diameter 6·6 cm (5·5-8·2), overall height 6·7 cm (5·8-10·2), depth of cup 4·3 cm (3·2-6·1), dry weight 24 g (14-51) (Groh 1975); overall height can be 15 cm (Makatsch 1976). Building: good evidence to suggest sometimes by both sexes, ♀ taking major part (Mildenberger 1940; Venables 1940; A and J Evans), but in extensive study in Pfälzerwald was by ♀ only, at times accompanied by ♂ (see also Social Pattern and Behaviour); at 11 nests took 4-6 days, building almost always in early morning (Groh 1975, which see for bouts and techniques); material collected from ground up to 200 m from nest; rootlets ripped up or grass pulled out with some effort (Brighouse 1954; Groh 1975; Federschmidt 1988). EGGS. See Plate 33. Sub-elliptical, smooth and very slightly glossy; greyish-white, often flushed pinkish or bluish, with many sepia spots, blotches, scrawls, and hairstreaks concentrated towards broad end, and pale grey undermarkings (Groh 1975; Harrison 1975; Makatsch 1976). 21·4 × 16·1 mm (19·2-24·0 × 14·5-18·0), n = 343; calculated weight 2·90 g (Schönwetter 1984). See Barba and López (1990) for 44 eggs from Valencia. Clutch 3-4 (2-5). In Pfälzerwald, of 57 complete clutches: 2 eggs, 9%; 3, 32%; 4, 58%; 5, 2%; average 3·53, no significant change over season (Groh 1975); for Rheinland, see Mild-

enberger (1984). In Algeria, of 50 clutches: 3 eggs, 16%; 4, 52%; 5, 32%; average 4·16 (Heim de Balsac and Mayaud 1962). In Valencia, average 3·81, n = 21 (Barba and López 1990); in north-east Spain, 3·88, n = 36 (Muntaner et al. 1983); in north-central France, 3·87, n = 52 (Labitte 1955a); in Devon, 3-4(-6) (A and J Evans). Eggs laid daily in early morning c. 1-3 days after nest completed (Groh 1975). 2 broods usual throughout range, often 3; in Valencia, 76% of 33 pairs double-brooded (Barba and López 1990); many records of ♀♀ laying 4-5 clutches including replacements (e.g. Groh 1975, A and J Evans); 2nd nest usually started (2-)6-8 days after independence of young or loss of clutch or brood, though sometimes about same time as fledging of 1st brood (Groh 1975). INCUBATION. 12-13 days (11-14, n = 11) from laying of last egg to hatching of last young; by ♀ only (Groh 1975). In Devon, 13-14 days (A and J Evans). In Pfälzerwald, 3 of 9 ♀♀ started fully with penultimate egg, 4 with last, and 2 started 1-2 days after last egg laid; ♀♀ had stints on eggs of 10-55 min and breaks of 5-40 min (Groh 1975). YOUNG. Fed and cared for by both parents; brooded by ♀ for up to c. 9 days (Haas 1943; Groh 1975, which see for brooding stints and feeding routines). See also Sitters (1991) for many details of brooding and feeding at several nests in Devon. FLEDGING TO MATURITY. Period from 1st hatching to 1st fledging 11-13 days (10-14, n = 10); young start to pick up food c. 6 days after fledging, independent after 8-16 days (Groh 1975, which see for development of young). Observations in Devon of young fed by parents 2-3 weeks after fledging (A and J Evans). Age of first breeding 1 year (Groh 1975). BREEDING SUCCESS. In Valencia, of 86 eggs, 65% hatched and 31% produced fledged young; hatching rate in April 84% (n = 19), June 56% (n = 18); fledging rate in April 50% (n = 16), June 70% (n = 10); 24% of eggs failed to hatch and 6% abandoned, 24% of nestlings starved and 14% taken by predators; 42% of 24 nests failed completely (Barba and López 1990). In Pfälzerwald, 37% of 73 nests failed; predators included domestic cat, Jay Garrulus glandarius, and Red-backed Shrike Lanius collurio; other losses caused by vineyard work and weather (Groh 1975). In Devon, of 10 nests, only 1 definitely produced fledged young, 1 other probably successful; nestlings in 4 died in heavy rain, 3 nests predated, and cause of failure of 1 unknown (Sitters 1991). Young thought to be vulnerable to predation because of very loud food-calls, audible at 70 m (A and J Evans). In Switzerland, one ♀ raised at least 9 young to fledging in 3 broods (Fuchs 1964).   BH

Plumages. ADULT MALE. In fresh plumage (August-December), nasal bristles small, black; forehead and crown olive-green, yellowish-green, or greyish, feathers with black shaft-streak, closest on side of crown. Supercilium yellow, rather broad near base of upper mandible, very narrow above eye, widening again behind eye. Lore and eye-stripe dark olive-grey, widening on upper ear-coverts. Upper cheek from base of upper mandible backwards to frontal part of ear-coverts yellow or pale yellow;

remainder of ear-coverts dark olive-grey with some yellow-green in centre, grey connected with rear of dark eye-stripe. Small yellow spot behind ear-coverts near distal part of supercilium. Eye-ring dark grey and yellow. Nape olive-green, yellowish-green, or greyish, sometimes with a few dark streaks. Feathers of mantle variable: chestnut with inconspicuous dark grey or black tips, or strongly marked with long and rather broad black central streak, bordered with narrow chestnut line; all feathers on mantle have pale yellowish, yellowish-green, or whitish edges. Scapulars similar, but dark centres of feathers absent or small and restricted to tip. Back and rump yellow-green or greenish-grey, back sometimes mixed with some chestnut, rump occasionally with narrow dark grey streaks. Upper tail-coverts like rump, but longest have chestnut tinge and usually narrow black shaft-streak. Stripe on lower cheek, chin, and upper throat sooty-black, all feathers with narrow pale yellow fringes, grey extending into bar at rear of cheek and connected to olive-grey on rear of ear-coverts. 3 bands across breast: upper (along border of lower throat, upper chest, and upper side of neck) bright yellow, central (across chest and lower side of neck) broadly yellowish-green or grey-green, lower (just below chest and on upper flank) chestnut but usually restricted to chestnut patch on each side. Belly pale or bright yellow; vent and under tail-coverts yellow or white; some dark grey of feather-bases sometimes shows through. Flank and thigh streaked yellow, olive, chestnut, and grey; longest under tail-coverts with narrow dark streak. Central pair of tail-feathers (t1) brown to grey-brown with greenish, brownish, yellowish, or whitish edges, widest on outer web; t2–t4 dark grey with narrow greenish, yellowish, or whitish edge, t4 occasionally with white tip; t5 dark grey with white wedge occupying about one-third of distal part of inner web, tip of outer web white; t6 as t5, but white wedge extends over two-thirds of distal part of inner web; outer web of t6 with narrow white fringe, dark grey on basal part extending to tip along shaft, widening on distal end. Flight-feathers and primary coverts dark brownish-grey; outer webs of flight-feathers with greenish, yellowish, or whitish edge, widest on basal three-quarters of p6–p8. Tertials chestnut with dark grey centres, most obvious on longest tertial; when very fresh, all fringed pale yellow or off-white. Bastard wing greyish-black with narrow chestnut and greenish fringe. Greater upper wing-coverts with grey centres, chestnut fringe along inner and outer web, and narrow pale greenish, yellowish, or whitish tip; inner greater coverts darker chestnut, outer paler; median coverts grey, fringed pale yellow or green and with chestnut hue on tip; lesser coverts grey with broad green fringes, similar in colour to upper chest. Greater under wing-coverts grey with broad white tips, under primary-coverts grey with lemon and white edges; marginal coverts mixed grey and yellow; axillaries mixed grey and bright yellow. *In worn plumage* (February–May), similar to fresh plumage, but pale feather-fringes on mantle and scapulars worn off, upperparts appearing darker; throat usually without pale feather-tips. ADULT FEMALE. In fresh plumage (August–December), forehead, crown, and nape greenish-olive with black shaft-streaks. Lore and supercilium pale yellow, faintly streaked dark brown behind eye; supercilium widest behind eye. Broad olive-brown eye-stripe, from eye to beyond upper ear-coverts. Ear-coverts pale yellow to olive-yellow with faint dark brown streaks and with chestnut hue, surrounded by brown eye-stripe above and at rear and by brown moustachial stripe below; pale yellow or whitish spot at rear of ear-coverts. Broad submoustachial stripe whitish or lemon-yellow with brown streaks, bordered below by faint malar stripe, not reaching base of bill. Hindneck less intensely streaked, with smaller streaks than crown and more spotted, especially on side. Feathers of mantle blackish-brown,

bordered with chestnut and fringed whitish, pale greenish-yellow, or olive, but sometimes almost no chestnut at all. Scapulars like mantle, but with less black and more chestnut; dark shaft-streaks variable, often widening at tip. Back, rump, and upper tail-coverts olive to brownish, usually with faint dusky streaks, but streaks occasionally absent or almost as strong as those on mantle. Chin and throat pale yellow, usually with at least some dark tips on feathers, appearing faintly spotted. Lower throat pale yellow or greyish-yellow with strong blackish-brown streaks, but these occasionally absent, yellow crescent then uniform, as in ♂. Side of breast chestnut, varying in depth and extent, sometimes absent. Centre of belly unstreaked, pale to strong yellow, dark grey feather-bases occasionally showing through, as in ♂. Flank yellow with strong dark brown streaks and with chestnut borders to feathers. Under tail-coverts lemon to yellow with dark brown streaks. Tail, flight-feathers, bastard wing, and wing-coverts as ♂. *In worn plumage* (February–May), pale edges on upperparts wear off, giving darker and more chestnut appearance; underparts more streaked due to wear. NESTLING. Down long and plentiful; grey, grey-brown, or grey-cinnamon (Witherby *et al.* 1938; Dementiev and Gladkov 1954; Groh 1975.) JUVENILE. Feathers on forehead, crown, neck, mantle, scapulars, and back dark grey with broad yellow or rusty-brown edges; edges broader on neck. Rump similar to upperparts, but feathers of loose structure and with much down on bases. Pattern of side of head as adult ♀. Chin, throat, chest, and flank yellow or buff-brown with dark grey-brown streaks, strongest on chest; sides of breast and flank with chestnut or rufous hue. Lesser upper wing-coverts dark grey with narrow buff-brown fringes, median and greater dark grey with broad buff-brown fringes. Flight-feathers, bastard wing, primary coverts, and tertials as adult. Under wing-coverts pale lemon with dark streaks and tips. FIRST ADULT. Like adult, but neck and rump perhaps more streaked. Tail-feathers usually more pointed than in adult, but see Moults (below); if tail unmoulted, t4 without white tip (Groh 1975).

Bare parts. ADULT, FIRST ADULT. Iris brown. Upper mandible dark horn, lower mandible light bluish to grey. Leg and foot brownish-flesh. NESTLING. Iris dark brown. Bill dark horn; inside of mouth pink or red, gape-flanges lemon. Leg and foot pale yellow. (Witherby *et al.* 1938; Dementiev and Gladkov 1954; Groh 1975.) JUVENILE. Upper mandible black-brown, lower light blue with pink tinge at base, cutting edges light horn-yellow. Leg and foot pale pink-flesh. (Heinroth and Heinroth 1931.)

Moults. ADULT POST-BREEDING. Complete, primaries descendent. Mid-July to mid-October. In Spain, June, single ♀ had 2 tertials new, remainder of plumage old. In Italy, single ♂ had primary moult score 15 on 1 August; in Greece, 2 ♂♂ and 1 ♀ had moult score 10, 14, and 15 in 1st half of August; 2 ♂♂ had moult score 15 and 33 on 16–17 September; all these birds had body in moult, but secondaries still old, except ♂ with primary score 33. Tail moulted centrifugally, starting with t1 and finishing with t5 and/or t6. One ♀ moulted virtually all greater coverts simultaneously. (RMNH, ZMA.) Single ♂ in Morocco, November, still moulting (Meinertzhagen 1940). Apparently no pre-breeding. POST-JUVENILE. Partial: body, lesser, median, and possibly all greater upper wing-coverts, some tertials, and sometimes a few tail-feathers. Starts *c.* 30 days after hatching, June (1st brood) to October (Groh 1975; Ginn and Melville 1983). However, a 1st-calendar-year ♂, November, Naples, had p9 (left), p8–p9 (right), bastard wing, some greater upper primary coverts, secondaries and tertials (s1 and s5–s9), all greater coverts, and complete tail new, and this perhaps happens more often, especially in early-hatched birds (Winkler and Jenni 1987).

Measurements. ADULT, FIRST ADULT. Whole geographical range, all year; skins (ZFMK, ZMA, RMNH: M Platteeuw, C S Roselaar). Bill (S) to skull, bill (N) to distal corner of nostril; exposed culmen on average 2·2 mm. shorter than bill (S).

WING	♂	81·0 (2·00; 45)	76–86	♀ 77·8 (1·95; 25)	74–81
TAIL		66·0 (2·68; 35)	60–71	63·9 (2·17; 23)	60–67
BILL (S)		14·0 (1·49; 43)	12·5–15·1	14·3 (1·39; 22)	12·2–15·3
BILL (N)		7·9 (0·35; 34)	7·1–8·7	8·0 (0·52; 23)	7·3–9·8
TARSUS		18·3 (0·68; 28)	17·2–19·7	18·3 (0·67; 10)	17·6–19·5

Sex differences significant for wing and tail.

(1) South-west Germany, all year, live birds (Groh 1975); (2) Makedonija, all year, skins (Stresemann 1920); (3) southern France, all year, live birds (G Olioso).

WING (1)	♂	79·8 (— ; 182)	74–85	♀ 75·8 (— ; 25)	73–79
(2)		81·9 (2·58; 54)	78–84	78·4 (1·78; 25)	75–81
(3)		82·4 (1·65; 39)	80–86	80·1 (1·06; 24)	77–82
TAIL (1)		69·8 (— ; 182)	63–81	66·6 (— ; 25)	61–73

Morocco, ♂ 79–84 (5), ♀ 76–82 (3) (Meinertzhagen 1940). Sicily, ♂♂ 81, 81·5; Sardinia, ♂♂ 80·5, 83 (Eck 1985b).

Weights. Nominate *cirlus*. South-west Germany: (1) March–July, (2) December–March (Groh 1975). (3) Southern France, all year (G Olioso).

(1)	♂	23·9 (— ; 173)	21–28	♀ 23·6 (— ; 16)	20–29
(2)		27·3 (— ; 11)	25–29	25·9 (— ; 11)	22–29
(3)		25·7 (1·26; 39)	23–28	25·5 (1·30; 24)	23–28

Sardinia: January, ♂♂ 27, 28, ♀ 26; August, ♀ 28 (Demartis 1987). Greece: December–January, ♂ 27, ♀♀ 24, 26, 27 (Mak-

atsch 1950). Turkey: May, ♂♂ 26, 26, ♀ 27; July, juvenile ♀ 23·9 (Rokitansky and Schifter 1971); August, juvenile 24·6 (Eyckerman *et al.* 1992). Weight at fledging, south-west Germany: 18–20 (Groh 1975).

Structure. Wing rather short, broad at base, tip bluntly pointed. 10 primaries: p7–p8 longest, p9 1–3 shorter, p6 0–3, p5 2–8, p4 7–12, p3 10–15, p2 11–18, p1 12–19; p10 strongly reduced, 52–55 shorter than p7–p8, 6–9 shorter than longest upper primary covert. P5–p8 clearly emarginated, p4 occasionally with slight emargination; p7–p8 (p6–p9) with slight notch. Tip of longest tertial falls between tips of (p1–)p3–p5. Tail long, slightly forked; 12 feathers, t4 longest, t1 2–6 shorter. Bill short, conical; similar to Yellowhammer *E. citrinella*, but slightly less deep at base and upper mandible slightly and evenly curved. Middle toe with claw 18·7 (12) 17·8–19·8; outer toe with claw *c.* 69% of middle with claw, inner *c.* 71%, hind *c.* 73%; hind claw 6·5 (8) 6–7.

Geographical variation. No races recognized. ♀♀ from Corsica and Sardinia slightly more heavily streaked black on underparts (especially flank) than those of France and Italy, and mantle slightly duller; sometimes separated as *nigrostriata* Schiebel, 1910, but not recognized here, as heavily streaked birds occasionally occur elsewhere in range (e.g. northern Algeria, Italy, Greece), and about half of birds from Corsica and Sardinia are not separable from birds of mainland; no difference in size. Birds from Portugal very slightly smaller and darker than elsewhere, sometimes separated as *portucaliae* Floericke, 1922, but difference very slight, and recognition not supported by specimens examined: wing of ♂ 79·8 (1·46; 10) 77–82, against 81·5 (1·86; 16) 77–83 in remainder of range (RMNH, ZMA).   GOK

Emberiza stewarti (Blyth, 1854) White-capped Bunting

FR. Bruant de Stewart   GE. Silberkopfammer

A central Asian species, breeding from Kazakhstan through Afghanistan to northern Pakistan, and wintering in Himalayan foothills and adjacent plains. ♀, showing no signs of captivity, was trapped in Herve (Belgium) on 9 August 1931, and kept in captivity until 10 April 1935. Not accepted by Commission pour l'Avifaune Belge (1967), but Lippens and Wille (1972) considered it likely to have been a wild bird, taking into account the state of its plumage and the fact that the species was not known to have been imported.

Emberiza cia Rock Bunting

PLATES 16 (flight) and 18
[between pages 256 and 257]

DU. Grijze gors   FR. Bruant fou   GE. Zippammer
RU. Горная овсянка   SP. Escribano montesino   SW. Klippsparv

Emberiza Cia Linnaeus, 1766

Polytypic. Nominate *cia* Linnaeus, 1766, central and southern Europe, North Africa east to Tunisia, north-west and western Asia Minor, and Levant, grading into *prageri* in southern and eastern Turkey east from Central Taurus mountains; *prageri* Laubmann, 1915, north-east Turkey, Transcaucasia, Caucasus, and Crimea; *par* Hartert, 1904, Iran, Afghanistan, western Pakistan (Baluchistan to Gilgit), and through central Asia to Tarbagatay and southern Altai, east to Tekes valley (Sinkiang, China) and Mongolia; probably this race in Iraq. Extralimital: *stracheyi* Moore, 1856, western Himalayas from Baltistan and Kashmir to Kumaon; *flemingorum* Martens, 1972, western Nepal.

Field characters. 16 cm; wing-span 21·5–27 cm. Slighter than Yellowhammer *E. citrinella* with rather short, fine bill, somewhat shorter and rounder wings, and narrower tail, though tail averages longer; 20% larger than House Bunting *E. striolata*. Medium-sized but relatively slim bunting, with long, thin tail contributing to more attenuated outline than any other congener. Shares rufous-buff ground-colour to plumage (particularly underparts) with 5 other *Emberiza*; best distinguished by rather small, lead-coloured bill, strong head pattern (particularly ♂) of blackish crown-stripes and complete black surround to ear-coverts on greyish ground, and strongly rufous rump. Terrestrial, rarely far from rocks. One call distinctive. Sexes dissimilar; some seasonal variation in ♂. Juvenile separable. 3 races in west Palearctic, poorly differentiated in colour but Caucasus birds somewhat larger. Only European race, nominate *cia* described here (see also Geographical Variation).

ADULT MALE. Moults: July–October (complete); full contrasts of head pattern produced by wear. Head, nape, and throat to upper breast pale grey, with supercilium, cheeks, and chin looking more whitish at some angles, strongly lined with close black streaks on sides of crown and quite deep black eye-stripe which turns down to outline ear-coverts and round to reach bill. When fresh, head washed yellowish-brown, and plumage contrasts obscured. Mantle and scapulars chestnut-buff, warmest on outer edges and near rump, and rather narrowly but distinctly streaked black. Long rump uniform chestnut, almost red on lateral tail-coverts, and eye-catching. Wings distinctly marked: lesser coverts mainly ashy-grey; median coverts black-centred and white-tipped, forming always striking upper wing-bar; greater coverts brown-black, buff-fringed and buffish-white-tipped, forming less obvious wing-bar, particularly when worn; tertials brown-black with buff margins; flight-feathers brown-black, fringed pale buff. Tail warm brown-black, with buff margins to inner feathers and fully white outer webs of outermost and decreasing white ends of next 2 pairs showing not just as outer edges but also as corners when fully spread. Underparts from lower breast to under tail-coverts uniform buffish-chestnut, looking brighter and warmer than back. Under wing-coverts white. Bill rather short and conical; lead-grey. Legs bright orange-flesh to reddish-brown. ADULT FEMALE. Less intensely coloured on head, back, and underparts than ♂, but with similarly vivid chestnut rump and well-marked wings except for duller, browner lesser coverts. When fresh, head basically buffish-grey, with black-streaked crown, dusky to black eye-stripe and broken surround to rather brown ear-coverts, small faint dusky spots on throat, and more striking but still fine black streaks on sides and centre of breast. When worn, head shows faintly spotted whitish supercilium and greyer cheeks, collar, throat, and upper breast, last 2 areas losing spots and most streaks. In shadow, may suggest Dunnock *Prunella modularis*, but tail

longer and with much white. Bill duller than ♂; legs often paler, reddish-straw. JUVENILE. Resembles newly moulted ♀ but lacks any grey on head or throat, while showing stronger black streaks on crown and back and from throat to breast and along flanks. Wing-bars less distinct than adult, with smaller and incomplete buffish-white tips to both median and greater coverts; lesser coverts greyish-brown. FIRST-YEAR MALE. Moult of head and body plumage and wing-coverts in early autumn. Resembles worn ♀ with only slightly greyer ground to head and warmer back. Retained juvenile flight- and tail-feathers lose pale fringes. FIRST-YEAR FEMALE. Duller than adult, with still brownish tinge to head; shows more fully streaked crown and breast; buff supercilium indistinct; flanks retain some streaks.

♂ unmistakable if seen well, but in glimpse or in shadow may suggest Cirl Bunting *E. cirlus* (black throat, yellow and olive ground-colour to head and breast, more yellow underbody) or *E. striolata* (much smaller, and lacks bright wing-bars and deep rufous rump, but shares grey throat and breast and, in western race *E. s. sahari*, black stripes on head). ♀ and immature troublesome, needing careful separation from Pine Bunting *E. leucocephalos* (slightly larger but less attenuated, with horn bill, buffish-white ground-colour to face, and dull white underparts, always well streaked on flanks), *E. citrinella* (also larger and fuller-tailed, with at least yellowish hue to face and underparts), and *E. striolata* (much smaller, even more uniform rufous-buff in appearance, with less marked head and unstreaked underparts). Important to recognize that *E. cia* is usually a bird of higher altitudes than any confusion species, mixing with *E. striolata* only during descent in winter to surroundings of arid mountains. Flight like *E. citrinella*, trailing tail, but wing-beats somewhat lighter making action more flitting; narrowness of tail except when spread contributes to less substantial silhouette. Escape-flight ascendant to boulders or tree canopy. Gait a hop, also creeping when feeding. Stance typically half-upright; tail sticks out but rarely drooped even when on ground. Unobtrusive, but ♂ flutters wings when singing; when disturbed, may flick and spread tail. Markedly terrestrial, as *E. cirlus*; ♂ easier to find than ♀. Widely scattered in breeding habitat but forms parties during descent in winter, when also associates loosely with other seed-eaters.

Song variously described: short high-pitched phrase, last rising note prolonged or repeated; suggests Reed Bunting *E. schoeniclus* in shortness and somewhat recalls Serin *Serinus serinus* in shrill but buzzing timbre, 'zi-zi-zi-zirr. . .'; alternative a fast, squeaky phrase of at least 10 notes (*c.* 2·5 s), last 2–3 prolonged and repeated and varied in pitch like Wren *Troglodytes troglodytes*, 'seut wit tell-tell wit drr weeay sit seeay'. Calls include sharp 'tzit' like *E. cirlus* or thin, weak 'zeet', one or other occasionally repeated at least 3 times; high, drawn-in 'seeee' or 'seea'; and fuller, rather bubbling 'tucc', which becomes extended into more twittering 'tootootooc' in flight.

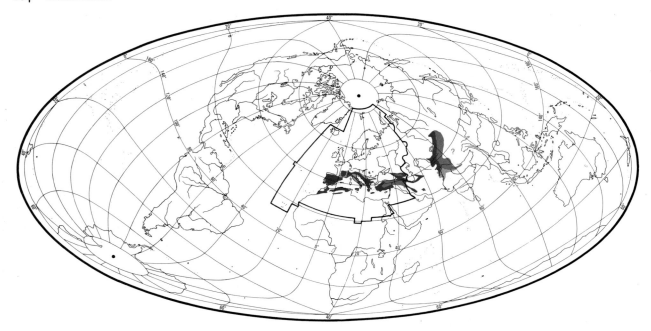

Habitat. Extends across lower middle latitudes of west Palearctic, from Mediterranean to Caucasus, from warm temperate to steppe climatic zones, within July isotherms of *c*. 20–30°C. Avoids most humid or wet situations, closed forest, and good agricultural land, preferring sunny semi-arid terrain, often stony or rocky, with more or less sparse shrub vegetation, and usually with no more than scattered trees. Often on slopes or hillsides, at 0–1900 m, and extra-limitally in Asia above 4000 m, frequenting mountain villages and gardens (Voous 1960*b*). Occupies open areas at upper forest limits, juniper *Juniperus* scrub, subalpine meadows with shrubs and screes, stone-walled cultivated areas, and vineyards on hillsides (Harrison 1982). In northern part of European range seeks out warm dry southern or south-west slopes with oak *Quercus* scrub or other bushy cover, hedges, or patches of young conifers. Slopes are often steep, with rocky outcrops, stony gullies, or quarries, and terrain must be open, with some bare earth and low vegetation, and preferably a few scattered trees. Favours lower mountain slopes (Hollom 1962). In Switzerland, breeds mostly at 500–1500 m but exceptionally higher, on sunny, dry, rocky or stony slopes with meagre vegetation, resorting to flat areas only for foraging. Spaced out shrubs or trees, surrounded by bare surfaces, are essential, but mosaics of cultivation, as in vineyards and on terraces, are acceptable, as are clumps of thorny bushes or hedges, and even fringes of pine *Pinus* or oak forests (Glutz von Blotzheim 1962). In USSR, breeds from sea-level to snowline, mostly above 500 m. Sometimes nests in ravines or on screes, but occasionally on level areas in foothills. For nesting, favours areas with plenty of boulders bordering spring or stream. In Siberia, open areas among patches of birch *Betula* and larch *Larix* on foothills are favoured (Dementiev and Gladkov 1954). In Afghanistan, very abundant in fir *Abies* and spruce *Picea* forests, especially in glades, but replaced by White-capped Bunting *E. stewarti* in neighbouring *Pinus gerardiana* forest at *c*. 3000 m (Paludan 1959).

Descends in winter to lower ground, and in India then found in semi-desert plains, cotton fields, and near canal cultivation (Ali and Ripley 1974). European wintering birds feed on ground close to hedges, bushes, copses, and wayside trees, to which they fly up when disturbed (Hollom 1962).

Distribution. A few changes recorded, at periphery of range.

FRANCE. Some range retraction in north since 1930s (Yeatman 1976). CZECHOSLOVAKIA. Breeding first recorded 1954, followed by limited spread (KH). HUNGARY. First recorded breeding in 1950s, in north; spread since, and by early 1980s present in suitable habitats throughout country (G Magyar). SYRIA. Presumed to breed in extreme north-west, but confirmation needed (Baumgart and Stephan 1987). ISRAEL. Breeding first recorded, Mt Hermon area, 1967 (Shirihai in press).

Accidental. Britain, Sweden, Poland, Malta, Libya, Canary Islands.

Population. No major changes; some increases, and one local decrease, in peripheral parts of range.

FRANCE. 1000–10 000 pairs (Yeatman 1976). GERMANY. Estimated 1500 pairs (Rheinwald 1992). In southern Schwarzwald, declined by at least 35–40% between 1950s

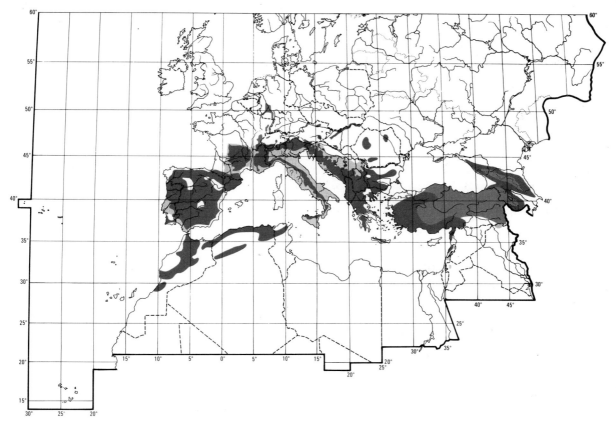

and 1988; main causes afforestation, agricultural changes, and disturbance; estimated *c.* 35 pairs 1986-8 (Mann *et al.* 1990). CZECHOSLOVAKIA. Probably some increase since 1954 (KH). AUSTRIA. Stable, or perhaps increasing (H-MB). HUNGARY. Estimated 200-300 pairs (G Magyar). ISRAEL. About 20 pairs in 1st half of 1980s; increasing during 2nd half of 1980s; now 50-100 pairs (Shirihai in press).

Movements. Most populations sedentary or dispersive, but partial migrant or migrant over short or medium distances from parts of range with coldest winters. Altitudinal movements in many high mountain areas. Review for western Europe and of scanty ringing data by Zink (1985). Migration said to be nocturnal (Shirihai in press), although unclear on what basis.

In western Europe, altitudinal movements reported from Pfalz (south-west Germany), Switzerland, France, and Spain (Mayaud 1936; Bernis 1954; Groh 1982, 1988; Zink 1985; Yeatman-Berthelot 1991). In France, mainly resident in or near breeding areas, but leaves Jura, parts of southern Massif Central, and central Pyrénées in winter; recorded wintering outside breeding range in central France, Bretagne, and Cotentin (Mayaud 1960; Yeatman-Berthelot 1991). In middle-Rhine region (western Ger-

many), formerly known only as summer visitor to breeding grounds, March-October, with only 2 winter records 1910-49; in more recent years seen regularly in winter, with flocks of more than 10 regular in 1978-88 (Macke 1980; Groh 1982, 1988; Zink 1985). Many birds leave German breeding areas October-November and return March-April; old suggestion that winter quarters are in north Africa (Niethammer 1937) unconfirmed and unlikely; Groh (1982, 1988) shows wintering sites in Pfalz at 5-7 km from breeding territories. In Hungary, near northern edge of breeding range, has been recorded regularly in winter since 1946, maximum 20 birds, wintering near breeding sites (Dandl 1959; Horváth 1975; Selley 1976). Near Graz (Austria), birds at winter feeding place were from nearby breeding sites (Wöhl 1980, 1981, 1985). In Switzerland, resident, scarce passage migrant, and winter visitor (Winkler 1984). Near Geneva (western Switzerland), most birds migratory, leaving from mid- to end of October, returning early to mid-March and probably into April (Géroudet 1954). At Col de Bretolet (western Switzerland), close to breeding areas, autumn migration starts beginning of October, reaches peak in 2nd half of October, ends November; spring migration from beginning of March to end of April (Dorka 1966; Winkler 1984); 142 recorded 1954-74 (Winkler 1975). The few ringing recoveries suggest that movements in western Europe not normally over great distances: nestling ringed Rüdesheim (western Germany) recovered in April 2 years later in Luxembourg (outside modern breeding range); another German bird moved 95 km to west; 4 distant recoveries of birds ringed in western Alps show movements between WSW (to Lot-et-Garonne, France) and south (to French Riviera); 2 Italian recoveries show movements south-east along coast in Genoa province; Spanish recovery 55 km south-east along coast (Zink 1985). In England and Wales, 5 records of vagrants: in February (2), June, August, and October (Dymond et al. 1989).

No evidence of movements in Sicily (Iapichino and Massa 1989). Regular autumn passage migrant at Gibraltar, reaching peak during second half of October (Finlayson 1992), but no true migration noted in north-west Morocco (Pineau and Giraud-Audine 1979), so perhaps only local movements involved.

In North Africa, at least partly leaves high ground in winter, when birds occur on plains and reaches edge of Sahara (Heim de Balsac and Mayaud 1962). In north-west Morocco, seen in winter on hills where breeding does not occur (Pineau and Giraud-Audine 1979). Single stragglers recorded on Jefren escarpment of Libya (Willcox and Willcox 1978), at Touggourt in Algeria (Dupuy 1969), and, in southern Sahara, at Abéché in Chad and in northern Sudan (Salvan 1963; Etchécopar and Hüe 1967; Mathiasson 1972).

In central Asia, par is partial migrant and altitudinal migrant in some regions, but apparently resident at low elevations. In some regions, ♀♀ arrive in breeding areas 7-10 days after ♂♂ (Dementiev and Gladkov 1954). Most birds leave Kazakhstan, although some winter in foothills and plains of south, but not in large numbers. Autumn migration fairly clearly expressed in foothills and hill regions of south and east Kazakhstan, with movements along river valleys and on mountain slopes, September-October. In Talasskiy Alatau, appears in cultivated regions from mid-September, but most migration is in mountains, where small groups pass throughout September; in 1st half of October is most numerous species in mountains; last records mid-October. Spring migration mostly March. In cultivated regions of Talasskiy Alatau, earliest records 7-15 March over 7 years. In Kirgizskiy Alatau, at 2000 m, appears 11-21 March. Other records also give dates in March and early April. At Alma-Ata, where individuals winter, arrival noted at end of February (perhaps wintering birds) and first half of March. Reaches breeding areas in mountains in April. (Korelov et al. 1974.) At Chokpak pass in western Tien Shan (Kazakhstan), regular on autumn passage although migration is mostly through mountains, beginning 10-18 October, mainly 2nd half of October; only occasional spring records (Gavrilov and Gistsov 1985). In western Tien Shan, arrives on breeding grounds at 1700-2000 m mid- to late March, ♂♂ before ♀♀ (Gubin 1980). In western Tadzhikistan, mainly

sedentary, but some birds make long movements having character of true migration, and others make altitudinal movements. Downward movement from higher elevations from 2nd half of September; on wintering grounds in foothills from end of October to end of February or March. Numbers wintering and dates of migration fluctuate strongly depending on weather, especially in spring; late snowfall drives back birds which have already arrived in breeding areas. Vertical migration in lower mountains of south-west Tadzhikistan is inconsiderable; birds wintering here leave breeding grounds only after heavy snowfall. (Abdusalyamov 1977.) E. c. stracheyi in eastern Tadzhikistan is mainly sedentary, but some depart for winter and others make altitudinal movements; spring arrival seen end of March and beginning of April; further study needed (Abdusalyamov 1977).

E. c. prageri resident in Crimea, but partial altitudinal migrant in Caucasus, where it leaves high ground in winter; winter flocks near Tbilisi occur mid-October to mid-March (Dementiev and Gladkov 1954). In Middle East, prageri and par are partial altitudinal migrants, leaving high ground in winter (Hüe and Etchécopar 1970; Paynter 1970; Hollom et al. 1988). Local winter visitor to Kurdish mountains of Iraq, recorded 12 November 1944 to 18 March 1945 (Moore and Boswell 1957). In Iran mainly resident, but wanders to lowlands in winter, including southern shore of Caspian (D A Scott). In eastern Afghanistan, summer visitor to breeding areas on wooded slopes, wintering in low-lying valleys of extreme east (S C Madge). Scarce winter visitor to Jordan, recorded 3 November to 2 March (Br. Birds 1990, 83, 230; I J Andrews). Fairly common winter visitor to mountainous parts of northern and central Israel with Mediterranean climate; arrivals November to mid-December, mainly second half of November; leaves end of February to end of March, mainly 1st half of March. Both immigration and emigration apparently by direct flights as not seen outside wintering areas, except single record outside normal winter range, at Beer Sheva in November. Small population breeding on Mt Hermon winters at 500-1200 m, lower than much of breeding range (900-1800 m) (Shirihai in press.) On Cyprus, described as very scarce and irregular passage migrant and perhaps also winter visitor (Flint and Stewart 1992), but total of only 6 records, mainly late March and April (Zink 1985; Flint and Stewart 1992), suggests it is only a straggler. Recorded as vagrant to Kuwait (2 records in March: V A D Sales; Hollom et al. 1988); reports from other Gulf states, Oman, and Saudi Arabia no longer accepted (F E Warr).

Western Himalayan populations of par are partial altitudinal migrants, wintering from lower part of breeding range at c. 2800 m down to foothills and plains, south in India to Bahawalpur, Haryana to Ambala, and Delhi; occurs in winter range November-March, a few until April (Ali and Ripley 1974). This race also winters in Baluchistan (Pakistan) (Roberts 1992). Himalayan race *stracheyi* (and intergrades with *par*) reported to spend November–March mainly below breeding range at *c*. 600–2100 m (Ali and Ripley 1974). However, in Pakistan many remain through winter in main valleys of far northern areas including Gilgit and Chitral; in Chitral, common down to 1200 m, early October to end of April (Roberts 1992). Likewise, in Nepal little altitudinal movement, breeding at 2440–4600 m, wintering down to 1800 m (Inskipp and Inskipp 1985). DTH

Food. Seeds, mainly of grasses (Gramineae), and other parts of plants; invertebrates in breeding season. Feeds principally on ground among rocks and scrubby vegetation, or in short grass in fields, at woodland edges, etc., but not infrequently in bushes or tall herbs taking both seeds and insects. Mostly picks seeds from ground, but will also stand on stems, sometimes several at a time, to bend them over and reach seed-head, reaches over to seed-head from neighbouring perch, or pulls seed-head down while standing on ground. Catches flying insects in short sallies just above ground. (Géroudet 1954; Dandl 1959; Glutz von Blotzheim 1962; Szabó 1962; Györgypál 1981; Groh 1988.) In France in winter, forages on wasteland, cultivated areas, and even in gardens (Yeatman-Berthelot 1991). Outside breeding season, feeds in flocks with other seed-eaters, though often noted in unmixed flocks; feeds for hours in one spot if undisturbed (Pricam 1957; Dandl 1959; Hollom 1962; Groh 1982; Shirihai in press). De-husks large seeds before swallowing (Groh 1988).

Diet in west Palearctic includes the following. Invertebrates: mayflies (Ephemeroptera: Siphlonuridae), grasshoppers (Orthoptera: Acrididae), bugs (Hemiptera: Pentatomidae, Aphidoidea), flies (Diptera), adult and larval Lepidoptera (Tortricidae, Pyralidae, Geometridae), ants (Hymenoptera: Formicidae), adult and larval beetles (Coleoptera: Cicindelidae, Carabidae, Cantharidae, Chrysomelidae, Curculionidae), spiders (Araneae), earthworms (Lumbricidae), snails (Pulmonata: Clausiliidae, Vertiginidae). Plants: seeds, etc., of knotgrass *Polygonum*, chickweed *Stellaria*, fleabane *Conyza*, woodrush *Luzula*, grasses (Gramineae: *Diplachne, Poa, Agropyron, Lolium, Deschampsia, Dactylis, Agrostis, Molinia, Stipa, Digitaria*), moss (Musci). (Dandl 1959; Jakobs 1959; Glutz von Blotzheim 1962; Szabó 1962; Rucner 1973*b*; Groh 1982, 1988; Guitián Rivera 1985.)

In Hungary in breeding season, 2 stomachs contained 23 snails (22 Vertiginidae), 5 adult and larval beetles (3 Curculionidae), 1 adult Diptera, 1 small wasp (Hymenoptera); in winter, main food seeds of grasses *Festuca* and *Diplachne*, also caterpillars (Dandl 1959). In Croatia, 2 stomachs in breeding season held only beetles (Carabidae, Curculionidae), Hemiptera (Pentatomidae), and ants (Rucner 1973*b*). In north-west Spain, of 16 faecal samples over 2 years, 75% contained beetles, 50% plant material, 25% eggs and larvae of insects, 19% spiders, 6% Diptera;

in spring, all samples contained animal and plant material; in summer, all contained animals and 56% plants (Guitián Rivera 1985). In Pfälzerwald (Pfalz, south-west Germany), feeds mostly on adult and larval insects in breeding season, but gradually moves from mainly animal to mainly plant diet over summer; in winter, eats almost only seeds and other parts of plants, only occasionally a few invertebrates. From July, principal food seeds of grass *Deschampsia flexuosa* in forest clearings; in winter, knotgrass and grass *Agropyron*. (Groh 1982, 1988.) In Switzerland, commonly eats seeds of grass *Molinia* and *Stipa* (Glutz von Blotzheim 1962). Many studies from south-central Asia. In Kazakhstan, contents of 8 stomachs, spring–winter, as follows: in April, Hemiptera and beetles (Curculionidae); in mid-May, 1 Hemiptera (Nabiidae), 1 caterpillar, 1 bee (Hymenoptera: Apoidea), 1 beetle (Curculionidae); end of June, Orthoptera; early July, 6 ants; mid-July (2 stomachs), Acrididae, 4 adult and larval Diptera, 4 beetles (Hydrophilidae, Curculionidae), seeds; mid-October, 2 beetles and seeds; mid-November, 500 seeds of grass *Setaria*, 35 seeds of *Amaranthus* (Kovshar' 1966); other stomachs held bulbils of bistort *Polygonum* (Korelov *et al.* 1974). In Tadzhikistan, of 23 stomachs, 57% contained seeds, 22% unidentified insects, 17% Orthoptera, 17% beetles, 9% other plant material; plants included Polygonaceae, Compositae, and grass *Poa* (Popov 1959); 7 stomachs, June–July, held Orthoptera, Hemiptera, adult and larval Lepidoptera, beetles (Chrysomelidae, Curculionidae), ants and other Hymenoptera, earthworms, plus seeds of Polygonaceae, Cruciferae, and grasses; in early spring, green parts of plants and milky seeds taken; winter stomachs contained only seeds (Abdusalyamov 1977). In Kirgiziya, often fed on spiders and snails; plant material included grains of wheat *Triticum* (Pek and Fedyanina 1961, which see for details of 45 stomachs). For 10 stomachs from Uzbekistan, see Salikhbaev and Bogdanov (1967). In Pakistan, recorded feeding on rosehips *Rosa* (Roberts 1992).

Diet of young almost wholly insects, with seeds fed later in season. In Pfälzerwald, size of invertebrates given increased with age of nestlings, and adult removed extremities of larger insects before feeding; commonest items in diet were caterpillars, especially of *Tortrix viridana* during infestations; adult Pyralidae also recorded in large numbers (Groh 1988). At one nest in Mosel valley (south-west Germany), principally caterpillars and Orthoptera, also 1 small earthworm (Jakobs 1959). In Switzerland, small green caterpillars important part of diet, as well as small adult Lepidoptera; seeds of *Stipa* brought to nest in some numbers (Géroudet 1954). Similarly in Hungary, young fed mostly green caterpillars, also Orthoptera, especially Acrididae (Dandl 1959; Szabó 1962; Györgypál 1981). Adults collect food up to *c*. 150–200 m from nest (Géroudet 1954; Mattes 1976; Gubin 1980). In Kazakhstan, nestlings also given small green caterpillars and adult Lepidoptera (Sphingidae, Noctuidae); several items brought on each visit by parents (Kovshar' 1966). Captive birds picked up larvae, beat them on ground, mandibulated them until soft and towards base of bill, then picked up another; took up to 10 at a time to nest; small nestlings given invertebrates in pieces, but later whole (Wunsch 1976).   BH

Social pattern and behaviour. Rather poorly known. Principal studies in Pfalz (south-west Germany) by Groh (1982, 1988) and Talasskiy Alatau (western Tien Shan) by Gubin (1980).

1. Outside breeding season, occurs singly or in groups, of young and adults, sometimes in larger flocks (e.g. Groh 1988, Yeatman-Berthelot 1991). In Switzerland, congregates in small parties after 2nd broods fledge in August and remains in groups on breeding grounds until emigration in mid- to late October (Géroudet 1954). Flocks of adults and juveniles form in USSR in late July and August (Dementiev and Gladkov 1954). European wintering flocks apparently seldom exceed 12 birds (e.g. Hollom 1962). Maxima of 20-30 recorded in Wallis and Tessin (Switzerland) (Winkler 1984), and of 19 in Pfalz (Groh 1982, 1988, which see for details of group sizes during 6 winters). On wintering grounds in Israel, flocks of more than 30 occur regularly; maxima recorded 80 and 170 (Shirihai in press). Mean winter densities (birds per km²) in Gibraltar area (southern Spain) 3·0 on South Terraces (Anon 1987), 6·1 in shrublands and 15·7 in cork oak *Quercus suber* forests (Arroyo and Tellería 1984). In Pfalz, individuals often reoccupy same wintering site in successive winters; one ♂ used same site for 5 winters (Groh 1982, 1988). In winter, occupies habitats frequented by other *Emberiza* and finches (Fringillidae) and often occurs in mixed flocks (see also Flock Behaviour, below) (e.g. Moore and Boswell 1957, Hollom 1962, Yeatman-Berthelot 1991, Roberts 1992). BONDS. Appears to be essentially monogamous; age of first breeding not known. Scanty data on duration of pair-bond, but marked pairs found together in 2 successive breeding seasons (twice) and 3 seasons (once) (Groh 1988). Nest-building and incubation by ♀ only (see Heterosexual Behaviour, below). Young fed by both sexes (see Relations within Family Group, below). In Pfalz, some territories occupied by unpaired ♂♂; during 1983-8 maximum 26·1% (n=23) in 1984, minimum 13·3% (n=15) in 1988 (Groh 1988). In Pakistan, at higher elevations, parties can still be seen in June (Roberts 1992), suggesting non-breeders involved. BREEDING DISPERSION. Territorial during breeding season. In Pfalz, territory size 1·1-4·3 ha, mean 2·55 ha (n=35) (Groh 1988). On Mt Hermon (Israel), occurs as scattered breeding pairs with a few tens or hundreds of metres between nests, although at a few places 3-5 pairs congregate with c. 50 m between nests (Shirihai in press). Nests 30-50 m apart in Khamar-Daban (south of Lake Baykal) (Vasil'chenko 1982). In western Tien Shan, nests usually 50-100 m apart; in one case, 2 nests containing young were separated by only 50 m, and a 3rd nest was built between these (Gubin 1980). Recorded flying up to 400-500 m from territory to feed and drink (Groh 1988). Adults may seek food for nestlings up to 150 m from nest (Géroudet 1954; Mattes 1976); 50-200 m recorded in study in Tien Shan (Gubin 1980). Nests for 2nd brood 8-65 m, mean 26 m (n=11) from 1st-brood nests (Groh 1988). ♂♂ often reoccupy same territory in successive breeding seasons: recorded for 2 years (6 ♂♂), 3 years (2), and 5 years (1); also 2 instances of ♀♀ occupying same territory in 2 successive years (Groh 1988). Similarly, in western Tien Shan, breeding ♀ ringed in 1972 nested c. 40 m away in 1973; several other cases of birds nesting in successive years at or close to same site, but these birds not

ringed (Gubin 1980). Breeding densities: 13-14 territories in 28 ha (equivalent to 48·2 per km²) at Morais (Tras-os-Montes, northern Portugal) (Mead 1975); 1·0 birds per km² in *Pinus radiata* plantations of Atlantic Basque country in Spain (Carrascal 1987); 5-34 pairs per km² (mean 9·8) in maquis in Morocco (Thévenot 1982); 7-8 pairs in 20 ha of rocky terrain in western Switzerland (Géroudet 1954); in Bulgaria (pairs per km²), 4 in *Carpinus orientalis* woodland, 8 in scrub dominated by *Juniperus communis*, 8 in *Pinus nigra* plantations (Petrov 1988), 6 in *Quercus pedunculiflorus* woodland, 14 in *Cercis siliquastrum*-dominated scrub (Simeonov and Petrov 1977), 32 in *Juniperus oxycedrus*-dominated communities (Petrov 1982). ROOSTING. In winter in Israel roosts communally in medium-sized trees and bushes, usually within a few hundred metres of both feeding and drinking places (Shirihai in press).

2. Not exceptionally shy, but often described as unobtrusive, and rather elusive outside breeding season (e.g. Géroudet 1954, Hollom 1962). ♀ much less conspicuous than ♂ on breeding territory (Groh 1988). In Kurdish mountains (Iraq), wintering birds flocked with Fringillidae and other *Emberiza*, but were tamer and remained behind when others flew off (Moore and Boswell 1957). In Pakistan, described as fairly tame and allowing close approach; some would regularly hop within 'a few feet' of observer (Roberts 1992). Tail constantly flicked in many contexts, exposing white outer feathers (Ali and Ripley 1974). FLOCK BEHAVIOUR. Flocks feed within easy reach of trees or bushes, into which they fly when disturbed, giving 'tzi-tzi' calls (see 2 in Voice); further disturbance may cause flights to several hundred metres distance or retreat through cover of vegetation (Pricam 1957; Dandl 1959; Hollom 1962; Roberts 1992). In winter, in mixed-species flocks (see above), *E. cia* maintain contact with each other and separate as group on taking flight (Hollom 1962). SONG-DISPLAY. Song (see 1 in Voice) used by ♂ only, to mark and defend territory (see also Antagonistic Behaviour, below) (Groh 1988). Song from ♀ reported once, from close to nest with young (Desfayes 1951), but not clear from published account that dull-plumaged ♂ was not involved. ♂ typically sings from bare spike or other exposed perch at top of tree, but also from vineyard stake or wire (Hollom 1962; Groh 1988; Roberts 1992); in western Tien Shan, mainly from top of bush (Gubin 1980). In Pfalz, begins to sing on return to territory (earliest 25 February). Song most frequent before nest-building, with decline during incubation and infrequent when young being fed (Groh 1988). In Pakistan, ♂ sings almost constantly during nest-building and incubation periods (Roberts 1992). In Pfalz, song scarce July and exceptional from August; rare at wintering sites even in March-April when birds on breeding territories in same region have started singing (Groh 1988). Quiet song reported in mild still weather in winter from area outside breeding range in Hungary (Dandl 1959). Unpaired ♂♂ sing more intensively than breeding birds and mostly continue later into summer (Groh 1988). In Switzerland, song-period early April to July or beginning of August, with some in October (Géroudet 1954). In USSR, song commences when ♂♂ arrive on territory in March and continues at least until late July; during breeding season ♂♂ sing all day (Dementiev and Gladkov 1954); song-period in western Tien Shan (1700-2200 m) 21 March to 10 July (Gubin 1980). Occasionally gives weak song with bill closed. Counter-singing against Dunnock *Prunella modularis* sometimes occurs, apparently due to similarity of songs. (Groh 1988.) ANTAGONISTIC BEHAVIOUR. (1) General. In Pfalz, territories occupied mostly in March, especially in 2nd half of month, exceptionally from late February (records of ♂♂ on territory from 23 February); leave territories 21 September to 11 November (Groh 1982, 1988); see also Movements. Song import-

ant in advertisement and defence of territory; ♂♂ can be captured for ringing using tape-recorded song to attract them (Groh 1988). In western Tien Shan, ♂ defends territory against intruding conspecifics, Red-headed Bunting *E. bruniceps*, and White-capped Bunting *E. stewarti* (Gubin 1980). In Pakistan, ♂♂ indulge in much chasing and territorial disputes with rival ♂♂ when they first reach nesting grounds and at onset of nesting (Roberts 1992). Both ♂ and ♀ reported to chase conspecific intruders from territory in study in Hungary (Györgypál 1981). (2) Threat and fighting. Little information (see above), but fights seen between territorial ♂♂ (Groh 1988). Heterosexual Behaviour. Few details of pair-formation, display, or mating. In USSR, reported to form pairs immediately on arrival of ♀♀ on breeding grounds (Dementiev and Gladkov 1954). On several occasions, ♂ (near ♀) seen continually tail-flirting (rapid opening and closing of tail: Fig A), calling with slight backward movement

A

of head (Géroudet 1954, 1957). Courtship-feeding during pair-formation not reported, but one record of ♂ carrying food to incubating ♀ (see below). Both partners involved in seeking nest-site (Groh 1988). Stem-display by ♂ occurs during nest-building, apparently much as in Cirl Bunting *E. cirlus* and Ortolan Bunting *E. hortulana*. Seen 4 times in German study; each time ♂ carried potential nest-material (stems, grasses) in bill for only short period and did not add it to nest (Groh 1988). In study in Tien Shan, ♂ once seen carrying nest-material which was eventually discarded (Gubin 1980). ♂ of captive pair often carried material and once sang with it in bill, but took no part in nest-building (Wunsch 1976). Nest-building by ♀ alone, mainly in early morning, commonly accompanied by ♂ (Géroudet 1954; Wunsch 1976; Gubin 1980; Györgypál 1981; Groh 1988). ♂ never seen with material by Géroudet (1954). While ♀ building, ♂ follows and sings (Dementiev and Gladkov 1954). Incubation by ♀ alone (Witherby *et al.* 1938; Géroudet 1954; Gubin 1980; Györgypál 1981; Groh 1988; Roberts 1992), although ♂ often seen near nest during incubation (Gubin 1980; Groh 1988). ♂ sometimes calls to incubating ♀, who answers softly from nest; once seen to carry food to incubating ♀ (Györgypál 1981). Relations within Family Group. ♀ eats eggshell (Gubin 1980). Young fed by both parents (Géroudet 1954; Mattes 1976; Groh 1988; initially mainly by ♀, although ♂ also brings some food (Roberts 1992). At nest studied in detail by Gubin (1980), share of visits by ♂ increased, e.g. from 9 out of 37 on day 1, to 24 out of 43 on day 7. Györgypál (1981) described similar observations at Hungarian nest. Young brooded by ♀ alone (Gubin 1980; Györgypál 1981; Groh 1988), for several days; more frequently and for longer stints when young small, e.g. for totals of 10 hrs 21 min on 1st day, 4hrs 21 min on 5th day, only at night on 7th day (Gubin 1980). Faeces of nestlings swallowed by ♀ for first 2–3 days, then removed by both parents and discarded 50–100 m from nest (Gubin 1980). Both parents

recorded swallowing faeces and carrying them shorter distances by Géroudet (1954). Development of young described by Gubin (1980): eyes open to slit on day 2–3, fully open on day 4; in Hungarian study, eyes reported to open at 5–6 days (Györgypál 1981). Wunsch (1976) described development of young bred in captivity, reporting that eyes begin to open on 8th day, but this probably erroneous. Young leave nest at 10–13 days; unable to fly for a few days after this, during which time they remain more or less close together in vicinity of nest (Géroudet 1954). According to Roberts (1992), young fledge on 8th–9th day but cannot fly until 14th. Gubin (1980) also recorded young leaving nest when disturbed on 8th–9th day. 3 young with feathers mainly in pin found scattered *c.* 2 m from nest (Jouard 1934). Young stay in thick ground vegetation for a few days after fledging, then move into bushes and saplings (Géroudet 1954; Groh 1988). Fledged young fed by both parents, mostly for 2–3 weeks, once for 27 days. When another nest is started, feeding period of fledglings shorter, especially by ♀. When no other nest started, young may remain longer in territory with parents, so that young of last brood can remain in or near territory with parents until October. (Groh 1988.) Adult recorded pecking at persistently begging fledgling (Géroudet 1954). Anti-predator Responses of Young. From 6th day onwards, nestlings crouch in nest when observer comes close; at 8 days, attempt to escape when handled (Gubin 1980). Nestlings fluttered from nest and hid in grass clumps nearby when observer approached nest closely (Roberts 1992). Recently fledged young, still unable to fly, hide on ground among stones and grass (Géroudet 1954), or in thick ground vegetation (Groh 1988). Young hidden near nest (with feathers still mainly in pin) crouched motionless with eyes closed when approached (Jouard 1934). Parental Anti-predator Strategies. (1) Passive measures. Incubating or brooding ♀ crouches low in nest on hearing alarm-call from ♂; sits very tightly, sometimes allowing herself to be touched, or may flush when man within *c.* 1–2 m (Gubin 1980). Adults feeding young land several metres from nest before final approach on ground, 'climbing slowly over the stones'; when leaving nest they fly directly from it (Géroudet 1954; Györgypál 1981). However, in Tien Shan, Gubin (1980) described ♀ flying to within *c.* 1 m of nest, looking about, then flying to nest. (2) Active measures: against birds. Mistle Thrushes *Turdus viscivorus* sometimes attacked if they perch close to nest (Gubin 1980). (3) Active measures: against man. ♀ flushed from nest may perform distraction-lure display of disablement-type, most assiduously when nest contains young: ♀ runs from nest, spreading wings occasionally, and appearing to somersault over small rocks in her path (Géroudet 1954; Gubin 1980).

(Fig A by E Dunn from drawing in Géroudet 1954.)   DTH

Voice. Used throughout year, but especially in breeding season. Songs varied, with possible geographical differences, but no detailed study. For additional sonagrams, see Bergmann and Helb (1982).

Calls of Adults. (1) Song of ♂. Likened to song of Dunnock *Prunella modularis* by many observers (e.g. Jakobs 1959, Hollom 1962, Groh 1988), although some variants suggest short song of Wren *Troglodytes troglodytes* (Bruun *et al.* 1986; P J Sellar). Very variable (e.g. Figs I–V), but typically fast, rather long (for *Emberiza*), with short, high-pitched units and musical phrases; often a squeaky timbre and somewhat jerky delivery (e.g. Bruun *et al.* 1986, Mild 1990, Jonsson 1992). Rendered as 'seut

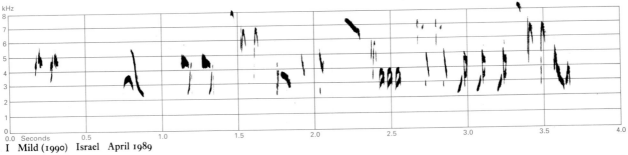

I   Mild (1990)   Israel   April 1989

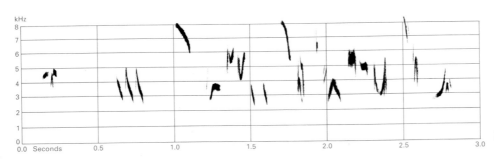

II   Mild (1990)   Israel   April 1989

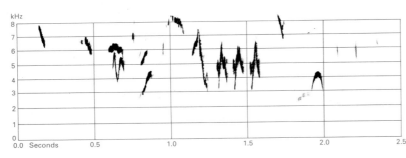

III   C Chappuis   France   April 1965

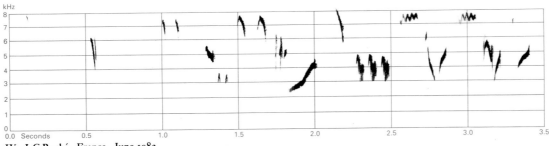

IV   J-C Roché   France   June 1983

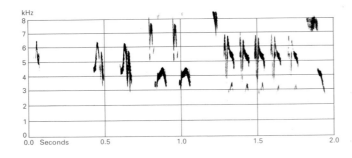

V   B N Veprintsev and V V Leonovich   Tadzhikistan   May 1974

wit tell-tell wit drr weeay sit seeay' (Bruun *et al.* 1986); shrill and in part rather suggestive of Serin *Serinus serinus* (Witherby *et al.* 1938). More complicated than song of Yellowhammer *E. citrinella* or Ortolan Bunting *E. hortulana*, but thin and feeble (Etchécopar and Hüe 1967). In Pfalz (south-west Germany), song incorporates mimicry of song and calls of other birds, including calls of Chaffinch *Fringilla coelebs* and parts of song of Redstart *Phoenicurus phoenicurus* (Groh 1988). Song in Morocco lower pitched than in France (Chappuis 1969). Recording from Tadzhikistan is of less musical song, recalling *P. phoenicurus* (Fig V: P J Sellar); possibly mimicry (see above). Recording from Kirgiziya is of unmusical, stumbling phrases consisting of mixture of dry and impure high-pitched units, 'zet drr sit drr set set drrr wit' (Mild 1987). Duration of song 1–3(–4) s (Géroudet 1954; Györgypál 1981), mean interval between songs 5 s (4–7) for short songs, 7 s (6–8) for longer songs (Györgypál 1981); in western Tien Shan, songs given at intervals of 3–4 s (Gubin 1980). Song in Pakistan (probably same as that in Europe: L Svensson), described by Roberts (1992) as quite melodious and bubbling, reminiscent of Goldfinch *Carduelis carduelis* but rather higher pitched and briefer; early in breeding season can be more prolonged with variations, but later becoming shorter and more stereotyped. Song opens with separate, emphasized, sibilant, thin, rising and falling units 'tsip-tsu' or 'tsu-tussip', followed by series of short rapid bubbling, twittering, rising and falling motifs; whole song-phrase lasts 1 s, given 8–9 times with minor variations. Generally very high-pitched and weak in volume, quite similar in tonal quality to song of Rufous-breasted Accentor *Prunella strophiata*. (Roberts 1992.) Further studies needed to determine whether individual ♂♂ sing more than one type of song, and extent to which the evidently wide variation in songs is due to racial differences, regional dialects, or individual differences. (2) Contact- and alarm-call. Thin, sharp 'tsi', often repeated (Fig VI); given as contact-call and when alarmed (Hollom

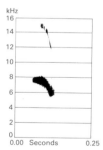

VI Mild (1990) Israel January 1989

1962). Also rendered a sharp 'tzit', hardly distinguishable from call 2 of Cirl Bunting *E. cirlus* (Witherby *et al.* 1938), 'zip' (Bergmann and Helb 1982), very high 'tsi', short and hard (P J Sellar), thin, weak 'zeet' (Bruun *et al.* 1986), short, hard, high-pitched 'tzi' and softer variants (Mild 1990), and 'tsee' or 'zie', the commonest calls (Jonsson

1992). A 'rattle' of 3 or more 'tit' type calls is given in alarm, occurring in flight or after landing (Andrew 1957b). Same call apparently repeated in longer, quick series in alarm by adults when fledged young are hidden, rendered 'si-titititi... titititi... zi-didididi...' (Géroudet 1954). (3) Long, high, but slightly descending 'tseeeee' (Fig VII), sometimes shorter (Fig VIII) when apparently passing

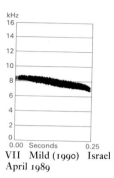

VII Mild (1990) Israel April 1989

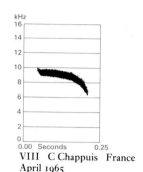

VIII C Chappuis France April 1965

into call 2. Rendered as a very high drawn-in whistle 'seeee', just a shade down-slurred or uniformly pitched (Bruun *et al.* 1986); high-pitched, drawn-in whistle, 'tzeeeee', very slightly falling in pitch (Mild 1990); 'zii' (Bergmann and Helb 1982); thin 'seea' (Peterson *et al.* 1983); distinctive, rather quiet, very thin, high-pitched, squeaky 'tswee'; given frequently on breeding grounds, as contact- or warning-call (Roberts 1992). According to Andrew (1957b), 'see' call is given to intruders near nest and clearly expresses strong fear; such calls sometimes approach 'seeoo', which is also given by solitary birds and in winter. (4) A 'tucc', fuller than call 2 and rather bubbling, becoming more of a twitter in flight, 'tootootooc' (Witherby *et al.* 1938). Short 'tüp', and on taking flight a variable 'chelut', 'chit vit', or slightly vibrant 'chee dee dee' (Jonsson 1992). A 'tu-tu-tu-uk' given in flight by winter flocks (Roberts 1992). (5) Rolling 'kierrrr' (Jonsson 1992). May be same as short hard 'trill', 'trr', commonly given by disturbed birds in flight and sometimes changed to 'zii trr' after landing (Bergmann and Helb 1982).

CALLS OF YOUNG. Food-calls of small young monosyllabic, becoming disyllabic when older (Wunsch 1976); piping hissing chirp (Györgypál 1981); conspicuously loud (Whistler 1923).   DTH

Breeding. SEASON. Rheinland and Pfälzerwald (Pfalz, western Germany): eggs laid mainly about 1st half of May (mid-April to July, rarely August) (Macke 1980; Groh 1988); for Bayern (southern Germany), see Model and Otremba (1986) and Berck and Berck (1976). Switzerland: eggs laid from late April or late May depending on altitude; 2nd clutches after mid-June, and eggs recorded mid-August (Géroudet 1954; Glutz von Blotzheim 1962). Hungary: early May to at least mid-July (Dandl 1959; Szabó 1962; Györgypál 1981). Greece: laying rarely before mid-May (Makatsch 1976). Caucasus (south-west Russia): young occasionally still in nest mid-September (Dementiev and Gladkov 1954). Algeria: full clutches found early April to mid-June (Heim de Balsac and Mayaud 1962). SITE. On or close to ground in cleft in rock or between boulders on slope, usually by bush, etc., generally hidden by vegetation though sometimes exposed; also in wall or earth bank, or low in dense tree or bush (Géroudet 1954; Dandl 1959; Hollom 1962; Szabó 1962; Model and Otremba 1986). In Caucasus, recorded under roofs (Dementiev and Gladkov 1954), and in Kashmir 4 m up in tree (Roberts 1992). In Pfälzerwald, 60% of 47 nests on ground, 26% in small coniferous trees, and 14% in dense creepers, shrubs, or herbs; average height above ground 51 cm (0-182) (Groh 1988); in Rheinland, 18% of 72 nests on ground (Mildenberger 1984). Nest: foundation of dry grass, stalks, and roots, occasionally leaves and bits of bark, lined with fine grasses, rootlets, and some hair (Géroudet 1954; Dandl 1959; Szabó 1962; Model and Otremba 1986). In Pfälzerwald, 12 nests had average outer diameter 12·9 cm (10·0-16·0), inner diameter 5·9 (5·5-6·5), overall height 7·2 (5·5-9·1), depth of cup 4·0 cm (2·9-5·2), dry weight 26 g (9-36) (Groh 1988). Building: by ♀ only, sometimes accompanied by ♂; takes c. 3-7 days, and material collected c. 10-30 m from site (Géroudet 1954; Gubin 1980; Györgypál 1981, which see for duration of building bouts; Groh 1988). EGGS. See Plate 33. Sub-elliptical, smooth, and faintly glossy; very pale greyish- to purplish-white, heavily and intricately scrawled with long and meandering dark violet to black hairstreaks, sometimes forming ring at broad end (Harrison 1975; Makatsch 1976; Wunsch 1976). Nominate *cia*: 20·6 × 15·8 mm (18·5-23·3 × 14·6-17·5), n = 109; calculated weight 2·57 g (Makatsch 1976); for 47 eggs from Catalonia (north-east Spain), see Muntaner *et al.* (1983), and for 29 from Pfälzerwald, see Groh (1988). North Africa: 20·8 × 16·0 mm (19·9-21·6 × 15·7-16·3), n = 7; calculated weight 2·73 g (Makatsch 1976); 19·0-21·0 × 15·0-16·5 mm, n = 36 (Etchécopar and Hüe 1967). See also Schönwetter (1984). Clutch: 4-5 (3-6). In Pfälzerwald, of 41 clutches: 2 eggs, 5%; 3, 22%; 4, 63%; 5, 10%; average 3·78 (April 4·25, n = 4; May 4·05, n = 18; June 3·75, n = 12; July 3·0, n = 5; August 2·5, n = 2); many pairs had 2 broods, a few had 3; replacement nest can be started 2 days after failure of clutch; eggs of 2nd clutch found 7-8 days after fledging of 1st brood; including replacements, up to 4 clutches can be laid by one ♀ (Groh 1988). See also Macke (1980). For timing of 3 clutches in captivity, see Wunsch (1976). Eggs laid daily in early morning 3-4 days after nest completed (Géroudet 1954; Györgypál 1981; Groh 1982); in Talasskiy Alatau (western Tien Shan) up to 15 days recorded between completion of nest and 1st egg (Gubin 1980). Average size in Switzerland 3·95, n = 20 (Glutz von Blotzheim 1962), and in Catalonia 3·87, n = 16 (Muntaner *et al.* 1983). In Algeria, of 42 clutches: 3 eggs, 17%; 4, 55%; 5, 26%; 6, 2%; average 4·14 (Heim de Balsac and Mayaud

1962). No evidence for 2nd broods in Israel (Shirihai in press) or Talasskiy Alatau (Gubin 1980). INCUBATION. 12-14 days, by ♀ only, starting with last egg (Géroudet 1954; Wunsch 1976; Shirihai in press); from laying of last egg to hatching of last young 13·4 days, n = 7 (Groh 1988). At one nest in Hungary, ♀ took breaks of maximum 20 min (Györgypál 1981); in Talasskiy Alatau one ♀ took breaks of 20-45 min, n = 9 (Gubin 1981). YOUNG. Fed and cared for by both parents; brooded only by ♀ (Géroudet 1954, which see for routine at 8 nests; Dandl 1959; Szabó 1962; Groh 1988). At one nest, ♀ brooded for 11 days (Györgypál 1981). For brooding stints and feeding rates, see also Gubin (1981) and Social Pattern and Behaviour. FLEDGING TO MATURITY. Fledging period 10-13 days, young leaving nest before able to fly (Géroudet 1954, which see for development of young). From 1st hatching to 1st fledging 11·3 days (10-13, n = 6); young fed for 2-3(-4) weeks after fledging (Groh 1988). BREEDING SUCCESS. In Pfälzerwald, of 155 eggs or nestlings, 47% produced fledged young, giving 3·5 per successful nest (n = 21), 1·8 overall (n = 41); of 20 nests which failed completely, 18 predated (predators included Jay *Garrulus glandarius*) and 2 broods died of cold (Groh 1988). In Rheinland, 41 eggs in 11 successful clutches resulted in 31 fledged young; predators included weasel and stoat *Mustela*, smooth snake *Coronella austriaca*, and Carrion Crow *Corvus corone*; some broods lost to heavy rain (Macke 1980; Mildenberger 1984). In India and Pakistan, frequently brood-parasitized by Cuckoo *Cuculus canorus* (Ali and Ripley 1974; Roberts 1992); also recorded in Switzerland (Ribaut 1954). BH

Plumages. (nominate *cia*). ADULT MALE. In fresh plumage (October-February), forehead black mottled with grey; crown basically light neutral grey, finely streaked black, tinged cinnamon or buff, especially towards nape; stripe at side of crown black, mottled with some grey and cinnamon-brown. Supercilium pale neutral grey, slightly washed buff; lore black, extending as a black eye-stripe through eye to rear of ear-coverts; eye-ring narrow, black. Upper cheek and ear-coverts light neutral grey, almost white near base of bill, mottled dark grey, partly suffused greyish-cinnamon, especially on ear-coverts; a narrow black bar at rear end. Lower cheek pale neutral grey, contrasting sharply with narrow well-defined black malar stripe, latter extending to lower side of neck where it reaches black bar behind ear-coverts. Nape and upper mantle light neutral grey, heavily mottled black and pale chestnut; lower mantle, scapulars, and back saturated rufous-cinnamon, marked with broad black streaks. Rump and upper tail-coverts dark cinnamon-rufous, upper tail-coverts with black shafts. Chin, throat, and chest light neutral grey, palest on chin. Side of breast, belly, and flank cinnamon-rufous, contrasting sharply with chest, merging into warm pink cinnamon or buff on thigh, vent, and under tail-coverts. Tail-feathers sooty black; both webs of central pair (t1) with broad cinnamon-rufous fringes, outer webs of t2-t5 with narrow cinnamon or buff fringe; t4 with narrow but well-defined white fringe along tip and sometimes with white blotch on tip of inner web; inner web of t5 with contrasting white wedge on tip (30% of length of feather), inner web of t6 with similar white wedge (about half of length of feather), outer web of t6 largely white. Flight-feathers sooty-black; primaries with narrow and sharply defined pale cinnamon or cream-buff fringes on outer webs, (whitish along emarginated parts); outer webs of secondaries with cinnamon to tawny fringes (widest on innermost). Tertials sooty black with broad rufous-cinnamon borders. Primary coverts and bastard wing greyish-black, outer webs narrowly margined greyish-buff or off-white. Greater upper wing-coverts with sooty black centre (mostly concealed) and broad rufous-cinnamon border along outer web, buff or cream-white on tip, tips sometimes forming indistinct pale wing-bar; median coverts sooty black with cream-white to off-white tips 2-3 mm wide, forming well-marked upper wing-bar; lesser coverts medium neutral grey with buff or pale cinnamon tips. Under wing-coverts and axillaries white; shorter coverts along leading edge of wing mottled dark grey. *In worn plumage* (March-May), grey of crown, nape, side of head and neck, from chin to chest, and lesser coverts purer, cinnamon or buff feather-tips completely worn off; broad black stripe on each side of crown, narrower black eye-stripe, black malar stripe, and black bar at rear border of ear-coverts more distinct, sharply contrasting with grey; centre of crown marked with sharp black shaft-streaks; cinnamon-rufous of mantle, scapulars, tertials, tail, and upper wing-coverts slightly paler, fringes and tips tinged pale grey-buff; tips of greater upper wing-coverts bleached to off-white, lower wing-bar better marked, but both bars largely worn off when heavily abraded. ADULT FEMALE. Like adult ♂, but black stripes on crown mixed brown, less pure black; grey feathers on centre of crown more extensively tipped buff-brown, black shaft-streaks slightly wider, grey hardly visible when plumage fresh (in contrast to adult ♂), grey tinged brown and more heavily streaked black than in adult ♂ when plumage worn, less contrasting with black-and-brown stripe at side of crown; black stripes on side of head slightly less sharply defined, tinged or mixed with some brown, grey duller and browner, less pure ash-grey; grey of throat and chest tinged buff and with faint dusky grey triangular marks when plumage fresh, more uniform buffish-grey when worn. NESTLING. Down long and fairly plentiful, restricted to upperparts, upper wing, vent, and thigh; dark grey (Ticehurst 1926; Witherby *et al.* 1938). JUVENILE. Forehead, crown, and nape evenly streaked pink-buff or cream-buff and sepia-black. Side of nape cream-buff to off-white, finely mottled grey, but almost uniform pale buff from nostril to above eye and on rear of lower cheek; broad but ill-defined sepia-black stripes run from below and behind eye over upper and lower ear-coverts; a broken sooty malar stripe. Mantle and scapulars evenly streaked pink-cinnamon (if fresh) or buff (if worn) and black, this gradually turning into tawny-cinnamon with some dull black spots on back and upper rump and this in turn to rufous-cinnamon on upper tail-coverts, latter with black shaft-streaks. Chin and throat pale greyish-buff to off-white, finely speckled dusky grey; side of breast, chest and upper flank warmer tawny-buff or cinnamon-buff with short and sharply-defined triangular black streaks; remainder of underparts slightly paler tawny-buff, palest on mid-belly, unmarked except for traces of black or grey dots on lower flank. Lesser upper wing-coverts tawny-buff with ill-defined dusky centres; median coverts black-brown with broad cream-buff to off-white tip, greater coverts black-brown with broad pink-cinnamon fringe along outer web and cream-buff tip, pale tips of median and greater coverts broken by black-brown point in middle, extending from centre. Tail as adult ♂, but t4 without white fringe along tip (Schuphan and Heseler 1965; Robert 1975). Flight-feathers, greater upper primary coverts, and bastard wing as in adult, but fringes slightly browner, less greyish-brown, especially along primary coverts; tertials as adult, but rufous-cinnamon fringes slightly narrower. FIRST ADULT. Like adult, but tail, flight-feathers, and greater upper primary coverts still juvenile, more worn than in adult at same time of year; tail-feathers often narrower and with more pointed tip than in adult, t4 without white fringe along tip (but fringe on t4 of adult liable to wear off by June-July); t1 sometimes contrastingly newer, more rufous, and with more rounded tip than t2-t6 when moulted in post-juvenile (all equally new in adult). Head and body of 1st adult ♂ approximately intermediate between adult ♂ and adult ♀; grey of top and side of head and neck on average browner than in adult ♂, but less so than in adult ♀; black stripes on top and side of head blacker and more distinct than in adult ♀; side of throat and all chest slightly washed buff when plumage fresh, marked with faint dusky spots (on average, less so than in adult ♀), this worn off in abraded plumage, but grey often still less pure than in adult ♂. Crown and nape of 1st adult ♀ rather evenly streaked dull black and buff-brown, black stripe on each side of crown scarcely broader than black streaks on central crown, no grey of feather-bases visible, unless plumage heavily worn; supercilium and stripe on lower cheek pale grey-buff, remainder of side of head and neck brownish-buff with faint and narrow grey streaks and specks; dark stripe behind eye, below and at rear of ear-coverts, as well as malar stripe generally indistinct (much more so than in adult ♀ and 1st adult ♂); grey of chin to chest extensively tinged buff when plumage fresh, marked with dusky grey triangular spots, sometimes coalescing into regular streaks; if worn, buff tinge and dusky specks or streaks on side of throat and all chest often still conspicuous, in contrast to adult ♀ and 1st adult ♂, which are predominantly (buffish-)grey.

Bare parts. ADULT, FIRST ADULT. Iris dark brown to red-brown. Upper mandible dark horn-grey, dark plumbeous-grey, or bluish-black, lower mandible light blue-grey to leaden-grey with slight yellow or pink tinge at base and bluish- or plumbeous-black tip. Leg and foot pale flesh-brown, red-brown, or dull orange-flesh, toes tinged brown or dull grey. (Richmond 1895*a*; Witherby *et al.* 1938; Schuphan and Heseler 1965.) NESTLING. Bare skin, including legs and bill, pink-flesh; mouth uniform yellow-pink, gape-flanges whitish (Witherby *et al.* 1938). JUVENILE, FIRST ADULT. Iris sepia or dull brown. Both mandibles bluish grey, sometimes tinged green at base of lower mandible and brown on tip. Leg and foot pale flesh to yellowish grey-brown. (Schuphan and Heseler 1965; Robert 1975; RMNH, ZMA.)

Moults. ADULT POST-BREEDING. Complete; primaries descendent. From late July until mid-September or October (Meinertzhagen 1940; Johansen 1944; Schuphan and Heseler 1965). In Caucasus area, small feathers replaced from late July, moult completed September, sometimes from late August (Dementiev and Gladkov 1954). In Transcaspia, *par* starts with small feathers in early August, plumage fresh from late September; all birds newly moulted by October (Dementiev and Gladkov 1954). In Iran, moult not started in single bird from 18 July, just started with tail 25 July, in full moult late July (Stresemann 1928); another had primary moult score *c.* 30 (p8-p9 still old) and active moult of tail as early as 6 July (RMNH). In Afghanistan, no moult up to late July (Paludan 1959). POST-JUVENILE. Partial: head, body, lesser and median upper wing-coverts, and sometimes greater coverts, tertials, and t1. August-October, young of late broods September-October (Géroudet 1954; Schuphan and Heseler 1965). In central Eurasia, *prageri* and *par* start moulting in July; some *par* still in moult September (Dementiev and Gladkov 1954).

Measurements. ADULT, FIRST ADULT. Nominate *cia*. Central and southern Europe, North Africa, western Turkey, and Israel, all year; skins (RMNH, ZFMK, ZMA). Bill (S) to skull, bill (N) to distal corner of nostril; exposed culmen on average 2·9 shorter than bill (S).

WING	♂	81·7 (2·84; 32)	74–87	♀	77·8 (3·96; 28)	72–83
TAIL		70·8 (2·83; 27)	65–75		69·1 (3·68; 26)	62–77
BILL (S)		13·6 (1·04; 31)	12·1–14·7		13·4 (1·14; 26)	11·1–15·3
BILL (N)		8·0 (0·58; 29)	6·8–8·8		7·9 (0·64; 25)	5·9–9·1
TARSUS		19·0 (1·09; 22)	16·8–21·1		19·2 (0·94; 21)	17·7–21·3

Sex difference significant for wing. Possibly some mis-sexed individuals included in ranges above, as usually less overlap in wing between sexes; thus, Europe, ♂ 78–87 (n=34), ♀ 74–80 (n=20) (Svensson 1992); Italy, ♂ mainly 77–85, ♀ mainly 75–79 (Cucco and Ferro 1988).

E. c. par. Iran and central Asia, all year; skins (RMNH, ZMA).

WING	♂	83·2 (4·03; 16)	73–90	♀	79·7 (0·88; 7)	78–84
TAIL		73·4 (4·34; 14)	67–81		68·6 (3·20; 7)	62–72
BILL (S)		11·7 (0·84; 12)	10·1–13·5		11·6 (0·44; 6)	11·2–12·5
BILL (N)		8·2 (0·84; 12)	7·0–9·8		8·4 (0·17; 6)	8·1–8·6
TARSUS		18·9 (1·03; 12)	17·2–20·9		18·9 (0·50; 6)	18·0–19·4

Sex differences significant for wing and tail.

Wing. Nominate *cia*. (1) North Africa (Jordans 1950). (2) Central and southern Spain (Jordans 1950; ZFMK, RMNH, ZMA). (3) South-east France (G Olioso). (4) Pfalz (south-west Germany), December–June (Groh 1988, which see for other details). (5) North-west Italy (Cucco and Ferro 1988). (6) Southern Yugoslavia and Greece (Stresemann 1920; Makatsch 1950; RMNH). Nominate *cia*, *prageri*, and intermediates between these. (7) Asia Minor (Jordans and Steinbacher 1948; Kumerloeve 1961, 1964a, 1967a, 1969a, 1970a; Rokitansky and Schifter 1971; RMNH, ZFMK). *E. c. prageri*. (8) Northern Caucasus (Vaurie 1956e). *E. c. par.* (9) Iran (mainly north) (Stresemann 1928; Paludan 1940; Schüz 1959; Diesselhorst 1962). (10) Iran (Vaurie 1956e). (11) Former Soviet Central Asia and Mongolia (Vaurie 1956e; Piechocki and Bolod 1972; RMNH, ZMA). (12) Eastern Afghanistan (Paludan 1959; see also Martens 1972). *E. c. stracheyi*. (13) Kashmir, Baltistan, and Ladakh (northern India) (Vaurie 1956e; Martens 1972; ZMA). (14) Northern Punjab to Kumaon (northern India) (Vaurie 1956e; Martens 1972; ZMA). *E. c. flemingorum*. (15) Western Nepal (Martens 1972).

(1)	♂	— (— ; 12)	79–86	♀	— (— ; 6)	76–79
(2)		82·2 (2·30; 14)	79–86		76·1 (1·80; 10)	74–79
(3)		84·1 (2·22; 13)	80–87		— (— ; —)	—
(4)		81·1 (2·50; 75)	75–87		75·3 (2·17; 56)	71–80
(5)		81·4 (2·66; 39)	76–87		77·1 (1·68; 44)	74–83
(6)		83·3 (2·05; 29)	79–87		77·8 (1·53; 11)	75–80
(7)		83·6 (2·17; 14)	80–87		77·6 (3·88; 11)	72–84
(8)		88·0 (— ; 5)	85–90		— (— ; —)	—
(9)		86·0 (2·17; 12)	82–89		78·3 (— ; 3)	78–79
(10)		89·0 (— ; 10)	85–92		— (— ; —)	—
(11)		86·1 (— ; 14)	82–91		79·8 (1·17; 14)	78–83
(12)		84·7 (— ; 18)	82–87		77·5 (— ; 8)	75–80
(13)		86·4 (— ; 22)	82–88		79·1 (1·62; 6)	77–81
(14)		84·0 (— ; 17)	81–90		77·0 (— ; 1)	—
(15)		81·5 (— ; 5)	79–82		74·0 (— ; 4)	72–76

Weights. Nominate *cia*. (1) Spain, Italy, Germany, Greece, and north-west Turkey, December and summer (Makatsch 1950; Rokitansky and Schifter 1971; ZFMK, ZMA). (2) South-east France, winter (G Olioso). Pfalz (south-west Germany): (3) April–June (Groh 1988; see also Groh 1982). Intermediates between nominate *cia* and *prageri*. (5) South-central and eastern Asia Minor, May–November (Kumerloeve 1961, 1964a, 1967a,

1969a, 1970a; ZFMK). *E. c. par*: (6) Northern Iran, March–June (Paludan 1940; Schüz 1959). (7) Eastern Afghanistan, March–July (Paludan 1959; Niethammer 1973). (8) Kazakhstan (Korelov *et al.* 1974).

(1)	♂	23·6 (1·52; 5)	21·5–25·0	♀	22·9 (0·85; 3)	22·0–23·7
(2)		24·3 (2·69; 13)	21·0–29·0		— (— ; —)	—
(3)		24·2 (1·62; 51)	21–29		22·7 (1·64; 38)	20–27
(4)		22·9 (1·18; 24)	20–25		21·1 (1·28; 18)	18–23
(5)		25·0 (2·19; 6)	23·0–27·0		24·0 (— ; 2)	23·0–25·0
(6)		26·5 (1·54; 4)	24·6–27·5		20·3 (— ; 1)	—
(7)		21·7 (7 ; 19)	17·0–26·1		20·7 (— ; 10)	18·9–22·5
(8)		23·4 (— ; 24)	20·3–27·8		22·1 (— ; 13)	19·5–24·5

Nominate *cia*. North-west Italy, August–November: 22·2 (1·61; 97) 19–28 (Cucco and Ferro 1988, which see for details). Belgium, August: juvenile (probably ♂) 22·5 (Robert 1975).

E. c. par. Mongolia, July: ♂♂ 22, 25; ♀ 23 (Piechocki and Bolod 1972).

E. c. stracheyi. Kashmir (India), August–September: ♂ 22·8 (2·38; 4) 19·2–24·3, ♀ 21·5, juvenile 21·7 (1·91; 6) 19·0–24·5 (P R Holmes and Oxford University Kashmir Expedition 1983). India, April–May: 19·3 (11) 18–21 (Ali and Ripley 1974).

Structure. Wing short, broad at base, tip bluntly pointed. 10 primaries: p7 longest, p9 3–6 shorter, p8 and p6 0–2, p5 1·5–4, p4 5–10, p3 9–14, p2 11–16, p1 12–18; p10 strongly reduced, narrow, 14.2 (10) 13–16 shorter than longest upper primary covert. Outer web of p5–p8 emarginated, inner web of p6–p9 with faint notch. Tip of longest tertial reaches to tip of p3–p4. Tail fairly long, tip slightly forked; 12 feathers, t1 and t6 1–5 shorter than t3–t4. Bill as in Yellowhammer *E. citrinella*, but slightly less deep and wide at base, more sharply pointed at tip. Middle toe with claw 16·6 (10) 15·3–18·5; outer toe with claw *c.* 79% of middle with claw, inner *c.* 71%, hind *c.* 86%. Remainder of structure as in *E. citrinella*.

Geographical variation. Based on specimens (BMNH, RMNH, ZFMK, ZMA) examined by C S Roselaar. Slight, strongly clinal. Of 5 races recognized here, 4 are fairly distinct: nominate *cia* in central Europe rather small with dark rufous-cinnamon upperparts; *par* in arid hills of west-central Asia is large, with sandy-cinnamon upperparts with narrow black streaks and more uniform pinkish-rufous-cinnamon chest to under tail-coverts; *stracheyi* in western Himalayas small, with dark rufous-brown or vinous-rufous ground-colour to upperparts and uniform deep rufous-cinnamon chest to under tail-coverts; *flemingorum* in western Nepal small, and pale rufous-cinnamon above. Typical birds of these races are connected by clinally intermediate populations; also, minor clines in colour run from darker birds in central Europe to slightly paler ones in south-west and south-east, with palest extremes in southern Spain and Bulgaria. Distinct cline of increasing paleness runs from Asia Minor through Caucasus, Iran, Transcaspia, and Tien Shan, ending in palest populations of Tarbagatay and southern Altai. Cline reversed again from Afghanistan to southeast, populations becoming increasingly darker from eastern Afghanistan and Pamirs through western Himalayas to Sutlej valley and Kumaon (area inhabited by typical *stracheyi*), but much paler again further east in western Nepal. All birds from central and southern Europe and North Africa here considered to belong to nominate *cia*, though many populations differ slightly in colour and have been named *hordei* C L Brehm, 1831 (south-east Europe), *africana* Le Roi, 1911 (North Africa), and *callensis* Ticehurst and Whistler 1938 (Portugal). From Germany to central Rumania, fairly dark on upperparts and dark below;

in France, Italy, and Yugoslavia, almost imperceptibly paler; in central Spain, Morocco, Algeria, Sicily, and Dalmatia, both upperparts and underparts fairly pale pinkish-cinnamon with buff cast, but this mainly due to more pronounced bleaching and abrasion there, as difference slight in fresh plumage; in Portugal, upperparts dark, but underparts pale; palest birds occur in southern Spain (Sierra Nevada), Tunisia, Bulgaria, and Greece; bill in North Africa slightly longer and heavier than elsewhere (C S Roselaar, RMNH, ZFMK, ZMA; see also Ticehurst and Whistler 1938, Meinertzhagen 1940, Jordans 1950, and Niethammer 1957). Birds from western and northern Asia Minor (east to at least Kastamonu) and Levant are referable to nominate *cia*, but those of north-east Turkey are similar in colour to *prageri* from Caucasus area and Transcaucasia, while birds from southern Turkey eastward from middle Taurus mountains are more or less intermediate between nominate *cia* and *prageri* in colour, though near nominate *cia* in size (Jordans and Steinbacher 1948; Kumerloeve 1961, 1964a, 1967a, 1969a, 1970a). *E. c. prageri* distinctly paler and larger than any population of nominate *cia*, but hardly distinguishable from *par*; rufouscinnamon of upperparts and belly brighter than in nominate *cia*, more pinkish, chin to chest paler grey; tips of median upper wing-coverts red-brown, not white as in nominate *cia* (but colour strongly dependent on bleaching and abrasion in both). Populations from Crimea and along shores of Black Sea in western Caucasus slightly darker than nominate *cia*, rump more intensely rusty, and lesser coverts greyer; darker and more intensely coloured than *par*, but tips of median coverts white; sometimes separated as *mokrzechyi* Molchanov, 1917, but here included in *prageri*, following Vaurie (1956e). North-east and eastern Iran, southern Zagros mountains of Iran, western and central Afghan-

istan, Turkmeniya, and mountainous areas of Central Asia east to Tarbagatay, southern Altai, and (isolated) in Mongolia inhabited by *par*, a pale and large race, which is rather similar to *prageri*, but slightly paler, especially in worn plumage and in extreme east of range; dark streaks on mantle and scapulars on average narrower, rump more tawny; in northern Iran (west from Gorgan) and in northern Zagros mountains, grades into *prageri*, but birds from Iran nearer *par* (Vaurie 1956e). In eastern Afghanistan, Pamirs, and extreme north-west Pakistan (Chitral, Gilgit), rufous colours deeper than in *par*, and these birds sometimes therefore separated as *lasdini* Zarudny, 1917, but best still included in *par*. Further east, birds from Baltistan, Ladakh, and Kashmir show increasing colour saturation (these closely similar to nominate *cia* in colour, but slightly larger), cline ending in dark *stracheyi* of upper Sutlej valley and Kumaon in northern India; *stracheyi* darker and more richly coloured than nominate *cia*, with sharper black-and-white marks on head, and with median and greater coverts tipped rufous (Hartert 1903-10; Hellmayr 1929; Vaurie 1956e). In western Nepal, *flemingorum* much paler than *stracheyi*, near *par*, but distinctly smaller (Martens 1972).

Forms superspecies with Eastern Rock Bunting *E. godlewskii* which differs mainly in chestnut rather than black stripes on head and is often included in *E. cia* (Hartert 1903-10, 1928; Vaurie 1956e, 1959). Ranges touch in Altai mountains without hybridization, pointing to competitive exclusion, and *E. cia* occurs locally within range of *E. godlewskii* in Mongolia without apparent interbreeding, thus behaving as separate species (Mauersberger and Portenko 1971; Mauersberger 1972). For races of *E. godlewskii*, see Hartert and Steinbacher (1932-8), Johansen (1944), and Vaurie (1956e, 1959).   MP

Emberiza cioides Brandt, 1843 Meadow Bunting

FR. Bruant des prés   GE. Wiesenammer

PLATES 15 and 16 (flight)
[between pages 256 and 257]

Breeds in open country and forest edges from central Asia east to eastern Siberia, Korea, Japan, and eastern China south to Kwangtung. Northern populations winter south of breeding range or at lower elevations within general breeding range; southern populations winter within breeding range. Record of singing ♂ in Finland, 20-27 May 1987, formerly accepted as wild bird but now presumed to be escape (Jännes 1992). Also unconfirmed record of 2 birds at Veneto (Italy), 1910 (Brichetti and Massa 1984).

Emberiza striolata House Bunting

DU. Huisgors   FR. Bruant striolé   GE. Hausammer
RU. Пестрая овсянка   SP. Escribano sahariano   SW. Hussparv

Emberiza striolata Lichtenstein, 1823

PLATES 16 (flight) and 19
[between pages 256 and 257]

Polytypic. *E. s. sahari* Levaillant, 1850, North Africa east to western Egypt, south to northern Niger and northern Chad; perhaps this race in Mauritania; nominate *striolata* (Lichtenstein, 1823), from Sinai, extreme south-east Egypt, eastern Sudan, and Eritrea east through southern Levant and Arabia (except south-west) to north-west India. Extralimital: *jebelmarrae* Lynes, 1920, west-central Sudan; *saturatior* Sharpe, 1901, central and southern Ethiopia to Somalia and northern Kenya, also Yemen; *sanghae* Traylor, 1960, southern Mali, and (perhaps this race) Sénégal.

Field characters. 13-14 cm; wing-span 21·5-26 cm. 15-20% shorter and much slighter than Rock Bunting *E. cia*, Cretzschmar's Bunting *E. caesia*, and Cinnamon-breasted Rock Bunting *E. tahapisi*, with rather stubby bill, somewhat small head, and narrow tail. Rather small, delicate, cryptically plumaged bunting, with dark rufous-edged tail shared only by *E. tahapisi*. Western race *sahari* tame and rather dull, with little-marked grey and rufous-buff plumage; larger eastern race nominate *striolata* shyer and more puzzling, with blackish stripes on head of typical ♂ suggesting *E. cia* and even more so *E. tahapisi*. Calls distinctive. Sexes rather similar; little seasonal variation. Juvenile resembles ♀. 2 races in west Palearctic; well marked ♂♂ separable.

(1) Middle East race, nominate *striolata*. ADULT MALE. Moults: June–October (complete). Ground of head, neck, and upper breast greyish, variably but distinctly striped black and white. Well-marked bird has narrow, almost white streak in centre of crown, black-streaked sides to crown, almost white supercilium, fully black eye-stripe, often pure white panel under eye-stripe, blackish moustachial stripe, whitish malar stripe, and black-speckled chin and throat, spreading onto breast; white stripes usually more extensive than intervening black ones, unlike *E. tahapisi*. Poorly-marked bird shows far less contrasting pattern, with reduced white on central crown-stripe and malar stripe, and dark stripes ashy-grey, not black; note that such pattern never recalls *E. tahapisi*. Ground of body and wings noticeably uniform sandy- to rufous-brown, little-marked except for ill-defined and discontinuous dusky streaks on mantle, stronger dusky marks on scapulars, and blackish-brown centres to rufous-edged tertials and larger coverts; note that all such marks are less discrete and contrasting than on *E. tahapisi*. Rump and upper tail-coverts concolorous with back but unstreaked, again unlike *E. tahapisi*. Tail is darkest part of plumage, being dull brownish-black with rufous (not white) edges to outer feathers providing instant separation from *E. cia* but not *E. tahapisi* which shows similar pattern. Bill small, shape recalling small *Carduelis* finch; dusky-brown on upper mandible but flesh to warm yellow on lower (again like *E. tahapisi*, not *E. cia*). Legs flesh, pale yellow, or brown. ADULT FEMALE. Typically duller and paler than ♂, with even more obscure head marks and streaks above, but shows quite strongly striated breast. Individual variation not as obvious as ♂ but sufficiently wide to make separation of brightest ♀ and dullest ♂ impractical. JUVENILE. Not studied in the field. Best distinguished from ♀ by uniformly dusky-horn bill. (2) North-west African race, *sahari*. Proportionately slightly shorter-winged but longer-tailed than nominate *striolata*. Unlike even poorly marked nominate *striolata*, head, neck, and breast of

adult ♂ look more wholly grey at distance and show far less contrasting dark and pale stripes at close range, with dark stripes finer and ill-defined; rest of plumage much more fully rufous than nominate *striolata* and even less streaked or appearing unstreaked so that at any distance may appear uniform. ♀ has faint dusky striations round neck, breast and fore-flanks.

Confusing. Where not commensal with man, difficult to observe due to small size and cryptic plumage producing distinctive 'will of the wisp' character. When size not apparent, can be confused in Middle East with migrant *E. caesia* and allies (pink bill, bright eye-ring, more streaked above and on wings, white outer tail-feathers); in north-west Africa, particularly in winter, with *E. cia* (grey bill, white bar on median coverts, rufous-chestnut rump, white outer tail-feathers); and with potential vagrant from north-east Africa and Arabia, *E. tahapisi* (also short-winged but noticeably longer tailed, more strongly marked head with fully black or blackish throat and bib, much darker streaks on back and centres to wing-coverts, dark-spotted rump, and different calls). Flight fast and flitting, with flickering wing-beats and light landing. Gait a hop. Stance low and rather level, with head raised above chest but showing little neck. In north-west Africa, locally markedly commensal with man, but wilder elsewhere. Seen in pairs and small parties, never flocks; consorts with *E. cia*.

Voice distinctive. Song a short to medium-length, rather feeble phrase, with accelerated delivery recalling Chaffinch *Fringilla coelebs*, 'wi-di-dji-du-wi-di-di' or 'witch witch a wee'. Calls include nasal 'tzswee', thin sharp 'tchiele', and 'sweee-doo'.

Habitat. Situation confused by existence of ecologically distinct races, one of which (*sahari* of north-west Africa) has long been largely adapted to commensalism with man in inhabited settlements, while nominate *striolata* in Asia has remained attached to natural rocky habitat apart from having colonized some ruins of forts and other buildings. In north-west Africa, replaces House Sparrow *Passer domesticus* in villages on edge of desert, even entering houses and shops. Also sometimes found in wild desolate places, at some altitude, but never in open desert (Etchécopar and Hüe 1967). In western Sahara, has become locally a village bird (Moreau 1966), and on edge of Grand Erg Occidental in northern Sahara birds were 'so tame that they fearlessly entered the hotel dining room through the kitchen and picked up crumbs at our feet' (Bannerman and Priestley 1952). By contrast, nominate *striolata* has no connection with houses at all in Eritrea (Smith 1955*b*).

Nominate *striolata* in Asia lives in arid areas on rocky slopes with sparse vegetation, but when found in desert and semi-desert is usually within reach of water. May

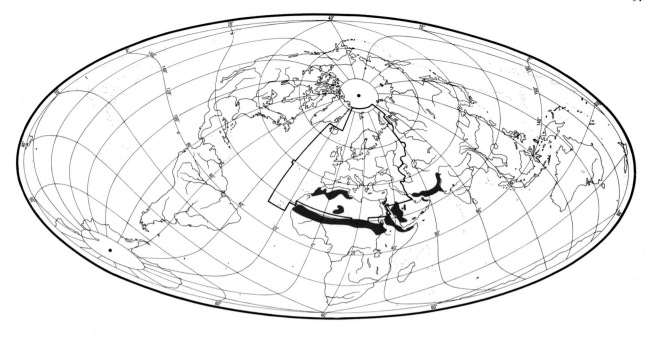

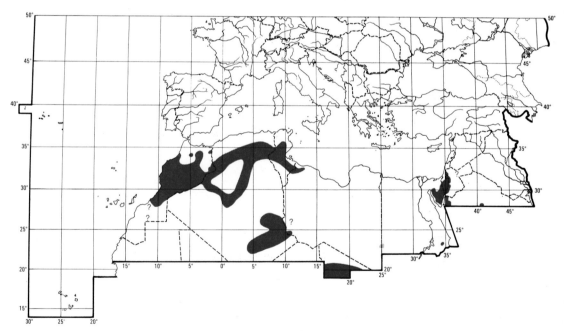

occur in lower and more open areas in winter. Confined to tropical and subtropical zones (Harrison 1982). In India and Pakistan, favours rocky or stony hillsides and nullahs bearing sparse scrub of *Euphorbia* or thorn bushes, but will occupy ancient hill forts and ruins. In winter, spreads to sandy plains, tamarisk scrub, and grass areas near canals, feeding on ground (Ali and Ripley 1974).

Distribution. Northward spread in north-west Africa, beginning in 19th century.

MOROCCO. Marked northward spread. Advance of more than 200 km along Atlantic coast recorded between 1867 and 1902. Northward spread continued, reaching Casablanca 1965; breeding first recorded Oujda 1980, Rabat 1983, Fes 1985. (Courteille and Thévenot 1988.) ALGERIA,

TUNISIA. Has spread north this century (Courteille and Thévenot 1988).

Accidental. Cyprus, Kuwait, Canary Islands.

Population. Has undoubtedly increased with northward expansion of range, but no data.

ISRAEL. Estimated *c.* 1000 pairs, with marked annual fluctuations; decreasing after dry winters (Shirihai in press). JORDAN. 21–40 pairs Petra, 1983 (Wittenberg 1987).

Movements. Essentially sedentary, but with short-distance movements in some populations.

In Morocco (*sahari*), some wandering after breeding season (Heim de Balsac and Mayaud 1962; Courteille and Thévenot 1988); in foothills at Errachidia, local breeding population supplemented by additional birds in non-breeding season (Destre 1984). Recorded north of breeding range in Morocco (which is expanding northwards; see Distribution) a few times, at Ceuta (September) and Tanger (September–October) (Giraud-Audine and Pineau 1973; Dubois and Duhautois 1977; Courteille and Thévenot 1988). Vagrants of *sahari* recorded Canary Islands and Egypt (Vaurie 1959; Armani 1985; Goodman and Meininger 1989). Old report from Andalucia (southern Spain) (Temminck 1835) perhaps also of vagrant (Courteille and Thévenot 1988).

Nominate *striolata* resident in Egypt (Goodman and Meininger 1989). In Israel, resident in some places but mostly makes local movements. Populations of Negev highlands and Judean Desert plateau move to lower elevations for winter. Movement of flocks from breeding grounds mostly October–November, return February–March. After particularly wet winters large numbers may invade desert areas such as in Negev and at Elat and remain to nest there. 2 records of wanderers to Mediterranean zone, both at Jerusalem (March and November–December). (Shirihai in press.) Described as resident at Dead Sea, Petra, and Wadi Rum in Jordan (I J Andrews) and in Iran (D A Scott). Reported once from Kuwait; other records there and at Qatar not accepted (F E Warr); reported or assumed to be resident in United Arab Emirates, Yemen (*saturatior*), and Saudi Arabia (Jennings 1981; F E Warr). In Oman, resident with seasonal movements; also scarce autumn migrant (northern Oman, Masirah) and winter visitor (Gallagher 1977; Gallagher and Rogers 1980; Gallagher and Woodcock 1980); visiting birds presumably not from far away. In northern Oman, makes local movements to coastal plains, where present early December to late February (Walker 1981a). In India and Pakistan, essentially resident but makes local and erratic movements in winter to plains and vicinity of water (Ali and Ripley 1974; Roberts 1992).

E. s. saturatior described as resident and wanderer in northern Kenya (Britton 1980).   DTH

Food. Seeds, mostly of grasses (Gramineae); invertebrates in breeding season. In North Africa, very dependent on man, feeding in streets, inside houses, restaurants, etc., and on rubbish tips and dung heaps, though also on ground in rocky country. In east of range, however, generally forages far from human habitation. (Whitaker 1894; Baker 1926; Meinertzhagen 1954; Jennings 1981c; Courteille and Thévenot 1988; Richardson 1990.) Frequently associates with flocks of domestic animals (Meinertzhagen 1940; Paz 1987), and feeds on spilled grain or open sacks in markets (Wallis 1912; Roux *et al.* 1990). Bird in Yemen seen standing on ground and pulling seeds off grass head (Brooks 1987).

Diet includes the following. Invertebrates: ants (Hymenoptera: Formicidae), spiders (Araneae). Plants: seeds, berries, flowers, etc., of toothbrush tree *Salvadora persica*, jujube *Ziziphus*, *Heliotropium*, Compositae, grasses (Gramineae, including reed *Phragmites*, *Phalaris*, *Bromus*, barley *Hordeum*, millet *Panicum*). (Heim de Balsac and Mayaud 1962; Laferrère 1968; Ali and Ripley 1974; Paz 1987; Roberts 1992.)

In North Africa, feeds on barley and other crops in spring and millet in autumn, but most seeds taken according to abundance (Heim de Balsac and Mayaud 1962; Laferrère 1968). In Tassili (southern Algeria), 2 stomachs obtained end of March and beginning of April contained only seeds (Roche 1958). In Pakistan, feeds on fallen jujube berries on ground and green unripe berries of toothbrush tree, but diet mostly seeds of grass and *Heliotropium*; often at water and wells (Roberts 1992). No further information.   BH

Social pattern and behaviour. Poorly known. Study in Casablanca (Morocco) and review by Courteille and Thévenot (1988); also studied in Marrakech (Morocco) by Roux *et al.* (1990).

1. In breeding season mainly in pairs; at other times often gathers in groups. In Israel, groups form as early as June; flocks consist of up to 30(–47) birds, average 8 (Shirihai in press). Maximum flock sizes elsewhere: 20 in western Sahara (Gaston 1970b), 30 in Oman (Gallagher and Rogers 1980), 10 in Saudi Arabia (Jennings 1980). In India and Pakistan likewise mainly singly or in pairs during breeding season, but in small flocks in winter (Ali and Ripley 1974; Roberts 1992); largest winter flock mentioned, 6–7 in Pakistan (Roberts 1992). In western Sahara, reported flocking in association with Desert Lark *Ammomanes deserti* and Trumpeter Finch *Bucanetes githagineus* (Gaston 1970b); in Dhofar (Oman) associates with Cinnamon-breasted Rock Bunting *E. tahapisi* (Walker 1981b); in Pakistan occurs alongside and feeds with Rock Bunting *E. cia* in some areas (Roberts 1992). BONDS. During breeding season pairs seem to be monogamous and stable (Courteille and Thévenot 1988; Roux *et al.* 1990). Pair-bond apparently year-long (Roux *et al.* 1990). Specimens indicate age of first breeding 1 year (C S Roselaar). Nest-building at least mainly by ♀; incubation by ♀ alone; nestlings and fledged young fed by both parents (see Heterosexual Behaviour, and Relations within Family Group, below). BREEDING DISPERSION. Territoriality may (Courteille and Thévenot 1988) or may not (Roux *et al.* 1990) be weak and rather fluid; perhaps undefended areas between territories (Pasteur 1956). Territory apparently serves for feeding and nesting, and may

include interior of occupied houses; defended by ♂ (Pasteur 1956; Courteille and Thévenot 1988). One pair flew 60 m to drink, crossing other territories; ♂ used regular place in territory (on wall top) for defecation; from single observation, territory thought to be chosen by ♂ alone (Pasteur 1956). In Israel, usually breeds in large separate territories, but occasionally in sparse colony-like groups, 10–30 m between nests; in hilly areas each pair usually on separate hillock (Shirihai in press). Nest-site often re-used, when same nest frequently rebuilt; majority of 2nd and 3rd clutches are in same nest as 1st; good proportion use same site year after year, one nest in Casablanca being used 11–12 times 1983–7 (Courteille and Thévenot 1988). In Marrakech, *c.* 2 pairs per ha (Roux *et al.* 1990). Transects 1 km in length in urban area of Casablanca recorded up to 10 singing birds during breeding season (Courteille and Thévenot 1988, which see for data on other times of year). At Petra (Jordan), pairs scattered in rocky habitat, with local concentrations (Wittenberg 1987). ROOSTING. Little information. ♀ roosts on nest until young 4 days old, but not on nest on 8th night (Courteille and Thévenot 1988). See also Roux *et al.* (1990).

2. *E. s. sahari* often or usually commensal with man (Heim de Balsac and Mayaud 1962); in Morocco, where regarded as sacred and receiving traditional protection, often very tame, feeding inside houses, shops, and mosques (Etchécopar and Hüe 1967; Dachsel 1975; Courteille and Thévenot 1988). Recorded tamely entering dining room, apparently regarding it as part of territory (Pasteur 1956). *E. s. striolata* not commensal, and in Israel not easily observed except when ♂♂ sing in breeding season (Paz 1987; Mild 1990); rather shy in Iran (D A Scott); in India and Pakistan, unobtrusive and easily overlooked according to Ali and Ripley (1974), but said by Roberts (1992) to be tame and confiding in Pakistan, even in remote regions, often allowing careful approach to within 3 m. FLOCK BEHAVIOUR. Little known. Flight-calls given when winter flocks disturbed (Roberts 1992). SONG-DISPLAY. Song (see 1 in Voice) used by ♂ to advertise territory (Pasteur 1956; see also Antagonistic Behaviour, below). In Morocco, song-posts on conspicuous perches, commonly on walls of houses and gardens, including occupied buildings in city of Casablanca; in early spring, ♂ spends much time on song-post, with few movements away from it (Pasteur 1956; Courteille and Thévenot 1988). At Petra (Jordan), sings only from rocks and not from trees or bushes (Wittenberg 1987). Song-posts in India and Pakistan include stones, bush-tops, and ruined buildings (Ali and Ripley 1974; Roberts 1992); singing from ground also recorded (Paige 1960). Sings all year in Morocco; in Casablanca, most song November to April or May (especially February–April), little in summer (Courteille and Thévenot 1988); in foothills, song strongly reduced in autumn and winter, but does not cease; more intense from January–February (Destre 1984). Just before beginning of breeding season, one ♂ sang for several hours each day, especially in morning; preferred one song-post, but used several others close by (Pasteur 1956). Song in Israel February–March (Shirihai in press). In northern Oman, song heard October–April (Gallagher 1977); in Saudi Arabia, most apparent February (Jennings 1980). In Pakistan, sings in spring and throughout breeding season; from mid-February onwards in Sind, but commencing about one month later in Punjab Salt Range (Roberts 1992). ANTAGONISTIC BEHAVIOUR. (1) General. No overt territorial defence seen by Pasteur (1956), but strong defence all year in Marrakech (Roux *et al.* 1990). In Casablanca, ♂♂ seen pursuing conspecifics (of uncertain sex) from October, and especially in winter, coincident with increase in singing (Courteille and Thévenot 1988). In Israel, 2 ♂♂ sang only a few metres apart, shortly after chasing and attacking each other (Mild 1990). (2) Threat and fighting.

In boundary disputes, resident ♂♂ face each other giving shrill calls and vibrating wings while moving sideways; ♀♀ also participate (Roux *et al.* 1990). Apparent threat directed at *A. deserti* on wall top involved hovering in flight 30 cm in front of it and moving slowly from side to side while hovering (Pasteur 1956). Fights not rare, but occur only when intruder penetrates far into territory; ♀ often watches ♂♂ fighting (Courteille and Thévenot 1988). HETEROSEXUAL BEHAVIOUR. (1) General. Little known. From single observation, thought that ♀ joins ♂ on territory at beginning of breeding season; no display seen; pair soon feeding together (Pasteur 1956). Pairs said to form in February in Saudi Arabia (Jennings 1980). Establishment of territory by ♂ and display said to take *c.* 1 week in the few cases observed, once 20 days; no details reported (Courteille and Thévenot 1988). (2) Pair-bonding behaviour. Chakir (1986) recorded pursuit-flights 5–10 days before nest-building; interpreted as display, but probably confused with territorial aggression (see Antagonistic Behaviour, above). In these, '♂' chases '♀' aggressively from place to place, sometimes leading to fighting in the air, or pecking, with sharp calls throughout, and birds falling to ground, after which chasing may continue. Both sexes reported to vibrate wings and press body close to ground in presence of mate (Chakir 1986). (3) Courtship-feeding. ♂ said to feed ♀ March–May at Tassili N' Ajjer (Algeria) (Heim de Balsac and Mayaud 1962), but no other reports, and confirmation needed that birds being fed were not juveniles. (4) Mating. On ground during nest-building. ♀ wing-shivers, raises tail slightly, and gives shrill calls; ♂ approaches and copulates quickly (Roux *et al.* 1990). (5) Nest-site selection and behaviour at nest. Not known which sex chooses site. Nest-building reported by Ali and Ripley (1974) to be by both sexes and this widely quoted; said to be by both sexes by Pasteur (1956), who did not see it. Both sexes also reported carrying nest-material by Courteille and Thévenot (1988). Chakir (1986) found building essentially by ♀, with ♂ participating only at start of building. Building was by ♀ alone at single nest in Saudi Arabia (Jennings 1980). More information needed on role of ♂ because some other *Emberiza* have nest-building by ♀ alone, but at beginning of building period ♂♂ perform Stem-display in which potential nest-materials carried but not added to nest. Incubation by ♀ alone (Courteille and Thévenot 1988). RELATIONS WITHIN FAMILY GROUP. Eggshells removed from nest by parent (Courteille and Thévenot 1988). Young fed by both parents, on regurgitated food (Ali and Ripley 1974; Courteille and Thévenot 1988; Roberts 1992). See Courteille and Thévenot (1988) for details of growth and feeding rates of young. ♀ broods young almost continually for first 4 days and roosts on nest; after this ♀ absent from nest more and more often, not roosting there on 8th night (Courteille and Thévenot 1988). Faecal sacs of young carried to distance from nest, sometimes eaten by ♀ (Courteille and Thévenot 1988). Fledglings fed by both sexes, but ♂ feeds them most often, apparently because ♀ soon occupied in starting another clutch. Fledglings beg with wings held low and vibrated and open bill, giving food-calls (see Voice). Fledglings fed for 9–20 days (*n* = 5), perhaps sometimes 22 days. 2 young roosted with parents more than 25 days after fledging. (Courteille and Thévenot 1988). ANTI-PREDATOR RESPONSES OF YOUNG. No information. PARENTAL ANTI-PREDATOR STRATEGIES. (1) Passive measures. ♂ very quiet and discreet when visiting nest (Courteille and Thévenot 1988). (2) Active measures: against man. Alarm-calls (not described) given by ♀ near fledglings (Courteille and Thévenot 1988). (3) Active measures: against other animals. No information. DTH

Voice. Birds often vocal. Song varied; extent of repertoire and geographical variation need further study. Wide variety of calls reported, but hardly any information on contexts or functions; apparently similar sounds are grouped together below, but more study needed.

CALLS OF ADULTS. (1) Song of ♂. Varied, short, musical 'wi-di-dji-du-wi-di-dii' or 'witch witch a wee', with delivery resembling Chaffinch *Fringilla coelebs* (Etchécopar and Hüe 1967; Hollom *et al.* 1988); like song of *F. coelebs*, but weaker and higher pitched (Jonsson 1992). In Israel, rendered as loud and rather monotonous 'weDJE-weDi-weDi-wiDJE' (Mild 1990). Recordings (Figs I–IV) show variety of songs, with different types likened to Scarlet Rosefinch *Carpodacus erythrinus* (Fig I), Cetti's Warbler *Cettia cetti* (Morocco) and Dunnock *Prunella modularis* (Israel, Algeria: Fig II) (P J Sellar). Song of Moroccan ♂ rendered 'tu-ip tyiè tyiè, tyiè tyiè tyiè, tyuit-it'; 2 phrases of 3 short units followed by 1 phrase of prolonged units. After pairing, often shortened by omission of last 1–2 units. Another song often heard after pairing comprised 2 phrases often repeated alternately: 'ta-ap tè-èp. . . tlu-tlu', 2nd phrase of 2–3 notes and fluty. 3rd song-type, heard only once, rendered 'tlu-tlu-trrru-tluu'. Song often repeated many times per minute over several hours. (Pasteur 1956.) Song in Jordan rather variable, sometimes recalling that of Black Redstart *Phoenicurus ochruros* (Wittenberg 1987). In Saudi Arabia, thin, sweet, whistled warble with some reedy notes (King 1978). In Oman, song rendered 'chui it cher lit tit' in Dhofar (Walker 1981*b*), 'cher-tit-cher-lit-twit' in north (Walker 1981*a*). In India and Pakistan, song a lively 'which-which-wheech-whichy-which', first 2 units short

and accent on final unit; also described as rich 'whee-chi-whee-wichee' ending in an extra, subdued 'chi' (Ali and Ripley 1974). In Pakistan, described as short and stereotyped, but quite pretty; lasts just under 2 s and consists of vehement short warbling notes similar in form to song of Grey-necked Bunting *Emberiza buchanani*: 'twetchu-u-twwitchu-teet-which-chu-u', the last part rising, with falling but emphasized note at end. Some variation in song of individual ♂♂; another song-type rendered 'trip-trip-te-tree-cha tre-chi-tah', last phrase repeated twice. (Roberts 1992.) (2) Nasal 'tszwee' (Hollom *et al.* 1988; Roberts 1992); nasal timbre and glissando suggest Rock Sparrow *Petronia petronia*; sometimes 'tswee tswee-ak', 2nd part prolonged (L Svensson); the usual call according to Etchécopar and Hüe (1967). Also rendered as reedy, scratchy 'skeeek' or 'skeeee-ek' (King 1978), 'dwee' (Paige 1960), and 'waeee' like Wayeek-call of Brambling *Fringilla montifringilla* when given singly, 'chawawee' suggesting Lapwing *Vanellus vanellus* when double (P J Sellar). Variable 'zweer' (Fig V), 'waeee', or 'wawee', rich in harmonics, giving nasal timbre (Figs VI–VIII). (3) Chirping-call. Flight-call, 'cherz', rather like sparrow *Passer* (Gallagher and Rogers 1980; Roberts 1992), quite loud when given by flocks disturbed in winter (Roberts 1992). Harsh, sparrow-like chirps (Fig IX). Characteristic, short, hard, harsh, slightly nasal 'chaw' (Mild 1990); harsh 'cheb' or 'crebb' (Walker 1981*a*, *b*). Abrupt 'chirrup' like lark (Alaudidae) (Phillips 1982) may be variant. (4) Other calls. (4a) Thin, sharp 'tchick' (King 1978; Hollom *et al.* 1988). 'Chip' (Paige 1960); calls given when feeding, rendered 'tip', 'tyièp tyiep', 'tyiet', etc.

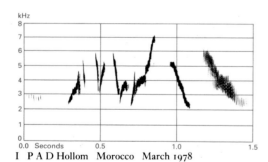

I   P A D Hollom   Morocco   March 1978

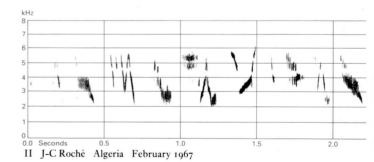

II   J-C Roché   Algeria   February 1967

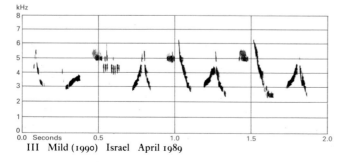

III   Mild (1990)   Israel   April 1989

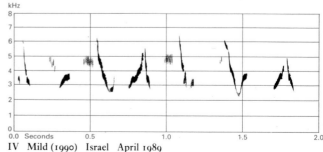

IV   Mild (1990)   Israel   April 1989

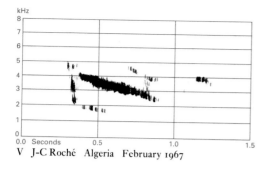

V J-C Roché Algeria February 1967

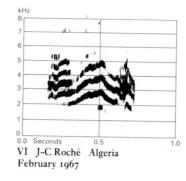

VI J-C Roché Algeria February 1967

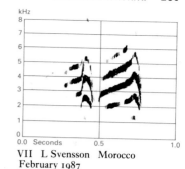

VII L Svensson Morocco February 1987

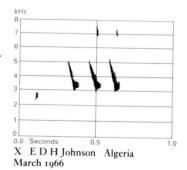

X E D H Johnson Algeria March 1966

VIII P A D Hollom Morocco March 1978

IX J-C Roché Algeria February 1967

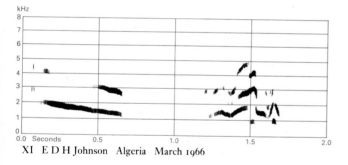

XI E D H Johnson Algeria March 1966

(Pasteur 1956), may be similar, as may rapid 'pliplipli' (Fig X) by ♂ calling to ♀ (E D H Johnson, P J Sellar). (4b) 'Sweee-doo' (Hollom *et al.* 1988); perhaps a variant of call 2. (4c) Drawn-out, descending 'pleep' (Fig XI, 1st unit) from ♀, given antiphonally in response to call 4d from ♂ (E D H Johnson, P J Sellar). (4d) A husky 'alala' from ♂ (Fig XI, 2nd unit), given antiphonally in response to call 4c from ♀ (E D H Johnson, P J Sellar). (4e) Single quiet 'cooi' (E D H Johnson, P J Sellar). Calls given when feeding, 'tiouip', 'tyuit', 'tuip', etc. (Pasteur 1956), may be similar.

CALLS OF YOUNG. Food-calls of nestlings far-carrying; fledglings beg incessantly with monosyllabic 'tchiù' (Courteille and Thévenot 1988). After day 13 (at least 4 days before fledging), young give very short 'tit' calls with bill closed; same calls used after fledging (Courteille and Thévenot 1988).   DTH

Breeding. SEASON. Marrakech (Morocco): eggs laid early March to beginning of August, with peaks mid-March, late May, and late July (Roux *et al.* 1990, which see for comparison between 2 urban habitats); another study found average date of start of 1st clutch 26 February, with range in Morocco late January to October, peak late February to early June (Courteille and Thévenot 1988). See also Thévenot *et al.* (1981, 1982). South-east Algeria: young in nest recorded April and October, even early December (Laferrère 1968). For review of north-west Africa, see Heim de Balsac and Mayaud (1962). Israel: eggs laid end of March to end of June (Paz 1987; Shirihai in press). Central Arabia: nest-building observed beginning of March and fledged young end of April and end of June (Jennings 1980). SITE. In North Africa, in or on buildings but also in rocky country away from settlements; in east of range, generally avoids human habitation. In Casablanca (Morocco), of *c.* 40 nests, 40% inside buildings on beam, ledge, or cornice, on water-heater, in crevice, etc., 25% on window-sill, balcony, etc., 25% in hole in wall or rock face, 10% in various places including mail-box, fuse-box, on shelf, and so on; almost always sheltered from above; usually 2–3 m above ground, although also recorded on soil below bush; sometimes in trees or shrubs, e.g. on trunk of palm. (Courteille and Thévenot 1988.) In Tafilalt (south-east Morocco), 9 of 31 nests were in rocks, remainder in or on buildings; also on sea-cliffs in western Morocco (Destre 1984). In Arabia, can be on ground among rocks, in ruins, or (e.g.) 20 m up on cliff ledge (Meinertzhagen 1954; Jennings 1980;

Richardson 1990), and in Pakistan also on ground or rock-face (Roberts 1992). Nest: small and cup-shaped; foundation of twigs, roots, grass stems, straw, etc., lined with hair, wool, plant down, and man-made material (Courteille and Thévenot 1988; Roberts 1992); 4 nests in Algeria had average outer diameter 15·0 cm (14–16), inner diameter 7·3 cm (7–8), overall height 5·3 cm (5–6), and depth of cup 1·8 cm (1·5–2·0) (Koenig 1896). Building: by ♀ only according to Roux et al. (1990); by both sexes according to Ali and Ripley (1974), and ♂ recorded helping in captivity (Preiss 1974); see also Courteille and Thévenot (1988) for similar observations, particularly in early stages; often re-uses old nest after refurbishment and same nest used for subsequent clutches; takes 7–10 days for new nest, 4–7 days to restore old nest (Courteille and Thévenot 1988). EGGS. See Plate 33. Sub-elliptical, smooth and faintly glossy; whitish, sometimes tinged faintly blue or green, speckled or spotted purplish or dark brown, markings increasing in intensity towards broad end (Harrison 1975; Courteille and Thévenot 1988; Roux et al. 1990). Nominate striolata: 19·3 × 14·2 mm (18·0–21·0 × 13·5–15·0), n = 15; calculated weight 2·0 g (Schönwetter 1984). E. s. sahari: 19·0–20·3 × 13·0–16·0 mm, n = 44 (Etchécopar and Hüe 1967); 19·5 × 14·1 mm (16·8–22·2 × 13·7–14·6 mm), n = 14 (Courteille and Thévenot 1988). Clutch: 2–4(–5). In Marrakech, of 35 clutches: 2 eggs, 37%; 3, 57%; 4, 6%; average 2·69; often 3 broods, 4 recorded; 1st clutch 2·75 (n = 16), 2nd 2·67 (n = 12), 3rd 3·14 (n = 7) (Roux et al. 1990, which see for comparison with other studies). Total of 156 clutches from Morocco, Tunisia, and elsewhere in north-west Africa as follows: 1 egg, 0·6%; 2, 10·2%; 3, 64·7%; 4, 23·7%; 5, 0·6%; average 3·13 (Heim de Balsac and Mayaud 1962; Etchécopar and Hüe 1967; Courteille and Thévenot 1988). In Israel, 1–2 broods (Shirihai in press), and in Pakistan, 2–3 (Roberts 1992). Eggs laid daily, generally in morning (Courteille and Thévenot 1988; Roux et al. 1990). INCUBATION. 13–14 (12–16) days, by ♀ only; at nest in Casablanca, all 3 eggs hatched on same day (Courteille and Thévenot 1988; Roux et al. 1990). In Israel, 12–14 days, n = 8 (Shirihai in press). YOUNG. Fed and cared for by both parents (Courteille and Thévenot 1988, which see for feeding routine; Roux et al. 1990; Roberts 1992). FLEDGING TO MATURITY. Fledging period 17–19 days; fed by parents, usually ♂, for 2–3 weeks after leaving nest (Courteille and Thévenot 1988, which see for development of young; Roux et al. 1990). Said to be 12–14 days (n = 8) in Israel (Shirihai in press). BREEDING SUCCESS. In Marrakech, 56% of 86 eggs in 32 nests hatched and 50% produced fledged young, giving 1·3 per nest; egg losses mostly due to human disturbance and interspecific competition (Roux et al. 1990). In Casablanca, 66% of 169 eggs in 58 nests hatched and 57% produced fledged young (1·7 per nest); human disturbance and (probably) domestic cats caused most losses (Courteille and Thévenot 1988).   BH

Plumages. (E. s. sahari). ADULT MALE. In fresh plumage (July–January), nasal bristles small, black; forehead, crown, and nape light ash-grey with darker grey shaft-streaks, feathers often with slight cinnamon hue on edges. Supercilium whitish, narrow and inconspicuous above lore and eye, broader behind eye, where speckled grey. Lore mottled grey and white; eye-stripe dark grey, narrow in front of eye, broader behind; eye-ring cream-white. Ear-coverts dark grey, stripe from gape to centre of ear-coverts mottled silvery-white and grey. Moustachial stripe grey, inconspicuous, connected with grey of lower ear-coverts; sub-moustachial stripe narrow and indistinct, mottled grey and white, reaching base of lower mandible, malar stripe dark grey. Mantle and scapulars rich cinnamon-rufous, sometimes with narrow and ill-defined grey shaft-streaks. Rump and upper tail-coverts uniform cinnamon-rufous. Chin, throat, and upper chest pale ash-grey, feathers with dull black centres, appearing mottled on chin, heavily blotched black on upper chest; lower chest cinnamon-rufous, sharply demarcated from grey upper chest; belly, thigh, flank, vent, and under tail-coverts similar or slightly paler cinnamon. Tail brownish-black, central pair (t1) with narrow cinnamon edge along both webs, t2–t5 edged cinnamon on outer web only, t6 with outer web and tip more broadly fringed cinnamon. Flight-feathers and tertials dark grey-brown to brown-black, outer web of primaries edged sandy-cinnamon, basal and middle portion of flight-feathers with broad cinnamon-rufous border; outer web of secondaries and both webs of tertials with broad cinnamon-rufous fringe. Greater upper primary-coverts, greater upper wing-coverts, and bastard wing cinnamon-rufous with largely concealed sooty stripe on centre, longest feather of bastard wing mainly greyish-black; median and lesser coverts uniform cinnamon-rufous. Underwing-coverts and axillaries pinkish-cinnamon, axillaries with grey bases. In worn plumage (November–April), dark shaft-streaks on head more conspicuous, light ash-grey partly or almost completely worn off; supercilium purer white, more contrasting; borders of feathers of upperparts bleached to ochre; cinnamon fringes on wing and tail bleached, partly worn off, especially on t1–t5 and tip of t6. ADULT FEMALE. In fresh plumage (July–January), similar to adult ♂, but top of head brown-cinnamon with greyish shaft-streaks, closely similar to remainder of upperparts; supercilium cinnamon-buff, lore greyish-cinnamon, eye-ring sandy-cinnamon, ear-coverts cinnamon-buff; moustachial stripe and lower part of ear-coverts greyish-brown, darker than ear-coverts; submoustachial stripe narrow and inconspicuous, greyish-white. Chin, throat, chest, and side of breast brown-grey, feathers with broad and ill-defined buff edges, less sharply demarcated from rufous-cinnamon underparts than in adult ♂. In worn plumage (November–April), head appears darker (brown edges worn off), closely streaked and mottled dark brown and buff, much less black than adult ♂; cinnamon-buff supercilium more conspicuous; cinnamon-rufous fringes of wing and tail bleached, partly worn off. NESTLING. Down white, on head and back (Courteille and Thévenot 1988, which see for development). JUVENILE. Like adult ♀, but general colour of head and body more diluted cinnamon, less saturated rufous; feathering distinctly shorter and looser; head, neck, and chest uniform cinnamon-buff or greyish-buff, almost completely unstreaked (especially in ♂) or with narrow and ill-defined grey streaks (in adult ♀, streaks more sharply defined, darker grey). Wing and tail as adult ♀, but rufous-cinnamon fringes of outer webs of tail-feathers and primaries extend along tips of feathers; greater upper primary and secondary coverts with narrower and less contrasting grey centres, often confined to faint shaft-streak. FIRST ADULT. Like adult, but juvenile flight-feathers, greater upper primary-coverts, and bastard wing retained; when plumage fresh, rufous fringes along tips of primaries and restricted amount of grey on centres of primary coverts useful for ageing; also, often a contrast in colour between fringes of newer tertials and older inner secondaries; if worn, best distinguished by contrast between abraded primary coverts and bastard wing (which show yellowish fringes) and newer bright rufous-cinnamon greater coverts. In some birds, some flight-feathers replaced, these showing contrast in colour and wear with retained older ones (unlike adult).

Bare parts. ADULT, FIRST ADULT. Iris brown or orange-brown. Upper mandible dark-horn or blackish-brown, lower flesh-coloured or yellow. Leg and foot flesh-colour or pale brown. (Meinertzhagen 1930; Ali and Ripley 1974; ZMA.) NESTLING. No information. JUVENILE. Bill dark horn (Hollom *et al.* 1988).

Moults. ADULT POST-BREEDING. Complete; primaries descendent. In birds examined, plumage of *sahari* from north Algerian Sahara still fresh up to March, worn but not yet moulting up to mid-June; in Beni-Abbès, 10–12 July, single ♀ not yet started, ♂ just started with inner primaries (moult score 9). In Ahaggar (southern Algeria), single ♂ from 10 June had primary moult score 28; tail just started (score 16), secondary moult just starting, wing-coverts and body mainly new, many feathers growing on head (ZMA). In 4 nominate *striolata* from Sinai, flight-feathers new but some feathers on head and body still growing 8–16 August (ZFMK). Active flight-feather moult occurs April–October (Ticehurst and Whistler 1938; Niethammer 1955; ZFMK, ZMA). Occasionally, apparently moults during breeding: single ♀ with brood-patch was in moult on 20 October (Ticehurst and Whistler 1938). Apparently no pre-breeding moult. POST-JUVENILE. Partial, but perhaps sometimes complete: head, body, lesser, median, and greater upper wing-coverts, usually 1–3 tertials, and often all tail (a few feathers sometimes retained); juvenile bastard wing, greater upper primary-coverts, and flight-feathers retained, but birds which retain some old inner or outer primaries or scattered secondaries are perhaps 1-year-olds. Birds in fresh juvenile plumage examined from mid-June (northern Algeria), late May and early June (Ahaggar), and early August (Sinai); birds in moult from early June (Ahaggar) and 7–23 August (Sinai), with moult just completed from 23 August (Sinai) but also as early as 20 April to 14 May in Ahaggar. (ZFMK, ZMA: C S Roselaar.)

Measurements. ADULT, FIRST ADULT. *E. s. sahari.* Morocco, Algeria, and Tunisia, all year; skins (ZFMK, ZMA). Bill (S) to skull, bill (N) to distal corner of nostril; exposed culmen on average 2·1 mm shorter than bill (S).

	♂			♀		
WING	79·6 (2·66; 32)	73–87		74·8 (1·98; 21)	72–78	
TAIL	60·6 (1·94; 9)	59–64		58·0 (2·0 ; 5)	55–60	
BILL (S)	12·8 (0·27; 11)	12·1–13·1		12·6 (0·62; 7)	12·0–13·0	
BILL (N)	7·2 (0·26; 13)	6·7–7·5		7·3 (0·55; 7)	6·7–8·3	
TARSUS	17·5 (0·79; 9)	16·4–19·2		17·5 (1·23; 7)	16·1–19·8	

Sex differences significant for wing and tail. Ages combined, though wing of juvenile and of 1st adult with retained juvenile flight-feathers shorter than of older birds. Thus, in ♂♂ from Algeria: adult 80·4 (2·69; 7) 76–84, juvenile 76·0 (2·39; 10) 73–79 (ZMA: C S Roselaar).

Nominate *striolata.* Sinai, Israel, and Sudan, all year; skins (ZFMK).

	♂			♀		
WING	77·0 (1·04; 11)	75–78		74·0 (2·19; 9)	71–79	
TAIL	56·4 (0·95; 9)	55–58		54·4 (2·17; 9)	52–57	
BILL (S)	11·4 (0·36; 9)	10·8–11·9		11·4 (0·44; 8)	10·9–12·3	
BILL (N)	6·6 (0·21; 10)	6·4–6·9		6·7 (0·30; 9)	6·2–7·1	

Sex differences significant for wing and tail.

E. s. jebelmarrae. Sudan, skins of 5 ♂♂ and 1 ♀: wing 81·6 (79–85) (Vaurie 1956).

E. s. saturatior. Somalia, skins: wing, 2 adult ♂♂ 74, 79 (Vaurie 1956).

E. s. sahari or *saturatior.* Chad, 3 ♂♂ and 2 ♀♀: wing 75–77 (Niethammer 1955).

Weights. *E. s. sahari* or *saturatior.* Algeria and Chad, November–January: ♂ 15·1 (2·17; 10) 12–18, ♀ 14·5 (1·08; 7) 13–16 (ZFMK; see also Niethammer 1955). Niger, July: ♂♂ 14·8, 15·2, 16·8 (Fairon 1975).

Structure. Wing rather short, broad at base, tip rounded. 10 primaries: p7–p8 longest, p9 0–3 shorter, p6 1–3, p5 2–7, p4 7–10, p3 9–13, p2 12–15, p1 14–18; p10 strongly reduced, 48–57 shorter than p7–p8, 4–7 shorter than longest upper primary covert. Outer web of p6–p8 clearly emarginated, p5 occasionally with slight emargination. Tip of longest tertial reaches tip of p3 (p2–p4). Tail rather short, tip slightly forked; 12 feathers, t6 longest, t1 1–3 shorter. Bill conical; rather similar to Yellowhammer *E. citrinella*, but less deep, less wide, and slightly shorter, and upper mandible more evenly curved. Middle toe with claw 16·0 (11) 14·7–16·8; outer toe with claw *c.* 66% of middle with claw, inner *c.* 64%, hind *c.* 63%; hind claw 5·1 (13) 4·5–5·7.

Geographical variation. Rather slight. Nominate *striolata* from Middle East slightly smaller than *sahari*, especially in bill and tail. Head pattern more prominent: cap grey with black streaks, but ground-colour of cap occasionally slightly pinkish-grey, less whitish, and black streaks sometimes less profuse; in worn plumage, prominent white stripe on central crown, supercilium and malar stripe whiter, eye-stripe and moustachial stripe blacker, sometimes a white streak across grey of cheeks and ear-coverts, upperparts grey with pinkish-cinnamon tinge rarely extending to upper tail-coverts; mantle and scapulars with more prominent black shaft-streaks; lower chin and chest paler grey, with slight sandy-brown suffusion (less pure medium grey), black streaks on chest narrower; remainder of underparts paler, pinkish-cinnamon, less rufous-cinnamon, and contrasting less with chest. Within *sahari*, populations not uniform in colour: when series of specimens from same time of year compared, birds from Beni-Abbès (western Algeria) more broadly streaked black on head, neck, and upper chest than typical *sahari* from northern fringe of Sahara in central and eastern Algeria, chin to throat almost uniform sooty when plumage worn, mantle and scapulars more clearly streaked, body darker rufous-cinnamon or rufous-brown, especially on underparts; birds from Tunisia and from Ahaggar in southern Algeria on average paler, streaks on head, neck, and upper chest narrower than in northern and north-east Algeria, mantle and scapulars virtually unstreaked in Tunisia, but fairly distinct in Ahaggar (though less so than in Beni-Abbès). Dark Beni-Abbès birds similar to population named *theresae* Meinertzhagen, 1939, from Morocco, but various populations within Morocco differ from each other, e.g. those of Tiznit and Ksar-es-Souk paler than those of Essaouira, Agadir, and Marrakech (Vaurie 1956e), and *theresae* thus not recognized. Birds from Ahaggar tend somewhat towards *jebelmarrae* from western Sudan, as do those of Aïr (northern Niger), Tibesti, Borkou, and Ennedi (northern Chad) (Niethammer 1955), but those of Ahaggar and Aïr at least are closer to *sahari* (ZMA, ZFMK: C S Roselaar). *E. s. saturatior* from north-east Africa larger and darker than nominate *striolata*. *E. s. jebelmarrae* from western Sudan larger and more richly coloured than *saturatior*, with bold blackish streaking on crown, mantle, scapulars, and chest, and with deep rufous belly (Vaurie 1956; Traylor 1960). *E. s. sanghae* from Mali is darker, more chestnut on body than *sahari*, but perhaps similar in size; differs from *jebelmarrae* in being more chestnut above, without heavy streaking, and from *saturatior* in being less streaked (Traylor 1960). Birds from south-east Iran, Pakistan, and north-west India said to be paler than those from Middle East, less rufous on body, and wing, with striping on average less conspicuous; these sometimes separated as *tescicola* Koelz, 1954, but not recognized here since

variation in colour is within range of nominate *striolata*; in fact, a few birds examined from south-eastern Iran and India were darker on head, neck, and chest than birds examined from Sinai and Middle East (ZFMK). See also Vaurie (1956*e*).

May form species-group with Cape Bunting *E. capensis*, Rock Bunting *E. cia*, Larklike Bunting *E. impetuani*, Socotra Mountain Bunting *E. socotrana*, and Cinnamon-breasted Rock Bunting *E. tahapisi* (Hall and Moreau 1970). For situation in area of overlap with *E. tahapisi*, see Walker (1981*b*) and Phillips (1982).   GOK

Emberiza tahapisi A Smith, 1836   Cinnamon-breasted Rock Bunting

PLATE 19
[between pages 256 and 257]

FR. Bruant cannelle   GE. Bergammer

A resident, mainly Afrotropical species, widespread in open, rocky habitats south of Sahara, also in Socotra and southern Arabia. Record of this species from Sinai (Egypt) 1984, formerly accepted by Goodman and Meininger (1989), now considered to relate to House Bunting *E. striolata* (M Beaman, P L Meininger).

Emberiza cineracea   Cinereous Bunting

PLATES 17 (flight) and 20
[between pages 256 and 257]

DU. Smyrnagors   FR. Bruant cendré   GE. Türkenammer
RU. Серая овсянка   SP. Escribano cinéreo   SW. Gulgrå sparv

Emberiza cineracea C L Brehm, 1855

Polytypic. Nominate *cineracea* C L Brehm, 1855, southern and western Asia Minor; *semenowi* Zarudny, 1904, south-west Iran, occurring on migration in Middle East. Not known what race breeds in south-east Turkey.

Field characters. 16–17 cm; wing-span 25–29 cm. Size between Yellowhammer *E. citrinella* and Black-headed Bunting *E. melanocephala*, looking rather more attenuated and longer-winged than partly sympatric Cretzschmar's Bunting *E. caesia*. Rather large, strangely featureless bunting, most resembling ♀ or immature *E. melanocephala* but showing white tail-feathers. At close range, shows faint plumage pattern converging with *E. caesia* and allies; pale grey bill and at least faintly yellow throat crucial in identification. Sexes closely similar; little seasonal variation. Juvenile separable. 2 races in west Palearctic, of which ♂♂ easily separable; Turkish race, nominate *cineracea* described here (see also Geographical Variation).

ADULT MALE BREEDING. Moults: early autumn (complete). Head basically yellow but strongly suffused ashy- or olive-grey so that yellow supercilium, conspicuous narrow yellow-white eye-ring, and pale yellow chin and throat stand out. Throat marked with ill-defined greyish malar streak connected to variable ashy-grey breast-band, collar, and nape. Mantle and scapulars pale ashy-brown, with soft dark streaks; rump plain grey. Wings ashy-brown, with greyish lesser coverts, blackish-centred and almost

white-tipped median coverts forming quite bright upper wing-bar, blackish-centred and pale greyish-buff-tipped greater coverts forming indistinct lower wing-bar, and blackish-centred and pale-buff-edged tertials. Tail brown-black with greyish margins to central feathers and bright white ends to 2 outermost. Lower breast tinged yellow, flanks washed grey, rest of underparts cream-white. In fresh plumage, head even greyer or more olive and fringes of upperpart feathers tinged rufous. Bill shape like *E. caesia*, dull horn-grey to blue-grey. Legs flesh-brown. ADULT FEMALE. Drabber than ♂, with more streaked plumage. Head browner, with yellowish tinge confined to crown, and streaked on crown, nape, and sides of neck; contrasts with indistinct pale buffish-white to buffish-yellow eye-ring, chin, and throat, with brownish-grey malar streak joined to ashy-brown breast and nape, both softly striped darker. Back slightly darker than ♂, with more distinct streaks particularly on lower mantle and scapulars. Rest of plumage as ♂ but upper wing-bar duller, while flanks even paler. JUVENILE. Distinctly buffier and less white below than adult, with only a tinge of yellow on throat. Streaking even more developed than

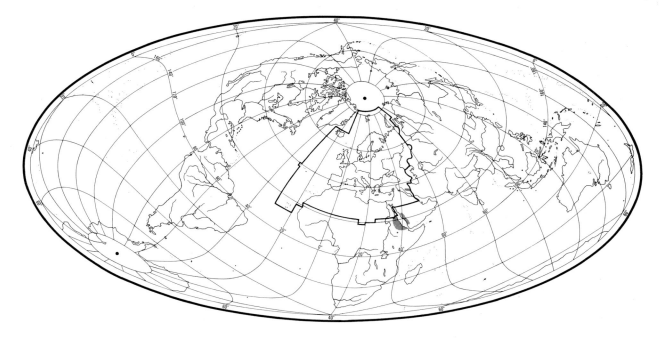

♀, with spots on throat, fully streaked breast and rump, and larger dark feather-centres on mantle and scapulars. Wings more distinctly patterned than adult, with broad rufous-buff margins to tertials.

Lacks obvious characters but, compared to much more colourful migrant congeners, its featurelessness is immediately striking. At close range, head of ♂ may suggest Ortolan Bunting *E. hortulana* but grey bill and predominantly dull greyish plumage, lacking any rufous tones on underparts, allow ready distinction. ♀ and immature even more nondescript than ♂ but their white outer tail-feathers instantly exclude *E. melanocephala* and Red-headed Bunting *E. bruniceps*. Note that Iranian race *semenowi* is distinctly yellow on belly and under tail, initially suggesting these species more strongly. Immature may also be confused with palest *E. hortulana* and particularly with Grey-necked Bunting *E. buchanani* but both these species show bright flesh- to reddish bill and at least buff, if not rufous-, tinged underparts. Flight, behaviour, and actions much as Rock Bunting *E. cia*. Unobtrusive and markedly wary. Terrestrial but perches in and sings from coniferous trees in mountains. Rare and local in breeding range; infrequently seen as migrant, usually travelling with *E. hortulana* and allies.

Song a simple, ringing, tuneful phrase of 5–6 notes: 'dir dir dir dli-di', 'deur deur deur-deur dree-do', or 'drip-drip-drip-drip-drie-drieh'. Commonest call a short, metallic 'kjip', 'kup', 'kleup', or 'cluff'.

Habitat. Imperfectly known, owing to scarcity of data from restricted and inaccessible areas of occurrence in south-east of west Palearctic and thinly inhabited winter quarters. Summer visitor to scrub-covered uplands in Turkey (Martins 1989), in warm temperate or Mediterranean climate. Dresser (1871–81) described it from limited data as living in rocky mountainous districts with scanty vegetation, and being found from base of mountains to high up in conifer region, usually resting on moderate-sized blocks of stone. Harrison (1982) described it as occurring on high slopes, dry and rocky with sparse shrubby vegetation, and wintering often in dry coastal areas. Seems also to occur on passage in lowland deserts.

Distribution. No changes reported.
GREECE. Breeds only on Lesbos and Chios (GIH).
Accidental. Tunisia.

Population. GREECE. Certainly less than 1000 pairs, perhaps not more than 100 (GIH).

Movements. Migrant, moving south or south-west to winter (apparently) in eastern Sudan, Eritrea, Yemen, and south-west Saudi Arabia. Migration nocturnal (Shirihai in press). Poorly known, with no ringing studies. For reviews, see Chappuis *et al.* (1973) and Knijff (1991).

Begins to leave breeding grounds in July, although juveniles sometimes stay as late as September (Knijff 1991). 2 main migration routes. Western route through southern Turkey via Syria, Lebanon, Jordan, Israel, and Egypt to Sudan and Eritrea used predominantly by nominate *cineracea*, but also by smaller numbers of *semenowi*. Eastern route follows coast of Arabian Gulf and is used

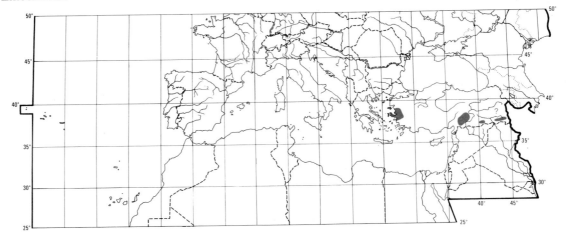

exclusively by *semenowi*; supposed wintering area for these is at south-west tip of Arabia, but no records from south-central Arabia (Knijff 1991).

Migrants recorded on both routes August–November (Knijff 1991). Apparently much scarcer on autumn passage than in spring (e.g. in Israel: Shirihai in press) and unrecorded in autumn in Cyprus (Flint and Stewart 1992) and Jordan (I J Andrews). Single August record outside breeding range in Rezaiye basin (Iran) (D A Scott). In Israel, recorded end of August to end of September, mainly 1st half of September; generally very rare, and none in some autumns, but maximum of 21 together recorded; chiefly in east of country and Judean highlands (Shirihai in press). Single autumn record in Egypt, late August (Goodman and Meininger 1989). In Gulf states, 3 autumn records late August and September (F E Warr).

Arrival in Sudan reported September (Prendergast 1985). Winter range very poorly known, involving only 11 records of 21 birds; definite records from eastern Sudan, Eritrea, northern Yemen, and south-west Saudi Arabia (Vaurie 1959; Chappuis *et al.* 1973; Knijff 1991; F E Warr). Not certain whether populations using eastern and western migration routes remain separate in winter or combine (Knijff 1991). 2 winter reports from United Arab Emirates unconfirmed (F E Warr).

Arrives on breeding grounds early April, although *semenowi* recorded in Zagros (Iran) as early as end of February. Uses same western and eastern migration routes as in autumn, with migrants recorded February–May. (Knijff 1991.) 3 spring records in Egypt, March–May; 2 of nominate *cineracea*, 1 *semenowi* (Bruun 1984; Goodman and Meininger 1989). In Israel, rare to scarce migrant March to mid-May, mainly end of March and 1st half of April; passage on broad front through most of country, especially eastern and central parts and mostly north of Negev; 70–80% of migrants are ♂♂; nominate *cineracea* predominant, with only 4 confirmed records of *semenowi* in north and central Israel but 30–40% of total in south at Elat; many ♂♂ occur that are intermediate in plumage

between the races (Shirihai in press). On Cyprus, very scarce passage migrant in spring (14 records), late March and April, all apparently nominate *cineracea* (Flint and Stewart 1992). In Jordan, 6 spring records of 9 birds, mostly late March and April (one 1 February) (I J Andrews). In Gulf states, very scarce and irregular spring migrant, mid-March to April, once mid-May (F E Warr).

Vagrant reported from Samarkand in central Asia (singing ♂, 10 May 1983: Knijff 1991), but report from Afghanistan in January 1972 (Smith 1974) probably erroneous (S C Madge).   DTH

Food. Little studied. Diet seems to be principally seeds and small invertebrates, which are probably main food in breeding season. On passage in Israel, feeds on seeds in stubble in flocks with Cretzschmar's Bunting *E. caesia* and Ortolan Bunting *E. hortulana*; also on rocky slopes and desert uplands with low vegetation and scrub. (Krüper 1875; Knijff 1991; Shirihai in press.) In winter in Eritrea, small parties foraged on rocky ground and short grass (Smith 1957). See also Prendergast (1985) for observations in Sudan. In Yemen, early October, 3 birds seen feeding in millet *Panicum* field (Phillips 1982).

In eastern Turkey, 3 stomachs collected late May contained 3–4 large caterpillars (Lepidoptera), adult flies (Diptera), adult beetles (Coleoptera), 28 small pupae, 1 small snail (Pulmonata), and many fragments of arthropods; no plant material present (Chappuis *et al.* 1973). In western Iran, April, of 8 stomachs, 3 contained only insect remains, 3 contained seeds, and 2 some of each (Paludan 1938). At Elat (southern Israel), migrating birds fed on beetles and spiders (Araneae) (Knijff 1991). No further information.   BH

Social pattern and behaviour. Little known; no detailed study.

1. On migration and in winter reported singly and in small groups. On spring migration in Israel groups up to 4 normal, sometimes more than 10, once 40 (Shirihai in press). On migration, often seen with flocks of Ortolan Bunting *E. hortulana* and

Cretzschmar's Bunting *E. caesia* (Prendergast 1985; Knijff 1991; Shirihai in press). BONDS. No details, although apparently in pairs during breeding season (Chappuis *et al.* 1973). At one nest, building by ♀, attended by ♂ (Krüper 1875). Roles of sexes in incubation and care of young not known (Harrison 1975). BREEDING DISPERSION. In Turkey, 4 pairs reported in *c.* 10 ha along ravine and 8 pairs elsewhere in *c.* 20 ha (Chappuis *et al.* 1973). ROOSTING. No information.

2. Very poorly known. Migrants at Elat (Israel) easy to approach, but shy on breeding grounds (Moore and Boswell 1957; Knijff 1991). Recorded mobbing viper *Vipera xanthina* in Greece (Dimitropoulos 1987). Song (see 1 in Voice) reported only from ♂ (Chappuis *et al.* 1973). Sings from bush, tree, or rock (Moore and Boswell 1957; P A D Hollom). Song reported from breeding grounds early May to mid-June (Moore and Boswell 1957; Chappuis *et al.* 1973; Martins 1989). No further information. DTH

Voice. Little known. See Chappuis *et al.* (1973) for additional sonagrams and evidence of similarity to Ortolan Bunting *E. hortulana* and Cretzschmar's Bunting *E. caesia*. Vocabulary presumably similar to those species, and thus wider than described in literature, but existing information insufficient to be sure of homologies in calls. Classification used here thus attempts to group similar sounds together, using similar numbering system to that for *E. hortulana*, in absence of information on their variability or contexts.

CALLS OF ADULTS. (1) Song of ♂. Somewhat resembles songs of *E. hortulana*, Black-headed Bunting *E. melanocephala*, and Grey-necked Bunting *E. buchanani*, with characteristic time pattern due to two brief pauses: thin, hoarse 'zrí zri-zri-zri-záh zrivé-zrivé', 'záh' often lower pitched (L Svensson); according to Knijff (1991), less sweet than above species, comprising 4-6 units, rising in pitch, followed by longer, descending couplet unit, 'dir-dir-dir-dir-dli-dlu'. Also described as 'dir dir dir dli-di' (Peterson *et al.* 1983); brief and simple, comprising 5-6 units of typical *Emberiza* character, 'drip-drip-drip-drip-drie-drieh' (Hollom *et al.* 1988). Recording from Turkey (Fig I), also analysed with sonagram in Chappuis *et al.* (1973), has songs of 1·0-1·2 s duration, separated by

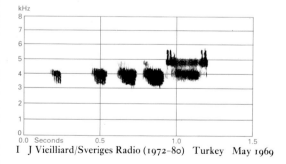

I J Vieilliard/Sveriges Radio (1972-80) Turkey May 1969

intervals of 5·5 s; 4 short units (duration *c.* 100-150 ms, at *c.* 3-5 kHz; last 2 with harmonics at *c.* 8 kHz, not shown in Fig I) are repeated at fairly regular intervals, followed

by longer unit at higher frequency (duration *c.* 360 ms, at *c.* 3·5-6 kHz); last unit said to sound lower pitched to human ear. This recording (Fig I) suggests 'ji-ji-jee' or 'ji-ji-j-jee'; another from Greece 'di-di-deerdi', both like quick abbreviated song of Yellowhammer *E. citrinella* (D T Holyoak). (2) Tsik-call. Described as the normal call, a soft 'tsik', very like call of *E. hortulana* (Knijff 1991). Rendered as short 'kip' (Peterson *et al.* 1983); short metallic 'kjip' or 'kyip' (Hollom *et al.* 1988; Jonsson 1992); sharp 'chip' (Moore and Boswell 1957). Call rendered 'chiff' (Bruun *et al.* 1986) is possibly similar. (3) Tieu-call. Sharply descending 'tieu' (Fig II, 1st unit, based on same recording as Fig 4A in Chappuis *et al.* 1973); 'chülp' or

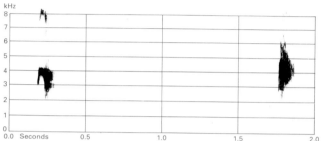

II J Vieilliard/Sveriges Radio (1972-80) Turkey May 1969

'chülup' (L Svensson). Calls rendered 'tyeeh' with clear brassy tone (Jonsson 1992) and 'dchu' or 'dju' (Mild 1990) appear similar. A 'chupe chip', with pause between units (Moore and Boswell 1957), is apparently combination of calls 2-3, such combinations being characteristic of *E. hortulana* species-group (L Svensson). (4) Vibrant, impure 'trri', very similar to call of *E. citrinella* (Fig II, 2nd unit, based on same recording as Fig 4A in Chappuis *et al.*

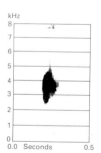

III Mild (1990) Israel April 1989

1973; Fig III). Also rendered as short, hard 'tshri' (Mild 1990).

CALLS OF YOUNG. No information. DTH

Breeding. SEASON. Turkey: in west, eggs recorded 2nd half of April and late May (Krüper 1875; Hüe and Etchécopar 1970); family parties seen last week of May (Kumerloeve 1961); in east, fresh full clutch and young 2-3 days old noted 23 May at *c.* 900 m (Chappuis *et al.* 1973); recently-fledged young mid-August (Beaman 1978). Western Iran: eggs laid towards end of April (Paludan 1938). SITE. On ground, on slope with sparse vegetation, though not usually arid, against rock partly hidden by over-hanging grass, etc. (Krüper 1875; Hüe and Etchécopar 1970; Chappuis *et al.* 1973; Hollom *et al.* 1988). Nest: one in eastern Turkey had foundation of stalks and stems, leaves of thistle, and grass-heads, lined with rootlets and hair; wall very thin where touching rock but thick and well-woven on opposite side; inner diameter 7 cm, depth of cup 3 cm (Chappuis *et al.* 1973). Building: at one nest by ♀, casually accompanied by ♂ (Krüper 1875). EGGS. Short sub-elliptical, smooth and slightly glossy; similar to other closely-related Emberizidae but perhaps more intensely blue, with fewer dark brown scrawls, and more spots and blotches rather evenly distributed over surface, with some concentration at broad end (Hüe and Etchécopar 1970; Chappuis *et al.* 1973, which see for comparison with other *Emberiza*; Harrison 1975). Nominate *cineracea*: 21·1 × 15·8 mm (20·0–23·0 × 15·0–17·0), $n = 5$; calculated weight 2·76 g (Schönwetter 1984); 3 eggs from eastern Turkey were 20·5 × 16·8 mm (20·2–20·7 × 16·5–17·0) (Hüe and Etchécopar 1970). *E. c. semenowi*: 20·3 × 15·5 mm (19·9–20·5 × 15·3–16·0), $n = 5$ (Chappuis *et al.* 1973). Clutch: (3-)4–6 (Hüe and Etchécopar 1970; Chappuis *et al.* 1973). No further information.   BH

Plumages. (nominate *cineracea*). ADULT MALE. Forehead, crown, and nape yellow, washed olive-grey to a variable extent; a very narrow and hardly contrasting yellow supercilium, extending from upper mandible to just behind eye. Lore pale yellow, mottled grey; eye-ring yellow; ear-coverts yellow with olive-grey wash. Hindneck smoke brown-grey with olive-green wash; remainder of upperparts drab-grey, centres of feathers of lower mantle and scapulars slightly browner and with narrow but distinct black-brown shaft-streaks. Lower cheek, chin, and throat sulphur-yellow, merging into light drab-grey of chest, latter with variable amount of green-yellow wash; malar stripe olive-grey, rather faint; side of breast and flank pale ash-grey with faint brown shaft-streaks merging into cream-white of belly, thigh, vent, and under tail-coverts. Tail-feathers dark sepia-brown, both webs of central pair (t1) and outer webs of others with narrow pale drab-grey fringe; t5–t6 with large white square-cut tip on inner web, occupying about one third of t5 and about half of t6; middle portion of outer web of t6 white. Flight-feathers dark drab-brown to greyish-black, outer webs of outer primaries with narrow off-white fringes (absent from the emarginated parts); fringes of outer webs of inner primaries and of secondaries pale drab-grey to off-white. Tertials dark drab-brown or sepia-brown, outer web with broad and con-trasting pink-brown to light drab-grey fringe. Primary coverts, bastard wing, and greater upper wing-coverts dark drab-brown, latter with narrow drab-grey tips (widest on outer web) and outer fringes; median upper wing-coverts black-brown with con-spicuous and rather broad pink-cream to grey-white tips (most distinct on outer webs); pale tips of greater and median coverts form a double wing-bar (most distinct on median coverts). Lesser upper wing-coverts drab-grey. Under wing-coverts and axillaries pale drab-grey or off-white, shorter coverts blotched black. *In fresh plumage* (autumn), upperparts and fringes of tertials and wing tinged pink-cinnamon or rufous-brown with vinous tinge, yellow of head tinged olive. ADULT FEMALE. Very similar to adult

male, differing mainly in colour of head, throat, and chest. Crown and nape pale drab-brown with green-yellow wash and narrow but distinct sooty shaft-streaks; yellow of side of head, chin, and throat less pure and uniform, paler or tinged buff, mottled drab-grey on side of head, only short stripe above eye and eye-ring more or less pure pale yellow or buff-yellow; lower cheek and side of chin and throat mottled grey, often with a broken sooty malar stripe; chest and side of breast washed cream-buff, upper chest with narrow sooty streaks. Remainder of body, tail, and wing as adult ♂, but upperparts sometimes more drab-brown, less grey, and with slightly bolder dark streaks on mantle and scapulars. Fringes of tertials and wing slightly more pinkish-drab, belly more buffish-cream. NESTLING. No information. JUVENILE. Forehead mottled brown-grey; crown and nape yellow-grey, streaked drab-brown, hindneck more uniform brown-grey. Lore grey-brown tinged yellow, eye-ring narrow, pale buff, ear-coverts brownish smoke-grey; no distinct head markings. Mantle and scapulars paler than in adult, marked with bold drab-brown streaks like crown and nape. Back, rump, and upper tail-coverts uniform drab-grey. Chin, throat, and chest buff; throat and (especially) chest rather heavily streaked dark brown. Belly and flank pale buff, marked with elongate drab-brown streaks, lighter and less streaked towards rear; vent and under tail-coverts pale buff, latter with well-marked black-brown feather-shafts. Tail as adult ♂, but white on inner webs of t5 and t6 tends to be V-shaped at base, less square-cut. Flight-feathers and tertials as adult, but fringes on outer webs more buffish, and fringes of tertials pale cinnamon-rufous; greater and median upper wing-coverts with broad, clear-cut fringes and tips, buff on greater, off-white on median, latter showing as distinct wing-bar. FIRST ADULT. Like adult, but juvenile tail-feathers, primary coverts and flight-feathers retained; tail-feathers often with more V-shaped white tips on inner webs of t5 and t6, less square-cut; tips of feathers often more pointed, less broadly truncate, more worn than those of adult at same time of year. Head of 1st adult ♀ suffused buff-grey, scarcely yellow (in adult ♀, clear yellow tinge on cap and pure pale yellow on lower cheek, chin, and throat apart from dusky malar stripe).

Bare parts. ADULT, FIRST ADULT. Iris brown. Bill pale blue-grey or horn-grey, darker grey on culmen, sometimes with flesh tinge. Leg and foot pale flesh or pale brown. (Hartert 1903–10; Meinertzhagen 1930; Hollom *et al.* 1988; BMNH, RMNH, ZFMK.) NESTLING, JUVENILE. No information.

Moults. ADULT POST-BREEDING. Complete; primaries descend-ent. No information on timing; moult not started in birds exam-ined from June; moult completed in birds from October. Apparently no pre-breeding moult. POST-JUVENILE. Partial; juvenile tail, flight-feathers, and greater upper primary coverts retained. 9 birds from Izmir area (Turkey), 12–25 August, either fully juvenile or with moult just started on mantle and side of breast (RMNH); moult generally completed by start of autumn migration.

Measurements. Nominate *cineracea*. Western Turkey, April–November; skins (RMNH, ZFMK, ZMA, ZMB). Bill (S) to skull, bill (N) to distal corner of nostril; exposed culmen on average 3·5 less than bill (S).

	♂		♀	
WING	90·3 (2·50; 41)	86–96	86·8 (2·32; 6)	84–90
TAIL	68·3 (2·46; 38)	63–73	67·0 (3·16; 5)	64–72
BILL (S)	13·0 (0·85; 30)	11·4–14·6	12·3 (0·85; 5)	11·0–13·2
BILL (N)	8·4 (0·43; 30)	6·9–9·0	8·2 (0·41; 5)	7·7–8·7
TARSUS	19·2 (0·65; 27)	18·0–20·4	19·3 (0·67; 4)	18·6–20·2

Differences between adults and 1st adults (with juvenile wing and tail) slight. Thus, in ♂: wing, adult 90·5 (2·43; 18) 87–95, 1st adult 90·2 (2·60; 23) 86–96; tail, adult 67·5 (2·15; 17) 64–71, 1st adult 69·0 (2·54; 21) 63–73 (C S Roselaar).

Wing. Israel, on migration, live birds: (1) nominate *cineracea*, (2) *semenowi* (Shirihai in press). (3) *E. c. semenowi*: Zagros mountains (south-west Iran), April (Paludan 1938; ZMB), and Syria (Kumerloeve 1969b).

(1)	♂	89·5 (— ; -)	83–95	♀	85·0 (— ; -)	82–91
(2)		93·5 (— ; -)	85–98		89·0 (— ; -)	84–95
(3)		93·3 (3·29; 7)	88–97		88·7 (— ; 3)	86–90

Weights. Nominate *cineracea*. Turkey: 20·6, 24·0 (BTO).
E. c. semenowi. Zagros mountains (Iran), April: ♂ 24·4 (3·06; 6) 21·1–29·7, ♀♀ 23·5, 24·8, and (containing egg) 27·5 (Paludan 1938).

Structure. Wing rather long, broad at base, tip bluntly pointed. 10 primaries: p8 longest, p9 0–1 shorter, p7 0·5–2, p6 2–5, p5 7–14, p4 12–17, p3 15–20, p2 18–22, p1 21–26; p10 strongly reduced, narrow, 57–65 shorter than p8, 15·4 (7) 13–17 shorter than longest upper primary covert. Outer web of p6–p8 emarginated, inner web of p6–p9 with faint notch. Tip of longest tertial reaches tip of p5. Tail fairly long, tip slightly forked; 12 feathers, t1 3–10 shorter than t4, t6 1–3 shorter. Bill slightly heavier than in Yellowhammer *E. citrinella*, c. 7 mm deep and wide at base. Tarsus and toes fairly short but stout. Middle toe with claw 18·1 (7) 17·9–19·2; outer toe with claw c. 69% of middle toe with claw, inner c. 66%, hind c. 69%. Remainder of structure as in *E. citrinella*.

Geographical variation. Marked, mainly involving colour and, to lesser extent, wing length. Compared with nominate *cineracea* of western Turkey, *semenowi* in south-west Iran and probably neighbouring Iraq has cap of ♂ brighter yellow, less greenish yellow, green tinge extending faintly over hindneck, mantle, and rump; entire underparts bright sulphur-yellow, tinged olive-grey on chest, side of breast, and flank (chest less grey and belly less white than in nominate *cineracea*); under tail-coverts greyish-white with narrow dusky shaft-streaks and grey centre (more uniform off-white in nominate *cineracea*). As in nominate *cineracea*, ♀ *semenowi* differs from ♂ by streaked greyish cap, similar to mantle and scapulars; upperparts of ♀ *semenowi* more greyish-olive than in ♀ nominate *cineracea*, less drab-brown, belly more yellowish, less buffish-cream. According to Shirihai (in press), many migrants in Israel are intermediate between typical birds of the 2 races; perhaps originate from south-central and south-east Turkey and (if still breeding) neighbouring Lebanon and Syria. Even in Izmir area, western Turkey, a few ♂♂ have pronounced suffusion of yellow just below grey of chest. (C S Roselaar.)

Based on characteristics of eggs and voice, *E. cineracea* seems to form species-group with Ortolan Bunting *E. hortulana*, Cretzschmar's Bunting *E. caesia*, and Grey-necked Bunting *E. buchanani*. Its range partly overlaps with *E. hortulana*, while bordering *E. caesia* in the west and *E. buchanani* in the east. Origin of this group probably in Asia Minor. (Chappuis *et al.* 1973.)

MP

Emberiza hortulana **Ortolan Bunting**

Du. Ortolaan Fr. Bruant ortolan Ge. Ortolan
Ru. Садовая овсянка Sp. Escribano hortelano Sw. Ortolansparv

Emberiza Hortulana Linnaeus, 1758

Monotypic

PLATES 17 (flight) and 21
[between pages 256 and 257]

Field characters. 16–17 cm; wing-span 23–29 cm. Only marginally smaller but noticeably more compact than Yellowhammer *E. citrinella*; slightly larger on average than Grey-necked Bunting *E. buchanani* and Cretzschmar's Bunting *E. caesia*. Relatively long-billed, rather round-headed, and rather plump bunting; epitome of trio which also includes *E. buchanani* and *E. caesia* and displays in all plumages common characters of bright eye-ring, pale sub-moustachial stripe contrasting with dark malar stripe, and rufous or at least warm buff underparts. ♂ shows diagnostic olive-toned head and breast isolating yellow throat; ♀ and immature less distinctive, requiring careful separation from allies. Sexes dissimilar; little seasonal variation. Juvenile separable.

ADULT MALE. Moults: July–September (almost complete); December–March (head and body). Head rather flat-toned olive-green, with variable greyish tinge, faint dark streaks on crown and nape, and pale lemon- to buffish-yellow lores, eye-ring and throat, last showing quite pronounced malar stripe coloured as head. Malar stripe joins olive-grey to olive-green breast which narrows at sides into similarly coloured or browner band round nape and contrasts strikingly with yellow throat. Mantle and scapulars fulvous-brown, tinged chestnut on fringes and with black feather-centres forming bold streaks. Long rump yellowish-brown, only finely streaked. Wings basically brown-black, with greyish lesser coverts, pale buff tips to median coverts and margins and tips to greater coverts (not forming striking double wing-bar), and broader, more chestnut fringes to tertials. Tail brown-black, with chestnut margins to central feathers and much white on outermost; tail does not, however, look as white-edged as *E. citrinella*. Underparts below breast warm orange-chestnut, palest on vent and under tail-coverts.

Underwing yellowish-white. Bill noticeably long, tapering to fine point; flesh- to orange-red, on some bright and shiny enough to suggest piece of sealing wax. Legs flesh. With wear, crown, nape, and upper breast become greyer; some lose all visible green tone. ADULT FEMALE. Resembles ♂ but head browner and rather more distinctly streaked; lores and throat always buffish-yellow, with more streaked malar stripe, and duller, less green breast which is also finely streaked, particularly in centre. Note that cream eye-ring can appear brighter than on ♂. Back, rump, wings, and tail as ♂ but duller; underparts less uniform, looking distinctly mottled at close range. JUVENILE. Lacks distinctive head and throat colours of adult but already shows whitish eye-ring. Upperparts darker and browner than adult, more heavily streaked, with rather brighter wing-bars; face and underparts whitish-buff, with more rufous centre to underbody and under tail-coverts difficult to see but with well-marked striations from throat to flanks, strong enough on breast to form loose gorget. At distance, can recall pipit Anthus. Bill grey-horn. FIRST-YEAR. Juvenile plumage moulted by October, and then resembles ♀ though duller, with still brownish head, less yellow throat, and greyer upper breast all noticeably streaked. Adult pattern of characteristic pale eye-ring and submoustachial stripe and dark-flecked malar stripe show well.

Adult ♂ and ♀ unmistakable if seen well, but can be confused with allies at distance when head colours obscured. Juvenile and 1st-winter far less distinctive, needing careful separation from allies. Note that amongst trio of similar rufous-bodied species, only E. hortulana shows extensive yellow on throat and underwing and green tone to head; also has much heavier streaks in all plumages than E. buchanani and dull brown, not rufous rump, unlike E. caesia. Flight recalls E. citrinella but action lighter and more fluent, recalling Tree Pipit Anthus trivialis in speed and undulations. Escape-flight (of migrant) usually short, often to ground cover. Gait a hop, also creeping when feeding. Stance half-upright to level with tail often cloaking ground. Actions include flick of wings and tail. Secretive except at times of mass passage; often quiet. Gregarious on migration and in winter.

Song variable, often with attractive ringing tone to 1st part and lower-pitched more melancholy 2nd part. Calls also vary: shrill 'tsee-up' and lower, fuller, more piping 'tseu' when breeding; shorter 'tsip' and incisive 'twick' in autumn, particularly from flushed diurnal migrant; clear, metallic 'sleee', repeated loosely by overflying nocturnal migrants.

Habitat. Very varied, lying within July isotherms of 15–30°C, from high boreal through temperate, Mediterranean, and steppe zones, and to montane zones at c. 1500–2500 m in south of range. Attracted to trees, even breeding in forest glades and clearings, as well as pine Pinus forests, tree plantations, forest steppe with birch Betula trees, slopes of low mountains overgrown with grass and small

pistachio Pistacia trees, and orchards. Contrastingly, occurs freely in steep ravines, on bare alluvial deposits, and on rocky ground scantily covered with prickly shrubs. A ground feeder and nester, often even singing from boulder or low bush, but at times from treetop. Favours regions of high sunshine and low rainfall, regardless of latitude, and where food is readily available will spread widely over cultivated open land. Does not avoid banks of rivers and lakes but shows little attraction to wetlands, or to human settlements, especially cities. (Voous 1960b; Dementiev and Gladkov 1954.)

Formerly ranged far north in Scandinavia, not only breeding at sea-level on islands in Gulf of Bothnia but up to c. 900 m on fjells of Norway, among junipers Juniperus and other shrub cover, reflecting apparent preference for high sunlit slopes (Bannerman 1953a). Avoids areas of high rainfall.

In Germany, has traditionally occupied wide range of habitats, especially open cultivated land with plenty of trees and bushes, both on fertile plains and in hilly country. Favours tree-lined country roads and farm tracks, lined with fruit trees or limes Tilia, but avoids low fruit trees and gardens, and occurs only thinly along forest edges. Local colonizations and desertions have taken place with no apparent change in habitat conditions, and apparently suitable sites may be left untenanted (Niethammer 1937). Decreasing numbers in Belgium are markedly attached to sunny dry situations on poor sandy soil with scattered trees, or to other dry sites such as neglected vineyards and rocky plateaux (Lippens and Wille 1972). Here perhaps is clearest indication of strength of attachment to continental rather than oceanic climates. Contrastingly, in Switzerland, expansion occurred from mid-20th century in dry sunny warm regions, on level or gently sloping and often cultivated ground with scattered trees. Mown and fallow fields, terraces, and vineyards with walls or rocky margins are preferred for breeding, normally no higher than 1000–1650 m, although non-breeders occupy forest edge above 2000 m. (Glutz von Blotzheim 1962.)

In southern France and Spain tends to avoid hot lowlands and to breed more in uplands and on mountains up to at least 2000 m. Even from these latitudes, however, generally migrates to subtropical savannas and cultivated areas, as well as steppes.

Distribution. Some range contraction in north-west, and expansion in east; no other major change reported except possibly recent colonization of Algeria.

FRANCE. Major contraction of breeding range between 1960 and 1990, disappearing from 17 départements in north of country; now almost confined to southern half (Claessens 1992). NORWAY. Now confined to a few scattered localities in south-east, where formerly more widespread (see Population). One pair bred Nordland 1987. (VR.) SWEDEN. Disappeared from southern Sweden

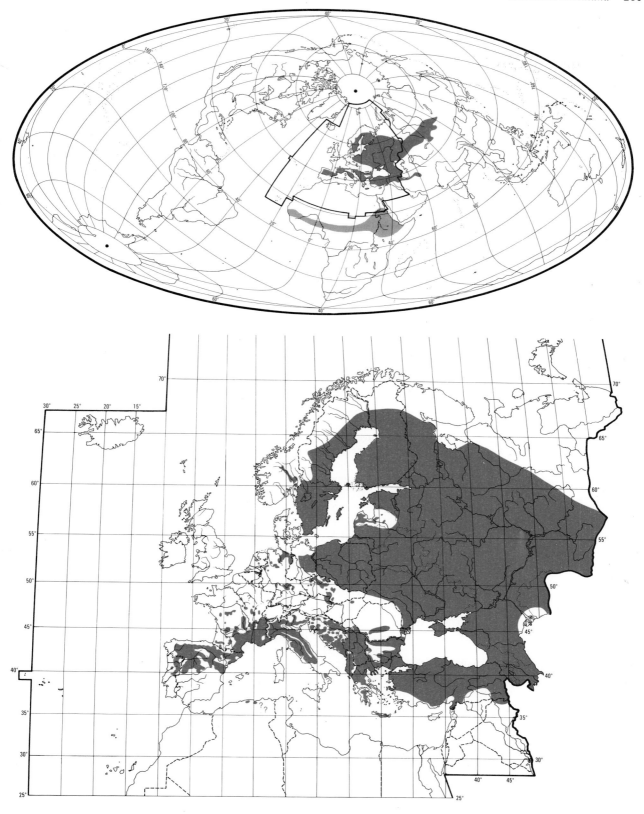

in 1950s and 1960s, due to effect of pesticides. Has since recovered but not re-occupied all former areas. During 1970s and 1980s has begun to breed in areas of intensive forestry, in clear-felled areas. (LR.) CZECHOSLOVAKIA. Breeding first recorded 1860; thereafter spread until 1950-60 when spread halted (KH). HUNGARY. Range contracted since 1960; only stable population now in Matra hills (G Magyar). LATVIA. Very rare in 19th century; from 1930s spread towards north-east, but expansion halted and now sporadic, patchily distributed (Vīksne 1983, 1989). SYRIA. Probably breeds in north-west, but confirmation needed (Baumgart and Stephan 1987). ALGERIA. Formerly regarded only as passage migrant. Several dozen pairs have bred regularly in Djurdjura since at least 1977; has certainly bred in Petite Kabylie for about same period, probably also in Aurès mountains and perhaps south-east of Sétif. (Burnier 1977; EDHJ.)

Accidental. Iceland, Faeroes, Canary Islands.

Population. Has decreased markedly in centre and west of range, but increased in north-east. Decrease in west probably due mainly to degradation of habitat caused by changing agricultural practices, and, in south-west, continued hunting (Claessens 1992). No evidence that population decline due to decreased breeding success (Stolt 1993).

FRANCE. 10 000-100 000 pairs (Yeatman 1976); estimated 10 000-23 000 pairs (Claessens 1992). SPAIN. Decreasing (AN). BELGIUM. Fewer than 100 pairs. Marked decrease since c. 1950, due to changing agricultural practices and habitat changes, perhaps also recent cold wet springs (Devillers et al. 1988). NETHERLANDS. Decreasing. 190-225 occupied territories 1975, 110-130 1978, 60-80 1983, 30-40 1986; actual breeding population smaller because some territories occupied by unpaired ♂♂. (SOVON 1987.) Population stabilized at 30-40 pairs 1985-90. Decline probably due to large-scale replacement of cereals by maize. (Noorden 1991.) Decrease continued in 1991, to c. 24 pairs (Br. Birds 1992, 85, 443-63). GERMANY. Estimated 3000 pairs (Rheinwald 1992). East Germany: 7000±3000 pairs (Nicolai 1993). NORWAY. Estimated 300-500 pairs. Dramatic decrease in recent decades (see Distribution), probably mainly due to changing agricultural practice. (VR.) SWEDEN. Estimated 40 000 pairs (Ulfstrand and Högstedt 1976). Decreased greatly in 1950s and early 1960s, due to effect of pesticides, followed by recovery (see Distribution) (LR). FINLAND. Estimated 100 000-200 000 pairs; long-term upward trend (Koskimies 1989). POLAND. Recently more numerous than in 19th century (Tomiałojć 1990). CZECHOSLOVAKIA. Marked decrease since 1960 (KH). AUSTRIA. Marked decline in recent decades (H-MB). About 200 pairs 1960-70, but censuses in 1981-6 showed decline to less than 10 pairs (Bauer et al. 1988). SWITZERLAND. Marked decline since 1950s (RW). Estimated c. 243 pairs 1978-9 (Biber 1984). HUNGARY. After peak in 1950-60 (several hundred

pairs), population gradually declined; now 10-12 pairs (G Magyar). LATVIA. Increase from 1930s, apparently now halted (see Distribution) (Vīksne 1983, 1989). LITHUANIA. Slight decrease since 1980 (Logminas 1991). ALGERIA. At least several dozen pairs (see Distribution) (EDHJ).

Movements. Long-distance migrant, wintering in sub-Saharan Africa, north of 5°N. More reported from eastern than western areas of Africa; small numbers winter in southern Arabia. Wintering birds use open upland habitats at 1000-3000 m. Migration and wintering areas not well documented, but see especially Stolt (1987) and Zink (1985).

Wintering in West Africa is probably mostly within 5-10°N. Reported by Brosset (1984) as common on broken ground above 1000 m on Guinea side of Mt Nimba, but not recorded from Ivory Coast (Thiollay 1987). Uncommon but annual winter visitor (December-January, perhaps later) to Loma and Tingi mountains of north-east Sierra Leone (G D Field). Few, if any, records in most other West African countries, e.g. 3 in Nigeria (north-central area), none in Ghana or Niger (Elgood 1982; Grimes 1987; Giraudoux et al. 1988). Further east, winters extensively in highland areas of Ethiopia (Urban and Brown 1971) and probably southern Sudan (Nikolaus 1987). Single record in mid-October almost at equator in Kenya (Britton 1980) and another on Seychelles in mid-November (Feare 1975). Small flocks reported in winter from northern Yemen (Brooks et al. 1987).

Stolt (1987) and Zink (1985) documented south-west (or SSW) direction of birds on passage from Fenno-Scandia and other parts of western Europe. Concentration of 17 recoveries in Landes (south-west France) probably due to trapping pressure, but 9 recoveries in south-west Iberia (including 3 from Switzerland and 1 from Italy) may indicate fattening area there. Other recoveries in north-east Spain (2), northern Morocco (2), and 2 from Finland to northern Italy. About 75% of recoveries are in autumn; none yet from south of Sahara. 2 recoveries from St Petersburg region (Russia) to Italy (Mal'chevski and Pukinski 1983). Birds from easternmost part of breeding range in western Siberia (45° further east than African wintering area) also head south-west.

Autumn migration mostly inconspicuous. In southern Sweden and Denmark, movement mainly mid-August to early or mid-September (Rabøl 1969; Edelstam 1972; Ulfstrand et al. 1974; Enquist and Pettersson 1986). Scarce on passage in Britain and Ireland (1320 records 1958-85); 70% of records are in autumn (mainly late August to September, latest early November); occurs mostly on south coast in autumn and east coast in spring (Dymond et al. 1989). Heavy passage at Col de Bretolet (Switzerland) from last week of August to mid-September (Dorka 1966). See also Stolt (1977) for west European migration dates. Scarce on passage in west and central Mediterranean region (e.g. Sultana and Gauci 1982, Iapichino and Massa

1989); most records in September (Stolt 1977; Finlayson 1992). Passage recorded late September to mid-October on west Moroccan coast (Smith 1965). Immediately south of Sahara, passage from late September to November or even December reported from Mauritania, Sénégal, Gambia, and Mali (Morel and Roux 1966; Lamarche 1981; Gee 1984; Gore 1990).

Spring migration much more conspicuous than autumn in most areas. Passage through west and central Mediterranean late March to mid-May with most in 2nd half of April (Zink 1985; Finlayson 1992). In Libya, scarce in north-west and probably common in north-east (Bundy 1976). At Col de Bretolet, passage peaks 26 April to 5 May (Winkler 1984). Passage (or vagrant) birds in Britain and Ireland, mainly on North Sea coast, peak 1st half of May (Dymond *et al.* 1989). Arrivals at breeding areas in Belgium and lower Rhein (Germany) from mid-April; main arrival in Denmark 2nd week of May, in northern Sweden 2nd half of May; at Helsinki (Finland) earliest birds averaged 2 May over 21 years. ♂♂ precede ♀♀ by 1-2 weeks. (Rabøl 1969; Tiainen 1979; Zink 1985.) In St Petersburg region, birds do not return until 2nd half of May and arrival prolonged into early June (Mal'chevski and Pukinski 1983).

In east of range, leaves western Siberia end of July to September (Johansen 1944; Gyngazov and Milovidov 1977). In Kazakhstan (breeds in north), passage via both north-west and south or south-east, chiefly mid-August to mid-September, beginning slightly earlier in north-west than in south (Korelov *et al.* 1974). In Ural floodlands in north-west, local birds leave by mid-September, but passage continues later (Levin and Gubin 1985). At Chokpak pass in south-east, movement peaks in early September, with adults before juveniles, but ♂♂ and ♀♀ together within each age-class (Gavrilov and Gistsov 1985). In Iran, fairly common passage migrant throughout north, but very scarce in south (D A Scott). Passage reported throughout Iraq, mostly inconspicuous, but locally common in north, chiefly to mid-September (Allouse 1953; Moore and Boswell 1957; Marchant 1963*a*). Regular in Turkey and Cyprus (Beaman 1978; Flint and Stewart 1992) and in much of Arabia (except central areas), mainly September-October (Bundy and Warr 1980; Jennings 1981*a*; F E Warr). Often numerous in south-west Saudi Arabia in autumn, but scarce in spring (Stagg 1985). Common in Israel (Shirihai in press), and regular in small numbers in eastern Egypt until mid-October (Goodman and Meininger 1989). Strong passage recorded Sudan and Eritrea, mostly September-October (Zink 1985). Arrives on wintering grounds near Addis Ababa (Ethiopia) from early October (Maréchal 1986).

Spring departure from Addis Ababa area during March, and apparently complete by end of month (Maréchal 1986). In Egypt, main passage mid-March to late April (Goodman and Meininger 1989). At Bahig on northern coast, over 6 years, 16-93 recorded 20 March to 26 April

(Horner 1977). In Levant and Middle East, passage chiefly mid-March to April (Bundy and Warr 1980; Flint and Stewart 1992; Shirihai in press; F E Warr). Migration through Chokpak pass end of April to mid-May, peaking in first 5 days of May (Gavrilov and Gistsov 1985). Breeding birds arrive in southern Volga-Kama region in first 10 days of May (Popov 1978) and in Kazakhstan in late April and May (earlier in west and north-west than south) (Korelov *et al.* 1974). In Ural valley, earliest arrivals recorded late April, and all have arrived within a few days (Levin and Gubin 1985). Arrives in western Siberia early to mid-May (Johansen 1944). Vagrant to India and Pakistan in spring (Ali and Ripley 1974; Roberts 1992). CJM

Food. Probably mainly invertebrates; also seeds, especially outside breeding season (Witherby *et al.* 1938; Glutz von Blotzheim 1962; Hölzinger 1987). In Westfalen (north-west Germany), May-June, forages on bare soil or sprouting crops, but also in deciduous trees, especially oak *Quercus*, for defoliating caterpillars (Lepidoptera), or in pine *Pinus* for seeds; after breeding season very often in harvested root-crop fields and in areas covered with bird's-foot *Ornithopus* (Conrads 1968). In Berlin area (Germany), early in season feeds on roads, field edges, and arable land; in windy weather, picked up caterpillars below trees (Garling 1941). On passage in Eritrea, forages in short grass, cereal fields, abandoned cultivation, and ploughed fields (Smith 1957), and in western Algeria and Mauritania, in April, fed on ground apparently taking seeds rather than insects (Heim de Balsac and Heim de Balsac 1949-50). In trees, forages up to topmost crown, and often observed hovering to pick off caterpillars dangling on silk or on leaves; commonly catches flying insects up to size of large beetles (Coleoptera), e.g. chafers (Scarabaeidae) (Durango 1948; Khokhlova 1960; Conrads 1969; Rymkevich 1977). On Isles of Scilly, October, observed methodically biting through stems of grass *Elymus*, using forceful sideways movement of head to snap stem over before moving bill along it towards head to extract pith as well as seeds (King 1971). Once seen extracting seeds from horse-droppings (Garling 1941). Pointed, cone-shaped bill ideal for handling both seeds and insects; seeds picked up in tweezer-like bill-tip and manoeuvred with tongue under central ridge of upper mandible, long seeds positioned across bill; de-husking effected by rapid up-and-down movement so husk split on central ridge and expelled via bill-tip; kernel then similarly crushed and swallowed. Freshly caught bird needed 2·3 s to de-husk seed of Canary grass *Phalaris* (0·3 s for positioning, 1 s for crushing, 0·5 s for de-husking, 0·5 s for crushing and swallowing kernel); practised bird required only 0·7 s; in preference trials, birds accepted only grass seeds. (Ziswiler 1965, which see for discussion and comparison with other seed-eaters.)

Diet in west Palearctic includes the following. Invertebrates: damsel flies and dragonflies (Odonata), grass-hoppers, etc. (Orthoptera: Gryllidae, Tettigoniidae, Acrididae), earwigs (Dermaptera), bugs (Hemiptera: Pentatomidae, Coreidae, Miridae, Cicadellidae, Psyllidae), lacewings, etc. (Neuroptera: Myrmeleontidae, Raphidiidae), adult and larval Lepidoptera (Pieridae, Tortricidae, Pyralidae, Tineidae, Notodontidae, Noctuidae, Geometridae), adult and larval flies (Diptera: Tipulidae, Asilidae, Muscidae), Hymenoptera (Pamphilidae, Tenthredinidae, Ichneumonidae, ants Formicidae, bees Apoidea), adult and larval beetles (Coleoptera: Carabidae, Histeridae, Scarabaeidae, Buprestidae, Elateridae, Tenebrionidae, Coccinellidae, Chrysomelidae, Bruchidae, Curculionidae), spiders (Araneae), millipedes (Diplopoda: Julidae), earthworms (Lumbricidae), snails (Pulmonata). Plants: seeds (etc.) of pine *Pinus*, hemp *Cannabis*, grasses (Gramineae, including *Elymus*, oats *Avena*, rye *Secale*, wheat *Triticum*, millet *Panicum*). (Zumstein 1921; Witherby *et al.* 1938; Durango 1948; Formozov *et al.* 1950; Tarashchuk 1953; Dementiev and Gladkov 1954; Mal'chevski 1959; Khokhlova 1960; Conrads 1968, 1969; King 1971; Popov 1978.)

In Ukraine, 34 stomachs, May–August, contained 332 invertebrates, of which 40·4% by number Hymenoptera (37% ants), 23·2% beetles (13% Curculionidae, 6% Elateridae), 15·7% Orthoptera (11% Acrididae, 4% Gryllidae), 12·1% Lepidoptera (9·7% larvae and pupae), 3·9% Diptera (3% Tipulidae), 2·7% Hemiptera (2% Coreidae), 1% spiders; remainder earwigs, millipedes, and snails; of *c.* 450 seeds, 84% of herbs, 11% wheat, and 4% millet (Tarashchuk 1953); another 34 stomachs, spring, contained 227 insects, of which 54·6% by number caterpillars (23% Geometridae, 21% Tortricidae, 8% Noctuidae), 33·0% beetles (18% Curculionidae, 7% Chrysomelidae, 6% Scarabaeidae), 11·8% Hymenoptera (mainly ants); plant material (in 7 stomachs) almost wholly cereal grain (Khokhlova 1960). In Volga-Kama region (east European Russia), of 16 stomachs, spring and summer, 94% contained animal food, 31% plant material (mostly seeds of grasses, including wheat); 81% held Curculionidae, 31% Orthoptera, 31% caterpillars, 19% Hemiptera, 13% Elateridae, 13% ants, and 13% spiders; June stomachs contained no plant food, which is first taken in July, becoming most important part of diet in autumn (Popov 1978). In Voronezh region (south European Russia), however, seeds taken regularly in small numbers throughout spring and summer (Mal'chevski 1959). For review of studies in European Russia, see Formozov *et al.* (1950). In Westfalen, probably eats mostly seeds on arrival in late spring, including large amounts of pine, then moves to invertebrates, (e.g.) defoliating caterpillars in trees, beetles, and Diptera larvae in leaves of root-crops; in autumn, very fond of oat grain (Conrads 1968).

Young fed invertebrates, especially caterpillars, so many of which are taken in oaks that this tree often considered necessary component of breeding habitat (e.g. Khokhlova

1960, Conrads 1968, Bülow 1990, Lang *et al.* 1990). In Voronezh region, 190 collar-samples contained 405 invertebrates, of which 61% by number adult and larval Lepidoptera (especially Geometridae during infestations), 18% beetles (mainly Scarabaeidae), 11% grasshoppers (Acrididae), and 8% Diptera (Tipulidae, Asilidae) (Mal'chevski 1959). In south-east European Russia, of 244 items, 48% (by number) larvae (21% Notodontidae, 7% Geometridae), 42% Orthoptera, 3% Tipulidae; proportion of larvae fell from 76% when nestlings 4–6 days old (*n* = 92 items) to 33% at 8–12 days (*n* = 152), while Orthoptera rose from 12% to 53%; young given exclusively thin, soft Notodontidae caterpillars when various larvae placed near nest in preference trial, others discarded by adults (Formozov *et al.* 1950). In Wertfalen, Scarabaeidae beetles brought to nest only after day 7; defoliating Tortricidae caterpillars often occur in large numbers only before nestling period and so to large extent unavailable, but young receive adult Lepidoptera also (Conrads 1969). At one site in Switzerland, Orthoptera main nestling food (Kunz 1950). Food usually collected within 200 m of nest (Kunz 1950; Conrads 1969), though up to 1500 m recorded (Formozov *et al.* 1950). Adults bring 2–3 large invertebrates each visit (e.g. beetles, earthworms, Orthoptera); caterpillars bundled in bill. Large beetles can either be fed whole or have wing-cases (etc.) removed. (Durango 1948; Formozov *et al.* 1950; Mal'chevski 1959; Conrads 1969.) One brood consumed total of 106 larvae per day (Formozov *et al.* 1950). BH

Social pattern and behaviour. Well known. Principal studies in Sweden by Durango (1948) and long-term study of marked birds in Westfalen (northern Germany) by Conrads (1969).

1. Regularly occurs singly as well as in small flocks when on passage (e.g. Witherby *et al.* 1938, Sultana and Gauci 1982, Finlayson 1992). In USSR, spring migrants mostly in small groups, sometimes singly, autumn migrants almost always in large flocks, which begin to form as soon as breeding finishes (Dementiev and Gladkov 1954). Migrants in Belgium mostly singles, less often 2(–5) together (Spaepen 1952), but in Westfalen often arrives in spring in groups as well as singly (Conrads 1969). Flocks on passage in Israel number scores of birds (Paz 1987); in Sicily occasionally 15 together (Iapichino and Massa 1989); on Cyprus in spring usually tens, maximum recorded 200 in mid-April (Flint and Stewart 1992). Migrants recorded in thousands in Sudan in September, visiting drinking place 08.00–09.00 hrs each morning (Prendergast 1985). Flocks, of both sexes, also occur in winter (Witherby *et al.* 1938); wintering birds in Sierra Leone in loose flocks (G D Field). Large flocks sometimes also recorded in spring when newly arrived migrants in northern Sweden encounter very cold weather (Durango 1948). Migrants passing through Belgium usually do not occur with other birds, except sometimes Tree Pipits *Anthus trivialis* (Spaepen 1952); flocks outside breeding season often in company with *A. trivialis* (Witherby *et al.* 1938). On passage in Syria noted in company with Cretzschmar's Bunting *E. caesia* and Black-headed Bunting *E. melanocephala* (Baumgart and Stephan 1987); migrant flocks in Jordan also often associated with *E. caesia* (Wittenberg 1987). Wintering birds in West Africa in

small groups, often in company with Cinnamon-breasted Rock Bunting *E. tahapisi* (Morel and Roux 1966; Smith 1967; G D Field). BONDS. Mainly monogamous, but record of polygamy in Sweden (Durango 1948). Polygamous ♂ sometimes accompanied 2nd ♀ but did not feed her young (Durango 1948). Pair-bond apparently lasts only for one breeding attempt, but information scanty. First breeding sometimes at 1 year old, but few details (Conrads 1969). Non-breeding ♂♂ are often present close to breeding birds (Durango 1948). In north-west Germany, 64% (*n*=11) of ♂♂ unpaired in 1st year, 50% (*n*=16) in 2nd year, 17% (*n*=6) in 3rd year; overall, 62% of ♂♂ in population were paired each year (Conrads and Quelle 1986). Bülow (1990) found 55% (*n*=38) of singing ♂♂ paired. Nest built by ♀. Incubation by ♀ alone. Young usually fed by both parents (see Relations within Family Group, below). BREEDING DISPERSION. In southern and central Sweden, often breeds in 'small colonies' of 2–4 pairs, and non-breeding ♂♂ tolerated in same areas. In northern Sweden, mainly occurs as isolated pairs. (Durango 1948.) Similarly, in southern Germany occurs both as widely separated pairs and in marked concentrations (Helb 1974). In German study, Schubert (1984) found 34–50% of isolated ♂♂, remainder in concentrations of 2–17 together. Concentrations of breeding birds also reported from Switzerland (Glutz von Blotzheim 1962) and Yugoslavia (Reid 1979). In Sweden, no strict territories defended, or strict territory is at most small and not rigorously defended against other ♂♂ (Durango 1948). In Sweden, song-posts of ♂♂ often only 25–50 m apart and non-breeding ♂♂ may also have song-posts in close proximity to those of breeding birds (Durango 1948; Géroudet 1951b). In Westfalen, however, song-posts are minimum 50 m apart, average 280 m (Conrads 1969). Several accounts suggest it does not defend territory close to nest or song-posts, and to some extent remains sociable during breeding season. 2 records of breeding birds with young in nests consorting with extra ♂ without apparent aggression. Often feed in groups near nests or song-posts, or some distance from them, occasionally at over 500 m from nest. (Niethammer 1937; Garling 1941; Durango 1948.) In Westfalen, however, Conrads (1969) found territory more actively advertised and defended. Paired ♂♂ dominate unpaired ♂♂. Breeding birds may collect food for nestlings up to 200 m from nest, but usually closer; unpaired ♂♂ often seen in vicinity of fledglings on breeding territories (Conrads 1969). In Yugoslavia, 8 ♂♂ found singing together in circular area of boulders no more than 80 m across (Reid 1979). Around Kamyshin (east European Russia), nests 100–200 m apart (Dementiev and Gladkov 1954). Territory size in Germany *c*. 3 ha (Mildenberger 1968), *c*. 2·1–4·4 ha (Conrads 1969). Overall breeding densities difficult to assess as breeding pairs commonly clustered; in Lozère (France), does not exceed 4·0 pairs per km² on steppe-like grasslands, 14–17 pairs per km² in cultivated hollows of limestone hills, and reaches 22–25 pairs per km² in stony, heavily grazed pastureland with low juniper *Juniperus* bushes (Lovaty 1991). On agricultural land in southern Finland, mean density increased from 3·2 pairs per km² in 1930s (on 655 ha) to 5·7 pairs per km² (on 476 ha) in 1979 (Tiainen and Ylimaunu 1984). In Poland, mean densities 0·10 and 0·11 pairs per km² in 2 areas (Górski 1988). In Bulgaria, mean densities (pairs per km²) range from 1–2 in shelter belts on agricultural land (B Ivanov) and 2 in woodland dominated by *Quercus pubescens* (Simeonov and Petrov 1977) to 114 in woodland dominated by Judas-tree *Cercis siliquastrum* (Simeonov and Petrov 1977) and 180 in plantations of *Pinus nigra* (Petrov 1988). Little evidence of site-fidelity, but ♂♂ re-found in later years at mean distance of 600 m (*n*=24) from ringing localities (Conrads and Quelle 1986). ROOSTING. On ground amongst grasses and herbs, at least when migrants first return in spring (Durango 1948).

During breeding season, several ♂♂ spent night in cornfields near nests; ♀♀ roosted on nest (Conrads 1969).

2. Rather quiet and secretive (Witherby *et al.* 1938). Migrants in Belgium particularly shy (Spaepen 1952). Becomes very inconspicuous during moult July–August (Géroudet 1951b); Conrads (1969) noted that they keep completely hidden then. Wintering bird in Nigeria remarkably tame, appearing on verandah of house and approaching to within a few feet (Smith 1967). Silent when raptors seen near nest, but attacked Cuckoo *Cuculus canorus* (see Parental Anti-predator Strategies, below). Crest often raised in fear (Andrew 1956b) and probably in varied other contexts, as in other *Emberiza*, but no details. FLOCK BEHAVIOUR. Calls often given in flocks include flight-call 'tjö' and 'jüp' (see 3b in Voice) and 'bit' call (see 4 in Voice), latter common also at upflights and as contact-call between perched birds (Conrads 1971). Migrants flying overhead in Belgium readily attracted by calls and song of caged conspecifics (Spaepen 1952). SONG-DISPLAY. Singing ♂ (Fig A) typically holds head slightly above horizontal, neck

A

somewhat extended, wings slightly hanging and rump feathers raised; bill is opened widely with each element of early part of song, but almost closed in later part of song (Conrads 1969). Often sings in resting attitude (Conrads 1969), like some other *Emberiza* (Andrew 1956b); description of singing posture of ♂ in Yugoslavia by Reid (1979) suggests resting attitude. Song-posts are on trees, rocks, bushes, telegraph wires or other elevated perches (Witherby *et al.* 1938; Durango 1948; Conrads 1969; Reid 1979), and territories normally include elevated song-posts (Durango 1948; Conrad 1969). See Schubert (1984) for details of tree species used as song-posts in Germany. Sometimes sings in Display-flight, when song may differ (see Antagonistic Behaviour, below). In Westfalen, seldom sings from ground (Conrads 1969), but in Sweden seen singing from ground regularly after sunset; sings from ground during day much less often, mainly when there is strong wind (Durango 1948). Occasional song widely reported from migrants on spring passage in North Africa, Middle East, and Mediterranean region (Witherby *et al.* 1938; Heim de Balsac and Mayaud 1962; Sultana and Gauci 1982; Baumgart and Stephan 1987), and in Switzerland (Géroudet 1951b). ♂♂ begin singing in spring as soon as they reach breeding grounds, before ♀♀ arrive, but song weak for first few days (Durango 1948). ♂♂ captured on spring passage through Belgium and held in aviaries sing little or not at all for a few days, then strongly after this (Spaepen 1952). Song reduced in cold weather, but one ♂ recorded singing in snowstorm in northern Sweden on 27 April. Song output declines after young hatch; except from non-breeding ♂♂, which continue to sing with great persistence (16 songs in 2 min, 640 in 2 hrs: Durango 1948). By August–September, occasional short songs

(Subsong according to Conrads 1969: see 1b in Voice) are rare. (Durango 1948; Géroudet 1951b; Conrads 1969.) See Conrads (1969) for details of annual and diurnal periodicity of song in Westfalen, influence of weather, and other factors and general information on diurnal activity pattern. Sometimes sings at night, not infrequently for much of night in northern Sweden (Witherby et al. 1938; Durango 1948). Sometimes sings with food in bill during period when nestlings being fed (Géroudet 1951b). ANTAGONISTIC BEHAVIOUR. (1) General. In Sweden, reported not to be quarrelsome on breeding grounds; one ♂ sometimes displaces another from song-post, but birds often feed in groups while breeding and tolerate conspecifics near nests (Durango 1948). In Westfalen, newly arrived ♂♂ seen in small groups in fields, with little sign of aggression, but later in breeding season considerable territorial rivalry evident (see below) (Conrads 1969). ♂♂ captured on spring migration through Belgium and held in aviaries generally tolerant of one another, although some fight repeatedly, to point where close to death (Spaepen 1952). Well-marked interspecific aggression ♂ recorded a few times in breeding season, including fight with ♂ Yellow Wagtail Motacilla flava and threatening A. trivialis with bill open (Conrads 1969). (2) Threat and fighting. Head-forward threat-posture used at close range (as in other Emberiza: Andrew 1957b), but weakly expressed; often accompanied by gaping (Conrads 1969). Threat by territorial ♂ commonly involves Tail-spreading display, in which white outer tail-feathers conspicuous; tail-spreading also prominent in Territorial Hopping-flights (see below) and Slow-chases (see below) (Conrads 1969). Rival ♂♂ sometimes give 'tjöi' call (see 3a in Voice) in various contexts (Conrads 1969, 1971). 3 types of flight-display are used in territorial advertisement or defence (Conrads 1969), here termed: Display-flights, Territorial Hopping-flights, and Territorial Round-flights. Display-flight ('Balzflug' of Hortling 1928, Homann 1959; 'Schauflug' of Conrads 1969) apparently represents advertisement of territory by ♂, given only over bird's own territory (Conrads 1969). ♂ flies steeply upwards from song-post in tree (rarely from ground), with undulating flight (Fig B) and descends abruptly. Characteristic song-phrases given (see Conrad 1969 for details) and 'hüit' variant of call 3a (see Voice) forms regular part of Display-flight. Display-flights seen rather infrequently: in German study recorded on 19 dates, 15 in May, 3 in June, 1 in August. (Conrads 1969.) Song in flight reported by Durango (1948), Géroudet (1951b), and Spaepen (1952)

apparently also refers to Display-flight. Territorial Hopping-flight very often used in territorial defence by ♂ at beginning of breeding season. ♂ flies with head and tail distinctly raised and rhythmically spreads tail with each wing-beat, conspicuously exposing white outer feathers; Subsong or full song may be given. (Conrads 1969.) Territorial Round-flights (Fig C) appear to represent lower-intensity display for territorial defence or advertisement, used during incubation and nestling periods; ♂ flies out over open field from perch and returns to near starting point (Conrads 1969). Territorial ♂ may chase another ♂ (sometimes 2 others) with characteristic Slow-chase (Fig D), accom-

D

panied by song and calls (Conrads 1969); Quick-chase sometimes also used; both Slow-chase and Quick-chase are much as described for some other Emberiza, with tail being spread in time with each wing-beat (Andrew 1957a, 1961; Conrads 1969). ♂♂ may fight over territory; fights commonly involve one ♂ pecking at tail of another; at higher intensity ♂♂ fight breast-to-breast in the air (as in other Emberiza: Andrew 1957a); 'chaa' call (see 5 in Voice) commonly given in territorial fights (Conrads 1969). Appeasement behaviour includes wing-quivering and adoption of soliciting posture (see Heterosexual Behaviour, below) by ♂ (Conrads 1969). HETEROSEXUAL BEHAVIOUR. (1) General. In spring, ♂♂ arrive on breeding areas some days before ♀♀ and commence singing. ♀♀ join ♂♂ as soon as they arrive (Durango 1948). On spring passage through Belgium, ♂♂ occur 8–14 days ahead of ♀♀ (Spaepen 1952). In Westfalen, first ♂♂ arrive up to 8 days before first ♀♀ (Conrads 1969). During breeding season pair commonly feed together and keep very close company when ♀ away from nest; whenever ♀ flies, ♂ follows closely (Conrads 1969). (2) Pair-bonding behaviour. ♂ may pursue ♀ in fast, twisting flights above fields, giving repeated dry 'tip tip tip' call (probably same as 4 in Voice). Other conspecifics often accompany these chases. (Durango 1948.) ♂ also reported to fly between song-posts in apparent Display-flights, with fluttering wing-beats during time when ♀ incubating (Durango 1948), but these flights may function in territorial defence rather than heterosexual display (see Antagonistic Behaviour, above). Bill-raised display of ♂ (Fig E) as in other Emberiza (Andrew 1957a).

B

C

E

Used both in sexual contexts and in encounters with rival ♂♂: head and bill lifted more or less high, with wings hanging slightly and tail sometimes partly spread; given in trees, and on ground where ♂ may follow ♀ with hopping gait; associated with mating, which often follows high point of display, but may also be used after mating. (Conrads 1969.) Géroudet (1951*b*) recorded instance of use of this display in early morning during laying period: ♂ flew down to ground at side of ♀, then raised head so bill pointed vertically up, exhibiting yellow throat to ♀. Conrads (1969) interpreted display as ritualized pre-mating movement of ♂ rather than exhibition of yellow throat, but these possibilities not exclusive. Wing-quivering display given in sexual contexts, most commonly in association with other sexual displays and mating; also in territorial encounters between ♂♂, when it may signify submission (see Antagonistic Behaviour, above). (Conrads 1969.) (3) Courtship-feeding. Apparently does not normally occur. ♀ once observed to beg food from ♂, with fluttering wings, when ♂ arrived at nest to feed young (Durango 1948). (4) Mating. Successful mating always preceded by Soliciting-display of ♀ (Fig F), in which head lowered and wings

F

vibrated. ♂ flies very close to ♀ prior to mating and once heard to give 'güb' call (see 3b in Voice). Mating seen mainly in trees, but sometimes on ground or near nest. (Conrads 1969.) ♀ normally gives 'bibibibibi. . .' call (see 6 in Voice) prior to mating (Conrads 1971). Very brief copulation seen in early evening on wire, at time when ♀ building replacement nest; ♀ gave 'bibibibibi. . .' call (Géroudet 1951*b*). (5) Nest-site selection. Nest-site flight appears to be display associated with nest-site selection (Conrads 1969). Typically, ♂ flies down from song-post or other high perch using characteristic 'hopping' flight, with short undulations and whirring wing-beats. ♀ usually follows ♂, but sometimes ♀ leads and ♂ follows. (Garling 1949; Conrads 1969.) Garling (1949) interpreted these flights as nest-site selection, choice probably being made by ♀. When ♀ nest-building, ♂ uses Stem-display ('Halmbalz' of Conrads 1969; 'Nest-site display' of Andrew 1957*a* for other *Emberiza*), carrying twig in bill, but without taking part in true nest-building (Durango 1948; Krampitz 1950; Conrads 1969). Display seen 7 times by Conrads (1969), and interpreted as helping to maintain pair-bond. Nest built by ♀ alone, mainly during morning (Durango 1948; Géroudet 1951*b*; Conrads 1969; Harrison 1975); ♂ frequently accompanies ♀ during building (Conrads 1969). Incubation by ♀ alone (Durango 1948; Géroudet 1951*b*; Harrison 1975); unconfirmed suggestion that ♂ may participate (Witherby *et al.* 1938). See Conrads (1969) for details of incubation routine. ♀ often spends time away from nest during incubation in company of ♂ (Conrads 1969). Durango (1948) notes that when ♀ leaves nest during incubation she joins ♂ and feeds with him. Relations within Family Group. Young usually fed by both parents, on food carried in bill (Durango 1948; Géroudet 1951*b*; Conrads 1969; Harrison 1975). Sometimes ♂ feeds young more often, sometimes ♀; occasionally, young fed by ♀ alone (Durango 1948; Conrads 1969). See Conrads (1969) for details of feeding rates. ♀ broods small young (Durango 1948), in dry weather mainly on 1st–4th day (Conrads 1969); ♀ recorded brooding 6-day-old young in cold weather and shading nestlings from hot sun (Géroudet 1951*b*). Faecal sacs of nestlings probably eaten by ♀ (and possibly also by ♂) for first 4 days; later removed (by both sexes) and

deposited in particular place, such as branch of bush (Durango 1948). Eyes of nestlings open on 3rd day; by 9th day almost, and on 10th fully feathered (Conrads 1969). Young leave nest several days early if disturbed, before able to fly, remaining concealed in vegetation nearby. One report of young leaving nest of their own accord before fully feathered, following ringing. (Durango 1948.) Géroudet (1951*b*) also reported young leaving nest before able to fly. Conrads (1969) recorded young leaving nest 4 days before able to fly; scattered in cornfields and usually fed by only one parent; after 4th day out of nest, probably self-feeding. Durango (1948) noted that after leaving undisturbed nest in morning, fledglings were fed same day by ♂ alone, moving up to 25 m from nest; following day young were 100 m from nest, still being fed by ♂ alone; 7 days later young not receiving food. Young may move up to 50 m from nest on day of fledging, although most remain near it (Conrads 1969). Fledged young beg for food with 'zië' or 'zi' food-calls (see Voice). 1st-years evidently migrate from breeding grounds before adults since they occur on passage through Belgium before adults (Spaepen 1952). Anti-predator Responses of Young. Not described for nestlings. Recently fledged young remain concealed in cover (Durango 1948; Géroudet 1951*b*; Conrads 1969). Parental Anti-predator Strategies. (1) Passive measures. Incubating ♀ sits tightly (e.g. Géroudet 1951*b*, Mildenberger 1968, Conrads 1969), allowing observer to approach within 6–7 m even when eggs fresh; when eggs well incubated or small young being brooded, may allow approach to within 1 m. (Durango 1948; Géroudet 1951*b*.) Adults feeding young often land several metres from nest, then perch there for some time (especially ♀) before making final approach on ground (Durango 1948; Conrads 1969). (2) Active measures: against birds. Calls when alarmed at nest similar to those given when human appears (see below). Silent when raptors seen; apparently no special alarm-call to warn of flying predators. Cuckoo *Cuculus canorus* flying near nest attacked on 3 occasions, once by both birds of pair. (Conrads 1969.) (3) Active measures: against man. 'Seeoo', 'tjöi', 'tjö', and 'jüp' calls given in alternation when alarmed at nest (see 2, 3a–b in Voice), while parents perch on vantage points nearby (Géroudet 1951*b*; Conrads 1969, 1971). Apparent distraction-displays recorded on 3 days at one nest with young, when parents surprised by sudden close appearance of observer: one bird (sex unclear) appeared to fall to ground from low flight, then flew off in hesitant manner; ♀ ran from nest like mouse, was pursued for 5 m by observer, then (3 times) flew up almost vertically, fell back to ground and continuing running; on last occasion ♀ ran off like mouse, without flying. (Géroudet 1951*b*.) ♀ removed 4 nestlings 3–4 days old from nest soon after visit by observer who ringed young, and close passage of cat; interpreted by observer as accidental removal of young due to rings being mistaken for droppings (Géroudet 1951*b*). Conrads (1969) also recorded 'injury-feigning' at nest-site from brooding ♀♀. Adults warn recently fledged young with 'zië' call (see 2 in Voice), given with raised crest and simultaneous upward flicking and spreading movements of tail (Conrads 1969).

(Figs by D Nurney: A from photograph in *Lintumies* 1991, **26**, 15; E from drawing in Géroudet 1951*b*; others from drawings in Conrads 1969.)   DTH

Voice. Frequently used in breeding season, less often at other times. Detailed studies (with sonagrams) in Westfalen (northern Germany) of calls by Conrads (1971) and of song by Conrads and Conrads (1971) and Conrads (1976). Regional variation in song well documented; also evidence

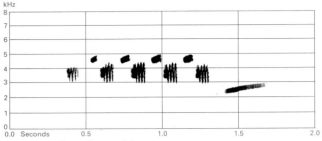

I   P A D Hollom   France   May 1974

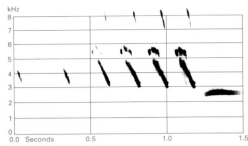

II   B N Veprintsev   Russia   May 1961

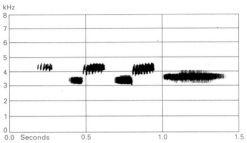

III   J-C Roché   France   June 1984

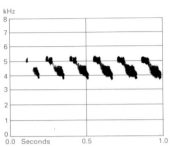

IV   J-C Roché   France   June 1984

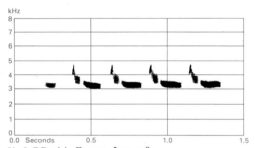

V   J-C Roché   France   June 1984

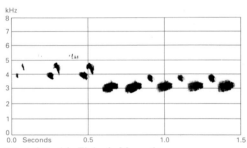

VI   J-C Roché   Finland   May 1964

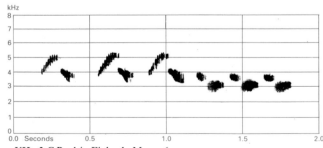

VII   J-C Roché   Finland   May 1964

that some calls vary between populations in northern and southern Germany (Helb 1981, 1986).

CALLS OF ADULTS. (1) Song of ♂. (1a) Full song. Varies individually and regionally (e.g. Witherby *et al.* 1938, Géroudet 1951*b*, Spaepen 1952, Conrads and Conrads 1971), but commonly has attractive ringing tone to 1st section, and lower-pitched, more melancholy 2nd part

(Bruun *et al.* 1986; Jonsson 1992). See Figs I–VII. Published sonagrams (Conrads and Conrads 1971; Conrads 1976; Helb 1986) show overall duration of song 1·5–2·0 s, overall frequency range 2–6 kHz; similar repeated units occur in 1–3(–4) groups, last group usually low pitched. In Westfalen, when singing strongly, song given 7–9 (5–11) times per min (Conrads 1969); single ♂ in Sweden sang average of 6 times per min (Durango 1948). See Conrads (1969) for data on number of songs per songbout. Song audible to man at 300–500 m (Spaepen 1952). Lower-pitched units sometimes form 1st and sometimes 2nd part of song; variants occur in which one or other unit predominates, or 3rd, tremulous unit, 'trrrull', softer and deeper than either, is added (Witherby *et al.* 1938). Usually 6–7 similar clear units with final unit lower or higher (Peterson *et al.* 1983). Songs of Swiss birds mainly have 2nd part of song restricted to single unit, usually of lower pitch than others, but sometimes higher pitched (Géroudet 1951*b*). Other renderings 'swee swee swee swee drü drü' or 'drü drü drü seea seea'; in southern Europe, 2nd part is usually only one hoarse, falling unit (Bruun *et al.* 1986; L Svensson); only 1 final unit usual in Greece and Turkey (Mild 1990); also 'tsie-tsie-tsie-tsie truh-truh-truh' (Jonsson 1992). Song dialects in Westfalen investigated in detail by Conrads and Conrads (1971) and Conrads (1976), which see for sonagrams. 5 regional dialects investigated were all very clearly differentiated, often separated by sharp boundary, with no mixed zone. End-phrases extraordinarily consistent in form and frequency and these differed most between dialects, but opening phrases also differed. Practically all ♂♂ in a dialect region sang same songs, but very small minority sang with 'foreign' dialects; geographical origins of such birds could often be deduced from song characteristics. Foreign dialect sometimes copied by neighbouring ♂♂, who learnt it in addition to usual song-type for area (which was apparently learnt when young). Most individual ♂♂ had repertoire of 2-3 (1-5) song-types. (Conrads 1969, 1976; Conrads and Conrads 1971.) See Helb (1986) for sonagrams of songs of different dialects in southern Germany; Reid (1979) for descriptions of songs in Austria; Balát (1963) for notes on dialects (using musical notation) in Czechoslovakia and Bulgaria; Stolt and Åström (1975) for analyses of songs (with sonagrams) in Sweden and Finland; Nonnenmühle and Nonnenmühle (1982) for details of songs in Bulgaria. For comparison with Grey-necked Bunting *E. buchanani*, see that species. (1b) Subsong. Described by Conrads (1969) as consisting of fewer units than full song. Heard in varied contexts including: soon after ♂♂ arrive in spring; in territorial conflicts; during Stem-display (see Social Pattern and Behaviour); when young are fledging; on ground before and after other activities; in cold weather and late in season (especially July). Single record of juvenile giving Subsong, in mid-July. Quite distinct type of song, a few chirping notes terminated by clear, piping 'chirrip-chip-hooo' like Bullfinch *Pyrrhula pyrrhula* (Witherby *et al.* 1938), may represent Subsong. Possible Subsong also described from bird heard in August by Géroudet (1951b). (2) Seeoo-call (Andrew 1957b). Varies from distinctly disyllabic 'psië', 'psili', or 'tsië' to almost monosyllabic 'psi' or 'tsi' (Conrads 1971), falling very slightly in pitch (L Svensson). Other renderings 'ziüh' (Niethammer 1937), shrill 'tsee-ip', or more curtailed 'tsip' (Witherby *et al.* 1938), 'tsjiü' (Géroudet 1951b; Spaepen 1952), clear, metallic 'SLEEe' (Bruun *et al.* 1986), 'sie' (Jonsson 1992); recordings suggest 'TSEEip' (Figs VIII-IX). 1st syllable is at *c.* 5 kHz, 2nd at 4·25 kHz; total duration of call 150-200 ms. Used by both sexes in variety of contexts: as 'social call' of solitary birds (Andrew 1957b), as introduction to song of ♂, but especially often as alarm-call at nest (when alternated with calls 3a-b). Apparently given with ambivalent motivation and developed from food-call of young. (Conrads 1971.) Given by migrants flying at night, heard from each bird 3-5 times as it flies over (Durango 1948; Bruun

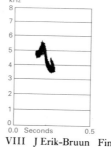
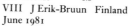

VIII J Erik-Bruun Finland
June 1981

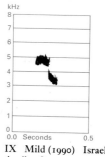

IX Mild (1990) Israel
April 1989

et al. 1986). (3) 'Tjöi', 'tjö', and 'jüp' calls. (3a) Disyllabic 'tjöi', 'jüi', or 'hüi' (Conrads 1971). Other renderings 'jühji' (Niethammer 1937), 'püit' (Géroudet 1951b), 'uwiet' (Spaepen 1952). Complete call begins at 3-3·5 kHz, falls slightly in pitch, then rises by 1-1·4 kHz; lower-pitched part of call shows 2-3 harmonics; whole call lasts 250-300 ms, but shorter versions are common (especially from ♀♀), and further shortening gives call 3b. Used by both sexes, mainly as alarm-call near nest, but sometimes given by rival ♂♂; seldom heard outside breeding season. The 'hüit' variant is part of Display-flight (see Social Pattern and Behaviour). (Conrads 1971.) (3b) 'Tjö' and 'jüp' calls; shortened versions of call 3a (Conrads 1971). Also rendered 'djöb', 'jük', 'güb' (Conrads 1971), 'jup' (Andrew 1957b), high whistling 'teu' (Witherby *et al.* 1938; Peterson *et al.* 1983), short 'chu' (Bruun *et al.* 1986); recordings suggest repeated 'ju' (Fig X) or 'jup' (Fig XI). Short versions fall quickly in pitch from *c.* 3 kHz to 2 kHz, often showing 2-3 harmonics and have total duration *c.* 70-95 ms (Conrads 1971). Used in many

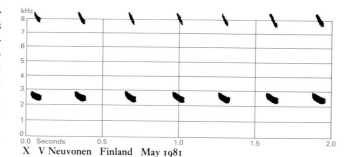

X V Neuvonen Finland May 1981

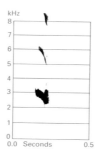

XI V Neuvonen Finland May 1981

contexts by both sexes, including as flight-call, by ♀ while nest-building or during pauses in incubation (along with call 2), and when nestlings or fledglings threatened. (Conrads 1971.) (4) Bit-call. Rendered 'hit', 'bit', 'büt', 'tjip', 'zück' (Conrads 1971), 'pfütt', 'tzück' (Niethammer 1937), incisive 'twick' (Witherby *et al.* 1938), 'tip' (Durango 1948), 'pit' (Géroudet 1951*b*), 'tit' (Andrew 1957*b*), muffled, dry 'plett' (Bruun *et al.* 1986), short 'plit' (Jonsson 1992). Rather variable in pitch, with higher versions at *c.* 4–5 kHz, lower-pitched ringing version at *c.* 2–4 kHz; brief, lasting only *c.* 50–70 ms. Used by both sexes, typically as flight-call, often on take-off; also as contact-call when perching, especially by paired birds. (Conrads 1971.) Heard from diurnal migrants (Bruun *et al.* 1986; Jonsson 1992). (5) Chaa-call (Andrew 1957*b*). Rendered 'chä' by Conrads (1971). Harsh call, with wide frequency range (*c.* 2–8 kHz), typically given in quick bursts with units 50–100 ms in duration, but sometimes single long unit (*c.* 400 ms). Given by ♂♂ during territorial fights. (Conrads 1969, 1971.) (6) A 'bibibibibibibi' by ♀ (Conrads 1971); 'tititi' (Andrew 1957*b*), 'zeuzeuzeu...' or 'zéézéé...' (Géroudet 1951*b*). Sonagram analysed by Conrads (1971) showed 17 similar units repeated in total duration of *c.* 2 s. Each unit falls quickly in frequency from *c.* 5 kHz to *c.* 3 kHz. Given by ♀ soliciting mating (see Social Pattern and Behaviour); some similarity to food-call of young. (Conrads 1971.)

CALLS OF YOUNG. Food-calls of nestlings a clear tinkling; at day 7, high pitched (from 6 kHz to above 8 kHz) and given *c.* 3 times per s (Conrads 1969, 1971). 10-day-old nestlings gave food-calls rendered 'tsé-tsé-tsé tsé-tsé-tsé' or 'chi-chi-chi-chi' when parents arrived; audible at more than 40 m (Durango 1948). After day 6, young give 'zip' call mixed with food-call (Conrads 1969). Fledged young beg for food with 'zië' or 'zi' call that may be precursor to call 2 of adults; sonagrams analysed by Conrads (1971) showed monosyllabic call with wide frequency range (*c.* 2·5–6·5 kHz). DTH

Breeding. SEASON. Sweden: see diagram; 85% of 117 clutches completed between 20 May and 20 June (Durango 1948); see also Stolt (1993). Southern Finland: eggs laid 2nd half of May (mid-May to early June) (Haartman 1969). St Petersburg region (north-west Russia): eggs laid from *c.* 20 May to mid-June (Rymkevich 1977); for Voronezh region (south European Russia), see Mal'chevski (1959). Northern Bayern (central Germany): eggs laid early May to late June, mostly late May or early June (Lang *et al.* 1990); for Westfalen, see Conrads (1977); for Baden-Württemberg, see Hölzinger (1986); for eastern Germany, see Garling (1941) and Eifler and Blümel (1983). Catalonia (north-east Spain): start of laying mid-April, eggs recorded early July (Muntaner *et al.* 1983). Israel: at 1500–1900 m, from beginning of May to end of July (Shirihai in press). For Algeria, see Burnier (1977). SITE. On ground; in north-west Europe, usually in cereals or other arable crop, often potatoes, frequently in depression in soil so top of nest-rim flush with ground; otherwise in vineyards, forest clearings, on rocky slopes, or in thick grass, heather (etc.), sheltered by overhanging rock or foliage (Ferguson-Lees 1950; Géroudet 1951*b*; Mal'chevski 1959; Conrads 1969; Popov 1978; Lovaty 1991). In Sweden, only 2 of 102 nests above ground (one 1·2 m up in juniper *Juniperus*), 50% on agricultural land (Durango 1948). In Niederrhein region (north-west Germany), 33% of 217 nests in oats *Avena*, 26% barley *Hordeum*, 22% in mixture of both, 13% in other cereals, 5% potatoes (Mildenberger 1984); for discussion, see Mildenberger (1968), Conrads (1968, 1969), and Bülow (1990). In northern Bayern, 83% of 23 nests in summer cereal; crop height probably most important factor in choice of site (Lang *et al.* 1990); in eastern Germany, 65% of 34 nests in cereals, 12% in potatoes (Eifler and Blümel 1983). In St Petersburg region, only 1 of 36 nests in cereal field, remainder in thick grass on hillside or by ditches (Rymkevich 1977). Nest: foundation of stalks, stems, roots, and leaves lined with fine grasses, rootlets, and hair (Zumstein 1921; Durango 1948; Géroudet 1951*b*; Rymkevich 1977). Sometimes when flush with soil, cup has no real foundation, and rough material arranged wreath-like on ground; 6–7 nests in Westfalen had average outer diameter 11–12 cm (9–14), inner diameter 6·5 cm (6–7), depth of cup 3·4 cm (3–4) (Conrads 1969); overall height of one nest in Sweden 7·5 cm (Durango 1948); see also Popov (1978). Building: by ♀ only, generally accompanied by ♂; sometimes excavates depression in soil by turning and scratching with feet (Mildenberger 1968; Conrads 1969); material gathered from ground rarely more than 50 m from nest; mostly done in morning and takes 2–3(–5) days (Géroudet 1951*b*; Conrads 1969; Popov 1978). EGGS. See Plate 34. Sub-elliptical, smooth and faintly glossy; bluish, greyish, purplish, or pinkish, sparsely but evenly marked with brownish-black speckles, blotches, and scrawls, sometimes forming ring at broad end; greyish undermarkings (Harrison 1975; Makatsch 1976). 19·9 × 15·4 mm

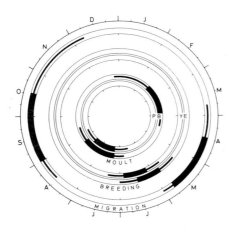

(17·2–22·1 × 14·1–17·1), $n = 637$; calculated weight 2·48 g (Schönwetter 1984). Clutch: 4–5 (3–6). In Sweden, of 89 clutches: 3 eggs, 13·5%; 4, 31·5%; 5, 46·1%; 6, 7·9%; 7, 1·1%; average 4·52; no significant change from south to north and no 2nd broods recorded (Durango 1948); see also Stolt (1993) for sizes of 167 broods. In Niederrhein, of 76 clutches: 3 eggs, 1·3%; 4, 26·3%; 5, 59·2%; 6, 13·2%; average 4·84; in May 5·0 ($n = 48$), June 4·65 ($n = 26$), July 3·5 ($n = 2$) (Mildenberger 1984); replacement clutches laid 9–11 days after loss ($n = 3$) (Mildenberger 1968). In Switzerland, replacement nest started 2 days after loss of young aged 3–4 days (Géroudet 1951b); in Finland, one pair probably had 3 replacements, but no record of genuine 2nd broods (Haartman 1969). Average size in Finland 4·93, $n = 27$ (Haartman 1969), St Petersburg region 4·32, $n = 28$ (Rymkevich 1977), Voronezh region 4·70, $n = 33$ (Mal'chevski 1959), Czechoslovakia 4·44, $n = 18$ (Hudec 1983), eastern Germany 4·31, $n = 16$ (Eifler and Blümel 1983), Catalonia 4·50, $n = 6$ (Muntaner *et al.* 1983). Only 1 brood recorded in Russia and Germany (Dementiev and Gladkov 1954; Eifler and Blümel 1983, Mildenberger 1984); for apparent 2nd broods in Germany, see Schuster (1926) and Garling (1943, 1949), but these accounts disputed by Conrads (1968, 1969). For possible 2nd broods in Switzerland, see Glutz von Blotzheim (1962). Eggs laid daily usually in early morning, 1–2 days ($n = 4$) after completion of nest (Conrads 1969). INCUBATION. 11–12(–13) days, by ♀ only; starts with last or penultimate egg (Géroudet 1951b; Mildenberger 1968; Conrads 1969). In Westfalen, average stint on eggs on day 6 at 2 nests was 28·2 min (15–45), average break 7·6 min (3–19) (Conrads 1969). YOUNG. Fed and cared for by both parents, ♀ frequently taking greater share (Durango 1948; Géroudet 1951b; Mildenberger 1968; Conrads 1969). For routine, see Zumstein (1921), Durango (1948), and Conrads (1969). Brooded by ♀ for *c.* 3–5 days (Mal'chevski 1959; Conrads 1969). FLEDGING TO MATURITY. Fledging period 12–13 days (9–14), usually leaving nest before able to fly (Durango 1948; Géroudet 1951b; Mildenberger 1968; Conrads 1969). Russian authors give lower figures: (7–)8–9 days in Voronezh region (Mal'chevski 1959), 9–10 (8–12) in St Petersburg region (Rymkevich 1977). Young still fed for 4–5 days and independent 8–12 days after leaving nest (Conrads 1969; Rymkevich 1977); see also Lang *et al.* (1990). For development of young, see Géroudet (1951b), Mal'chevski (1959), and Conrads (1969). Age of first breeding in ♂ 1 year (Conrads and Quelle 1986). BREEDING SUCCESS. In Voronezh region, of 155 eggs in 33 clutches, 65% hatched (32% predated), 51% produced fledged young (2·4 per nest overall); other nests destroyed by agricultural activity (Mal'chevski 1959). In Westfalen, of 19 breeding attempts, 74% successful, 21% lost brood, and 5% outcome unknown; where known, losses due to agricultural activity and human disturbance (Bülow 1990); predators have very little effect on success (Conrads 1977, which see for discussion). In

Niederrhein, 10–12% of eggs lost, mostly to heavy rain or agricultural work (Mildenberger 1968). Heavy downpours not uncommon in Germany at this time of year, sometimes killing all broods in study area (Conrads 1969; Helb 1974; Lang *et al.* 1990). In northern Bayern, nests destroyed by grass-mowing; in best year, 92% of 50 chicks fledged (Lang *et al.* 1990). Also in Sweden, many losses caused by rain; recorded predators were cats, dogs, and rats *Rattus* (Durango 1948); see Stolt (1993) for discussion of whether population decline is due to reduced breeding success. In Belgium, 75% of 36 pairs raised at least 1 fledged young; nests lost to rain and egg collectors (Maréchal 1988; Maes 1989). In St Petersburg region, losses caused by crows (Corvidae) and human interference (Rymkevich 1977). BH

Plumages. ADULT MALE BREEDING. Top and side of head and neck down to upper cheek and ear-coverts medium grey with distinct olive-green tinge, crown sometimes with faint dusky shaft-streaks, sometimes with slightly paler olive-grey supraloral stripe and short dusky stripe below eye; eye-ring yellow-white, markedly contrasting. Mantle and scapulars rather variable, rather dull olive-brown, sometimes with brighter rufous-cinnamon outer scapulars, or more uniform warm cinnamon-brown, marked with dull black shaft-streaks 1–2 mm wide, but streaks sometimes narrow and inconspicuous, especially on scapulars. Back to upper tail-coverts dull cinnamon-brown, feather-tips slightly greyer, feather-centres sometimes with dusky brown dot or streak. Lower cheek pale yellow, forming contrasting pale stripe, widening on lower side of neck, bordered below by olive malar stripe; chin and throat lemon-yellow or pale yellow, forming contrastingly pale bib. Chest and side of breast olive-grey, centre often with green-yellow tinge; remainder of underparts rufous-cinnamon, brighter and almost rusty-rufous at border of chest, paler pink-cinnamon or buff-cinnamon on flank, thigh, and under tail-coverts; feather-tips narrowly fringed pale buff when plumage fresh, partly concealing rufous. Tail dull black, slightly browner on central pair (t1); both webs of t1 and outer web of t2–t3(–t4) narrowly fringed cinnamon-brown, bleaching to grey-white when worn, tips of (t3–)t4–t6 narrowly fringed pink-buff to white, fringes soon wearing off; inner web of t5 with contrasting white wedge 20–30 mm long on tip, inner web of t6 with wedge of 25–40 mm long; inner web of t4 sometimes with short white wedge or elongated blotch; basal and middle portion of outer web of t6 white, shaft black, widening into dull black patch on tip. Flight-feathers, greater upper primary coverts, and bastard wing greyish-black, darkest on outer web and tip of primaries and on bastard wing; outer webs of primaries (except emarginated parts) and outer webs and tips of primary coverts narrowly and indistinctly fringed dull cinnamon or olive-brown, bleaching to a more contrasting off-white on outer primaries in abraded plumage; outer webs of secondaries fringed dull cinnamon, widest on middle portion of each feather. Tertials black, outer webs and tips broadly fringed rufous-cinnamon, rufous invading black at middle portion of outer web of longer tertials. Greater and median upper wing-coverts black, tips broadly fringed pink-cinnamon, bleaching to pale grey-buff in worn plumage, outer webs of greater coverts fringed duller cinnamon or grey-buff. Lesser upper wing-coverts dull black with olive-brown or-grey fringes. Under wing-coverts and axillaries grey, paler on longer feathers which have bases partly yellow-white, mixed pale yellow on marginal coverts. In some birds (of any age or sex), yellow pigments sometimes absent, head, neck, and chest then pure bluish-grey and eye-ring, stripe on lower cheek, and bib pure white; birds like these may occur in any part of range (examined from, e.g. western Europe and Turkey) and are rather similar to Grey-necked Bunting *E. buchanani*, differing mainly from that species by grey instead of rufous chest. ADULT FEMALE BREEDING. Like adult ♂ breeding, but grey of head, neck, and chest less clearly washed olive-green, more gradually merging into brown of mantle, crown with narrow but distinct dull black shaft-streaks, lore speckled cream, ear-coverts washed olive-brown; ground-colour of remainder of upperparts slightly duller brown, less rufous on mantle or outer scapulars; malar stripe and upper chest often marked with dull black spots or short streaks; underparts down from breast paler pink-cinnamon or buff-cinnamon, less deep rufous at border of chest. ADULT MALE NON-BREEDING. Like adult ♂ breeding, but plumage entirely freshly moulted (except frequently for some secondaries, which differ mainly then in paler colour of outer fringe); top and side of head and neck as well as chest clear olive-green, less grey, feathers on cap, nape, and ear-coverts often with dusky tips or shaft-streaks, malar stripe and upper chest or entire chest often marked with dull black triangular dots, chest sometimes appearing streaked black and green-yellow; stripe on lower cheek as well as bib on throat bright pale yellow, sometimes less sharply defined from malar stripe and chest than in breeding plumage; body as in adult ♂ breeding, ground-colour of upperparts rather variable, dull cinnamon-brown to olive-brown with distinct grey cast, mantle and scapulars distinctly streaked black; underparts below chest as adult ♂ breeding, belly deep rufous-cinnamon, partly concealed by narrow buff-yellow feather-tips. ADULT FEMALE NON-BREEDING. Like adult ♀ breeding, but plumage fresh (as adult ♂ non-breeding), top and side of head as well as chest more olive-green; malar stripe and chest spotted and streaked black; stripe on lower cheek and bib often buff-yellow, less pale yellow. Rather like adult ♂ non-breeding, but top and side of head as well as malar stripe and chest more heavily spotted and streaked black and underparts below chest paler cinnamon-buff, belly less deep rufous. NESTLING. Down fairly long and plentiful, pearl-grey with slight buff or brown tinge; confined to top of head, shoulder and back, with traces on belly (Heinroth and Heinroth 1924-6; Witherby *et al.* 1938; Durango 1948). JUVENILE. Entire upperparts closely streaked black-brown and buff, black streaks wider on cap, mantle, and scapulars, narrower and with buff more predominant on nape and rump. Side of head and neck mottled brown and buff, darkest on ear-coverts, which show darker surround, purer buff on lore, supercilium, upper side of neck, eye-ring, and in stripe over lower cheek; underparts cinnamon-buff (when fresh) or pale buff (when worn), paler on chin, mid-belly, and on woolly vent and under tail-coverts, heavily marked with short black streaks or spots on malar stripe, chest, side of breast, and flank. Tail as in fresh adult ♂ non-breeding, but tips of feathers ending in blunt point, less rounded; ground-colour more sepia, less black; no white on tip of inner web of t4 (absent in some adults, too), white wedges on t5 and t6 sometimes shorter and less sharply defined. Wing as adult, but tips of greater and median coverts buff, bleached to white by September, when fringes along tips of adult pink-cinnamon or pink-buff; black of centres of coverts sharply contrasting with tip (more so than adult), gradually tapering to sharp point towards feather-tip (less bluntly pointed than in adult); lesser coverts dark grey-brown, fringed buff; under wing-coverts and axillaries grey, virtually without yellow; all secondaries equally new. FIRST ADULT MALE NON-BREEDING. Rather like juvenile, but ground-colour less buff, less contrasting. Entire upperparts including cap dull cinnamon-brown or olive-brown, heavily marked with black streaks 1-2 mm wide, reduced to narrow dusky shaft-streaks on rump and upper tail-coverts. Lore mottled buff and brown, rather ill-defined supercilium and bar along upper side of neck grey-olive or grey-buff with fine dusky specks, rather indistinct, bordering cinnamon-brown, buff-brown, or olive-brown ear-coverts. Eye-ring prominent, buff-white. Stripe over lower cheek warm buff, pale buff, or buff-yellow, bordered by distinct black malar stripe below. Chin and throat pale yellow or yellow-buff, partly marked with fine dusky specks; remainder of underparts pale buff-cinnamon, often mixed with some olive-grey on chest and deep rufous-cinnamon on belly; chest and side of breast marked with short dull black streaks, forming gorget, flank and much of belly marked with narrow dusky streaks. Tail, flight-feathers, primary coverts, bastard wing, many or all greater coverts, and sometimes tertials still juvenile (see Juvenile for characters), tips of greater coverts and fringes of tertials bleached to cream or white, contrasting in colour and pattern with tips of new median coverts (colour and pattern uniform in adult non-breeding). Often closely similar to adult ♀ non-breeding, but tail and much of wing still juvenile. FIRST ADULT FEMALE NON-BREEDING. Like 1st adult ♀ non-breeding, but ground-colour of stripe on lower cheek and of entire underparts warm buff or pale buff, black malar stripe narrower but more contrasting, black streaks on chest, flank, and belly more contrasting, throat usually without yellow tinge, chest without grey, belly without deep rufous. Part of juvenile feathering retained, as in 1st adult ♂ non-breeding. FIRST ADULT MALE BREEDING. Like adult ♂ breeding, and often indistinguishable. Head, neck, and chest on average greyer than in adult, less tinged green, throat and stripe on lower cheek paler yellow, less bright lemon-yellow. Primary coverts and flight-feathers still juvenile, on average more abraded on tips than in adult; some or all tail-feathers and outer greater coverts sometimes still juvenile, but sometimes replaced in winter quarters and then sometimes contrastingly newer than 1st adult feathers replaced at an earlier stage of moult. FIRST ADULT FEMALE BREEDING. Like adult ♀ breeding, but cap on average more heavily streaked, less tinged green, malar stripe and chest more copiously spotted and streaked, throat pale buff-yellow, less pure yellow, ground-colour of chest light buffish-grey, less green-grey; primary coverts and flight-feathers still juvenile, tips on average more pointed and more worn at same time of year than in adult; all or much of tail and occasionally greater coverts still juvenile, characters as in Juvenile. Head and body rather similar to adult ♀ non-breeding and 1st adult ♂ non-breeding.

Bare parts. ADULT, FIRST ADULT. Iris dark brown or black-brown. Bill pink-flesh, reddish-flesh, or brownish-flesh, leg and foot pink-flesh or pale flesh-horn, sometimes with slight grey tinge. (Hartert 1903-10; Heinroth and Heinroth 1924-6; Witherby *et al.* 1938; RMNH, ZMA.) NESTLING. Bare skin, leg, and bill pale pink-yflesh or yellowish-flesh; mouth pink- or flesh-red; gape-flanges pale cream-yellow (Heinroth and Heinroth 1924-6; Witherby *et al.* 1938; Durango 1948). JUVENILE. Iris dark brown. Bill flesh-brown or grey-horn. Leg and foot brownish pink-flesh. (Heinroth and Heinroth 1924-6.)

Moults. Based mainly on Stresemann and Stresemann (1969*b*). ADULT POST-BREEDING. Almost complete; primaries descendent. In breeding area in late summer; duration 4-6 weeks. Starts between early and late July, completed late July to early September, before start of autumn migration (most often s4-s5), these apparently replaced in winter quarters. (Stresemann; 1920, 1928*b*; Diesselhorst 1962; Stresemann and Stresemann 1969; Piechocki and Bolod 1972; Brokhovich 1990.) ADULT PRE-

BREEDING. Partial: head and body. Starts with underparts, followed by mantle and back; head and neck last. In winter quarters, late December to late March; many in heavy moult March, usually completed late March. POST-JUVENILE. Partial: head, body, lesser and median upper wing-coverts, and (sometimes) tertials or innermost greater coverts. Starts at age of 20–30 days; depending on hatching date, starts mid-July to early September, generally completed on start of autumn migration. FIRST PRE-BREEDING. Partial, as in adult pre-breeding, but in part a continuation of post-juvenile, including variable number of greater coverts and tail-feathers; starts between early November and early March, completed late January to mid-April. (Wallgren 1954; Stresemann and Stresemann 1969; RMNH, ZMA.)

Measurements. Whole geographical range, April–September; skins (RMNH, ZMA). Bill (S) to skull, bill (N) to distal corner of nostril; exposed culmen on average 3·1 less than bill (S).

	♂		♀	
WING	87·5 (2·81; 53)	83–96	83·9 (3·72; 28)	77–92
TAIL	63·1 (2·88; 53)	56–71	61·5 (2·51; 27)	57–69
BILL (S)	14·0 (0·96; 55)	11·7–15·3	14·2 (0·87; 26)	12·7–15·5
BILL (N)	8·1 (0·43; 55)	7·1–9·0	8·1 (0·47; 26)	7·0–9·1
TARSUS	18·5 (0·75; 50)	17·0–20·3	18·6 (0·86; 24)	16·8–19·7

Sex differences significant for wing.

Wing. (1) France, Netherlands, and Germany (ZMA). (2) Camargue (France), live birds, spring (Isenmann 1992). (3) Greece, Turkey, Armeniya, and Iran (Stresemann 1928b; Paludan 1938, 1940; Niethammer 1943; Makatsch 1950; Schüz 1959; Nicht 1961; Kumerloeve 1963, 1964a, 1970b; Rokitansky and Schifter 1971; Vauk 1973; RMNH, ZMA). (4) Southern Yugoslavia (Stresemann 1920).

	♂		♀	
(1)	87·1 (2·10; 8)	83–89	84·2 (2·71; 6)	81–87
(2)	90·2 (3·27; 19)	83–96	87·5 (3·41; 15)	82–93
(3)	88·0 (3·34; 29)	83–96	83·3 (1·93; 8)	81–87
(4)	89·8 (1·96; 40)	86–96	85·1 (1·55; 8)	82–87

Camargue, sexes combined, live autumn and spring migrants: 88·1 (3·13; 103) 81–97 (Isenmann 1992). Southern Spain: ♂♂ 86, 91 (Niethammer 1957a). Kazakhstan: ♂ 84–91 (42), ♀ 77–88 (23) (Korelov *et al.* 1974). Mongolia: ♂ 91·0 (3) 90–93 (Piechocki and Bolod 1972). See also Mauersberger (1982).

Weights. ADULT, FIRST ADULT. Belgium: (1) August and early September, (2) April and early May (Spaepen 1952). (3) Camargue (France), spring migrants (Isenmann 1992). Whole geographical range: (4) April, (5) May, (6) June, (7) August–September (Krohn 1915; Paludan 1938, 1940; Niethammer 1943; Makatsch 1950; Schüz 1959; Nicht 1961; Kumerloeve 1963, 1964a, 1969b, 1970a, b; Rokitansky and Schifter 1971; Piechocki and Bolod 1972; Vauk 1973). (8) Morocco, spring migrants (Ash 1969). (9) Kazakhstan, summer (Korelov *et al.* 1974).

	♂		♀	
(1)	22·9 (— ; 19) 20·8–27·8		22·7 (— ; 10) 19·9–26·7	
(2)	22·1 (— ; 17) 20·3–24·6		19·1 (— ; 1)	
(3)	24·4 (2·26; 19) 18·8–28·1		23·4 (2·13; 15) 19·0–27·0	
(4)	23·8 (3·13; 5) 19·3–27·0		22·1 (— ; 2) 21·3–23·0	
(5)	23·2 (3·83; 11) 16·0–30·0		27·5 (3·11; 4) 24·0–31·0	
(6)	21·4 (3·13; 6) 17·0–24·7		26·1 (— ; 2) 25·0–27·2	
(7)	24·0 (4·53; 5) 16·0–27·0		21·7 (— ; 2) 19·9–23·5	
(8)	20·1 (— ; 8) 17·3–26·3		18·4 (— ; 3) 17·0–20·5	
(9)	22·8 (— ; 29) 19·6–30·0		21·9 (— ; 6) 20–24·8	

Unsexed birds or sexes combined. (10) Belgium, August and early September (Spaepen 1952). Camargue: (11) late August and September, (12) April and early May (Isenmann 1992). (13) Helgoland (Weigold 1926). (14) Germany, May–June (♂ only) (Eck 1985b). (15) Central Nigeria, March–April (Smith 1966b).

(10)	23·2 (— ; 50) 19·9–27·0	(13)	24·6 (— ; 26) 20–30
(11)	23·8 (2·20; 44) 19·7–30·5	(14)	22·8 (1·20; 9) 21·5–25·0
(12)	23·6 (2·13; 59) 18·8–28·1	(15)	30·9 (5·03; 5) 22·0–36·0

Structure. Wing rather long, broad at base, tip bluntly pointed. 10 primaries: p8 longest, p9 (0–)1–4 shorter, p7 0–2 shorter, p6 2–6, p5 8–12, p4 14–18, p3 16–21, p2 20–25, p1 20–28; p10 strongly reduced, a tiny pin, concealed below reduced outermost upper primary covert, 55–63 shorter than p8, 8–12 shorter than longest upper primary covert. Outer web of (p6–)p7–p8 emarginated, inner web of p9 with slight notch (p7–p8 sometimes faintly also). Tip of longest tertial reaches tip of p5–p6 in fresh plumage. Tail rather long, tip slightly forked; 12 feathers, t4–t5 longest, t1 2–6 shorter, t6 0–3. Bill as in Yellowhammer *E. citrinella*, but tip slightly more attenuated; depth at base 6·2 (10) 5·8–6·7, width at base 7·1 (10) 6·6–7·5. Tarsus and toes rather short and stout. Middle toe with claw 18·1 (10) 16–20; outer toe with claw *c.* 69% of middle with claw, inner *c.* 71%, hind *c.* 77%. Remainder of structure as in *E. citrinella*.

Geographical variation. Slight, if any; mainly in colour. Birds from south of range slightly paler than elsewhere, probably due mainly to more intense bleaching. Variation mainly individual or due to age, not geographical, and thus no races recognized (Vaurie 1956e, 1959), e.g. *antiquorum* (southern Europe), *shah* (Iran), and *elisabethae* (Mongolia) not valid. Birds from northern Turkey and Armeniya perhaps inclined to have deeper rufous scapulars and underparts, purer blue-grey head, neck, and chest, and whitish throat (Nicht 1961; ZMA), but similar birds occur occasionally elsewhere. For variation, see also Johansen (1944), Niethammer (1957a), and Eck (1985b).

For relationships, see Cretzschmar's Bunting *E. caesia*.

Recognition. See Recognition of *E. buchanani* and *E. caesia*.

MP, CSR

Emberiza buchanani Grey-necked Bunting

PLATES 17 (flight) and 20
[between pages 256 and 257]

Du. Steenortolaan   Fr. Bruant à cou gris   Ge. Steinortolan
Ru. Скальная овсянка   Sp. Escribano cabecigrís   Sw. Bergortolan

Emberiza Buchanani Blyth, 1844

Polytypic. *E. b. cerrutii* De Filippi, 1863, south-east Turkey, southern Transcaucasia, Iran (east to Kuh-e-Taftan), and southern Turkmeniya in Bol'shoy Balkhan and Kopet-Dag; *obscura* Zarudny and Korejev, 1903 (or *neobscura* Paynter, 1970, if name *obscura* already occupied), hills of northern Kazakhstan at 47–51°N, from Mugodzhary plateau and Ulutau east to western Altai, from Dzhungarskiy Alatau, Tarbagatay, and southern Altai through Dzhungaria to Goviʾaltay and eastern Hangay in Mongolia, and in Tien Shan west to Fergana area, Tashkent, and Talasskiy Alatau; also isolated in hills of southern Kazakhstan (Chu-Ili mountains, Karatau) and (perhaps this race) Nuratau and hills of Kyzyl-Kum desert; vagrant Orenburg (south-east European Russia). Extralimital: nominate *buchanani* Blyth, 1844 (synonym: *huttoni* Blyth, 1849), Tadzhikistan from Zeravshan and Alay mountains south through southern Uzbekistan, Afghanistan, and western Pamirs to Quetta area (western Pakistan).

Field characters. 15–16 cm; wing-span 24–27 cm. Slightly smaller than Ortolan Bunting *E. hortulana*, with somewhat shorter wings but proportionately longer tail. Counterpart of Cretzschmar's Bunting *E. caesia* but with slightly more pointed bill and rather shorter tail and characteristically faded, softly streaked plumage. Adult lacks grey breast-band of *E. caesia*, having wholly vinous underbody. At all ages, pale brownish fringes to flight-feathers add to washed-out appearance. Sexes rather similar; little seasonal variation. Juvenile separable.

ADULT MALE. Moults: June–September (complete). Head, malar stripe, and surround to shoulder ashy-grey, marked by narrow, bright cream eye-ring and pale greyish-cream to white submoustachial stripe and throat. Note lack of pale lores. In good light, contrasts of head pattern are as striking as in *E. hortulana* and much stronger than in *E. caesia*, but other plumage noticeably paler, more washed-out or faded than either close ally. Mantle dusky brownish-grey, rather narrowly and usually inconspicuously streaked dull grey-brown and showing rusty tone only on scapulars which show as warm band above wing. Back and rump as mantle, lacking visible streaks though these reappear on upper tail-coverts. Wings generally as dull as back, but at close range show (a) grey lesser coverts, (b) dark brown bases and whitish-buff tips to median coverts forming indistinct upper wing-bar, and (c) broad dull cream to faintly rufous-brown fringes to greater coverts and tertials which both lack more than dusky-brown centres and are far less striking than on close allies. Flight-feathers black but look dull brown when folded, with fringes coloured as those of greater coverts. Tail brown-black, with brownish fringes to central feathers and bright white edges and corners. Underparts from breast to belly dull vinous-buff, darkest from breast to rear flanks but copiously mottled by whitish to buff-grey tips and fringes, particularly when fresh; vent and under tail-coverts pale buff or almost white. Underwing white. Bill and legs pale pink-straw to reddish-brown; colourful compared with plumage. ADULT FEMALE. Differs from ♂ less than in close allies, but in comparison ♀ slightly paler and even duller than ♂; crown slightly streaked, short supra-loral stripe mottled cream, and underparts more cream-buff than vinous-buff. JUVENILE. Differs from adult as in close allies, but already shows rather dull, softly streaked plumage pattern. Best told from *E. caesia* by dull brownish (not rufous) rump, pale, at most buff (not rufous) vent, less sharp striations on mantle and breast, and pale, uniform flight-feathers. Distinguished from *E. hortulana*

by last 2 characters and lack of yellow on throat. FIRST-YEAR. See Plumages.

Adult instantly distinguished from *E. hortulana* and *E. caesia* by lack of dark breast-band. Juvenile more difficult to separate but has typical faded appearance of species; easily distinguished from *E. caesia* but not from paler *E. hortulana* (see above). Flight, behaviour, and action as *E. caesia*. Approachable. Sociable, forming flocks on migration and in winter. Strictly montane in breeding season; always markedly terrestrial.

Song a quite long, loud, rich 'trill', ascending towards higher-pitched and emphasized penultimate note, 'dze dze dze dzee-oo' or 'di-di-dew de-dew'; recalls *E. hortulana* more than *E. caesia*. Calls include 'chep', 'tcheup', and 'choup' from perched bird and 'tip', 'tsip', 'sip', 'sik', and 'tsik-tsik' in flight.

Habitat. A counterpart in south-east of west Palearctic to Cinereous Bunting *E. cineracea* further west. A more montane temperate to warm temperate species, occurring on dry rocky slopes of foothills and mountains, and on screes and rocky outcrops, as well as in ravines, favouring arid and barren terrain (Harrison 1982). In former USSR, avoids tree-grown slopes, in some regions favouring barren low mountains in intermediate mountain zone at *c.* 900–1400 m but elsewhere readily breeds higher, even up to 2500 m. Essential requirement seems to be patchy cover of grass and other xerophytic vegetation, including scattering of bushes, or stony desert. Steep slopes are not avoided, but open rocky ground is also chosen. Numerous in Afghanistan, especially among tussock grass on slopes at *c.* 3000 m, and elsewhere to 3200 m (Paludan 1959).

Ecological distinctions from Rock Bunting *E. cia* remain obscure; Ortolan Bunting *E. hortulana* is mainly a counterpart at lower elevations among taller and denser shrub growth, but not uncommonly overlaps at higher altitudes. Explanation of relations may lie rather in past evolutionary history and responses to glacial changes than in present-day ecology.

Wintering birds in India live on stony ground with sparse shrubs and on *Euphorbia*-covered broken hillsides: sometimes also on stubble (Ali and Ripley 1974).

Distribution and population. No changes reported.

Movements. Migratory, wintering mainly in India. West Palearctic populations mainly move south-east to wintering grounds.

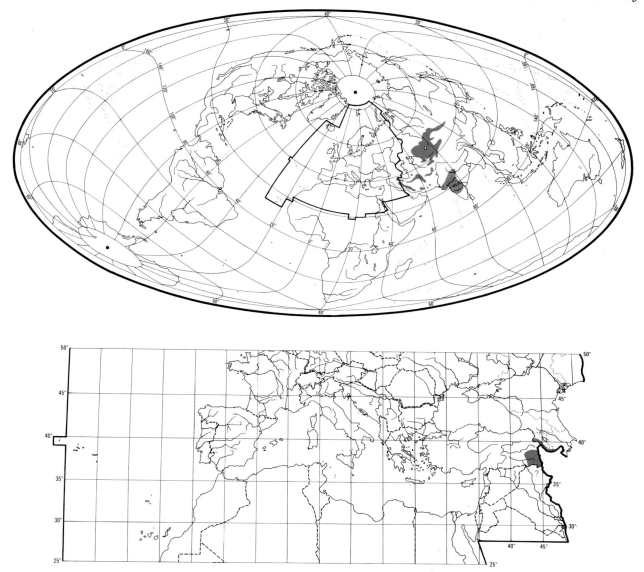

Main departure from breeding grounds in Iran before mid-September (D A Scott); latest record from Afghanistan 24 September (Hüe and Etchécopar 1970). Autumn migration in Turkmeniya mid-August to September (Dementiev and Gladkov 1954; Rustamov 1958); in Tadzhikistan, chiefly September, continuing to early October (Abdusalyamov 1977); begins late August in Pamir-Alay (Tadzhikistan) (Ivanov 1969) and birds become numerous on passage there by mid-September (Dementiev and Gladkov 1954). In Kazakhstan, migration apparently begins in 1st half of August. In foothills of Dzhungarskiy Alatau, flocks noted 29 August to 12 September in 1940; in 1941, migration weaker and later, with small flocks noted 30 September. In Talasskiy Alatau, noticeable but not intensive migration in some years but not in others, via both

hills and foothills, mid-August to mid-September; in 1933 and 1935 seen in 1st half of August with sharp drop in numbers by end of August; in 1959–65 not encountered at all in foothills and last observation in mountains was 26 August. (Korelov *et al.* 1974.) At Chokpak pass (western Tien Shan, Kazakhstan), regular migrant in very small numbers, mostly 21 August to 15 September (Gavrilov and Gistsov 1985). Leaves Baluchistan (Pakistan) in September, when strong passage implies more than local and Afghan populations involved; occurs on passage in Pakistan during September and abundant on passage through Kutch (India) in same month (Ali and Ripley 1974; Roberts 1992).

Winters in western and central India from northern Gujarat east and south to central Uttar Pradesh, eastern

Maharashtra, northern Andhra Pradesh and central Karnataka; also sparsely in Sind (southern Pakistan), where perhaps regular only in years with good monsoon rain; recorded in wintering range 8 October to 29 April (Ali and Ripley 1974; Roberts 1992). All races are presumed to winter in same region, in view of strong passage through Baluchistan; races apparently not identifiable in winter (Ali and Ripley 1974). No winter records of the species elsewhere.

Spring passage through Kutch (abundant March) and Pakistan (mainly April, one as late as 25 May); arrives on breeding grounds in Baluchistan in March (Ali and Ripley 1974; Roberts 1992). Arrives Afghanistan end of April (Hüe and Etchécopar 1970). In Iran, arrives southern Baluchestan early March and on breeding grounds in north by last week of April; recorded on passage in southeast Iran (D A Scott). Arrives eastern Turkey end of April (Jonsson 1992). Appears Turkmeniya 2nd half of April (Rustamov 1958). Migration mid-April in Tadzhikistan; at Badakhshan most seen 23-29 April (Abdusalyamov 1977). Migration in Kazakhstan April-May; in western Tien Shan appears in 2nd half of April (Korelov et al. 1974). At Chokpak pass, regular migrant in very small numbers, mostly 1st half of May (Gavrilov and Gistsov 1985). See Dementiev and Gladkov (1954) for additional dates of spring migration and arrival in former USSR.

Populations breeding as far east as western Mongolia and Sinkiang (China) are tentatively assumed to share same winter range as other forms (Ali and Ripley 1974; Schauensee 1984). These may migrate around western end of Himalayas, since no passage or winter records from central Himalayas; however, confirmation of winter range needed because races apparently not identifiable in winter (Ali and Ripley 1974).

Vagrants away from normal range in USSR recorded Derbent, Orenburg region, and perhaps Crimea (Dementiev and Gladkov 1954). Vagrant also reported Kuwait (F E Warr). DTH

Food. All information extralimital. Diet mainly seeds and other parts of plants, plus invertebrates in breeding season. Feeds on ground on rocky slopes with scattered scrubby vegetation, foraging slowly and methodically, often remaining motionless for long periods. (Salikhbaev and Bogdanov 1967; Armani 1985; Roberts 1992.) In winter quarters or on passage, also in cultivated areas such as fields of *Sorghum*, lucerne *Medicago*, stubble, pasture, or weedy places (Ticehurst 1922; Mauersberger 1960; Piechocki et al. 1982; Roberts 1992). In April, in Tadzhikistan, feeds in morning until 10.00-11.00 hrs then flies to water, feeding again in evening (Abdusalyamov 1977).

Diet includes the following. Invertebrates: grasshoppers (Orthoptera: Acrididae), bugs (Hemiptera: Reduviidae, Cicadidae), ants (Hymenoptera: Formicidae), beetles (Coleoptera: Buprestidae, Curculionidae), snails (Pulmonata). Plants: seeds, buds, shoots (etc.) of bistort

Polygonum, Ranunculaceae, spurge *Euphorbia*, *Halimodendron*, buckthorn *Rhamnus*, Compositae, grasses (Gramineae, including millet *Panicum*, wild barley *Hordeum*, *Bromus*, *Stipa*). (Pek and Fedyanina 1961; Kovshar' 1966; Salikhbaev and Bogdanov 1967; Korelov et al. 1974; Abdusalyamov 1977; Piechocki et al. 1982; Roberts 1992.)

In Kazakhstan, July-August, 13 stomachs contained mainly seeds (buckthorn, spurge, grasses); invertebrates included grasshoppers and cicadas; 4 stomachs in June contained mainly insects (grasshoppers, cicada nymphs, beetles); plant material included bulbils of *Polygonum viviparum* (Kovshar' 1966; Korelov et al. 1974). In Uzbekistan, 1st half of July, 19 stomachs contained only insects; grasshoppers in 17, beetles in 8 (Salikhbaev and Bogdanov 1967). In Kirgiziya, 10 summer stomachs held mainly invertebrates (Reduviidae, Curculionidae, ants, snails), plus seeds and green parts of Ranunculaceae and other plants (Pek and Fedyanina 1961). On passage in Mongolia, 2nd half of May, groups of up to 25 picked up previous year's *Halimodendron* seeds below shrubs (Piechocki et al. 1982).

Young are fed insects, adults bringing several at each visit (Abdusalyamov 1977). BH

Social pattern and behaviour. Little known; no detailed studies.

1. Reported to occur in pairs and flocks in former USSR, migrating singly and in groups; flocks form from beginning of July in Transcaspia, in August further east, and flocking more evident on autumn migration than in spring (Dementiev and Gladkov 1954; Korelov et al. 1974; Flint et al. 1984). In Kazakhstan, Uzbekistan, and Tadzhikistan, occurs singly or in small flocks (3-6) on migration (Salikhbaev and Bogdanov 1967; Korelov et al. 1974; Abdusalyamov 1977). In Kazakhstan, flocks of up to 20 seen in Karzhantau mountains on 30 April; small flocks (of juveniles only) noted on 10 July, but flocking not seen in 1st half of August in Talasskiy Alatau (Korelov et al. 1974). Flock of 6 or more seen in Iran, April (P A D Hollom). In India, gregarious in winter, when usually in small scattered flocks of 8-20, sometimes in larger flocks; apparently seen singly or in pairs during spring migration (Ali and Ripley 1974). Largest flock reported of over 50 in Sind (Pakistan) (Roberts 1992). Migrants in Uzbekistan sometimes occur in flocks mixed with other species (Salikhbaev and Bogdanov 1967); in Kazakhstan, sometimes flocks with Ortolan Bunting *E. hortulana* (Korelov et al. 1974) and migrants in Tadzhikistan flock with Pine Bunting *E. leucocephala* and Scarlet Rosefinch *Carpodacus erythrinus* (Abdusalyamov 1977). In India in winter, sometimes in company with other *Emberiza* (Ali and Ripley 1974); in Chitral (Pakistan) recorded in flock with Rock Bunting *E. cia* and White-capped Bunting *E. stewarti* (Roberts 1992). BONDS. No details of type of pair-bond, its duration, or age at 1st breeding. Nest-building and incubation mainly or entirely by ♀ (see Heterosexual Behaviour, below); young fed by both parents (see Relations within Family Group). BREEDING DISPERSION. In USSR, usually nests as separate pairs, but in some places (e.g. Transcaspia) may form neighbourhood groups (Dementiev and Gladkov 1954; Rustamov 1958). In Uzbekistan, nests as solitary pairs, but small flocks and family parties occur on feeding grounds and at watering places (Salikhbaev and Bogdanov 1967). In Tadzhikistan,

breeding pairs 500–600 m apart in Mogoltau mountains, 1000–2500 m apart in Aktau mountains (Abdusalyamov 1977). Near Quetta (Pakistan) 12–16 pairs breeding within *c.* 64 ha, but usually more widely separated (Roberts 1992). ROOSTING. Under plants, shrubs or rocks (Abdusalyamov 1977).

2. Very little information. Behaviour in general typical of so-called 'long-winged' *Emberiza* buntings (E N Panov); behaviour and general habits very similar to *E. hortulana* (Korelov *et al.* 1974). In Pakistan, not very shy, but extremely difficult to detect once settled on ground (Roberts 1992). FLOCK BEHAVIOUR. No information. SONG-DISPLAY. ♂♂ in Elburz (Iran) sing (see 1 in Voice) from stones, clods of earth, and low stems such as those of Umbelliferae, in habitat with no bushes (Martens 1979). In Kazakhstan, ♂ sings from cliff-tops or rocks, not from bushes (Korelov *et al.* 1974), but from rocks or bushes in Tadzhikistan (Abdusalyamov 1977) and reported to sing from upper branches of bushes by Dementiev and Gladkov (1954) and Hüe and Etchécopar (1970). In Mongolia, ♂ sang from top of cliff (Kozlova 1930). In Pakistan, ♂ sings from quite low *Artemisia* bushes or stones or even on bare open ground, not seeking prominent perches (Roberts 1992). In USSR often sings on spring migration (Dementiev and Gladkov 1954; Korelov *et al.* 1974). In Kazakhstan, ♂♂ continued singing at one site up to mid-July (Korelov *et al.* 1974); latest date for song 23 July (Kovshar' 1966). In Uzbekistan, song noted early May (Salikhbaev and Bogdanov 1967); in Tadzhikistan, from end of April (Abdusalyamov 1977). In Mongolia, song first noted 16 May, ceased mid-July (Kozlova 1930); in Mongolian Altai, singing 11–14 July (Potapov 1986). ♂ sang vigorously at dawn and dusk, with head held high (Kozlova 1930); song commences as soon as day starts to warm up (Abdusalyamov 1977, which see for other information on diurnal activity rhythm at beginning of breeding season). ANTAGONISTIC BEHAVIOUR. No information. HETEROSEXUAL BEHAVIOUR. After arrival on breeding grounds in Kazakhstan, birds pair up more or less straight away and start nest-building (Korelov *et al.* 1974). Peak courtship period end of May in Uzbekistan (Salikhbaev and Bogdanov 1967). Nest-hollow said to be sometimes excavated by bird (Dementiev and Gladkov 1954; Abdusalyamov 1977), but confirmation needed. Nest-building reported to be by both sexes, with ♀ doing most (Dementiev and Gladkov 1954; Abdusalyamov 1977), but confirmation needed that ♂ takes real part in building rather than giving 'stem-display' as in (e.g.) *E. hortulana*. Incubation by ♀ (Harrison 1975; Abdusalyamov 1977), but ♂ stays close by and sings (Abdusalyamov 1977). Kovshar' (1966) reported most ♂♂ do not incubate, but that 1 of 3 shot had brood-patch. RELATIONS WITHIN FAMILY GROUP. Little information. Young fed by both sexes (Kovshar' 1966; Abdusalyamov 1977), with up to 10 small insects brought on each visit (Kovshar' 1966). Family remains together for *c.* 7–8 days after fledging, then breaks up and young independent (Abdusalyamov 1977). Fledged young in family parties fed by parents (Salikhbaev and Bogdanov 1967). ANTIPREDATOR RESPONSES OF YOUNG. No information. PARENTAL ANTI-PREDATOR STRATEGIES. ♀ sits very tightly, allowing man to come right up to nest (Korelov *et al.* 1974); ♀ once touched on nest (Kovshar' 1966). DTH

Voice. Little known. Studies of song in Elburz (Iran) by Martens (1979) and in Mongolia by Wallschläger (1983), which see for additional sonagrams. Information on calls too scanty to allow more than listing of similar sounds together.

CALLS OF ADULTS. (1) Song of ♂. Resembles song of

Ortolan Bunting *E. hortulana*, but varies within (Wallschläger 1983) and between individual ♂♂ (Martens 1979; Wallschläger 1983), resulting in varied descriptions in literature. Figs I–III show different song-types from 3

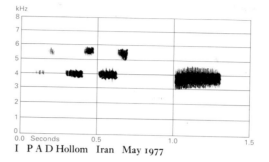

I P A D Hollom Iran May 1977

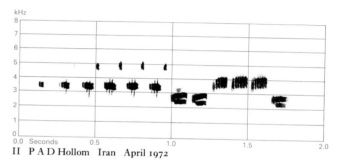

II P A D Hollom Iran April 1972

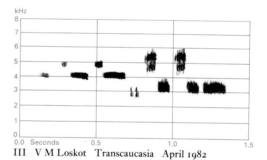

III V M Loskot Transcaucasia April 1982

different ♂♂. Rendered 'dze dze dze dzee-oo', ascending to last unit which then descends (Heinzel *et al.* 1972); quite loud and rich, typically 'di-di-dew de-dew'; penultimate unit higher pitched and more emphasized; delivery can be fast or slurred and more like *E. hortulana* than Cretzschmar's Bunting *E. caesia* (Hollom *et al.* 1988); song in Turkey weak, somewhat suggesting *E. hortulana* or short version of Yellowhammer *E. citrinella*, typically with higher-pitched and more emphasized unit followed by 2 lower-piched units, 'sresre-sreSRIH sre-sre' (L Svensson); recording from Mongolia (Schubert 1982) is similar (see Mauersberger *et al.* 1982 for other descriptions from Mongolia). Wallschläger (1983) gave additional descriptions and sonagrams; 2 ♂♂ studied each gave 2 different song-types, repeated in apparently rather irregular alternation. In Pakistan, song rather brief and stereotyped with little variation between birds; lasts just over 1 s and typically repeated at rate of 7–8 songs per min for 2–3 min; rather metallic in timbre, but quite a pleasant little jingle, rendered 'tswee-tswee-tswee-tsweee' (rapidly repeated and on same high metallic note), 'dzwe-e e-h-dul' (more grating and drawn-out), or 'tsi-tsi-tsi-tsi dzu-u-dzwid-dul' (2nd sub-phrase rasping or buzzing and lower pitched); easily recognized as distinct from song of *E. hortulana* (Roberts 1992). Also described as apparently identical to song of *E. hortulana*, a pleasant melancholy ditty of *c.* 5 units, first 4 at same pitch, last lower (Ali and Ripley 1974). Study of songs of 7 ♂♂ in Elburz (Martens 1979) revealed 2 main song-types with additional minor variants; only 1 type reported from each ♂. Total duration of song 1·1–1·5 s; overall frequency range mainly 3–5 kHz. All songs consisted of 2 sub-phrases: 1st with 2–7 units (either all similar or of 2 types given alternately), 2nd with 2–3 pairs of alternating units (1st in each pair high-pitched, 2nd low-pitched). (Martens 1979.) (2) Flight-call. A 'tsip'. Other renderings: 'sip' (Heinzel *et al.* 1972); 'tip', 'tsip', or 'sik' (Hollom *et al.* 1988); soft 'tsik-tsik' (Flint *et al.* 1984); rather soft, sibilant, disyllabic 'trip-trip' or 'tsik-tsik' given in flight in winter in Pakistan (Roberts 1992). Recordings suggest short sharp 'tsi' followed by 'chi' (Fig IV). 'Squeaking alarm note' (Meiklejohn 1948) might be

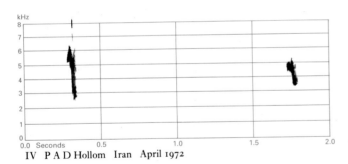

kHz

IV P A D Hollom Iran April 1972

this or another call. (3) Tcheup-call. Rendered 'choup' (in flight: Heinzel *et al.* 1972), 'chep', or 'tcheup' (perched: Hollom *et al.* 1988). Calls 2–3 often given alternately, perched or in flight, as in other members of *E. hortulana* species-group.

CALLS OF YOUNG. No information.   DTH

Breeding. SEASON. Almost all information extralimital. Armeniya: newly-fledged young and family parties seen 6–10 June (Lyaister and Sosnin 1942). Iran: eggs recorded from end of April to July (Blanford 1876; Grote 1934a; Paludan 1938; Hüe and Etchécopar 1970). Southern Turkmeniya: in Kopet-Dag mountains, newly-fledged young recorded end of May (Rustamov 1958). Kazakhstan: eggs recorded late May to early July, fledglings from late June; peak period for completed clutches about end

of June (Dementiev and Gladkov 1954; Kovshar' 1966; Korelov *et al.* 1974). Afghanistan: fresh clutch found late May at 2700 m (Whistler 1945); see also Paludan (1959). SITE. On ground, usually well concealed in shelter of rock, shrub, or overhanging grass on stony slope (Dementiev and Gladkov 1954; Ivanov 1969; Hüe and Etchécopar 1970; Korelov *et al.* 1974). Sometimes readily visible (Meinertzhagen 1920). One nest in south-east Iran 30 cm above ground in thorny bush; ♀ obtained for confirmation (Blanford 1876). Nest: rather fragile foundation of coarse stalks and grass stems, occasionally some twigs, neatly and smoothly lined with fine grass, wool, and hair; on slope, rear wall of foundation sometimes lacking and front built up (Yanushevich *et al.* 1960; Ivanov 1969; Hüe and Etchécopar 1970; Korelov *et al.* 1974). 3 nests in Kazakhstan and Tadzhikistan had outer diameter 10·0–13·0 cm, inner diameter 5·7–7·0 cm, overall height 5·9–7·0 cm, depth of cup 4·0–5·0 cm (Kovshar' 1966; Abdusalyamov 1977); outer diameter can be 18 cm (Dementiev and Gladkov 1954). Building: by ♀ only (Dementiev and Gladkov 1954; Korelov *et al.* 1974); according to Abdusalyamov (1977), ♂ also builds but confirmation needed. ♀ apparently excavates scrape in soil before building (Dementiev and Gladkov 1954; Ivanov 1969; Abdusalyamov 1977). EGGS. Sub-elliptical, smooth and slightly glossy; white, faintly tinged blue, green, or buff, sparsely speckled faint purplish-grey and with scattered spots and hairstreaks of purplish-black concentrated towards broad end; greyish-violet undermarkings (Harrison 1975; Schönwetter 1984); hairstreaks sometimes absent (Hüe and Etchécopar 1970). Nominate *buchanani*: 19·7 × 15·4 mm (19·0–20·3 × 14·9–15·9), *n* = 3; calculated weight 2·45 g (Schönwetter 1984); for 22 eggs from Semirech'e (south-east Kazakhstan), see Schlegel (1920). Clutch: 4–5 (3–6); in Turkmeniya, 3 clutches were of 3, 5, and 6 eggs (Rustamov 1958). Possibly 2 broods but requires confirmation; eggs laid daily in early morning (Dementiev and Gladkov 1954; Hüe and Etchécopar 1970; Korelov *et al.* 1974). INCUBATION. No information on period. By ♀ only (Dementiev and Gladkov 1954; Abdusalyamov 1977), but see Kovshar' (1966) for report of ♂ with brood-patch. YOUNG. Fed and cared for by both parents (Kovshar' 1966; Abdusalyamov 1977). FLEDGING TO MATURITY. Family breaks up 7–8 days after young leave nest (Abdusalyamov 1977). No further information.   BH

Plumages. (*E. b. cerrutii*). ADULT MALE. Top and side of head and neck medium grey, slightly tinged brown on hindneck; eye-ring contrastingly white, almost 1 mm wide behind and below eye. Mantle and scapulars to rump pale brown-grey, upper tail-coverts sepia-brown; lower mantle and scapulars with narrow sepia shaft-streaks, each streak narrowly bordered by dull brown at sides on central mantle, extensively cinnamon-brown or rusty-cinnamon on outer mantle and outer scapulars, forming contrasting patch, which is sometimes partly concealed below grey feather-fringes; feathers of rump and tail-coverts fringed grey when plumage fresh. Sharply defined white stripe backwards from lower mandible over lower cheek, often slightly tinged cinnamon. Chin white with slight cinnamon suffusion; side of chin grey, forming malar stripe, throat off-white, feather-tips of side of chin and throat often slightly tinged cinnamon. Underparts backward from chest vinous-cinnamon, feather-tips whitish or pale buff-grey when plumage fresh, partly concealing cinnamon; side of breast medium grey, upper flank light grey; cinnamon of lower belly, rear of flank, and vent paler, broadly fringed pale cream-buff; under tail-coverts pale cream-buff to off-white, longer coverts sometimes with dusky shafts. Tail black, both webs of central pair (t1) and outer webs of others fringed light cinnamon-brown, soon bleaching to off-white; base and middle portion of outer web of t6 white, tip of inner web of t6 with large white wedge (30–45 mm long); inner web of t5 with white wedge 20–35 mm long on tip. Flight-feathers, tertials, greater upper primary coverts, bastard wing, and greater and median upper wing-coverts black, outer webs of primaries and tips of secondaries narrowly fringed pale grey-buff or off-white, primary coverts and bastard wing faintly edged grey; outer webs of secondaries and outer webs and tips of greater and median coverts broadly fringed cinnamon-brown or rufous-cinnamon when fresh, tips paler cinnamon-buff when worn. Lesser upper wing-coverts light brown-grey. Under wing-coverts and axillaries white, grey of feather-bases exposed on primary coverts and shorter coverts. *In fresh plumage* (autumn), pale brown-grey feather-fringes on upperparts and whitish fringes on underparts more extensive, more fully concealing cinnamon on mantle and scapulars and partly concealing cinnamon of chest and belly (pale fringes mainly 2–3 mm wide, wearing gradually off in mainly April–May: Mauersberger 1982); if worn, cinnamon more fully exposed, white of stripe on cheek and of throat-patch purer, more contrasting, fringes of tertials, secondaries, and greater and median upper wing-coverts paler and partly worn off, and black of wing and tail browner. ADULT FEMALE. Upperparts and side of head and neck drab-brown-grey (head and neck less ash-grey than in adult ♂), marked with fine dark shaft-streaks on crown, lower mantle, and scapulars (in ♂, crown without streaks, and scapulars often more extensively cinnamon, especially when worn); short stripe above lore mottled cream-buff and light grey, less uniform grey than in adult ♂; eye-ring white, conspicuous. Stripe on lower cheek light cream-grey or cream-white, chin and throat isabelline-white, occasionally partly suffused cinnamon; malar stripe dark grey, mottled cream-grey, more distinct than in adult ♂. Chest rufous-cinnamon, each feather broadly fringed cream-white when plumage fresh, partly concealing cinnamon, often marked with short black shaft-streak or spot (in ♂, usually no black marks); remainder of underparts extensively cream-buff (belly less cinnamon than in ♂). Wing and tail as adult ♂, but fringes of secondaries, tertials, and greater and median upper coverts sometimes less bright. Upperparts and side of head and neck greyer when plumage worn, but crown and chest usually still narrowly streaked (unlike ♂) and mantle, scapulars, and belly still less cinnamon than in ♂. NESTLING. No information. JUVENILE. Rather like adult ♀, but upperparts and side of head and neck more diluted pale grey-brown, on upperparts marked with rather sharply defined dark grey streaks 2–4 mm broad, extending to rump; cinnamon on mantle and scapulars absent; tail-coverts with narrower streaks, fringes rufous. Eye-ring off-white, fairly contrasting; lower cheek often with distinct cream-white stripe. Ground-colour of underparts pink-buff on chest, pale buff on flank and vent; chest, side of breast, and flank marked with ill-defined dark grey streaks. Tail and flight-feathers as adult, but ground-colour dark brown, less blackish; tertials and greater and median upper wing-coverts dark brown, rather narrowly fringed pink-buff, soon bleaching to contrasting

cream-white. Rather like juvenile *E. hortulana*, but that species has ground-colour of upperparts darker buff-brown, of underparts deeper ochre-buff, and dark streaks on body darker, broader, and more contrasting. See also Recognition. FIRST ADULT MALE. Like adult ♂, but juvenile flight-feathers, primary coverts, variable number of tertials and greater upper wing-coverts, and t2–t6 retained; outer tail-feathers, flight-feathers, and primary coverts browner than in adult at same time of year, tips more distinctly frayed, less smoothly rounded; new t1 contrasts in shape and abrasion with worn neighbouring juvenile ones; old tertials and outer greater coverts (if any) with rather narrow and frayed light vinous-grey fringes, contrasting with neighbouring ones which show broader and more cinnamon fringes. Head and body as in adult ♂, but upperparts browner, less grey, dusky streaks broader, especially those of upper tail-coverts (but not as sharply defined as in juvenile), rump faintly spotted dusky; underparts extensively cinnamon, like adult ♂, but pale feather-fringes slightly wider, cinnamon more concealed in fresh plumage; in some birds, throat and chest marked with dull black specks. If worn (spring and early summer), indistinguishable from adult ♂, except by browner and more heavily worn primaries and primary coverts and (sometimes) by showing contrast in colour and abrasion within greater coverts or tail-feathers. FIRST ADULT FEMALE. Like adult ♀, but part of juvenile feathering retained, as in 1st adult ♂. Head and body as in adult ♀, but upperparts slightly more extensively marked with ill-defined grey-brown streaks, extending faintly to rump, chest to belly cream-buff, cinnamon on centres of feathers of chest more restricted; streaks on chest slightly broader and darker, extending to throat.

Bare parts. ADULT, FIRST ADULT. Iris brown or dark brown. Bill pale brown-yellow, brownish orange-yellow, red-brown, light brown-flesh, or yellow-flesh, sometimes with darker brown or grey culmen. Mouth yellowish-flesh. Leg and foot flesh-brown, red-brown, or reddish-flesh, toes often slightly browner or greyer, tarsus sometimes yellowish-flesh. (Ali and Ripley 1974; BMNH, RMNH, ZFMK, ZMA, ZMB.) NESTLING, JUVENILE. No information.

Moults. ADULT POST-BREEDING. Complete; primaries descendent. According to specimens examined and data from literature, moult starts with p1 between mid-June and late July, completed mid-August to late September; body and tail start from primary moult score *c.* 13–18; tail-feathers moult centrifugal, all new at primary score *c.* 40–45, tertials new at score *c.* 35, body new at about same time as outer primaries (Hellmayr 1929; Kozlova 1933; Paludan 1940, 1959; Piechocki and Bolod 1972; BMNH, ZFMK, ZMB). Starts early to late July (earliest in south of range), completed mid- or late August (Mauersberger 1982). No pre-breeding moult (Ticehurst 1922; Svensson 1992) or limited moult of cheek and perhaps t1 (Mauersberger 1982). POST-JUVENILE. Partial: head, body, lesser, median, and variable number (none to all) of greater upper wing-coverts and tertials, and usually t1. Starts soon after fledging, and timing thus highly variable: some fully juvenile up to early September, some in moult mid-July to late August, others in 1st adult plumage from late July onwards (Mauersberger 1982; BMNH, ZMB). Moult mainly when birds still in or near breeding area, but if t1, some or all tertials and greater coverts, and part of body still old during autumn migration, these sometimes replaced after arrival in winter quarters, November–April (Mauersberger 1982, Svensson 1992); some birds have all tail-feathers, greater coverts, and tertials still juvenile in late spring (BMNH, ZMA).

Measurements. ADULT, FIRST ADULT. *E. b. cerrutii*. South-east Turkey, Iran, and Kopet-Dag (Turkmeniya), March–August; skins (BMNH, ZFMK, ZMB). Bill (S) to skull, bill (N) to distal corner of nostril; exposed culmen on average 3·4 less than bill (S).

WING	♂	89·3 (0·96; 12)	87–92	♀ 84·8 (2·48; 6)	80–87
TAIL		70·4 (1·52; 12)	68–73	66·0 (2·51; 6)	63–70
BILL (S)		15·1 (0·55; 12)	14·4–16·2	14·3 (0·65; 6)	13·4–15·2
BILL (N)		8·8 (0·72; 12)	8·4–9·6	8·3 (0·68; 6)	7·8–9·4
TARSUS		20·4 (0·54; 12)	19·3–21·5	19·5 (0·57; 6)	18·9–20·4

Sex differences significant.

E. b. obscura. Uzbekistan and Kazakhstan to Tien Shan, April–September; skins (BMNH, RMNH, ZFMK, ZMB, ZMA).

WING	♂	88·1 (1·35; 16)	86–90	♀ 86·3 (1·26; 3)	85–88
BILL (S)		14·6 (0·55; 14)	13·9–15·5	13·8 (0·41; 4)	13·4–14·3

Wing. (1) *E. b. cerrutii*, south-east Turkey, Armenia, and Iran (Paludan 1938, 1940; Nicht 1961; Kumerloeve 1969a). (2) *E. b. obscura*, Mongolia (Piechocki and Bolod 1972; Piechocki *et al.* 1982). (3) Nominate *buchanani*, Afghanistan (Paludan 1959). (4) All races combined (Mauersberger 1982).

(1)	♂	88·1 (1·94; 14)	83–91	♀ 80·9 (2·96; 8)	76–85
(2)		88·2 (1·60; 11)	85–91	84·0 (— ; 1)	—
(3)		86·0 (— ; 11)	82–89	81·7 (2·07; 6)	79–84
(4)		82·6 (1·51; 89)	79–87	82·0 (1·53; 38)	—

See also Dementiev and Gladkov (1954), Ali and Ripley (1974), Korelov *et al.* (1974), and Svensson (1992).

Weights. ADULT, FIRST ADULT. *E. b. cerrutii*: (1) south-east Turkey, Armenia, and Iran, May–July (Paludan 1938, 1940; Nicht 1961; Kumerloeve 1969a; Desfayes and Praz 1978; ZFMK). *E. b. obscura*: (2) Kazakhstan (Korelov *et al.* 1974); (3) Mongolia, late May to mid-July (Piechocki and Bolod 1972; Piechocki *et al.* 1982). Nominate *buchanani*: Afghanistan, (4) late May to mid-July, (5) September (Paludan 1959). Race unknown: (6) India, October (Ali and Ripley 1974).

(1)	♂	21·3 (2·08; 17)	18·8–26·0	♀ 20·9 (1·99; 8)	17·2–24·3
(2)		20·7 (— ; 15)	18·9–24·0	19·9 (— ; 15)	17·0–22·5
(3)		20·6 (1·12; 11)	19–22	23·0 (— ; 1)	—
(4)		20·2 (1·30; 5)	19–22	20·0 (— ; 4)	20–20
(5)		22·6 (— ; 6)	18–26	20·7 (2·89; 3)	19–24
(6)		20·1 (— ; 8)	17–22	20·9 (— ; 10)	20–22

Structure. Wing long, broad at base, tip bluntly pointed. 10 primaries: p7–p8 longest, p9 0–2 shorter, p6 1–4, p5 6–10, p4 13–16, p3 17–19, p2 19–22, p1 20–24; p10 strongly reduced, 57–64 shorter than p7–p8, 7–11 shorter than longest upper primary covert. Outer web of (p6–)p7–p8 emarginated, inner web of p8 (p7–p9) with faint notch. Tip of longest tertial reaches to tip of p3–p5. Tail rather long, tip slightly forked; 12 feathers, t3–t4 longest, t1 3–7 shorter, t6 0–4. Bill as in *E. hortulana*, but distinctly longer, lower mandible relatively less deep; depth of bill at base 5·8 (5) 5·4–6·1, width 6·1 (5) 5·2–6·7; bill longer, lower mandible less deep, cutting edges more clearly toothed, and nostrils less broadly bordered by feathering than in Yellowhammer *E. citrinella*. Middle toe with claw 18·0 (5) 17·5–18·5 mm; outer and inner toe with claw both *c.* 65% of middle with claw, hind *c.* 77%. Remainder of structure as in *E. citrinella*.

Geographical variation. Slight; involves colour only. *E. b. cerrutii* from southern Transcaucasia and south-east Turkey to Iran and neighbouring parts of Turkmeniya a rather pale sandy-coloured race. Nominate *buchanani* from Quetta area (Pakistan) and Afghanistan north to western Pamir and Zeravshan mountains (Alay range) darker, upperparts dark grey with more extensive rusty-cinnamon on mantle and scapulars (less extensively sandy-grey), deeper cinnamon on underparts. *E. b. obscura* from northern Kazakhstan to Mongolia and south through Tien Shan and Kara Tau to Fergana area rather dark and grey on upperparts, usually more clearly streaked blackish on mantle and scapulars than both other races, less extensively rusty; upperparts less sandy-brown and underparts deeper cinnamon than in *cerrutii*; pale stripe on lower cheek as well as chin and throat greyish-white or cream-white, usually less tinged cinnamon than in *cerrutii*, grey of malar stripe more contrasting. See also Kozlova (1933), Vaurie (1956e, 1959), Mauersberger (1960, 1982), Vuilleumier (1977), and Desfayes and Praz (1978).

Forms species-group with Cinereous Bunting *E. cineracea*, Cretzschmar's Bunting *E. caesia*, and *E. hortulana*.

Recognition. Upperparts of adult and 1st adult closely similar to *E. hortulana*, but top and side of head and neck grey, not yellowish-green (some *E. hortulana* lack yellow pigment on head and neck, however, and these rather similar to *E. buchanani*); streaks on mantle and scapulars much narrower and ground-colour greyer than in *E. hortulana*; rufous on upperparts restricted to some rather dull brown-rufous on (mainly) outer scapulars; unlike *E. caesia*, no bright rufous on rump. Rufous on underparts more vinous than in *E. hortulana* and *E. caesia*, fringed pale grey, less uniform cinnamon, extending up to lower throat, without broad greyish chest-band of *E. hortulana* and *E. caesia*. In all plumages, tail distinctly longer than *E. hortulana* and *E. caesia*. In all plumages, tail distinctly longer than *E. hortulana* and *E. caesia* (but about equal in length to that of *E. cineracea*), wing/tail ratio 1·28 (21) 1·23–1·34 (mainly 1·32–1·42 in *E. hortulana* and *E. caesia*); bill longer, 13·4–16·2 to skull (in *E. hortulana*, 10·1–13·9; in *E. caesia* and *E. cineracea*, mainly 10·5–14·0); tip of inner web of t6 on average more extensively white, but some overlap in extent and pattern.   CSR

Emberiza caesia   **Cretzschmar's Bunting**   PLATES 17 (flight) and 21
[between pages 256 and 257]

DU. Bruinkeelortolaan   FR. Bruant cendrillard   GE. Grauortolan
RU. Красноклювая овсянка   SP. Escribano ceniciento   SW. Rostsparv

Emberiza caesia Cretzschmar, 1826

Monotypic

Field characters. 16 cm; wing-span 23–26·5 cm. Averages slightly smaller than Ortolan Bunting *E. hortulana*. Close counterpart of *E. hortulana*, with similar structure and almost identical plumage pattern; adult differs most in pure grey head and breast-band and orange-chestnut throat. Sexes dissimilar; little seasonal variation. Juvenile separable.

ADULT MALE. Moults: July–August (complete). Looks more immaculate than *E. hortulana*. Head, nape, neck, malar stripe, and breast grey, sometimes appearing ashy but in good light showing strong bluish suffusion; bright orange-chestnut lores, sub-moustachial stripe and throat and bright cream eye-ring. Note that contrast between head and throat is less than in *E. hortulana* and Grey-necked Bunting *E. buchanani*. Upperparts, wings, and tail less fulvous, more rufous, in general tone than *E. hortulana*, with mantle and scapulars more heavily streaked than *E. buchanani*. Lesser coverts mainly grey; tips of median and greater coverts and fringes of tertials bright orange-chestnut, contrasting with black bases or centres. Underparts intensely rufous-chestnut, somewhat darker and less bright than throat and distinctly rustier than *E. hortulana*. Underwing rufous-white to pale grey. Bill bright reddish-flesh in breeding season, fading to horn-grey in winter. Legs pale pink to reddish-brown. For slight differences in fresh plumage, see Plumages. ADULT FEMALE. Resembles ♂, but at close range shows browner, streaked crown and nape, ashy-brown face, and paler, buff-white throat with dark-flecked malar stripe. Sides of neck and breast suffused grey but at least centre of breast finely streaked dark brown. Underparts paler than ♂, particularly in fresh plumage, with pinkish-buff mottling. JUVENILE. Differs from adult as in *E. hortulana* and closely resembles that species, but in fresh plumage shows warmer, more rufous-orange ground-colour to plumage, with grey (not green) tinge to crown, white (not yellow) ground-colour to throat, and white underwing. FIRST-YEAR. ♂ shows at least some pure grey on back of head and breast and full rusty colour on belly, but ♀ has only hints of these while retaining much heavier streaks on throat, breast, and flanks.

Adult unmistakable if true colours, including blue-grey of breast-band, are apparent; at distance or in incomplete view, requires care. Important to remember that *E. caesia* is darker-bodied than *E. hortulana* and much more so than *E. buchanani*, with typically the darkest, most orange or rusty throat and vent of trio. Note however that one British vagrant (June) showed yellowish chin and sub-moustachial stripe (Holloway 1984). Juvenile difficult to distinguish from *E. hortulana*; see Recognition. Separation of juvenile from *E. buchanani* easier, since *E. caesia* much more heavily streaked above and on breast, with darker flight-feathers, rufous rump and vent, and shorter tail. Head may appear bulbous, even peaked on rear crown. Flight, behaviour, and actions as *E. hortulana* but less shy, with migrants allowing much closer approach. Markedly

terrestrial. Escape-flight often to trees. Gregarious, but not in such large parties as *E. hortulana*.

Song similar to *E. hortulana* but shorter and thinner-sounding, lacking pleasant ringing tone; usually 3–4 notes, last one longer, 'dzee-dzee-dzree'. Contact-call also like *E. hortulana* but sharper, 'tchipp'.

Habitat. Breeds in east Mediterranean region in warm temperate climate, mainly not far from sea. Occurs on rocky hillsides and islands among sparse herbage, with some shrub or tree growth, usually below *c.* 1300 m. While overlapping with Ortolan Bunting *E. hortulana*, tends to spread more onto drier and more barren rocky slopes, being ecologically intermediate between those buntings inseparable from vegetation cover and those preferring bare open ground or rocks (Harrison 1982). Found within July isotherms of 24–30°C. In some areas recorded inhabiting lower levels than *E. hortulana*, while in others the reverse appears to be the case. Occurs in maquis with kermes oak *Quercus coccifera* (Voous 1960b). In Palestine, said to replace Yellowhammer *E. citrinella*, occupying bare hillsides and rocky wadis with scrub, but retreating to thicket when disturbed. In Greece, on arrival in spring, frequents hillocks and dried-up shores of mountain streams near sea. Generally on ground, never perching on stones or walls, and avoiding buildings, but occasionally briefly calling from low bush. On passage in Egypt, generally frequents borders of desert and cultivated ground, as well as dunes and heaps of rubbish. (Dresser 1871–81.) Breeds in Lebanon and Cyprus up to *c.* 1700 m on slopes sheltered by *Cistus* and on forest land (Vere Benson 1970; Bannerman and Bannerman 1971). In Jordan, found in variety of habitats from slopes with bushes above Zarqa river and wadi cliffs near Karak, to rubbish dump on approach to Petra and to camel compound deep in desert at Wadi Rum (Mountfort 1965).

Winters in dry savanna, steppe, cultivated areas, and gardens within arid regions (Harrison 1982).

Distribution. No changes reported.

ALBANIA. May breed, but no proof (EN).

Accidental. Britain, France, Netherlands, Germany, Sweden, Finland, Poland, Austria, USSR, Malta, Algeria, Libya, Kuwait, Canary Islands.

Population. No information on trends.

ISRAEL. A few thousand pairs (Shirihai in press).

Movements. Migrant, completely vacating breeding areas to move mainly southwards to winter in north-east Africa and perhaps west Arabia. Movement mainly nocturnal.

Winters in Sudan, south to *c.* 11°N (Nikolaus 1987), and in Eritrea (Urban and Brown 1987). Also recorded Abéché (Chad) in January (Etchécopar and Hüe 1967). No firm evidence of occurrence south of Sudan, though

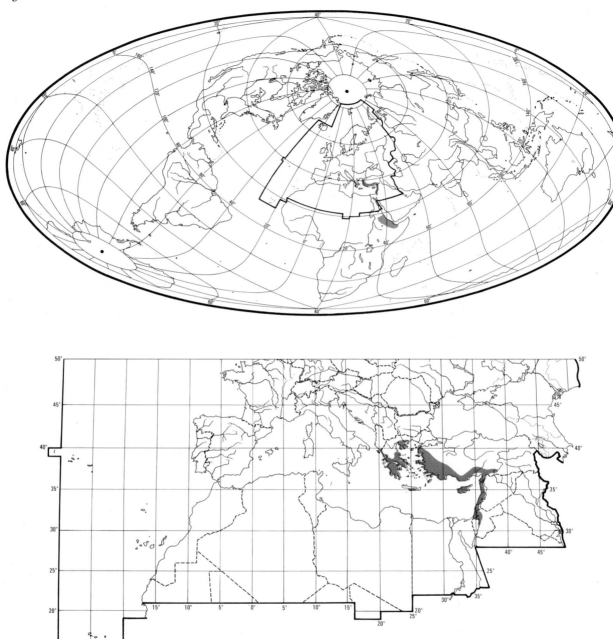

said to have occurred in Kenya (Moreau 1972; Britton 1980). Also recorded in south-west Saudi Arabia in winter (Meinertzhagen 1954); one December sighting in north-west Saudi Arabia (Baldwin and Meadows 1988). According to Meinertzhagen (1930), a few remain in Egypt in some winters; no recent confirmation (Goodman and Meininger 1989). Record of ♀ flying round ship in Red Sea soon after midday on 13 February 1953 (Phillips 1953)

might give hint of more regular movements during winter, but these otherwise unreported.

Apparently migrates on broad front across and around east Mediterranean, in both autumn and spring. In Turkey, occurs outside breeding areas on passage (Martins 1989). In Israel, autumn migration mainly over north, west, and centre of country; in spring, mainly in south and east; in Sinai, common in spring, less so in autumn

(Shirihai in press). In Jordan, uncommon on passage September–October and March–May (I J Andrews). In south-west Saudi Arabia, uncommon migrant west of Hejaz and Asir mountains (Jennings 1981a). In north-west Saudi Arabia, spring migrant in small numbers at Tabuk (Stagg and Walker 1982) and regular passage migrant at Yanbu (F E Warr). Fairly common on autumn and spring passage in Egypt (Goodman and Meininger 1989).

In Israel, juveniles leave breeding areas immediately on independence in June–July(–August), dispersing up to a few tens of km from breeding site. Adults leave gradually July–August. Common passage migrant August to mid-October (rarely end of October), with peak 3–16 September. (Shirihai in press.) Leaves Cyprus mainly mid-August to mid-September, stragglers to mid-November; autumn passage obscured by departure of local breeders (Flint and Stewart 1992). In Egypt, passage mid-August to late September, exceptionally until early November (Goodman and Meininger 1989). Migrants recorded arriving in Sudan in September (Prendergast 1985).

In spring, recorded at Tabuk (north-west Saudi Arabia) early March to early April (Stagg and Walker 1982). In Egypt, passage mid-March to late April, exceptionally from mid-February (Goodman and Meininger 1989). At Bahig (north coast of Egypt), 5–35 birds recorded 8 March to 22 April during 1967 and 1969–73 (Horner 1977). Very common on spring passage through much of Israel, mid-February to mid-May, mainly March; 2 peaks of passage, *c.* 8–15 March (mainly ♂♂) and 19–26 March (mainly ♀♀); overall, numbers mostly greater than in autumn. Arrives on breeding grounds end of February or March, mostly 2nd half of March, coincident with massive passage of migrants. (Shirihai in press.) Paz (1987) also noted that spring passage of ♂♂ precedes that of ♀♀ by *c.* 10 days. Arrives on breeding grounds in Cyprus early or mid-March (twice late February), common by late March or early April; also fairly common or common on passage on low ground and at northern capes, mainly mid- or late March to early or mid-April (Flint and Stewart 1992).

Apparently rare visitor to Arabian Gulf states, with a few sightings reported Kuwait in September and April–May and in eastern Saudi Arabia (F E Warr). Vagrant record also in south-west Iran (Hüe and Etchécopar 1970; Hollom *et al.* 1988). 　　　　　　　　　　DTH

Food. Seeds and small invertebrates. Feeds almost exclusively on ground and said to be probably most terrestrial of Emberizidae of region. (Meinertzhagen 1954; Hüe and Etchécopar 1970; Reid 1979; Armani 1985; Paz 1987.) On passage in Arabia and Israel forages in flocks of a few hundred with Cinereous Bunting *E. cineracea* and Ortolan Bunting *E. hortulana* on rocky slopes, in stubble, and in other cultivated areas bordering desert; also recorded in gardens (Meinertzhagen 1954; Paz 1987; Knijff 1991; Shirihai in press). In Jordan, mid-April, searched for food in groups of 3–8 among semi-arid herbs and scrub (Wit-

tenberg 1987); flock observed feeding on rubbish dump (Mountfort 1965). In winter in Eritrea, forages on coastal plains of short grass and in cereal fields in large flocks; appears regularly in February coinciding with maturing of grain (Smith 1957). See also Prendergast (1985) for observations in Sudan. Vagrant on Fair Isle preferred to forage on bare earth and apparently fed on seed-heads, while another fed skulking in short vegetation in field of growing oats *Avena* (Dennis 1969; Oddie 1981). In Arabia, fed on small seeds (not cereals) and small insects, including flying ants (Formicidae) and pupae (Meinertzhagen 1954). No further information. 　　BH

Social pattern and behaviour. Very poorly known; no detailed study.

1. In Israel, concentrations of tens or hundreds occur at main drinking spots after breeding season and before migration. Often migrates in groups of a few tens; large gatherings occur on passage, with maxima counted in small areas of 220 in September, 290 in March; a few hundred may gather in one small field in Arava desert. (Shirihai in press.) Flocks numbering scores seen on spring passage (Paz 1987). Similarly, in Cyprus usually occurs in flocks of tens on spring passage, maximum 180 in mid-March (Flint and Stewart 1992). Passage migrants in Syria noted a few times associating with Ortolan Bunting *E. hortulana* and Black-headed Bunting *E. melanocephala* (Baumgart and Stephan 1987). Migrants noted associating with *E. hortulana* also in Jordan (Wallace 1982b; Wittenberg 1987). However, migrant *E. caesia* and *E. hortulana* flocks in Cyprus reported to remain separate (Reid 1979). BONDS. Nature and duration of pair-bond and age at first breeding unknown. Shares of sexes in nest-building unknown. Incubation by ♀ (Paz 1987), although incubation by ♂ reported once from Cyprus (Flint and Stewart 1992). Young apparently fed by both parents (see Relations within Family Group, below). BREEDING DISPERSION. Information very scanty. In Israel often occurs in concentrations of breeding pairs, with tens of pairs in some areas while in adjacent areas it is absent or a few pairs are widely spread; average distance between active nests 400 m, minimum *c.* 200 m (Shirihai in press). In Cyprus, also occurs in breeding concentrations, 18 singing birds being recorded along 500 m of track (Reid 1979). No data on breeding densities or territory. ROOSTING. No information.

2. Very tame; mainly terrestrial, perching on bushes only when alarmed (Meinertzhagen 1954; Kumerloeve 1961; Etchécopar and Hüe 1967). Seeks shelter among tree branches; also sings from elevated perches that include bushes and trees (Paz 1987). FLOCK BEHAVIOUR. Contact-calls heard from nocturnal migrants (Paz 1987). No other information. SONG-DISPLAY. In Israel, ♂♂ use song-posts on trees or bushes (Paz 1987). In Jordan, sings from top of rock or bush, also on stony ground; sits on same perch for long periods (Hollom 1959). In Cyprus, noted singing from low shrubs, lower branches of conifers, pile of stones, or on ground in middle of road (Ashton-Johnson 1961); also from tip of weed 1·3 m high (Reid 1979). Song-period in Cyprus at least early April to June, once late August (Ashton-Johnson 1961; Reid 1979; Flint and Stewart 1992). ANTAGONISTIC BEHAVIOUR, HETEROSEXUAL BEHAVIOUR. No information (for roles of sexes in incubation, see Bonds, above). RELATIONS WITHIN FAMILY GROUP. Roles of sexes in caring for young little known: ♀ seen carrying food (Hollom 1959) and ♂ seen carrying food to nest and flying away with faeces (Ashton-Johnson 1961). Young leave nest before capable of flight (Paz 1987). ♂ seen feeding 'strong-flying' young (Ashton-Johnson 1961). ANTI-PREDATOR

RESPONSES OF YOUNG. No information. PARENTAL ANTI-
PREDATOR STRATEGIES. ♀ made last 4 m of approach on ground
to nest (with eggs). Incubating ♀ sat very tightly; when flushed,
performed 'rodent run injury display', giving 'chip-chip' call (see
4 in Voice). (Ashton-Johnson 1961.)   DTH

Voice. Vocal, at least in breeding season and on migration.
Detailed studies only of song (Bergmann 1983a); for additional sonagrams of song and calls, see Bergmann and Helb
(1982) and Bergmann (1983a). Fragmentary information
suggests similar variety of calls to Ortolan Bunting *E.
hortulana*, but no detailed studies. Most calls are short,
sharp sounds; similar structure of vocabulary to *E. hortulana* is presumed here.

   CALLS OF ADULTS. (1) Song of ♂. Resembles song of
south European and Middle East populations of *E. hortulana*, but shorter. Typically consists of 4 units, final one
differing from others (never has 2 or more final units
differing from others as often in *E. hortulana*); thus, 'dzee-
dzee-dzee-dzree', sometimes a clearer more fluty 'dee-
dee-dee-dree'; occasionally 1st unit omitted, and

sometimes a whole phrase is sung clearer and higher
pitched or with congested, squeaky timbre; averages 8
songs per min. (Hollom 1959; Peterson *et al.* 1983; L
Svensson.) Song on the whole thinner than that of *E.
hortulana*; sometimes described as thin and scratchy, sometimes with noticeable musical content, but still thinner
than most *E. hortulana* songs (Kumerloeve 1961; Reid
1979); somewhat mournful (Paz 1987). 2 song-types used
alternately according to Bruun *et al.* (1986), and Mild
(1990) recorded one bird giving 2 types of song (Figs
I–II). Individual ♂ in Cyprus gave 4 widely differing
song-types, each consisting of 4–5 units (diads in 2 of the
song-types); total duration of song *c.* 1 s; overall frequency
range *c.* 1·8–6·0 kHz; last unit (or diad) longer than others
in 3 of 4 song-types (Bergmann 1983a). Recordings reveal
considerable variety of songs (Figs I–VI), individual ♂♂
evidently having (at least) 2 or 3 types, which may be sung
alternately or almost alternately (P J Sellar, D T Holyoak).
Nearly all recordings are of songs consisting of 2 or 3
short units followed by longer unit, but wide variety

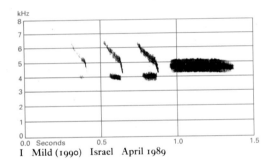

I   Mild (1990) Israel April 1989

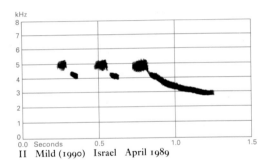

II   Mild (1990) Israel April 1989

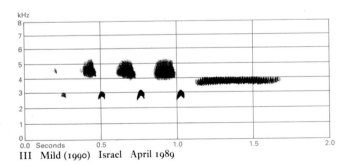

III   Mild (1990) Israel April 1989

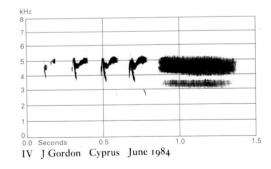

IV   J Gordon Cyprus June 1984

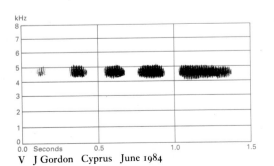

V   J Gordon Cyprus June 1984

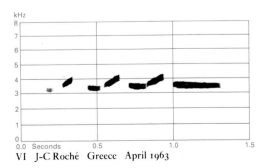

VI   J-C Roché Greece April 1963

occurs within this format, suggestive of complexity similar to that revealed by studies of *E. hortulana* and Cirl Bunting *E. cirlus* (D T Holyoak). Song in Cyprus does not show any difference of dialect from continental populations (Bergmann 1983a; Flint and Stewart 1992). (2) Contact-call. Similar to call 2 of *E. hortulana* but monosyllabic, harder and sharper, slightly less metallic, 'tchipp' or 'tchitt' (Mild 1990); can recall Yellowhammer *E. citrinella* (L Svensson). Also rendered as 'tsip' or 'djüb' (Bergmann and Helb 1982, which see for sonagrams), 'tsip', as anxiety call of breeding pair (Hollom 1959), liquid 'pwit pwit' (P J Sellar), sharp 'chi' (Fig VII), 'zi', or 'tsi'

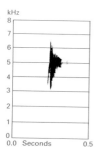

VII   P A D Hollom   Israel   May 1979

(D T Holyoak), metallic 'spit', sharper than in *E. hortulana* (Bruun *et al.* 1986), and rather harsh, sometimes grating 'zie' or 'chit' (Jonsson 1992). Insistent 'styip', less soft than call of *E. hortulana* (Nisbet and Smout 1957; Peterson *et al.* 1983), and 'tyup' (Fig VIII: P J Sellar) imply somewhat disyllabic variants as often occur in *E. hortulana*. Given in flight (Mild 1990) and when perching. (3) 'Teu' calls. Piping 'teu' or 'teup' call, heard from anxious breed-

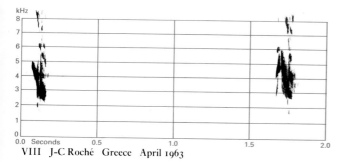

VIII   J-C Roché   Greece   April 1963

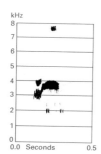

IX   P A D Hollom   Israel   May 1979

ing pair (Hollom 1959), also rendered 'tchu' (Mild 1990); 'chü' or 'cheü' (L Svensson); recording (Fig IX) suggests 'tuu' (D T Holyoak). (4) Short 'plet', quiet and dry, given in flight (Mild 1990). 'Flight-call' rendered 'pt', 'but', or 'zit' by Bergmann and Helb (1982, which see for sonagrams) is apparently the same. Probably analogous to call 4 of *E. hortulana*. 'Chep', described as alarm-call from anxious breeding pair (Hollom 1959), and 'chip' or 'chipchip' when disturbed at nest (Ashton-Johnson 1961), are probably also the same.

CALLS OF YOUNG. No information.   DTH

Breeding. SEASON. Greece: eggs laid from April, probably mainly 2nd half of month (Steinfatt 1954; Makatsch 1976); fresh eggs found mid-June (Krüper 1875); see also Reiser (1905) and Peus (1954). Cyprus: eggs laid from April at low altitude, from end of April at 700-1300 m; fledged young observed early May (Bannerman and Bannerman 1958; Ashton-Johnson 1961; Flint and Stewart 1992). Turkey: near south coast, eggs laid in one nest around beginning of May (Beaman 1978). Israel: eggs laid late March to mid-July (Pitman 1922; Shirihai in press). SITE. On ground, often in depression, sheltered by rocks and vegetation, at times within roots of shrub, though frequently quite visible; usually on slope, and commonly under rock-rose *Cistus* (Reiser 1905; Pitman 1922; Bannerman and Bannerman 1958; Mauersberger 1960; Paz 1987; Flint and Stewart 1992). In Greece, recorded in wall (Banzhaf 1937b). Nest: foundation of stalks, roots, and grass, thickly lined with rootlets and hair; on slope, rear wall can be formed by rock or earth (Pitman 1922; Bannerman and Bannerman 1958; Ashton-Johnson 1961; Makatsch 1976). Building: ♀ seen carrying material in Greece and Cyprus (Steinfatt 1954; Took 1971). EGGS. See Plate 34. Sub-elliptical, smooth and slightly glossy; very pale yellowish-, bluish-, purplish-, or greyish-white with fine purplish-black speckling and hairstreaks over whole surface though mostly at broad end (Pitman 1922; Bannerman and Bannerman 1958; Ashton-Johnson 1961; Harrison 1975). For comparison with Ortolan Bunting *E. hortulana*, see Reiser (1905) and Makatsch (1976). 19·6 × 15·0 mm (17·5-21·9 × 13·5-16·7), $n=116$; calculated weight 2·31 g (Schönwetter 1984); see also Pitman (1922) for 8 eggs from Israel. Clutch: 4-5(-6); Cyprus 4-5, $n=6$ (Bannerman and Bannerman 1958); Israel 4-5, $n=12$ (Shirihai in press); 6 recorded in Turkey (Krüper 1875). Often 2 broods (Makatsch 1976; Shirihai in press). INCUBATION. 12-14 days (Shirihai in press); 19 days according to Paz (1987). ♀ sits very tightly (Pitman 1922; Ashton-Johnson 1961). One observation of incubation by ♂ (Bennett 1976). YOUNG. Fed and cared for by ♂ at least (Ashton-Johnson 1961); ♀ seen carrying food in Jordan (Hollom 1959). FLEDGING TO MATURITY. Fledging period 12-13 days (Paz 1987; Shirihai in press); young leave nest unable to fly (Paz 1987). BREEDING SUCCESS. No information.   BH

Plumages. ADULT MALE BREEDING. Top and side of head and neck medium grey, forehead narrowly mottled brown; lore mottled rufous and grey, bordered above by narrow uniform rufous-cinnamon stripe. Eye-ring narrowly but conspicuously pale rufous-cream to cream-white. Broad and contrasting stripe from base of lower mandible over lower cheek to below or just behind lower ear-coverts deep rufous-cinnamon, bordered by narrow medium grey malar stripe below, which ends in grey of lower side of neck. Mantle, scapulars, and back rufous-cinnamon to buff-brown, often brighter and deeper rufous on outer scapulars, slightly paler and more greyish-buff along fringes of feathers; lower mantle and scapulars contrastingly marked with narrow black streaks. Rump bright rusty-cinnamon or rufous-chestnut, upper tail-coverts slightly paler rufous-cinnamon. Chin and throat rufous-cinnamon, forming contrasting patch, often faintly bordered by pale buff rim; upper side of breast and all chest contrastingly medium grey. Belly deep rufous-chestnut, gradually paler rufous-cinnamon to lower side of breast, flank, vent, and under tail-coverts, feathers narrowly tipped pale pinkish-grey when plumage fresh. Tail black; both webs of central pair (t1) broadly fringed rufous-cinnamon; t2–t5 narrowly fringed pink-buff; inner web of t4 with broad pink-buff to white border along tip; inner web of t5–t6 with white wedge on tip, occupying *c.* 25% of feather-length of t5, *c.* 30% of t6. Flight-feathers, greater upper primary coverts, and bastard wing black-brown; outer webs of primaries narrowly fringed pale buff (except emarginated parts), outer webs of secondaries with broader cinnamon fringes, primary coverts narrowly fringed brown-grey on outer webs and tips. Tertials and greater and median upper wing-coverts brown-black or black, outer webs and tips broadly fringed pink-cinnamon or rufous-cinnamon, cinnamon bleaching to pink-buff on tip when plumage worn; lesser coverts greyish-buff with darker grey centres. Under wing-coverts and axillaries buff-grey or whitish-grey, smaller coverts along leading edge of wing with dark grey bases. ADULT MALE NON-BREEDING. Like adult ♂ breeding, but crown sometimes with faint dusky shaft-streaks, rump and mid-belly sometimes less deep rufous-chestnut, more rufous- or tawny-cinnamon; tertials, greater and median upper wing-coverts, and t1 all equally fresh, fringes and tips rufous-cinnamon (in breeding plumage, median coverts and often inner secondaries and some or all tertials new and with rufous-cinnamon fringes, fringes of remainder of feathers bleached to tawny or pink-buff, but colour strongly dependent on timing and extent of pre-breeding moult). ADULT FEMALE BREEDING. Like adult ♂ breeding, but grey of forehead and crown slightly tinged brown, hindneck completely buff-brown, less grey, all with narrow dusky shaft-streaks; mantle, scapulars, and back often slightly paler and more olive, less warm cinnamon, rump markedly paler, buff-brown or olive-brown, not as deep rusty as adult ♂ breeding. Lore, stripe over upper cheek, and bar behind ear-coverts pale tawny-cinnamon or buff, remaining side of head and neck greyish-buff, darker and greyer at rear of ear-coverts; head pattern far less contrasting than in adult ♂. Chin and throat tawny-cinnamon, malar stripe broken up into grey or dull black spots, sides of throat and lower throat with fine black streaks; side of breast and all chest mixed grey and rufous-cinnamon with fine black shaft-streaks (ground-colour and amount of streaks depending on extent of pre-breeding moult: greyer and less streaked when moult extensive, but never as uniform grey as adult ♂ breeding); belly rufous-cinnamon, grading to pink-cinnamon on flank and under tail-coverts (all paler than in adult ♂ breeding). Wing and tail as adult ♂ breeding, but fringes of t1, tertials, and greater and median upper wing-coverts paler, pink-cinnamon (when fresh) or pale tawny-buff (when worn),

colour of fringes and contrast in colour between fringes of coverts or tertials in part depending on extent of pre-breeding moult. ADULT FEMALE NON-BREEDING. Like adult ♀ breeding, but top and side of head pale buff-brown or grey-brown, finely streaked dull black, sometimes with limited amount of pure grey just visible on crown or nape; side of head without distinct pattern, except for narrow but clear pale buff eye-ring and for pale tawny-buff or cream-buff stripe backwards from lower mandible, turning upward behind ear-coverts, which is bordered above and below by indistinct and broken black moustachial and malar stripe. Chin and throat pale cream-buff to off-white, partly speckled grey, bordered at side and below by gorget of pink-buff and dull black streaks over lower side of neck and chest. Belly pink-cinnamon, grading to cream-buff on flank and vent. Remainder of body and wing and tail as in adult ♀ breeding; fringes of wing-coverts and tertials all equal in colour. NESTLING. No information. JUVENILE. Forehead, crown, and hindneck closely mottled or streaked pale buff and dark brown, mantle and scapulars to upper tail-coverts more clearly streaked buff and brown. Side of head and neck pale cream-buff, finely speckled dark brown (especially on supercilium and side of neck, less so on lore and in narrow eye-ring), upper cheek and ear-coverts slightly darker grey-brown. Ground-colour of entire underparts pale pink-buff (when fresh) or cream-white (when worn), finely mottled dull black or grey on side of chin (forming broken malar stripe) and all throat; chest, side of breast, flank, and (sometimes) part of belly and under tail-coverts with narrow dusky streaks. Tail as adult, but fringes rather pale pink-cinnamon or pink-buff, narrow; tips of feathers more pointed than in adult, less broad and truncate; tertials as adult, but fringes paler, less deep rufous, tips rapidly bleaching to cream-white; lesser, median, and greater upper wing-coverts dark sepia-brown, extending into blunt point towards tip of each covert, side of each broadly and contrastingly pale cream or off-white. Feathering rather short and loose, especially on head, neck, and underparts. FIRST ADULT MALE NON-BREEDING. Closely similar to adult ♀ non-breeding, top and side of head and neck grey-brown with limited amount of pure grey at feather-bases (less extensively and uniformly grey than adult ♂ non-breeding), a narrow but distinct pale eye-ring, broad buff streak over lower cheek, bordered above by indistinct mottled black moustachial stripe and below by more distinct black-and-grey mottled malar stripe; chin and throat buff, streaked gorget and chest as in adult ♀ non-breeding. In contrast to adult ♀ non-breeding, juvenile outer tail-feathers, usually outer greater upper wing-coverts, and occasionally a few tertials retained, contrasting in colour, shape, and wear with neighbouring new feathers (in adult non-breeding, all coverts equally new); also, flank and belly often a mixture of pale buff and rufous-cinnamon feathers, partly streaked dull black (in adult ♀, uniform pink-cinnamon without streaks, or with some fine streaks on flank only). FIRST ADULT FEMALE NON-BREEDING. Like 1st adult ♂ non-breeding, but top of head heavily streaked black, similar to mantle and scapulars; side of head and neck brown or grey-brown with fine dusky specks or streaks and with contrasting pale eye-ring and broad cream stripe over lower cheek (ear-coverts almost uniform brown); chin and throat cream-buff (when fresh) or off-white (when worn), gradually merging into tawny-buff to cream-buff on remainder of underparts; sides of throat, side of breast, and chest fairly heavily marked black, central throat, upper belly, flank, and under tail-coverts with fine black shaft-streaks. Wing and tail as 1st adult ♂ non-breeding, less uniform than adult ♀ non-breeding. FIRST ADULT MALE BREEDING. Like adult ♂ breeding, distinguishable only by more worn and pointed primaries and (especially) greater upper primary coverts; on average, hindneck browner (less grey); rump paler, more buff-brown (rarely as rufous-chestnut as adult ♂ breeding); grey of side of head and chest faintly tinged buff or brown, less pure; grey of malar stripe sometimes mottled black; belly on average more rufous-cinnamon, less deep rufous-chestnut. FIRST ADULT FEMALE BREEDING. Like adult ♀ breeding, but juvenile flight-feathers, greater upper primary coverts, and occasionally some tail-feathers retained, tips more pointed and more worn than adult ♀ at same time of year. Also, streaks on crown, malar stripe, and streaks on chest on average more distinct than in adult ♀ breeding (and thus quite different here from unstreaked 1st adult ♂ non-breeding), and belly paler, more pinkish-cinnamon.

Bare parts. ADULT, FIRST ADULT. Iris brown or black-brown. In spring and summer, bill reddish-flesh, or pink-brown, culmen brown-horn to dark slate-grey; in autumn and winter, more extensively flesh-grey or horn-grey. Leg and foot pink-flesh, pale flesh-brown, or purplish red-brown, often tinged grey on toes. (Hartert 1903–10; Dennis 1969; RMNH, ZMA.) NESTLING, JUVENILE. No information.

Moults. ADULT POST-BREEDING. Complete; primaries descendent. Little information on timing; single ♀ from western Turkey heavily worn by late July, moult not yet started; in Greece, single ♂ from late July had p1–p7 new and p8–p9 growing, another from early September had moult completed (RMNH). Single ♂ and ♀ in heavy moult 29 July and 1 August, Turkey (Eyckerman *et al.* 1992). ADULT PRE-BREEDING. Partial: head, neck (sometimes except nape), variable amount of scapulars and rump, chin to upper chest, variable amount of remainder of underparts, median upper wing-coverts, often inner greater coverts (tertial coverts) and tertials, and frequently t1; moult in ♂ more extensive than in ♀, in latter often restricted to head, side of neck, chin to upper chest, median coverts, and limited scattered feathering elsewhere on body. In winter quarters, but timing unknown; usually completed on arrival on breeding grounds, but some ♀♀ on Cyprus had part of feathers of body still growing in March (ZMA). POST-JUVENILE. Partial: head, body, lesser, median, and usually all greater upper wing-coverts, usually all tertials, usually t1 and (in ♂) often all tail. Starts shortly after fledging; in small sample from western Turkey, July, some fully juvenile, others in full moult, starting with side of breast, upper belly, upper flank, and side of mantle (RMNH, ZMA); probably usually completed at start of autumn migration. FIRST PRE-BREEDING. Either as adult pre-breeding or more restricted, especially in ♀♀; in ♂, head, neck, much of underparts, outer scapulars, scattered feathers of rump, median coverts, and usually inner greater coverts, tertials, and t1 new (only flight-feathers and greater upper primary coverts still juvenile, remainder of wing and tail in 1st non-breeding); in ♀, 20–70% of feathering of head, neck, and throat new, 0–50% of remainder of upperparts, 10–80% of underparts, often t1, and highly variable number of median and greater coverts and tertials (t2–t6 or all tail, flight-feathers, and primary coverts still juvenile, remainder in 1st non-breeding). (C S Roselaar.)

Measurements. ADULT, FIRST ADULT. Whole geographical range, all year; skins (RMNH, ZMA). Bill (S) to skull, bill (N) to distal corner of nostril; exposed culmen on average 3·2 less than bill (S).

	♂		♀	
WING	84·5 (2·27; 34)	79–88	80·5 (1·76; 13)	77–83
TAIL	61·9 (2·86; 34)	55–70	59·8 (2·23; 13)	56–63
BILL (S)	13·8 (0·59; 34)	12·7–15·0	13·4 (0·69; 12)	11·8–14·6
BILL (N)	7·5 (0·51; 34)	6·2–8·9	7·7 (0·56; 12)	7·1–8·9
TARSUS	18·2 (0·65; 31)	16·9–19·1	18·3 (0·82; 10)	16·8–19·8

Sex differences significant for wing, tail, and bill (S).

Wing. Whole range, summer: ♂ 84·1 (2·35; 14) 78·5–87, ♀♀ 81·5, 84·5 (Niethammer 1943; Jordans and Steinbacher 1948; Makatsch 1950; Kumerloeve 1961, 1964*a*; G O Keyl).

JUVENILE. Unsexed birds, Turkey and Syria, June–July: wing 82·0 (2·28; 10) 78–86, tail 59·4 (2·33; 10) 56–64; tarsus 17·8 (0·74; 8) 16·3–18·9 (RMNH, ZMA).

Weights. Greece, western Turkey, and Israel, late March to July: ♂ 21·2 (1·16; 8) 19·8–23·0, ♀♀ 19·8, 22·2 (Niethammer 1943; Makatsch 1950; Kumerloeve 1964*a*; Rokitansky and Schifter 1971; G O Keijl). Israel: 20–25 (Paz 1987). Turkey, 29 July and 1 August, adult: ♂ 23·5, ♀ 18·5 (Eyckerman *et al.* 1992). Fair Isle (Britain), June: 23 (Dennis 1969).

Structure. Wing rather short, broad at base, tip bluntly pointed. 10 primaries: p8–p9 longest or either one 0–2 shorter than other; p7 0–1 shorter, p6 2–5, p5 8–11, p4 12–16, p3 15–18, p2 18–21, p1 20–24; p10 reduced, 10·2 (10) 9–13 shorter than longest upper primary covert. Tip of longest tertial reaches halfway to tip of p3 and p4. Bill rather finer at base than in Yellowhammer *E. citrinella*, longer and more sharply pointed; as in Ortolan Bunting *E. hortulana*, but base more slender and culmen and gonys almost straight. Middle toe with claw 18·2 (9) 17·7–19·8; outer toe with claw *c.* 73% of middle with claw, inner *c.* 65%, hind *c.* 76%. Remainder of structure as in *E. hortulana*.

Geographical variation. Very slight. When series of skins compared, slight cline discernible of increasingly pale ground-colour and gradually fewer dark markings, running from Greece through Asia Minor to Levant (Böhr 1962).

Forms species-group with *E. hortulana*, Grey-necked Bunting *E. buchanani*, and Cinereous Bunting *E. cineracea*. Where *E. caesia* and *E. hortulana* occur side-by-side on mainland, competitive exclusion results in altitudinal and/or habitat segregation; on most islands in eastern Mediterranean, only one species occurs, usually *E. caesia*, sometimes *E. hortulana* (e.g. on Crete), but both on Samothraki and Rhodes (Niethammer 1942; Mauersberger 1960; Kumerloeve 1962*a*; Kinzelbach and Martens 1965).

Recognition. Juvenile and 1st adult non-breeding often hard to distinguish from *E. hortulana*. Characters cited by Dennis (1969) and difference in tail pattern given by Svensson (1992) not always valid. Often, cap of *E. hortulana* tinged green (grey in *E. caesia*), throat pale yellow (sometimes on buff ground; in *E. caesia*, while with buff or cinnamon suffusion), axillaries and shorter under wing-coverts partly pale yellow (white in *E. caesia*).

MP

Emberiza chrysophrys **Yellow-browed Bunting**

PLATES 17 (flight) and 22
[between pages 256 and 257]

Du. Geelbrauwgors   Fr. Bruant à sourcils jaunes   Ge. Gelbbrauenammer
Ru. Желтобровая овсянка   Sp. Escribano cejigualdo   Sw. Gulbrynad sparv

Emberiza chrysophrys Pallas, 1776

Monotypic

Field characters. 14–15 cm; wing-span 21·5–25 cm. Slightly but distinctly larger than Little Bunting *E. pusilla*, with longer and deeper bill, rather heavier head, and slightly plumper body; only slightly smaller than Rustic Bunting *E. rustica* and Yellow-breasted Bunting *E. aureola* and allies. Rather small bunting, with somewhat sparrow-like profile to bill and head; bill proportionately larger than any other *Emberiza* of similar size. Plumage of body, wings, and tail recalls *E. pusilla*, but head colours and pattern quite distinct, with at least partly yellow supercilium and white median crown-stripe contrasting strongly with blackish or black crown-sides and ear-coverts, latter distinctly spotted white at rear. Sexes closely similar; little seasonal variation. Juvenile inseparable.

ADULT MALE BREEDING. Moults: probably July–August (almost complete); winter (head and body). Head heavily striped black from forehead along sides of crown, through eye, and around or across ear-coverts; dark stripes isolate long, narrow white median crown-stripe (easiest seen from behind on nape), mainly yellow, terminally white supercilium (narrow before eye, not reaching bill; broader behind eye, reaching almost to nape) and small white spot on rear ear-coverts. Striped head contrasts with white throat, which is sharply divided by sharp black malar stripe. Note that ♂ *E. rustica* in moult from 1st-winter to breeding plumage may show not dissimilar set of contrasts but these differ in producing more isolated pale spot on nape (not continuous central stripe), pure white supercilium, and only broken or vestigial lower malar stripe; beware particularly presence of similar white spot on rear ear-coverts; look at body for diagnostic difference in breast marks, banded rufous in *E. rustica*, streaked blackish-brown on no more than warm buff ground-colour in *E. chrysophrys*. Confusion of head pattern with *E. pusilla* once considered possible, but that species has thinner bill, predominantly rufous-buff ground-colour to head, noticeable narrow black edges only to rear ear-coverts, and narrow but distinct cream eye-ring which is completely absent on *E. chrysophrys*. Back tawny-brown, streaked black but centre of mantle, scapulars, and wings rather more generally rufous; rump more chestnut than back. Wings show grey lesser coverts, almost white tips to black-based median coverts (forming noticeable upper wing-bar), pale buff tips to greater coverts (forming less distinct lower wing-bar), and strongly chestnut edges to tertials and inner secondaries; wing markings altogether

brighter than *E. pusilla*. Tail brown-black, with rufous edges to central feathers and white outer tail-feathers. Centre of breast and rest of underparts white, with black-brown streaks with diffuse buff edges continuing from breast along flanks. In fresh plumage, shows brown tinge to crown and area below eye and less vividly yellow supercilium; appearance converges with ♀ and 1st-winter ♂. Bill dark blackish-horn to grey-horn on upper mandible, with striking pale yellowish-pink to flesh-horn cutting edges and lower mandible made even more obvious by basal depth. Legs pale flesh. ADULT FEMALE. Closely resembles ♂ but black crown-stripes tinged brown, fore-supercilium rather more distinct, and yellow part of supercilium duller and paler, while all ear-coverts dull brown except for complete black edge and white spot at rear. FIRST-YEAR. Closely resemble adult but a few retain apparently wholly whitish supercilium of juvenile plumage (see Plumages).

♂ unmistakable if seen well, but in glimpse could be confused with 'tan-striped' morph of White-throated Sparrow *Zonotrichia albicollis* (larger with longer tail lacking white edges, grey ear-coverts and collar, and little-streaked underparts), *E. rustica*, and *E. pusilla* (see above). ♀ and immature more troublesome, inviting confusion with *E. rustica* and *E. pusilla* and with ♀ and immature Black-faced Bunting *E. spodocephala* which has however much duller head pattern, having no black on sides of crown, only pale buff supercilium, greyish cheeks and neck-sides, less discrete pale spot on rear ear-coverts, and less white, more diffusely streaked underparts. Flight, behaviour, and actions much as other smaller *Emberiza*. In breeding range, frequently consorts with *E. pusilla* and shows in comparison slightly less sprightly mien.

Commonest call a short, distinct 'zit'; likened to *E. pusilla* but softer and shriller; also resembles that of *E. rustica*.

Habitat. Breeds in taiga forests of deep interior of south-east Siberia, inhabiting pines *Pinus* and larches *Larix*, and also thickets in shrub layer, apparently in lowlands, shifting to similar habitats elsewhere for winter (Dementiev and Gladkov 1954).

Distribution. Breeds in south-east Siberia, from Irkutsk east to Barguzin mountains and Stanovoy range. Winters in central and south-east China.

Accidental. Britain: Norfolk, 19 October 1975 (Holman

1990); Fair Isle, 12–23 October 1980 (Kitson and Robertson 1983). Netherlands: Schiermonnikoog, 19 October 1982 (Vonk and IJzendoorn 1988). Ukraine: 1 January 1983 (Davydovich and Gorban' 1990).

Movements. Migratory, but little information.

Birds migrate via south-east Russia (Transbaykalia to Ussuriland), north-east China (Heilungkiang, eastern Inner Mongolia, Hopeh, Shantung) and Korea to winter in eastern China from Kiangsu south to Kwangtung, and west along Yangtze valley to Red Basin (eastern Szechwan) (Vaurie 1959; Gore and Won 1971; Schauensee 1984). Vagrant as far south as Hong Kong (Chalmers 1986). Annual but scarce on passage both seasons on islands off western Japan, and sometimes recorded in winter (Brazil 1991).

Time of autumn departure from breeding grounds uncertain; perhaps begins August with main movement 1st half of September (Dementiev and Gladkov 1954; Reymers 1966), though recorded at Ulan-Ude (south-east of Lake Baykal) to 20 October in one year (Shkatulova 1979). Passage in north-east China and Korea September–November (Hemmingsen and Guildal 1968; Gore and Won 1971).

Spring passage in north-east China (mid-March–)April–May (Hemmingsen 1951; Hemmingsen and Guildal 1968; Williams 1986); latest record in Hong Kong 26 April (Chalmers *et al.* 1991). In southern Ussuriland (south-east Russia), recorded (very rarely) in mid-May (Panov 1973a). In area east of Lake Baykal, over 2 years, various passage records 25 April to 11 June (Shkatulova 1979); further north, on Lena river, migrants observed mid- to late May (Reymers 1966; Larionov *et al.* 1991). No reports of vagrancy in western Siberia (Johansen 1944; Ravkin 1973) or Kazakhstan (Korelov *et al.* 1974).

For autumn vagrancy to west Palearctic, see Distribution. British records coincided with arrival of other Asian vagrants (Kitson and Robertson 1983; Holman 1990); for discussion of weather conditions and possible passage routes, see Kitson and Robertson (1983) and Elkins (1985). DFV

Voice. See Field Characters.

Plumages. ADULT MALE. In fresh plumage (autumn), forehead and crown deep black, marked in middle by narrow and sharply-defined white stripe, often concealed on forehead and forecrown, more distinct at rear. Supraloral stripe yellow, short and narrow, sometimes indistinct, extending into broad and sharply-defined bright yellow supercilium from above eye backwards. Narrow eye-ring pale yellow, broadly broken by black in front and behind. Feather-tufts at lateral base of upper mandible dark grey; lore, upper cheek, ear-coverts, and narrow line at upper rear of eye black; centre of ear-coverts mottled brown and white, upper rear with uniform white spot. Sharply defined cream-white stripe over lower cheek, bordered below by distinct black single or double malar stripe, widening into mottled black patch on lower side of neck. Hindneck and side of neck white with black

streaks, forming mottled white shawl, connected with white on lower cheek. Mantle and scapulars to upper tail-coverts buff-brown or olive-brown, each feather of mantle and scapulars with sharp black central streak, narrower and sometimes less sharp black streaks on remainder; inner border of feathers of mantle and of inner scapulars and tips of feathers of rump brighter cinnamon-rufous. Chin and throat cream-white, marked with some black triangular spots at sides and lower border; chest and upper flank sharply and evenly marked with black streaks on cream-buff ground, side of breast and lower flank warmer buff-brown and with broader and duller black streaks. Remainder of underparts cream-white. Central pair of tail-feathers (t1) rusty-rufous with ill-defined dark grey centre; t2–t6 black, outer webs bordered by hardly contrasting dark olive-brown fringes; tip of inner web of t4 sometimes with small white spot, inner web of t5 with sharply contrasting white wedge 2–3 cm long on tip; middle portion of outer web of t6 white, inner web of t6 with white wedge *c.* 3·5–5 cm long along shaft, appearing largely white. Flight-feathers, greater upper primary coverts, and bastard wing greyish-black, outer webs of secondaries narrowly and rather faintly fringed rusty-brown, primary coverts and bastard wing fringed olive-brown, primaries more contrastingly bordered by narrow pink-brown fringes along outer webs. Tertials black with broad rufous-cinnamon outer fringe and tip, rufous extending into notch on black of middle portion of outer web of longer tertials. Greater upper wing-coverts black with contrasting pink-buff fringes *c.* 1 mm wide along outer web, extending into pale pink-buff fringe on tip; median coverts black with contrasting buff-white or pure white tips 1–2 mm long, forming narrow but distinct wing-bar; lesser coverts olive-grey with dark grey centres. Axillaries and under wing-coverts white, longest coverts grey, marginal coverts dotted black. *In worn plumage* (spring), white median crown-stripe often more distinct; ground-colour of upperparts greyer (less brown), black streaks more contrasting; ground-colour of lower cheek and of entire underparts bleached to white, less cream or buff; rufous-cinnamon of t1 and tertials bleached, partly worn off; tips of greater coverts whiter, but these (and those of median coverts) sometimes almost completely worn off. Some variation in plumage, especially in autumn, when some ♂♂ (although adult according to tail-shape) show brown fringes along black feathers of crown (not as pure black as described above), brown central ear-coverts with black-brown rim above and below and black border at rear, somewhat less contrasting with white or pale buff spot on upper rear corner of ear-coverts, more buff-yellow supercilium, and more numerous black triangular spots at border of cream-white belly. Supercilium usually bright yellow, but occasionally rather pale yellow, even in adult ♂ in spring. ADULT FEMALE. Like adult ♂, but black lateral crown-stripes with prominent brown fringes, black partly concealed, especially in autumn; mid-crown stripe buff, less pure white; supercilium variable, deep yellow, buff-yellow, pale yellow, or white, apparently independent of age, hardly contrasting with buff-brown lore; upper cheek and ear-coverts buff-brown, ear-coverts with dark brown or dull black surround and with pale buff or off-white spot on upper rear corner; black-mottled white shawl across hindneck and side of neck sullied brown and grey, pale stripe of lower cheek and ground-colour of chest and flank washed buff, less conspicuously pale cream or white than adult ♂, unless plumage abraded. JUVENILE. Lateral crown-stripe dark brown, median crown-stripe and supercilium pale buff or buff-white. Lore buff, upper cheek and ear-coverts buff-brown mottled grey and pale buff, bordered by darker brown rim at upper and lower border and at rear of ear-coverts. Mantle to upper tail-coverts tawny-buff, more rufous on rump, marked with narrow dull black streaks (faint on rump); outer borders of feathers of mantle and of scapulars paler buff. Ground-colour of lower cheek and underparts cream-yellow, almost white on throat and mid-belly, side of throat, lower throat, side of breast, chest, and flank closely marked with narrow black streaks widening into triangle on feather-tip. Tail as adult, but no white on tips of t4 and tips of all feathers pointed, less rounded than in adult. Wing as adult, but lesser coverts sepia-brown with buff fringe, median dull black with white tip 2–3 mm wide, greater black-brown with narrow buff fringe along outer web and tip, fringe on tip broken by dull black along shaft. FIRST ADULT MALE. Like adult ♀, but juvenile tail, flight-feathers, greater upper primary coverts, and outer greater coverts retained, tips of tail-feathers and primary coverts more pointed, more worn at same time of year. Black of crown mottled brown, but on average less so than in adult ♀, mid-crown stripe white with black spots, less clearly defined than in adult ♂. Supercilium variable, as adult ♀ (in spring, usually bright yellow); lore and upper cheek mottled brown, buff, and grey, contrasting rather with supercilium; ear-coverts buff-brown to dark brown, spot at upper rear corner buff-white, less contrasting than in adult ♂, borders of ear-coverts spotted black-brown or dull black. Ground-colour of upperparts rufous-buff or tawny, less olive than in adult, paler buff 'braces' on mantle and scapulars. Lower cheek, chin, chest, side of breast, and flank suffused buff-brown (darker than in adult ♂), lower side of neck, chest, and flank more closely marked with narrower, less even, and duller black streaks than in adult. Tips of greater and median coverts cream-buff (in fresh adult, tips of median coverts whiter, of greater coverts warmer pink-buff, but soon bleached to white at all ages). FIRST ADULT FEMALE. Like 1st adult ♂, but crown in autumn browner, black (if any) more concealed; central crown-stripe buff, inconspicuous; front part of supercilium buff-yellow, rear part pale yellow or white, ear-coverts buff-brown with dark brown surround, pale buff spot on rear less contrasting, supercilium hardly contrasting with lore; chin and throat partly marked with dull black triangular spots.

Bare parts. ADULT, FIRST ADULT. Iris dark brown. Bill flesh-pink, pale flesh-horn, or greyish-flesh with broad dark grey, steel-grey or horn-grey stripe on culmen and on tip of gonys. Leg and foot pale flesh-pink, dull pink-flesh, or brownish-flesh. (Hartert 1903–10; Kitson and Robertson 1983; Vonk and IJzendoorn 1988; BMNH, ZMB.) JUVENILE. No information.

Moults. ADULT POST-BREEDING. Complete; primaries descend-

ent. None examined in moult. In Siberia, plumage heavily abraded July, fresh in September, and moult thus probably August; some apparent 1-year-olds in fresh plumage by late July (Dementiev and Gladkov 1954). ADULT PRE-BREEDING. Partial; in winter quarters, but no information on timing and extent. In some birds, includes at least head and front of body (BMNH). POST-JUVENILE. Partial: head, body, lesser and median upper wing-coverts, and usually at least innermost greater coverts. Late July to early August (Dementiev and Gladkov 1954), but some September migrants had body apparently still partly juvenile (BMNH). FIRST PRE-BREEDING. Partial; in winter quarters. As adult pre-breeding, but frequently t1 as well.

Measurements. ADULT, FIRST ADULT. Whole geographical range (but mainly eastern China), September–June; skins (BMNH, RMNH, ZFMK, ZMB). Bill (S) to skull, bill (N) to distal corner of nostril; exposed culmen on average 3·5 less than bill (S).

	♂		♀	
WING	80·3 (1·86; 20)	78–84	74·8 (2·10; 18)	71–78
TAIL	62·2 (2·14; 16)	58–65	59·1 (2·28; 12)	55–62
BILL (S)	14·2 (0·45; 16)	13·4–14·8	14·1 (0·61; 12)	13·4–15·0
BILL (N)	8·5 (0·41; 16)	7·9–9·2	8·5 (0·43; 12)	7·8–9·3
TARSUS	20·0 (0·72; 16)	19·2–21·5	19·7 (0·42; 12)	19·2–20·8

Sex differences significant for wing and tail.

Weights. Fair Isle (Scotland), 1st adult ♂, October: 19·9 (Kitson and Robertson 1983).

Structure. Wing short, broad at base, tip rounded. 10 primaries: p7–p8 longest, p9 1–3·5 shorter, p6 0–1, p5 3–4, p4 7–12, p3 11–16, p2 14–18, p1 15–20; p10 strongly reduced, 49–57 shorter than p7–p8, 7–9 shorter than longest upper primary covert. Outer web of p5–p8 emarginated, inner web of p7–p8 (p6–p9) with notch, sometimes faint. Tip of longest tertial reaches tip of p1–p3. Tail rather short, tip slightly forked; 13–14 longest, t1 3–5 shorter, t6 1–2 shorter. Bill strong, conical, 7·2 (5) 7·0–7·5 deep at base, 7·7 (5) 7·2–8·3 wide; heavier than in other stripe-headed congeners with white in tail; culmen straight or slightly concave, ending in sharply pointed tip. Tarsus and toes rather short and thick. Middle toe with claw 17·6 (8) 16·5–18·5 mm; outer toe with claw c. 73% of middle with claw, inner c. 69%, hind c. 81%.

Geographical variation. None.   CSR

Emberiza rustica **Rustic Bunting**

PLATES 17 (flight) and 22
[between pages 256 and 257]

DU. Bosgors   FR. Bruant rustique   GE. Waldammer
RU. Овсянка-ремез   SP. Escribano rústico   SW. Videsparv

Emberiza rustica Pallas, 1776

Polytypic. Nominate *rustica* (Pallas, 1776), northern Eurasia except Anadyrland and Kamchatka. Extralimital: *latifascia* Portenko, 1930, Anadyrland and Kamchatka.

Field characters. 14·5–15·5 cm; wing-span 21–25 cm. Slightly shorter than Reed Bunting *E. schoeniclus* due to shorter tail; 10% larger than Little Bunting *E. pusilla*.

Medium-sized, rather upstanding, perky bunting, with rather square or peaked crown. All plumages show silky-white underparts with bold pattern of spots across breast and on flanks. ♂ beautiful: has black head with white lines and spots, strongly rufous breast, flanks, and rump, and bright white upper wing-bar and tail-edges. ♀ and immature less distinctive, lacking as distinctive a head pattern but still showing wide rufous streaks on flanks and mottled warm rufous-buff rump. Restless, popping up and down. Sexes dissimilar; some seasonal variation in ♂. Juvenile separable, being least rufous below.

ADULT MALE. Moults: June–September (complete); March–May (most of head and throat); full breeding plumage acquired by wear. Head jet-black, usually with narrow broken greyish-white line in centre of crown, white spot on nape (most visible from behind), bold white supercilium (from just in front of eye to hindneck), and small white spot on rear ear-coverts. Note that all white head markings may have yellowish tinge. From hindneck, past ear-coverts and around breast, variable rufous-chestnut shawl-cum-band produces striking 'scarf' which deepens on centre of breast where dark spot may show as 'tie-pin'. Above 'scarf', throat pure white, with some black on chin and short dark rufous-chestnut lower malar stripe (not reaching bill); below, underparts shining, almost silvery-white, with bold lines of rufous-chestnut marks from sides of breast along flanks. Back chestnut, sharply but not heavily streaked buff and less completely black, least so on yellow-fringed scapulars; rump rufous-chestnut, faintly mottled buff. Wings brightly patterned: lesser coverts chestnut and buff; median coverts mainly black with quite broad white tips (forming obvious upper wing-bar); greater coverts black, with buff edges and narrow white tips (forming less distinct lower wing-bar); tertials black, with rufous to buff margins; flight-feathers brown-black with pale brown fringes. Tail brown-black with bright white outer tail-feathers. Bill dark horn to lead-grey, with flesh base to lower mandible. Legs pale reddish-flesh. Fresh plumage less contrasting than that of spring, as fresh rufous-brown and buff tips overlay both black and white areas on head, most obviously on sides of crown, lores, and ear-coverts which appear almost wholly chestnut, chestnut and black, or distinctly mottled, thus contrasting less with buffier supercilium and spot on nape. Elsewhere, fresh buff tips show as scalloping on chestnut 'scarf', flanks, and back (underparts thus look more spotted, less lined), and both wing-bars have buff tinge. ADULT FEMALE. Plumage pattern and colours similar to fresh ♂ but duller with nape-spot rarely visible, 'scarf' less developed, malar stripe more pronounced (reaching bill), and breast and flanks with dark brown striations in chestnut patches. In fresh plumage, fore-supercilium and lore rufous-buff and ear-coverts dark brown, last emphasizing cream lower eye-crescent and pale buff spot on rear ear-coverts, but most striking feature are buffish-cream rear supercilium and wide pale buff to white sub-moustachial stripe which runs back under ear-coverts. Rump more mottled than ♂ and upper wing-bar always less pure white. JUVENILE. Differs from ♀ in even duller,

buffier head, warmer back, buffier upper wing-bar, and (most distinctively) blackish spots on breast-centre and tawny-brown streaks on sides of breast and within rufous-buff-washed flanks. FIRST-YEAR. Doubtfully distinguishable from adult, but retained juvenile flight-feathers may show at close range. At least some birds also keep heavier, dark breast markings including blackish 'tie-pin' spot.

♂ showing rufous breast and flanks unmistakable, but note that head and wing markings are similar in pattern to Yellow-browed Bunting *E. chrysophrys* (in close view, latter's yellow supercilium, grey lesser coverts, and striped breast dispel confusion). ♀ and immature much more liable to confusion with *E. chrysophrys*, *E. schoeniclus*, *E. pusilla*, and even Pine Bunting *E. leucocephalos* and Meadow Bunting *E. cioides*. In western Europe, *E. schoeniclus* will present most frequent trap, but it lacks rufous on rump and along flanks and shows only dull wing-bars, while calls quite different (see that species). In east of west Palearctic, *E. schoeniclus*, *E. pusilla*, and (marginally) *E. leucocephalos* will require separation. *E. pusilla* gives similar call but is noticeably smaller, with wholly rufous-buff median crown-stripe and face (not matched by *E. rustica* in any plumage) and dull, greyish rump. *E. leucocephalos* recalls *E. rustica* in showing whitish ground to underparts and rufous breast, flanks, and rump but is 15% larger and proportionally longer tailed, with grey lesser coverts, duller wing-bar, diffuse supercilium, often darker throat, more evenly streaked underparts (except in ♂) and quite different calls. Joint occurrence of *E. rustica* and *E. cioides* extremely unlikely; latter shows rufous rump but is somewhat larger and proportionally longer tailed, with dark uniform crown and nape, dark cheeks showing white supra-moustachial streak, grey collar, no discrete streaks on underparts, and repetitive calls. Flight light and fluttering, lacking uneven rhythm of *E. schoeniclus*; action closer to Yellow-breasted Bunting *E. aureola* than *E. pusilla*. Escape-flight usually short to dense cover but also stands ground, ascending ground plant to see better. Gait a perky hop, often (unlike larger buntings) with tail held up. Stance fairly level except when singing, but head usually held up, exaggerating depth of chest and displaying markings. Occasionally flicks wings and tail. Restless and active, feeding quickly and often moving on. Fond of damp habitats, even when vagrant. Gregarious where population dense, but usually solitary as migrant in western Europe.

Song a fairly short phrase, clear, mellow and melodious but varied, even irresolute in delivery (in last characteristic recalling Dunnock *Prunella modularis* or broken song of Robin *Erithacus rubecula*); can sound mournful, 'dudeleu-deluu-delee'. Commonest call much like Song Thrush *Turdus philomelos* but more distinct and higher pitched, 'zit'; also likened to louder of *E. pusilla*'s 2 common monosyllables. Also has disyllabic 'tic tic', recalling *E. rubecula* but with 2nd note sometimes lower pitched.

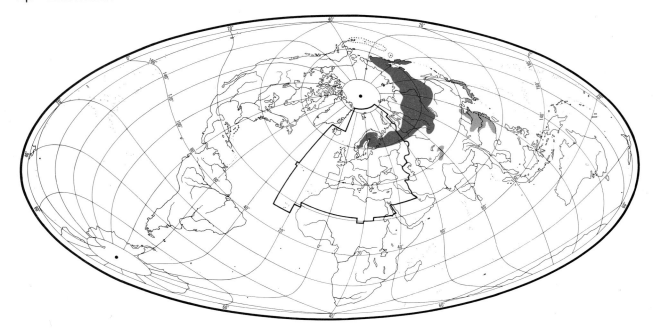

Habitat. Breeds in north Palearctic, in boreal zone between July isotherms of 12° and 22–23°C, having expanded in modern times from more easterly regions. Favours moist and wooded lowland situations, especially growth of willow *Salix*, birch *Betula*, and poplar *Populus* on margins of coniferous taiga forest by fens or river banks, or moist mosses (Voous 1960*b*). In northern Russia, prefers marshy forests, such as margins of spruce *Picea* or birch forests and extensive mosses, as well as river banks. Extralimitally ascends to 600 m in Altai. (Dementiev and Gladkov 1954.) In Finland, favoured breeding site is very damp pine swamp, pines being stunted, almost dead and covered with moss; a few willows, and birches of varying height, usually present, with, characteristically, ground vegetation of *Sphagnum*, cloudberry *Rubus chamaemorus*, and horsetail *Equisetum sylvaticum* (Bannerman 1953*a*). Normal habitat requirements for breeding are scattered or marginal trees, thickets or dense undergrowth, moors or heaths and often streams or pools.

Winters in cool temperate climates, on open as well as wooded terrain.

Distribution. Has expanded to west and south in 20th century.

NORWAY. Recently spread south-west (VR). SWEDEN. Has expanded range. Only 2 records before 1897, when first recorded breeding. In 1950s rather common north of 64°N along coast and in west; subsequently recorded with increasing frequency further south, and expansion apparently still in progress. (LR.) FINLAND. Has expanded to west and south in 20th century (OH). LATVIA. First recorded breeding 1985 (Vīksne 1989).

Accidental. Iceland, Britain, Portugal, France, Belgium, Netherlands, Denmark, Germany, Poland, Czechoslovakia, Switzerland, Austria, Italy, Malta, Yugoslavia, Greece, Bulgaria, Syria, Iraq, Kuwait, Egypt.

Population. Has increased in Fenno-Scandia in 20th century.

NORWAY. Estimated 100–300 pairs 1988 (VR); presumably increasing with spread to south-west. SWEDEN. Estimated 50 000 pairs (Ulfstrand and Högstedt 1976); increasing (LR). FINLAND. Increase since 19th century in connection with range expansion; reasons obscure. 500 000 to 1 million pairs. (Koskimies 1989.)

Movements. All populations migratory. Western birds head east then south, and eastern birds head south or south-west, to reach winter quarters via south-east Russia and north-east China.

Winters in eastern China from southern Hopeh and Shantung south to Fukien, and west along Yangtze valley to southern Shensi and eastern Szechwan; perhaps also in Tien Shan in north-west China (Schauensee 1984; Cheng 1987); vagrant as far south as Hong Kong (Chalmers 1986). Abundant in winter in lowlands of Korea, especially in south (Austin 1948; Gore and Won 1971). Common and locally abundant in Japan; winters from southern Hokkaido south to Kyushu, with highest numbers in coastal regions of central and northern Honshu; very uncommon or rare in Nansei islands in south (Brazil 1991). In Ussuriland (south-east Russia), winters irregularly in small numbers (Panov 1973*a*).

Migration prolonged in both seasons. In autumn,

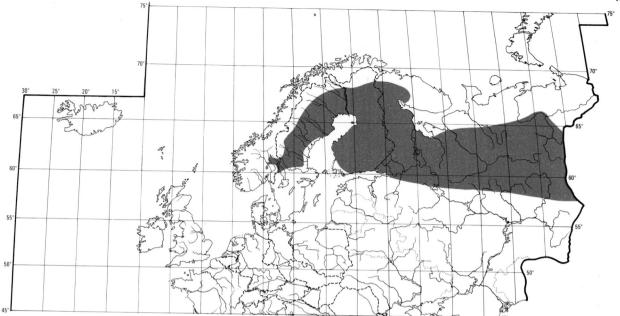

begins in August in northern Europe. Birds leave Sweden end of August to September; occurs in large numbers on east coast, mainly immediately east of breeding areas, but also further south (south to *c.* 58°45′N) (Staav 1976; SOF 1990). On Kola peninsula (north-west Russia), average latest report 12 September (3-20 September) over 10 years (Semenov-Tyan-Shanski and Gilyazov 1991). In St Petersburg region, movement probably begins in last third of August, and peaks early or mid-September; main migration ends in early October, with stragglers throughout month (Rymkevich 1979). Apparently now regular as far south as Ryazan' region (south-east of Moscow) in autumn (first reported 1953), and rare there in spring (Ivanchev 1988). Also a few recent records both seasons in Voronezh region further south (Vorobiev and Likhatski 1987). Birds migrate via forest regions, skirting (to north) European steppes and central Asian deserts (Dementiev and Gladkov 1954). Regular but rare as far south as Ural valley in north-west Kazakhstan (Korelov *et al.* 1974). In western Siberia, passage September to mid-October, sometimes from end of August; movement intensive through southern taiga region (e.g. at Tomsk), but very scarce further south, e.g. in Zaysan depression (north-east Kazakhstan); latest records at Tomsk 11-17 October over 3 years (Johansen 1944; Gyngazov and Milovidov 1977; Vartapetov 1984). In Mongolia, reported only in extreme north on passage (Zink 1985), and occasionally occurs in south-west Transbaykalia (Kozlova 1933). Further north, in southern Yakutiya, movement very intensive; starts end of July and continues to beginning of October, with stragglers to late October (Noskov and Gaginskaya 1977). Leaves lowlands of north-east Siberia late August to early September, and

scarce by mid-September (Krechmar *et al.* 1991). Similarly on Kamchatka peninsula, main autumn migration begins 20-25 August, continuing mostly to 13-16 September; a few late records (to 10 December); common and widespread, especially on woodland edges (Lobkov 1986). Passage through Ussuriland usually begins only in early October (occasionally late September), peaking in mid-October (Panov 1973*a*). On Sakhalin island, movement mid- to late October (Gizenko 1955). Arrives in north-east China and Japan from mid-October (Hemmingsen and Guildal 1968; Brazil 1991). Highest numbers in Korea from mid-November (Gore and Won 1971); many birds ringed there retrapped in later years, and several recoveries in southern and central China of birds ringed Korea (McClure 1974).

In spring, migrates through north-east China from late February, mostly to end of March, with small numbers to mid-April; markedly earlier than most other Emberizidae (Williams 1986). Begins to leave southern Korea in mid-March, with only stragglers left by early April (Austin 1948). Departs from Japan late March to April, with a few remaining until early May (Brazil 1991). In southern Ussuriland, earliest records 10-24 March over 3 years, and peak passage in 2nd half of April (Panov 1973*a*). Arrives on southern Sakhalin in 2nd half of April; bird ringed Japan, November, recovered in Sakhalin in May (Gizenko 1955; Zink 1985). Occasionally recorded in late April on Kamchatka peninsula, usually not until mid-May, with peak in last third of May; widespread from seashore to subalpine zones (Lobkov 1986). Reaches southern Yakutiya from early May (Vorobiev 1963), and middle reaches of Anadyr' in extreme north-east of range in last third of May (Krechmar *et al.* 1991). Main passage through central Siberia 25 April to 12 May (Reymers 1966). Reaches south of western Siberia in April (Johansen 1944), but arrives in north-east Altai in 1st half of May (Ravkin 1973), and northern taiga zones not until end of May or beginning of June (Vartapetov 1984). In St Petersburg region, passage usually begins in late April, but local birds arrive with main movement, in early May (Rymkevich 1979). First arrivals in southern Finland in early May, occasionally late April (Tiainen 1979). In Finnish Lapland, 1974–85, average earliest record 14 May (26 April to 29 May) (Pulliainen and Saari 1989), and on Kola peninsula, over 16 years, average earliest 19 May (6–31 May) (Semenov-Tyan-Shanski and Gilyazov 1991). In Sweden, spring records south of breeding range are far more widespread than autumn, reaching many inland areas (Staav 1976).

Although in autumn most birds change to southward heading only in south-east Russia, a few head south at earlier stage of migratory route. Thus, in Kazakhstan, sometimes encountered in south-east in winter, especially at Alma-Ata (Korelov *et al.* 1974), and also reported in Amu Dar'ya valley, Uzbekistan (Salikhbaev and Bogdanov 1967). 2 records (December, March) from extreme east of Afghanistan (S C Madge); in Iran, probably a scarce passage migrant, with 6 recent records mid-February to mid-March and late October to late November (D A Scott); also February record from Iraq (Marchant and Macnab 1962); vagrant to eastern Arabia (F E Warr). In Israel, 1–9 individuals occur annually in autumn (once in March) at Elat in south, with rare reports elsewhere (Shirihai in press).

Vagrant west and south of range both seasons in Europe. In Britain, now occurs annually; recorded late March to June (mostly May), and end of August to early November; 34 records before 1958, and 210 in 1958–91 (including Ireland); 55% of records 1958–85 were in autumn; occurs chiefly on east coast, especially Shetland (Scotland); bird ringed Fair Isle, June 1963, was recovered on Chios island off western Turkey, October 1963 (Dymond *et al.* 1989; Rogers *et al.* 1992). In Netherlands, 32 records in all, of which 25 in September–October, 1 in November, and 6 in March–May (Berg *et al.* 1992). In Denmark, 19 records 1965–90 (Frich and Njordbærg 1992). In western Germany, recorded mostly on Helgoland; 16 reports 1977–90 south to Nordrhein-Westfalen (Bundesdeutscher Seltenheitenausschuss 1989, 1992). In France, 8 autumn records in 20th century and 1 in spring, mostly in south and west (Dubois and Yésou 1992). In Malta, records (rare) chiefly October–November, once in March (Sultana and Gauci 1982). For vagrancy elsewhere in Europe, see Distribution.   DFV

Food. Mainly seeds, plus insects and spiders in breeding season. Feeds on ground and in bushes, commonly in damp and marshy places, and much invertebrate prey associated with water. In forest, tosses leaves aside while foraging. (Neufeldt 1961; Rymkevich 1979; Kishchinski 1980; Bergmann and Helb 1982; Armani 1985; Bentz and Génsbøl 1988; Semenov-Tyan-Shanski and Gilyazov 1991.) In China and Japan in winter, most often seen feeding on ground and low vegetation, sometimes in flocks of hundreds; in woodlands, clearings, forest edges, open spaces including urban parks, etc., but especially in rice stubble, reedbeds, and on riverbanks (Cheng 1964; Brazil 1991). In north-west Russia, August–September, fed on grain in fields of barley *Hordeum* and oats *Avena*; also in stubble and other arable crops (Mewes and Homeyer 1886; Mal'chevski and Pukinski 1983). Vagrants on coast of south-east England, November, fed among weeds and grass (Romer 1947); on Fair Isle (northern Scotland) and Helgoland (north-west Germany) foraged in rough grass and vegetable fields, and on oat stacks (Bannerman 1953*a*).

Diet in west Palearctic includes the following. Invertebrates: damsel flies and dragonflies (Odonata), stoneflies (Plecoptera), grasshoppers, etc. (Orthoptera), aphids (Aphidoidea), adult and larval Lepidoptera (Tortricidae, Geometridae), adult and larval caddis flies (Trichoptera), flies (Diptera: Tipulidae, Culicidae, Bibionidae), adult and larval Hymenoptera (Tenthredinidae), beetles (Coleoptera: Cerambycidae, Chrysomelidae, Curculionidae), spiders (Araneae). Plants: seeds (etc.) of cloudberry *Rubus*, grasses (Gramineae, including barley *Hordeum*, oats *Avena*), sedge *Carex*. (Mewes and Homeyer 1886; Dementiev and Gladkov 1954; Neufeldt 1961; Mal'chevski and Pukinski 1983; Semenov-Tyan-Shanski and Gilyazov 1991.)

On Kola peninsula (north-west Russia) in summer, 41% of diet (presumably by frequency of occurrence) Curculionidae, 37% caterpillars, 27% Diptera, 20% spiders, 17% aphids, 11% Hymenoptera; proportion of animal material declines over summer and by September diet mostly seeds (Dementiev and Gladkov 1954); see also Semenov-Tyan-Shanski and Gilyazov (1991). In Kareliya (north-west Russia), of 20 stomachs, 18 contained invertebrates and 5 plant material; invertebrates were Odonata, stoneflies, caterpillars (Tortricidae, Geometridae), caddis flies, Diptera, and beetles (Cerambycidae, Chrysomelidae, Curculionidae) (Neufeldt 1961). In Yakutiya (eastern Russia), July–August, 6 stomachs contained caterpillars, Diptera, adult and larval beetles, spiders, large seeds of trees, and grass seeds (Larionov *et al.* 1991). In South Korea, October, 12 stomachs held 279 items, of which only 0·4% by number animal prey (Diptera); 41% seeds of grass, 23% barberry *Berberis*, 13% millet *Panicum*, 11% *Callicarpa* (Verbenaceae), 6% mouse-ear *Cerastium*, 2% rice *Oryza* (Won 1961). In Japan, winter, takes seeds of Coniferae (including Japanese cedar *Cryptomeria*), Moraceae, Polygonaceae, Leguminosae, Oxalidaceae, Rubiaceae, Caprifoliaceae, Gramineae, and insects (Orthoptera, Lepidoptera, Diptera, Coleoptera); one stomach contained 100 grass seeds (Cheng 1964).

Diet of young probably mostly invertebrates. In St Petersburg region (north-west Russia), adults seen to feed young insects, including caterpillars, Tipulidae, and sawflies (Hymenoptera: Symphyta) (Mal'chevski and Pukinski 1983). At nests in Finnish Lapland and Volga-Kama region (east European Russia), nestlings given small caterpillars (Davidson 1951; Popov 1978).   BH

Social pattern and behaviour. Some aspects little known. Important study, using marked birds, in St Petersburg region (north-west Russia) by Rymkevich (1979). Following account also based heavily on extralimital information, eastern Russia except where otherwise stated.

1. Gregarious outside breeding season. In St Petersburg region, from beginning of July, family parties start to form small nomadic flocks which move rapidly through forests and river valleys (Rymkevich 1979). On autumn migration, Kamchatka, seen singly and in flocks of up to 40 birds; at this time up to 470 birds per km² (Lobkov 1986); migrating flocks of 3–10, often with some Olive-backed Pipits *Anthus hodgsoni* (Sokolov and Lobkov 1985). By November, in Ussuriland, birds are encountered singly in flocks of other buntings (Emberizidae), including Yellow-throated Bunting *E. elegans*; apparently a few birds overwinter in south of Ussuriland, typically in flocks of up to 7 birds (Panov 1973a). Wintering birds in Japan gather in flocks of up to 30 birds, but in late winter up to 400 together (Brazil 1991). Occasionally also flocks of several hundred on spring migration (Yamashina 1982). On spring migration, Ussuriland, birds migrate singly, in twos, or in small flocks (mostly up to 6 birds) (Panov 1973a). In Kamchatka, during peak spring migration (along forest edge, with many thawed patches), up to 1000 birds per km² (Lobkov 1986). In Finland, after spring arrival, initially in flocks of up to 10–20 (Carpelan 1932). In Yakutiya, mid-May, small flock exlusively ♂♂ (♀♀ probably not yet arrived) (Vorobiev 1963). Flocks later break up into pairs. BONDS. No evidence for other than monogamous mating system. No information on duration of pair-bond. Nest built apparently by ♀ (Rymkevich 1979). Both sexes incubate (Haftorn 1971; Popov 1978), at least for 1st clutch, although ♂ has no brood-patch and tends to do shorter stints (Rymkevich 1979). Young are fed by both sexes (Popov 1978), more by ♂ towards end of nestling period. When ♀ starts incubating 2nd clutch (*c.* 1 month apart after 1st) ♂ takes over feeding 1st-brood fledglings. Parents feed young for 9–15 days after leaving nest. (Rymkevich 1979; see also Relations within Family Group, below.) Age of first breeding not known. BREEDING DISPERSION. Solitary and territorial. Little known about size or detailed function of territory, but 'nesting area' of pair in Kamchatka was 0·6 ha (Lobkov 1986). In Finnish peatland ('mires'), 1963–84, average *c.* 3·6 pairs per km² (Kouki and Häyrinen 1991); in Värriö (northern Finland), 1985–87, 0·9 pairs per km² in spruce *Picea* forest, 2·0 in pine *Pinus* forest with undergrowth (Pulliainen and Saari 1989); in Sompio, 1982–85, 0·2 pairs per km² in virgin pine forest, 2·2 in virgin spruce (Virkkala 1987); in same region, 1983–89, 1·0 in virgin coniferous forest (Virkkala 1991). In Sweden, average 1 pair per km² (Ulfstrand and Högstedt 1976). In Hedmark (Norway) 0·03–0·4 territories per km² (Sonerud and Bekken 1979). In European Russia: 166 birds per km² in pine forest, Vologda region (Vtorov and Drozdov 1960); in St Petersburg region, over 2 years, maximum 167 pairs per km² in damp forest, maximum 28 in mosaic of dry pine forest and wetter boggy areas (Rymkevich 1979); in Kareliya, 1·2 pairs per km² in pine forest, 10·1 in bogs with and without trees (Ivanter 1962). In northern Urals, 1979–86, mean 7·0 ± 2·0 pairs per km² (Shutov 1990). In Kamchatka, highest densities in riverine woods of alder *Alnus*, willow *Salix*, and poplar *Populus*: up to 190 pairs per km²; in birch *Betula ermani*, 28·9–103·4; in mixed forest, average 18·2, locally up to 44 (Lobkov 1986). For densities in different habitats in central Siberia, see Rogacheva (1988) and Larionov *et al.* (1991). In St Petersburg region, average 30–40 m between nests for 1st and 2nd broods (Rymkevich 1979). ROOSTING. During incubation, ♀ sits on nest at night (Rymkevich 1979). No information on other roost-sites. On migration, small flocks rest in bushes or marshland (Staav and Fransson 1987). No further information.

2. Fairly unobtrusive in breeding season (Staav and Fransson 1987). In northern Urals, more secretive than Yellow-breasted Bunting *E. aureola*, with which shares same habitat (Portenko 1937). According to Sushkin (1938), however, very confiding in Altai (southern Russia), allowing close approach compared with other Emberizidae. When excited, flicks tail, raises crown feathers (including in Song-display: see below), and gives Contact-alarm calls (see 2 in Voice) (Carpelan 1932; Etchécopar and Hüe 1983). FLOCK BEHAVIOUR. Migrants fly up into trees when disturbed (Panov 1973a). On arrival on breeding grounds, flocks move about in forest, feeding on ground, notably in thawed patches (Carpelan 1932; Rymkevich 1979). SONG-DISPLAY. ♂ sings (see 1 in Voice) from elevated perch, with crown raised (Fig A). Singing starts in flocks on spring migration, e.g. heard

A

in March (Panov 1973a). In Fenno-Scandia, song starts beginning of May, continuing for only relatively short period, after which little heard till young fledge in June (Staav and Fransson 1987). In St Petersburg region, resurgence of song in 2nd half of June coincides with dispersal of 1st broods and incubation of 2nd clutches (Rymkevich 1979). In northern Urals, song heard up to beginning of August (Portenko 1937). Some song heard on autumn migration, including quiet song from young ♂♂ (Panov 1973a; Lobkov 1986). For singing by both sexes in apparent precopulatory display, see Heterosexual Behaviour, below; for singing during disturbance at nest, see Parental Anti-predator Strategies (below). ANTAGONISTIC BEHAVIOUR. No information. HETEROSEXUAL BEHAVIOUR. Little known. ♂♂ usually start arriving on breeding grounds ahead of ♀♀ (Rymkevich 1979; Lobkov 1986). In Kamchatka, pair-formation after *c.* 18 May, territorial establishment and courtship mainly 23 May–2 June, copulation 2–17 June, nest-building up to 20 June (Lobkov 1986). In St Petersburg region, where double-brooded, whole breeding cycle takes *c.* 90 days (Rymkevich 1979). Little known about intimate courtship. During nest-building, copulation regularly seen, quite often followed by ♂ chasing ♀; in possible soliciting-display, ♀ with raised tail and wing-shivering moved slowly around in tree canopy, alternating quiet song with ♂ who followed close behind. Several records of ♂ collecting and carrying nest material but no confirmation that he helps to build. (Rymkevich 1979.) Both sexes approach nest furtively low over ground (Kapitonov and Chernyavski 1960). In most cases, incubating bird leaves nest on hearing call of approaching mate, less often when mate reaches nest if approach is silent (Rymkevich 1979). RELATIONS WITHIN FAMILY GROUP. For physical development of young see Rymkevich (1979). Young brooded up to day 5. Faecal sacs are swallowed (presumably when young are small) or carried away. (Rymkevich 1979.) For role of sexes in feeding young up to independence, see Bonds (above). In evening peak feeding period, parents tend to forage close to nest, at other times quite far away. Young leave nest at 9–10 days and disperse over territory; flight-powers still not fully developed at 14–15 days, but once able to fly well, family party reunites, initially tending to stay within c. 200 m of nest; one record of family party still intact when young 24 days old (i.e. 14–15 days after fledging). If laying 2nd clutch, ♀ starts new nest almost immediately after 1st brood leaves nest, but continues to help ♂ with feeding fledglings for a few days till she starts incubating. After young independent, family party leaves territory and joins up with others to form small nomadic flocks. (Rymkevich 1979.) ANTI-PREDATOR RESPONSES OF YOUNG. Young newly out of nest (10 days old) tend to crouch when approached; at 12 days, more often fly short distance before seeking cover (Rymkevich 1979). PARENTAL ANTI-PREDATOR STRATEGIES. Only reactions known are those to man. In Finland, parents disturbed at nest tail-flicked, called, etc., and hopped from twig to twig a few metres from intruder; ♀ leaves nest when observer c. 2–4 m away and performs impeded flight (distraction-lure display of disablement type), calling the while; ♂ joins in calling; if nest contains eggs, pair often move away and return only after ½ hr or more (Carpelan 1932). Also in Finland, impeded flight described by Davidson (1954) at nest with week-old young: parents kept fluttering around intruders, just above ground as if hawking insects, with tail depressed and extended, legs dangling and sometimes brushing ground; then both would alight, raise their crests and keep up Alarm-calls (see 3 in Voice). In St Petersburg region, both sexes perform distraction-display on ground when they have eggs, nestlings, and sometimes recently-fledged young: adult will run, wings spread and raised, tail sometimes fanned (this perhaps

display reported by Witherby et al. 1938); if not followed by intruder, bird will return, 'collapse' at intruder's feet, then purposefully run off again; when highly excited will also run along branches or glide down from a height with widely spread tail (Rymkevich 1979). In northern Urals, when man approaches nest, owners call excitedly (2–3 in Voice) and ♂ once started singing; sometimes run along low branches, head bowed low, and gradually move off into cover (Portenko 1937). In Altai, bird disturbed at nest seen fluttering around observer, almost to within touching distance, but no records of injury-feigning (Sushkin 1938).

(Fig by D Nurney from photograph in Sonerud and Bekken 1979.) EKD

Voice. Freely used throughout year, though song confined mainly to breeding season. Repertoire not fully known. For additional sonagrams see Bergmann and Helb (1982).

CALLS OF ADULTS. (1) Song. Clear, melodious, and rather short, with melancholy timbre (like Lapland Bunting Calcarius lapponicus) and irregular patterning (like Dunnock Prunella modularis), e.g. 'dúdely díi-dah delúú-deli' (L Svensson, H Delin). Charming, melodious, liquid song, not like any other Emberiza (Staav and Fransson 1987; Bentz and Génsbøl 1988) except perhaps Rock Bunting Emberiza cia, albeit phrases (of E. rustica) shorter, lower-pitched, and lack repetition of units; can also recall Garden Warbler Sylvia borin (Haftorn 1971; Bergmann and Helb 1982; Etchécopar and Hüe 1983; P J Sellar). In recordings analysed by J Hall-Craggs, average phrase-length 1·8 s (1·5–2·5, n = 26 phrases from 4 birds). Most distinctive feature of phrase is its melodic line which, when quite slowly delivered, may be said to undulate (e.g. Fig I); but when delivered rapidly (e.g. Figs II–

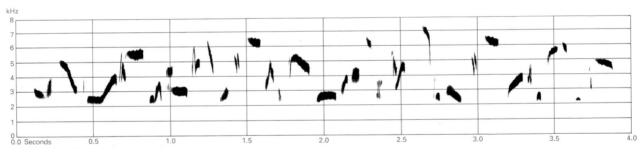

I J-C Roché Finland June 1968

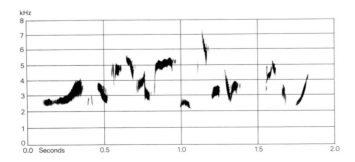

II P A D Hollom Finland June 1988

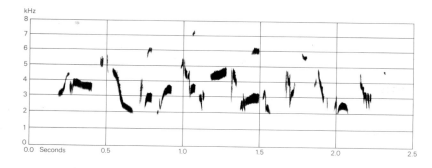

III B N Veprintsev Russia June 1977

III) individual units (or notes) zig-zag between low band (fairly constant range of less than 2 kHz to 3 kHz) and medium/high bands; middle and upper notes often describe ascending and descending melodies which, in unabridged phrases, peak at 4·0-6·5 kHz, then descend to low range at end when, occasionally, high note added (e.g. Fig IV). In one unusually long phrase (Fig I) this rise and fall occurs twice within 3·9 s. First and last notes of phrase may be very quiet. Songs of some birds have characteristic introductory note (recalling Bullfinch

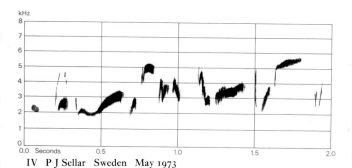

IV P J Sellar Sweden May 1973

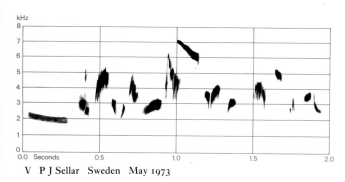

V P J Sellar Sweden May 1973

Pyrrhula pyrrhula, e.g. 'hooEEE' in Fig I, 'pyooo' in Fig V: P J Sellar) or notes (e.g. Fig IV shows introductory subphrase of 5 units ending in rising 'hooEEE', common to all phrases in recording of this bird). (J Hall-Craggs.) According to Bergmann and Helb (1982), ♂ has several phrase-types, each starting the same, after which may vary, some phrases being lengthened, others abbreviated. Quiet snatches of song are heard from some 1st-year ♂♂ in autumn (Lobkov 1986). (2) Contact-alarm call. Thin, high-pitched 'tsip, tsip, tsip' (Yamashina 1982), most often compared to Tsip-call of Song Thrush *Turdus philomelos*: 'zit', very like *T. philomelos* but slightly higher and more distinct (Bruun *et al.* 1986), hard 'tzit' or 'tsik' (Bentz and Génsbøl 1988; Jonsson 1992), 'zik(k)' (Haftorn 1971; Bergmann and Helb 1982). In recordings, calls are quiet, even at close quarters; Fig VI depicts series of 'tsik-' or 'zik-' calls at half speed to accommodate top of pitch range in excess of 8 kHz (J Hall-Craggs). This call similar to call 2a of Little Bunting *E. pusilla*, in which introduction of 'z' sound denotes some agitation (Brood and Söderquist 1967), but *E. rustica* lacks metallic quality of Meadow Bunting *E. cioides* (Wild Bird Society of Japan 1982). (3) Alarm-call. Soft drawn-out 'tsie' like Penduline Tit *Remiz pendulinus* (Jonsson 1992); drawn- out 'tsiee', rather like Spotted Flycatcher *Muscicapa striata* (Bentz and Génsbøl 1988). High-pitched 'twüit' (Bergmann and Helb 1982) perhaps the same. (4) Other calls. Short 'trill', similar to Yellow-throated Bunting *E. elegans*, sometimes heard in autumn (Panov 1973a: Ussuriland, eastern Russia).

CALLS OF YOUNG. No information.   EKD

Breeding. SEASON. Sweden: start of laying mid-May to 1st half of June (Ehrenroth and Jansson 1966; Haftorn 1971). Finnish Lapland: average date of 1st egg 6 June (22 May to 5 July, $n=24$); nests above ground started *c.*

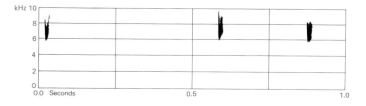

VI J-C Roché Finland June 1968

10 days earlier than those on ground; nests for 2nd broods started late June or early July (Pulliainen and Saari 1989); see also Carpelan (1932). St Petersburg region (north-west Russia): eggs laid beginning of May to beginning of July (Rymkevich 1979; Mal'chevski and Pukinski 1983). Kola peninsula (north-west Russia): at 4 nests, clutches started 25 May to 16 June (Semenov-Tyan-Shanski and Gilyazov 1991). SITE. Generally on ground, often near water, in tussocky grass, *Sphagnum* moss cushions, among thick roots, etc., beside bush or tree or under overhanging grass, sedge, etc., or (rarely) in depression in level ground; sometimes above ground in tree or on stump (Carpelan 1932; Haartman 1969; Rymkevich 1979; Krechmar *et al.* 1991; Semenov-Tyan-Shanski and Gilyazov 1991). In Finnish Lapland, of 50 nests, 48% on ground in tussock or among grasses and herbs, 20% under tree or bush, 16% just above ground in roots, 16% in juniper *Juniperus* tree; average height of last 2 categories 0·8 m (0·1–3·0, $n = 16$) (Pulliainen and Saari 1989). In Kamchatka (eastern Russia), of 46 nests, 54% on ground, mostly by tree, 20% in roots of fallen tree or on stump, 17% in hollow birch *Betula* or alder *Alnus* at 2·9 m (1·5–4·5) above ground, 9% in dwarf pine *Pinus pumila* or alder bush at 0·8 m (Lobkov 1986). Nest: foundation of grass, sedge, horsetail *Equisetum*, moss, leaves, needles, lichen (etc.), lined with fine plant material, hair, and sometimes feathers (Carpelan 1932; Swanberg 1951; Rymkevich 1979; Krechmar *et al.* 1991; Semenov-Tyan-Shanski and Gilyazov 1991). In Kamchatka, 31 nests had average outer diameter 12·0 cm (9·1–16·0), inner diameter 6·7 cm (5·0–9·5), overall height 7·5 cm (5·6–9·5), depth of cup 5·0 cm (4·0–6·5) (Lobkov 1986); see also Popov (1978), Larionov *et al.* (1991), and Semenov-Tyan-Shanski and Gilyazov (1991). Building: probably by ♀ only (Rymkevich 1979). EGGS. See Plate 34. Sub-elliptical, smooth and faintly glossy; pale bluish-green or greenish-white, densely covered with fine olive-brown or greyish spots, blotches, and scrawls, though these can be absent; pale grey undermarkings (Dementiev and Gladkov 1954; Harrison 1975; Makatsch 1976); see also Carpelan (1932) for discussion of egg types. Nominate *rustica*: 20·3 × 15·2 mm (18·5–22·3 × 13·8–16·0), $n = 253$; calculated weight 2·46 g (Schönwetter 1984). Clutch: 4–5 (3–6). In Finnish Lapland, average 5·03 ($n = 31$); before mid-June 5·20 ($n = 25$), after 4·33 ($n = 6$); *c.* 20–75% of pairs throughout Finland have 2 broods (Pulliainen and Saari 1989); of 11 clutches from Sweden and Finland, 3 of 4 eggs, 5 of 5, and 3 of 6; average 5·0 (Makatsch 1976); see also Haartman (1969) for Finland; in Finland, 1st clutch commonly 5–6 eggs, 2nd 4–5 (Carpelan 1932). In St Petersburg region, of 24 clutches: 4 eggs, 58%; 5, 33%; 6, 8%; average 4·5, but no difference between early and late clutches noted (Mal'chevski and Pukinski 1983); in 2 cases, 28 and 32 days between hatching of 1st and 2nd clutches (Rymkevich 1979). In Kamchatka, 41 1st clutches averaged 4·2, 8 replacement clutches 3·0 (Lobkov 1986). In Yakutiya (eastern Russia), of 19 clutches: 4 eggs, 37%;

5, 58%; 6, 5%; average 4·7 (Larionov *et al.* 1991). Eggs laid daily (Makatsch 1976; Pulliainen and Saari 1989). INCUBATION. 11(–13) days; by both sexes (Popov 1978; Mal'chevski and Pukinski 1983; Lobkov 1986; Krechmar *et al.* 1991). In 21 observations of adult on nest in Finnish Lapland, 71% ♀; in one case, hatching spread over 2 days (Pulliainen and Saari 1989). In St Petersburg region, adults changed over regularly during day; ♀ had stints of 30–90 min, ♂ 30–60 min (Rymkevich 1979). YOUNG. Fed and cared for by both parents; brooded by ♀ for *c.* 5 days (Davidson 1951; Rymkevich 1979; Mal'chevski and Pukinski 1983). FLEDGING TO MATURITY. Average fledging period in one study 8·5 days (7–10, $n = 7$) (Pulliainen and Saari 1989). In St Petersburg region, 9–10 days, and young leave nest before able to fly; fed by adults for 9–15 days after leaving nest (Rymkevich 1979, which see for development of young; Mal'chevski and Pukinski 1983). BREEDING SUCCESS. In Finnish Lapland, 82% of 39 breeding attempts survived egg stage, and 67% nestling stage, giving overall success of 55% (47% when partial losses included), resulting in 1·7 fledged young per nest overall ($n = 26$), or 3·5 per successful nest ($n = 13$); 20% of eggs produced fledged young in 1974–82, 68% in 1983–8 (difference significant); variation perhaps due to weather. Clutches started in May had success rate of 36%, June–July 72%; no difference in success recorded between different habitats or nest-sites. Of 13 losses, 8 caused by predators, 2 by weather, 2 by starvation, and 1 by desertion; 2 young killed by Diptera larvae. (Pulliainen and Saari 1989.) Also in Kamchatka, nestlings killed by warble-fly (Oestridae) larvae (Lobkov 1986). BH

Plumages. (nominate *rustica*). ADULT MALE BREEDING (March–August). Nasal bristles black. Forehead, crown, and side of nape jet-black with yellowish-white edges to feathers, latter widest on centre of forehead and crown, but fully worn off in abraded plumage. Spot on nape white to yellowish-white, varying in size and sometimes hardly visible. Supercilium white or yellowish-white, narrow or absent in front of eye, broad behind eye, running to end of black cap. Lore and ear-coverts black, centre of ear-coverts with faint light specks, rear with yellowish-white spot, but spot sometimes hardly visible. Submoustachial stripe white, bordered by black of ear-coverts above and by narrow rufous-chestnut malar stripe below; malar stripe rather inconspicuous, not reaching base of bill. Eye-ring black, white above eye. Broad rufous band across hindneck, feathers with traces of grey-buff fringes when plumage fresh. Centres of feathers of mantle and scapulars rufous-chestnut, fringes pale yellowish or greenish-buff, widest on side of mantle, forming yellow stripe; mantle sometimes with black shaft-streaks, scapulars occasionally with narrow black streaks. Rump and upper tail-coverts rufous-chestnut, feathers of rump with pale yellowish or buff edges, longest tail-coverts occasionally with narrow dark central streak. Chin and throat white; feathers at base of lower mandible black. Gorget rufous-chestnut, rather narrow, pointing downward on central chest, joining rufous of band across neck; flank and thigh rufous-chestnut, feathers fringed white when fresh. Remainder of underparts white, except sometimes for scattered rufous feathers on belly. Central pair of tail-feathers (t1) pale brown with dark-brown shaft-streak or dark-brown spot at tip, edged buffish or slightly rufous, especially on outer web; t2–t4 blackish-brown with yellowish or buffish edge, t4 sometimes with small white tip to inner web; t5 black with white wedge on inner web, covering one-third to three-quarters of distal part; t6 white, except for black spot on tip and black basal quarter of inner web. Flight-feathers, bastard wing, and greater upper primary-coverts dark grey-brown; outer web of primaries fringed whitish, orange-brown, or chestnut (palest on p6–p9); secondaries broadly fringed orange-brown or chestnut; primary-coverts and bastard wing with dull brown edge. Tertials black with broad chestnut, pale brown, or orange-brown fringe on outer web, extending onto tip of inner web. Greater upper wing-coverts black, fringed buff, brown, or buffish-brown on outer web, tipped pure white, tips forming wing-bar; innermost subterminally tinged chestnut. Median coverts black with broad white tip and small subterminal chestnut spot; lesser coverts rufous-chestnut with narrow yellowish fringe and grey base. Under wing-coverts white; axillaries white with grey bases; marginal coverts grey-brown with white tip. ADULT MALE NON-BREEDING (August–March). Like adult ♂ breeding, but centre of feathers on forehead, crown, nape, and ear-coverts black, chestnut, or black with chestnut edge, broadly fringed yellowish-white, buff, or rufous, widest on forehead and along central part of crown; black centres on feathers of crown and nape sometimes narrow, edges buffish, and chestnut or rufous absent, but this perhaps in 1st non-breeding ♂ only. Super-cilium, submoustachial stripe, and throat with yellowish tinge, occasionally with narrow black tips on throat-feathers. Feathers on neck, mantle, gorget, and flank with broader buff, pale yellow, or whitish fringes than in adult ♂ breeding. Inner primaries and all secondaries with narrow greyish tips; tips of greater and median upper wing coverts buffish-white. ADULT FEMALE BREED-ING (March–August). Feathers on forehead and crown brown, blackish-brown, or black, bordered with chestnut and edged yellowish-white, or with narrow black shaft-streak, bordered chestnut and edged buff; buff or yellowish edges sometimes absent; feathers in centre of crown with broad buff or yellowish edges, forming ill-defined crown-stripe. Nape with white spot. Supercilium narrow and brown or off-white in front of eye, broad and white or yellowish-white above and behind eye. Lore brown or buff. Eye-stripe light brown, chestnut or blackish, starting behind eye. Ear-coverts yellowish-rusty brown, buff, chocolate-brown, or blackish, feathers edged buffish when fresh; rear of coverts with pale buff or whitish spot, partly bordered with dark brown. Moustachial stripe brownish-black, incon-spicuous. Submoustachial stripe pale buff or white, malar stripe narrow, but usually more conspicuous than in ♂, blackish-brown or rufous-chestnut. Rufous band across hindneck, feathers fringed yellowish when fresh. Mantle and scapulars like adult ♂ breeding, but greenish-buff fringes broader, largely concealing chestnut and black, scapulars usually with more black along shaft. Remainder of body like adult ♂ breeding, but rufous centres of feathers of gorget and flank usually narrower, less extensive; sometimes, a few black-brown feathers or black spots on ventral part of gorget. Wing and tail as in adult ♂ breeding, but grey bases of lesser wing-coverts sometimes more extensive. ADULT FEMALE NON-BREEDING (August–March). Similar to adult ♀ breeding, but feathers on crown, nape, hindneck, upperparts, gorget, and flank with broad greenish-buff, pale yellow, or whit-ish fringes, widest on forehead and along central part of crown. Inner primaries and secondaries with narrow greyish tips; tips on greater and median upper wing-coverts buffish-white. NESTLING. Down on upperparts dense; dark grey (Rymkevich 1979). JUVEN-ILE (July–September). Feathers of forehead, crown, and mantle with black central streak and broad rufous fringes; fringes of feathers of central crown sometimes partly yellowish or buffish,

forming ill-defined crown-stripe. Rump and upper tail-coverts buff-brown with rufous tinge and narrow dusky shaft-streaks. Lore and front part of supercilium rich buff or whitish, rear part of supercilium from above eye backwards broad, yellowish-white or white. Chin and upper throat buff or buffish-white; lower throat, chest, side of breast, and flank with brownish-black tips, bordered rufous on side of breast and flank, edged buff; feathers on central chest with strong yellow-buff wash. Belly and under tail-coverts pure white. Wing and tail as in adult ♀ non-breeding, but tips of greater coverts rich buff, of median coverts buff-white, each tip divided by brown in middle; lesser coverts grey with rather narrow rufous-chestnut fringes. FIRST ADULT MALE. Like adult ♂, but juvenile flight-feathers, tail, and greater upper prim-ary coverts retained, more worn and sometimes more pointed on tip than in adult at same time of year. Birds with narrow black shaft-streaks on feathers of crown, less or no black on ear-coverts, less rufous in feathers of hindneck, mantle, scapulars, and rump, black shaft on longest upper tail-coverts, black tips on feathers of throat, and narrower rufous marks on gorget and flanks, which are more concealed below by broad yellowish-buff edges, are perhaps 1st adult ♂. FIRST ADULT FEMALE. Like adult ♀, but part of juvenile feathers retained, as in 1st adult ♂. Birds with little or no rufous on neck, mantle, scapulars, and rump, with blackish subterminal spots on feathers of hindneck, with broad black central streak on scapulars, or with little or no rufous on gorget and flank are perhaps 1st adult ♀.

Bare parts. ADULT, FIRST ADULT. Iris dark brown. Base of bill pinkish-grey, remainder dark leaden-grey. Leg and foot light red-brown or pinkish-flesh, tinged grey on toes. NESTLING. Skin pale pink; bill and claws almost white at hatching. Inside of bill pale pink at hatching, later becomes fleshy-red; flanges whitish-yellow at hatching, later bright yellow (Rymkevich 1979). JUVEN-ILE. No information.

Moults. ADULT POST-BREEDING. Complete, primaries descend-ent. In north-west Russia, starts late June to early August (mainly mid-July), completed mid-August to late September. Total moulting period *c.* 50–70 days (Rymkevich 1976, 1990). ADULT PRE-BREEDING. Partial: forehead, crown, ear-coverts, chin, and upper throat, but in some individuals breeding plumage might be acquired by wear. Single ♂, Manchuria, with light moult on head in late March (Meise 1934a). Moult (March–)April–May (Witherby *et al.* 1938). POST-JUVENILE. Starts mid-June to mid-August (mainly mid-July), completed mid-August to late September (mainly early September). Birds from early broods start when juvenile plumage just completed, later-hatched birds start moult when juvenile flight-feathers and tail not quite completed; moult in early-hatched birds takes *c.* 50 days, in later-hatched birds *c.* 60 days (Rymkevich 1976, 1990). For influence of light on moult, see Rymkevich (1976).

Measurements. Races combined. Whole geographical range, all year; ages combined; skins (RMNH, ZMA). Bill (S) to skull, bill (N) to distal corner of nostril; exposed culmen on average 3·4 mm shorter than bill (S).

	♂		♀	
WING	80·7 (1·56; 18)	78–83	77·5 (2·33; 10)	73–81
TAIL	55·2 (2·99; 16)	48–59	54·1 (2·79; 7)	49–57
BILL (S)	13·7 (0·79; 19)	11·6–14·9	13·6 (0·88; 12)	12·0–15·0
BILL (N)	8·3 (0·44; 19)	7·4–9·1	8·0 (0·47; 12)	7·2–8·8
TARSUS	19·4 (0·54; 13)	18·8–20·3	19·1 (0·48; 15)	18·4–19·7

Sex differences significant for wing.

Wing. Northern Europe: ♂ 74–83 (73), ♀ 71–79 (53) (Svens-son 1992). Russian Lapland: ♂♂ 76–81 (4), ♀♀ 70–80 (7)

(Semenov-Tyan-Shanski and Gilyazov 1991). Manchuria, ♂♂ 77, 77, 80, ♀♀ 74, 75 (Meise 1934a).

Wing. Japan, live birds: (1) adult, (2) 1st adult (Dornberger 1983).

(1)	♂ 80·8 (1·90; 53) 77·0–85·0	♀ 77·0 (1·20; 35) 74·0–79·0
(2)	79·5 (1·69; 55) 76·0–82·0	75·3 (1·92; 45) 72·0–79·5

See also Kaneko (1976). For growth of nestling, see Rymkevich (1979).

Weights. Nominate *rustica*. Yamal peninsula (north-west Siberia): May–June, ♂♂ 18·1, 19·8; ♀ 17·8 (Danilov *et al.* 1984). Russian Lapland: ♂ 19·7 (4) 17·5–20·8, ♀ 18·4 (7) 17·1–19·9 (Semenov-Tyan-Shanski and Gilyazov 1991). Kazakhstan: ♂ 20·2 (10) 15·5–22·0 (Korelov *et al.* 1974). Poland: 1st adult, August–September, 17·9 (1·85; 4) 16·2–20·5 (Busse 1968). Netherlands: 1st adult, September–October, 18·1, 18·9 (Winkelman 1981; Free University Amsterdam). Skokholm (Wales): 19·2 (Browne and Browne 1956). For weight development of nestling, see Rymkevich (1979).

E. r. latifascia. Japan, October–November: (1) adult, (2) 1st adult (Dornberger 1983).

(1)	♂ 21·0 (1·64; 54) 17·7–24·3	♀ 20·1 (1·31; 35) 18·0–22·9
(2)	20·8 (1·72; 54) 17·4–25·8	18·8 (1·43; 45) 16·0–21·2

See also Nechaev (1969).

Structure. Wing rather long, broad at base, tip bluntly pointed. 10 primaries: p8 longest, p7 0–1 shorter, p6 1–2, p5 5–7, p4 11–14, p3 15–16, p2 17–18, p1 19–20, p9 0·5–2; p10 greatly reduced, 50–57 shorter than p8, 9–11 shorter than longest upper primary covert; p6–8 emarginated; emargination on p5 hardly visible. Tip of longest tertial reaches tip of p3–p5. Tail slightly forked; 12 feathers, t4–t5(–t6) longest; t1 4–6 mm shorter. Bill conical; similar to that of Yellowhammer *E. citrinella*, but upper mandible slightly thicker and lower mandible thinner; depth at feathering near base 6·3 (16) 5·9–6·7, width 6·7 (16) 6·2–7·1; bill slightly longer and more sharply pointed, culmen straight or slightly concave, tip occasionally slightly convex. Middle toe with claw 18·2 (8) 17·5–18·7; outer and inner toe with claw both *c.* 71% of middle with claw, hind *c.* 80%; hind claw 7·5 (11) 7·1–8·2.

Geographical variation. Rather slight, if any. 2 races sometimes recognized, nominate *rustica* from northern Europe to central Siberia, and *latifascia* in eastern Siberia; latter said to differ in black rather than brown-black cap and ear-coverts and broader rufous chest-band (*c.* 12 mm instead of *c.* 7·5 mm) (Portenko 1930). Also, bill of eastern birds slightly longer and more sharply pointed: in Japan, to skull 14·0 (17) 12·0–15·0, further west 13·3 (14) 11·6–14·9 (G O Keijl). Boundary between races uncertain, *latifascia* either occurring east from Lena or Yana river (Portenko 1930; Paynter 1970) or restricted to Kamchatka and Anadyrland (Vaurie 1959; Portenko and Stübs 1971). *E. r. latifascia* recognized here, following Vaurie (1959) and Paynter (1970), though characters poorly defined in specimens examined and no races recognized by Dementiev and Gladkov (1954), Vaurie (1956e), and Stepanyan (1978). GOK

Emberiza pusilla **Little Bunting**

PLATES 17 (flight) and 23 [between pages 256 and 257]

Du. Dwerggors   Fr. Bruant nain   Ge. Zwergammer
Ru. Овсянка-крошка   Sp. Escribano pigmeo   Sw. Dvärgsparv

Emberiza pusilla Pallas, 1766

Monotypic

Field characters. 13–14 cm; wing-span 20–22·5 cm. Distinctly less bulky than Reed Bunting *E. schoeniclus*; usually 10% smaller, with sharply pointed bill, flat sloping forehead, little or no neck, shorter, straight-edged tail, and shorter legs; close in size to but with less ample tail than Pallas's Reed Bunting *E. pallasi*; somewhat shorter and distinctly smaller-billed than Yellow-browed Bunting *E. chrysophrys*. Smallest bunting of west Palearctic, with delicate but compact form (lacking obviously long tail of larger *Emberiza*) and terrestrial behaviour recalling Linnet *Carduelis cannabina* and Dunnock *Prunella modularis*. At all ages, bright chestnut face with narrow but distinct cream eye-ring and dark stripes on crown-sides and around rear of ear-coverts (not reaching bill) distinctive. Pale double wing-bar most obvious feature of buff- to grey-brown upperparts; fine, discontinuous streaks on underparts also characteristic. Flight recalls small finch. Call useful in separation from *E. schoeniclus*. Sexes similar; no seasonal variation. Juvenile separable.

Adult. Moults: July–September (complete); February–April (face). Plumage basically buff- to grey-brown above and clean buffish-white below, with bright, warm-coloured, and quite strongly marked head, more rufous, pale-barred wings, white-edged tail, and finely streaked breast and flanks. Tone of face and upperparts varies, breeding ♂ having most rufous lores and chin and darkest head and back-stripes; sex of occasional birds with coppery mantle and back not certainly known. Separation from other small buntings, including unusually small *E. schoeniclus*, requires close observation of: (a) relatively quite long, pointed bill, with straight, even slightly concave culmen; (b) narrow but distinct, usually rufous-toned median crown-stripe, emphasized by (c) dark brown to black lateral crown-stripes; (d) contrasting rufous to cream supercilium, merging in front of eye with rufous lores and central ear-coverts to create (e) bright chestnut face; (f) narrow but distinct cream eye-ring; (g) usually small, pale cream spot on rear ear-coverts; (h) almost black rear eye-stripe, broadening at end and turning down to form dark border to lower ear-coverts but not extending forward of eye (unlike *E. schoeniclus* which has complete moustachial stripe); (i) dull grey-brown lesser coverts; (j) whitish tips to median coverts forming bright upper wing-bar; (k) pale buff tips to greater coverts forming less striking lower wing-bar; (l) sharp black malar stripe, like moustachial stripe not reaching bill and broadening into (m) fine, blackish streaks at sides of breast, across it, and along flanks, rarely forming continuous lines. Bill dusky on tip and culmen, flesh-pink to flesh-grey on lower mandible and cutting edges. Legs flesh-pink to rather pale bright pinkish-brown. When fresh, stripes on crown and mantle obscured by buff fringes; when very worn, lower wing-bar lost and underbody becomes whiter. Important to note that tail is only narrowly edged white (lacking apparent white corners of *E. schoeniclus*). For distinction of some ♀♀, see Plumages. JUVENILE. Ground-colour of upperparts paler, more yellowish-buff than fresh adult; stripes on head less distinct but spots and streaks on underparts heavier and more extensive. After moult in July–August, assumes adult-like head and body plumage so that 1st-winter not distinguishable from ♀, except by grey-brown (not reddish-brown) eye.

Combination of small size, bill shape, pale eye-ring, and bright head pattern allow diagnosis, but *E. pusilla* nevertheless subject to persistent confusion with smallest *E. schoeniclus* (particularly those with bright plumage), Rustic Bunting *E. rustica*, *E. pallasi*, and *E. chrysophrys*. *E. schoeniclus* and *E. pallasi* easily separated on call and close observation of head and underpart patterns; *E. rustica* and *E. chrysophrys* have similar calls but *E. rustica* shows much more obvious, dark rufous streaks on breast and flanks and intensely rufous-brown rump (dull, even greyish on *E. pusilla*), while *E. chrysophrys* has distinctly black and yellow or white head markings, including much more distinct white spot on rear ear-coverts, and lacks rufous face. All confusion species also lack striking eye-ring. *E. pusilla* may also suggest small Nearctic emberizid sparrows but all species recorded in west Palearctic lack fully white outer tail-feathers. Flight light and fast, with silhouette suggesting small finch (Carduelinae) until longer, straight-edged tail apparent; action recalls *C. cannabina* in tight spaces, but less obviously undulating over distance, while never showing erratic wing-beats and tail-spreading of *E. schoeniclus*. Gait a hop, sometimes apparently creeping or shuffling; when feeding in crouch, can suggest small pipit *Anthus*. Perches perkily, dropping directly to ground. Stance usually level, emphasized by neckless, hunched posture, but raises head in excitement or alarm. Twitches wings and occasionally flicks tail (but tail action lacks frequent slight spreading so characteristic of *E. schoeniclus*). Slower gait, dumpiness, and wing movement all evocative of *P. modularis*. Unobtrusive, often stays deep in ground cover but also visits tree canopies. Sociable rather than gregarious, with migrants travelling

with congeners and often joining other small terrestrial passerines to feed. Increasing winter occurrences in western Europe usually associated with damp, seed-bearing habitats.

Song a fairly quiet, sweet, and varied phrase of unusual timbre, with tonal sounds, buzzy units, and clicks; phrase structure can suggest *E. rustica*, Ortolan Bunting *E. hortulana* and *E. schoeniclus*. Subsong gentle and twittering. Calls include 2 short, flat monosyllables given by migrants: hard, sharp, clicking 'zik', 'tik', 'tzik', or 'pwick', recalling Hawfinch *Coccothraustes coccothraustes*, *E. rubecula*, and *E. rustica*; quieter, lower, dry 'tick', 'tip', 'tsih', 'twit', or 'stip', suggesting *E. rubecula* or Song Thrush *Turdus philomelos*; both often quickly repeated 2–3 times in alarm. Breeding birds also utter a short, slightly hoarse 'tse', like Spotted Flycatcher *Muscicapa striata*, and quiet 'tsee', like *E. rubecula* and other buntings.

Habitat. In west Palearctic, breeds only in boreal and arctic continental climatic zones, from 7°C July isotherm in north to 18–20°C in south, having more northerly range than any other Emberizidae except Lapland Bunting *Calcarius lapponicus* and Snow Bunting *Plectrophenax nivalis*. Ecologically it comes between *C. lapponicus* and Rustic Bunting *E. rustica*, being a bird of moister and shrubbier tundra than *C. lapponicus* but less wooded situations than *E. rustica*. Favours willow *Salix* zone along rivers through northern taiga, and open forest by river mouths. (Voous 1960b.) Towards west of range shows preference for undergrowth of dwarf birch *Betula nana* or willow among taller trees, which may be birch, spruce *Picea*, or other species. In forest country, selects open types with clearings close by, and favours lowland or sometimes hilly tundra with much moss or swampy ground cover, especially near stream or river bank (Mikkola and Koivunen 1966). In northern Russia, shows preference for dwarf birch, alder *Alnus*, and willow, and concentrates in river valleys for breeding, although occupying southern mountains and foothills in winter (Dementiev and Gladkov 1954).

Winters in eastern Asia, even in warm climates, as in northern Burma, where it favours bracken and short grass on hillsides or along mule tracks; on plains, frequents stubble and visits patches of flowers in rest-house gardens (Smythies 1986).

Distribution. Has spread west into Fenno-Scandia since 1930s, and nort-east into Kola peninsula (north-west Russia).

NORWAY. Sporadic records of singing ♂♂ outside area of proved breeding (VR). SWEDEN. Few records, but may well breed annually (LR). FINLAND. First recorded breeding 1935; then spread rapidly westward. In several years in 1980s, when particularly numerous, recorded breeding far beyond main area occupied. (OH.) RUSSIA. No records for Kola peninsula in early 20th century. First recorded

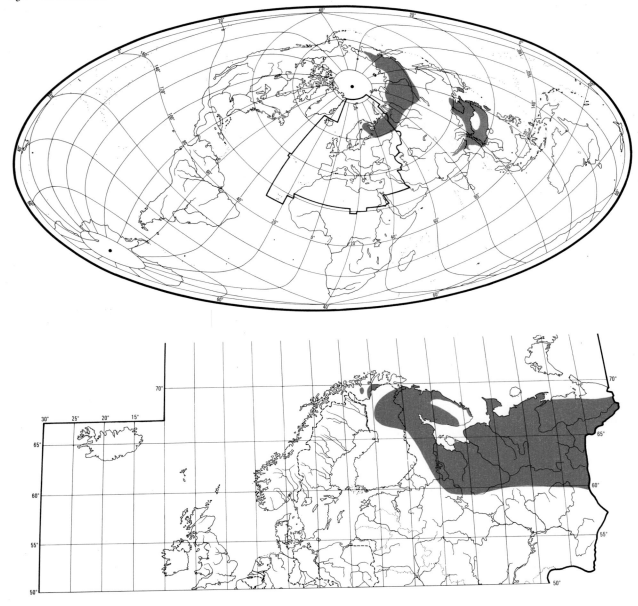

Khibin, and has now probably spread along whole coast to Ponoy estuary. (Mikhaylov and Fil'chagov 1984.)

Accidental. Iceland, Britain (annual), Ireland, Channel Islands, Netherlands (annual), France, Spain, Belgium, Germany, Denmark, Poland, Austria, Hungary, Switzerland, Italy, Malta, Yugoslavia, Greece, Bulgaria, Turkey, Egypt, Lebanon, Jordan, Kuwait, Canary Islands.

Population. Has increased in west with colonization of Fenno-Scandia, and probably still increasing.

NORWAY. Estimated 100–300 pairs (VR). SWEDEN. Probably fewer than 10 pairs (LR). FINLAND. Annual

fluctuations, but recent trend seems to be upwards. Estimated about 1000 pairs in 1988. (Koskimies 1989; OH.)

Movements. All populations migratory. Western birds head east from breeding grounds then south or south-east, and eastern birds head south, to reach winter quarters via Mongolia, south-east Russia, and north-east China.

Winters in southern China, mainly south of Yangtze river, west to Yunnan and south-east half of Szechwan, also north to southern Shensi and Hopeh (Schauensee 1984; Cheng 1987). Fairly common and widespread in Hong Kong (Chalmers 1986); rare in Taiwan (Chang 1980). Uncommon on passage and in winter in Korea

(Gore and Won 1971). In Japan, regular in small numbers on passage in western islands, with occasional widespread winter records (Brazil 1991). Further south, winters in north-west and north-east Thailand, northern Vietnam and northern Laos (King *et al.* 1975). In Burma, winters in large numbers in northern hills, and fairly common in central areas; scarce further south, and absent from southern Tenasserim (Smythies 1986). Common in north-east India and Sikkim (Ali and Ripley 1974), and fairly common in Nepal (Inskipp and Inskipp 1985). Rare vagrant to Borneo (Smythies 1960) and Philippines (Dickinson *et al.* 1991).

Autumn migration August–November. Birds leave breeding grounds from mid-August to early or mid-September, e.g. Yamal peninsula (Danilov *et al.* 1984), west Taymyr peninsula (Krechmar 1966), lowlands of north-east Siberia (Krechmar *et al.* 1991), Koryak highlands (Kishchinski 1980). In north-east Finland also, movement begins in August, with some birds still present in September; most birds migrate east or south-east, but observations at bird stations show that small numbers move gradually south-west August–November (only 3 spring records from this direction) (Koivunen *et al.* 1975). Rare in St Petersburg region (north-west Russia), and lack of records in central and south European Russia shows that western populations move directly eastward (Dementiev and Gladkov 1954; Mal'chevski and Pukinski 1983). In Volga-Kama region, occurs (rarely) only in north (Popov 1978); similarly, movement through western Siberia is most intensive in northern taiga zones, with smaller numbers further south (Johansen 1944; Vartapetov 1984). Change to south-east heading apparently begins at *c.* 75°E (much earlier than in Rustic Bunting *E. rustica*); in Kazakhstan, regular (though in very small numbers) in Irtysh valley in north-east, but rare elsewhere (Korelov *et al.* 1974). Occurs on passage throughout northern Mongolia, September to early October (Kozlova 1933; Mauersberger 1980; Mey 1988), and common south of Lake Baykal (Vasil'chenko 1982). Movement through central Siberia and Yakutiya chiefly in September (Vorobiev 1963; Reymers 1966). Numerous in Manchuria in September (Piechocki 1959), but rare as far east as Ussuriland (Panov 1973*a*), and very rare on Sakhalin (Gizenko 1955). In Korea, passage is mostly in north, with few records from peninsula (Austin 1948). In Hopei (north-east China), passage end of September to beginning of November (Wilder and Hubbard 1938; Hemmingsen and Guildal 1968), and reaches Hong Kong chiefly from mid-November (earliest 11 October) (Chalmers 1986). Present in Burma and north-east India from October (Ali and Ripley 1974; Smythies 1986).

Spring migration late March to June. Leaves India March–April (Ali and Ripley 1974), and last records in Burma in early May (Smythies 1986). In Hong Kong, evidence of passage late March to early April, with latest birds in early May (Chalmers 1986). At Beidaihe in north-east China, very common in spring (though not as abundant as formerly), peaking in 2nd week of May; in one year, recorded 28 March to 20 May, with few before 21 April or after mid-May (Williams 1986). In Mongolia, movement chiefly May, continuing to early June (Piechocki and Bolod 1972; Mauersberger *et al.* 1982). Main passage in Yakutiya in 2nd half of May (Vorobiev 1963), and reaches lowlands of north-east Siberia in last third of May, when 30–50% of forest tundra is snow-free (Krechmar *et al.* 1991), and Yamal and Taymyr peninsulas at end of May and beginning of June (Krechmar 1966; Danilov *et al.* 1984). One of latest migrants to reach north-east Finland; at Kuusamo, 1964–72, average earliest bird 6 June, and never earlier than 30–31 May (Koivunen *et al.* 1975).

Occasional winter or passage records in Asia, southwest of usual range, show that some individuals change to south-east heading earlier than mainstream; reported from Iran (D A Scott), Afghanistan (S C Madge), south-east Kazakhstan (Korelov *et al.* 1974), south-west Sinkiang (western China) (Schauensee 1984), western Himalayas (Ludlow and Kinnear 1933; Delany *et al.* 1982), and northern Pakistan (Roberts 1992). North-east of range, rare vagrant to Alaska (Kessel and Gibson 1978).

Widespread records in west Palearctic, mostly in autumn (see also Distribution). 3 birds have reached Iceland (Pétursson *et al.* 1991). Annual in Britain, with 93 before 1958, and 457 in 1958–91 (including Ireland). Autumn records chiefly in Shetland (especially Fair Isle), on British east coast and in Isles of Scilly; spring records well scattered, and include a number of inland localities, but with few in Shetland. Reports span September–May, with 85% of 1958–85 records in autumn (mostly late September to October) and noticeable spring peak April–May. (Dymond *et al.* 1989; Rogers *et al.* 1992.)

Several reports midwinter, from Scotland south to Jersey (Channel Islands); in Merseyside (north-west England), bird remained from January to early April (Eades 1984; Paintin 1991); 3 birds present March–April at different sites in southern Britain (Brucker *et al.* 1992) may also have overwintered. On Finnish coast also, a few midwinter records. In Sweden, up to 1986, 209 records April–November, chiefly May to early July; autumn records mainly in south (compare south-west heading of some Finnish birds, above), spring records mainly in north. (Breife *et al.* 1990.) In Denmark, 28 records 1965–90 (Frich and Njordbjærg 1992). In Netherlands, 59 records in all, of which 53 in September–November and 6 in February–May (Berg *et al.* 1992). In western Germany, 19 records 1977–90, mostly in autumn, south to Rheinland-Pfalz (Bundesdeutscher Seltenheitenausschuss 1989, 1992). In France, far more records than of *E. rustica*; in 19th century, a few caught annually in autumn at Marseille in south-east. Of 35 records (involving 37 birds) in 20th century, 80% in west and south-west France, and 14% in south-east; recorded in all months September–November and January–April, chiefly October–November. (Dubois and Yésou 1992.) In Israel, occurs in very small numbers; regular at Elat, with 1–6 individuals each year, and 7 birds reported 1979–89 in Jerusalem hills; most records late October to mid-November (Shirihai in press).   DFV

Food. Much information extralimital; diet seeds, also invertebrates in breeding season (Witherby *et al.* 1938; Dementiev and Gladkov 1954; Haftorn 1971; Staav and Fransson 1987). In northern Sweden and Norway, summer, forages for defoliating caterpillars (especially Geometridae) a few metres above ground in birches *Betula*, also in bilberry *Vaccinium* and similar low vegetation; also observed feeding repeatedly in short grass in meadow (Swanberg 1954; Curry-Lindahl 1962; Sudhaus 1969b). On migration, most often feeds in crops, on turned soil, paths, and roads, almost wholly on ground (Wallace 1976b). In winter quarters, in short grass, scattered woodland, marshy places, river banks, and (particularly) stubble and paddy fields (Cheng 1964; Smythies 1986; Roberts 1992). In Mongolia, end of May, passage birds fed on ground in grassy places and woodland clearings, but also in willow *Salix* bushes; also noted in park 4 m up in larch *Larix* picking at bases of needle clusters (Mauersberger 1980). Vagrant in north-west England seen to forage most often in stubble with mixed flock of other seed-eaters, and also in garden hedge (Eades 1984). In Denmark, spring, and Brandenburg (eastern Germany), autumn, only observed to feed on ground in mixed flocks by water, on bare soil, arable fields, and weedy areas (Læssøe and Nørregaard 1968; Dittberner *et al.* 1969). Distribution or numbers possibly influenced by occurrence of defoliating caterpillar infestations (Mikkola and Koivunen 1966; Sudhaus 1969b).

Diet includes the following. Invertebrates: mayflies (Ephemeroptera), stoneflies (Plecoptera), bugs (Hemiptera: Psyllidae), adult and larval Lepidoptera (Geometridae), caddis flies (Trichoptera), adult and larval flies (Diptera: Tipulidae, Culicidae, Chironomidae, Rhagionidae, Syrphidae, Calliphoridae), Hymenoptera (Tenthredinidae, Ichneumonidae), adult and larval beetles (Coleoptera: Carabidae, Chrysomelidae, Curculionidae), spiders (Araneae: Lycosidae), earthworms (Lumbricidae). Plant material: seeds of cowberry *Vaccinium*, crowberry *Empetrum*, grasses (Gramineae, including millet *Panicum*), sedges (Cyperaceae). (Kapitonov and Chernyavski 1960; Vorobiev 1963; Cheng 1964; Krechmar 1966; Brood and Söderquist 1967; Sudhaus 1969b; Verzhutski *et al.* 1979; Danilov *et al.* 1984; Chernov and Khlebosolov 1989; Larionov *et al.* 1991.)

On Yamal peninsula (northern Russia), August, of 10 stomachs, 9 contained 48 invertebrates, and 5 contained plant material (seeds of sedges and grasses); invertebrates were 48% by number Diptera (42% Calliphoridae, 4% Tipulidae), 42% beetle larvae, 6% spiders, 4% caterpillars; 2 spring stomachs held Diptera, beetles, remains

of cowberries and grass seeds (Danilov *et al.* 1984). In Yakutiya (eastern Russia), early spring, 4 stomachs contained 31% by volume beetles (16% Carabidae, 15% Curculionidae, of which 2% larvae), 28% caddis flies, 25% Psyllidae (Chernov and Khlebosolov 1989; E I Khlebosolov); another 4 stomachs, July–August, contained Hemiptera, Diptera, Ichneumonidae, small beetles, spiders, earthworms, and some grass seeds (Larionov *et al.* 1991). On Taymyr peninsula (northern Russia), June, 7 stomachs held adult and larval Diptera (Tipulidae, Chironomidae, Rhagionidae) and beetles (Carabidae, Chrysomelidae) (Krechmar 1966). In Hunan (southern China), spring and autumn, of 57 stomachs, 79% contained plant material (grass and cereal seeds) and 42% insects (Hemiptera, Lepidoptera, Hymenoptera, beetles); in Hopeh (northern China), spring and autumn, contents of 35 stomachs were 82% plant food and 18% insects (Cheng 1964). On lower Ob' river (northern Russia), June–August, 28 stomachs (probably of adult birds) contained 584 items, of which 75% by number seeds and 25% invertebrates, including 7% beetles, 2% Hemiptera, 2% Hymenoptera, 1% Diptera, and 1% spiders (Verzhutski *et al.* 1979). ♂ in Norway ate 6 caterpillars in quick succession (Sudhaus 1969b).

Young fed on invertebrates, mostly Diptera and larval Lepidoptera. In Yakutiya, 32 collar-samples contained *c.* 280 items, of which 81% by volume (62% by number) Diptera, including 51% (21%) Tipulidae adults, 19% (12%) Tipulidae pupae, 3% Syrphidae, 1% Chironomidae, and 0·5% (11%) Culicidae; also 6% (15%) spiders, 5% (23%) Tenthredinidae, 2% caddis flies, and 0·3% beetles (Chernov and Khlebosolov 1989; E I Khlebosolov). On Yamal peninsula, 10 collar-samples contained 102 invertebrates, of which 93% by number Diptera (89% Chironomidae, 2% Tipulidae), 5% beetle larvae, and 2% spiders (Danilov *et al.* 1984). Several observations in Fenno-Scandia of caterpillars, commonly Geometridae, being given to young (Swanberg 1954; Mikkola and Koivunen 1966; Sudhaus 1969b; I and M Hills).   BH

Social pattern and behaviour. Most aspects reasonably well known, but more information needed on courtship and other display.

1. Gregarious outside breeding season. In Burma, typically in flocks of *c.* 12, sometimes up to several hundreds (Smythies 1986). In India, usually encountered singly or in small flocks, often in company with Grey-headed Bunting *E. fucata* and Tree Pipit *Anthus trivialis* (Ali and Ripley 1974). In Japan, often found among other *Emberiza*, especially Rustic Bunting *E. rustica*, and occasionally Redpoll *Carduelis flammea* (Brazil 1991). In Szechuan (China), generally associates with other *Emberiza* and also with Great Tit *Parus major* (Morrison 1948). Birds wintering in Britain typically mix with (variously) Meadow Pipit *A. pratensis*, Tree Sparrow *Passer montanus*, Linnet *Carduelis cannabina*, Yellowhammer *E. citrinella*, and Reed Bunting *E. schoeniclus* (e.g. Stokoe 1949, Harris 1957, Eades 1984). Also migrates in flocks, e.g. south of Lake Baykal in flocks of 3–100 or more (Vasil'chenko 1982). During autumn migration, northern

Urals (Russia), mixes with Arctic Warbler *Phylloscopus borealis*, tits *Parus*, and redpolls *Carduelis* (Portenko 1937); in August, lower Kolyma (eastern Siberia), mixes with Red-throated Pipit *A. cervinus* (E R Potapov). BONDS. No evidence for other than monogamous mating system. No polygyny found in 3-year study of marked birds on Yamal peninsula, northern Siberia (Danilov *et al.* 1984). Both sexes incubate (Danilov *et al.* 1984; Krechmar *et al.* 1991) though mostly ♀ (Brood and Söderquist 1967). Study by Mikkola and Koivunen (1966) suggested ♂'s contribution perhaps biased towards later in incubation. Both sexes also brood and feed young, though more brooding by ♀ (Mikkola and Koivunen 1966). Of 2 nesting pairs studied in Pasvik (Norway), ♂ helped to feed young in one case, but apparently not in the other (I and M Hills). ♂ took over care of 1st brood when ♀ started 2nd (Koivunen *et al.* 1975). Young still fed by parents at *c.* 4–5 days after leaving nest, but not when checked again *c.* 12 days later (Brood and Söderquist 1967). No information on age at first breeding. BREEDING DISPERSION. Solitary and territorial. In Kuusamo (eastern Finland), 2–3 ♂♂ often settle near one another (Koivunen *et al.* 1975); not known whether this due to aggregation in suitable habitat or tendency to form neighbourhood groups. In northern Norway, territory size *c.* 0·3–0·7 ha, within which singing area *c.* 0·4 ha (Sudhaus 1969*b*). On Yamal peninsula, average 0·3 ha (0·2–0·5, $n=10$); in territories along watercourse, average length 85 m (62–107) (Danilov *et al.* 1984). In Fenno-Scandia, distance between nests (or singing ♂♂) typically 100–200 m (Mikkola and Koivunen 1966; Brood and Söderquist 1967; Sudhaus 1969*b*). In preferred habitat (bushy tundra) in Russia, sometimes only 10 m between nests (Portenko 1937: northern Urals); up to 8–10 pairs per ha, though such densities very sporadic (Kishchinski 1980: Koryak highlands). Territory serves for courtship, nesting, and feeding young. Parents feed young both in and outside territory (residents not hostile to foraging neighbours) (Danilov *et al.* 1984). In Vorkuta (north-east European Russia) up to 126 birds per km² (Morozov 1986). In northern Urals, 1979–86, average 76 pairs per km² (Shutov 1990). Densities in extralimital Russia include: on Yamal peninsula, 2·6–100 pairs per km² (Danilov *et al.* 1984, which see for variation with location and year); in central Siberia, 79–101 birds per km² in bushy tundra, 68–69 in forest tundra, average 22 (up to 200 locally) in relatively open coniferous forest (sometimes after fire) with sphagnum moss floor, average 17 (rarely up to 175) in rich riverine meadows (Rogacheva 1988). In western Siberia, density ranged from 0·7 birds per km² in bogs to 121 in pine *Pinus* forest (Ravkin 1984; see also Vartapetov 1984). For Yakutiya, see Shmelev and Brunov (1986) and Larionov *et al.* (1991). Perhaps little site-fidelity from year to year: of 13 colour-ringed ♂♂ and 7 ♀♀, only 1 ♂ was seen in study area the next year (Danilov *et al.* 1984). No known nesting associates, but territories in Pasvik held several breeding pairs of *E. schoeniclus* (I and M Hills). ROOSTING. In breeding season, Finland, birds apparently slept in late evening before starting to sing from 01.00 hrs (see Song-display, below); at 22.00 hrs unpaired ♂ was flushed from roost-site among dried twigs on ground (Palmgren 1936). No information for other times of year.

2. Has reputation for being tame and approachable outside breeding season (Morrison 1948; Wallace 1976*b*), e.g. September vagrant (not tired) allowed approach to *c.* 60 cm (Ruttledge 1954). Also has remarkable ability to vanish in minimal vegetation, tending to go to ground rather than cover (Wallace 1976*b*). When approached, flushes from ground like some other *Emberiza*: dives instantly behind next bush a few paces away, and, if followed, repeats manoeuvre (Ali and Ripley 1974). In breeding season, much tamer than (e.g.) *E. rustica*, but gets very alarmed

if sitting on nest (Johansen 1944; see Parental Anti-predator Strategies, below). In study on Finnish breeding grounds, ♂ generally gave alarm-calls (see 2a in Voice) less than ♀, also flew higher into trees, and raised crown feathers more prominently and more often (Mikkola and Koivunen 1966). Tail-flicks in excitement, rather like flycatcher (Muscicapidae) (Palmgren 1936). FLOCK BEHAVIOUR. No information. SONG-DISPLAY. ♂ sings (see 1 in Voice) typically from exposed tree-top, with bill raised and throat feathers ruffled (Fig A), body vibrating with

A

effort (Swanberg 1954; Sudhaus 1969*b*; I and M Hills) also slight wing-twitching (Sudhaus 1969*b*). Singers fly towards playback-song (Mikkola and Koivunen 1966), suggesting at least territorial function. During laying period, ♂ sang usually less than 10 m from nest (Mikkola and Koivunen 1966). ♂ continues singing during incubation, nestling period (when ♀ brooding) (Swanberg 1954; Brood and Söderquist 1967), and also heard after young fledged (Sudhaus 1969*b*; I and M Hills). However, seasonal pattern varies with, perhaps, number of broods: e.g. at nest in Kuusamo, ♂ did not sing after 7 July when clutch was near completion (Mikkola and Koivunen 1966), lack of song thereafter perhaps indicating no subsequent brood. Birds occurring Britain heard singing in April (Eades 1984; N J Hallam). In Russia, song heard during spring migration; in Timanskaya tundra (north European Russia) song-period 10 June to early July (Dementiev and Gladkov 1954); in Anadyr' region (north-east Russia) 27 May to 20 July (Portenko 1939). No details of diurnal pattern other than in Roosting (above). ANTAGONISTIC BEHAVIOUR. ♂ guards territory, often helped by ♀ (who more often confronts other ♀♀) (Danilov *et al.* 1984). When another bird (presumably conspecific) flew through territory, resident ♂ chased it for *c.* 50 m through cover, giving 'zick' calls (Sudhaus 1969*b*: see 2a in Voice). Sketches in Danilov *et al.* (1984) depict display of forward-threat type to mirror image; threat to distant conspecific began with raising one wing before take-off. No further information. HETEROSEXUAL BEHAVIOUR. Little information. In study in Kiruna (northern Sweden) courtship-display apparently occurred many times in mornings (also later in day) of 3–5 July; sequence included the following: ♂ sang loudly in tree-top, ♀ skulking in scrub just beneath, then ♂ dived into scrub, changing at same time to low-intensity song; ♀ took up position on bare patch on ground; then ♂, with wings extended vertically (Wing-raising), and with no audible sound, 'danced on the ground in S-shaped curves in a semi-circle' in front of ♀. Copulation followed 'in the normal bunting manner'. Once, after copulating, Reverse-mounting occurred, ♀ jumping onto back of ♂ who pressed himself, immobile, to ground for a few moments. After separation, ♀ perched in scrub and preened her rump. (Swanberg 1954.) Adults typically approach nest furtively, low under cover (Mikkola and Koivunen 1966). Sudhaus (1969*b*) described one case of flight rather like display-flight of *A. trivialis*: bird flew diagonally downwards for *c.* 25 m to ground; observers expected to find nest where it landed, but none found. ♀ sometimes gave faint 'siuu' (see 4 in Voice) when settling on eggs or young (Brood and Söderquist 1967, which see for description of nest-relief). No further information. RELATIONS WITHIN FAMILY GROUP. Both sexes feed young. Provisioning by parents studied closely at nest in Sweden: one adult would fly in from surrounding terrain, approaching by degrees, till nest-tussock reached, then walked to nest; various sequences then followed, including silently feeding 1 or rarely 2 nestlings, or awaiting arrival of mate which would then feed young while 1st parent departed without feeding them. Towards end of nestling period, both parents flew more directly to nest, communicating with 'tuck' and 'tjurrrk' calls (see 2a–b in Voice). (Brood and Söderquist 1967.) Parents seen carrying off faecal sacs (I and M Hills). Young leave nest as yet unable to fly, initially clambering into surrounding vegetation, and only just able to fly c. 4–5 days later (Brood and Söderquist 1967). Once capable of flight, young scattered widely in wood and were fed in bouts: parent would feed one repeatedly, then fly to another (sometimes 200 m away) and feed it 3–5 times in succession, and so on (Swanberg 1954). See Bonds (above) for age at independence. ANTI-PREDATOR RESPONSES OF YOUNG. No information. PARENTAL ANTI-PREDATOR STRATEGIES. (1) Passive measures. In 2 studies at nests in Swedish Lapland, incubating ♀ a tight sitter, allowing approach to c. 1–1·5 m, but less passive once off nest (see below) (Curry-Lindahl 1962; Brood and Söderquist 1967). (2) Active measures: against man. Vary from mild demonstration to injury-feigning. In Sweden, when disturbed off nest, did not allow such close approach as when sitting, constantly and vigorously hopping and flitting around nest-site, calling the while (Curry-Lindahl 1962; Brood and Söderquist 1967). More demonstrative and more vocal after hatching, yet much more subdued than *E. schoeniclus*; unlike *E. rustica*, *E. pusilla* never strayed far from nest-site when disturbed (Brood and Söderquist 1967). ♀ gave c. 2 calls per s (see 2a in Voice) when far from nest, increasing to c. 5 per s as she flew slowly low over nest, decreasing again as she passed beyond it (Mikkola and Koivunen 1966). Rapid series of 'zick' calls developed without interruption into song as bird flew in alarm around nest-area (Sudhaus 1969b). Mobile distraction-lure display also occurs, e.g. once (no details) in northern Norway (Sudhaus 1969b). In Kiruna (Sweden) when young bird was flushed while ♂ was feeding it, ♂ feigned injury in manner resembling *E. schoeniclus* (Swanberg 1954). Display performed when young are in nest, described by Danilov *et al.* (1984, Yamal peninsula), is presumably the same: adult runs with head and both wings raised and spread tail lowered (Fig B).

B

Occasionally attacks or mock-attacks man (Danilov *et al.* 1984), e.g. will hover in front of man's face (E R Potapov). (3) Active measures against other animals. Sometimes attacks voles (Microtinae, nest predator), striking with bill and legs, and usually forcing retreat (Danilov *et al.* 1984).

(Figs by D Nurney: A from photograph in Bentz and Génsbøl 1988; B from drawing in Danilov *et al.* 1984.)   EKD

Voice. Freely used throughout year, though song confined to breeding season. For additional sonagrams see Bergmann and Helb (1982). In the following scheme, it is not always known to what extent sounds distinguished as different calls may simply be variants of a smaller basic repertoire.

CALLS OF ADULTS. (1) Song. Short phrase of 1·4–2·2 s (up to 2·4 s in available recordings: W T C Seale) with pauses of c. 4–12 s between phrases (Sudhaus 1969b); in full song c. 10 phrases per min (Swanberg 1954). Often begins with segment (series of similar motifs), followed by other such segments and/or by assortment of different unit-types creating greater variation, e.g. 'zwi zwi zwi zwiehdidi'. Construction of phrases roughly intermediate between Ortolan Bunting *E. hortulana* and Yellow-breasted Bunting *E. aureola*; introductory segment can start either high pitched like *E. hortulana* or low like *E. aureola*; phrase can end either high or low, and in latter case with almost inaudible final units. ♂ has several different phrase-types, and phrases are not uncommonly curtailed. (Sudhaus 1969b; Bergmann and Helb 1982.) Song also described as not unlike Rustic Bunting *E. rustica* but more 'bitty' and slightly ticking in character, something between Reed Bunting *E. schoeniclus* and *E. rustica*, and can also recall section from song of Tree Pipit *Anthus trivialis*; often 3–4 sections distinguishable, e.g. 'titti-chup chup-sturriep' or 'pie pie-sturi sturi-tulee-tchee' (Jonsson 1992). In recordings of birds from various regions (analysis and renderings, below, by W T C Seale), constituent sounds highly diverse in structure and timbre, variously pure, buzzy, vibrant, clicking, and also include short tremolos, transients, and diads; opening segment of phrases commonly comprises repetition (2–7 times) of 2-unit motif; e.g. Fig I shows 7 'zree' motifs, followed respectively by high-pitched 'zee', diadic 'paa', subdued high-pitched 'si' and vibrant 'zrrreee'. Fig II (another phrase from same bird as Fig I) has shorter introduction, whole phrase 'srri zee srri zee sip sip sip zrree dz-oo dz-oo'. In 2 different

TREVOR BOYER

PLATE 12. *Pipilo erythrophthalmus erythrophthalmus* Rufous-sided Towhee (p. 77): 1–2 ad ♂, 3 ad ♀. *Chondestes grammacus grammacus* Lark Sparrow (p. 80): 4 ad. *Ammodramus sandwichensis* Savannah Sparrow (p. 80). *A. s. princeps*: 5 ad. *A. s. labradorius*: 6 ad. *Zonotrichia iliaca iliaca* Fox Sparrow (p. 84): 7 ad summer. *Zonotrichia melodia melodia* Song Sparrow (p. 86): 8 ad summer. (IL)

PLATE 13. *Zonotrichia leucophrys* White-crowned Sparrow (p. 90). Nominate *leucophrys*: 1 ad summer, 2 1st ad winter. *Z. l. gambelii*: 3 ad summer. *Zonotrichia albicollis* White-throated Sparrow (p. 94): 4 ad white-striped morph, 5 ad tan-striped morph. *Junco hyemalis hyemalis* Dark-eyed Junco (p. 97): 6 ad ♂, 7 ad ♀, 8 1st ad ♂ winter. (IL)

PLATE 14. *Calcarius lapponicus* Lapland Bunting (p. 101): **1–2** ad ♂ breeding, **3–4** ad ♂ non-breeding, **5** ad ♀ breeding, **6–7** ad ♀ non-breeding, **8** 1st ad ♂ breeding, **9** 1st ad ♂ non-breeding, **10** juv. (TB)

PLATE 15. *Emberiza spodocephala spodocephala* Black-faced Bunting (p. 138): **1** ad ♂ breeding, **2** ad ♂ non-breeding, **3** ad ♀, **4** 1st ad ♀ non-breeding. *Emberiza cioides weigoldi* Meadow Bunting (p. 195): **5** ad ♂ winter, **6** ad ♀ winter. (TB)

PLATE 16. *Emberiza spodocephala spodocephala* Black-faced Bunting (p. 138): **1** ad ♂ non-breeding. *Emberiza leucocephalos leucocephalos* Pine Bunting (p. 142): **2** ad ♂ winter. *Emberiza citrinella citrinella* Yellowhammer (p. 153): **3** ad ♂ winter, **4** ad ♀ winter. *Emberiza cirlus* Cirl Bunting (p. 170): **5** ad ♂ winter, **6** ad ♀ winter. *Emberiza cia cia* Rock Bunting (p. 182): **7** ad ♂ winter, **8** ad ♀ winter. *Emberiza cioides weigoldi* Meadow Bunting (p. 195): **9** ad ♂ winter. *Emberiza striolata sahari* House Bunting (p. 195): **10** ad ♂ winter. (TB)

PLATE 17. *Emberiza cineracea cineracea* Cinereous Bunting (p. 204): **1** ad ♂ winter. *Emberiza hortulana* Ortolan Bunting (p. 209): **2** ad ♂ non-breeding, **3** 1st ad ♂ non-breeding. *Emberiza buchanani cerrutii* Grey-necked Bunting (p. 223): **4** ad ♂ winter. *Emberiza caesia* Cretzschmar's Bunting (p. 230): **5** ad ♂ non-breeding. *Emberiza chrysophrys* Yellow-browed Bunting (p. 238): **6** ad ♂ winter. *Emberiza rustica rustica* Rustic Bunting (p. 240): **7** ad ♂ non-breeding. *Emberiza pusilla* Little Bunting (p. 250): **8** ad ♂ non-breeding. *Emberiza rutila* Chestnut Bunting (p. 261): **9** ad ♂ winter, **10** 1st ad ♂ winter. (TB)

PLATE 18. *Emberiza cia* Rock Bunting (p. 182). Nominate *cia*: **1** ad ♂ spring, **2** ad ♂ winter, **3** ad ♀ spring, **4** ad ♀ winter, **5** 1st ad ♂ winter, **6** juv, **7** ad ♂ spring '*africana*'. *E. c. prageri*: **8** ad ♂ spring. (TB)

PLATE 19. *Emberiza striolata* House Bunting (p. 195). *E. s. sahari*: **1** ad ♂ spring, **2** ad ♂ winter, **3** ad ♀, **4** juv. Nominate *striolata*: **5** ad ♂ spring, **6** ad ♀. *Emberiza tahapisi septemstriata* Cinnamon-breasted Rock Bunting (p. 204): **7** ad ♂, **8** ad ♀, **9** juv. (TB)

PLATE 20. *Emberiza cineracea* Cinereous Bunting (p. 204). Nominate *cineracea*: **1** ad ♂ spring, **2** ad ♂ winter, **3** ad ♀, **4** juv. *E. c. semenowi*: **5** ad ♂ spring, **6** ad ♀. *Emberiza buchanani cerrutii* Grey-necked Bunting (p. 223): **7** ad ♂ spring, **8** ad ♂ winter, **9** ad ♀, **10** juv. (TB)

PLATE 21. *Emberiza hortulana* Ortolan Bunting (p. 209): **1** ad ♂ breeding, **2** ad ♂ non-breeding, **3** ad ♀, **4** 1st ad ♂ non-breeding, **5** juv. *Emberiza caesia* Cretzschmar's Bunting (p. 230): **6** ad ♂ breeding, **7** ad ♂ non-breeding, **8** ad ♀, **9** 1st ad ♂ non-breeding, **10** juv. (TB)

PLATE 22. *Emberiza chrysophrys* Yellow-browed Bunting (p. 238): **1** ad ♂ spring, **2** ad ♂ winter, **3** ad ♀, **4** 1st ad ♂ winter.   *Emberiza rustica rustica* Rustic Bunting (p. 240): **5** ad ♂ breeding, **6** ad ♂ non-breeding, **7** ad ♀ non-breeding, **8** juv. (TB)

PLATE 23. *Emberiza pusilla* Little Bunting (p. 250): **1** ad ♂ breeding, **2** ad ♀ breeding, **3** ad non-breeding, **4** 1st ad non-breeding, **5** juv.   *Emberiza rutila* Chestnut Bunting (p. 261): **6** ad ♂ spring, **7** ad ♂ winter, **8** ad ♀, **9** 1st ad ♂ winter. (TB)

PLATE 26. *Emberiza aureola* Yellow-breasted Bunting (p. 264): **1** ad ♂ non-breeding, **2** 1st ad ♂ non-breeding. *Emberiza schoeniclus schoeniclus* Reed Bunting (p. 276): **3** ad ♂ winter, **4** 1st ad ♂ winter. *Emberiza pallasi polaris* Pallas's Reed Bunting (p. 295): **5** ad ♂ winter, **6** 1st ad ♂ winter. *Emberiza bruniceps* Red-headed Bunting (p. 303): **7** ad ♂ non-breeding. *Emberiza melanocephala* Black-headed Bunting (p. 313): **8** ad ♂ non-breeding, **9** ad ♀ non-breeding. (TB)

PLATE 27. *Miliaria calandra* Corn Bunting (p. 323). Nominate *calandra*: **1** ad spring, **2–3** ad winter, **4** juv. *M. c. buturlini*: **5** ad spring, **6** 1st winter (TB)

PLATE 28. *Spiza americana* Dickcissel (p. 339): **1** ad ♂ spring, **2** ad ♀ spring, **3** 1st ad ♂ non-breeding, **4** ad ♂ winter. *Pheucticus ludovicianus* Rose-breasted Grosbeak (p. 342): **5-6** ad ♂ breeding, **7** 1st ad ♂ non-breeding, **8-9** ad ♀ spring, **10** ad ♂ non-breeding. (CETK)

PLATE 29. *Guiraca caerulea* Blue Grosbeak (p. 346): **1** ad ♂ summer, **2** ad ♂ winter, **3** 1st ad ♂ non-breeding, **4** ad ♀ summer. *Passerina cyanea* Indigo Bunting (p. 349): **5** ad ♂ breeding, **6** ad ♂ non-breeding, **7** ad ♀ breeding, **8** 1st ad ♂ non-breeding. *Passerina amoena* Lazuli Bunting (p. 352): **9** ad ♂ breeding, **10** ad ♂ non-breeding, **11** ad ♀ summer, **12** 1st ad ♀ winter. (CETK)

PLATE 30. *Dolichonyx oryzivorus* Bobolink (p. 355): **1** ad ♂ breeding summer, **2** ad ♂ breeding spring, **3** ad ♀ breeding summer, **4** ad ♂ non-breeding, **5** 1st ad ♂ non-breeding.   *Molothrus ater ater* Brown-headed Cowbird (p. 359): **6** ad ♂, **7** ad ♀, **8** juv.   *Quiscalus quiscula* Common Grackle (p. 361). *Q. q. versicolor*: **9** ad ♂, **10** ad ♀. *Q. q. stonei*: **11** ad ♂. (CETK)

PLATE 31. *Xanthocephalus xanthocephalus* Yellow-headed Blackbird (p. 364): **1–2** ad ♂ summer, **3** ad ♂ winter, **4** ad ♀ summer, **5** 1st ad ♂ winter.   *Icterus galbula galbula* Northern Oriole (p. 367): **6** ad ♂, **7** ad ♀, **8–9** 1st ad ♂ non-breeding. (CETK)

TREVOR BOYER

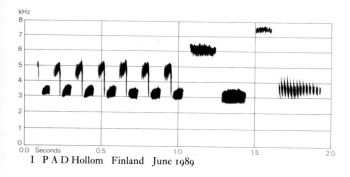

I   P A D Hollom   Finland   June 1989

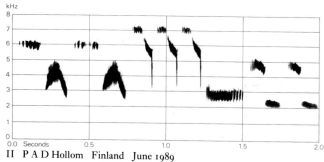

II   P A D Hollom   Finland   June 1989

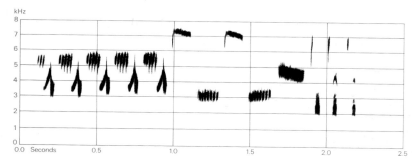

III   I Hills   Norway   June 1978

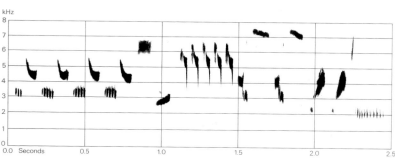

IV   I Hills   Norway   June 1978

songs (Figs III–IV) of one bird, introduction sounds rather like Yellowhammer *E. citrinella* (P J Sellar), while overall segmental structure also somewhat reminiscent of fragment of rather low-pitched song of Wren *Troglodytes troglodytes* (W T C Seale). After 2 introductory 'dzoo' motifs, song shown (half speed) in Fig V continues with series of 4 'dzrip' motifs remarkable for harmonic rela-

PLATE 32. *Emberiza schoeniclus* Reed Bunting (p. 276). Nominate *schoeniclus*: **1** ad ♂ spring, **2** ad ♂ winter, **3** ad ♀ spring, **4** ad ♀ winter, **5** 1st ad ♂ spring, **6** 1st ad ♂ winter, **7** 1st ad ♀ spring, **8** juv, **9** variant with pale crown. *E. s. pallidior*: **10** ad ♂ spring, **11** 1st ad ♂ winter. *E. s. incognita*: **12** ad ♂ spring, **13** 1st ad ♂ winter. *E. s. reiseri*: **14** ad ♂ spring. *E. s. intermedia*: **15** ad ♂ spring. *E. s. pyrrhuloides*: **16** ad ♂ spring. *E. s. witherbyi*: **17** ad ♂ spring.

Emberiza pallasi polaris Pallas's Reed Bunting (p. 295): **18** ad ♂ spring, **19** ad ♂ winter, **20** ad ♀ spring, **21** ad ♀ winter, **22** 1st ad ♂ winter, **23** juv. (TB)

tionship between units, producing curious jarring, rather metallic timbre. Introductory motifs each consist of 2 units in Figs I–V; however, in phrase (Fig VI) from Finnish bird, introduction (roughly 'strileeleelee') made up of single rather drawn-out compound unit of considerable complexity; in Fig VII (same bird as Fig VI) the 3 introductory motifs are of 2 units, each approximately 'zee-rrae' in which 'rrae' is diadic and vibrant-sounding. (W T C Seale.) Song bout is built up of several different phrase-types, each repeated 10–20 times (without change, or with addition or subtraction of 1 or 2 motifs, rarely more: W T C Seale) before switching to next phrase-type; one ♂ studied had 4 phrase-types: 1st one with opening 'tsi-' units like those of *E. citrinella*, 2nd with rapidly rattling units as in song of *A. trivialis*, 3rd resembling *E. hortulana* (as near in pitch as if mimicking though not considered to be actually doing so), 4th a more twittering phrase suggestive of Linnet *Carduelis cannabina* or Lapland Bunting *Calcarius lapponicus* (Swanberg 1954, which see for renderings). Song sometimes preceded by series of call 2a

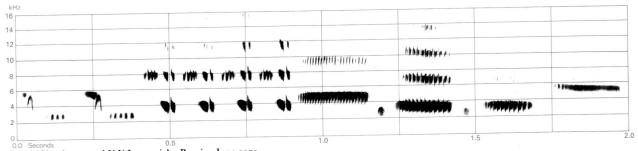

V B N Veprintsev and V V Leonovich Russia June 1979

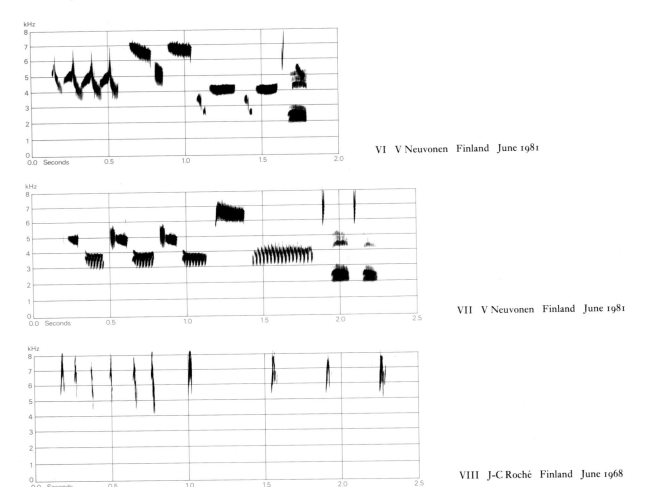

VI V Neuvonen Finland June 1981

VII V Neuvonen Finland June 1981

VIII J-C Roché Finland June 1968

(Sudhaus 1969*b*). Low-intensity song (or Subsong) heard from ♂ during intimate courtship shortly before copulation (Swanberg 1954). (2) Contact-alarm and alarm-calls. (2a) Commonest call (heard throughout the year) a hard sharp clicking 'tzik' like reduced version of Hawfinch *Coccothraustes coccothraustes* (Svensson 1975; Bruun *et al.* 1986). Also described as a metallic piercing 'tsik' or 'tik' (Jonsson 1992); also recalls Tic-call of Robin *Erithacus*

rubecula (Wallace 1976), often given in series (Fig VIII). Wallace (1976) distinguished 'zik' (or 'tic', 'tzik' etc.) from quieter, lower-pitched dry 'tip' (or 'stip', 'tsip', 'tsitt', etc.). Although probably a continuum, the harder sounds indicate greater agitation, e.g. typically when nest is closely approached. Thus alarmed bird introduces 's' or 'z' sound into normal contact-call to produce 'ksitt' (compared with 'kitt') (Swanberg 1954); likewise normal 'tick'

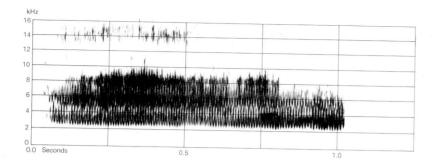

or 'sitt' becomes 'tzick' in alarm (Brood and Söderquist 1967). When alarmed by man at nest, ♀ gave (maximum) more than 5 'tsik' (or aspirated 'pt') calls per s, while ♂'s rate did not exceed *c.* 1 per s (Mikkola and Koivunen 1966). Rattling 'tett-rett-rett-tett', like *C. cannabina*, but fainter, given when disturbed at nest (Brood and Söderquist 1967), probably belongs here. Smacking 'tack' when disturbed at nest, and clicking 'tuck' between mates at nest-relief (Brood and Söderquist 1967) perhaps variants of basic call. (2b) Low buzzing 'tzurrrk' heard at nest-relief (Brood and Söderquist 1967). Recording of song includes harsh rasping buzz with odd rattling timbre, 'dzzzaair' (Fig IX, shown half speed to clarify rapid, rather wide-band, frequency modulation) (W T C Seale). (3) Brief, slightly hoarse 'tse', like Spotted Flycatcher *Muscicapa striata*, sometimes heard on breeding grounds (Svensson 1975). Function not known. (4) Drawn-out, very thin, quiet 'tseeee', similar to *E. rubecula* but quieter, heard rarely on breeding grounds (Svensson 1975). Intermittent drawn-out, quiet 'siiih' with bill barely open, given when nervous (Palmgren 1936). Also rendered drawn-out 'ziih' rather like *E. citrinella* (Sudhaus 1969*b*), which may be alarm-call for aerial predator (Bergmann and Helb 1982). Faint 'siuu', not unlike *E. schoeniclus*, sometimes heard from ♀ settling on nest for incubation or brooding (Brood and Söderquist 1967).

Calls of Young. Food-call of young 'sitt' or 'tick', just like adult call 2a but weaker; audible from *c.* 10 m (Brood and Söderquist 1967). No further information.   EKD

Breeding. Season. North-central Finland: eggs laid early June to mid-July; laying starts average 11 June (*n* = 19); July clutches replacements and 2nd broods (Koivunen *et al.* 1975). Swedish Lapland: eggs laid 1st half of June (Swanberg 1954; Curry-Lindahl 1962; Brood and Söderquist 1967). Vorkuta area (north European Russia): fresh eggs found beginning of July (Gladkov 1962). Northern Russia: eggs laid about mid-June (Krechmar 1966; Danilov *et al.* 1984). Site. Usually on ground, on grass tussock or moss cushion sheltered by overhanging grass or twigs of (e.g.) alder *Alnus*, birch *Betula*, willow *Salix*, *Vaccinium*, etc., also on tree stump (Dementiev and Gladkov 1954; Krechmar 1966; Koivunen *et al.* 1975). On Yamal peninsula (northern Russia), 44% of ground nests

on even surface, 40% on, and 16% at side of tussock or cushion; some in trees up to 1·5 m above ground, particularly during flooding (Danilov *et al.* 1984). In north-central Finland, 57% of 21 nests on such hummocks (Koivunen *et al.* 1975). Nest: foundation of thin twigs, stalks of herbs, grass, sedge, horsetail *Equisetum*, moss, lined with fine grass, lichen, and sometimes hair (Dementiev and Gladkov 1954; Curry-Lindahl 1962; Makatsch 1976; Danilov *et al.* 1984; Krechmar *et al.* 1991). In north-central Finland, 12–17 nests had average outer diameter 7·9 cm (7·0–10·0), inner diameter 6·3 (5·5–7·2), overall height 4·8 (4·0–7·7), depth of cup 4·0 cm (2·8–5·5); replacement nests flimsier (Koivunen *et al.* 1975, which see for details of material). For dimensions of 18 nests in northern Russia, see Gladkov (1962) and Krechmar *et al.* (1991). Building: presumably by ♀ only as in other *Emberiza* of region. Eggs. See Plate 34. Sub-elliptical, smooth and slightly glossy; very pale green, olive, grey, or pink with sparse spots and blotches of purplish-black or dark brown, or scrawls and hairstreaks, sometimes forming vague circle at broad end; undermarkings of grey or violet (Harrison 1975; Makatsch 1976). See Koivunen *et al.* (1975) for discussion of occurrence of 2 distinct types, light and dark, in north-central Finland. 18·2 × 14·0 mm (16·4–20·2 × 13·2–15·0), *n* = 181; calculated weight 1·89 g (Schönwetter 1984). See also Danilov *et al.* (1984) and Krechmar *et al.* (1991). Clutch: 4–6 (3–7). In north-central Finland, of 21 clutches: 3 eggs, 10%; 4, 14%; 5, 29%; 6, 48%; average 5·14; in June 5·5 (*n* = 16), July 4·3 (*n* = 5); some pairs have 2 broods (Koivunen *et al.* 1975). In north-east Russia, average 4·6 (*n* = 15); 2 replacement clutches each of 3, one 2nd replacement of 1 egg (Krechmar 1966; Krechmar *et al.* 1991); average on Yamal peninsula 5·03, *n* = 32 (Danilov *et al.* 1984). Eggs laid daily (Mikkola and Koivunen 1966; Makatsch 1976). Incubation. 11–12 days, by both sexes (Dementiev and Gladkov 1954; Brood and Söderquist 1967; Makatsch 1976). Starts with 3rd–5th egg, and young hatch over 1–3 days (Mikkola and Koivunen 1966; Danilov *et al.* 1984). At one nest in Swedish Lapland ♀ sat for 4 hrs (Curry-Lindahl 1962). Young. Fed and cared for by both parents (Swanberg 1954; Makatsch 1976; I and M Hills). At one nest, casually brooded by ♂ but mostly by ♀ (Mikkola and Koivunen 1966). Fledging to Maturity. Fledging period 5–8(–11) days. At one nest in Finland 5·5–7·5 days (Mikkola and Koivunen 1966). In general, leaves nest unable to fly at 6–8 days, fully fledged *c.* 3–5 days later (Brood and Söderquist 1967; Danilov *et al.* 1984; Bentz and Génsbøl 1988). Fed by parents for at least 5 and at most *c.* 16 days after leaving nest (Brood and Söderquist 1967). BREEDING SUCCESS. In north-central Finland, 76% of 107 eggs produced fledged young; 6% failed to hatch; 14% of 21 clutches destroyed and 5% of broods (Koivunen *et al.* 1975). Voles (Microtinae) included among nest predators (Danilov *et al.* 1984). BH

Plumages. ADULT MALE BREEDING. Centre of forehead, crown, and hindneck rufous-chestnut, forming broad median crown-stripe, bordered on each side by broad black lateral crown-stripe. Broad cinnamon-buff supercilium extending backwards from just above and behind eye. Eye-ring broad and conspicuous, cream-buff to off-white. Front part of supercilium, lore, upper cheek, and ear-coverts rufous-chestnut, often with paler cinnamon spot at upper rear of ear-coverts; upper, lower, and rear of ear-coverts bordered by black line, varying in extent: sometimes complete, but often restricted to a stripe behind eye over upper ear-coverts, widening at rear, or to bar behind ear-coverts. Bar along upper side of neck pale cream-buff to off-white, often rather contrasting with chestnut or black at rear of ear-coverts and with grey-olive of remaining side of neck. Lower cheek and chin rufous-cinnamon, sometimes with black malar stripe, partly separating cinnamon of lower cheek from that of chin, black not reaching base of lower mandible. Upperparts backwards from lower mantle greyish-olive, marked with ill-defined sooty streaks on upper mantle, upper rump, and tail-coverts, virtually uniform grey-olive on lower rump, broadly streaked deep black on lower mantle and scapulars; black on lower mantle and scapulars partly bordered rufous-chestnut, sides of feathers cinnamon-buff, forming pale lines, less grey-olive than tips of feathers. Throat white, often rather contrasting with rufous-cinnamon of chin; lower side of neck with cream-mottled black triangular patch, sometimes extending into incomplete malar stripe, extending into a gorget of short black streaks on pale cream-buff to off-white background across chest. Side of breast and flank pale cream-grey to off-white, marked with sooty streaks (sooty sometimes partly bordered by cinnamon), remainder of underparts including under tail-coverts uniform cream-white or off-white. Tail greyish-black, outer web and tip of central 4 pairs (t1–t4) narrowly fringed pale cinnamon (when fresh) or pale grey (when worn); inner web of t5 with narrow white stripe along shaft; basal and middle portion of outer web of t6 and long wedge on tip of inner web of t6 white. Flight-feathers, greater upper primary coverts, and bastard wing greyish-black, outer webs narrowly fringed cinnamon-brown, but fringes of primaries and shorter feathers of bastard wing paler, pink-buff (when fresh) to off-white (when worn). Tertials deep black, broad fringes along outer web and tip rufous-chestnut, black on middle of outer web often with chestnut notch. Greater upper wing-coverts black, outer webs and tips fringed rufous-cinnamon (soon bleaching to pink-cinnamon or pink-buff on tips); median coverts black with broad and contrasting pink-cinnamon tips, soon bleaching to pink-buff or cream-white; lesser coverts dark grey with grey-olive fringes. Under wing-coverts and axillaries white, marginal coverts dotted sooty, longer primary coverts grey. *In fresh plumage* (early spring), feathers of cap fringed pale grey-olive, chestnut and black crown-stripes less distinct; feathers of mantle and scapulars more broadly fringed grey-olive, chestnut and cin-

namon more concealed; ground-colour of side of neck and underparts more cream-buff, less whitish. *In worn plumage* (when nesting), ground-colour of side of neck and from throat downwards whiter, black streaks of chest, side of breast, and flank more sharply contrasting; fringe along tips of tertials and greater and median coverts and along outer webs of flight-feathers and tail paler, virtually white on median coverts, but sometimes largely worn off. ADULT FEMALE BREEDING. Head pattern rather variable; often indistinguishable from adult ♂ breeding; some ♀♀ have narrow black streaks on chestnut of crown (absent in ♂), or have entire supercilium, lore, and cheek rufous-cinnamon to cinnamon-buff, with deeper rufous-chestnut restricted to ear-coverts; chin sometimes pale cream-buff or off-white instead of rufous-cinnamon, not contrasting with cream-white throat. ADULT MALE NON-BREEDING. Like adult ♂ breeding in fresh plumage, but grey-olive fringes along tips of feathers of cap broader, chestnut and black crown-stripes more concealed; rufous-chestnut of front part of supercilium, lore, upper cheek, and ear-coverts finely speckled or streaked pale buff, ear-coverts more fully bordered by black; lower cheek and upper chin cinnamon-buff, less rufous, often with pronounced black malar stripe (not reaching bill-base), chin contrasting less with throat. ADULT FEMALE NON-BREEDING. Like adult ♀ breeding in fresh plumage, but chestnut and black of crown-stripes more concealed below broader olive-grey feather-fringes, chestnut in particular often hard to detect; supercilium, lore, and cheek rather pale cinnamon-buff, chestnut restricted to ear-coverts, often with pronounced black moustachial and malar stripes, neither reaching bill-base. Chin and throat cream-buff. NESTLING. Down long, plentiful, on upperparts and vent only; dark brown (Witherby *et al.* 1938). JUVENILE. Feathers streaked dull black and pale buff on mid-crown, dull black and cinnamon at each side, stripes on cap rather indistinct. Side of head cinnamon-buff or rufous-buff, paler buff on indistinct supercilium, mottled black or dark grey on tips of longer ear-coverts, sometimes with faint dark border along upper or (rarely) lower ear-coverts. Lower cheek tawny-buff, bordered below by short and irregular black malar stripe. Hindneck mottled buff and dark grey, side of neck whitish and dark grey. Mantle, scapulars, and back closely streaked black and light cinnamon-buff, rump and upper tail-coverts black and warmer cinnamon. Ground-colour of underparts pale cream-buff, darker buff on side of breast, chest, and flank, almost white on vent; throat, chest, side of breast, upper belly, and flank closely marked with bold dull black triangular spots (broader than black streaks of adult, less deep black, less sharply defined, and reaching higher up on throat). Lesser upper wing-coverts dark grey with light buff-cinnamon fringe; median coverts dull black with pale cream-buff tip, latter interrupted in middle by narrow black shaft-streak; greater coverts dull black, short fringe along tip of outer web cream-white, along tip of inner web cream-buff, black of centre extending to feather-tip along shaft. Tertials, tail, flight-feathers, and greater upper primary coverts as in adult, but tips of tail-feathers and primary coverts more pointed, less rounded, fresh at same time as those of adult are worn or in moult. FIRST ADULT NON-BREEDING. Rather like adult non-breeding, but juvenile tail, flight-feathers, greater upper primary coverts, usually tertials, and variable number of greater upper wing-coverts retained. Tips of tail-feathers, primary coverts, and outer primaries more pointed and frayed, less rounded and smoothly edged than in adult at same time of year. Sexes generally indistinguishable; much variation in head pattern, and birds with much chestnut on mid-crown stripe, front of supercilium, and lores mainly ♂, birds with chestnut restricted to ear-coverts and with narrow dark streaks on mid-crown mainly ♀, but some overlap in characters between sexes. FIRST ADULT

BREEDING. Like adult breeding, but part of juvenile non-breeding retained, as in 1st adult non-breeding. Head pattern as in adult breeding, but some birds of either sex show less chestnut than adult, amount about as in adult non-breeding. Ageing often impossible.

Bare parts. ADULT, FIRST ADULT, JUVENILE. Iris red-brown or dark brown, grey-brown in 1st autumn. Bill dark leaden-grey or greyish-black, cutting edges, extreme base of upper mandible, and basal half of lower mandible flesh-pink (in autumn) to dull flesh-grey or light grey with pink tinge (in spring). Leg and foot pink-flesh, reddish-flesh, deep flesh, dark flesh, or pale red-brown, tinged lilac or grey on toes. (Wallace 1957, 1982a; Mills 1982; ZMA.) NESTLING. No information.

Moults. ADULT POST-BREEDING. Complete; primaries descendent. On breeding grounds, starting when feeding young. Starts with shedding of p1 5–30 July (on average, *c.* 21 July), completed with regrowth of p10 or inner secondaries after 39–55 days, mid-August to early September (on average, *c.* 23 August). No difference between sexes in timing of moult. Autumn migration may start before all body moult finished, but moult of flight-feathers completed. (Ryzhanovski 1986.) ADULT PRE-BREEDING. Partial: side of head, chin, and throat. In winter quarters, February–April. (Witherby *et al.* 1938.) POST-JUVENILE. Partial: head, body, lesser and median upper wing-coverts, sometimes innermost greater coverts (tertial coverts) or (occasionally) some more greater coverts or tertials. In breeding area, starting at age of 17–23 days, duration 40–45 days. (Ryzhanovski 1986.) FIRST PRE-BREEDING. Apparently as adult pre-breeding.

Measurements. ADULT, FIRST ADULT. Mainly breeding area, northern Europe and north-west Siberia, summer, some from western Europe, autumn; skins (BMNH, RMNH, ZFMK, ZMA). Bill (S) to skull, bill (N) to distal corner of nostril; exposed culmen on average 3·2 less than bill (S).

WING	♂	72·4 (1·59; 30) 69–76	♀	68·8 (0·88; 13) 67–70
TAIL		54·9 (2·26; 11) 51–58		51·6 (2·57; 9) 46–55
BILL (S)		12·4 (0·49; 11) 11·6–13·4		11·6 (0·61; 9) 11·2–12·6
BILL (N)		7·2 (0·46; 11) 6·3–8·1		7·0 (0·34; 9) 6·2–7·3
TARSUS		17·5 (0·60; 9) 16·5–18·2		17·7 (0·44; 7) 17·0–18·4

Sex differences significant for wing and tail.

Wing. Yamal peninsula (Russia): ♂ 72·9 (90) 69–78, ♀ 69·3

(87) 64–75 (Danilov *et al.* 1984). Manchuria (northern China): 69–75 (22), ♀ 65–70 (11) (Meise 1934a; Piechocki 1959).

Weights. Yamal peninsula, summer: ♂ 15·7 (85) 13·8–19·3, ♀ 15·1 (79) 12·9–18·4 (Danilov *et al.* 1984). Lake Chany (south-west Siberia), September: 14·3, 15·8 (Havlín and Jurlov 1977). Slovenia, October: ♂ 14·5 (Bračko 1992). Netherlands, autumn: 14·1 (1·42; 7) 12·7–16·5 (ZMA). Mongolia, May: ♂ 16, ♀ 13·5 (0·58; 4) 13–14 (Piechocki and Bolod 1972). Manchuria (northern China), late August and early September: ♂♂ 14, 15; ♀♀ 13, 13 (Piechocki 1959).

Structure. Wing rather long, broad at base, tip bluntly pointed. 10 primaries: p7 and p8 longest, p9 1–2 shorter, p6 0–2, p5 2·5–7, p4 10–12, p3 13–16, p2 15–18, p1 17–21 (see also Königstedt and Müller 1988); p10 strongly reduced, a tiny pin, 49–55 shorter than p7–p8, 7–10 shorter than longest upper primary covert. Outer web of p6–p8 emarginated (occasionally p5 slightly also); inner web of p7–p9 with notch. Tip of longest tertial reaches to tip of p2–p4(–p5). Tail rather long, tip slightly forked; 12 feathers, t4–t5 longest, t1 4–7 shorter, t6 1–3. Bill short, slender, conical; depth at base 5·1 (5) 4·9–5·4, width 5·4 (5) 5·0–5·7 wide; tip sharply pointed. Culmen straight or slightly concave. Leg and foot short and slender. Middle toe with claw 15·4 (10) 14·5–16·5; outer and inner toe with claw both *c.* 70% of middle with claw, hind *c.* 81%; hind claw 6·4 (13) 6·0–7·0. Remainder of structure as in Yellowhammer *E. citrinella*.

Geographical variation. None.

Recognition. In non-breeding plumages, rather similar to ♀ and non-breeding Reed Bunting *E. schoeniclus*, but usually smaller; p5 (almost) without emarginations, 2·5–7 mm shorter than wing-tip (p7–p8) (in *E. schoeniclus*, 1–3 mm); emarginations on primaries less deep (see Svensson 1992); ear-coverts almost uniform rufous-chestnut (mixed brown, black, and grey-white in *E. schoeniclus*); eye-ring usually prominent, whitish; lesser coverts grey with grey-olive tips (in *E. schoeniclus*, rufous-cinnamon or rufous-chestnut), tips of median coverts paler (but colour in both species affected by bleaching); culmen straight or slightly concave (but some birds of arctic populations of *E. schoeniclus* similar). See also Field Characters, and Svensson (1975, 1992), Wallace (1976b), Robel (1985), Königstedt and Müller (1988), and Olsen (1989).   GOK, CSR

Emberiza rutila   **Chestnut Bunting**

DU. Rosse gors   FR. Bruant roux   GE. Rötelammer
RU. Рыжая овсянка   SP. Escribano herrumbroso   SW. Rödbrun sparv

Emberiza rutila Pallas, 1776

Monotypic

Field characters. 14–15 cm; wing-span 21–23·5 cm. Slightly larger and noticeably less delicate than Little Bunting *E. pusilla*, with proportionately larger head, deeper chest, and longer wings; close in size to Yellow-breasted Bunting *E. aureola* but with relatively shorter tail. In all plumages, shows yellow underparts, unstreaked chestnut rump, and little or no white on outer tail-feathers. ♂ bright chestnut on hood, back, and inner wing-feathers; ♀ and immature less distinctive, recalling *E. aureola* but with less striped head and less sharply streaked underparts. Sexes dissimilar; some seasonal variation. 1st winter separable.

ADULT MALE. Moults: August–October (complete). Head, hood down to breast, back, rump, wing-coverts, tertials, and inner secondaries bright rufous-chestnut, marked only by black centres to tertials. Flight-feathers and tail brown-black, tail lacking all but vestigial white patch on outer feathers. Underbody bright yellow, washed and streaked olive and grey on flanks. Bill spiky, with flesh-horn to ochre-horn base and dusky culmen. Legs rather short; yellowish-brown. When fresh, off-white fringes to chestnut tracts create mottled appearance, most noticeably on throat and breast (breast sometimes appearing banded). ADULT FEMALE. Noticeably yellower than ♂, with pattern recalling *E. aureola* but differing in (a) duller head with less contrasting, streaked chestnut and black crown-stripes, less clean, dusky yellow supercilium, dull, rather uniform dusky-olive ear-coverts; (b) pale surround to face restricted to yellow submoustachial stripe, emphasized by short, blackish malar stripe; (c) less boldly striped back, lacking yellowish ground-colour; (d) duller wings, sometimes with chestnut on lesser and median coverts and buff tips to larger coverts (not showing as bright wing-bars); (e) clouded breast, dusky at sides and more or less mottled chestnut in centre, with soft dusky-olive streaks extending along upper flanks; (f) unstreaked, bright rufous rump; (g) pale yellow under tail-coverts; (h) little or no white in tail. Noted as showing fairly obvious whitish eye-ring (Bradshaw 1992). FIRST-WINTER MALE. Some similar to winter ♂ but lack rufous throat, while showing more olive fringes to mantle and wing-coverts and much narrower rufous margins to retained juvenile tertials. Others similar to brighter ♀♀ and not distinguishable in the field. Note that any retained juvenile median and greater coverts may show pale, whitish tips, creating broken wing-bars. FIRST-WINTER FEMALE. Resembles adult but lacks any rufous feathers except on rump; lesser wing-coverts grey-brown.

♂ unmistakable. ♀ and 1st-winter tricky when rufous rump hidden, closely resembling dull *E. aureola*. Distinction requires close observation of head, back, wings, and flanks (see above); *E. aureola* also shows distinct pale central crown-stripe, less distinct malar stripe, more distinctly streaked flanks, and narrow white edges to tail. ♀ and immature may also suggest Yellowhammer *E. citrinella* but that species much larger and longer, with bold white outer tail-feathers. Flight action and silhouette much as *E. aureola* but general character on ground differs in distinctly dumpier form and unobtrusive behaviour. Escapes to tree canopy.

Commonest call similar to *E. pusilla*, a short monosyllabic 'zic'; also a thin high 'teseep'.

Habitat. Breeds in east Palearctic in temperate forest zone of Siberia, in open forests of larch *Larix* and also broad-leaved trees such as alder *Alnus* and birch *Betula*, apparently favouring rich ground-cover of herbaceous plants such as *Ledum*, and dense grass. Frequents mountain slopes and lake shores, and during spring migration also fields and gardens near villages (Dementiev and Gladkov 1954).

Wintering birds in India frequent rice stubbles and bushes in cultivation and forest clearings, feeding on ground and flying up into trees and bushes when disturbed. Recorded, probably on migration, up to 2100 m and in Ladakh at *c.* 4500 m (Ali and Ripley 1974).

Distribution. Breeds eastern Siberia, from north-west Irkutsk region east to Sea of Okhotsk, south to Baykal region and probably northern Mongolia and northern Manchuria. Winters from Assam east to south-east China, south to northern Burma, Thailand, and northern Indochina.

Accidental. 4 autumn records probably true vagrants: Netherlands, 1st-winter ♀, 5 November 1937; Norway, juvenile, 13–15 October 1974; Yugoslavia, 1st-winter ♂, 10 October 1987; Malta, 1st-winter ♂, November 1983. 4 records (June) for Britain, from Scottish and Welsh islands, of doubtful status; some or all may have been escapes from small numbers regularly imported as cage-birds. (Alström and Colston 1991.)

Movements. Migratory. Little information.

Winters in south-east China from Fukien west to Hunan and south to Hainan (Schauensee 1984); also in Vietnam (rather uncommon in south) and southern Laos; fairly common in western Thailand; regular in Burma (except north-east and Tenasserim), and scarce winter visitor to extreme north-east of India (west to *c.* 89°E) (Wildash 1968; Ali and Ripley 1974; King *et al.* 1975; Lekagul and Round 1991). A few winter records from west Japanese islands (Brazil 1991).

Migrates via Ussuriland (south-east Russia), eastern China (west to Shensi) and Korea (Vaurie 1959; Schauensee 1984). Birds ringed Korea recovered Burma (3 in January–March) and Taiwan (1 in April) (McClure 1974). Autumn migration apparently long drawn-out, August–December. Reported 10 August and 10 September in different years in central Siberia (Reymers 1966), early August to mid-October in Ussuriland and adjoining areas (Dementiev and Gladkov 1954; Panov 1973a), and September–October in north-east China and Korea (Hemmingsen and Guildal 1968; Gore and Won 1971). In Hong Kong, passage in small numbers October–December, especially in 2nd and 3rd weeks of November (Chalmers 1986).

In spring, passage in Hong Kong chiefly March–April (fewer than in autumn) (Chalmers 1986). Main movement late, however; almost entirely in May in north-east China (Williams 1986), and 2nd half of May in Ussuriland

(Panov 1973*a*). Reaches breeding areas May–June (Dementiev and Gladkov 1954).

Rare records March–June of vagrancy west of winter range, in Sikkim, Nepal, Ladakh (north-west India), and Chitral (northern Pakistan) (Ali and Ripley 1974). No reports in western Siberia (Johansen 1944; Ravkin 1973) or Kazakhstan (Korelov *et al.* 1974).

For vagrancy to west Palearctic, see Distribution.

DFV

Voice. See Field Characters.

Plumages. ADULT MALE. Upperparts, side of head and neck, and chin to chest deep rufous-chestnut, side of breast and flank mixed olive and yellow, streaked olive-grey on lower flank, remainder of underparts bright sulphur-yellow. Tail greyish-black, uniform or with faint olive-brown edge along central feathers; (t5–)t6 with white border *c.* 1 mm wide along tip of inner web, sometimes extending into small white wedge 6–12 mm long. Flight-feathers, greater upper primary coverts, and bastard wing greyish-black, outer web narrowly fringed light olive-grey (except longest feather of bastard wing), tips of primary coverts sometimes chestnut; basal and middle portion of inner webs bordered light grey. Tertials greyish-black (longest) or deep black (shorter ones), outer webs largely or fully rufous-chestnut. Upper wing-coverts uniform rufous-chestnut, greyish-black inner webs of greater coverts concealed. Under wing-coverts and axillaries pale yellow, shorter coverts brighter sulphur-yellow, dotted dark grey, primary coverts mainly grey. In fresh plumage (autumn), feathers of upperparts with even pale green-grey or off-white fringes 0·5–1 mm wide, chest with sharply contrasting off-white fringes *c.* 1 mm wide, partly concealing rufous-chestnut; pale fringes fully worn off in spring. In some birds, mantle and scapulars faintly streaked black, these perhaps 2-year-olds or advanced 1-year-olds. ADULT FEMALE. In fresh plumage (autumn), cap, mantle, scapulars, and back brownish-olive; forehead, crown, lower mantle, and scapulars finely streaked black; side of crown intensely suffused chestnut, forming distinct lateral crown-stripe; black streaks of lower mantle and scapulars partly bordered by chestnut. Brown-olive of back gradually shades into deep rufous-cinnamon of rump and upper tail-coverts; longest upper tail-coverts with faint black shaft-streak and brown-grey fringe along tip. Side of head and neck brown-olive, mottled chestnut on lore, upper cheek, and ear-coverts, purer and paler olive-buff on supercilium, on centres of ear-coverts and on bar behind ear-coverts; stripe over lower cheek from base of lower mandible backwards pale buff, bordered by faint, mottled black moustachial stripe above and by short black malar stripe below. Narrow and indistinct pale grey eye-ring. Chin and throat pale yellow-buff or buff-white; chest buff-yellow or yellow-buff with ill-defined dark olive spots. Flank yellow, heavily streaked olive, especially on lower flank, remainder of underparts bright pale yellow, often paler contrasting with chest and flank. Tail uniform dusky grey, t6 sometimes with small white spot on tip of inner web. Flight-feathers, greater upper primary coverts, and bastard wing dusky grey; outer webs of primaries narrowly fringed pale buff-yellow, outer webs of secondaries fringed olive-grey; basal and middle portions of inner webs of flight-feathers bordered light pink-grey. Tertials and greater and median upper wing-coverts greyish-black; fringes along tip of median coverts and along outer web and tip of greater coverts and tertials rufous-cinnamon, *c.* 2 mm wide; lesser coverts brown-olive with rufous-chestnut centres (mainly con-

cealed), sometimes with dusky shaft-streak. Under wing-coverts and axillaries as in adult ♂. *In worn plumage* (spring), ground-colour of upperparts more grey-olive, less brownish, dark shaft-streaks more distinct, more rufous-chestnut of feather-centres exposed, especially on side of crown, nape, mantle, and scapulars; supercilium, bar behind ear-coverts, and submoustachial stripe on lower cheek paler, face and ear-coverts often mottled rufous; black malar stripe conspicuous; chin and throat cream-white, chest yellow (less buff), often with some partly chestnut feathers admixed; rarely, whole throat mixed yellow and chestnut. JUVENILE. Upperparts and side of head and neck dark olive-brown with broad ill-defined black-brown shaft-streaks; each side of crown with dark red-brown lateral stripe, paler yellow-brown mid-crown stripe and supercilium; rump and upper tail-coverts red-brown. Underparts dirty pale green-yellow, tinged white on chin and throat, ochre on chest; chest with short blackish streaks, side of breast and flank with dark grey streaks. Wing and tail as adult ♀, but no chestnut on lesser coverts, fringes along outer webs of tail-feathers paler grey-olive, fringes and tips of tertials and median and greater coverts paler rufous-cinnamon, tawny-buff on outer coverts. (Hartert 1903–10; Dementiev and Gladkov 1954; RMNH.) FIRST ADULT MALE. Top and side of head and neck rufous-cinnamon or pale chestnut, slightly less deep than in adult ♂, feathers broadly fringed olive-grey when plumage fresh, largely concealing rufous, especially on supercilium, side of neck, and lower cheek; mantle, scapulars, and back olive-grey with broad black shaft-streaks, black bordered by limited amount of rufous. Rump and upper tail-coverts rufous-cinnamon or pale chestnut, feathers fringed grey, but rufous well-visible. Underparts as adult ♂, but pale fringes of feathers of chin to chest wider, less even, almost fully concealing rufous-chestnut; both chestnut and yellow slightly less deep than in adult ♂. Tertials fringed chestnut (less extensive than in adult ♂); greater coverts fringed dull pale brown, chestnut restricted to innermost. Lesser and median upper wing-coverts rufous-chestnut with dusky central mark and olive-grey fringe. *In worn plumage* (spring), chestnut of upperparts, side of head and neck, chin to chest, and upper wing-coverts more exposed, but less deep and uniform than in adult ♂, variegated olive-grey to varying extent; mantle and scapulars mixed rufous and olive-grey, usually heavily streaked black; supercilium, centre of ear-coverts, side of neck, and lower cheek rufous-chestnut with traces of grey or yellow-buff fringes; chin and throat mixed rufous-chestnut and cream-yellow or off-white; ♂♂ with relatively little amount of chestnut closely similar to those adult ♀♀ which have much of it. Juvenile tail, flight-feathers, and greater upper primary coverts retained, slightly greyer and more worn than those of adult at same time of year, and fringes of tail and primary coverts greyer olive, less dull brown or chestnut, black on inner tertials less deep, less sharply defined, but difference sometimes hard to see. FIRST ADULT FEMALE. Rather like adult ♀, but virtually without rufous, except on rump and upper tail-coverts; elsewhere, rufous (if any) restricted to traces on side of crown and nape; lesser coverts brown-olive with dark grey bases, without rufous-chestnut sub-terminally. Forehead to scapulars and back olive-brown with dull black shaft-streaks (faint on hindneck, darkest on each side of crown), mid-crown stripe and long supercilium paler brown-buff, mottled with some olive-grey. Lore, upper cheek, and ear-coverts olive-brown with grey mottling, ear-coverts with rather ill-defined dull black surround; distinct submoustachial stripe on lower cheek pale buff. Chin and throat pale buff; malar stripe, side of throat, and chest pale tawny-buff with short black streaks or spots. Side of breast olive-brown; flank mottled buff and white, marked with narrow dull black streaks; remainder of underparts pale yellow or yellow-white, not sharply demarcated from chest and flank. Lesser coverts grey-olive, without chestnut. Part of juvenile feathering retained, as in 1st adult ♂; tail more sharply pointed than in adult, inner web of outer feathers concave at tips rather than straight or convex. See also Svensson (1992).

Bare parts. ADULT, FIRST ADULT. Iris brown or dark brown. Upper mandible brownish- or bluish-horn, lower mandible paler. Leg and foot light brown. (Hartert 1903-10; RMNH). JUVENILE. No information.

Moults. ADULT POST-BREEDING. Complete; primaries descendent. Specimens from USSR moulting body on 16 August and in intense moult of wing, tail, and body on 21 and 23 August; some moult of body up to at least 20 September (Dementiev and Gladkov 1954). Moult completed in birds on autumn migration, China, late September and October (BMNH, ZFMK). Apparently no pre-breeding moult (Svensson 1992). POST-JUVENILE. Partial; involves head, body, lesser and median upper wing-coverts, and sometimes a number of greater coverts and a few tertials. Starts shortly after fledging; moult completed in ♂ from 18 August, Manchuria (China) (Piechocki 1959); one in USSR still in moult 11 September (Dementiev and Gladkov 1954). Occasionally perhaps a partial pre-breeding moult of head and chest in spring (Svensson 1992).

Measurements. ADULT, FIRST ADULT. Eastern Siberia, summer, and eastern China, on autumn and spring migration; skins (RMNH, ZFMK, ZMA). Bill (S) to skull, bill (N) to distal corner of nostril; exposed culmen on average 3·2 less than bill (S).

WING	♂	74·1 (1·67; 14)	71-77	♀ 71·4 (1·90; 8)	69-75
TAIL		55·3 (2·51; 11)	51-59	52·3 (1·83; 8)	49-55
BILL (S)		13·2 (0·83; 13)	12·3-14·5	13·0 (0·75; 8)	12·0-14·3
BILL (N)		7·7 (0·66; 11)	6·9-8·7	7·6 (0·55; 8)	6·9-8·6

TARSUS 19·0 (0·41; 12) 18·4-19·9   18·9 (0·31; 8) 18·4-19·4
Sex differences significant for wing and tail.

Wing. USSR: ♂ 72·2 (12) 71-75, ♀ 69 (6) 68-70 (Dementiev and Gladkov 1954). Whole geographical range: ♂ 72-79 (n=23), ♀ 67-73(-76) (n=18) (Svensson 1992). Mongolia: 65-78 (Piechocki and Bolod 1972).

Weights. ADULT, FIRST ADULT. USSR: ♂ 17·6 (3) 16·8-18·9, ♀♀ 16·5, 16·6 (Dementiev and Gladkov 1954). Mongolia, August-September: ♂ 12, unsexed 14 (Piechocki and Bolod 1972). Manchuria (China), August: ♂ 15 (Piechocki 1959). Thailand, January-February: ♂♂ 15·2, 16·2; ♀ 15·8 (Melville and Round 1984). Malta, November: ♂ 18 (Sultana and Gauci 1985a). Netherlands, November: ♀ 16·3 (RMNH).

Structure. Wing rather short, broad at base, tip bluntly pointed. 10 primaries: p7-p8 longest, p9 0-3 shorter, p6 1-2·5, p5 6-8, p4 10-12, p3 12-14, p2 14-16, p1 15-19; p10 strongly reduced, 48-55 shorter than p7-p8, 8-11 shorter than longest upper primary covert. Outer web of p6-p8 emarginated, faint notch on inner web of p7-p9. Tip of longest tertial reaches tip of secondaries. Tail rather short, tip shallowly forked; 12 feathers, each with rather pointed tip, t1 6-10 shorter than t6. Bill short, conical, culmen and gonys straight or faintly convex; 6·0 (7) 5·8-6·4 mm deep at base, 6·4 (7) 6·0-6·8 mm wide. Leg and foot rather short, slender. Middle toe with claw 17·2 (7) 16-18; outer and inner toe with claw both c. 65% of middle with claw, hind c. 77%. Remainder of structure as in Yellowhammer E. citrinella.

Recognition. Smaller than other Emberiza except Black-faced Bunting E. spodocephala, Little Bunting E. pusilla, and Pallas's Reed Bunting E. pallasi; differs from these by almost uniformly dark tail without long white wedge on inner web of outer 2 feathers (see Plumages), rufous rump and yellow-tinged belly, pointed tail-feathers, and (from E. spodocephala and E. pallasi only) by absence of emargination of p5.   CSR

Emberiza aureola Yellow-breasted Bunting

PLATES 24 and 26 (flight)
[between pages 256 and 257]

Du. Wilgengors   Fr. Bruant auréolé   Ge. Weidenammer
Ru. Дубровник   Sp. Escribano aureolado   Sw. Gyllensparv

Emberiza aureola Pallas, 1773

Polytypic. Nominate *aureola* Pallas, 1773, north-east Europe, east to Lake Baykal and through Yakutiya to Anadyrland, south to Mongolia. Extralimital: *ornata* Shulpin, 1928, south-east Siberia, from Chita region (eastern Transbaykalia) through northern Manchuria and Amurland to Sakhalin and Japan; also Kamchatka.

Field characters. 14-15 cm; wing-span 21·5-24 cm. Slightly smaller and noticeably shorter tailed than north-western race of Reed Bunting E. schoeniclus schoeniclus; slightly larger and longer tailed than Chestnut Bunting E. rutila; up to 10% larger than Little Bunting E. pusilla. Rather small but robust bunting, with spiky bill and rather short tail contributing to flight silhouette suggestive of Fringilla. In all plumages, ground-colour of little-streaked underparts yellowish; outer tail-feathers show white.

Adult ♂ has narrow black-brown necklace, rich brown back, white lesser and median coverts (again recalling Fringilla), and whitish wing-bar; front of head and throat black when breeding. ♀ and immature have heavily striped head, broadly streaked back, double whitish wing-bar, and streaked flanks. Call assists separation from other large Emberiza. Sexes dissimilar; some seasonal variation. Juvenile separable.

ADULT MALE BREEDING. Moults: August-October

(complete); March–May (mainly face). Vivid plumage contrasts come mainly through wear. Forehead, forecrown, lores, ear-coverts, and deep throat jet-black, combining with dark chestnut rear-crown and nape to give extremely dark-headed appearance. Bright, rich yellow half-collar separated from similarly coloured underbody by black-edged, chestnut necklace around breast which breaks up into chestnut streaks on sides of breast and down upper flanks. Mantle and scapulars dark chestnut, showing only vestigial (if any) blackish streaks; rump similarly coloured and unstreaked; at some angles, these tracts show pinkish wash. Wings more boldly patterned than any other *Emberiza* of the region, with white blaze on longer lesser and median coverts as obvious as on Chaffinch *F. coelebs* and set off by blackish-centred, rufous greater coverts whose whitish tips form narrower but still distinct lower wing-bar; tertials black-centred, with bright rufous-buff margins; flight-feathers brown-black, with narrow buff fringes most obvious on secondaries. Tail brown-black, with rufous-brown fringes and much white on outer feathers. Vent and under tail-coverts buffish-white. Under wing-coverts mainly white. Bill pale pinkish to brown-horn, with dusky culmen and tip. Legs pale flesh-brown to orange-brown. ADULT FEMALE BREEDING. Strikingly different from ♂ in plumage pattern, but, if seen well, no less distinctive. Ground-colour of plumage noticeably pale, with warm yellowish-buff tone on head, upperparts, and wings and somewhat richer yellow tints on half-collar and from throat to breast and along flanks. Main plumage features: (a) dark blackish-brown lateral crown-stripes, isolating narrow median crown line and long, bold supercilium which is broad and almost white behind eye; (b) black-brown rear eye-stripe, further emphasizing supercilium and contrasting with whitish spot on upper rear ear-coverts; (c) black-brown border to lower ear-coverts extending forward to bill; (d) rather pale buff nape; (e) bold black streaks on lower mantle, contrasting particularly with broad pale stripes on either side of mantle-centre; (f) bright white bar across median coverts, some forming small white blaze; (g) narrower, buffier bar across greater coverts; (h) buff-brown (sometimes rufous) rump; (i) dusky-brown streaks from sides of breast to rear flanks; (j) orange tinge to upper breast, where ill-defined chestnut necklace or gorget sometimes shows. Effect is to make ♀ more heavily marked than other *Emberiza*, even *E. rutila*. ADULT MALE NON-BREEDING. Fresh plumage has similar basic pattern to breeding ♂ but black throat lost and head becomes dusky-brown with dull buffish crown-stripe, supercilium, and centre to black-edged cheeks. All chestnut tracts show paler, buffish tips, with soft black streaks on mantle and scapulars. Striking wing markings unchanged. ADULT FEMALE NON-BREEDING. Appearance after moult changes much less than ♂ but ground-colour of plumage buffer, rump lacking obvious rufous, and necklace always concealed. Note that most of breast unstreaked, unlike 1st winter and juvenile. JUVENILE.

Resembles ♀ but face plainer, with dark lower border to cheeks restricted to rear spot, all of breast finely streaked brown-black on rather darker buff ground, wing-bars whiter, and rump quite distinctly streaked. FIRST-WINTER MALE. Resembles ♀ but shows chestnut on crown and rump and at least vestige of necklace. FIRST-SUMMER MALE. Lacks immaculate appearance of adult, with less uniform black face and throat, still heavily streaked mantle, narrower necklace, less white on wing-blaze, and less rich yellow below. FIRST-YEAR FEMALE. Appearance intermediate between juvenile and adult, retaining fine breast-streaks on paler yellow underparts. Note that autumn vagrants may retain wholly juvenile plumage and young ♂♂ sing in ♀-like plumage.

At close range, adult ♂ unmistakable but at distance or in poor view may briefly suggest ♂ Black-headed Bunting *E. melanocephala* (10% larger, with proportionately larger bill and more attenuated build; lacks necklace, flank-streaks, and white wing-blaze) and on ground even Black-headed Wagtail *Motacilla flava feldegg* (10% larger, with much finer bill and longer tail and legs; lacks black throat and has green upperparts). Female and immature distinctive compared to other large *Emberiza*, lacking fully streaked underparts and particularly lemon-yellow hue of Yellowhammer *E. citrinella*, but needing careful separation from *E. rutila* (same size but duller, more olive-toned above, with far less distinct head pattern and wing-bars, much more rufous rump and yellow under tail-coverts), Black-faced Bunting *E. spodocephala* (same size but darker, more brown-toned above, with greyish shawl, cheeks, and collar, distinct dusky malar stripe—vestigial in *E. aureola*—duller wing-bars, and greyish rump), Red-headed Bunting *E. bruniceps* (up to 10% larger, with more uniform, less streaked plumage and especially yellow rump and under tail-coverts), and even Bobolink *Dolichonyx oryzivorus* (plumage pattern and colours remarkably convergent except for lack of white on outer tail-feathers but at least 15% larger, with deeper, more conical bill). *E. aureola* thus has pivotal position within bunting identification and worth full study. Flight light and fast, but, due to relatively compact silhouette, appearance and action may suggest *Fringilla* finch as much as typical bunting (particularly in adult ♂ showing white wing-blaze); does not trail tail as *E. citrinella*, landing tidily without hint of collapse. Escapes to thick cover. Gait a hop, also creeping when intent on feeding. Stance variable, often quite upright on perch but level (with more hunched posture) on ground. Undemonstrative but occasionally flicks wings and tail. Unobtrusive; often shy on migration but frequently tame on breeding grounds. Addicted to damp thickets on breeding grounds but enters cereal crops or skulks along hedges and ditches on migration; markedly gregarious when undertaking mass movement.

Song recalls Ortolan Bunting *E. hortulana* but includes jingling or chiming also suggestive of Lapland Bunting *Calcarius lapponicus* and *E. schoeniclus*; typical phrase ter-

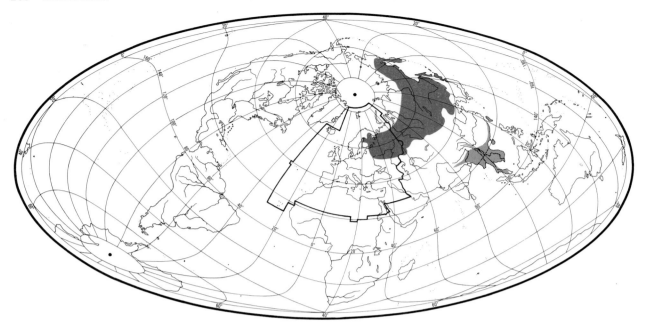

minally higher pitched and faster than *E. hortulana* but with similar melodious timbre, 'tu tu-li tu-li-ti he-li li-lu-li' or 'treeu-treeu-trteeu-huhuhu-treea-treea trip-treeh'. Calls include short 'tik', 'zipp', or 'tzip', reminiscent of short call of Spotted Flycatcher *Muscicapa striata* and *E. pusilla*, and soft trilling 'trssit'; south Siberian birds call 'tick' or 'zick', sounding louder than *E. pusilla* and *E. spodocephala*.

Habitat. In west Palearctic breeds in boreal zone between July isotherms of 12°C and 23–24°C, overlapping with Rustic Bunting *E. rustica* and sharing with it a recent westward expansion over northern Europe. Favours moist or wet, often riverside or floodland sites, with willow *Salix* and birch *Betula* thickets or spinneys, interspersed with tall grass (Voous 1960*b*). Occurs in low-lying wet or dry meadows with tall herbage and scattered shrubs, riverside thickets and shrubby willows in river flood zones, fields and boggy areas with secondary scrub, open burnt forest, mountain meadows with scrub and scattered trees, birch forest edge, and sparse growth of young forest (Harrison 1982). In European USSR, typically occupies relatively dry river valley meadows covered with dense tall grasses and scattered bushes of alder *Alnus* and willow. In taiga belt, inhabits also peat bogs, scorched areas with remains of birch and spruce *Picea* trees, and alpine meadows with thick osier growths and solitary spruces (Dementiev and Gladkov 1954). In Siberia, common in forest-steppe intersected by ravines, margins of sparse forest, wet valleys, and floodplains, preferring areas with shrubby vegetation to those with many trees, and avoiding dry open steppes (Shkatulova 1962). Further east, on Amur, occupies both

dry and damp meadows with rich or sparse shrub undergrowth and sometimes trees, as well as sedge and grass (Barancheev 1963). While most records of habitat refer to valleys or low-lying places, with emphasis on wet or swampy conditions, a few relate to dry open steppes and some to higher elevations up to *c.* 1700 m (Dresser 1871–81).

In Indian winter quarters, frequents grassland and farms, in or around small settlements, and also hedgerows and gardens in hills up to 1500 m, feeding on ground but flying up to trees when disturbed; roosts communally in bushes (Ali and Ripley 1974).

Distribution. Westward spread began in 19th century. At least 2 recent waves of expansion, first in 1960–5, then 1974 onwards (Mal'chevski and Pukinski 1983).

NORWAY. Adults feeding young at Pasvik (near border with Finland) 1967, may have bred in Norway (VR). FINLAND. First bred in 1920s; breeding areas in central Finland not occupied until 1940s and 1950s (Koskimies 1989).

Accidental. Iceland, Britain, Ireland, France, Belgium, Netherlands, Germany, Spain, Denmark, Sweden, Poland, Latvia, Estonia, Czechoslovakia, Italy, Malta, Greece, Cyprus, Turkey, Israel, Jordan, Egypt.

Population. Probably increasing slowly in west of range.

FINLAND. Rapid increase in middle of 20th century; since then levelling off, but seems still to be increasing slowly. Estimated 300 pairs. (Koskimies 1989; OH.)

Movements. All populations migratory. Western birds head east from breeding grounds then south, and eastern

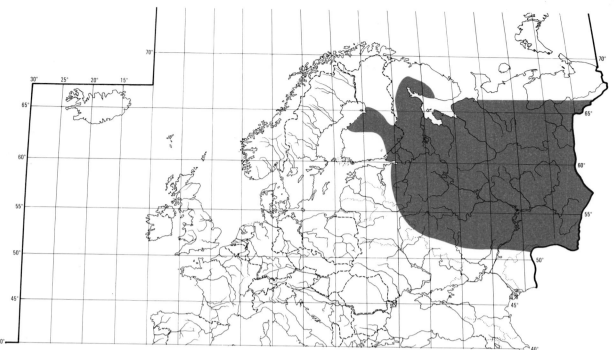

birds south or south-west, to reach winter quarters via Mongolia, south-east Russia, and China.

Winters throughout south-east Asia (apparently not in south-east Thailand) (King *et al.* 1975; Lekagul and Round 1991), with small numbers on Malay peninsula south to Malacca (*c.* 2°50′N) (Medway and Wells 1976); also in southern China from Yunnan east to Kwangtung, south to Hainan (Schauensee 1984). Uncommon in Taiwan (Chang 1980). Abundant winter visitor to plains of Burma (occurring in hills chiefly on passage), reaching extreme south of Tenasserim (Smythies 1986). Also winters in north-east India and Bangladesh, Sikkim, and (chiefly eastern) Nepal (Ali and Ripley 1974; Inskipp and Inskipp 1985). Vagrant to Philippines (Kennedy 1984) and Borneo (Smythies 1981), also to Arabia (F E Warr). 2 long-distance recoveries: bird ringed at Oulu (northern Finland), June, recovered in southern Thailand, December, *c.* 9600 km south-east; and bird ringed in central Thailand, November, recovered in Yakutiya (north-east Russia), July, *c.* 6720 km north-east (McClure 1974).

Leaves breeding grounds early and returns late. Departure begins July, adults earlier than juveniles. Movement reported from late July in Finland (Hildén 1969*a*); in St Petersburg region (north-west Russia), adults begin to leave at end of July, with latest records (juveniles) late August or early September (Mal'chevski and Pukinski 1983). In Volga-Kama region (west of Urals), main departure and passage by end of August, though some juveniles still present at beginning of September (Popov 1978). In western Siberia, adults present until end of August, and juveniles until mid-September (Johansen 1944). In Kazakhstan, no evidence of passage except in Ural valley in north-west, so presumably birds breeding in northern Kazakhstan head directly eastwards (Korelov *et al.* 1974). Many pass through north-east Altai, at low levels and in foothills; passage especially intensive at end of August, but continues to mid-October (Ravkin 1973); timing similar in Mongolia (Mey 1988). Most birds leave breeding grounds in north-east Russia (e.g. Yakutiya, Kamchatka peninsula) in August, with a few stragglers in 1st week of September (Vorobiev 1963; Lobkov 1986; Larionov *et al.* 1991). However, birds breeding on Sakhalin island (*ornata*) depart late (following moult), vacating northern and central areas by end of September, southern areas not until end of October (Gizenko 1955). In Ussuriland (south-east Russia), intensive and widespread passage in 1st half of August, with stragglers sometimes present to mid-October (Panov 1973*a*), and passage in north-east China also August–October (Hemmingsen 1951). Many migrate through northern Korea (Austin 1948; Gore and Won 1971). Japanese birds (breeding in Hokkaido and extreme north of Honshu) present until late September, and migrate to winter quarters via mainland; fairly common on passage also on west Japanese islands, but rare in Honshu (Brazil 1991). Abundant both seasons in lower Yangtze valley (China) (Gee and Moffett 1917). Many birds stop over to moult in southern China, especially in Yangtze valley (Portenko and Stübs 1971). In Hong Kong, passage October–November, with large numbers in concentrated movement mid-October to mid-November (Chalmers 1986).

Arrives in India, Nepal, and Burma in October (Ali and Ripley 1974; Inskipp and Inskipp 1985; Smythies 1986); in south of range (Malay peninsula) not until November (Medway and Wells 1976).

Spring migration April–June; present in winter quarters until May, but most birds leave in April. In Hong Kong, passage early April to early May, but less regular and numerous than in autumn (Chalmers 1986). In north-east China, common end of April to end of May or beginning of June (Williams 1986). Breeding birds reach Japan early or mid-May (even as late as June in northern areas), occasionally late April (Brazil 1991). In north-east of range, average earliest record 19 May (16–26 May) over 13 years in Lena valley at Yakutsk (Larionov *et al.* 1991), and 28 May (10 May to 3 June) over 10 years on Kamchatka peninsula (Lobkov 1986). Reaches Anadyr' region in extreme north-east of range only in early June, later than Little Bunting *E. pusilla* and Rustic Bunting *E. rustica* (Portenko 1939). Arrives in northern Mongolia and south-west Transbaykalia in 2nd half of May (Kozlova 1933), passage in north-east Altai continuing throughout June (Ravkin 1973). In Kazakhstan, migrates in large numbers through Irtysh valley in north-east, inconspicuous elsewhere (Korelov *et al.* 1974). In western Russia (west of Urals) arrivals late May to June (Ptushenko and Inozemtsev 1968; Popov 1978; Mal'chevski and Pukinski 1983). In Karelia (eastern Finland), birds usually arrive at beginning of June or exceptionally in last days of May (Karttunen *et al.* 1970); north-west direction of spring migration is reflected in breeding range, lying in zone 300 km wide extending from south-east Finland to northern Gulf of Bothnia (Koskimies 1989).

Most records of vagrancy in Europe are in Britain, Scandinavia, and Italy (Alström and Colston 1991). Annual in Britain in recent years; almost all records in autumn, especially September (suggesting reverse migration); 10 before 1958 and 145 in 1958–91 (including Ireland); steady increase since 1950s probably genuine and not due to increase in observers, as most are from Fair Isle (Scotland) where coverage consistently high (Dymond *et al.* 1989; Rogers *et al.* 1992). In Sweden, 23 records up to 1986, of which 19 in May–July (chiefly June) in north and south-east, suggesting overshooting, and 4 in August in south-east (Breife *et al.* 1990). For vagrancy elsewhere in Europe, see Distribution.   DFV

Food. Almost all informatiom extralimital. Diet seeds and other plant material; invertebrates in breeding season (Promptov 1934; Dementiev and Gladkov 1954; Portenko and Stübs 1971; Armani 1985). On arrival on river meadows south of Moscow, feeds on grass and herbs exposed by receding water collecting insects and seeds, and also in rose *Rosa* bushes (Promptov 1934). In Volga-Kama region (east European Russia), foraged on stems of tall vegetation, rarely on ground (Popov 1978). In Mongolia, May, fed tamely in groups by roads, tracks, and streams (Piechocki and Bolod 1972); in summer, foraged in grass far from trees and also in bushes; sometimes in larch *Larix* trees nibbling at twigs and cones (Mauersberger 1980); in October, in snow, recorded in larches searching for insects (Mey 1988). In eastern Hokkaido (Japan), summer, 94% of 198 feeding observations were on ground (almost all in tangled tall grass, reeds, sedges, etc., *c.* 3% on pasture or bare soil), 6% on vegetation (3% rose bushes, 3% herbs); not seen to move up and down stems (Kidono 1977, which see for comparison with Reed Bunting *E. schoeniclus*). In India and Burma, winter, feeds on grass seeds, including bamboo *Bambusa*, but as rice crop ripens descends on paddyfields in huge flocks to feed on it and also later in stubble; moves into villages when this supply exhausted; occasionally makes sallies after flying insects (Ali and Ripley 1974; Smythies 1986).

Diet includes the following. Invertebrates: damsel flies and dragonflies (Odonata), stoneflies (Plecoptera), grasshoppers, etc. (Orthoptera: Tettigoniidae, Acrididae), bugs (Hemiptera: Pentatomidae, Fulgoridae, Cicadellidae, Aphididae), adult and larval Lepidoptera (Tortricidae, Pyralidae, Noctuidae), lacewings (Neuroptera: Chrysopidae), caddis flies (Trichoptera), adult and larval flies (Diptera: Tipulidae, Culicidae, Stratiomyidae, Tabanidae, Syrphidae, Tachinidae, Muscidae), adult and larval Hymenoptera (Tenthredinidae, Ichneumonidae, ants Formicidae), beetles (Coleoptera: Cicindelidae, Carabidae, Dytiscidae, Staphylinidae, Scarabaeidae, Buprestidae, Elateridae, Cantharidae, Coccinellidae, Chrysomelidae, Curculionidae), spiders (Araneae), snails (Pulmonata). Plants: seeds (etc.) of rose *Rosa*, grasses (Gramineae, including wheat *Triticum*, rice *Oryza*, *Sorghum*, bamboo *Bambusa*, *Milium*). (Ferguson-Lees 1959; Chia and Li 1973; Ali and Ripley 1974; Korelov *et al.* 1974; Popov 1978; Kishchinski 1980; Larionov *et al.* 1991.)

In Volga-Kama region, summer, 100% of 33 stomachs contained animal material and 6% plant material (seeds of grass *Milium*); 60% contained Curculionidae, 39% caterpillars, 36% Diptera, 18% Orthoptera, 15% Elateridae, 15% Buprestidae, 12% spiders, 6% Chrysomelidae, 3% Staphylinidae, 3% Hemiptera (Popov 1978). In Yakutiya (eastern Russia), 8 gullets and stomachs in breeding season contained Odonata, stoneflies, caddis flies, Diptera, various Hymenoptera, and beetles (Cantharidae, Chrysomelidae, Curculionidae); herb seeds present in 3 samples (Larionov *et al.* 1991). In Ural valley (north-west Kazakhstan), 2 stomachs held 4 Noctuidae, 4 beetles (2 Chrysomelidae, 2 Curculionidae), and 1 Hymenoptera (Korelov *et al.* 1974). In China, over year, 859 stomachs contained 89% by volume (81% by number) plant material, of which cereal grain 70% (44%) and grass seeds 19% (37%), and 11% (16%) insects, mostly beetles, Hemiptera, and Diptera; in 34 stomachs, July, insects were 80% by volume (53% by number), but less than 6% by volume in every other month (Chia and Li 1973). Also in China, summer, of 25 stomachs, 5 contained wheat grains and 9 other plant material, but most of diet invertebrates; on migration, causes great damage in paddyfields (Cheng 1964). In Ussuriland (eastern Russia) in autumn, feeds on grass seeds (Panov 1973a).

Young fed on invertebrates. In Yakutiya, 211 items from nestling collar-samples comprised 27% (by number) Diptera (15% adult Tabanidae, 11% adult Tipulidae, 1% larval Stratiomyidae), 26% Orthoptera, 23% caterpillars, 13% Hymenoptera (10% sawfly Symphyta larvae, 1% ants), 4% adult beetles (2% Chrysomelidae, 1% Curculionidae), 4% spiders, 1% Odonata, 1% snails (Larionov et al. 1991). In Volga-Kama region, 16-day-old young were given Orthoptera, adult and larval Lepidoptera, Diptera, and Chrysomelidae beetles; most food collected within 10-25 m of nest and adults not seen to forage more than 70 m away (Popov 1978). In Finland, adults at one nest brought lacewings, Diptera (Tipulidae, Tabanidae, Syrphidae), and caterpillars (Ferguson-Lees 1959). At other nests in Russia and Kazakhstan, young received Tettigoniidae, adult and larval Lepidoptera, and Tipulidae (Promptov 1934; Ptushenko and Inozemtsev 1968; Korelov et al. 1974; Kishchinski 1980). For food in captivity, see Stadie (1983).   BH

Social pattern and behaviour. Most aspects quite well known, but only detailed study of courtship is for *ornata* in Hokkaido (Japan) (Masatomi and Kobayashi 1982).

1. Gregarious outside breeding season, in flocks of up to hundreds and sometimes thousands. In Amur region (eastern Russia), after young independent, family parties initially form flocks of 25-30, these becoming larger prior to autumn migration (Barancheev 1963). In Nepal, flocks of 400 regular in east, November-April, and enormous flocks reported flying to roost in March-April 1982, e.g. 3500 at Chitwan and over 7000 at Kosi Tappu (Inskipp and Inskipp 1985). In Hong Kong, often in flocks of up to 500 on autumn passage (Chalmers 1986, which see for other concentrations). In Burma, 'immense' flocks recorded feeding in fields of grain (Smythies 1986). In spring, migrates in typically small flocks but may congregate in enormous numbers at favourable feeding or drinking sites (Kozlova 1930; Dementiev and Gladkov 1954). In Mongolia, seen feeding together with Little Buntings *E. pusilla* during spring migration (Piechocki and Bolod 1972). In winter, often associated with munias (Estrildidae) (Ali and Ripley 1974). BONDS. Mating system mainly monogamous, but ♂♂ bigamous exceptionally: 2 cases recorded in which ♂ simultaneously fed young at nests of 2 ♀♀ and showed anxiety at both nests (Mal'chevski and Pukinski 1983). Nest built by ♀ alone (for apparent contribution by ♂ in captive pair, see Timmis 1972); both sexes incubate, but ♂ has no brood patch (Mal'chevski and Pukinski 1983) and makes only minor contribution (Masatomi and Kobayashi 1982). According to Popov (1978) ♀ incubates alone for first 6-8 days, after which sexes share incubation. Both sexes brood (Ferguson-Lees 1959) and feed young (e.g. Promptov 1934, Masatomi and Kobayashi 1982). If ♀ dies, ♂ takes over responsibility for chick-rearing (Popov 1978). After leaving nest, young fed by both parents for up to 7-10 days (Ptushenko and Inozemtsev 1968), 10-12 days (Mal'chevski and Pukinski 1983). No information on age at first breeding. BREEDING DISPERSION. Typically in neighbourhood groups, with relatively small territories leading to locally high

densities. In Volga-Kama region (east European Russia), nests only 8-10 m apart in some well-vegetated sites, but 200 m or more if there is little cover (Popov 1978). In St Petersburg region (Russia), 'semi-colonial', with 20-30 m between nests (Mal'chevski and Pukinski 1983). In central-southern Siberia, territory size in valleys (optimal habitat) c. 0·25 ha, nests 50 m apart, but in forests 70-100 m apart (Reymers 1966). Territory serves for courtship, feeding and nesting, but some feeding occurs outside territory. In Volga-Kama region (where density very high: see above) feeds usually 10-25 (-70) m from nest. In Hokkaido, both sexes will forage outside territory, ♀♀ staying out longer than ♂♂ (Masatomi and Kobayashi 1982). In study by Nakamura et al. (1968) in Hokkaido, density relatively low (17·8 pairs per km²), and 'home ranges' large and less overlapping than sympatric Reed Buntings *E. schoeniclus*; each ♂ *E. aureola* had 2 singing areas, one central (around nest-area), other (used for song-duels with neighbours) 100-200 m from nest-site. Also in Hokkaido, over 3 years, maximum 143 pairs per km² in grassland (Fujimaki and Takami 1986). Only other known densities are for Russia (both western Palearctic and extralimital) as follows: in Komi (north-east European Russia) 25-34 pairs per km² in raised bogs (Guriev and Bashkova 1986). In Moscow region, minimum 18 pairs per ha in 1930s, but, with irrigation, distribution became uneven and density had dropped to 6-12 pairs per km² by 1959-61 (Ptushenko and Inozemtsev 1968; see also Promptov 1934). In west Siberian taiga, 8-11 birds per km² in northern bogs, 265 in south (Vartapetov 1984). In central Siberia, up to 500 birds per km² in some areas but mostly 60-120; average 13 in shrubby sub-zone of mountain foothills, 43 in mesotrophic bogs, 89 in boggy areas around Kansk settlement, 197 in river valleys in southern taiga, up to 422 birds per km² (2·1 pairs per ha) along Yenisey River (Rogacheva 1988). In Lena-Amga interfluve (Yakutiya) up to 18·8 birds per km² in meadows (Larionov et al. 1991). In Kamchatka, 7-32 pairs per km² on riverine willow *Salix* scrub, 11·8 in scrubby riverine meadows, 53·8 on riverine terraces with sparse birch *Betula* woodland; occupation generally very patchy, e.g. can be 3 pairs on 0·5-0·9 ha, then none on apparently same habitat elsewhere (Lobkov 1986). ROOSTING. During laying, ♀ starts sitting on nest at night from 2nd-3rd egg (Popov 1978); no other information for breeding season. Outside breeding season, roosts communally in bushes (Ali and Ripley 1974). In Burma, large numbers seen at dusk flying to roost in chosen clump of bushes fringing stream (Smythies 1986). For nocturnal activity see Song-display (below).

2. At least in winter quarters, not very shy (Roberts 1992). Flies up into trees when disturbed during ground-feeding (Ali and Ripley 1974). Will also crouch on ground when danger threatens (Piechocki and Bolod 1972). Tail erect when perching (Cooke 1960). FLOCK BEHAVIOUR. See introduction to part 1 (above) for flock associates, and Roosting (above) for roosting flights. No other information. SONG-DISPLAY. ♂ sings (see 1 in Voice) from exposed tree-tops and bushes (several such song-posts used within territory: Masatomi and Kobayashi 1982), also from grass tussocks (Piechocki 1959). Starts in winter quarters in March-April, i.e. shortly before spring migration (Ali and Ripley 1974), and heard on arrival (or soon after) on breeding grounds (e.g. Kozlova 1930, Masatomi and Kobayashi 1982). For role of song in territorial defence and courtship see Antagonistic Behaviour and Heterosexual Behaviour (below). In Moscow region, song-activity most intense 8-11 June (Ptushenko and Inozemtsev 1968). In Amur region, song-display intense early June, sudden reduction after June 15, and by July 18 there was only low-intensity singing from 3 ♂♂ (Barancheev 1963); on 12 June, combined output of ♂♂ in study area was 943 songs per

day (max. 105 per hr); on 25 June 994, 95; on 1 July 923, 78; on 7 July 609, 66 (Denisova and Gomolitskaya 1967, calculated from Barancheev 1963). End of song-period in Chita region (south-east of Lake Baykal, Russia) 19 July (Shkatulova 1962), in Altai end of July (Dementiev and Gladkov 1954). Diurnal rhythm studied in Russia: in Amur region, song begins 45–50 min before sunrise, peaks at *c.* 04.30–05.00 hrs, then gradual reduction to low around mid-day, building up again from *c.* 15.00 hrs to peak *c.* 19.00 hrs, then decline to finish *c.* 1 hr after sunset (Barancheev 1963, which see for influence of weather on song-activity). In Novosibirsk region (western Siberia) 35% of song in morning, 24% later in day, 30% evening, 11% night (Podarueva 1979). Singing also heard in early June at *c.* 22.20 hrs in full moon, Ulan-Bator, Mongolia (Mauersberger *et al.* 1982). Captive ♂♂ also often sing at night (Dornberger 1982; Stadie 1983). ANTAGONISTIC BEHAVIOUR. Little information. ♂♂ defend territory against other ♂♂ (Ptushenko and Inozemtsev 1968). In captive-study, both pair-members expelled intruders from nest area (Timmis 1972). After pair-formation, ♂♂ defend territories 'more definitely', staying there for most of day and singing at several song-posts on top of bushes, etc. (Masatomi and Kobayashi 1982). In Hokkaido, chasing and physical fighting frequently seen at some territorial borders (Nakamura *et al.* 1968). In study in Moscow region, little aggression between neighbouring ♂♂, at least during human disturbance near nest, at which time neighbouring ♂♂ sometimes encroached with impunity on nest-area (Promptov 1934). During fighting (Hokkaido), both sexes (i.e. ♂-♂ fights, and ♀-♀) show presumably submissive posture similar to Soliciting-display of ♀ (Masatomi and Kobayashi 1982; see Heterosexual Behaviour, below). HETEROSEXUAL BEHAVIOUR. (1) General. Older ♂♂ typically arrive on breeding grounds ahead of ♀♀ and younger ♂♂ (Promptov 1934; Dementiev and Gladkov 1954; Masatomi and Kobayashi 1982). (2) Pair-bonding behaviour. Following account based entirely on detailed observations of *ornata* by Masatomi and Kobayashi (1982). ♂♂ sing on arrival (see Song-display, above) and establish territories. When ♀ arrives in or near territory, neighbouring ♂♂ often together chase and court her. During chases, ♂ commonly seen Tail-fanning (tail spread and lowered). When ♀ returns to territory and joins singing ♂, he commonly performs Wing-raising (Fig A) in which, at high intensity, one

B

A

wing fully extended and raised vertically, while other remains folded. Unlike *E. schoeniclus*, rarely raises both wings together. ♀ sometimes Wing-raises near mate during nest-building, incubation, and brooding; see also subsection 4 (below). Initial stages of pair-formation also include Material-carrying by ♂ (same as Stem-display in some other *Emberiza*): picks up and holds dead grass stem in bill and tries to fly far from ♀ who often follows him; ♂ later drops material while flying or perched; when perching (Fig B) or walking with material, body rather erect so that

tail touches ground, and head plumage somewhat ruffled (even when displaying in flight). Material-carrying typically final display in mating sequence (see subsection 4, below) and also recurs if nest destroyed (though ♂ never takes any material to nest). ♀ occasionally gives Chi-call (see 4 in Voice) when she approaches ♂. (3) Courtship-feeding. None. (4) Mating. Account based entirely on Masatomi and Kobayashi (1982). Takes place in territory, usually on bare patch on ground, occasionally on small branch of tree or shrub. Both ♂ and ♀ often Wing-raise and sometimes Wing-shiver when they approach mate for mounting (♂ on ♀ and, in Reverse-mounting described below, ♀ on ♂) and copulation. Precopulatory behaviour highly variable from simple to complex. In 'Simple type', ♂ approaches (sometimes in flight) ♀ perched on branch who then performs Soliciting-display (Fig C, showing ♀ soliciting thus on ground): lowers her forebody,

C

holding head and tail high, wings slightly lifted (sometimes to extent of Wing-raising) and vibrated; feathers of underside are markedly ruffled; ♂ then sidles up, mounts, and copulates (head down and wings beating) for a few seconds. Simple type also occurs on ground: ♀ solicits (as above), attracting ♂ to walk or run towards her, mount and copulate; ♀ may then Reverse-mount ♂ (Fig D), beating her wings and flicking closed tail

D

sideways several times; sometimes, ♀ and ♂ walk about on ground for a while, then she solicits, but even though several copulation attempts may occur in relatively short time, there are seldom any elaborate preliminaries of type described below. In other cases of copulation on ground, 'Complex type' of precopulatory behaviour occurs: ♂ solicits by approaching ♀ with Tail-fanning-walk, and ♀ invites with Soliciting-display; in ♂'s Tail-fanning-walk, body and head held horizontally, sometimes accompanied by prominent Wing-raising; alternatively, wings may just be lifted slightly or lowered; plumage on head (not rest of body) markedly ruffled. Often ♂ approaches ♀ in full circle or semi-circle; before mounting, ♂ approaching ♀ from behind in low posture, occasionally pulls her tail a couple of times, eliciting no obvious reaction. After dismounting (usually obliquely in front of ♀) ♂ often stays alongside ♀ and performs Prone-display ('head-up-lie-flat'): lies flat and motionless but with bill vertical and wings slightly lifted, with wing-posture varying from sideways extension (Fig E) to full Wing-raising (Fig F); sometimes

E

F

raised wings are slightly shivered, highlighting white patches. Prone-display, similar to soliciting by ♂ Yellowhammer *E. citrinella*, followed mounting in more than half the cases seen and therefore highly ritualized post-copulatory behaviour in *E. aureola*. When ♂ performs Prone-display, ♀ mounts and ♂ then extricates himself by walking forwards and often crouching. ♂ sometimes tries to copulate repeatedly, but after successive copulations Reverse-mounting less likely. ♂ finally (but not always) ends complex sequence with Material-carrying, sometimes followed by Twee-call (see 3 in Voice). In variants, ♂ may perform Prone-display before ♀ lands; occasionally, before copulation, ♀ performs Reverse-mounting and Tail-fanning-walk. (Masatomi and Kobayashi 1982, which see for other sequences and relative frequency of different types.) (5) Nest-site selection and behaviour at nest. In captive study, ♂ apparently took initiative in selecting site; after much prospecting on her own and with mate, ♀ finally settled on site chosen by ♂ (Timmis 1972). For role of sexes in incubation, see Bonds (above). No further information. RELATIONS WITHIN FAMILY GROUP. For 3 days after hatching, both parents share continuous brooding of young (see

Ferguson-Lees 1959 for photograph of ♂ brooding), thereafter (mostly ♀) at night only (Popov 1978). Eyes of young start to open at 3–4 days, fully open 6–7 days (Shkatulova 1962). Both sexes feed young and apparently both participate in nest-hygiene; faecal sacs may be swallowed (presumably early on) or taken away (Ferguson-Lees 1959). In captive study, faecal sacs eaten by both parents for first few days, later carried away (Timmis 1972). Mates often perform mutual low-intensity Wing-raising when meeting at nest to feed young (Masatomi and Kobayashi 1982). Young leave nest unable to fly, yet quite mobile by jumping and flapping; initially seek refuge in cover, whence they call to summon provisioning parents (Promptov 1934; Barancheev 1963). After independence (see Bonds, above) family parties congregate in flocks (Barancheev 1963). ANTI-PREDATOR RESPONSES OF YOUNG. Young may explode from nest prematurely from 7 days if disturbed (Ptushenko and Inozemtsev 1968). PARENTAL ANTI-PREDATOR STRATEGIES. (1) Passive measures. No information. (2) Active measures. Only known responses are to man. When nest disturbed, both parents fly around nest with Alarm-calls (see 5 in Voice), sometimes attracting neighbouring ♂♂ to join them (Promptov 1934). Incubating bird (♂ or ♀) flushed from nest makes short flight and walks about with spread tail (typical of distraction-display, notably impeded flight, in *Emberiza*: Andrew 1956); when young (in or out of nest) are threatened, parents often spread tail repeatedly, closing it slowly, and giving Alarm-calls (Masatomi and Kobayashi 1982). In Finland, ♀ flushed from nest in twilight performed 'injury-feigning' (Ferguson-Lees 1959) but no details. According to Portenko (1939), distraction-display in Anadyr' region (north-east Russia) is as in *E. pusilla*. In captive study, ♀ feigned injury by running along ground dragging both wings (Timmis 1972).

(Figs by D Nurney from photographs in Masatomi and Kobayashi 1982.)   EKD

Voice. Song confined to breeding season, but call 2 heard throughout year. There are probably calls other than those listed below yet to be described. For additional sonagrams see Bergmann and Helb (1982) and (song) Wallschläger (1983).

CALLS OF ADULTS. (1) Song. Short melodious phrase (1–2 s) most closely resembling Ortolan Bunting *E. hortulana* in timbre and syntax (segmental structure) but 1st segment of *E. hortulana* longer and higher-pitched (Bergmann and Helb 1982, which see for other differences). Tempo of *E. aureola* song slightly slower than that of *E. hortulana* (Staav and Fransson 1987; Bentz and Génsbøl 1988; L Svensson). Renderings include 'tyy-tyy-tsyy-tsyy-tsitsi-tuu' (Haftorn 1971), 'tse tuee-tuee tsiu-tsiu zeeu' and 'tsiu-tsiu-tsiu vüe-vüe tsia-tsia trip-trip' (Jonsson 1992). Figs I–II illustrate diversity of unit-types, Fig I a fine whistling song with long tonal units, Fig II more varied. Most extensive analysis is by Wallschläger (1983) from study in eastern Russia and Mongolia; relative simplicity of rather short phrase from Russian recording (Fig III) shows to advantage several basic features described by Wallschläger (1983) as follows: song-phrase shows quite considerable variation on basic structure of usually 6–8 units (5–11); units are often double (with marked pitch change from 1st to 2nd syllable), and arranged in 3 short segments (roughly 2 units each) then coda which is

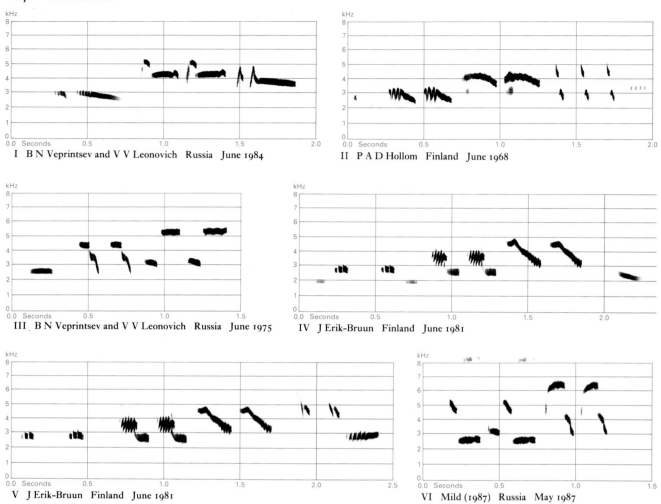

I   B N Veprintsev and V V Leonovich   Russia   June 1984

II   P A D Hollom   Finland   June 1968

III   B N Veprintsev and V V Leonovich   Russia   June 1975

IV   J Erik-Bruun   Finland   June 1981

V   J Erik-Bruun   Finland   June 1981

VI   Mild (1987)   Russia   May 1987

highly variable in length and pitch (or, as in Fig III, no coda). Overall, pitch rises stepwise through the 3 segments, though there is sometimes an alternation between high- and low-pitched segments. Where populations in Mongolia were at high density (described as 'colonial') there were unit-types common to all individuals, especially in 3rd segment and coda (e.g. all ♂♂ in Ulan-Bator had trill in coda), creating local dialects reminiscent of *E. hortulana*. In addition, locally shared units can also occur in other quite distant populations. (Wallschläger 1983). Individuals have repertoire of song-types, as found (e.g.) by J Bjørn Anderson in southern Siberia: 1 ♂ usually varied number of units from one song to next, often adding one unit each time so that successive phrases had (e.g.) 5, then 6, 7, and 8 units; after this maximum, song reverted to few units and same pattern was repeated in following songs; major changes in unit-structure and segment type often took place just after returning to short song-type. Song sometimes compressed, e.g. in middle section (Berg-

mann and Helb 1982). Figs IV–V, representing consecutive songs in bout from Finnish bird, illustrate common source of variation in that song remains constant except for variable coda; each of these phrases starts with 4 identical units but first 2 too quiet to depict on Figs IV–V; next phrase in this bout reverted to song-type shown in Fig IV. Figs VI–IX, depicting 4 consecutive phrases from Russian bird, show more elaborate manipulation; sequence starts with short phrase which is effectively middle section on its own (Fig VI), then an opening added (Fig VII), next (Fig VIII) switches to different opening and appends coda (single unit showing rapid frequency modulation), finally in 4th phrase (Fig IX) retains same coda but reverts to opening used in 2nd (Fig VII) phrase. (E K Dunn.) Occasional mimicry of other *Emberiza* songs may occur (Bergmann and Helb 1982) but not proven beyond doubt, and care needed in view of intrinsic similarity to *E. hortulana*; Finnish bird said to have imitated *E. hortulana* perfectly and Yellowhammer

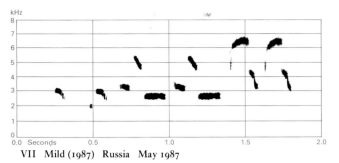

VII Mild (1987) Russia May 1987

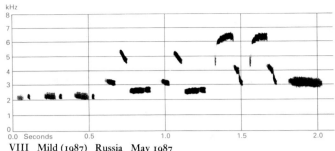

VIII Mild (1987) Russia May 1987

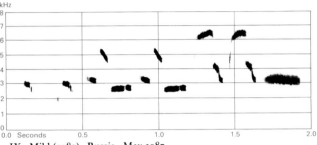

IX Mild (1987) Russia May 1987

E. citrinella almost perfectly, as well as singing *E. aureola* song (Hongell 1977). Dornberger (1982, which see for sonagrams) recorded song that began with call 2 (below) or sparrow-like 'tschilp'. (2) Contact-alarm call. Commonest call a short hard piercing 'tsik' or 'zick' or 'zip', very like Robin *Erithacus rubecula* (Bergmann and Helb 1982; Jonsson 1992), heard in any situation from mild excitement to alarm (then given in rapid succession), and also recorded as preliminary to song (see 1, above). Other renderings include thin 'tsip' (thinner and more nasal than *E. citrinella*) from September vagrant, Northern Ireland (Cooke 1960). Calls which are probably variants: soft 'tsii' (Bentz and Génsbøl 1988) or 'tsee' (Jonsson 1992); soft (not hard) 'trssit' or 'triist' (Haftorn 1971); low-pitched 'tk' (Bergmann and Helb 1982). (3) Twee-call of ♂. Drawn-out 'twee' sometimes heard from ♂ at end of sequence of courtship, e.g. at end of post-copulatory display (Masatomi and Kobayashi 1982: *ornata*). See also call 5, below. (4) Soliciting-call of ♀. Short soft 'chi' or 'chichi', occasionally given by ♀ when she approaches ♂; homologous to call 6 of *E. citrinella* (Masatomi and Kobayashi 1982: *ornata*).

CALLS OF YOUNG. Little known. Food-contact call of well-grown young apparently repeated 'zick' (Mauersberger *et al.* 1982) which (as in e.g. Little Bunting *E. pusilla*) resembles call 2 of adult. EKD

Breeding. SEASON. Late and short throughout range. Finland: eggs laid around 2nd half of June to early July (Ferguson-Lees 1959; Casén and Hildén 1965; Haartman 1969). St Petersburg region (north-west Russia): eggs laid

from mid-June (Mal'chevski and Pukinski 1983). South of Moscow: most clutches laid 1st half of June (Promptov 1934). Volga-Kama region (east European Russia): eggs laid 2nd half of June (Popov 1978). Northern Kazakhstan: 2nd half of June (late May to end of June) (Korelov *et al.* 1974). SITE. On ground, either on tussock or, where dry, in depression, nest-rim at times flush with ground, sheltered by scrub, commonly birch *Betula* or willow *Salix*, or in tree roots; where wet, often slightly above ground in bush or stout herb (Johansen 1944; Shkatulova 1962; Barancheev 1963; Korelov *et al.* 1974). In St Petersburg region, 14 of 22 nests were on ground in grass, mostly in small hollows, remainder 10-15 cm up in vegetation (Mal'chevski and Pukinski 1983). On river meadows south of Moscow often under tall docks *Rumex* (Promptov 1934). Nest: foundation of dry grass and stalks lined with soft grass, rootlets, and sometimes hair (Ferguson-Less 1959; Shkatulova 1962; Barancheev 1963; Korelov *et al.* 1974). In Volga-Kama region, 12 nests had average outer diameter 9·0 cm (8·5-10·5), inner diameter 6·4 cm (5·5-7·5), and depth of cup 4·3 cm (2·0-6·0) (Popov 1978). See also Lobkov (1986) and Larionov *et al.* (1991). Building: by ♀ only; takes 3-5 days (Ptushenko and Inozemtsev 1968; Korelov *et al.* 1974; Masatomi and Kobayashi 1982; Mal'chevski and Pukinski 1983); ♀ sometimes creates hollow in ground (Dementiev and Gladkov 1954; Korelov *et al.* 1974). In captivity, one ♂ did most of work for first 3 days then ♀ took over (Timmis 1972). EGGS. See Plate 34. Sub-elliptical, smooth and slightly glossy; greyish or greenish with olive or purplish-grey undermarkings, sparsely to heavily marked with brown to purplish-black blotches and scrawls (Harrison 1975; Makatsch 1976). Nominate *aureola*: 20·3 × 15·2 mm (18·0-22·0 × 14·0-16·0), *n* = 144; calculated weight 2·46 g (Schönwetter 1984). Clutch: 4-5 (3-7). In Moscow region, of 7 clutches, 1 of 3 eggs, 1 of 4, and 5 of 6; average 5·28 (Ptushenko and Inozemtsev 1968). In St Petersburg region, 84% of 19 clutches 4-5 (Mal'chevski and Pukinski 1983); average in Volga-Kama 5·25 (Popov 1978). In Yakutiya (eastern Russia), of 57 clutches: 2 eggs, 2%; 3, 23%; 4, 26%; 5, 49%; average 4·23 (Larionov *et al.* 1991). Average size increases with ♀ age (Kilin 1983). Very probably 2 broods at times, but more often 1 (Johansen 1944; Dementiev and Gladkov 1954; Shkatulova 1962). Eggs laid daily in early morning (Korelov *et al.* 1974; Popov 1978). INCUBATION. 13–14 days, by both sexes, starting properly with last or penultimate egg, hatching over 1–3 days (Barancheev 1963; Ptushenko and Inozemtsev 1968; Korelov *et al.* 1974). ♂ takes only minor part, having no brood-patch (Makatsch 1976; Mal'chevski and Pukinski 1983); in Volga-Kama, ♀ incubated for first 6–8 days, then both (Popov 1978). In Chita region (east of Lake Baykal, Russia), total of 46–52 min per hr spent on eggs (Shkatulova 1962). For routine in captivity, see Stadie (1983). YOUNG. Fed, cared for, and brooded by both parents (Promptov 1934; Ferguson-Lees 1959; Shkatulova 1962; Popov 1978). FLEDGING TO MATURITY. Fledging period (9–)11–14 days; leaves nest before able to fly but periods of 8–9 days perhaps caused by disturbance (e.g. Shkatulova 1962, Mal'chevski and Pukinski 1983); other authors give more than 10 days (Ptushenko and Inozemtsev 1968), 11–12 (Barancheev 1963, which see for development of young), 13–14 (Dementiev and Gladkov 1954; Korelov *et al.* 1974), and 13–15 days (Harrison 1975; Makatsch 1976); longest periods are probably time to first flight. Independent 7–14 days after fledging (Ptushenko and Inozemtsev 1968; Makatsch 1976; Mal'chevski and Pukinski 1983). BREEDING SUCCESS. In Yakutiya, 21 nests contained 79 eggs, of which 71% hatched, 24% lost to predation, and 5% infertile (Larionov *et al.* 1991). In Amur region (eastern Russia), 50% of eggs in 40 nests produced fledged young; in one area, 4 of 9 nests destroyed at egg stage, 2 at nestling stage; number of infertile eggs very low (Barancheev 1963). In Volga-Kama, only 15% of nests successful, 85% deserted due to predation by dogs and fox *Vulpes* (Popov 1978). On Ob' river (central Russia), nest losses twice as high in years of high flood levels; earlier clutches always more successful (Ananin and Ananina 1983); in river meadows along Oka river south of Moscow, many nests lost during hay-making (Promptov 1934). Success greater when brood fed by older ♂ (Kilin 1983).   BH

Plumages. (nominate *aureola*). ADULT MALE BREEDING (March–August). Nasal bristles small, black. Forehead, lore, narrow line above eye, eye-ring, ear-coverts, and cheek black; black of forehead sometimes extending to forecrown above eye. Crown to upper tail-coverts rich maroon-chestnut; feathers of mantle, rump, and longest upper tail-coverts with narrow black centres near tip, but black sometimes almost absent; in latter case, black restricted to a few black subterminal shaft-streaks on centre of mantle. Occasionally, longest upper tail-coverts with grey centre. Chin and upper throat black, bordered by yellow bar extending from side of neck across lower throat. Each side of breast maroon-chestnut, connected by narrow clear-cut maroon-chestnut band across upper chest; in some birds, some black feathers between maroon-chestnut ones, or feathers yellow at base, broadly tipped maroon-chestnut, with narrow black subterminal band. Remainder of underparts pale or bright yellow, often with some white feather-fringes; flank with maroon-chestnut or black drop-like spots, sometimes forming black patch on upper flank; under tail-coverts white, yellow-white, or buff-white. Tail-feathers grey-brown or greyish-black; central pair

(t1) with chestnut edges, t2–t4 with narrow brown edges, t4 sometimes with white tip; t5 with narrow white wedge on tip of inner web, sometimes absent; outer pair (t6) with white wedge on tip of inner web and with white outer web with blackish-brown tip. Flight-feathers and greater upper primary-coverts greyish-black, outer webs with light brown edges; secondaries usually with broader maroon-chestnut fringe on base of outer web; outermost (reduced) primary (p10) and outermost upper primary-covert brown with white outer web. Tertials brownish-black or black, outer web and tip of inner web with broad rufous-chestnut fringe and yellowish-brown or earth-brown edge. Greater upper wing-coverts black with broad maroon-chestnut or rufous-chestnut fringe along outer web, broadly tipped brown, buffish, or white; innermost 2 coverts tipped maroon-chestnut or rich brown on inner web, maroon-chestnut, brown, or white on outer web. Median and longer lesser coverts white with narrow yellow or brownish tips; remainder of lesser coverts grey-black or black, sometimes tipped white. Under wing-coverts white, sometimes with yellowish tinge; greater under primary-coverts grey, broadly tipped white. ADULT FEMALE BREEDING (March–August). Feathers of side of forehead and side of crown brownish-black, edged grey-brown, sandy-brown, or chestnut, feathers of central forehead and central crown grey-brown or buff with black shaft-streaks, forming pale crown-stripe. Supercilium yellowish or buffish-white, narrow in front of eye, broad and off-white behind eye; lore and ear-coverts silvery-grey, grey-brown, yellow-brown, or buffish. Stripe behind eye sandy-grey, ending in black spot on upper rear of ear-coverts and extending in narrow black line along rear and lower border of ear-coverts, merging with malar stripe. Eye-ring buff or sandy-brown. Narrow and inconspicuous buffish or greyish-buff submoustachial stripe, not reaching base of lower mandible. Black centres of feathers of lower nape occasionally with some chestnut at border. Upper mantle and side of neck greyish, sometimes with small black feather-tips. Centres of feathers of lower mantle black, bordered chestnut and with broad grey, grey-brown, or yellowish-buff fringes; scapulars like mantle, but black centres narrower and merging into chestnut and grey on edge. Rump brown, slightly or strongly tinged chestnut, with ill-defined grey-black feather-centres and buff fringes. Longest upper tail-coverts grey-black with grey-buff fringes. Centre of chin and throat pale yellow or white, side buff-yellow. Side of chest buff-brown or grey-brown with black shaft-streaks and broad grey fringes. Central chest pale yellow with narrow ill-defined brown shaft-streaks, forming inconspicuous gorget. Lower chest and belly pale yellow. Flank yellow-white, marked with long black or brown streaks, latter faintly bordered brown. Under tail-coverts off-white or buff-white. Tail- and flight-feathers as adult ♂ breeding. Tertials brown-black with buff-brown fringes, rarely with some maroon-chestnut in between. Greater upper wing-coverts brown-black or black, edged whitish-grey, outer 3–4 with white or off-white on tip of outer web; innermost greater coverts with some chestnut on inner web. Median upper wing-coverts black or brownish-black, tipped white. Lesser upper wing-coverts grey or grey-black with grey-white edge. Underwing and axillaries as adult ♂ breeding. ADULT MALE NON-BREEDING (September–March). Like adult ♂ breeding, but maroon-chestnut of crown extending onto forehead, reaching base of bill. Feathers of side of crown tipped black, feathers of central crown broadly edged greyish-buff or yellowish-buff, sometimes with black spot on tips, giving impression of pale crown-stripe. Supercilium buff or straw-yellow behind eye. Eye-ring buff. Ear-coverts buff, streaked brown, largely concealing blackish bases of feathers. Feathers of mantle, back, rump, and scapulars with greyish-buff or yellowish-buff fringes. Chin and throat whitish-yellow. Band across chest narrow, very dark maroon, black, or predominantly black on side of chest and with maroon feathers on central chest. Remainder of plumage as adult ♂ breeding. ADULT FEMALE NON-BREEDING. As adult ♀ breeding. NESTLING. Down brown, on upperside only (Stresemann and Stresemann 1969a). JUVENILE. Head and upperparts like adult ♀ breeding, but supercilium less whitish; rump with black shaft-streaks, edged pale brown, occasionally with yellowish fringes, black on central and lower rump bordered with chestnut; longest upper tail-coverts grey with black shaft-streaks. Chin and throat pale yellow; feathers of side of throat with small brown subterminal spots. Lower throat and lower chest with slight orange hue. Feathers of centre of chest greenish-yellow with grey-brown or brownish-black shaft-streaks; side of breast and flank tinged greyish, brownish, or greenish. Belly pale yellow or yellowish-white; vent and under tail-coverts white, usually with pale yellow tinge. Flight-feathers and greater upper primary coverts grey-black with pale brown or greenish-yellow edges; inner secondaries edged brown. Tertials black, bordered with some chestnut on outer web and on tip of inner web, broadly edged buff, brown, or pale chestnut. Greater upper wing-coverts black with orange-brown or yellow fringes, broadly tipped buff; median coverts similar, but tips yellowish-white; lesser coverts grey-black with greyish-buff edges. FIRST ADULT MALE BREEDING. Similar to adult ♂ breeding, but forehead with narrow black band; crown and nape maroon-chestnut, feathers sometimes with black central streaks or black tips; when fresh, feathers with broad buff edges. Lore and ear-coverts black with some buffish tips. Eye-ring buff above eye, black below. Supercilium black with buff spots above eye, broader and buff to off-white behind eye, but in some birds uniform black. Neck maroon-chestnut or mixed greyish and maroon-chestnut. Mantle as adult ♀ breeding, but black feather-centres bordered by some maroon-chestnut edges or upper mantle fully maroon-chestnut. Scapulars maroon-chestnut with black shaft-streaks and buff edges. Longest upper tail-coverts greyish with blackish-brown centres. Chin and throat black with some yellow feather-tips, black usually somewhat less extensive than in adult ♂. Underparts similar to adult ♂ breeding, but paler yellow and with some white feathers on breast and belly; band across chest perhaps narrower, mixed with yellow feathers. Greater upper wing-coverts black, bordered narrowly maroon-chestnut or chestnut-brown on outer web, edged greyish-white; median coverts black with white tips; lesser coverts greyish-black, tipped grey-buff or white. FIRST ADULT FEMALE BREEDING. As adult ♀ breeding.

Bare parts. ADULT, FIRST ADULT. Iris brown. Upper mandible plumbeous or dark-grey, cutting edges and lower mandible pink-flesh, pale greyish-flesh, or brownish-flesh. Leg and foot pale flesh-horn, flesh-brown, or light brown. (Niethammer 1937; Ali and Ripley 1974; ZMA.) NESTLING. No information. JUVENILE. In autumn, upper mandible plumbeous-black to steel-black, cutting edges pale greyish-flesh below nostril, pale horn-yellow at gape; lower mandible pale pink-flesh, tinged grey on side, cutting edges at gape pale horn-yellow. Leg and foot dark grey with flesh tinge, rear of tarsus and sole pale purplish-flesh. (ZMA.)

Moults. Based mainly on Stresemann and Stresemann (1969a). ADULT POST BREEDING. Complete; primaries descendent. Nominate *aureola* moults completely between late August and early October at migration stop-over sites in China; *ornata* has complete moult late July to mid-September on breeding grounds, before start of autumn migration. ADULT PRE-BREEDING. Partial; March–May. Involves forehead, crown, lore, area round eye,

ear-coverts, chin, throat, and possibly some or most feathers of band across chest. Moult in ♂ probably more extensive than in ♀. (Stresemann and Stresemann 1969a; RMNH, ZFMK, ZMA.) POST-JUVENILE. Partial: body, tertials, and wing-coverts. In both races, timing largely similar to adult post-breeding. In Netherlands, August–September, 2 birds still largely in juvenile plumage with restricted amount of 1st-adult feathering on throat, side of crown, and mantle (ZMA).

Measurements. ADULT, FIRST ADULT. Races combined. Whole geographical range, all year; skins (RMNH, ZMA). Bill (S) to skull, bill (N) to distal corner of nostril; exposed culmen on average 4·1 less than bill (S).

	♂		♀	
WING	77·8 (2·12; 19)	73–81	73·5 (1·17; 8)	72–76
TAIL	55·4 (1·95; 18)	53–59	53·0 (1·77; 8)	51–56
BILL (S)	14·4 (0·47; 19)	13·6–15·2	14·0 (0·42; 8)	13·5–14·5
BILL (N)	8·0 (0·36; 19)	7·6–8·7	7·8 (0·39; 7)	7·6–8·5
TARSUS	20·9 (0·59; 12)	20·0–21·5	20·3 (0·53; 6)	19·5–20·8

Sex differences significant for wing and tail.

Wing. Nominate *aureola*. South-west Siberia, August: juvenile 73·2 (2·12; 39) 67–76 (Havlín and Jurlov 1977). Northern Russia: ♂ 76·5 (148) 73–80, ♀ 72·6 (55) 68–76 (Dementiev and Gladkov 1954). Mongolia, May–August: ♂ 78·6 (1·89; 24) 75–82, ♀ 74·2 (2·25; 8) 71–78 (Piechocki and Bolod 1972; Piechocki et al. 1982).

E. a. ornata. Amurland and Ussuriland (south-east Siberia): ♂ 75·7 (36) 72–79, ♀ 71·3 (12) 69–74 (Dementiev and Gladkov 1954). Manchuria (northern China): ♂ 73–79, ♀ 69·5–75·5 (Meise 1934a; see also Piechocki 1959). Kuril Islands: ♂ 79·7 (3) 78–81 (Nechaev 1969).

Weights. Nominate *aureola*. South-west Siberia, August: juvenile 19·8 (1·54; 41) 17·2–24·3 (Havlín and Jurlov 1977). Kazakhstan: ♂ 21·2 (8) 19·5–22·9, ♀ 22·0 (Korelov et al. 1974). Northern Russia: ♂ 22·3 (15) 20·5–28·8, ♀ 20·0 (5) 17·5–28 (Dementiev and Gladkov 1954). Mongolia, May–August: ♂ 22·1 (1·78; 24) 19–26, ♀ 20·8 (2·44; 8) 17–24 (Piechocki and Bolod 1972; Piechocki et al. 1982).

E. a. ornata. Manchuria, May: ♂ 20·7 (3) 19–23, ♀ 18 (Piechocki 1959). Kuril Islands: ♂ 23·6 (3) 23·1–24·0, ♀ 24·0 (Nechaev 1969). South-east Siberia: ♂ 21·6 (8) 20·5–22·5, ♀ 21·3 (5) (Dementiev and Gladkov 1954).

Structure. Wing rather short, broad at base, tip bluntly pointed. 10 primaries: p8 longest, p9 0–1 shorter, p7 0–2, p6 1–3, p5 7–9, p4 11–14, p3 13–17, p2 15–19, p1 18–21; p10 strongly reduced, 50–52 shorter than p8, 7–10 shorter than longest upper primary covert. Outer web of p6–p8 clearly emarginated. Tip of longest tertial reaches tip of p5 (p2–p6). Tail rather short, tip forked; 12 feathers, t4–t5 longest, t1 4–7 shorter. Bill conical, depth at base in nominate *aureola* 6·1 (9) 5·8–6·8, width 6·8 (11) 6·0–7·7, depth in *ornata* 6·3 (7) 6·0–6·9, width 7·1 (7) 6·0–7·8; culmen virtually straight. Middle toe with claw 18·7 (9) 16·7–20·9; outer toe with claw c. 72% of middle with claw, inner c. 70%, hind c. 81%; hind claw 6·9 (15) 6·4–7·8. Remainder of structure as in Yellowhammer *E. citrinella*.

Geographical variation. Slight, clinal; involving depth of colour only, not size (Johansen 1944). ♂ *ornata* from Chita region (east of Lake Baykal) east to Sakhalin and northern Japan darker chestnut on upperparts than ♂ nominate *aureola* from northern Eurasia, deeper yellow on underparts, streaks on mantle and flank heavier, blacker, bar on chest more often black; ♀ darker, more heavily streaked, deeper yellow below; juvenile browner.

Birds darkest on Sakhalin and Japan (named *insulanus* Portenko, 1960), paler towards north and west where intergrading with nominate *aureola*, but differences between *ornata* and nominate *aureola* too slight and birds show too much individual variation to warrant recognition of more than 2 races (including intermediate

'*kamtschatica*' Stanchinsky, 1929, on Kamchatka and along Sea of Okhotsk, and '*suschkini*' Stanchinsky, 1929, in Mongolia and western Transbaykalia), following Vaurie (1956*e*), contra Timoféeff-Ressovsky (1940).

Forms species-group with Chestnut Bunting *E. rutila*. GOK

Emberiza schoeniclus **Reed Bunting**

PLATES 26 (flight) and 32
[between pages 256 and 257]

DU. Rietgors   FR. Bruant des roseaux   GE. Rohrammer
RU. Камышовая овсянка   SP. Escribano palustre   SW. Sävsparv

Fringilla Schoeniclus Linnaeus, 1758

Polytypic. *SCHOENICLUS* GROUP. Nominate *schoeniclus* (Linnaeus, 1758), western and northern Europe from Britain to Urals, south to western and central France, northern Switzerland, north-west Austria, western Czechoslovakia, western and northern Poland, northern Belorussiya, and Smolensk, Ryazan', Kazan, and Perm' in European Russia, grading into *stresemanni* in Czechoslovakia south of Brno and Tatra mountains and into *ukrainae* in south-east Poland, southern Belorussiya, and at 53–55°N in European Russia (Tula to Ul'yanovsk regions); *passerina* Pallas, 1771, north-west Siberia from Urals east to lower Khatanga river, south to 62°N on Ob' and Yenisey rivers, grading into nominate *schoeniclus* in basin of Pechora river and perhaps further west, into *pallidior* in western Siberia at 59–62°N, and into *parvirostra* east from basin of middle Yenisey (Yeloguy to Turukhansk) and probably east of Khatanga; *stresemanni* Steinbacher, 1930, Carpathian basin from eastern Austria and southern Slovakia through Hungary east to north-west Rumania and south to Sava and lower Morava basin in northern Yugoslavia, probably grading into *tschusii* in south-east Rumania and north-west Bulgaria; *ukrainae* (Zarudny, 1917), northern Moldavia, northern Ukraine, and south European Russia, north to Kiev, Orel, and Tambov regions, south to Poltava, Khar'kov, and Voronezh, east to Saratov and Kuybyshev (Samara); *pallidior* Hartert, 1904, Ufa region of northern Bashkiriya (east European Russia) and south-west Siberia north to Sverdlovsk, Tobol'sk, Tara, and Tomsk, east to foot of Altai (Minusinsk, Biysk, Semipalatinsk), south to plains on northern border of Kazakhstan. Extralimital: *parvirostra* Buturlin, 1910, east from basin of Khatanga and middle Yenisey to Lena and Olekma valleys, south to Lake Baykal and Stanovoy mountains; *pyrrhulina* (Swinhoe, 1876), south-east Siberia from Lake Baykal to Sakhalin and Kamchatka, south to Manchuria (north-east China) and Japan. *PYRRHULOIDES* GROUP. *E. s. tschusii* Reiser and Almásy, 1898, north-east Bulgaria, eastern Rumania, southern Ukraine, and Crimea east to lower Volga between Saratov and Sarpa; *incognita* (Zarudny, 1917), east from Volga (Orenburg, Orsk) through uplands of northern Kazakhstan to regions of Karaganda and Karkaralinsk; *witherbyi* Von Jordans, 1923, Iberia, Balearic Islands, Mediterranean coast of France, and perhaps North Africa; *intermedia* Degland, 1849, Corsica, Italy, and Dalmatian coast of Yugoslavia; *reiseri* Hartert, 1904, eastern Albania, Makhedonija (southern Yugoslavia), Greece, and Turkey; *caspia* Ménétries, 1832, eastern Transcaucasia, north-west, northern, and south-west Iran, and (perhaps this race) Syria; *korejewi* (Zarudny, 1907) Seistan and Baluchestan (south-east Iran); *pyrrhuloides* Pallas, 1811, northern shore of Caspian Sea (mouth of Terek, Volga, and Ural rivers) east through southern Kazakhstan and Turkmeniya to foot of Tien Shan and Lake Alakol', north to basins of Irgiz and Turgay rivers and to Lake Balkhash. Extralimital: *harterti* Sushkin, 1906, Zaysan basin, Dzhungaria (northern Sinkiang, China), and western Mongolia; *centralasiae* Hartert, 1904, Tarim basin (southern Sinkiang, China); *zaidamensis* Portenko, 1929, Tsaidam basin, China.

Field characters. 15–16·5 cm; wing-span 21–28 cm. Size varies considerably between 2 main racial groups, being smallest in group containing nominate *schoeniclus* of north-west Europe and allied northern races which average 10% shorter and noticeably slighter than Yellowhammer *E. citrinella*; intermediate in south European races, and largest in east European and Asian group containing *pyrrhuloides* and allies in which birds overlap with *E. citrinella* and have as long a tail; bulbous shape, and size, of bill also increase from north to south and east. Medium-sized to rather large bunting, seemingly large-

headed and thick-necked (in ♂), with fairly lengthy form and distinctive voice. Breeding ♂ instantly recognized by black head and bib and white collar, shared only by Pallas's Reed Bunting *E. pallasi*; non-breeding ♂, ♀, and juvenile essentially brown above and buffish below, quite heavily streaked but lacking striking diagnostic character (since neither convex culmen nor diagnostic reddish tone of lesser wing-coverts easily seen). Paler eastern birds, rare small individuals, and (in ♀ and juvenile) occasional aberrant head plumage present serious pitfalls. Has habit of nervously spreading tail. Sexes dissimilar; marked seasonal variation in ♂. Juvenile separable. 13 races in west Palearctic in 2 groups, with extreme western and eastern races distinguishable in the field. Only northern race, nominate *schoeniclus*, described fully here (see also Plumages and Geographical Variation).

(1) Western and north European race, nominate *schoeniclus*. ADULT MALE BREEDING. Moults: July–November (complete); March–May (face). Full contrasts of summer plumage produced by wear. Head and deep bib jet-black, separated by broadening white submoustachial stripe which joins broad white collar as it extends forward towards bib. Collar often seemingly expanded into deep band on hindneck, forming obvious blaze which contrasts with head and dark back (and 'shines' over several hundred metres). Mantle and scapulars greyish-brown, noticeably streaked black and less obviously pale buff (particularly on central lines of mantle) and rufous (particularly on scapulars). Wings form most warmly coloured area of plumage, with distinctly rufous-chestnut lesser coverts and similarly coloured margins and tips to black-centred median and greater coverts and tertials. Important to note that tips of coverts do not form obvious wing-bars and (in nominate *schoeniclus*) overlay of pale rufous to buff fringes to flight-feathers do not create pale wing-panel except in worn plumage. Dull grey rump distinctive, being little streaked and becoming brown only just above tail, but difficult to see except from behind. Tail brown-black, with buff margins and brilliant white panels on outermost pair and ends of next pair which 'flash' when spread, particularly on take-off. Underparts below whitish breast sullied with grey, particularly on flanks, and rather indistinctly streaked blackish-brown from lower sides of breast to rear flanks. White belly, vent, and under tail-coverts rarely show. Under wing-coverts white. Bill proportionately shorter than *E. citrinella* and somewhat bulbous, with convex culmen (except in some northern birds) visible at close range; looks less pointed than those of smaller buntings; dusky-black, with horn tinge to lower mandible. Legs flesh-grey to dark brown. ADULT MALE NON-BREEDING. Much less contrasted than breeding plumage due to pale-tipped fresh feathers. Black head obscured by buff and rufous-brown tips; shows buffish supercilium and much less clean white submoustachial stripe. Narrow buff eye-ring does not catch eye. Nape-band lost under brown feather-tips but white collar usually evident on sides of neck. Black bib indistinct but still visible as mottled patch on throat and upper breast. Head thus has characteristically swarthy, speckled appearance, always more darkly patterned than ♀. Rest of plumage more like that of summer, but back rather darker, with fresh rump feathers looking less grey due to buff-brown tips, and underparts often buffish-tinged and less distinctly streaked on flanks. Bill paler, with brown-horn base. ADULT FEMALE BREEDING. Differs from ♂ in head and underpart pattern and colours. Head never wholly black, with (a) more chestnut sides to crown, at least

trace of pale cream to buff rear supercilium and sometimes similarly coloured supraloral streak, (b) more mottled chestnut to brown ear-coverts almost wholly bordered dark brown or black, (c) distinctive dark moustachial stripe reaching bill, and (d) pale buff, rarely white surround to ear-coverts not forming full collar but ending in distinctly pale submoustachial stripe. Submoustachial stripe contrasts strongly with bold black malar stripe, which also joins dark rufous to blackish streaking of entire breast and flanks, often forming dark blob on side of neck. Variations include buffer or greyer ground-colour to head and dusky lores. Rump much browner than ♂. Bill usually as non-breeding ♂. ADULT FEMALE NON-BREEDING. Relative contrasts of breeding plumage obscured by fresh brown tips to head feathers; beware swarthier birds suggesting ♂♂. Any hint of collar lost and obvious markings on head often restricted to dull buff supercilium, buff-white submoustachial stripe, dark edges of (usually) mottled brown cheeks, and black malar stripe and blob on side of neck. Underparts much duller than ♂ and most breeding ♀♀. JUVENILE. Resembles ♀ but has more yellowish or pale buff appearance to body looking boldly streaked above and on breast. Head pattern obscure, with short supercilium and only dark malar stripe as obvious as ♀. Beware dull buff-tipped, black lesser wing-coverts which may suggest other *Emberiza* until lost in 1st moult. FIRST-WINTER MALE. Appearance intermediate between adult ♂ and ♀ but most already show partly white collar and patchy bib (not extending as low on upper breast as in adult ♂). Complete buff supercilium is clear sign of immaturity. FIRST-SUMMER MALE. Retains dull, even spotted collar and shorter bib. FIRST-YEAR FEMALE. Crown browner and breast and flanks more streaked than adult. Puzzling variations (perhaps also of 1st-winter ♂) include pale yellowish-buff central crown-stripe, clear rufous crown-sides, and cream supercilium. Importantly, dark malar stripe and its contrast with pale chin and particularly with pale submoustachial stripe remain constant features. (2) Other races. In *schoeniclus* group, *passerina* of north-west Siberia potentially a troublesome vagrant: although similar in size to nominate *schoeniclus*, its paler mantle colours, finer streaking, paler grey rump, and (often) bill shape all converge with characters of *E. pallasi*; since some ♀ *passerina* are over 10% smaller than ♂ *passerina* and nominate *schoeniclus*, they clearly constitute hitherto unnoted pitfall. In intermediate birds and *pyrrhuloides* group, increases in size and increasingly bulbous bill become obvious in south-east Europe and thence eastward. Plumage tones vary, however, with dark birds in Carpathian basin, southern Balkans, Ukraine, and Turkey but strikingly pale ones north of Black Sea (where *pallidior* may show almost white rump in worn plumage) and eastwards across steppes of Asia. For further details, see Measurements and Geographical Variation.

Breeding adult ♂ almost unmistakable, since only ♂ *E. pallasi* shows similar plumage but is instantly distinguishable by different call, typically smaller size, always whitish rump, and (when visible) grey lesser wing-coverts. In all other plumages (and remembering its quite marked geographical variation), adult ♂, ♀, and immature *E. schoeniclus* are troublesome, particularly when silent, and, in these plumages, potential and known confusion species include *E. pallasi* (smaller, with usually pale rump, less marked head, usually more striped back, paler fringes to tertials and secondaries, finer streaks on underparts, and greyish or grey lesser coverts), Rustic Bunting *E. rustica* (somewhat smaller and more compact, with bright rufous, not grey-brown rump), Little Bunting *E. pusilla* (usually noticeably smaller, with cream eye-ring, uniformly rufous, not mottled cheek), Black-faced Bunting *E. spodocephala* (somewhat smaller and noticeably more compact, with similar contrast between pale submoustachial and dark malar stripes but greyish face and half-collar), Yellow-browed Bunting *E. chrysophrys* (intermediate in size between *E. schoeniclus* and *E. pusilla*, with yellow and white supercilium and white rear cheek-spot), Song Sparrow *Melospiza melodia* (similar in size but lacks white outer tail-feathers), and 1st-winter White-throated Sparrow *Zonotrichia albicollis* (up to 10% larger and lacks white outer tail-feathers). Happily, all the above have different calls which quickly dispel confusion. Inexperienced observer might also confuse *E. schoeniclus* with Lapland Bunting *Calcarius lapponicus* (much bulkier with long wings, plainer face, terrestrial habits, and different call). Occurrence of eastern races of *E. schoeniclus* in western Europe problematic; 3 early 20th century English specimens considered fraudulent, but sight records of 2 bulbous-billed birds, one pale-striped above, as large as *E. citrinella*, and resembling *pallidior*, and another matching *passerina*, in late autumn on east coast of England (D I M Wallace), are suggestive of true vagrancy. Flight like *E. citrinella*, being most similar in fluttering, direct, fast take-off and trailing of sometimes bulbous-ended tail, but action actually less strong, more often recalling small pipit *Anthus*, with erratic bursts of sometimes uneven wing-beats producing jerky, seemingly hesitant progress. At distance, flight silhouette also suggests pipit. Escape-flight usually short, direct, and level to denser cover (or in fields to hedge or bush); migrant flushed from open ground may tower. Gait a hop; also creeps when feeding. Stance as *E. citrinella* but carries tail up more often when on ground. Habitually flicks and spreads tail, exposing white on outer feathers. Territorial ♂ often as conspicuous as ♂ *E. citrinella*, perching upright on top of reed or other tall plant, but ♀ (and both sexes in winter) unobtrusive, often remaining hidden in ground cover. Sociable, particularly at roost, but on migration and in winter normally seen in scattered parties rather than flocks.

Song rather short and disjointed, even staccato in rather slow phrasing, with timbre variably described as tinkling, metallic, and squeaky; certainly more monotonous than most other buntings, irritatingly repeated: 'tweek-tweek-tweek-tititick', 'zink-zink-zi-zi-zi-zinck', or 'tsee tsee tseea tsisirrr'. Commonest calls: soft, whistling 'seeoo', falling smoothly in pitch; short, hoarse, even-pitched 'bzy', 'chü', or 'bzree', often in flight or from autumn migrant. Absolutely no record of short ticking call.

Habitat. Most widespread in range of west Palearctic breeding Emberizidae, inhabiting oceanic islands and peninsulas, and continental plains from arctic through boreal, temperate, and Mediterranean to steppe and even desert climatic zones, between July isotherms as low as 10–11°C in north to above 32°C in south (Voous 1960b). Yet within this vast range, choice of occupied sites is ecologically restricted to particular types of dense and prolific fairly low vegetation, mainly associated with intense soil moisture. Avoids both closed forest and typical open country, as well as bare, rocky, or frozen surfaces, steep or broken ground, and areas of immediate human disturbance or settlement. Apparent attachment to marshes, fens, bogs, riversides, and inland waters occurs indirectly, rather through dependence on their associated vegetation types than being linke with any special need for water.

Occupies tall herbage and small shrubs found in marshy and swampy areas bordering fresh or brackish water of all kinds, normally in valleys and lowlands. Even in Switzerland, generally lives below 700 m, with only local exceptions (Glutz von Blotzheim 1962). Other common habitats include peat bogs, wet meadows with tall herbage, reedbeds, shrub tundra, swampy areas in grass steppes, and wet grassy clearings in forests (Harrison 1982). These tend to occur as transitions between reed or willow marshes and herbaceous or woody fens; normally absent from extensive flooded reedbeds. Abundance of suitable plant or animal food and unsuitability of terrain for competitors also influence habitat choice. In parts of range this has recently expanded to cover drier situations such as young conifer plantations, hawthorn scrub on waterless chalk downlands, and even middle of large fields of barley or other tall crops. Breeding densities on farmland, however, are less than 10% of those typical for wetlands, and no more than 2% of those on modern sewage farm. Thus able to maintain in Britain a national population higher than that of many more generally known birds. (Bell 1969; Sharrock 1976.)

The more specialized habitats in Britain include middle and upper zones of salt-marshes on low coasts, open areas and moist slacks of coastal sand-dunes, gravel pits, and upland rivers. Woodlands may be occupied where there is water, or where clearings or young plantations create suitable conditions (Fuller 1982). For use of farmland, moorland, or ditches distant from breeding area in foraging, see Howard (1929).

Habitat throughout west Palearctic broadly conforms to above description, although in certain regions (e.g. southern Norway) *E. schoeniclus* becomes an upland bird, and in southern USSR thick-billed birds are said to

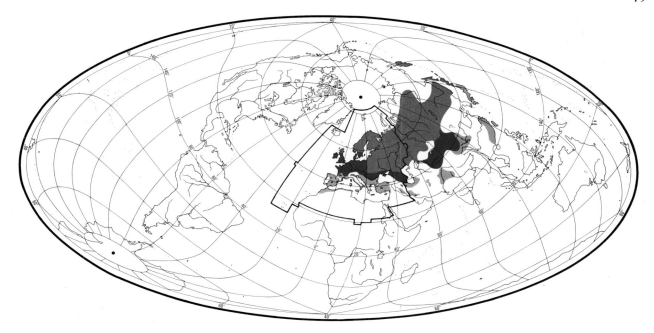

inhabit huge reed thickets, preferring flooded reed growth (Dementiev and Gladkov 1954). Throughout range, whether resident or migratory, there is a shift after breeding season to drier and more open situations, such as fields, lake shores, and marram grass *Ammophila* on sand-dunes, more similar to habitats of other Emberizidae at this season.

Distribution. No changes reported, except for probable recent spread into Kola peninsula.

FAEROES. One breeding record, 1972; probably regular summer visitor (Bloch and Sørensen 1984). RUSSIA. First recorded breeding north of forest zone in Kola peninsula in late 1950s, spreading north along Voron'ya and Iokan'ga valleys to north coast (Mikhaylov and Fil'chagov 1984).

Accidental. Iceland, Egypt, Kuwait.

Population. Apparently no general trends, but local and regional changes, attributed to various causes. Reported increases outnumber decreases.

BRITAIN, IRELAND. Probably in excess of 600 000 pairs, 1968–72 (Sharrock 1976). Britain: steep decline 1975–83, following high levels; now stable at lower level; changes probably related to severity of winters (Marchant *et al.* 1990); *c.* 400 000 pairs (Hudson and Marchant 1984). Ireland: increasing, and apparently expanding into drier habitats (Hutchinson 1989). FRANCE. 10 000 to 100 000 pairs; increasing in north, with adaptation to breeding in cultivated land (Yeatman 1976). BELGIUM. Estimated 8000–11 000 pairs; probably decreasing, due to habitat changes (Devillers *et al.* 1988). NETHERLANDS. Estimated 40 000–

70 000 pairs (SOVON 1987). GERMANY. 400 000 pairs (G Rheinwald). East Germany: 150 000 ± 60 000 pairs (Nicolai 1993). SWEDEN. Estimated 700 000 pairs (Ulfstrand and Högstedt 1976); 800 000 pairs (Arvidsson *et al.* 1992). FINLAND. Estimated 500 000 to 1 million pairs; marked increase this century, due to eutrophication of waters and discontinuation of cattle grazing promoting reedy and shrubby growth along lake shores (Koskimies 1989). BULGARIA. 100–1000 pairs; no change (TM).

Survival. Britain: average annual mortality of ♂♂ 47 ± SE5%, ♀♀ 51 ± SE7% (Dobson 1987); sexes combined 43 ± SE2·3% (Prŷs-Jones 1977). Switzerland: sexes combined 47 ± SE4·4% (Prŷs-Jones 1977). Oldest ringed bird 10 years 8 months (Rydzewski 1978).

Movements. Northern group *schoeniclus* sedentary to migratory; southern *pyrrhuloides* group of races chiefly sedentary. Winters in areas with little or no snow cover (except *pyrrhuloides* group) making mid-winter flights if snowfall persists. In Europe, nominate *schoeniclus* migratory in north-east, increasingly sedentary towards southwest; migrants head between SSW and west. Account based on Prŷs-Jones (1984), which see for discussion of environmental factors; for review, see also Zink (1985).

Chiefly sedentary in Britain. Winter distribution widespread and similar to summer, but birds withdraw from upland areas. Ringing data show that *c.* 40% of ♀♀ and *c.* 80% of ♂♂ move no more than 5 km between summer and winter; of those which disperse further, less than 20% recorded at over 100 km from breeding areas, showing movement mainly towards mild south-west; birds leave Britain only exceptionally. Winter visitors (in small num-

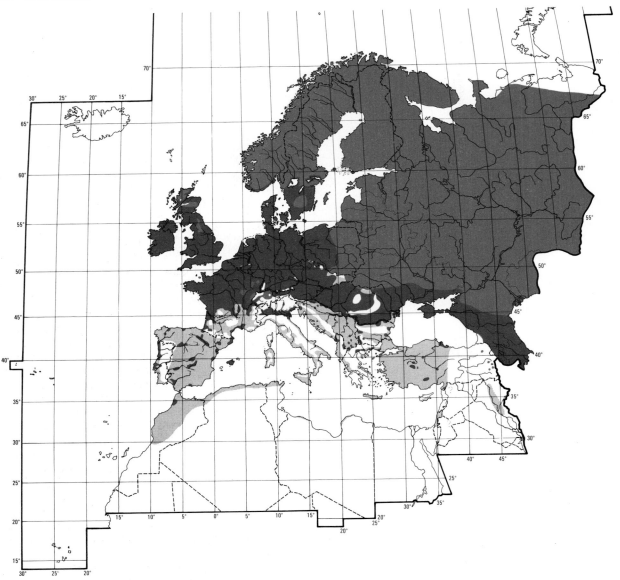

bers) originate chiefly in western Scandinavia, with a few from Low Countries and north-west Germany. Some west Scandinavian birds migrate along east coast of Britain to wintering grounds further south in Europe. (Prŷs-Jones 1984; Lack 1986.) Ringing recoveries up to 1987 (including passage-migrants) show movements to or from Low Countries (17), France (11), Norway (9), Sweden (5), Germany (1), Spain (1) (Mead and Clark 1987).

West Scandinavian birds (west of 20°E) almost entirely migratory, wintering chiefly in southern France, with smaller numbers in Britain, Low Countries, northern France, and as far as southern Iberia; main route via Low Countries, but perhaps some reach south-east France via Alpine passes. East Scandinavian and Finnish birds

wholly migratory, wintering chiefly in northern and central Italy; some migrate via Swiss Alpine passes, but probably most between eastern end of main Alps and Illyrian Alps; some winter in south-east France, and a few further west, overlapping with west Scandinavian birds. 2 recoveries of birds ringed near Baltic coast show similar pattern. North-central European birds chiefly migratory, especially in east; winter recoveries are between west and south-west of breeding areas, in northern, south-west, and southern coastal France, and in north-east and south-west Iberia; tend to be progressively further east dependent on origin. Relatively few birds winter in Switzerland, mainly near Lake Geneva; recoveries of birds ringed on passage or (fewer) in breeding season in Switzerland are chiefly in south-east France, with others in south-west France, Spain, and northern Italy. In Netherlands, *c.* 25% local birds remain to winter, and *c.* 5% in Belgium; recoveries of birds ringed in Low Countries (chiefly passage birds from north-west Germany and western Scandinavia) are mainly in south-west France, with some reaching Spain. (Prŷs-Jones 1984.) Partially migratory in France, and many present all year throughout country; in southern half, higher numbers (including immigrants) winter than summer; recovery data from Camargue in south-east suggest that birds usually arrive via northern Italian plain in autumn, but return north-east more directly in spring, via eastern France and western Switzerland. Yugoslavian birds tend to move west across northern Italy. (Prŷs-Jones 1984; Yeatman-Berthelot 1991.)

Some birds winter south of range in Mediterranean region. In Strait of Gibraltar (where few breed), locally common in winter, and regular on passage (Finlayson 1992). In Corsica (where none breed), regular in small numbers (Thibault 1983). Uncommon on passage and in winter in Sicily (Iapichino and Massa 1989), and a few records each year in Malta (Sultana and Gauci 1982). Widespread winter visitor in small numbers in north-west Africa, reported to edge of Sahara (Heim de Balsac and Mayaud 1962). In Libya, scarce and local and perhaps regular in coastal Tripolitania, not recorded elsewhere (Bundy 1976). Common in winter throughout Greece (breeds in north) (G I Handrinos). Local birds resident in Turkey; immigrants winter in western two-thirds, generally in moderate numbers (Martins 1989). Common winter visitor to reedbeds in Cyprus (Flint and Stewart 1992). Regular in Iraq (Moore and Boswell 1957). In Israel, winters in small but very varying numbers (Shirihai in press). Vagrant to Egypt and Arabia (Goodman and Meininger 1989; F E Warr).

Autumn movement chiefly mid-September to mid-November. For short-distance post-breeding dispersal in Finland, see Haukioja (1971*b*). Leaves western Scandinavia end of September to early November (Rendahl 1959), southern Finland and St Petersburg region (Russia) September to mid-October (Haukioja 1969; Mal'chevski and Pukinski 1983), and north-central Europe mainly in October (Niethammer 1937). Passage in Switzerland peaks 13–22 October (Winkler 1984). Highest numbers in Mediterranean region mid-November to February (Brosset 1961; Blondel and Isenmann 1981; Flint and Stewart 1992; Shirihai in press). Winter site-fidelity shown in Rhône-Alpes (south-east France), with birds retrapped up to 4 years after ringing (Olioso 1987).

Spring movement chiefly mid-February to April. In Switzerland, passage from last third of February to mid-April, peaking in 1st half of March (Winkler 1984). At Chertal (Belgium), earliest arrivals 18 February to 10 March over 6 years, and end of passage early to mid-April; in first 5 days of passage, 90% ♂♂, and passage of ♀♀ continues *c.* 10 days later than ♂♂ (Collette 1972).

Reaches west Scandinavia mid-March to May according to latitude (Rendahl 1959). Arrives in southern Finland late March to April (Haukioja 1969); at Helsinki, average earliest record 3 April (22 March to 20 April) over 23 years (Tiainen 1979). Reaches St Petersburg region end of March or beginning of April in early springs, more usually in 2nd week of April, with main movement last third of April (Mal'chevski and Pukinski 1983).

In east of breeding range also, vacates northern areas, but present all year further south, and also winters south of range. *E. s. passerina* winters in Iran, Kazakhstan, and north-west Mongolia, *pallidior* from Caucasus east to north-west India, *pyrrhulina* in Japan and eastern China. *E. s. pyrrhuloides* mostly resident in Kazakhstan and adjoining regions. (Vaurie 1959.) In Iran, very common in south Caspian lowlands, less common in wetlands in west and in Tehran area (D A Scott). Widely dispersed in small numbers in Pakistan (Roberts 1992), and locally common in north-west India, November–March (Ali and Ripley 1974). Reported from most of China in winter (Schauensee 1984). Vagrant to Nepal and Hong Kong (Inskipp and Inskipp 1985; Chalmers 1986). In Japan (breeds chiefly in Hokkaido), locally common October–March from northern Honshu to southern Kyushu, mainly in coastal marshes (Brazil 1991). For Kazakhstan, where several races occur, see Korelov *et al.* (1974). Passage chiefly September–November and March–April (e.g. Johansen 1944, Korelov *et al.* 1974), but continues in spring to May or early June in far east (Lobkov 1986; Williams 1986). DFV

Food. Seeds and other plant material; invertebrates in breeding season, and also opportunistically during remainder of year. Takes plant and animal material on ground among sedges, rushes, reeds, etc., in pasture and marshy grasslands, and also low in waterside bushes and trees (e.g.) willow *Salix*, alder *Alnus*, or on stems of reed *Phragmites*; outside breeding season, more often on ground in open countryside and cultivated fields, weedy areas, woodland clearings, uplands, etc., well away from water, often in flocks with other seed-eaters. (Witherby *et al.* 1938; Glutz von Blotzheim 1962; Ptushenko and Inozemtsev 1968; Górski 1976; Prŷs-Jones 1977; Eifler and Blümel 1983.) When foraging, probably most agile of Emberizidae of region, particularly in spring and summer; when feeding in tall grass, bushes, etc., seen to use all positions employed by finches (Fringillidae) except hanging upside-down (Prŷs-Jones 1977). In scrubby grassland in southern England, perches on twigs of shrubs or on stout herbs, less often on grass stems, to reach grass seed-heads; of 71 observations, 76% in long grass, 11% in mixed grass and herbs, 7% on ground, and 6% on bushes; in winter, more often on ground and in bushes (Beven 1964, which see for comparison with other species). In Oxfordshire (southern England), fed extensively in grass in late summer and early autumn taking seeds from heads of grasses and herbs; landed on lower half of stem then moved upward, bending it over; stems also gripped in one foot from neighbouring perch; in winter picked small invertebrates from trees, fences, etc. (Prŷs-Jones 1977). Seed-heads also reached by standing on ground and stretching up (Blümel 1982). Commonly catches flying insects, especially Odonata and Diptera, in sallies from perch (Tucker and Rowcliffe 1950; Summers-Smith and Summers-Smith 1952; Beven 1964; Prŷs-Jones 1977). In winter, frequently at bird-tables in gardens, as well as in farmyards (Blümel 1982; Eifler and Blümel 1983; Zimmerli 1986); in Devon (south-west England), when feeding on nut feeder, always had one foot on mesh, the other on perch (Stewart-Hess 1992). In Oxfordshire, winter, when feeding on short grass, 53% of 250 pecking actions were level or upwards, significantly less than in Yellowhammer *E. citrinella*, suggesting that *E. schoeniclus* took less invertebrates in this habitat (Prŷs-Jones 1977, which see for comparison of both species). For discussion of bill morphology, diet, and comparison with other seed-eaters, see Eber (1956). In Cornwall (south-west England), very often fed on rotting potatoes in harvested field, slowly chewing pieces before swallowing (King 1985a). In Mallorca, March, birds made holes in *Phragmites* stems and extracted larvae 1–2 cm long (Isenmann 1990, which see for discussion of bill size and diet). In Ili delta (south-east Kazakhstan), winter, thin-billed (*passerina*) and thick-billed (*pyrrhuloides*) birds showed marked feeding differences: thin-billed birds fed almost exclusively on seeds of Chenopodiaceae occurring in huge expanses there; thick-billed birds remained in reedbeds feeding only on insects all year round, concentrating on Diptera larvae in winter, snipping off *Phragmites* leaves to get at them inside the stems, removing all from one stem then moving on to next; heavy bills apparently necessary to remove tough leaves and open stems (Stegmann 1948). See also Schüz (1959) for similar findings from stomachs from south Caspian Sea area.

Diet in west Palearctic includes the following. Invertebrates: springtails (Collembola), mayflies (Ephemeroptera), adult and larval damsel flies and dragonflies (Odonata: Agriidae, Coenagriidae, Platycnemididae, Lestidae, Libellulidae), stoneflies (Plecoptera), grasshoppers, etc. (Orthoptera: Tettigoniidae, Acrididae, Tetrigidae), bugs (Hemiptera: Notonectidae, Auchenorrhyncha, Aphidoidea), lacewings, etc. (Neuroptera), adult and larval Lepidoptera (Satyridae, Pieridae, Hepialidae, Zygaenidae, Tortricidae, Pyralidae, Notodontidae, Noctuidae, Lymantriidae, Lasiocampidae, Sphingidae, Geometridae), caddis flies (Trichoptera), adult and larval flies (Diptera: Tipulidae, Culicidae, Chironomidae, Stratiomyidae, Tabanidae, Muscidae), adult and larval Hymenoptera (Tenthredinidae, ants Formicidae), beetles (Coleoptera: Buprestidae, Elateridae, Cantharidae, Chrysomelidae, Curculionidae, Scolytidae), spiders (Araneae: Theridiidae, Araneidae, Tetragnathidae, Lycosidae, Pisauridae,

Clubionidae, Thomisidae, Salticidae), harvestmen (Opiliones), ticks (Acari), Crustaceae, snails (Pulmonata: Succineidae, Planorbidae, Lymnaeidae, Cochlicopidae, Vertiginidae, Zonitidae), Bivalvia (Sphaeriidae). Plants: seeds, shoots, etc., of spruce *Picea*, birch *Betula*, alder *Alnus*, nettle *Urtica*, knotgrass *Polygonum*, dock *Rumex*, orache *Atriplex*, goosefoot *Chenopodium*, amaranth *Amaranthus*, chickweed *Stellaria*, mouse-ear *Cerastium*, buttercup *Ranunculus*, rape, etc. *Brassica*, shepherd's purse *Capsella*, gold of pleasure *Camelina*, pennycress *Thlaspi*, meadowsweet *Filipendula*, lupin *Lupinus*, flax *Linum*, willow-herb *Epilobium*, cowberry *Vaccinium*, scarlet pimpernel *Anagallis*, forget-me-not *Myosotis*, basil *Clinopodium*, potato *Solanum*, plantain *Plantago*, mugwort *Artemisia*, bur-marigold *Bidens*, hawksbeard *Crepis*, sowthistle *Sonchus*, *Galinsoga*, mare's-tail *Hippuris*, water plantain *Alisma*, sedges (Cyperaceae), rushes (Juncaceae), reedmace *Typha*, grasses (Gramineae, including maize *Zea*, millet *Panicum*, reed *Phragmites*, *Lolium*, *Poa*, *Deschampsia*, *Alopecurus*, *Molinia*, *Setaria*, *Agrostis*, *Festuca*, *Echinochloa*). (Witherby *et al.* 1938; Kovačević and Danon 1952; Dementiev and Gladkov 1954; Turček 1961; Glutz von Blotzheim 1962; Beven 1964; Ptushenko and Inozemtsev 1968; Górski 1976; Prŷs-Jones 1977; Popov 1978; Blümel 1982; Kostin 1983; Mal'chevski and Pukinski 1983; King 1985a; Okulewicz 1991; Semenov-Tyan-Shanski and Gilyazov 1991.)

In Oxfordshire, over year, animal material accounted for *c*. 30% by corrected volume of contents of 108 stomachs; 100% in May (no data for June) declining to minimum of less than 5% November–December; present in 100% of stomachs March–May, 60% in December. Over year, 103 stomachs contained 23 647 seeds, of which 67·5% by number (82·1% by weight) grass, 18·0% (8·8%) Caryophyllaceae, 5·0% (4·8%) Chenopodiaceae, and 3·4% (1·5%) Polygonaceae. Grasses fairly constant over year, cereals at maximum in March when 52% by weight of all seeds; at one site in September, 85% by weight of 1531 seeds in 7 stomachs were Chenopodiaceae and 11% grass. Animal prey September–February small numbers of spiders, springtails, Hemiptera, Diptera, and beetles, many as larvae or pupae; in March, springtails and especially Diptera increased greatly, mainly due to emergence of midges (Chironomidae), picked up from damp open ground at rate of 1–2 per s, and larger Diptera which are taken in air; from mid-April through May, caterpillars from trees and hedges main component of invertebrate diet, and in June–July spiders and Odonata become important. (Prŷs-Jones 1977, which see for many details.) In Volga-Kama region (east European Russia), spring-summer, of 36 stomachs, 78% contained animal material, 39% plant material; 39% contained Curculionidae, 25% Hemiptera, 22% caterpillars, 17% Buprestidae, 14% Diptera, 11% Orthoptera, 11% spiders, and 6% Elateridae; 8% contained seeds of Polygonaceae, 6% Chenopodiaceae, 6% Labiatae, and 6% grass *Echinochloa* (Popov

1978). In Ukraine, breeding season, 5 stomachs contained 81 invertebrates, of which 43% by number Curculionidae, 30% caterpillars, and 27% Diptera (Kistyakovski 1950). In Crimea (southern Ukraine), October–January, of 6 stomachs, 5 contained only seeds of herbs and grasses (*Polygonum*, *Amaranthus*, *Phragmites*), and 1 contained Orthoptera and larvae (Kostin 1983). In Poznań area (western Poland), winter, of 193 feeding observations, 31% on goosefoot, 17% lupin, 16% rape, 11% amaranth, 8% gold of pleasure, 5% pennycress, 5% *Echinochloa*, 3% cereals, 2% shepherd's purse, 2% *Galinsoga*, 1% flax (Górski 1976). On Kola peninsula (north European Russia), 10 spring stomachs contained seeds of birch, cowberry, sedges, and grasses, as well as caddis flies and other insects (Semenov-Tyan-Shanski and Gilyazov 1991). In Moscow and St Petersburg regions (and generally elsewhere), diet changes from mainly plant to animal food in spring, then back again around late July (Ptushenko and Inozemtsev 1968; Mal'chevski and Pukinski 1983). In autumn in eastern Germany often on maize flowers taking stamens (Blümel 1982). For proportion of food weight and energy reserves accounted for by items carried to roost in Oxfordshire, and comparison with *E. citrinella*, see Prŷs-Jones (1977).

Nestlings fed only invertebrates. In Wrocław area (south-west Poland), 641 collar-samples from 51 broods contained 1835 invertebrates, of which 24·6% by number spiders (17% Araneidae, 2% Clubionidae, 2% Thomisidae), 18·8% Diptera (14·2% adults, 4·6% larvae; 8% Tipulidae, 2% Chironomidae, 2% Tabanidae), 12·6% mayflies, 10·3% Hymenoptera (10% larval Tenthredinidae), 8·9% Lepidoptera (7·9% larvae, 1·0% adults), 4·9% snails, 4·7% Hemiptera, 4·7% beetles (4% Chrysomelidae), 4·0% Orthoptera (2% Tettigoniidae, 2% Acrididae), 3·0% Odonata (2·5% adults, 0·5% larvae), 2·4% caddis flies. Adult Diptera and beetles decrease as proportion of diet towards end of season, while Orthoptera, adult Lepidoptera, and snails increase; spiders present in 80–90% of samples during July–August; in mid-season (June), mayflies, caterpillars, and larval Hymenoptera commonest prey. In samples from grassland habitats, main component of diet was adult Diptera, especially crane-flies Tipulidae, and in those from areas with more trees and bushes, caterpillars, larval Hymenoptera, and adult beetles; diversity of prey greatest early in season. (Okulewicz 1991.) In Oberlausitz (eastern Germany), end of May, 50% of diet by observation at one nest was *Tortrix* caterpillars, collected in oak *Quercus* tree, 25% crane-flies, and 25% other invertebrates. At end of June and beginning of July, almost all food was damsel flies (Lestidae, Agriidae) taken on vegetation by water, most of them newly emerged and unable to fly, and brought to nest in large amounts at each visit; *c.* 50% of adult Odonata given to young had wings removed first. (Blümel 1982.) In St Petersburg region, diet of young mostly larval Odonata; large adults had wings removed, small ones fed with wings

attached (Mal'chevski and Pukinski 1983). In Oxfordshire, nestling collar-samples held only invertebrates, which at this time (July) accounted for only 40% by volume of adult diet; young switched to grass seeds as soon as independent (Prŷs-Jones 1977). BH

Social pattern and behaviour. Well known, mainly from substantial studies by Andrew (1956*b*, 1957*a*) and O'Malley (1993) in England, Ghiot (1976) in Belgium, and Blümel (1982) in Germany. See O'Malley (1993) for detailed analysis of mating behaviour and reproductive success.

1. Regularly in flocks outside breeding season, often with other species (especially Fringillidae and other Emberizidae); numbers of *E. schoeniclus* in such flocks very variable, up to 1500 recorded. Post-breeding flocks form in September, Belgium. Flocks begin to break up in early spring, in Belgium mainly February, in central Europe in March or, if weather cold, early April. ♂♂ usually leave flocks before ♀♀, visiting prospective territories in early morning; leave territory at *c.* 09.00 hrs to rejoin feeding flock, not returning to territory until following morning (see also Heterosexual Behaviour, below). (Ghiot 1976; Blümel 1982.) BONDS. Monogamous mating system, but extra-pair paternity common, only 50% of offspring being result of within-pair mating and 69% of broods containing extra-pair offspring (O'Malley 1993). Polygamy occasional. In first study, involving 4 proven cases of bigyny and other probable cases, and one suspected case of trigyny, one bigamous ♂ was 2 years old, 2 were 3 years old, and 1 at least 4 years old, suggesting polygyny may be feature of birds that have bred at least once before. High densities of breeding population probably a main predisposing factor for polygyny; in several cases, 2 nests of bigynous ♂♂ only a few metres apart. High density probably also responsible for one case of 2 ♀♀ sharing a nest, and another suspected case. (Bell and Hornby 1969.) In more recent study, involving 8 cases, polygyny not associated with individual ♂ quality, but mainly followed death of ♂, neighbouring ♂ subsequently obtaining widowed ♀ as 2nd mate (O'Malley 1993). ♀ alone builds nest, and incubates (but see Breeding); both sexes feed nestlings and care for and feed fledged young (Ghiot 1976; Blümel 1982). In some cases of bigamy (see above), ♂♂ fed young in both nests, but usually did not (Bell and Hornby 1969). Age of first breeding 1 year (Hornby 1971). BREEDING DISPERSION. Solitary and territorial. ♂♂ usually re-ocupy territory held in previous year; ♀♀ tend also to do so, but less strongly (Bell 1968; Hornby 1971; O'Malley 1993). Territory size (area occupied and used) varies in course of breeding cycle, with averages in Belgian study as follows: before ♂ is paired 710 m², during pair-formation 1950 m², during nest-building 2720 m², during egg-laying 1890 m², during incubation 840 m², when feeding young 1810 m². In densely populated areas, when territories are at largest stages, overlaps between neighbouring territories may occur and conflicts increase. (Ghiot 1976; Blümel 1982.) In Netherlands study, territory sizes very variable: average size in 1982 (*n*=26) 1659 m² (96–3264), in 1983 (*n*=28) 1535 m² (500–3470) (Hut 1985, which see for changes in course of breeding cycle and relation between territories of *E. schoeniclus* and other species breeding in same habitat). Territory used mainly for pair-formation, maintenance of pair-bond, and nesting. During pair-formation, some food obtained in territory, but generally obtained on neutral ground elsewhere, neighbouring pairs meeting without conflict. Territory very seldom capable of providing enough food for young. (Ghiot 1976; Blümel 1982.) Breeding densities very variable. Can be high in small areas of suitable habitat, e.g. 148 pairs per km² on island with willow *Salix* thickets, Finland (Haukioja

1968), 140 per km² in marshland, eastern Germany (Blümel 1982), 182 per km² in flooded meadowland, northern Germany (Harms 1975). Examples of more usual densities in areas of mixed habitats, where proportion of suitable habitat small: 11·0 pairs per km² in mixed bog, heath, and peripheral woodland, southern England (Glue 1973); c. 8 pairs per km² in mountain birch *Betula* woodland, Sweden (Ulfstrand and Högstedt 1976); 0·98 pairs per km² in farmland, Poland (Górski 1988); average (8 years) 0·7 pairs per km² in mixed farmland, southern England (Benson and Williamson 1972). ROOSTING. Outside breeding season, roosts communally in thick vegetation, e.g. reeds *Phragmites*, growth fringing lake shore, conifer stands; usually up to 30 birds together, occasionally 100, once 500 in large reedbed (Blümel 1982). 2216 birds counted arriving at exceptionally large roost in reedbed 50 × 30 m near Hamburg (Germany) (Bruster 1988). During pair-formation, birds continue to go to communal roost; begin to roost in territory when first eggs laid (Ghiot 1976).

2. Not shy; easy to observe, and generally not very sensitive to disturbance by man (Ulbricht 1975). Where coexisting with Yellow-breasted Bunting *E. aureola* in northern Scandinavia, neighbouring ♂♂ of the 2 species stimulate each other to sing, often singing alternately, but do not come into conflict (G Åström). FLOCK BEHAVIOUR. Not studied in detail, but apparently similar to Yellowhammer *E. citrinella*, winter flocks being essentially feeding aggregations, often with other species, not closely coordinated. Flocks generally small in winter, much larger during migration, especially spring migration, when flocks of several hundred may form, up to recorded maximum of 1500 (Mecklenburg, 27 March). (Blümel 1982.) SONG-DISPLAY. Song usually given from high point within territory, such as bush or reed stem; occasionally from ground. Singer adopts relaxed posture (Fig A) with ruffled plumage, notably on head and rump.

A

No other special display associated with song, but singing bird often mounts higher on perch between each song (Ulbricht 1975). Tempo of song changes according to pairing status of ♂. 'Rapid song' (see 1 in Voice), with short inter-unit intervals and long inter-song intervals, restricted to unpaired ♂♂. 'Slow song', with long inter-unit intervals and short inter-song intervals (so that in extreme cases song may sound continuous), mainly restricted to paired ♂♂. Intergrades occur, especially during pair-formation. ♂♂ which lose mates during breeding season revert from slow to rapid song. (E Nemeth.) Song-period from taking up of territories in early spring to end of breeding: late January to early August, central England (Bell 1967); beginning of March to beginning of August, Germany (Blümel 1982); early April to

early July, Sweden at 60°N (G Åström); occasional autumn song after moult (Blümel 1982). Diurnal pattern of singing varies greatly according to latitude. In Germany at c. 51°N, song begins before sunrise and ends after sunset; occasional at night (Blümel 1982). At 60°N, Sweden, morning song begins c. 30 min before sunrise in April, c. 2 hrs before in May, and c. 2·5 hrs before in June; ends c. 1 hr after sunset in April, and earlier relative to sunset in May and June. In northern Norway (Finnmark) and northern Finland, song begins about midnight, and is most intense 00.00–02.00 hrs. Maximum duration of song at 60–65°N, 20–21 hrs. (Åström 1976, which see for detailed analysis of song activity in relation to light and other environmental factors.) In relation to breeding cycle of individual pairs, song activity is intense when ♂ unpaired, and falls off to very low level during and immediately after pair-formation, followed by resurgence during nesting. Song loud and long when ♀ incubating, with peaks at 10.00–11.00, 15.00–15.30, and 18.30–19.00 hrs. (Ghiot 1976.) Main functions of song are to advertise territory to rival ♂♂ and potential mates (Ghiot 1976); also probably to synchronize activities of pair throughout breeding season (Ewin 1978). ANTAGONISTIC BEHAVIOUR. Aggressive behaviour in winter flocks not studied in detail; involves Head-forward threat display; pivoting (see below) does not occur (Andrew 1957a). In threat display to intruder in territory, main display is Head-forward threat: ♂ faces intruder with body horizontal and head aligned with body, head feathers sleeked; commonly gives aggressive calls (see 3 in Voice), occasionally gapes; may repeatedly lower bill, or lower fore-part of body with bill horizontal; may pivot, swinging body from side to side. In early stages of pair-formation (see below), when ♀♀ are visiting territories, ♂♂ especially aggressive to one another and fights often develop. ♂♂ fly up together, sometimes to nearly 1 m, face to face, clawing and pecking, then descend still face to face or one following the other. Aggressive and fear calls given (see 3 in Voice). (Andrew 1957a; Ghiot 1976.) If one ♂ begins to retreat, 'slow chase' may follow, birds flying with slow wing-beats of normal amplitude, sometimes separated by brief glides; tail held spread, both in flight and in brief perching between flights. Between encounters, 'moth flights' (rapid wing-beats of small amplitude) sometimes occur, which occasionally pass into wing-quivering on landing (Andrew 1957a). HETEROSEXUAL BEHAVIOUR. (1) Pair-bonding behaviour. Takes place in ♂'s territory, after break-up of winter flocks. On arrival of ♀ in territory, ♂ stops singing and approaches ♀; may engage in conflict with other ♂♂ who approach ♀ from neighbouring territories (Ghiot 1976). Another account describes ♂ as approaching incoming ♀ with plumage ruffled; ♂ often picks up piece of dry grass, flies short distance with it, then drops it (Ulbricht 1975). ♀ at first avoids ♂, giving 'see' calls (see 3 in Voice); visits a number of territories, at first apparently interested in site, not in ♂. ♀ then begins to re-visit one ♂ more often, for longer periods, and finally associates with him in all activities. At this stage, ♀ does not know limits of territory; ♂ follows her continually, trying to keep her within territory; ♀ gradually learns limits and stays within them. ♂ makes frequent sexual approaches, sometimes leading to ♂ gripping ♀; pair may roll on ground, then separate; ♀ tries to avoid him. During period until ♀ ready to mate (c. 20 days), ♂ accompanies ♀ closely. Towards end of this period, ♀ begins to attack other ♀♀ which approach. (Ghiot 1976; O'Malley 1993, which see for detailed analysis of mate-guarding.) 'Fluffed run' (Fig B) is the only common courtship display. With body horizontal or nearly so, bill somewhat lowered, legs slightly flexed, tail drooped and spread, head and body feathers (especially those of rump) markedly ruffled, wings (largely hidden by flank feathers) often trailed or (at higher intensity) raised, ♂ makes series of swift

B

runs, at first away from ♀. Later, may run just behind her, assuming more erect posture and raising wings higher, then leap onto her back and attempt to copulate. Sites chosen for display are open spaces on ground, providing scope for running. (Andrew 1957*a*.) Displays take place mainly during pair-formation, when often followed by attempts to mate (see above), and as prelude to mating; pairs normally show no marked sexual behaviour between end of pair-formation and beginning of breeding. (Andrew 1957*a*.) (2) Courtship-feeding. None recorded. (3) Mating. Begins 19 days before laying of 1st egg and continues to day of laying of 2nd egg (O'Malley 1993). May take place either on perch or on ground. Before mating, both members of pair may pick up pieces of dry vegetation and let them drop; ♂ may fly with rapid wing-beats ('whirring' flight) towards or over ♀ (Ulbricht 1975). Soliciting ♀ adopts crouching posture, with legs flexed, body more or less horizontal, bill pointed obliquely upwards, tail raised and commonly spread; wings raised, usually with little extension, and rapidly quivered; cloaca exposed. ♀ at first gives rapidly repeated 'seeoo'-type calls, replaced at high intensity by usual 'tititi...' soliciting calls (see 4 in Voice). ♂ usually flies straight to ♀, then hovers over her or lands directly on her back, where he stands with beating wings, rather erect posture, and bent legs. Copulation behaviour same as *E. citrinella* (see that species). (Andrew 1957*a*.) (4) Nest-site selection and behaviour at nest. ♀ apparently chooses nest-site, accompanied by ♂ (Blümel 1982). ♀ alone builds. ♂ may pick up material; twice seen to place it in alternative site, as if indicating preference (Ghiot 1976). ♀ alone incubates, but ♂ may get onto nest when ♀ leaves it to feed. ♂ approaches nest, calling; ♀ leaves, and ♂ goes onto nest. When ♀ returns she calls, and ♂ (if still there) leaves nest. If weather fine, ♂ stays only a few minutes on nest, then sings on nearby perch, or joins ♀. RELATIONS WITHIN FAMILY GROUP. Both parents feed young, and remove faeces, either swallowing them or dropping them in flight. (Blümel 1982.) When nestlings are small, parents do not go to nest together; if one is at nest, other waits. From 6th day, both may visit nest at same time. Both parents feed young after fledging, but mainly ♂; ♀, if nesting again, begins to build 2–3 days after young have fledged. (Ghiot 1976.) After fledging, while still unable to fly, young hide in vegetation near nest; located by parents by their calls (see Voice). From age of 15 days, young can fly well enough to leave ground vegetation and move into reeds, bushes, and trees. ♂ then usually tolerated by neighbouring ♂♂ if he trespasses to feed them, with food in bill. (Ghiot 1976; Blümel 1982.) Young independent *c.* 20 days after leaving nest (Hermann 1983). Length of post-fledging care apparently not precisely recorded. ANTI-PREDATOR RESPONSES OF YOUNG. Young cower in nest-cup in response to warning call of parent (see below) (Ulbricht 1975). PARENTAL ANTI-PREDATOR STRATEGIES. Incubating or brooding ♀ sits tight, leaving nest only when (human) intruder 1–2 m away (Blümel 1982). ♀ gave excited warning calls (see 2c in Voice) at approach of Marsh Harrier *Circus aeruginosus* to nest with young, causing young to cower in nest (Ulbricht 1975). Both parents perform distraction-lure display of disablement type. Display silent or ♀ gives 'dschiep' calls (Blümel 1982: see 2 in Voice); usually begins with short flight down from nest with tail widely

spread; then series of short flights, alternating with movement on ground, carry bird away from intruder; tail spread throughout. During flights, wing-beats are of small amplitude and short glides often occur, so that bird moves only slowly forward. In moving on ground, wings may be fluttered, held by side, or raised to almost full elevation and extension; bill sometimes slightly lowered; rump feathers markedly raised (Howard 1929; Ghiot 1976), or body feathers may be either normal or all slightly ruffled. (Andrew 1956*b*.) ♂ executed 'frenzied dance', with ruffled plumage and drooped wings, in response to stuffed Goshawk *Accipiter gentilis* placed near nest, moving round it in decreasing circles (Ulbricht 1975).

(Figs by D Nurney: A from photograph in Ghiot 1976; B from drawing in Andrew 1957*a*.)   DWS

Voice. Song well studied (Ewin 1976, 1978; Ghiot 1976; Gailly 1982, 1988; G Åström, E Nemeth). For calls, most detailed study is by Andrew (1957*b*), on which following account mainly based. For additional sonagrams, see Arendt and Schweiger (1982) and Bergmann and Helb (1982).

CALLS OF ADULTS. (1) Song of ♂. Short series of rather unmusical, tinkling units, of variable tempo; 'tweek tweek tweek tititick' (Witherby *et al.* 1938) gives good idea of typical song. Frequency range mostly 3–7·5 kHz (Ghiot 1976). Tempo varies greatly, depending on stage of breeding (see Social Pattern and Behaviour), with striking distinction between 'rapid' and 'slow' song. In rapid song (Figs I–II), intervals between units typically less than 0·15

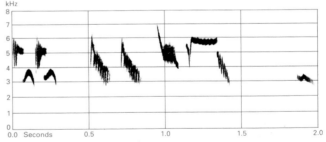

I E Nemeth Austria March 1991

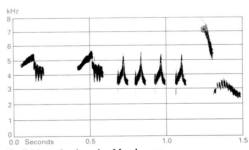

II E Nemeth Austria March 1991

s, and song-phrases usually well spaced, with intervals between phrases longer than length of phrase; in slow song (Figs III–IV), intervals between units more than

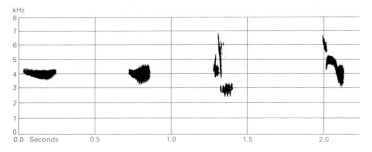

III E Nemeth   Austria   April 1991

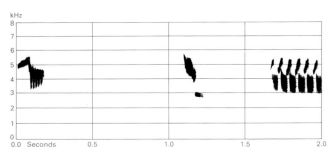

IV E Nemeth   Austria   April 1991

0·15 s, often much more (so that song sounds halting and hesitant), and intervals between phrases often not much longer than intervals between units (so that song series may be more or less continuous). (Ewin 1976; E Nemeth.) Sonagraphic analysis shows more complex song structure than detectable by human ear, with 4 classes of unit distinguishable: introductory units, 'note-complexes', 'multiple note-complexes', 'trills' (Ewin 1976). Introductory units comparatively simple in structure, distinctive and invariable in each individual (and apparently main component of song used in individual recognition by conspecifics: Ghiot 1976). Song-phrases generally begin with repetition of introductory units (2–4, seldom more: G Åström), followed by one or more note- complexes, with trills tending to come at end; sequence very variable (Ewin 1976; Ghiot 1976; Gailly 1982); total usually 3–7 units (Arendt and Schweiger 1982). Usual duration of song 1–3 s; averages for 22 ♂♂ ranged from 1·22 to 2·74 s. Intervals between songs very variable (see above); when very short, songs may be given at rate of *c*. 20 per min. (G Åström.) Song-phrases show well-marked development within song-bout: bird usually begins with monotonous series of contact-calls (see 2, below), gradually introducing elements of song; bill at first closed, increasingly opened as song-bout proceeds (Bergmann *et al.* 1984, which see for detailed analysis). Individual ♂♂ have repertoires of 20–30 different units, mostly note-complexes (Ewin 1978); *c*. 30 (Gailly 1982). ♂♂ usually share much of repertoire with neighbours, especially close neighbours, and when counter-singing use closely similar songs; exceptions perhaps result from displacement of ♂ to another territory after song repertoire established (Ewin 1976, 1978). Similarity of repertoires of neighbouring ♂♂, southern

England, suggests formation of local dialects (Ewin 1976, 1978), but no evidence for dialects in Belgian population (Ghiot 1976). Some individuals in local populations (perhaps related to one another) have very similar introductory units (G Åström). Imitations of other species occasionally incorporated into songs, e.g. Redstart *Phoenicurus phoenicurus*, Bluethroat *Luscinia svecica*, Dipper *Cinclus cinclus* (Blümel 1982), but this not reported in detailed studies of song (see references above). (2) Contact-alarm and alarm-calls. (2a) Commonest call (Figs V–VII), used all

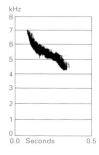

V C J Hazevoet   Netherlands   April 1982

VI C J Hazevoet   Netherlands   May 1984

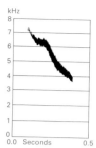

VII P A D Hollom England March 1984

year, when feeding or moving around in loose flocks with conspecifics, characteristic 'seeoo' (Andrew 1957*b*), 'tseep' (Witherby *et al.* 1938), or 'zieh' (Blümel 1982); soft, whistling timbre; typical frequency range 3·5–7 kHz, falling in pitch; lasts *c.* 0·2 s (Andrew 1957*b*; L Svensson). Usually given in slight alarm (e.g. Fig VII, call of ♀ at nest with young), and by solitary birds to make contact with others; in breeding territory may serve as warning to mate or young (Blümel 1982). ♂♂ give series of 'seeoo' calls before singing; as tendency to sing increases, 'seeoo' deepens and approaches 'cheep' (see below) in form (Andrew 1957*b*). (2b) Short, hoarse, even-pitched 'bzy', 'chü', or 'bzree' heard mostly outside breeding season, often in flight and often from autumn migrants (L Svensson). (2c) Deeper, briefer 'cheep' described by Andrew (1957*b*) is possibly same as 2b, but said to fall in pitch even more sharply than call 2a; frequency range 2–4 kHz, duration less than 0·2 s; apparently mainly a contact-call; given by solitary birds (chiefly juveniles in autumn), and occasionally in apparent alarm; very similar call also used as copulation-call. Higher-pitched 'see', with frequency range 6–8 kHz and falling little in pitch (less than 'seeoo'), duration also *c.* 0·2 s, expresses strong fear or alarm; given by both sexes to human intruder at nest (Andrew 1957*b*), and apparently same as anxious-sounding 'dschiep', given by ♀ during distraction-lure display of disablement type (Blümel 1982). Warning against flying predator overhead, a very high-pitched call, apparently similar to calls used in same context by many other passerines (E Nemeth). (3) Calls associated with aggression. Head-forward threat display (see Social Pattern and Behaviour) usually accompanied by 'tchaa' call, loud and vibrant (Fig VIII), or less loud

'tchu'; 'eee' indicates fear, or less confidence (Andrew 1956*d*; Ghiot 1976). Call accompanying fighting, characteristic 'tchap-tchap', given when combatants separate. ♀, when attacked by ♂ in early stages of pair-formation, gives typical fear call, 'see'. (Ghiot 1976.) (4) Soliciting-call. ♀ soliciting mating gives rapidly repeated 'seeoo' calls (contact-alarm call), replaced with increased excitement by 'tititi...' (typical *Emberiza* soliciting call) (Andrew 1956*d*; Ghiot 1976). (5) Other calls. ♂ gives husky 'shee' as nest-site call; also (rather uncommonly), 'chu' calls during flights about territory and occasionally as prelude to song; 'chu' call occasionally given by ♀. (Andrew 1957*b*.)

CALLS OF YOUNG. First audible calls, 'sieh sieh', given from 6th day; become louder and shriller (Blümel 1982). This evidently same as close begging-call, 'surr', very rapidly repeated series of very short syllables ranging from 1 to *c.* 8 kHz, calls consisting typically of *c.* 5 syllables lasting *c.* 0·1 s (Andrew 1957*b*, which see for sonagram); first heard on 8th day and almost disappeared by 18th day (Andrew 1956*d*). From 10th day, young begin to give 'distant begging-call', 'zit... zit' (Blümel 1982), 'tip' (Andrew 1957*b*, which see for sonagram), short pulse of sound lasting *c.* 0·03 s with main energy at 7–8 kHz. Distant begging-call continues after fledging, when it also indicates position of young to parents; audible (to man) for considerable distance, but difficult to locate (Blümel 1982); still used by young *c.* 40 days old (Andrew 1957*b*).

DWS

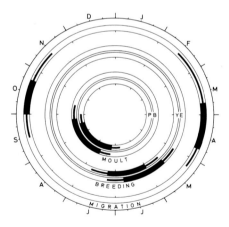

Breeding. SEASON. Nottinghamshire (central England): eggs laid 1st half of May (late April to mid-July) (Hornby 1971, which see for factors influencing time). England (from nest record cards): mean laying date 21 May (14 April to 6 August, *n* = 529); mean date of *c.* 106 1st clutches 30 April (Yom-Tov 1992, which see for comparison with Yellowhammer *E. citrinella* and Corn Bunting *Miliaria calandra*). Finland: see diagram. Kola peninsula (north European Russia): eggs laid from about beginning of June (Semenov-Tyan-Shanski and Gilyazov

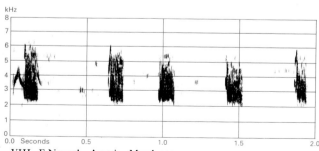

VIII E Nemeth Austria March 1991

1991). For Swedish Lapland, see Lennerstedt (1964). Wrocław region (south-west Poland): eggs laid mid-May (early April to mid-July) (Okulewicz 1989). Eastern Germany: peak egg-laying May (late April to early July) (Ulbricht 1975; Blümel 1982). For Netherlands, see Hut (1985, 1986). Catalonia (north-east Spain): season end of April to end of August (Muntaner *et al.* 1983). SITE. Usually well hidden on ground or on sedge tussocks, heaps of dead rushes *Juncus*, reeds *Phragmites*, etc., by water; also up to 4 m above ground in (e.g.) willow *Salix* or alder *Alnus* (Witherby *et al.* 1938; Haukioja 1970; Blümel 1982; Hermann 1983). Also recorded in cavity in wall or earth bank near water (Glutz von Blotzheim 1962). In Wrocław region, of 263 nests, *c.* 70% less than 10 cm above ground or water; 74% on sedge or grass tussock, 13% on ground in tangle of vegetation, 8% in or under bush, 4% on pile of dead vegetation, 2% in depression in ground (Okulewicz 1989). On Kola peninsula, of 39 nests, 54% on tussocks, 31% on ground, 15% in bushes, including juniper *Juniperus* (Semenov-Tyan-Shanski and Gilyazov 1991). In Swedish Lapland, nests perhaps more often above ground in years of high population of predators, especially rodents (Arheimer and Enemar 1974); in St Petersburg region (north-west Russia), 1st nests tend to be closer to ground than later ones (Mal'chevski and Pukinski 1983). Increasingly on ground in typical *E. citrinella* habitat, principally agricultural landscapes, some distance from water (Bell 1969; Eifler and Blümel 1983; Diesselhorst 1986, which see for comparison with *E. citrinella*; Lieder 1987, which see for review). Nest: foundation of stems and blades of sedges, grasses, and other waterside plants, occasionally small twigs, lined with finer plant material, moss, rootlets, and sometimes hair or feathers (Ulbricht 1975; Blümel 1982; Hermann 1983; Mal'chevski and Pukinski 1983; Okulewicz 1989). On Kola peninsula, 39 nests had average outer diameter 8·4 cm, inner diameter 5·2 cm, depth of cup 4·5 cm (Semenov-Tyan-Shanski and Gilyazov 1991); 10 nests in Oberlausitz (eastern Germany) had outer diameter 10–14 cm, overall height 5–12 cm, and dry weight 15–24 g (Blümel 1982). Building: by ♀ only, occasionally accompanied by ♂; material gathered near nest, though ♀ will fly up to 50 m while collecting (Ptushenko and Inozemtsev 1968; Ulbricht 1975; Blümel 1982; Hermann 1983). In Wrocław region, took average 2·6 days (1–4, $n=20$); in eastern Belgium 2–7 days (Ghiot 1976); in eastern Germany, *c.* 5 days for 1st nest and 3 for replacement (Ulbricht 1975, which see for details). EGGS. See Plate 34. Sub-elliptical, smooth and slightly glossy; very pale purplish, lilac-grey, or olive-brown, rarely buffish or greenish, with scrawls, spots, and blotches of purplish-brownish-black irregularly distributed over whole surface, or concentrated at either end; occasionally unmarked (Harrison 1975; Makatsch 1976). Nominate *schoeniclus*: 19·5 × 14·6 mm (17·4–22·6 × 13·6–16·0), $n=859$; calculated weight 2·19 g. *E. s. intermedia*: 20·7 × 14·8 mm (19·0–22·2 × 14·0–16·1), $n=25$; calculated weight 2·38 g.

E. s. reiseri: 20·2 × 15·3 mm, $n=5$; calculated weight 2·45 g. *E. s. pyrrhuloides*: 20·3 × 15·1 mm (19·4–21·1 × 14·7–16·1), $n=32$; calculated weight 2·43 g. *E. s. pallidior*: 19·3 × 14·6 mm (16·5–20·0 × 13·5–15·2), $n=14$; calculated weight 2·17 g. (Schönwetter 1984.) Clutch: 4–5 (3–7). In eastern Germany, of 321 clutches: 2 eggs, 1%; 3, 4%; 4, 18%; 5, 73%; 6, 4%; average 4·76; in April 5·06 ($n=64$), May 4·85 ($n=196$), June 4·37 ($n=51$), July 3·10 ($n=10$) (Blümel 1982). In St Petersburg region, of 116 clutches: 2 eggs, 2%; 3, 3%; 4, 25%; 5, 57%; 6, 14%; average 4·78; in May 5·08 ($n=61$), June 4·91 ($n=45$), July 4·29 ($n=7$) (Mal'chevski and Pukinski 1983). Average in Wrocław region 4·72 ($n=269$); in April 4·92 ($n=76$), July 3·56 ($n=16$) (Okulewicz 1989); Swedish Lapland 6·1 ($n=11$) (Lennerstedt 1964); south-west Finland 5·14 ($n=203$) (Haukioja 1970, which see for discussion); Nottinghamshire 4·3 ($n=430$), 4·1–4·8 over 7 years (Hornby 1971); England 4·46 ($n=529$) (Yom-Tov 1992); Munich area (southern Germany) 4·70 ($n=44$) (Hermann 1983, which see for review of size in Europe); Catalonia 5·25 ($n=4$) (Muntaner *et al.* 1983). Clutches of 7 recorded in Finland; 2nd broods probably only reared by ♀♀ older than 1 year (Haartman 1969; Haukioja 1970). In Munich area, 47% of 19 pairs had 2 broods (Hermann 1983); in Netherlands, 43% of pairs in 14 territories (Hut 1986), but in Wrocław region, only 4 double broods noted (Okulewicz 1989); in Catalonia 2, perhaps 3, broods (Muntaner *et al.* 1983). Replacement clutch laid *c.* 6 days after loss of 1st (Hornby 1971; Hermann 1983); in St Petersburg region, 2nd brood started after 1st fledged (Mal'chevski and Pukinski 1983). Eggs laid daily in early morning 1–2 days after nest completed (Haukioja 1970; Ulbricht 1975; Blümel 1982; Hermann 1983). For records of more than one ♀ laying in same nest, see Bell and Hornby (1969) and Okulewicz (1989). INCUBATION. 13 days (12–15); proper incubation probably by ♀ only, though ♂♂ recorded covering eggs for considerable periods; starts with last or penultimate egg (Makatsch 1976; Mal'chevski and Pukinski 1983; Okulewicz 1989). In Brandenburg area (eastern Germany), average time spent on eggs was 54 min per hr; usually only 1 break per hr, sometimes 5–6 shorter ones; in 4 pairs, ♂♂ took stints of 7–35 min per day (Ulbricht 1975). In Wrocław region, ♂ on eggs for 6–10 min (3–42) up to 12 times per day (Okulewicz 1989). In Munich area, some ♂♂ took stints of 1–3 hrs in afternoon (Hermann 1983). For exact measurements of duration, and temperature fluctuations, see Schäfer (1968). YOUNG. Fed and cared for by both parents; brooded by ♀ continuously for *c.* 4 days (Ptushenko and Inozemtsev 1968; Ghiot 1976; Okulewicz 1989, which see for details). FLEDGING TO MATURITY. Fledging period 10–12 days (9–13); leaves nest 3–5 days before able to fly (Ulbricht 1975; Blümel 1982; Hermann 1983; Mal'chevski and Pukinski 1983; Okulewicz 1989). Perhaps less in Fenno-Scandia: 8–9 days in Swedish Lapland (Lennerstedt 1964); average 9·2 days (9–10, $n=21$) in south-west Finland (Haukioja 1970, which see for development of young). However, 10–12 days on Kola peninsula (Semenov-Tyan-Shanski and Gilyazov 1991). Young independent *c.* 20 days after leaving nest (Hermann 1983). Age of first breeding 1 year (Hornby 1971). BREEDING SUCCESS. In Nottinghamshire, of 315 nests (annual average from 6 years' data) started by 85 pairs, 20% produced fledged young; 93% of total failures caused by predation, 7% by floods and weather. 230 young left 63 successful nests, but only 60% of them fledged, giving 1·6 fledged young per breeding pair (0·4 per nest started, 2·2 per successful nest). Of 1846 eggs, 67% hatched (51–87% over 7 yrs), 50% (32–81%) produced young which left nest (38% lost to predation, 6% failed to hatch, 3% lost to floods or weather, 1% starved), but only *c.* 30% produced flying young. Predators included Carrion Crow *Corvus corone*, Magpie *Pica pica*, weasel *Mustela nivalis*, fox *Vulpes*, rat *Rattus*, and man. (Hornby 1971, which see for discussion of methods.) In Munich area, of 207 eggs laid by 19 pairs, 64% hatched and 42% produced fledged young (29% in one year, $n = 93$, 52% in another, $n = 114$); this gave 4·5 fledged young (2·7 and 6·2) per pair. Poor success due to flooding and loss of 2nd broods; predators included grass snake *Natrix natrix*, stoat *Mustela erminea*, slug *Arion*, and probably *P. pica*. (Hermann 1983.) In eastern Germany, of 116 nests in Oberlausitz, eggs hatched in 78% and young fledged in 69%, giving 2·5 per nest (Eifler and Blümel 1983); in Brandenburg area, 60% of 98 eggs in 20 nests hatched, and 20% produced fledged young, or just 1·0 per nest; almost all losses due to predation (Ulbricht 1975). In Wrocław region, 67% of 96 total losses caused by predation, 13% high water, 6% fire, 5% rain, and 2% human disturbance; maximum predation mid-June when 16% of 49 nests destroyed, mostly by White Stork *Ciconia ciconia*, *Corvus corone*, grass snake, and vole *Arvicola* (Okulewicz 1989). In eastern Belgium, up to 83% of eggs and young lost to weasels, cats, and Corvidae (Ghiot 1976). In south-west Finland, 53% of 90 clutches and 21% of 53 broods lost, mostly to predators; 9% of 585 eggs failed to hatch (Haukioja 1970, which see for productivity of different clutch sizes). For decreased success in Swedish Lapland caused by higher hatching failure by lake compared with nests away from water, possibly due to pollution, see Nyholm and Myhrberg (1977). For nest parasitism by Cuckoo *Cuculus canorus* in Norway, see Moksnes and Róskaft (1987).   BH

Plumages. (nominate *schoeniclus*). ADULT MALE. In fresh plumage (autumn), forehead and crown buff-brown or rufous-brown, broad deep black centres of feathers often partly visible. Supraloral stripe and supercilium buff-brown or yellow-buff, narrow above eye, wider above ear-coverts, often partly speckled black, in particular at rear end. Narrow eye-ring cream-buff or pale brown. Lore, upper cheek, and ear-coverts rufous-brown to black-brown, narrowly streaked buff below eye, more broadly buff or buff-brown on centre of ear-coverts, forming pale spot with darker stripe above and below. Hindneck and side of neck white, feathers with olive tips, sometimes bordered black subterminally; each feather white for about two-thirds of length, olive tip sharply contrasting, latter concealing white when plumage quite fresh (least so on lower side of neck), but white generally visible as mottled shawl from October–November onwards. Lower cheek white with slight cream or buff wash, forming distinct stripe, connected with white on side of neck. Upper mantle olive-brown or olive-grey with ill-defined sooty spots; feathers of lower mantle and scapulars with broad black centres and rufous-pink or ochre sides, latter warmer rufous-cinnamon at border of black. Back, rump, and upper tail-coverts buff-brown or olive-brown, some medium grey of feather-centres partly visible; back with narrow black streaks, longer upper tail-coverts (and sometimes shorter coverts and rump) with ill-defined dark brown centres. Chin and throat cream-buff, each feather with extensive black centre, partly exposed on side of chin and on entire throat, forming dark malar stripe and mottled black bib. Side of breast olive-grey with brown feather-tips; side of throat, chest, and flank white, feather-tips suffused cream-buff, marked with ill-defined grey- or rufous-brown dots on side of throat and on mid-chest, more sharply streaked with grey-brown, rufous, or black on side of chest and flank. Belly, vent, and under tail-coverts white, feather-tips with cream suffusion. Central pair of tail-feathers olive-brown, shading to blackish at shaft, outer web broadly fringed rufous; t2–t4 black with narrow rufous- or olive-grey outer fringe; t5 black with narrow white outer fringe and with white wedge on tip of inner web *c.* 20–40 mm long; t6 white with black shaft widening to dark grey blob on tip and with black wedge at extreme base, white wedge on inner web at least 40 mm long. Flight-feathers greyish-black, darkest on outer webs and tips; outer web of p9 and emarginated parts of p5–p8 narrowly fringed off-white, remainder of outer webs of primaries and tips of inner primaries contrastingly fringed pink-grey, outer webs and tips of secondaries more broadly vinous-brown or rufous-pink. Tertials black with broad and contrasting rufous-cinnamon fringes, cinnamon forming notch at centre of outer web of longest 2 feathers. Greater upper primary coverts and bastard wing greyish-black, coverts and shorter feathers of bastard wing with dull rufous fringes. Greater upper wing-coverts deep black, outer webs and tips broadly fringed rufous-cinnamon; lesser and median coverts deep rufous-cinnamon or rufous-chestnut, black of bases of median coverts partly visible. Under wing-coverts and axillaries white, shorter primary coverts mottled grey. *Bleaching and wear* have marked effect: rufous- or buff-brown fringes of top and side of head and of upperparts gradually bleach to yellow-ochre or cream, fringes of head, neck, and throat gradually wear off, fully exposing underlying black-and-white pattern (see Fennell and Stone 1976 for gradual change in appearance). By about March, before a partial moult of head, cap as well as throat to central chest mainly black with traces of cream feather-tips, bar on hindneck and side of neck and stripe on lower cheek fully white, but side of head and chin sometimes still largely brown, with contrastingly pale buff supercilium and stripe from below eye to centre of ear-coverts; after partial pre-breeding moult in spring, top and side of head as well as chin to centre of chest deep black, with narrow buff feather-tips when freshly-moulted, strongly contrasting with white of neck, lower cheek, and side of throat. By about April, mantle and scapulars broadly streaked black and grey-buff or yellow-ochre, back to upper tail-coverts grey with black streaks on back and olive-brown spots on rump and tail-coverts; later on, mantle and scapulars mainly black with narrow cream or off-white streaks, back to upper tail-coverts dusky grey, underparts (below black bib) dirty white with narrow sharply contrasting black or brown streaks on side of breast and flank;

rufous fringes of tail, flight-feathers, tertials, and greater coverts bleached to pale buff or off-white, partly worn off, but basal outer webs of secondaries as well as lesser and median upper wing-coverts still deep rufous, colour in sharp contrast with largely black mantle and scapulars. ADULT FEMALE. In fresh plumage, rather like adult ♂ in fresh plumage and sometimes hard to distinguish. Centre of forehead and crown olive-grey or buff-grey, sides warmer rufous-brown; centres of feathers either rather narrowly marked dull black, showing as short dark streaks on crown, but black on feather-centres sometimes as broad and extensive as in adult ♂. Side of head as adult ♂, in some birds with black specks on rear of supercilium and with some black exposed on upper cheek and on border of ear-coverts, as in adult ♂, but supercilium in others uniform cream-buff and lore, upper cheek, and ear-coverts without black. Stripe on lower cheek and bar at side of neck white, but feather-tips often more extensively suffused olive than in ♂, white less conspicuous. Hindneck either uniform olive-brown or-grey, without concealed white, but some birds have basal half of part of feathers of hindneck white (less extensive and contrasting than in ♂). Chin and throat buff, bordered at each side by black malar stripe (often not reaching base bill); malar stripe broader and mottled buff towards rear, merging into buff-and-sooty streaked gorget across chest; centres of feathers of throat and central chest sometimes black, sometimes partly visible. Remainder of body, tail, and wing as in adult ♂. *In worn plumage* (spring), head and upperparts markedly darker, but head highly variable; in darkest birds, top and side of head as well as throat to mid-chest black, cap with traces of brown fringes, eye-ring, supercilium, and central ear-coverts mottled black and buff-brown (in ♂, uniform black), stripe on lower cheek and bar at side of neck cream-white, narrower, less uniform and less contrasting than in adult ♂, chin mainly buff, central throat and chest mottled black-and-buff; black of throat and mid-chest extends into narrow black streaks on off-white ground on side of chest, side of breast, and upper belly (in ♂, virtually unstreaked); hindneck mottled off-white, black, and olive (in ♂, fully white); in paler birds, mid-crown closely streaked black-and-grey, side black-and-brown; super-cilium and eye-ring buff, lore, stripe on upper cheek, and stripe on upper ear-coverts sepia-brown, patch from below eye to centre of ear-coverts mottled buff and pale brown; stripe on lower cheek and bar at side of neck cream-white; ground-colour of underparts off-white, marked with bold mottled black malar stripe and sharp dark brown or rufous-brown streaks on chest, side of breast, and flank; in all birds, worn upperparts, tail, and wing as in worn adult ♂. NESTLING. Down sooty or dull black, rather long, restricted to sparse tufts on cap and upperparts, with traces on wing, thigh, and vent. For development, see Blümel (1982). JUVENILE. Rather like adult in fresh plumage, but forehead more evenly streaked black and buff or dull brown; ground-colour of remainder of upperparts paler, cream-buff or ochre, less dull rufous than in fresh adult, streaks bolder, blacker, more strongly contrasting; streaks on rump sharp, deep black, *c.* 1 mm broad, not as grey and obscure as in adult. Side of head and neck as adult ♀, but upper cheek and rear of ear-coverts darker, more uniform deep black; malar stripe heavy, uniform black; black pattern sharply contrasting with buff or cream-buff supercilium, lower cheek, and bar along side of neck. Ground-colour of underparts yellow-ochre or buff, merging into buff-white on belly and vent (in adult, warmer buff on chest, purer white on belly), marked with sharp and narrow deep black streaks on chest, side of breast, flank, upper belly, and side of belly (in adult, streaks more diffuse, less deep black, bordered by rufous or black partly replaced by rufous-brown); also, adult usually with a black-spotted gorget across upper chest, in contrast to

juvenile). Lesser, median, and greater upper wing-coverts black, fringes contrastingly buff, *c.* 1 mm wide, black ending in point at tip of each covert (in adult, lesser coverts uniform deep rufous, median and greater with broad rufous-cinnamon fringes). For tail, see 1st adult, below. Sexes virtually similar, but hindneck of ♂ buff with dull black mottling, paler than crown and upper mantle, hindneck of ♀ more heavily streaked black and brown, similar to crown and mantle; also, side of neck of ♀ speckled black (virtually uniform buff in ♂), and ground-colour of side of head and neck and of underparts tends to be warmer buff in ♂, paler buff or grey-buff in ♀. FIRST ADULT MALE. Like adult ♂, but juvenile tail, flight-feathers, greater upper primary coverts, and sometimes a few outer greater coverts or tertials retained; if still present, juvenile outer greater coverts or tertials contrast in colour, shape, and degree of abrasion with neighbouring new feathers; juvenile tail-feathers pointed at tip, less broadly rounded than in adult, more worn at same time of year, t1 sometimes contrastingly newer (unlike adult); occasionally, entire juvenile tail replaced, and ageing then difficult; when on breeding grounds, from about April, tail and primaries become heavily worn and ageing often impossible. Head and body as in adult, but brown fringes of feathers of head and neck sometimes slightly broader, black and white more concealed. FIRST ADULT FEMALE. Like adult ♀, but part of juvenile feathering retained, as in 1st adult ♂. Head and body as in adult ♀ but concealed black on cap and throat and white on neck generally restricted, reduced to narrow black streaks on cap and virtually absent on neck and throat, but a few are as dark on head and throat in worn plumage as some adult ♀♀.

Bare parts. ADULT, FIRST ADULT. Iris brown or dark brown. Upper mandible leaden-grey, cutting edges and lower mandible pink-horn or whitish-horn with variable grey tinge, flanges of gape yellow in 1st autumn; in ♂, bill darkens to black February-March, paler again August-September. Leg and foot pink-brown, flesh-grey, dull flesh-brown, horn-brown, or dark brown, darkest on toes (not as invariable dark brown as suggested by Mills 1982). (Poulsen 1950; RMNH, ZMA; R Reijnders.) NEST-LING. Mouth pale red or deep pink, spurs and tip of tongue whitish; gape-flanges yellow-white or pale yellow. Bare skin flesh-pink at hatching, gradually darkening to grey; bill and leg pale flesh. (Heinroth and Heinroth 1924-6; Witherby *et al.* 1938; Blümel 1982). JUVENILE. Bill flesh-grey with dark grey culmen and tip. Leg and foot flesh-grey, darker on joints and toes. (RMNH, ZMA.)

Moults. ADULT POST-BREEDING. Complete; primaries descendent. In England, starts with p1 mainly 10 July to 15 August, completed with regrowth of p9 mainly late August to late September; duration of moult *c.* 60 days, median dates 17 July to 15 September in ♂, 21 July to 19 September in ♀ (Bell 1970); in larger sample from Britain generally, moult late June to early October, starting mainly early July to mid-August, duration of primary moult *c.* 60 days, or 50 days in earlier birds and *c.* 33 days in later ones (Ginn and Melville 1983). In Britain, southern Germany, and Austria, primary moult mainly mid-July to late September; body mainly from early August onwards, most intense at primary moult score 15–35, completed at same time as regrowth of p9 (moult score 45); head starts at primary moult score 5–20, most intense at 25–40, often completed after regrowth of p9; secondaries start with shedding of s1 at primary score 20·0 (4·29; 725), mainly 13–26, completed with s6 at same time as p9 or slightly later; tail starts with t1 at primary score 10·2 (5·10; 682), mainly 6–14, fully regrown at *c.* 34–45 (on average, *c.* 39); tertials start at score 8·2 (5·78; 625), mainly 0–20 (Kasparek

1979, which see for other details). In central Sweden, p1 shed 1-25 July (rarely, from 20 June or up to 15 August), primary moult completed with regrowth of p9 late August to mid-September (Sondell 1977, which see for moult of other tracts). Small sample of moulting birds examined from Netherlands fell within British data (RMNH, ZMA). Primary moult duration in south of Britain c. 45 days, in north c. 46 days, on average starting 25 July; in western France c. 45 days, average start c. 15 July; in eastern Austria c. 42 days, average start 27 July; in central Sweden 55-60 days, average start 12 July; in southern Finland 56 days in ♂, 54 days in ♀, average start 8 July; in northern Finland (66-70°N) 50 days in ♂, 43 days in ♀, average start 15 July (Kasparek 1980, which see for many details); see also Haukioja (1971c) and Lehikoinen and Niemelä (1977). ADULT PRE-BREEDING. Partial: mainly confined to forehead, side of head, chin, and mid-throat, sometimes cap, chin to upper chest, and part of neck also, rarely some feathering of mantle or a few tail-feathers; mainly February-March (Witherby 1916; Witherby et al. 1938; Poulsen 1950; Bell 1970; RMNH, ZMA). POST-JUVENILE. Partial: head, body, lesser, median, and often many or all greater upper wing-coverts, many or all tertials, frequently t1, sometimes all tail. Starts shortly after fledging, lesser coverts first. (RMNH, ZMA.) Early-hatched birds most likely to moult tail, those hatched May in moult August-September; moult of head sometimes up to October (Bell 1970). In sample from Netherlands (RMNH, ZMA), moult just started in juveniles examined late June to mid-August, in full moult late July to mid-September, almost completed early August to early October (rarely, late October). Moult perhaps occasionally complete (Ginn and Melville 1983).

Measurements. ADULT, FIRST ADULT. All data taken by W R R de Batz and C S Roselaar (BMNH, RMNH, ZMA, with extra Italian birds supplied by C Violani). Nominate *schoeniclus*. Wing, bill to nostril, and depth of bill at base from: (1) southern Sweden, eastern Germany, Poland, and south-west Belorussia, breeding; (2) Netherlands, breeding; (3) Netherlands and Germany winter; (4) Iberia, North Africa, and Balkans, winter; (5) Italy and Sardinia, September-April. Other measurements combined; skins. Bill (S) to skull, width is width of bill at base; exposed culmen on average 3·2 less than bill (S).

WING (1)	♂	81·4 (2·79; 16)	77-85	♀	75·4 (2·96; 7)	71-79
(2)		79·8 (1·99; 44)	76-82		74·3 (2·72; 24)	70-78
(3)		80·7 (2·38; 73)	76-85		75·8 (2·73; 36)	71-80
(4)		82·0 (2·81; 19)	77-87		76·3 (2·43; 28)	72-81
(5)		82·2 (3·39; 24)	78-86		76·7 (2·31; 42)	73-81
TAIL		63·8 (2·32; 34)	60-69		60·6 (1·64; 18)	58-64
BILL (S)		12·2 (0·66; 158)	10·9-13·5		11·8 (0·77; 130)	10·4-13·2
BILL (1)		7·3 (0·37; 16)	6·7-7·8		6·8 (0·29; 6)	6·4-7·2
(2)		7·4 (0·32; 28)	7·0-8·1		7·1 (0·47; 20)	6·6-7·6
(3)		7·1 (0·38; 42)	6·4-7·8		6·9 (0·30; 26)	6·4-7·4
(4)		7·0 (0·35; 19)	6·4-7·6		6·9 (0·33; 28)	6·3-7·4
(5)		7·0 (0·39; 24)	6·2-7·6		6·8 (0·35; 42)	6·0-7·3
DEPTH (1)		5·3 (0·18; 16)	5·1-5·6		5·1 (0·26; 6)	4·8-5·2
(2)		5·3 (0·28; 28)	4·9-5·7		5·1 (0·27; 20)	4·8-5·5
(3)		5·2 (0·31; 42)	4·9-5·8		5·2 (0·31; 26)	4·8-5·6
(4)		5·3 (0·23; 19)	4·9-5·5		5·2 (0·22; 28)	4·9-5·5
(5)		5·4 (0·27; 24)	4·7-5·8		5·2 (0·34; 42)	4·6-5·6
WIDTH		5·2 (0·24; 154)	4·6-5·6		5·1 (0·24; 129)	4·6-5·6
TARSUS		20·5 (0·87; 162)	18·7-22·0		20·1 (0·94; 130)	18·0-21·7

Sex differences significant, except for bill (4) and bill depth (1), (3), and (4).

Bill to nostril and bill depth of ♂ from (1) Lapland, (2) south of Sweden and Finland, (3) St Petersburg area (Matoušek 1969).

(1) BILL (N)	6·8 (0·2; 32) 6·4-7·2	DEPTH 5·3 (0·2; 32) 4·7-5·8

(2)	6·9 (0·2; 17) 6·4-7·3	5·3 (0·2; 17) 5·0-5·6
(3)	7·1 (0·2; 25) 6·9-7·6	5·4 (0·3; 25) 5·1-6·0

Wing; live birds. (1) England, winter (Bell 1970). (2) Belgium, spring migrants (Collette 1972). (3) Hamburg area, Germany, migrants (Eck 1990; see also Eck 1985b). (4) Switzerland, spring migrants (Christen 1984). (5) South-east France, winter (G Olioso). Finland: (6) adult, (7) 1st adult (Haukioja 1969).

(1)	♂	78·0 (— ; 404)	—	♀	72·8 (— ; 209)	—
(2)		78·6 (2·08; 275)	71-83		73·2 (1·96; 225)	67-78
(3)		80·5 (2·15; 1538)	—		74·6 (1·93; 1437)	—
(4)		79·4 (— ; 253)	73-85		73·6 (— ; 300)	68-81
(5)		81·3 (2·04; 217)	75-86		75·9 (1·75; 159)	71-80
(6)		80·9 (1·38; 36)	78-84		75·1 (1·31; 22)	72-77
(7)		78·0 (1·47; 32)	74-81		73·3 (1·24; 53)	71-76

Western France: bill to nostril, ♂ 7·6 (0·34; 15) 7·0-8·0, ♀♀ 6·8, 7·0 (Mayaud 1933a). See also Schekkerman (1986b).

E. s. passerina. Wing, bill depth, and bill width of birds from (1) lower Yenisey valley, and (2) middle Ob' valley, latter probably in part tending to *pallidior* from further south; other measurements combined; skins. Bill (N) is bill to nostril.

WING (1)	♂	82·3 (1·46; 10)	80-84	♀	78·1 (0·96; 4)	77-79
(2)		81·2 (3·50; 5)	78-84		77·7 (2·52; 3)	75-81
BILL (S)		11·5 (0·44; 15)	10·4-12·5		11·4 (0·60; 7)	10·4-12·5
BILL (N)		6·6 (0·30; 15)	6·2-7·1		6·4 (0·51; 7)	5·4-7·1
DEPTH (1)		5·0 (0·24; 10)	4·8-5·3		5·0 (0·24; 4)	4·8-5·2
(2)		5·2 (0·17; 5)	5·1-5·4		5·0 (— ; 3)	5·0-5·0
WIDTH (1)		4·9 (0·18; 10)	4·6-5·2		4·8 (0·22; 4)	4·6-5·1
(2)		5·2 (0·18; 5)	4·9-5·4		5·2 (0·21; 3)	5·0-5·4
TARSUS		20·0 (0·51; 15)	19·4-20·8		19·7 (0·83; 7)	18·7-20·4

Sex differences significant for wing.

Yamal peninsula (north-west Siberia), wing: ♂ 82·9 (23) 79-84, ♀ 77·1 (21) 75-79 (Danilov et al. 1984). North-west Siberia, bill depth: ♂ 5·2 (17) 5·0-6·0, ♀ 5·0 (3) 4·5-5·3 (Dementiev and Gladkov 1954; apparently including intermediates with *pallidior*).

E. c. parvirostris. North-central Siberia, bill depth: ♂ 5·2 (6) 5·0-5·3, ♀ 4·8 (4) 4·5-5·0 (Dementiev and Gladkov 1954).

E. c. stresemanni. Eastern Austria, Hungary, and northern Yugoslavia, breeding; skins.

WING	♂	82·5 (2·49; 23)	78-86	♀	74·8 (0·84; 5)	74-76
BILL (S)		12·8 (0·73; 23)	11·4-13·5		11·8 (0·57; 5)	11·4-12·5
BILL (N)		7·6 (0·34; 23)	7·1-8·3		7·2 (0·40; 5)	6·8-7·8
DEPTH		6·0 (0·25; 23)	5·7-6·4		5·9 (0·16; 5)	5·8-6·3
WIDTH		5·7 (0·28; 23)	5·2-6·3		5·6 (0·14; 5)	5·5-6·3
TARSUS		22·1 (0·83; 23)	20·8-23·9		21·8 (1·58; 5)	20·8-25·0

Sex differences significant for wing and bill. In another sample of ♂♂, bill (N) 7·7 (0·5; 36) 6·2-8·5, bill depth 6·0 (0·3; 36) 5·2-6·5 (Matoušek 1968). See also Jirsík (1951) and Beretzk et al. (1962). For Austrian winter sample which includes nominate *schoeniclus* and for sample from Greece, October, probably consisting of mixture of nominate *schoeniclus*, *strèsemanni*, *ukrainae*, and *tschusii*, see Dornberger (1979).

E. s. ukrainae. Northern Ukraine to western side of lower Volga; skins (including ♂ from Kumerloeve 1967b): wing, ♂ 81·0 (2·67; 5) 77·5-84, ♀ 75·9 (3) 74-78; bill (S) 12·5 (0·78; 8) 11·5-13·5, bill (N) 7·0 (0·42; 7) 6·7-7·7, bill depth 6·0 (0·33; 8) 5·6-6·5, bill width 5·6 (0·13; 7) 5·5-5·8, tarsus 21·1 (0·79; 7) 20·0-22·0 (W R R de Batz, C S Roselaar). In sample from Khar'kov, Ryazan', Voronezh, and Minsk (western Russia and Belorussiya), ♂: bill (N) 7·2 (0·4; 37) 6·6-8·2, bill depth 5·7 (0·3; 37) 5·2-6·5 (Matoušek 1968).

E. s. pallidior. Tarim basin (Sinkiang, China), winter: skins.

WING	♂	83·5 (2·73; 13)	79-88	♀	79·6 (3·00; 4)	76-83
TAIL		65·2 (2·53; 11)	62-70		62·8 (2·08; 3)	60-65

BILL (S)	12·1 (0·43; 13) 11·6–12·8	12·2 (0·52; 4)	11·7–12·6
BILL (N)	7·0 (0·33; 13) 6·4–7·6	7·2 (0·30; 4)	6·8–7·7
DEPTH	5·4 (0·19; 13) 5·1–5·7	5·5 (0·05; 4)	5·4–5·8
WIDTH	5·2 (0·26; 13) 4·8–5·6	5·5 (0·15; 4)	5·3–5·8
TARSUS	20·2 (0·67; 13) 19·4–21·3	20·0 (0·52; 4)	19·7–20·6

Sex differences significant for wing.

Probably this race. Iraq, winter: wing, ♂ 82·6 (1·90; 7) 79–85, ♀ 75·7 (1·96; 59) 72–80 (Čtyroky 1987). Lake Chany (south-west Siberia), breeders and migrants: wing, ♂ 80·4 (2·09; 145) 74–85, ♀ 75·1 (1·84; 114) 70–83, juvenile 77·1 (3·20; 48) 72–84 (Havlín and Jurlov 1977). South-west Siberia: bill depth, ♂ 5·3 (18) 5·0–5·8, ♀ 5·2 (8) 4·7–5·5 (Dementiev and Gladkov 1954).

E. s. pyrrhulina. Japan, live birds: (1) adult, (2) 1st adult (Dornberger 1983).

(1)	♂	84·0 (1·95; 18)	78–86	♀ 78·9 (1·30; 15) 76–81
(2)		82·6 (2·15; 30)	77–86	77·4 (1·98; 29) 73–82

Japan: wing, ♂♂ 76, 86, ♀ 77·6 (1·11; 5) 76–79; bill (S) 11·8 (0·57; 7) 11·2–12·5, bill (N) 7·1 (0·20; 7) 6·7–7·3, bill depth 5·9 (0·23; 7) 5·6–6·2, bill width 5·7 (0·16; 7) 5·5–6·0, tarsus 20·9 (0·40; 7) 20·3–21·8 (W R R de Batz, C S Roselaar). South-east Siberia, bill depth: ♂ 5·5 (8) 5·2–5·8 (Dementiev and Gladkov 1954). See also Kaneko (1976).

E. s. tschusii. North-east Rumania, southern Ukraine (including Crimea), and Sarpa area (west of lower Volga); wing, ♂ 85·0 (1·20; 5) 83–86, ♀ 80·7 (3) 78–84; bill (S) 12·9 (0·53; 8) 12·5–13·5, bill (N) 7·4 (0·34; 8) 6·9–8·0, bill depth 6·9 (0·35; 8) 6·4–7·4, bill width 6·6 (0·35; 8) 6·0–6·9, tarsus 21·8 (0·96; 8) 20·8–22·9 (W R R de Batz, C S Roselaar). Moldavia, southern Ukraine and Crimea: bill depth, ♂ 6·2 (6) 6·1–6·8, ♀ 6·0 (4) 5·8–6·0 (Dementiev and Gladkov 1954; likely to include *ukrainae*).

E. s. incognita. Central Kazakhstan: wing, ♂ 76–87, ♀ 75–85; bill (N), ♂ 7·5–10·6, ♀ 8·2–10·0; bill depth, ♂ 5·5–7·5, ♀ 5·0–7·0 (n = 28 in ♂, 20 in ♀) (Korelov et al. 1974); bill depth, ♂ 6·1 (19) 5·1–6·8, ♀ 5·8 (8) 5·1–5·9 (Dementiev and Gladkov 1954; likely to include some *pallidior*).

E. s. witherbyi. Mallorca, wing: ♂ 78–80 (n = 11), ♀ 70–74 (n = 4) (Jordans 1924); ♂ 80·3 (7) 79–82 (Vaurie 1958a). Camargue (France), winter: wing, ♂ 78·3 (2·31; 46) 73–84, ♀ 72·5 (1·71; 25) 69–76 (G Olioso). Portugal, wing: ♂ 75·8 (3·30; 4) 72–79, ♀ 74·5 (Steinbacher 1930a; BMNH, ZMA).

E. s. intermedia. Italy, April–September; skins.

WING	♂	83·7 (2·06; 14) 80–87	♀ 79·5 (1·84; 9) 76–82
BILL (S)		13·6 (0·86; 14) 12·3–14·5	12·6 (0·34; 9) 12·0–13·5
BILL (N)		7·8 (0·52; 14) 7·3–8·6	7·2 (0·32; 9) 6·9–7·7
DEPTH		7·4 (0·44; 14) 7·0–7·8	7·2 (0·42; 9) 6·9–7·7
WIDTH		7·1 (0·44; 14) 6·6–7·9	7·1 (0·25; 9) 6·6–7·5
TARSUS		22·5 (1·12; 14) 21·3–24·4	22·0 (0·62; 9) 20·8–22·9

Sex differences significant for wing and bill length. Dalmatia (western Yugoslavia), ♂: bill (N) 7·5 (0·3; 18) 6·9–8·3, bill depth 7·0 (0·3; 18) 6·6–7·6 (Matoušek 1968).

E. s. reiseri. Southern Yugoslavia and Greece: wing, ♂ 86·7 (11) 82–93 (Vaurie 1958a); wing, ♂ 86·7 (3·77; 9) 82–92, ♀ 79·7 (3) 78–83 (Stresemann 1920; Makatsch 1950). Turkey, May–June, ♂: wing 85·3 (2·32; 9) 82·5–89; exposed culmen 11·0 (0·84; 9) 9·5–12·3, bill depth 8·1 (0·27; 9) 7·7–8·5 (Kumerloeve 1964b, 1969b). Perhaps this race, Cyprus, winter: wing, ♂ 81·0 (14) 77–86·5, ♀ 77·8 (54) 73–81 (Flint and Stewart 1992).

E. s. caspia. Northern Iran, skins: wing, ♂♂ 85, 86, ♀ 80·7 (3) 78–83; bill (S) 13·7 (0·45; 5) 13·3–14·5, bill (N) 7·7 (0·54; 5) 7·3–8·6, bill depth 7·2 (0·16; 5) 7·0–7·4, bill width 7·1 (0·28; 5) 6·8–7·4, tarsus 22·5 (0·60; 5) 21·5–23·4 (W R R de Batz, C S Roselaar).

E. s. korejewi. Seistan, south-west Iran: wing 80, 84; bill (S) 13·5, 13·5; bill (N) 7·8, 8·1; bill depth 7·2, 7·6; bill width 7·2, 7·4; tarsus 22·4, 22·9 (W R R de Batz, C S Roselaar).

E. s. pyrrhuloides. Volga delta eastward through southern Kazakhstan and Turkmeniya to Lake Balkhash; skins.

WING	♂ 90·0 (2·11; 16) 86–94	♀ 81·7 (1·42; 6)	79–84
BILL (S)	13·0 (0·65; 16) 11·9–13·5	13·0 (0·57; 6)	12·5–13·5
BILL (N)	7·5 (0·28; 16) 7·0–8·0	7·2 (0·24; 6)	6·9–7·4
DEPTH	8·0 (0·54; 16) 7·7–8·4	7·8 (0·57; 6)	7·6–8·4
WIDTH	7·9 (0·46; 16) 7·1–8·7	7·5 (0·31; 6)	7·1–7·8
TARSUS	23·1 (0·89; 16) 21·8–25·0	22·0 (1·22; 6)	20·7–23·5

Sex differences significant for wing, bill (N), and tarsus.

Southern Kazakhstan (probably including some *harterti*): bill depth, ♂ 7·3–9·3 (n = 74), ♀ 7·2–9·0 (n = 25) (Korelov et al. 1974).

E. s. harterti. Zaysan basin, skins of ♂: wing 87·6 (1·65; 5) 85–90, bill (S) 13·1 (0·57; 5) 12·5–13·5, bill (N) 7·6 (0·25; 5) 7·2–7·8, bill depth 7·5 (0·47; 5) 7·2–7·8, bill width 7·0 (0·20; 5) 6·8–7·2, tarsus 21·8 (1·04; 5) 20·5–23·0 (W R R de Batz, C S Roselaar). Mongolia, wing: ♂ 88·7 (3) 87–92, probably ♀ 85·2 (2·63; 4) 83–88 (Piechocki and Bolod 1972).

E. s. centralasiae. Tarim basin east to Lop Nor (Sinkiang, China), skins.

WING	♂ 84·7 (3·17; 16) 80–90	♀ 78·6 (2·62; 8)	75–83
TAIL	69·6 (1·55; 8) 67–73	70·0 (— ; 3)	64–75
BILL (S)	11·8 (0·84; 16) 11·0–13·5	11·9 (0·78; 8)	11·6–13·3
BILL (N)	7·0 (0·31; 16) 6·6–7·5	6·8 (0·42; 8)	6·4–7·4
DEPTH	7·3 (0·36; 16) 6·8–7·6	7·1 (0·37; 8)	6·6–7·4
WIDTH	6·9 (0·33; 16) 6·3–7·4	6·7 (0·23; 8)	6·4–7·1
TARSUS	22·5 (0·92; 16) 21·2–23·4	22·2 (0·95; 8)	20·3–23·9

Sex differences significant for wing.

Tarim basin, wing: ♂ 84·2 (3·01; 10) 81–91 (Hellmayr 1929).

E. s. zaidamensis. Tsaidam basin (China), wing: ♂ 90 (5) 87–92, ♀♀ 81, 83 (Vaurie 1972); ♀, bill (N) 6·5, bill depth 6·9, bill width 6·5 (W R R de Batz).

Weights. ADULT, FIRST ADULT. Nominate *schoeniclus*. Netherlands, southern Sweden, eastern Germany, and Czechoslovakia, combined: (1) April–September, (2) October–March (Scott 1965b; Havlín and Havlínová 1974; Eck 1985b; ZMA). (3) Netherlands, exhausted, all year (ZMA). South-east France: (4) October–November, (5) December–January, (6) February (G Olioso). (7) Belgium, spring migrants (Collette 1972). *E. s. passerina.* (8) Yamal (north-west Siberia), summer (Danilov et al. 1984). *E. s. pallidior* (probably mainly this race). (9) Northern Kazakhstan (Korelov et al. 1974). (10) Lake Chany (south-west Siberia), August–September (Havlín and Jurlov 1977). *E. s. pyrrhulina.* (11) Japan (Dornberger 1983). *E. s. incognita* (probably mainly this race). (12) Central Kazakhstan (Korelov et al. 1974). *E. s. witherbyi.* (13) Camargue (France), winter (G Olioso). *E. s. reiseri.* (14) Greece and Turkey, May–June (Makatsch 1950; Kumerloeve 1964b, 1969b). (15) Probably this race, Cyprus, winter (Flint and Stewart 1992). *E. s. pyrrhuloides* (probably including *harterti*). (16) Southern Kazakhstan (Korelov et al. 1974).

(1)	♂	19·7 (1·77; 27) 16·5–21·5	♀ 17·9 (1·75; 18) 15·5–21·6
(2)		20·2 (1·90; 19) 16·5–23·1	18·3 (2·55; 5) 16·2–22·0
(3)		13·6 (0·45; 8) 13·0–14·2	11·6 (1·74; 6) 10·0–14·5
(4)		20·6 (1·31; 100) 18·0–25·0	18·4 (1·32; 93) 15·5–22·5
(5)		21·9 (1·47; 91) 18·0–25·0	19·2 (1·28; 50) 17·0–22·5
(6)		21·1 (1·67; 26) 18·5–24·5	18·1 (1·24; 16) 16·0–20·0
(7)		21·2 (— ; 307) 16·6–26·7	18·4 (— ; 232) 14·5–23·6
(8)		20·7 (— ; 18) 18·5–24·3	18·9 (— ; 20) 17·0–22·6
(9)		20·9 (— ; 21) 16·5–22·8	18·3 (— ; 16) 16·0–20·8
(10)		19·7 (1·20; 153) 16·1–24·4	17·2 (0·99; 117) 14·2–20·6

(11)	22·5 (1·47; 47) 19·8–26·5	20·3 (1·53; 42) 17·3–23·9	
(12)	23·9 (— ; 8) 20·0–27·0	20·5 (— ; 1) —	
(13)	20·1 (1·34; 46) 17·0–24·0	17·1 (0·89; 25) 15·5–18·5	
(14)	26·1 (1·88; 12) 24·0–30·0	— (— ; —) —	
(15)	21·7 (— ; 13) 19·0–24·5	20·5 (— ; 51) 18·2–26·0	
(16)	26·0 (— ; 20) 22·3–30·5	21·1 (— ; 9) 18·6–23·6	

Nominate *schoeniclus*. In Finland, average per 10-day period fairly constant throughout March–October in ♂, 19·8–20·2 (occasionally, 19·1–20·5); in ♀, average 17·6–18·5 11 April to 10 May, 18·7–19·8 11 May to 19 June (when laying), 17·6–18·2 19 August to 27 September, 18·8 late September and early October (when migrating) (Haukioja 1969, which see for graphs and table). In England, rather constant April–August (♂ on average 20·0, ♀ 18·7, *n*=72); heavier in winter, ♂ *c.* 21·5, ♀ *c.* 19·3, heavier when colder (Fennell and Stone 1976); average before moult, ♂ 20·9 (5), ♀ 18·7 (11); during moult, ♂ 21·2 (34), ♀ 19·7 (21) (Bell 1970); see also Scott (61), Reynolds (1974), and Warrilow *et al.* (1978).

E. s. stresemanni or *tschusii*. North-west Bulgaria, July: ♂ 22 (Niethammer 1950). *E. s. intermedia* or *stresemanni*. Greece, winter: ♂ 22 (3) 21–23, ♀ 18 (Makatsch 1950).

E. s. pyrrhuloides. Iraq, February: ♀ 21·8 (Čtyroký 1987, 1989).

For various thin- and thick-billed races in winter quarters, see also Schüz (1959), Kumerloeve (1967*b*), Dornberger (1979), and Flint and Stewart (1992).

NESTLING. For growth curve, see Blümel (1982).

JUVENILE. For weight in relation to age, Finland, see Haukioja (1969). Juvenile *pallidior*, Lake Chany, 18·5 (1·54; 48) 16·1–22·5 (Havlín and Jurlov 1977); juvenile *harterti*, Mongolia, 22·3 (1·70; 7) 20–24 (Piechocki and Bolod 1972).

Structure. Wing rather short, broad at base, tip rounded. 10 primaries: in nominate *schoeniclus*, p7 longest, p8 0(–1) shorter, p9 2·7 (10) 1·5–3·5, p6 0–0·5(–1), p5 1–2·5, p4 5–8, p3 8–14, p2 11–17, p1 15·0 (10) 13–19; p10 strongly reduced, 2 shorter to 2 longer than reduced outermost upper primary covert, 7–11 shorter than longest primary covert, 48–59 shorter than p7. In *pallidior* and *pyrrhulina*, p9 2·9 (15) 1–6 shorter than p7, p1 16·0 (15) 14–19; in *stresemanni*, *tschusii*, and *intermedia*, p9 2·6 (15) 1·5–4 shorter than p7, p1 14·2 (15) 12–17; in *pyrrhuloides*, p9 3·3 (10) 1·5–5 shorter, p1 16·2 (10) 14–18; in *centralasiae*, p9 4·3 (6) 2–6 shorter, p1 12·5 (6) 11–14. Outer web of p5–p8 emarginated, inner web of p6–p9 with notch (often faint). Tip of longest tertial reaches to tip of about p4 when not worn. Tail rather long, tip slightly forked; 12 feathers, t4–5 longest, t1 4·5 (20) 3–9 shorter, t6 0–3 shorter. Bill short, less than half head length. Bill shape highly variable, depending on race: base relatively slender, culmen (almost) straight or slightly curved, and lower mandible deeper than upper mandible in northern races, nominate *schoeniclus*, *passerina*, and *parvirostra*; slightly wider and deeper and culmen gently decurved in *stresemanni*, *ukrainae*, *pallidior*, and *pyrrhulina*; still heavier with both mandibles about equal in depth in *tschusii* and *incognita*; very heavy with strongly arched gonys and deep upper mandible in remaining southern races, most markedly in *reiseri* and *pyrrhuloides*, with tip slightly pointed in most but markedly blunt (bill to nostril equal to or less than bill depth at base) in *zaidamensis*, *harterti*, *centralasiae*, *pyrrhuloides*, and *reiseri*; see Measurements for depth and width of bill. Cutting edges, nostril, and bristles as Yellowhammer *E. citrinella*. Leg and foot rather short and slender. Middle toe with claw 18·6 (20) 16·5–20 (all races combined); outer toe with claw in nominate *schoeniclus* *c.* 74% of middle with claw, inner *c.* 73%, hind *c.* 78%; outer toe relatively longer in races of *pyrrhuloides* group.

Geographical variation. Marked, but largely clinal; involves depth of ground-colour and width of dark streaking on upperparts and flanks of both sexes and on chest of ♀, as well as depth and width of bill at base, relative depth of upper and lower mandible, shape of culmen, and (to some extent) size (wing, tail, tarsus, weight) and wing formula. Thoroughly investigated by Portenko (1929) and Steinbacher (1930*a*, *b*), followed by Hartert and Steinbacher (1932–8) and Johansen (1944), with important studies in part of species' range by Jirsík (1951), Vaurie (1956*d*, 1958*a*), Beretzk *et al.* (1962), Matoušek (1968, 1969, 1971), and Matoušek and Jablonski (1969). Races and groups recognized here based on data of these authors as well as on *c.* 600 specimens examined (BMNH, RMNH, ZMA, ZFMK: C S Roselaar). 2 subspecies-groups recognized: thin-billed *schoeniclus* group in north, thick-billed *pyrrhuloides* group in south. Among birds examined, very few had bill depth 6·6–6·7 mm and bill width 6·3–6·4 mm, but many birds above and below this, hence division into 2 groups rather than 3 as advocated by (e.g.) Hartert (1921–2) and Vaurie (1959) (however, very few *incognita* examined, a race which appears to be intermediate between groups). Within each main group, 2 sub-groups recognizable, also based on bill size (see also Structure), but these less well-defined than main groups, apparently grading into each other. Sub-groups of thick-billed group comprise: (1) nominate *schoeniclus* in western and northern Europe, *passerina* in north-west Siberia, and *parvirostra* in north-east Siberia, with bill depth mainly 4·7–5·7 and bill width 4·5–5·6 mm; (2) *stresemanni* in Carpathian basin, *ukrainae* from northern Moldavia and northern Ukraine east to Saratov and Kuybyshev on middle Volga, *pallidior* in south-east European Russia (east of middle Volga) and south-west Siberia east to Yenisey and foot of Altai, and *pyrrhulina* from Lake Baykal east to lower Amur, Manchuria, and Japan, as well as Kamchatka, with bill depth mainly 5·8–6·5 mm and bill width 5·2–6·2 mm. In thick-billed group, sub-groups comprise: (1) *tschusii* from north-east Bulgaria and eastern Rumania through southern Moldavia and Ukraine to Crimea and Sarpa area on lower Volga (*c.* 48–50°N), *incognita* in zone from Orenburg area (south-east European Russia) through central Kazakhstan between *c.* 49° and *c.* 51°N to *c.* 76°E, *witherbyi* in Iberia, Balearic Islands, southern France, and (at least formerly) North Africa and Sardinia, *intermedia* on Corsica, Italy, and Dalmatian coast of Yugoslavia, *caspia* in eastern Transcaucasia and northern Iran and (perhaps this race) south-west Iran and Syria, *korejewi* in south-east Iran, *centralasiae* in Tarim basin (China), and *harterti* in Zaysan area (eastern Kazakhstan), Dzhungaria (China), and Mongolia, with bill depth mainly 6·8–7·7 mm and bill width mainly 6·5–7·5 mm; (2) very thick-billed *reiseri* in southern Balkans, Greece, and Turkey, *pyrrhuloides* from western and northern shore of Caspian Sea (north from Terek river) east through Transcaspia to Alakol' in eastern Kazakhstan, and *zaidamensis* in Tsaidam basin (China), with bill depth 7·8 mm or more and bill width mainly 7·5 or more. In each sub-group, colour gradually paler towards east, and birds inhabiting same longitude are similar in colour, independent of bill size; variation in tone of ground-colour sometimes hard to assess because of marked effect of bleaching and wear.

Nominate *schoeniclus* a dark small-billed race, sometimes split into 5 races: *steinbacheri* in Lapland and north European Russia with fine bill, virtually straight culmen, bill to nostril less than 7·0 mm and often black spots in white of hindneck of ♂ (see Steinbacher 1930*a* and Dementiew 1937), nominate *schoeniclus* with intermediate bill size and somewhat variable culmen shape in southern Fenno-Scandia, *turonensis* from much of Britain, Netherlands, and western France through central Europe to northern Poland and Baltic states with slightly thicker bill, slightly curved culmen, and bill to nostril generally over 7·0 mm, *mackenziei* in western Scotland and western Ireland, which is similar to *turonensis* but shows broader black streaks to upperparts (see Meinertzhagen 1953), and *goplanae* with still thicker bill from northern Belorussiya, Kaliningrad, and St Petersburg east to northern Perm' area in east European Russia; these trends discernible when large series examined, but overlap in characters is large and all forms included in nominate *schoeniclus* here, following Lundevall (1950), Vaurie (1956*d*), Matoušek (1969), and Matoušek and Jablonski (1969). Nominate *schoeniclus* grades into *stresemanni* in southern Moravia and central Slovakia (Jirsík 1951) and into *ukrainae* in southern Belorussiya and in European Russia from Tula to Ul'yanovsk regions, but in fact continuous cline of increasing bill depth probably exists in north and east in sequence *steinbacheri*, nominate *schoeniclus*, *turonensis*, *goplanae*, *ukrainae*, and (perhaps) *tschusii* of thick-billed group; intergradation zone of nominate *schoeniclus* and *stresemanni* narrow and secondary in character, probably because *stresemanni* well-isolated by mountain ranges. *E. s. passerina* from north-west Siberia has rather fine bill, but within range of nominate *schoeniclus*; differs mainly in paler ground-colour of upperparts (more ochre rather than dull rufous), narrower black streaks of mantle, scapulars, flank, and (in ♀) chest, and paler grey rump; birds from Pechora basin (north-east European Russia) similar in size to *passerina*, but dark streaks heavier, apparently forming part of cline of decreasing colour saturation running from 'steinbacheri' eastward. *E. s. parvirostris* from north-east Siberia similar in size to *passerina*, but paler still, black streaks on cap of ♀ and on flank of both sexes virtually absent, streaks on mantle and scapulars short and narrow; for intergradation zone of *passerina* and *parvirostra* with *pallidior* further south, see Johansen (1944) and Stepanyan (1990). *E. s. stresemanni* from Carpathian basin similar in colour to nominate *schoeniclus*, differing only in heavier bill; may grade into *tschusii* of eastern Rumania along Danube of north-west Bulgaria and south-west Rumania (Niethammer 1950). *E. s. ukrainae* similar in size to *stresemanni* (Matoušek 1968), but paler, ground-colour of upperparts more buff, dark streaks narrower, rump paler olive-grey, less suffused brown, and lesser coverts tawny-cinnamon, less deep rufous-chestnut. *E. s. pallidior* and *pyrrhulina* both similar in size to *stresemanni* and *ukrainae*; *pallidior* very pale, ground-colour of upperparts sandy-buff, of underparts buff-white; rump grey-white when worn; dark streaks on upperparts and flanks rather narrow, tinged black-brown on lower mantle and scapulars, brown and reduced in extent on cap, back, and flank; *pyrrhulina* has warmer brown ground-colour of upperparts than *pallidior*, cap of ♀ more rufous, chest and flank warmer buff; streaking reduced, rather as in *pallidior*; for populations of *pyrrhulina* in Manchuria, see Dementiev (1934) and Vaurie (1956*d*).

E. s. intermedia of thick-billed *pyrrhuloides* group similar in colour to nominate *schoeniclus*, but ground-colour of rump paler and flank profusely streaked rufous-brown, streaks less black. *E. s. witherbyi* from Balearic Islands, eastern Spain, and Mediterranean France has bill as heavy as *intermedia*, but slightly more pointed; wing shorter, colour similar to nominate *schoeniclus* or slightly darker; birds formerly breeding Sardinia intermediate between *witherbyi* and *intermedia* (Steinbacher 1930*a*; Hartert and Steinbacher 1932–8). Position of *lusitanica* Steinbacher, 1930, breeding Portugal, unclear; included in *witherbyi* by Vaurie (1959), but very different (Ticehurst and Whistler 1933); streaks on upperparts and flank in a few birds examined broader,

blacker, and more extensive than in *witherbyi*, and bill finer and more sharply pointed; both colour and size rather like nominate *schoeniclus* from western Scotland, but bill slightly deeper (near *stresemanni*), though culmen less curved, and wing shorter; maintained in *witherbyi*, pending further study of south-west European populations, though better positioned in thin-billed group near nominate *schoeniclus*. *E. s. reiseri* from southern Balkans, Greece, and Turkey large and dark (like *stresemanni*), and with markedly heavy bill (see Vaurie 1958*a* and Kumerloeve 1964*b*, 1969*b*). *E. s. tschusii* (synonym: *volgae* Stresemann, 1919) similar in colour to *ukrainae*, but bill deeper and wider at base and wing longer. *E. s. incognita* similar in size to *tschusii*, similar in colour to *pallidior*; *haermsi* Zarudny, 1911, perhaps an older name available for *incognita*. *E. s. pyrrhuloides* large and with markedly thick and blunt bill; like *reiseri*, but much paler, even more so than *pallidior*, ground-colour of upperparts pink-buff (when fresh) to cream-buff (when worn), of underparts cream-buff to white, streaks on lower mantle and scapulars black-brown, short, reduced in extent, streaking on underparts virtually absent. Isolated *zaidamensis* of Tsaidam basin in China is similar, but ground-colour of upperparts slightly more cinnamon. Remainder of thick-billed races in Asia have bill less wide and deep at base than *pyrrhuloides*, *zaidamensis*, or *reiseri*, but heavier than *incognita*. *E. s. caspia* of eastern Transcaucasia and northern Iran and *korejewi* of south-east Iran both rather dark and greyish on upperparts (particularly *caspia*), darker and less buffy than *pyrrhuloides* with streaks heavier and blacker, rump darker grey, and lesser coverts deeper rufous, but paler than *reiseri*; birds intermediate between *caspia* and *korejewi* occur or formerly occurred Syria and south-west Iran (Vaurie 1956*d*). *E. s. centralasiae* and *harterti* both pale, like *pyrrhuloides* or slightly warmer ochre-buff, wing-tip of at least *centralasiae* markedly blunt, *harterti* doubtfully separable, being more or less intermediate between *centralasiae*, *pyrrhuloides*, and *incognita*.

Relationships of thick- and thin-billed groups not fully clarified. Races of the 2 groups said to intergrade in eastern Europe and Asia, *tschusii* and *incognita* providing link between them (Portenko 1929; Steinbacher 1930*a*; Johansen 1944; Dementiev and Gladkov 1954). However, groups largely segregated spatially in Kazakhstan, though some overlap occurs locally (Korelov *et al.* 1974), apparently not leading to extensive interbreeding, and marked differences between groups in structure of palate (Portenko 1929) and in feeding ecology (Stegmann 1948, 1956) probably prevent mixing. In southern Europe, where races largely separated by mountain ranges, no intermediate races occur, thinner-billed nominate *schoeniclus* and *stresemanni* being abruptly replaced by thick-billed *witherbyi*, *intermedia*, and *reiseri* further south without intergradation. Locally, thick- and thin-billed races said to breed in same area without any known intermediates (e.g. northern Greece: Makatsch 1950). On Sicily, thick-billed *intermedia* bred in past, as on Italian mainland (Hartert 1903–10; Priolo 1988), but thin-billed nominate *schoeniclus* reported to breed at present, far south of remainder of range (Massa 1976); see also Brichetti and Cova (1976). Many nominate *schoeniclus* present in mainland Italy in winter, within breeding range of thick-billed *intermedia*, and a few *stresemanni* present in north-east then also, but none present May–August according to specimens examined, while those lingering into April were mainly ♀♀ (C S Roselaar). Further data from southern Europe required.   CSR

295

Emberiza pallasi **Pallas's Reed Bunting**

Du. Pallas' Rietgors   Fr. Bruant de Pallas   Ge. Pallasammer
Ru. Полярная овсянка   Sp. Escribano de Pallas   Sw. Dvärgsävsparv

Cynchramus Pallasi Cabanis, 1851

Polytypic. *E. p. polaris* Middendorff, 1851, north-east European Russia and north-west Siberia east to Lena valley, south to *c.* 60°N. Extralimital: nominate *pallasi* (Cabanis, 1851), eastern Tien Shan north through Altai and north-west Mongolia to Sayan mountains and south-west Transbaykalia; *minor* Middendorff, 1851, Lena valley east to Chukotskiy peninsula, south to northern and eastern Transbaykalia, north-east Mongolia, northern Manchuria, and (possibly) Sikhote-Alin'; *lydiae* Portenko, 1929, Orog Nuur (southern Mongolia), Tuva area (southern Siberia), Buyr Nuur (eastern Mongolia), and neighbouring parts of Argun river and Russian Transbaykalia.

Field characters. 13–14 cm; wing-span 20·5–23 cm. Over 10% smaller than Reed Bunting *E. schoeniclus* but with similar form except for straight culmen, flatter crown, and slimmer body; only marginally larger than Little Bunting *E. pusilla*, with similarly pointed bill. 2nd smallest bunting of west Palearctic, closely resembling *E. schoeniclus* in general character and plumage pattern but having diagnostic calls. At close range, compared to western *E. schoeniclus*, both sexes show much more streaked upperparts (due to distinctively paler fringes to back and wing-feathers emphasizing black-brown centres), usually pale greyish rump, and only lightly marked underparts; identification confirmed by dull (never rufous) lesser wing-coverts, bright double wing-bar, and pale panel on folded wing (but only 2nd of these obvious on juvenile). Adult ♂ further distinguished by yellowish to buffish tinge to rear collar, white rump, and virtually unstreaked underparts; adult ♀ and immature by much more uniform head (lacking obvious dark borders to crown and ear-coverts of *E. schoeniclus*) and heavy black malar stripes turning into throat (and not breaking up to form obvious streaks on underparts). Sexes dissimilar; marked seasonal variation in ♂. Juvenile separable.

ADULT MALE. Moults: August–September (complete). In worn breeding plumage, instantly recalls *E. schoeniclus* but lacks brown ground to back and rufous margins to wing-feathers, these being instead pale yellowish-grey to white. Pale ground-colour contrasts vividly with virtually black stripes down mantle and scapulars, black bases of median coverts (creating narrow white wing-bar across their tips), black centres to greater coverts (forming pale buffish wing-bar across their tips), and black centres to tertials whose pale yellowish fringes merge with those of inner secondaries to form conspicuous pale panel. When wings drooped or in flight, greyish-white rump and white, faintly streaked upper tail-coverts diagnostic, as are bluish-grey lesser wing-coverts when visible. Bill short and spiky, with straight culmen; dusky-black. Legs pale, bright flesh-brown. In fresh autumn plumage, also recalls *E. schoeniclus*, but already shows diagnostic pale rump and bluish-grey lesser wing-coverts. If those characters hidden, further distinguished in autumn by: (a) less heavily marked head and throat, lacking dark blackish crown-sides, rear eye-stripe, and patchy, heavily underlined ear-coverts typical of *E. schoeniclus*, last being largely uniform rufous-buff on *E. pallasi*; (b) warm yellowish-buff nape and rear collar; (c) more evenly pale buff- and blackish-striped back; (d) buff tinge to sides of upper breast and unmarked flanks; bill becomes bi-coloured, with pinkish-horn lower mandible. ADULT FEMALE. Recalls *E. schoeniclus* less than ♂. Except when heavily worn, lacks strongly marked head which resembles non-breeding ♂ but shows also rufous crown, more distinct cream supercilium, and even duller border to lower ear-coverts. Further distinguished from non-breeding ♂ by duller greyish-brown lesser wing-coverts, more rufous on tips of coverts and particularly tertial fringes, paler buff margins to more streaked rump and upper tail-coverts, and buffier underparts with striking black malar stripe turning into throat and faint rufous-brown streaks on sides of breast and discontinuously along flanks. When heavily worn, crown becomes streaked blackish, lower border to ear-coverts sharpens into blackish rear patch and narrow line reaching bill (as in *E. schoeniclus*, not *E. pusilla*), and pale panel on tertial-edges and inner secondaries is much reduced; isolated black malar stripe and its extension into throat remain obvious (not breaking up into lines of dark streaks on breast and flanks as on *E. schoeniclus* and *E. pusilla*). Bill as non-breeding ♂. JUVENILE. Much closer in appearance to *E. schoeniclus* than adult, with dark-streaked rump, less even pattern of stripes on buffier back, more rufous on wings, and (in particular) full dusky streaks over breast and at least forepart of flanks. May be distinguished at close range by dull head pattern (like worn ♀ but with indistinct supercilium), dull brown lesser wing-coverts, pale buff to almost white tips to median and greater coverts (forming most contrasting wing-bars of any plumage), somewhat paler, rufous-buff (not chestnut) tertial panel, and paler margins to upper tail-coverts; of these, double near-white wing-bar most trustworthy.

♂ showing pale rump unmistakable. ♂ hiding rump, ♀, and immature all require careful separation from *E. schoeniclus*, and also from *E. pusilla*, ♀ and immature Black-faced Bunting *E. spodocephala*, and even small

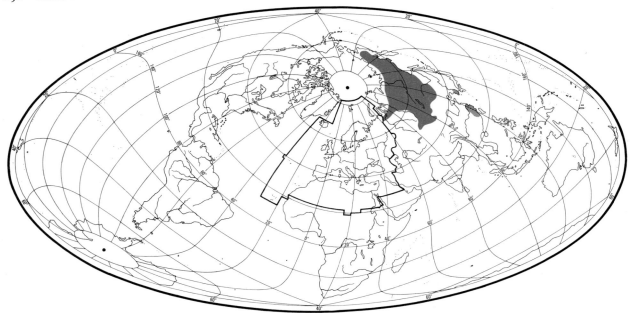

Nearctic emberizid sparrows. For distinction from *E. sch-oeniclus*, see above (but note that some far-eastern races of *E. schoeniclus* also show pale plumage and little streaking on underparts); see also below for diagnostic calls. Separation from other small buntings also made easy by call and following plumage differences: in *E. pusilla*, rufous head with dark lateral crown-stripes and dark surround to cheeks not reaching bill, more striking cream eye-ring, duller ground-colour to upperparts and finely lined underparts; in *E. spodocephala*, greyer face and upper mantle, duller browner upperparts, more rufous feather-margins on wings, buffier wing-bars, and fully streaked underparts (but beware particularly its greyish lesser wing-coverts); in small Nearctic sparrows, lack of white outer tail-feathers, yellowish or greyish supercilium, and usually fully streaked underparts. Flight lighter, more flitting than *E. schoeniclus*; flying bird looks even smaller and more delicate than when perched. Gait a hop. Stance and actions as *E. schoeniclus*; has similar flicking and nervous spreading of tail. Unobtrusive, spending much time on ground. Northern races *polaris*, *minor*, and nominate *pallasi* adapted not to reeds but to dwarf trees, thickets, and nearby open ground in tundra; Mongolian race *lydiae* (possibly separate species) prefers dry steppe by lakes. Sociable, often in small parties in normal range.

Song (of immature ♂ *lydiae*) short and pronounced: 'cheep-preep-deet-preep'. Diagnostic calls do not suggest *E. schoeniclus*; transcriptions for *polaris* include 'tsleep', uttered when perched and recalling Tree Sparrow *Passer montanus*, 'chirrup'; 'pseeoo(p)' when flushed, 'tsee-see', and another note in flight somewhat like Richard's Pipit *Anthus novaeseelandiae* but very much fainter; for *lydiae*, 'chreep' or 'preep'.

Habitat. Breeds mainly in east Palearctic, in drier and cooler situations than overlapping Reed Bunting *E. schoeniclus*, occupying tundra with tall herbage and shrubs, but also shrubs and grass areas of steppes and semi-desert and, in south, mountain tundra (Harrison 1982; Flint *et al.* 1984). Northern races inhabit river valleys with thickets of willow *Salix* and alder *Alnus* in lowland tundra, though *minor* occurs up to 1800 m. Further south, breeds on high plateaux up to 2200–2500 m, in dwarf birch *Betula nana* and other shrub growth. Also inhabits steppe (*lydiae*, in Mongolia) and semi-deserts with bushes (Dementiev and Gladkov 1954; Loskot 1986*b*).

Winters in plains, preferring irrigated areas with shrubs and stands of reeds near rivers and lakes (Dementiev and Gladkov 1954).

Distribution. Has bred since at least 1981 in eastern Bol'shezemel'skaya tundra (north-east European Russia). Western limit of range previously thought to be Taz basin. Occasionally recorded in southern Yamal peninsula, and sporadic breeding confirmed near Kharp. In Bol'shezemel'skaya tundra, breeds north to upper reaches of Kara river, west to source of Seyda river. (Morozov 1987.)

Accidental. Britain: adult ♀, Fair Isle, 29 September to 11 October 1976; juvenile, Fair Isle, 17–18 September 1981 (Alström and Colston 1991).

Population. No information.

Movements. Northern populations long-distance migrants; southern populations short-distance and altitudinal migrants (perhaps dispersive rather than migratory in south-west). Limited information, and some aspects

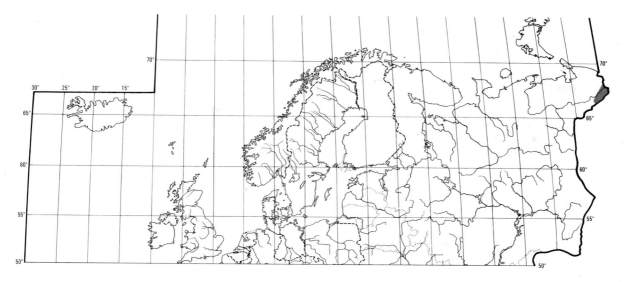

unclear, as division and range of races differently described by different authors; wintering grounds incompletely known. For review, see Loskot (1986*b*).

E. p. polaris and *minor* winter from central Manchuria, central and southern Korea, and (in small numbers) extreme south of Ussuriland (south-east Russia) south through north-east and east China to Chekiang, Kiangsi, and Fukien; also possibly annual in small numbers in south-west Kyushu (Japan). Probably *polaris* occupies southern part of range, and *minor* more northern areas. (Austin 1948; Vorobiev 1954; Vaurie 1959; Loskot 1986*b*; Kennerley 1987; Brazil 1991.) Nominate *pallasi* winters in north-west China (western Sinkiang, and further east in oases of Nan Shan and Ala Shan), probably also in north-west Mongolia (Kobdo, and Haar Us Nuur lake), where recorded November to 17 December; *lydiae* winters in Nan Shan, and probably at Haar Us Nuur lake (where recorded 17 November to 4 December) (Schauensee 1984; Loskot 1986*b*); perhaps both these races also winter in intermediate areas.

E. p. polaris (and *minor*). In autumn, birds leave northern breeding areas August–September. In Bol'shezemel'skaya tundra (north-east European Russia), over 2 years, movement 14 August to 29 September (Morozov 1987). Recorded until early September in Koryak highlands and Kamchatka peninsula (north-east Siberia) (Kishchinski 1980; Lobkov 1986). Passage near Tomsk (in south of western Siberia) mid-September to beginning of October (Johansen 1944), continuing to about mid-October at Krasnoyarsk (south-central Siberia); recorded to 20 October at Irkutsk (west of Lake Baykal). In Transbaykalia, Amur region and Ussuriland, passage last third of September and throughout October. (Loskot 1986*b*.) Inconspicuous in Ussuriland, and not recorded in some years (Panov 1973*a*). In north-east China, main arrival

and passage in October (La Touche 1920), and present in Japan from mid-October (Brazil 1991). First-ever record in Hong Kong, 8–14 December 1991 (Leader 1992).

Spring migration begins March. Present in Japan until early March (Brazil 1991), and in Korea until April (Gore and Won 1971). Passage and departures in coastal Hopei (north-east China) from early March to about mid-May, with highest numbers in late April in one year (La Touche 1920; Williams 1986). In Ussuriland, passage recorded mid-March to April (Vorobiev 1954; Panov 1973*a*). Arrives on Kamchatka peninsula 19–26 May (Lobkov 1986). Reaches Yakutiya and lowlands of extreme north-east Siberia in last third of May or early June, slightly later than Little Bunting *E. pusilla* (Vorobiev 1963; Krechmar *et al.* 1991). One record 14 June 1983 on Wrangel Island (Stishov *et al.* 1991), presumably due to overshooting. Vagrant to Alaska (Kessel and Gibson 1978). At Krasnoyarsk, passage mainly in 2nd half of April and early May, continuing to late May or early June (Loskot 1986*b*); at Tomsk, in one year, reported from 15 April, with heavy passage 20–25 April (Gyngazov and Milovidov 1977); movement there continues in May, with stragglers to beginning of June (Johansen 1944). Recorded exceptionally in north-east Kazakhstan (Korelov *et al.* 1974). On Putorana plateau (north-central Siberia), where very rare, noted 3 June in one year (Morozov 1984). At Noril'sk lakes (western Taymyr), earliest records 7–10 June over 2 years (Krechmar 1966). Presumably birds skirt north of Urals to reach extreme north-west of range, Bol'shezemel'skaya tundra, where earliest records 6–10 June over 3 years (Morozov 1987).

E. p. pallasi disperse early from high latitudes after breeding; recorded outside breeding areas late July or beginning of August, though movement continues until at least September (Loskot 1986*b*). Also recorded in south of known winter range (Ala Shan and Nan Shan) as early as August (Vaurie 1959). Spring migration is late, from last third of April continuing as late as early June, due to snow cover on high-level breeding grounds (Loskot 1986*b*). At Orog Nuur lake (south-west Mongolia), earliest record 25 April in one year, and passage observed in last third of May and beginning of June in central and northern Mongolia; in Tes-Khem valley near Erzin (south-central Russia, near Mongolian border), recorded on passage with other Emberizidae in last third of May (Kozlova 1930; Loskot 1986*b*).

E. *p. lydiae* still present at or near breeding grounds in late August and early September (recorded 8 September in Transbaykalia), but reported in Tuul Gol valley (north-central Mongolia) 4 August (apparently on passage), in southern Gobi (southern Mongolia) in September, and Kansu (northern China) in August. In spring, seen at Orog Nuur lake 4 April (first day of survey) in one year (earlier than nominate *pallasi*), with first small flock 14 April. Recorded in southern Transbaykalia 1–2 April (2 years' observations). (Kozlova 1930; Loskot 1986*b*.) At Ulan Bator in north-central Mongolia (race not given) reported from 30 March in one year, one of earliest spring migrants (Kitson 1979).

For vagrancy to west Palearctic, see Distribution.

DFV

Food. Little information, all extralimital. Diet seeds and other plant material, invertebrates in breeding season (Dementiev and Gladkov 1954; Armani 1985). In Chukotka (north-east Siberia), when collecting food for young, foraged in willow *Salix* scrub and on wet meadows, and also on muddy shore of lake (Kishchinski *et al.* 1983). In northern China, forages in reedbeds, rice fields, meadows, etc., in large flocks outside breeding season (Chong 1938); also in willows, energetically collecting insects and larvae (Shaw 1936); seen perching in large numbers on stems of tall *Stipa* grass taking seeds from heads (Piechocki 1958). When feeding on such grasses, foraging actions acrobatic, reminiscent of Redpoll *Carduelis flammea*, and often hangs upside down (Kolthoff 1932).

Diet includes the following. Invertebrates: adult Lepidoptera, adult and larval flies (Diptera: Tipulidae, Chironomidae), Hymenoptera (larval sawflies Symphyta), beetles (Coleoptera: Curculionidae). Plants: seeds, shoots (etc.) of alder *Alnus*, crowberry *Empetrum*, grass *Stipa*. (Piechocki 1958; Kapitonov and Chernyavski 1960; Kishchinski 1980; Danilov *et al.* 1984; Krechmar *et al.* 1991.)

In north-east Siberia, early June, stomach of one ♂ contained 10 crowberry seeds, 3 adult Lepidoptera, and beetles (Krechmar *et al.* 1991); stomachs in another study held beetles, seeds, and shoots (Uspenski *et al.* 1962); in September, birds on passage perched in bush de-husking alder seeds (Kishchinski 1980). In south-east Siberia, 5 May stomachs contained seeds and insects, including beetles (Dementiev and Gladkov 1954). On Yamal peninsula (north-west Siberia), summer, stomachs contained Tipulidae, Chironomidae, sawfly larvae, and Curculionidae (Danilov *et al.* 1984).

In lower Lena valley (north-east Siberia), young were fed large adult Tipulidae (Kapitonov and Chernyavski 1960).

BH

Social pattern and behaviour. Many aspects little known. For summary of main sources, Russia and Mongolia, see (primarily taxonomic) review by Loskot (1986*b*).

1. In China, winter, occurs in sometimes huge flocks in reed-beds, paddyfields, and grassy fields near swamps (La Touche 1925–30; Kolthoff 1932; Chong 1938); at Beidaihe (north-east China), mobile flock of up to 150 present March–May and presumed to have wintered there (Williams 1986). Flock of *c.* 100–120 also recorded lower Amur (eastern Russia) in March (Vorobiev 1954). Migrates in both spring and autumn in generally small flocks, sometimes singly (e.g. Dementiev and Gladkov 1954, Spangenberg 1965, Lobkov 1986); largest flock in spring *c.* 30–40, Ussuriland (eastern Russia) (Panov 1973*a*). In Amur region, mid-May, recorded in small mixed flocks with other Emberizidae (Loskot 1986*b*), and in lower Lena valley (Yakutia, north-east Russia), around time of arrival on breeding grounds, single birds associated with flocks of Redpolls *Carduelis flammea* (Kapitonov and Chernyavski 1960). BONDS. Little information, but nothing to suggest other than monogamous mating system. No information on duration of pair-bond. Studies by Gladkov and Zaletaev (1962) and Krechmar (1966) suggested only ♀ incubates, but at 2 nests in Yakutia ♂ on nest when discovered (Vorobiev 1963), and ♂ recorded apparently incubating in other studies (Sokolov 1986*a*; Krechmar *et al.* 1991, which see for photograph). Young (nestlings and fledglings) fed by both sexes (Kishchinski 1980; Morozov 1984); at one nest, lower Lena, well-grown nestlings seen being fed by ♀, fledglings apparently by both parents (Kapitonov and Chernyavski 1960). Age of independence and of first breeding not known. BREEDING DISPERSION. Solitary or sometimes more clustered. Presumably territorial, as in other *Emberiza*, but no details. In Chita region (south-east of Lake Baykal), generally in clusters of 3–8 pairs, very rarely solitary (Sokolov 1986*a*); near Zabaykal'sk (Chita), 2 nests *c.* 60 m apart (Loskot 1986*b*). In Kanchalan basin (Chukotskiy peninsula, north-east Russia), breeding pairs locally 100–200 m apart in willow *Salix* scrub with sedge *Carex* meadows, bogs, and lakes (Kishchinski *et al.* 1983). One ♀, lower Lena river, seen to forage near nest for young close to fledging (Kapitonov and Chernyavski 1960); no other information indicating function of presumed territory. On Bol'shezemel'skaya tundra (north-east European Russia), July, density over 3 years 3·4–17·3 birds per km² (Morozov 1987), or up to 33·6 per km², and unaffected by varying levels of pollution (from heavy industry in Vorkuta) (Morozov 1986). Examples of density in Russia east of Urals as follows. In dense, low willow and birch *Betula* scrub along river, Koryak highlands (north of Kamchatka), locally 3–4 pairs in 0·5 ha (600–800 per km²); up to 300 pairs per km² in slightly taller thickets (Kishchinski 1980). Elsewhere in Kamchatka, up to 300–400 pairs per km², preferred habitat comprising dense willow scrub abutting open, meadow-like patches on flat lowlands (Lobkov 1986, which see for further densities in different habitats). In central Siberia, 32 birds per km² recorded in bare summit zone and willow tundra, western Sayan mountains, 69–76 in wooded tundra along Yenisey river (Rogacheva 1988). In Chita region, average 5 birds per km² in steppe with fairly dense, tall vegetation (Sokolov 1986*a*). ROOSTING. No information.

2. Koryak study found birds quiet and secretive in post-

breeding period (Kishchinski 1980). In Mongolia, birds frequently flicked up tail and simultaneously slightly fanned it (Kitson 1979). Vagrant on Fair Isle (Shetland) generally wary; spread tail when perched on wire, but lacked nervousness of Reed Bunting *E. schoeniclus* (also present) and wing-flicked less (Riddiford and Broome 1983). FLOCK BEHAVIOUR. No information. SONG-DISPLAY. ♂ sings (see 1 in Voice) while perched on side of plant stem (e.g. tall grass) (Sokolov 1986a), or from top of shrub, flying repeatedly from one song-post to another (Kishchinski 1980). In Mongolia, *lydiae* ♂♂ recorded ascending for Song-flights (Piechocki and Bolod 1972), but no details. No song noted from wintering birds in eastern China (Kolthoff 1932), but will sing on spring migration (e.g. Spangenberg 1965 for Iman river, Ussuriland); at Lake Khanka (southern Ussuriland), song noted over several days from late April, presumably from passage birds (Dementiev and Gladkov 1954; Vorobiev 1954). Sings otherwise from arrival on breeding grounds (e.g. Krechmar *et al.* 1991 for north-east Russia). On Bol'shezemel'skaya tundra, over 3 years, noted from 8, 16, and 19 June (Morozov 1987). In northern Russia east of Urals, also in Altai, mid- to late June and (Yakutiya) up to 22 July (Sushkin 1938; Dementiev and Gladkov 1954; Vorobiev 1963; Danilov *et al.* 1984; Morozov 1984). ANTAGONISTIC BEHAVIOUR. In Chita region, frequent and persistent ♂-♂ chases apparently associated with pairing; at end of bout of chasing, one ♂ would land by and escort ♀ (Sokolov 1986a). No further information. HETEROSEXUAL BEHAVIOUR. (1) General. On Bol'shezemel'skaya tundra, breeding pairs first noted over 4 years between 13 and 22 June (Morozov 1987). Further east, on lower Lena river, pair-formation from early June (Kapitonov and Chernyavski 1960). In Chita study, birds in pairs from late May (Sokolov 1986a). Apparently only very short period between arrival and start of laying (Kishchinski 1980 for Koryak highlands). (2) Pair-bonding behaviour. Includes sexual chases, noted Chita as late as 25–29 July, sometimes 5–6 ♂♂ in pursuit of one ♀ (Sokolov 1986a). (3) Courtship-feeding. In Lena valley, ♂ several times recorded carrying food to nest to feed incubating mate (Kapitonov and Chernyavski 1960). No confirmation of this by other authors and no further information on other aspects. RELATIONS WITHIN FAMILY GROUP. In Koryak study, young brooded by ♀, also briefly by ♂; brood finally left nest at *c.* 10 days, remaining close by for further *c.* 5 days (Kishchinski 1980). Tend to stay well hidden in dense vegetation during this initial period after fledging (Sokolov 1986a). ANTI-PREDATOR RESPONSES OF YOUNG. No information. PARENTAL ANTI-PREDATOR STRATEGIES. (1) Passive measures. In Yakutiya, ♀ flushed from nest several times, each time made fairly rapid and furtive return to nest, moving low through vegetation (Vorobiev 1963). (2) Active measures. On lower Lena, ♀ flew about and called (see 2d in Voice) when man near nest; also flicked tail after landing (Kapitonov and Chernyavski 1960). Other studies found both parents showing alarm when nest or young threatened, flying about, perching prominently, and calling (see 2a in Voice) (Portenko 1939; Vorobiev 1963), though some variation between pairs (Sokolov 1986a). MGW

Voice. Characteristic calls generally distinct from Reed Bunting *E. schoeniclus* (Kitson 1979; Alström and Colston 1991); see, however, 2c (below) for report of call similar to that species. Song (also distinct from *E. schoeniclus*) apparently varies geographically, but no detailed study. Little information on use of calls through the year.

CALLS OF ADULTS. (1) Song of ♂. Simple, short, some-

what monotonous phrase comprising series of repeated similar notes. Recording from northern Yakutiya (eastern Russia: presumably *polaris*) contains song comprising 6 slightly shrill and ringing '(s)rrie' notes, each a complex unit with very gentle onset and offset (J Hall-Craggs, M G Wilson: Fig I). Other descriptions of *polaris* (Yakutiya):

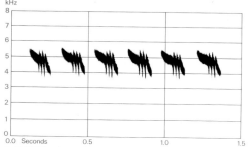

I B N Veprintsev and V V Leonovich Russia July 1976

monotonous 'chi-chi-chi-chi' (Vorobiev 1963); bell-like 'trill' (Loskot 1986b). Song of *lydiae* apparently of similar structure (Loskot 1986b); like quiet stridulation of bush-cricket (Tettigoniidae) 'tsisi-tsisi-tsisi-tsisi' (Kozlova 1930, 1932). Song of nominate *pallasi* apparently significantly different (Loskot 1986b); clear, ringing phrase likened to song of Willow Tit *Parus montanus* (Kozlova 1932) or Yellow-breasted Bunting *E. aureola* (Sushkin 1938). (2) Contact- and alarm-calls. Further study needed to determine how many different calls exist and true extent of variation within them (calls perhaps also vary geographically). Following scheme provisional, especially in respect of calls 2a–c which may well intergrade according to context. (2a) Often likened to sparrow *Passer* and thus distinct from *E. schoeniclus*. Slightly variable 'tsleep' recalling Tree Sparrow *P. montanus*, usually given when perched (Alström and Colston 1991). Recording of calls given when disturbed (Fig II) suggests sparrow-like

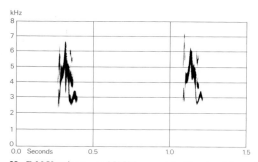

II B N Veprintsev and V V Leonovich Russia July 1976

'tsiup' or 'pseeup' (M G Wilson), or 't' si' (J Hall-Craggs). Further descriptions: 'tsirk' by alarmed birds of both sexes in Yakutiya when man near nest (Vorobiev 1963); trisyllabic 'peeseeoo' recalling Yellow-browed Warbler

Phylloscopus inornatus humei noted in Mongolia (Kitson 1979); 'chulp', clear 'chee-ulp', or 'tschee-ulp' rather like House Sparrow *P. domesticus* (Riddiford and Broome 1983; Jonsson 1992), harsh and intermediate between *P. domesticus* and Richard's Pipit *Anthus novaeseelandiae* according to Olsen (1989), so could equally well be in following category. (2b) At least some calls listed here are perhaps only harsher variants of call 2a (above). Frequently given in flight and suggesting very faint call of *A. novaeseelandiae* (Alström and Colston 1991). In Mongolia, 'ch-reep' recalling both *P. domesticus* and Tawny Pipit *A. campestris* (Kitson 1979). Vagrant ♀ temporarily held captive gave anxiety-call intermediate between *P. domesticus* and Yellow Wagtail *Motacilla flava* (Broad and Oddie 1980). Frequently-uttered flight-call of wintering birds in China rendered 'pitch pitch' (Kolthoff 1932). (2c) Calls grouped here perhaps only quieter variants or alternative renderings of some described above. In China, in winter, faint 'se se' said by Kolthoff (1932) to be typical call; apparently similar are quiet 'sip sip' given by vagrant ♀ when first flushed (Riddiford and Broome 1983), and 'psieh' (resembling call of *E. schoeniclus*) by ♂ running about on ground in Mongolia (Piechocki *et al.* 1982). (2d) In Yakutia, abrupt 'vcha vcha vcha' given by ♀ when man near nest with young (Kapitonov and Chernyavski 1960); no obvious relationship with calls in above scheme, though perhaps close to 'chulp' (see 2a, above).

CALLS OF YOUNG. No information.   MGW

Breeding. Almost all information extralimital. SEASON. North-east European Russia: eggs laid from late June or early July; young generally fledged by end of July (Morozov 1987). Northern Siberia: eggs laid from 1st half of June into July (Gladkov and Zaletaev 1962; Krechmar 1966). North-east Siberia: nest-building throughout June once habitat dried out after thaw; eggs laid from beginning of June, those in July probably replacements (Kapitonov and Chernyavski 1960; Kishchinski 1980; Loskot 1986*b*; Krechmar *et al.* 1991). Chita region (south-east of Lake Baykal): extended, perhaps because of high predation rate; eggs laid early June to mid-July (Sokolov 1986*a*). Mongolia: also extended, early June to late July (Kozlova 1930, 1933). Eastern Tien Shan and Altai: clutches found from mid-June to end of July; nestlings recorded towards end of August (Loskot 1986*b*, which see for review). SITE. Well hidden on ground or tussock, or in depression in moss, lichen, etc., sheltered by shrub (e.g. *Ledum*, *Vaccinium*, *Salix*) or grass; also less than *c.* 50 cm above ground in bush or small tree (e.g. birch *Betula nana*, pine *Pinus pumila*) (Vorobiev 1963; Kishchinski 1980; Lobkov 1986; Loskot 1986*b*, which see for review; Krechmar *et al.* 1991). In Chita region, of 16 nests, 10 in tangle of wormwood *Artemisia*, 3 in sow-thistle *Sonchus*, and 2 in grass *Stipa* (Sokolov 1986*a*). See Loskot (1986*b*) for *lydiae* in dry steppe habitat, where perhaps more often above ground. Nest: rather flimsy foundation of dry stems and blades of grass and sedge, lined with similar but finer material, hair, and sometimes dry needles of larch *Larix* (Vorobiev 1963; Krechmar 1966; Kishchinski 1980; Lobkov 1986). In Chita region, 7 nests had average outer diameter 9·0 cm (8·0–10·7), inner diameter 5·1 cm (4·6–5·6), overall height 8·0 cm (6·4–9·5), and depth of cup 4·6 cm (4·2–5·4) (Sokolov 1986*a*). See also Kishchinski (1980) and Loskot (1986*b*). Can be simple depression in moss cushion lined with hair (Krechmar *et al.* 1991). Nest of *lydiae* described as being more similar to Japanese Reed Bunting *E. yessoensis* than to other races of *E. pallasi* (Loskot 1986*b*). Building: no information. EGGS. See Plate 34. Sub-elliptical, smooth and glossy; creamy-pink to reddish-brown, sometimes darker towards broad end, with scattered blackish-brown spots, small blotches, and hair-streaks and greyish-brown undermarkings and scrawls (Gladkov and Zaletaev 1962; Schönwetter 1984; Loskot 1986*b*). *E. p. polaris*: 18·4 × 14·1 mm (18·0–19·4 × 14·0–14·2), *n* = 11 (Gladkov and Zaletaev 1962; Vorobiev 1963); weight of 1 egg 1·67 g (Kishchinski 1980, which see for dimensions of 4 eggs); see also Loskot (1986*b*). *E. p. lydiae*: 17·9 × 13·4 mm (16·1–18·8 × 12·8–13·7), *n* = 28 (Sokolov 1986*a*); see also Loskot (1986*b*). Clutch: 4–5 (3–6) (Krechmar 1966; Loskot 1986*b*; Morozov 1987). In north-east Siberia, of 16 clutches, 6 were of 4 eggs, 9 of 5, and 1 of 6; average 4·7 (Loskot 1986*b*; Krechmar *et al.* 1991). In Chita region, 9 clutches comprised 2 of 3 eggs, 6 of 4, and 1 of 5; average 3·9 (Sokolov 1986*a*). Perhaps 2 broods in Mongolia (Kozlova 1933). INCUBATION. About 11 days, by both sexes (Vorobiev 1963; Krechmar *et al.* 1991); in Chita region, almost wholly by ♀ at one nest, ♂ only recorded once on eggs (Sokolov 1986*a*). YOUNG. Fed and cared for by both parents (Kishchinski 1980; Sokolov 1986*a*); in Koryak highlands (north-east Siberia), brooded by ♀ and also briefly by ♂ (Kishchinski 1980). FLEDGING TO MATURITY. In one case, young left nest at 10 days old (first left at 8 days when disturbed) (Kishchinski 1980). BREEDING SUCCESS. Predation rate thought to be high in Chita region (Sokolov 1986*a*). No further information.

BH

Plumages. (*E. p. polaris*). ADULT MALE. In fresh plumage (autumn), top and side of head and neck buff-brown, forehead and crown with round-ended black centres of feathers partly visible, centres of feathers of hindneck white (fully concealed); stripe on lore through eye as well as ear-coverts more black-brown, but this largely concealed by broad buff feather-tips, except on lower rear end of ear-coverts; supercilium warm buff, only slightly paler than crown and stripe through eye; stripe over lower cheek backwards from lower mandible and broad bar along side of neck distinct, paler buff or ochre, some white of feather-centres partly visible. Mantle and scapulars buff-brown, feathers broadly streaked black; black on some feathers partly bordered by chestnut. Back to upper tail-coverts pale buff-brown, white of feather-centres partly visible, back and tail-coverts with faint grey shaft-streaks. Chin and throat black, feathers with broad warm buff fringes (about equal amount of black and buff visible); remainder of underparts white with warm buff suffusion on feather-tips, side of breast and flank faintly streaked brown or grey. Central pair of tail-feathers dark grey-brown, broadly fringed warm brown along both webs; t2-t4 black, faintly edged grey-buff along outer web, more broadly so along tip and inner web; t5 black with white outer fringe and white tip to inner web 20-30 mm long; t6 white except for black wedge at base, dark shaft, and narrow dusky shaft-streak on tip. Flight-feathers greyish-black, fringe along outer web of primaries and inner secondaries light pink-buff, *c.* 0·5-1 mm wide, similar but whiter fringes along outer webs of outer secondaries, narrow grey-white edges along emarginated parts of outer primaries. Tertials black, outer web and tip bordered by contrastingly warm buff fringe 2-4 mm wide, black of tertials ending rounded subterminally; greater upper wing-coverts similar, fringes 2-3 mm wide; median coverts black with broad warm buff tip bordered white subterminally; lesser coverts dark grey with broad ash-grey to off-white fringes. Greater upper primary coverts and bastard wing greyish-black, outer web with narrow ill-defined pale buff-grey edge (except on longest feather of bastard wing). Under wing-coverts and axillaries pale grey to off-white with slight buff tinge, marginal ones with black base. Influence of bleaching and wear enormous. During autumn, buff tinges bleach to pale ochre, pale yellowish grey-buff, or (by mid-winter) greyish- or creamy-white. From mid-winter onwards, pale fringes wear off, top and side of head as well as chin and throat becoming deep black (last on supercilium and on centre of ear-coverts), contrasting sharply with broad pale buff-yellow to pure white collar round neck, white stripe on lower cheek, white rump and upper tail-coverts, and white underparts; initially, mantle, scapulars, and back black with broad cream-white streaks ('tramlines'), but by July pale fringes largely worn off here, too, much uniform black remaining; heavily abraded hindneck, cheek-stripe, rump, and upper tail-coverts then often with ill-defined grey mottling or streaks. Buff fringes of t1, flight-feathers, tertials, and outer webs of greater coverts and tips of median and greater coverts become paler and narrower through bleaching and wear, only traces of grey-white fringes remaining by June-July. ADULT FEMALE. In fresh plumage (autumn), rather like adult ♂ in fresh plumage, but forehead and crown narrowly streaked black on centres of feathers (in ♂, centres broadly black); concealed white on hindneck absent, pattern of side of head even less contrasting, bases of feathers of eye-stripe browner, less blackish; eye-ring narrow, buff-white, broken in front and behind; stripe on lower cheek and bar on side of neck buffier, without white admixture; chin and throat buff, without black, bordered at side by broad black malar patch, not reaching bill-base, often extending into narrow black-spotted gorget along lateral upper chest. As in ♂, buff tinges become paler due to bleaching and abrasion during winter and spring, pale fringes partly worn off, but much less black exposed then than in adult ♂ in spring. By June, cap of ♀ black with narrow mottled pale buff streaks; supercilium more conspicuously buff-white, broad above ear-coverts but hardly extending to above lore; lore, ear-coverts, and upper cheek mottled buff, brown, and black, without black surround to ear-coverts, except for dark patch at lower rear of ear-coverts, sometimes with black line at border of upper cheek from gape backwards to below eye; hindneck mixed grey and buff (no white), side of neck dirty cream-buff with fine dusky mottling; underparts dirty white, throat, chest, side of breast, and flank washed cream-buff; side of breast and flank with faint brown streaks; throat spotted black, forming ill-defined patch (not extending to chin), or buff with mottled black gorget across lower throat, connected with conspicuous black malar patches. Wing and tail worn, as in worn adult ♂, but fringes and tips usually still pink-cinnamon, less whitish, especially towards feather-bases. NESTLING. No inform-

ation. JUVENILE. Entire upperparts closely streaked buff-white and dull black or grey-black, slightly warmer buff on hindneck, paler and more extensively buff-and-white on rump and upper tail-coverts. Supercilium broad (narrower above front of eye), pale buff; eye-ring conspicuous, cream-white. Upper cheek and ear-coverts dark brown, tinged buff on centre of ear-coverts, brown merging into pale buff side of neck, which is speckled dusky grey. Lower cheek and underparts dirty buff-white; a mottled black malar stripe, merging into gorget of mottled black streaks across chest; side of breast and flank with faint black streaks. Tail and wing as adult, but fringes of tail, flight-feathers, and tertials dull cinnamon, less pale than adult; all upper wing-coverts black on centre, broadly fringed pale buff, black extends into sharp point on tip of covert. FIRST ADULT. Like adult, but juvenile tail, flight-feathers, greater upper primary coverts, bastard wing, and variable number of greater upper wing-coverts retained, ground-colour dark brown-grey, less greyish-black, tips of tail-feathers and primary coverts more sharply pointed than in adult, more frayed than in adult at same time of year, juvenile greater coverts (if any) with black on centres extending into point towards tip, less broadly rounded. In ♂, head and body like adult ♂, but all feather-tips more broadly fringed buff when plumage fresh, more fully concealing black on head and mantle and white on neck and rump; black on centres of feathers of crown extends into small point (in adult, broadly rounded); black on centres of mantle and scapulars narrower, more broadly bordered by rufous-chestnut; chin and throat more extensively mottled white. When worn, head and body of 1st adult ♂ more similar to adult ♂, but black of side of head and of chin and throat less uniform, partly mottled buff-white on supercilium and throat; white of hindneck mottled dusky; mantle and scapulars more clearly streaked white. Head and body of 1st adult ♀ as in adult ♀, but cap more extensively pink-cinnamon, black streaks narrower; upper cheek and ear-coverts more cinnamon, less spotted black or brown, except at rear end, usually no black spots on central throat, and black gorget more broadly broken in middle.

Bare parts. ADULT, FIRST ADULT. Iris brown or dark brown. In breeding season, bill plumbeous- or greyish-black with slightly paler dark grey-horn base of lower mandible; in winter, bill dark horn-grey to blackish-horn, lower mandible pink-grey or pale horn-grey with slight pink tinge. Leg and foot flesh-brown or light brown, tinged grey on toes. (BMNH, ZMA.) NESTLING. No information. JUVENILE. Iris brown. Bill dark horn, paler at base of lower mandible. Leg and foot pale flesh. (BMNH.)

Moults. ADULT POST-BREEDING. Complete; primaries descendent. Heavily worn but moult not started in 6 birds in eastern Tien Shan (Sinkiang, China), 4-6 August (BMNH), in 3 from Mongolia on 6-12 August (Piechocki and Bolod 1972), or in 2 from Manchuria on 25 August (Piechocki 1959), but another from Mongolia had just started 12 August, primary moult score *c.* 2 (Piechocki and Bolod 1972). Moult just starting mid-August in Transbaykalia and from 25 July in Altai (Dementiev and Gladkov 1954). Plumage fresh on 20-25 September (BMNH, ZFMK). POST-JUVENILE. Partial: head, body, lesser and median upper wing-coverts, and variable number of greater coverts (sometimes none). In Manchuria, 2 birds fully juvenile 16 September, another in moult then (Meise 1934a). In birds examined, plumage fully juvenile 5-7

August, moult completed in birds from mid-September and later (BMNH, ZFMK). Moult starts August; in some already completed in 2nd half of August (Dementiev and Gladkov 1954).

Measurements. ADULT, FIRST ADULT. *E. p. polaris.* Central Siberia (mainly along Yenisey), April–October, and migrants north-east China, March–April and October–November; skins (BMNH, ZFMK, ZMA). Bill (S) to skull, bill (N) to distal corner of nostril; exposed culmen on average 2·5 less than bill (S).

WING	♂	73·6 (2·05; 10)	71–77	♀ 69·9 (1·47; 9)	68–73
TAIL		60·8 (2·82; 10)	57–65	58·9 (1·70; 9)	57–62
BILL (S)		11·5 (0·67; 10)	10·8–12·5	11·4 (0·44; 9)	10·6–12·0
BILL (N)		6·7 (0·31; 10)	6·3–7·1	6·6 (0·24; 9)	6·3–7·0
TARSUS		18·2 (0·49; 10)	17·5–19·0	18·2 (0·56; 9)	17·3–18·9

Sex differences significant for wing.

Yamal peninsula (north-west Siberia), wing: ♂ 71 (8) 70–73, ♀ 68·0 (3) 64–70 (Danilov *et al.* 1984). Exposed culmen: 8·1–9·2 (12) (Olsen 1989). See also Loskot (1986*b*).

Nominate *pallasi.* Eastern Tunisia (China): wing, 75·5 (3) 74–78, ♀ 71·7 (3) 70–75; bill (S), ♂ 12·5 (3) 11·9–12·8, ♀ 12·1 (3) 11·9–12·5; bill (N), ♂ 7·3 (3) 7·1–7·5, ♀ 7·1 (3) 7·0–7·2; tarsus, ♂ 18·6 (3) 18·1–18·9, ♀ 18·6 (3) 18·6–18·7 (BMNH). Exposed culmen: 8·9–9·5 (13) (Olsen 1989).

E. p. minor. Manchuria (northern China), wing: ♂ 69–74, ♀ 66·5–71 (Meise 1934*a*).

E. p. lydiae. Orog Nuur (southern Mongolia), wing: ♂♂ 67, 69; ♀ 70 (Piechocki and Bolod 1972).

Weights. ADULT, FIRST ADULT. *E. p. polaris.* Yamal peninsula (north-west Siberia), summer: ♂ 14·6 (8) 13·9–15·3, ♀ 14·2 (3) 13·6–16·4 (Danilov *et al.* 1984). USSR: ♂ 12·6 (5) 11·8–14 (Dementiev and Gladkov 1954).

Nominate *pallasi.* Mongolia: May, ♂♂ 13, 18; August, ♂♂ 16, 16, ♀♀ 15, 16 (Piechocki and Bolod 1972). USSR: ♀♀ 13, 13·4 (Dementiev and Gladkov 1954).

E. p. minor. Manchuria (north-east China), ♀♀: August, 11·3; September, 12, 14 (Piechocki 1959). See also Loskot (1986*b*).

E. p. lydiae. Mongolia, early June, ♂♂ 13, 13, ♀ 13 (Piechocki and Bolod 1972).

Structure. Wing rather long, broad at base, tip bluntly pointed. 10 primaries: p6–p8 longest, p9 2–3·5 shorter, p5 1–2·5, p4 5–7, p3 8–11, p2 10–12, p1 12–15 (*polaris*; for other races, see Loskot 1986*b*); p10 reduced, 47–54 shorter than p6–p8, 6–9 shorter than longest upper primary covert. Outer web of p5–p8 emarginated; inner web of p6–p9 with faint notch. Tip of longest tertial between tips of secondaries and p2. Tail rather long, tip forked; 12 feathers, t4 longest, t1 2–6 shorter, t6 1–2. Bill short, rather fine, conical; depth and width at base each 5·0 (10) 4·7–5·5 mm;

culmen and gonys straight or slightly curved (less so than in Reed Bunting *E. schoeniclus*), ending in sharply pointed tip. Tarsus and toes rather short, slender. Middle toe with claw 17·2 (7) 16–18 mm; outer toe with claw *c.* 71% of middle with claw, inner *c.* 67%, hind *c.* 76%. Remainder of structure as in *E. schoeniclus*.

Geographical variation. Slight; involves size (wing, bill, or tarsus length) and general colour; as colour strongly dependent on time of year also, differences between various races often difficult to establish. Account based mainly on Loskot (1986*b*). Generally, birds in north smaller, and paler there in west than in east; birds in south larger, paler in east than in west. Wing of *polaris* in north-west mainly 64–73 (Sushkin 1925; Johansen 1944; Dementiev and Gladkov 1954; Danilov *et al.* 1984); palest of all races, especially in west of range. *E. p. minor* of north-east (east of Lena valley, where it grades into *polaris*) on average darker, more heavily streaked black on rump, and slightly smaller. In south, nominate *pallasi* from mountains of central Asia (eastern Tien Shan through Mongolia to south-west Transbaykalia, north to Sayan mountains) is largest race, wing mainly 70–78; upperparts in fresh plumage buff-brown, underparts warm buff, less pale than in *polaris*; black streaks on upperparts much broader (though still largely concealed in autumn), but mantle and scapulars mainly black in spring, pale fringes ('tramlines') almost completely worn off. Apparently grades into *polaris* in north, as birds collected as spring migrants in Krasnoyarsk along Yenisey are rather large, close to nominate *pallasi* (see Measurements), though pale, as *polaris*; perhaps *polaris* is not always small (see Mauersberger 1983). Nominate *pallasi* is replaced by *minor* east from northern end of Lake Baykal and eastern Transbaykalia. Position of southern race *lydiae* problematic: occurs locally within range of nominate *pallasi* but at lakes in plains, not in mountains; smaller and paler than nominate *pallasi*, tail relatively longer, wing shorter and more rounded; streaks on upperparts and wing brown instead of black. Here considered a race of *E. pallasi*, following Loskot (1986*b*) and Stepanyan (1990), but sometimes not recognized (Dementiev and Gladkov 1954) or perhaps a separate species (Vaurie 1959); further study required.

Recognition. Rather like *E. schoeniclus*, but ground-colour of upperparts and of fringes of tail- and flight-feathers distinctly paler, mantle and scapulars black with pale 'tramlines' when plumage worn; rump whitish; sides of head indistinctly patterned, except for pale stripe on lower cheek; lesser upper wing-coverts ash-grey or grey-brown (in *E. schoeniclus*, rufous-cinnamon to chestnut), median and greater coverts black with tips buff-grey (when fresh) or white (when worn), forming distinct pale wing-bars; tail more extensively white; culmen straight or virtually so. See also Kitson (1979, 1982), Broad and Oddie (1982), Olsen (1989), Bradshaw (1991, 1992), and Svensson (1992).

CSR

Emberiza bruniceps **Red-headed Bunting**

Du. Bruinkopgors   Fr. Bruant à tête rousse   Ge. Braunkopfammer
Ru. Желчная овсянка   Sp. Escribano carirrojo   Sw. Stäppsparv

Emberiza bruniceps Brandt, 1841. Synonym: *Emberiza icterica.*

Monotypic

Field characters. 16 cm; wing-span 24·5–28 cm. Slightly smaller and more compact than Black-headed Bunting *E. melanocephala*, with somewhat shorter, more rounded wings and slightly stubbier bill. Rather large bunting; structure and plumage of ♀ and immature much as *E. melanocephala*. ♂ has variable golden and chestnut head and bib, distinctly greenish mantle, bright yellow-green rump, and yellow underparts. ♀ resembles *E. melanocephala* but many show greenish-olive or grey tone on crown and back, while a few have buff-chestnut on forecrown, lower throat and upper breast. Juvenile plumage overlaps with *E. melanocephala* but typically less buff below. Call distinctive. Sexes dissimilar; individual and seasonal variation in ♂. Juvenile separable. Hybridizes with *E. melanocephala* (♀ hybrids indistinguishable from that species).

ADULT MALE. Moults: September–December (complete). Crown chestnut to golden, when fresh tipped brown and on some streaked; when golden, forms pale cap above dark face. Sides of head, throat, and centre of breast usually drenched deep chestnut; feathers tipped buff when fresh, but some birds have throat golden and many have stripe under ear-coverts golden. Nape, back, and upper tail-coverts distinctly greenish, quite strongly streaked dark brown and contrasting with long yellowish rump. Wings and tail much as *E. melanocephala* but lesser coverts fringed green (as back). Rest of underparts yellow, with golden to lemon tone. Bill blue-grey; legs brownish-flesh. Note that individual variation or hybridization can produce chestnut spots on mantle and rump. ADULT FEMALE. Differs from ♂ as in *E. melanocephala* which it closely resembles except for (a) somewhat darker more sandy green- or grey-brown head and upperparts, only showing tinge of chestnut, (b) usually sharper dark brown streaks above, especially on scapulars, and (c) typically paler greenish-yellow rump. When worn in spring, often shows traces of pale chestnut on forecrown, face, throat, and breast. JUVENILE, FIRST-WINTER. Usually considered indistinguishable from *E. melanocephala*, but typically has paler underparts, sometimes with more obvious and extensive dark brown spots and streaks on breast and flanks (see also Plumages).

♂ virtually unmistakable, but in glimpse on ground, beware confusion with Chestnut Bunting *E. rutila* (15% smaller, with short tail and wholly rufous upperparts). ♀

and immature much more difficult, and in fresh plumage may be inseparable from *E. melanocephala*. Flight, actions, and behaviour much as *E. melanocephala*.

Song very similar to *E. melanocephala*. Commonest call more penetrating than *E. melanocephala*, a brisk musical 'pwip', 'tweet', 'pweek', 'tliip', or 'tlyp', somewhat like House Sparrow *Passer domesticus* but with liquid tone; other calls include 'chuh' (suggesting Yellowhammer *E. citrinella*), 'chip' or rather harsh 'jyp', and more metallic 'ziff'.

Habitat. Adjoining and complementary to that of Black-headed Bunting *E. melanocephala* in south Palearctic, but mainly in warmer, drier, and more open country, with less vigorous vegetation, in steppe, semi-desert, and desert oasis situations (Harrison 1982). Prefers thickets where available, and sings from top of bush or telephone wire (Flint *et al.* 1984). Occupies all kinds of shrubby and herbaceous thickets, scattered in thin patches over relatively open countryside, but is highly typical of cultivated areas, seeking out water. Often nests close to human habitations, and ascends mountains freely to *c.* 2000 m, in places nesting up to 3000 m. Flies readily, in preference to covering even short distances on ground (Dementiev and Gladkov 1954).

In Indian winter quarters associates with *E. melanocephala*, sharing habitat and habits (Ali and Ripley 1974).

Distribution. Breeds central Asia, from Kazakhstan east to Altai mountains, south to eastern Iran, Afghanistan, northern Pakistan, and western Sinkiang. Winters in peninsular India.

Accidental. Iceland, Faeroes, Britain, Ireland, France, Spain, Netherlands, Belgium, Germany, Denmark, Norway, Sweden, Czechoslovakia, Switzerland, Italy, Israel, Kuwait.

Movements. Migratory, all birds moving south-east to winter in India. Movement diurnal, at least in part (Korelov *et al.* 1974).

Widespread on Indian plains in winter, from Haryana, Rajasthan, and Gujarat east and south to north-east Bangladesh, eastern Maharashtra, western Andhra Pradesh, southern Karnataka, and western Tamil Nadu (Ali and

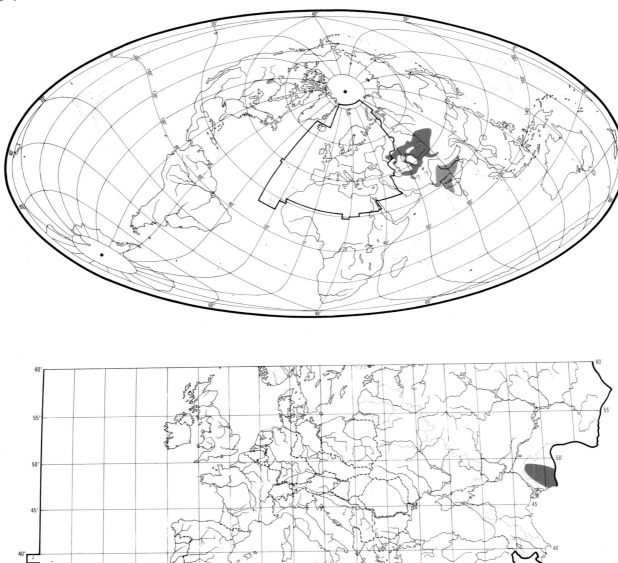

Ripley 1974). Exceptionally recorded from Nepal (Inskipp and Inskipp 1985). Winter quarters thus more extensive than in Black-headed Bunting *E. melanocephala*.

Autumn migration July–September. In Kazakhstan, birds leave Barsa-Kel'mes island (Aral Sea) late July to mid-August, and Kurgal'dzhino in north by *c*. 20 August; movement most conspicuous in southern Kazakhstan, and many birds migrate through foothills of Talasskiy Alatau. In Talasskiy Alatau, local birds vacate breeding grounds at high levels by about the middle of August, and occur at lower levels in cultivated areas until early or mid-September; over 5 years, latest records (juveniles) 4–11

September. (Kovshar' 1966, 1985; Korelov *et al.* 1974; Eliseev 1985; Kovshar' *et al.* 1986.) In Pamir-Alay mountains, passage mid-August to mid-September; apparently regular at high levels, e.g. recorded at Yashil'kul' and Rangkul' lakes, *c*. 4000 m (Ivanov 1969). Most birds leave Turkmeniya in August; at Badkhyz in south, latest records 5 August to 8 September over 6 years; exceptionally recorded 27 September, once 6 December (Rustamov 1958). In Iran (breeds in north-east), recorded only until early August, and apparently migrates directly south-east, as not recorded outside breeding areas (D A Scott). In north-east Afghanistan, leaves some breeding areas in July (when crops harvested), and large flocks gather in central mountains (where crops not yet harvested) in mid- to late July (Reeb 1977). Passage through Pakistan to India mostly August–September. In Pakistan, migrates mainly in northern Sind and southern Punjab; uncommon in northern Punjab, and absent from southern Sind where *E. melanocephala* common. In Jhang district (Punjab), over 2 years, recorded 22 July to 14 September, chiefly from 2nd week of August to 1st week of September. (Whistler 1922*a*; Ali and Ripley 1974; Roberts 1992.)

In spring, migration begins in March, and passage through Pakistan is late March to May; in Kohat and Kurram districts, occurs in large numbers mid-April to mid-May (Whitehead 1909; Ali and Ripley 1974; Roberts 1992). Recorded in Iran from 4th week of April (D A Scott); in north-east Afghanistan, earliest records 20–21 April over 2 years (Reeb 1977). Usually reaches southern Turkmeniya early, by mid-April; at Badkhyz, earliest 8–15 April over 6 years; arrives in northern and western Turkmeniya in late April and early May; some birds continue along east shore of Caspian Sea, and both east and west shores of Aral Sea (Rustamov 1958). Arrives in Uzbekistan and Tadzhikistan from last third of April, with intensive passage in early May, e.g. 3–8 May in one year in Kafirnigan valley south of Gissarskiy mountains (Salikhbaev and Bogdanov 1967; Ivanov 1969; Kovshar' *et al.* 1986). Marked easterly passage recorded in Nuratau mountains (Uzbekistan) in May. In mountainous areas, birds apparently migrate chiefly via foothills in spring. (Dementiev and Gladkov 1954; Kovshar' *et al.* 1986.) Arrives on breeding grounds in southern Kazakhstan from early May (occasionally late April), in northern Kazakhstan and at high levels not before mid-May (Korelov *et al.* 1974); in Talasskiy Alatau foothills, over 19 years, average earliest record 6 May (21 April–16 May) (Kovshar' *et al.* 1986). On Barsa-Kel'mes island, over 3 years, first ♂♂ arrived 3–4 May, and first ♀♀ 10–13 May (Eliseev 1985); at Kurgal'dzhino, earliest arrivals 17–24 May over 3 years (Kovshar' 1985).

Records of vagrancy to west Palearctic (see Distribution) tend not to be published in detail, as attributed mostly to escaped cagebirds, especially in northern Europe, where imported in large numbers until fairly recently (Alström and Colston 1991). In Britain (and probably elsewhere, e.g. France: Dubois and Yésou 1992), reports tend to be at migration seasons and in coastal areas, thus coinciding with vagrants of other species; this not surprising, however, since escaped individuals of migratory species are likely to show seasonal movements in appropriate directions, and to be observed especially at well-studied sites; also, observations inland not necessarily reported, owing to dubious origin (Ferguson-Lees 1967; Hudson 1967). Some genuine vagrancy probably occurs, however (Alström and Colston 1991; Dubois and Yésou 1992); this supported by rapid colonization northwestwards in Kazakhstan, from *c.* 57°30′E along Emba river in *c.* 1900, to north-west border of Kazakhstan and beyond to *c.* 45°E along Volga river in *c.* 1960–70 (Korelov *et al.* 1974).

DFV

Food. All information extralimital. Diet seeds (especially cereals) and other plant material, invertebrates in breeding season. Adults apparently eat much plant food throughout summer, though diet of young almost wholly invertebrates. Little information from summer quarters on foraging sites, techniques, etc.; feeds mostly on ground often in or near areas of cultivation, also in shrubs and bushes. (Dementiev and Gladkov 1954; Armani 1985.) Most feeding observations are of wintering or passage birds since it occurs in considerable flocks causing damage to crops; in Pakistan, feeds in fields of *Sorghum*, millet *Pennisetum*, etc., taking grain from seed-heads as well as from ground; in April, flock fed in green fodder picking up caterpillars (Roberts 1992). In India, diet and habits identical to Black-headed Bunting *E. melanocephala* (Ali and Ripley 1974). Vagrant (or escape) in Belgium, early September, fed with House Sparrows *Passer domesticus* on fruits of goosefoot *Chenopodium* and cereals (Lefever and Hublé 1955).

Diet includes the following. Invertebrates: damsel flies and dragonflies (Odonata), grasshoppers, etc. (Orthoptera: Gryllidae, Tettigoniidae, Acrididae), mantises (Dictyoptera: Mantidae), bugs (Hemiptera: Pentatomidae, cicadas, etc. Auchenorrhyncha), lacewings, etc. (Neuroptera), adult and larval Lepidoptera (Noctuidae), flies (Diptera), Hymenoptera (ants Formicidae, wasps Vespidae, bees Apoidea), beetles (Coleoptera: Carabidae, Geotrupidae, Scarabaeidae, Buprestidae, Tenebrionidae, Coccinellidae, Chrysomelidae, Curculionidae), spiders (Araneae: Lycosidae), ticks (Acari), snails (Pulmonata). Plants: seeds, etc., of Polygonaceae, goosefoot *Chenopodium*, Leguminosae, grasses (Gramineae, including millet *Pennisetum*, *Panicum*, oats *Avena*, wheat *Triticum*, *Sorghum*, rice *Oryza*). (Dubinin 1953; Lefever and Hublé 1955; Rustamov 1958; Popov 1959; Pek and Fedyanina 1961; Kovshar' 1966; Salikhbaev and Bogdanov 1967; Ali and Ripley 1974; Korelov *et al.* 1974; Eliseev 1985; Roberts 1992.)

In Ural valley (north-west Kazakhstan), 13 stomachs, late June to August, contained 49 insects, of which 41% (by number) Orthoptera (12% Acrididae, 6% Gryllidae), 31% caterpillars, 18% beetles (6% Curculionidae, 4% Carabidae, 4% Chrysomelidae, 4% Coccinellidae), 4% adult and pupal Diptera, 4% Hymenoptera, 2% Odonata; some seeds in 3 stomachs (Dubinin 1953); in Talasskiy Alatau (southern Kazakhstan), 3 stomachs in May contained Orthoptera, Hemiptera, caterpillars, beetles (mostly Chrysomelidae and Curculionidae), and spiders; 1 stomach in June held Dermaptera, caterpillars, beetles, and plant material, and 5 stomachs in July contained Orthoptera, beetles, and cereal grains (Kovshar' 1966); see also Korelov *et al.* (1974) and Eliseev (1985). In Tadzhikistan, summer, 40 stomachs contained 83 items, of which 69% by number invertebrates (including 22% beetles, 19% Orthoptera, 6% caterpillars, 2% Hemiptera and Neuroptera, 1% spiders) and 31% plant material (including 10% seeds and 2% berries) (Popov 1959). In Uzbekistan, summer, of 20 stomachs, 19 contained invertebrates and 3 contained seeds; Orthoptera were in 6, Scarabaeidae in 6, Lepidoptera in 5, Tenebrionidae in 4, snails in 4, and Buprestidae in 3 (Salikhbaev and Bogdanov 1967); of 17 summer stomachs, Diptera were in 14, caterpillars in 14, Hemiptera in 8, Curculionidae in 8, Hymenoptera in 6, and 3 contained seeds (Sagitov and Bakaev 1980). In Turkmeniya, 5 stomachs in breeding season contained 13 insects, of which 9 were beetles (6 Carabidae, 2 Curculionidae) and 3 were Orthoptera (Rustamov 1958). In Kirgiziya, summer, 40 stomachs contained insects and plant material, 20% of which was wheat grains; remainder included oats and seeds of Polygonaceae and Leguminosae (Pek and Fedyanina 1961). On breeding grounds, feeds in flocks in cereal fields late July and August before migrating (Grote 1934b; Dementiev and Gladkov 1954). Captured wild bird accepted only seeds, refusing green material and seeds still attached to heads; fed only on aviary floor (Vauk 1961).

Young are fed invertebrates. In Turkmeniya, collar-samples contained 196 invertebrates, of which 80% by number Orthoptera, 18% Lepidoptera (17% caterpillars), and 3% Hymenoptera (Rustamov 1958). On Barsa-Kel'mes island (Aral Sea), 131 collar-samples contained Orthoptera, mantises, adult and larval Lepidoptera, beetles, and *Tarentula* spiders (Eliseev 1985); see also Stepanyan and Galushin (1962). In Talasskiy Alatau, 18 collar-samples held 12 Orthoptera, 4 adult Noctuidae, and 2 caterpillars (Kovshar' *et al.* 1987). In captivity, fed by parents with small insects and regurgitated seeds at first; after 3 days given larvae and seeds (Timmis 1973). BH

Social pattern and behaviour. Most aspects quite well known, but largely from extralimital sources. Main study by Panov (1989) includes some useful comparisons with similar Black-headed Bunting *E. melanocephala* (with which *E. bruniceps* hybridizes in sympatric zone). Also important study by Andrew (1957a) of displays of captive birds (2 pairs), and comparison with other *Emberiza*. In view of ♂'s apparently minimal role in all nesting activities, information needed on possibility that mating system polygamous.

1. Gregarious for much of year, and even in breeding season sometimes forages in groups. Regular flocking starts as soon as young fledge, 10–15(–50) birds per flock. (Grote 1934b; Kovshar' 1966.) In immediate post-breeding period, flocks are of juveniles (Korelov *et al.* 1974). In winter quarters, flocks of 2–300; migrates in mostly large flocks, e.g. 50–100 in April, but autumn flocks smaller and more dispersed (Roberts 1992). Smaller flocks also indicated by the time birds reach breeding grounds, e.g. ♂♂ (which migrate ahead of ♀♀) in tight-knit flocks of 5–7(–30) birds (Panov 1989); in Turkmeniya, in year when both sexes arrived together, flocks of 2–8, mid-April, containing both sexes (Rustamov 1958). In early May, southern Uzbekistan, migrant flocks of 20–50 ♀♀ (including very few ♂♂) (Stepanyan 1970). Com-

monly associates with other birds outside breeding season. In Afghanistan, early August, flocks of hundreds mixed with House Sparrow *Passer domesticus* and Tree Sparrow *P. montanus* (Smith 1974). Said not to mix on passage with *E. melanocephala* (Roberts 1992) though see Erard and Etchécopar (1970) for ♂ accompanying *E. melanocephala* in Iran. In winter quarters, India, often associates with *E. melanocephala* 'in enormous numbers' to ravage crops (Ali and Ripley 1974; see also Roosting, below). BONDS. Mating system not adequately studied; lack of involvement by ♂ in nesting activities and brood-care (see below) suggests likelihood of polygyny and promiscuity (Panov 1989), but no empirical evidence available. Widespread agreement that only ♀ builds nest, incubates, and rears (broods and feeds) young, at least to fledging (Rustamov 1958; Ivanov 1969; Sagitov and Bakaev 1980; Kovshar' *et al.* 1987). For suggestion that both sexes build nest and rear young, see Korelov *et al.* (1974), but all other sources agree that only ♀ builds. In captive study, ♂ took very little part in feeding young till *c.* 5 days, thereafter took about equal role with ♀ (Timmis 1973). In the wild, however, even after fledging, ♂ made no contribution; when brood part-fledged, ♀ fed chick which had left nest as well as those remaining in nest, and apparently fed the entire brood once fledged (Kovshar' *et al.* 1987). Parental roles require further study in cases where new clutch started just as brood leaves nest; in one case, ♀ seen building new nest while still caring for young near to fledging, and 4 days later new nest contained 2 eggs; in another case, ♀ with nest containing 3 eggs seen feeding fledglings nearby (Rustamov 1958). Although no involvement of ♂ indicated in such cases of overlap, difficult to see how ♀ alone could adequately feed fledged brood while incubating new clutch. No information on age at independence in the wild, but in captive study parents continued feeding young for *c.* 2 weeks after leaving nest (Timmis 1973). Age of first breeding not known. *E. bruniceps* hybridizes with *E. melanocephala* in contact zone in northern Iran (Paludan 1940; Haffer 1977). In 3 study sites, proportion of hybrids ranged from 8–57% (Haffer 1977). For isolating mechanisms between the 2 species see Panov (1989). In Suffolk (south-east England) apparent ♂ *E. bruniceps* bred with ♀ Yellowhammer *E. citrinella* and 1 out of 3 eggs hatched (Benson 1967). BREEDING DISPERSION. Territorial (apparently all-purpose type) and solitary or in neighbourhood groups. Size and density of groups vary with region and year. Where population high, groups typically contain 8–10 territories, with distance between (main) song-posts of neighbours not exceeding 15 m; where population lower, e.g. 50–100 m apart. (Panov 1989.) However, in Turkmeniya, no 'colonies' formed even where local density high; record of 54 nests (of which 12 active) in 2·4 ha of Kushka valley; in another area, not more than 3–4 singing ♂♂ in 2 km² (Rustamov 1958). Likewise in southern Uzbekistan, no 'colonies' although nests as little as 10–30 m apart (typically *c.* 40–50 m) (Sagitov and Bakaev 1980). In contact zone of *E. bruniceps* and *E. melanocephala* in northern Iran, both species dispersed in groups of *c.* 5–10 pairs, with ♂♂ spaced *c.* 100 m apart (Haffer 1977); also in northern Iran, in irrigated fields with scattered fruit trees, average 1 singing ♂ (*E. bruniceps*) every 60 m (Erard and Etchécopar 1970). Other (mostly extralimital) densities of *E. bruniceps* as follows. In Kazakhstan, 6–8·9 pairs along 10 km; also 80 pairs along 10 km of lower Ural river (Poslavski 1974); 1·0–6·5 birds along 10 km of lower Emba river (Neruchev and Makarov 1982). In Turkmeniya, 20–22 birds per km² (Vronski 1986); also 173 birds per km² on agricultural land, Murgab valley (Drozdov 1968). In Uzbekistan, 10 singing ♂♂ along 4 km of lower Amu-Dar'ya river (Paevski *et al.* 1990). In Tadzhikistan, 10–20 pairs per km² in tall grass semi-savanna; 200–300 in open grassy habitat with scattered shrubs and trees

(Tolstoy 1986). 2nd clutches laid in new nest built near 1st (Rustamov 1958). No information on site-fidelity between years. ROOSTING. Little information, and none for adults in breeding season. Young recorded returning to nest to roost for 3 nights after fledging (Kovshar' *et al.* 1987). In winter, India, forms communal roosts of tens to hundreds with, variously, wagtails *Motacilla*, Rose-coloured Starling *Sturnus roseus*, mynas *Acridotheres*, *P. domesticus*, weavers *Ploceus*, munias (Estrildidae), Scarlet Rosefinch *Carpodacus erythrinus*, and *E. melanocephala* (Gadgil and Ali 1975; Dhindsa and Toor 1981).

2. In alarm, flicks tail (first up, then down) and raises crown feathers (Andrew 1957a). For alarm reactions in winter flocks mixed with *E. melanocephala*, see that species. Advertising behaviour (song-display, etc.) generally similar to *E. melanocephala* (Panov 1989). FLOCK BEHAVIOUR. Flocks on spring passage keep up constant soft twittering when they settle in tree (Roberts 1992). In compact flocks of ♂♂ on spring migration, and when newly arrived on breeding grounds, individual distance sometimes not more than 1 m, and practically no mutual aggression; during the week following arrival on breeding grounds, ♂♂ occupy territories but remain to some extent sociable, resulting in close breeding groups, even in relatively uniform habitat (Panov 1989). SONG-DISPLAY. ♂'s song (see 1 in Voice) serves for territorial advertisement and mate-attraction. Sings typically from top of tall shrub (Fig A, right), hummock, or any available vantage point including large stones and overhead wires along road (Ivanov 1969; Haffer 1977; Piechocki *et al.* 1982), also in 2 kinds of Song-flight (see below). Observations in Iran apparently established that *E. bruniceps* sang much more in flight than when perched, whereas *E. melanocephala* gave opposite impression (Erard and Etchécopar 1970), but hard evidence lacking for this difference. Once territory established, *E. bruniceps* sings for long spells from favoured song-posts in territory, e.g. 1 ♂ spent most time singing from 2 perches 2–40 m from nest, average 78 songs per hr (54–117, *n* = 8 hrs over various days), and continued singing during chick-rearing (by ♀) (Kovshar' *et al.* 1987). Where nests in groups (see Breeding Dispersion, above) Song-duels occur, with 2–3 ♂♂ audible at any time and place (Panov 1989). From time to time, ♂ performs Moth-flight (Fig A: middle), strongly reminiscent of Dangling-legs display of Corn Bunting *Miliaria calandra* (Erard and Etchécopar 1970): flies, singing (or calling: see 2–3 in Voice) from one song-post to another, low over ground, with shallow, quivering wing-beats; especially shallow

when excited, bird then appearing scarcely to progress; also at high intensity, tail slightly raised and plumage markedly ruffled (rendering yellow rump highly conspicuous: Meinertzhagen 1938); at times, Moth-flight interspersed with short glides. Moth-flight continues right up to when bird lands on next song-post, and if flying thus towards rival, displaying bird keeps wings raised briefly on landing (Fig A, left). (Panov 1989.) Wing-quivering of Moth-flight begins when still perched (Wing-quivering display: Fig B); rump may be raised, and bill may be

B

held lowered during wing-quivering or repeatedly lowered during it (Bill-lowering); also tail may occasionally be drooped and spread (Andrew 1957a, homologous with 'fluffed run display' of other *Emberiza*). Elements of Wing-quivering display may be seen in various combinations during and after pair-formation in all sorts of social interactions, territorial and heterosexual (Panov 1989; see Heterosexual Behaviour, below). In another, less common type of Song-flight at peak of courtship period, ♂ flies high up and circles over territory (Panov 1989); ascends rapidly, singing, followed by parachute-descent recalling song-flight of Tree Pipit *Anthus trivialis* (Grote 1934b; Erard and Etchécopar 1970; Korelov *et al.* 1974). ♂♂ start singing 2–3 days after arrival on breeding grounds (Sagitov and Bakaev 1980; see also Flock Behaviour, above). In Zeravshan mountains (Pamir-Alay), vocal activity low on first arrival, songs being truncated (see 1 in Voice) and confined to morning and evening; activity intensifies markedly with arrival of ♀♀ (at which time territories also become firmly established), ♂♂ now singing more-or-less throughout day (during May) except for slight decrease around hottest time of day; gradual decline from end of May, with reversion to singing truncated songs restricted to morning and evening; song-period ends mid-July (Glushchenko 1986), coinciding with start of moult (Ivanov 1969). Few data on song-period elsewhere, but in

A

southern Uzbekistan song heard from 29 April and intense up to c. 20 May (Sagitov and Bakaev 1980). ANTAGONISTIC BEHAVIOUR. (1) General. ♂ establishes territory before arrival of ♀♀, and is responsible for defence (Sagitov and Bakaev 1980; Glushchenko 1986). (2) Threat and fighting. ♂ will chase conspecific intruders out of territory (Salikhbaev and Bogdanov 1967). In competition for newly-arrived ♀♀ (see Heterosexual Behaviour, below), conflicts arise between ♂♂, usually brief but sometimes ending in fight which starts in the air and finishes on ground (Panov 1989). Following description of threat-display based on captive study (Andrew 1957a): threat is of kind seen in E. citrinella ('head forward posture'; see 5 in Voice for associated call): crouches horizontally, with head (which may be somewhat sleeked) drawn into shoulders, then thrust forward at opponent; rivals face each other gaping, often simultaneously raising and lowering heads; threat sometimes leads to attack (forward rush) or retreat. One ♂ was subordinate to mate and often gave body-cooling responses (variously bill-lowering and gaping, wing-raising, feather-sleeking) after attacking her; also sang in disputes with her. HETEROSEXUAL BEHAVIOUR. (1) General. ♂♂ usually arrive on breeding grounds c. 1 week before first ♀♀ (Eliseev 1985; Glushchenko 1986; Panov 1989), but gap varies from sometimes almost simultaneous arrival of the sexes (Rustamov 1958) to 10–15 days apart (Korelov et al. 1974). Typically ♂♂ already on territory and singing when ♀♀ arrive (see above) and nest-building starts within 4–5 days (Rustamov 1958). (2) Pair-bonding behaviour. When there are still few ♀♀, and territorial structure still loose, ♀ entering area tends to provoke collective chase by up to 6–7 ♂♂ (and skirmishing among them: see Antagonistic Behaviour, above); later on, when territories better-established, usually only 2–3 ♂♂ within immediate vicinity (and response-distance) of any newly-arrived ♀ (Panov 1989). ♂ attracts mate initially by singing, including Song-flights (see Song-display, above). When ♀ first appears in or near ♂'s territory he attempts to approach her, flying quickly towards her in normal flight, but singing vigorously. Unreceptive ♀ attempts to hide in cover or flees, ♂ (sometimes with others) in pursuit. If ♀ receptive, ♂ approaches her in Moth-flight (see Song-display, above) and, if he succeeds in close approach (within 20–30 cm), displays while perched (Fig C) or creeping around her. (Panov

C

1989.) According to Andrew (1957a), this display not different from Wing-quivering display (see Song-display, above), but Panov (1989) apparently distinguishes it from Fig B in that no mention in this context of Wing-quivering or Bill-lowering: ♂ ruffles plumage (notably nape and rump), droops wings, lowers and half-spreads tail (Fig C). At close quarters, ♂ sometimes pivots in front of ♀, suggesting approach–retreat conflict. In one-sixth of all Wing-quivering displays, Bill-lowering was used by ♂ to pick up nest material, which was sometimes then carried in flight, with Wing-quivering on landing. (Andrew 1957a.)

After short interaction with ♀, ♂ flies to next song-post, usually in Moth-flight. ♀ also may use quivering flight, accompanied by call 3, when approaching ♂. (Panov 1989.) (3) Courtship-feeding. None recorded. (4) Mating. ♂ flies to ♀ (who adopts Soliciting-posture (not described, but see 3 in Voice for associated call), mounts, and flies off again (Panov 1989). No further information. (5) Nest-site selection and behaviour at nest. No information on which sex selects nest-site. In captive study, ♂ picked up nest material, but for display purposes only: performed Nest-site display, almost invariably in a dark hollow, sometimes preceded by collecting nest material during Wing-quivering display with Nest-site calls (see 4 in Voice); in Nest-site display, ♂ occasionally shuffles backwards (nest-shaping movement) and gives rapid series of Nest-site calls, intensity increasing if ♀ approaches; unlike in similar display of M. calandra, ♂ very rarely picks up nest-material, although repeated Bill-lowering may occur, and he does not usually Wing-quiver. ♀ gives 'tic' calls when entering nest-site, these becoming almost continuous during her nest-shaping movements or if ♂ approached (Andrew 1957a; see 3 in Voice). In another captive study, when ♀ collected material near nest, ♂ perched high on nearby branch and called very softly (not described); ♂ generally accompanied ♀ when she went further from nest (Timmis 1973). In the wild, during nest-building phase, ♂ seen following ♀ when she left nest but did not accompany her assiduously, instead waiting on song-post for her return (Kovshar' et al. 1987). RELATIONS WITHIN FAMILY GROUP. Young brooded till 8 days old, after which no brooding even in wet weather; all feeding by ♀, and ♂ never seen to visit nest (n = 40 hrs at 6 nests) (Kovshar' et al. 1987). For presumably exceptional feeding of brooding ♀ and young by ♂ in captive study, see Timmis (1973). Eyes of young open not earlier than 4–5 days (Rustamov 1958, which see for further details of physical development). In captive study (in which ♂, probably exceptionally, was involved in care of young), nest-sanitation by both sexes, though apparently far more by ♀; sometimes on feeding visits to nest, faecal sacs were swallowed, but mostly discarded some distance from nest (Timmis 1973). When chick ready to leave nest, ♀ flew there with food (♂ without) and perched quietly nearby until chick left; ♀ then alternately fed fledgling and remaining nestlings; after whole brood fledged, ♀ started foraging near them; young roosted in nest for 3 nights after fledging (Kovshar' et al. 1987). In Kazakhstan, juveniles form flocks in last third of July (Korelov et al. 1974). ANTI-PREDATOR RESPONSES OF YOUNG. No information. PARENTAL ANTI-PREDATOR STRATEGIES. (1) Passive measures. ♀ a tight sitter (Kovshar' 1966; Korelov et al. 1974). (2) Active measures: against man. In Talasskiy Alatau, parents very agitated when young left nest, ♂ singing intensely and mock-preening (Kovshar' et al. 1987). On Barsa-Kel'mes island (Aral Sea), when man approaches nest containing eggs, ♀ leaves silently whereas ♂ flies towards intruder singing; when nestlings or fledglings threatened, both parents flutter around giving Alarm-calls (see 6 in Voice); 'injury-feigning' performed by 5% of ♀♀ (Eliseev 1985). In Kazakhstan, ♂ seen 'injury-feigning' (♀ passive) when young threatened (Korelov et al. 1974).

(Figs by D Nurney based on drawings in Panov 1989.)   EKD

Voice. Freely used in breeding season, but birds relatively silent at other times of year. Calls differ from those of Black-headed Bunting E. melanocephala, but, like that species, often described as sparrow-like (Erard and Etchécopar 1970). Structural differences between the 2 species show in comparison of recordings; all calls of E. bruniceps (except 'chip' and 'tnk': see 2a, below) ascend to the end following brief transient-like clicks (Fig V), while all in *E. melanocephala* descend to the end, some preceded by brief ascending transient (see that species for details) (J Hall-Craggs). However, songs of the 2 species hard to differentiate, at least aurally. For additional sonagrams see Andrew (1957*b*) and especially Panov (1989).

CALLS OF ADULTS. Song of ♂. Rather harsh (yet quite melodious), fairly low-pitched, ringing, short, rapidly delivered phrase of uneven rhythm, usually beginning with a few (e.g. 2-4) 'zrit' units (call-type 2a) (Schüz 1959; Alström and Colston 1991; L Svensson). ♂ sings from perch, typically raised, occasionally on ground, also in display-flight (see Social Pattern and Behaviour); in flight, several songs given in quick succession, and delivery also at times faster (Timmis 1973). Whole song rendered 'trit-trit-chri-chri-cheuh-cheuh-ah', generally repeated monotonously at intervals of 2-3 s, sometimes with minor variations at end of phrase (Roberts 1992, based on Tadzhikistan recording by B N Veprintsev). In Iran, 'tsit-tsit-titeriteri' (Erard and Etchécopar 1970); introductory 'ik ik' followed phrase very similar to Whitethroat *Sylvia communis* but, unlike that species, fading towards end (Schüz 1959). In Tadzhikistan, 665 full songs of 40 ♂♂ recorded in various localities were rather uniform in length (1·4-2 s) but demonstrated clear geographical and individual variation in volume and pitch; songs of birds from different altitudes audibly different, being more flowing at high altitude (Salmanova 1986). Song of any individual fairly stereotyped but sometimes truncated (see below), and number of preparatory call-type units varies (sometimes completely absent) (Panov 1989). Song generally truncated (missing final unit) at start of breeding season before ♀♀ arrive, given in full when ♀♀ arrive, and truncated again towards end of breeding season (Glushchenko 1986). In 7 recordings of song (3 from Kirgiziya, 2 from Iran, 1 from Tadzhikistan, 1 from Netherlands) analysed by J Hall-Craggs, all phrases were preceded by 'zit' (or 'zhu', etc.) calls; Figs I-IV show, respectively, 2, ½, 1, and 4 such calls; song-phrase characteristics include

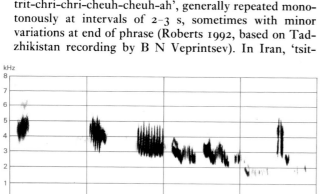

I P A D Hollom Iran May 1977

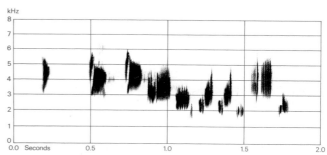

II L Svensson Tadzhikistan June 1986

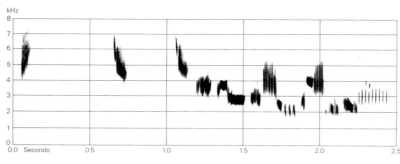

III Mild (1987) Kirgiziya June 1987

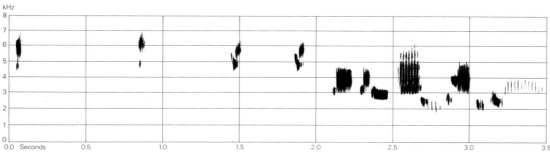

IV L Svensson Kirgiziya June 1983

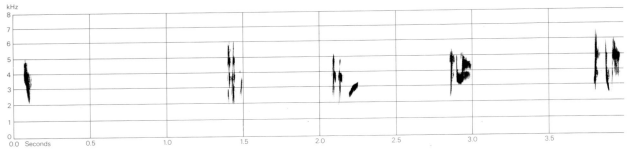

V   L Svensson   Kirgiziya   June 1983

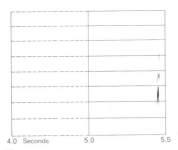

V *cont.*

repeated unit in 1st part (in 4 out of 7 recordings), lower-pitched voice in 2nd part (in all 7), and final buzzy unit (3 of 7) or rattle (3 of 7); in sum, these characteristics offer no simple, useful distinction from *E. melanocephala*, although preparatory 'zit' (etc.) calls are variable (compare Figs I–IV) in *E. bruniceps* whereas those in *E. melanocephala* are quite stereotyped; however, this difference not helpful in the field since 'zit' calls, while visually different, are too brief and high-pitched to differentiate aurally. In northern Iran (where *E. bruniceps* and *E. melanocephala* overlap), *E. bruniceps* described as identical to *E. melanocephala* (Paludan 1940; Schüz 1959); however, according to Erard and Etchécopar (1970), main part of song (i.e. after introductory units) seemed longer, more variable, more agreeable, and less shrill in *E. melanocephala* than in *E. bruniceps*. According to Panov (1989), comparison of songs and sonagrams (not from zone of overlap) revealed the following differences between the 2 species: (a) motifs of vibrato type more predominant in song of *E. bruniceps* than *E. melanocephala*; (b) less change in pitch between consecutive motifs in *E. bruniceps*; (c) less clear temporal pattern in *E. bruniceps*. None of these features absolutely constant or diagnostic, but together they impart more hurried quality to song of *E. bruniceps*. (Panov 1989.) (2) Contact-alarm calls. (2a) Subdued 'zrit' serves as flight- and flight-intention call, and (in series) as song-introduction (see above). Other renderings (not in song) include 'chip' (Andrew 1957*b*), rather harsh 'jyp' (Alström and Colston 1991), 'chzhik' (Korelov *et al.* 1974); brief nasal 'zhu', used in various contexts, sometimes singly (e.g. in flight on migration) but more often in series

(e.g. quite often in ♂'s Moth-flight: see Social Pattern and Behaviour), when may be mixed with call 2b (Panov 1989, which see for sonagram showing both calls alternating). Similar series (Fig V) analysed by J Hall-Craggs suggests description given by Panov (1989) may be oversimplified in that 6 different call-types identifiable, of which probably only 1st ('chip') is call 2a (see 2b, below, for others); recording consists of bout of 45 calls, of which Fig V shows first 5 call-types delivered consecutively (rendered 'chip t-t-wi t-t-ui t-de-li ch-tui'); final call ('tnk', with tonal component) in Fig V is not 6th in sequence of delivery but 25th (appended here for convenience); although call-types 1–5 (Fig V) are delivered consecutively twice at start of bout, commonest associations between pairs of the 6 call-types in whole bout (45 calls) is 1–4–5–1–2–3–6–4–2, i.e. no immediate repetition of any particular call-type. (J Hall-Craggs.) Rather high-pitched variant of call 2a sometimes heard in alarmed flight, probably tending towards call 6 (Andrew 1957*b*). (2b) Chink-call. A 'chink' often given in same contexts as call 2a (including allegedly preceding song, although introductory calls in recordings of song are different from 2b: J Hall-Craggs, E K Dunn); higher pitched than 2a although every intermediate between the two is heard (Andrew 1957*b*); 'pli' (Panov 1989) and musical 'tweet' (Roberts 1992: 'flight-call') apparently this call, perhaps also 'tliip' and 'tlyyp' (Alström and Colston 1991). Sometimes given in twittering series, mixed or not with call 2a (Panov 1989); middle 4 calls in Fig V represent call 2b or variants (J Hall-Craggs; see description of 2a for further explanation). 'Chipz-chew', with pause between units, much like sound of rusty gate swinging, given by captive ♂ from elevated perch (Timmis 1973) perhaps sequence 2b–2a. (3) Excitement-call. Quiet tremolo 'che che che che' or hoarse 'ke ke ke ke' with notable harmonic structure, used rather rarely in high-intensity excitement; like calls 2a–b, can be given by ♂ in Moth-flight, sometimes also by ♀ trying to approach (in quivering flight) ♂ (Panov 1989). Thus Andrew (1957*b*) designated this call ('chichichi...') as 'soliciting-call of ♀'. Variant ('tictic...') heard from ♀ entering nest-site, call repeated almost continuously when making nest-shaping movements or if ♂ approached her (Andrew 1957*a*, *b*). (4) Nest-site call of ♂. A 'si' given singly or (at high intensity) rapidly repeated (Andrew 1957*b*). May be given sporadically in Wing-quivering display, but repeatedly during Nest-site display, repetition-rate and volume increasing if ♀ approaches ♂ in nest-site; then sounds not unlike call 3 of soliciting ♀ (Andrew 1957*a*). (5) Threat-call. A 'chaa' heard during threat and fighting (Andrew 1957*a*, *b*: 'aggressive call'). (6) Alarm-call. Rattling series of 'tit' sounds given typically during escape-flight (Andrew 1957*b*). June vagrant, Denmark, gave harsh rattle from the ground when alarmed by observers (Hansen and Birkholm-Clausen 1968).

CALLS OF YOUNG. In Talasskiy Alatau, young very silent up to fledging, but oldest chick started calling (not described) shortly before fledging (Kovshar' *et al.* 1987). No further information.   EKD

Breeding. All information extralimital. SEASON. Transcaspia and Iran: eggs laid from 2nd half of May (Grote 1934*b*; Erard and Etchécopar 1970). Turkmeniya: eggs laid early May to last week of June (Rustamov 1958; Panov 1989). Northern Uzbekistan: eggs laid from 2nd half of May (Paevski *et al.* 1990). Kazakhstan: mid-May to late June (Stepanyan and Galushin 1962; Kislenko 1974; Korelov *et al.* 1974; Berezovikov and Kovshar' 1992). Eggs recorded into July throughout range (Dementiev and Gladkov 1954). SITE. Low and well hidden in dense or thorny shrub, vine, fruit tree, etc., or very close to ground in thick grass (Ticehurst 1926–7; Dementiev and Gladkov 1954; Sagitov and Bakaev 1980; Paevski *et al.* 1990). In Talasskiy Alatau (southern Kazakhstan), 69% of 452 nests were on bushes or stout herbs (36% on rose *Rosa*, 13% honeysuckle *Lonicera*, 12% thistles, 8% juniper *Juniperus*); of 364 nests, 31% were 1–10 cm above ground in grass tussocks, 54% 10–50 cm, 9% 50–100 cm, 7% more than 100 cm above ground; highest was 4 m up in juniper (Kovshar' *et al.* 1986). On Barsa-Kel'mes island (Aral Sea), 82 nests were 6–160 cm above ground (Eliseev 1985). Nest: rather loose and untidy foundation of stems of cereals, rough grasses, Umbelliferae, Cruciferae, etc., often with flowers attached, sometimes pieces of bark or leaves; lined with fine grass, plant fibres, rootlets, and hair (Grote 1934*b*; Dementiev and Gladkov 1954; Hüe and Etchécopar 1970; Makatsch 1976; Roberts 1992). In Talasskiy Alatau, 232 nests had average outer diameter 10·0 cm (8·5–18), inner diameter 6·5 cm (5·0–7·8), overall height 7·0 cm (4·6–10·4), depth of cup 5·0 cm (2·7–6·7), and dry weight 4–38 g (Kovshar' *et al.* 1986). For 74 nests from Barsa-Kel'mes, see Eliseev (1985). Building: by ♀ only, accompanied by ♂; takes 2–3(–6) days (Rustamov 1958; Timmis 1973; Sagitov and Bakaev 1980; Kovshar' *et al.* 1987). See Korelov *et al.* (1974) for suggestion of ♂ participation. EGGS. See Plate 34. Sub-elliptical, smooth and slightly glossy; white or pale bluish-white, finely and sparsely spotted purplish-grey to brown, concentrated at broad end; no hairstreaks or scrawls (Harrison 1975; Makatsch 1976). For discussion and comparison with Black-

headed Bunting *E. melanocephala*, see Etchécopar (1961); for discussion of colour types in Talasskiy Alatau, see Kovshar' *et al.* (1987). 21·2 × 15·8 mm (19·0–23·3 × 14·4–17·0), $n = 108$; calculated weight 2·77 g (Schönwetter 1984); 66 eggs from Barsa-Kel'mes had fresh weights 2·0–2·3 g (Eliseev 1985). For 111 eggs from Talasskiy Alatau, see Kovshar' *et al.* (1987); for 64 eggs from Turkmeniya, see Rustamov (1958). Clutch: 4–5 (3–6). In Talasskiy Alatau, of 85 clutches: 2 eggs, 4%; 3, 33%; 4, 58%; 5, 6%; average 3·66 (Kovshar' *et al.* 1987). In Turkmeniya, mostly 3–5, replacement 3–4; average of 52 clutches 3·15, but some of 1 egg so presumably incomplete and average brood size was 3·38, $n = 29$ (Rustamov 1958). In Uzbekistan, of 12 clutches, 4 of 4 eggs and 8 of 5; average 4·67 (Salikhbaev and Bogdanov 1967). 2 broods recorded in Turkmeniya; in 2 cases, ♀ apparently built nest and laid eggs while still caring for 1st brood (Rustamov 1958); probably 2 broods in Transcaspia and Iran (Grote 1934*b*). Eggs laid in morning, (1–)2–3 days after nest completed (Sagitov and Bakaev 1980; Kovshar' *et al.* 1987; Berezovikov and Kovshar' 1992). Captive birds laid eggs daily (Timmis 1973). INCUBATION. 12–13 days (10–14); by ♀ only (Rustamov 1958; Korelov *et al.* 1974; Kovshar' *et al.* 1987); hatching started 11 days after laying of last egg (Eliseev 1985). Some authors give 16–17 days (Salikhbaev and Bogdanov 1967; Sagitov and Bakaev 1980). In captivity took 14 days, all 3 eggs hatching in one day (Makatsch 1976). In Talasskiy Alatau, in 29 hrs of observation at 3 nests, average 86% (80–94%) of time spent on eggs (Kovshar' *et al.* 1987; see also Berezovikov and Kovshar' 1992); in Uzbekistan, ♀ took stints of 5–120 min and breaks of 5–30 min (Sagitov and Bakaev 1980). YOUNG. Fed and cared for by ♀ only (Rustamov 1958; Sagitov and Bakaev 1980; Kovshar' *et al.* 1987). In captivity, ♂ did some feeding when young *c.* 5 days old; ♀ brooded continuously for 3–4 days (Timmis 1973). FLEDGING TO MATURITY. Fledging period (9–)12–13 days; often leaves nest before able to fly (Eliseev 1985, which see for development of young; Kovshar' *et al.* 1987; Berezovikov and Kovshar' 1992). Captive young fledged after 15 days; fed by parents for further 2 weeks (Timmis 1973). BREEDING SUCCESS. On Barsa-Kel'mes, of 246 eggs in 71 nests, 70% hatched and 58% produced fledged young, giving 2·0 per nest overall; 24% of eggs deserted or predated, 5% failed to hatch, 10% taken as chicks by Great Grey Shrike *Lanius excubitor*, 2% deserted as chicks, 1% died (Eliseev 1985). In Talasskiy Alatau, of 118 eggs in 40 nests, 17% hatched and 14% produced fledged young; only 5 nests successful, producing 3·4 fledged young per successful nest, 0·4 per nest overall; of 35 nests destroyed, 83% predated at egg stage, 14% deserted, and 3% predated at nestling stage (Kovshar' *et al.* 1987). For south-east Kazakhstan, see Berezovikov and Kovshar' (1992).   BH

Plumages. ADULT MALE BREEDING (November–July). Nasal bristles small, black. Cap down to upper cheek and ear-coverts rich rufous-brown, orange-brown, or yellow-brown, darkest and slightly glossy at base of bill, round eye, and on cheek, usually lighter, more golden-orange, on crown; when fresh, feathers with broad greenish-buff tips, partly concealing brown, tips whitish when worn. Eye-ring rufous-brown. Neck yellow or rufous-brown, feather-tips broadly greenish-buff when fresh; side of neck rufous-brown or yellow. Occasionally a faint yellow malar stripe. Mantle greenish-yellow or greenish, marked with faint narrow blackish or broader dark brown centres (latter sometimes faint) and with yellow or greenish-yellow edges, latter broadly fringed grey-buff when fresh. Scapulars grey-brown, outer web with broad buff or chestnut outer border, inner web with dark brown, buff, or yellowish-green border. Rump and shorter upper tail-coverts yellow with brownish or buffish-green tips; longest upper tail-coverts grey-brown with darker centres, greyish or yellowish-green fringes, and (occasionally) with narrow chestnut tips. Chin, throat, and chest rich rufous-brown or orange-brown, occasionally with golden-yellow tinge, forming large bib; feathers tipped greenish-buff when fresh, whitish when worn. Belly, flank, and under tail-coverts bright yellow, feathers with narrow white fringes when fresh. Tail dark brown with greenish, yellowish, or chestnut edges on outer webs, central pair (t1) slightly paler; outer pair (t6) with dark brown shaft, pale brown to buff inner web and tip of outer web, and with whitish edge on basal half of outer web and near tip of inner web; rarely, inner web with buffish-white or white wedge on distal part of inner web and with white on tip of outer web. Flight-feathers brown; outer web of outer primaries fringed whitish, outer webs of inner primaries greenish, yellowish-green, or buff-brown; outer webs of secondaries fringed pale chestnut. Tertials dark brown with broad chestnut or buffish-chestnut fringes, widest on outer web and near tip. Greater upper primary coverts, greater upper wing-coverts, and bastard wing dark brown, primary coverts with narrow chestnut or brownish edges, greater coverts broadly fringed buff, chestnut, or white with narrow chestnut borders. Median coverts dark brown with broad white, buff, or chestnut tips; tips of greater and median coverts bleached to white when worn, outer fringes paler. Basal half of lesser coverts brown, distal half yellow-green with chestnut and green edges; rarely, outer lesser coverts chestnut with green edges and some chestnut present on median coverts. Under wing-coverts and axillaries yellowish-white; marginal under wing-coverts bright yellow. ADULT FEMALE BREEDING (November–July). Forehead, crown, and nape brown with sandy-grey, greyish-buff, or yellowish-green fringes, feather-centres occasionally with rufous-chestnut spots. Supercilium and lore whitish-grey or buff, ear-coverts sandy-grey or buff. Neck sandy-grey or greyish-green, with little or no streaking; mantle sandy-grey or greenish, feathers with dark brown centres, feathers sometimes with rufous-chestnut hue when fresh; scapulars similar to mantle, but brown centres wider and less sharply demarcated. Rump pale yellowish, yellow, greyish, or green, with faint shaft-streaks; longest upper tail-coverts drab-brown, fringed pale grey when fresh. Chin and throat off-white or buff; faint band across chest greyish-buff, often marked with narrow brown shaft-streaks or small brown spots. Belly off-white or pale yellow; flank greyish-yellow or buff-yellow, occasionally with faint shaft-streaks. Under tail-coverts yellow. Tail and wing as in adult ♂ breeding, but with little or no yellow or chestnut; lesser upper wing-coverts brown-grey with whitish edges. Axillaries as adult ♂, but with narrow shaft streaks. ADULT MALE NON-BREEDING (August–November). Rather like adult ♀ breeding. If new, feathers of forehead, crown, nape, neck, mantle, and scapulars greyish-buff or greenish with dark brown shaft-streaks, often with rufous-chestnut tinge on forehead and chestnut hue on mantle. Lore and patch below eye

yellowish-chestnut, supercilium faint, greenish-grey with rufous-chestnut tinge; ear-coverts greenish-grey. Chin and centre of chest rufous-chestnut, feathers with broad yellowish-white edges, throat and side of neck yellowish-white, or chin, throat, and chest whitish-yellow with scattered rufous-chestnut feathers. Side of breast dirty yellowish-buff. Underparts yellow, mixed with scattered yellowish-white feathers when partly new. Rump, under tail-coverts, tail, and wing as adult ♂ breeding, but feathers heavily abraded. ADULT FEMALE NON-BREEDING (August–December). As adult ♀ breeding, but feathers of upperparts with broad greyish-buff tips, rump with faint greyish-green tinge, and feathers on underparts with pale yellowish-white edges. NESTLING. Down cinnamon according to Harrison (1975), dirty white according to Sagitov and Bakaev (1980); on head and back (Harrison 1975). JUVENILE. Rather like adult ♀, but feathers of forehead, crown, neck, mantle, and back, as well as all upper wing-coverts brown with clear-cut buff edges, rump without yellow or green, and longest upper tail-coverts with dark brown centres. Chin, throat, chest, and belly whitish, buffish-white or yellowish-buff. Faint band across chest, formed by dark brown shaft-streaks. Side of chest, lower cheek, belly, and flank more strongly tinged buff than in adult. Under tail-coverts pale yellow. Tail-feathers similar to adult ♂ breeding, but narrower, tips pointed, with little or no white, and edged and tipped more buffish. Flight-feathers dark brown with narrow buff edges and tips. Under wing-coverts without yellow, whitish or buffish-white, with dark brown centres; axillaries dark brown with buff edges. FIRST ADULT MALE NON-BREEDING. Similar to adult ♂ non-breeding, but in early autumn with variable number of juvenile feathers on mantle, rump, underparts, and all feathers of tail and wing retained. Ear-coverts buff. FIRST ADULT MALE BREEDING. Similar to adult ♂ breeding. A few may be recognized on less extensive rufous-chestnut on head and breast. FIRST ADULT FEMALE NON-BREEDING. Similar to 1st adult ♂ non-breeding. FIRST ADULT FEMALE BREEDING. Similar to adult ♀ breeding.

Bare parts. ADULT, FIRST ADULT. Iris dark brown. Bill light blue-grey, bluish-horn, or leaden-grey, slightly darker on tip, sometimes faintly tinged pink at base. Leg and foot pale flesh-brown, greyish flesh-horn, or reddish-horn (Vauk 1961; ZFMK). NESTLING. Bare skin of throat yellowish-red, of underparts whitish-yellow (Harrison 1975). JUVENILE. No information.

Moults. ADULT POST-BREEDING. Partial: in ♂, involves part of crown, nape, mantle, back, upper tail-coverts, chin, throat, and most tertials and upper wing-feathers, but extent highly variable, and some birds moult only a few feathers. In ♀, moult restricted to a few or many feathers on upperparts, but often no moult at all. Moult may start as early as late June, and most birds usually have at least some feathering new at start of migration in August. Single ♂♂ from 30 June and 24 July had much of upperparts new and many feathers growing on head and throat; another from 17 July had feathers of winter plumage on head, mantle, and underparts. Single ♂ from late September in heavy moult of ear-coverts, chin, throat, chest, and neck, remainder of plumage old. (Stresemann 1924; Paludan 1959; Stresemann and Stresemann 1969a; RMNH, ZMA.) ADULT PRE-BREEDING. Complete or largely so; primaries descendent. Starts on arrival in winter quarters, sometimes as early as late August; duration c. 60 days. Feathers moulted in post-breeding are not replaced again, except for those which show different non-breeding pattern. Primaries start with p1; up to 4 primaries moulted simultaneously. Tail moult rapid, centrifugal. Secondaries, upper primary-coverts, and tail moulted at same time as primaries. Due to rather restricted post-breeding moult in ♀, pre-breeding is more extensive than in ♂; feathers moulted in late summer are not replaced again in winter. (Stresemann and Stresemann 1969a; RMNH, ZMA.) POST-JUVENILE. Partial: head, body, and upper wing-coverts. Starts immediately after leaving nest. FIRST PRE-BREEDING. Partial: in winter quarters; involves body feathers and wing-coverts, but no information on extent and timing.

Measurements. Adult. Whole geographical range, all year; skins (RMNH, ZMA). Bill (S) to skull, bill (N) to distal corner of nostril; exposed culmen on average 3·4 less than bill (S).

WING	♂	88·7 (2·67; 12)	84–92	♀ 85·0 (2·45; 7)	81–88
TAIL		67·2 (3·36; 13)	59–71	63·3 (1·03; 6)	62–65
BILL (S)		15·8 (0·68; 12)	15·1–17·2	14·9 (1·34; 6)	12·8–16·2
BILL (N)		9·2 (0·32; 11)	8·6–9·7	9·5 (0·28; 4)	9·2–9·8
TARSUS		20·8 (0·95; 10)	19·6–22·3	20·6 (0·64; 6)	19·5–21·4

Sex differences significant for wing, tail, and bill (S). Juvenile tail 62·7 (2·31; 3) 60–64, shorter than in adult.

North-east Iran, ♂: wing 87·5 (1·50; 10) 86–90, tail 69·9 (2·04; 18) 67–72·5, bill (N) 9·3 (0·40; 10) 8·7–10·0 (Haffer 1977, which see for measurements of intermediates with *E. melanocephala*). Afghanistan: wing, ♂ 86·5 (1·63; 11) 84–89, ♀ 81·8 (1·50; 4) 80–83 (Paludan 1959). Mongolia: wing, ♂ 87·5 (1·38; 6) 85–89 (Piechocki and Bolod 1972; Piechocki *et al.* 1982). Whole range, ♂: wing 88·2 (10) 85–90, tail 67·5 (10) 65–72, tarsus 18·9 (10) 18·0–19·5 (Vaurie 1956e). Wing: ♂ 82–93 (75), ♀ 78–87 (52) (Svensson 1992).

Weights. (1) Afghanistan, April–October (Paludan 1959). (2)

Kazakhstan, summer (Korelov *et al.* 1974). (3) Mongolia, May–July (Piechocki and Bolod 1972; Piechocki *et al.* 1982). (4) Kirgiziya (Yanushevich *et al.* 1960). (5) USSR (Dementiev and Gladkov 1954).

(1)	♂	24·4 (1·38; 12)	22–27	♀	23·2 (1·26; 4)	22–25
(2)		25·3 (— ; 27)	23·0–30·5		24·2 (— ; 19)	18·0–28·5
(3)		24·7 (0·82; 6)	24–26		— (— ; —)	—
(4)		— (— ; 25)	20–33		— (— ; 9)	23·5–30·2
(5)		24·9 (— ; 4)	21–27·3		29 (— ; 1)	—

Helgoland (Germany), April and June: ♂ 30, ♀ 34 (Vauk 1961).

Structure. Wing rather long, broad at base, tip bluntly pointed. 10 primaries: p8 longest, p9 0–5 shorter, p7 0–1, p6 1–4, p5 6–10, p4 12–16, p3 15–20, p2 19–25, p1 23–29; p10 strongly reduced, 61–64 shorter than p9, 6–11 shorter than longest upper primary covert. Outer web of p6–p8 clearly emarginated. Tip of longest tertial reaches tip of p3–p5. For slight differences from Black-headed Bunting *E. melanocephala* in wing formula, see Vaurie (1956e) and Svensson (1992). Tail of average length, tip slightly forked; 12 feathers, t5(–t4) longest, t1 4–7 shorter. Bill long, strong, conical, as in *E. melanocephala*; depth at feathering 7·0 (9) 6·5–7·5, width 7·2 (9) 6·5–7·7. Middle toe with claw 21·7 (9) 20·5–23·3; outer toe with claw *c.* 67% of middle with claw, inner *c.* 66%, hind *c.* 65%; hind claw 7·1 (11) 6·5–7·6.

Geographical variation. None.
For relationships, see *E. melanocephala*. GOK

Emberiza melanocephala **Black-headed Bunting**

PLATES 25 and 26 (flight)
[between pages 256 and 257]

DU. Zwartkopgors   FR. Bruant mélanocéphale   GE. Kappenammer
RU. Черноголовая овсянка   SP. Escribano cabecinegro   SW. Svarthuvad sparv

Emberiza melanocephala Scopoli, 1769

Monotypic

Field characters. 16–17 cm; wing-span 26–30 cm. Looks noticeably larger than Yellowhammer *E. citrinella*, with proportionately rather longer bill, more obvious neck, rather longer wings, and distinctly longer legs, but similar tail length; marginally larger than Red-headed Bunting *E. bruniceps*, but no visible structural difference. 2nd largest bunting of west Palearctic, with rather long, tapering bill, rather long body, and noticeably long legs combining into characteristically heavy but sleek form shared only by *E. bruniceps*. Combination of uniformly pale, unstreaked underparts and lack of white outer tail-feathers excludes all other buntings except *E. bruniceps*. ♂ distinctive, with black head, chestnut back, and yellow underparts; ♀ and immature lack obvious characters and may not be separable from *E. bruniceps*. Sexes dissimilar; some seasonal variation in ♂. Juvenile separable. Hybridizes with *E. bruniceps*.

ADULT MALE. Moults: October–December (complete); June–July (some upperpart feathers and tertials). Whole of head except throat black, when fresh tipped buff-brown particularly on crown and nape. Diffuse nuchal collar mixed yellow and chestnut, only obvious when plumage worn and neck stretched. Back to upper tail-coverts mainly bright chestnut, when fresh fringed buff-brown and on long rump variably mixed with yellow, often creating paler area between wings. Chestnut of shoulder continues onto side of breast, sometimes forming patch. Wings black-brown with chestnut and brown lesser coverts and (when fresh) sharp buff to whitish fringes and tips, broad enough to form obvious pale bar across black-centred median coverts, but similar marks across greater coverts and along edges of tertials and secondaries less eye-catching. Almost black bastard wing and dusky primaries often appear as long dark panel on lower edge of wing. Tail black-brown, narrowly fringed buff-brown. Entire underparts including wing-coverts and axillaries yellow, tinged buff on chin and throat and sulphur elsewhere but some birds much paler, a few even appearing yellowish-white especially under tail; a few may have speckled or wholly black throat. Bill deep-based but rather long, with mandibles tapering to point; lead-blue, with greyer lower mandible. Legs dark flesh, brighter when sunlit. ADULT FEMALE. Most appear colourless compared to ♂, with even more featureless appearance than *E. bruniceps*. When fresh, head and upperparts dull brown, slightly ashy, with slight to strong chestnut tinge on mantle, usually faint yellow or chestnut mottling on rump, and indistinct black or brown streaks on crown and back. When worn, mantle more chestnut and virtually unstreaked and rump paler. Head shows faint greyish-buff eye-ring and hint of similarly coloured supercilium but only obvious contrast is with pale whitish-buff throat. Wings and tail patterned as ♂ but duller. Underparts and underwing pale buff, only faintly tinged yellowish except on under tail-coverts; some birds have faint dusky malar stripe and are more greyish or warmer buff on sides of or across breast; when worn, underparts may look merely dull yellow or white. Bill as ♂, but duller; legs usually paler flesh than ♂. A few acquire noticeably dusky head, showing ♂'s pattern. JUVENILE. Somewhat resembles ♀ but generally buffier above and below, with more distinct streaks, even on sandy rump; streaked also on sides of breast and rarely flanks. Wings and tail less dark than adult, due to visibly broader sandy-buff fringes and tip; both median and greater coverts show obvious wing-bars. Obvious yellow tone restricted to under tail-coverts. Bill greyish-horn, paler than adult. FIRST-WINTER. Assumes adult plumage from June onwards, and after December ♂ usually indistinguishable from adult, but ♀ may retain signs of immaturity, particularly more streaked back (see Plumages).

♂ virtually unmistakable, but, in glimpse of bird on ground, beware confusion with similarly patterned ♂ Yellow-breasted Bunting *E. aureola* (20% smaller, wholly black head, chestnut necklace, darker upperparts, wing markings like Chaffinch *Fringilla coelebs*, and streaked flanks) and ♂ of black-headed race of Yellow Wagtail *Motacilla flava feldegg* (as long but much slimmer, with fine bill, greenish back, and white-edged tail). ♀ and immature always initially confusing and hard to separate not only from *E. bruniceps* but also from Cinereous Bunting *E. cineracea*. Distinction from similarly sized *E. bruniceps* not always possible, but that species typically greyer or greener above, with brighter rump, and paler below, sometimes with more pronounced streaking on breast and flanks. *E. cineracea* 15% smaller, with white outer tail-feathers. Flight as other *Emberiza* but noticeably strong, lacking rather erratic rhythm to wing-beats of smaller species; since flight silhouette lacks full end to tail of *E. citrinella*, form and undulating progress may also suggest

large pipit *Anthus*. ♂ leaving song perch or in display may dangle legs in manner of Corn Bunting *Miliaria calandra*. Gait a hop. Stance usually quite upright, with leg length usually obvious below sleek, unruffled flanks. Gregarious, forming huge flocks in winter range, but west Palearctic breeding population only locally dense.

Song quite short but more musical than usual in *Emberiza*, usually beginning with rather grating twitter, then 2-3 accelerating, far-carrying, warbled phrases; suggests warbler and (partly) Lapland Bunting *Calcarius lapponicus*: 'zip-zip-zip', 'chit-chit-chit', or 'sitt süt süt', then 'süterEE-süt-süte-ray'. Calls include 'cheuh' or 'styu' recalling *E. citrinella*, and clicking 'plüt' or 'chup' recalling Ortolan Bunting *E. hortulana*.

Habitat. Breeds in south-west Palearctic in warm temperate, Mediterranean, and steppe zones, between July isotherms of 23–32°C, generally in lowlands, avoiding both drier and wetter extremes. Favours fairly dense and tall bushy and scrub vegetation, including open maquis, wooded steppes, orchards, olive groves, and vineyards, and groves or thickets along streamsides, roadsides, or field borders (Voous 1960b). Also found in open forest with undergrowth, in open lowland grassland with scrub, especially thorn scrub, and on mountain slopes (Harrison 1982). Fond of perching on fairly high branches or other commanding features, especially as song-posts. In Cyprus, breeds in mountains up to at least 1200 m (Bannerman and Bannerman 1971). In southern parts of former USSR occupies hilly steppes covered with dense tall grassy or herbaceous plants, shrubs, and isolated low trees, and open plains broken with gullies and low hills; also orchards and cornfields. Avoids damp localities; locally ascends foothills and lower mountain ranges, in Armeniya to 1500–2000 m, exceptionally higher (Dementiev and Gladkov 1954).

Wintering birds in India feed in flocks and cultivated fields, sometimes causing serious damage to standing crops, and proving very difficult to scare away. Also occupy scrub jungle, roosting in enormous concentrations with other species in thorn scrub and thickets (Whistler 1941; Ali and Ripley 1974).

Distribution. Has recently spread north along Black Sea coast.

ITALY. Locally common in scattered areas of south, east, and west coasts; has bred as far north as southern Lombardia (Brichetti 1976; Bocca and Maffei 1984). RUMANIA. Spread north to Danube delta areas in 1960s (Tălpeanu and Paspaleva 1979). BULGARIA. Now breeding along whole Black Sea coast; inland, Stara Planina forms northern limit (Simeonov 1970; TM).

Accidental. Iceland, Britain, Ireland, Netherlands, Belgium, France, Spain, Denmark, Norway, Sweden, Finland, Poland, Czechoslovakia, Switzerland, Austria, Malta, Morocco, Algeria, Tunisia.

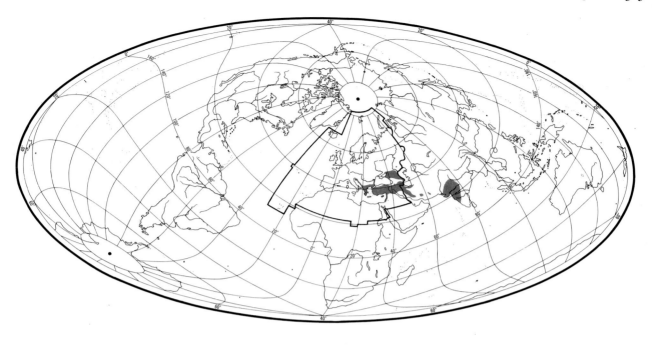

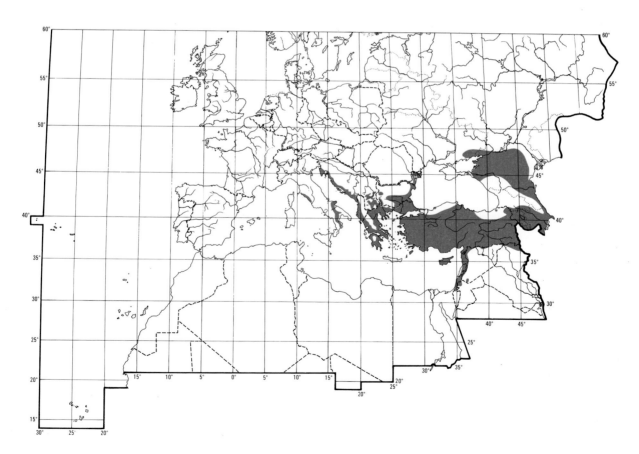

Population. BULGARIA. Increasing, with spread along Black Sea coast; now 5000–50 000 pairs (TM).

Movements. Migratory, all birds moving south-east or ESE to winter in western and central India: Rajasthan, Gujarat, Madhya Pradesh (east to Saugor), Maharashtra (east to Nagpur and Nander), and Karnataka (south to Hiriyur). Exceptionally recorded from Nepal (Inskipp and Inskipp 1985). 3 ringing recoveries: bird ringed Kutch (*c.* 23°N 69°E, western India), September, recovered *c.* 3700 km north-west in Krasnodar region (*c.* 45°N 40°E, south European Russia), May; bird ringed Kathiawar (*c.* 22°N 71°E, western India), September, recovered *c.* 4500 km WNW in Cyprus, May (Ali and Ripley 1974); also bird ringed Yugoslavia, June, recovered Cyprus in following April (Flint and Stewart 1992).

Leaves breeding grounds early, and returns late. Departure (inconspicuous) late July to August; autumn route presumably same as spring (see below). Apparently most birds leave Cyprus in July, with only a few records August–September, exceptionally later (Flint and Stewart 1992); also on islands of Aegean Sea, departures late July and early August (Watson 1964). Vacates Turkey August to mid-September (Beaman *et al.* 1975; Beaman 1978). In Israel, movement (chiefly through central and eastern areas) August–September, with stragglers present until 3rd week October; most local birds leave by mid-August (Shirihai in press). Some birds reach Sinai; at St Katherine in south, over 2 years, 22 individuals recorded 9 August to 19 September (Goodman and Meininger 1989). Recorded irregularly in Gulf states of Arabia mid-August to early October, and annually in very small numbers in Oman, July–December (Oman Bird List; F E Warr). In Iran, few records after late August and none after end of September (D A Scott). At Karachi (southern Pakistan), recorded 12 August to 22 September over 7 years (Roberts 1992). Arrives in India August–September (Ali and Ripley 1974). Occasional midwinter records from breeding range or intermediate areas, e.g. Israel (Shirihai in press) and Kuwait (Bundy and Warr 1980).

In spring, leaves winter quarters March–April (Ali and Ripley 1974). Passage through Pakistan is on fairly narrow front, mainly through Sind and southern Baluchistan; rare as far north as Punjab; at Karachi, over 7 years, passage recorded 2 March to 22 April, ♂♂ preceding ♀♀. Many birds fly along Makran coast; others move west along Ras Koh foothills further north, then turn north or north-west along Tahlab river bordering Iran. (Roberts 1992.) In Iran, abundant on migration in south and south-east, less common on central plateau; arrives in south-east at end of March, and in north in 3rd week of April (D A Scott). In Transcaucasia, earliest ♂♂ arrive in last third of April, and earliest ♀♀ in 1st third of May; dates similar in western Azerbaydzhan; in Armeniya, ♂♂ reported 3–6 May, and ♀♀ 3–5 days later (Panov 1989). Recorded in Voronezh region (south European Russia) 28 March in one year

(Vorobiev and Likhatski 1987). In Iraq, most migration is in north; evidently birds travel north-west along foothills of Zagros and Kurdish mountains; common at Kirkuk in April (Marchant 1963a). Reaches Turkey mostly from late April, southern coast sometimes mid-April (Kasparek 1992). Sometimes reported in Cyprus as early as March, but usually arrives in early or mid-April (sometimes not until late April), with movement continuing to mid-May (Flint and Stewart 1992). Earlier observations in Cyprus and southern Turkey than in central Turkey suggest that ongoing passage from Iraq may be along southern foothills of eastern Turkey. Arrives on Aegean islands late April and early May (one of latest migrant breeders) (Watson 1964). Reaches Makedonija (Yugoslavia) very suddenly at end of April or beginning of May, with no singing ♂♂ present one day and many the next; ♀♀ arrive a few days later (Makatsch 1950).

Some birds take more southerly route. Recorded in Azraq (north-east Jordan) in small numbers mid-April to mid-May, with ♂♂ preceding ♀♀ (Wallace 1982b). In one year, common at Badanah in northern Saudi Arabia, 20–24 April (Jennings 1981a), presumably including birds bound for breeding areas in Israel; arrives in Israel during 2nd half of April and 1st half of May (Shirihai in press).

Vagrancy west of range is mostly in spring, suggesting overshooting. Some records may involve escaped cagebirds. In Britain and Ireland, 9 records prior to 1958, and 83 in 1958–91; of 55 records 1958–85, 60% in spring, chiefly in May; distribution well scattered, though 25% of spring and 41% of autumn records were from Shetland (Dymond *et al.* 1989; Rogers *et al.* 1992). In France, 30 records in 20th century, chiefly May–June; most are in south-east, but a few also in north and west; in 1988, 6 singing ♂♂ reported in south (Dubois and Yésou 1992). In Sweden, up to 1986, 11 records May–June and 2 in August, all in south (Breife *et al.* 1990). Only 2 records (June, August) in western Germany, 1977–90 (Bundesdeutscher Seltenheitenausschuss 1991, 1992), and 10 in Denmark since 1965 (Frich and Nordbjærg 1992). 3 birds have reached Iceland (Pétursson *et al.* 1992). Despite breeding in southern Italy, only 9 reports from Malta, 1867–1988 (Sultana and Gauci 1982; Coleiro 1989), and 11 from Sicily since 1847 (Iapichino and Massa 1989). DFV

Food. Much information extralimital. Diet seeds and other plant material; invertebrates in breeding season (Witherby *et al.* 1938; Dementiev and Gladkov 1954; Armani 1985). In summer quarters, forages principally in cultivated areas: cereal or sunflower *Helianthus* fields, vineyards, orange groves, etc., feeding both on ground and in shrubs or low in trees (Maštrović 1942; Makatsch 1963; Armani 1985; Shirihai in press). In Yugoslavia, often feeds in maize *Zea* stubble covered with weedy herbs (Bannerman 1953a). Most foraging observations concern migrant birds or winter visitors, since species occurs then in huge numbers often causing considerable crop damage. In Iraq and Israel, large flocks feed in cereals such as millet *Panicum* (Moore and Boswell 1957; Paz 1987; Shirihai in press), and in Sind (Pakistan) 'vast clouds' settled in cereal fields (Ticehurst 1922). Also in Pakistan, fed on maize flowers, clinging to stems (Roberts 1992); in India, commonly in rice *Oryza* fields (Ali and Ripley 1974). Vagrant in Britain, beginning of June, fed mostly on dry dusty track; pulled 2·5-cm crane-fly (Tipulidae) larva from grassy bank and swallowed it (Jennings 1981*d*).

Diet includes the following. Invertebrates: larval mayflies (Ephemeroptera), crickets (Orthoptera: Tettigoniidae), earwigs (Dermaptera), cicadas, etc. (Hemiptera: Auchenorrhyncha), larval Lepidoptera, larval flies (Diptera: Tipulidae), Hymenoptera (ants Formicidae, wasps Vespidae), beetles (Coleoptera: Carabidae, Tenebrionidae, Cerambycidae, Chrysomelidae, Curculionidae), harvestmen (Opiliones). Plants: seeds, etc., of bugloss *Anchusa*, *Salvadora*, grasses (Gramineae, including millet *Pennisetum*, *Panicum*, wheat *Triticum*, maize *Zea*, *Sorghum*, rice *Oryza*). (Witherby *et al.* 1938; Kovačević and Danon 1952, 1959; Moore and Boswell 1957; Ali and Ripley 1974; Pätzold 1975; Jennings 1981*d*; Kostin 1983; Paz 1987; Roberts 1992.)

In Crimea (southern Ukraine), 9 stomachs (May–June) contained 94% insects and 6% seeds of bugloss and other herbs; insects included 71% beetles (39% Tenebrionidae, 16% Chrysomelidae, 10% Curculionidae, 6% Carabidae and Cerambycidae), 19% wasps, 2% Auchenorrhyncha, 1% Orthoptera, and 1% caterpillars (Kostin 1983). In Pakistan, feeds on unripe berries of *Salvadora* as well as various cereals (Roberts 1992). Captive bird refused all seeds except millet, Canary grass *Phalaris*, and rolled oats *Avena*; all seeds de-husked before eating; was also fond of small Orthoptera and larvae (Rokitansky 1969, which see for discussion of bill shape and diet). See also Maštrović (1942) and Pätzold (1975) for diet in captivity

In Bulgaria, young at one nest over 8-hr period were fed almost only bush crickets *Tettigonia viridissima*; 1 large cricket brought for all 4 nestlings (Pätzold 1975). For diet of young in captivity, see Maštrović (1942). BH

Social pattern and behaviour. Some aspects (e.g. dispersion, song-display) well known but, with no comprehensive studies, remarkably little known about others. In particular, no details of postures and displays in antagonistic and heterosexual contexts, and some confusion in literature (not known if from observer error or genuine diversity) about role of sexes in nest duties. In these respects, *E. melanocephala* less well known than closely related Red-headed Bunting *E. bruniceps* which has similar behavioural repertoire (certainly so in case of song-display) (Panov 1989) and which hybridizes with *E. melanocephala* in sympatric zone.

1. Gregarious outside breeding season. Huge flocks reported wintering in north-west and central India (Blanford 1876). Spring migration usually in flocks: in Iran, flocks of 10–30 (Erard and Etchécopar 1970); in Pakistan, a dozen up to 40–50 birds (Roberts 1992), first flocks typically ♂♂ which migrate ahead of

♀♀ (Ticehurst 1922, 1926–7; Ali and Ripley 1974; Roberts 1992), though see Flock Behaviour (below) for mixed flock, early May; no association on passage with *E. bruniceps*, the two species only converging on wintering grounds (Roberts 1992) where large mixed flocks ravage crops (Ali and Ripley 1974); see Roosting (below) for other associations. In Israel, at end of breeding season, mainly mid-June to mid-July, juveniles and adults form flocks of a few tens to hundreds which concentrate in good areas for drinking, feeding, and roosting (see below) (Shirihai in press). BONDS. No study of mating system. As in *E. bruniceps*, ♂ may play minimal role in nesting activities (at least up till fledging), suggesting possibility of polygyny. However, sources differ in contribution of sexes, suggesting that ♂'s role may vary with local conditions. According to Makatsch (1979: Greece), ♀ alone builds nest, incubates, broods, and rears young; in Cyprus also, ♀ alone responsible for nest-building and incubating (Ashton-Johnson 1961). However, in Balkans, ♂ takes equal share in building according to Reiser and Führer (1896). In Croatia, ♂ said to share building, make minor contribution to incubation (taking over for short periods 2–3 times per day) and share equally with ♀ in feeding and caring for young (Maštrović 1942). In close study of nest in southern Bulgaria, ♂ did not begin to share feeding of young till they left nest, but then contributed intensively (Pätzold 1975). No information on age at independence in the wild, or age at first breeding. One captive chick took food independently at 19 days (Pätzold 1975), though uncertain fledging period (see Breeding) makes dependent period hard to evaluate. For hybridization with *E. bruniceps*, see that species. BREEDING DISPERSION. Territorial (apparently all-purpose type) and solitary or in neighbourhood groups. Territory relatively small (see below), with requisite elevated song-posts (e.g. Rokitansky 1969; see Song-display, below). In Israel, tends to form neighbourhood groups, e.g. *c.* 25 pairs in field of sunflowers *Helianthus*, each pair having territory of *c.* 200–400 m², minimum *c.* 15 m between nests (Shirihai in press). According to Panov (1989), quite often solitary, tending not to form compact groups like *E. bruniceps*; where local population high (as in Armeniya), average distance between nests 101 ± 17 m (*n* = 9); in 8 cases, distance to nearest neighbour 75–200 m, once 30 m (Panov 1989). In agricultural area in north-west Rhodos (Greece), 10–30 singing ♂♂ per km² (Scharlau 1989). In Makedonija, density remarkably high in optimal habitat, e.g. 3 nests in 30 m (Makatsch 1950). In Albania and Bulgaria, sometimes only 30–40 m between song-posts of neighbouring ♂♂; in Albania, *c.* 40 pairs per km²; in Bulgaria *c.* 37·5–50 pairs per km², also 9 pairs along 2300 m; in both countries, highest density on coastal hill slopes (Kolbe and Kolbe 1981). Also in Bulgaria, 137 pairs per km² in habitat dominated by *Paliurus aculeati* (Milchev 1990), 48 in Christ's thorn *P. spina-christi* (Petrov 1982), 19 in wheat *Triticum* and barley *Hordeum* (Milchev 1990). In Azerbaydzhan, 80 birds per km² in settlements and adjoining plantations, 4 in riverine tamarisk *Tamarix*, 5 in arable land, 1 in semi-desert (Drozdov 1965). ROOSTING. No information for breeding season. In winter, India, forms mixed roosts with Yellow Wagtail *Motacilla flava*, Rose-coloured Starling *Sturnus roseus*, House Sparrow *Passer domesticus*, and *E. bruniceps* (Gadgil and Ali 1975). In Bampur (southern Iran), early April, migrants 'roosted for the night in such numbers . . . that a shot, almost at hazard, into the tree, brought down ten or a dozen birds' (Blanford 1876).

2. In breeding season, singing ♂♂ can be approached to a few metres before they fly off a short distance, perch, and resume singing (Frank 1943; Rokitansky 1969). In winter quarters, not shy in presence of man, and not easily flushed from crops by shouting, flocks merely rising and settling (readily taking to trees)

in furthest corner of field (Ali and Ripley 1974; Roberts 1992). General demeanour and display repertoire much as in *E. bruniceps* (see that species, and below). FLOCK BEHAVIOUR. Little information. On Syrian/Turkish border, early May, flock of *c.* 20 birds (both sexes) rested together on rocky hillside, with a little preening, feeding, and song (see below), before departing in groups (Hollom 1959). Song also heard from April migrant flocks in Iran (Erard and Etchécopar 1970). SONG-DISPLAY. In breeding season, ♂ sings indefatigably almost throughout day (Frank 1943: see 1 in Voice), with bill raised and crown slightly so, mostly from any convenient elevated song-post, typically tree, tall bush, vineyard post, telegraph pole or wire (Makatsch 1950; Rokitansky 1969). In south-west Iran, sang from tops of rocks, bushes, somtimes clods of earth and stones, also regularly (but less intensely) from ground while feeding. Song-posts mostly used early morning and evening, but were resorted to at irregular intervals throughout the day. One bird sang 6 times per min for 3 min, being answered (Song-duel) by neighbour *c.* 100 m away; when 2nd ♂ stopped answering, 1st ♂ continued singing less intensely (*c.* 4–5 times per min) and began preening between songs. (Almond 1946.) Although song from perch predominates, *E. melanocephala*, like *E. bruniceps*, also sings during sexual chases (see Heterosexual Behaviour, below) and has 2 types of Song-flight. In 1st, more common type of Song-flight, ♂ *E. melanocephala* sings in level flight, with shallow quivering wing-beats and dangling legs (Moth-flight: Fig A), similar to Corn

A

Bunting *Miliaria calandra*; displays thus regularly when flying between song-posts (occasionally when disturbed from song-post), also occasionally just before landing on song-post after feeding (Almond 1946; Rokitansky 1969). Less common, towering type of Song-flight occurs, associated in *E. bruniceps* with peak courtship and (with 'parachute' descent) likened to song-flight of Tree Pipit *Anthus trivialis*; in *E. melanocephala*, seen only once in south-west Iran, bird on song-post flying up almost vertically to *c.* 6–7 m, song starting at top of ascent and continuing on steep descent; as soon as song finished, bird 'flattened out' and flew to another perch *c.* 40 m away (Almond 1946). In Lebanon, when bird (originally giving alarm-calls: see 2d in Voice) was flushed, it flew up almost vertically and, at *c.* 10–15 m, broke into song with a fluttering, parachute descent (Hollom 1959). ♂♂ sing from early morning till twilight, including hottest time of day (Makatsch 1979; Kolbe and Kolbe 1981). Captive ♂ sang only once at night, contra view gained from literature that *E. melanocephala* sings regularly at night (Rokitansky 1969). Song-period relatively short. Song sometimes heard in winter quarters just before spring departure (Ali and Ripley 1974), also on spring migration (Roberts 1992; see also Flock behaviour, above). In Makedonija, song starts immediately upon arrival at

end of April, and stops at end of June (Makatsch 1950). In north-west Turkey, song said to begin at start of June, and finish by end of July (Rokitansky 1969). In south-west Iran, all birds seen had stopped singing by June 29, but 2 heard on July 13, *c.* 190 km further north (Almond 1946). ANTAGONISTIC BEHAVIOUR. Little information. In nesting groups, proximity of song-posts leads to continual territorial disputes between neighbours, sometimes involving 3 ♂♂ at once (Kolbe and Kolbe 1981). No information on displays but presumably similar to *E. bruniceps*. ♂ shared use of tree as song-post and look-out with ♂ Woodlark *Lullula arborea* (Pätzold 1975). HETEROSEXUAL BEHAVIOUR. With relatively late arrival in spring, and prompt departure in summer, effectively short time spent on breeding grounds, estimated at 17 weeks in Peloponnisos (southern Greece) (Makatsch 1979). ♂♂ typically arrive on breeding grounds a few days ahead of ♀♀ (e.g. Makatsch 1950, 1979; see also start of part 1, above). Nest-building begins only a couple of days after ♀♀ arrive (Maštrović 1942). When ♀♀ arrive they keep fairly quiet but any ♀ flying from one bush to another is immediately pursued by 5–6 ♂♂ giving intense song-fragments (Rokitansky 1969); same sexual chasing seen in *E. bruniceps*, to which rest of courtship in *E. melanocephala* apparently similar (Panov 1989). No further information. RELATIONS WITHIN FAMILY GROUP. For discussion of role of sexes in care of young, see Bonds (above). Otherwise little known. In captive study, ♀ swallowed *c.* 50% of faecal sacs, and carried off remainder. At 8–9 days old, 4 young (possibly disturbed by observer) all left nest very quietly within 20 min, and disappeared in thorny nest-bush; at this stage, ♂ (who had not provisioned young as nestlings), started feeding them intensively. (Pätzold 1975.) In Peloponnisos, as soon as young fledge, family party roams through olive *Olea* orchards (etc.), starts dispersing further by end of July, and all gone by mid- to late August (Makatsch 1979). ANTI-PREDATOR RESPONSES OF YOUNG. See Relations within Family Group (above) for possible premature fledging. Chick (which left nest at 8–9 days) hopped off when placed on bare soil (Pätzold 1975). No further information. PARENTAL ANTI-PREDATOR STRATEGIES. (1) Passive measures. ♀ very quiet and discreet during early nesting phase: e.g. in one study, no nests found till eggs hatched, when ♀♀ (for first time) conspicuous with alarm-calls (Kolbe and Kolbe 1981: see 2d in Voice). In Crete, ♀ found to sit quite tightly on nest, slipping off quietly (Meiklejohn 1936). (2) Active measures. In Lebanon, towering Song-flight by ♂ (see Song-display, above), after giving alarm-calls, possibly also an alarm response. No further information.

(Fig by D Nurney from painting in Matthews *et al.* 1988.)   EKD

Voice. Freely used in breeding season. In winter, apart from song sometimes heard just before spring departure, the only call heard was musical 'tweet' (Ali and Ripley 1974), i.e. perhaps call 2b or 2c (below). Medley of calls of contact-alarm type given in various circumstances, and further study needed to elucidate function and context of constituent vocalizations; following scheme therefore tentative. Also, no information on calls used specifically in heterosexual contexts. Like Red-headed Bunting *E. bruniceps*, some calls quite sparrow-like, but recordings show structural differences from that species: in *E. melanocephala* all calls of ♂♂ descend, though some incorporate brief ascending transient (Figs IV–V); 2 calls of ♀ (Fig VI) show relatively weak ascent followed by long strong descent; see Voice of *E. bruniceps* for contrasting tendency of calls, after transient-like click, to ascend to the end. No detailed published studies of differences from *E. bruniceps* except for song which is scarcely distinguishable in the field. For additional sonagrams see Bergmann and Helb (1982) and Panov (1989: song only).

CALLS OF ADULTS. (1) Song of ♂. (1a) Full song. Variable but characteristic phrase, usually introduced (as in *E. bruniceps*) by short harsh call-type units (see 2a, below), and ending with a few mellow sounds with distinctive tinny ring, e.g. whole song 'pzt pzt pzt pzt pzt chirri chirri chirrli chürrlü chürrlü' (Jonsson 1992). Also described as melodious loud phrase beginning with quieter single units, followed (with accelerando and crescendo for the most part) by rapid sequence of full-sounding, rather throaty units showing frequency modulation and varying in pitch (Bergmann and Helb 1982). Very similar to *E. bruniceps* (see that species for detailed comparison, also below). Overall structure of *E. melanocephala* song apparently the same throughout geographical range, though recordings show marked variation in details, and dialects presumably occur (as in *E. bruniceps*). 6 recordings from 5 countries (2 Turkey, 1 each Israel, Cyprus, Bulgaria, and Transcaucasia) analysed by J Hall-Craggs show that structure of preparatory 'zit' calls is quite stereotyped in *E. melanocephala* (see Figs I–III) but variable in *E. bruniceps*; like *E. bruniceps*, song-phrase characteristics of *E. melanocephala* further include: repeated unit in 1st part (5 out of 6 recordings); lower-pitched voice in 2nd part (all 6), *E. melanocephala* sounding more assured (or practised) in this than *E. bruniceps*; distinctive final unit which may be, variously, buzzy (1 of 6), rattle (2 of 6), tremolo (2 of

6) or trill (1 of 6). Fig I shows lovely mellifluous phrase preceded by 2 'zit' calls (also a 3rd but too quiet to depict). Only 1 of many introductory 'zit' calls is shown in Fig II which, as in Fig I, ends with quiet low rattle. Typically lower-pitched 2nd part of song illustrated well in Fig III (2nd part a distinct ascending arpeggio); this recording features 2 ♂♂ counter-singing with same song. Renderings of songs from different regions as follows. In Peloponnisos (Greece) rather short 'di-di-drüli TRA', 'better and purer' than in, e.g., Makedonija (Makatsch 1979). Bulgarian song described as 'riri ri-re-dri-dri-dri-ür-üra' (Schubert and Schubert 1982); for Bulgaria, see also Balát (1963). In north-west Turkey, sweet 'tsiliritsilirup', almost like oriole *Oriolus* in quality and temporal pattern, quite often introduced with stammering 'zit-zit' (Rokitansky 1969). In northern Iran, song consists of 2 noisy 'ticking' notes followed by phrase recalling Whinchat *Saxicola rubetra*, roughly 'dridri didä-düdadi' (Schüz 1959); also in northern Iran, 'psitt-psitt-teutittiteuriterrit' and 'psitt-psitt-titerueutirit' (Erard and Etchécopar 1970). For song description in Ukraine see Frank (1943). Same song often repeated monotonously for long spells, e.g. over 1 hr (Paz 1987). In south-west Iran, one bird sang (from elevated perch) 6 times per min during song–duel, appreciably less often from ground while foraging (Almond 1946). Intense song-fragments given by ♂♂ in sexual chases at start of breeding season (Rokitansky 1969). (1b) Subsong. Quiet song with bill closed and throat vibrating heard in breeding season, Turkey (Rokitansky 1969). (2) Contact- and alarm-calls. When disturbed in territory, gives (just as does *E. bruniceps*) persistent series (medley), apparently in no fixed order (but see below), of short single calls of

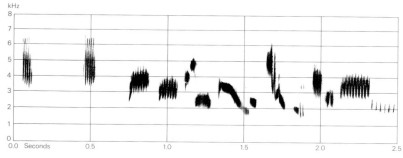

I   Mild (1990)   Israel   April 1989

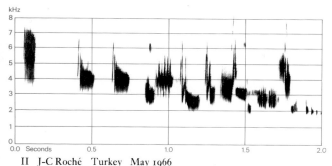

II   J-C Roché   Turkey   May 1966

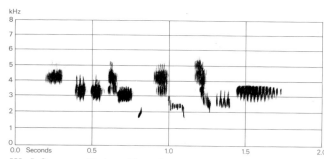

III   L Svensson   Turkey   May 1989

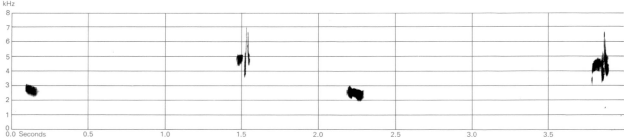

IV Mild (1990) Israel April 1989

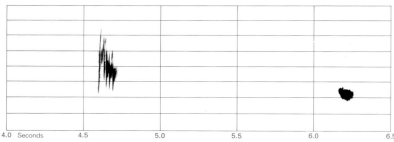

IV *cont.*

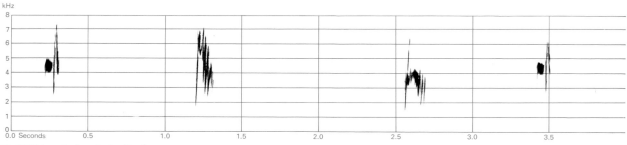

V Mild (1990) Israel April 1989

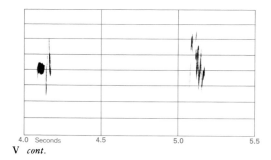

V *cont.*

several types, some short (e.g. 'pit'), some barely disyllabic (Bergmann and Helb 1982). Recordings include such medleys, but call-sequence not random; Fig IV shows representative selection (6 calls) of series of 16 calls containing 4 disyllabic call-types as follows: 'siu (call-type 1), si-tik (2), pee-oo (3), si-tik (2), zriik (4), siu (1)'; order of 4 call-types in whole bout 1 2 1 2 3 2 4 2, 1 2 1 2 3 2 4 1.

Fig V similarly shows, from series of 9 calls, 6 consecutive calls (different from those in Fig IV) as follows: 'si-td (1) tziurk (2) tziurk (3) si-td (1) si-td (1) tziirk (2)'. (J Hall-Craggs.) Only some of these calls easily reconciled with those from published sources; latter include: (2a) Short 'pit', listed as flight-call (Bergmann and Helb 1982); somewhat metallic 'pit' (Schüz 1959); sharp 'zitt' (Witherby *et al.* 1938; Rokitansky 1969); 'zitt' listed as warning-call of ♀ (Frank 1943). Apparently higher pitched (5–6kHz) than equivalent call of *E. bruniceps* (Panov 1989). Series of presumably this call serves as song-introduction (see above). (2b) Usual flight-call a sombre 'dripping' 'pchlü' or 'plüt', not unlike some deeper flight-calls of Goldfinch *Carduelis carduelis* (Jonsson 1992). (2c) Lively 'pyiup' and 'tyilp' (Jonsson 1992). In recording of ♀ (Fig VI), 'seup tyup' (J Hall-Craggs). The following also apparently the same or related: sharp 'tjipp' (Took 1973); 'tje teje' of apparent warning given by ♀♀ (Frank 1943); quiet 'djb djb' from captive ♂ (Rokitansky 1969). (2d)

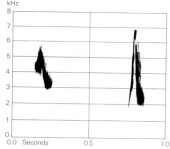

VI P S Hansen Greece June 1975

Zee- and Tsip-calls. Low (presumably low-pitched) 'zee' of warning or anxiety (Witherby *et al.* 1938). 'Tsip' calls often alternate with a softer call: in Greece, 'tsip-jüp' thought to be contact-call (Makatsch 1963); in Lebanon, repeated 'teu tseep teu tseep', perhaps of alarm, from ♂ which, when then flushed, performed towering song-flight (Hollom 1959); recordings (Fig IV, 1st 2 calls; Fig VII)

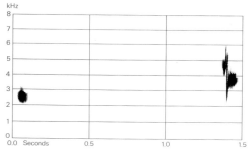

VII P S Hansen Turkey July 1987

of medley of calls also include apparently this combination: soft low-pitched 'siu' (recalling Bullfinch *Pyrrhula pyrrhula*: P J Sellar) followed by higher-pitched 'si-tik' (Fig IV) or 'ch-di' (Fig VII) (J Hall-Craggs). In Ukraine, 'tjzü-de(t) hju' and 'tsjü-he tsje', with stress on 1st and 3rd units (Frank 1943) presumably much the same.

CALLS OF YOUNG. Captive young first heard giving food-calls (not described) at 9–10 days; louder and harsher by start of independence at 19 days (Pätzold 1975). Curious sizzling sound in recording by P S Hansen (Turkey) perhaps from nestlings.   EKD

Breeding. SEASON. Short throughout region. Western Italy: nest-building recorded 10 May (Allavena 1970). Croatia: eggs laid from mid-May, mostly 1st half of June (Maštrović 1942). Greece: eggs laid mid-May to end of June (Meiklejohn 1936; Makatsch 1950, 1979). Cyprus: laying starts mid-May (Ashton-Johnson 1961; Flint and Stewart 1992). Turkey: eggs found early May to mid-June (Beaman *et al.* 1975). Levant: eggs laid early May to 1st half of July, peak May (Pitman 1922; Moore and Boswell

1957; Shirihai in press). Nakhichevan' (south of Caucasus): eggs laid mainly late May and early June (Panov 1989). SITE. Low down in dense, often thorny shrub (e.g. bramble *Rubus*, rose *Rosa*, rockrose *Cistus*, Christ's thorn *Paliustrus*), and commonly on vine *Vitis*; usually 0·5–1·0 m above ground (Makatsch 1963; Took 1971; Pätzold 1975; Kolbe and Kolbe 1981); sometimes on ground (Dementiev and Gladkov 1954), and recorded 6 m up in ash *Fraxinus* (Reiser 1939). On Croatian coast, almost all of 100 nests were on vines (Maštrović 1942); on Crete, 11 of 20 nests on thistles, against stem supported by shoots and leaves (Meiklejohn 1936). In Greece, nests recorded attached to stems of wheat *Triticum* (Makatsch 1963). Nest: loose, untidy foundation of stalks of herbs, grass (including reed *Phragmites*), and leaves, lined with fine grasses, stems, rootlets, hair, and sheep's wool (Meiklejohn 1936; Makatsch 1963, 1976; Pätzold 1975). Fairly often with brightly-coloured flower-heads on outside (Pitman 1922; Lucas and Took 1974). In Pilos area (southern Greece), 16 nests had average outer diameter 14·2 cm, inner diameter 6·9 cm, overall height 7·2 cm, depth of cup 4·9 cm, dry weight 23·4 g (Makatsch 1979, which see for plants used); see Meiklejohn (1936) for 11 nests on Crete. Building: by ♀ only, according to most authors, taking 10–14 days (Ashton-Johnson 1961; Lucas and Took 1974; Makatsch 1979; Paz 1987), but see Reiser and Führer (1896) and Maštrović (1942) for confident statements that ♂ takes share, equal to ♀'s according to Reiser and Führer (1896); ♀ in Cyprus added material to nest containing 5 eggs (Lucas and Took 1974). EGGS. See Plate 34. Sub-elliptical, smooth and slightly glossy; very pale blue or greenish-blue, rarely buff, rather sparsely speckled purplish-grey to olive-brown, usually concentrated towards broad end; sometimes unmarked (Harrison 1975; Makatsch 1976). Lack of hairstreaks and scrawls give appearance quite unlike those of almost all other *Emberiza* except Red-headed Bunting *E. bruniceps* (e.g. Pitman 1922, Took 1971, Makatsch 1979). For discussion and for comparison with *E. bruniceps*, see Etchécopar (1961). 22·4 × 16·2 mm (19·0–26·2 × 14·0–18·2), *n* = 297; calculated weight 3·08 g (Schönwetter 1984). Clutch: 4–5 (3–7). In Croatia, of 34 broods: 2 young, 9%; 3, 24%; 4, 44%; 5, 18%; 6, 3%; 7, 3%; average 3·9 (Maštrović 1942). In Bulgaria, 3 clutches were all of 4 (Pätzold 1975). Apparently slightly smaller in Israel: 17 clutches were 3–5, mostly 4 (Shirihai in press); perhaps increases from west to east as 6–7 more common in former USSR (Grote 1934*b*; Korelov *et al.* 1974). Short season throughout range means more than 1 brood unlikely (Maštrović 1942; Makatsch 1979; Panov 1989; Shirihai in press). Eggs laid daily (Makatsch 1979). INCUBATION. About 14 days; by ♀ only (Meiklejohn 1936; Ashton-Johnson 1961; Makatsch 1979; Kolbe and Kolbe 1981); 10 days recorded in Cyprus (Cole and Flint 1970); 14–16 in Israel (Shirihai in press). According to some authors, by both sexes: ♂ takes 2–3 stints per day (Maštrović 1942), or takes over at midday and evening (Reiser and Führer 1896). Probably starts with last or penultimate egg (Makatsch 1979). YOUNG. Fed and cared for by ♀ only (Makatsch 1979), though ♂ fed young in Bulgaria study once they had left nest (Pätzold 1975, which see for routine); by both parents throughout according to Maštrović (1942). FLEDGING TO MATURITY. Fledging period 13-16 days, $n = 17$, in Israel, according to Shirihai (in press). In Bulgaria, normally 10 days (Kolbe and Kolbe 1981), 8-9 days when disturbed by observer (Pätzold 1975). BREEDING SUCCESS. In Greece, predation levels very low, and main predators thought to be owls (Strigidae); principal danger to nests probably agricultural work in vineyards; some eggs probably taken by lizard *Lacerta viridis* (Makatsch 1963, 1979). Brood-parasitism by Cuckoo *Cuculus canorus* recorded in Bulgaria (Robel 1983). BH

Plumages. ADULT MALE BREEDING (November-July). Nasal bristles small, black. Cap down to upper cheek and ear-coverts glossy black, feathers with grey-buff fringes when plumage fresh. Hindneck, mantle, scapulars, and back rufous-chestnut to tawny-brown, sometimes with slight yellow tinge, feathers with grey-buff fringes when fresh; occasionally, feathers of upper mantle with dusky tips, lower mantle with dark shaft-streaks, or hindneck yellow; longest scapulars grey-brown on outer web, with dark brown shaft-streak near tip, and with greenish-yellow fringe on inner web. Rump variable, either uniform yellowish-green, orange-yellow, or tawny-brown with broad lemon-yellow fringes. Upper tail-coverts similar to rump; longest upper tail-coverts brown with dark brown shaft-streaks and chestnut or yellowish fringes. Upper chin white, side of breast chestnut, remainder of underparts bright yellow, feathers with narrow white fringes when fresh; occasionally, whole underparts pale yellow; rarely, a variable amount of black feathers in irregular pattern on throat, extending onto chest in some birds; flank occasionally deep orange or with some chestnut. Tail-feathers greyish-black, central pair (t1) slightly paler, all with narrow buff or white edges; t6 paler than t1-t5, inner web with off-white edge and pale grey wedge along shaft, basal half of outer web pale grey, but t6 sometimes with a grey tip on inner web only. Flight-feathers, greater upper primary-coverts, and bastard-wing greyish-black with off-white, yellow-white, or buff edges along outer web. Tertials black, outer webs and tips bordered by broad but ill-defined white and buff fringes on outer web and tip, narrower pale fringes along inner web. Greater and median upper wing-coverts black, tips of median and tips and outer webs of greater broadly fringed buff (when fresh) or white to off-white (when worn). Lesser coverts chestnut with yellow fringes. Greater under wing-coverts grey with white and yellow tips and narrow dark brown shaft streak; axillaries, remainder of under wing-coverts, and marginal coverts white with yellow edges. ADULT FEMALE BREEDING (November-July). Feathers of forehead and crown pale brown with grey or grey-black centre; lore, cheek, and ear-coverts pale brown with slight yellow hue. Eye-ring pale sandy-brown. Feathers of neck chestnut with some yellow on edge and with grey tips. Mantle and scapulars brown with some chestnut and yellow on centre and with dark brown shaft-streak, or plain chestnut (as in adult ♂ breeding). When fresh, feathers on head, mantle, scapulars, and rump with broad grey-buff fringes. Rump and shorter upper tail-coverts yellow, yellow-green, chestnut with green, or greyish-green; longest upper tail-coverts browner, sometimes with faint dark brown

shaft-streak. Chin, throat, chest, belly, vent, and under tail-coverts pale yellow, broad grey tips of feathers largely concealing yellow when plumage fresh, chin sometimes buffish-yellow; rarely, chest with fine streaks, forming faint gorget. Flank dirty grey, sometimes with inconspicuous shaft-streaks. Tail and wing as in adult ♂, but lesser upper wing-coverts grey-brown with only a hint of chestnut and yellow. ADULT MALE NON-BREEDING (September-December). New feathering as in adult ♀ breeding. Feathers of cap black with distal third brown, but lore and upper cheek black. Hindneck to scapulars and back brown or grey-brown, feathers with broad grey-buff fringes; mantle, scapulars, and back with broad, ill-defined dark brown shaft-streaks. Chin, throat, and breast white, pale yellow, or mixed pale yellow and white; remainder of underparts yellow, sometimes with some white feathers, all yellow feathers with broad white fringes. Tail and wing as adult ♂ breeding. ADULT FEMALE NON-BREEDING (September-December). Like adult ♀ breeding, but some old abraded tertials and greater and median upper wing-coverts retained. Throat, chest, and belly occasionally with many whitish feathers, bright yellow restricted to under tail-coverts. NESTLING. Down cinnamon, restricted to cap and back. JUVENILE. Like adult ♀, but feathers on head rusty-brown or cinnamon with dark shaft-streak; upperparts greyish, distal half of feathers light cinnamon; lower back and rump pale cinnamon; chin and breast pale cinnamon, remainder of underparts whitish (Stresemann and Stresemann 1969a; Pätzhold 1975, which see for development). FIRST ADULT NON-BREEDING (July-November). Forehead, crown, mantle, and scapulars as in adult ♀, but dark centres of feathers less clear-cut, without any chestnut tinge. Supercilium, ear-coverts, lore, edges of crown feathers, mantle, scapulars, and lesser upper wing-coverts dark sandy-brown or buffish. Stripe over lower cheek off-white, inconspicuous, bordered above by pale brown ear-coverts and below by buff malar stripe. Rump and upper tail-coverts with faint dark brown streaks; rump greenish, buffish, or slightly chestnut (latter perhaps in ♂ only). Underparts from chin to vent off-white with slight buff wash, side of breast and sometimes chest more strongly tinged buff, chest occasionally with narrow brown shaft-streaks extending onto lower breast, flank, and thigh. Under tail-coverts pale lemon-yellow or deep yellow. Tail black-brown with buff edges, widest on both webs of t5; t6 like adult, but with dark brown shaft-streak. Fringes of tertials buff, sharply demarcated from black-brown centres. Greater and median upper wing-coverts as well as bastard wing black-brown, broadly fringed buff; lesser coverts grey with faint buff edges. By October-November, some yellow feathers appear on underparts, and buff fringes and tips of tertials and greater and median coverts fade to white. FIRST ADULT MALE BREEDING. Like adult ♂ breeding, but juvenile flight-feathers and greater upper primary-coverts retained, browner and more worn than those of adult at same time of year; occasionally, some juvenile tertials retained. ♂♂ with ♀-type feathers on crown, mantle, or back and pale yellow underparts perhaps in 1st breeding plumage. FIRST ADULT FEMALE BREEDING. Like adult ♀ breeding, but part of juvenile feathering retained, as in 1st adult ♂ breeding.

Bare parts. ADULT, FIRST ADULT. Iris dark brown. Bill leaden-grey or steel-grey, lower mandible with paler cutting edge. Leg and foot light flesh-brown or horn-brown, tinged grey on toes. NESTLING. No information. JUVENILE. Bill greyish-horn.

Moults. ADULT POST-BREEDING. Partial; in ♂, involves part of crown, nape, mantle, scapulars, back, upper tail-coverts, and most tertials and upper wing-coverts, but extent of moult varies individually, and some birds replace only a few feathers; in

♀♀, no moult, or (occasionally) a few or even many feathers of upperparts replaced: e.g. on 1 August, single ♀ had many feathers of head, mantle and scapulars in moult, but another not yet moulting (Stresemann 1928b). In Makedonija (southern Yugoslavia), 5 ♂♂ and 4 ♀♀ had body moult just starting 1–10 July (Stresemann 1920). In northern Iran, 22–29 July, 3 ♂♂ in heavy moult to non-breeding plumage, 3 ♀♀ just started (Paludan 1940). Single ♂ had not yet started 9 June, another had body moult completed on 25 August (Diesselhorst 1962). Birds in full moult of body in southern Yugoslavia or northern Greece on 25 June, 27 July, and 30 July (Stresemann 1928b). In western Turkey, in heavy moult of body on 10 July (Rokitansky and Schifter 1971, which see for moult of bird in captivity). Moult may start as early as June, and usually most birds have at least some feathering new at start of migration in August. ADULT PRE-BREEDING. Complete or almost so; primaries descendent. Starts after arrival in winter quarters, from early October onwards; duration *c.* 60 days. Primaries start with p1, and up to 4 primaries are moulted simultaneously; tail moult rapid, centrifugally; secondaries, upper primary-coverts, and tail are moulted during primary moult. As ♀ has very restricted post-breeding moult, pre-breeding moult is more complete than in ♂. Body feathers which have been moulted to non-breeding in late summer are not replaced during pre-breeding moult, except differently coloured ones (Stresemann and Stresemann 1969). Some birds retain non-breeding feathers during summer, e.g. single bird from June, Dalmatia, had scattered non-breeding feathers on head and mantle (RMNH). POST-JUVENILE. Completed within a few weeks after fledging; involves head, body, and lesser and median wing-coverts. Some birds had freshly moulted innermost greater upper wing-coverts between early July and late August (RMNH). In Afghanistan, 22–29 July, single ♂ and 2 ♀♀ not yet started, another ♂ in full moult (Paludan 1940). In northern Iran, 26 July, single bird had body moult almost completed (Stresemann 1928b). In southern Yugoslavia, moult of 4 birds just started 2–10 July, more advanced in 2 from 10–16 July (Stresemann 1928b). FIRST PRE-BREEDING. Involves head, body, and wing-coverts; in winter quarters, but no information on extent and timing. (Paludan 1940; Stresemann and Stresemann 1969; Pätzold 1975; RMNH, ZMA).

Measurements. ADULT, FIRST ADULT. Whole geographical range, all year; skins (RMNH, ZMA). Bill (S) to skull, bill (N) to distal corner of nostril; exposed culmen on average 4·2 less than bill (S).

WING   ♂ 96·1 (1·81; 16) 94–101   ♀ 89·4 (2·31; 12) 86–94
TAIL   67·7 (2·71; 15) 63–73   66·2 (1·85; 12) 64–70

BILL (S)   17·6 (0·57; 16) 16·4–18·4   17·6 (0·63; 11) 16·2–18·5
BILL (N)   10·3 (0·50; 16) 9·1–11·0   10·3 (0·43; 12) 9·3–11·0
TARSUS   22·3 (0·71; 14) 21·3–24·0   22·2 (0·56; 10) 21·2–23·0

Sex differences significant for wing. For data on hybrids with Red-headed Bunting *E. bruniceps*, see Paludan (1940).

Wing. Makedonija (southern Yugoslavia), May–July, skins: ♂ 95·1 (2·19; 97) 90–101, ♀ 87·4 (1·62; 12) 84–89 (Stresemann 1920). Whole geographical range, skins: ♂ 93·5 (2·24; 19) 89–97, ♀ 88·3 (2·56; 7) 84–92 (Stresemann 1928b; Paludan 1938; Schüz 1959; Nicht 1961; Diesselhorst 1962; Rokitansky and Schifter 1971). Live birds, Turkey: ♂ 95·3 (4) 94–97, ♀ 88·8 (4) 85–91 (Vauk 1973). Iran, skins of ♂: wing 93·5 (1·04; 18) 92–95·5, tail 72·7 (1·74; 18) 69·5–75, bill (N) 10·4 (0·45; 18) 10·0–11·1 (Haffer 1977). See also Vaurie (1956e).

Weights. Breeding area. Adult and 1st adult: (1) April–May, (2) June–July. Juvenile: (3) July–August (Banzhaf 1931; Paludan 1938, 1940; Schüz 1959; Kumerloeve 1961; Nicht 1961; Diesselhorst 1962; Rokitansky and Schifter 1971; Vauk 1973; Desfayes and Praz 1978; Eyckerman *et al.* 1992). (4) Intermediates with *E. bruniceps*, adult, northern Iran, June (Paludan 1940).

(1)   ♂ 29·7 (2·58; 20) 25·0–33·0   ♀ 27·0 (— ; 9) 23·0–31·0
(2)   29·9 (1·24; 6) 27·7–30·7   26·9 (3·56; 4) 24·0–32·0
(3)   30·2 (0·84; 3) 29·7–31·2   25·6 (1·17; 5) 24·4–27·5
(4)   27·8 (1·81; 22) 25·2–31·5   27·0 (1·94; 3) 24·4–27·0

For effect of light on weight, see Kumar and Tewary (1983).

Structure. Wing rather long, broad at base, tip pointed. 10 primaries: p8 longest, p9 0–2 shorter, p7 0–3, p6 4–7, p5 10–15, p4 15–20, p3 18–24, p2 21–27, p1 24–29; p10 greatly reduced, 62–68 shorter than p8, 8–12 shorter than longest upper primary covert. Outer web of p6–p8 clearly emarginated. Tip of longest tertial reaches tip of p2–p5. Tail fairly long, tip slightly forked; 12 feathers, t3(–t4) longest, t1 3–6 shorter. Bill strong, conical, 7·7 (10) 7·4–8·2 mm deep at base, 8·3 (10) 7·5–8·7 wide; much longer and heavier than bill of Yellowhammer *E. citrinella*, upper mandible deeper, cutting edges virtually straight, but more heavily toothed at base. Middle toe with claw 22·2 mm (9) 20·0–24·4; outer toe with claw *c.* 68% of middle with claw, inner *c.* 67%, hind *c.* 69%; hind claw 7·8 (18) 7·0–9·0.

Geographical variation. None. Forms superspecies with *E. bruniceps*, with which it hybridizes in areas where breeding ranges meet (north and south of Caspian Sea). For discussion of hybrid zone, description of hybrids, and taxonomic status, see Paludan (1940), Schüz (1959), and Haffer (1977). GOK

Miliaria calandra   Corn Bunting

PLATE 27
[between pages 256 and 257]

Du. Grauwe gors   Fr. Bruant proyer   Ge. Grauammer
Ru. Просянка   Sp. Triguero   Sw. Kornsparv

Emberiza calandra Linnaeus, 1758

Polytypic. Nominate *calandra* (Linnaeus, 1758), Canary Islands, North Africa, and Europe, east to Caucasus area, Asia Minor (except south-east) and coastal Levant; *clanceyi* (Meinertzhagen, 1947), western Ireland and western Scotland;

buturlini (Hermann Johansen, 1907), interior of Levant and south-east Turkey, east through Iraq, Iran, and southern Transcaspia to western China and Afghanistan.

Field characters. 18 cm; wing-span 26–32 cm. Noticeably larger, bulkier, but proportionately shorter-tailed than Yellowhammer *Emberiza citrinella*; close in size to Skylark *Alauda arvensis*. ♂ is largest bunting of west Palearctic with heavily streaked buff-brown plumage; ♀ 10% smaller but still bulky, sharing ♂'s heavy bill and stout legs. Recalls ♀ sparrow *Passer* or *A. arvensis* far more than other buntings. No white in tail. Size, flight, and voice all more important to identification than plumage details. Sexes similar; no seasonal variation. Juvenile separable at close range.

ADULT. Moults: July–November (complete). Ground-colour of upperparts and wings yellowish- to greyish-brown (palest when worn), of face, breast, and flanks buff, and of rest of underparts pale buff to dull white (when worn). Uniformity of appearance emphasized by copious, regular black-brown streaking of upperparts, sides of neck, breast, flanks, and most wing-coverts. Tail also uniformly coloured, dark brown, with only buff fringes and tips; lacks white outer feathers so characteristic of most *Emberiza* and *A. arvensis*. Where overlying streaks do not obscure ground-colour, plumage shows the following pale marks: on head, diffuse supercilium, indistinct eye-ring, central ear-coverts, obvious surround to ear-coverts, chin, faint double wing-bar, tertial fringes, rump, belly, and vent. Strength or proximity of streaks form the following dark marks: dark edge to rear ear-coverts, prominent malar stripe joining clustered streaks on upper breast, and dark centres to coverts and tertials. Bare parts as striking as any plumage character. Bill large and heavy, sparrow-like with strongly developed tooth on lower mandible; usually yellowish-horn with dark culmen. Legs and heavy feet bright flesh- to straw-yellow, paler on feet. A few birds show patches of warmer colours (brown to rusty) on head and fore-upperparts (see Plumages). JUVENILE. Up to September or October, ground-colour of plumage paler and brighter than adult and streaks broader and heavier. Head pattern differs distinctly in paler crown centre contrasting with dark lateral stripes, broader supercilium, paler cheeks, and only vestigial malar stripe. Elsewhere, paler tips to coverts form brighter wing-bars and flanks show fewer streaks.

Within Emberizidae, large size, lack of obvious marks, flight, and voice form unmistakable combination. Inexperienced observer must, however, beware similarity of appearance to other streaked passerines which may share habitat of *M. calandra* or use it on migration. Commonest confusions are with larger streaked larks *Melanocorypha* and *Alauda*, ♀ and immature sparrows *Passer* and *Petronia*, ♀ and immature rosefinches *Carpodacus*, Rose-breasted Grosbeak *Pheucticus ludovicianus*, and Bobolink *Dolichonyx oryzivorus*. Flight recalls rock sparrow *Petronia*,

differing from that of other buntings in loose, surging take-off (during which ♂ in breeding season often dangles and trails legs and bunched feet) and steeper, less undulating, and faster escape-flight. Over longer distance, alternation of strong wing-beats and closed-wing attitudes produces powerful undulations, particularly noticeable in descent to feeding area or roost. Throughout flight, relatively large dark wings catch eye as much as dull, not proportionately long tail. Gait a hop; bird appears reluctant to climb within ground cover or around perch. Stance often noticeably upright, particularly on song-post when tail may hang down; at other times half-erect, rarely level except on ground. Occasionally flicks wings and tail; raises head when alarmed on ground. Beware similarity of ♂'s fluttering courtship to hover of *A. arvensis*. Most easily found in spring when ♂♂ perch and sing openly; in autumn and winter, flocks become secretive among cover of fields and scrub. Gregarious, usually in unmixed flocks where common but joining other buntings and finches on migration.

Song diagnostic, a short, rushed jangle of ticking and discordant units: 'tük tük zik-zee-zrrississ', recalling rattle of bunch of small keys; often given repeatedly for long periods, forming characteristic sound of (particularly) barley fields in spring and summer. Calls varied but only 2 important for identification: on take-off and in flight, quite loud, initially slightly liquid but terminally clicked 'quit' or 'quick', often run together in accelerating series 'quit-it-it' and forming distinctive chorus when given by flock; rather harsh 'chip' as contact-call in breeding season.

Habitat. Breeds in middle latitudes of south-west Palearctic, in cool and warm temperate, Mediterranean, and steppe climatic zones, within July isotherms of 17–32°C, including extremes of both oceanic and continental types (Voous 1960*b*). Mainly in lowlands, preferably undulating or sloping rather than level, and with pronounced liking for vicinity of sea coasts. Avoids forest, wetlands, rocky and broken terrain, and, in most regions, mountains or high plateaux, as well as built-up areas. Apart from need for perches to overlook territory and to serve as song-posts (low trees, bushes, overhead cables, fences, or walls), is at home in fully open country, and has minimal demands for cover, except to some extent for roosting, e.g. in reeds.

General requirements for habitat appear plain, but no explanation yet found for arbitrary tendency to occupy at high density specific areas while other seemingly suitable sites are left vacant, and also to abandon areas, apparently unchanged, after years of successful breeding there.

Survey in farming area of north-east Scotland revealed strong association with arable land particularly linked with barley, and in breeding season with peas grown for animal

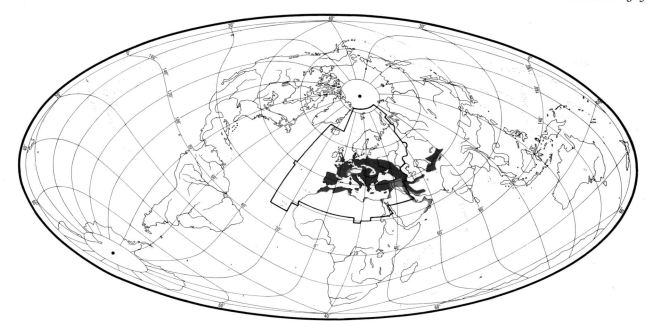

fodder, whereas vegetables grown for human consumption were avoided. Here, at extreme north of range, favours maritime grass and raised bogs in winter, but earlier habit of sheltering from bad weather in straw and hay ricks has been frustrated by recent changes in farm practice. This is one of many areas where diminution of numbers and contraction of breeding range is reported (Buckland *et al.* 1990).

Thorough national reviews of habitat factors in Britain (Parslow 1973; Fuller 1982) have failed to account for widespread declines and regional or local fluctuations, but changes in farming practice are implicated, e.g. reductions in cultivated area of barley, switch towards autumn growing of cereals, replacement of hay by silage, and decline in traditional rotations and mixed farming practices; increased use of pesticides and removal of hedges may have reduced food supply, and climate perhaps also involved (Donald *et al.* in press). Nesting within growing crops is common, but some indication that survival of local stocks may benefit where gorse *Ulex* and bramble *Rubus* offer readily accessible alternatives (Sharrock 1976). Marked differences in many habits in different regions add to difficulties of interpretation, as do variations in practice of polygamy and group territories (Bannerman 1953*a*). Habit of roosting on ground has made it vulnerable to trapping locally.

In USSR, habitat generally similar, but some overspill occurs into edges of forest clearings, plantations, or orchards, and even to grassy sections of oases in sandy deserts. Where conditions suitable, nesting recorded in subalpine meadows up to 1200–2600 m, or even higher in isolated cases, but not really a montane species, and not often found at any distance from cultivation (Dementiev and Gladkov 1954).

Distribution. Range has contracted in north-west.

BRITAIN. Range contraction, especially in north and west, since *c.* 1930 (Sharrock 1976). Has virtually disappeared from Wales and south-west England, and in Scotland now mainly in coastal areas (Thom 1986; Marchant *et al.* 1990). IRELAND. Formerly widespread; range contracted throughout 20th century (Hutchinson 1989). BELGIUM. Spread over large part of Lorraine since 1974 (Jacob 1982). Has spread, due to changing agricultural practice (very large fields) (Devillers *et al.* 1988). NORWAY. Formerly bred in south-west, but (except for possible breeding attempts 1971 and 1977: Olsen 1984) no breeding records since 1928; now accidental visitor (VR). CZECHOSLOVAKIA. Slow spread 1850–1960; marked contraction after 1965 (KH).

Accidental. Faeroes, Norway (see above), Finland.

Population. Has decreased in many parts of northern, western, and central Europe, probably due mainly to changing agricultural practice and land use. See review by Donald *et al.* (in press).

BRITAIN, IRELAND. Decline since *c.* 1930; estimated 30 000 pairs 1968–72 (Sharrock 1976). Marked decline north-east Scotland 1988–92 compared with 1944–8 (Watson 1992*a*). Now on verge of extinction as breeder in Ireland, Wales, western Scotland, and south-west England (Donald *et al.* in press). FRANCE. 100 000 to 1 million pairs (Yeatman 1976). BELGIUM. Estimated 7100 pairs (Devillers *et al.* 1988). NETHERLANDS. Has declined, prob-

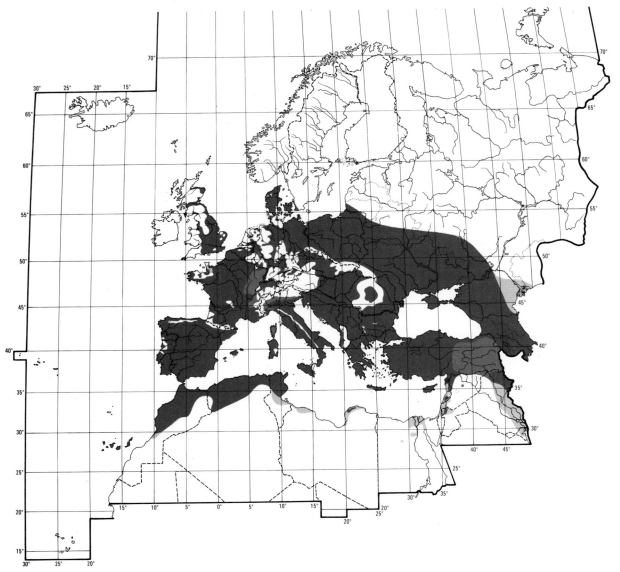

ably from before 1970. Estimated 1100-1250 territories 1975, 450-700 1980, 250-375 1985, 100-200 1989. (Hustings *et al.* 1990.) GERMANY. Estimated 27 000 pairs (Rheinwald 1992). East Germany: decreased, cause uncertain (Gliemann 1973); 20 000 ± 8000 pairs (Nicolai 1993). Schleswig-Holstein: drastic decrease, due to intensive farming methods, from 3000-4000 pairs in 1955 to 40 pairs in 1987. Decline began 1960-5 and steepened rapidly after 1975. Wintering population (formerly considerable) now practically extinct. (Busche 1989.) Bayern: decline in Danube valley (Regensburg-Vilshofen), from 152 singing ♂♂ in 1975 to 43 in 1988-91. Main causes probably modern farming methods and increase in wet springs and early summers after 1945. (Vidal 1991.) SWEDEN. Decrease since beginning of 20th century, probably due to changing agri-

cultural practice; *c.* 100 birds 1982 (Jönsson 1982), *c.* 20 1988 (Jönsson 1989); decrease continuing to 1991, and population perhaps no longer viable (Jönsson 1992). CZECHOSLOVAKIA. Marked decrease after 1965 (KH). AUSTRIA. Decrease in recent decades, mainly due to loss of habitat (H-MB). SWITZERLAND. Has recently declined (Schmid 1991; Christen 1991). BULGARIA. No recent change; 100 000 to 1 million pairs (TM). LITHUANIA. Common until 1970-80, when numbers suddenly fell (Logminas 1991).

Survival. Germany: estimated annual mortality 42 ± SE4% (Møller 1983c). Oldest ringed bird 10 years 6 months (Dejonghe and Czajkowski 1983).

Movements. Resident to partially migratory. Winters

chiefly within breeding range. Western migrants head mostly south-west or SSW, and some southern birds move west; at least some eastern birds head south or east of south. Data suggest central European birds migrate more than north European ones, but confirmation required from more extensive ringing. Resident birds roam in flocks in winter, resulting in absence from some breeding localities. Account based chiefly on Zink (1985).

Resident in Britain; of birds ringed up to 1976, only 8 recoveries at 15 km or more; such movements probably represent wanderings of winter feeding flocks (Boddy and Blackburn 1978). Longest distance within Britain 74 km east from Surrey to Kent in south-east (Mead and Clark 1987). 1 recovery outside Britain: bird ringed Kent, February, recovered May (presumably in breeding area) in Pas de Calais in north-east France, c. 200 km south-east (Spencer 1964).

In France, absent from some breeding localities in winter, but little known of movements of local birds; immigrants arrive from further east and north-east, wintering mainly in southern France (Zink 1985; Yeatman-Berthelot 1991). Resident in Sweden (SOF 1990), and chiefly resident in Belgium (Lippens and Wille 1972) and Denmark, though Bornholm vacated in winter; up to 1979, furthest distance recorded for Danish-ringed bird 35 km (Møller 1978a; Zink 1985). Overwinters in Mecklenburg (north-east Germany), with no reports of passage (Klafs and Stübs 1987); in Rheinland (western Germany), many overwinter in Aachen and Köln area, but very few further north in Kleve and Wesel area (Mildenberger 1984). Observations fewer in southern Germany in winter, and flock-size smaller, than in Denmark and northern Germany; in Bayern (southern Germany), where many breed, occurs on passage but winters only in small numbers; exceptional in winter in Baden-Württemberg (Zink 1985). Overwinters in Switzerland in fluctuating numbers in mild years, and widespread on passage, mostly at lower levels (Schifferli *et al.* 1982; Winkler 1984).

Almost all long-distance recoveries involve birds ringed south of 51°N, and chiefly south of 50°N. Migrants from central Europe head south-west or SSW to winter chiefly in south-east France (especially in lower Rhône area), also on French Atlantic coast and in north-east Spain (with 1 recovery in south-west Spain); birds ringed in Belgium and central France recovered in south-west France and north-east Spain, and bird ringed Poland recovered in north-west Italy. Some northern Italian birds head WSW to winter in south-east France. 3 records of birds moving south from central to southern Spain, of which 1 ringed in breeding season. Individuals recorded wintering in different areas in different years: e.g. birds ringed in upper Rhine area (western Germany), winter, recovered in later winters in eastern France (1) and north-east Spain (1); also in winter, bird ringed Czechoslovakia recovered in north-east Italy, and 3 birds ringed in Camargue recovered in north-west Italy. (Zink 1985; Bendini and Spina 1990.)

Some birds cross Strait of Gibraltar to Morocco, where wintering of local birds or immigrants extends slightly south of breeding range (Heim de Balsac and Mayaud 1962; Finlayson 1992). In eastern Morocco, high numbers (at least in some years) probably include many European immigrants (Brosset 1961). Bird ringed northern Tunisia, March, recovered WNW at 37°33′N 5°04′W in south-central Spain, July. In western Algeria, recorded south to Beni-Abbès (30°05′N). (Zink 1985.) Recorded exceptionally as far south as Mauritania and Sénégal (Morel 1980; Lamarche 1988). Fairly common on passage in Malta (where breeds), and some migrants probably winter (Sultana and Gauci 1982). Locally common in winter in northern Libya (where no confirmed breeding) (Bundy 1976). In Turkey, widespread resident or partial migrant, recorded only from western two-thirds in winter; some evidence of immigration in autumn, with larger numbers on south coast (Beaman 1978); in Cyprus (where also breeds), common on passage, and winters on low ground in variable numbers (Flint and Stewart 1992). In Jordan, winter flocks occur mainly in northern highlands and on desert fringe (I J Andrews). In Israel (breeds in north and centre), very common passage migrant, and abundant and widespread winter visitor; large flocks move south along Mediterranean coastal plain (Shirihai in press). In Egypt, regular winter visitor in small flocks to cultivated parts of Nile delta and valley (south to 27°44′N), Suez Canal area, Faiyum, and northern Sinai; irregular south to Kharga (25°26′N) (Goodman and Meininger 1989). Winters in small numbers in Gulf states (Bundy and Warr 1980) and Oman (Oman Bird List); reported exceptionally in western Arabia, but not in centre or south (F E Warr). In Iraq and Iran, distribution extends further south winter than summer (Moore and Boswell 1957; D A Scott), and in northern Afghanistan reported further west in winter than summer (S C Madge).

Autumn migration protracted overall, August–December, and timing varies; formation of winter flocks (from August) masks departure and passage. Long-distance recoveries reported from October. Passage mostly September–November in central Europe (Lippens and Wille 1972; Schifferli *et al.* 1982; Zink 1985). At Bodensee (southern Germany), leaves Rhine delta chiefly by late July, but present in cornfields until October (Zink 1985). In Camargue (southern France), post-breeding flocks joined by immigrants from October (Blondel and Isenmann 1981), and passage at Strait of Gibraltar October–November (Finlayson 1992). Vacates breeding area in western Sardinia as early as end of August (Mocci Demartis 1973), and passage in Malta begins August (Sultana and Gauci 1982). At Bosporus (western Turkey), in one year, small movement eastward recorded 3–24 October (Porter 1983). In Israel, 2 influxes occur, late October to mid-November (involving chiefly passage birds) and mid-November to early December (chiefly winter visitors) (Shirihai in press). Present Libya September–April

(Bundy 1976), and in Gulf states mainly mid-November to mid-March (Bundy and Warr 1980).

Spring movement also protracted, (January-)February-May. In Strait of Gibraltar, passage late February to May (Finlayson 1992), and in Malta small return passage February-March (Sultana and Gauci 1982). At Bahij on north Egyptian coast, recorded 19 February to 18 March over 2 years (Horner 1977), and passage in Cyprus mid-February to early May (mainly March to mid-April) (Flint and Stewart 1992), in Israel mid-February to mid-April, mostly in 2nd half of March (Shirihai in press). Returns to breeding grounds in western Sardinia at end of March (Mocci Demartis 1973). Arrives in south-west Germany end of February to end of April (Zink 1985), and returns to breeding grounds in Rheinland in 1st third of April (Mildenberger 1984). Movement in Denmark March-May (Møller 1978a). In southern England, reported returning to local breeding sites as early as January, more usually late February to March (Br. Birds 1950, 43, 83). January movement also reported in Belgium (Lippens and Wille 1972).

In east of range (south-east Kazakhstan and adjoining areas of south-central Asia), present in much of breeding area all year, but vacates northern and mountainous regions September-October, e.g. Talasskiy Alatau and Zailiskiy Alatau; fairly common in Tashkent area and Golodnaya steppe in winter, with fewer in Chimkent area. Exceptionally recorded from Zaysan depression, well north of breeding range. In spring, returns to Talasskiy Alatau from mid-February. (Korelov et al. 1974.) In eastern Uzbekistan, in one year, passage reported 28 March to 20 April (Ostapenko et al. 1980).　　　　DFV

Food. Seeds (often of cereals), other plant material, and invertebrates, especially in breeding season. Feeds almost wholly on ground in arable fields, damp meadows, short rough grass, etc.; in autumn, commonly in stubble and fields where root crops have been harvested or dung spread; forages in farmyards, grain depots, etc., only in harsh winters, and much less so than, e.g. Yellowhammer *Emberiza citrinella*. (Bannerman 1953a; Glutz von Blotzheim 1962; Gliemann 1973; Eifler and Blümel 1983; Lack 1986; Haack and Schmidt 1989.) In north-west England, cereal fields held higher densities of animal and plant food than did root and horticultural crops (Thompson and Gribbin 1986). Rarely feeds in bushes and trees at field edges, moving rather clumsily along twigs toward tips searching for invertebrates, then flying to next twig when it bends; much less agile in trees than *E. citrinella*, though clambers fairly skilfully in cereals to reach seed-heads (Gliemann 1973). In Turkmeniya, noted in January feeding in wormwood *Artemisia* shrubs (Rustamov 1958). In Germany in August, 3-4 together seen making repeated sallies from perches to catch flying insects (probably Diptera) above cereal field, at times hovering briefly but awkwardly (Dathe 1961). Pulls up cereal seedlings, squeezes

out milky endosperm, and discards remainder (Goodbody 1955). In north-east Scotland, seen to de-husk grains of barley *Hordeum*, wheat *Triticum*, and oats *Avena* by striking them on road, post, dry earth, etc. (Watson 1992a). In Kazakhstan, very often observed feeding on dry cow dung, though not clear whether on (e.g.) dung beetles (Scarabaeidae) or seeds (Kovshar' 1966). For discussion of bill and tongue morphology in connection with diet, and comparison with other seed-eaters, see Eber (1956).

Diet in west Palearctic includes the following. Invertebrates: dragonflies and damsel flies (Odonata), grasshoppers, etc. (Orthoptera: Gryllidae, Tettigoniidae, Gryllotalpidae, Acrididae), earwigs (Dermaptera: Forficulidae), bugs (Hemiptera: Aphidoidea), adult and larval Lepidoptera (Pieridae, Tortricidae, Noctuidae), flies (Diptera: Tipulidae, Rhagionidae, Tabanidae), ants (Hymenoptera: Formicidae), adult and larval beetles (Coleoptera: Scarabaeidae, Curculionidae), spiders (Araneae), millipedes (Diplopoda), earthworms (Lumbricidae), snails (Pulmonata). Plants: seeds, buds, etc., of sycamore *Acer*, ivy *Hedera*, Virginia creeper *Parthenocissus*, knotgrass *Polygonum*, dock *Rumex*, rape *Brassica*, pear *Pyrus*, broom *Cytisus*, grasses (Gramineae, including wheat *Triticum*, barley *Hordeum*, oats *Avena*, rye *Secale*). (Rey 1908; Collinge 1924-7; Ryves and Ryves 1934; Witherby et al. 1938; Kovačević and Danon 1952; Turček 1961; Glutz von Blotzheim 1962; Macdonald 1965; Gliemann 1973; Hegelbach 1980; Haack and Schmidt 1989.)

In England, all year, 37 stomachs contained 28·5% by volume animal material (22·5% insects, 1·5% earthworms, 1·5% snails, 3% other invertebrates, including spiders and millipedes) and 71·5% plant material (59·5% seeds of herbs, 7·5% cereals, 4·5% miscellaneous); excluding period August-October, only 2% of diet cereal grains (Collinge 1924-7). In Aberdeenshire (north-east Scotland), 16 stomachs, March-June, contained 126 identifiable items among mass of vegetable matter, of which 65·9% by number seeds (61% cereals), 32·5% small beetles, and 1·6% snails; no beetles taken in March, but in June comprised 79% of 33 items in 3 stomachs, though almost all of these were in one stomach (Goodbody 1955). In Schleswig-Holstein (northern Germany), stomachs obtained in August contained many fragments of insects and snail shells; thought possibly to eat more animal than plant material over year (Eber 1956). In Kirgiziya, 83 stomachs (time of year unknown) contained Orthoptera (Tettigoniidae, Acrididae), Hemiptera (Pentatomoidea, Reduviidae), adult Lepidoptera, Diptera (including Bibionidae), adult and larval beetles (Carabidae, Scarabaeidae, Chrysomelidae, Curculionidae), snails, and seeds (etc.) of cereals and other grasses, juniper *Juniperus*, Polygonaceae, Ranunculaceae, and Leguminosae (Pek and Fedyanina 1961). In western Tien Shan (Kazakhstan), 7 stomachs, June-August, held Orthoptera (Tettigoniidae, Acrididae), Hemiptera (Pentatomidae, Cicadellidae), ants, beetles (including Carabidae, Curculionidae), and only a few seeds (Kovshar' 1966). In Schleswig-Holstein, wheat grains important when available; in autumn, seeds of weeds and root crops preferred but later forages mainly in stubble and dung-spread pasture (Haack and Schmidt 1989). In Aberdeenshire, main winter food was grains of stacked cereal, but as this traditional practice is disappearing may have some difficulty finding food at this time of year; takes much grain during sowing (Goodbody 1955). For effect on food supply of changing methods in agriculture, see also Lack (1986), Thompson and Gribbin (1986), Watson (1992a), and Donald *et al.* (in press).

Young fed insects and their larvae, and seeds (mostly cereal). In Zurich region (northern Switzerland), 86 collar-samples from 9 nests contained 324 invertebrates and cereal grains; until end of June young given mostly adult Diptera (in cool weather mostly Tabanidae and in warm weather Rhagionidae because of different activity patterns); after early July, Diptera disappeared from diet and Orthoptera became main component; cereals not important, less than 1 ear of grain being brought per nest over nestling period (Hegelbach 1980). In eastern Germany, collar-samples from 4 broods contained 354 items, of which 42·3% by number Lepidoptera (great majority of which Noctuidae: 30% caterpillars, 11% pupae, 1% adults), 13·6% Orthoptera (mainly Tettigoniidae and Acrididae), 9·9% beetles (almost all Scarabaeidae), 8·5% earwigs, 4·8% spiders, 3·7% Diptera, also Hemiptera, earthworms, and snails; 16·1% by number wheat grains. Adults crush larvae in bill before feeding to young, and large insects have extremities removed and are beaten and chewed; virtually all cereal grain brought is almost ripe, and nestlings receive more as they get older. (Gliemann 1973.) In Cornwall, south-west England, adult butterflies (Pieridae) brought to nest with wings attached; main nestling food probably small green caterpillars of this species (Ryves and Ryves 1934), and similarly in northern Scotland dominant items in nestling diet were green caterpillars, also Tipulidae and seeds; ♀ collected food usually within 200 m of nest (20–300 m) (Goodbody 1955; Macdonald 1965). In north-east Scotland, chicks seemingly fed mainly pulp of cereal grain, even those freshly-hatched, particularly in cold or wet weather when insect activity low; almost all grain brought (barley, wheat, and oats) was greenish and partly ripe, not wholly unripe; season when this food available starts much earlier where winter cereals grown: winter barley taken late May to mid-July, winter wheat mid-June to mid-September, and spring barley and oats late June to September (Watson 1992b). In study in eastern England, ♀ rarely brought less than 5 caterpillars per visit (Robertson 1954). BH

Social pattern and behaviour. Major studies have concentrated on breeding behaviour, e.g. Ryves and Ryves (1934) in Cornwall (England); MacDonald (1965) in Sutherland (Scotland); Gliemann (1973) in Germany; Hartley (1991) and Shepherd (1992) in North Uist (Outer Hebrides, Scotland), recently using DNA-fingerprinting techniques (Hartley *et al.* 1993).

Studies frequently address influence of agricultural practices (e.g. Hegelbach and Ziswiler 1979, Thompson and Gribbin 1986). Pattern of ♂'s singing behaviour also well described (e.g. Gyllin 1965a, 1967, Møller 1983c). Antagonistic- and courtship-display compared with *Emberiza* buntings by Andrew (1957a).

1. Generally gregarious outside breeding season; flocks typically very variable in size and composition, both geographically and in relation to weather. Flocking starts at end of breeding season; in northern Jylland (Denmark), early flocks consisted of ♀♀ and juveniles; later, 'family' groups may aggregate, flock size increasing till December, then decreasing (Møller 1983c). In West Midlands (England) flock size typically up to 60–100 (Harrison *et al.* 1982). In north-east Scotland, flocks (from late October) typically of 20–40 birds but up to 300 recorded in severe weather; in some years sizeable flocks still present late April and early May (Buckland *et al.* 1990). Winter flocks, composed of ♂♂ and ♀♀, will associate with House Sparrow *Passer domesticus*, Twite *Carduelis flavirostris*, Yellowhammer *Emberiza citrinella*, and Reed Bunting *E. schoeniclus* in England (P K McGregor) and western Scotland (I R Hartley); also with Tree Sparrow *P. montanus*, Greenfinch *C. chloris*, Goldfinch *C. carduelis*, Linnet *C. cannabina*, Chaffinch *Fringilla coelebs*, Brambling *F. montifringilla*, and Snow Bunting *Plectrophenax nivalis* in north-east Scotland (A Watson). In winter, ♂♂ may sing in flocks (north-east Scotland: A Watson) or leave flocks to sing on nearby, previously defended territories (P K McGregor, M Shepherd) or move much further to defend territories, though still returning to flock later (D G C Harper); see also Andrew (1956c) and Song-display (below). ♂♂ start defending territories from November in Kincardineshire, north-east Scotland (A Watson), from January in Cambridge, eastern England (Andrew 1956c) and Sussex, southern England (D G C Harper), and from March in North Uist (Shepherd 1992). In Cambridge, ♀♀ did not remain on territories for any length of time until April; before then (and to lesser extent until end of May) ♀♀ continuously in large flocks (Andrew 1956c). BONDS. Pair-bond not close: ♂ and ♀ hardly ever together on or off territory unless sexually or aggressively motivated (Andrew 1956c; Møller 1983c). Mating system complex and typically variable within and between areas; numbers of ♂♂ unmated, monogamous and successively (or occasionally simultaneously) polygynous in approximate ratio 1:2:1, e.g. 23·5%, 41·2%, and 33·3% respectively in North Uist (Hartley *et al.* 1993). In Lancashire (northern England), of 75 ♂♂, *c.* 72% monogamous, 19% polygynous (2–3 ♀♀ breeding per territory), 9% unmated; ♀♀ (unlike in study of Møller 1983c) not found to be promiscuous (Thompson and Gribbin 1986). Polyandry reported only by MacDonald (1965). Extent of (successive) polygyny variable: in North Uist, generally 2 ♀♀ (maximum 3 ♀♀) per polygamous ♂ (Hartley 1991; Shepherd 1992; see also Walpole-Bond 1938, Humphrey 1967, and Hegelbach 1984; but compare Ryves and Ryves 1934); however, up to 18 ♀♀ (6 of these simultaneous) in Sussex (D G C Harper), also 6 ♀♀ simultaneously in north-east Scotland (A Watson). In Sussex, probability (for ♂♂) of pairing related to date of arrival on territory: arrival before 1 April gives pairing probability of 0·79, after 1 April 0·3 (D G C Harper). However, in north-east Scotland, ♂ and ♀ present in same area throughout year (A Watson). Extent of polygyny unrelated to territory quality in North Uist (Hartley 1991) but correlated with this in Lancashire (Thompson and Gribbin 1986) and apparently in Sussex (D G C Harper); occasionally associated with polyterritoriality in Sussex (D G C Harper) and North Uist (M Shepherd). Rarely does either sex move territory after nest failure, although unmated ♂♂ most likely to move territory within a year (Shepherd 1992). Agricultural practice (e.g. silage cutting) in north-east Scotland is

major cause of nest failure, after which pair will move to other suitable habitat, although ♂ and ♀ will also move to new territory (available because of agricultural practice) after successful attempt (A Watson). ♂♂ arrive on territories considerably before ♀♀. ♀♀ seldom seen prior to settlement (Ryves and Ryves 1934; Shepherd 1992), may arrive in groups (D G C Harper). ♀♀ construct nest over 2–4 days (North Uist) and incubate alone (Ryves and Ryves 1934; MacDonald 1965; Shepherd 1992). Nestlings provisioned mostly by ♀, contribution of ♂ varying from 0–50% (Hegelbach 1984; Hartley 1991). One instance of ♂ rearing brood alone after ♀ disappeared at 4 days after hatching (I R Hartley, M Shepherd). In North Uist, 19% (n=43) of broods received some provisioning from ♂ up to 4 days after hatching, proportion increasing with nestling age; 8 days after hatching 82% (n=36) of nests received feeding visits from ♂, but food loads brought by ♂ smaller than by ♀ throughout (Hartley 1991). Broods of primary and secondary ♀♀ provisioned at similar rates by ♂ (Hartley 1991), but differed when more than 3 ♀♀ bred simultaneously with 1 ♂ in north-east Scotland (A Watson). Juveniles seen to help with provisioning, but relationship to nestlings unclear (D G C Harper, A Watson); other ♀♀ of polygamous ♂ also seen helping to provision young (A Watson). Young independent after 3–4 weeks old, but may still be fed after this time (M Shepherd, A Watson). Data sparse on age of first breeding, but unmated ♂♂ more likely to be 1-year-olds and all 1st-year ♂♂ hold territories for at least part of 1st season (Shepherd 1992). Sex-ratio 1:1 in marked populations (Gyllin 1965a; Gliemann 1972; Hegelbach and Ziswiler 1979; Hegelbach 1974; Hartley 1991; Shepherd 1992; Hartley et al. 1993), previous reports of skewed sex ratios erroneous because based on nests per ♂ (e.g. Ryves and Ryves 1934). BREEDING DISPERSION. Solitary and territorial, but often aggregated (e.g. Møller 1983c). Size of territory difficult to ascertain (rival ♂♂ fighting only at places where they had adjacent song-posts). In Denmark, over 2 years, size declined from 2·8 ± SD1·2 ha to 1·3 ± SD0·5 ha due to disappearance of some large marginal territories in less preferred areas (i.e. preferred territories smaller than less preferred ones) (Møller 1983c). In Cambridge, mean maximum width of territory 140 m, average area c. 1 ha (0·7–1·1, n=4). In Lancashire, song-posts up to 140 m from nest (Thompson and Gribbin 1986). Data on shift of territory from year to year in North Uist (see below) suggest territory c. 170 m across. Territory of all-purpose type. In Scotland, territory-holding ♂♂ commonly allow access to ♀♀ for feeding (Hartley 1991; A Watson). Polyterritoriality rare: 3 instances in 3 years in North Uist; 2 involved takeovers at distance of 2–3 territories from original territory (with single incubating ♀) and included ♀ with very recently-failed 1st nesting attempt; 3rd case involved unmated male and previously-unoccupied territory, 12 territories away from original territory (Shepherd 1992). No data on distance between replacement nests, although usually within same territory (see above). In North Uist, c. 60% ♂♂ (40% ♀♀) returned to breed on previous year's territory; ♂♂ changing territories moved mean 3·6 ± SE1·0 territories, median 600 m (200–2400, n=16), ♀♀ moved mean 2·8 ± SE0·6 territories, median 500 m (200–1300, n=13), and unsuccessful ♂♂ moved significantly further; mean natal dispersal of ♂♂ was 6·2 ± SE1 territories, median 1100 m (200–4500, n=23); for ♀♀, mean 7·2 ± SE1·4 territories, median 1250 m (200–4000, n=16), positively related to population density in year of hatching and (in ♂ only) to natal brood size (Shepherd 1992). In Sussex, modal dispersal distance of ♂♂ 200–600 m (n=18), of ♀♀ 400–600 m (n=23) (McGregor et al. 1988). Estimations of density affected by aggregated dispersion (giving importance to size of area sampled) and, in many areas, by recent declines. In Britain,

decline well-documented (e.g. Brown et al. 1984, Marchant 1984, Thom 1986, Williams et al. 1986, Marchant and Whittington 1988), probably reflecting agricultural practices and their intensification (Harrison et al. 1982; O'Connor and Shrubb 1986; Thompson and Gribbin 1986; Thompson et al. 1992). Density estimates based on singing ♂♂ (e.g. Terry 1986) though sometimes (misleadingly for a polygamous species) converted into 'pairs'. Course (1941) recorded 6·3 ♂♂ per km² on arable land in Cambridgeshire, while Sharrock (1976) reported 2·4 'pairs' per km² for British farmland. In Lothians (Scotland), very low density of 1 ♂ per 11·9 km² reflects decline (Brown et al. 1984). In Outer Hebrides, 5·0–5·4 ♂♂ per km² in preferred habitats (sand-dune and machair) (Williams et al. 1986). Other densities as follows. At Lac de Grand-Lieu (western France), 16·7–20 ♂♂ per km² (Marion and Marion 1975). In Reusstal (Switzerland) over 4 years, 29–74 ♂♂ per km² (Schifferli et al. 1982). In Vorarlberg (Austria), 16 territories per km² locally in Rhein valley, 7 over larger area (Kilzer and Blum 1991). In Germany: 50 censuses 1948–75 gave mainly 0·3–2·6 ♂♂ per km in Mecklenburg (Klafs and Stübs 1987); in Niedersachsen, 60 ♂♂ along 63 km (Schoppe 1986); in Danube valley (Bayern), decline from 0·36 ♂♂ per km² in 1975, to 0·10 in 1988–91 (Vidal 1991); for Brandenburg, see Rutschke (1983c); for Oberlausitz, see Eifler and Blümel (1983). In farmland in Damnica plateau (Poland), 2·83 'pairs' per km² (Górski 1988). In Bulgaria, 2–155 'pairs' per km² in different habitats (T Michev, P Iankov, B Ivanov, and L Profirov). In Morais (northern Portugal), 62·5 territories per km² (Mead 1975). In Extremadura (western Spain), 33 ♂♂ per km² (Hellmich 1985). In Morocco, 12·8 'pairs' per km² in maquis (Thévenot 1982). ROOSTING. Outside breeding season, roosts communally in groups up to few hundred, commonly in reedbeds (e.g. Taverner 1958, Link and Ritter 1973a, Boddy and Blackburn 1978, Bibby and Lunn 1982), but in bushes and on open marshy ground in small groups of 1–9 (M Shepherd, I R Hartley). In West Midlands, maximum 500 reported roosting together at end of December (Harrison et al. 1982). Continual turnover of marked individuals at roosts (Boddy and Blackburn 1978; C Thomas). Often gathers in staging posts c. 1 hr before dark, close to site of roosting (I R Hartley). In North Uist during this period, loafs diurnally in groups of less than 20 on old farm machinery, also on bales of wire close to sources of grain (C Thomas). In north-east Scotland, recorded taking shelter in straw and hay ricks in severe weather (Buckland et al. 1990). For roosting by fledglings in nest, see Relations within Family Group (below).

2. Relatively approachable, especially on fence-lines alongside frequently used paths and by houses; calls 2a and 2c (see Voice) commonly elicited by close approach of human. Often tail-flicks (with spreading to highlight white edges) and wing-flicks after landing; also wing-flicks with or without intense tail-flicking when alarmed (Andrew 1956a). Remains still on perch or drops into cover when hawk Accipiter flies over (Andrew 1956c). Rap-

A

B

tors also elicit flight to wire (telephone wire, fence, etc.) (A Watson). See also Parental Anti-predator Strategies (below). FLOCK BEHAVIOUR. Little hard information. Individual distance not less than 40 cm when foraging, but as little as 10 cm when resting on wire (A Watson). See introduction to part 1 (above) for singing in winter flocks, and 2b in Voice for characteristic calls. SONG-DISPLAY. For functions of song see 1 in Voice and Møller (1983*c*). ♂ sings from exposed perch (Fig A), less commonly in flight, and frequently on alighting. Perch-sites vary from ground-level, small thistles (Compositae) and grass clumps to tree-tops, but most commonly human artefacts (fence-wires and posts, abandoned machinery, hedges, stone walls, telegraph poles, pylons, and overhead wires) (P K McGregor, A Watson). On leaving and approaching song-post, ♂, often while singing, commonly performs characteristic Dangling-legs display (Fig B): head lowered, feet dangled and legs pointed out slightly, accompanied by rapid, shallow wingbeats (e.g. Jonsson 1992, E N Panov). ♀ does not sing. During breeding season, high level of song output maintained throughout day, even through midday; for detailed data on diurnal and seasonal variation in song output, see Gyllin (1965*a*, 1967), Gliemann (1973), and Møller (1983*c*). Singing (including Subsong: P K McGregor) occurs sporadically throughout winter in flocks, at roosts, and occasionally on territory (Møller 1983*c*; Shepherd 1992) (see also introduction to part 1, above), but main song-period April to September (e.g. Ryves and Ryves 1934, MacDonald 1965, Hegelbach and Ziswiler 1979, Møller 1983*c*, Shepherd 1992). ♂ continues singing and displaying to strange ♀♀ on territory despite almost continuous presence of resident ♀ (Andrew 1956*c*), and (in Denmark) successful (polygynous) ♂♂ continue singing late into breeding season (Møller 1983*c*). In Lancashire also, polygynous ♂♂ sang more intensively and for longer than other ♂♂ (Thompson and Gribbin 1986). Rate of song-production early in season not associated with subsequent polygyny in Scotland (Shepherd 1992), but may be in Sussex (D G C Harper). ANTAGONISTIC BEHAVIOUR. (1) General. ♂♂ will attack conspecific intruders throughout territorial period, less commonly at end of flocking period (P K McGregor, A Watson); see introduction to part 1 (above) for sporadic territorial defence during winter. ♀ takes no part in defence of territory, but for apparent ♀-♀ chase see Andrew (1956*c*). In peak period of territorial defence, most attacks directed at intruding neighbouring territory-holders ('floaters' rare, observed only once in study in North Uist: I R Hartley, M Shepherd), focus of defence equally ♀ and territory; attacks on ♀♀ were attempted extra-pair copulations (see subsection 4 of Heterosexual Behaviour, below). (2) Threat and fighting. Detailed descriptions of displays, postures (with illustrations), and associated calls in Andrew (1957*a*). Head-forward threat-posture similar to that in *Emberiza*: legs flexed, body horizontal with head in line, bill opened wide (Fig

1 in Andrew 1957*a*), but may be closed during head-thrust. In Breast-to-breast fighting, ♂♂ fly vertically face to face, clawing and pecking, up to 8 m; if one ♂ seizes other, both fall to ground where fighting may continue (Andrew 1957*a*; P K McGregor, M Shepherd). In Swoop-attack (typically seen in boundary disputes: Andrew 1956*c*), ♂ flies over perched rival, sometimes striking with claws; ♂♂ commonly switch roles (attacked bird slowly following attacker). Breast-to-breast fighting may follow if ♂ lands after Swoop-attack (Andrew 1957*a*). Attacks on other bird species noted (Andrew 1956*c*), especially *E. citrinella* (A Watson). HETEROSEXUAL BEHAVIOUR. (1) General. ♂♂ arrive on territories considerably before ♀♀, latter up to 2 months after first ♂♂ in Scotland (MacDonald 1965; Hartley 1991; Shepherd 1992), but compare Kincardineshire where arrival simultaneous (A Watson); ♀♀ 2–3 weeks later in Switzerland (Hegelbach 1984). Laying dates positively correlated with arrival date. Mean delay 27 days between settlement and laying date of first egg, but delay negatively correlated with settlement date, i.e. late-arriving ♀♀ lay sooner after arrival (M Shepherd). (2) Pair-bonding behaviour. ♂ recorded chasing ♀♀ and, at close quarters, settles on ground close to ♀ (also on ground) and makes repeated upward flights, each *c*. 1 m, singing and calling, and lands beside her each time (Walpole-Bond 1932). Detailed description of intimate courtship-display by Andrew (1957*a*): Wing-quivering display usually accompanies courtship; body horizontal, bill lowered or repeatedly lowered and raised, wings raised and somewhat extended, vibrated in bursts, occasionally tail spread (Fig 6 in Andrew 1957*a*); ♂ displaying thus may approach ♀; associated call ('titi') similar to flight-call (see 2b in Voice). Bill-gaping may accompany Wing-quivering display. Collection and presentation of nest-material (A Watson; see also subsection 5, below) sometimes occur as part of courtship. No antagonism observed during pair-formation, but ♀♀ removed for ringing sometimes attacked by mate after return (I R Hartley); also ♂♂ attempting to copulate often rebuffed (M Shepherd; see subsection 4, below). No particular mate-guarding behaviour (I R Hartley, M Shepherd), though Thompson and Gribbin (1986) recorded ♂♂ escorting ♀♀ to and from nest. (3) Courtship-feeding. Extremely rare; none observed in North Uist (I R Hartley, M Shepherd), only by one ♂ in north-east Scotland (A Watson); therefore no data on associated display or timing in relation to nesting cycle. (4) Mating. Copulation readily observed; occurs on ground, fence-posts and wire fences; no apparent concealment (Hartley 1991; M Shepherd, A Watson). Soliciting-display by ♀ as follows: body horizontal, bill and tail raised somewhat, wings raised and rapidly quivered (Fig 2 in Andrew 1957*a*); perched ♀ may wing-quiver and gape aggressively at approaching ♂. ♂ gives 'kwaa' call (see 3 in Voice) as he flies to ♀, hovers above her, calling more excitedly (may land directly on her), mounts with rather erect posture and beating

wings, lowers hind body, pushing ♀'s tail to one side, wings ceasing beating at cloacal contact. (Andrew 1957a.) Repeated cloacal contact not observed. ♀ raises tail, and vent feathers noticeably spread (A Watson). Copulation lasts 1–4 s; no particular post-copulatory behaviour, but ♀ may shake body (A Watson). Copulation rate of 1–2 matings per clutch (Hartley et al. 1993). Occurs at any time of day (I R Hartley, M Shepherd, A Watson), but data scarce. Observed from 1–10 days before first egg laid (Hartley et al. 1993) and from 1–2 days after ♀ begins nest-building (A Watson). Attempted extra-pair copulations quite common but rarely successful (I R Hartley, M Shepherd) and not concentrated in ♀'s fertile period (Shepherd 1992). Apparently low risk of cuckoldry confirmed by rarity of extra-pair paternity (determined by DNA-fingerprinting) in same population: only 2 out of 44 offspring (1 of 15 broods) not fathered by territory-holding ♂, although probable that the one nest with apparent extra-pair paternity was not assigned to correct ♂. Single extra-pair copulation attempted 5 days after 1st egg laid. (Hartley et al. 1993.) (5) Nest-site selection and behaviour at nest. ♀ selects nest-site. ♂'s Nest-site display is variant of Wing-quivering display with associated soft 'sisi' calls, given on ground; nest-material may be picked up when near ♀ (Andrew 1957a). ♀ builds and incubates. ♀♀ rarely fly directly into nest when incubating (I R Hartley), sometimes arriving and leaving at different points (A Watson). RELATIONS WITHIN FAMILY GROUP. ♂ feeds young directly but less frequently than ♀ (for relationship with nestling age see Bonds, above). Faecal sacs removal by ♂ and ♀ for older broods (I R Hartley, A Watson). Begging occurs in and out of nest with associated calls of young (see Voice). Eyes open at day 3–4 (I R Hartley, A Watson). Young leave nest 10–13 days after hatching (North Uist), but can be flushed by predator at earlier stage (I R Hartley, M Shepherd: see Anti-predator Responses of Young, below). Can only fly weakly on first leaving nest (M Shepherd). Move up to 20 m on 1st day (I R Hartley), often hiding in crops in vicinity of nest (M Shepherd: North Uist). Unlikely to return to roost in nest in North Uist (I R Hartley), but frequently do so in north-east Scotland for 1–2 days after fledging (A Watson). Family bonds maintained for c. 2 weeks after leaving nest (I R Hartley), longer for final brood (A Watson), families often forming small flocks (M Shepherd). ANTI-PREDATOR RESPONSES OF YOUNG. Young, alerted by parental calls (see 2a in Voice), crouch in nest and become silent when approached; will explode from nest from c. 5 days after hatching (I R Hartley, M Shepherd). PARENTAL ANTI-PREDATOR STRATEGIES. (1) Passive measures. ♀♀ sit very tightly, flushing at last moment, probably only when vegetation in contact with nest is touched (M Shepherd). (2) Active measures: against birds. Close approach of raptor to nestlings causes parents to fly around, ♀ apparently attempting to lead predator away (A Watson). (3) Active measures: against man. ♀ flushed from nest flutters over vegetation for some distance before flying away, perhaps laboured flight serving dis-

traction function (I R Hartley). No special display on leaving nest according to Andrew (1956a) but Walpole-Bond (1932) states ♀ may flutter along ground. ♂ and ♀ will mob during ringing of nestlings; ♀ will also sit 10–25 m away calling quietly (I R Hartley: 2a see Voice). No further information.

(Figs by D Nurney: A from artist's own material: B from drawing by E N Panov.)   PKM

Voice. Both sexes call throughout year, and ♂♂ in winter flocks may sing sporadically. Song variation and singing behaviour well studied in Britain, notably in Oxfordshire, Sussex, and Cornwall (England), north-east Scotland, and Outer Hebrides (north-west Scotland), yielding numerous published sonagrams (e.g. McGregor 1980, McGregor et al. 1988; see also Pellerin 1981). The few studies from continental Europe and elsewhere suggest there may be differences from Britain in pattern of local song variation (see below) (e.g. Czikeli 1980, 1982, Sultanov and Gumbatova 1986; but see Pellerin 1981). Comparative study of Emberiza calls by Andrew (1957a, b) included M. calandra and traced sonagrams.

CALLS OF ADULTS. (1) Song of ♂. Basically segmented in structure. Begins with accelerating series (trill) of variable number of discrete units (e.g. 'pit pitpitpit': Bergmann and Helb 1982) with very similar structure (Figs I–II), followed by number of very rapidly executed groups of tightly packed units (rattles) containing variety of unit-types with predominant form being rapid frequency modulation over wide frequency range. Rattles have jangling timbre, likened to bunch of keys being shaken (Witherby et al. 1938), or glass-splintering 'tük-tük-tük tri-rililiririlililee' (Jonsson 1992) which at distance takes on tinkling quality. Central section lasts for 30–50% of song-phrase and is followed by more widely-spaced units, com-

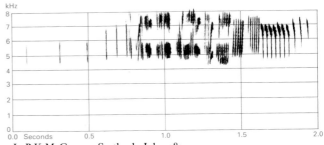

I P K McGregor Scotland July 1989

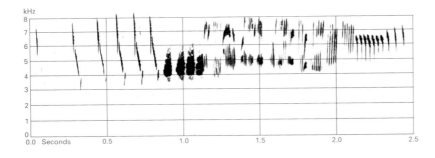

II P K McGregor Scotland July 1989

monly falling in pitch in decelerating 'trill', giving impression of mechanical device coming to halt. This pattern widespread throughout Britain and other parts of Europe. Commonness of song throughout breeding season strongly suggests that it functions in similar manner to other passerine song, i.e. in territory defence and mate attraction or stimulation. Certainly ♂♂ respond aggressively to song playback in the field (e.g. McGregor 1983, 1986, Pellerin 1982, 1983), whereas no response noted from ♀♀ (which do not sing). Song accompanies approach of potential mate in the field and captivity (Andrew 1957*a*). Although generally given from low perches, song can be given in flight (often accompanied by Dangling-legs display) and regularly given on alighting. 2 important aspects of ♂'s song variation. Firstly, individuals often sing incomplete song-phrases (McGregor 1981; McGregor *et al.* 1988; Constable 1989; Shepherd 1992). Extent of song completeness varies from a few units of accelerating introductory trill to starting 2nd song-phrase immediately after 1st has ended, with no noticeable pause. Commonness of partial songs is related to song-type (see below). Partial songs may have decelerating 'trill' appended. 2nd form of within-♂ variation is repertoire of song-types. Differences between song-types are concentrated almost exclusively in 1st third of song-phrase and are usually readily apparent to human ear in accelerating introductory trill. Song-types are sung in bouts, a number of one type being followed by another. Number of song-types varies between areas, e.g. 2 in Oxfordshire (McGregor 1980, 1981, 1986) and Outer Hebrides (P K McGregor) and 3 in Sussex (McGregor *et al.* 1988), Cornwall (McGregor 1980, 1981), and north-east Scotland (P K McGregor); also 3 in Azerbaydzhan (Sultanov and Gumbatova 1986). In Britain, all ♂♂ in a local population sing same number and form of song-types (rare exception are mixed dialect singers discussed below). Field playback experiments have shown ♂♂ capable of distinguishing between song-types (McGregor 1986). In study in central England, song-types sung approximately equally and used in same contexts (McGregor 1981, 1986). However, in a Hebridean population, one song-type (type 2, with more complex introductory units and longer overall duration: Fig II) was 3 times more likely to be sung as partial song than other song-type (Fig I: Constable 1989; Shepherd 1992); ratio of complete to incomplete songs was 3:1 for song-type 1 and 1:4 for song-type 2 (Constable 1989). Mated ♂♂ sang higher proportion of song-type 2 and sang more incomplete song-type 2 (Shepherd 1992), suggesting signal value in communication with mates. Different pattern of song-type variation has been described by Czikeli (1982), with an association between population density and form of introductory units; in low population densities (e.g. those of central Europe moist meadowland) birds sang introductory units with harmonic overtones whereas these lacking in songs in high density populations of dry, partially cultivated steppe habitats of southern Europe

(Fig 1 in Czikeli 1982). Playback elicited 'high-density' song-type from all populations, perhaps because it simulated high population densities (Czikeli 1982). Further research necessary to reconcile these findings with those in British populations (see above). Geographical variation in song known for some time (e.g. Berndt and Frieling 1944, Duchrow 1959). Most striking feature of between-♂ variation is existence of local song dialects, characterized by mosaic distribution of song variation, prerequisite of which is very little variation between ♂♂ in same local dialect, i.e. all ♂♂ must sing same song (or songs). Such a pattern is found in Oxfordshire (McGregor 1980, 1981; McGregor and Thompson 1988), Cornwall (McGregor 1980, 1981), Sussex (McGregor *et al.* 1988), and Outer Hebrides (P K McGregor). Playback shows ♂♂ are able to discriminate between local dialects (McGregor 1983; Pellerin 1982, 1983). Adjacent dialects have distinct songs with no intergrading at dialect boundaries. In detailed studies in Oxfordshire, Sussex, and Outer Hebrides, only one instance of ♂ singing songs intermediate in structure between 2 populations ('mixed dialect singer': McGregor and Thompson 1988); also, though rarely, ♂♂ near common boundary sang 2 local dialects. In Hebrides, ♂♂ singing 2 dialects sang local dialect of area in which they held territory 3 times more often than local dialect of adjacent population (J L Kitwood). In Britain there is also widespread change in local dialects from year to year (McGregor and Thompson 1988), marked individuals changing fine details of songs from year to year (P K McGregor). Factors stimulating these changes and resulting in ♂♂ of all ages adopting changes are unknown. Sultanov and Gumbatova (1986) failed to find local dialects in Azerbaydzhan, but dialects described in Pelopónnisos (Greece) (Kroymann and Kroymann 1992). Song quite distinct from that of other Emberizidae, consistent with placing *M. calandra* in its own genus. However, there are a number of reports of *M. calandra* apparently singing song of *Emberiza*, including Ortolan Bunting *E. hortulana* (Schumann 1956), but particularly Yellowhammer *E. citrinella* (Fig III); such individuals were isolated from other *M. calandra*, interacted with *E. citrinella*, and rarely sang *M. calandra* song (Donovan 1978, 1984; Richards 1981; Stirrup and Eversham 1984), and situation thus best not described as mimicry, but

III A K Pearce Wales July 1978

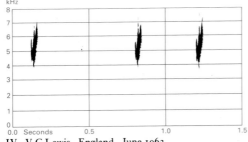

IV   V C Lewis   England   June 1963

instead probably exemplifies importance of social inter-action in song learning (Pepperberg 1985). (2) Contact-alarm and alarm-calls. (2a) Simple, brief 'tic' or click (Fig IV). Also rendered as loud abrupt 'quit', or when run together (as in flight) as 'quit-it-it' (Witherby *et al.* 1938), 'tit' (Fig 1 in Andrew 1957*b*), 'tipit' (M G Wilson). Given in variety of contexts, notably excitement or alarm (e.g. on close approach of recordist to singer or nest-site); also commonly precedes song, and may be produced by both sexes apparently as contact-call (Jonsson 1992), including as flight-intention signal. Rate of repetition seems to reflect level of excitement (etc.), with highest rates commonly preceding song by ♂. (2b) Similar, but simpler, call given by birds in winter feeding flocks sounds like hot fat sput-tering in pan (P J Sellar: Fig V). (2c) Persistent, brief,

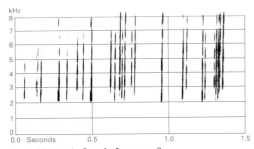

V   Mild (1990)   Israel   January 1989

VI   P K McGregor   Scotland   July 1989

single rasping calls (Fig VI) variously rendered as 'jüir' (Bergmann and Helb 1982), 'chirr' or 'jurr' (Witherby *et al.* 1938), 'weep' (Fig 5 in Andrew 1957*b*), and soft rolling 'dchrrut' (Jonsson 1992). Occurs in much the same context

as call 2a, often being interspersed in series of such calls; possibly restricted to disturbance in nesting territory (Bergmann and Helb 1982). May express fear (Andrew 1957*b*), and also function as warning-call. Series of call 2c given by singing ♂ (known to have recently fledged young) when Peregrine *Falco peregrinus* approached; ♂ stopped singing, turned to face raptor, and began calling at inter-vals of *c.* 0·5 s, continuing to call and track raptor until it disappeared from view, whereupon ♂ relaxed and resumed singing. (2d) Rippling 'cheer' (P J Sellar), which may

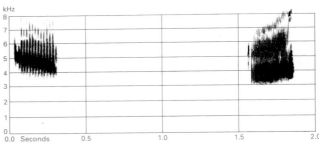

VII   J-C Roché   France   June 1966

ascend or descend (Fig VII); rendered as prolonged grat-ing 'zeep' (Witherby *et al.* 1938: 'scolding note'), and plaintive, nasal 'zeeaa', recalling Greenfinch *Carduelis chl-oris*, sometimes combined with call 2a (M G Wilson); thus, creaky 'zieh... tzek' (Jonsson 1992). (2e) A 'sow' or 'see-ow', somewhat like Reed Bunting *E. schoeniclus*, used espe-cially when nest contains young (Witherby *et al.* 1938). (3) Other calls. A 'kwaa' (Fig 8 in Andrew 1957*b*) given by ♂ during copulation.

CALLS OF YOUNG. Soft, indistinct, high tinkling food-

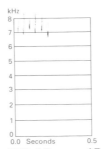

VIII   J Burton and D J Tombs/BBC   Spain   1983

call from nestlings (Fig VIII), very similar to 'close begging-call' of *E. citrinella* (Fig 7 in Andrew 1957*b*). Hand-reared birds at 10 days gave faint 'sipsipsip', at 20 days a more vibrant 'surr' (Andrew 1957*b*). Call similar to adult call 2a given after fledging (Fig IX). Fledged young after a few days also give calls (Fig X) intermediate between adult calls 2a and 2c and which seem to function in begging; these vary in mean frequency by *c.* 0·5 kHz

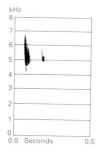

IX V C Lewis England June 1967

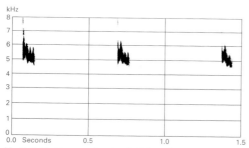

X P K McGregor Scotland July 1989

and to some extent in structure (P K McGregor). Little known of song acquisition, although pattern of local dialects and widespread change in dialects from year to year (see above) suggest that song is acquired early in each breeding season. Single individual raised in acoustic isolation produced song considered normal (Thorpe 1958), though no sonagram presented; see above (Fig III) for apparent importance of social tutoring.   PKM

Breeding. SEASON. Britain: in Sutherland (north-west Scotland), eggs laid early June to mid-July (mid-May to August) (Macdonald 1965); in England (from nest record cards), mean date for *c.* 62 1st clutches 25 May (Yom-Tov 1992, which see for comparison with Yellowhammer *E. citrinella* and Reed Bunting *E. schoeniclus*); in Cornwall (south-west England), mainly July (early June to early August) (Ryves and Ryves 1934); for north-west England,

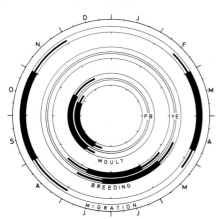

see Thompson and Gribbin (1986); eggs recorded September (Witherby *et al.* 1938). Southern Sweden: peak egg-laying July (Jönsson 1992). Germany: see diagram; in Oberlausitz (eastern Germany), eggs laid mainly June (late April to end of July) (Eifler and Blümel 1983); for northern Germany, see Haack and Schmidt (1989). Israel: in lowlands, 2nd half of February to 2nd half of May; at higher altitude, mid-March to end of June (Shirihai in press). Canary Islands: eggs laid mid-March to mid-June (Martín 1987). SITE. Generally on ground, in thick tangled grass or shrub, in depression in soil of arable field (sometimes perched 'awkwardly' on bare ground), or in pasture, often in clump of thick weeds, e.g. thistle or other stout Compositae (Walpole-Bond 1932; Robertson 1954; Macdonald 1965; Gliemann 1973; Jönsson 1992). Recorded up to 1·5 m above ground in bush (Witherby *et al.* 1938; Woods 1950); in Morocco, often in low branches of tree (Heim de Balsac and Mayaud 1962; Etchécopar and Hüe 1967). In Cornwall, 89% of 54 nests in dense bramble *Rubus* or gorse *Ulex*, 9% in crops, and 2% in rough grass (Ryves and Ryves 1934); of 35 nests in Sutherland, all but one in crops, 22 in cereals (Macdonald 1965); see also Hayhow (1984) for Sheffield area, central England, and Thompson and Gribbin (1986) for north-west England. In Oberlausitz, 36% of 39 nests in meadows, 31% rough grassland, 18% in broom *Cytisus*, *Rubus*, etc., 15% in arable fields (Gliemann 1973). Nest: fairly large loose construction of stalks, grass stems (which can be green), and roots, lined with fine grass, rootlets, and sometimes hair (Walpole-Bond 1932; Ryves and Ryves 1934; Dementiev and Gladkov 1954; Robertson 1954). In Czechoslovakia, 10 nests had average outer diameter 12·6 cm (9·5–16), inner diameter 7·6 (5·5–9), overall height 7·4 (5–11), and depth of cup 6·1 cm (4–8) (Hudec 1983). Building: by ♀ only, accompanied by ♂; usually finished in 1–3(–4) days, ♀ working all day into evening (Walpole-Bond 1932; Ryves and Ryves 1934; Gyllin 1965a; Macdonald 1965; Gliemann 1973). Said to take 14–18 days by Woods (1950) but this considered highly exceptional by other authors. All material collected from ground usually within *c.* 50 m; sometimes pulls up fresh grass (Macdonald 1965; Gliemann 1973; see these sources for bouts and technique). See also Ryves and Ryves (1934). EGGS. See Plate 34. Very variable in shape and colour; generally sub-elliptical, smooth and slightly glossy; whitish, often tinged with blue, purple, or buff, sparsely but boldly marked with blotches and meandering scrawls and hairstreaks of brownish-black or purple, though some with hardly any pattern or only large pale brown blotches; greyish-violet undermarkings often present (Gliemann 1973; Harrison 1975; Makatsch 1976). Nominate *calandra*: 24·1 × 17·6 mm (20·8–28·6 × 15·8–19·2), *n* = 844; calculated weight 3·91 g (Schönwetter 1984). *M. c. buturlini*: 23·4–17·5 mm (21·9–25·1 × 16·3–18·3), *n* = 38; average fresh weight 3·62 g, *n* = 33 (Kovshar' 1966). See also Makatsch (1976) for *c.* 400 eggs from various regions. Clutch: 4–6 (1–7). In Cornwall, of 78 clutches: 2 eggs, 1%; 3, 23%; 4, 54%; 5, 21%; 6, 1%; average 3·97 (Ryves and Ryves 1934); average in England 3·82, n = 310 (Yom-Tov 1992), and in Sutherland 4·0, n = 14 (Macdonald 1965). Clutches of 1 and 7 recorded in Britain (Witherby *et al.* 1938). In north-west Africa, of 131 clutches: 4 eggs, 36%; 5, 40%; 6, 21%; 7, 3%; average 4·92 (Heim de Balsac and Mayaud 1962). Average in Oberlausitz 4·63, n = 38 (Eifler and Blümel 1986), and in Czechoslovakia 4·20, n = 15 (Hudec 1983). In southern England, 2 broods common, occasionally 3 (Walpole-Bond 1932); in Cornwall, c. 5 of 14 ♀♀ had 2 broods (Ryves and Ryves 1934); in north-west England, 2 of 27 ♀♀ (Thompson and Gribbin 1986); only replacement clutches noted in Sutherland (Macdonald 1965). Very few 2nd broods recorded in Oberlausitz (Gliemann 1973), and extremely unusual in Israel (Shirihai in press). In Cornwall, 2nd broods started 8–14 days (n = 5) after fledging of 1st (Ryves and Ryves 1934). Eggs laid daily c. 1–2 days after nest completion (Ryves and Ryves 1934; Dementiev and Gladkov 1954; Gliemann 1973). INCUBATION. 12–14 days, by ♀ only (Witherby *et al.* 1938; Makatsch 1976); in Cornwall, 12·5 days (11–13, n = 8) from laying of last egg to hatching of last young; incubation proper probably started with penultimate egg; stints on eggs were 40–90 min, breaks 15–40 min (Ryves and Ryves 1934). In Sutherland study, stints averaged 60 min (30–117), breaks 12·5 min (2–29) (Macdonald 1965). Incubation can apparently start with any egg (Makatsch 1976). YOUNG. Fed and cared for mainly by ♀, ♂ helping occasionally late in nestling period (Ryves and Ryves 1934, which see for feeding routine; Macdonald 1965; Makatsch 1976). ♂♂ with more mates more likely to feed young; some juveniles feed chicks (D G C Harper); see Hartley (1991) for details, also Social Pattern and Behaviour. Brooded by ♀ for 3–4(–7) days (Ryves and Ryves 1934; Gliemann 1973). FLEDGING TO MATURITY. Fledging period 9–13 days, often leaving nest before able to fly (Ryves and Ryves 1934; Glutz von Blotzheim 1962; Macdonald 1965). In East Sussex (southern England), young from early nests fed by parents after fledging for 7–10 days, later ones for much longer (D G C Harper); in Oberlausitz, fed for c. 2 weeks after leaving nest (Gliemann 1973). For age of first breeding, see Social Pattern and Behaviour. BREEDING SUCCESS. In Cornwall, c. 150 young flew from 54 nests fathered by 24 ♂♂ (c. 6 fledged young per ♂, 2·8 per nest); of 32 eggs in 8 nests, 94% hatched and 66% produced fledged young; smallest chick often dies (Ryves and Ryves 1934). In East Sussex, c. 75% of eggs hatched and 56% produced fledged young (sample size unknown); ♀'s success increases with number of other ♀♀ her ♂ is paired to, though data may be distorted by extra-pair copulations; 3·14 fledged young produced per ♀ (n = 51), 4·21 per ♂ (n = 38); 42% of ♂♂ raised no young (D G C Harper). In north-east England, average 3·3 young fledged per nest (n = 9), slightly more in barley (3·5), but sample size small (Thompson and Gribbin 1986). In Sutherland, 88% of

56 eggs hatched and probably 50–60% of 91 nestlings flew from 27 nests; ♀♀ liable to desert if flushed early in incubation (Macdonald 1965). In North Uist (Outer Hebrides, Scotland), monogamous ♂♂ raised 2·0 fledged young each, bigamous ♂♂ 5·5; primary and secondary ♀♀ had similar rates of success, both greater than monogamous ♀♀, whose chicks starved more often because smaller food loads brought to nest (Hartley 1991; I R Hartley). In Oberlausitz, c. 12 of 37 nests failed completely; 173 eggs produced c. 114 fledged young (c. 4·6 per successful nest, 3·1 overall); most losses caused by agricultural machines (Gliemann 1973); average brood size 4·1 (n = 49), 88·6% of average clutch (n = 38) (Eifler and Blümel 1983). In eastern England, and presumably elsewhere, many nests destroyed by harvesting machines, etc. (Robertson 1954). In southern Sweden, most successful clutches tend to be later ones and probably high proportion of 1st clutches lost (Jönsson 1992).   BH

Plumages. (nominate *calandra*). ADULT MALE. Forehead black-brown (if fresh) to grey-brown (if worn), crown, hindneck, and upper side of neck grey-brown streaked black. Lore and narrow eye-ring cream-buff or tawny, often mottled grey in front of eye. Supercilium and ear-coverts like crown but dark streaks less intense, especially at border with eye-ring. Mantle, scapulars and back dark grey-brown to warm brown with slight rusty tinge, boldly streaked black; rump and upper tail-coverts slightly more buffish-grey (if fresh) or greyish-olive (if worn), dark streaks much finer. Lower cheek, chin, and throat pale cream-pink or buff-white, centre of throat usually with clear-cut triangular black spots; lower cheek and side of throat separated by contrasting black malar stripe, sometimes broken into short streaks. Lower side of neck, side of breast, chest, and flank warm cream-buff (when fresh) to dirty off-white (when worn), feathers with dark brown or black shaft-streaks of varying width, broadest forming an interrupted black gorget across upper chest; remainder of underparts pale cream-buff, cream-white, or off-white, marked with narrow dark shaft-streaks on side of belly; bases of feathers of side of breast and of under tail-coverts pale drab-grey, but these concealed unless plumage heavily worn. Tail-feathers dark sepia-brown on upper surface, pale brown-grey below; sides and tips narrowly fringed buff (almost white along outer web of outer feather). Flight-feathers, greater upper primary coverts, and bastard wing dark grey-brown; outer primaries narrowly fringed pale cream or yellowish-white on outer web, remainder of feathers less contrastingly fringed pale buff-brown or greyish-buff on outer webs, greyish-white or pinkish-white on inner webs and tips. Tertials and median and greater upper wing-coverts dark brown or black-brown on centres, outer webs of tertials and greater coverts broadly fringed buff-brown (soon fading to whitish on tip); tips of median coverts broadly pink-cinnamon, extreme tips bleaching rapidly to off-white. Lesser upper wing-coverts grey-brown with slightly paler buff tips. Under wing-coverts and axillaries pink- or cream-white, shorter coverts with dark brown bases. *In worn plumage* (about February–May), brown of upperparts becomes greyer, underparts become whiter; some individuals may show an ill-defined buff-white supercilium from above eye backwards. Apart from influences of wear on colour, also some individual variation, and sometimes 2 morphs recognized; in darker birds, ground-colour of fresh upperparts more dark grey-brown with dull rusty-brown tinge, underparts more cream-buff; in paler birds fresh upperparts more brownish-grey, underparts more yellowish-cream; both distinctly greyer above and whiter below when plumage worn. ADULT FEMALE. As ♂, but see Measurements. NESTLING. Upperparts covered with long and plentiful yellow-buff down, mainly on head, shoulder, and back; otherwise naked (Heinroth and Heinroth 1924-6); at *c.* 2 weeks fully feathered, but cap and shoulder still downy. JUVENILE. Rather like adult, but ground-colour more yellowish-buff, less grey; best distinguished by clear-cut and sharply contrasting pale buff to off-white fringes along tertials, primary coverts, and greater, median, and lesser upper wing-coverts; pale fringes on upper wing-coverts narrower than in adult. Indistinguishable from adult once complete post-juvenile moult finished from about middle of October; last recognizable juvenile feathers are white-fringed outer greater upper primary coverts.

Bare parts. ADULT, FIRST ADULT. Iris brown or dark brown. Bill pinkish-, yellowish-, greenish-, or greyish-horn, more yellowish at base below nostril, plumbeous-grey to black-brown on culmen and extreme tip. Leg and foot flesh-brown with pink, yellow, or grey tinge, ochre-brown, yellow-brown, or light horn-brown, rear of tarsus and soles paler, yellowish. (Hartert 1903-10; Gliemann 1973; RMNH, ZMA.) NESTLING. Bare skin, including leg and bill, pink-flesh. Mouth flesh, gape flanges yellow. (Heinroth and Heinroth 1924-6.) JUVENILE. Iris brown. Bill light greyish-horn, tinged yellow at base. Leg and foot pale flesh-brown or pale yellow-brown. (RMNH, ZMA.)

Moults. ADULT POST-BREEDING. Complete; primaries descendent. In Britain, starts with p1 between mid-July and mid- or late August; primaries completed with regrowth of p9-p10 after *c.* 80 days early October to early November; some tail-feathers or secondaries sometimes still growing after completion of primaries (Newton 1968; Ginn and Melville 1983). In Belgium and Germany, occasional birds (perhaps juveniles) still in last stages of moult up to December (Gliemann 1973). On Malta, starts mid- or late June, apparently immediately after termination of breeding, ♂ perhaps slightly earlier than ♀; primary moult completed after *c.* 78 days, finished about mid-August to mid-September (Gauci and Sultana 1983). In Algeria, some still in moult 15 October (Ticehurst and Whistler 1938). In Turkey, 5 birds in heavy moult late July and early August (Eyckerman *et al.* 1992). In Iran, 3 not yet moulting 8-24 July, another just started 25 July (Stresemann 1928; Paludan 1940). POST-JUVENILE. Unusual among Emberizidae in having complete moult. Starts at age of *c.* 5 weeks (Heinroth and Heinroth 1924-6). In Britain, timing and duration apparently as adult post-breeding. On Malta, duration as adult; starts with p1 between late June and late July, all primaries full-grown between late August and early October (Gauci and Sultana 1983).

Measurements. ADULT, FIRST ADULT. Nominate *calandra*. Britain, Netherlands, and Germany south to Spain, North Africa, northern Italy, Greece, and north-west Asia Minor, whole year; skins (RMNH, ZMA). Bill (S) to skull, bill (N) to distal corner of nostril; exposed culmen on average 3·3 less than bill (S).

	♂			♀		
WING	101·4 (2·40; 76)	96-107		92·8 (2·39; 36)	87-98	
TAIL	69·7 (3·02; 49)	62-74		62·8 (4·15; 27)	54-70	
BILL (S)	16·5 (1·02; 48)	15·0-18·2		15·6 (0·88; 27)	14·4-17·2	
BILL (N)	9·4 (0·47; 48)	8·2-10·3		8·7 (0·48; 27)	7·5-9·4	
TARSUS	25·3 (0·98; 47)	23·8-27·1		24·7 (0·72; 27)	23·3-26·0	

Sex differences significant.

Wing. Nominate *calandra*. Skins, unless otherwise noted. (1) Britain, mainly live birds (Prŷs-Jones 1976; Boddy and Blackburn 1978). (2) Netherlands (RMNH, ZMA). (3) Central Europe (mainly eastern Germany) (Eck 1985*b*). (4) Sweden (Svensson 1992). (5) Portugal and Spain, (6) Morocco, Algeria, and Tunisia (RMNH, ZMA). (7) Mallorca (Jordans 1924). (8) Corsica (Stresemann; 1920). (9) Sardinia (Stresemann 1920; Eck 1985*b*; ZMA). (10) Italy, mainly live birds (Montecristo island and Sicily) (Baccetti *et al.* 1981; RMNH). (11) Malta, live birds, whole year (Gauci and Sultana 1983). (12) Yugoslavia and Greece (Stresemann 1920; ZMA). (13) Turkey (except south-east) and Caucasus area (Stresemann 1920; Jordans and Steinbacher 1948; Kumerloeve 1961, 1963; Nicht 1961; Rokitansky and Schifter 1971; Eyckerman *et al.* 1992; RMNH, ZMA). (14) Tenerife, Gran Canaria, and Lanzarote (Canary Islands) (BMNH, RMNH, ZMA). *M. c. buturlini.* (15) South-east Turkey, Iran, and Transcaspia (Hartert 1921-2; Paludan 1938; Diesselhorst 1962; Kumerloeve 1970*b*; RMNH, ZMA).

	♂		♀	
(1)	99·4 (2·2 ; 93)	94-105	90·3 (2·0 ; 218)	85-97
(2)	101·5 (2·23; 48)	96-107	92·7 (2·54; 25)	87-98
(3)	102·7 (2·65; 53)	96-109	92·5 (1·96; 10)	90-96
(4)	— (—; 27)	96-105	— (—; 9)	88-96
(5)	101·6 (2·13; 8)	9-105	93·5 (—; 2)	92-95
(6)	100·3 (2·98; 15)	97-107	92·2 (1·34; 7)	90-94
(7)	— (—; 20)	95-104	— (—; 5)	90-92
(8)	98·6 (1·67; 9)	96-101	88·3 (—; 3)	87-89
(9)	103·3 (—; 3)	102-105	95·0 (—; 3)	93-98
(10)	— (—; —)		90·5 (2·48; 12)	86-95
(11)	98·4 (2·59; 24)	94-105	90·1 (2·12; 55)	86-96
(12)	102·2 (2·10; 48)	97-109	93·2 (1·74; 30)	90-98
(13)	101·5 (2·24; 17)	98-106	91·6 (2·19; 23)	88-96
(14)	97·4 (1·89; 19)	93-100	89·4 (1·90; 21)	86-93
(15)	101·8 (2·55; 9)	98-106	93·7 (1·03; 7)	92-95

Nominate *calandra*. Algeria (*n*=11), ♂ 96-105, ♀ 92-97. Canary Islands (*n*=33), ♂ 92-100, ♀ 87-91. (Jordans 1924.) For unsexed birds, see also Görnitz (1922).

JUVENILE. Wing and tail distinctly shorter than in adult. In Netherlands, both on average *c.* 5-6 shorter (samples small) (RMNH); on Malta, wing *c.* 7 shorter: ♂ 90·8 (1·66; 14) 88-94, ♀ 82·9 (2·74; 26) 78-87 (Gauci and Sultana 1983).

Weights. Nominate *calandra*. Whole geographical range: (1) November-February, (2) April-May, (3) June-July (Krohn 1915; Bacmeister and Kleinschmidt 1920; Paludan 1940; Kumerloeve 1961, 1963; Nicht 1961; Rokitansky and Schifter 1971; Eck 1985*b*; Demartis 1987; Eyckerman *et al.* 1992; G O Keijl, RMNH, ZMA). (4) England, winter (Boddy and Blackburn 1978, which see for monthly variations). Malta: (5) breeding adult, March to mid-June; (6) juvenile, April-September, (7) adult and 1st adult, mid-June to February (Gauci and Sultana 1983). *E. c. buturlini.* (8) South-east Turkey and Iran, March-May (Paludan 1938; Kumerloeve 1970*b*). (9) Kazakhstan (Korelov *et al.* 1974).

	♂		♀	
(1)	55·6 (8·68; 4)	44-64	52·7 (—; 1)	—
(2)	55·7 (5·61; 7)	46-63	39·6 (3·15; 7)	35-45
(3)	52·4 (8·60; 15)	34-62	40·9 (3·10; 11)	37-47
(4)	53·6 (4·1 ; 68)	45-63	43·9 (3·1 ; 202)	36-53
(5)	47·8 (3·57; 11)	44-56	38·7 (2·73; 21)	35-43
(6)	44·6 (3·74; 14)	38-50	36·1 (2·75; 26)	31-42
(7)	47·3 (2·60; 13)	43-52	38·6 (3·17; 34)	34-49
(8)	48·9 (2·39; 4)	45-51	40·6 (0·84; 5)	39-41
(9)	49·1 (—; 20)	40-58	42·2 (—; 9)	38-51

In Belgium, range 37-64·5 (*n*=70) (Gliemann 1973); in Britain, February, 35-56 (*n*=103) (Follows 1969); in south-east France, winter, 48·8 (7·43; 14) 36-61 (G Olioso); in Balearic Islands, September-October, 43·4 (23) 37-53 (Mester 1971); on

Montecristo island (near Elba, Italy), April, possible ♀♀, 32·3 (3·7; 11) 25–36 (Baccetti *et al.* 1981).

Structure. Wing rather long, broad at base, tip bluntly pointed. 10 primaries: p7–p8 longest; p9 0–2 shorter, p6 2–5, p5 9–15, p4 15–21, p3 19–25, p2 21–27, p1 23–30; p10 reduced, 60–72 shorter than p7–p8, 9–15 shorter than longest upper primary covert. Outer web of p6–p8 emarginated, inner of p7–p9 sometimes with faint notch. Tip of longest tertial reaches tip of p5–p6. Tail rather short, tip square or slightly notched; 12 feathers, t1 1–7 shorter than t6. Bill basically rather like that of Yellowhammer *Emberiza citrinella*, but much heavier at base, depth at base 9·0 (10) 8·3–9·7, width 9·8 (10) 9·0–10·8; culmen and gonys curved; 'teeth' on middle of palate and near base of cutting edge of lower mandible strongly developed. Tarsus and toes rather short, strong. Middle toe with claw 23·7 (10) 20–28; outer toe with claw *c.* 69% of middle with claw, inner *c.* 67%, hind *c.* 87%. Remainder of structure as in *E. citrinella*.

Geographical variation. Slight and largely clinal, but nevertheless many races sometimes recognized (e.g. in review by Görnitz 1921, 1922), of which only extremes considered valid here. Size markedly constant throughout much of range, but birds from Britain, Canary Islands, and North Africa tend to be slightly smaller than those on continent, as are a number of populations on islands in western Mediterranean (Corsica, Malta, Sicily, and apparently Balearic Islands as well as Montecristo island near Elba, but not Sardinia); more information on these required. Birds from Corsica also slightly darker than nominate *calandra* from continent, upperparts dull dark grey, underparts more heavily marked black, and these as well as other west Mediterranean populations of small size perhaps separable as *parroti* Görnitz, 1921. Birds from Tenerife sometimes separated as *thanneri* (Tschusi, 1903), and perhaps this race also on other Canary Islands, but see Bannerman (1912, 1919). Variation clinal throughout Britain, continental Europe, and Asia, depth of ground-colour and width and extent of dark streaks on body decreasing from west to east; dark brownish western end of cline

(*clanceyi* in western Scotland and western Ireland) and pale greyish eastern end (*buturlini* from inland Syria, Jordan, southeast Turkey, and southern and eastern Iran throughout Transcaspia east to Dzhungarskiy Alatau mountains) fairly distinct, but nominate *calandra* in between highly variable, in many areas occurring in 2 morphs, browner one predominating in west and greyer one in east: grey exceptional in (e.g.) Britain or Netherlands; both morphs common in Spain, eastern Germany, or Kiev area (Ukraine) (Jordans 1950; Gliemann 1973). In *clanceyi*, ground-colour of upperparts deep warm brown, of underparts yellow-buff or tawny, fringes of feathers of upperparts, tail, and wing tinged rufous when plumage fresh (Meinertzhagen 1947, 1953); in *buturlini*, ground-colour of upperparts sandy-grey, of underparts pale cream to white, ear-coverts pale, fringes of flight-feathers almost white, chin and throat almost unspotted (Johansen 1907). As not all birds from Asia are pale or show reduced spotting, and some birds from Tien Shan inseparable from some of Italy, Hungary, or North Africa (Hartert 1903–10; Meinertzhagen 1954; Keve and Rokitansky 1966), those from Iran inseparable from Denmark (Paludan 1938), and birds with *clanceyi* characters occur east into Germany, both *buturlini* and *clanceyi* are often included in nominate *calandra*, and species thus monotypic. Following Vaurie (1956e, 1959), *buturlini* recognized here, as birds from Asia are constantly paler and more sandy-olive-grey on upperparts than nominate *calandra* from Europe, less dull grey, and whiter and less streaked on underparts. *M. c. clanceyi* recognized here as birds examined from Outer Hebrides and western Ireland are constantly darker brown above and darker yellow and more heavily streaked below than brown birds of west and central continental Europe. An alternative is to recognize no races at all, as done by Paynter (1970).

M. calandra often included in *Emberiza*, but separated here on account of marked sexual dimorphism in size (in contrast to most other *Emberiza*, which usually show sexual dimorphism in colour instead), complete post-juvenile moult (partial in *Emberiza*), and differences in bill structure and behaviour (Andrew 1956a, 1957a; Gliemann 1973). For possible relationships, see Andrew (1956a, 1957a) and Lees-Smith and Madge (1982).   MP

Subfamily CARDINALINAE cardinal-grosbeaks and allies

Small to medium-sized emberizids, variously known as 'cardinals', 'grosbeaks', 'saltators', and even 'buntings', and collectively also as 'cardinal finches'. Mainly arboreal, feeding on seeds (which they crush in their strong bills), fruits, and insects. Except as stragglers, found in New World only from central Canada to northern Argentina but only *c.* 9 species in North America with remainder in tropics. Mostly sedentary but some migratory. About 47 species in 15 genera, of which 5 species of 4 genera accidental in west Palearctic (all from North America).

Sexes closely similar in size. Bill mainly large and stout, in some species recalling that of grosbeaks of subfamily Carduelinae (Fringillidae); differs internally in a number of features from bill of emberizines (and also from that of cardueline grosbeaks). Wing fairly long to rather short,

bluntly pointed; 9 primaries (p10 minute and hidden). Tail short to medium, graduated or somewhat forked; 12 feathers. Leg rather long; tarsus scutellate, booted, and ridged as in Fringillinae. Head-scratching by indirect method as far as known (Simmons 1957b for *Passerina*). Little information on bathing and sunning (but see Hauser 1957 and Potter and Hauser 1974 for Cardinal *Cardinalis cardinalis*). Active (direct) anting reported from at least 7 species in genera *Cardinalis*, *Pheucticus*, *Saltator*, and *Passerina* (Whitaker 1957; Simmons 1961b, 1963; Potter 1970; Hauser 1973; Potter and Hauser 1974).

Plumages markedly bright, with large and contrasting patches of different colours (red, yellow, blue, green, white, black, etc.), being especially multicoloured in *Passerina*; generally not so streaky as in Emberizinae, adult ♂

usually lacking contrasting head-stripes. Sexes usually very different, with ♂ much brighter than ♀; no marked seasonal differences in most species. Juvenile and 1st non-breeding plumage close to adult ♀.

Spiza americana Dickcissel

PLATE 28
[between pages 256 and 257]

Du. Dickcissel   Fr. Dickcissel d'Amérique   Ge. Dickzissel
Ru. Американская спиза   Sp. Dickcissel   Sw. Dickcisselsparv

Emberiza americana Gmelin, 1789

Monotypic

Field characters. 13–17·5 cm; wing-span 22·5–27·5 cm. Up to 10% larger than House Sparrow *Passer domesticus* but with rather slimmer form except for square tail with pointed feathers and rather long legs; 5–10% smaller than Bobolink *Dolichonyx oryzivorus*. Sparrow-like, with yellow on lores, supercilium, and submoustachial stripe, sharp blackish malar stripe, and bright bay-chestnut forewing-coverts in every plumage except 1st winter ♀. Adult ♂ has bluish bill (when breeding), grey crown and hind cheeks, and black bib contrasting with yellow breast. ♀ and immature less colourful but similarly patterned, lacking only bib. Sexes dissimilar; no seasonal variation. Juvenile separable.

ADULT MALE. Moults: June–October (complete); January–February (mainly head and underparts). Crown, nape, and rear ear-coverts clean grey, faintly and finely streaked black on yellow-tinged forehead; supercilium yellow, fading to white with wear and broadening behind eye; lores and lower eye-crescent mottled grey and white; submoustachial stripe yellow. Narrow black malar stripe runs into quite deep black bib which contrasts with white chin and panel on side of neck over breast. Mantle, greater coverts, and tertials tawny-brown, with black feather-centres creating obvious streaks. Most of scapulars, and lesser and median coverts red-chestnut, with black central streaks only on scapulars. Rump mottled greyish-brown. Flight- and tail-feathers dark brown with pale grey to buff fringes. Area around bib to centre of belly yellow, contrasting with grey fore-flanks and grey-buff rear flanks and whitish belly and under tail-coverts. When fresh, colours of upperparts show both greeer tones on head and browner wash to back; black bib obscured by pale cream tips. Bill of breeding bird pale blue-grey, with dusky culmen; fades to pinkish-horn in winter. Legs brown. ADULT FEMALE. Patterned as ♂ but head and upperparts duller buff- to grey-brown, yellow patches less clean, no or only incomplete mottled bib, and fine streaks on breast and flanks. FIRST-WINTER. Resembles adult ♀ but shows much less obvious yellow on head and breast which (like flanks) are always quite finely but distinctly streaked. ♂ usually shows chestnut at shoulder but ♀ lacks this, showing instead chestnut-buff-tipped, blackish-centred median coverts; ♀ is also less streaked on brèast.

♂ virtually unmistakable; only ♂ Dead Sea Sparrow *Passer moabiticus* has convergent plumage colours and pattern. ♀ and immature confusing, suggesting *D. oryzivorus*, other Nearctic sparrows, and west Palearctic buntings (Emberizidae.) Important to realize that *S. americana* is ground-haunting and has distinctive tail shape and call (see below); in addition, all but immature ♀ shows restricted but vivid chestnut on forewing, lacking in all possible confusion species. Direct flight and other actions recall *D. oryzivorus*, also sparrows, and even small lark (Alaudidae). ♂ more conspicuous in behaviour than ♀.

Commonest call a low buzzing 'br-r-r-r-rt', rather like similar call of Long-tailed Tit *Aegithalos caudatus*; given in flight and during nocturnal migration.

Habitat. Breeds in Nearctic temperate open lowlands, either prairies and other grassland bearing tall grasses, herbs, or shrubs, or crops of grass, clover, or alfalfa (Johnsgard 1979); also in meadows, pastures, weed patches, and grain fields (Terres 1980). A ground bird, but uses raised perches such as roadside fences or overhead wires for singing, and will sometimes nest in shrub or tree.

Winters in South America in forest clearings, forest edges, llanos, open country, and cultivation such as rice-fields at low altitude, in Venezuela up to 600 m (Schauensee and Phelps 1978).

Distribution. Breeds in North America, from eastern Montana and south-east Saskatchewan east to Massachusetts, south to central Colorado, southern Texas, southern Louisiana, central Mississippi, and South Carolina; formerly also in Atlantic lowlands from Massachusetts to North Carolina. Winters in Mexico, Central America, and northern South America, from Michoacán south to northern Colombia, Venezuela, and the Guianas; locally also in small numbers in coastal lowlands of USA from southern New England to southern Texas.

Accidental. Norway: adult, Måløy (Sogn og Fjordane), 29 July 1981 (Michaelsen 1985).

Movements. Migratory; erratic in distribution both winter and summer (Fretwell 1980). Populations fluctuate greatly; breeding range east of Appalachians abandoned 1860–95, perhaps due to loss of winter habitat, but partially reoccupied 1928–54; has declined in most of range since then (Bent 1968; Fretwell 1980). Autumn records in eastern North America, and attempted wintering where artifical food available, have increased greatly since 1949 (Bent 1968; Squires 1976; Tufts 1986).

Timing of autumn departure from breeding areas is variable, partly due to often irregular and unsuccessful nesting in recent decades. Some birds leave in late July (with first passage records in Texas), but most in August–September (e.g. Oberholser 1974, Hall 1983, Janssen 1987). Main passage in southern states, often involving large flocks, continues chiefly to mid-October, with frequent stragglers into early November (Sutton 1967; Bent 1968; Oberholser 1974; James and Neal 1986). Earliest records in winter range late August, with main arrival September–October, but in most years few remain in Central America then (Fretwell 1980; Hilty and Brown 1986; Binford 1989; Stiles and Skutch 1989). Initial winter destination in Venezuela is reached in October (Schauensee and Phelps 1978), but in December and January, perhaps due to drought, many depart, moving gradually east to Trinidad or west to Central America (ffrench 1967; Fretwell 1986). Migratory route (both seasons) between Mexico and West Indies (Peterson and Chalif 1973; Bond 1985), but main corridor apparently chiefly overland and quite narrow, through Texas, coastal Mexico and western Gulf to Central America and eventually to Trinidad (Stevenson 1957; ffrench 1967; Bent 1968). Winter ringing in Trinidad showed one case of site-fidelity, with return a year later (Loftin 1977).

In spring, leaves winter range late March to early May (Bent 1968; Fretwell 1986); in Trinidad, birds increase weight by up to 50% before heading westward in mid-April (ffrench 1967); presumably this fattening is preparation for passage across less favourable land areas, as few cross sea east of Honduras and Yucatán (eastern Mexico) (Stevenson 1957; Bond 1985). Main migration, in large flocks as in autumn, occurs in 2nd half of April throughout Central America and southern states (Monroe 1968; Lowery 1974; Imhof 1976; Binford 1989). Most birds reach breeding areas by mid- or late May (Sutton 1967; Bent 1968).

Individuals frequently recorded north of usual range, but north only to 52°N; in Canada, nearly all east coast records are in autumn (Bent 1968; Sealy 1971; Godfrey 1986). AJE, DFV

Voice. See Field Characters.

Plumages. ADULT MALE. In spring, forehead and crown yellow-green, merging into medium grey on nape and hindneck, marked with narrow black shaft-streaks. Narrow but distinct supercilium bright yellow, extending from nostril to above rear of ear-coverts, bordered below by mottled pale grey and white patch on lore and just below eye. Upper cheek and shorter ear-coverts yellow-green, merging into medium grey on longer ear-coverts, latter with a slightly darker grey border at rear. Upper side of neck medium grey. Lower cheek mainly bright yellow, except for small white patch at border of lower mandible, yellow at rear rather sharply defined from white lower side of neck. Mantle and scapulars rufous-cinnamon to buff-brown (paler on side of mantle, where forming pale stripes), contrastingly streaked deep black on lower mantle, more narrowly and less contrastingly streaked sooty on scapulars. Back, rump, and upper tail-coverts greyish drab-brown. Chin white; throat black, forming rather narrow patch ending in bib on central chest, sharply contrasting with yellow of lower cheek and with white on lower side of neck. Side of breast medium grey with some bright yellow suffusion, chest to upper mid-belly bright yellow, often with small and irregular black patch on mid-breast (occasionally, connected with black of bib), chest often with olive bar when plumage fresh. Flank light grey with buff or yellow-green suffusion, side of belly, vent, and under tail-coverts white. Tail dark sepia, feathers narrowly edged pale grey or off-white. Flight-feathers, tertials, greater upper primary coverts, and bastard wing brown-black or black, outer webs of primaries narrowly fringed pale grey to off-white, outer webs of primary coverts and secondaries rather narrowly fringed pale cinnamon-buff to light grey-buff, tertials more broadly fringed and tipped rufous-cinnamon to light grey-buff. Lesser and median upper wing-coverts bright rufous-chestnut, greater coverts brown-black or black with broad and contrasting rufous-chestnut to pale grey-buff outer fringe and tip (colour dependent on abrasion and extent of pre-breeding moult: when not moulted, fringes bleached and rather narrow; when moulted, fringes broader and more rufous, darkest on newer outer coverts, paler and more cinnamon on slightly older inner coverts). Under wing-coverts and axillaries yellow, brightest along leading edge of wing, tipped greyish-white on longer feathers. *In fresh plumage* (autumn), forehead and crown brighter yellow-green; hindneck, upper side of neck, and ear-coverts washed green; mantle and scapulars brighter rufous-cinnamon; back to upper tail-coverts olive-brown or buff-brown, less greyish; black of throat-patch partly concealed by narrow pale cream-yellow or white feather-tips; yellow of chest tinged buff or orange, light grey of flanks and white of belly downward washed pale yellow or cream; fringes of greater coverts and tertials cinnamon-rufous, those of primary coverts, flight-feathers, and tail pink-brown. In early spring, when plumage partly fresh again after pre-breeding moult, similar to fresh autumn plumage, but fringes of throat-patch narrow, and fringes of non-moulted parts of plumage (tail, flight-feathers, primary coverts, sometimes tertials, greater coverts, and back to upper tail-coverts) bleached, less brown and rufous, distinctly greyer. *In worn plumage* (June–July), top and side of head mainly grey, supercilium and stripe over lower cheek white (green and yellow tinges lost), ground-colour of mantle and scapulars as well as entire back to upper tail-coverts greyish-drab; black of throat partly mottled white; yellow of chest to mid-belly rather pale, remainder of underparts rather dirty white. ADULT FEMALE. Ground-colour of entire upperparts and side of neck buff-brown (when fresh) to dull greyish-drab (when worn), forehead and crown with slightly broader black streaks than in adult ♂, virtually without green; hindneck virtually without grey (unless heavily worn). Side of head grey-brown to grey-buff, less grey than adult ♂; supercilium buff-yellow, bright yellow, or pale buff (depending on wear), as distinct as adult ♂, but shorter, tinged buff below ear-coverts and merging into buff-brown (if fresh) or light grey (if worn) on lower side of neck; a distinct black malar stripe (unlike adult ♂). Chin and throat white, often bordered by mottled black gorget at lower end, connecting both ends of malar stripe (black throat patch absent). Side of breast and flank tawny-buff (if fresh) to pale buff-grey (if worn), chest to centre of upper belly deep ochre-yellow (if fresh) to sulphur-yellow (if worn), chest and flank with narrow but distinct dull black or dark grey shaft-streaks (unstreaked in adult ♂), remainder of underparts cream to off-white. Wing and tail as adult ♂ (equally liable to bleaching and wear), fresh lesser and median upper wing-coverts less chestnut than in adult ♂, more rufous-cinnamon, some dull black of centres of coverts partly visible. See also Measurements. JUVENILE. Upperparts buff-brown, shading to sepia on crown, feathers of mantle and scapulars fuscous-black with cinnamon-buff tips and fringes. Supercilium and malar stripe orange-buff or ochre-buff; chest and upper flank greyish-buff, tinged yellow when plumage fresh, chin to throat light buff, remainder of underparts pale buff to off-white; no streaks on underparts. Flight-feathers grey to brown-black, narrowly fringed pale grey; upper wing-coverts olive-black with broad cream fringes. (Gross 1921, which see for development.) FIRST ADULT NON-BREEDING. Like adult ♀, but head-pattern often less pronounced, supercilium and stripe over lower cheek cream-buff with more restricted amount of bright yellow, black malar stripes less deep black, more mottled; dark spots on lower throat less conspicuous or absent, yellow on chest and belly deep but rather restricted, chest, breast, and flank narrowly but distinctly marked with dusky streaks (more so than adult ♀). New lesser upper wing-coverts drab-brown or drab-grey with ill-defined black streak (often less rufous than adult ♀, and with less black at centre); median coverts variable, either all or partly juvenile, drab-grey with black shaft-streaks and off-white tip (showing as pale wing-bar), or all or partly new, rufous-cinnamon with black centre. Greater upper wing-coverts, primary coverts, tertials, flight-feathers, and tail still juvenile, fringes distinctly paler and more abraded than those of adult at same time of year, pink-buff to pale drab-brown, distinctly paler and less rufous than scapulars (in adult non-breeding, scapulars about equal in tinge and abrasion to tertials and greater coverts); juvenile tail-feathers sharply pointed, tip frayed (in adult, bluntly pointed); t1 occasionally new, contrasting in wear and shape with other tail-feathers, unlike any adult; see Pyle *et al.* (1987) for shape of tail-tip; juvenile tertials rounded at tip (in adult, more truncate). Sexes similar, but see Measurements, and new median coverts of ♂ show less black on centre than those of ♀, black of ♂ usually concealed; some advanced ♂♂ have black streaks on underparts reduced or virtually absent, and these may show some black spots at side of throat or on lower throat, as well as bright chestnut median coverts and broad rufous-cinnamon fringes to new greater coverts, differing from adult ♂ in white mid-throat and juvenile primary coverts, tertials, flight-feathers, and tail. FIRST ADULT BREEDING. Like adult, but bastard wing, primary coverts, flight-feathers, and tail still juvenile, centres distinctly browner and tips more worn than those of adult at same time of year.

Bare parts. ADULT, FIRST ADULT. Iris dark brown. Culmen and tip of gonys slate-black to black, remainder of bill pale pink-horn in winter, pale horn with grey tinge to pale bluish-grey in breed-

ing. Leg and foot dull flesh-brown, purplish-brown, or dark horn-brown, paler horn on soles. (Ridgway 1901-11; Gross 1921; RMNH, ZMA.) JUVENILE. Bill and leg pale flesh, gradually darker when older (Gross 1921).

Moults. ADULT POST-BREEDING. Complete; primaries descendent. On breeding grounds; starts between late June and early August, completed August to early October, but some feathering of head, neck, and chest still growing in migrant ♀♀ late October and November (RMNH, ZMA). In small samples from Illinois (USA), ♂♂ moult from mid-July, but no ♀ in moult up to early August (Gross 1921; Zimmerman 1965). ADULT PRE-BREEDING. Partial: head, neck, underparts, lesser and median upper wing-coverts, and variable number of feathers of mantle to rump and greater upper wing-coverts (Pyle *et al.* 1987; RMNH, ZMA). In Panama, starts from late January or early February onwards, completed from late February onwards (Zimmerman 1965). On Trinidad, some moult by time of arrival in December, peak in February, completed in some by early or mid-February but mainly in March (♂) or early April (♀) (ffrench 1967). POST-JUVENILE. Partial: head, body, lesser and variable number of median upper wing-coverts, and occasionally some or all greater upper wing-coverts, variable number of tertials, and some feathers of bastard wing (RMNH, ZMA). Starts shortly after fledging, when on or near breeding grounds; starts between late June and early August, completed late July to late August, moult slower in birds from early broods than in those from later ones (Gross 1921; RMNH, ZMA). FIRST PRE-BREEDING. As adult pre-breeding; juvenile flight-feathers, tail, and greater primary coverts retained as well as variable amount of 1st non-breeding plumage from mantle to upper tail-coverts, bastard wing (if replaced in post-juvenile), and variable number of tertials and greater upper wing-coverts. (Pyle *et al.* 1987; RMNH, ZMA.)

Measurements. ADULT, FIRST ADULT. USA, spring and summer, and Central America and northern South America, autumn and winter; skins (BMNH, RMNH, ZMA). Bill (S) to skull, bill (N) to distal corner of nostril; exposed culmen on average 3·2 less than bill (S).

	♂		♀	
WING	84·0 (2·07; 28)	81–88	76·4 (2·13; 14)	73–79
TAIL	57·0 (2·45; 14)	53–61	50·9 (1·37; 14)	49–54
BILL (S)	17·0 (0·55; 14)	16·1–17·7	16·1 (0·61; 14)	15·3–16·8
BILL (N)	10·3 (0·62; 14)	9·5–11·2	9·5 (0·42; 14)	9·0–10·2
TARSUS	23·3 (0·62; 14)	22·5–24·5	22·6 (0·69; 13)	21·7–23·4

Sex differences significant. See also ffrench (1967).

Weights. Illinois (USA): mid-July to early August, adult, ♂ 33·5 (2·27; 6) 30·7–36·7, ♀ 26·7 (2·62; 7) 23·4–30·6; juvenile and 1st adult, ♂ 33·8 (3·45; 4) 29·5–36·7, ♀ 30·3 (3·52; 5) 27·1–36·0 (Gross 1921); breeders, mid-June to early August, ♂ 28·5 (0·66; 22), ♀ 25·2 (0·77; 9) (Zimmerman 1965). Kentucky (USA): May, ♂ 28·3; September, ♂ 27·5 (1·51; 4) 25·5–29·0, ♀ 23·1 (1·58; 4) 21·0–24·5 (Mengel 1965). Coastal New Jersey (USA), autumn: 27·7 (2·08; 4) 25·0–29·5 (Murray and Jehl 1964). Kansas (USA), early October: 35·6 (7) 33·8–38·4, ♀ 29·2 (29) 25·6–32·8 (Tordoff and Mengel 1956).

Belize, April: ♂ 27·2 (7·04; 4) 20·0–35·1 (Russell 1964). Panama: October, 24·4 (Rogers 1965); wintering birds January to mid-February, ♂ 29·8 (1·85; 5);, ♀ 24·2 (1·73; 5); migrants late February to mid-April, ♂ 27·9 (1·23; 23), ♀ 24·4 (1·85; 7) (Zimmerman 1965).

On Trinidad, evening weights more or less constant December to early March, average of ♂ *c.* 31, ♀ *c.* 25, increasing slightly late March and early April and markedly mid- and late April, when ♂♂ reach average *c*. 47·5 and ♀♀ *c*. 38·5 (ffrench 1967, which see for details).

Structure. Wing rather long, broad at base, tip pointed. 10 primaries: p8–p9 longest, p7 0·5–2·5 shorter, p6 3·5–7, p5 7–11, p4 11–15, p3 13–17, p2 16–20, p1 18–23. P10 strongly reduced, a minute pin concealed below reduced outermost upper primary covert; 50–61 shorter than p8–p9, 7–10 shorter than longest upper primary covert, 0–3 shorter than reduced outer primary covert. Outer web of p7–p8 emarginated, inner web of p8(–p9) with very faint notch. Tip of longest tertial variable in length, reaching between tips of p1 and p6. Tail short, tip square or slightly forked; 12 feathers, each with pointed tip, especially in juvenile. Bill short, conical, somewhat swollen at base; visible part of culmen *c*. of head-length; *c*. 8·5–10·5 mm deep and wide at base; culmen gently decurved, cutting edges straight or upper with a blunt indistinct 'tooth' in middle; tip slightly compressed laterally; bill relatively slightly deeper at base than in Indigo

Bunting *Passerina cyanea*, but distinctly less so than in grosbeaks *Pheucticus ludovicianus* and *Guiraca caerulea*. Nostrils small, rounded, exposed. About 6 short bristles at each side of base of upper mandible, projecting obliquely downward. Tarsus and toes rather long, thick. Middle toe with claw 21·1 (10) 20·0–23·0; outer toe with claw *c*. 70% of middle with claw, inner *c*. 68%, hind *c*. 78%.

Geographical variation. None.

May have formed superspecies with Townsend's Bunting *S. townsendii*, a smaller bird with dark slate-grey head, neck, and underparts, white supercilium and malar stripe, white bib surrounded by narrow black gorget, grey-brown upperparts with narrow blackish streaks on mantle and scapulars, pale yellow lower belly, and white vent and under tail-coverts; known from 1 specimen taken Pennsylvania (USA) in 1833 and now apparently extinct.

CSR

Pheucticus ludovicianus **Rose-breasted Grosbeak**

PLATE 28
[between pages 256 and 257]

Du. Roodborstkardinaal   Fr. Cardinal à poitrine rose   Ge. Rosenbrust-Kernknacker
Ru. Красногрудый дубоносовый кардинал   Sp. Candelo tricolor   Sw. Brokig cardinal

Loxia ludoviciana Linnaeus, 1766

Monotypic

Field characters. 18–21·5 cm; wing-span 30–32·5 cm. Head and body similar in size to Corn Bunting *Miliaria calandra* but bill even larger and square-ended tail longer; at least 20% larger than medium-sized rosefinches *Carpodacus*. Rather large, stout-billed, and quite long-tailed Nearctic passerine, with colourful underwing and double wing-bar common to all plumages. Breeding ♂ black and white, with rose-pink breast. ♀ and immature recall *Carpodacus* finch but have more strongly striped head including pale crown-centre, more heavily marked breast, and proportionately larger and paler bill. Flight strong, recalling Hawfinch *Coccothraustes coccothraustes* over long distance. Call distinctive. Sexes dissimilar; marked seasonal variation in ♂. Juvenile separable.

ADULT MALE BREEDING. Moults: July–September (complete). January–May (mostly head and neck). Head, breast, back, wings, and tail glossy black, strikingly marked by mainly white median coverts forming strong upper wing-bar, white tips to outer webs of greater coverts forming rather broken lower wing-bar (fading out on outer feathers), white bases to primaries forming bold patch by primary coverts, partially white tips to middle primaries and tertials, and striking white panel on outer tail-feathers. Most of chest and centre of forebelly rose-pink. Rest of underparts, and rump and tail-coverts white, cross-hatched black on upper tail-coverts and lateral rump feathers. Underwing rose-pink on lesser median and bases of

great coverts. Bill deep and quite short, whitish to pinkish-horn. Legs bluish-grey. ADULT MALE NON-BREEDING. Appearance much closer to adult ♀ but quite readily distinguished by blacker wings retaining bold white patch at bases of primaries, tail with obvious white panels, dark-barred but still basically white rump, splashes of pink on breast, and retention of pink underwing. ADULT FEMALE. Dark brown above and buffish-white below, streaked blackish on back and from malar area, across breast to rear flanks. Head pattern strikingly marked with pale buff median crown-stripe and lores, and dull white supercilium, submoustachial stripe, and border to ear-coverts, all contrasting with blackish-edged, dark brown lateral crown-stripes and ear-coverts. Wing pattern also distinctive but far less eye-catching than in ♂, with only white wedges on tips of median and greater coverts (forming rows of spots rather than continuous wing-bars) and only narrow white band at bases of primaries. Ground-colour of breast and upper flanks yellow-buff, of underwing yellow through yellow and pink to (rarely) pure pink. With wear, wing-spots become abraded and may even be lost completely from greater coverts, and white underparts look colder. Bill pinkish-horn with dusky tip. FIRST-WINTER MALE. Resembles ♀ but breast less streaked, often patched pink, upperparts more darkly marked, new coverts blacker on base and more boldly tipped white, and patch at bases of primaries rather broader and purer white.

Underwing wholly pink as adult ♂. FIRST-WINTER FEMALE. Difficult to distinguish from adult but may show bright new coverts, while underwing never pink, usually pure yellow. FIRST-SUMMER MALE. At close range, separable from adult by retained worn 1st-year outer wing and flight-feathers.

♂ unmistakable. ♀ and immature recall *Carpodacus* finch, particularly heavily marked Purple Finch *C. purpureus* which might also cross Atlantic to western Europe, but *P. ludovicianus* much larger, with size of bill, pale crown-centre, colourful underwing, form of wing-bars, and pattern and unforked shape of tail adding to distinctive appearance. Important to recognize that *P. ludovicianus* is no less bulky than Redwing *Turdus iliacus* or Pine Grosbeak *Pinicola enucleator*. Flight strong, with beats of quite long wings obvious in bursts which produce both shooting acceleration, and marked undulations; in open flight, fore-silhouette and action can recall *C. coccothraustes*, but in glimpsed passes between cover general character can also suggest tanager *Piranga* or small thrushes *Turdus* and *Catharus*. Gait a hop. Stance usually half-upright with relaxed tail, sometimes more level with raised tail. Bold and excitable at times, especially in autumn parties, but often quiet and unobtrusive as vagrant. Closely associated with deciduous trees and thickets.

Call a single, rather sharp, even squeaky, metallic 'peek', 'kick', or 'kink'; loud and far-carrying.

Habitat. Breeds in temperate Nearctic wooded lowlands, deciduous or mixed, strongly favouring edges between stands of large tall trees and thickets of tall shrubs, especially where an opening is made by a stream, pond, or marsh. Also in parks, on wooded farmland, and even in villages and large gardens which provide required edge effect. (Pough 1949.) On Great Plains, associated with relatively open deciduous forest on floodplains, slopes, and bluffs, seemingly unaffected by state of development of understorey (Johnsgard 1979).

Winters in South America in open rain and cloud forest, rarely below 1000 m, and in secondary growth, brush, and cultivated land, especially near streams, foraging from medium levels to treetops (Schauensee and Phelps 1978).

Distribution. Breeds in North America, from north-east British Columbia east to Nova Scotia, south to south-east Alberta, North Dakota, eastern Nebraska, central Kansas, central Oklahoma, southern Missouri, eastern Tennessee, and Maryland. Winters from central Mexico south through Central America to north-west South America.

Accidental. Britain, Ireland, Channel Islands, France (Ouessant), Norway, Sweden, Spain, Yugoslavia, Malta.

Movements. Migrant. Whereas Canadian breeding range extends *c.* 5200 km east to west (60–125°W), and winter range *c.* 3800 km from Venezuela to Mexico (65–

100°W), migration through southern USA is mostly concentrated within *c.* 1600 km (80–95°W) (Bent 1968; Rappole *et al.* 1983).

In autumn, movement within breeding areas begins early August, and by late August main migration is underway across range; few remain in west by early September (Palmer 1949; Bent 1968; Salt and Salt 1976; Tufts 1986). Peak movement early to mid-September in northernmost states, mid- to late September in mid-latitudes, and October near Gulf of Mexico coast (e.g. Imhof 1976, Hall 1983, Janssen 1987). In November, stragglers reported on east coast, and more frequently on Gulf coast, where a few winter rarely (Stewart and Robbins 1958; Lowery 1974; Fisk 1979). Few birds reach winter range before late September, with most arrivals mid-October to December (Bent 1968; Hilty and Brown 1986; Binford 1989). Migration front narrows southward and most birds cross Gulf between eastern Texas and western Florida, with fewer following east coast of Mexico or traversing West Indies (Bent 1968; Peterson and Chalif 1973; Bond 1985). Birds ringed Alabama recovered Michigan and Guatemala (Imhof 1976). Winter site-fidelity shown by ringing in Belize, Honduras, and Panama, with returns up to 2 years later (Loftin 1977).

Spring movement within winter range poorly documented, but apparently begins late March, continuing chiefly to late April (Bent 1968; Hilty and Brown 1986; Stiles and Skutch 1989). Early-April movements include exceptionally early arrivals north to Wisconsin and Nova Scotia (Bent 1968; Tufts 1986). Peak migration starts mid- or late April in southern states (Lowery 1974; Oberholser 1974; Potter *et al.* 1980), reaching southern Canada by mid-May (Squires 1976; Knapton 1979; Francis and Cooke 1990). Arrives in northernmost areas as late as 25 May (Munro and Cowan 1947; Erskine 1985), and stragglers remain in most southern states to mid-May (e.g. Sutton 1967, Imhof 1976). Route followed is reverse of autumn, crossing rather than skirting Gulf (Stevenson 1957).

Accidental in most western states (most frequently in California), and in southern British Columbia, both spring and autumn; also recorded north of range in northern Ontario and Newfoundland, and in Greenland (Garrett and Dunn 1981; American Ornithologists' Union 1983; Godfrey 1986).

Rare autumn vagrant to west Palearctic. 21 records from Britain and Ireland up to 1990, all October, excepting 1 in December–January, all concerned immature individuals (Dymond *et al.* 1989; Rogers *et al.* 1989, 1991).

AJE, PRC

Voice. See Field Characters.

Plumages. ADULT MALE BREEDING. Forehead to back, side of head and neck, and chin to throat and side of breast deep black; frequently some non-breeding feathers retained, these duller black and with traces of buff fringes (mantle and scapulars), or with white (hindneck, upper mantle, supercilium, side of neck) or pink-and-white (chin, throat) base partly visible. Rump and shorter upper tail-coverts white (rarely, pink), upper rump with some variegation; longer upper tail-coverts black with large white blotch on tip. Chest contrastingly poppy-red or pink-red, sharply divided from white at border of side of breast, narrowly extending into a pink-red median stripe to centre of belly. Remainder of underparts white, sometimes with some black spots at border of side of breast and on flank, sometimes (if non-breeding plumage partly retained) with some cream-buff feathers with dusky spot or streak on flank. Central 3 pairs of tail-feathers (t1-t3) black, often with faint off-white fringe on tip; t4 black, usually with large white subterminal patch on inner web; terminal half of inner web of t5-t6 white. Flight-feathers, tertials, and all upper wing black; bases of primaries extensively white (except on shafts and on outer web of p9), showing as large white wing patch extending c. 7 mm beyond primary coverts on p8 to c. 20 mm on p1; tips of median coverts broadly white, forming wing-bar c. 3 mm wide on outer coverts to c. 10 mm on inner coverts; tips of secondaries, tertials, and sometimes inner or all greater coverts with rather small triangular white spot on tip of outer web, sometimes forming narrow wing-bar on greater coverts but soon affected by abrasion, especially on tertials. Under wing-coverts and axillaries pink-red; small coverts along leading edge of wing black, longer under primary coverts dark grey to dull black with faint pink suffusion. Much individual variation, especially in depth of colour, size and shape of red chest-patch, and in white in wing (Whittle 1938; Smith 1966a), in part due to variation in extent of pre-breeding moult. ADULT MALE NON-BREEDING. Like adult ♂ breeding, but tips of black feathers of cap, mantle, scapulars, side of head, cheek, and lesser upper wing-coverts fringed buff, partly concealing black (especially on centre of crown, where showing as buff median crown-stripe, and on mantle and scapulars); feather bases of hindneck, lore, and side of neck partly visible; chin and throat pink-red, bases of feathers white (mainly hidden), tips with variable amount of black (thus, throat mainly pink like chest, not contrastingly black); tips of white feathers of rump, upper tail-coverts, underparts, and of median upper wing-coverts washed cream-buff, sometimes partly with small dusky subterminal speck. ADULT FEMALE. In fresh plumage (autumn), forehead and crown dull black, each feather narrowly fringed olive-brown; a distinct pale cream-buff median crown-stripe, narrow on mid-forehead, broader but partly dotted brown on mid-crown. A distinct long supercilium, pale cream-buff from nostril to above front of eye, almost white further back, continued into white-and-brown mottled band across nape. Mantle cream-buff, heavily spotted or streaked dark olive-brown or black-brown, scapulars and back greyish-olive with broad but ill-defined dark olive-brown streaks. Rump and upper tail-coverts greyish-olive, feathers with ill-defined olive-brown centres, longest upper tail-coverts with off-white fringe or blotches on tip. Lore black, eye-ring off-white, broken by black in front and behind; upper cheek and ear-coverts olive-brown with indistinct cream shaft-streaks, rear border of ear-coverts black-brown. Lower cheek, chin, and throat pale cream-buff or cream-white, each feather with small dusky brown triangular mark on tip, marks sometimes forming indistinct broken malar stripe; side of neck, chest, and side of breast warm cream-buff, densely spotted with dark olive-brown triangular marks; remainder of underparts gradually paler cream-buff to pink-cream, marked with narrower and more elongate olive-brown streaks, mid-belly, vent, and under tail-coverts almost cream-white, streaks obsolete. Tail dark olive-brown; t4-t6 with off-white border (1-2 mm wide) along tip of inner web. Flight-

feathers dull black or brown-black, outer webs tinged olive-brown (especially on secondaries), bases of primaries white, of which 1-4 mm (RMNH, ZMA) or 3-10 mm (Pyle et al. 1987) just visible beyond tips of primary coverts; greater upper primary coverts and bastard wing dull black with faint olive-brown edges. Inner webs of tertials and greater upper wing-coverts dull black or olive-black, outer webs olive-brown with cream-white spot on tip (faint on outermost). Median upper wing-coverts black or black-brown, c. 5 mm of tip cream-white (except for dark shaft and small brown triangle at extreme tip), lesser coverts black to black-brown with olive-brown fringes. Under wing-coverts and axillaries pale or light yellow, salmon, orange, yellow with red-pink fringes, yellow with scattered red-pink feathers, or (exceptionally) entirely bright red-pink (Moyer 1930; Goodpasture 1972; Leberman 1984); marginal coverts cream with black spot, longer primary coverts grey. Some individual variation in shade of brown, in amount and intensity of streaks on chest, in ground-colour of chest (white, yellow, or buff), and in colours of underwing (Smith 1966a). *In worn plumage* (late spring and early summer), ground-colour of head, neck, and underparts off-white, more sharply contrasting with dark marks on head and with dark spots and streaks on underparts; often an off-white spot on central ear-coverts, a more distinct white bar behind black rear of ear-coverts; upperparts from mantle backwards duller greyish-olive with less contrasting olive-brown marks; a small mottled grey and off-white patch just below eye; cream tips of median coverts, cream spots on tips of greater coverts and tertials, and off-white fringes along tips of outer tail-feathers bleached to white, largely worn off. JUVENILE. Upperparts and side of head olive-brown, feathers fringed tawny-cinnamon and white, centres blackish; median crown-stripe and supercilium white with buff tinge, stripe at each side of crown black-brown; underparts pure white, usually with a few olive-brown streaks on side of chin and throat; tips of median and greater upper wing-coverts pale cinnamon (if fresh) to white (if worn), forming distinct wing-bars; sexes similar, but ♂ usually shows 4-8 mm of white at bases of primaries (♀ 0-4 mm), under wing-coverts and axillaries of ♂ mixed pink and yellow (♀ yellow only) (Pyle et al. 1987), and some ♂♂ have primaries partly tinged black, especially at border of white primary patch (Cannell et al. 1983; ZMA). FIRST ADULT MALE NON-BREEDING. Like adult ♀, but flight-feathers, primary coverts, bastard wing, and tail still juvenile, slightly browner and more worn than those of adult ♀ at same time of year (but primaries sometimes blackish at border of white base, unlike ♀), tail-feathers narrower, tips more pointed (less broad and rounded than adult), off-white fringe along terminal part of inner web faint or absent. Under wing-coverts and axillaries variable (always containing at least some pink), as adult ♀; greater under primary coverts pale pink-grey. From August-September, many ♂♂ start to show some pink fringes along tips of lesser upper wing-coverts, as well as some pink suffusion or pure pink feathers on throat or chest (unlike adult ♀). FIRST ADULT MALE BREEDING. Highly variable; during autumn and early winter, ♀-like non-breeding plumage gradually replaced by a plumage similar to adult ♂ non-breeding, or intermediate in character between ♀ and ♂ non-breeding, later by plumage similar to adult ♂ breeding or intermediate in character between non-breeding ♂ and breeding ♂; in any plumage, differs from adult ♂ by retention of most or all juvenile flight-feathers and primary coverts, these distinctly grey-brown (black in adult) contrasting with new deep black greater coverts and tertials (see Moults). In typical 1st breeding plumage, rather similar to adult ♂ non-breeding (upperparts and side of head mainly black), but feathers of mid-crown and of mantle and scapulars broadly fringed olive; rump and upper tail-coverts tinged buff, partly spotted black; supercilium, central hindneck, and side of neck mottled black-and-white; chin, throat, and chest pink with buff suffusion, remainder of underparts white with buff suffusion, profusely spotted black; tail often a mixture of brown juvenile and black or black-and-white adult feathers (if new, t4 more often fully black than in adult); flight-feathers, primary coverts, and bastard wing juvenile, brown. In 2nd non-breeding and breeding plumage, when in 2nd year of life, plumage as adult, but some birds retain a few contrastingly brown and worn juvenile secondaries (Cannell *et al.* 1983). FIRST ADULT FEMALE. Like adult ♀, but flight-feathers, primary coverts, and variable number of feathers of tail and bastard wing still juvenile, more worn than those of adult, tips of tail-feathers more pointed; colour of old feathers not much contrasting with new ones (unlike 1st adult ♂). Under wing-coverts and axillaries yellow or orange-yellow, no pink (unlike adult ♀ and 1st adult ♂); greater under wing-coverts pale yellow-grey, a limited amount of ill-defined white at base of primaries or none at all. In spring, much individual variation in colour and abrasion, due partly to differences in extent of pre-breeding moult, partly to variation in bleaching and wear; in some birds, tail, tertials, and greater upper wing-coverts still juvenile, heavily worn, white spots on tips worn off, body and lesser and median coverts largely in fairly worn non-breeding; in others, body, tail, tertials, and upper wing-coverts of fresh breeding plumage, only flight-feathers and primary coverts distinctly worn.

Bare parts. ADULT, FIRST ADULT. Iris dark brown. Bill pink-white to slate-grey, palest on base of lower mandible; tip of culmen darkest, light horn-grey to slate-black. Leg and foot leaden-grey, slate-brown, or slate-black, fissures between scutes and soles dull flesh-grey to light grey. (RMNH, ZMA.) JUVENILE. Bill pale pink-horn, tip of culmen grey. Leg and foot grey with flesh fissures between scutes; soles flesh. (Farrand 1983).

Moults. ADULT POST-BREEDING. Complete; primaries descendent. Starts in breeding area, mainly late July or August, completed from September, either in breeding area, during migration, or in winter quarters. Of 57 adults killed during autumn migration, 91% in active moult, mainly of flight-feathers, often of body feathers, occasionally of tail; most birds in last stage of moult, with head and body virtually complete non-breeding, (p8-)p9 and some secondaries growing (in 58%), or primaries just completed, but secondaries still growing (in 28%); the few remaining birds less advanced (Cannell *et al.* 1983). Some old secondaries, in particular, frequently retained; either replaced in winter quarters or during next post-breeding moult. ADULT PRE-BREEDING. Partial; in winter quarters, but frequently still in active moult on spring migration (Tordoff and Mengel 1956). Moult January to early May (Ivor 1944; West 1962; Pyle *et al.* 1987). Extent rather variable, and some birds (especially ♀♀) have no moult at all. In birds with moult, includes much or all feathering of head and neck and variable amount of feathering (0–95%) of mantle, scapulars, chin to chest, or entire underparts, sometimes a few tertials or wing-coverts (RMNH, ZMA). POST-JUVENILE. Partial: head, body, lesser and median upper wing-coverts, and variable number of greater coverts. On breeding grounds, July–September; some body feathers often still growing during autumn migration. FIRST PRE-BREEDING. Partial. In winter quarters, October–April. Some birds moult variable number of greater coverts and tertials in early winter, but this probably a continuation of post-juvenile, as these are replaced for 2nd time later in winter during pre-breeding. Includes head, much of body, many or all upper wing-coverts and tertials, and variable number of tail-feathers. In spring, mainly in 1st breeding plum-

age, but flight-feathers, greater upper primary coverts, some or all feathers of bastard wing, and often some tail-feathers still juvenile (a few scattered secondaries sometimes new, probably after accidental loss), some other feathers still 1st non-breeding (mainly scattered feathers of, especially, rear half of body, and variable number of tertials and wing-coverts). (RMNH, ZMA.) Of 90 birds from spring, 48% retained some juvenile tail-feathers, especially t2–t5 (Cannell *et al.* 1983). In ♂, feathers moulted early in winter (e.g. t1 and inner greater coverts, gc4–gc9) browner than those moulted later (e.g. t2–t6 and gc1–gc3 and, moulted for 2nd time, gc6–gc9); these appear juvenile (e.g. t1 and gc4–gc5 in our sample), but are not.

Measurements. ADULT, FIRST ADULT. Eastern North America, May–September, Central America, and north-west South America, September–May; skins (RMNH, ZMA). Juvenile wing and tail include retained juvenile tail and wing of 1st adult; adult tail includes that of 1st adults with (partial) new tail. Bill (S) to skull, bill (N) to distal corner of nostril; exposed culmen on average 4·4 shorter than bill (S).

	♂		♀	
WING AD	104·9 (2·55; 18)	100–109	102·4 (2·01; 7)	98–105
JUV	104·0 (3·15; 14)	100–112	101·1 (2·50; 8)	98–104
TAIL AD	73·8 (2·61; 26)	69–78	72·0 (1·38; 7)	70–74
JUV	72·4 (3·45; 9)	69–79	70·8 (2·73; 8)	67–75
BILL (S)	20·5 (0·67; 34)	19·3–21·8	20·6 (0·94; 15)	19·0–21·8
BILL (N)	12·1 (0·49; 34)	11·4–13·2	12·2 (0·58; 15)	11·3–13·3
TARSUS	23·3 (0·71; 16)	22·4–24·8	22·9 (0·81; 15)	22·0–23·8

Sex differences significant for wing.

Weights. Pennsylvania (USA): adult and 2nd calendar year, (1) July–August, (2) September–October, (3) May; 1st calendar year, (4) July–August, (5) September, (6) October (Clench and Leberman 1978).

	♂		♀	
(1)	41·9 (3·16; 20)	36·3–49·2	41·7 (2·80; 11)	35·4–46·8
(2)	47·9 (3·27; 25)	40·5–55·4	46·2 (5·60; 22)	38·7–62·4
(3)	45·5 (4·65; 92)	37·3–57·5	46·2 (4·13; 67)	38·3–56·1
(4)	40·5 (3·17; 25)	32·8–48·6	41·6 (2·16; 18)	37·7–45·3
(5)	45·0 (3·63; 88)	37·5–53·4	45·4 (3·68; 70)	33·8–56·0
(6)	54·5 (6·76; 7)	46·3–63·0	47·9 (3·14; 7)	43·3–52·2

Coastal New Jersey (USA), autumn: 45·1 (4·7; 13) 38·4–53·9 (Murray and Jehl 1964). See also Baldwin and Kendeigh (1938), Tordoff and Mengel (1956), Rand (1961b), Graber and Graber (1962), Mengel (1965), Grant (1968), and Kuenzi *et al.* (1991).

Curaçao, October: ♀ 26 (ZMA). El Salvador, November: ♀ 40 (ZMA). Belize, April: ♂ 29·2 (Russell 1964). Mexico, April: ♂ 58·9 (RMNH). Isles of Scilly (England), October: single ♀ 37·0–41·4 (Grant 1968).

Structure. Wing rather long, broad at base, tip bluntly pointed. 9–10 primaries: p8 longest, p9 1·5–4 shorter, p7 0–1, p6 1–3, p5 7–11, p4 13–18, p3 17–21, p2 19–25, p1 22–28. P10 strongly reduced, a tiny pin hidden beneath outermost upper primary coverts, apparently absent in 6 of 11 birds; 67–76 shorter than p8, 10–13 shorter than longest upper primary covert. Outer web of p6–p8 emarginated, inner web of p7–p9 with faint notch. Tip of longest tertial reaches to tip of p1–p2. Tail rather long, tip square; 12 feathers. Bill short, exposed culmen *c.* 70% of head length; markedly thick and swollen at base, depth at base 13·5–15·5 mm, width at base 11·0–12·0 mm; culmen gently decurved, cutting edges almost straight except for sharp deflection at base, corresponding with tooth on lower mandible. Bill superficially as in Hawfinch *Coccothraustes coccothraustes*, but in that species much wider at base and less compressed laterally at tip, with more broadly flattened culmen, deeper base of upper mandible but shallower base of lower, longer gonys, virtually straight cutting edges, and no fine hook at tip of upper mandible. Nostril fairly large, rounded, partly covered by membrane at base. Many short but distinct bristle-like hairs below nostril and at lateral base of upper mandible. Tarsus and toes short, rather stout. Middle toe with claw 21·0 (10) 19·0–23·0; outer toe with claw c. 73% of middle with claw, inner c. 69%, hind c. 74%.

Geographical variation. None.

Forms superspecies with Black-headed Grosbeak *P. melanocephalus* from western North America (Mayr and Short 1970); in Great Plains area, where ranges overlap, some hybridization occurs, and thus sometimes considered single species, but interbreeding appears non-random and best considered a superspecies pair. See West (1962), Short (1969), Paynter (1970), Anderson and Daugherty (1974), and Kroodsma (1974*a*, *b*) for situation in and extent of zone of hybridization.

Recognition. For differentiation of ♀ from ♀ of *P. melanocephalus*, see Pyle *et al.* (1987) and Marlan (1991).   CSR

Guiraca caerulea Blue Grosbeak

PLATE 29
[between pages 256 and 257]

Du. Blauwe Bisschop   Fr. Guiraca bleu   Ge. Azurbischof
Ru. Голубая гуирака   Sp. Cardenal azul   Sw. Blåtjocknäbb

Loxia caerulea Linnaeus, 1758. Synonym: *Passerina caerulea*.

Polytypic. Nominate *caerulea* (Linnaeus, 1758), south-east USA, west to central Kansas and Texas; *interfusa* Dwight and Griscom, 1927, dry interior of USA from central South Dakota and eastern Nebraska south to western Texas, south-east California, and neighbouring north-central Mexico; *salicaria* Grinnell, 1911, western USA from north-central California, western Nevada, and western Arizona south to north-west Baja California (Mexico); 3 races Mexico to Costa Rica.

Field characters. 14–17 cm; wing-span 26–28·5 cm. About 15% larger than Scarlet Rosefinch *Carpodacus erythrinus* but with rather similar though stockier form; about 15% smaller than Rose-breasted Grosbeak *Pheucticus ludovicianus*, with proportionately shorter tail. Rather large, robust but shy Nearctic finch, with two rusty wing-bars in all plumages. ♂ deep blue, with blacker flight- and tail-feathers; ♀ and immature recall lightly marked *Carpodacus* finch but show some blue on wing and tail. Sexes dissimilar; no seasonal variation. Immature similar to ♀.

ADULT MALE. Moults: July–October (complete). Head and ground-colour of back, rump, and body quite deep azure to violet-blue; black lore and area round bill. In fresh plumage at close range, shows darker feather-centres and rufous tips on mantle and scapulars and buff or whitish tips on rump and underparts. Wings blue on lesser coverts, blue-black across greater coverts and bastard wing, and black fringed pale cinnamon on tertials and (less obviously) other flight-feathers; rusty median coverts form striking broad upper wing-bar and buff fringes and rusty tips of greater coverts create 2nd lower wing-bar. Under wing-coverts grey. Tail black with blue edges. Bill grey-white, with dark culmen and tip. Legs dusky-black. ADULT FEMALE. Head and back warm greyish-brown, with buff supercilium, pale buff eye-ring, buff surround to ear-coverts, buff-white throat with dusky malar stripe, and soft dark brown streaks on mantle and scapulars. Lower back and rump greyer than mantle with tinge of blue and hardly streaked except on upper tail-coverts. Wings patterned as ♂ but both wing-bars are paler, cream- to rusty-buff. Tail duller than in ♂, with only blue tinge to edges. Underparts buff, with paler whitish-buff belly and under tail-coverts; when worn, softly streaked brown across breast and along flanks. FIRST-WINTER. Resembles ♀ but retains juvenile wing and tail. Young ♂♂ may show bluish fringes on wing and tail, rarely on head and body. FIRST-SUMMER MALE. Oddly variegated due to mix of blue adult and grey and brown immature tracts. Even when fully worn, never as blue as adult ♂.

♂ unmistakable, being over 20% larger than main Nearctic confusion species, Indigo Bunting *Passerina cyanea*, which lacks heavy bill and wing-bars. ♀ and 1st-winter far less distinctive since appearance of bill, head, and wings converges with *Carpodacus* finches; essential to confirm presence of blue feathers and depth and tone of upper wing-bar. Flight, gait, and behaviour all recall *Carpodacus* finch; occasionally flicks tail. Likes thick cover.

Commonest call a sharp 'chink'.

Habitat. A Nearctic forest species, like Rose-breasted Grosbeak *Pheucticus ludovicianus*, firmly linked with edges between dense woodland and more open lower vegetation, as naturally occurs by streamsides, swamps, or where opened-up patches are undergoing second growth. Human intervention has partly replicated such terrain in roadside plantings, along farm hedgerows and ditches, and in weedy fields, into which range has recently expanded, although attachment to thick cover and to moist places persists and open ground is avoided (Pough 1949). Weedy pastures are, however, acceptable (Johnsgard 1979). In west found at base of mountains, and in California to *c.* 1200 m, occurring in mesquite and often in willows along irrigation ditches; also in young pines, orchards, and gardens, and old fields overgrown with brambles; tops of small trees, bushes, or overhead cables are used as song-posts (Terres 1980). In south, in Louisiana, favours mixed woods or oak and pine, particularly where there are stands of scrub oak, but shifts in autumn to brush and briar thickets, old cornfields, and tall grass (Lowery 1974).

Distribution. Breeds in North America, from central California, Nevada, and the Dakotas east to New Jersey, south through Central America to central Costa Rica. Winters from northern Mexico south to central Panama; rarely in southern USA and Cuba.

Accidental. Norway: ♂, Øyeren, 14 June 1970 (Haftorn 1971); ♂, Rogaland, 22 November 1987 (Bentz 1989).

Movements. Migrant (northern races) to resident (southern races). Migrants winter within and slightly south of residents' range; northern birds migrate up to 6000 km (American Ornithologists' Union 1983).

Autumn migration is inconspicuous, except north-east of breeding range where vagrants appear annually mid-August to October (Bailey 1955; Bull 1974; Tufts 1986). Breeding birds depart from northern areas chiefly by mid-September (Stewart and Robbins 1958; Bent 1968), with passage further south continuing to mid- or late October (Oberholser 1974; Imhof 1976; Potter *et al.* 1980). Further west, most leave by end of September (Grinnell and Miller 1944; Phillips *et al.* 1964). Arrives in West Indies (where sparse) and Central America from September, chiefly October to early November (Bond 1985; Rogers *et al.* 1986a; Stiles and Skutch 1989). Migration evidently on broad front both seasons, but little studied (Bent 1968). Winter site-fidelity shown by ringing in Belize, with one return a year later (Loftin 1977).

Northward movement begins late March, with last records in winter range mid- or late April (Ridgely 1981; Rogers *et al.* 1986a; Stiles and Skutch 1989). A few birds appear in southern states, from Texas to South Carolina, in early April, and individuals have reached Massachusetts and Nova Scotia by mid-April in some years (Bailey 1955; Bent 1968; Tufts 1986; Kale and Maehr 1990). Main passage in USA starts mid-April (Oberholser 1974; Potter *et al.* 1980; Garrett and Dunn 1981), and most breeding areas are occupied by mid-May (Stewart and Robbins 1958; Bailey and Neidrach 1967; Bohlen 1989). Along north-east coasts, many migrants overshoot range, reaching Nova Scotia annually April–May (e.g. Bull 1974, Tufts 1986).

Vagrant north to Washington state, and from Saskatchewan to Nova Scotia, with single records in Newfoundland (autumn) and nearby St Pierre-et-Miquelon

(Etcheberry 1982; Maunder and Montevecchi 1982; American Ornithologists' Union 1983).   AJE

Voice. See Field Characters.

Plumages. (nominate *caerulea*). ADULT MALE. Top and side of head and neck glistening violet-blue, lore and strip of feathering at base of lower mandible sooty-black. Mantle, scapulars, and back slightly darker and duller violet-blue, feathers with rufous-cinnamon tips, partly concealing blue; rump and upper tail-coverts brighter violet-blue, feathers with narrow buff or pale brown tips. Entire underparts violet-blue, tinged sooty on upper chin; feathers with traces of pale buff to off-white tips, especially on lower flank, belly, thigh, vent, and under tail-coverts. Tail black, feathers narrowly fringed violet-blue along outer webs, tip of outer 3–4 feathers with sharply defined grey or off-white border. Flight-feathers, tertials, greater upper primary coverts, and bastard wing black, outer webs narrowly fringed violet-blue (except for emarginated parts of outer primaries), fringes of tertials and sometimes secondaries partly pale cinnamon. Lesser upper wing-coverts glistening violet-blue; median coverts rufous-chestnut, forming broad wing-bar; greater coverts black, tip of outer web rather narrowly fringed cinnamon (forming paler and narrower wing-bar than that on median coverts), outer fringe partly violet-blue. Longer under wing-coverts and axillaries dark grey, shorter coverts violet-blue. *In fresh plumage* (about August–March), feathers of crown, neck, rump, upper tail-coverts, and entire underparts narrowly tipped red-brown to buff (wider and paler on rump, vent, and tail-coverts), partly concealing blue; mantle and scapulars more broadly tipped rufous, blue more fully hidden. *In worn plumage* (June–July), pale brown to off-white feather-fringes worn off, blue fully exposed, except for traces of fringes on lower mantle, scapulars, vent, and under tail-coverts; some dark grey or dull black of feather-bases visible amidst blue of head and underparts; blue of head and chest slightly less violet, more azure-blue; fringes of tail, flight-feathers and tertials paler, more greyish, partly worn off; cinnamon tips of greater coverts bleached to buff or cream, partly or (on outer coverts) almost fully worn off; some black of bases sometimes visible on median coverts. ADULT FEMALE. In fresh plumage, top and side of head and neck dark brown, feathers with faint rufous tips, lore buff-white with fine grey specks. Dark brown of hindneck shades into buff-brown on mantle, scapulars, and back, and this in turn to grey-brown on rump and upper tail-coverts; feathers of rump with buff-brown tips, tail-coverts with narrow off-white tips. Chin backwards to upper belly and upper flank saturated tawny-buff, gradually paler buff on vent and under tail-coverts. Tail brown-black; fringes along outer webs (except outermost feather, t6) cerulean-blue, fringe along distal part of outer web of t4–t5 off-white. Flight-feathers, tertials, and upper wing-coverts dark sepia-brown; flight-feathers narrowly fringed buff along outer webs (almost white along emarginated parts of primaries), tertials more broadly fringed cinnamon; lesser upper wing-coverts with blue-grey fringes, median and greater coverts with broad rufous-cinnamon tips, forming wing-bars. Longer under wing-coverts and axillaries grey, shorter coverts buff. Some birds show a variable amount of blue feathers mixed within brown ones, mainly on head, rump, and chest. *In worn plumage* (spring and early summer), upperparts dark olive-brown, dusky brown feather-centres visible on crown, lower mantle, and scapulars; back to upper tail-coverts tinged olive-grey; underparts less saturated and more restricted tawny, side of breast with dark brown-grey patch, chin, belly, vent, and under tail-coverts buff, some faint dark grey-brown streaks on flank and (sometimes) belly; fringes of tail largely worn off, grey rather than blue, those of tertials and median and greater upper wing-coverts bleached to buff-white and partly worn off. JUVENILE. Rather like adult ♀, but less warm buff-brown and tawny; upperparts greyish-buff, underparts buff with some rufous-buff on feather-tips, much grey of bases of feathers visible on underparts. Under tail-coverts short, loose, and fluffy. Tips of median and greater upper wing-coverts pink-buff (white if worn), *c.* 2–3 mm long (narrower than in adult ♀). No bluish tinge to buff-grey fringes of tail-feathers and lesser upper wing-coverts, except on tail of some ♂♂. FIRST ADULT NON-BREEDING. Like adult ♀, but juvenile tail, flight-feathers, tertials, greater upper primary coverts, and sometimes a few outer greater upper wing-coverts or feathers of bastard wing retained; tail-feathers rather narrow, tips rounded or slightly pointed, without sharply-defined pale border along tip (in adult, feathers wider, tip more truncate and with sharp off-white fringe). Sexes similar, but fringes of tail-feathers, flight feathers, and lesser coverts often bluish in ♂, and some ♂♂ start to show scattered partly or fully blue feathers on head or body in course of 1st winter. FIRST ADULT BREEDING MALE. Forehead, forecrown, lore, ear-coverts and cheeks blue, as in adult ♂, part of feathers with narrow brown tip. Hindcrown, hindneck, side of neck, mantle, and scapulars dark grey with variable amount of blue tinge (generally limited), feathers with broad red-brown or cinnamon tips, partly concealing grey. Back to upper tail-coverts blue, partly concealed by ill-defined buff-brown feather-tips on back, pale buff fringes on rump, and grey fringes on upper tail-coverts. Chin and throat a variable mixture of blue and buff-white feathers, remainder of underparts dark grey, feather centres variably tinged blue, largely concealed by broad buff-brown fringes on side of breast and chest and by grey-buff fringes elsewhere, appearing heavily mottled. Tail as adult ♀; flight-feathers and greater upper primary coverts as in juvenile, but variable number of outer ones frequently new, strongly contrasting with old and frayed browner inner ones; fringes blue-grey, generally less deep blue than in adult ♂. Lesser upper wing-coverts blue with buff fringes, median and greater coverts blue-grey to blue, median with cinnamon tips *c.* 4 mm long (narrower and less deep than in adult ♂), greater with tips 2–3 mm long (narrower, less deep, and less sharply-defined). *In worn plumage*, brown and buff fringes partly wear off, more blue exposed, but generally much less uniform blue than in adult ♂, strongly mixed and washed grey and brown on body. FIRST ADULT BREEDING FEMALE. Like adult ♀, and sometimes indistinguishable; as in 1st breeding ♂, some tail-feathers, flight-feathers, and greater upper primary coverts sometimes new, these contrasting markedly in colour and abrasion with neighbouring retained juvenile ones; when flight-feathers and primary coverts all still juvenile, these browner and more worn than those of adult at same time of year, less dark than non-breeding newer median or greater coverts.

Bare parts. ADULT, FIRST ADULT. Iris dark brown to black-brown. Bill plumbeous-grey to black, darkest on culmen; extreme base of upper mandible, cutting edges, and entire lower mandible bluish-white to pale greyish-horn, palest at undersurface of lower mandible; in 1st autumn, traces of yellow at gape. Leg and foot dark grey to black, soles grey. (RMNH.) JUVENILE. No information.

Moults. ADULT POST-BREEDING. Complete; primaries descendent. Starts July–August, completed September–October; on breeding grounds, but sometimes suspended during autumn migration and then completed after arrival in winter quarters.

No pre-breeding moult. POST-JUVENILE. Partial: head, body, lesser and median upper wing-coverts, sometimes variable number of greater coverts, tertials, or tail-feathers. On breeding grounds, July–October, but continued after arrival in winter quarters, where merging into 1st pre-breeding moult. FIRST PRE-BREEDING. Partial or almost complete; in winter quarters, November–April. Involves head, body, lesser and median upper wing-coverts, some or all tail-feathers, tertials, and greater coverts, frequently 3–8 outer or all primaries, and occasionally some greater upper primary coverts or secondaries. (Dwight 1900; Pyle *et al.* 1987; RMNH.)

Measurements. Nominate *caerulea*. Eastern USA, summer, and eastern part of Central America, winter and spring; skins (BMNH, RMNH, ZMA). Bill (S) to skull, bill (N) to distal corner of nostril; exposed culmen on average 4·5 less than bill (S).

	♂		♀	
WING	89·1 (2·57; 16)	86–94	84·2 (1·81; 10)	81–87
TAIL	63·6 (1·37; 10)	61–66	62·0 (3·07; 10)	57–67
BILL (S)	19·2 (0·66; 11)	18·2–20·0	19·3 (0·44; 10)	18·7–20·1
BILL (N)	12·4 (0·46; 9)	11·8–13·2	12·2 (0·54; 10)	11·5–12·9
TARSUS	21·1 (0·31; 10)	20·6–21·7	21·1 (0·65; 10)	20·2–22·0

Sex differences significant for wing.

Weights. New Jersey (USA), autumn: 27·5 (Murray and Jehl 1964). Georgia and South Carolina (USA), July to 1st half of September: ♂ 29·0 (0·64; 8), ♀ 25·8 (1·06; 4) (Norris and Johnston 1958). Belize: March, ♂ 28·5 (Russell 1964); see also Rogers and Odum (1966).

Structure. Wing rather short, broad at base, tip bluntly pointed. 10 primaries: p7–p8 longest, p9 3–5 shorter, p6 0·5–2, p5 3–6, p4 8–12, p3 12–16, p2 14–19, p1 17–21. P10 reduced; 55–65 shorter than p7–p8, 7–10 shorter than longest upper primary covert. Outer web of p6–p8 emarginated, inner of p7–p9 with faint notch. Tip of longest tertial reaches to tip of p2–p3. Tail rather short, tip square or slightly forked; 12 feathers. Bill short, very strong, conical; closely similar to bill of Rose-breasted Grosbeak *Pheucticus ludovicianus*, but even shorter and somewhat wider at base, tip more compressed laterally. Tarsus and toes rather short and stout. Middle toe with claw 20·8 (5) 19·5–22; outer toe with claw *c.* 76% of middle with claw, inner *c.* 69%, hind *c.* 77%. Remainder of structure as in *P. ludovicianus*.

Geographical variation. Rather slight, involving size and colour. *G. c. interfusa* of interior USA larger than nominate *caerulea* from south-east USA, especially tail and bill; blue of ♂ less purplish, more greyish, median coverts paler chestnut, tips of greater coverts broader, paler pinkish-cinnamon; ♀ paler (Dwight and Griscom 1927; Storer and Zimmerman 1959). *G. c. salicaria* from south-west USA has wing and tail longer than nominate *caerulea*, similar to *interfusa*, but bill markedly smaller and less swollen than both of these; colour pale, as *interfusa*. Remaining 3 races from Central America larger, colour varying between as dark as nominate *caerulea* (in north) to even paler than *interfusa* (in south); see Ridgway (1901–11), Dwight and Griscom (1927), and Storer and Zimmerman (1959).

G. caerulea often included in *Passerina*, with which it agrees in voice, nest, bill colour, and behaviour (Phillips *et al.* 1964; Blake 1969), and perhaps nearer Indigo Bunting *P. cyanea* and Lazuli Bunting *P. amoena* than are some other species at present included in *Passerina* (Blake 1969). However, maintained in *Guiraca* following American Ornithologists' Union (1983). CSR

Passerina cyanea Indigo Bunting

PLATE 29
[between pages 256 and 257]

Du. Indigo-gors   Fr. Passerin indigo   Ge. Indigofink
Ru. Индиговый овсянковый кардинал   Sp. Pape azulejo   Sw. Indigofink

Tanagra cyanea Linnaeus, 1766

Monotypic

Field characters. 13–14·5 cm; wing-span 19·5–21·5 cm. Close in size and form to Linnet *Carduelis cannabina* but with more slender tail; 25% smaller and much daintier than Blue Grosbeak *Guiraca caerulea*. Small, active Nearctic bunting, with quite strong bill. Lacks obvious wing-bars except in juvenile. Breeding ♂ almost wholly deep blue. Winter ♂, ♀, and immature mainly buff-brown, often showing little blue and with faint plumage pattern suggesting *Carduelis* or *Carpodacus* finch. Sexes dissimilar; marked seasonal variation in ♂. Juvenile separable. Hybridizes with Lazuli Bunting *P. amoena*.

ADULT MALE BREEDING. Moults: July–September (complete), December–April (mainly head, body, and wing-coverts). Whole plumage indigo-blue, looking very dark at distance. At close range, shows black lores and centres to median and greater coverts and tertials and mainly black flight-feathers. Bill dusky black on upper mandible, pale grey, almost whitish on lower. Legs brown-black. ADULT MALE NON-BREEDING. Although blue always visible on tail and outer wing, rest of plumage basically warm buff-brown on head and back and buff-white below, with pure white under tail-coverts. With wear, patches of blue feathers show on head, back, chin, and breast. Warm buff wing-bars formed by tips to median and greater coverts less obvious than those of *G. caerulea*. Bill duller horn than in spring. ADULT FEMALE. Against ♂, remarkably plain buffish-brown above and buff and dull white below, with patches of pale greyish-blue restricted to lesser wing-coverts and fringes of outer wing, secondaries, and tail. At close range, shows faint face pattern of pale buff marks above and behind eye, and pale cream throat with buff malar stripe leading into soft brownish streaks over breast, fading out on flanks. On wings, quite marked contrast between dark brown centres and warm buff edges and tips to tertials and greater and median coverts, but no obvious wing-bars. Bill flesh-horn. FIRST-WINTER. Close to adult ♀, but ♂ shows much darker malar stripe and often blue feathers in head and body plumage. Due to retention of juvenile tertials and greater coverts, may show paler edges to tertials and sharp whitish bar across wing. For fuller details of this and 1st-summer plumage, see Plumages.

♂ unmistakable in good view if size well judged. ♀ and immature nondescript, inviting confusion with ♀ and immature *Carduelis* and *Carpodacus* finches; important to note that *P. cyanea* lacks strong streaks typical of those species and also bulbous bill of *Carpodacus*. Flight and behaviour much as *Carduelis* finch. ♀ noticeably shyer than ♂.

Commonest call a loud 'pwit', sometimes repeated.

Habitat. Breeds in temperate and warmer Nearctic lowlands, from coniferous forest zone southwards, including upland fringes but excluding most intensively cultivated and grazed areas, deserts (except around shrubs near water), and closed-canopy forest, as well as human residential areas. Inhabits brushy and weedy fringes of cultivated lands, roads, railways, and rivers, as well as woods, where they are open, deciduous, and broken by clearings; also found in abandoned fields, weedy cultivated land, and swamps. On migration and in winter, frequents weedy fields and cropland, especially when recently harvested, citrus orchards, savanna, and low second growth, then foraging on ground, although in breeding season also in trees to 15 m. (Payne 1991.) Requires dense ground cover of brushy growths with an occasional tree or telephone wire for song-post; favours forest edges by fields, streams, and lakes (Pough 1949). Often on river floodplains and in successional situations such as second growth; even in orchards (Johnsgard 1979).

Distribution. Breeds in North America, from south-east Saskatchewan east to New Brunswick and Maine, south to southern New Mexico, south-east Texas, Gulf of Mexico coast, and central Florida. Winters from central Mexico south through Central America with a few in Colombia and north-west Venezuela, also in southern Florida and Caribbean (chiefly Greater Antilles).

Accidental. Iceland, Britain, Ireland, Netherlands, Denmark, Sweden, Finland. Some records, at least, may involve escapes (e.g. Dymond *et al.* 1989).

Movements. Migratory, all birds wintering south of breeding range. Along western edge of breeding range, hybridizes regularly, but in limited area, with Lazuli Bunting *P. amoena*; migration picture for such hybrids is unstudied. (American Ornithologists' Union 1983; Rappole *et al.* 1983.)

Although a few birds stray south-west in July, e.g. to Arizona, main autumn departure from north of range begins late August, with peak movement September in northern states, continuing to October further south, and to early November along Gulf of Mexico (Palmer 1949; Phillips *et al.* 1964; Bent 1968; James and Neal 1986).

By early October, birds are arriving throughout Central America and West Indies (Monroe 1968; Brudenell-Bruce 1975; Ridgely 1981; Rogers *et al.* 1986*a*). Migration on broad front, including movements to West Indies, across Gulf of Mexico (most birds), and probably through eastern Mexico (Bent 1968; Bond 1985; Payne 1991). Winter site-fidelity has been shown from ringing e.g. in Jamaica, Belize, and Guatemala, with returns up to 8 years later (Loftin 1977). Long-distance ringing recoveries include movements between Belize and Ohio, Costa Rica and Michigan, and Jamaica and Maine (Payne 1991).

In spring, some movement occurs in March, mostly in areas where wintering is sporadic, and late March airstreams have carried stray birds north to Arkansas, Maryland, and even Nova Scotia (Stewart and Robbins 1958; Bent 1968; James and Neal 1986; Tufts 1986). Chief departure from winter quarters in April (e.g. Brudenell-Bruce 1975, Binford 1989), and main movement northward April to early May, with passage and arrivals from early April in southern states, late April in Kentucky and Missouri, and early May in Michigan and southern Ontario (Mengel 1965; Bent 1968; Sprague and Weir 1984; Payne 1991). Most birds are in northernmost breeding areas soon after mid-May (Speirs 1985). As in autumn, most migration is across Gulf of Mexico (Stevenson 1957; Moore *et al.* 1990); reconstruction from weather maps suggests that some mid-April records involve direct flight from Central America to New England (Bent 1968).

Vagrant north to Newfoundland, Hudson Bay coast, central Alberta, and southern British Columbia, and west to most western states (American Ornithologists' Union 1983; Speirs 1985). AJE

Voice. See Field Characters.

Plumages. ADULT MALE BREEDING. Lore black, remainder of head and chin to chest bright smalt- or violet-blue, on hindneck gradually merging into cobalt-blue or cerulean-blue of remainder of upperparts; breast and mid-belly bright blue with violet tinge, gradually merging into turquoise-blue or cerulean-blue of side of neck and breast, flank, vent, and under tail-coverts. Tail black, both webs of central pair (t1) and outer web of t2–t5 fringed blue (widest towards base). Flight-feathers, tertials, greater upper primary coverts, and bastard wing black, primaries and coverts narrow but sharply fringed cerulean-blue along outer webs (hardly so on emarginated part of primaries), secondaries, tertials, and middle feather of bastard wing more broadly fringed with darker cobalt-blue. Upper wing-coverts cobalt-blue, contrasting black on centres and inner webs largely concealed. Under wing-coverts and axillaries dark grey, axillaries and shorter coverts tipped cerulean- or turquoise-blue, coverts along leading edge of wing tinged violet. *In fresh plumage* (late winter), feathers of body partly tipped brown, tips soon bleaching to buff and worn off by February–March; in worn plumage (late spring and early summer), blue slightly less intense, violet more restricted, on body more extensively cerulean- or turquoise-blue; some black at base of lower mandible; dark grey of feather-bases on body partly visible, especially on mantle and belly; fringes of tail and flight-feathers bleached to light grey-blue, partly worn off.

Occasionally, some partly brown or buff non-breeding feathers retained on body or upper wing-coverts, exceptional birds being even browner than bluest breeding ♀, independent of age (primary coverts of these ♂♂ fringed blue, in contrast to those of 1st adult ♂, which are partly or fully brown) (Payne 1991). ADULT FEMALE BREEDING. Upperparts umber-brown, dark grey feather-centres partly visible, especially on crown and mantle; feathers of rump and upper tail-coverts with paler buff fringes, centres in some birds partly washed light blue-grey. Side of head buff-brown; slightly paler, more buffish, on spot above dull grey lore and on eye-ring; ear-coverts narrowly streaked buff; side of neck umber-brown. Ground-colour of underparts buff, slightly browner on side of breast and flank, paler on chin, throat, and belly, almost white on vent; all feathers except those of chin, mid-belly, and vent with short dark grey shaft-streaks, usually rather ill-defined and faint, but sometimes distinct on (especially) throat, chest, and belly. Tail dark grey, t1 and narrow outer fringes of t2–t5 light grey-blue or a mixture of buff and blue. Flight-feathers, tertials, primary coverts, and bastard wing greyish- or brownish-black, fringes along outer webs brown, but especially those of primaries or primary coverts sometimes partly or fully light grey-blue. Upper wing-coverts black or brownblack with drab-brown fringes; lesser coverts washed light greyblue to varying extent, median coverts with broad rufous-brown tips, greater with narrower pale rufous tips, these tips forming 2 wing-bars. Under wing-coverts and axillaries grey-brown, tips pale buff to dirty white (most extensively so on axillaries and longer coverts). Occasionally, varying amount of feathers of body with light grey-blue wash. *In worn plumage* (late spring and early summer), upperparts distinctly duller and greyer brown; ground-colour of side of head and of underparts paler dirty buff-white, short dull grey streaks and spots more extensive; tips of median and greater coverts and tip of outer fringe of tertials bleached to off-white, more prominent, but frequently fully worn off when heavily abraded, especially on greater coverts and tertials. ADULT MALE NON-BREEDING. Like adult ♂ breeding, but all blue feathers of head, body, and wing are broadly tipped rufous-brown on upperparts, pale buff to white on underparts, except for tail, flight-feathers, and primary coverts (which are as blue as in breeding); blue of feathers restricted to centres, largely concealed below brown feather-tips when plumage fresh (in early autumn), partly exposed when worn (by mid-winter); brown tips to median and greater upper wing-coverts form double wing-bar, but these indistinct unless plumage worn. ADULT FEMALE NON-BREEDING. Like adult ♀ breeding; streaks on underparts sometimes obsolete, tips of median and greater coverts umber-brown to buff-brown, sometimes scarcely paler than remainder of upper wing. JUVENILE. Upperparts and side of head and neck dark brown, feathers with ill-defined paler buff-brown fringes and more sooty subterminal centres. Underparts buff to isabellinewhite, palest on chin and from mid-belly to under tail-coverts; underparts extensively marked with narrow and sharp dark brown streaks, least so on mid-belly to under tail-coverts. Tail and flight-feathers sooty-grey, feathers with rather faint and narrow rufous edges, narrower, paler, and greyer on primaries. Tertials and upper wing-coverts sooty brown, tertials with narrow buff-brown fringe along outer web and tip, median and greater coverts with broader and more strongly contrasting buff or pink-buff fringe along tip (narrowest at shaft), latter forming 2 distinct bars on wing. In worn plumage, rufous and buff of fringes of flight-feathers bleached to off-white, partly worn off. FIRST ADULT MALE NON-BREEDING. Rather like adult ♀, being mainly brown; blue-grey fringes along tail- and flight-feathers similar to those of adult ♀ or partly replaced by pale buff or light buff-grey, those of primary coverts mainly buff-brown; variable amount of grey-blue on rump or lesser upper wing-coverts (as in adult ♀); tips of tail-feathers pointed, less broadly rounded or truncate than in adult ♀; a distinct dark brown-grey malar stripe (faint in adult ♀); tips of greater upper wing-coverts cream-buff (rufous in adult ♀), underparts sometimes with broad dark streaks and spots (faint in adult ♀); axillaries sometimes bluish (Johnston 1967a). During autumn and winter, new feathers grow which are similar to those of adult ♀ non-breeding plumage, showing (largely concealed) blue centres; up to December, generally less than 10% of plumage blue (Johnston 1967a; Payne 1991). FIRST ADULT FEMALE NON-BREEDING. Like adult ♀, but no blue tinge on body or wing-coverts, and fringes of tail and flight-feathers fully buff to pale grey-buff; tips of tail-feathers pointed, as in 1st non-breeding ♂; plumage usually worn, especially wing-coverts and tail (fresh in adult), streaks on underparts more distinct. Similar to 1st non-breeding ♂, but no traces of blue or blue-grey, and malar stripe indistinct. (Johnston 1967a.) FIRST ADULT MALE BREEDING. Like adult ♂, but variable amount of non-breeding plumage retained, showing as buff-brown patches (on upperparts, flank, and/or wing-coverts) or pale buff to white variegation (on breast to under tail-coverts), and secondaries, inner or all primaries, and usually all primary coverts still juvenile (primaries, if partly new, showing a strong contrast between older brown-fringed inner ones and newer blue-fringed outers); brown tips of greater coverts (if any) bleach to pale buff, showing as pale wing-bar. If worn, brown tips of retained non-breeding feathers partly or fully worn off, plumage more extensively blue, but less deep and uniform than adult ♂ breeding, and at least some inner primaries and primary coverts worn and brown-fringed. See Moults (below) or Payne (1991) for variation in amount of blue and brown. FIRST ADULT FEMALE BREEDING. Like adult ♀; rump and lesser coverts of some birds partly grey-blue, others uniform brown, fringes of tail sometimes partly grey; in some birds, outer primaries new, fringes light blue-grey, contrasting with worn brown-fringed inner ones (unlike adult ♀); primary coverts worn, fringed brown-buff; tips of median and greater coverts often paler than in adult, wing-bars more distinct, but more liable to wear off; underparts on average more distinctly and sharply spotted and streaked with grey.

Bare parts. ADULT. Iris brown or dark chestnut-brown. In breeding ♂, upper mandible and gonys ridge plumbeous-black or black, cutting edges and lower mandible light blue-grey, gape black; in breeding ♀, upper mandible and gonys ridge dark horn-brown to brown-black, cutting edges and lower mandible pale grey or pale pink-horn, gape horn. In non-breeding plumage, bill light horn-brown, pink-horn, grey-horn, or slate-grey, culmen and tip of lower mandible darker dusky grey-horn or black; gape yellow. Leg and foot purple-brown, dark slate-grey, or plumbeous-black, paler slate-grey on soles. FIRST ADULT. Like adult; gape yellow until May of 2nd calendar year. JUVENILE. Iris olive-brown. Bill dusky horn-brown, base and middle portion of lower mandible paler pink-horn or yellowish. Leg and foot purple-flesh or dull flesh-grey. (Kaufman 1989; Payne 1991; RMNH, ZMA.)

Moults. ADULT POST-BREEDING. Complete; primaries descendent. On breeding grounds; ♂ starts July–August, ♀ when young have fledged, August or early September; completed from August (♂) or September (♀), in late-starting ♀♀ not until October (Blake 1969; Pyle *et al.* 1987; RMNH, ZMA); single ♀ from Kansas in last stages of moult 1 October (Tordoff and Mengel 1956). ADULT PRE-BREEDING. Partial; in winter quarters, December to early April, sometimes already completed February (Johnston and Downer 1968; RMNH, ZMA). Includes head and body (some-

times excluding up to 30% of feathering, especially nape, scattered feathers of mantle and scapulars, or some feathers of side of breast, flank, and vent), lesser and many or all median upper wing-coverts, usually 3–9 greater upper wing-coverts (mostly 6–8) and 1–3 tertials (mostly 1), occasionally some tail-feathers or feathers of bastard wing (RMNH, ZMA). POST-JUVENILE. Partial: head, body, lesser and median upper wing-coverts, occasionally some inner greater coverts, tertials, or t1. On breeding grounds: June–October. Starts at age of *c.* 16 days, before flight-feathers full-grown (Sutton 1935; Blake 1969), completed from late July onwards. Moult largely suspended during autumn migration, but resumed immediately after arrival in winter quarters, November–January, merging into 1st pre-breeding moult which spans February–May. Early winter moult thus partly a continuation of post-juvenile or an early pre-breeding (C S Roselaar), but as much of head and body moulted for 2nd time in November–January, and for a 3rd time in pre-breeding, (December–)February–April, early winter moult better considered an extra moult which leads to a supplementary plumage (Rohwer 1986), in which feathering shows mixed features of breeding and non-breeding plumage. See Rohwer (1986) for details, and Johnston and Downer (1968), Rohwer *et al.* (1992), and Willoughby (1992) for further details and interpretation of immature moults. The extra moult includes a varying amount of feathering of head and body and of lesser and median upper wing-coverts; also tail, tertials, greater upper wing-coverts, bastard wing (as far as not moulted in post-juvenile), often outer primaries (starting descendently somewhere in middle, mainly with p4 or p5), but occasionally all flight-feathers or none at all, and occasionally some greater upper primary coverts (RMNH, ZMA). In sample from North Carolina (USA), 5·06 (0·6; 37) 4–7 outer primaries new in spring (Blake 1969); rarely, some primaries already new or moulting in autumn (Johnston 1967a). For moult of outer primaries, see also Johnston and Downer (1968) and Rohwer (1986). FIRST PRE-BREEDING. Partial; involves 50–95% of feathering of head and body (occasionally, down to 20% or up to 100%), often all lesser upper wing-coverts and some or all median coverts, occasionally some tertials or greater coverts (RMNH, ZMA). Mainly completed before spring migration (occasionally already by early March), but some scattered feathers of (especially) head and neck frequently still growing just after arrival in breeding area (Blake 1969; RMNH, ZMA).

Measurements. Whole geographical range, all year; skins (RMNH, ZMA). Juvenile wing includes wing of 1st-year birds with new outer primaries; adult tail includes tail of 1st-year birds with new tail; bill (S) to skull, bill (N) to distal corner of nostril; exposed culmen on average 2·3 less than bill (S).

WING AD	♂	69·2 (1·42; 28)	67–72	♀	66·5 (0·71; 5)	65–68
JUV		68·8 (1·34; 29)	66–71		64·6 (2·05; 8)	62–68
TAIL AD		50·6 (1·65; 27)	48–53		47·7 (1·20; 10)	45–49
JUV		49·0 (1·90; 5)	46–51		47·2 (1·60; 4)	46–49
BILL (S)		13·3 (0·51; 56)	12·4–14·1		13·0 (0·51; 12)	12·4–13·8
BILL (N)		7·9 (0·43; 34)	7·3–8·5		7·6 (0·31; 12)	7·2–8·1
TARSUS		17·8 (0·44; 18)	17·2–18·7		17·5 (0·44; 11)	17·0–18·3

Sex differences significant for wing, adult tail, and bill (N).

For wing of migrants in Florida (USA), see Johnston (1967a).

Weights. (1) Eastern USA, breeding season (Payne 1991). Pennsylvania (USA): July–August, (2) adult, (3) juvenile and 1st adult; September, (4) adult, (5) juvenile and 1st adult; October, (6) adult, (7) 1st adult; (8) May–June (Clench and Leberman 1978). (9) Kansas and Georgia (USA), early October (Tordoff and Mengel 1956; Johnston and Haines 1957). (10) East Ship

Island (off south-east Mississippi, USA), spring (Kuenzi *et al.* 1991, which see for details).

(1)	♂	15·0 (0·73; 1111) 12·5–17·5	♀	14·4 (0·96; 398) 11·9–18·5
(2)		14·8 (0·84; 76) 12·8–16·8		13·9 (1·28; 89) 11·2–18·0
(3)		14·3 (1·15; 32) 12·0–16·5		13·8 (0·96; 34) 12·2–16·7
(4)		16·0 (1·76; 77) 13·3–21·4		14·8 (1·19; 39) 12·7–20·1
(5)		14·6 (1·05; 65) 12·5–19·0		14·2 (1·26; 101) 11·7–20·0
(6)		16·5 (1·38; 14) 14·4–18·5		15·7 (1·68; 13) 13·7–18·5
(7)		15·2 (1·48; 8) 13·2–17·8		15·9 (2·30; 16) 12·9–19·9
(8)		14·6 (0·96; 291) 12·3–20·7		13·9 (1·14; 198) 11·4–18·6
(9)		16·3 (— ; 23) 13·9–19·1		16·0 (— ; 16) 12·7–19·0
(10)		13·9 (1·7 ; 332) —		12·5 (1·6 ; 254) —

North Carolina (USA): April–May, 13·9 (19) 12·1–15·4; September–October, 14·4 (34) 12·6–18·9 (Blake 1969).

Belize: late August to mid-November, 13·1 (1·3; 31) (Mills and Rogers 1990); February, ♂ 12·7, ♀ 13·1; April, ♂ 17·7 (Russell 1964); see also Rogers and Odum (1966) for spring data and fat-free weight. Bonaire (Netherlands Antilles), January: ♀ 14·8 (ZMA). Panama, March–April: 14·4 (1·3; 23) (Leck 1975). For many data Florida and Jamaica, see Johnston and Downer (1968). For fat-free weights, see Connell *et al.* (1960) and Odum *et al.* (1961).

Structure. Wing rather short, broad at base, tip bluntly pointed. 10 primaries: p7–p8 longest, p9 1·5–3 shorter, p6 0–1, p5 3–6, p4 6–10, p3 8–12, p2 10–13, p1 12–16. P10 strongly reduced, a narrow pin-like feather laying along outer base of p9; 45–53 shorter than p7–p8, 4–8 shorter than longest upper primary covert. Outer web of p6–p8 emarginated, inner of p7–p9 with notch. Tip of longest tertial reaches tip of p3–p4. Tail rather short, tip square or slightly forked; 12 feathers. Bill short, straight, conical; exposed culmen *c.* 50% of length of head; culmen and cutting edges slightly decurved, cutting edges strongly angled downward at extreme base; lower mandible rather deep at base, but upper rather shallow, depth of bill (at feathering of forehead) 6·2–6·8 mm, width at base 6·2–7·2 mm. Nostrils small, rounded, broadly bordered by loral feathering at rear, some short soft bristles project obliquely downward from basal side of upper mandible. Tarsus and toes rather short and strong. Middle toe with claw 14·9 (5) 14–16 mm; outer toe with claw *c.* 70% of middle with claw, inner *c.* 66%, hind *c.* 77%.

Geographical variation. None in plumage. For variation in song, see Thompson (1970), Shiovitz and Thompson (1970), Emlen (1971), Payne (1983), and Payne *et al.* (1981, 1988).

Hybridizes extensively with Lazuli Bunting *P. amoena* where ranges overlap in Great Plains (Sibley and Short 1959; Short 1969; Kroodsma 1975), but as interbreeding is not random, both better considered full species, together comprising superspecies (Paynter 1970). For hybridization with Painted Bunting *P. ciris* of south-east USA and neighbouring north-east Mexico, see Taylor (1974). *P. ciris* and Varied Bunting *P. versicolor* from Mexico and border of southern USA form species-group with superspecies formed by *P. cyanea* and *P. amoena* (Mayr and Short 1970). CSR

Passerina amoena **Lazuli Bunting**

PLATE 29
[between pages 256 and 257]

Du. Lazuli-gors   Fr. Passerin azuré   Ge. Lazulifink
Ru. Лазурный овсянковый кардинал   Sp. Paserina mariposa   Sw. Lazulifink

Emberiza amoena Say, 1823

Monotypic

Field characters. 12–13·5 cm; wing-span 20–23 cm. Slightly smaller than Indigo Bunting *P. cyanea*. Small Nearctic bunting of similar form to *P. cyanea* but showing double white wing-bar in all plumages. ♂ pale blue above, with pale rufous and white underbody in breeding plumage. Winter ♂, ♀, and immature mainly dull buff-brown, often showing blue only on rump and tail; all lack any streaks above and below. Sexes dissimilar; marked seasonal variation. Juvenile separable. Hybridizes with *P. cyanea*.

ADULT MALE BREEDING. Moults: August–December (complete). Brightness of breeding plumage produced by wear. Hood down to mantle and breast and long rump bright turquoise-blue, divided by dusky blue, softly streaked back. Wings basically black, with turquoise-blue lesser coverts, white median coverts with rufous-tinged tips forming bold upper wing-bar, pale rufous tips to greater coverts forming less distinct, narrow lower wing-bar, dull rufous-brown fringes to tertials and secondaries, and bluish fringes to outer flight-feathers. Tail dusky blue. Chest-band and most of flanks pale rufous; belly and under tail-coverts white. Bill black to dark grey, with base of lower mandible paler blue-grey. Legs dusky-brown. ADULT MALE NON-BREEDING. From December to May, tawny fringes and tips to hood and upperparts conceal blue bases to feathers, initially almost completely. Rufous colour of tips to wing-coverts and tertials at brightest. ADULT FEMALE. Compared with ♂, even more nondescript than *P. cyanea* and similar to ♀ of that species in fresh plumage, when upperparts, fore-underparts, and fringes and tips of tertials and larger wing-coverts all have rufous tone. No distinct wing-bars in winter, but by spring wear and bleaching produce almost white double wing-bar and bluish tinge to rump and tail. At close range, face shows pale rufous to whitish eye-ring. Virtual lack of streaks above and below is helpful. FIRST-WINTER. Resembles adult ♀ but worn flight- and tail-feathers retained from juvenile plumage, while plumage lacks obvious blue. Bird showing blue on head, upperparts, and lesser wing-coverts likely to be ♂. For 1st-summer, see Plumages.

Breeding ♂ unmistakable. In other plumages, subject to same confusions as *P. cyanea* but separation made easier by whiteness of wing-bars in worn plumage and lack of streaks. Flight and behaviour as *P. cyanea*.

Call a sharp 'tsip'.

Habitat. In western temperate Nearctic lowlands and dry uplands, in similar environments to Indigo Bunting *P. cyanea*. Needs abundance of shrubs, low trees, and herbaceous vegetation, which in much of its range involves riverside situations (Johnsgard 1979). Found in thickets of willow, rose, and other scrub or tall *Artemisia*, and in shrub-grass stage following fire in chapparal or woodlands. Also occupies old forest clearings and coastal sagebrush woodland, but liable to leave sites which have grown up too tall or densely, and to colonize newly emerging ones. Forages in grass or other herbs and in low shrub layers (Pough 1957). In Washington State, hillsides and brushy gullies are especially favoured (Larrison and Sonnenberg 1968). Habitat includes dry bushy hillsides and wooded valleys, with aspen, willow, and alder, as well as wild rose thickets along mountains streams. Ascends to c. 2800 m in Rockies and to 3500 m in Californian Sierras. (Terres 1980.)

Distribution. Breeds in North America, from southern British Columbia east to southern Saskatchewan and north-east South Dakota, south to northern Baja California, southern California, central New Mexico, and central Texas. Winters in southern Arizona and Mexico south to Guerrero and Veracruz.

Accidental. Faeroes: ♂, Skálavík, 14–18 June 1981 (Bloch and Sørensen 1984).

Movements. Short-distance migrant; breeding range separated by less than 700 km from winter range. In autumn, some birds migrate without stopovers; others (perhaps most) interrupt migration for completion of moult, at stopovers where feeding at that season seems more favourable than in breeding range (Young 1991). Hybridizes with Indigo Bunting *P. cyanea* locally along eastern edge of breeding range, and some eastern records may be of undetected hybrids or intergrades.

Post-breeding groups start wandering in late July, and visit areas at higher altitude than those used for breeding (Bent 1968). Numbers in most breeding areas dwindle through August, and most birds have departed from north by mid-September (Jewett *et al.* 1953; Burleigh 1972; Cannings *et al.* 1987). Museum specimens show that adults in moult concentrate during August–September at 2 stopover sites, on border of southern Arizona and northern Mexico, and in southern Baja California (Mexico); juveniles linger on breeding grounds until September, and moult on aver-

age a month later, at same stopover sites as adults; birds move on to western Mexico October–November, after completing moult (Young 1991). Scattered sight records support that pattern, with better documentation in peripheral areas (Sutton 1967; Bent 1968; Oberholser 1974). Migration apparently on broad front (both seasons), but presumably avoiding forest and alpine areas in mountains (Bent 1968).

Spring migration begins in March; movement starts in west, reaching California in early April, and Arizona and Texas a week or two later (Bent 1968; Oberholser 1974; Garrett and Dunn 1981). Earliest birds reach north-west states by late April, with arrivals later inland and in mountains (Bent 1968; Burleigh 1972). Few reach Canadian breeding areas before late May (Munro and Cowan 1947; Sadler and Myres 1976; Cannings *et al.* 1987).

Vagrants have been detected north to Mackenzie District (61°N, in July) and northern Ontario (51°N, in May), as well as in eastern states (American Ornithologists' Union 1983; Godfrey 1986).   AJE

Voice. See Field Characters.

Plumages. ADULT MALE BREEDING. Head, entire upperparts, chin, throat, and side of neck bright cerulean- or turquoise-blue, feathers narrowly tipped rufous-brown when plumage fresh, narrowest on chin and forehead, widest on mantle and scapulars, especially on latter partly concealing blue. Lore dark grey or dull black. Chest and side of breast rufous-brown or tawny, sharply demarcated from blue of throat (especially in worn plumage) and from cream-white breast, belly, vent and under tail-coverts, gradually merging into pale cinnamon-rufous of flank. Thigh mixed grey, blue, and white. Tail black, outer webs fringed cerulean- or turquoise-blue, widest on central pair (t1), hardly so on outer pair (t6). Flight-feathers, greater upper primary coverts, and bastard wing black, outer webs fringed cerulean- or turqoise-blue (except for p9 and for emarginated parts of outer primaries), but terminal part of inner secondaries partly rufous-brown. Tertials and greater upper wing-coverts black, fringes along outer webs and tips rufous-brown or tawny, often indistinctly bordered by blue submarginally. Median upper wing-coverts white, tips suffused rufous-cinnamon when plumage fresh, showing as broad rufous, white, or mixed rufous-and-white wing-bar; lesser upper wing-coverts bright cerulean-blue. Under wing-coverts and axillaries blue, but longer feathers mainly or fully dark grey. *In worn plumage* (May–July), brown and tawny fringes of blue head and upperparts worn off (except sometimes for traces on mantle and back), blue more glistening, more turquoise, especially on face; rufous-brown or tawny of chest and of fringes of greater coverts and tertials paler pink-cinnamon, on coverts sometimes almost white and largely worn off; median coverts all-white, tips partly worn off, wing-bar narrower; white of underparts purer and more extensive (unless contaminated by dirt); blue fringes of tail and flight-feathers paler, sometimes light sky-blue or greyish-white. ADULT MALE NON-BREEDING. Like fresh adult ♂ breeding, but tawny fringes of head, upperparts, chin and throat slightly broader, especially blue of mantle and scapulars virtually concealed; chest, median upper wing-coverts, and fringes of greater coverts and tertials brighter and more saturated tawny-brown. ADULT FEMALE. In fresh plumage (autumn), upperparts saturated cinnamon-brown, sometimes slightly tinged buff on hindneck and olive on rump, some pale grey-blue on feather-centres sometimes visible, rump and upper tail-coverts often mixed cerulean-blue. Side of head and neck paler cinnamon-brown, lore, cheek, and front part of ear-coverts variegated pale buff-grey; narrow pink-cinnamon eye-ring. Chin and throat pale buff-brown, some light grey-buff of feather-bases sometimes visible; often a faint, short, dark grey malar stripe. Chest, side of breast, and flank saturated cinnamon-brown, merging into paler cinnamon-brown on side of belly and under tail-coverts and into cream-pink on mid-belly and vent; occasionally, some very faint grey streaks on chest or side of breast. Tail- and flight-feathers and greater upper primary coverts greyish-black, outer webs narrowly fringed pale grey-blue (faintly edged brown on t6 and p9). Tertials and greater upper wing-coverts black, faintly tinged grey-blue on outer webs, these as well as tips broadly fringed rufous-cinnamon. Lesser and median upper wing-coverts cinnamon-brown, centres of coverts with variable amount of grey-blue wash, median coverts broadly and contrastingly tipped rufous-cinnamon. Under wing-coverts and axillaries pale grey with cream-white tips. Closely similar to fresh adult ♀ Indigo Bunting *P. cyanea*, but colour of upperparts including upper wing-coverts more cinnamon or rufous, less olive, throat more cinnamon, less pale; chest, side of breast, and flank much warmer cinnamon, less buff, without dark streaks or with obsolete streaks only; fringes of tertials and tips of median and greater coverts broader, more deeply rufous, not strongly contrasting with upperparts and rest of coverts. *In worn plumage* (spring and early summer), markedly less rufous-cinnamon on upperparts, chin to chest, and upper wing; upperparts grey-brown (less greyish-olive than *P. cyanea* in worn plumage), rump and upper tail-coverts more distinctly pale grey- or green-blue, contrasting with remainder of upperparts; chin to chest and upper flank pale cinnamon to buff, virtually unstreaked, remainder of underparts buff to cream-white; wing-bars formed by tips of median and greater coverts bleached to pale cinnamon or off-white, sharply contrasting with remainder of wing (but colour and contrast depending on extent of pre-breeding moult, median coverts sometimes being new and rufous-tipped, and on individual variation in amount of bleaching). JUVENILE. Like juvenile *P. cyanea*, but upperparts more diluted cinnamon-buff, less brown-buff, and underparts warmer buff with less distinct dark marks. FIRST ADULT. 1st adult non-breeding like adult ♀, but rump, upper tail-coverts, lesser and median upper wing-coverts, and fringes of flight- and tail-feathers without blue or virtually so; tail, flight-feathers, and greater upper primary coverts still juvenile, more worn at tips than those of adult at same time of year, tips of tail-feathers slightly pointed (in adult, broadly rounded). Colour and contrast of wing-bars depends on extent of post-juvenile moult (in fresh plumage, median coverts often broadly rufous, greater narrowly cream). Sexes similar, but ♂ on average larger (see Measurements), and perhaps more often with traces of grey-blue on fringes of tail- and flight-feathers; from November, many ♂♂ attain some blue on head, upperparts, or lesser wing-coverts. In 1st breeding similar to adult, but outer primaries frequently contrastingly newer than inner ones (all equally new in adult), these fringed blue in ♂, pale grey-blue or green-blue in ♀ (old

inner ones and primary coverts contrastingly brown), tail either still juvenile (tail-tips pointed) or (often) all or partly new (when all new, relatively newer than in adult); in ♂, variable amount of brown retained among blue in upperparts and often some white on throat (unlike adult ♂).

Bare parts. ADULT, FIRST ADULT. Iris brown or dark brown. Upper mandible and gonys plumbeous-grey, slate-black, or black, cutting edges and rest of lower mandible pale grey-blue to light pinkish-grey. Leg and foot dark horn-brown, dark purplish-brown, or brown-black, soles dull slate-grey. Gape probably as in *P. cyanea*. (Ridgway 1901-11; RMNH, ZMA.) JUVENILE. No information.

Moults. Rather similar to *P. cyanea* in sequence, but some differences in timing and extent. Adult post-breeding involves only a limited amount of feathering on breeding grounds, most birds moving to stopover sites in south-central USA and north-west Mexico (including southern Baja California), where they have a complete post-breeding moult during August-December; on these stopover sites, immatures have a partial supplemental moult, involving head, body, outer 4-7 primaries, tertials, sometimes inner secondaries, and some or all tail-feathers; this moult follows an earlier partial post-juvenile moult from July to early August on breeding grounds; some birds of all ages delay moult entirely until reaching winter quarters. In contrast to *P. cyanea*, pre-breeding moult limited to a few feathers of head and throat (during February-April), bright breeding plumage mainly obtained by wearing off of brown feather-tips. (Young 1991.) See Rohwer *et al.* (1992) and Willoughby (1992) for interpretation of supplemental moult.

Measurements. ADULT, FIRST ADULT. Western USA (mainly California), May-September; skins (BMNH, RMNH, ZMA). Bill (S) to skull, bill (N) to distal corner of nostril; exposed culmen on average 2·2 less than bill (S).

WING	♂	73·6 (1·30; 20) 71-76	♀	70·4 (1·37; 21)	67-73
TAIL		53·6 (1·47; 10) 51-57		51·9 (2·40; 11)	48-55
BILL (S)		12·3 (0·53; 10) 11·8-13·0		12·6 (0·42; 9)	12·2-13·1
BILL (N)		7·1 (0·33; 10) 6·6-7·4		7·2 (0·45; 9)	6·6-8·0
TARSUS		17·1 (0·64; 10) 16·2-18·0		17·3 (0·39; 11)	16·6-17·8

Sex differences significant for wing. See also Sibley and Short (1959) and Kroodsma (1975).

Weights. Northern Great Plains (USA), summer: average 15·3 ($n=38$), 0·5 heavier than Indigo Bunting *P. cyanea* (Kroodsma 1975). Northern California (USA), June: ♂♂ 14, 16; ♀ 15 (Grinnell *et al.* 1930).

Structure. 10 primaries: p7-p8 longest, p9 2-3 shorter, p6 0-2, p5 3-6, p4 6-10, p3 9-13, p2 11-15, p1 13-18; p10 reduced, 47-52 shorter than p7-p8, 6-9 shorter than longest upper primary covert. Tip of longest tertial reaches to p2-p3. Middle toe with claw 15·7 (5) 15-16·5. Remainder of structure as in *P. cyanea*.

Geographical variation. None.

For relationships, see Indigo Bunting *P. cyanea*.   CSR

Passerina ciris (Linnaeus, 1758) **Painted Bunting**

Fr. Passerin nonpareil   Ge. Papstfink

A North American species, breeding in south-east USA and northern Mexico and wintering in Florida, Bahamas, Greater Antilles, and Central America. British records formerly thought to be of escapes, but 6 records (April–July) now under review for acceptance as wild birds (*Ibis* 1991, **133**, 222). 2 records from Norway are thought to be escapes (VR).

Family ICTERIDAE New World blackbirds, orioles, and allies

Rather small to fairly large 9-primaried oscine passerines (suborder Passeres), known variously also as 'grackles', 'troupials', 'oropendolas', 'caciques', 'hangnests', etc. Most species arboreal but some terrestrial. Diet varied; includes insects and seeds (and even small mammals, reptiles, and fish in case of certain larger species) but many frugivorous to greater or lesser extent, some also taking nectar. Except as vagrants, icterids occur only in New World, from Alaska to Tierra del Fuego but mainly in tropics; those breeding in North America mostly migratory. About 100 species in 23 genera of which 5 migrant North American species accidental in west Palearctic.

Sexes often markedly dissimilar in size, with ♂ the larger bird. Bill variable: conical and finch-like in some species, including cowbirds *Molothrus* and Bobolink *Dolichonyx oryzivorus*; quite slender and decurved in many others, including 'orioles' of genera *Quiscalus* and *Icterus*; straight and pointed in meadowlarks *Sturnella* and most 'blackbirds' *Euphagus*, *Agelaius*, and *Xanthocephalus*; long, heavy, and pointed in oropendolas and caciques *Cacicus* (etc.), with frontal casque and (in some) fine terminal hook. Bill used for forceful prying (open-billed probing; see

Sturnidae). Conspicuous rictal bristles lacking. Wing usually long and pointed; 9 primaries (p10 minute and concealed). Tail usually quite long, often rounded. Leg typically quite long and strong. Headscratching by indirect method only so far as known (see Simmons 1957*b*, 1961*a* for *Icterus* and *Psarocolius*). No detailed information on bathing or sunning behaviour; lateral sunning observed in Common Grackle *Q. quiscula* (Hauser 1957; Potter and Hauser 1974; Simmons 1986*a*). Direct (active) anting reported in at least 14 species (*c.* 5 in the wild) of genera *Dolichonyx* (1 species), *Molothrus* (1), *Quiscalus* (3), *Sturnella* (1), *Agelaius* (1), and *Icterus* (7) (Whitaker 1957; Simmons 1961*b*, 1966; Potter 1970).

Plumages of many species black, often with metallic gloss, or partially so, combined with yellow, orange, green, red, etc. Sexes usually dissimilar with ♂ brighter than ♀; no marked seasonal differences.

Close relationship to other American 9-primaries oscines confirmed by DNA evidence (Sibley and Ahlquist 1990); see Fringillidae and Emberizidae for further discussion.

Dolichonyx oryzivorus **Bobolink**

Du. Bobolink   Fr. Goglu des prés   Ge. Bobolink
Ru. Боболинк   Sp. Charlatán   Sw. Bobolink

Fringilla oryzivora Linnaeus, 1758

Monotypic

PLATE 30
[between pages 256 and 257]

Field characters. 17–19 cm; wing-span 26–32 cm (♂ up to 20% larger than ♀). Size and wing length close to Corn Bunting *Miliaria calandra*, but with more pointed bill, proportionately smaller head, narrower neck when extended, plumper body, and shorter rounded tail with narrow pointed ends to feathers. Distinctive Nearctic bird of

ground cover, with plumage of all but breeding ♂, and structure, suggesting strange weaver (Ploceidae), and also Quail *Coturnix coturnix*. Head and mantle of winter ♂, ♀, and immature dramatically striped black-brown and golden-buff. Breeding ♂ black with yellow nape and white scapulars and rump. Call distinctive. Sexes dissimilar in breeding season; marked seasonal variation in ♂. Juvenile separable at close range.

ADULT MALE BREEDING. Moults: June–August (complete), February–June (complete). When worn, head, mantle, wings, underparts, and tail black, with yellowish-white fringes on mantle, tertials, and inner greater coverts. Rear and side of neck bright pale yellow-buff; scapulars and long area from back to upper tail-coverts white. Bill deep-based but rather long and sharply pointed; black-horn with paler grey-blue lower mandible. Legs bright brown. ADULT MALE NON-BREEDING. After autumn moult, indistinguishable from ♀, except by size and sometimes obvious black on bases of flight-feathers and black speckles on fore-underparts. Bill paler than in spring, grey to flesh-brown, with dusky culmen. ADULT FEMALE. Ground-colour of plumage yellow-buff above and buffish-white below. Head pattern recalls Aquatic Warbler *Acrocephalus paludicola*: long black-brown lateral crown-stripes isolate pale yellow median stripe and yellower supercilium; greyish or whitish lore and eye-ring; almost black rear eye-stripe. Blackish lines of streaks on back, broadest in centre and there delineating 2 pale yellowish lines forming pale braces or V-mark). Wings show contrast between black-brown feather-centres and pale buff fringes and tips, with overlap of secondaries forming pale panel; no obvious wing-bars. Along breast-sides, flanks, and vent, 2–3 rows of dusky spots form discontinuous lines. Bill pinkish-horn with dusky-grey culmen. FIRST-YEAR. In winter resembles adult ♀ but most may be aged by even buffier appearance, whiter tips to median coverts (bright enough to form jagged bar) and usually retained juvenile tertials which look darker and whiter-edged than adult. Bill as ♀ or even paler. See also Plumages.

Breeding ♂ unmistakable. Winter ♂, ♀, and 1st-winter distinctive but subject to confusion with other striped, buff passerines. Of these, greatest chance of mistake lies with escaped ♀ or immature weaver, since that family shares similar bill and tail shape and behaviour on ground. Of wild confusion species, needs to be separated from Rose-breasted Grosbeak *Pheucticus ludovicianus* (10% larger, with dark ear-coverts, fully streaked breast, pale patch at base of primaries, and long, square-ended tail) and Dickcissel *Spiza americana* (15% smaller, with no discrete black stripes on crown and face and diagnostic chestnut median and lesser wing-coverts). Flight quite rapid and direct, with action recalling *M. calandra* but silhouette lacking as long and full a tail. Hops, also jumps, and climbs quite nimbly on ground plants. Habitually raises head to look around, then recalling lark or quail. Gregarious in Nearctic, but vagrants do not readily join flocks of other seed-eating passerines. Usually found in tall, seed-bearing ground cover; will perch on bushes.

Commonest call distinctive, a soft but metallic 'pink', 'chink', or 'pint'.

Habitat. Breeds in temperate Nearctic open lowland country, originally in valley grasslands between forests and on coastal marshes, but, since widespread clearance for farming, has extensively colonized fields of grain, hayfields, and meadows (Pough 1949; Forbush and May 1955). On Great Plains, inhabits tall-grass prairies, ungrazed to lightly grazed mixed-grass prairies, wet meadows, hayfields, and retired croplands (Johnsgard 1979). Further west, in Washington State, prefers low-lying moist meadows of considerable size with some water nearby; fields of hay, alfalfa, or wild grass are sought after (Larrison and Sonnenberg 1968). In autumn, shifts to river and coastal marshes and grain fields, and in south to ricefields.

Distribution. Breeds in North America from south-east British Columbia east to Nova Scotia, south to north-east California, northern Utah, north-east Kansas, central Illinois, West Virginia, and New Jersey. Winters in South America, south to Bolivia and northern Argentina.

Accidental. Britain and Ireland (see Movements), France, Norway, Italy, Gibraltar.

Movements. Long-distance migrant, breeding in temperate North America and wintering beyond equator in South America.

Post-breeding birds begin flocking in late July; they appear then in local non-breeding areas, and there is some southward drift along Atlantic coast during July (Stewart and Robbins 1958). Migration becomes general after mid-August, when thousands per day reported locally on passage (Trautman 1940; Bailey 1955; Bull 1974). Most birds have left Canada by early September, and northern USA by late September (Bull 1974; Squires 1976; Knapton 1979; Janssen 1987). Only stragglers remain in USA after mid-October (Burleigh 1958; Imhof 1976). Migrants cross West Indies and Caribbean Sea mostly September–October, passage in South America continuing through November (Bent 1958; Brudenell-Bruce 1975; Ridgely 1981; Hilty and Brown 1986). Main route for all populations is via south-east states (notably Florida), whence migration crosses Gulf of Mexico and West Indies (chiefly east of Mexico, west of Puerto Rico), some birds probably flying direct to South America. Passage within South America mainly east of Andes, but poorly known. (Bent 1958; Peterson and Chalif 1973; Raffaele 1989; Ridgely and Tudor 1989.) Regular (chiefly autumn) in Galapagos (Harris 1974).

Spring migration in South America begins late February, with first arrivals in Colombia and West Indies in early March (Bond 1985; Hilty and Brown 1986). During April, migration sweeps swiftly northward; a few early flocks, comprising only conspicuous ♂♂, reach southern Canada before May (Sprague and Weir 1984; Tufts 1986; Cannings *et al.* 1987). Main passage in south-east states from mid-April (Imhof 1976; Kale and Maehr 1990), but reaches New England and Great Lakes by early May (Bailey 1955; Sprague and Weir 1984; Janssen 1987). Northward advance is slower further west, and southern prairies are not occupied until May, with birds reaching Manitoba and Alberta after mid-May (Sadler and Myres 1976; Knapton 1979). Most have passed northward everywhere by late May, although stragglers remain in many southern states into June (e.g. Oberholser 1974, Imhof 1976). The sparse movement west of Rockies is mostly in May (Bailey and Niedrach 1967; Garrett and Dunn 1981; Alcorn 1988). Spring route is reverse of autumn, main passage going from Colombia to Florida and radiating thence (Stevenson 1957; Bent 1958).

Vagrant northward to arctic Alaska, Hudson Bay and southern Labrador, also to Greenland (American Ornithologists' Union 1983; Godfrey 1986).

Rare autumn vagrant to Atlantic seaboard of west Palearctic. 14 records from Britain and Ireland up to 1990, September–October, of which 8 from Isles of Scilly (Dymond *et al.* 1989; Rogers *et al.* 1991). The only spring record is from Gibraltar: ♂, 11–16 May 1984 (Holliday 1990).
AJE, PRC

Voice. See Field Characters.

Plumages. ADULT MALE BREEDING. In fairly worn plumage (about June), entire head black, bordered by contrasting buff-yellow band across hindneck and side of neck. Mantle black, some feathers at centre with pale buff or cream-white fringe at side. Scapulars, back, rump, and upper tail-coverts white, tips of feathers sometimes washed pale buff, concealed centres of feathers black, showing through as grey wash on (especially) inner scapulars and back. Sides of breast partly white (largely hidden below wing-bend at rest), remainder of underparts black, feathers of lower flank, belly, vent, thigh, and under tail-coverts with traces of pale buff fringe along tip. Tail black, sides of feathers with traces of off-white edge, tip with dull grey fringe on inner web. Flight-feathers, tertials, and all upper wing-coverts black, distal part of inner web of outer primaries grey, outer webs of outer primaries, outer greater primary coverts, and longest feather of bastard wing narrowly edged cream or white, outer webs and tips of tertials more broadly fringed buff-grey; some traces of off-white fringes sometimes visible on other wing feathers. Under wing-coverts and axillaries black. *In fresh plumage* (April and early May), black of feathers of head narrowly fringed buff on tips (browner on each side of crown, sometimes giving hint of crown-stripes), on underparts broadly so, black on (especially) chest, vent, and thigh sometimes largely concealed; band on nape warmer ochre or rufous-buff, feather-tips dusky grey; mantle-feathers broadly fringed rufous-buff; scapulars, back, rump, and upper tail-coverts cream with broad buff-brown feather-tips, largely concealing cream; tail- and flight-feathers narrowly but distinctly edged pale buff; tertials with broad buff-grey fringes; tips of median and greater upper wing-coverts with distinct pale buff fringe, showing as 2 narrow wing-bars. *When heavily worn* (July), band on nape bleached to pale cream; head, mantle, underparts, tail, and wing (except tertials) uniform black; white tips of inner scapulars, back, and rump largely worn off, scapulars streaked black, grey, and cream, back and rump virtually completely grey. ADULT FEMALE BREEDING. In fresh plumage (April–May), stripe along middle of crown

pale buff, narrowing towards forehead, marked with short black streaks; this pale median crown-stripe bordered at each side by broad black stripe at side of forehead and crown, these in turn bordered laterally by broad and distinct pale buff supercilium. Eye-ring, lore, and feathering on front of upper cheek off-white, rear of upper cheek and all ear-coverts greyish-buff with narrow off-white shaft-streaks. A narrow but distinct black stripe behind eye separates buff of supercilium from greyish-buff of ear-coverts. Hindneck and side of neck warm buff or olive-buff, hindneck with short black streaks. Mantle and scapulars broadly streaked black and warm buff, fringes of feathers of outer mantle and of inner scapulars partly pale buff or cream, showing as double pale V-mark. Back, rump, and upper tail-coverts pale greyish-olive or buff-olive, feathers with black marks on centre. Ground-colour of lower cheek, side of breast, and chest warm-buff, gradually merging into white on remainder of underparts; all buff and white slightly suffused pale yellow; side of breast blotched black, chest often with narrow black marks, feathers of flank, thigh, and under tail-coverts with black streak on centre, those of upper flank and tail-coverts ending in point, those of lower flank into a black blob. Tail dull olive-grey; central feathers (t1–t2) with contrastingly black central stripe, t3–t4 with black wash on centre. Flight-feathers, greater upper primary coverts, and bastard wing black-brown on outer webs, dark grey-brown on inner webs; fringes olive-grey to pale cream along outer webs (faint on primary coverts), pale grey along tip. Tertials and remainder of upper wing-coverts black, fringes along outer webs of median and greater coverts pale cream, those along inner webs of these and along lesser coverts and tertials warm buff or buffish-olive. Under wing-coverts and axillaries yellow-white, longer feathers pale grey at base, smaller coverts along leading edge of wing spotted black. *In worn plumage* (June–July), buff and olive fringes on upperparts duller, greyer, black of lateral crown-stripes and on mantle, scapulars, and back more prominent; buff of sides of head and neck and of all underparts bleached, chin and belly more extensively white, but entire underparts usually contaminated grey by dirt; fringes of tail- and flight-feathers, tertials, and upper wing-coverts bleached to dirty pale grey, partly worn off. ADULT MALE NON-BREEDING. Rather like adult ♀ breeding; plumage fresh in autumn, worn by mid-winter. Best distinguished by greater size (see Measurements); also, part of flight-feathers often with extensive though variable black wash on bases, and some feathers of (in particular) chin, cheek, or chest frequently partly or fully black. ADULT FEMALE NON-BREEDING. Like adult ♀ breeding; plumage fresh in autumn, worn by mid-winter. JUVENILE. Upperparts and upper wing dull brown-black (less black than in adult ♀), median crown-stripe, supercilium, band on hindneck, and fringes of feathers of mantle to upper tail-coverts and of upper wing buff, darkest on hindneck. Side of head and neck and entire underparts rich buff, paler on chin, with dusky brown stripe behind eye and grey-brown spots on side of throat; no distinct streaks on flank and under tail-coverts. Tail, flight-feathers, primary coverts, and bastard wing grey-brown, paler than in adult ♀, fringed greyish-white. FIRST ADULT MALE AND FEMALE NON-BREEDING. Rather like adult ♀ non-breeding, but juvenile tail, flight-feathers, greater upper primary coverts, bastard wing, and a variable number of tertials retained; tail-feathers sharply pointed, as in adult, but both webs of each feather gradually tapering to sharp tip (in adult, inner web with marked bulge subterminally, especially on outer feathers; see also Pyle *et al.* 1987); tips of greater primary coverts and feathers of bastard wing with narrow and even contrasting off-white fringe (more buffish and less even in width in adult). Also, ground-colour of fresh feathering of head and body more yellowish, on upperparts mainly yellowish-buff (in adult, more cinnamon-buff and olive), underparts strongly tinged pale yellow, less warm buff and white; chest usually without small dark spots; black streaks on lower flank ending in a point (in adult, in a rounded blob). Sexes similar, but ♂ larger (see Measurements), and some ♂♂ develop a few partly or fully black feathers on cheek, throat, or chest in course of autumn and winter. First Adult Male Breeding. Like adult ♂ breeding, but some birds retain variable number of worn brown non-breeding secondaries, primary coverts, or (occasionally) some feathers elsewhere (see First Pre-breeding Moult); tips of primaries sometimes more extensively grey-brown than adult, tail less extensively black; buff and brown fringes on head and body on average broader when plumage fresh. First Adult Female Breeding. Generally indistinguishable from adult ♀ breeding, but birds which show some old and worn secondaries or greater upper primary coverts contrasting with newer ones are probably 1st-years.

Bare parts. Adult, First Adult. Iris brown to black-brown; sepia-brown in some 1st-autumn birds. Upper mandible dark plumbeous-grey to slate-black in adult ♂ breeding, lower mandible dark horn-grey to slate-blue, sometimes with flesh tinge, gape black; bill turns paler from July, during post-breeding moult, after which it is grey, purplish-grey, or flesh-brown with dark grey or horn-black culmen ridge. In 1st non-breeding and in ♀, culmen and tip of upper mandible dark horn-brown to slate-black, remainder of upper mandible reddish-horn to pale flesh-grey, entire lower mandible pink-white, flesh-pink, or pale horn, gape pink or flesh-pink. Leg and foot pale brown, pink-brown, or light horn-brown, sometimes with flesh, bluish, or grey tinge, especially on toes. Juvenile. Iris grey-brown or sepia-brown. Bill pale flesh-pink, upper mandible with slate-black tip and brown base. Leg and foot light brown or flesh-horn. (Dwight 1900; Parslow and Carter 1965; RMNH, ZMA.)

Moults. Adult Post-breeding. Complete or virtually so; primaries descendent. In breeding area; starts with p1 late June (some ♂♂) to mid-August (some ♀♀); some ♂♂ not yet in flight-feather moult mid-July, but body moult just started; moult completed at onset of autumn migration, but occasionally some flight-feathers still old in migrants from October, with moult suspended (RMNH, ZMA). Adult Pre-breeding. Complete; February–June (Pyle et al. 1987). Only a very few from winter quarters examined; single ♂♂ from December and January just started moult on body (some black feathers growing), wing and tail still in fresh non-breeding plumage (RMNH, ZMA). Post-juvenile. Usually partial: head, body, lesser, median, and greater upper wing-coverts (occasionally excluding outermost greater), and a variable number of tertials. Starts shortly after fledging, completed or virtually so upon start of autumn migration. Moult advanced in 4 birds from end of July (Dwight 1900); partial body moult generally completed August (RMNH, ZMA), but 2 exceptional birds from New York, early September, had body and outer primaries in moult and remainder of flight-feathers apparently new (Parkes 1952). First Pre-breeding. Complete or almost so. Unknown proportion of birds arrive on breeding grounds with moult completed (these indistinguishable from older birds), others retaining part of juvenile greater upper prim-

ary coverts and secondaries; some of latter also retain scattered non-breeding feathers on body (e.g. 30% of underparts in one June ♂), and birds with flight-feathers and primary coverts new but body feathering partly in worn non-breeding are probably also 1st-years. (RMNH, ZMA.)

Measurements. Adult, First Adult. Eastern North America, May–August, Caribbean islands, September–November and May, and a few from South America, winter; skins (RMNH, ZMA). Bill (S) to skull, bill (N) to distal corner of nostril; exposed culmen on average 1·8 less than bill (S).

WING	♂	98·6 (2·59; 24)	94–103	♀ 89·5 (2·14; 13)	86–95
TAIL		64·7 (2·52; 22)	60–69	60·0 (2·17; 13)	57–64
BILL (S)		16·4 (0·57; 18)	15·5–17·2	15·8 (0·50; 11)	15·0–16·4
BILL (N)		9·9 (0·43; 18)	9·3–10·4	9·2 (0·39; 11)	8·6–9·7
TARSUS		27·8 (0·74; 19)	26·9–29·3	26·1 (1·04; 11)	24·9–27·6

Sex differences significant.

Weights. Eastern USA, spring: ♂ 50·7 (5) 48·2–51·7, ♀ 39·8 (14) 34·2–45·6 (Nice 1938). Kentucky (USA), May: ♂♂ 33·8, 36·1 (Mengel 1965). Coastal New Jersey (USA), autumn: 28·5 (10) 20·4–34·5 (Murray and Jehl 1964). Illinois (USA), September: ♂ 50·7 (10), ♀ 39·9 (7) (Graber and Graber 1962). Kentucky and Kansas (USA), September and early October: ♂ 49·5, ♀ 39·0 (3·81; 7) 31·6–42·9 (Tordoff and Mengel 1956; Mengel 1965).

Surinam, October: ♀ 27·1 (RMNH). Aruba and Bonaire (off Venezuela), October and May: ♂♂ 23·3, 23·9, ♀ 16·2 (all more or less exhausted) (ZMA). Belize, early June; ♀ 31·3 (Russell 1964). For fat-free weights, see Connell et al. (1960) and Odum et al. (1961), for fat-content Caldwell et al. (1963).

Isles of Scilly (England), September: probable ♂, 39·5 (Parslow and Carter 1965). Lucca (Italy), September: 32·2 (Massi et al. 1991).

Structure. Wing rather long, broad at base, tip pointed. 10 primaries: p9 longest, p8 0–1 shorter, p7 2–6, p6 7–10 (♀) or 9–12 (♂), p5 12–14 (♀) or 13–19 (♂), p1 26–30 (♀) or 28–34 (♂). P10 strongly reduced, a tiny pin at outer base of p9; 57–60 (♀) or 64–70 (♂) shorter than p9, 8–12 shorter than longest upper primary covert. Outer web of p7–p8 emarginated, inner of (p7–) p8–p9 with indistinct notch. Tip of longest tertial reaches tip of p1–p4. Tail rather long, tip slightly rounded; 12 feathers, each with sharply pointed tip (most markedly so in juvenile), t6 3–10 shorter than t1. Bill short, straight, conical; visible part of culmen c. 70% of head length; depth at base 8·5–10·5 mm, width at base 8·0–10·0 mm. Culmen slightly decurved or almost straight, with slightly elevated broad rounded ridge at base in cross-section. Cutting edges straight, those of upper mandible sometimes with faint blunt tooth in middle, edges strongly sloped downwards at extreme base. Bill-tip sharply pointed. Nostril rather small, partly covered by small operculum at proximal upper corner. Bristles at base of mandibles soft, short, indistinct. Tarsus rather long, toes long; fairly slender. Middle toe with claw 26·7 (10) 25–28; outer toe with claw c. 66% of middle with claw, inner c. 69%, hind c. 83%.

Geographical variation. None.   CSR

Molothrus ater Brown-headed Cowbird

PLATE 30
[between pages 256 and 257]

Du. Bruinkopkoevogel   Fr. Vacher à tête brune   Ge. Kuhstärling
Ru. Буроголовый коровий трупиал   Sp. Tordo-cuco canadiense   Sw. Brunhuvad kostare

Oriolus ater Boddaert, 1783

Polytypic. Nominate *ater* (Boddaert, 1783), eastern North America, west to central Ontario, eastern Minnesota, central Iowa, eastern Nebraska, north-west Kansas, south-east Colorado, and north-central Texas, south to south-central Louisiana, and central Alabama; *artemisiae* Grinnell, 1909, western North America, west of nominate *ater* (with which it intergrades), south to north-east and east-central California, southern Nevada, Utah, north-east and east-central Arizona, and western New Mexico; *obscurus* (Gmelin, 1789), southern USA and northern Mexico, south of nominate *ater* and *artemisiae*, from southern Louisiana west to north-west California.

Field characters. 17·5–21 cm; wing-span 29–35 cm. Nearly 10% larger than Starling *Sturnus vulgaris*, with deeper-based, rather short, noticeably conical bill, and often raised tail. Quite small but robust Nearctic icterid, with relatively dark and uniform plumage. ♂ dusky-black, with umber-brown hood extending to mantle and breast. ♀ and immature dusky mouse-brown, with paler throat. Sexes dissimilar; no seasonal variation. Juvenile separable.
ADULT MALE. Moults: July–October (complete). Head, neck, throat, and breast wholly dark umber-brown; rest of plumage dusky-black, faintly glossed silver-blue and green on wings and purple on mantle and breast. Bill and legs blackish-grey. ADULT FEMALE. Paler and more uniform in appearance than ♂. Plumage slightly mottled mouse-grey to grey-brown, with faint face pattern most noticeable in dark crown and cheeks, greyish chin and throat, and slightly paler edges to inner wing-feathers. FIRST-WINTER. Resembles adult but usually retains some juvenile feathers on wing; ♂ less glossed than adult. Bill sometimes brown.
Unmistakable, with finch-like bill, stocky, bustling character, and raised tail. Flight rapid, with silhouette and action recalling both *S. vulgaris* or thrush *Turdus* more than grackle *Quiscalus*, but with more abrupt undulations. Gait includes walk and run, varied by occasional leap; habitually carries tail up when on the move. Feeds on ground in open country, habitually associating with cattle, etc.
Commonest calls: 'chuck'; in flight, high whistled 'phee de de' or 'weee-titi' with accent on 1st note and then drop in pitch.

Habitat. In temperate to subtropical Nearctic, linked with prairies and pastures through ancient adaptation to feeding with bison *Bison*, and later cattle and horses, which stir up insect food. Range enlarged by modern forest clearance. (Pough 1949.) Habit of brood-parasitism has, however, led to attachment to woodland edges, thickets, and places where low or scattered trees are interspersed with grassland (Johnsgard 1979). Also found in pastures, meadows, willows *Salix*, orchards, and suburbs; mainly a lowland bird but at times ascends to *c.* 1200 m in mountains (Larrison and Sonnenberg 1968).

Distribution. Breeds in North America from north-east British Columbia and southern Mackenzie District east to central Quebec and Nova Scotia, south to Mexico. Winters in southern and eastern North America, north to north-central California, Texas, Michigan, southern Ontario, New York, and Massachusetts.
Accidental. Norway: ♀, Telemark, Jomfruland, 1 June 1987 (Bentz 1989).

Movements. Status varies from fully migratory (populations breeding north of 40°N in west—*artemisiae*, and north of 45°N in east—nominate *ater*), through partly migratory, to sedentary (most birds breeding south of 35°N, including most *obscurus* and southern populations of other races) (American Ornithologists' Union 1957, 1983; Bent 1958). Wintering north of snow-line has increased greatly since 1950, especially in north-east, and now involves flocks around grain elevators and where cattle fed (e.g. Tufts 1986); birds wintering in north comprise only small percentage of summer numbers, but tend to be overemphasized in regional summaries (A J Erskine).
Adults of this parasitic species have no parental ties, so they are found in wandering flocks from July onwards, and these are often reported as migrants. This, combined with wide breeding range, blurs detection of migration. Records suggest a gradual departure from northern breeding areas late August to mid-September, with few remaining in Canada and north-west USA by October (Jewett *et al.* 1953; Houston and Street 1959; Butler and Campbell 1987). From Great Lakes eastward movement is later, extending through October, with larger concentrations southward towards winter range (Palmer 1949; Bull 1974; Bohlen 1989). By mid-November, main migration is over, although movement continues later, especially in severe winters. Ringing recoveries show south-west heading from east of Great Lakes to Carolinas, Alabama, and Arkansas (Stewart and Robbins 1958; Bull 1974; Imhof 1976; James and Neal 1986). In Arkansas, recoveries show movements

north to Ontario, and south to Mexico, with many on coast of Texas and Louisiana (James and Neal 1986). Birds ringed on prairies (*artemisiae*) migrated south from Alberta, Saskatchewan, and Dakotas to Texas and Mexico (Bray *et al.* 1974). Long-distance recoveries of birds ringed in northern California August–December were almost entirely north of study-site, west of Cascade mountains and north to southern British Columbia; other recoveries showed altitudinal movements (Crase *et al.* 1972).

Spring migration becomes general from early March, though sometimes reported in February (Stewart and Robbins 1958; Fawks and Petersen 1961; Mengel 1965). By late March or early April, first flocks (all ♂♂) are in eastern Canada and Great Lakes states (Squires 1976; Sprague and Weir 1984; Janssen 1987). Many reach eastern breeding areas during April, but flocks are still evident in northern Ontario after mid-May (Todd 1963; A J Erskine). Movement on prairies and inter-mountain plateaux in west is later, reaching Canada after mid-April, with full numbers only in May (Burleigh 1972; Sadler and Myres 1976; Cannings *et al.* 1987), though birds reach coastal British Columbia from late March (Butler and Campbell 1987).

Vagrant to western and northern Alaska (mostly late summer), Bathurst Island (75°N), and southern Labrador and Cuba (Kessel and Gibson 1978; American Ornithologists' Union 1983; Godfrey 1986).   AJE

Voice. See Field Characters.

Plumages. (nominate *ater*). ADULT MALE. Head, neck, and chin to chest fuscous-brown, feather-tips sometimes faintly glossed purple when plumage fresh. Remainder of body deep black, glossed blue-green or green (depending on light and abrasion), more purplish on mantle and breast, purple sometimes rather poorly demarcated from fuscous-brown of neck and chest. Vent and thigh virtually without gloss, sooty-black. Tail, flight-feathers, and all upper wing-coverts black with green gloss, like body; outer webs of primaries often slightly duller, inner webs more greyish-black or sooty. Under wing-coverts and axillaries deep black, feather-tips glossed blue-green or purple-blue. *In worn plumage* (late spring and early summer), head and neck slightly less dark fuscous-brown, more drab, some pale grey of feather-bases sometimes partly visible; gloss on remainder of body and on tail and wing slightly less intense, more oil-green; some sooty-black of feather-centres of breast and belly exposed; primaries and greater upper primary coverts virtually without gloss, sooty-black. ADULT FEMALE. Upperparts dark brown-grey, feathers with slightly paler drab-grey fringes; ill-defined streaks on feather-centres of mantle, scapulars, and sometimes cap sooty-black, sometimes with slight green or purple gloss. Side of head and neck, side of breast, chest, belly, vent, and under tail-coverts paler brown-grey, fawn, or drab-grey, flanks slightly darker drab-grey, all feathers on underparts with narrow dark brown-grey shaft-streaks; lore and cheek with pale grey or off-white mottling, ear-coverts with fine pale grey or off-white shaft-streaks. Chin and throat pale grey bordered at side by rather ill-defined brown-grey malar stripe. Tail black-brown, faintly glossed green, fringed drab-grey. Flight-feathers, greater upper primary and secondary coverts, tertials, and bastard wing black-brown; outer webs of primaries narrowly and sharply fringed pale grey to white, of others rather ill-defined drab-grey (paler towards tips on secondaries and greater coverts). Lesser and median upper wing-coverts like scapulars. Under wing-coverts and axillaries dark drab-grey with small paler grey tips. *In worn plumage*, entire plumage darker grey-brown, especially on cap, mantle, scapulars, and from chest backwards, mottled paler grey due to abraded feather-tips; chin and throat whiter, dark malar stripe more distinct; fringes of flight-feathers, tertials, and greater coverts paler, but partly worn off. JUVENILE. Upperparts dark brown-grey, each feather with pale buff fringe, appearing scalloped; centres of feathers of hindneck, rump, and upper tail-coverts slightly browner than more blackish ones of cap, mantle, scapulars, and back. Side of head and neck mottled buff and dark grey, upper cheek purer buff, lower cheek with faint dusky malar stripe. Chin buff-white; remainder of underparts pale buff or cream-white, chest, side of breast, and upper flank with large dark grey mark on centre of feathers (appearing spotted), breast and lower flank with dark streaks or arrow-marks, showing as short streaks; dark marks on centres of feathers of vent and on under tail-coverts mainly concealed. Tail as adult ♀, but feather-tips more pointed; flight-feathers and greater upper primary coverts as adult ♀, but slightly browner. Tertials and all upper wing-coverts black-brown with sharply contrasting buff fringes, latter slightly wider and paler cream or white along tips of median and greater coverts, forming 2 indistinct wing-bars. In worn plumage, fringes of upperparts and upperwing bleached to off-white, partly worn off (especially on nape, upper mantle, coverts, and tertials). FIRST ADULT MALE. Like adult ♂, but gloss on average slightly less intense, and usually some juvenile feathering retained, especially among under wing-coverts and tertials (see Post-juvenile in Moults), dull grey-brown, markedly contrasting with glossy black of remainder of feathering and with new black-fringed glossy grey greater under wing-coverts. FIRST ADULT FEMALE. Like adult ♀, but some juvenile feathering retained, as in 1st adult ♂. Contrast between paler drab-grey and more worn juvenile feathering and new darker fuscous-drab of remainder of feathering often difficult to see, especially in spring and summer, when entire plumage worn.

Bare parts. ADULT, FIRST ADULT. Iris brown to black-brown. Bill slate-black or brownish-black, base and middle portion of lower mandible of ♀ slightly paler brown or grey. Leg and foot black, soles dark grey to dull black. JUVENILE. Iris brown. Bill horn-brown, darkest on culmen and tip, slightly tinged flesh at base. Leg and foot dark horn-brown to brown-black, soles dull flesh-colour. (RMNH, ZFMK, ZMA.)

Moults. ADULT POST-BREEDING. Complete; primaries descendent. On breeding grounds, July–October (Pyle *et al.* 1987). In a few ♀♀ examined, moult started with p1 mid- or late July, completed with p9 late September or early October (RMNH, ZFMK). ADULT PRE-BREEDING. Partial, mainly head and neck, March (Phillips *et al.* 1964). POST-JUVENILE. Complete or virtually so. In sample of 39 ♂♂ and 10 ♀♀ wintering in Texas (USA), only 1 ♂ had moult complete; 97% of ♂♂ and 100% of ♀♀ retained juvenile greater under primary coverts, 92% of ♂♂ and 100% of ♀♀ retained juvenile greater under secondary coverts, 66% of ♂♂ and 80% of ♀♀ median under primary coverts, 37% of ♂♂ and 50% of ♀♀ median under secondary coverts, 82% of ♂♂ and 70% of ♀♀ some tertials, 42% of ♂♂ and 0% of ♀♀ some smaller inner under wing-coverts; a very few retained some juvenile feathers elsewhere on wing or body or (once) a single primary (Selander and Giller 1960). In Rhode Island, north-east USA, a greater amount of juvenile feathering is retained: most often some to many under wing-coverts (in 71% of many ♂♂ of all ages examined), in decreasing frequency also some scapulars, some feathers of sides of back and rump, feathers on mid-belly or round eye, some flight-feathers (especially s3-s5), or a few feathers elsewhere (Baird 1958). Likewise, in sample examined from eastern USA, many juvenile under wing-coverts retained, as well as some odd tertials, tertial coverts, a few scapulars, or feathers on belly (RMNH, ZMA). Timing of moult as in adult post-breeding; for sequence of replacement, see Friedmann (1929) and Baird (1958).

Measurements. Nominate *ater*. ADULT, FIRST ADULT. Eastern North America, mainly April–October; skins (RMNH, ZFMK, ZMA). Bill (S) to skull, bill (N) to distal corner of nostril; exposed culmen on average 2·0 less than bill (S).

	♂		♀	
WING	111·8 (2·48; 21)	108–116	99·8 (1·56; 16)	97–102
TAIL	72·7 (2·68; 15)	69–76	65·3 (1·35; 9)	63–68
BILL (S)	19·1 (0·71; 16)	18·1–20·2	17·0 (0·78; 9)	16·2–18·2
BILL (N)	11·8 (0·79; 15)	10·8–12·8	10·2 (0·55; 9)	9·6–11·0
TARSUS	26·7 (0·70; 16)	25·6–27·8	24·6 (0·55; 9)	23·8–25·5

Sex differences significant.

JUVENILE. Wing and tail on average 5·0 shorter than in adult.
M. a. artemisiae. For wing of migrants in Kansas (USA), see Hill (1976).

Weights. Nominate *ater*. Pennsylvania (USA): (1) March, (2) April–May (Clench and Leberman 1978).

		♂		♀	
(1)		50·3 (2·82; 15)	45·5–54·3	40·4 (2·18; 7)	37·9–43·9
(2)		48·9 (3·18; 733)	32·4–58·0	38·8 (2·93; 671)	30·5–51·2

Averages Ohio (USA): April–May, ♂ 46·3 (24), ♀ 38·1 (34); June–July, adult ♂ 46·4 (10), adult ♀ 39·6 (39), juvenile 37·0 (248); August–September, juvenile and 1st adult, 37·0 (34) (Baldwin and Kendeigh 1938). Pennsylvania: July–August, mainly juvenile, 34·4 (4·08; 18) 30·4–45·4; September–November, sexes and ages combined (mainly ♀), 38·3 (5·01; 12) 30·5–47·6 (Clench and Leberman 1978). Coastal New Jersey (USA),

autumn: 40·9 (4) 38·1–44·3 (Murray and Jehl 1964). On ship off Florida, March: ♂ 46 (ZMA).
For development of juvenile, see Friedmann (1929).

Structure. Wing rather long, broad at base, tip bluntly pointed. 10 primaries: p8–p9 longest, p7 0–2 shorter, p6 2–6 (♀) or 4–7 (♂), p5 8–13 (♀) or 12–15 (♂), p4 12–18 (♀) or 17–21 (♂), p1 25–29 (♀) or 30–35 (♂). P10 strongly reduced 67–70 (♀) or 75–80 (♂) shorter than p8–p9, 11–15 shorter than longest upper primary covert. Outer web of p6–p8 emarginated, inner of (p7-)p8–p9 with slight notch. Tip of longest tertial reaches to about tip of p3. Tail rather long, tip square or slightly rounded; 12 feathers, t6 0–6 mm shorter than t1. Bill short, straight, conical; depth at base 10–13 mm in nominate *ater*, 19·5–11·5 in *artemisiae*, 8·5–10·5 in *obscurus*, width at base 8–11 mm. Visible length of culmen *c.* 72% of length of head. Culmen slightly decurved in nominate *ater*, base slightly swollen, almost straight in *artemisiae* and *obscurus*; bill-tip sharp. Cutting edges, nostrils, and bristles as in Common Grackle *Quiscalus quiscula*. Tarsus and toes rather short and thick. Middle toe with claw 22·2 (10) 19·5–25·0; outer and inner toe with claw both *c.* 76% of middle with claw, hind *c.* 83%.

Geographical variation. Rather slight, mainly involving size. *M. a. artemisiae* of western North America on average larger than nominate *ater* from east (wing on average *c.* 3·8 mm longer, tail *c.* 1·8, tarsus *c.* 1·3), bill proportionally longer and more slender (on average, *c.* 0·5 mm longer, but *c.* 0·4 mm less deep at base). *M. a. obscurus* from southern USA (east to Louisiana) and northern Mexico distinctly smaller (wing on average *c.* 7·8 mm shorter than in nominate *ater*, tail *c.* 4·9, tarsus *c.* 2·2), bill even finer at base than in *artemisiae*, and gape-flanges of nestling differently coloured (Ridgway 1901–11; Rothstein 1978). For history of distribution of races and zones of intergradation, see Rothstein (1978), Rothstein *et al.* (1980), Fleischer and Rothstein (1980), and Fleischer *et al.* (1991).   CSR

Euphagus carolinus (Müller, 1776) **Rusty Blackbird**

FR. Quiscale rouilleux   GE. Roststärling

A North American species, breeding widely in Alaska and Canada, south to southern Canada and in USA to New York, Massachusetts, and Maine; winters from southern Alaska and southern Canada south to Gulf of Mexico coast. 2 recorded in Britain: one near Cardiff (Wales), 4 October 1881, was perhaps a genuine drift-migrant; another in St James's Park (London), July–August 1938, was probably an escape (Alexander and Fitter 1955).

Quiscalus quiscula **Common Grackle**

PLATE 30
[between pages 256 and 257]

DU. Glanstroepiaal   FR. Quiscale bronzé   GE. Purpurgrackel
RU. Обыкновенный гракл   SP. Zánate común   SW. Mindre båtstjärt

Gracula Quiscula Linnaeus, 1758

Polytypic. *Q. q. versicolor* (Vieillot, 1819), Canada and north-east and central USA, south to eastern Connecticut, central and western New York, western Pennsylvania and West Virginia, Kentucky, western Tennessee, north-west Mississippi, and western Louisiana; *stonei* Chapman, 1935, south of *versicolor*, from south-west Connecticut and south-east and central-south New York south-eastward through Virginia and northern Georgia and Alabama to central Louisiana; nominate *quiscula* (Linnaeus, 1758), south-east USA, south of *stonei*, from extreme south-east Virginia through eastern parts of North and South Carolina and southern parts of Alabama and Mississippi to south-east Louisiana, south to southern Florida.

Field characters. ♂ 27·5–34, ♀ 25–30 cm; wing-span ♂ 39–45, ♀ 36–41 cm. Structure of bill, head, and body somewhat suggest large starling *Sturnus* but with long wedge-shaped tail; Tristram's Grackle *Onychognathus tristramii* is the only west Palearctic species at all similar in form. Strange-looking, long-stepping, and long-tailed Nearctic icterid, with pale eye and all-dark plumage with astonishing iridescence. Sexes closely similar; no seasonal variation. Juvenile separable.

ADULT MALE. Moults: July–October (complete). Black, but with such intense gloss that head, mantle, and breast appear violet, bluish-green, or steel-blue, back, rump, and flanks look bronze or brass, inner wing bronze or reddish-purple, and flight-feathers and tail red, bronze, and violet (races differ: see Geographical Variation). Eye straw-yellow to yellow-white. Bill steely-black. Legs dusky-black. ADULT FEMALE. Smaller and slightly duller than ♂, with browner underparts and shorter tail. FIRST-YEAR. Both sexes resemble adult ♀. Some may show dull eye or retained dull brown feathers from juvenile plumage.

Unmistakable. Occurrence within range of *O. tristramii* virtually impossible; larger Boat-tailed Grackle *Q. major* and Great-tailed Grackle *Q. mexicanus* are not known to stray from North American ranges. Flight suggests strange thrush rather than starling: rapid and usually level, lacking rather abrupt undulations of other Nearctic relatives; ends in bouncing landing. Gait a long-striding walk (slower than jaunt of *S. vulgaris*) or loping run. Stance usually level on ground, with tail raised; more upright on perch. Occasionally flicks wings and jerks tail in excitement. References in field guides to tail having keel shape are unlikely to be relevant to observations of vagrants as this is only associated with ♂'s courtship display. Feeds on ground in open country or woodland.

Calls hoarse and grating, 'chuck', 'chack', and 'check'; also shrill 'cheer'.

Habitat. Breeds in Nearctic from cool to warm temperate regions, foraging on ground, especially where wet, in open places such as fields, pastures, lawns, golf courses, shores, marshes, or open wet woodlands. Nests in open woods, parks, groves, and shade trees and bushes, in Canada preferably in conifers (Godfrey 1979). Invades cities to roost at night in shade trees, feeding on lawns and nesting in parks. Also attracted to wet areas, feeding along water's edge or wading out from it. Sometimes nests in cavities in buildings. (Pough 1949.) Generally frequents woodland edges, residential areas, parks, shelterbelts, farms with planted trees, and tall shrub thickets (Johnsgard 1979).

Distribution. Breeds throughout North America east of Rocky Mountains, except extreme north, south to southern Texas and Florida. Winters in southern parts of breeding range.

Accidental. Denmark: Roskilde (Zealand), March–May 1970 (Schelde 1970).

Movements. Variable status; winters in southern half of breeding range. Most populations breeding north of 45°N (*versicolor*) largely migratory, moving 1000 km or more to winter range, with only small proportion wintering north to Canada–USA border. Mid-latitude populations (*versicolor* and *stonei*) partly migratory, and south of 35°N most birds (of all 3 races) are sedentary. (Bent 1958; American Ornithologists' Union 1957, 1983.)

After breeding, birds assemble in progressively larger flocks, which move between feeding areas and roosts for up to 3 months before general southward migration is evident. Thus, large areas are vacated from July, which may suggest early migration, especially when birds are recovered south of ringing-site then (e.g. in Dakota study: Bray *et al.* 1973). On broader scale, migration begins only in mid- to late September in southern Canada (Sadler and Myres 1976; Speirs 1985; Tufts 1986), and becomes general early in October, with widespread passage reported across east by late October (Bailey 1955; Bull 1974; Janssen 1987). Most birds are in winter range by late November (Bent 1958). Breeding densities decrease rapidly west of 100°W, even in Canada where breeding recorded to 120°W (Erskine 1971), and movements in west involve few birds. Migration often involves daytime movements of vast flocks, and is widespread from Atlantic coast to Great Plains (Bent 1958). Ringing recoveries show concentration southward, birds which bred at 65–115°W wintering mostly from Carolinas to eastern Texas, 75–95°W (Bray *et al.* 1973; Bull 1974; Imhof 1976; James and Neal 1986). Ringing in many areas shows winter site-fidelity, with recoveries up to 11 years later (e.g. James and Neal 1986).

Spring migration becomes general in March; February records partly involve individuals that wintered north of main range. Most birds have left winter roosts by early April, but local breeding occurs throughout winter range, so only general patterns are evident. Ringing confirms northward movement mainly March–April (e.g. Bray *et al.* 1973). Arrives in northern states and eastern Canada mid- to late March (Palmer 1949; Sprague and Weir 1984), and by late April or early May most birds are in breeding areas (Bent 1958; Houston and Street 1959).

Vagrant west to Pacific, from California to Alaska, and north to tundra, from Alaska to Hudson Bay and northern Newfoundland (American Ornithologists' Union 1983; Godfrey 1986). AJE

Voice. See Field Characters.

Plumages. (*Q. q. versicolor*). ADULT MALE. Head, neck, upper mantle, and chest strongly glossed violet-blue or green-blue, depending on angle of light; forehead, ear-coverts, and cheeks mostly violet, side of neck and chest mostly green-blue, but with much variation; some sooty-black of feather-centres sometimes just visible on throat. Remainder of body glossy bronze-green to brass-olive, sharply demarcated from blue of lower mantle and chest, appearing black-brown in dull light but strongly lustrous when exposed to sun; some sooty black of feather-bases just visible, vent mainly black, under tail-coverts glossed violet and bronze. Tail black with bronze-violet lustre. Flight-feathers, greater upper primary coverts, and bastard wing black, inner webs slightly grey, outer webs of primaries, greater coverts, and bastard wing slightly glossed violet, of secondaries strongly violet-bronze or coppery-red, depending on light; remainder of upper wing-coverts and tertials strongly greenish-bronze, violet-bronze, or coppery-red, depending on light and abrasion. Under wing-coverts and axillaries sooty black, shorter ones tipped glossy violet or green-blue. For individual variation in colour of gloss, see Snyder (1937). *In worn plumage* (late spring and early summer), gloss on head and neck more variable, between blue-green and reddish-purple (in part, depending on light); body slightly less strongly lustrous bronze, especially from breast downwards; tail and primaries duller black, gloss on remainder of upperwing more violet or coppery-red. ADULT FEMALE. Like adult ♂, but head to upper mantle and chest on average less intensely glossy, some more black-brown of feather-bases of chin and throat visible; bronze-green on remainder of body less lustrous, especially much black-brown of feather-bases exposed on underparts; gloss on upper wing less intense, upper wing-coverts greyish-bronze to violet. Best distinguished by size (see Measurements). JUVENILE. Entirely uniform dark drab-grey to sooty-grey, slightly paler on underparts, where sometimes with faint darker streaks (especially in ♀); slightly darker, blackish-grey or black-brown, round eye and on wing. FIRST ADULT MALE. Like adult ♂, but gloss often less intense, in some almost as in adult ♀ (but size larger). Usually differs from adult ♂ and ♀ by retention of some juvenile feathering, especially greater under wing-coverts, sometimes a variable number of tertials, and occasionally some feathers of body, these contrastingly greyer than neighbouring new black feathers; new tail-feathers on average slightly narrower and with more rounded feathers than in adult, less broad and truncate, more worn at same time of year. FIRST ADULT FEMALE. Like adult ♀, but part of juvenile plumage retained (as in 1st adult ♂), and less glossy throughout; much dark grey-brown of feather-bases visible on head and body, in particular rump, ear-coverts, chin, throat, and from flank and breast to under tail-coverts virtually uniform dark brown to sooty black, virtually or fully without gloss.

Bare parts. ADULT, FIRST ADULT. Iris straw-yellow to yellow-white. Bill dark purplish-black, greyish-black, or deep black. Leg and foot dusky purple-black or deep black; soles grey. JUVENILE. Iris dark brown once eye open, passing through brown-grey and grey to greyish-green in late summer; bare skin on head yellowish. Bill, leg, and foot sepia-brown, purple-brown, or dusky purple-black, soles purplish-grey or flesh-grey. (Dwight 1900; Ridgway 1901–11; Gillespie and Gillespie 1932; Bent 1958; RMNH.)

Moults. ADULT POST-BREEDING. Complete; primaries descendent. In breeding area, July–October (Wood 1945; Pyle *et al.* 1987). In 6 moulting birds from north-east USA examined, moult started with p1 early July to early August, completed late August to early October; all tail-feathers simultaneously growing in one from late September (primary moult score 35), (virtually) completed in 2 others from 8 August (score 34) and 4 October (score 42) (RMNH, ZMA). For sequence of moult, see Stone (1937) and Wood (1945). No pre-breeding moult. POST-JUVENILE. Complete or virtually so; some juvenile under wing-coverts and/or tertials usually retained. Mainly in breeding area, July–October (Laskey 1940; Pyle *et al.* 1987). Some moulting birds cited under adult post-breeding above perhaps in post-juvenile moult. A limited 1st pre-breeding moult of head and chest in at least some spring ♂♂.

Measurements. *Q. q. versicolor*. ADULT, FIRST ADULT. North-central and north-east USA and Ontario (Canada), April–October; skins (BMNH, RMNH, ZFMK, ZMA). Bill (S) to skull, bill (N) to distal corner of nostril; exposed culmen on average 3·0 less than bill (S).

	♂	♀
WING	144·3 (4·09; 37) 135–152	131·2 (4·13; 12) 123–137
TAIL	127·2 (5·44; 21) 118–136	105·4 (7·02; 12) 95–116
BILL (S)	33·4 (1·28; 22) 31·5–35·8	30·2 (1·45; 12) 28·3–32·5
BILL (N)	22·4 (1·13; 22) 20·7–24·3	19·6 (0·79; 12) 18·1–20·7
TARSUS	37·3 (1·12; 21) 35·8–39·2	34·5 (0·76; 12) 33·5–35·5

Sex differences significant. Ages combined above, though new wing and tail of 1st adult shorter than in older birds. Thus: ♂, 1st adult wing 142·2 (3·33; 7) 138–147, adult wing 147·1 (2·98; 15) 143–152; 1st adult tail 122·2 (3·83; 7) 118–125, adult tail 129·8 (4·29; 14) 122–136.

Weights. *Q. q. versicolor*. Pennsylvania (USA): (1) October, (2) April–July (Clench and Leberman 1978).

(1) AD ♂ 118·2 (— ; 3) 107·5–126·0 ♀ 96·2 (— ; 2) 94·9–97·6
 1ST AD 117·3 (10·37; 13) 90·4–136·4 93·4 (5·06; 14) 83·2–101·0
(2) 119·9 (8·80; 108) 90·4–144·8 97·1 (9·14; 39) 82·1–124·7

Michigan (USA): ♂ 120·2 (23) 103·7–132·8, ♀ 101·3 (9) 96·4–107·8 (Amadon 1944; see also Rand 1961a, b). Averages Ohio (USA), April–July: ♂ 117·7 (4), ♀ 98·9 (2) (Baldwin and Kendeigh 1938). Kentucky (USA): April, ♂ 115·1; September–October, ♀ 91·6 (3) 88·1–93·4 (Mengel 1965). Averages, Canada: ♂ 131·4 (99), ♀ 100·8 (105) (Snyder 1937).

Nominate *quiscula*. Georgia and South Carolina (USA), July: ♂ 107, ♀ 76·6 (Norris and Johnston 1958). Florida (USA), late March and April: ♂ 108·1 (8) 97–114, ♀ 79·8 (6) 74–86 (Amadon 1944).

Structure. Wing rather short, broad at base, tip bluntly pointed. 10 primaries: p7 longest, p6 and p8 0–3 shorter, p9 3–9, p5 4–10, p4 12–20, p3 19–26, p2 23–30, p1 28–36. P10 reduced, a tiny pin, 88–105 shorter than p7, 13–20 shorter than longest upper primary covert, 0–3 shorter than reduced outermost upper primary covert, beneath which it is concealed. Outer web of (p5–) p6–p8 emarginated, inner web of p6–p9 with rather faint notch. Tip of longest tertial reaches to tip of p1–p2. Tail rather long, tip rounded or graduated; tail can be folded vertically, showing V-shape in transverse section; 12 feathers, each gradually broader towards bluntly rounded or truncate tip, t6 15–35 mm shorter than t1. Bill long, visible part of culmen *c.* 90% of head-length, thus almost equal; 11–14·5 mm deep at base, 10–12·5 mm wide; culmen with bluntly rounded ridge in transverse section, almost straight over entire length, but slightly elevated at base and gently decurved at tip, ending in fine hook; cutting edges almost straight, but flexed with oblique hook downwards at extreme base. Nostril rounded-triangular, partly covered by operculum above, bordered by plush-like feathering of lore at base; some short fine bristles project obliquely downward at each side of base of upper mandible. Tarsus and toes rather short, strong. Middle toe with claw 30·3 (10) 27·5–32·0; outer toe with claw *c.* 75% of middle with claw, inner *c.* 74%, hind *c.* 85%.

Geographical variation. Marked, involving size (as expressed in measurements and weight), relative bill depth, and colour of gloss on body. 2 main races: (1) large *versicolor*, occurring west from Nova Scotia, northern New England, and Appalachians, with head, neck, and chest mainly greenish-blue or purple, body and upper wing-coverts uniform bronze-green, more violet or purplish on longer coverts and on wing and tail; (2) small nominate *quiscula* from south-east USA (extreme south-east Virginia south-west to south-east Louisiana, south to southern tip of Florida), which has head, neck, and chest dark purplish-bronze or violet, remainder of body and upper wing-coverts extensively and strongly glossed purple, bluish-purple, purple-bronze, or violet, feathers with greener bases and fringes (not as uniform as *versicolor*), flight-feathers and tail purple-blue to violet-purple. Both extreme races connected by *stonei* (occupying wide zone from southern New England to central Louisiana) which combines large size of *versicolor* with purplish body colour and barred wing-coverts of nominate *quiscula*, though plumage generally less uniform than in nominate *quiscula* and varying more between individuals. For characters and distribution of hybrid forms, see Oberholser (1919), Chapman (1935, 1939, 1940), Huntington (1952), Bent (1958), Yang and Selander (1968), and Zink *et al.* (1991c).

Perhaps forms superspecies with Antillean Grackle *Q. niger* of Greater Antilles, Carib Grackle *Q. lugubris* of Lesser Antilles and northern South America, and Nicaraguan Grackle *Q. nicaraguensis* of Nicaragua (Yang and Selander 1968; Mayr and Short 1970), but see Zink *et al.* (1991c).   CSR

Sturnella magna (Linnaeus, 1758) **Eastern Meadowlark**

Fʀ. Sturnelle des prés   Gᴇ. Lerchenstärling

A North American species, breeding in south-east Canada, eastern and central USA, Cuba, Mexico, Central America, and northern South America; winters throughout breeding range except for northernmost parts. 4 records in England in 19th century have been dismissed as escapes, but the fact that 2 of the 3 dated records were in October, the most likely month for drift-migrants, suggests they may have been wild birds (Alexander and Fitter 1955).

Agelaius phoeniceus (Linnaeus, 1766) **Red-winged Blackbird**

Fʀ. Carouge à épaulettes   Gᴇ. Rotschulterstärling

A North American species, breeding throughout Canada and USA except extreme north, also in Bahamas, Cuba, and Isla de Pinos; winters from southern Canada south through breeding range. Alexander and Fitter (1955) listed 16 dated records from 19th century, 15 from Britain and 1 from Italy, and mentioned 2 undated British records. They considered that some at least of the records must be escapes as the species has been a favourite cage-bird, all records were of ♂♂, and concentration of records in south-east England in 1863–6 suggests escapes from consignment of birds imported at that time; also, bird recorded in Scotland was almost certainly one of several released not far away 20 days before.

Xanthocephalus xanthocephalus **Yellow-headed Blackbird**

PLATE 31
[between pages 256 and 257]

Dᴜ. Geelkoptroepiaal   Fʀ. Carouge à tête blanche   Gᴇ. Brillenstärling
Rᴜ. Желтоголовый кассик   Sᴘ. Tordo cabecidorado   Sᴡ. Gulhuvad trupial

Icterus xanthocephalus Bonaparte, 1825

Monotypic

Field characters. ♂ 24–27·5, ♀ 20–25 cm; wing-span ♂ 38–45, ♀ 35–38 cm. ♂ noticeably larger than Blackbird *Turdus merula* but ♀ somewhat smaller; form differs in deep-based, long, pointed bill, stockier body, and looser tail. Large, lengthy, but also stocky Nearctic icterid; very dark plumage with yellow foreparts; adult ♂ has mainly white carpal patches. Sexes dissimilar; no seasonal variation. Juvenile resembles ♀.

ADULT MALE. Moults: July–October (complete); January–April (mainly head and forebody). Head, hindneck, throat, and breast deep yellow, with black surround to bill, lores, and eye-patch and dark steel-blue bill. Rest of plumage black, with white bases to bastard wing and primary coverts which form obvious carpal patch. When fresh, crown and hindneck spotted black. Legs grey-black. ADULT FEMALE. Noticeably smaller than ♂. Rather oily, dusky plumage marked only by yellow on face, whitish throat, and yellow breast which abuts mottled patch of black and white lines on forebelly. Often a dark malar stripe. FIRST-YEAR. Resembles adult ♀ but shows whitish tips to bastard wing, primary coverts, and some outer greater coverts. ♂ best distinguished from ♀ by larger size, more distinct eye-stripe and supercilium, cleaner chest, and (after winter moult) yellower head. Legs initially grey-flesh.

Unmistakable. Flight somewhat recalls starling *Sturnus*: heavy and somewhat laborious, with marked abrupt undulations, but quite fast and direct over distance; flutters slowly with drooped tail before entering ground cover. Walks and runs but apparently does not hop; adept at clambering on marsh plants. Stance usually half upright with head often sunk into shoulders and tail often hanging down when perched. Gregarious in Nearctic wetlands but vagrants have shown no habitat preference.

Voice harsh and rasping, with guttural notes recalling suckling pigs and low monosyllabic 'krack' or 'kack'.

Habitat. Breeds in temperate Nearctic lowlands, favouring deep marshes fringing lakes and shallow river impoundments, where there are stands of cattail *Typha*, bulrush *Scirpus*, or reed *Phragmites*. Shows preference for emergent vegetation in fairly deep water for nesting. (Johnsgard 1979.) Forages in meadows and fields and nearby marshes, and locally by highways; found on ocean coast in September (Larrison and Sonnenberg 1968). In Canada, forages also on grainfields, freshly ploughed land, and in barnyards (Godfrey 1979).

Distribution. Breeds in North America from central British Columbia and northern Alberta east through central Manitoba to north-west Ohio, south to southern California, southern New Mexico, north-east Mexico, south-west Missouri, and central Illinois. Winters from southern USA south to southern Mexico.

Accidental. Iceland: adult ♂, 23–24 July 1983 (Pétursson and Ólafsson 1985). There are also other records,

not fully accepted as wild birds, from France (23 August–15 September 1979: Dubois and Yésou 1986), Denmark (2 October 1918: Schiøler 1922), and Norway (30 May 1979: Ree 1980).

Movements. Migrant, winter range overlapping with extreme south of breeding range. Breeding is in fertile, deep-water marshes, with most foraging then occurring in this habitat; records outside breeding habitat in mid-July involve post-breeding dispersal rather than migration (Bent 1958; A J Erskine).

Southward migration begins early or mid-August (Jewett *et al.* 1953; Royall *et al.* 1971), and most birds depart from northern breeding areas by late September or early October (Cannings *et al.* 1987; Janssen 1987). Interior and plateau areas are vacated chiefly by end of October (Bent 1958; Royall *et al.* 1971; Alcorn 1988). Birds ringed in Dakotas recorded in Texas and Mexico from 2nd week of September; recoveries show movements of 2000–3000 km in less than 30 days (Royall *et al.* 1971). Stragglers appear regularly in east coast areas August–October, with most in September (e.g. Bailey 1955, Bull 1974, Tufts 1986). The few that winter annually in south-east states (Bent 1958; Potter *et al.* 1980; Kale and Maehr 1990) arrive gradually; not known if they include birds that strayed to north-east earlier in season. Migration, mainly by day and in flocks, probably widespread except in western mountains, but movements are channelled by need to forage near water in open land (A J Erskine).

In spring, limited movement starts in late February, and individuals appear right across range in March (Bent 1958; Burleigh 1972; Janssen 1987; Alcorn 1988). Main migration much later, with most birds leaving winter range in April, some in 1st half of May (Royall *et al.* 1971; Oberholser 1974). Main arrival begins early April in intermountain west (Burleigh 1972; Cannings *et al.* 1987), and late April or early May on prairies (Royall *et al.* 1971; Bohlen 1989), not reaching northern range limits until mid-May (Houston and Street 1959; A J Erskine). Far fewer vagrants to east in spring than autumn.

Vagrant to west and north coasts of Alaska (July–October), and to northern Canada, as well as to most eastern states, and also to Greenland. Recorded at sea in Atlantic Ocean *c.* 480 km north-east of New York city. (Kessel and Gibson 1978; American Ornithologists' Union 1983; Godfrey 1986.)   AJE

Voice. See Field Characters.

Plumages. ADULT MALE. Head, neck, and throat to chest bright golden-yellow, often tinged orange or tawny-gold, especially on top and side of head. Lore, small patch round eye, feathering at base of lower mandible, and upper chin black. Lower chest and upper mid-belly yellow with some black bars or spots; remainder of body backwards from mantle, side of breast, and centre of belly uniform black, shorter under tail-coverts with some concealed yellow at bases. Tail black. Entire wing including tertials, most wing-coverts, and under wing black, but greater upper primary coverts contrastingly white (except for some black on tips, especially on outer ones), 2 outer greater upper wing-coverts with one or both webs extensively white (some white also on inner webs of middle greater coverts, mainly concealed), and outer web of middle feather of bastard wing largely or fully white. *In fresh plumage* and in non-breeding (September to about April–May), top of head and hindneck tawny-yellow with numerous dull black feather-tips, latter partly or largely concealing yellow. *In worn plumage* (June–July), yellow paler, especially on those parts of head, neck, or chest which were not affected by pre-breeding moult; tips of primaries, tertials, and tail-feathers brown. ADULT FEMALE. Upperparts uniform black-brown, fringes of feathers slightly greyer if plumage fresh. A distinct yellow supercilium, extending from forehead to above ear-coverts. Lore and upper cheek mottled off-white, yellow, and brown; ear-coverts dark brown, showing as dark smudge on side of head; lower cheek yellow, sometimes bordered below by indistinct brown-mottled malar stripe; side of neck mottled yellow, off-white, and pale brown, bordered at rear by dark brown bar extending from mantle down to side of lower throat. Chin, throat, and chest yellow, mixed off-white to variable extent, lower chest with dark brown spots of varying size. Breast and upper mid-belly off-white, spotted or streaked dark brown, remainder of underparts dark brown, feathers with slightly paler brown or grey-brown fringes when plumage fresh and with some grey of feather-bases exposed when plumage worn. Tail and entire wing black-brown (dark sepia-brown when plumage worn), no white (except for some concealed white on inner webs of primary coverts). Much individual variation in amount of yellow on side of head and neck and from chin to breast, especially in spring and early summer, depending on extent of pre-breeding moult. In autumn, side of head and neck and chin to breast mainly pale yellow (mottled white on lore, with dark smudge on ear-coverts, and with varying amount of dark mottling or streaking on chest and breast); this yellow bleached to off-white by spring, but part of it replaced by bright yellow during spring moult. Yellow acquired in spring often darker than that of autumn, often tinged tawny-yellow or even rufous on supercilium and cheek, less inclined to show dark malar stripe or brown spots on side of neck and chest, often forming contrasting bright yellow patches amidst dirty pale yellow or off-white on remaining front part of body (e.g. brighter yellow chest contrasting with off-white throat and breast, or tawny-yellow front part of supercilium contrasting with dirty pale yellow rear supercilium); some birds acquire only limited amount of new feathering, showing (e.g.) off-white chest with dark brown triangular spots. JUVENILE. Head, neck, and chest cinnamon-buff to buff-yellow, chin and throat buff-white. Mid-crown, ear-coverts, upperparts back from mantle, underparts back from side of breast and belly, tertials, and wing-coverts dark brown, longer feathers fringed cinnamon or tawny along tips; breast streaked dull white, thigh and middle of breast and belly off-white. FIRST ADULT MALE. Like adult ♀, thus top of head mainly brown (less yellow than adult ♂), body dark brown (not as black as adult ♂), and without extensive white on outer wing-coverts (many greater primary coverts as well as middle feather of bastard wing showing broad but ill-defined fringe of white on tip only). Differs from adult ♀ in larger size, more distinct dark eye-stripe (extending from base of bill to ear-coverts), contrasting more with long and distinct yellow supercilium; more uniform yellow chest, darker breast (black-brown with off-white spots, less extensively streaked off-white), more tapering tips of tail-feathers (less truncate; see Pyle *et al.* 1987), and white-tipped greater upper primary coverts (in ♀, black-brown with concealed white bases).

In spring, new feathering on top of head, mantle, eye-stripe (including lore), side of neck, and from side of breast backwards blacker, forehead and hindneck partly or fully tawny or golden-orange, chin to chest brighter and more uniform yellow, but much individual variation in colour and in extent of 1st pre-breeding moult. FIRST ADULT FEMALE. Like adult ♀, but upperparts, wing, and tail browner, less black-brown, especially when worn; side of head and neck and chin to chest buff-white to off-white instead of yellow, partly mottled brown on supercilium, cheek, and chest, and with distinct dark brown malar stripe; tail-feathers more tapering at tip than in adult ♀, similar to 1st adult ♂; some greater upper wing-coverts with poorly defined white fringes along tips (as 1st adult ♂, unlike adult ♀). In spring and early summer, new feathers on side of head and/or chest yellow, as in adult ♀; tips of primaries and tail-feathers browner and more heavily worn than in adult ♀ at same time of year; tips of inner greater primary coverts with traces of mottled white fringes. On average, less new plumage acquired in spring than in adult ♀, supercilium and cheek more often cream-yellow rather than orange-yellow. See also Crawford and Hohman (1978).

Bare parts. ADULT, FIRST ADULT. Iris brown or dark brown. Bill dark leaden-grey to greyish-black (darkest in ♂ and in breeding plumage). Leg and foot dark leaden-grey, greyish-black, or black. JUVENILE. Bill dark horn-brown. Leg and foot dull greyish-flesh with dark slate or blackish scutes. (BMNH, RMNH.)

Moults. ADULT POST-BREEDING. Complete; primaries descendent. On breeding grounds, July to early October; in birds examined (sample small), late July to mid-October (RMNH, ZMA). ADULT PRE-BREEDING. Partial; extent rather variable; in some birds, involves head, neck, chest, and scattered feathers of remainder of body, in others scattered feathers of head (mainly on side of head) and/or chest only, rarely t1 (RMNH, ZMA). In winter quarters, January–April. POST-JUVENILE. Partial; starts shortly after fledging, in moult July–August. Involves head, body, and lesser, median, and greater upper wing-coverts. FIRST PRE-BREEDING. Like adult pre-breeding, equally variable or more limited in extent. (Pyle *et al.* 1987; RMNH.)

Measurements. ADULT, FIRST ADULT. Whole geographical range, all year; skins (RMNH, ZMA). Juvenile wing and tail are retained juvenile wing and tail of 1st adult. Bill (S) to skull, bill (N) to distal corner of nostril; exposed culmen on average 2·3 shorter than bill (S).

	♂	♀
WING AD	146·4 (3·68; 28) 140–153	118·5 (2·72; 7) 115–122
JUV	135·5 (2·43; 7) 132–139	115·5 (1·80; 4) 114–118
TAIL AD	100·8 (3·46; 19) 94–106	81·1 (2·42; 7) 77–85
JUV	88·3 (2·69; 7) 84–91	76·3 (2·75; 3) 74–80
BILL (S)	24·9 (1·76; 27) 22·3–27·9	20·7 (0·76; 10) 19·7–21·9
BILL (N)	16·5 (1·26; 21) 14·0–18·3	13·2 (0·51; 10) 12·7–14·1
TARSUS	37·2 (1·14; 21) 35·3–39·5	31·1 (0·91; 11) 29·5–32·2

Sex differences significant.

Birds from central-north USA (Iowa, Wisconsin) in sample above smaller than those from southern and western USA (mainly California, some Texas); e.g., adult ♂: central-north, wing 143·2 (2·57; 11) 140–147, bill (S) 23·2 (0·68; 7) 22·5–24·5; south and west, wing 148·3 (3·03; 16) 143–153, bill (S) 26·1 (1·30; 16) 24·0–27·9. Wing of live birds, Washington (north-west USA): adult ♂ 144·6 (2·81; 90), 1-year-old ♂ 136·0 (2·46; 42), adult ♀ 115·9 (3·54; 28) (Searcy 1979, which see for influence of age on wing-length). For ♀, see also Crawford and Hohman (1978).

Weights. Average of adults, Washington (north-west USA): ♂ 92·8 (*n*=90), ♀ 51·1 (*n*=28) (Searcy 1979). Iowa (USA), May-July: ♀ 46·6 (3·57; 30) 41·1–56·7 (Crawford and Hohman 1978). Oaxaca (Mexico), October: ♂ 82·7, ♀ 47·9 (Binford 1989).

Structure. Wing rather long, broad at base, tip bluntly pointed. 10 primaries: p8–p9 longest, p7 0–2 shorter, p6 2–5, p5 9–13, p4 16–21, p1 26–31 (♀) or 31–38 (♂). P10 strongly reduced, a tiny pin; 78–98 shorter than p8–p9, 12–18 shorter than longest upper primary covert, 2 shorter to 2 longer than reduced outermost upper primary covert. Outer web of p6–p8 emarginated, inner of p7–p9 with notch. Tip of longest tertial reaches to tip of p3. Tail rather long, tip square or slightly rounded; 12 feathers. Bill rather long, visible length of culmen *c.* 80% of head length, rather conical in some birds, more elongate in others; *c.*

11–13 mm deep at base and *c.* 8·5–11 mm wide. Culmen straight, but base slightly elevated, ending in small shield on mid-forehead; cutting edges straight, but kinked downwards at extreme base. Nostrils oval, covered by operculum above. Bristles at base of mandibles short, soft, reduced. Tarsus and toes rather short, strong. Middle toe with claw 28·2 (8) 25–31; outer and inner toe with claw both *c.* 80% of middle with claw, hind *c.* 86%.

Geographical variation. Slight. Birds from north-east of range slightly smaller than those from west (see Measurements). Head, neck, and chest in populations from east of range (Mississippi valley) perhaps paler yellow (less orange) than in those west of Rockies (Ridgway 1901–11).

CSR

Icterus wagleri Sclater, 1857   **Black-vented Oriole**

Fr. Oriole cul-noir   Ge. Waglertrupial

A Mexican and Central American species, resident from northern Mexico south through highlands to north-central Nicaragua. One in Rogaland (Norway), 10 July to 7 November 1975, may have been an escape (VR).

Icterus galbula   **Northern Oriole**

PLATE 31
[between pages 256 and 257]

Du. Baltimore-troepiaal   Fr. Oriole du Nord   Ge. Baltimoretrupial
Ru. Северный цветной трупиал   Sp. Bolsero norteño   Sw. Baltimoretrupial

Coracias Galbula Linnaeus, 1758

Polytypic. GALBULA GROUP. Nominate *galbula* (Linnaeus, 1758), eastern Canada and eastern USA, west to a line from central Alberta to north-east Texas, vagrant to west Palearctic. BULLOCKII GROUP (extralimital). *I. g. bullockii* (Swainson, 1827), extreme south-west Canada, western USA (except part of California), and northern Mexico, hybridizing with *galbula* in south-east Alberta, North Dakota, eastern Colorado, western Nebraska, western Oklahoma, and north-central Texas; *parvus* Van Rossum, 1945, west-central and southern California and neighbouring southern Nevada, western Arizona, northern Baja California, and north-west Sonora. ABEILLEI GROUP (extralimital). *I. g. abeillei* (Lesson, 1839), central plateau of Mexico, grading into *bullockii* in Durango.

Field characters. 17–20 cm; wing-span 28–32 cm. Approaches size of Starling *Sturnus vulgaris* with rather similar bill and head shape but much longer body and rather long, slightly rounded tail. Colourful, arboreal icterid, with long pointed bill. Yellow-orange below at all ages, but tone varies. Adult ♂ striking: black hood, back, wings, and tail-centre, orange forewing, rump, underparts, and tail-edges, and white wing-bar. ♀ duller, with double white wing-bar. Sexes dissimilar; no seasonal variation. Juvenile separable. Only eastern race described here.

ADULT MALE. Moults: June–September (complete). Head, breast, back, wings, and base and centre of tail black; wings show orange lesser and median coverts, white bar across tips of greater coverts, and narrow white edges to tertials. Long rump, underparts, and rest of tail vivid orange. Bill blue-grey with dusky culmen. Legs blue-grey. ADULT FEMALE. Resembles ♂ in having dark upperparts contrasting with paler underparts but few approach his vividness, most being only mottled brown on head and back and orange-yellow below. Head variably marked: dark birds have only smudges of yellow by base of bill; on others, warm orange or yellow extends over lore, into cheek and chin so that either throat appears only partly black and does not extend into full bib or whole head appears more or less dusky-orange and lacks bib. Wings duller than ♂: no wholly orange fore-coverts, showing instead broad whitish bar across tips of median coverts and narrow whitish bar across tips of greater. Tail more yellow or green than ♂, having only dusky centre. Bill duller. FIRST-WINTER MALE. Resembles typical adult ♀ but again variable, some with markedly orange head and warm back but others with dusky crown and cheeks and even olive- to greyish-brown back. FIRST-WINTER FEMALE. Duller than any adult ♀, never showing black on throat.

Unmistakable. Plumage colours and head and wing pattern may recall Brambling *Fringilla montifringilla*, but general character and behaviour distinctive. Note however that closely related Orchard Oriole *I. spurius* could conceivably cross Atlantic (slightly smaller than *I. galbula*, with slighter bill and relatively shorter wings and longer tail; underparts brick-red, not orange, in ♂, and greenish-yellow, not orange or warm yellow, in ♀ and immature). Flight free and rapid, with easy wing-beats; suggests large long-tailed warbler. Stance usually rather level; often raises tail above wing-tips and occasionally flicks it. As vagrant, often skulking and difficult to relocate.

Calls include rich, fluted whistle, 'pew-li', nasal 'ucht', and hard rattling 'cher-r-r-r-r'.

Habitat. Breeds in temperate Nearctic lowlands, favouring wooded river bottoms, upland forest, shelterbelts, and partially wooded residential areas and farmsteads. Absent from pure coniferous forest but after their clearance colonizes ensuing deciduous growth. (Johnsgard 1979.) Lives in scattered tall trees, particularly elms *Ulmus*, as well as more open deciduous woodland (Godfrey 1979). Also occurs along country roads and streams (Pough 1949), in cottonwoods, willows *Salix*, and sycamores *Acer* lining streams and irrigation ditches of open country, in farmyards, and in trees on ranches; in Arizona, in semi-arid mesquite *Prosopis* groves; ascends into lower mountain canyons (Terres 1980).

Distribution. Breeds in North America from central British Columbia east to Nova Scotia, south to central Mexico. Winters mainly from Mexico to north-west South America.

Accidental. Iceland, Britain, Netherlands, Norway.

Movements. Status varies between migratory and sedentary from north to south of range; most birds breeding north of 35°N are long-distance migrants. Western populations (*bullockii* group of races) winter south to Guatemala, eastern population (nominate *galbula*) south to north-west South America; birds wintering in extreme southern USA include both *bullockii* group and nominate *galbula*. (Bent 1958; American Ornithologists' Union 1983.)

Some birds leave breeding areas in July, but main migration evident only from mid-August (Bent 1958). Last records early September in northern breeding areas (Palmer 1949; Houston and Street 1959; Sprague and Weir 1984). By mid-September, most have vacated mid-latitude states (Stewart and Robbins 1958; Mengel 1965;

Alcorn 1988), and some have reached main wintering range in Central America (Monroe 1968; Stiles and Skutch 1989); peak movement crosses southern states and reaches Central America and Colombia in October (Ligon 1961; Lowery 1974; Hilty and Brown 1986). Autumn migration on broad front, but chiefly west of Florida, Cuba, and Yucatán (Mexico) (Peterson and Chalif 1973; Bond 1985; Kale and Maehr 1990). Arrives in South America (where uncommon) later than in Central America. Long-distance ringing recoveries show movements between Manitoba and Guatemala, Ontario and Guatemala or Honduras, northern USA (Minnesota east to Massachusetts) and Mexico or Central America (south to El Salvador). Ringing in Honduras has shown winter site-fidelity, with returns up to 6 years later. (Sealy 1985.) Since *c.* 1950, numbers wintering in temperate North America (especially on Atlantic seaboard) have increased greatly, notably where artificial food available; ringing in North Carolina shows that birds return in successive winters (Erickson 1969; Root 1988).

Northward migration starts late March, and birds gradually leave winter range during April (Oberholser 1974; Hilty and Brown 1986; Stiles and Skutch 1989). Arrivals become general in early April, from California to Florida (Grinnell and Miller 1944; Lowery 1974; Kale and Maehr 1990). Main migration moves north evenly, reaching Canadian border from British Columbia to New Brunswick from 1st week of May (Sadler and Myres 1976; Squires 1976; Cannings *et al.* 1987). Peak movement in early to mid-May across north of range, and migration is over by end of May (Palmer 1949; Jewett *et al.* 1953; Janssen 1987). Spring route mostly reverse of autumn, but birds more common in Florida in spring (Stevenson 1957; Kale and Maehr 1990).

Vagrants have reached south-east Alaska, Churchill (Manitoba, 59°N), and south-west Greenland (65°N), and many others wander shorter distances beyond breeding range (Bent 1958; American Ornithologists' Union 1983).

Nominate *galbula* is rare vagrant to Atlantic seaboard of west Palearctic, mainly in autumn. Of 17 records in Britain up to 1990, 13 in September–October, 2 in May, and 1 in December, plus 1st-winter ♀ found wintering at Roch, Dyfed from 2 January to 23 April 1989 (Dymond *et al.* 1989; Rogers *et al.* 1990, 1991).   AJE, DFV, PRC

Voice. See Field Characters.

Plumages. (nominate *galbula*). ADULT MALE. Entire head and neck deep black, on throat extending downwards into bib on upper chest; some grey of feather-bases sometimes visible on lower cheek, chin, and throat. Mantle, scapulars, and back black, feathers narrowly fringed greyish-orange when fresh (fringes sometimes faint, sometimes completely worn off from October, but traces occasionally present until March), some orange-yellow of feather-bases sometimes visible on mantle when worn. Rump and upper tail-coverts rich orange, feather-tips slightly suffused olive-grey when plumage fresh in autumn, some yellow of feather-bases sometimes visible when worn in late spring and summer. Entire underparts from chest downwards fiery reddish-orange, grading to yellowish-orange on mid-belly and vent; tinge deeper and more reddish when plumage fresh, more yellow when worn. Basal and terminal thirds of tail bright yellow or orange-yellow, middle portion of tail black, forming broad black band (*c.* 20–30 mm wide on t2, *c.* 15–20 mm on t6), but middle and terminal portion of central pair (t1) black, except for greyish-green fringe along tip; tips of t2–t6 often slightly washed dusky grey. Flight-feathers black, p6–p8 with white border along outer web (except on emarginated part; a trace sometimes on p5 and p9 also), outer web of secondaries with white border along terminal part; base of inner web of flight-feathers broadly fringed whitish. Tertials black with broad white outer fringes and (if not worn off) narrow greyish tip; greater upper primary coverts and bastard wing uniform black. Greater upper wing-coverts black, terminal part of outer webs white, forming broken bar *c.* 5–10 mm wide across wing; in worn plumage, white fringes and tips of flight-feathers, tertials, and greater coverts partly worn off. Lesser and median upper wing-coverts and small coverts along leading edge of wing cadmium-yellow to fiery orange. Axillaries and under wing-coverts bright yellow or orange-yellow. ADULT FEMALE. Rather variable. Forehead, crown, and side of head dull black, each feather with olive-yellow or buff-yellow fringe; fringes sometimes narrow, cap and side of head appearing black with pale mottling, or black may be restricted, cap and side of head then appearing olive-yellow or buff-yellow with some fine black spots (mainly on rear crown, upper and rear of ear-coverts, and cheek). Hindneck and side of neck more uniform olive-yellow, faintly speckled dark grey, often markedly less black than cap and mantle; lore and area below eye mottled dark and pale grey. Head and neck rarely all-black (Dwight 1900). Mantle, scapulars, and back dull black, each feather broadly fringed buff-yellow or olive-yellow on mantle, more olive-grey towards lower scapulars and back, appearing black-spotted with paler scalloping. Rump and upper tail-coverts buffish-orange to yellow, tinge partly concealed by olive-grey suffusion on feather-tips (least so on shorter upper tail-coverts). Some birds have chin and throat black, usually mixed with much white or yellow; in others, chin grey, throat buff-orange, mixed with black on central throat or with mottled black T-mark on lower throat and central upper chest. Chest and side of breast buffish-orange or yellow-orange (sometimes slightly suffused olive-grey, especially at side), grading into paler orange-yellow on upper flank and breast and this in turn to yellow on mid-belly and vent and into mixture of yellow and pale grey on lower flank; under tail-coverts bright yellow or buffish-yellow. Tail olive-green to greenish-yellow, slightly purer yellow towards outer feathers; in some birds, middle and tip of t1 and middle portion of some other feathers partly suffused black. Wing as in adult ♂, but greyish-black rather than pure black, and white fringes slightly less pure and less contrasting. Tip of outer web of greater upper wing-coverts broadly fringed white, forming broken pale wing-bar (as in ♂, but slightly narrower); outer webs of greater coverts fringed grey. Median upper wing-coverts black, entire tip broadly white or yellowish-white, forming broad full wing-bar (unlike ♂). Longer lesser upper wing-coverts buffish-yellow or orange-yellow with black centres, shorter ones largely orange-yellow; under wing-coverts yellow (less bright than adult ♂), longer coverts and axillaries grey. JUVENILE. Upperparts olive-brown with slight orange tinge, brightest on crown and upper tail-coverts; paler and greyer than adult ♀, without black feather-centres. Underparts ochre-yellow or buffish-olive, sometimes tinged orange, palest (pale grey) on chin, throat, and mid-belly, brightest on chest and under tail-coverts.

Tail dark olive, slightly more greenish-olive or yellowish-olive towards outer feathers. Wing as adult ♀, but brownish-black rather than greyish-black; lesser upper wing-coverts olive-grey, median and greater brownish-black with broad pale buff tips (browner and with off-white tips when worn). Plumage looser and shorter than in 1st non-breeding, especially on lower flank, where *c.* 8–13 barbs along terminal 10 mm of each web (*c.* 13–19 in 1st adult) (Rohwer and Manning 1990). FIRST ADULT MALE NON-BREEDING. Rather like adult ♀, but head and neck virtually without black. Cap to upper mantle and side of head and neck olive-green, forehead, short line above eye, ear-coverts, and upper cheek tinged buff-yellow, side of crown and band across upper mantle often slightly brighter yellowish-green. Lower mantle, scapulars, and back olive-green, each feather with dull black or greyish-black centre, appearing mottled, but not as distinctly and contrastingly scalloped as adult ♀. Rump olive-green, upper tail-coverts brighter yellowish-green. Lore grey. Chin to chest buffish-yellow with paler yellow feather-bases, remainder of underparts bright yellow, tinged olive on side of belly, upper flank, and thigh, partly washed light grey on lower flank. Tail and wing still juvenile, except for lesser upper wing-coverts (dull black with yellow fringe) and often inner or all median upper wing-coverts (black with broad yellow-white tip); white fringes of flight-feathers and tertials often largely worn off, white tips of outer webs of juvenile greater coverts and white tips of outer juvenile median coverts (if any) heavily abraded, wing-bars sometimes indistinct. Retained juvenile feathering distinctly worn, contrasting with neighbouring fresh feathers; tips of tail-feathers pointed (in adult ♀, all wing and tail uniformly fresh, tips of tail-feathers rounded). From October, some uniformly black feathers appear on lower cheek or throat. FIRST ADULT FEMALE NON-BREEDING. Like 1st adult ♂ non-breeding, but no black on mantle and scapulars and entire plumage more olive-grey, less yellowish. Forehead to back olive-green, slightly yellow on forehead, olive with ill-defined grey fringes on mantle and scapulars; rump and upper tail-coverts olive-grey, tinged yellow-green on shorter upper tail-coverts. Side of head as 1st adult ♂ non-breeding. Chin and central throat yellow-white, side of throat and chest deep buff-yellow, remainder of underparts pale yellow, tinged white on vent and extensively tinged grey on flank and side of belly. Tail and wing largely juvenile, as in 1st adult ♂ non-breeding, but lesser upper wing-coverts dull black with olive-grey fringes, new median coverts as 1st adult ♂ non-breeding. FIRST ADULT MALE BREEDING. Variable. Black and orange feathering similar to adult ♂ appears from mid-winter (in ♀, new feathers tipped olive). In May, advanced birds are similar to adult ♂, but black of head and neck sometimes mottled yellow, and fiery-orange of underparts mixed with orange-yellow; also, flight-feathers, greater upper primary coverts, bastard wing, and sometimes outer greater coverts or some tertials and tail-feathers still juvenile, old feathers on wing much browner and more abraded than neighbouring fresh black feathers, white tips and fringes largely worn off; lesser and median coverts often still 1st non-breeding, mixed black and yellow. Less advanced birds have variable combination of old olive feathering above and dull yellow below, mixed with fresh deep black and bright yellow feathers; wing and tail still juvenile; differ from adult ♀ in lack of olive fringes to new black feathers, absence of evenly scalloped mantle and scapulars, and heavily worn wing and tail. FIRST ADULT FEMALE BREEDING. Some similar to adult ♀, showing black and buff-yellow mottled cap and side of head, and scalloped mantle and scapulars (though black feather-centres smaller and less deep black than adult ♀); chin and throat less black than adult ♀, mainly yellow; chest buff-yellow with less fiery-orange tone than in adult ♀, lower flank and vent greyer. Much of wing and usually all tail still juvenile, heavily abraded. Some birds retain much of 1st non-breeding plumage, only part of feathering of head, mantle, and chest new; new feathers of cap and mantle black with broad olive-green fringe (in 1st breeding ♂, new feathers uniformly black); tail and much of wing juvenile.

Bare parts. ADULT, FIRST ADULT. Iris dark brown or black-brown. Bill pale grey-blue or blue-grey, culmen dark plumbeous-grey or black. Leg and foot blue-grey to dark plumbeous-grey. JUVENILE. Iris grey-brown to dark brown. Bill mauve, pale flesh-grey, light greyish-blue, or steel-blue, culmen slate-grey or dark plumbeous-grey. Mouth flesh-pink. Leg and foot bright light blue. (Workman 1963; Broad 1978; BMNH, RMNH, ZMA).

Moults. Nominate *galbula*. ADULT POST-BREEDING. Complete; primaries descendent. In or near breeding area, about mid-June to mid-September, 1-year-olds start slightly earlier than adults (Sealy 1979, which see for details). ADULT PRE-BREEDING. Restricted in extent or no moult at all; scattered feathers of head or body only. Of adult ♂♂ examined by Rohwer and Manning (1990), none in moult October–December, 9–14% of birds in moult January–March (*n* = 22 in each month), 33% in April (*n* = 9); of ♀♀, a few in moult February–March. POST-JUVENILE. Partial: head, body, lesser and many or all median upper wing-coverts, occasionally a few tertial coverts or inner greater coverts. Starts (June–)July–August, shortly after fledging, completed August–September(–October), usually before start of migration (Pyle *et al.* 1987; Rohwer and Manning 1990; RMNH, ZMA). FIRST PRE-BREEDING. In ♂, starts from October, immediately after arrival in winter quarters; black feathers of throat acquired first. From December–February onwards, variable amount of remainder of non-breeding body feathering of ♂ replaced; moult most intense February–April and continued in May shortly after arrival on breeding grounds and possibly also during spring migration. By May, much of head, body, and wing-coverts of ♂ new, as well as often some tertials (mainly s8–s9) and occasionally bastard wing and central or all tail-feathers. In ♀, moult starts from February–March, extent variable; in some birds, all head and neck new by May, as well as much or all of mantle and scapulars and chin to breast; in others, only scattered feathers of head or front part of body new. (Rohwer and Manning 1990; RMNH, ZMA.)

I. g. bullockii. Unlike nominate *galbula*, adult post-breeding and post-juvenile both in September–October, during migration stopover in south-west USA, 1st adult ♂ acquiring black throat-patch during post-juvenile moult. No pre-breeding moult, 1st adults retaining non-breeding plumage throughout 1st summer. (Rohwer and Manning 1990.)

Measurements. Nominate *galbula*. Eastern North America, May–September, and eastern Mexico to Costa Rica, October–April; skins (RMNH, ZMA). Bill (S) to skull, bill (N) to distal corner of nostril; exposed culmen on average 2·2 less than bill (S). 1st adult wing is retained juvenile wing of birds in 1st adult non-breeding or breeding plumage; 1st adult tail includes birds with juvenile tail as well as new 1st breeding tail.

WING AD ♂	97·3 (1·84; 21)	94–101	♀	92·2 (3·40; 5)	88–97
1ST AD	93·9 (2·36; 11)	89–98		92·6 (1·81; 9)	89–96
TAIL AD	71·7 (2·17; 15)	68–75		70·1 (1·92; 5)	67–72
1ST AD	70·9 (2·63; 10)	65–74		68·3 (2·30; 9)	64–71
BILL (S)	21·0 (1·04; 25)	19·2–23·3		21·2 (0·60; 10)	20·5–22·2
BILL (N)	13·3 (0·53; 25)	12·0–14·1		13·3 (0·51; 10)	12·4–14·0
TARSUS	24·0 (0·64; 24)	23·0–25·3		23·9 (0·72; 10)	22·9–25·1

Sex differences significant for adult wing only.

I. g. bullockii. Western USA (summer) and western Mexico (winter); skins (RMNH); ages combined.

WING	♂ 103·1 (1·91; 11)	99–106	♀ 95·8 (1·99; 6)	93–98
BILL (S)	21·4 (1·04; 11) 19·8–23·0		21·0 (1·30; 6)	19·9–23·2

Sex differences significant for wing.

Weights. Nominate *galbula*. Ohio (USA): May, ♀ 34·2; average of immatures, July 33·3 (3), August 32·6 (3) (Baldwin and Kendeigh 1938). Belize, November–April: ♂ 33·2 (3) 30·1–37·0 (Russell 1964). Mexico and El Salvador: October–January, ♂ 33·2 (2·43; 6) 30·5–36, ♀ 30·5; April, ♂ 39·4 (RMNH, ZMA). Britain: Lundy (Devon), October, single bird on various dates 26·9–29·6 (Workman 1963). Netherlands: Vlieland, October, single bird 22·8–29·0 (E B Ebels).

Intermediates between nominate *galbula* and *bullockii*. Oaxaca (Mexico), October–March: ♂♂ 36·2, 37·7 (Binford 1989).

Structure. Wing rather long, tip bluntly pointed. 9–10 primaries: p7–p8 longest or either one 0–1 shorter than other; p9 1–5 shorter than longest, tip equal to tip of p6 or slightly shorter; p6 1–2 shorter than p7–p8, p5 4–8, p4 6–14, p3 10–20, p1 20–27. A tiny p10 usually present, 58–73 shorter than p7–p8, 9–14 shorter than longest primary covert; 2 shorter to 1 longer than reduced outermost upper primary covert, under which it is hidden. Outer web of (p5–)p6–p8 distinctly emarginated, inner web of p7–p8 (p6–p9) with rather faint notch. Tip of longest tertial reaches to about tip of p2. Tail rather long, tip slightly rounded; 12 feathers, t6 4–11 shorter than t1–t2. Bill slightly shorter than head, straight or slightly decurved; deep at base (especially lower mandible), gradually tapering to acutely pointed tip. Nostril oval, covered by narrow membrane above. Many short fine hairs at base of bill, no bristles. Leg and toes rather short, but strong. Middle toe with claw 19·1 (5) 18·0–20·0; outer and inner toe with claw both *c.* 77% of middle with claw, hind *c.* 91%.

Geographical variation. Complex, marked. Comprises 3 morphologically distinct groups: *galbula* group (nominate *galbula* only, Baltimore Oriole) in eastern North America, *bullockii* group (*bullockii* and *parvus*, Bullock's Oriole) in western North America, and *abeillei* group (*abeillei* only, Abeille's Oriole) in north-central Mexico, each formerly often considered to be separate species, but in view of hybridization in contact zones now each considered semispecies within a single polytypic species. Apart from marked difference in colour and pattern of (especially) adult ♂♂, nominate *galbula* and *bullockii* also differ in measurements (e.g. Ridgway 1901–11), voice (Saunders 1951), physiology (Rising 1969), social behaviour in breeding season (Rising 1970), moults, and time needed for 1st adult ♂ to acquire adult characters (Rohwer and Manning 1990). Secondary intergradation of characters of *galbula* and *bullockii* probably enabled by habitat changes brought about by man in Great Plains area (Anderson 1971); in some localities (in particular, apparently, where expansion is in an early phase) meeting populations hybridize extensively, but in others (perhaps after stabilization) *galbula* and *bullockii* overlap completely and assortative mating occurs, so hybridization limited. For hybrid zone and hybrid characters, see Sutton (1938, 1968), Sibley and Short (1964), Misra and Short (1974), Rising (1970, 1973, 1983), Corbin and Sibley (1977), and Rohwer and Manning (1990). Differences in moult and migration strategy are perhaps source of selection against hybrids (Rohwer and Manning 1990). Secondary contact zone between *bullockii* and *abeillei* in Durango (Mexico) less well investigated; see Miller (1906) and Rising (1970, 1973).

Adult ♂ *bullockii* differs from adult ♂ *galbula* in (1) restricted black on head and neck, forehead (narrowly) orange, side of head orange except for black stripe through eye, cheek orange, chin and throat with narrow black bib; (2) extensive white on wing, black on median and greater upper wing-coverts restricted to inner webs, outer webs largely white; white outer fringes of tertials, secondaries, and outer primaries broader than in nominate *galbula*; (3) t1 mainly black, as in nominate *galbula*, but t2–t6 with black tips 5–15 mm wide (in nominate *galbula*, band *c.* 2 cm wide over middle of tail instead). In adult ♀ *bullockii*, lower mantle and scapulars either narrowly streaked black, appearing greenish-grey with black streaks, or uniform grey (in nominate *galbula*, black with yellow-buff or orange-buff scalloping); supercilium more distinct, yellow, bordered below by grey eye-stripe; cap yellowish- or greyish-green, rump grey, upper tail-coverts, underparts down to chest, and under wing bright yellow (upper tail-coverts and chest in particular much less orange-buff than adult ♀ nominate *galbula*); narrow throat-patch mottled grey and black or almost uniform black; flank, belly, and vent extensively pale grey, more so than some 1st adult ♀♀ nominate *galbula*. Juvenile and 1st adult non-breeding *bullockii* are yellowish-olive on cap, olive-grey on mantle, scapulars, back, and rump, and

(greenish-)yellow on upper tail-coverts, from throat to chest, and on under tail-coverts; yellow supercilium and dusky eye-stripe sometimes less distinct than in adult ♀; flank and belly greyish-white; as in adult ♀, upper tail-coverts, cheek and chest more lemon- or sulphur-yellow, less buff-orange than 1st adult non-breeding nominate *galbula*. First non-breeding ♂ *bullockii* shows narrow black throat-patch, but (unlike nominate *galbula*) does not show pure orange and black feathers from mid-winter onwards. Californian *parvus* differs from *bullockii* only in smaller size. Adult ♂ of Mexican *abeillei* differs from *bullockii* in uniform black upperparts, broad black eye-stripe and throat-patch (orange-yellow supercilium short, malar stripe rather narrow), and black side of neck, side of breast, and flank; for ♀, juvenile, and non-breeding plumages, see Ridgway (1901–11) and Rohwer and Manning (1990).

Recognition. In ♀ and non-breeding plumages, differs from Orchard Oriole *I. spurius* in heavier bill and in relatively longer wing and shorter tail, tail/wing ratio 0·75 (37) 0·71–0·79, against 0·86 (10) 0·81–0·92 in *I. spurius*.   CSR

ADDITIONAL SPECIES
and changes to taxonomy

The following list contains brief details of species which are new to the west Palearctic list since the various earlier volumes were published. Also included are notes on a few species which have been affected by subsequent, and now widely accepted, changes to taxonomy.

Shy Albatross *Diomedea cauta*. Israel: Elat, 20 February to 7 March 1981, when found dead (Shirihai 1987).

Southern Giant Petrel *Macronectes giganteus*. France: record off Ouessant, November 1967, now accepted on French list only as *Macronectes*, not specifically identified (Dubois and Yésou 1986).

Cape Petrel *Daption capense*. Italy: captured off Sciacca (Sicily), September 1964, often considered to be of doubtful origin (Bourne 1967), but accepted on Italian list (Brichetti and Massa 1984).

Atlantic Petrel *Pterodroma incerta*. Czechoslovakia: Zolinski (Zips), 1870 (Godman 1907-10; Bourne 1992). Israel: Elat, 31 May 1982 (Shirihai 1987); Elat, 19 April 1989 (Schot 1989).

Jouanin's Petrel *Bulweria fallax*. Italy: 3 seen (1 obtained) Cinadolmo (Treviso), 2 November 1953; discussed by Olson (1985) and Zonfrillo (1988) and accepted provisionally on Italian list (Brichetti and Massa 1991).

Streaked Shearwater *Calonectris leucomelas*. Israel: 2 or 3, Elat, 21 June to mid September 1992 (Morgan and Shirihai 1992).

Flesh-footed Shearwater *Puffinus carneipes*. Israel: Elat, 15 August 1980 (Paz 1987; Goodman and Storer 1987).

Wedge-tailed Shearwater *Puffinus pacificus*. Egypt: 24 November 1983 (Bezzel 1987); Port Said, 10 March 1988 (Everett 1988; Goodman and Meininger 1989).

Manx Shearwater *Puffinus puffinus* (monotypic) to be split from Mediterranean Shearwater *P. yelkouan* (comprising nominate *yelkouan* and *mauretanicus*) (*Br. Birds* 1988, 81, 306-19; 1990, 83, 299-319).

Audubon's Shearwater *Puffinus lherminieri*. Israel: Tel-Aviv, January 1984 (*Br. Birds* 1988, 81, 14); Ma'agan Mikhael, 31 December 1989 (*Br. Birds* 1990, 83, 222).

White-bellied Storm-petrel *Fregatta grallaria*. Near Cape Verde Islands, 17 August 1986 (Haase 1988).

Swinhoe's Petrel *Oceanodroma monorhis*. Britain and Ireland: Tynemouth, at least three birds 1989-1992 (*Ibis* 1994, 136). France: Brittany 1989, Madeira: 1983, 1988, and 1991 (*Ardea* 1985 73, 105-6, *Birding World* 1992, 5, 438-42).

Red-footed Booby *Sula sula*. Cape Verde Islands: Cima, 24 August 1986 (Hartog 1987). Norway: Molen, Vestfold, 29 June 1985 (Bentz 1988).

Gannet *Sula bassana*. Name changed to *Morus bassanus* (*Notornis* 35, 35-57; *Bull. Br. Orn. Club* 108, 9-12).

Masked Booby *Sula dactylatra*. Spain: adult, Puerto Sotogrande, Cádiz, 10 October 1985; Torremolinos, 14 December 1985 (Juana *et al.* 1989).

Double-crested Cormorant *Phalacrocorax auritus*. Britain: Billingham (Cleveland), 11 January to 29 April 1989 (*Ibis* 1992, 134, 211-14). Azores: Mosteiros (São Miguel), 24-26 October 1991 (Duin 1992).

Chinese Pond Heron *Ardeola bacchus*. Norway: adult, Hellesylt (Møre og Romsdal), autumn 1973 (Folkstad 1978).

Little Blue Heron *Hydranassa caerulea*. Azores: 28 November 1964, ringed in USA (Denis 1981).

Tricoloured Heron *Hydranassa tricolor*. Azores: Pico, 22-24 October 1985 (Parrott *et al.* 1987).

Black Heron *Hydranassa ardesiaca*. Cape Verde Islands: 1985 (C. Madge and C. J. Hazevoet, *Birds of the Cape Verde Islands*, BOU checklist, in press).

Snowy Egret *Egretta thula*. Iceland: dead on ship just south of Iceland, 1974; specimen, 6 June 1983; May or June 1985 (G Pétursson). Azores: 2, Santa Cruz (Flores), 11 October 1988 (*Dutch Birding* 1988, 193).

Great Blue Heron *Ardea herodias*. Azores: c. 10, of which 3 found dead, São Miguel, Pico, and Faial, 4 April to 24 June 1984 (Le Grand 1986).

African Spoonbill *Platalea alba*. Spain: Ebro delta, 20 February 1989; Mallorca, 1 May to 24 September 1989 (Juana *et al.* 1991). But large numbers recently imported into Europe.

White-faced Whistling Duck *Dendrocygna viduata*. Spain: Mallorca, end December 1973 (Ximenes 1975); Canary Islands: one captured Los Rodeos (Tenerife) between 1967 and 1969 (Estarriol Jiménez 1974).

Lesser Whistling Duck *Dendrocygna javanica*. Israel: Ma'agan Mikhael, 15 November 1966 to mid-March 1967, photographed and considered unlikely to be an escape (Shirihai in press).

Black Swan *Cygnus atratus*. Slovenia: feral breeder (*Br. Birds* 1992, 85, 289).

Bar-headed Goose *Anser indicus*. Accidental in Russia (1985) and Hungary (*Br. Birds* 1992, 85, 289). Feral breeding in Czechoslovakia, Germany (*Br. Birds* 1992, 85, 289), and (apparently) Norway (*Br. Birds* 1989, 82, 322).

Ross's Goose *Anser rossii*. Netherlands: Santpoort and Assendelft (Noordholland), 30 November to 1 December 1985 (Berg and Cottaar 1986); Zeeland, 19 January to 28 February 1988 (Berg 1990). Britain: several of unknown origin (*Birding World* 1991, 4, 138-40).

Spur-winged Goose *Plectropterus gambensis*. Morocco: Oued Massa, 8-10 June 1984 (Bouwan 1985).

Cotton Pygmy Goose *Nettapus coromandelianus*. Iraq: 2 live ♀♀ in Basrah market, 19 November 1975, were said to have been taken in Hammar marshes (*c*. 30°30′N 47°35′E) the previous day (Kainady 1976), but their origin cannot be certainly known; the specimens are in Basrah Natural History Museum.

Wood Duck *Aix sponsa*. Iceland: 26-27 April 1984 (Pétursson and Ólafsson 1986).

Red-billed Duck *Anas erythrorhyncha*. Israel: Ma'agan Mikhael, 20 June to 12 July 1958 (Hovel 1987).

Cape Shoveler *Anas smithii*. Morocco: near Agadir, 26 April 1978 (Duff 1979).

Canvasback *Aythya valisineria*. Iceland: ♀ collected, Gull, 11 April 1977 (G Pétursson).

Lesser Scaup *Aythya affinis*. Britain and Ireland: Chasewater (West Midlands), 8 March to 26 April 1987 (Holian and Fortey 1992); Co. Down and Co. Armagh, 13 February to 14 April 1988 and subsequent seasons.

African Fish Eagle *Haliaeetus vocifer*. Formerly included on the strength of a record from Sinai (Egypt) in 1967, but this was not accepted by Hovel (1987). However, Goodman and Meininger (1989) have given an Egyptian record of an adult collected near Abu Simbel on 1 November 1947.

Gabar Goshawk *Micronisus gabar*. Egypt: several old undated records (Goodman and Meininger 1989).

Swainson's Hawk *Buteo swainsoni*. Norway: 2nd-year, Rost (Nordland), 6 May 1986 (Bentz 1988).

California Quail *Callipepla californica*. Feral breeding in Corsica (France), Italy (Cruon and Nicolau-Guillaumet 1985), Mecklenburg (Germany) (Klafs and Stübs 1977), and Denmark (*Br. Birds* 1992, **85**, 289).

Daurian Partridge *Perdix dauurica*. Feral breeding in Italy (*Riv. ital. Orn.* **54**, 3-87), Russia, and Ukraine (*Br. Birds* 1992, **85**, 289).

Reeves's Pheasant *Syrmaticus reevesii*. Czechoslovakia: feral breeding (Pokorny and Pikula 1987). France: feral breeding (Cruon and Nicolau-Guillaumet 1985).

Wild Turkey *Meleagris gallopavo*. Germany: feral breeding (Niethammer *et al.* 1964).

Hooded Crane *Grus monacha*. Russia (*Br. Birds* 1992, **85**, 289).

Oriental Pratincole *Glareola maldivarum*. Britain: Dunwich (Suffolk) and Old Hall marshes (Essex), 22 June to *c*. 11 October 1981; Harty and Elmley (Kent), 21 or 22 June to 3 October 1988 (Burns 1993).

Three-banded Plover *Charadrius tricollaris*. Egypt: Gebel Asfar (Cairo), 5 March to at least 24 March 1993 (*Birding World* 1993, **6**, 100).

Lesser Golden Plover *Pluvialis dominica* now split into American Golden Plover *P. dominica* and Pacific Golden Plover *P. fulva*. (Knox 1987*a*).

Spoonbill Sandpiper *Eurynorhynchus pygmeus*. Ukraine: 2 records (*Br. Birds* 1992, **85**, 289).

Hudsonian Godwit *Limosa haemastica*. Britain: adult, Blacktoft (Humberside), 10 September to 3 October 1981; Devon, 22 November 1981 to 4 January 1982; Humberside, 26 April to 6 May 1983 (Grieve 1987).

Grey-tailed Tattler *Heteroscelus brevipes*. Britain: Dyfi estuary (Dyfed/Gwynedd), 13 October to 17 November 1981 (*Ibis* 1988, **130**, 334-7).

Willet *Catoptrophorus semipalmatus*. Finland: Kemi (Kuivnanuoro), 21 September 1983 (Solonen 1985). Norway: Molen (Vestfold), 14 October to December 1992 (Sondbø 1993).

South Polar Skua *Stercorarius maccormicki*. Faeroes: specimen, September 1889 (Alström and Colston 1991). Israel: Elat, 28 June 1992 (*Br. Birds* 1993, **86**, 41).

Herring Gull/Lesser Black-backed Gull *Larus argentatus/L. fuscus* complex. The problems with the taxonomy of this difficult group were discussed in Vol III of BWP (p. 836). The relationships remain unclear, but among the changes recommended by some since the publication of Volume III are the following:

L. f. heuglini and *L. f. taimyrensis* may be races of *L. argentatus*, not *L. fuscus* (Grant 1986, Poyser).

L. a. thayeri may be treated as a separate species, Thayer's Gull, *L. thayeri* (*BWP*, Vol. **III**, 840; *Check-list of North American Birds*, 6th Edition, 1983), but is considered by others to be a race of Iceland Gull, *L. glaucoides* (Sibley and Monroe 1990, Yale University Press). This is of relevance due to possible recent sightings of this form in Ireland.

L. a. cachinnans, *L. a. michahellis*, *L. a. atlantis*, *L. a. omissus*, *L. a. barabensis*, and *L. a. mongolicus* are now considered to be a separate species, the Yellow-legged Gull, *L. cachinnans* (e.g. Yésou, P. 1991, *Ibis* **133**, 256-63; Sibley and Monroe 1990, Yale University Press).

L. a. armenicus is considered to be a separate species, Armenian Gull, *L. armenicus* (Dubois 1985, *Alauda* **53**, 226-28).

Brown-headed Gull *Larus brunnicephalus*. Israel: Elat, 12 May 1985 (Shirihai *et al.* 1987, but identification disputed by Hoogendoorn 1991).

Elegant Tern *Sterna elegans*. Ireland: Carlingford Lough (Down), 22 June to 3 July 1982; Ballymacoda (Cork), 1 August 1982 (O'Sullivan and Smiddy 1988). France: Gironde, June 1974 to 1988 (Dubois *et al.* 1989). Belgium: Zeebrugge 12 June to 15 July 1988 (Boesman 1992).

Little Tern *Sterna albifrons*. The New World form of this species is considered by some to be a separate species, Least Tern *S. antillarum* (Sibley and Monroe 1990, Yale University Press; Chandler and Wilds 1994, *Br. Birds* **87**, 60-6). A bird apparently of this form occurred in Sussex in the summers of 1983-1992, and there were other sightings, possibly of the same bird.

Saunders's Tern *Sterna saundersi*. Egypt: dried carcass in nest of Osprey *Pandion haliaetus*, Gezira Mahabis, 21

May 1982 (Goodman and Storer 1987). Israel: Elat, June–July 1988 (*Br. Birds* 1989, 19).

Ancient Murrelet *Synthliboramphus antiquus*. Britain: Lundy (Devon), 27 May to 26 June 1990, 14 April to 20 June 1991 (*Ibis* 1992, **134**, 211–14), and 1992.

Mourning Dove *Zenaida macroura*. Britain: Calf of Man (Isle of Man), 31 October 1989 (*Ibis* 1993, **135**, 220–2).

Monk Parakeet *Myiopsitta monachus*. Belgium: feral breeding (Schaezen and Jacob 1985). Italy: feral breeding (Spanò and Truffi 1986). Spain: established feral breeding (Battlori and Nos 1985).

Chimney Swift *Chaetura pelagica*. Britain: 2, Porthgwarra (Cornwall), 21–27 October 1982 (Willams 1986); Isles of Scilly, 4–9 November 1986; Grampound (Cornwall), 18 Ocober 1987.

Eastern Phoebe *Sayornis phoebe*. Britain: Lundy (Devon), April 1990 (*Ibis* **135**, 220).

Chestnut-headed Finch Lark *Eremopterix signata*. Israel: Elat, 1 March 1983 (H Shirihai).

Hume's Lark *Calandrella acutirostris*. Israel: Elat, 4–14 February 1986 (Shirihai and Alström 1990).

Banded Martin *Riparia cincta*. Egypt: Aswan, 15 November 1988 (Clements 1990).

Tree Swallow *Tachycineta bicolor*. Britain: St Marys (Isles of Scilly) 6–10 June 1990 (*Ibis* 1992, **134**, 380–1).

Ethiopian Swallow *Hirundo aethiopica*. Israel: trapped, Bet She'an, 22 March 1991 (Bear 1991).

Cliff Swallow *Hirundo pyrrhonota*. Britain: juvenile, Isles of Scilly, 10–27 October 1983 (Crosby 1988); juvenile, South Gare (Cleveland), 23 October 1988 (Little 1990).

Rock/Water Pipit complex (formerly united in *Anthus spinoletta*) to be split as: Rock Pipit *A. petrosus* (comprising nominate *petrosus*, *meinertzhageni*, *kleinschmidti*, and *littoralis*), Water Pipit *A. spinoletta* (nominate *spinoletta*, *coutellii*, and *blakistoni*), and Buff-bellied Pipit *A. rubescens* (nominate *rubescens*, *japonicus*, *pacificus*, and *alticola*) (Knox 1988a).

Northern Mockingbird *Mimus polyglottos*. Netherlands: Schiermonnikoog (Friesland), 16–23 October 1988 (Ebels 1991). Britain: Saltash (Cornwall) 30 August 1982; Hamford Water (Essex) 17–23 May 1988 (*Ibis* **135**, 496).

Varied Thrush *Zoothera naevia*. Britain: 1st-winter, Nanquidno (Cornwall), 14–23 November 1982 (Madge *et al.* 1990).

Greenish Warbler *Phylloscopus trochiloides*. The following species, *P. plumbeitarsus*, and the Green Warbler, *P. nitidus*, are now considered to be conspecific with Greenish Warbler (*Ibis* **135**, 221).

Two-barred Greenish Warbler *Phylloscopus plumbeitarsus*. Britain: Gugh (Isles of Scilly), 22–27 October 1987 (*Ibis* 1993, **135**, 220–2). Netherlands: Castricum (Noordholland), 17 September 1990 (Schekkerman 1992).

Treecreeper *Certhia* species. The status of the North American representatives of this genus is unclear. These birds have often been included (as in this work) as races of *Certhia familiaris*, but because some think that they might be more closely related to *C. brachydactyla*, the American Ornithologists Union currently lists them as a separate species *C. americana* (*Check-list of North American Birds*, 6th Edition, 1983).   TI, DJB

REFERENCES

ABBOTT, W M (1931) *Irish Nat. J.* 3, 191-2. ABDULALI, H (1947) *J. Bombay nat. Hist. Soc.* 46, 704-8. ABDUSALYAMOV, I A (1973) *Fauna Tadzhikskoy SSR* 19 *Ptitsy* 2; (1977) 3. Dushanbe. ÅBRO, A (1964) *Sterna* 6, 81-5. ABS, M (1961) *Falke* 8, 370-1; (1964) *Vogelwarte* 22, 173-6; (1966) *Ostrich* suppl. 6, 41-9. ABSHAGEN, K (1963) *Beitr. Vogelkde.* 8, 325-38. ACHARYA, H G (1953) *J. Bombay nat. Hist. Soc.* 50 (1951), 169-70. ACKERMANN, A (1967) *J. Orn.* 108, 430-73. ADAMS, A L (1864) *Ibis* (1) 6, 1-36. ADAMS, R (1989) *Devon Birds* 42, 40-43. ADAMS, R G (1948) *Br. Birds* 41, 210-11. ADAMYAN, M S (1965) *Zool. Zh.* 44, 569-77. ADKISSON, C S (1977a) *Avic. Mag.* 83, 195-8; (1977b) *Wilson Bull.* 89, 380-95; (1981) *Condor* 83, 277-88. ADOLPH, P A (1943) *Br. Birds* 37, 134. ADRET-HAUSBERGER, M (1983) *Z. Tierpsychol.* 62, 55-71; (1984) *Ibis* 126, 372-8; (1989) *Bioacoustics* 2, 137-62. ADRET-HAUSBERGER, M, and GÜTTINGER, H R (1984) *Z. Tierpsychol.* 66, 309-27. ADRET-HAUSBERGER, M, and JENKINS, P F (1988) *Behaviour* 107, 138-56. AERTS, M A P A and SPAANS, A L (1987) *Limosa* 60, 169-74. AHARONI, I (1931) *Beitr. Fortpfl. Vögel* 7, 161-6, 222-6; (1942) *Bull. zool. Soc. Egypt* 4, 13-19. AICHHORN, A (1966) *J. Orn.* 107, 398-9; (1969) *Verh. dt. zool. Ges.* 32, 690-706; (1970) *Ber. Nat.-Med. Ver. Innsbruch* 58, 347-52; (1989) *Egretta* 32, 58-71. AKHMEDOV, K P (1957) *Uchen'iye Zap. Stalinabad Zh. pedagog. Inst.* 1, 101-13. ALATALO, R (1975) *Lintumies* 10, 1-7. ALBERNY, J-C, TANGUY LE GAC, J, and VENANT, H (1965) *Oiseaux de France* 44, 18-25. ÅLBU, T (1983) *Fauna norv.* (C) *Cinclus* 6, 53-6. ALCORN, J R (1988) *The birds of Nevada*. Fallon. AL-DABBAGH, K Y and JIAD, J H (1988) *Int. Stud. Sparrows* 15, 22-43. ALDRICH, J W (1940) *Ohio J. Sci.* 40, 1-8; (1984) *Orn. Monogr. AOU* 35. ALEKSEEVA, N S (1986) *Ornitologiya* 21, 145. ALERSTAM, T (1988) *Anser* 27, 181-218; (1990) *Bird migration*. Cambridge. ALERSTAM, T and ULFSTRAND, S (1972) *Ornis scand.* 3, 99-139. ALEX, U (1985) *Zool. Abh. Staatl. Mus. Tierkde. Dresden* 41, 200. ALEXANDER, B (1898a) *Ibis* (7) 4, 74-118; (1898b) *Ibis* (7) 4, 277-85. ALEXANDER, C J (1917) *Br. Birds* 11, 98-102. ALEXANDER, W B (1933) *J. Anim. Ecol.* 2, 24-35. ALEXANDER, W B and FITTER, R S R (1955) *Br. Birds* 48, 1-14. ALI, S (1963) *J. Bombay nat. Hist. Soc.* 60, 318-21; (1968) *The book of Indian birds*. Bombay. ALI, M H, RAO, B H K, RAO, M A, and RAO, P S (1982) *J. Bombay nat. Hist. Soc.* 79, 201-4. ALI, S and RIPLEY, S D (1972) *Handbook of the birds of India and Pakistan* 5; (1973a) 8; (1973b) 9; (1974) 10. Bombay. ÅLIND, P (1991) *Calidris* 20, 94-8. AL-JOBORAE, F F (1979) D Phil Thesis. Oxford. ALLARD, H A (1939) *Science* 90, 370-1. ALLAVENA, S (1970) *Riv. ital. Orn.* 40, 460-1. ALLIN, E K (1968) *Br. Birds* 61, 541-5. ALLISON, G W (1975) *Glos. nat. Soc. J.* 26 (7), 84. ALLOUSE, B E (1953) *Iraq nat. Hist. Mus. Publ.* 3. ALLSOPP, K (1976) *Br. Birds* 69, 532-4. ALLSOPP, K and NIGHTINGALE, B (1991) *Br. Birds* 84, 137-45. ALMOND, W E (1946) *Br. Birds* 39, 315. ALONSO, J A, MUÑOZ-PULIDO, R, BAUTISTA, L M, and ALONSO, J C (1991) *Bird Study* 38, 35-43. ALONSO, J C (1984a) *J. Orn.* 125, 209-23; (1984b) *J. Orn.* 125, 339-40; (1984c) *Ardeola* 30, 3-21; (1985a) *Ardeola* 32, 31-8; (1985b) *Ardeola* 32, 405-8; (1985c) *J. Orn.* 126, 195-205; (1986a) *Intl. studies on sparrows* 13, 35-43; (1986b) *Ekol. Polska* 34, 63-73. ALSOP, F J (1973) *Wilson Bull.* 85, 484-5. ALSTRÖM, P and COLSTON, P (1991) *A field guide to the rare birds of Britain and Europe*. London. ALTEVOGT, R, and DAVIS, T A (1980) *J. Bombay nat. Hist. Soc.* 76, 283-90. ALTNER, H (1957) *Orn. Mitt.* 9, 115. ALTNER, H and REGER, K (1959) *Anz. orn. Ges. Bayern* 5, 224-34. ALTRICHTER, K (1974) *Anz. orn. Ges. Bayern* 13, 231-9. ALVAREZ, F (1975) *Doñana Acta Vert.* 1 (2), 67-75. ALVAREZ, F and AGUILERA, E (1988) *Ardeola* 35, 269-75. ALVAREZ, F and ARIAS DE REYNA, L (1975) *Doñana Acta Vert.* 1 (2), 77-95. AMADON, D (1943) *Wilson Bull.* 55, 164-77; (1944) *Auk* 61, 136-7; (1967) *Linnaean News-Letter* 20, 2-3. AMANOVA, M A (1977) *Ekologiya* 1, 99-101. AMAT, J A and OBESO, J R (1989) *Ardeola* 36, 219-24. AMBEDKAR, V C (1972) *J. Bombay nat. Hist. Soc.* 69, 268-82. AMERICAN ORNITHOLOGISTS' UNION (1957) *Checklist of North American birds*, 5th ed. Baltimore; (1983) *Checklist of North American birds*, 6th edn. Lawrence. AMLANER, C J and BALL, N J (1983) *Behaviour* 87, 85-119. AMMERSBACH, R (1960) *Gef. Welt* 84, 81-85. ANANIN, A A and ANANINA, T L (1983) In Kuchin, A P (ed.) *Ptitsy Sibiri*, 163-4. Gorno-Altaysk. ANANIN, A A and FEDOROV, A V (1988) In Korneeva, T M (ed.) *Fauna Barguzinskogo Zapovednika*, 8-33. Moscow. ANDELL, P, EBENMAN, B, EKBERG, B, LARSSON, P-G, NILSSON, L, PERSSON, O, and ÖHRSTRÖM, P (1983) *Anser*, Suppl. 15. ANDERSEN, H H (1989) *Br. Birds* 82, 380-1. ANDERSEN-HARILD, P, BLUME, C A, KRAMSHJ, E, and SCHELDE, O (1966) *Dansk orn. Foren. Tidsskr.* 60, 1-13. ANDERSON, A (1961) *Scott. Nat.* 70, 60-74. ANDERSON, B W (1971) *Condor* 73, 342-7. ANDERSON, B W and DAUGHERTY, R J (1974) *Wilson Bull.* 86, 1-11. ANDERSON, T R (1977) *Condor* 79, 205-8; (1978) *Occ. Pap. Mus. nat. Hist. Kansas* (70), 1-58; (1980) *Proc. int. orn. Congr.* 17, 1162-70; (1984) *Ekol. Polska* 32, 693-707; (1991) In Pinowski, J and Summers-Smith, J D (eds) *Granivorous birds in the agricultural landscape*, 87-93. Warsaw. ANDREEV, A V (1977) *Zool. Zh.* 56, 1578-81; (1982a) *Ornitologiya* 17, 72-82; (1982b) In Gavrilov, V M and Potapov, R L (eds) *Ornithological studies in the USSR* 2, 364-76. Moscow. ANDRÉN, H (1985) *Vår Fågelvärld* 44, 261-8. ANDREW, D G (1969) *Br. Birds* 62, 334-6. ANDREW, R J (1956a) *Behaviour* 10, 179-204; (1956b) *Br. J. anim. Behav.* 4, 125-32; (1956c) *Ibis* 98, 502-5; (1956d) *Br. Birds* 49, 107-11; (1957a) *Behaviour* 10, 255-308; (1957b) *Ibis* 99, 27-42; (1961) *Ibis* 103a, 315-48. ANDRIESCU, C, and ANDRIESCU, I (1972) *Muzeul de Ştiinţele Naturii Dorohoi, Botoşani, Studii şi comunicări 1972*, 205-10. ANDRIESCU, C and CORDUNEANU, V (1972) *Muzeul de Ştiinţele Naturii Dorohoi, Botoşani, Studii şi comunicări 1972*, 199-204. ANGELL-JACOBSEN, B (1980) *Ornis scand.* 11, 146-54. ANGWIN, E (1977) *Bull. E. Afr. nat. Hist. Soc.* Nov-Dec, 131. ANIKIN, V I (1963) *Ornitologiya* 6, 463. ANON (1976) *BTO News* 79, 3; (1985) *Reader's Digest complete book of New Zealand birds*. Sydney; (1986) *Israel Land Nat.* 12, 37; (1987) *Strait of Gibraltar Bird Observatory Spec. Rept.* 1; (1990) *Ardeola* 37 (2), 325-52. ANTAL, L, FERNBACH, J, MIKUSKA, J, PELLE, I, and SZLIVKA, L (1971) *Larus* 23, 73—127. ÁNTIKAINEN, E (1978) *Savonia* 2, 1-45; (1981) *Ornis fenn.* 58, 72-7. ANTIKAINEN, E, SKARÉN, U, TOIVANEN, J, and UKKONEN, M (1980) *Ornis fenn.* 57, 124-31. APLIN, O V (1911) *Zoologist* (4) 15, 112-13. ARAUJO, J (1975)

Ardeola **21** (Espec.), 469-85. ARCHER, G and GODMAN, E M (1961) *Birds of British Somaliland and the Gulf of Aden* **4**. Edinburgh. ARDAMATSKAYA, T B (1968) In Ivanov, A I (ed.) *Migratsii Zhivotnykh* **5**, 146-52. St Petersburg. ARENDT, E and SCHWEIGER, H (1982) *Publ. Wiss. Film. Sekt. Biol.* **15/34**, 3-10. ARFF, E (1962) *Falke* **9**, 390. ARHEIMER, O and ENEMAR, A (1974) *Fauna och Flora* **69**, 153-64. ARIAS DE REYNA, L M, RECUERDA, P, CORVILLO, M, and CRUZ, A (1984) *Doñana Acta Vert.* **11** (1), 79-92. ARMANI, G C (1985) *Guide des passereaux granivores: Embérizinés*. Paris. ARMITAGE, J (1927) *Br. Birds* **21**, 117-19; (1932) *Br. Birds* **26**, 206-7; (1933) *Br. Birds* **27**, 153-7; (1937) *Br. Birds* **31**, 98-100. ARMSTRONG, E A (1954) *Ibis* **96**, 1-30. ARNHEM, R and VAN LOMMEL, J (1964) *Gerfaut* **54**, 458-65. ARNOLD, E L and ELLIS, J C S (1957) *Br. Birds* **50**, 347. ARNOLD, M A (1955) *Br. Birds* **48**, 91. ARNOLD, P (1962) *Birds of Israel*. Haifa. ARRIGONI DEGLI ODDI, E (1929) *Ornitologia italiana*. Milan; (1931) *Riv. ital. Orn.* (2) **1**, 100-4. ARROYO, B and TELLERÍA, J L (1984) *Ardeola* **30**, 23-31. ARTHUR, R W (1963) *Br. Birds* **56**, 49-51. ARVIDSSON, B L, BOSTRÖM, U, DAHLÉN, B, DE JONG, A, KOLMODIN, U, and NILSSON, S G (1992) *Ornis svecica* **2**, 67-76. ASBIRK, S and FRANZMANN, N-E (1978) In Green, G H and Greenwood, J J D (Eds) *Joint biological expedition to north east Greenland 1974*, 132-42. Dundee; (1979) *Dansk orn. Foren. Tidsskr.* **73**, 95-102. ASCHOFF, J and HOLST, D VON (1960) *Proc. int. orn. Congr.* **12**, 55-70. ASENSIO, B (1984) *Ardeola* **31**, 128-34; (1985a) *Ardeola* **32**, 173-8; (1985b) *Ardeola* **32**, 179-86; (1986a) *Ardeola* **33**, 176-83; (1986b) *Doñana Acta Vert.* **13**, 103-10; (1986c) *Suppl. Ric. Biol. Selvaggina* **10**, 375-6. ASENSIO, B and ANTÓN, C (1990) *Ardeola* **37**, 29-35. ASENSIO, B and CARRASCAL, L M, (1990) *Folia Zool.* **39**, 125-30. ASH, J (1949) *Br. Birds* **42**, 289. ASH, J S (1964) *Br. Birds* **57**, 221-41; (1969) *Ibis* **111**, 1-10; (1980) *Proc. Pan-Afr. orn. Congr.* **4**, 199-208; (1988) *Sandgrouse* **10**, 85-90. ASH, J S and NIKOLAUS, G (1991) *Bull. Br. Orn. Club* **111**, 237-9. ASHTON-JOHNSON, J F R (1961) *Ool. Rec.* **35**, 49-55. ÅSTRÖM, G (1976) *Zoon Suppl.* **2**. ATKINSON, C T, and RALPH, C J (1980) *Auk* **97**, 245-52. ATTLEE, H G (1949) *Br. Birds* **42**, 85. AUBIN, T (1987) *Behaviour* **100**, 123-33. AUBRY, J (1970) *Ann. Zool.-Écol. animale* **2**, 509-22. AUEZOVA, O N (1982) In Gvozdev, E V (ed.) *Zhivotnyi mir Kazakhstana i problemy ego okhrany*, 9-11. Alma-Ata. AUSTIN, O L (1948) *Bull. Mus. comp. Zool.* **101** (1). AVERY, G R (1991) *Glos. Nat. Soc. J.* **42**, 2-5. AXELL, H (1989) *Acrocephalus* **10**, 36-7. AXELSSON, P, KÄLLANDER, H, and NILSON, S (1977) *Anser* **16**, 241-6.

BABENKO, V G (1984a) *Ornitologiya* **19**, 171-2; (1984b) *Ornitologiya* **19**, 172 BACCETTI, N, FRUGIS, S, MONGINI, E, and SPINA, F (1981) *Riv. ital. Orn.* **51**, 191-240. BACCHUS, J G (1941) *Br. Birds* **35**, 17; (1943) *Br. Birds* **37**, 38. BACHKIROFF, Y (1953) *Le moineau steppique au Maroc*. Rabat. BACMEISTER, W and KLEINSCHMIDT, O (1920) *J. Orn.* **68**, 1-32. BAEGE, L (1968) *Beitr. Vogelkde.* **14**, 81-3. BAER, W (1910) *Orn. Monatsschr.* **35**, 401-8. BAEYENS, G (1979) *Ardea* **67**, 28-41; (1981a) *Ardea* **69**, 69-82; (1981b) *Ardea* **69**, 125-39; (1981c) *Ardea* **69**, 145-66. BAGG, A M (1943) *Auk* **60**, 445. BÄHRMANN, U (1937) *Mitt. Verein. sächs. Orn.* **5**, 115-18; (1942) *Beitr. Fortpfl. Vögel* **18**, 203-4; (1950) *Syllegomena biologica 1950*, 41-9; (1958) *Vogelwelt* **79**, 129-35; (1960a) *Abh. Ber. Mus. Tierkde. Dresden* **25**, 71-9; (1960b) *Anz. orn. Ges. Bayern* **5**, 510-13; (1960c) *Anz. orn. Ges. Bayern* **5**, 573-7; (1963) *Zool.*

Abh. Staatl. Mus. Tierkde. Dresden **26**, 187-218; (1964) *Zool. Abh. Staatl. Mus. Tierkde. Dresden* **27**, 1-9; (1966) *Zool. Abh. Staatl. Mus. Tierkde. Dresden* **28**, 221-34; (1967) *Beitr. Vogelkde.* **12**, 363-6; (1968a) *Die Elster*. Wittenberg Lutherstadt; (1968b) *Zool. Abh. Staatl. Mus. Tierkde. Dresden* **29**, 177-90; (1968c) *Beitr. Vogelkde.* **14**, 8-28; (1970a) *Beitr. Vogelkde.* **15**, 434-6; (1970b) *Beitr. Vogelkde.* **15**, 454-5; (1971a) *Beitr. Vogelkde.* **17**, 180-1; (1971b) *Beitr. Vogelkde.* **17**, 413-4; (1972) *Beitr. Vogelkde.* **18**, 89-122; (1973) *Beitr. Vogelkde.* **19**, 153-69; (1976) *Zool. Abh. Staatl. Mus. Tierkde. Dresden* **34**, 1-37; (1978a) *Zool. Abh. Staatl. Mus. Tierkde. Dresden* **34**, 199-228; (1978b) *Zool. Abh. Staatl. Mus. Tierkde. Dresden* **35**, 223-52. BÄHRMANN, U and ECK, S (1975) *Zool. Abh. Staatl. Mus. Tierkde. Dresden* **33**, 237-43. BAILEY, A M and NIEDRACH, R J (1967) *Pictorial checklist of Colorado birds*. Denver. BAILEY, A M, NIEDRACH, R J, and BAILY, A L (1953) *Publ. Denver Mus. Nat. Hist.* **9**. BAILEY, W (1955) *Birds in Massachusetts: when and where to find them*. South Lancaster. BAIRD, J C (1958) *Bird-Banding* **29**, 224-8; (1967) *Bird-Banding* **38**, 236-7. BAIRLEIN, F (1979) *Vogelwarte* **30**, 1-6; (1988) *Vogelwarte* **34**, 237-48. BAKER, A J (1980) *Evolution* **34**, 638-53. BAKER, A J, DENNISON, M D, LYNCH, A, and LE GRAND, G (1990b) *Evolution* **44**, 981-99. BAKER, A J, PECK, M K, and GOLDSMITH, M A (1990a) *Condor* **92**, 76-88. BAKER, C E (1927) *Br. Birds* **20**, 200. BAKER, E C S (1926) *The fauna of British India: Birds* **3**. London; (1932) *The nidification of birds of the Indian Empire*. **1**. London; (1934) *The nidification of birds of the Indian Empire*. **3**. London. BAKER, H R and INGLIS, C M (1930) *The birds of southern India*. Madras. BAKER, M C (1975) *Evolution* **29**, 226-41. BALANÇA, G (1984a) *Gibier Faune sauvage* **2**, 45-78; (1984b) *Gibier Faune sauvage* **3**, 37-61; (1984c) *Gibier Faune sauvage* **4**, 5-27. BALÁT, F (1963) *Larus* **15**, 141-4; (1971) *Zool. Listy* **20**, 265-80; (1974) *Zool. Listy* **23**, 123-35; (1976) *Zool. Listy* **25**, 39-49. BALDA, R P (1980) *Z. Tierpsychol.* **52**, 331-46. BALDWIN, P J and MEADOWS, B S (1988) *Birds of Madinat Yanbu Al-Sinaiyah and its hinterlands*. Riyadh. BALDWIN, S P and KENDEIGH, S C (1938) *Auk* **55**, 416-67. BALDWIN, S P, OBERHOLSER, H C, and WORLEY, L G (1931) *Sci. Publ. Cleveland Mus. Nat. Hist.* **2**, 1-165. BALFOUR, D (1976) *Sterna* **15**, 169-73. BALFOUR, E (1968) *Scott. Birds* **5**, 89-104. BALPH, M H (1975) *Bird-Banding* **46**, 126-30. BALTVILKS, J (1970) *Mat. 7. Pribalt. orn. Konf.* **2**, 23-8. BANGJORD, G (1986) *Vår Fuglefauna* **9**, 251. BANKIER, A M (1984) *Br. Birds* **77**, 121. BANKS, K W, CLARK, H, MACKAY, I R K, MACKAY, S G, and SELLERS, R M (1989) *Ring. Migr.* **10**, 141-57; (1991a) *Bird Study* **38**, 10-19; (1991b) *Scott. Birds* **16**, 57-65. BANKS, R C (1959) *Condor* **61**, 96-109; (1964) *Univ. Calif. Publ. Zool.* **70**, 1-123. BANNERMAN, D A (1911) *Ibis* (9) **5**, 401-2; (1912) *Ibis* (9) **6**, 557-627; (1919) *Ibis* (11) **1**, 84-131; (1936) *The birds of tropical West Africa* **4**; (1948) *The birds of tropical West Africa* **6**; (1949) *The birds of tropical West Africa* **7**. London; (1953a) *The birds of the British Isles* **1**; Edinburgh; (1953b) *The birds of west and equatorial Africa* **2**. Edinburgh; (1956) *The birds of the British Isles* **5**. Edinburgh; (1963) *Birds of the Atlantic islands* **1**. Edinburgh. BANNERMAN, D A and BANNERMAN, W M (1958) *Birds of Cyprus*. London; (1965) *Birds of the Atlantic islands* **2**; (1966) **3**; (1968) **4**. Edinburgh; (1971) *Handbook of the birds of Cyprus and migrants of the Middle East*. Edinburgh;; (1983) *The birds of the Balearics*. London. BANNERMAN, D and PRIESTLEY, J (1952) *Ibis* **94**, 654-82.

BANZHAF, E (1937a) *Vogelzug*, **8**, *114-8*. BANZHAF, W (1937b) *Ver. orn. Ges. Bayern* **21**, 123-36. BAPTISTA, L F (1977) *Condor* **79**, 356-70; (1990) *Vogelwarte* **35**, 249-56. BARANCHEEV, L M (1963) *Ornitologiya* **6**, 173-6. BARANOV, L S and MARGOLIN, E A (1983) *Ornitologiya* **18**, 186. BARBA, E and LÓPEZ, J A (1990) *Mediterránea Ser. Biol.* **12**, 79-88. BÁRDHARSON, H R (1986) *Birds of Iceland*. Reykjavík. BARDIN, A V (1990) *Trudy Zool. Inst. Akad. Nauk SSSR* **210**, 18-34. BARIŞ, S, AKÇAKAYA, R, and BILGIN, C (1984) *Birds of Turkey* **3**. BARLOW, J C (1973) *Ornith. Monogr. AOU* **14**, 10-23; (1980) *Proc. int. orn. Congr.* **17**, 1143-9. BARLOW, J C and POWER, D M (1970) *Canad. J. Zool.* **48**, 673-80. BARNARD, C (1979) *New Scientist* **83**, 818-20. BARNARD, C J (1980a) *Anim. Behav.* **28**, 295-309; (1980b) *Anim. Behav.* **28**, 503-11; (1980c) *Behaviour* **74**, 114-27. BARNARD, C J and SIBLY, R M (1981) *Anim. Behav.* **29**, 543-50. BARNES, J A G (1941) *Br. Birds* **35**, 17. BARRAUD, E M (1956) *Br. Birds* **49**, 289-97. BARREAU, D, BERGIER, P, and LESNE, L (1987) *Oiseau* **57**, 307-67. BARRETT, J H (1947) *Ibis* **89**, 439-50. BARRIETY, L (1965) *Bull. Centr. Étud. Sci. Biarritz* **5**, 267-71. BARROWCLOUGH, G F (1980) *Auk* **97**, 655-68. BARROWS, W B (1889) *US Dept. Agric. Div. Econ. Orn. and Mamm. Bull.* **1**. BARTA, Z (1977) *Aquila* **83**, 308. BARTELS, M (1931) *Beitr. Fortpfl. Vögel* **7**, 129-30. BARTELS, M and BARTELS, H (1929) *J. Orn.* **77**, 489-501. BARTH, R and MORITZ, D (1988) *Beitr. Naturkde. Niedersachs.* **41**, 118-29. BARTHEL, P H, HANOLDT, W, HUBATSCH, K, KOCH, H-M, KONRAD, V, and LANNERT, R (1992) *Limicola* **6**, 265-86. BARTLETT, E (1976) *Br. Birds* **69**, 312. BÄSECKE, K (1950) *Vogelwelt* **71**, 53; (1955) *Vogelwelt* **76**, 187-8; (1956) *Vogelwelt* **77**, 190. BASSINI, E (1970) *Ric. Zool. appl. Caccia* **47**. BATES, D J (1979) *Scott. Birds* **10**, 276-7. BATES, G G and BATES, I M (1989) *Scott. Birds* **15**, 132. BATES, G G and WHITAKER, D S (1980) *Scott. Birds* **11**, 85-7. BATES, G L (1930) *Handbook of the birds of West Africa*. London; (1934) *Ibis* (13) **4**, 685-717; (1936) *Ibis* (13) **6**, 531-56; (1937) *Ibis* (14) **1**, 786-830. BATES, R S P (1938) *J. Bombay nat. Hist. Soc.* **40**, 183-90. BATES, R S P and LOWTHER, E H N (1952) *Breeding birds of Kashmir*. Oxford. BATT, L (1993) *Br. Birds* **86**, 133. BATTLORI, X and NOS, R (1985) *Misc. Zool.* **9**, 407-11. BAU, A (1902-3) *Z. Ool.* **12**, 81-6. BAUER, C-A (1976) *Anser* **15**, 221. BAUER, K (1953) *Orn. Mitt.* **5**, 224-5. BAUER, K, DVORAK, M, KOHLER, B, KRAUS, E, and SPITZENBERGER, F (1988) *Artenschutz in Österreich*. Vienna. BAUER, K and ROKITANSKY, G (1951) *Arbeiten aus der Biologischen Station Neusiedler See* **4** (1). BAUER, M (1961) *Vogelwelt* **82**, 118-19. BAUER, W, HELVERSEN, O VON, HODGE, M, and MARTENS, J (1969) *Catalogus Faunae Graeciae* **2**. Thessaloniki. BAUGNEE, J-Y (1988) *Aves* **25**, 59-61. BAUMGART, W (1967) *J. Orn.* **108**, 341-5; (1978) *Falke* **25**, 372-85; (1980) *Falke* **27**, 78-85; (1984) *Beitr. Vogelkde.* **30**, 217-42. BAUMGART, W and KASPAREK, M (1992) *Zool. Middle East* **6**, 13-19. BAUMGART, W and STEPHAN, B (1974) *Zool. Abh. Staatl. Mus. Tierkde. Dresden* **33**, 103-38; (1987) *Mitt. zool. Mus. Berlin* **63**, Suppl. *Ann. Orn.* **11**, 57-95. BÄUMER-MÄRZ, C and SCHUSTER, A (1991) In Glandt, D (ed.) *Der Kolkrabe (Corvus corax) in Mitteleuropa. Metelener Schriftenreihe Naturschutz* **2**, 69-81. BAUR, P (1981) *Vögel Heimat* **51**, 209. BAUWENS, P, BAUWENS, L, and RAHDER, J (1976) *Limosa* **49**, 201-3. BAWTREE, R F (1950) *Br. Birds* **43**, 16. BAXTER, E V and RINTOUL, L J (1953) *The birds of Scotland*. BAZELY, D R (1987) *Condor* **89**, 190-2. BAZIEV, D K (1976) *Int. Stud. Sparrows* **9**, 30-4. BEAMAN, M (ed.) (1978) *Orn. Soc. Tur-*

key Bird Rep. 1974-5. Sandy; (1986) *Sandgrouse* **8**, 1-41. BEAMAN, M, PORTER, R F, and VITTERY, A (eds) (1975) *Orn. Soc. Turkey Bird Rep. 1970-3*. Sandy. BEAR, A (1991) *Torgos* **9** (2), 41-2, 73. BEAUD, M (1991) *Nos Oiseaux* **41**, 249-50. BEAUD, M and SAVARY, L (1987) *Nos Oiseaux* **39**, 40-1. BEAUD, P and MANUEL, F (1983) *Nos Oiseaux* **37**, 39-41. BEAUDOIN, J-C (1976) *Alauda* **44**, 77-90. BEAUFORT, L F DE (1947) *Ardea* **35**, 226-30. BECHER, O M (1949) *Country Life* **106**, 120. BECHSTEIN, J M (1853) *Cage and chamber-birds*. London. BECHTOLD, I (1967) *Aquila* **73-4**, 161-70. BECKER, G B and STACK, J W (1944) *Bird-Banding* **15**, 45-68. BEECHER, W J (1978) *Bull. Chigago Acad. Sci.* **11**, 269-98. BEHLE, W H (1950) *Condor* **52**, 193-219. BEHLE, W H and ALDRICH, J W (1947) *Proc. Biol. Soc. Washington* **60**, 69-72. BEKLOVÁ, M (1972) *Zool. Listy* **21**, 337-46. BELCHER, C F (1930) *The birds of Nyasaland*. London. BELIK, V P (1981) *Ornitologiya* **16**, 151-2. BELL, B D (1967) *Br. Birds* **60**, 139; (1968) *Br. Birds* **61**, 529-30; (1969) *Br. Birds* **62**, 209-18; (1970) *Bird Study* **17**, 2169-81. BELL, B D and HORNBY, R J (1969) *Ibis* **111**, 402-5. BELL, D G (ed.) (1978) *County Cleveland Bird Rep. 1977*, 36. BELL, T H (1962) *The birds of Cheshire*. Altrincham. BELOPOL'SKI, L (1962) *Loodus. Seltsi Aastaraamat* **55**, 227-39. BELOUSOV, Y A (1986) *Tez. Dokl. 1. S'ezda Vsesoyuz. orn. Obshch. 9. Vsesoyuz. orn. Konf. 1*, 75-6. BEL'SKAYA, G S (1963) *Ornitologiya* **6**, 45-7; (1974) In Rustamov, A K (ed.) *Fauna i ekologiya ptits Turkmenii* **1**, 34-54. Ashkhabad; (1987) *Izv. Akad. Nauk Turkmen. SSR Ser. biol. Nauk* (5), 41-9; (1989) *Izv. Akad. Nauk Turkmen. SSR Ser. biol. Nauk* (3), 53-9. BENEDEN, A VAN (1946) *Alauda* **14**, 70-86. BENDER, R O (1949) *Bird-Banding* **20**, 180-2. BENDINI, L and SPINA, F (eds) (1990) *Boll. dell' Attività Inanellamento* **3**. BENECKE, W (1907) *Falke* **17**, 268-9. BENKMAN, C W (1987a) *Ecol. Monogr.* **57**, 251-67; (1987b) *Wilson Bull.* **99**, 351-68; (1988a) *Ibis* **130**, 288-93; (1988b) *Auk* **105**, 370-1; (1988c) *Auk* **105**, 578-9; (1988d) *Auk* **105**, 715-19; (1988e) *Behav. Ecol. Sociobiol.* **23**, 167-75; (1989a) *Ornis scand.* **20**, 65-8; (1989b) *Auk* **106**, 483-5; (1990) *Auk* **107**, 376-86; (in press) White-winged Crossbill. In Poole, A, Stettenheim, P, and Gill, F (eds) *The birds of North America*. Philadelphia. BENNETT, C J L (1976) *Cyprus Orn. Soc. (1957) Rep.* **20**, 1-63. BENSON, C W (1946) *Ibis* **88**, 444-61. BENSON, G B G and WILLIAMSON, K (1972) *Bird Study* **19**, 34-50. BENSON, G B G (1967) *Br. Birds* **60**, 343. BENSON, H (1890) *Zoologist* (3) **14**, 17-18. BENT, A C (1946) *Bull. US natn. Mus.* **191**; (1950) *Bull. US natn. Mus.* **197**; (1953) *Bull. US natn. Mus.* **203**; (1958) *Bull. US Nat. Mus.* **211**; (1968) *Bull. US natn. Mus.* **237**. BENTZ, P-G (1987) *Vår Fuglefauna* **10**, 91-5; (1988) *Vår Fuglefauna* **11**, 87-93; (1989) *Vår Fuglefauna* **12**, 101-10; (1990) *Fauna och Flora*, 1-9. BENTZ, P-G and GÉNSBØL, B (1988) *Norsk fuglehåndbok*. TÅnder. BERCK, K-H (1961-2) *Vogelwelt* **82**, 129-73; **83**, 8-26. BERCK, K-H and BERCK, U (1976) *Anz. orn. Ges. Bayern* **15**, 95-6. BÉRESS, J and MOLNÁR, P (1964) *Aquila* **69-70**, 57-70. BERETZK, P and KEVE, A (1971) *Lounais-Hämeen Luonto* **42**. BERETZK, P, KEVE, A, and MARIÁN, M (1962) *Acta zool. Acad. Sci. Hung.* **8**, 251-71. BERETZK, P, KEVE, A, and MARIÁN, M (1969) *Bonner zool. Beitr.* **20**, 50-9. BEREZOVIKOV, N N (1983) *Ornitologiya* **18**, 187. BEREZOVIKOV, N N and KOVSHAR', A F (1992) *Russ. Orn. Zh.* **1**, 221-6. BERG, A B VAN DEN (1982a) *Dutch Birding* **4**, 60-2; (1982b) *Dutch Birding* **4**, 136-9; (1988) *Dutch Birding* **10**, 92-3; (1989) *List of Dutch bird species 1989*; (1990) *Vogels nieuw in Nederland*

1990. BERG, A B VAN DEN, BY, R A DE, and CDNA (1991) *Dutch Birding* 13, 41-57; (1992) *Dutch Birding* 14, 73-90. BERG, A B VAN DEN, and BLANKERT, J J (1980) *Dutch Birding* 2, 33-5. BERG, A B VAN DEN and COTTAAR, F (1986) *Dutch Birding* 8, 57-9. BERG, A B VAN DEN and ROEVER, J W DE (1984) *Dutch Birding* 6, 139-40. BERGER, S B (1968) *Sterna* 8, 157, 159. BERGH, L M J VAN DEN, JAARSVELD, B VAN, and DRIEL, F VAN (1989) *Limosa* 62, 91-2. BERGIER, P (1989) *Newsl. Bahrain nat. Hist. Soc.* 1. BERGMAN, G (1952) *Ornis fenn.* 29, 105-7; (1953) *Acta Soc. Faun. Flor. Fenn.* 69(4), 1-15; (1956) *Ornis fenn.* 33, 61-71. BERGMAN, S (1935) *Zur Kenntnis nordostasiatischer Vögel.* Stockholm. BERGMANN, H-H (1983a) *Cyprus orn. Soc. (1969), Rep.* 8, 41-54; (1983b) *Vogelkdl. Ber. Niedersachs.* 15, 1-4; (1991) *Gef. Welt* 115, 426-8. BERGMANN, H-H, FABREWITZ, S, GRAUPNER, B, HINRICHS, K, and ZUCCHI, H (1982) *Math.-naturwiss. Unterricht* 35, 172-81. BERGMANN, H-H and HELB, H-W (1982) *Stimmen der Vögel Europas.* Munich. BERGMANN, H-H, ROY, A, and SCHRÖDER, H (1988) *Gef. Welt.* 112, 280-4. BERGMANN, H-H, ZIETLOW, S, and HELB, H-W (1984) *J. Orn.* 125, 59-67. BERG-SCHLOSSER, G (1978) *J. Orn.* 119, 111-13. BERNASEK, O (1985) *Gef. Welt* 109, 124-5. BERNDT, R (1960) *Orn. Mitt.* 12, 181. BERNDT, R and DANCKER, P (1960) *Proc. int. orn. Congr.* 12, 97- 109. Helsinki. BERNDT, R and FRANTZEN, M (1987) *Vogelkdl. Ber. Niedersachs.* 19, 93. BERNDT, R and FRIELING, F (1939) *J. Orn.* 87, 593-638. BERNDT, R and FRIELING, H (1944) *Beitr. Fortpfl. Vögel* 22, 68. BERNDT, R and MOELLER, J (1960) *Braunschweig. Heimat* 46, 119-24. BERNDT, R and WINKEL, W (1987) *Vogelwelt* 108, 98-105. BERNHOFT-OSA, A (1956) *Vår Fågelvärld* 15, 245-7; (1959) *Stavanger Mus. Årb. 1959,* 139-42; (1960) *Vår Fågelvärld* 19, 220-3; (1965) *Stavanger Mus. Årb.,* 109-18; (1978) *Vår Fuglefauna* 1, 93-5. BERNIS, F (1933) *Bol. Soc. Esp. Hist. Nat.* 33, 377-84; (1945) *Bol. Soc. Esp. Hist. Nat.* 43, 93-145; (1954) *Ardeola* 1, 11-85; (1989a) *Commun. INIA* 53 Ser. *Recursos naturales;* (1989b) *Commun. INIA* 54 Ser. *Recursos naturales.* BERROW, S D, KELLY, T C, and MYERS, A A (1991) *Irish Birds* 4, 393-412. BERRUTI, A and NICHOLS, G (1991) *Birding in Southern Africa* 43, 52-7. BERTHOLD, P (1964) *Vogelwarte* 22, 236-75; (1971) *Vogelwelt* 92, 141-7. BERTHOLD, P and BERTHOLD, H (1987) *Bocagiana* 110, 1-8. BERTHOLD, P and GWINNER, E (1972) *Vogelwarte* 26, 356-7; (1978) *J. Orn.* 119, 338-9. BERTHOLD, P and SCHLENKER, R (1982) *Dutch Birding* 4, 100-2. BESSON, J (1968) *Alauda* 36, 292-3; (1982) *Nos Oiseaux* 36, 289-90. BÉTHUNE, G DE (1961) *Gerfaut* 51, 387-98; (1986) *Oriolus* 52, 38. BETTMANN, H (1969) *Orn. Mitt.* 21, 26-7. BEVEN, G (1946) *Br. Birds* 39, 23; (1947) *Br. Birds* 40, 308-10; (1964) *Lond. Nat.* 43, 86-109. BEZEMER, K W L (1979) *Vogeljaar* 27, 128-30. BEZZEL, E (1957) *Anz. orn. Ges. Bayern* 4, 589-707; (1972) *Vogelwarte* 26, 346-52; (1985) *J. Orn.* 126, 434-9; (1987) *Vogelwelt* 108, 71-2; (1988) *J. Orn.* 129, 71-81. BEZZEL, E and BRANDL, R (1988) *Anz. orn. Ges. Bayern* 27, 45-65. BEZZEL, E and LECHNER, F (1978) *Die Vögel des Werdenfelser Landes.* Greven. BEZZEL, E, LECHNER, F, and RANFTL, H (1980) *Arbeitsatlas der Brutvögel Bayerns.* Greven. BHARUCHA, E K (1989) *J. Bombay nat. Hist. Soc.* 86, 450. BIBBY, C J (1977) *Ring. Migr.* 1, 148-57. BIBBY, C J and CHARLTON, T D (1991) *Açoreana* 7, 297-304. BIBBY, C J, CHARLTON, T D, and RAMOS, J (1992) *Br. Birds* 85, 677-80. BIBBY, C J and LUNN, J (1982) *Biol. Conserv.* 23, 167-186. BIBER, J-P and LINK, R (1974) *Nos Oiseaux* 32, 273-4. BIBER, O (1984) *Orn. Beob.* 81, 1-28. BIDDULPH, C H (1954) *J. Bombay*

nat. Hist. Soc. 52, 208-9. BIEDERMANN-IMHOOF, R (1913) *Orn. Monatsber.* 21, 4-6. BIELFELD, H (1980) *Kanarien.* Stuttgart; (1981) *Zeisige, Kardinäle und andere Finkenvögel.* Stuttgart. BIERI, W (1945) *Orn. Beob.* 42, 140-2. BIGGER, W K (1931) *Ibis* (13) 1, 584-5. BIGNAL, E and CURTIS, D J (eds) (1989) *Choughs and land-use in Europe.* Scottish Chough Study Group. BIGNAL, E, MONAGHAN, P, BENN, S, BIGNAL, S, STILL, E, and THOMPSON, P M (1987) *Bird Study* 34, 39-42. BIGOT, L (1966) *Terre Vie* 113, 295-315. BIJLSMA, R G (1979) *Limosa* 52, 53-71; (1982) *Ardea* 70, 25-30. BIJLSMA, R G and MEININGER, P L (1984) *Gerfaut* 74, 3-13. BIJLSMA, R G, RODER, F E DE, and BEUSEKOM, R VAN (1988) *Limosa* 61, 1-6. BILLE, R-P (1978) *Nos Oiseaux* 34, 261; *Nos Oiseaux* 35, 227-31. BILLETT, A E (1989) *Br. Birds* 82, 81. BINFORD, L C (1971) *Californian Birds* 2, 1-10; (1989) *Orn. Monogr. AOU* 43. BIRCH, A (1990) *Birding World,* 3, 308-9. BIRD, E G (1935) *Ibis* (13) 5, 438-41. BIRKHEAD, T R (1972) *Br. Birds* 65, 356-7; (1974a) *Br. Birds* 67, 221-9; (1974b) *Ornis scand.* 5, 71-81; (1979) *Anim. Behav.* 27, 866-74; (1982) *Anim. Behav.* 30, 277-83; (1989) *Br. Birds* 82, 583-600; (1991) *The magpies.* London. BIRKHEAD, T R, ATKIN, L, and MØLLER, A P (1987) *Behaviour* 101, 101-38. BIRKHEAD, T R and CLARKSON, K (1985) *Behaviour* 94, 324-32. BIRKHEAD, T R, EDEN, S F, CLARKSON, K, GOODBURN, S F, and PELLATT, J (1986) *Ardea* 74, 59-68. BIRKHEAD, T R and GOODBURN, S F (1989) In Newton, I (ed.) *Lifetime reproduction in birds,* 173-82. London. BIRKHEAD, T R and MØLLER, A P (1992) *Sperm competition in birds.* London. BISWAS, B (1950) *J. zool. Soc. India* 2 (1). BJORDAL, H (1983a) *Fauna norv.* (C) *Cinclus* 6, 105-8; (1983b) *Vår Fuglefauna* 6, 34-6; (1984) *Fauna norv.* (C) *Cinclus* 7, 21-3. BJÖRKLUND, M (1989a) *Behav. Ecol. Sociobiol.* 25, 137-40; (1989b) *Anim. Behav.* 38, 1081-83; (1989c) *Ornis scand.* 20, 255-64; (1990a) *Auk* 107, 35-44; (1990b) *Ibis* 132, 613-17. BJØRNSEN, B (1988) Cand. Sci. Anim. Ecol. Thesis. Bergen Univ. BLAIR, C M G (1961) *Ibis* 103a, 499-502. BLAIR, H M S (1936) *Ibis* (13) 6, 280-308. BLAIR, R H and TUCKER, B W (1941) *Br. Birds* 34, 206-15. BLAKE, C H (1956) *Bird-Banding* 27, 16-22; (1962) *Bird-Banding* 33, 97-9; (1964) *Bird-Banding* 35, 125-7; (1967) *Bird-Banding* 38, , 234; (1969) *Bird-Banding* 40, 133-9. BLANA, E (1970) *Charadrius* 6, 23-5. BLANCHARD, B D (1941) *Univ. Calif. Publ. Zoöl.* 46 (1). BLANCHARD, B D and ERICKSON, M M (1949) *Univ. Calif. Publ. Zoöl.* 47, 255-318. BLANCO, G, CUEVAS, J A, and FARGALLO, J A (1991) *Ardeola* 38, 91-9. BLANCOU, L (1939) *Oiseau* 9, 410-85. BLANFORD, W T (1876) *Eastern Persia. An account of the journeys of the Persian Boundary Commission, 1870-2* 2: *Zoology and Geology.* London. BLASER, P (1970) *Orn. Beob.* 67, 297-9; (1973) *Orn. Beob.* 70, 186-7; (1974) *Orn. Beob.* 71, 322-3. BLASIUS, R (1886) *Ornis* 2, 437-550. BLATHWAYT, F L (1903) *Zoologist* (4) 7, 26-7. BLEDSOE, A H (1988) *Wilson Bull.* 100, 1-8. BLEM, C R (1975) *Wilson Bull.* 87, 543-9; (1981) *Condor* 83, 370-6. BLOCH, D and SØRENSEN, S (1984) *Yvirlit yvir Føroya fuglar.* Tórshavn. BLOK, A A and SPAANS, A L (1962) *Limosa* 35, 4-16. BLOMGREN, A (1964) *Lavskrika.* Stockholm; (1971) *Br. Birds* 64, 25-8; (1983) *Vår Fågelvärld* 42, 343-6. BLONDEL, J (1963) *Alauda* 31, 22-6; (1969) *Synécologie des passereaux résidents et migrateurs dans le Midi Méditerranéen français.* Marseille; (1979) *Biogéographie et écologie.* Paris. BLONDEL, J and ISENMANN, P (1981) *Guide des oiseaux de Camargue.* Neuchâtel; 750e 764e 768e 687e. BLUME, C A (1963) *Dansk orn. Foren. Tidsskr.* 57, 19-21. BLUME, D (1967) *Aus-*

drucksformen unserer Vögel. Wittenberg Lutherstadt. BLU-MEL, H (1976) *Der Grünling*. 2nd edition. Wittenber Lutherstadt; (1982) *Die Rohrammer*. Wittenberg Lutherstadt; (1983*a*) *Der Grünling*. Wittenberg Lutherstadt; (1983*b*) *Abh. Ber. Naturkundemus. Görlitz* **56** (4); (1986) *Abh. Ber. Naturkdemus. Görlitz* **59** (3). BLYTH, E (1867) *Ibis* (2) **3**, 1-48. BOAG, P T and RATCLIFFE, L M (1979) *Condor* **81**, 218-9. BOBRETSOV, A V and NEUFELD, N D (1986) *Tez. Dokl. I. S'ezda Vsesoyuz. orn. Obshch. 9. Vsesoyuz. orn. Konf.* **1**, 85-6. BOCCA, M and MAFFEI, G (1984) *Gli uccelli della Valle d'Aosta*. Aosta. BOCK, W J and MARONY, J J (1978) *Bonn. zool. Beitr.* **29**, 122-47. BOCHEŃSKI, Z (1970) *Acta Zool. Cracov.* **15**, 1-59. BOCHEŃSKI, Z and OLEŚ, T (1977) *Acta Zool. Cracov.* **22**, 319-71; (1981) *Acta Zool. Cracov.* **25**, 3-12. BOCK, C E and LEPTHIEN, L W (1976) *Amer. Nat.* **110**, 559-71. BOCK, W J and MORONY, J J (1978) *Bonn. zool. Beitr.* **29**, 122-47. BODDY, M (1979) *Ringers' Bull* **5**, 87; (1981) *Ring. Migr.* **3**, 193-202; (1983) *Ornis scand.* **14**, 299-308; (1984) *Ring. Migr.* **5**, 91-100. BODDY, M and BLACKBURN, A C (1978) *Ring. Migr.* **2**, 27-33. BODDY, M and SELLERS, R M (1983) *Ring. Migr.* **4**, 129-38. BODENSTEIN, G (1953) *Orn. Mitt.* **5**, 72. BOECKER, M (1970) *Bonn. zool. Beitr.* **21**, 183-236. BOEHME, R L (1958) *Uchen. zap. Severo-Osetinsk. gos. ped. Inst.* **23** (1), 111-83. BOERTMANN, D, OLSEN, K M, and PEDERSEN, B B (1986) *Dansk orn. Foren. Tidsskr.* **80**, 35-57. BOERTMANN, D, SØRENSEN, S, and PIHL, S (1986) *Dansk. orn. Foren. Tidsskr.* **80**, 121-30. BOESMAN, P (1992) *Dutch Birding* **14**, 161-9. BOURNE, W R P (1967) *Ibis* **109**, 141-67. BOGLIANI, G (1985) *Riv. ital. Orn.* **55**, 140-50. BOGLIANI, G and BRANGI, A (1990) *Bird Study* **37**, 195-8. BOGORODSKI, Y V (1981) *Ornitologiya* **16**, 153. BOHAC, D (1967) *Angew. Orn.* **2**, 151-2. BOHLEN, H D (1989) *The birds of Illinois*. Bloomington. BÖHMER, A (1973) *Orn. Beob.* **70**, 103-12; (1974) *Orn. Beob.* **71**, 279-82; (1976*a*) *Orn. Beob.* **73**, 109-36; (1976*b*) *Orn. Beob.* **73**, 136-40. BÖHNER, J, CHAIKEN, M L, BALL, G F, and MARLER, P (1990) *Hormon. Behav.* **24**, 582-94. BÖHR, H-J (1962) *Bonn. zool. Beitr.* **13**, 50-114. BOIE, F (1866) *J. Orn.* **14**, 1-4. BOIE, H (1922) *Biol. Centralbl.* **42**, 87-93. BOLAM, G (1912) *Birds of Northumberland and the eastern borders*. Alnwick; (1913) *Wildlife in Wales*. London. BOLLE, C (1856) *J. Orn.* **4**, 17-31; (1857) *J. Orn.* **5**, 305-51; (1858*a*) *J. Orn.* **6**, 125-51; (1858*b*) *Naumannia* **8**, 369-93. BOLTON, M (1986) *Distribution of breeding birds in the Algarve in relation to habitat*. In Pullan, R (ed.) *A Rocha Bird Report 1986*, 5-8. BOND, J (1938) *Can. Field-Nat* **52**, 3-5; (1985) *Birds of the West Indies*. London. BOND, L M G (1946) *Bird Notes News* **22**, 32. BONHAM, P F (1970*a*) *Br. Birds* **63**, 28-32; (1970*b*) *Br. Birds* **63**, 262-4. BONHAM, P F and SHARROCK, J T R (1969) *Br. Birds* **62**, 550-52. BOOTH, C J (1979) *Scott. Birds* **10**, 261-7; (1986) *Scott. Birds* **14**, 51. BOOTH, C, CUTHBERT, M, and REYNOLDS, P (1984) *The birds of Orkney*. Stromness; (1986) *Orkney Bird Rep. 1985*, 3-57. BOOTH, C J and REYNOLDS, P (1992) *Br. Birds* **85**, 245-6. BORGVALL, T (1952) *Vår Fågelvärld* **11**, 11-15. BOROS, I and HORVÁTH, L (1955) *Acta Zool. Acad. Sci. Hung.* **1**, 43-51. BORRAS, A and SENAR, J C (1986) *Misc. Zool.* **10**, 403-6; (1991) *J. Orn.* **132**, 285-9. BORROR, D J (1961) *Ohio J. Sci.* **61**, 161-74. BORTOLI, L (1973) *Productivity, population dynamics and systematics of granivorous birds*, 249-52. Warsaw. BORTOLI, L and BRUGGERS, R (1976) PNUD/Recherche pour la lutte contre les oiseaux granivores 'Quelea quelea'. SR 258. Dakar. BOS, G, SLIJPER, H J, and TAAPKEN, J (1945) *Limosa* **18**, 56-68. BÖSENBERG, K

(1958) *Falke* **5**, 58-61. BOSSEMA, I (1967) *Levende Nat.* **70**, 86-92; (1979) *Behaviour* **70**, 1-117; (1980) *Corvid Newsl.* **1** (2). BOSSEMA, I and BENUS, R F (1985) *Behav. Ecol. Sociobiol.* **16**, 99-104. BOSSEMA, I and POT, W (1974) *Levende Nat.* **77**, 265-79. BOSSEMA, I, RÖELL, A, BAEYENS, G, ZEEVALKING, H, and LEEVER, H (1976) *Levende Nat.* **79**, 149-66. BOSWALL, J (1970) *Br. Birds*, **63**, 256-7; (1985*a*) *BBC Wildl.* **3**, 174-9; (1985*b*) In Campbell and Lack (1985), 599. BOSWELL, C and NAYLOR, P (1957) *Iraq nat. Hist. Mus. Publ.* **13**, 16. BOUARD, R (1983) *Bull. UCAGO* 3-4, 6-15. BOURDELLE, E and GIBAN, J (1950-51) *Bull. stations françaises de Baguage* **7**. Paris. BOURNE, W R P (1953) *Br. Birds* **46**, 381-2; (1955) *Ibis* **97**, 508-56; (1957) *Ibis* **99**, 182-90; (1966) *Ibis* **108**, 425-9; (1967) *Ibis* **109**, 141-67; (1986) *Bull. Br. Orn. Club* **106**, 163-70. BOURNE, W R P (1992) *Dutch Birding* **7**, 21-2. BOURNE, W R P and NELDER, J A (1951) *Br. Birds* **44**, 386-7. BOURRILLON, P (1961) *Oiseau* **31**, 247. BOUTET, J-Y and PETIT, P (1987) *Atlas des oiseaux nicheurs d'Aquitaine 1974-1984*. Bordeaux. BOXBERGER, L VON (1930) *Beitr. Fortpfl. Vögel* **6**, 211. BOWDEN, C G R (1987) *Sandgrouse* **9**, 94-7. BOWDEN, C G R and BROOKS, D J (1987) *Sandgrouse* **9**, 111-14. BOYARCHUK, V P (1990) *Vestnik Zool.* 1990 (2), 52-7. BOYD, A W (1932) *Br. Birds* **25**, 278-85; (1933) *Br. Birds* **26**, 273-4; (1934) *Br. Birds* **27**, 259-60; (1935) *Br. Birds* **28**, 347-9; (1949) *Br. Birds* **42**, 213-4; (1951) *A country parish*. London. BOYLE, G (1966) *Br. Birds* **59**, 342. BOZHKO, S I (1974) *Acta Orn.* **14**, 39-57; (1980) *Der Karmingimpel*. Wittenberg Lutherstadt. BOZSKO, S (1977) *Aquila* **83**, 289-90, 305. BRAAE, L (1975) *Dansk orn. Foren. Tidsskr.* **69**, 41-53. BRACK, H (1977) *Bird-Banding* **48**, 370. BRAČKO, F (1992) *Acrocephalus* **13**, 57-8. BRADER, M (1989) *Egretta* **32**, 18-20. BRADSHAW, C (1991) *Birding World* **4**, 354-5; (1991) *Br. Birds* **84**, 310-11; (1992) *Br. Birds* **85**, 653-65. BRANDL, R and BEZZEL, E (1988) *Zool. Anz.* **221**, 411-17; (1989) *Orn. Beob.* **86**, 137-43. BRANDNER, J (1991) *Egretta* **34**, 73-85. BRANDT, H (1960) *Anz. orn. Ges. Bayern* **5**, 597-8; (1962) *Vogelwelt* **83**, 81-2. BRAY, O E, DE GRAZIO, J W, GUARINO, J L, and STREETER, R G (1974) *Inland Bird Banding News* **46**, 204-9. BRAY, O E, ROYALL, W C, GUARINO, J L, and DE GRAZIO, J W (1973) *Bird-Banding* **44**, 1-12. BRAZIER, H, DOWDALL, J F, FITZHARRIS, J E, and GRACE, K (1986) *Irish Birds* **3**, 287-336. BRAZIL, M A (1991) *The birds of Japan*. London. BRAUN, P (1989) *Falke* **36**, 154-5. BRAUNBERGER, C W (1990) *Orn. Mitt.* **42**, 15-16. BREIFE, B, HIRSCHFELD, E, KJELLÉN, N, and ULLMAN, M (1990) *Vår Fågelvärld Suppl.* **13**. BRÉMOND, J-C (1962) *Angew. Orn.* **1**, 49-63. BRENCHLEY, A (1986) *J. Zool. Lond.* **210**, 261-78. BRENNECKE, H-E (1953) *Orn. Mitt.* **5**, 149. BREWER, D (1990) *Br. Birds* **83**, 289-90. BRIAN, M V and BRIAN, A D (1948) *Trans. Herts. nat. Hist. Soc.* **23**, 30-6. BRICHETTI, P (1973) *Riv. ital. Orn.* **43**, 519-649; (1976) *Atlante ornitologico italiano* **2**. Brescia; (1982) *Riv. ital. Orn.* **52**, 3-50; (1983) *Riv. ital. Orn.* **53**, 101-44; (1986) *Riv. ital. Orn.* **56**, 3-39. BRICHETTI, P and COVA, C (1976) *Gli Ucc. Ital.* **1**, 28-31. BRICHETTI, P, CAFFI, M, and GANDINI, S (1992) *Nat. Bresciana* **28**. BRICHETTI, P and MASSA, B (1984) *Riv. ital. Orn.* **54**, 3-37; (1991) *Riv. ital. Orn.* **61**, 3-9. BRICHOVSKY, A (1968) *Falke* **15**, 170-3. BRIDGMAN, C J (1962) *Br. Birds* **55**, 461-70. BRIEN, Y, BESSEC, A, and LESOUEF, J-Y (1982) *Oiseau* **52**, 87-9. BRIGGS, F S and OSMASTON, B B (1928) *J. Bombay Nat. Hist. Soc.* **32**, 744-61. BRIGHOUSE, U W (1954) *Devon Birds* **7**, 38-41. BRINGELAND, R (1964) *Sterna* **6**, 4-6. BRITISH ORNITHOLOGISTS' UNION (1992) *Ibis* **134**, 211-14. BRIT-

TON, P L (1970) *Ostrich* **41**, 145-90; (ed.) (1980) *Birds of East Africa*. Nairobi. BRITTON, P L, and DOWSETT, R J (1969) *Ostrich* **40**, 55-60. BROAD, R A (1974) *Br. Birds* **67**, 297-301; (1978) *Scott. Birds* **10**, 58-9; (1981) *Br. Birds* **74**, 90-4. BROAD, R A and ODDIE, W E (1980) *Br. Birds* **73**, 402-8; (1982) In Sharrock, J T R and Grant, P J (eds) *Birds new to Britain and Ireland*, 212-16. Calton. BROCKMANN, H J (1980) *Wilson Bull.* **92**, 394-98. BROD, G (1988) *Beih. Veröff. Nat. Land. Bad.-Württ.* **53**, 83-90. BRODIE, E (1985) *Br. Birds* **78**, 244. BROKHOVICH, S A (1990) *Ornitologiya* **24**, 168-9. BROMLEY, F C (1947) *Br. Birds* **40**, 114. BROOD, K and SÖDERQUIST, T (1967) *Vår Fågelvärld* **26**, 266-8. BROOKE, R K (1962) *Ostrich* **33** (1), 23-5; (1973) *Auk* **90**, 206; (1976) *Bull. Br. Orn. Club* **96**, 8-13. BROOKS, D J (1987) *Sandgrouse* **9**, 115-20. BROOKS, D J, EVANS, M I, MARTINS, R P, and PORTER, R F (1987) *Sandgrouse* **9**, 4-66. BROOKS, W S (1968) *Wilson Bull.* **80**, 253-80; (1973) *Bird-Banding* **44**, 13-21. BROSSET, A (1956) *Alauda* **24**, 266-71; (1957a) *Alauda* **25**, 43-50; (1957b) *Alauda* **25**, 224-6; (1961) *Trav. Inst. sci. chérifien Ser. Zool.* **22**, 7-155; (1984) *Alauda* **52**, 81-101. BROSSET, A and ERARD, C (1986) *Les oiseaux des régions forestières du nord-est du Gabon* 1. Paris. BROUN, M (1971) *Auk* **88**, 924-5. BROWN, A W, LEVEN, M R, and PRATO, S R D DA (1984) *Scott. Birds* **13**, 107-111. BROWN, B J (1985) *Br. Birds* **78**, 244. BROWN, J L (1987) *Helping and communal breeding in birds*. Princetown. BROWN, L H (1967) *Ibis* **109**, 275. BROWN, R G B (1959) *Br. Birds* **52**, 98; (1963) *Ibis* **105**, 63-75. BROWN, R H (1924) *Br. Birds* **18**, 122-8; (1942) *Naturalist*, 39-45. BROWN, R N (1974) M Sc Thesis. Alaska Univ. BROWNE, K and BROWNE, E (1956) *Br. Birds* **49**, 241-57. BROWNE, P W P (1960) *Br. Birds* **53**, 575-7; (1981) *Bull. Br. Orn. Club* **101**, 306-10. BROWNING, M R (1976) *Bull. Br. Orn. Club* **96**, 44-7. BROYD, S J (1985) *Br. Birds* **78**, 647-56. BRUCH, A, ELVERS, H, POHL, C, WESTPHAL, D, and WITT, K (1978) *Orn. Ber. Berlin (West)* 3 suppl. Berlin. BRUCH, A and LÖSCHAU, M (1960) *Orn. Mitt.* **12**, 31. BRUCKER, J W (ed.) (1985) *Oxford Orn. Soc. Rep. 1984.* BRUCKER, J W, GOSLER, A G, and HERYET, A R (eds) (1992) *Birds of Oxfordshire*. Newbury. BRUDENELL-BRUCE, P G C (1975) *The birds of New Providence and the Bahama Islands*. London. BRUGGERS, R L and BORTOLI, L (1976) *Terre Vie* **30**, 521-7. BRUSTER, K-H (1988) *Hamburger avifaun. Beitr.* **21**, 189. BRUUN, B (1984) *Courser* 1, 44-6. BRUUN, B, DELIN, H, and SVENSSON, L (1986) *Birds of Britain and Europe*. Twickenham. BRYSON, D K (1947) *Br. Birds* **40**, 209. BUB, H (1953) *Orn. Mitt.* 5, 6; (1955) *Ibis* **97**, 25-37; (1962) *Falke* **9**, 164-71; (1969) *Vogelwarte* **25**, 134-41; (1976a) *Orn. Mitt.* **28**, 6-12; (1976b) *J. Orn.* **117**, 461; (1977) *Orn. Mitt.* **29**, 55-60; (1978) *Sterna* **17**, 21-3; (1985) *Beitr. Vogelkde.* **31**, 189-213; (1986) *Beitr. Vogelkde.* **32**, 249-65; (1987) *Beitr. Vogelkde.* **33**, 313-25; (1989) *Falke* **36**, 41. BUB, H, HEFT, H, and WEBER, H (1959) *Falke* **6**, 3-9, 48-54. BUB, H and HINSCHE, A (1982) *Hercynia* (NF) **19**, 322-62. BUB, H and KUMERLOEVE, H (1954) *Orn. Mitt.* **6**, 205-12, 225-31. BUB, H and PANNACH, G (1988) *Verh. orn. Ges. Bayern* **24**, 411-65; (1991) *Beitr. Naturk. Niedersachs.* **44**, 272-90; (1992) *Beitr. Naturk. Niedersachs.* **45**, 192-215 BUB, H and PRÄKELT, A (1952) *Beitr. Naturkde. Niedersachs.* 1, 10-2 BUB, H and VRIES, R de (1973) *Das Planberingungs-Programm am Berghänfling (Carduelis f. flavirostris) 1952-70*. Wilhelmshaven. BUBNOV, M A (1956) *Zool. Zh.* **35**, 316-18. BÜCHEL, H P (1974) *Mitt. nat.-forsch. Ges. Luzern* **24**, 73-94; (1983) *Orn. Beob.* **80**, 1-28. BUCKLAND, S T, BELL, M

V, and PICOZZI, N (1990) *The birds of north-east Scotland*. Aberdeen. BUCKLAND, S T and KNOX, A G (1980) *Br. Birds* **73**, 360-1. BUCKNALL, R H (1983) *Scott. Birds* **12**, 191-3. BÜHLER, A (1968) *Orn. Beob.* **65**, 26-8. BUITRON, D (1983a) *Anim. Behav.* **31**, 211-20; (1983b) *Behaviour* **87**, 209-36; (1988) *Condor* **90**, 29-39. BULL, J (1974) *Birds of New York State*. New York. BULLOCK, and DEL-NEVO, A (1983) *Peregrine* **5**, 226-9. BULLOCK, I (1985) *Br. Birds* **78**, 247-8. BULLOCK, I D, DREWETT, D R, and MICKLE-BURGH, S P (1983a) *Br. Birds* **76**, 377-401; (1983b) *Irish Birds* **2**, 257-71; (1983c) *Peregrine* **5**, 229-37; (1986) *Nature in Wales* (N.S.) **4** (1985), 46-57. BULLOUGH, W S (1942a) *Ibis* (14) **6**, 225-39; (1942b) *Phil. Trans. Roy. Soc.* (B) **231**, 165-241. BULLOUGH, W S and CARRICK, R (1940) *Nature* **145**, 629. BÜLOW, B von (1990) *Charadrius* **26**, 151-89. BUNDESDEUTSCHER SELTENHEITENAUSSCHUSS (1989) *Limicola* **3**, 157-96; (1991) *Limicola* **5**, 186-220; (1992) *Limicola* **6**, 153-177. BUNDY, G (1976) *The birds of Libya*. London. BUNDY, G, CONNOR, R J, and HARRISON, C J O (1989) *Birds of the Eastern Province of Saudi Arabia*. London. BUNDY, G and MORGAN, J H (1969) *Bull. Br. Orn. Club* **89**, 139-44, 151-9. BUNDY, G and WARR, E (1980) *Sandgrouse* **1**, 4-49. BUNYARD, P F (1932) *Bull. Br. Orn. Club* **52**, 83-4. BURG, G DE (1911) *Les oiseaux de la Suisse* **13**. Berne. BURKE, T and BRUFORD, M W (1987) *Nature* **327**, 149-52. BURKITT, J P (1935) *Br. Birds* **28**, 322-6; (1936) *Br. Birds* **29**, 334-8. BÜRKLI, W (1972) *Orn. Beob.* **69**, 183-4. BURLEIGH, T D (1958) *Georgia birds*. Norman; (1960) *Auk* **77**, 210-5; (1972) *Birds of Idaho*. Caldwell. BURLEIGH, T D and PETERS, H (1948) *Proc. Biol. Soc. Washington* **61**, 111-26. BURNETT, B W (1965) *Countryman* **65**, 130. BURNIER, E (1977) *Alauda* **45**, 238-9. BURNS, D W (1993) *Br. Birds* **86**, 115-20. BURNS, P S (1957) *Bird Study* **4**, 62-71. BURT, E H and HAILMAN, J P (1978) *Ibis* **120**, 153-70. BURTON, J A (1974) *The naturalist in London*. Newton Abbot. BURTON, M (1976) *Daily Telegraph* London 30 Oct 1976; (1979) *Wildlife* **21** (10), 19. BURTON, M and BURTON, R (1977) *Inside the animal world*. London. BUSCHE, G (1970) *Corax* **3**, 51-70; (1980) *Vogelbestände des Wattenmeeres von Schleswig-Holstein*. Heide; (1989) *Vogelwarte* **35**, 11-20; (1991) *Vogelwelt* **112**, 162-76. BUSCHE, G, BOHNSACK, P, and BERNDT, R K (1975) *Corax* **5**, 114-26. BUSCHING, W-D (1988) *Falke* **35**, 42-7. BUSSE, P (1962) *Acta Orn.* **6**, 209-30; (1963) *Acta Orn.* **7**, 189-220; (1965) *Ekol. Polska* (A) **13**, 491-514; (1968) *Notatki Orn.* **9** (1-2), 24-6; (1969) *Acta orn.* **11**, 263-328; (1970) *Notatki Orn.* **11**, 1-15; (1976) *Acta Zool. Cracov.* **21**, 121-261. BUSSE, P and HALASTRA, G (1981) *Acta Orn.* **18** (3), 1-122. BUTLER, A G (1899) *Foreign species in captivity*; (1905) *Ibis* (8) **5**, 301-401. BUTLER, A S (1989) *Br. Birds* **82**, 474-5. BUTLER, R W and CAMPBELL, R W (1987) *Can. Wildl. Serv. Occas. Pap.* **65**. BUTLIN, S M (1959) *Br. Birds* **52**, 387-8. BUTTERFIELD, R (1906) *Zoologist* (4) **10**, 31-3. BUTURLIN, L (1929) *Sistematicheskie zametki o ptitsakh Severnogo Kavkaza*. Makhachkala. BUTURLIN, S A (1906) *Ibis* (8) **6**, 407-27; (1907) *Orn. Monatsber.* **15**, 8-9. BUXTON, E J M (1960) *Ibis* **102**, 127-9. BUZZARD, G G (1989) *Br. Birds* **82**, 620-1. BYARS, T and GALBRAITH, H (1980) *Br. Birds* **73**, 2-5. BYKOVA, L P (1990) In Kurochkin, E N (ed.) *Sovremennaya ornitologiya*, 98-116. Moscow. BYRD, G V (1979) *Elepaio* **39**, 69-70.

CABOT, D (1965) *Irish Nat. J.* **15**, 95-100. CADE, T J (1952) *Condor*, **54**, 363; (1953) *Condor* **55**, 43-4. CADMAN, M D, EAGLES, P F J, and HELLEINER, F M (1987) *Atlas of the* breeding birds of Ontario. Waterloo. CADMAN, W A (1947) Br. Birds 40, 209-10. CAIRNS, J (1952) Malayan Nat. J. 7, 106-7. CALDWELL, J A (1949) Br. Birds 42, 288. CALDWELL, L D, ODUM, E P, and MARSHALL, S G (1963) Wilson Bull. 75, 428-34. CALHOUN, J B (1947a) Auk 64, 305-6; (1947b) Amer. Nat. 81, 203-28. CALVERT, M (1988) Br. Birds 81, 531-2. CAMBI, D and MICHELI, A (1986) Nat. Bresc. Ann. Mus. Civ. Sc. Nat. 22, 103-78. CAMPBELL, B (1953) Finding nests. London; (1968) Countryman 70 (2), 306-9; (1973) Forest Comm. Forest Rec. 86. CAMPBELL, B and FERGUSON-LEES, J (1965) A field guide to birds' nests. London. CAMPBELL, B and LACK, E (eds) (1985) A dictionary of birds. Calton. CAMPBELL, P O (1972) M Sc Thesis. Massey Univ. CAMPBELL, J W (1936) Br. Birds 29, 306-9; (1936) Br. Birds 30, 209-18. CAMPBELL, R W, SHEPARD, M G, and MACDONALD, B A (1974) Vancouver birds in 1972. Vancouver. CAMPINHO, F, LOURENÇO, J, and RODRIGUES, P (1991) Airo 2, 21. CANNELL, P F, CHERRY, J D, and PARKES, K C (1983) Wilson Bull. 95, 621-7. CANNINGS, R A, CANNINGS, R J, and CANNINGS, S G (1987) Birds of the Okanagan valley, British Columbia. Victoria. CANO, A and KÖNIG, C (1971) J. Orn. 112, 461-2. CARACO, T and BAYHAM, M C (1982) Anim. Behav. 30, 990-6. CARLO, E A DI (1991) Sitta 5, 35-47. CARLOTTO, L (1991) Riv. ital. Orn. 61, 48-9. CARLSON, T (1946) Vår Fågelvärld 5, 37-38. CARPELAN, J (1929a) Beitr. Fortpfl. Vögel 5, 60-3; (1929b) Beitr. Fortpfl. Vögel 5, 198-201; (1932) Beitr. Fortpfl. Vögel 8, 56-8. CARR, D (1969) Br. Birds 62, 238. CARRASCAL, L M (1987) Ardeola 34, 193-224; (1988) Donãna Acta Vert. 15, 111-31. CARROLL, C J (1916) Br. Bird 9, 293-4. CARRUTHERS, D (1910) Ibis (9) 4, 436-75. CARY, P (1973) A guide to birds of southern Portugal. Lisbon. CASÉN, R and HILDÉN, O (1965) Ornis fenn. 42, 33-5. CASTELLI, M (1988) Cahiers Éthol. appl. 8, 501-82. CASTLE, M E (1977) Scott. Birds 9, 327-34. CASTO, S D (1974) Wilson Bull. 86, 176-7. CATLEY, G P and HURSTHOUSE, D (1985) Br. Birds 78, 482-505. CATZEFLIS, F (1975) Nos Oiseaux 33, 64-5. CAVE, F O and MACDONALD, J D (1955) Birds of the Sudan. Edinburgh. CAWKELL, E M (1947) Br. Birds 40, 212; (1949) Br. Birds 42, 85; (1951) Br. Birds 44, 36. CEDERHOLM, G, FLODIN, L-Å, FREDRIKSSON, S, GUSTAFSSON, L, JACOBSSON, S, and PATERSSON, L (1974) Fauna och Flora 69, 134-45. CELLIER, M (1992) D. Phil. Thesis. Oxford Univ. CERNY, W (1938) Alauda 10, 76-90. ČERNÝ, W (1946) Sylvia 8, 13-18. CHAIKEN, M (1986) Ph D Thesis. Rutgers University; (1990) Develop. Psychobiol. 23, 233-46; (1992) Behaviour 120, 139-50. CHAKIR, N (1986) Ecologie du Bruant striolé, contribution à la biologie et à la dynamique de population à Marrakech. C.E.A. de biologie générale, Faculté des Sciences de Marrakech.. CHALMERS, M L (1986) Annotated checklist of the birds of Hong Kong. Hong Kong. CHALMERS, M L, TURNBULL, M, and CAREY, G J (1991) Hong Kong Bird Rep. 1990, 4-63. CHAMBERS, W T H (1867) Ibis (2) 3, 97-104. CHANDLER, C R and MULVIHILL, R S (1988) Ornis scand. 19, 212-16; (1990) Condor 92, 54-61; (1992) Auk 109, 235-41. CHANG, JAMES WAN-FU (1980) A field guide to the birds of Taiwan. Tai-pei. CHAPIN, J P (1954) Bull. Amer. Mus. nat. Hist. 75B (4). CHAPMAN, E A and McGEOCH, J A (1956) Ibis 98, 577-94. CHAPMAN, F M (1935) Auk 52, 21-9; (1939) Auk 56, 364-5; (1940) Auk 57, 225-233. CHAPPATTE, B (1980) Nos Oiseaux 35, 345. CHAPPELL, B M A (1946) Br. Birds 39, 352; (1949) Br. Birds 42, 84. CHAPPELLIER, A (1932) Oiseau (NS) 2, 535-42 CHAPPUIS, C (1969) Alauda 37, 59-71; (1976) Alauda 44, 475-95. CHAPPUIS, C, HEIM DE

BALSAC, H, and VIELLIARD, J (1973) Bonn. zool. Beitr. 24, 302-16. CHARLES, J K (1972) Ph D Thesis. Aberdeen Univ. CHARMAN, K (1965) Ph D Thesis. Durham Univ. CHARVOZ, P (1953) Nos Oiseaux 22, 137. CHAVIGNY, J DE and MAYAUD, N (1932) Alauda 4, 304-48, 416-41. CHEESMAN, R E (1919) Bull. Br. Orn. Club 40, 59. CHEKE, A S (1966) Ibis 108, 630-1; (1967) Ringers Bull. 3 (2), 7-8; (1973) In Kendeigh, S C and Pinowski, J (eds) Productivity, population dynamics and systematics of granivorous birds, 211-12. Warsaw. CHENG, T-H (1964) China's economic fauna: birds. Washington; (1987) A synopsis of the avifauna of China. Peking. CHERNOV, Y I and KHLEBOSOLOV, E I (1989) In Chernov, Y T (ed.) Ptitsy v soobshchestvakh tundrovoy zony, 39-51. Moscow. CHESNEY, M C (1986) Bird Study 33, 196-200; (1987) Tay Ringing Group Rep. 1984-6, 26-32. CHETTLEBURGH, M R (1952) Br. Birds 45, 359-64. CHEYLAN, G (1973) Alauda 41, 213-26. CHIA HSIANG-KANG and LI SHI-CHUN (1973) Acta zool. Sinica 19, 190-7. CHILGREN, J D (1977) Auk 94, 677-88; (1978) Condor 80, 222-9. CHIPLEY, R M (1980) In Keast, A and Morton, E S (eds) Migrant birds in the Neotropics: ecology, behavior, distribution, and conservation, 309-17 Washington, DC. CHITTENDEN, D E (1973) Br. Birds 66, 121. CHONG, L T (1938) Contr. biol. Lab. Sci. Soc. China 12 (Zool. Ser. 9), 183-373. CHRISTEN, W (1984) Orn. Beob. 81, 227-31. CHRISTENSEN, H Ø (1957) Dansk orn. Foren. Tidsskr. 51, 168-75. CHRISTENSEN, N H and ROSENBERG, N T (1964) Dansk orn. Foren. Tidsskr. 58, 13-35. CHRISTIANSEN, A (1935) Dansk orn. Foren. Tidsskr. 29, 22-9. CHRISTIE, D A (1983) Br. Birds 76, 462-3. CHRISTIE, D S (1963) Saint John Nat. Club Bull. 4, 1. CHRISTIE, G H (1927) Scott. Nat. 167, 158-9. CHRISTISON, A F P (1940) Birds of Northern Baluchistan. Quetta; (1941) Ibis (14) 5, 531-56. CHRISTOPHERS, S M (1984) Birds in Cornwall 1983, 11-96. CHURCHER, P B and LAWTON, J H (1987) J. Zool. Lond. 212, 439-55. CICHOKI, W (1988) Notatki Orn. 28, 97-8. CINK, C L (1976) Condor 78, 103-4. CLANCEY, P A (1938) Ibis (14) 2, 746-54; (1940) Ibis (14) 4, 91-9; (1943) Bull. Br. Orn. Club 64, 27-31; (1945) Bull. Br. Orn. Club 66, 20-1; (1946) Ibis 88, 518-19; (1947) Bull. Br. Orn. Club 67, 76-7; (1948a) Br. Birds 41, 115-16; (1948b) Bull. Br. Orn. Club 68, 92-4; (1948c) Bull. Br. Orn. Club 68, 132-7; (1948d) Bull. Br. Orn. Club 68, 137-41; (1948e) Ibis 90, 132-4; (1953) Bull. Br. Orn. Club 73, 72; (1954) Ibis 96, 317-18; (1989) Cimbebasia 11, 111-33; (1964) Bull. Br. Orn. Club 84, 110. CLARK, C C and CLARK L (1990) Wilson Bull. 102, 167-9. CLARK, J H (1903) Auk 20, 306-7. CLARK, L and MASON, J R (1988) Oecologia 77, 174-80. CLARKE, G C W (1949) Country Life 105, 1131. CLARKSON, K (1984) Ph D Thesis. Sheffield Univ. CLARKSON, K and BIRKHEAD, T R (1987) BTO News 151, 8-9. CLARKSON, K, EDEN, S F, SUTHERLAND, W J, and HOUSTON, A I (1986) J. Anim. Ecol. 55, 111-21. CLEGG, T M (1962) Br. Birds 55, 88-9. CLEMENTS, F A (1990) Sandgrouse 12, 55-6. CLENCH, M H (1973) Orn. Monogr. AOU 14, 32-33. CLENCH, M H and LEBERMAN, R C (1978) Bull. Carnegie Mus. nat. Hist. 5. CLERGEAU, P (1989) Oiseau 59, 101-15; (1990) J. Orn. 131, 458-60. CLESSE, B, DEWITTE, T, and FOUARGE, J-P (1991) Aves 28, 57-74. CLUNIE, F (1976) Notornis 23, 77. COCHET, P and FAURE, R (1987) Bièvre 9, 83-5. CODD, R B (1947) Br. Birds 40, 210. CODOUREY, J (1966) Nos Oiseaux 28, 177; (1968) Nos Oiseaux 29, 338-41. COE, M and COLLINS, N M (eds) (1986) Kora: an ecological inventory of the Kora National Reserve, Kenya. London. COHEN, E (1963) Birds of Hampshire and the Isle *of Wight.* Edinburgh. COLE, A (1990) *Devon Birds* **43**, 63–71. COLE, L R and FLINT, P R (1970) *Cyprus Orn. Soc. (1957) Rep.* **17**, 9–94. COLEIRO, C (1989) *Il-Merill* **26**, 1–26. COLEMAN, J D (1972) *Notornis* **19**, 118–39; (1973) *Notornis* **20**, 324–9; (1977) *Proc. New Zealand Ecol. Soc.* **24**, 94–109. COLEMAN, J D and ROBSON, A B (1975) *Proc. New Zealand Ecol. Soc.* **22**, 7–13. COLLAR, N J and STUART, S N (1985) *Threatened birds of Africa and related islands.* Cambridge. COLLETTE, P (1972) *Aves* **9**, 226–40. COLLETTE, P and FOUARGE, J (1978) *Aves* **15**, 19–29. COLLIAS, N E and COLLIAS, E C (1964) *Univ. Calif. Publ. Zool.* **73**, 1–239. COLLINGE, W E (1924–7) *The food of some British wild birds.* York; (1930) *J. Min. Agric.* (May), 151–8. COMTE, A (1926) *Bull. Soc. Zool. Genève* **3**, 34–8. CONDER, P J (1947) *Br. Birds* **40**, 212–3; (1948) *Ibis* **90**, 493–525. CONGREVE, W M (1936) *Ool. Rec.* **16**, 73–8. CONNELL, C E, ODUM, E P, and KALE, H (1960) *Auk* **77**, 1–9. CONNER, R N (1985) *Condor* **87**, 379–88. CONNOR, R J (1965) *Ool. Rec.* **39** (4), 4–9. CONRAD, E (1979) *Regulus* **13**, 83. CONRADS, K (1968) *Vogelwelt Suppl.* **2**, 7–21; (1969) *J. Orn.* **110**, 379–420; (1971) *Vogelwarte* **26**, 169–75; (1976) *J. Orn.* **117**, 438–50; (1977) *Vogelwelt* **98**, 81–105; (1984) *J. Orn.* **125**, 241–4. CONRADS, K and CONRADS, W (1971) *Vogelwelt* **92**, 81–100. CONRADS, K and QUELLE, M (1986) *Limosa* **59**, 67–74. CONSUL, C and ALVAREZ, F (1978) *Doñana Acta Vert.* **5**, 73–88. COOK, A (1975) *Bird Study* **22**, 165–8. COOKE, C H (1947) *Br. Birds* **40**, 308. COOKE, F, ROSS, R K, SCHMIDT, R K, and PAKULAK, A J (1975) *Can. Fld.-Nat.* **89**, 413–22. COOKE, H (1960) *Br. Birds* **53**, 229. COOMBS, C J F (1945) *Br. Birds* **38**, 154; (1946) *Cornwall Bird Watching Pres. Soc. Ann. Rep.* **16**, 49–50; (1960) *Ibis* **102**, 394–419; (1961) *Bird Study* **8**, 32–7, 55–70; (1978) *The crows.* London. COOPER, J and UNDERHILL, L G (1991) *Ostrich* **62**, 1–7. COOPER, J E S (1985) *Ring. Migr.* **6**, 61–5; (1987) *Sussex Bird Rep.* **39**, 88–91. COOPER, J E S and BURTON, P J K (1988) *Ring. Migr.* **9**, 93–4. COOPER, W W (1847) *Zoologist* **5**, 1775. COPETE, J L (1990) *Butll. GCA* **7**, 19–20. CORBIN, K W and SIBLEY, C G (1977) *Condor* **79**, 335–42. CORDERO, P J (1990) *Butll. GCA* **7**, 3–6; (1991) In Pinowski, J, Kavanagh, B P, and Górski, W (eds) *Proc. Int. Symp. Working Group Granivorous Birds, INTECOL*, 111–20. Warsaw. CORDERO, P J and RODRIGUEZ-TEIJEIRO, J D (1990) *Ekol. Polska* **38**, 443–52. CORDERO, P J and SALAET, M A (1987) *Publ. Dept. Zool. Barcelona* **13**, 111–16. CORDERO, P J and SUMMERS-SMITH, J D (1993) *J. Orn.* **134**, 69–77. CORKHILL, P (1973) *Bird Study* **20**, 207–20. CORNISH, A V (1947) *Br. Birds* **40**, 115. CORNWALLIS, L and PORTER, R F (1982) *Sandgrouse* **4**, 1–36. CORNWALLIS, R K and SMITH, A E (1964) *The bird in the hand.* Oxford (BTO guide 6). CORTÉS, J E (1982) *Alectoris* **4**, 26–9. CORTÉS, J E, FINLAYSON, J C, MOSQUERA, M A J, and GARCIA, E F J (1980) *The birds of Gibraltar.* Gibraltar. CORTI, U A (1935) *Bergvögel.* Bern; (1939) *Orn. Beob.* **36**, 121–40; (1949) *Einführung in die Vogelwelt des Kantons Wallis.* Chur; (1952) *Die Vogelwelt der schweizerischen Nordalpenzone.* Chur; (1959) *Die Brutvögel der deutschen und österreichischen Alpenzone.* Chur; (1961) *Die Brutvögel der französischen und italienischen Alpenzone.* Chur. CORTOPASSI, A J and MEWALDT, L R (1965) *Bird-Banding* **36**, 141–69. COULSON, J C (1960) *J. Anim. Ecol.* **29**, 251–71. COUNSILMAN, J J (1971) MA Thesis. Auckland Univ; (1974a) *Notornis* **21**, 318–33; (1974b) *Emu* **74**, 135–48; (1977) *Babbler* **1**, 1–13. COURSE, H A (1941) *Br. Birds* **35**, 154–5. COURTEILLE, C and THÉVENOT, M (1988) *Oiseau* **58**, 320–49. COVERLEY, H W (1933) *Ibis* (13) **3**, 782–5. COWDY,

S (1962) *Br. Birds* **55**, 229–33; (1973) *Bird Study* **20**, 117–20; (1976) *Birds* **6** (4), 30–32. COWIE, R J and HINSLEY, S A (1988) *Bird Study* **35**, 163–8. COX, J (1981) *Wielewaal* **47**, 322–5. COX, S (1984) *A new guide to the birds of Essex.* Ipswich. CRAGGS, J D (1967) *Bird Study* **14**, 53–60; (1976) *Bird Study* **23**, 281–4. CRAIG, A J F K (1983) *Ibis* **125**, 346–52; (1988) *Bonn. zool. Beitr.* **39**, 347–60. CRAMB, A P D (1972) *Br. Birds* **65**, 167. CRAMP, S (1969) In Gooders, J (ed.) *Birds of the world*, 2715–16; (1971) *Ibis* **113**, 244–5; (1985) *The birds of the western Palearctic* **4**. Oxford. CRAMP, S, PARRINDER, E R, and RICHARDS, B A (1957) In London Natural History Society *The birds of the London area since 1900*, 106–17. CRAMP, S and TEAGLE, W G (1952) *Br. Birds* **45**, 433–56. CRASE, F T, DE HAVEN, R W, and WORONECKI, P P (1972) *Bird-Banding* **43**, 197–204. CRAWFORD, R D and HOHMAN, W L (1978) *Bird-Banding* **49**, 201–7. CRAWHALL, E W (1952) *Country Life* **111**, 506. CREUTZ, G (1949) *Zool. Jahrb.* **78**, 133–72; (1953) *Beitr. Vogelkde.* **3**, 91–103; (1961) *Vår Fågelvärld* **20**, 302–18; (1962) *Orn. Mitt.* **14**, 64–6; (1967) *Falke* **14**, 93–6; (1970) *Falke* **17**, 426; (1981) *Der Graureiher.* Wittenberg Lutherstadt; (1988) *Beitr. Vogelkde.* **34**, 61. CREUTZ, G and FLÖSSNER, D (1958) *Beitr. Vogelkde.* **6**, 234–51. CROCQ, C (1974) *Alauda* **42**, 39–50; (1990) *Le casse-noix moucheté.* Paris. CROMBRUGGHE, S DE (1980) *Aves* **17**, 48. CROOK, J H (1963a) *J. Bombay nat. Hist. Soc.* **60**, 1–48; (1963b) *Ibis* **105**, 238–62; (1964a) *Proc. zool. Soc. Lond.* **142**, 217–55; (1964b) *Behaviour Suppl.* **10**, 1–178; (1969) In Hinde, R A (ed.) *Bird vocalizations*, 265–89. CROOK, J H and ALLEN, P M (1960) *J. E. Afr. nat. Hist. Soc.* **23**, 246. CROOK, S (1921) *Br. Birds* **15**, 10–15. CROSBY, M J (1988) *Br. Birds* **81**, 449–52. CROSSNER, K A (1977) *Ecology* **58**, 885–92. CROUCH, D J (1948) *Br. Birds* **41**, 149. CROUSAZ, G DE and LEBRETON, P (1963) *Nos Oiseaux* **27**, 46–61. CROWE, T M, BROOKE, R K, and SIEGFRIED, W R (1980) Abstr. 5th Pan-Afr. Orn. Congr. Malawi. CROZE, H (1970) *Z. Tierpsychol. Suppl.* **5**, 1–85. CRUMB, D W (1985) *Kingbird* **35**, 238–40. CRUON, R, ÉRARD, C, LEBRETON, J-D, and NICOLAU-GUILLAUMET, P (1992) *Alauda* **60**, 57–63. CRUON, R and NICHOLAU-GUILLAUMET, P (1985) *Alauda* **53**, 34–63. CRUZ, C DE LA, LOPE, F DE, and SILVA, E DA (1990) *Ardeola* **37**, 179–95. CRUZ SOLIS, C DE LA (1989) Thesis. Badajoz Univ. CRUZ SOLIS, C DE LA, LOPE REBOLLO, F DE, and SANCHEZ, J M (1992) *Ring. Migr.* **13**, 27–35. CRUZ SOLIS, C DE LA, LOPE REBOLLO, F DE, and SILVA RUBIO, E DA (1991a) *Ardeola* **38**, 101–15. CRUZ SOLIS, C DE LA, LOPE REBOLLO, F DE, and SILVA RUBIO, E DA (1991b) *Ring. Migr.* **12**, 86–90. CSIKI, E (1913) *Aquila* **20**, 375–96; (1914) *Aquila* **21**, 210–29; (1919) *Aquila* **26**, 76–104. ČTYROKÝ, P (1987) *Beitr. Vogelkde.* **33**, 141–204; (1989) *Zpravy MOS* **47**, 107–24. CUCCO, M and FERRO, M (1988) *Sitta* **2**, 99–103. CUDWORTH, J (1979) *Br. Birds* **72**, 291–3. CUGNASSE, J-M (1975) *Alauda* **43**, 478–9. CUGNASSE, J-M and RIOLS, C (1987) *Nos Oiseaux* **39**, 57–65. CULLEN, J P (1989) In Bignal, E and Curtis, D J (eds) *Choughs and land-use in Europe*, 19–22. Scottish Chough Study Group. CULLEN, J M, GUITON, P E, HORRIDGE, G A, and PEIRSON, J (1952) *Ibis* **94**, 68–84. CULLEN, P (1978) *Peregrine* **4**, 264–73. CUMMING, I G (1979) *Br. Birds* **72**, 53–9. CURRY-LINDAHL, K (1962) *Vår Fågelvärld* **21**, 161–73. CURTIS, S (1969) *Passenger Pigeon* **31**, 151–9. CUSTER, T W, OSBORN, R G, PITELKA, F A, and GESSAMAN, J A (1986) *Arctic Alpine Res.* **18**, 415–27. CUSTER, T W and PITELKA, F A (1975) *Condor* **77**, 210–12; (1977) *Auk* **94**, 505–25; (1978) *Condor* **80**, 295–301. CUTHILL, I and HINDMARSH, A M (1985)

Anim. Behav. **33**, 326-8. CVITANIĆ, A (1980) *Larus* **31-2**, 385-419; (1986) *Larus* **36-7**, 249-52. CVITANIĆ, A and TOLIĆ, R (1988) *Larus* **40**, 129-36.

DAAN, S (1972) *Flora och Fauna* **67**, 211-14. DAANJE, A (1941) *Ardea* **30**, 1-42. DACHSEL, M (1975) *Gef. Welt* **99**, 168. DAHLÉN, B (1988) *Vår Fågelvärld* **47**, 150. DALE, E M S (1924) *Can. Fld.-Nat.* **38**, 119-20. DALE, I H (1980) *Br. Birds* **73**, 480. DALMON, J (1932) *Oiseau* **2** (NS), 339-72. DALZIEL, L, FITCHETT, A, and WYNDE, R M (1986) *Shetland Bird Rep. 1985*, 3-35. DAMSTÉ, P H (1947) *J. exp. Biol.* **24**, 20-35. DANCKER, P (1956) *J. Orn.* **97**, 430-7. DANDL, J (1959) *Aquila* **65**, 175-88. DANFORD, C G (1877-8) *Ibis* (4) **1**, 261-74, (4) **2**, 1-35. DANIELS, D and EASTON, A (1985) *Devon Birds* **38**, 3-8. DANILOV, N N, NEKRASOV, E S, DOBRINSKI, L N, and KOPEIN, K I (1969a) *Int. Stud. Sparrows* **3**, 21-7. (1969b) *Ekol. Polska A* **17**, 489-501. DANILOV, N N, RYZHANOVSKI, V N, and RYABITSEV, V K (1984) *Ptitsy Yamala*. Moscow. DARE, P J (1986) *Bird Study* **33**, 179-89. DAROM, L (1979) *Israel Land Nat.* **5**, 23. DARWIN, C (1872) *The expressions of the emotions in man and animals*. London. DATHE, H (1962) *Orn. Mitt.* **14**, 56. (1971) *Beitr. Vogelkde.* **17**, 83-4. (1983) *Falke* **30**, 204. (ed.) (1986) *Handbuch des Vogelliebhabers* **2**. Berlin. DAVIDSON, A (1951) *Br. Birds* **44**, 346-8. DAVIDSON, A (1954) *A bird watcher in Scandinavia*. London. DAVIDSON, C (1985) *Trans. Norfolk Norwich Nat. Soc.* **27** (2), 98-102 705ghi. DAVIES, M (1988) In Cadbury, C J and Everett, M (eds) *RSPB Cons. Rev.* **2**, 91-4. DAVIS, D E (1959) *Ecology* **40**, 136-9. (1960) *Bird-Banding* **31**, 216-9. DAVIS, G (1928) *Avic. Mag.* **6**, 241-7. (1930) *Avic. Mag.* **8**, 289-94. DAVIS, J (1953) *Condor* **55**, 117-20. (1954) *Condor* **56**, 142-9. (1957) *Condor* **59**, 195-202. DAVIS, M (1951) *Auk* **68**, 529-30. DAVIS, P (1954) *Br. Birds* **47**, 21-3. (1963) *Bird Migration* **2**, 260-4. (1964) *Br. Birds* **57**, 477-501. DAVIS, P and DENNIS, R H (1959) *Br. Birds* **52**, 419-21. DAVIS, P E and DAVIS, J E (1986) *Nat. Wales* **3** (NS) (1984), 44-54. DAVIS, P G (1976) *Ring. Migr.* **1**, 115-6. (1977) *Bird Study* **24**, 127-9. DAVYDOV, A F (1976) *Dokl. Uchast. 2 Vsesoyuz. Konf. Poved. Zhiv.* 87-9. Moscow. DAVYDOVICH, L I and GORBAN', I M (1990) *Ornitologiya* **24**, 147 DAWSON, A (1991) *Ibis* **133**, 312-6. DAWSON, R (1975) *Br. Birds* **68**, 159-60. DAY, D H (1975) *Ostrich* **46**, 192-4. DEAN, F (1947) *Br. Birds* **40**, 191. DEBRU, H (1958) *Oiseau* **28**, 112-22. (1961) *Oiseau* **31**, 100-10. DE BRUN, N (1988) *Ornis Flandriae* **7**, 10. DECKERT, G (1962) *J. Orn.* **103**, 428-86. (1968a) *Beitr. Vogelkde.* **14**, 97-102. (1968b) *Der Feldsperling*. Wittenberg Lutherstadt. (1969) *Beitr. Vogelkde.* **15**, 1-84. (1980) *Beitr. Vogelkde.* **26**, 305-34. DEELDER, CL (1949) *Ardea* **37**, 1-88. (1952) *Ardea* **40**, 63-6. DEHN, W (1990) *Gef. Welt* **114**, 356. DEIGNAN, H G (1945) *US natn. Mus. Bull.* **186**. DEJONGHE, J-F (1984) *Les oiseaux de montagne*. Maisons-Alfort. DEJONGHE, J F and CZAJKOWSKI, M A (1983) *Alauda* **51**, 27-47. DEKEYSER, P L and VILLIERS, A (1950) *Bull. IFAN* **12** (3), 660-99. DELACOUR, J (1929) *Ibis* (12) **5**, 403-29. (1935) *Oiseau* **5** (NS), 377-88. (1943) *Zoologica* **28**, 69-86. (1947) *Birds of Malaysia*. New York. DELAMAIN, J (1912) *Rev. fr. Orn.* **2**, 298-302, 322-5. (1929) *Alauda* **1**, 59-63. DELANY, S, CHADWELL, C, and NORTON, J (1982) In *Univ. Southampton Ladakh Exped. Rep.* (1980), 5-153. Southampton Univ. DELAVELEYE, R (1964) *Gerfaut* **54**, 12-15. DELBOVE, P and FOUILLET, F (1986) *Oiseau* **56**, 77. DELESTRADE, A (1989) In Bignal, E, and Curtis, D J (eds) *Choughs and land-use in Europe*, 70-71.

Scottish Chough Study Group. DELESTRADE, A (1991) In Curtis, D J, Bignal, E M, and Curtis, M A (eds) *Birds and pastoral agriculture in Europe*, 72-5. Scottish Chough Study Group. DELIN, H and SVENSSON, L (1988) *Photographic guide to the birds of Britain and Europe*. London. DELMOTTE, C (1981) *Bull. Rech. Agron. Gembloux* **16**, 99-110. DELMOTTE, C, and DELVAUX, J (1981) *Aves* **18**, 108-18. DELVAUX, J (1983) *Aves* **20**, 174-5. DELVINGT, W (1961) *Gerfaut* **51**, 53-63. (1962) *J. Orn.* **103**, 260-5. DEMARTIS, A M (1987) *Gli uccelli Ital.* **12**, 15-26. DÉMENTIEFF, G (1934) *Oiseau* **4**, 525-9. DÉMENTIEFF, G P (1935) *Alauda* **7**, 153-69. DEMENTIEV, G P and GLADKOV, N A (1954) *Ptitsy Sovietskogo Soyuza* **5**. Moscow. DEMENTIEV, G P, KARTASHEV, N N, and SOLDATOVA, A N (1953) *Zool. Zh.* **32**, 361-75. DEMENTIEV, G P, KARTASHEV, N N, and TASHLIEV, A O (1956) *Trudy Inst. Biol. Akad. Nauk Turkmen. SSR* **4**, 77-119. DEMENTIEV, G P and PTUSHENKO, E (1939) *Ibis* (14) **3**, 507-12. DEMENTIEV, G P (1937) *Orn. Monatsber.* **45**, 86-7. DENDALETCHE, C (1991) In Curtis, D J, Bignal, E M and Curtis, M A (eds) *Birds and pastoral agriculture in Europe*, 68-9. Scottish Chough Study Group. DENDALETCHE, C and SAINT-LÈBE, N (1988) *Acta biol. mont.* **8** (no. spéc.), 147-70. (1991) *Acta biol. mont.* **10**, 45-50. DENIS, J V (1981) *N. Amer. Bird Band.* **6**, 88-96. DENISOVA, M N and GOMOLITSKAYA, R D (1967) *Ornitologiya* **8**, 344-5. DENNEMAN, W D (1981) *Vogeljaar* **29**, 194-203. DENNIS, R H (1969) *Br. Birds* **62**, 144-8. DENNY, J (1950) *Br. Birds* **43**, 333. DENSLEY, M (1990) *Br. Birds* **83**, 195-201. DE RUWE, F, DE PUTTER, G, and VANPRAET, J (1990) *Mergus* **4**, 166-70. DESBROSSE, A and ETCHEBERRY, R (1986) *Oiseau* **56**, 291-4. DESFAYES, M (1951) *Nos Oiseaux* **21**, 132. (1969) *Oiseau* **39**, 21-7. DESFAYES, M and PRAZ, J C (1978) *Bonn. zool. Beitr.* **29**, 18-37. DESMET, J (1981) *Gerfaut* **71**, 627-57. DESTRE, R (1984) Doctoral Thesis. Languedoc Univ. DEUNERT, J (1981) *Beitr. Vogelkde.* **27**, 125-6. (1989) *Abh. Ber. Naturkundemus. Görlitz* **63** (2). DEVILLERS, P, ROGGEMAN, W, TRICOT, J, MARMOL, P DEL, KERWIJN, C, JACOB, J-P, and ANSELIN, A (1988) *Atlas des oiseaux nicheurs de Belgique*. Brussels. DEWOLFE, B B (1967) *Condor* **69**, 110-32. DEWOLFE, B B, WEST, G C, and PEYTON, L J (1973) *Condor* **75**, 43-59. DHARMAKUMARSINHJI, R S (1955) *Birds of Saurashtra, India*. Bhavnagar. DHINDSA, M S (1983) *Ibis* **125**, 243-5. (1986) *Indian Rev. Life Sci.* **6**, 101-40. DHINDSA, M S and BOAG, D A (1989a) *Ibis* **132**, 595-602. (1989b) *Ornis scand.* **20**, 76-9. DHINDSA, M S, KOMERS, P E, and BOAG, D A (1989) *Can. J. Zool.* **67**, 228-32. DHINDSA, M S and TOOR, H S (1981) *Indian J. Ecol.* **8**, 156-62. (1990) In Pinowski, J and Summers-Smith, J D (eds) *Granivorous birds in the agricultural landscape*, 217-36. Warsaw. DHONDT, A A, and SMITH, J N M (1980) *Can. J. Zool.* **58**, 513-20. DIAMOND, A W, LACK, P, and SMITH, R W (1977) *Wilson Bull.* **89**, 456-66. DIAMOND, A W and SMITH, R W (1973) *Bird-Banding* **44**, 221-4. DÍAZ, J A and ASENSIO, B (1991) *Bird Study* **38**, 38-41. DI CARLO, E A (1956) *Riv. ital. Orn.* **26**, 55-61. DI CARLO, E A and LAURENTI, S (1991) *Gli uccelli Ital.* **16**, 81-96. DICE, L R (1918) *Condor* **20**, 129-31. DICK, W (1973) *Beitr. Vogelkde.* **19**, 397-405. DICKINSON, E C, KENNEDY, R S, and PARKES, K C (1991) *The birds of the Philippines*. Tring. DICKINSON, B H B and DOBINSON, H M (1969) *Bird Study* **16**, 135-46. DICKINSON, J C (1952) *Bull. Mus. comp. Zool.* **107**, 271-352. DICKSON, R C (1969) *Br. Birds* **62**, 497. (1972) *Br. Birds* **65**, 221-2. DIEN, J (1965) *Hamburger avifaun. Beitr.* **2**, 120-94. DIERSCHKE, J (1989) *Limicola* **3**, 246—51. DIESING, P (1984) *Beitr.*

Naturk. Niedersachs. **37**, 196-7. (1987) *Beitr. Naturk. Niedersachs.* **40**, 302. DIESSELHORST, G (1949) *Orn. Ber.* **2**, 1-31. (1950) *Orn. Ber.* **3**, 69-112. (1956) *Vogelwelt* **77**, 190. (1962) *Stuttgarter Beitr. Nturkde* **86**. (1968) *Khumbu Himal* 2. Innsbrück. (1971a) *Anz. orn. Ges. Bayern* **10**, 38-42. (1971b) *Vogelwelt* **92**, 201-26. (1986) *Mitt. Zool. Mus. Berlin* **62**, Suppl. *Ann. Orn.* **10**, 3-23. DIESSELHORST, G and POPP, K (1963) *Vogelwelt* **84**, 184-90. DIJKSEN, A J (1976) *Limosa* **49**, 204-6. DIJKSEN, L J (1989) *Limosa* **62**, 48. DIJKSEN, L J and KLEMANN, M (1992) *OSME Bull.* **28**, 21. DILGER, W C (1960) *Wilson Bull.* **72**, 115-32. DIMITROPOULOS, A (1987) *Herptile* **12**, 72-81. DIMOVSKI, A, and MATVEJEV, S (1955) *Arhiv Biol. Nauka Srpsko Biol. Društvo* **7**, 121-138. DINETTI, M and ASCANI, P (1990) *Atlante degli uccelli nidificanti nel comune di Firenze.* Florence. DIRECTIE NMF (1989) *Limosa* **62**, 11-14. DITTBERNER, H and DITTBERNER, W (1971) *Falke* **18**, 418-23. (1992) *Orn. Mitt.* **44**, 123-7. DITTBERNER, H, DITTBERNER, W, and LENZ, M (1969) *Vogelwelt* **90**, 225-33. DITTBERNER, H, DITTBERNER, W, and SADLIK, J (1979) *Falke* **26**, 296-8. DITTBERNER, H and KAGE, J (1991) *Beitr. Vogelkde.* **37**, 239-49. DITTRICH, W (1981) *J. Orn.* **122**, 181-5. DOBBEN, W H VAN (1949) *Ardea* **37**, 89-97. DOBBRICK, L (1931) *Abh. Westfäl. Mus. Naturk.* **2**, 27-33. DOBBRICK, W (1933) *Orn. Monatsber.* **41**, 55-6. DOBRYNINA, I N (1981) *Tez. Doklad. 10 Pribalt. orn. Konf.* **1**, 107-10. (1982) *Ornitologiya* **17**, 181. DOBRYNINA, I N (1986) In Sokolov, V E and Dobrynina, I N (eds) *Kol'tsevanie i mechenie ptits v SSSR, 1979-82, gody*, 35-167. Moscow. DOBSON, A P (1987) *Ornis Scand.* **18**, 122-8. DODSWORTH, P T L (1911) *J. Bombay nat. Hist. Soc.* **21**, 248-9. DOERBECK, F (1963) *Vogelwelt* **84**, 97-114. DOHERTY, P (1992) *Br. Birds* **85**, 595-600. DOHRN, H (1871) *J. Orn.* **19**, 1-10. DOÏCHEV, R L (1973) *Ann. Univ. Sofia Fac. Biol.* **65**, 1-10. DOLGUSHIN, I A (1968) *Trudy Inst. Zool. Akad. Nauk Kaz. SSR* **29**, 15-18. DOLGUSHIN, I A, KORELOV, M N, KUZ'MINA, M A, GAVRILOV, E I, GAVRIN, V F, KOVSHAR', A F, BORODIKHIN, I F, and RODIONOV, E F (1970) *Ptitsy Kazakhstana* **3**. Alma-Ata. DOL'NIK, V R (1975) *Zool. Zh.* **54**, 1048-56. DOL'NIK, V R (1980) *Zool. Zh.* **59**, 91-9. DOL'NIK, V R (1982) *Populyatsionnaya ekologiya zyablika.* Leningrad. DOL'NIK, V R and BLYUMENTHAL, T I (1967) *Condor* **69**, 435-68. DOL'NIK, V R and GAVRILOV, V M (1974) *Ornitologiya* **11**, 110-25. (1975) *Ekol. Polska* **23**, 211-26. (1979) *Auk* **96**, 253-64. (1980) *Auk* **97**, 50-62. DONALD, P F, WILSON, J D, and SHEPHERD, M (in press) *Br. Birds.* DONČEV, S (1958) *Bull. Inst. zool. Acad. Sci. Bulgarie* **7**, 269-313. (1981) *Aquila* **87**, 27-9. DONNER, E (1908) *Orn. Monatsschr.* **33**, 30-8. DONOVAN, J W (1978) *Nat. Wales* **16**, 142-3. (1984) *Br. Birds* **77**, 491. DONTSCHEV, S (1986) *Suppl. alle Ric. di Biologia della Selvaggina* **10**, 117-21. DORKA, U (1986) *Orn. Jahresh. Bad.-Württ.* **2**, 57-71. DORKA, V (1966) *Orn. Beob.* **63**, 165-223. (1973) *Anz. orn. Ges. Bayern* **12**, 114-21. DORKA, V, PFAU, K, and SPAETER, C (1970) *J. Orn.* **111**, 495-6. DORN, J L (1972) M Sc Thesis. Wyoming. DORNBERGER, W (1977) *Auspicium* **6**, 163-74. (1978) *Anz. orn. Ges. Bayern* **17**, 335-7. (1979) *Vogelwarte* **30**, 28-32. (1982) *Voliere* **5**, 59-61. (1983) *Verh. orn. Ges. Bayern* **23**, 501-9. DORNBUSCH, M (1972) *Apus* **2**, 286. (1973) *Falke* **20**, 193-5. (1981) *Beitr. Vogelkde.* **27**, 73-99. DOROGOY, I V (1982) *Ornitologiya* **17**, 119-24. DORSCH, H (1970) *Beitr. Vogelkde.* **15**, 437-51. DOS SANTOS, J R, DOS SANTOS, J N, and PEREIRA, A DE J (1985) *Cyanopica* **3**, 269-308. DOST, H (1957) *Falke* **4**, 101-3. DOUHAN, B (1978) *Vår Fågelvärld* **37**, 33-6. Dow,

D D (1966) *Ontario Bird-Banding* **2**, 1-14. DRESSER, H E (1871-81) *A history of the birds of Europe.* London. DRIVER, R (1957) *Br. Birds* **50**, 397-8. DROST, R (1930) *Vogelzug* **1**, 69-72. (1940a) *Orn. Monatsber.* **48**, 61-2. (1940b) *Vogelzug* **11**, 65-70. DROZDOV, N N (1965) *Ornitologiya* **7**, 166-99. (1968) *Ornitologiya* **9**, 345-7. DROZDOV, N N and ZLOTIN, R I (1962) *Ornitologiya* **5**, 193-207. DRURY, W H (1961) *Bird-Banding* **32**, 1-46. DRURY, W H and KEITH, J A (1962) *Ibis* **104**, 449-89. DUBALE, M S and PATEL, G (1975) *Pavo* **10**, 8-20. DUBERY, P (1983) *Birds* **9** (6), 71. DUBININ, N P (1953) *Trudy Inst. Lesa Akad. Nauk SSSR* **18**. DUBOIS, P J and COMITÉ D'HOMOLOGATION NATIONAL (1989) *Alauda* **57**, 263-94; (1990) *Alauda* **58**, 245-66. DUBOIS, P and DUHAUTOIS, L (1977) *Alauda* **45**, 285-91. DUBOIS, P J and YÉSOU, P (1986) *Inventaire des espèces d'oiseaux occasionnelles en France.* Paris. (1992) *Les oiseaux rares en France.* Bayonne. DUCHROW, H (1959) *Orn. Mitt.* **11**, 12. DUFF, A G (1979) *Alauda* **47**, 216-17. DUCKWORTH, E (1983) *Bird Life* (May-June), 43. DUIN, G VAN (1992) *Dutch Birding* **14**, 173-6. DUNAJEWSKI, A (1938) *Acta Orn. Mus. Zool. Polonici* **2**, 145-56. DUNLOP, E B (1917) *Br. Birds* **10**, 278-9. DUNN, E H (1990) *Ontario Birds* **7**, 87-91. DUNN, P J (1985) *Br. Birds* **78**, 151-2. (1990) *Br. Birds* **83**, 123-4. DUNN, P O and HANNON, S J (1989) *Auk* **106**, 635-44. DUNNET, G M (1955) *Ibis* **97**, 619-62. (1956) *Ibis* **98**, 220-30. DUNNET, G M, FORDHAM, R A, and PATTERSON, I J (1969) *J. appl. Ecol.* **6**, 459-73. DUPOND, C (1939) *Gerfaut* **29**, 185-203. DUPONT, P-L (1944) *Gerfaut* **34**, 23-8. DUPUY, A (1966) *Oiseau* **36**, 256-68. (1969) *Oiseau* **39**, 225-41. DUPUY, A and JOHNSON, E D H (1967) *Oiseau* **37**, 143. DUQUET, M (1984) *Nos Oiseaux* **37**, 331-40. (1986) *Nos Oiseaux* **38**, 263-8. DUQUET, M and PÉPIN, D (1987) *Nos Oiseaux* **39**, 170-1. DURAND, A L (1963) *Br. Birds* **56**, 157-64. DURANGO, S (1948) *Alauda* **16**, 1-20. DUREL, J (1927) *Oiseau* **8**, 235-43. DURHAM, M E and SELLERS, R M (1984) *Gloucs. Bird Rep.* **22**, 57-60. DURNEV, Y A, SONIN, V D, and SIROKHIN, I N (1984) *Ornitologiya* **19**, 177-8. DURNEV, Y A, SONIN, V D, LIPIN, S I, and SIROKHIN, I N (1991) In: Tsyrenov, V Z (ed.) *Ekologiya i fauna ptits Vostochnoy Sibiri*, 45-54. Ulan-Ude. DWENGER, R (1989) *Die Dohle.* Wittenberg Lutherstadt. DWIGHT, J (1897) *Auk* **14**, 259-72. (1900) *Ann. New York Acad. Sci.* **13**, 73-360. DWIGHT, J and GRISCOM, L (1927) *Amer. Mus. Novit.* **257**. DYBBRO, T (1976) *De danske ynglefugles udbredelse.* Copenhagen. DYER, M I, PINOWSKI, J, and PINOWSKA, B (1977) In Pinowski, J and Kendeigh, S C (eds) *Granivorous birds in ecosystems*, 53-105. London. DYMOND, J N (1991) *The birds of Fair Isle.* Edinburgh. DYMOND, J N, FRASER, P A, and GANTLETT, S J M (1989) *Rare birds in Britain and Ireland.* Calton. DYRCZ, A (1966) *Acta Orn.* **9**, 227-40.

EADES, R A (1984) *Br. Birds* **77**, 616-17. EAGLE CLARKE, W (1904) *Ibis* (8) **4**, 112-42. EARLÉ, R and GROBLER, N J (1987) *First atlas of bird distribution in the Orange Free State.* Bloemfontein. EAST, M (1988) *Ibis* **130**, 294-9. EASTWOOD, E, ISTED, G A, and RIDER, G C (1962) *Proc. Roy. Soc. Lond.* (B) **156**, 242-67. EATON, S W (1957a) *Auk* **74**, 229-39; (1957b) *Sci. Studies St. Bonaventura Univ.* **19**, 7-36. EBELS, E B (1991) *Dutch Birding* **13**, 86-9. EBER, G (1956) *Biol. Abh.* **13-14**, 1-60. ECK, S (1975) *Zool. Abh. Staatl. Mus. Tierkde. Dresden* **33**, 277-302; (1977) *Falke* **24**, 114-7; (1981) *Zool. Abh. Staatl. Mus. Tierkde Dresden* **37**, 183-207; (1984) *Zool. Abh. Staatl. Mus. Tierkde Dresden* **40**, 1-32; (1985a) *Zool. Abh. Staatl. Mus. Tierkde*

Dresden **40**, 79-108; (1985*b*) *Zool. Abh. Staatl. Mus. Tierkde Dresden* **41**, 1-32; (1990) *Zool. Abh. Staatl. Mus. Tierkde Dresden* **46**, 1-55. ECK, S and PIECHOCKI, R (1988) *Zool. Abh. Staatl. Mus. Tierkde Dresden* **43**, 135-41. EDDINGER, C R (1967) *Elepaio* **28**, 1-5, 11-18. EDELSTAM, C (1972) *Vår Fågelvärld*, supp. 7. ÉDEN, S F (1985) *J. Zool. Lond.* (A) **205**, 325-34; (1987*a*) *Ibis* **129**, 477-90; (1987*b*) *Anim. Behav.* **35**, 608-10; (1987*c*) *Anim. Behav.* **35**, 764-72; (1989) *Ibis* **131**, 141-53. EDGAR, A T (1972) *Notornis* **19** suppl. 89; (1974) *Notornis* **21**, 349-78. EDHOLM, M (1979) *Vår Fågelvärld* **38**, 106-7; (1980) *Vår Fågelvärld* **39**, 102. EDQVIST, T (1945) *Fauna och Flora* **40**, 92-3. EDWARDS, K D and OSBORNE, K C (1972) *Br. Birds* **65**, 203-5. EELLS, M M (1980) *Murrelet* **61**, 36-7. EENS, M (1992) Ph D Thesis. Antwerp Univ. EENS, M and PINXTEN, R (1990) *Ibis* **132**, 618-19. EENS, M, PINXTEN, R, and VERHEYEN, R F (1989) *Ardea* **77**, 75-86. EENS, M, PINXTEN, R, and VERHEYEN, R F (1990) *Bird Study* **37**, 48-52. EENS, M, PINXTEN, R, and VERHEYEN, R F (1991*a*) *Behaviour* **116**, 210-38. EENS, M, PINXTEN, R, and VERHEYEN, R F (1991*b*) *Belg. J. Zool.* **121**, 257-78. EENS, M, PINXTEN, R, and VERHEYEN, R F (1992*a*) *Ibis* **134**, 72-6. EENS, M, PINXTEN, R, and VERHEYEN, R F (1992*b*) *Anim. Behav.* EGGELING, W J (1960) *The Isle of May*. Edinburgh. EGGERMONT, D (1956) *Gerfaut* **46**, 17-21. EHRENROTH, B and JANSSON, B (1966) *Vår Fågelvärld* **25**, 97-105. EHRLICH, P R, DOBKIN, D S, and WHEYE, D (1986) *Auk* **103**, 835. EIFLER, G (1990) *Abh. Ber. Naturkundemus. Görlitz* **64** (2). EIFLER, G and BLÜMEL, H (1983) *Abh. Ber. Natkundemus. Görlitz* **57** (2). EISENHUT, E and LUTZ, W (1936) *Mitt. Vogelwelt* **35**, 1-14. EISENMANN, E (1969) *Bird-Banding* **40**, 144-5. EISFELD, D, STRÖDE, P, and OPHOVEN, E (1991) In Glandt, D (ed.) *Der Kolkrabe (Corvus corax) in Mitteleuropa. Metelener Schriftenreihe Naturschutz* **2**, 41-3. Metelen. EKELÖF, O and KUSCHERT, H (1979) *Corax* **7**, 37-9. EKSTRÖM, S (1952) *Vår Fågelvärld* **11**, 36. ELCAVAGE, P and CARACO, T (1983) *Anim. Behav.* **31**, 303-4. ELEY, C C (1991) D Phil Thesis. Sussex Univ. ELGAR, M A and CATTERALL, C P (1982) *Emu* **82**, 109-11. ELGOOD, J H (1982) *The birds of Nigeria*. London. ELGOOD, J H, FRY, C H, and DOWSETT, R J (1973) *Ibis* **115**, 1-45, 375-411. ELISEEV, D O (1985) In Prokofieva, I V (ed.) *Èkologiya ptits i reproduktivnyi period*, 3-10. St Petersburg. ELISEEVA, V I (1961) *Zool. Zh.* **40**, 583-91. ELKINS, N (1985) *Br. Birds* **78**, 51-2 ELLIOT, R D (1985) *Anim. Behav.* **33**, 308-14. ELLIOTT, C C H and JARVIS, M J F (1970) *Ostrich* **41**, 1-117; (1973) *Ostrich* **44**, 34-78. ELLIOTT, S (1991) *Wildlife Sound* **6** (6), 28-31. ELLIS, C R (1966) *Wilson Bull.* **78**, 208-24. ELLISON, A (1910) *Br. Birds* **3**, 300-2. ELMBERG, J (1991*a*) *Br. Birds* **84**, 344-5; (1991*b*) *Vår Fågelvärld* **50** (2), 38-9; (1992) *Birding World* **5**, 193. ELSACKER, L VAN, PINXTEN, R, and VERHEYEN, R F (1988) *Behaviour* **107**, 122-30. ELVERS, H, PFEIFFER, K, and WESTPHAL, D (1974) *Orn. Mitt.* **26**, 83-6. ELY, C A, LATAS, P J, and LOHOEFENER, R R (1977) *Bird-Banding* **48**, 275-6. ELZEN, R VAN DEN (1983) *Girlitze: Biologie, Haltung and Pflege*. Baden-Baden. ELZEN, R VAN DEN, and KÖNIG, C (1983) *Bonn. zool. Beitr.* **34**, 149-96. ELZEN, R VAN DEN, KÖNIG, C, and WOLTERS, H E (1978) *Bonn. zool. Beitr.* **29**, 323-59. EMLEN, S T (1971) *Anim. Behav.* **19**, 407-8. EMMERSON, K, MARTÍN, A, and BACALLADO, J J (1982) *Doñana Acta Vert.* **9**, 408-9. ENA, V (1984*a*) *Alytes* **2**, 144-59; (1984*b*) *Ibis* **126**, 240-9. ENA ALVAREZ, V (1979) Doctoral Thesis. Oviedo Univ. ENCKE, F-W (1965) *Beitr. Vogelkde.* **11**, 153-84. ENEMAR, A (1963) *Acta Univ. Lund* **2** (58), 1-21; (1964) *Flora och Fauna* **59**, 1-23; (1969) *Vår Fågelvärld* **28**, 230-5. ENEMAR, A, HANSON, S Å, and SJÖSTRAND, B (1965) *Acta Univ. Lund* **2** (5), 1-11. ENEMAR, A and NYSTRÖM, B (1981) *Vår Fågelvärld* **40**, 409-26. ENEMAR, A and SJÖSTRAND, B (1970) *Bull. ecol. Res. Comm. Lund* **9**, 33-7. ENGGIST-DÜBLIN, P VON (1988) *Beih. Veröff. Natursch. Landschaftsplege Bad.-Württ.* **53**, 175-82 ENGLAND, M D (1945*a*) *Br. Birds* **38**, 274; (1945*b*) *Br. Birds* **38**, 315; (1951) *Br. Birds* **44**, 386; (1970) *Br. Birds* **63**, 385-7; (1974) *Br. Birds* **67**, 218. ENNION, E A R and ENNION, D (1962) *Ibis* **104**, 158-68. ENQUIST, M and PETTERSSON, J (1986) *Spec. Rep. Ottenby Bird Obs.* 8. EPPRECHT, W (1965) *Orn. Beob.* **62**, 118. ERARD, C (1964) *Alauda* **32**, 105-28; (1966) *Alauda* **34**, 102-19; (1970) *Alauda* **38**, 1-26. ERARD, C (1968) *Bulletin du Centre de Recherches sur les Migrations des Mammifères et des Oiseaux* **19**, 3-62. ERARD, C and ETCHÉCOPAR, R-D (1970) *Mém. Mus. natn. Hist. nat.* (A) **66**. ERARD, C, and LARIGAUDERIE, F (1972) *Oiseau* **42**, 81-169. ERDMANN, E (1972) *Vogelk. Ber. Niedersachs.* **4**, 13-14. ERDMANN, G (1985) *Falke* **32**, 84-7. ERIKSEN, J (1990) *Oman Bird News* **9**, 10-13. ERIKSSON, K (1970*a*) *Ann. zool. Fenn.* **7**, 273-82; (1970*b*) *Ornis fenn.* **47**, 52-68; (1970*c*) *Sterna* **9**, 77-90. ERIKSSON, M and HANSSON, J-Å (1973) *Vår Fågelvärld* **32**, 11-22. ERIKSTAD, K E, BLOM, R, and MYRBERGET, S (1982) *J. Wildl. Mgmt* **46**, 109-14. ERLANGER, C F VON (1899) *J. Orn.* **47**, 449-532. ERNST, S (1983*a*) *Falke* **30**, 150-6; (1983*b*) *Naturschutzarb. Sachsen* **25**, 22-6; (1986) *Falke* **33**, 28-9; (1988) *Mitt. Zool. Mus. Berlin* **64**, Suppl. *Ann. Orn.* **12**, 3-50; (1990) *Beitr. Vogelkde.* **36**, 65-108. ERNST, S and THOSS, M (1977) *Falke* **24**, 48-53. ERPINO, M J (1968*a*) *Condor* **70**, 91-2; (1968*b*) *Condor* **70**, 154-65. ERSKINE, A J (1964) *Murrelet*, **45**, 15-22; (1971) *Wilson Bull.* **83**, 352-370; (1977) *Can. Wildl. Serv. Rep. Ser.* **41**; (1985) *Can. Field-Nat.* **99**, 188-95; (1992) *Atlas of breeding birds of the Maritime Provinces*. Halifax. ERSKINE, A J and DAVIDSON, G S (1976) *Syesis* **9**, 1-11. ERZ, W (1968) *Anthus*, **5**, 4-8. ESPMARK, Y (1972) *Fauna och Flora* **67**, 250-3. ESTARRIOL JIMÉNEZ, M (1974) *Ardeola* **20**, 330-1. ESTEN, S R (1931) *Auk* **48**, 572-4. ETCHEBERRY, R (1982) *Les oiseaux de St. Pierre et Miquelon*. Saint-Pierre. ETCHÉCOPAR, R-D (1961) *Oiseau* **31**, 158-9. ETCHÉCOPAR, R-D and HÜE, F (1967) *The birds of North Africa*. Edinburgh; (1983) *Les oiseaux de Chine, de Mongolie et de Corée: passereaux*. Paris. EVANS, A D (1992) *Bird Study* **39**, 17-22. EVANS, G H (1965) *Br. Birds* **58**, 457-61. EVANS, P G H (1980) D Phil Thesis. Oxford; (1986) *Ibis* **128**, 558-61; (1988) *Anim. Behav.* **36**, 1282-94. EVANS, P R (1966) *Ibis* **108**, 183-216; (1969*a*) *Condor* **71**, 316-30; (1969*b*) *J. Anim. Ecol.* **38**, 415-23; (1971) *Ornis fenn.* **48**, 131-2. EVANS, P R, ELTON, R A, and SINCLAIR, G R (1967) *Ornis fenn.* **44**, 33-41. EVANS, S M (1970) *Anim. Behav.* **18**, 762-7. EVERETT, M (1988) *Orn. Soc. Middle East Bull.* **20**, 3-5. EWERT, D N, and LANYON, W E (1970) *Auk* **87**, 362-3. EWIN, J P (1976) *Ibis* **118**, 468-9; (1978) Ph D Thesis, Univ. London. EWINS, P J (1979) *Ornithological observations in Morocco*. Unpubl; (1986) *Scott. Bird News*, 2; (1989) *Br. Birds* **82**, 331. EWINS, P J and DYMOND, J N (1984) *Shetland Bird Rep. 1983*, 48-57. EWINS, P J, DYMOND, J N, and MARQUISS, M (1986) *Bird Study* **33**, 110-6. EXCELL, J, KORKOLAINEN, V, and LINKOLA, P (1974) *Lintumies* **9**, 40-4. EYBERT, M-C (1980) *Oiseau* **50**, 295-7; (1985) Thèse Univ. Rennes I. EYCKERMAN, R, LOUETTE, M, and BECUWE, M (1992) *Zool. Middle East* **6**, 29-37. EYGELIS, Y K (1958) *Vestnik Leningrad. Univ.* **3**, *Biol.* 1, 108-15; (1961) *Zool. Zh.* **40**, 888-

99; (1964) *Zool. Zh.* **43**, 1517–29; (1965) *Zool. Zh.* **44**, 95–100; (1970) *Zool. Zh.* **49**, 892–7.

FAABORG, J and ARENDT, W J (1984) *J. Fld. Orn.* **55**, 376–8. FAIRHURST, A R (1970) *Br. Birds* **63**, 387; (1974) *Br. Birds* **67**, 215. FAIRON, J (1971) *Gerfaut* **61**, 146–61; (1975) *Gerfaut* **65**, 107–34; (1972) *Gerfaut* **62**, 325–30. FALCONER, D S (1941) *Br. Birds* **35**, 98–104. FALLET, M (1958a) *Schrift. naturwiss. Ver. Schleswig-Holstein* **29**, 39–46; (1958b) *Zool. Anz.* **161**, 178–87. FARINHA, J C (1989) In Bignal, E and Curtis, D J (eds) *Choughs and land-use in Europe*, 89–93. Scottish Chough Study Group. FARNER, D S, DONHAM, R S, MOORE, M C, and LEWIS, R A (1980) *Auk* **97**, 63–75. FARRAND, J (ed.) (1983) *The Audubon Society Master Guide to Birding* 3. New York. FARRAR, R B (1966) *Auk* **83**, 616–22. FASOLA, M and BRICHETTI, P (1983) *Avocetta* **7**, 67–84. FASOLA, M, PALLOTTI, E, CHIOZZI, G, and BALESTRAZZI, E (1986) *Riv. ital. Orn.* **56**, 172–80. FATIO, V and STUDER, T (1889) *Catalogue des Oiseaux de la Suisse*. Genève. FAWKS, E and PETERSEN, P, JR (1961) *A field list of birds of the tri-city region*. Davenport. FAXÉN, L (1945) *Vår Fågelvärld* **4**, 18–26. FEARE, C J (1974) *J. appl. Ecol.* **11**, 897–914; (1975) *Bull. Br. Orn. Club* **95**, 48–50; (1976) *J. Bombay nat. Hist. Soc.* **73**, 525–7; (1978) *Ann. appl. Biol.* **88**, 329–34; (1980) *Proc. int. orn. Congr.* **17** (2), 1331–6; (1981) In Thresh, J M (ed.) *Pests, pathogens and vegetation*, 393–400. London; (1984) *The Starling*, Oxford; (1986) *Gerfaut* **76**, 3–11; (1991) *Ibis* **133**, 75–9. FEARE, C J (1993) *Wilson Bull.* **105**. FEARE, C J and BURHAM, S E (1978) *Bird Study* **25**, 189–91. FEARE, C J and CONSTANTINE, D A T (1980) *Bird Study* **27**, 119–20. FEARE, C J, FRANSSU, P D DE, and PERIS, S J (1992) *Proc. Vert. Pest Contr. Conf.* **15**, 83–8. FEARE, C J, GILL, E L, McKAY, H V, and BISHOP, J D (in press) *Ibis*. FEARE, C J and INGLIS, I R (1979) *Ornis scand.* **10**, 42–7. FEARE, C J and McGINNITY, N (1986) *Bird Study* **33**, 164–7. FEARE, C J and MUNGROO, Y (1989) *Bull. Br. Orn. Club* **109**, 199–201. FEARE, C J, DUNNET, G M, and PATTERSON, I J (1974) *J. appl. Ecol.* **11**, 867–96. FEDERSCHMIDT, A (1988) *Ökol. Vögel* **10**, 151–64. FEDIUSCHIN, A V (1927) *J. Orn.* **75**, 490–5. FEDYUSHIN, A V and DOLBIK, M S (1967) *Ptitsy Belorussii*. Minsk. FEENEY, P P, ARNOLD, R W, and BAILEY, R S (1968) *Ibis* **110**, 35–86. FEIJEN, H R (1976) *Limosa* **49**, 28–67. FELLENBERG, W (1986) *Charadrius* **22**, 199–215; (1988a) *Charadrius* **24**, 85–7; (1988b) *Charadrius* **24**, 92–5. FELLENBERG, W and PFENNIG, H G (1986) *Charadrius* **22**, 216–20. FELTON, C (1969a) *Br. Birds* **62**, 80; (1969b) *Br. Birds* **62**, 445–6. FENK, R (1911a) *Orn. Monatsschr.* **36**, 233–44; (1911b) *Orn. Monatsschr.* **36**, 429–38; (1914) *Orn. Monatsber.* **22**, 85–90. FENNELL, J F M, SAGAR, P M, and FENNELL, J S (1985) *Notornis* **32**, 245–53. FENNELL, J F M and STONE, D A (1976) *Ring. Migr.* **1**, 108–14. FERDINAND, L (1991) *Bird voices in the North Atlantic*. Torshavn. FERGMAN, U (1988) *Torgos* **7** (2), 74–9, 101–2. FERGUSON-LEES, I J (1956) *Br. Birds* **49**, 398–400; (1957) *Br. Birds* **50**, 200; (1958) *Br. Birds* **51**, 99–103; (1959) *Br. Birds* **52**, 161–3; (1967) *Br. Birds* **60**, 344–7; (1971) In Gooders, J (ed.) *Birds of the world*, 2753–5. London. FERNANDEZ-CRUZ, M, ARAUJO, J, TEIXEIRA, A M, MAYOL, J, MUNTANER, J, EMMERSON, K W, MARTIN, A, and LE GRAND, G (1985) *Situacion de la avifauna de la Península Ibérica, Baleares y Macaronesia*. Madrid. FERRER, X (1987) *Ardeola* **34**, 110–3. FERRER, X, MOTIS, A, and PERIS, S J (1991) *J. Biogeography* **18**, 631–6. FERRY, C and FROCHOT, B (1970) *Terre Vie* **24**, 153–250. FEY, A

(1982) *Gef. Welt* **106**, 343–4. FFRENCH, R P (1967) *Living bird* **6**, 123–40; (1976) *A guide to the birds of Trinidad and Tobago*. Valley Forge, Pa; (1991) *Glos. nat. Soc. J.* **42**, 16. FICKEN, M S (1965) *Wilson Bull.* **77**, 71–5. FICKEN, M S and FICKEN, R W (1965) *Wilson Bull.* **77**, 363–75. FICKEN, R W, FICKEN, M S, and HAILMAN, J P (1978) *Z. Tierpsychol.* **46**, 43–57. FIEBIG, J (1983) *Mitt. zool. Mus. Berlin* **59**, *Suppl. Ann. Orn.* **7**, 163–87. FIEBIG, J and JANDER, G (1987) *Mitt. zool. Mus. Berlin* **63** *Suppl. Ann. Orn.* **11**, 123–35. FINLAYSON, C (1992) *Birds of the Strait of Gibraltar*. London. FINLAYSON, J C and CORTÉS, J E (1987) *Alectoris* **6**. FIRTH, F and FIRTH, F M (1945) *Br. Birds* **38**, 235–6. FISCHER, S, MAUERSBERGER, G, SCHIELZETH, H, and WITT, K (1992) *J. Orn.* **133**, 197–202. FISHER, J (1948) *Agriculture* **55**, 20–3. FISHER, J and HINDE, R A (1949) *Br. Birds* **42**, 347–57. FISHER, R H (1969) *Animals* **12**, 362. FISK, E J (1979) *Bird-Banding* **50**, 224–43, 297–303. FITTER, R S R (1948) *Br. Birds* **41**, 343; (1949) *London's Birds*. London. FITTER, R S R and LOUSLEY, J E (1953) *The natural history of the City*. London. FITTER, R S R and RICHARDSON, R A (1951) *Br. Birds* **44**, 16. FITZHARRIS, J and GRACE, K (1986) *IWC News* **47**, 13. FITZPATRICK, J (1978) *Br. Birds* **71**, 134. FITZWATER, W D (1967) *J. Bombay nat. Hist. Soc.* **64**, 111. FIUCZYNSKI, D (1961) *J. Orn.* **102**, 96–8. FJELD, P E and SONERUD, G A (1988) *Ornis scand.* **19**, 268–74. FJELDSÅ, J (1972) *Norwegian J. Zool.* **20**, 147–55; (1976) *Sterna* **15**, 133–5; (1981) *Dansk orn. Foren. Tidsskr.* **75**, 31–9. FLAXMAN, E W (1983) *Br. Birds* **76**, 352. FLEGG, J J M (1974) *Br. Birds* **67**, 517. FLEGG, J J M, and MATTHEWS, N J (1980) *Ringers' Bull.* **5**, 95. FLEISCHER, R C and JOHNSTON, R F (1984) *Can. J. Zool.* **62**, 405–10. FLEISCHER, R C, LOWTHER, P E, and JOHNSTON, R F (1984) *J. Fld. Orn.* **55**, 444–56. FLEISCHER, R C and ROTHSTEIN, S I (1988) *Evolution* **42**, 1146–58. FLEISCHER, R C, ROTHSTEIN, S I, and MILLER, L S (1991) *Condor* **93**, 185–9. FLEMING, R L, SR, FLEMING, R L, JR, and BANGDEL, L S (1976) *Birds of Nepal*. Kathmandu. FLIEGE, G (1984) *J. Orn.* **125**, 393–446. FLINT, P R and STEWART, P F (1992) *The birds of Cyprus*. London. FLINT, V E, BOEHME, R L, KOSTIN, Y V, and KUZNETSOV, A A (1984) *A field guide to birds of the USSR*. Princeton. FLOHART, G (1985) *Héron* **3**, 51. FLOWER, W U, WEIR, T, and SCOTT, D (1955) *Br. Birds* **48**, 133–4. FLUX, J E C (1978) *Notornis* **25**, 350–2. FOCKE, E (1966) *Veröff. Übersee Mus. Bremen* A, **3**, 259–64. FOELIK, R F (1970) *Zool. Jhrb. Anat.* **87**, 523–87. FOG, M (1963) *Danish Rev. Game Biol.* **4**, 63–110. FOKKEMA, J, BAKKER, A G, HOLLENGA, D, JUKEMA, J, and RIJPMA, U (1987) *Vanellus* **31**, 130–5. FOLK, Č (1966) *Zool. Listy* **15**, 273–83; (1967a) *Zool. Listy* **16**, 61–72; (1967b) *Zool. Listy* **16**, 379–93; (1968) *Zool. Listy* **17**, 221–36. FOLK, Č and BEKLOVÁ, M (1971) *Zool. Listy* **20**, 357–63. FOLK, Č, HAVLÍN, J, and HUDEC, K (1965) *Zool. Listy* **14**, 143–50. FOLK, Č and NOVOTNÝ, I (1970) *Zool. Listy* **19**, 333–42. FOLK, Č and TOUŠKOVÁ, I (1966) *Zool. Listy* **15**, 23–32. FOLKESTAD, A O (1967) *Sterna* **7**, 343–4; (1978) *Cinclus* **1**, 8–11. FOLLOWS, G (1969) *Ringers' Bull.* **3** (5), 11–12. FONSTAD, T (1981) *Fauna norv.* (C) *Cinclus* **4**, 89–96; (1984) *Oikos* **42**, 314–22. FORBUSH, E H and MAY, J B (1955) *A natural history of American birds*. New York. FORMOSOF, A N (1933) *J. Anim. Ecol.* **2**, 70–81. FORMOZOV, A N, OSMOLOVSKAYA, V I, and BLAGOSKLONOV, K N (1950) *Ptitsy i vrediteli lesa*. Moscow. FORNAIRON, F (1977) *Alauda* **45**, 341–2. FOUARGE, J (1980) *Nos Oiseaux* **35**, 373–5. FOUARGE, J and RAPPE, A (1966) *Aves* **3**, 52–9. FOX, A D, FRANCIS, I S, MADSEN, J, and STROUD, J M (1987) *Ibis* **129**, 541–52.

FOX, A D, FRANCIS, I S, McCARTHY, J P, and McKAY, C R (1992) *Dansk orn. Foren. Tidsskr.* **86**, 155-62. FRANCIS, C M and COOKE, F (1990) *J. Fld. Orn.* **61**, 404-12. FRANCIS, I S, FOX, A D, McCARTHY, J P, and McKAY, C R (1991) *Ring. Migr.* **12**, 28-37. FRANCIS, C M and WOOD, D S (1989) *J. Fld. Orn.* **60**, 495-503. FRANDSEN, J (1982) *Birds of the south western Cape.* Cape Town. FRANK, F (1943) *Orn. Monatsber.* **51**, 138-9; (1951) *J. Orn.* **93**, 61. FRANZ, D, HAND, R, and KAMRAD-SCHMIDT, M (1987) *Anz. Orn. Ges. Bayern* **26**, 237-50. FRANZ, J (1949) *Z. Tierpsychol.* **6**, 309-29. FRATICELLI, F (1984) *Riv. ital. Orn.* **54**, 98. FRATICELLI, F and GUSTIN, M (1987) *Avocetta* **11**, 161. FRAZIER, J G, SALAS, S S, and SALEH, M A (1984) *Courser* **1**, 17-27. FREDRIKSSON, S, JACOBSSON, S, and SILVERIN, B (1973) *Vår Fågelvärld* **32**, 245-51. FREEDMAN, B and SVOBODA, J (1982) *Can. Fld.-Nat.* **96**, 56-60. FREITAG, F (1978) *Orn. Mitt.* **30**, 255. FRENCH, N R (1954) *Condor* **56**, 83-5. FRENDIN, H (1943) *Fauna och Flora* **38**, 116-22. FRETWELL, S (1980) In Keast, A and Morton, E S (eds) *Migrant birds in the Neotropics: ecology, behavior, distribution, and conservation,* 517-27. Washington, D C; (1986) In Johnston, R F (ed.) *Current ornithology* **4**, 211-42. FREUCHEN, P and SALOMONSEN, F (1959) *The Arctic year.* London. FREY, M (1989a) *Orn. Beob.* **86**, 265-89; (1989b) *Orn. Beob.* **86**, 291-305. FRICH, A S and NORDBJÆRG, L (1992) *Dansk orn. Foren. Tidsskr.* **86**, 107-22. FRIEDMANN, H (1929) *The cowbirds.* Springfield; (1950) *The breeding habits of the weaverbirds: a study in the biology of behaviour patterns.* Washington. FRIEDRICH, W (1974) *Vogelwarte* **27**, 223-4. FRINGS, H, FRINGS, M, JUMBER, J, BUSNEL, R-G, GIBAN, J, and GRAMET, P (1958) *Ecology* **39**, 126-31. FRITH, J H (1957) *Emu* **57**, 287-8. FRITZ, Ö (1989) *Calidris* **18**, 143-64; (1991) *Calidris* **20**, 48-57. FROCHOT, B and PETITOT, F (1964) *Jean le Blanc* **3**, 32-40. FRÖDING, L (1987) BSC Thesis. Lund Univ. FROST, M P (1985) *Br. Birds* **78**, 50. FROST, R A (1979) *Br. Birds* **72**, 595-6; (1986) *Br. Birds* **79**, 508-9. FROST LARSEN, K and AAGAARD ANDERSEN, P (1965) *Feltornithologen* **7**, 24-5. FRUGIS, S, PARMIGIANI, S, PARMIGIANI, E, and PELLONI, C (1983) *Avocetta* **7**, 13-24. FRY, C H (1970) *Ostrich Suppl.* **8**, 239-63. FUCHS, E (1964) *Orn. Beob.* **61**, 132-7. FUCHS, W (1984) *Vögel Heimat* **55**, 42-3. FUGLE, G N, ROTHSTEIN, S I, OSENBERG, C W, and McGINLEY, M A (1984) *Anim. Behav.* **32**, 86-93. FUGLE, G N and ROTHSTEIN, S I (1985) *J. Fld. Orn.* **56**, 356-68. FUJIMAKI, Y and TAKAMI, M (1986) *Jap. J. Orn.* **35**, 67-73. FULLER, R J (1982) *Bird habitats in Britain.* Calton. FULTON, H T (1906) *J. Bombay nat. Hist. Soc.* **16**, 44-64. FUYE, M DE LA (1911) *Rev. fr. Orn.* **3**, 147-51.

GABRIELSON, I N and LINCOLN, F C (1959) *The birds of Alaska.* Harrisburg. GADGIL, M and ALI, S (1975) *J. Bombay nat. Hist. Soc.* **72**, 716-27. GAGINSKAYA, E R (1969) *Vopr. ekol. biotsenol* **9**, 37-48. GAILLY, P (1982) *Aves* **19**, 13-21, 99-102; (1988) *Alauda* **56**, 404. GAINES, D (1988) *Birds of Yosemite and the East Slope.* Lee Vining. GAIT, R P (1947) *Br. Birds* **40**, 341-2. GALEA, R (1987) *Il-Merill* **24**, 16. GALLACHER, H (1978) *De Spreeuw.* Utrecht. GALLAGHER, M (1989a) *Oman Bird News* **5**, 7-8; (1989b) *Oman Bird News* **7**, 10-11. GALLAGHER, M and WOODCOCK, M W (1980) *The birds of Oman.* London. GALLAGHER, M D (1977) *J. Oman. Stud. spec. Rep.* **1**, 27-58. GALLAGHER, M D and ROGERS, T D (1978) *Bonn. zool. Beitr.* **29**, 5-17; (1980) *J. Oman Stud. spec. Rep.* **2**, 347-85. GALLARDO, M (1986) *Faune de Provence* **7**, 18-29. GALLEGO, S and

BALCELLS, E (1960) *Ardeola* **6**, 337-9. GALLOWAY, D (1972) *Br. Birds* **65**, 522-6. GAMBLE, R and HAYCOCK, R J (1989) In Bignal, E and Curtis, D J (eds) *Choughs and land-use in Europe,* 39-41. Scottish Chough Study Group. GANGULI, U (1975) *A guide to the birds of the Delhi area.* New Delhi. GANSO, M (1960) *Egretta* **3**, 26-31. GANZHORN, J U (1986) *Ökol. Vögel* **8**, 49-56. GARCÍA DORY, M A (1983) *Alytes* **1**, 411-47. GARLING, M (1941) *Beitr. Fortpfl. Vögel* **17**, 51-8; (1943) *Beitr. Fortpfl. Vögel* **19**, 165; (1949) *Vogelwelt* **70**, 101-4. GARRETT, K, and DUNN, J (1981) *Birds of southern California.* Los Angeles. GARZÓN, J (1969) *Ardeola* **14**, 97-130. GASSMANN, H (1989-90) *Voliere* **12**, 324-8, **13**, 17-20. GASTON, A J (1968) *Ibis* **110**, 17-26; (1970) *Bull. Br. Orn. Club* **90**, 53-60, 61-6; (1978) *J. Bombay nat. Hist. Soc.* **75**, 115-28. GATEHOUSE, A G and MORGAN, M J (1973) *Nat. Wales* **13**, 267. GÄTKE, H (1891) *Die Vogelwarte Helgoland.* Braunschweig. GATTER, W (1974) *Vogelwarte* **27**, 278-89; (1976) *Vogelwarte* **28**, 165-70; (1977) *Verh. orn. Ges. Bayern* **23**, 61-9. GATTER, W, KLUMP, G, and SCHÜTT, R (1979) *Vogelwarte* **30**, 101-7. GAUCI, C and SULTANA, J (1983) *Il-Merill* **22**, 12-16. GAUHL, F (1984) *Vogelwelt* **105**, 176-87. GAVRILO, M V (1986) *1. S'ezda Vsesoyuz. orn. Obshch. 9. Vsesoyuz. orn. konf.* **1**, 140-1. St Petersburg. GAVRILOV, E I (1962a) *Ibis* **104**, 416-17; (1962b) *Trudy nauch.-issled. Inst. Zashch. Rast.* **7**, 459-528; (1963) *J. Bombay nat. Hist. Soc.* **60**, 301-17; (1965) *Bull. Br. Orn. Club* **85**, 112-14; (1968) *Ornitologiya* **9**, 343-4; (1972) *Int. Stud. Sparrows* **6**, 11-23. GAVRILOV, E I and GISTSOV, A P (1985) *Sezonnye perelety ptits v predgor'yakh zapadnogo Tyan'-Shanya.* Alma-Ata. GAVRILOV, E I, GUBIN, B M, and LEVIN, A S (1984) In Shukurov, E D (ed.) *Migratsii ptits v Azii* **7**, 74-96. Frunze. GAVRILOV, E I and KORELOV, M N (1968) *Byull. Mosk. Obshch. Ispyt. Prir. Otd. Biol.* **73** (4), 115-22. GAVRILOV, E I and KOVSHAR', A F (1968) *Trudy Inst. Zool. Akad. Nauk Kazakh. SSR* **29**, 41-9. GAVRILOV, E I, NAGLOV, V A, FEDOSENKO, A K, SHEVCHENKO, V L, and TATARINOVA, O M (1968) *Trudy Inst. Zool. Akad. Nauk. Kazakh. SSR* **29**, 153-207. GAVRILOV, V M (1979) *Ornitologiya* **14**, 158-63. GEBHARDT, E (1944) *Beitr. Fortpfl. Vögel* **20**, 98; (1955) *Vogelwelt* **76**, 188. GEE, J P (1984) *Malimbus* **6**, 31-66. GEE, N G and MOFFETT, L I (1917) *A key to the birds of the lower Yangtse valley.* Shanghai. GEILER, H (1959) *Beitr. Vogelkde.* **6**, 359-66. GEISTER, I (1974) *Biol. Vestn. Ljubljana* **22**, 71-3. GÉNARD, M and LESCOURRET, F (1986) *Vie Milieu* **36**, 27-36; (1987) *Bird Study* **34**, 52-63. GEORG, P V (1969) *Bull. Iraq. nat. Hist. Mus.* **4**, 21; (1971) *Bull. Iraq nat. Hist. Mus.* **5**, 45. GEORG, P V and VIELLIARD, J (1970) *Bull. Iraq nat. Hist. Mus.* **4**, 61-85. GEORGE, J C (1976) *J. Bombay nat. Hist. Soc.* **71**, 394-404. GEORGE, W G (1968) *Wilson Bull.* **80**, 496-7. GÉNSBØL, B (1964) *Feltornithologen* **6**, 59-63. GENT, C J (1949) *Br. Birds* **42**, 242. GENTZ, K (1970) *Beitr. Vogelkde.* **16**, 109-18; (1971) *Falke* **18**, 112-18. GERBER, R (1955) *Beitr. Vogelkde.* **5**, 36-45; (1956) *Die Saatkrähe.* Wittenberg Lutherstadt; (1963) *Beitr. Vogelkde.* **8**, 341-8. GERMOGENOV, N I (1982) In Labutin, Y V (ed.) *Migratsii i ekologiya ptits Sibiri,* 74-87. Novosibirsk. GÉROUDET, P (1951a) *Les passereaux 1.* Neuchâtel; (1951b) *Nos Oiseaux* **21**, 1-6, 23-31; (1952) *Nos Oiseaux* **21**, 160-8; (1954) *Nos Oiseaux* **22**, 145-56; (1955) *Nos Oiseaux* **23**, 89-95; (1957) *Les passereaux 3.* Neuchâtel; (1961) *Les passereaux 1* 2nd edn. Neuchâtel; (1964a) *Nos Oiseaux* **27**, 251; (1964b) *Nos Oiseaux* **27**, 299-303; (1991) *Nos Oiseaux* **41**, 119-36. GEUENS, A (1968) *Wielewaal* **34**, 357-8. GEYER, C (1985)

Aves **22**, 53-4. Geyr von Schweppenburg, H (1918) *J. Orn.* **66**, 121-76; (1920) *Falco* **16**, 17-26; (1930) *Orn. Monatsber.* **38**, 118-21; (1939) *Beitr. Fortpfl. Vögel* **15**, 198-9; (1942a) *Beitr. Fortpfl. Vögel* **18**, 27; (1942b) *Beitr. Fortpfl. Vögel* **18**, 1-5. Ghabbour, S I (1976) *Int. Stud. Sparrows* **9**, 17-29. Ghigi, A (1932) *Ann. Mus. Civ. Stor. nat. Genova* **55**, 268-92. Ghiot, C (1976) *Gerfaut* **66**, 267-305. Giban, J (1947) *Ann. des Épiphyt.* **13** (NS), 19-41. Gibb, J (1951) *Ibis* **93**, 109-27. Gibbons, D W (1987) *J. Anim. Ecol.* **56**, 403-14. Gibson-Hill, C A (1950) *Malayan Nat. J.* **5**, 58-75. Giebing, M (1992) *Voliere* **15**, 209-13. Gierow, P and Gierow, M (1991) *Orn. Svec.* **1**, 103-11. Gil, A (1927) *Bol. real Soc. española Hist. nat.* **27**, 81-96. Gil-Delgado, J A (1981) *Mediterránea* **5**, 97-114. Gil-Delgado, J A and Catalá, M C (1989) *Mediterránea Ser. Biol.* **11**, 121-32. Gil-Delgado, J A and Gómez, J A (1988) *Doñana Acta Vert.* **15**, 201-14. Gill, E H N (1923) *J. Bombay nat. Hist. Soc.* **29**, 757-68. Gill, E L (1919) *Br. Birds* **13**, 23-5. Gillespie, M and Gillespie, J A (1932) *Auk* **49**, 96. Gilroy, N (1922) *Ool. Rec.* **2**, 76-80. Ginn, H B and Melville, D S (1983) *Moult in birds.* Tring. Ginn, P J, McIlleron, W G, and Milstein, P le S (1989) *The complete book of southern African birds.* Cape Town. Giraud-Audine, M and Pineau, J (1973) *Alauda* **41**, 317. Giraudoux, P, Degauquier, R, Jones, P J, Weigel, J, and Isenmann, P (1988) *Malimbus* **10**, 1-140. Gistsov, A P and Gavrilov, E I (1984) *Int. Stud. Sparrows* **11**, 22-33. Gizenko, A I (1955) *Ptitsy Sakhalinskoy oblasti.* Moscow. Gladkov, N A and Zaletaev, V S (1962) *Ornitologiya* **5**, 31-4. Gladwin, T W (1985) *Br. Birds* **78**, 109-10. Glandt, D (ed.) (1991) *Der Kolkrabe (Corvus corax) in Mitteleuropa. Meteлener Schriftenreihe Naturschutz* **2**. Metelen. Glandt, D and Jansen, M (1991) In Glandt (1991), 113-16. Glas, P (1960) *Arch. Néerl. Zool.* **13**, 466-72. Glatthaar, R and Ziswiler, V (1971) *Rev. suisse Zool.* **78**, 1222-30. Glaubrecht, M (1989) *J. Orn.* **130**, 277-92. Glause, J (1969) *Vogelwelt* **90**, 66. Glayre, D (1970) *Nos Oiseaux* **30**, 230-4; (1979) *Nos Oiseaux* **35**, 71-4. Glayre, D and Magnenat, D (1984) *Oiseaux nicheurs de la haute vallée de l'Orbe.* Geneva. Gliemann, L (1973) *Die Grauammer.* Wittenberg Lutherstadt. Gloe, P (1982) *Ökol. Vögel* **4**, 209-11. Glück, E (1978) *J. Orn.* **119**, 336-8; (1980) *Ökol. Vögel* **2**, 43-91; (1982) *Vogelwarte* **31**, 395-422; (1983) *J. Orn.* **124**, 369-92; (1984) *Voliere* **7**, 7-12; (1985) *Ibis* **127**, 421-9; (1986) *Oecologia* **71**, 149-55; (1987) *Oecologia* **71**, 268-72; (1988) *Experientia* **44**, 537-9. Glück, E and Massoth, K (1985) *Voliere* **8**, 193-226. Glue, D E (1973) *Br. Birds* **66**, 461-72. Glushchenko, F P (1986) *Ornitologiya* **21**, 158-9. Glutz von Blotzheim, U N (1956) *Orn. Beob.* **53**, 36-40; (1962) *Die Brutvögel der Schweiz.* Aarau; (1987) *Orn. Beob.* **84**, 249-74. Gnielka, R (1978) *Orn. Mitt.* **30**, 81-90; (1986) *Beitr. Vogelkde.* **32**, 235-44. Godfrey, W E (1965) *Auk* **82**, 510-11; (1979) *The birds of Canada.* Ottawa; (1986) *The birds of Canada* rev. ed. Ottawa. Godin, J, Degauquier, R, and Tonnel, R (1977) *Héron* **4** (4), 1-26. Godman, F du C (1870) *Natural history of the Azores, or Western Islands.* London; (1872) *Ibis* (3) **2**, 209-24; (1907-10) *A monograph of the Petrels.* London. Gooch, S, Baillie, S R, and Birkhead, T R (1991) *J. appl. Ecol.* **28**, 1068-86. Goodacre, M J (1959) *Bird Study* **6**, 180-92. Goodbody, I M (1955) *Scott. Nat.* **67**, 90-7. Goodfellow, D J and Slater, P J B (1990) *Bioacoustics* **2**, 249-51. Goodfellow, P (1977) *Birds as builders.* Newton Abbot. Goodman, S M (1984) *Bonn. zool. Beitr.* **35**, 39-

56. Goodman, S M and Ames, P L (1983) *Sandgrouse* **5**, 82-96. Goodman, S M and Abdel Mowla Atta, G (1987) *Gerfaut* **77**, 3-41. Goodman, S M and Meininger, P L (eds) (1989) *The birds of Egypt.* Oxford. Goodman, S M, Meininger, P L, and Mullié, W C (1986) *Misc. Publ. Mus. Zool. Univ. Michigan* **172**. Goodman, S M and Storer, R W (1987) *Gerfaut* **77**, 109-45. Goodman, S M and Watson, G E (1983) *Bull. Br. Orn. Club* **103**, 101-6. Goodpasture, K A (1963) *Bird-Banding* **34**, 191-9; (1972) *Bird-Banding* **43**, 136. Goodwin, D (1949) *Br. Birds* **42**, 278-87; (1951) *Ibis* **93**, 414-42, 602-25; (1952a) *Behaviour* **4**, 293-316; (1952b) *Br. Birds* **45**, 113-22; (1952c) *Br. Birds* **45**, 364; (1953a) *Br. Birds* **46**, 113; (1953b) *Ibis* **95**, 147-9; (1955a) *Br. Birds* **48**, 181-3; (1955b) *Bull. Br. Orn. Club* **75**, 97-8; (1956) *Ibis* **98**, 186-219; (1960) *Avic. Mag.* **66**, 174-99; (1962) *Ibis* **104**, 564-6; (1964) *Br. Birds* **57**, 82-3; (1965a) *Avic. Mag.* **71**, 76-80; (1965b) *Domestic birds.* London; (1971) *Avic. Mag.* **77**, 88-93; (1975) *Br. Birds* **68**, 484-8; (1976) *Estrildid finches of the world.* London; (1985) *Avic. Mag.* **91**, 143-56; (1986) *Crows of the world.* London; (1987) *Avic. Mag.* **93**, 38-50. Gordon, P, Morlan, J, and Roberson, D (1989) *Western Birds* **20**, 81-7. Gordon, S (1949) *Country Life* **106**, 1239. Gore, M E J (1990) *Birds of The Gambia.* 2nd rev. edn. Tring. Gore, M E J and Won, P-O (1971) *The birds of Korea.* Seoul. Görner, M (1971) *Beitr. Vogelkde.* **17**, 173-4. Görnitz, K (1921) *Falco* **17**, 3; (1922) *Verh. orn. Ges. Bayern* **15**, 134-46; (1927) *J. Orn.* **75**, 58-60. Górski, W (1976) *Acta Orn.* **16**, 79-116; (1982) *Notatki Orn.* **23**, 3-13; (1988) *Acta Orn.* **24**, 29-62. Gosnell, H T (1932) *Br. Birds* **26**, 196-7; (1947) *The science of birdnesting.* Wirral. Gothe, H (1954) *Vogelwelt* **75**, 204-5. Göthel, H (1969) *Falke* **16**, 410-16. Götmark, F (1981) *Vår Fågelvärld* **40**, 47-56; (1982) *Vår Fågelvärld* **41**, 315-22. Götmark, F and Åhlund, M (1984) *J. Wildl. Management* **48**, 381-7. Götmark, F, Wallin, K, Jacobsson, S, and Alström, P (1979) *Vår Fågelvärld* **38**, 201-20. Göttgens, F, Göttgens, H, and Kollibay, F-J (1985) *Beitr. Naturkde. Niedersachs.* **38**, 233-8. Göttgens, H (1989) *Beitr. Naturkde. Niedersachs.* **42**, 148-57. Gotzman, J and Wisiński, P (1965) *Przeglad Zool.* **9**, 280-3. Göwert, R (1978) *Gef. Welt* **102**, 84-5, 224-5. Graber, R R and Graber, J W (1962) *Wilson Bull.* **74**, 74-88. Grabovski, V I (1983) *Zool. Zh.* **62**, 389-98. Graczyk, R (1961) *Przeglad Zool.* **5**, 241-5. Graczyk, R and Michocki, M (1975) *Roczniki Akad. Roln. Poznan.* **87**, 79-87. Grahn, M (1990) *Ornis scand.* **21**, 195-201. Gramet, P (1956) *Bull. Soc. zool. Fr.* **81**, 207-17; (1973) In Kendeigh, S C and Pinowski, J (eds) *Productivity, population dynamics and systematics of granivorous birds*, 181-94. Warsaw. Gramet, P and Dubaille, E (1983) Académie d'agriculture de France, Séance du 13 Avril, 455-64. Grampian and Tay Ringing Group (1981) *Rep. Grampian Ring. Group* **3**, 61-6. Grande, J L G (1986) *Ronda*, March 1986, 37-44. Grant, C H B and Mackworth-Praed, C W (1944) *Bull. Br. Orn. Club* **64**, 35-40. Grant, G S (1982) *Elepaio* **42**, 97-8. Grant, G S and Quay, T L (1970) *Bird-Banding* **41**, 274-8. Grant, P J (1968) *Br. Birds* **61**, 176-80; (1970) *Br. Birds* **63**, 153-5. Grant, P R (1972) *System. Zool.* **21**, 23-30; (1976) *Proc. int. orn. Congr.* **16**, 603-15; (1979) *Biol. J. Linn. Soc.* **11**, 301-32; (1980) *Bonn. zool. Beitr.* **31**, 311-17. Granvik, H (1916) *J. Orn.* **64**, 371-8; (1923) *J. Orn. Suppl.*, 1-280. Grätz, D and Grätz, H-P (1985) *Falke* **32**, 301-2. Gray, A P (1958) *Bird hybrids.* Farnham Royal. Green, G H and Summers, R W (1975) *Bird Study* **22**, 9-17. Green, P T

(1980) *Br. Birds* **73**, 358-60; (1981) *Ring. Migr.* **3**, 203-12; (1982*a*) *Ibis* **124**, 193-6; (1982*b*) *Ibis* **124**, 320-4. GREENHALGH, M (1965) *Br. Birds* **58**, 511. GREENING, M (1992) *J. Gloucs. Nat. Soc.* **43**, 34. GREENSLADE, D W (1979) *Br. Birds* **72**, 553. GREIG-SMITH, P W (1977) *Bull. Niger. orn. Soc.* **13**, 3-14; (1982) *Ibis* **124**, 529-34; (1985) In Sibly, R M and Smith, R H *Behavioural ecology*, 387-92. Oxford; (1987) *Behaviour* **103**, 203-16. GREIG-SMITH, P W and CROCKER, D R (1986) *Anim. Behav.* **34**, 843-59. GREIG-SMITH, P W and DAVIDSON, N C (1977) *Bull. Br. Orn. Club* **97**, 96-9. GREIG-SMITH, P W and WILSON, G M (1984) *J. appl. Ecol.* **21**, 401-22. GREIG-SMITH, P W and WILSON, M F (1985) *Oikos* **44**, 47-54. GRENQUIST, P (1947) *Ornis fenn.* **24**, 1-10. GREVE, K (1983) *Vogelkde. Ber. Niedersachs.* **15**, 5-10; (1990) *Beitr. Naturkde. Niedersachs.* **43**, 28-37. GREVE, K, and DORNIEDEN-GREVE, R (1982) *Beitr. Naturkde. Niedersachs.* **35**, 127-8. GRIEVE, A (1987) *Br. Birds* **80**, 466-73. GRIMES, L G (1987) *The birds of Ghana.* London. GRIMM, H (1954) *J. Orn.* **95**, 306-18; (1989) *Acta ornithoecol.* **2**, 100-2. GRIMSBY, A and ROER, J E (1992*a*) *Fauna norv.* (C) *Cinclus* **15**, 17-24; (1992*b*) *Vår Fuglefauna* **15**, 90-1. GRINNELL, J (1908) *Univ. Calif. Publ. Zool.* **5**, 1-170; (1928) *Condor* **30**, 185-9. GRINNELL, J, DIXON, J, and LINSDALE, J M (1930) *Univ. Calif. Publ. Zool.* **35**. GRINNELL, J and MILLER, A H (1944) *Pacific Coast Avifauna* **27**. GRINNELL, L I (1943) *Wilson Bull.* **55**, 155-63; (1944) *Auk* **61**, 554-60. GRISCOM, L (1937) *Proc. Boston Soc. nat. Hist.* **41**, 77-209. GRITTNER, I (1941) *Vogelzug* **12**, 56-73. GROBE, D W (1983) *Orn. Mitt.* **35**, 159. GRODZIŃSKI, Z (1971) *Acta Zool. Cracov.* **16**, 735-72; (1976) *Acta Zool. Cracov.* **21**, 465-500; (1980) *Acta Zool. Cracov.* **24**, 375-410. GROEBBELS, F (1960) *Vogelwelt* **81**, 94. GROH, G (1975) *Mitt. Pollichia* **63**, 72-139; (1982) *Mitt. Pollichia* **70**, 217-34; (1988) *Mitt. Pollichia* **75**, 261-87. GROMADZKA, J (1980) *Acta Orn.* **17**, 227-55. GROMADZKA, J and GROMADZKI, M (1978) *Acta Orn.* **16**, 335-64. GROMADZKA, J and LUNIAK, M (1978) *Acta Orn.* **16**, 275-85. GROMADZKI, M (1969) *Ekol. Polska* **17**, 287-311; (1980) *Acta Orn.* **17**, 195-224. GROMADZKI, M and KANIA, W (1976) *Acta Orn.* **15**, 279-321. GROŠELJ, P (1983) *Acrocephalus* **4**, 56-8. GROSS, A O (1921) *Auk* **38**, 1-26, 163-84. GROTE, H (1934*a*) *Beitr. Fortpfl. Vögel* **10**, 20-6; (1934*b*) *Orn. Monatsber.* **42**, 17-21; (1943*a*) *Beitr. Fortpfl. Vögel* **19**, 98-104; (1943*b*) *J. Orn.* **91**, 136-43; (1937) *Beitr. Fortpfl. Vögel* **13**, 150-1; (1940) *Vogelzug* **11**, 127-9; (1947) *Orn. Beob.* **44**, 84-90. GROTH, J G (1988) *Condor* **90**, 745-60; (1992*a*) *Auk* **109**, 383-5; (1992*b*) *Western Birds* **23**, 35-7. GRÜLL, A (1981) *Egretta* **24** Suppl., 39-63; (1988) *Beih. Veröff. Nat. Land. Bad.-Württ.* **53**, 65. GRÜN, G (1975) *Int. Stud. Sparrows* **8**, 24-103. GRÜNKORN, T (1991) In Glandt (1991), 9-15. GRUYS-CASIMIR, E M (1965) *Arch. Néerl. Zool.* **16**, 175-279. GUBIN, B M (1979) *Ornitologiya* **14**, 211-13; (1980) *Ornitologiya* **15**, 111-16. GUBLER, W (1978) *Orn. Beob.* **75**, 279-80. GUÉNIAT, E (1948) *Orn. Beob.* **45**, 81-98. GUERMEUR, Y, HAYS, C, L'HER, M, and MONNAT, J-Y (1973) *Ar Vran* **6**, 199-260. GUERRERO, J, LOPE, F DE, and CRUZ, C DE LA (1989) *Alauda* **57**, 234. GUEX, M-L (1986) *Nos Oiseaux* **38**, 343. GUICHARD, K M (1955) *Ibis* **97**, 393-424. GUILLORY, H D and DESHOTELS, J H (1981) *Wilson Bull.* **93**, 554. GUILLOU, J J (1964) *Alauda* **32**, 196-225; (1981) *Oiseau* **51**, 177-88. GUITIÁN, J (1987) *Ardeola* **34**, 25-35; (1989) *Ardeola* **36**, 73-82. GUITIÁN RIVERA, J (1985) *Ardeola* **32**, 155-72. GULAY, V I (1989) In Konstantinov, V M and Klimov, S M (eds) *Vranovye ptitsy v estestvennykh i antropogennykh landshaftakh* **1**, 53-5. Lipetsk. GURIEV, V N and BASHKOVA, E N (1986) *Tez. Dokl. S'ezda 1. Vsesoyuz. orn. Kongr. 9. Vsesoyuz. orn. Konf.* **1**, 182-3. St Petersburg. GUSH, G H (1975) *Br. Birds* **68**, 342; (1978) *Br. Birds* **71**, 40-1; (1980) *Devon Birds* **33**, 75-80. GÜTTINGER, H R (1970) *Z. Tierpsychol.* **27**, 1011-75; (1974) *J. Orn.* **115**, 321-37; (1976) *Bonn. zool. Beitr.* **27**, 218-44; (1977) *Behaviour* **60**, 304-18; (1978) *J. Orn.* **119**, 172-90; (1981) *Gef. Welt* **105**, 210-22; (1985) *Behaviour* **94**, 254-78. GÜTTINGER, H R and CLAUSS, G (1982) *J. Orn.* **123**, 269-86. GÜTTINGER, H R, WOLFFGRAMM, J, and THIMM, F (1978) *Behaviour* **65**, 241-62. GUYOT, A, LIGOR, J-L, and ROSE, R (1991) *Nos Oiseaux* **41**, 195. GWINNER, E (1964) *Z. Tierpsychol.* **21**, 657-748; (1965*a*) *J. Orn.* **106**, 145-78; (1965*b*) *Vogelwarte* **23**, 1-4; (1966) *Vogelwelt* **87**, 129-33. GWINNER, E and KNEUTGEN, J (1962) *Z. Tierpsychol.* **19**, 692-6. GYLLIN, R (1965*a*) *Fauna och Flora* **60**, 225-74; (1965*b*) *Vår Fågelvärld* **24**, 420-1; (1967) *Vår Fågelvärld* **26**, 19-29. GYLLIN, R and KÄLLANDER, H (1976) *Ornis scand.* **7**, 113-25; (1977) *Fauna Flora* **72**, 18-24. GYLLIN, R, KÄLLANDER, H and SYLVÉN, M (1977) *Ibis* **119**, 358-1. GYNGAZOV, A M and MILOVIDOV, S P (1977) *Ornitofauna Zapadno-Sibirskoy ravniny.* Tomsk. GYÖRGYPÁL, Z (1981) *Aquila* **87**, 71-8.

HAACK, W and SCHMIDT, G A J (1989) *Vogelkundl. Tageb. Schleswig-Holstein* **16**, 509-29. HAAPANEN, A (1965) *Ann. zool. Fenn.* **2**, 153-96; (1966) *Ann. zool. fenn.* **3**, 176-200. HAAR, H (1975) *Mitt. Abt. Zool. Landesmus. Joanneum* **4**, 105-14. HAARHAUS, D (1968) *Oecologia* **1**, 176-218. HAARTMAN, L VON (1952) *Ornis fenn.* **29**, 73-6; (1969) *Comm. Biol. Soc. Sci. Fenn.* **32**; (1972) *Ornis fenn.* **49**, 15; (1973) *Ornis fenn.* **50**, 49. HAARTMAN, L VON and NUMERS, M VON (1992) *Ornis fenn.* **69**, 65-71. HAAS, G (1939) *Beitr. Fortpfl. Vögel* **15**, 52-62; (1943) *Beitr. Fortpfl. Vögel* **19**, 43-6. HAASE, B J M (1988) *Ardea* **76**, 210. HAASE, E (1975) *Gen. comp. Endocrin.* **26**, 248-52. HACHFELD, B (1979) *Voliere* **2**, 7-11, 79-84; (1983) *Voliere* **6**, 146-9; (1984) *Voliere* **7**, 67. HAENSEL, J (1967) *Beitr. Vogelkde.* **13**, 1-28; (1970) *Beitr. Vogelkde.* **16**, 169-91. HAFFER, J (1977) *Bonn. zool. Monogr.* **10**; (1989) *J. Orn.* **130**, 475-512. HAFSTEINSSON, H T and BJÖRNSSON, H (1989) *Bliki* **8**, 47-9. HAFTORN, S (1952) *Fauna* **5**, 105-141. Oslo; (1971) *Norges Fugler.* Oslo. HAGEN, Y (1942) *Archiv Naturgesch.* NF **11**, 1-132; (1956) *Nytt Magasin for Zoologi* **4**, 107-8. HAGGER, C H E (1961) *Br. Birds* **54**, 291. HAGMANN, J and DAGAN, D (1992) *Orn. Beob.* **89**, 72. HAILA, Y, JÄRVINEN, O, and VÄISÄNEN, R A (1979) *Ornis scand.* **10**, 48-55. HAILA, Y, TIAINEN, J, and VEPSÄLÄINEN, K (1986) *Ornis fenn.* **63**, 1-9. HAILS, C J (1985) *Studies of problem bird species in Singapore: 1 Sturnidae.* Rep. Parks and Recreation Dept., Ministry of National Development, Singapore. HÁJEK, V (1969) *Vertebr. Zprávy* **1969** (2), 73-6; (1974) *Sylvia* **19**, 145-56. HÁJEK, V and BAŠOVÁ, D (1963) *Zool. Listy* **12**, 115-20. HAKALA, A V K and NYHOLM, E S (1973) *Ornis fenn.* **50**, 46-7. HAKE, M and EKMAN, J (1988) *Ornis scand.* **19**, 275-9. HÅLAND, A (1980) *Viltrapport* **10**, 104-6. HALD-MORTENSEN, P (1970) *Ibis* **112**, 265-6. HALL, B P (1953) *Bull. Br. Orn. Club* **73**, 2-8; (1957) *Bull. Br. Orn. Club* **77**, 44-6. HALL, B P and MOREAU, R E (1970) *An atlas of speciation in African passerine birds.* London. HALL, G A (1981) *J. Fld. Orn.* **52**, 43-9; (1983) *West Virginia birds.* Pittsburgh; (1985) *American birds* **39**, 911-4. HAM, I and ŠOTI, J P (1986) *Larus* **36-37**, 297-303. HAMILTON, S and JOHNSTON, R F (1978) *Auk* **95**, 313-23. HAMILTON, T H (1958) *Wilson Bull.* **70**, 307-46. HAMILTON-HUNTER,

R (1909) *Br. Birds* 3, 188-9. HAMMER, H (1958) *Falke* 5, 141. HAMMER, M (1948) *Danish Rev. Game Biol.* 1 (2), 1-59. HANCOX, M (1985) *Scott. Nat.* 1985, 37-40. HANDTKE, K (1975) *Naturk. Jber. Mus. Heineanum* 10, 33-41. HANDTKE, K and WITSACK, W (1972) *Naturk. Jber. Mus. Heineanum* 7, 21-41. HANF, B (1887) *Ornis* 3, 267. HANFORD, D M (1969) *Br. Birds* 62, 158. HANMER, D B (1989) *Safring News* 18, 19-30. HANN, H W (1937) *Wilson Bull.* 49, 145-237. HANSCH, A (1938) *Vogelzug* 9, 110. HANSEN, L (1950) *Dansk orn. Foren. Tidsskr.* 44, 150-61. HANSEN, O and BIRKHOLM-CLAUSEN, F (1968) *Dansk orn. Foren. Tidsskr.* 62, 97. HANSEN, P (1975) *Biophon* 3 (3), 2-5; (1981a) *Nat. Jutlandica* 19, 107-20; (1981b) *Nat. Jutlandica* 19, 121-38; (1984) *Ornis scand.* 15, 240-7; (1985) *Natura Jutlandica* 21, 209-19. HANSEN, W and OELKE, H (1976) *Beitr. Naturkde. Niedersachs.* 29, 85-158. HANSKI, I K and HAILA, Y (1988) *Ornis fenn.* 65, 97-103. HANTZSCH, B (1905) *Beitrag zur Kenntnis der Vogelwelt Islands*. Berlin. HARASZTHY, L (1988) *Magyarország madárvendégei*. Budapest. HARBARD, C (1989) *Songbirds*. London. HARBER, D D (1945a) *Br. Birds* 38, 211; (1945b) *Br. Birds* 38, 296. HARDENBERG, J D F (1965) In Busnel, R-G and Giban, J (eds) *Colloque. Le problème des oiseaux sur les aérodromes*, 121-6. Paris. HÄRDI, M (1989) *Orn. Beob.* 86, 209-17. HARDING, K C (1931) *Auk* 48, 512-22. HARDY, E (1932) *Br. Birds* 25, 301; (1946) *A handlist of the birds of Palestine*. Unpubl. MS, Edward Grey Inst., Oxford Univ; (1971) *Br. Birds* 64, 77-8. HARLOW, R C (1922) *Auk* 39, 399-410. HARMS, W (1975) *Hamburger avifaun. Beitr.* 13, 133-44. HARPER, D G C (1984) *Ring. Migr.* 5, 101-4. HARPUM, J R (1985) *Glos. nat. Soc. J.* 36, 115-7. HARRIS, A, TUCKER, L, and VINICOMBE, K (1989) *The Macmillan field guide to bird identification*. London. HARRIS, G J (1957) *Br. Birds* 50, 206-8. HARRIS, G J, PARSLOW, J L F, and SCOTT, R E (1960) *Br. Birds* 53, 513-8. HARRIS, M P (1962) *Br. Birds* 55, 97-103; (1974) *A field guide to the birds of Galapagos*. New York. HARRIS, M P, NORMAN, F I, and McCOLL, R H S (1965) *Br. Birds* 58, 288-94. HARRISON, C J O (1956) *Avic. Mag.* 62, 128-41; (1962a) *Proc. zool. Soc. Lond.* 139, 261-82; (1962b) *Bull. Br. Orn. Club* 82, 126-32; (1962c) *J. Orn.* 103, 369-79; (1964) *Ibis* 106, 462-8; (1965a) *Ardea* 53, 57-72; (1965b) *Bull. Br. Orn. Club* 85, 26-30; (1965c) *Behaviour* 24, 161-209; (1967) *Wilson Bull.* 79, 22-7; (1975) *A field guide to the nests, eggs and nestlings of European birds*. London; (1976) *Bird Study* 23, 59; (1978a) *A field guide to the nests, eggs and nestlings of North American birds*. Glasgow; (1978b) *Avic. Mag.* 84, 80-7; (1982) *An atlas of the birds of the western Palearctic*. London; (1983) *Avic. Mag.* 89, 163-9. HARRISON, G R, DEAN, A R, RICHARDS, A J, and SMALLSHIRE, D (1982) *The birds of the West Midlands*. Studley. HARRISON, H H (1984) *Wood Warblers' World*. New York. HARRISON, J M (1928) *Br. Birds* 22, 36-7; (1934) *Ibis* (13) 4, 396-8; (1937) *Bull. Br. Orn. Club* 57, 64-5; (1938) *Vogelzug* 9, 36; (1947a) *Ibis* 89, 411-8; (1947b) *Ibis* 89, 664; (1954) *Bull. Br. Orn. Club* 74, 105-12; (1955) *Bull. Br. Orn. Club* 75, 6-12, 17-21; (1958) *Bull. Br. Orn. Club* 78, 9-14, 23-8; (1961) *Bull. Br. Orn. Club* 81, 96-103, 119-124. HARRISON, J M and PATEFF, P (1933) *Ibis* (13) 3, 494-521. HARRISON, R (1970) *Br. Birds* 63, 302-3. HARROP, A (1988) *Bull. Oriental Bird Club* 8, 31. HARROP, H (1992) *Birding World* 5, 133-7. HARROP, J M (1970) *Nature in Wales* 12 (2), 65-9. HARS, D (1991) *Héron* 24, 289-92. HARTBY, E (1968) *Dansk orn. Foren. Tidsskr.* 62, 205-30. HARTERT, E (1903-10) *Die Vögel der paläarktischen Fauna* 1. Berlin; (1913) *Novit. Zool.* 20, 37-76; (1915) *Novit. Zool.* 22, 61-79; (1918a) *Bull. Br. Orn. Club* 38, 58-60; (1918b) *Novit. Zool.* 25, 327-337; (1918c) *Novit. Zool.* 25, 361; (1921-2) *Die Vögel der paläarktischen Fauna* 3. Berlin; (1928) *Novit. Zool.* 34, 197. HARTERT, E and OGILVIE-GRANT, W R (1905) *Novit. Zool.* 12, 80-128. HARTERT, E and STEINBACHER, F (1932-8) *Die Vögel der paläarktischen Fauna, Ergänzungsband*. Berlin. HARTLEY, I R (1991) Ph D Thesis. Leicester Univ. HARTLEY, I R, SHEPHERD, M, ROBSON, T, and BURKE, T (1993) *Behav. Ecol.* 4 in press. HARTOG, J C DEN (1987) *Zool. Meded. Leiden* 61 (28), 405-19; (1990) *Cour. Forsch.-Inst. Senckenberg* 129, 159-90. HARTWIG, W (1886) *J. Orn.* 34, 452-86. HARVEY, P (1990) *Birding World* 3, 266-7; (1991) *Fair Isle Bird Observatory Rep.* 43, 15-57. HARWIN, R M (1959) *Ostrich* 30, 97-104. HARWOOD, N (1959) *Br. Birds* 52, 166. HASSE, H (1961) *Regulus* 41, 115-9; (1962) *Vogelwelt* 83, 173-7; (1963) *Die Goldammer*. Wittenberg Lutherstadt; (1965) *Beitr. Vogelkde.* 10, 406-7. HAUKIOJA, E (1968) *Ornis fenn.* 45, 105-13; (1969) *Ornis fenn.* 46, 171-8; (1970) *Ornis fenn.* 47, 101-35; (1971a) *Ornis fenn.* 48, 25-32; (1971b) *Ornis fenn.* 48, 45-67; (1971c) *Rep. Kevo Subarctic Res. Stat.* 7, 60-9. HAUKIOJA, E, and REPONEN, J (1969) *Orn. Soc. Pori Ann. Rep.* 2, 49-51. HAURI, R (1956) *Orn. Beob.* 53, 28-35; (1957) *Orn. Beob.* 54, 41-4; (1966) *Orn. Beob.* 63, 77-85; (1988a) *Orn. Beob.* 85, 1-79; (1988b) *Orn. Beob.* 85, 305-7. HAUSBERGER, M and GUYOMARCH, J C (1981) *Biol. Behav.* 6, 79-98. HAUSBERGER, M and BLACK, J M (1991) *Ethol. Ecol. Evol.* 3, 337-44. HAUSER, D C (1957) *Wilson Bull.* 69, 78-90; (1973) *Chat* 37, 91-5. (1974) *Auk* 91, 537-63. HAVERSCHMIDT, F (1934) *Beitr. Fortpfl. Vogel* 10, 73; (1937) *Beitr. Fortpfl. Vögelkde.* 13, 228-30. HAVLÍN, J (1957) *Zool. Listy* 6, 247-56; (1976) *Zool. Listy* 25, 51-63; (1988) *Fol. zool.* 37, 59-66. HAVLÍN, J and FOLK, C (1965) *Zool. Listy* 14, 193-208. HAVLÍN, J and HAVLÍNOVÁ, S (1974) *Sylvia* 19, 89-116. HAVLÍN, J and JURLOV, K T (1977) *Acta Sci. Nat. Brno* 11 (2), 1-50. HAYCOCK, R J and BULLOCK, I D (1982) *Br. Birds* 75, 91-2. HAYHOW, S J (1984) *Magpie* 3, 22-9. HAYMAN, R W (1949) *Br. Birds* 42, 84-5; (1953) *Br. Birds* 46, 378; (1958) *Br. Birds* 51, 275. HAZELWOOD, A and GORTON, E (1953) *Bull. Br. Orn. Club* 73, 1-2; (1955) *Bull. Br. Orn. Club* 75, 98. HAZEVOET, C J and HAAFKENS, L B (1989) *Nature reserve development and ornithological research in the Republica de Cabo Verde*. ICBP Rep. Amsterdam. HEATHERLEY, F (1910) *Br. Birds* 3, 234-42. HECKE, P VAN and VERSTUYFT, S (1972) *Gerfaut* 62, 245-72. HEEB, P A (1991) D Phil Thesis. Oxford Univ. HEGELBACH, J (1980) *Orn. Beob.* 77, 60; (1984) Ph D Thesis. Zurich Univ. HEGELBACH, J and ZISWILER, V (1979) *Orn. Beob.* 76, 119-132. HEGNER, R E and WINGFIELD, J C (1986) *Horm. Behav.* 20, 294-312; (1987) *Auk* 104, 470-80. HEIDEMANN, J and SCHÜZ, E (1936) *Mitt. Vogelwelt* 35, 37-44. HEIJ, C J (1986) *Int. Stud. Sparrows* 13, 28-34. HEIJ, C J and MOELIKER C W (1991) In Pinowski, J and Summers-Smith, J D (eds) *Granivorous birds in the agricultural landscape*, 59-85. Warsaw. HEIM DE BALSAC, H (1929) *Alauda* 1, 68-77; (1948) *Alauda* 16, 75-96. HEIM DE BALSAC, H and HEIM DE BALSAC, T (1949-50) *Alauda* 17-18, 206-21. HEIM DE BALSAC, H and MAYAUD, N (1962) *Les oiseaux du nord-ouest de l'Afrique*. Paris. HEIMERDINGER, M A and PARKES, K C (1966) *Br. Birds* 59, 315-6. HEINIGER, P H (1991) *Orn. Beob.* 88, 193-207. HEINRICH, B (1988a) *Condor* 90, 950-2; (1988b) *Behav. Ecol. Sociobiol.* 23, 141-56; (1988c) In Slobodchikoff, C N (ed.) *The ecology of social behaviour*,

285-311. London; (1990) *Ravens in winter*. London. HEIN-RICH, B and MARZLUFF, J (1992) *Condor* **94**, 549-50. HEIN-ROTH, O and HEINROTH, M (1924-6) *Die Vögel Mitteleuropas* 1. Berlin-Lichterfelde; (1931) *Die Vögel Mitteleuropas. Ergänzungsband.* HEINZE, J and KROTT, N (1979) *Vogelwelt* **100**, 225-7. HEINZEL, H, FITTER, R, and PARSLOW, J (1972) *The birds of Britain and Europe with North Africa and the Middle East.* London. HEINZEL, H and WOLTERS, H E (1970) *J. Orn.* **111**, 497-8. HEISE, G (1970) *Orn. Mitt.* **22**, 144-5. HELB, H-W (1974) *Verhandl. Ges. Ökol.* (1974), 55-8; (1981) *J. Orn.* **122**, 325; (1985) *Behaviour* **94**, 279-323; (1986) In Wüst, W (ed.) *Avifauna Bavariæ* 2. Munich. HELL, P and SOVIŠ, B (1958) *Zool. Listy* **7**, 38-56; (1959) *Aquila* **65**, 145-60. HELLMAYR, C E (1929) *Field Mus. nat. Hist. Publ.* **263**, Zool. Ser. **17**, 27-144; (1935) *Catalogue of birds of the Americas and the adjacent islands* 8; (1938) 11. Chicago. HELLMICH, J (1985) *Orn. Mitt.* **37**, 178-81. HELM, F (1894) *Orn. Monatsschr.* **19**, 239. HELMS, C W (1959) *Wilson Bull.* **71**, 244-53. HELMS, C W, AUSSIKER, W H, BOWER, E B, and FRETWELL, S D (1967) *Condor* **69**, 560-78. HELMS, C W and DRURY, W H (1960) *Bird-Banding* **31**, 1-40. HÉMERY, G and PASCAUD, P-N (1981) *Oiseau* **51**, 1-16. HEMMINGSEN, A M (1951) *Spolia zool. Mus. Haun.* **11**; (1958) *Vidensk. Medd. dansk nat. Foren.* **120**, 189-206. HEMMINGSEN, A M and GUILDAL, J A (1968) *Spolia zool. Mus. haun.* **28**. HENDERSON, I G (1990) *Proc. int. orn. Congr.* **20** suppl., 377-8; (1991) *Ring. Migr.* **12**, 23-7. HENDY, E W (1939) *Br. Birds* **33**, 162; (1943) *Br. Birds* **35**, 37; (1943) *Somerset birds.* London. HENRICI, P (1927) *Beitr. Fortpfl. Vögel* **3**, 7-13. HENRIKSEN, K (1989) *Dansk orn. Foren. Tidsskr.* **83**, 55-9. HENRY, G M (1971) *A guide to the birds of Ceylon.* London. HENS, P A (1931) *Proc. int. orn. Congr.* **7**, 439-64. HENS, P A and MARLE, J G VAN (1933) *Org. Club. Nederl. Vogelk.* **6**, 49-58. HENTY, C J (1975) *Br. Birds* **68**, 463-6; (1979) *Bird Study* **26**, 192-4. HENZE, O (1975) *Gef. Welt* **99**, 31-2; (1979) *Falke* **26**, 13-20. HEPWORTH, N M (1946) *Br. Birds* **39**, 84-5. HERMANN, H (1983) *Verh. orn. Ges. Bayern* **23**, 459-77. HERREMANS, L (1973) *Wielewaal* **39**, 185-7. HERREMANS, M (1977) *Wielewaal* **43**, 133-41; (1982) *Gerfaut* **72**, 243-54; (1987) *Oriolus* **53**, 149-53; (1988) *Gerfaut* **78**, 243-60; (1989) *Dutch Birding* **11**, 9-15; (1990a) *Ardea* **78**, 441-58; (1990b) *Ring. Migr.* **11**, 86-9. HERRERA, C M (1980) *Ardeola* **25**, 143-80. HERRLINGER, E (1966) *Egretta* **9**, 55-60. HERRMANN, J (1977) *Orn. Ber. Berlin (West)* **2** (2), 121-38. HERROELEN, P (1962) *Gerfaut* **52**, 173-205; (1967) *Giervalk* **57**, 81-3; (1974) *Wielewaal* **40**, 69-74; (1980) *Ornis Brabant* **85**, 13-15; (1983) *Ornis Flandriae* 1983, 92-3; (1987) *Gerfaut* **77**, 99-104. HESELER, U (1966) *Luscinia* **39**, 69-71. HESS, R (1975) *Orn. Beob.* **72**, 120. HESSE, E (1915) *Orn. Monatsber.* **23**, 112-8. HEUER, J (1986) *Gef. Welt* **110**, 59. HEUGLIN, M T VON (1869-74) *Ornitologie Nordost-Afrika's* 1. Kassel. HEWSON, R (1957) *Br. Birds* **50**, 432-4; (1981) *Br. Birds* **74**, 509-12; (1984) *J. appl. Ecol.* **21**, 843-68. HEWSON, R and LEITCH, A F (1982) *Bird Study* **29**, 235-8. HICKLING, R (1983) *Enjoying Ornithology.* Calton. HICKS, L E (1934) *Bird-Banding* **5**, 103-18. HIETT, J C and CATCHPOLE, C K (1982) *Anim. Behav.* **30**, 568-74. HILDÉN, O (1969a) *Ornis fenn.* **46**, 93-112; (1969b) *Ornis fenn.* **46**, 179-87; (1972) *Ornis fenn.* **49**, 14-15; (1974a) *Lintumies* **9**, 45-51; (1974b) *Ornis fenn.* **51**, 10-35; (1977) *Ornis fenn.* **54**, 170-9; (1988) *Sitta* **2**, 21-57. HILDÉN, O and NIKANDER, P J (1986) *Lintumies* **21**, 88-93; (1988) *Lintumies* **23**, 80-5; (1991) *Lintumies* **26**, 104-117. HILL, A (1986) *Orn. Mitt.* **38**, 72-84. HILL, D,

TAYLOR, S, THAXTON, R, AMPHLET, A, and HORN, W (1990) *Bird Study* **37**, 133-41. HILL, R A (1976) *Bird-Banding* **47**, 112-4. HILPRECHT, A (1964) *Beitr. Vogelkde.* **10**, 177-83; (1965) *Auspicium* **2**, 91-118. HILTY, S L and BROWN, W L (1986) *A guide to the birds of Colombia.* Princeton. HIMMER, K H (1967) *Gef. Welt* **91**, 188-92. HINDE, R A (1947) *Br. Birds* **40**, 246-7; (1952) *Behaviour Supp.* **2**; (1953) *Behaviour* **5**, 1-31; (1954) *Behaviour* **7**, 207-32; (1954) *Proc. Roy. Soc. B* **142**, 306-31, 331-58; (1955-6) *Ibis* **97**, 706-45; **98**, 1-23; (1958a) *Anim. Behav.* **6**, 211-18; (1958b) *Proc. zool. Soc. Lond.* **131**, 1-48; (1959) *Bird Study* **6**, 15-19; (1960) *Proc. Roy. Soc. B* **153**, 398-420. HINDE, R A and STEEL, E (1972) *Anim. Behav.* **20**, 514-25. HINDMARSH, A M (1984) *Behaviour* **90**, 302-24; (1986) *Behaviour* **99**, 87-100. HIRALDO, F and HERRERA, C M (1974) *Doñana, Acta Vertebr.* **1** (2), 149-70. HIRSCHFELD, E (1988) *Birding World* **1**, 380. HIRSCHFELD, E, HOLST, O, KJELLEN, N, PERSSON, O, and ULLMAN, M (1983) *Anser* Suppl. **14**. HIRSCHFELD, E and SYMENS, P (1992) *Sandgrouse* **14**, 48-51. HIRSCHI, W (1986) *Orn. Beob.* **83**, 145-6. HISS, J-P (1979) *Ciconia* **3**, 184-5. HJORT, C and LINDHOLM, C-G (1978) *Oikos* **30**, 387-92. HOBBS, J N (1955) *Emu* **55**, 202. HOEHL, O (1939) *Jber. vogelkd. Beob. Stat. 'Untermain', Frankfurt,* 24-6. HOFER, H (1935) *Verh. zool. bot. Ges. Wien* **85**, 60-87. HOFFMANN, E C (1930) *Bird-Banding* **1**, 80-1. HOFFMANN, B (1925) *Mitt. Ver. sächs. Orn.* **1** suppl., 37-43; (1928) *Ver. orn. Ges. Bayern* **18**, 75-107. HOFFMANN, L (ed.) (1955) *Premier compte rendu 1950-54* Station Biol. de la Tour du Valat; (1956) *Deuxième compte rendu 1955* Station Biol. de la Tour du Valat. HOFSHI, H (1985) *M Sc Thesis.* Ben Gurion Univ. HOFSHI, H, GERSANI, M, and KATZIR, G (1987a) *Ostrich* **58**, 156-9; (1987b) *Ibis* **129**, 389-90. HÖGLUND, J (1985) *Ornis fenn.* **62**, 19-22. HOGSTAD, O (1967) *Sterna* **7**, 255-60; (1969) *Nytt Mag. Zool.* **17**, 81-91; (1975) *Norw. J. Zool.* **23**, 223-34; (1977) *Sterna* **16**, 19-27; (1982) *Fauna norv. (C) Cinclus* **5**, 59-64; (1985) *Ornis fenn.* **62**, 13-18; (1988) *Fauna norv. (C) Cinclus* **11**, 27-39. HOGSTAD, O and RØS-KAFT, E (1986) *Fauna norv. (C) Cinclus* **10**, 7-10. HÖGS-TEDT, G (1980a) *Ornis scand.* **11**, 110-15; (1980b) *Nature* **283**, 64-6; (1981) *J. Anim. Ecol.* **50**, 219-29. HOHLT, H (1956) *Vogelwelt* **77**, 194. HOLGERSEN, H (1982) *Sterna* **17**, 85-123. HOLIAN, J J and FORTEY, J E (1992) *Br. Birds* **85**, 370-6. HÖLLER, C and TEIXEIRA, A M (1983) *Vogelwarte* **32**, 81-2. HOLLIDAY, S T (1990) *Alectoris* **7**, 49-57. HOLLOM, P A D (1940) *Br. Birds* **34**, 86-7; (1955) *Ibis* **97**, 1-17; (1959) *Ibis* **101**, 183-200; (1962) *Br. Birds* **55**, 158-64; (1971) *The popular handbook of British birds.* London. HOLLOM, P A D, PORTER, R F, CHRISTENSEN, S, and WILLIS, I (1988) *Birds of the Middle East and North Africa.* Calton. HOLLOWAY, J (1984) *Fair Isle's garden birds.* Lerwick. HOLLYER, J N (1970) *Br. Birds* **63**, 353-73; (1971) *Br. Birds* **64**, 196-7. HOLMAN, D (1990) *Br. Birds* **83**, 430-2. HOLMAN, D and KEMP, J (1991) *Birding World* **4**, 353-4. HOLMAN, D J (1981) *Br. Birds* **74**, 203-4. HOLMAN, D J and MADGE, S C (1982) *Br. Birds* **75**, 547-53. HOLMES, D A and WRIGHT, J O (1969) *J. Bombay nat. Hist. Soc.* **66**, 8-30. HOLMES, P (1982) *Birds of Oxfordshire 1981*, 36-8. HOLMES, P R, HOLMES, H J, and PARR, A J (eds) (1985) *Rep. Oxford Univ. Exped. Kashmir 1983.* Oxford. HOLMS-TRÖM, C T (ed.) (1959-62) *Våra Fåglar i Norden.* Stockholm. HOLSTEIN, V (1934) *Dansk orn. Foren. Tidsskr.* **28**, 116-18. HOLTMEIER, F-K (1966) *J. Orn.* **107**, 337-45. HOLYOAK, D (1967a) *Br. Birds* **60**, 52; (1967b) *Bird Study* **14**, 61-2; (1967c) *Bird Study* **14**, 153-68; (1968) *Bird*

Study 15, 147-53; (1970a) *Ibis* 112, 397-400; (1970b) *Bull. Br. Orn. Club.* 90, 40-2; (1971) *Bird Study* 18, 97-106; (1972a) *Bird Study* 19, 59-68; (1972b) *Bird Study* 19, 215-27; (1974a) *Bird Study* 21, 15-20; (1974b) *Bird Study* 21, 117-28. HOLYOAK, D and RATCLIFFE, D A (1968) *Bird Study* 15, 191-7. HOLYOAK, D T and SEDDON, M B (1991) *Alauda* 59, 55-7, 116-20. HOLTMEIER, F K (1966) *J. Orn.* 107, 337-45. HÖLZINGER, J (1987) *Die Vögel Baden-Württembergs* 1. Karlsruhe; (1992a) *Kart. Med. Brutvögel* 7, 3-8; (1992b) *Kart. Med. Brutvögel* 7, 15; (1992c) *Kart. Med. Brutvögel* 7, 17-25. HOMANN, J (1959) *Beitr. Naturkde. Niedersachs.* 12, 58-62. HONGELL, H (1977) *Ornis fenn.* 54, 138. HOOGENDOORN, W B (1991) *Dutch Birding* 13, 104-6. HOPE JONES, P (1980) *Br. Birds* 73, 561-8. HOPKINS, J R (1985) *Br. Birds* 78, 597. HOPPE, R (1976) *Falke* 23, 29-33. HOPSON, A J (1964) *Bull. Nigerian orn. Soc.* 1 (4), 7-15. HORDOWSKI, J (1989) *Notatki orn.* 30 (1-2), 21-36. HORNBUCKLE, J (1984) *Derbyshire Bird Rep.* 1983, 62-3. HORNBY, R J (1971) Ph D Thesis. Nottingham Univ. HORNER, K O (1977) Ph D Thesis. Virginia Polytechnic Inst. State Univ. HORTLING, I (1928) *Ornis fenn.* 5, Supp; (1929) *Ornitologisk handbok.* Helsingfors; (1938) *Mitt. Ver. sächs. Orn.* 5, 219-27. HORTLING, I and BAKER, E C S (1932) *Ibis* (13) 2, 100-27. HORVÁTH, L (1975) *Aquila* 80-1, 310-1; (1976) *Aquila* 82, 37-47; (1977) *Aquila* 83, 91-5. HORVÁTH, L and HÜTTLER, B (1963) *Acta Zool. Sci. Hung.* 9, 271-6. HORVÁTH, L and KEVE, A (1956) *Bull. Br. Orn. Club* 76, 92-5. HOSONO, T (1966a) *Misc. Rep. Yamashina Inst. Orn.* 4, 327-47; (1966b) *Misc. Rep. Yamashina Inst. Orn.* 4, 481-7; (1967a) *Misc. Rep. Yamashina Inst. Orn.* 5, 34-47; (1967b) *Misc. Rep. Yamashina Inst. Orn.* 5, 177-93; (1967c) *Misc. Rep. Yamashina Inst. Orn.* 5, 278-86; (1969) *Misc. Rep. Yamashina Inst. Orn.* 5, 659-75; (1971) *Misc. Rep. Yamashina Inst. Orn.* 6, 231-49; (1973) *Misc. Rep. Yamashina Inst. Orn.* 7, 56-72; (1975) *Misc. Rep. Yamashina Inst. Orn.* 7, 533-49; (1983) *Misc. Rep. Yamashina Inst. Orn.* 15, 63-71; (1989) *Jap. J. Orn.* 37, 103-27. HOUSTON, C S (1963) *Bird-Banding* 34, 94-5. HOUSTON, C S and STREET, M G (1959) *Sask. Nat. Hist. Soc. Spec. Publ.* 2. HOUSTON, D (1974) *Unpubl. Rep., Dept. Forestry nat. Res.,* Edinburgh Univ; (1977a) *J. appl. Ecol.* 14, 1-15; (1977b) *J. appl. Ecol.* 14, 17-29; (1978) *Ann. appl. Biol.* 88, 339-41. HOVEL, H (1960) *Bull. Br. Orn. Club* 80, 75-6; (1987) *Check-list of the birds of Israel with Sinai.* Tel-Aviv. HOWARD, H E (1929) *An introduction to the study of bird behaviour.* HOWARD, D V (1968) *Bird-Banding* 39, 132. HOWARD, D V and DICKINSON HENRY, D (1966) *Bird-Banding* 37, 123. HOWARD, H E (1929) *An introduction to the study of bird behaviour.* Cambridge; (1930) *Territory in bird life.* London. HOWARD, R and MOORE, A (1980) *A complete checklist of the birds of the world.* Oxford. HOWELL, T R, PAYNTER, R A, and RAND, A L (1968) Carduelinae, in *Check-list of birds of the world* 14. Cambridge (Mass.). HOWELLS, V (1956) *A naturalist in Palestine.* London. HUBÁLEK, Z (1976) *Vertebr. Zprávy* 1975-6 (1), 82-6; (1978a) *Vest. Česk. Spol. Zool.* 42, 1-14; (1978b) *Vest. Česk. Spol. Zool.* 42, 15-22; (1980) *Acta orn.* 16, 535-53; (1983) *Acta Sc. nat. Brno (NS)* 17 (1), 1-52. HUBÁLEK, Z and HORÁKOVÁ, M (1988) *Acta Sci. Nat. Brno* 22 (5), 1-44. HUBBARD, J P (1969) *Auk* 86, 393-432; (1970) *Wilson Bull.* 82, 355-69; (1980) *Nemouria* 25, 1-9. HUBER, B (1991) In Glandt (1991), 45-59. HUCKRIEDE, B (1969) *Vogelwarte* 25, 23-5. HUDEC, K (1983) *Ptáci CSSR* 3 (1). Prague. HUDEC, K and FOLK, C (1961) *Zool. Listy* 10, 305-30. HUDSON, R (1967) *Br. Birds* 60, 423-6; (1969)

Br. Birds 62, 13-22. HUDSON, R and MARCHANT, J H (1984) *BTO Res. Rep.* 13. HUDSON, W H (1915) *Birds and man.* London. HÜE, F and ETCHÉCOPAR, R-D (1958) *Terre Vie* 105, 186-219; (1970) *Les oiseaux du proche et du moyen orient.* Paris. HUGHES, J (1976) *Br. Birds* 69, 273. HUGHES, S W M (1972) *Br. Birds* 65, 445; (1986) *Br. Birds* 79, 342. HUI, P A and HUI, M (1974) *Vögel Heimat* 45, 122-4. HULTEN, M (1967) *Regulus* 47, 27-31; (1972) *Regulus* 10, 463-5. HUME, A O (1874a) *Stray Feathers* 2, 29-324; (1874b) *Stray Feathers* 2, 467-84. HUME, R A (1975) *Br. Birds* 68, 515-16; (1980) *Br. Birds* 73, 478-9; (1983) *Br. Birds* 76, 90. HUMPHREY, D (1967) *Ool. Rec.* 41, 69-71. HUMPHREYS, G R (1928) *Bull. Br. Ool. Assoc.* 2, 51-7. HUND, K and PRINZINGER, R (1981) *Ökol. Vögel* 3, 261-5. HUNTINGTON, C E (1952) *Syst. Zool.* 1, 149-70. HURRELL, A G (1951) *Br. Birds* 44, 88-9. HURRELL, H G (1956) *Br. Birds* 49, 28-31. HUSAIN, K Z (1964) *Bull. Br. Orn. Club* 84, 9-11. HUSBY, M (1986) *J. Anim. Ecol.* 55, 75-83. HUSSELL, D J T (1972) *Ecol. Monogr.* 42, 317-64; (1985) *Ornis scand.* 16, 205-12. HUSSELL, D J T and HOLROYD, G L (1974) *Can. Fld.-Nat.* 88, 197-212. HUSTINGS, F, POST, F, and SCHEPERS, F (1990) *Limosa,* 63, 103-11. HUT, R M G VAN DER (1985) *Graspieper* 5, 43-64, 103-16. HUT, R M G VAN DER (1986) *Ardea* 74, 159-76. HUTCHINSON, C D (1989) *Birds in Ireland.* Calton. HUTSON, H P W (1945) *Ibis* 87, 456-9; (1954) *The birds about Delhi.* Kirkee. HYTÖNEN, O (1972) *Ardeola* 16, 277.

IANKOV, P N (1983) Ph D Thesis. Inst. Zool. Acad. Sci. Belorussiya Minsk. IAPICHINO, C and MASSA, B (1989) *The birds of Sicily.* Tring. IDZELIS, R F (1986) *Tez. Dokl. 1. S'ezda Vsesoyuz. orn. Obshch. 9. Vsesoyuz. orn. Konf.* 1, 261-2. IJZENDOORN, A L J VAN (1950) *The breeding-birds of the Netherlands.* Leiden. ILANI, G and SHALMON, B (1983) *Israel Land Nat.* 9, 39. IL'YASHENKO, V Y (1986) *Trudy Zool. Inst. Akad. Nauk SSSR* 150, 77-81. IL'YASHENKO, V Y, KALYAKIN, M V, SOKOLOV, E P, and SOKOLOV, A M (1988) *Trudy Zool. Inst. Akad. Nauk SSSR* 182, 70-88. ILYINA, T A (1990) In Kurochkin, E N (ed.) *Sovremennaya ornitologiya 1990,* 48-54. Moscow. IMHOF, T A (1976) *Alabama birds.* Alabama Univ. IMMELMANN, K (1966) *Ostrich Suppl.* 6, 371-9; (1973) *Der Zebrafink.* Wittenberg Lutherstadt. IMMELMANN, K, STEINBACHER, J, and WOLTERS, H E (1965) *Vögel in Käfig und Voliere: Prachtfinken* 1. Aachen. INBAR, R (1971) *Birds of Israel.* Tel Aviv; (1979) *Israel Land Nat.* 5, 20-2. INDYKIEWICZ, P (1988) *Przeglad Zool.* 32, 281; (1990) In Pinowski, J and Summers-Smith, J D (eds) *Granivorous birds in the agricultural landscape,* 95-121. Warsaw. INGLIS, I R, FLETCHER, M R, FEARE, C J, GREIG-SMITH, P W, and LAND, S (1982) *Ibis* 124, 351-5. INGRAM, C (1965) *Bull. Br. Orn. Club* 85, 20. INOZEMTSEV, A A (1962) *Ornitologiya* 5, 101-4. INSKIPP, C and INSKIPP, T (1985) *A guide to the birds of Nepal.* London. INSLEY, H and WOOD, J B (1973) *Nat. Wales* 13, 165-73. ION, I (1971) *Muz. Stiint. Nat. Bacau, Stud. Communic.* 263-76. ION, I and SARACU, S (1971) *Studii și Comunicări Științele Naturii, Muz. Județ. Suceava* 2, 271-8. IOVCHENKO, N P (1986) *Trudy Zool. Inst. Akad. Nauk SSSR* 147, 7-24. IRISOV, E A (1972) *Ornitologiya* 10, 334. IRVING, L (1960) *Bull. US natn. Mus.* 217. IRWIN, M P S (1981) *The birds of Zimbabwe.* Salisbury. ISAKOV, Y A and VOROBIEV, K A (1940) *Trudy Vsesoyuz. orn. zapoved. Gassan-Kuli,* 1, 5-159. ISENMANN, P (1990) *Nos Oiseaux* 40, 308; (1992) *Alauda* 60, 109-11. ISHIMOTO, A (1992) *J. Yamashina Inst. Orn.* 24, 1-12.

IVANAUSKAS, T L (1961) *Ekol. Migr. Ptits Pribaltiki*, 329-31. IVANCHEV, V P (1988) *Ornitologiya* 23, 209-10. IVANITSKI, V V (1984) *Zool. Zh.* 63, 1374-87; (1985a) *Nauch. Dokl. vyssh. Shk. Biol. Nauki* (9), 50-5; (1985b) *Zool. Zh.* 64, 1213-23; (1986) *Zool. Zh.* 65, 387-98; (1991) *Zool. Zh.* 70, 104-17. IVANOV, A I (1969) *Ptitsy Pamiro-Alaya.* Leningrad. IVANOV, B E (1987) *Ekol. Polska* 35, 699-721. IVANTER, E V (1962) *Ornitologiya* 5, 68-85. IVASHCHENKO, A A and KOVSHAR', A F (1972) *Ornitologiya* 10, 333-4 IVOR, H R (1944) *Wilson Bull.* 56, 91-104. IZMAYLOV, I V and BOROVITSKAYA, G K (1967) *Ornitologiya* 8, 192-7.

JACK, J (1974) *Br. Birds* 67, 356. JÄCKEL, A J (1891) *Systematische Übersicht der Vögel Bayerns.* Munich. JACKSON, F J (1938) *Birds of Kenya colony and the Uganda protectorate* 2. London. JACKSON, H D (1989) *Bull. Br. Orn. Club* 109, 100-6. JACOB, J-P (1982) *Aves* 19, 37-45; (1984) *Aves* 21, 261. JACOBSEN, J R (1963) *Dansk orn. Foren. Tidsskr.* 57, 181-220. JACOBY, H, KNÖTZSCH, G, and SCHUSTER, S (1970) *Orn. Beob.* 67 suppl. JAHN, H (1942) *J. Orn.* 90, 7-302. JAHNKE, W (1955) *Orn. Mitt.* 7, 51. JAKOBS, B (1959) *Orn. Mitt.* 11, 121-5. JAKOBSEN, O (1986) *Zool. Middle East* 1, 32-3. JAKUBIEC, Z (1972a) *Ekol. Polska* 20, 609-35; (1972b) *Ochrona Przyrody* 17, 135-52. JALIL, A K (1985) BSc Hons Thesis. National Univ. Singapore. JAMES, D A and NEAL, J C (1986) *Arkansas birds.* Fayetteville. JAMES, P (1986) *Br. Birds* 79, 299-300. JAMES, P C, BARRY, T W, SMITH, A R, and BARRY, S J (1987) *Ornis scand.* 18, 310-12. JANES, S W (1976) *Condor* 78, 409. JÄNNES, H (1992) *Lintumies* 27, 240-7. JÄNNES, H, NUMMINEN, T, NIKANDER, P J, and PALMGREN, J (1990) *Lintumies* 25, 254-71. JANSSEN, R B (1983) *Loon* 55, 64-5; (1987) *Birds in Minnesota.* Minneapolis; (1990) *Loon* 62, 69-71. JARDINE, D (1991) *Birds in Northumbria 1990*, 103-6. JARDINE, D C (1992a) *Br. Birds* 85, 619; (1992b) *Scottish Bird Rep. 1990*, 65-9. JARRY, C (1976) *Passer* 12, 64-75. JARRY, C and LARIGAUDERIE, F (1974) *Oiseau* 44, 62-71. JÄRVI, E and MARJAKANGAS, A (1985) *Ornis fenn.* 62, 171. JÄRVINEN, A and PIETIÄINEN, H (1982) *Mem. Soc. Fauna Flora Fenn.* 58, 21-6. JÄRVINEN, O and VÄISÄNEN, R A (1976) *Ornis fenn.* 53, 115-18; (1978) *J. Orn.* 119, 441-9; (1979) *Oikos* 33, 261-71. JEDRASZKO-DABROWSKA, D, and SZEPIETOWSKA, S B (1987) *Falke* 34, 337-8. JENNER, H E (1947) *Br. Birds* 40, 176. JENNI, L (1980) *Orn. Beob.* 77, 62; (1982) *Orn. Beob.* 79, 265-72; (1983) *Orn. Beob.* 80, 136-7; (1985) *Vogelwarte* 33, 53-63; (1986) *Orn. Beob.* 83, 267-8; (1987) *Ornis scand.* 18, 84-94; (1991) *Ornis scand.* 22, 327-34. JENNI, L and JENNI-EIERMANN, S (1987) *Ardea* 75, 271-84. JENNI, L and NEUSCHULZ, F (1985) *Orn. Beob.* 82, 85-106. JENNI, L and SCHAFFNER, U (1984) *Orn. Beob.* 81, 61-7. JENNING, W (1959) *Vogelwarte* 20, 35-6. JENNINGS, M C (1980) *Sandgrouse* 1, 71-81; (1981a) *The birds of Saudi Arabia: a check-list.* (1981b) *Birds of the Arabian Gulf.* London; (1981c) *J. Saudi Arab. nat. Hist. Soc.* 2 (1), 8-14; (1986a) *Phoenix* 3, 1-2; (1986b) *Phoenix* 3, 8-9; (1987) *Phoenix* 4, 7-8; (1988a) *Phoenix* 5, 3-4; (1988b) *Fauna of Saudi Arabia* 9, 457-67; (1992) *Sandgrouse* 14, 27-33. JENNINGS, M C and AL SALAMA, M I (1989) *Nat. Commiss. Wildl. Cons. Dev. Tech. Rep.* 14. JENNINGS, M C, AL SALAMA, M I, and FELEMBAN, H M (1988) *Nat. Commiss. Wildl. Cons. Dev. Tech. Rep.* 4. JENNINGS, M C, AL SHODOUKHI, S A, AL ABBASI, T M, and COLLENETTE, S (1990) *Nat. Commiss. Wildl. Cons. Dev. Tech. Rep.* 19. JENNINGS, M C, ABDULLA, I A, and MOHAMMED, N K (1991) *Nat. Commiss. Wildl. Cons. Dev. Tech.*

Rep. 25. JENNINGS, P (1981d) *Peregrine* 5 (3), 114-19; (1984) *Peregrine* 5 (6), 282. JENTZSCH, M (1988) *Acta ornithoecol.* 1, 415. JEPSON, P R (1987) *Br. Birds* 80, 19. JERDON, T C (1877) *The birds of India* 2 (1). Calcutta. JERZAK, L (1988) *Notatki Orn.* 29, 27-41. JERZAK, L and KAVANAGH, B (1991) *Br. Birds* 84, 441-3. JESPERSEN, P (1945) *Dansk orn. Foren. Tidsskr.* 39, 92-8. JESSE, W (1902) *Ibis* (8) 2, 531-66. JEWETT, S G, TAYLOR, W P, SHAW, W T, and ALDRICH, J W (1953) *Birds of Washington state.* Seattle. JIAD, J H and BUNNI, M K (1965) *J. biol. Sci. Res. Baghdad* 16, 5-15; (1988) *Bull. Iraq nat. Hist. Mus.* 8, 145-8. JIRSIIK, J (1951) *Sylvia* 13, 125-32. JOENSEN, A H and PREUSS, N O (1972) *Medd. Grøn.* 191 (5). JOHANNSEN, O F (1974) *Vögel Heimat* 45, 85. JOHANSEN, H (1907) *Orn. Jahrb.* 18, 198-203; (1944) *J. Orn.* 92, 1-105. JOHANSSON, L and LUNDBERG, A (1977) *Vår Fågelvärld* 36, 229-37. JOHN, A W G and ROSKELL, J (1985) *Br. Birds* 78, 611-37. JOHNSGARD, P A (1979) *Birds of the Great Plains.* Lincoln. JOHNSON, L R (1958) *Iraq nat. Hist. Mus. Publ.* 16, 1-32. JOHNSON, N K and ZINK, R M (1985) *Wilson Bull.* 97, 421-35. JOHNSON, N K, ZINK, R M, and MARTEN, J A (1988) *Condor* 90, 428-45. JOHNSTON, D W (1962) *Auk* 79, 387-98; (1963) *Wilson Bull.* 75, 435-46; (1967a) *Bird-Banding* 38, 211-14. JOHNSTON, D W and DOWNER, A C (1968) *Bird-Banding* 39, 277-93. JOHNSTON, D W, and HAINES, T P (1957) *Auk* 74, 447-58. JOHNSTON, R F (1967b) *Int. Stud. Sparrows* 1, 34-40; (1967c) *Auk* 84, 275-7; (1969a) *Condor* 71, 129-39; (1969b) *Auk* 86, 558-9; (1969c) *Syst. Zool* 18, 206-31; (1972) *Boll. Zool.* 39, 351-62; (1973a) *Syst. Zool.* 22, 219-26; (1973b) *Orn. Monogr. AOU*, 14, 24-31; (1976) *Occ. Pap. Mus. nat. Hist. Univ. Kansas* 56, 1-8; (1981) *J. Fld. Orn.* 52, 127-33. JOHNSTON, R F and FLEISCHER, R C (1981) *Auk* 98, 503-11. JOHNSTON, R F and KLITZ, W J (1977) In Pinowski, J and Kendeigh, S C (eds) *Granivorous birds in ecosystems*, 15-51. Cambridge. JOHNSTON, R F and SELANDER, R K (1964) *Science* 144, 548-50; (1966) *System. Zool.* 15, 357-8; (1971) *Evolution* 25, 1-28; (1973a) In Kendeigh, S C and Pinowski, J (eds) *Productivity, population dynamics, and systematics of granivorous birds*, 301-26. Warsaw; (1973b) *Amer. Nat.* 107, 373-90. JOLLET, A (1984) *Oiseau* 54, 109-30; (1985) *Alauda* 53, 263-86. JOLLIE, M (1985) *J. Orn.* 126, 303-5. JONASSON, H (1960) *Ornis fenn.* 37, 46-51. JONES, A E (1947-8) *J. Bombay nat. Hist. Soc.* 47, 117-25, 219-49, 409-32. JONES, H (1955) *Br. Birds* 48, 91. JONES, M B (1950) *Field* 195, 761. JONES, P J and WARD, P (1977) *Ibis* 119, 200-3. JONES, R, DAVIS, P G, and MEAD, C J (1975) *Ringers' Bull.* 4, 99-100. JONG, C DE and SCHILTHUIZEN, M (1987) *Vogeljaar* 35, 344-5. JONG, F DE (1981) *Vogeljaar* 29, 261. JONIN, M and LE DEMEZET, M (1972) *Penn ar bed* 8 (NS), 214-18. JONKERS, D A (1983) *Vogeljaar* 31, 37. JONSSON, L (1978a) *Birds of lake, river, marsh and field.* Harmondsworth; (1978b) *Birds of sea and coast.* Harmondsworth; (1979) *Birds of mountain regions.* Harmondsworth; (1982) *Birds of the Mediterranean and Alps.* London; (1992) *Birds of Europe with North Africa and the Middle East.* London. JÖNSSON, P E (1982) *Anser* 21, 213-22; (1989) *Anser* 28, 17-24; (1992) *Anser* 31, 101-8. JONSSON, P N (1949) *Fauna och Flora* 44, 76-84. JORDANS, A VON (1923) *Arch. Naturgesch.* 89A (3), 1-147; (1924) *J. Orn.* 72, 381-410; (1935) *Mitt. Vogelwelt* 34, 81-5; (1950) *Syllegomena biol.*, 65-181. JORDANS, A VON and STEINBACHER, J (1948) *Senckenbergiana* 28, 159-86. JOST, K (1951) *Orn. Mitt.* 3, 140. JOUARD, H (1930) *Orn. Monatsber.* 38, 137-9; (1934) *Alauda* 6, 396-9. JOURDAIN, F C

R (1906) *The eggs of European birds*. London; (1915) *Ibis* (10) **3**, 133-69; (1930) *Orn. Monatsber.* **38**, 155; (1935) *Br. Birds* **29**, 148; (1936) *Ibis* (13) **6**, 725-63. JOURDAIN, F C R and LYNES, H (1936) *Ibis* (13) **6**, 39-47. JÓZEFIK, M (1960) *Acta Orn.* **5**, 307-24; (1976) *Acta Orn.* **15**, 339-482. JUANA, A E DE and COMITÉ IBÉRICO DE RAREZAS DE AL SEO (1989) *Ardeola* **36**, 111-23; (1991) *Ardeola* **38**, 149-66. JUKEMA, J (1992*a*) *Dutch Birding* **14**, 12-14. JUKEMA, J (1992*b*) *Limosa* **65**, 30-1. JUKEMA, J and FOKKEMA, J (1992) *Limosa* **65**, 67-72. JUNG, E (1966) *Falke* **13**, 408-11; (1968) *Falke* **15**, 238-9. JUNG, K (1955) *Beitr. Naturkde. Niedersachs.* **8**, 44-6. JUNG, N (1975) *Falke* **22**, 194; (1983) *Beitr. Vogelkde.* **29**, 249-73. JUNGE, G C A (1942) *Ardea* **31**, 19-22.

KAATZ, C (1986) *Falke* **33**, 328-31. KAATZ, C and OLBERG, S (1975) *Int. Stud. Sparrows* **8**, 107-16. KADHIM, A-H H, AL-DABBAGH, K Y, MAYSOON, M A-N, and WAHEED, I N (1987) *J. Biol. Sci. Res. Baghdad* **18**, 1-9. KADOCHNIKOV, N P and EYGELIS, Y K (1954) *Zool. Zh.* **33**, 1349-57. KAINADY, P V G (1976) *Bull. Basrah nat. Hist. Mus.* **3**, 107-9. KAISER, W (1983) *Falke* **30**, 17-23. KALCHREUTER, H (1969*a*) *Auspicium* **3**, 437-57; (1969*b*) *Anz. orn. Ges. Bayern* **8**, 578-92; (1970) *Vogelwarte* **25**, 245-55; (1971) *Jh. Ges. Naturkde. Württ.* **126**, 284-338. KALDEN, G (1983) *Vogelkdl. Hefte Edertal* **9**, 91-2. KALE, H W, II and MAEHR, D S (1990) *Florida's birds*. Sarasota. KÄLIN, H (1983) *Orn. Beob.* **80**, 296-7. KALINOSKI, R (1975) *Condor* **77**, 375-84. KALITSCH, L VON (1943) *Beitr. Fortpfl. Vögel* **19**, 116-17. KÄLLANDER, H (1982) *Vår Fågelvärld* **41**, 268; (1988) *Ökol. Vögel* **10**, 113-14. KÄLLANDER, H, NILSSON, S G, and SVENSSON, S (1978) *Vår Fågelvärld* **37**, 37-46. KALMBACH, E R (1940) *US Dept. Agric. Tech. Bull.* **711**. KALOTÁS, Z (1986*a*) *Aquila* **92**, 162-70; (1986*b*) *Aquila* **92**, 175-239; (1988) *Beih. Veröff. Nat. Land. Bad.-Württ.* **53**, 67-74. KAMIŃSKI, P (1983) *Notatki orn.* **24**, 167-75; (1986) *J. Orn.* **127**, 315-29. KAMIŃSKI, P and KONARZEWSKI, M (1984) *Ekol. Pol.* **32**, 125-39. KANE, C P (1960) *Br. Birds* **53**, 223. KANEKO, Y (1976) *Misc. Rep. Yamashina Inst. Orn.* **8**, 206-12. KANG, N (1989) Ph D Thesis. National Univ. Singapore; (1991) *Malaysiana* **16**, 98-103; (1992) In Priede, I G and Swift, S M (eds) *Wildlife telemetry, remote monitoring and tracking of animals*, 633-41. Chichester. KANIA, W (1981) *Acta Orn.* **18**, 375-418. KANISS, M (1970) *Anz. orn. Ges. Bayern* **9**, 173-4. KAŇUŠČÁK, P (1979) *Zprávy MOS* **37**, 69-97; (1988) *Orn. Mitt.* **40**, 227-9. KAŇUŠČÁK, P and ŠNAJDAR, M (1972) *Ochrana Fauny* **6**, 128-32. KAPITONOV, V I and CHERNYAVSKI, F B (1960) *Ornitologiya* **3**, 80-97. KAREILA, R (1958) *Ornis fenn.* **35**, 140-50. KARLSSON, J (1983) Dissertation. Lund Univ. KARLSSON, L (1969) *Vår Fågelvärld* **28**, 252. KARPLUS, M (1952) *Ecology* **33**, 129-34. KARR, J R (1976) *Bull. Br. Orn. Club* **96**, 92-6. KARTTUNEN, L, LAAKSONEN, A, and LAPPI, E (1970) *Ornis fenn.* **47**, 30-4. KASPAREK, M (1979) *Ring. Migr.* **2**, 158-9; (1980) *Ökol. Vögel* **2**, 1-36; (1981) *Die Mauser der Singvögel Europas: ein Feldführer*; (1992) *Die Vögel der Türkei: eine Übersicht*. Heidelberg. KATZIR, G (1981) *Ardea* **69**, 209-10; (1983) *Ibis* **125**, 516-23. KAUFMAN, K (1989) *Amer. Birds* **43**, 385-8. KAVANAGH, B P (1987*a*) *Irish Birds* **3**, 387-94; (1987*b*) *Br. Birds* **80**, 383; (1988) *Ring. Migr.* **9**, 83-90. KAZAKOV, B A (1976) *Ornitologiya* **12**, 61-7. KAZAKOV, B A and LOMADZE, N K (1984) *Ornitologiya* **19**, 179-80. KAZANTSEV, A N (1967) *Ornitologiya* **8**, 356-7. KEAR, J (1960) *Ibis* **102**, 614-16; (1962) *Proc. zool. Soc. Lond.* **138**, 163-204. KEAST,

A and MORTON, E S (eds) (1980) *Migrant birds in the Neotropics: ecology, behavior, distribution, and conservation*. Washington DC. KECK, W N (1934) *J. Exp. Zoöl.* **67**, 315-47. KEIL, W (1973) In Kendeigh, S C and Pinowski, J (eds) *Productivity, population dynamics and systematics of granivorous birds*, 253-62. Warsaw. KEKILOVA, A F (1978) *Tez. Soobshch. 2. Vsesoyuz. Konf. Migr. Ptits.* **1**, 29-30. Alma-Ata. KELLER, M (1979) *Notatki orn.* **20**, 1-16. KELM, H (1936) *Vogelzug* **7**, 67-8. KELM, H and ECK, S (1985) *Zool. Abh. Staatl. Mus. Tierkde. Dresden* **42**, 1-40. KEMPER, H (1964) *Z. angew. Zool.* **51**, 31-47. KEMPER, T (1959) *Auk* **76**, 181-9. KEMPPAINEN, J and KEMPPAINEN, O (1991) *Lintumies* **26**, 20-9. KENDRA, P E, ROTH, R R, and TALLAMY, D W (1988) *Wilson Bull.* **100**, 80-90. KENNEDY, R J (1969) *Br. Birds* **62**, 249-58. KENNEDY, R S (1984) *Bull. Br. Orn. Club* **104**, 149-50. KENNERLEY, P R (1987) *Hong Kong Bird Rep. 1984-5*, 97-111. KEPLER, A K, KEPLER, C B, and DOD, A (1975) *Condor* **77**, 220-1. KERÄNEN, S and SOIKKELI, M (1985) *Ornis fenn.* **62**, 23-4. KESSEL, B (1951) *Bird-Banding* **22**, 16-23; (1957) *Amer. Midl. Nat.* **58**, 257-331. KESSEL, B and GIBSON, D D (1978) *Stud. avian Biol.* **1**. KESSEL, B and SPRINGER, H K (1966) *Condor* **68**, 185-95. KETTERSON, E D, and NOLAN, V (1978) *Auk* **95**, 755-8; (1983) *Wilson Bull.* **95**, 628-35. KEULEMANS, J G (1866) *Nederland. Tijds. Dierk.* **3**, 363-74. KEVE, A (1943) *Anz. Akad. Wiss. Wien, math.-naturwiss.* **80**, 16-20; (1958*a*) *Bull. Br. Orn. Club* **78**, 88-90; (1958*b*) *Bull. Br. Orn. Club* **78**, 155-7; (1960) *Proc. int. Orn. Congr.* **12**, 376-95; (1966*a*) *Lounais-Hämeen Luonto* **23**, 49-52; (1966*b*) *Riv. ital. Orn.* **36**, 315-23; (1966*c*) *Aquila* **71-2**, 39-65; (1967*a*) *Bull. Br. Orn. Club* **87**, 39-40; (1967*b*) *Ibis* **109**, 120-2; (1969) *Der Eichelhäher*. Wittenberg Lutherstadt; (1970) *Riv. ital. Orn.* **40**, 37-42; (1973) *Zool. Abh. Staatl. Mus. Tierkde. Dresden* **32**, 175-98; (1976*a*) *Emu* **76**, 152-3; (1976*b*) *Il-Merill* **17**, 25-6; (1978) *Zool. Abh. Staatl. Mus. Tierkde. Dresden* **34**, 245-73; (1985) *Der Eichelhäher*. Wittenberg Lutherstadt. KEVE, A and DONČEV, S (1967) *Zool. Abh. Staatl. Mus. Tierkde. Dresden* **29**, 1-16. KEVE, A and KOHL, Ş (1978) *Nymphaea* **6**, 583-606. KEVE, A and ROKITANSKY, G (1966) *Ann. naturhist. Mus. Wien* **69**, 225-83. KEVE, A and STERBETZ, I (1968) *Falke* **15**, 184-7, 230-3. KEVE-KLEINER, A (1942) *Aquila* **46-9**, 146-224; (1943) *Aquila* **50**, 369-70. KEYMER, I F (1975) *Br. Birds* **68**, 49. KHAIRALLAH, N H (1986) *Bull. Orn. Soc. Middle East* **16**, 16-17. KHAKHLOV, V A (1991) *Ornitologiya* **25**, 214-15. KHAN, R (1986) *Devon Birds* **39**, 63-4; (1987) *Devon Birds* **40**, 41. KHOKHLOV, A N and KONSTANTINOV, V M (1983) In Kuchin, A P (ed.) *Ptitsy Sibiri*, 224-5. Gorno-Altaysk. KHOKHLOVA, N A (1960) *Ornitologiya* **3**, 259-69. KHOKHLOVA, T Y, SAZONOV, S V, and SUKHOV, A V (1983) In Ivanter, E V (ed.) *Fauna i ekologiya ptits i mlekopitayushchikh severo-zapada SSSR*, 41-52. Petrozavodsk. KIDONO, H (1977) *Misc. Rep. Yamashina Inst. Orn.* **9**, 271-9. KIIS, A (1985) *Dansk. orn. Foren. Tidsskr.* **79**, 107-12; (1986) *Ornis scand.* **17**, 80-3. KIIS, A and MØLLER, A P (1986) *Anim. Behav.* **34**, 1251-5. KILHAM, L (1989) *The American Crow and the Common Raven*. Texas. KILIÇ, A and KASPAREK, M (1989) *Birds of Turkey* **8**. KILIN, S V (1983) In Kuchin, A P (ed.) *Ptitsy Sibiri*, 140-1. KILLPACK, M L (1986) *Utah Birds* **2**, 23-4. KILZER, R and BLUM, V (1991) *Atlas der Brutvögel Vorarlbergs*. Wolfurt. KING, B (1969) *Br. Birds* **62**, 201; (1971) *Br. Birds* **64**, 423; (1976*a*) *Bristol Orn.* **9**, 159; (1976*b*) *Br. Birds* **69**, 507; (1985*a*) *Br. Birds* **78**, 401; (1985*b*) *Br. Birds* **78**, 512-13. KING, B (1978) *J. Saudi*

Arab. nat. Hist. Soc. 1 (21), 3-24. KING, B and KING, M (1968) *Br. Birds* 61, 316. KING, B and ROLLS, J C (1968) *Br. Birds* 61, 417-18. KING, B, WOODCOCK, M, and DICKINSON, E C (1975) *A field guide to the birds of south-east Asia*. London. KING, J R, BARKER, S, and FARNER, D S (1963) *Ecology* 44, 513-21. KING, J R and FARNER, D S (1959) *Condor* 61, 315-24; (1961) *Condor* 63, 128-42; (1966) *Amer. Nat.* 100, 403-18. KING, J R, FARNER, D S, and MORTON, M L (1965) *Auk* 82, 236-52. KING, J R and MEWALDT, L R (1987) *Condor* 89, 549-65. KING, R (1976c) *N. Am. Bird-Bander* 1, 172-3. KING, W B and KEPLER, C B (1970) *Auk* 87, 376-8. KINGTON, B L (1973) *Br. Birds* 66, 231. KINSEY, R N (1972) IIIB Project. Auckland Univ. KINZELBACH, R (1962) *Vogelwelt* 83, 187; (1969) *Bonn. zool. Beitr.* 20, 175-81. KINZELBACH, R and MARTENS, J (1965) *Bonn. zool. Beitr.* 16, 50-91. KIPP, F A (1978) *Vogelwelt* 99, 185-9. KIRNER, O (1964) *Gef. Welt* 88, 29-31, 50-3. KIRSCH, K-W (1992) *Beitr. Naturkde. Niedersachs.* 45, 89-122. KISHCHINSKI, A A (1980) *Ptitsy Koryakskogo nagor'ya*. Moscow. KISHCHINSKI, A A, TOMKOVICH, P S, and FLINT, V E (1983) In Flint, V E and Tomkovich, P S (eds) *Rasprostranenie i sistematika ptits*, 3-76. Moscow. KISLENKO, G S (1974) *Ornitologiya* 11, 381-2. KISS, J B and RÉKÁSI, J (1983) *An. Baustului Stüntele nat. Timişoara* 15, 133-40; (1986) *A Magyar Madártani Egyesület* 2 *Tudományos Ülése*, 75-82. KISS, J B, RÉKÁSI, J, and STERBETZ, I (1978) *Avocetta* 1 (2), 3-18. KISTYAKOVSKI, O B (1950) *Trudy Inst. Zool. Akad. Nauk USSR* 4, 3-77. KITSON, A R (1979) *Br. Birds* 72, 94-100; (1982) *Br. Birds* 75, 40. KITSON, A R and ROBERTSON, I S (1983) *Br. Birds* 76, 217-25. KIUCHI, K (1988) *Strix* 7, 296. KIZIROĞLU, I, ŞIŞLI, M N, and ALP, Ü (1987) *Vogelwelt* 108, 169-75. KLAAS, C (1967) *Natur Mus. Frankfurt* 97, 29-32. KLAFS, G and STÜBS, J (1977) *Die Vogelwelt Mecklenburgs*. Jena; (1987) *Die Vogelwelt Mecklenburgs* 3rd edn. Jena. KLEHM, K (1967) *Falke* 14, 328-33. KLEIN, R (1988) Diss. Saarbrücken Univ; (1989) *J. Orn.* 130, 361-5. KLEINER, A (1938) *Oiseau* 8, 149-50; (1939a) *Aquila* 42-5, 79-140; (1939b) *Aquila* 42-5, 141-226; (1939c) *Aquila* 42-5, 542-9; (1939d) *Bull. Br. Orn. Club* 59, 70-1; (1939e) *Orn. Beob.* 36, 117-18; (1939f) *Bull. Br. Orn. Club* 60, 11-14. KLEINSCHMIDT, A (1938) *Falco* 34, 49-52. KLEINSCHMIDT, O (1893) *Orn. Jahrb.* 4, 167-219; (1906) *J. Orn.* 54, 78-99; (1909-11) Berajah, Zoographia infinita. Corvus Nucifraga. Halle; (1919) *Falco* 14, 15-17; (1922) *Abh. Ber. Zool. Mus. Dresden* 15 (3), 1; (1940) *Falco* 36, 22-25. KLICKA, J and WINKER, K (1991) *Condor* 93, 755-7. KLIJN, H B (1975) *Bijdr. Dierkde.* 45, 39-49. KLIMANIS, A (1987) *Gef. Welt* 111, 184-6. KLITZ, W J (1973) *Orn. Monogr. AOU* 14, 34-8. KLUIJVER, H N (1933) *Versl. Meded. Plantenziektekd. Dienst* 69; (1935) *Versl. Meded. Plantenziektekd. Dienst* 81; (1945) *Limosa* 18, 1-11. KNAPTON, R W (1979) *Sask. Nat. Hist. Soc. Spec. Publ.* 10. KNAPTON, R W and FALLS, J B (1982) *Can. J. Zool.* 60, 452-9. KNECHT, S (1960) *Anz. orn. Ges. Bayern* 5, 525-56. KNECHT, S and SCHEER, U (1968) *Z. Tierpsychol.* 25, 155-69. KNEIS, P (1977) *Falke* 24, 132-3; (1987) *Beitr. Vogelkde.* 33, 56-8. KNEUTGEN, J (1969) *J. Orn.* 110, 158-60. KNIEF, W (1988) *Beih. Veröff. Nat. Land. Bad.-Württ.* 53, 31-54. KNIGHTS, J C and WALKER, D (1989) *Adjutant* 19, 8-12. KNIJFF, P DE (1977) *Vogeljaar* 25, 268; (1991) *Birding World* 4, 384-91. KNOLLE, F (1990) *Vogelkdl. Ber. Niedersachs.* 22, 70-2; (1991) *Vogelkdl. Ber. Niedersachs.* 23, 48-53. KNORRE, D VON, GRÜN, G, GÜNTHER, R, and SCHMIDT, K (1986) *Die Vogelwelt Thüringens*. Jena.

KNOWLES, R K (1972) *Bird-Banding* 73, 114-17. KNOX, A G (1976) *Bull. Br. Orn. Club* 96, 15-19; (1987a) *Br. Birds* 80, 482-7; (1987b) In Cameron, E (ed.) *Glen Tanar: its human and natural history*, 64-74; (1988a) *Br. Birds* 81, 206-11; (1988b) *Ardea* 76, 1-26; (1990a) *Br. Birds* 83, 89-94; (1990b) *Ibis* 132, 454-66; (1990c) *Scott. Birds* 16, 11-18; (1992) *Biol. J. Linn. Soc.* 47, 325-35. KNYSTAUTAS, A (1987) *The natural history of the USSR*. London. KOBUS, D (1967) *Orn. Mitt.* 19, 259. KOCH, W (1914) *Orn. Monatsschr.* 39, 241-57, 273-88. KOELZ, W (1937) *Ibis* (14) 1, 86-104; (1948) *Auk* 65, 444-5; (1949) *Auk* 66, 208-9. KOENIG, A (1888) *J. Orn.* 36, 121-298; (1890) *J. Orn.* 38, 257-488; (1895) *J. Orn.* 43, 113-238, 257-321, 361-457; (1896) *J. Orn.* 44, 101-216; (1905) *J. Orn.* 53, 259-60; (1920) *J. Orn.* 68 suppl., 83-148; (1924) *J. Orn.* 72 suppl; (1926) *J. Orn.* 74 suppl., 152. KOES, R F (1989) *Blue Jay* 47, 104-6. KÖHLER, F (1943) *Fauna och Flora* 38, 3-7. KÖHLER, K-H (1990) *Orn. Mitt.* 42, 129. KOIVUNEN, P, NYHOLM, E S, and SULKAVA, S (1975) *Ornis fenn.* 52, 85-96. KOKHANOV, V D (1982) In Zabrodin, V A (ed.) *Ekologiya i morfologiya ptits na kraynem severo-zapade SSSR*, 124-37. Moscow. KOKHANOV, V D and GAEV, Y G (1970) *Trudy Kandalaksh. gos. Zapoved.* 8, 236-74. KOLBE, H and KOLBE, E (1981) *Falke* 28, 228-31. KOLBE, U (1982) *Falke* 29, 197-201, 209. KOLLIBAY, P (1913) *J. Orn.* 61, 612-17. KOLLMANSPERGER, F (1959) *Bonn. zool. Beitr.* 10, 21-67. KOLTHOFF, K (1932) *Medd. Göteborgs Mus. Zool. Avd.* 59. KOMEDA, S, YAMAGISHI, S, and FUJIOKA, M (1987) *Condor* 89, 835-41. KOMERS, P E (1989) *Anim. Behav.* 37, 256-65. KOMERS, P E and BOAG, D A (1988) *Can. J. Zool.* 66, 1679-84. KOMERS, P E and DHINDSA, M S (1989) *Anim. Behav.* 37, 645-55. KÖNIG, D (1966) *Corax* 1 (17), 203-9. KÖNIGSTEDT, D and MÜLLER, H E J (1988) *Zool. Abh. Staatl. Mus. Tierkde. Dresden* 43, 143-8. KÖNIGSTEDT, D and ROBEL, D (1977) *Zool. Abh. Staatl. Mus. Tierkde. Dresden* 34, 301-18; (1983) *Mitt zool. Mus. Berlin* 59, Suppl. Ann. Orn. 7, 127-49. KÖNIGSTEDT, D, ROBEL, D, and GOTTSCHALK, W (1977) *Beitr. Vogelkde.* 23, 347-50. KONRADT, H-U (1968) *Falke* 15, 278-9. KONSTANTINOV, V M, BABENKO, V G, and BARYSHEVA, I K (1982) *Zool. Zh.* 61, 1837-45. KONSTANTINOV, V M, MARGOLIN, V A, and BARANOV, L S (1986) *Tez. Dokl. I. S'ezda Vesesoyuz. orn. Obshch.* 9. *Vsesoyuz. orn. Konf.* 1, 312-13. KONTOGIANNIS, J E (1967) *Auk* 84, 390-5. KOOIKER, G (1991) *Vogelwelt* 112, 225-36. KORBUT, V V (1981) *Zool. Zh.* 60, 115-25; (1982) *Nauch. dokl. vyssh. Shk. Biol. nauki* (4) 29-33; (1985) *Ornitologiya* 20, 186-9; (1989a) *Zool. Zh.* 68 (11), 125-34; (1989b) *Zool. Zh.* 68 (12), 88-95. KORELOV, M N, KUZ'MINA, M A, GAVRILOV, E I, KOVSHAR', A F, GAVRIN, V F, and BORODIKHIN, I F (1974) *Ptitsy Kazakhstana* 5. Alma-Ata. KORENBERG, E I, RUDENSKAYA, L V, and CHERNOV, Y I (1972) *Ornitologiya* 10, 151-60. KORHONEN, K (1981) *Ann. zool. fenn.* 18, 165-7. KORPIMÄKI, E (1978) *Ornis fenn.* 55, 93-104. KORODI GÁL, I (1958) *Orn. Mitt.* 10, 66-9; (1965) *Zool. Abh. Staatl. Mus. Tierkde. Dresden* 28, 113-25; (1968) *Falke* 15, 296-301; (1969) *Revista Musedor* 3 (6), 251-5; (1972) *Trav. Mus. Hist. nat. 'Grigore Antipa'* 12, 355-83 KORTE, J de (1972) *Beaufortia* 20, 23-58. KORZYUKOV, A I (1979) *Ornitologiya* 14, 216. KOSHKINA, T V and KISHCHINSKI, A A (1958) *Trudy Kandalaksh. gos. Zapoved.* 1, 79-88. KOSKIMIES, P (1985) *Lintumies* 20, 302-6; (1989) *Distribution and numbers of Finnish breeding birds*. Helsinki. KOSTIN, Y V (1983) *Ptitsy Kryma*. Moscow. KOUKI, J and HÄYRINEN, U (1991) *Ornis fenn.* 68, 170-7. KOVAČEVIĆ, J and DANON, M (1952) *Larus* 4-5, 185-217; (1959) *Larus* 11, 111-30. KOVÁCS, G (1981) *Aquila* 87, 49-70. KOVÁTS, L (1973) *Nymphaea* 1, 71-85. KOVSHAR', A F (1966) *Ptitsy Talasskogo Alatau.* Alma-Ata; (1979) *Pevchie ptitsy v subvysokogor'e Tyan'-Shanya.* Alma-Ata; (1981) *Osobennosti razmnozheniya ptits v subvysokogor'e.* Alma-Ata; (ed.) (1985) *Ptitsy Kurgal'dzhinskogo Zapovednika.* Alma-Ata. KOVSHAR', A F, IVASHCHENKO, A A, and KOVSHAR', V A (1986) *Vestnik Zool.* (5), 36-40. (1987) *Vestnik Zool.* (1), 59-64. KOWSCHAR, A F (1966) *Falke* 13, 48-53. KOZLOVA, E V (1930) *Ptitsy yugo-zapadnogo Zabaykal'ya severnoy Mongolii i tsentral'noy Gobi.* St Petersburg; (1932) *Trudy Mongol. Komiss. Akad. Nauk SSSR* 3; (1933) *Ibis* (13) 3, 59-87; (1939) *Byull. Mosk. Obshch. Ispyt. prir. Otd. Biol.* 48, 63-70; (1975) *Ptitsy zonal'nykh stepey i pustyn' Tsentral'noy Azii.* St Petersburg. KRAATZ, S (1979) *Falke* 26, 299-306. KRABBE, N (1980) *Checklist of the birds of Elat.* Unpubl. KRÄGENOW, P (1986) *Der Buchfink.* Wittenberg Lutherstadt. KRAMER, G (1941) *J. Orn.* 89 suppl., 105-31. KRAMPITZ, H-E (1950) *Vogelwelt* 71, 7-9. KRÄTZIG, H (1936) *Vogelzug* 7, 1-16. KRAUS, M and GAUCKLER, A (1970) *Vogelwelt* 91, 18-23. KRECHMAR, A V (1966) *Trudy Zool. Inst. Akad. Nauk SSSR* 39, 185-312. KRECHMAR, A V, ANDREEV, A V, and KONDRATIEV, A Y (1978) *Ekologiya i rasprostranenie ptits na severo-vostoke SSSR.* Moscow. KRECHMAR, A V, ANDREEV, A V, and KONDRATIEV, A Y (1991) *Ptitsy severnykh ravnin.* St Petersburg. KREIBIG, K (1957) *Falke* 4, 63. KREMENTZ, D G, NICHOLS, J D, and HINES, J E (1989) *Ecology* 70, 646-55. KREUTZER, M (1979) *Behaviour* 71, 291-321; (1983) *C. R. Acad. Sci. Paris* (3) 297, 71-4; (1985) *Thèse Doc. Etat. Univ. Pierre et Marie Curie, Paris;* (1987) *J. comp. Psychol.* 101, 382-6; (1990) *Terre Vie* 45, 147-64. KREUTZER, M and GÜTTINGER, H R (1991) *J. Orn.* 132, 165-77. KRICHER, J C and DAVIS, W E (1986) *J. Fld. Orn.* 57, 48-52. KRIŠTÍN, A (1988) *Folia Zool.* 37, 343-56. KROHN, H (1915) *Orn. Monatsber.* 23, 147-51. KROODSMA, D E (1974a) *Auk* 91, 54-64; (1974b) *Wilson Bull.* 86, 230-6; (1975) *Auk* 92, 66-80; (1981) *Auk* 98, 743-51. KROYMANN, B (1965) *Orn. Mitt.* 17, 231-2; (1967) *Vogelwelt* 88, 170-3. KROYMANN, B and GIROD, R (1980) *BUND Information* 9, 37-40. KROYMANN, B and KROYMANN, H (1992) *Kart. Med. Brutvögel* 7, 47-8. KRÜGER, C (1944) *Dansk. orn. Foren. Tidsskr.* 38, 105-14. KRÜGER, S (1976) *Falke* 23, 283-4; (1979) *Der Kernbeisser.* Wittenberg Lutherstadt. KRULL, F, DEMMELMEYER, H, and REMMERT, H (1985) *Naturwiss.* 72, 197-203. KRÜPER, T (1875) *J. Orn.* 23, 258-85. KRUSEMAN, G (1942) *Ardea* 31, 302. KÜBEL, M and ULLRICH, B (1975) *J. Orn.* 116, 323-4. KUENZEL, W J and HELMS, C W (1974) *Auk* 91, 44-53. KUENZI, A J, MOORE, F R, and SIMONS, T R (1991) *Condor* 93, 869-83. KUHK, R (1931 *J. Orn.* 79, 269-78. KULCZYCKI, A (1973) *Acta Zool. Cracov.* 18, 583-666. KULCZYCKI, A and MAZUR-GIERASIŃSKA, M (1968) *Acta Zool. Cracov.* 13, 231-50. KUMAR, V and TEWARY, P D (1983) *Indian J. Zool.* 13, 25-31. KUMARI, E (1958) *J. Orn.* 99, 32-4; (1972) *Soobshch. Pribalt. Kom. Izuch. Migr. Ptits* 7, 58-83. KUMARI, E V (1960) *Trudy prolemn. tematich. Soveshch. Zool. Inst. Akad. Nauk SSSR* 9, 119-28. KUMERLOEVE, H (1957) *Orn. Mitt.* 9, 133; (1961) *Bonn. zool. Beitr.* 12 spec. vol; (1962a) *Bonn. zool. Beitr.* 13, 327-32; (1962b) *Iraq nat. Hist. Mus. Publ.* 20, 1-36; (1963) *Alauda* 31, 110-36, 161-211; (1964a) *Istanbul Üniv. Fakült. Mecmuasi* (B) 27, 165-228; (1964b) *J. Orn.* 105, 307-25; (1965a) *J. Orn.* 106, 112; (1965b) *Alauda* 33, 257-64; (1967a) *Istanbul Üniv. Fakült. Mecmuasi* (B) 32, 79-213; (1967b) *Alauda* 35, 1-

19; (1969a) *Ibis* 111, 617-18; (1969b) *Istanbul Üniv. Fen. Fakült. Mecmuasi* (B) 34, 245-312; (1969c) *Orn. Mitt.* 21, 84-5; (1969d) *Alauda* 37, 43-58, 114-34, 188-205; (1970a) *Istanbul Üniv. Fen. Fakült. Mecmuasi* (B) 35, 85-160; (1970b) *Beitr. Vogelkde.* 16, 239-49; (1974) *Orn. Mitt.* 26, 235. KUMMER, J (1983) *Beitr. Vogelkde.* 29, 244-5. KUNKEL, P (1959) *Z. Tierpsychol.* 16, 302-50; (1961) *Z. Tierpsychol.* 18, 471-89; (1967a) *Behaviour* 29, 237-61; (1967b) *Bonn. zool. Beitr.* 18, 139-68; (1969) *Z. Tierpsychol.* 26, 277-83. KUNZ, H (1950) *Orn. Beob.* 47, 1-4. KUNZ-PLÜSS, H (1969) *Orn. Beob.* 66, 22-3. KUUSISTO, A P (1927) *Luonnon Ystävä* 31, 106-7. KUZNETSOV, A A (1962) *Ornitologiya* 5, 215-42; (1967) *Ornitologiya* 8, 262-6. KUZYAKIN, A P and VTOROV, P P (1963) *Ornitologiya* 6, 184-94. KYDYRALIEV, A (1972) *Ornitologiya* 10, 352-6.

LABITTE, A (1937) *Oiseau* 7, 85-104; (1953) *Oiseau* 23, 247-60; (1954) *Oiseau* 24, 197-210; (1955a) *Alauda* 23, 212-16; (1955b) *Oiseau* 25, 57-8. LACK, D (1942-3) *Ibis* (14) 6, 461-84, 85, 1-27. LACK, D (1940) *Condor* 42, 239-41; (1946) *Br. Birds* 39, 258-64; (1948) *Evolution* 2, 95-110; (1954a) *The natural regulation of animal numbers.* Oxford; (1954b) *Br. Birds* 47, 1-15; (1955) *Proc. int. orn. Congr.* 11, 176-8; (1957) *Br. Birds* 50, 10-19. LACK, D and SOUTHERN, H N (1949) *Ibis* 91, 607-26. LACK, P (ed.) (1986) *The atlas of wintering birds in Britain and Ireland.* Calton. LACK, P C (1988) *Sitta* 2, 3-20. ŁACKI, A (1959) *Biul. Inst. Ochr. Ros.* 5, 239-49; (1962) *Acta Orn.* 6, 195-207. LÆSSØE, and NØRREGAARD, K (1968) *Dansk orn. Foren. Tidsskr.* 62, 95-6. LAFERRÈRE, M (1968) *Alauda* 36, 260-73. LAKHANOV, D L (1977) *Trudy Samarkand gos. univ.* NS 324, 33-49. LAKHANOV, Z L (1967) *Ornitologiya* 8, 364-66. LAM, C Y and COSTIN, R (1991) *Hong Kong Bird Rep. 1990,* 123-4. LAMARCHE, B (1981) *Malimbus* 3, 73-102; (1988) *Études sahariennes et ouest-africaines* 1 (4). LAMBA, B S (1963a) *J. Bombay nat. Hist. Soc.* 60, 121-33; (1963b) *Res. Bull.* (NS) *Panjab Univ.* 14, 11-20; (1969) *J. Bombay nat. Hist. Soc.* 65, 777-8. LAMBERT, K (1965) *Falke* 12, 318. LAMBERTINI, M (1981) *Avocetta* 5, 65-86. LANCUM, F H (1928) *Br. Birds* 21, 264. LAND, H C (1970) *Birds of Guatemala.* Wynnewood. LAND, R and LEWIS, P (1986) *Trans. Norfolk Norwich Nat. Soc.* 27, 256-9. LANE, C (1957) *Ibis* 99, 116. LANE, M (1984) *Notornis* 31, 283-4. LANG, E M (1939) *Orn. Beob.* 36, 141-5; (1946a) *Orn. Beob.* 43, 33-43; (1946b) *Orn. Beob.* 43, 117-18; (1948) *Orn. Beob.* 45, 197-205. LANGE, G (1960) *J. Orn.* 101, 360. LANG, M, BANDORF, H, DORNBERGER, W, KLEIN, H, and MATTERN, U (1990) *Ökol. Vögel* 12, 97-126. LANGRAND, O (1990) *Guide to the birds of Madagascar.* New Haven. LANNER, R M and NIKKANEN, T (1990) *Ornis fenn.* 67, 24-7. LANSDOWN, P and CHARLTON, T D (1990) *Br. Birds* 83, 240-2. LANSDOWN, P, RIDDIFORD, N, and KNOX, A (1991) *Br. Birds* 84, 41-56. LANZ-WÄLCHLI, H (1953) *Orn. Beob.* 50, 12-20. LAPOUS, E (1988) *Alauda* 56, 437-8. LARIONOV, G P, DEGTYAREV, V G, and LARIONOV, A G (1991) *Ptitsy Leno-Amginskogo mezhdurech'ya.* Novosibirsk. LARIONOV, P D (1959) *Zool. Zh.* 38, 253-60. LARIONOV, V F (1927) *Trudy Lab. exper. Biol. Mosk. Zoo.* 3, 119-137. LARRISON, E J, and SONNENBERG, K G (1968) *Washington birds.* Seattle. LARSEN, T and TOMBRE, I (1988) *Vår Fuglefauna* 11, 68-70; (1989) *Fauna norv.* (C) *Cinclus* 12, 3-10. LASKEY, A R (1940) *Bird Lore* 42, 25-30. LASSEY, P A and WALLACE, D I M (1992) *Bird Watching* 79, 84-5. LA TOUCHE, J D D (1912) *Bull. Br. Orn. Club* 29, 124-60; (1920) *Ibis* (11) 2, 629-71; (1923) *Ibis* (11) 5, 300-332;

(1925-30) *A handbook of the birds of eastern China* 1. London. LATSCHA, H (1979) *Orn. Mitt.* **31**, 225. LATZEL, G (1968) *Vogelwelt* **89**, 231-2. LATZEL, G and WISNIEWSKI, H-J (1971) *Vogelkde. Ber. Niedersachs.* **3**, 79-81. LAUBMANN, A (1912a) *Orn. Jahrb.* **23**, 81-8; (1912b) *Verh. Orn. Ges. Bayern* **11**, 164-5. LAUDAGE, C and SCHROETER, W (1982) *Orn. Mitt.* **34**, 168. LAURENT, G and MOUILLARD, B (1939) *Alauda* **11**, 104-74. LAURENT, J-L (1986) *Oiseau* **56**, 263-86. LAVAUDEN, L (1930) *Alauda* **2**, 133-5. LAVIN CASTANEDO, J (1978) *Ardeola* **24**, 245-8. LAWTON, J (1959) *Br. Birds* **52**, 433-4. LAY, H G (1970) *Br. Birds* **63**, 38-9. LEA, D and BOURNE, W R P (1975) *Br. Birds* **68**, 261-83. LEADER, P J (1992) *Hong Kong Bird Rep. 1991*, 127-30. LEBEDEV, V G (1986) *Tez. Dokl. 1. s'ezda Vsesoyuz. orn. Obshch. 9. Vsesoyuz. orn. Konf.* **2**, 16-17. LEBERMAN, R C (1984) *J. Fld. Orn.* **55**, 486-7. LEBEURIER, E (1955) *Oiseau* **25**, 102-43. LEBEURIER, E and RAPINE, J (1937) *Oiseau* **7**, 583-93; (1939) *Oiseau* **9**, 219-32. LEBRETON, J D (1975) *Oiseau* **45**, 65-71; (1976) *Trav. scient. Parc. nat. Vanoise* **7**, 157-61. LEBRETON, P (ed.) (1977) *Atlas ornithologique Rhône-Alpes.* Lyon. LEBRETON, P, TOURNIER, H, and LEBRETON, J D (1976) *Trav. scient. Parc. nat. Vanoise* **7**, 163-243. LECK, C F (1973) *Auk* **90**, 888; (1975) *Bird-Banding* **46**, 201-3. LE DU, R (1935) *Alauda* **7**, 198-209. LEESSMITH, D T and MADGE, S C (1982) *Bull. Orn. Soc. Middle East* **9**, 5-6. LEEVER, J J (1982) *Roek en landbouw.* Zeist. LEFEVER, H and HUBLÉ, J (1955) *Gerfaut* **45**, 299. LEFRANC, N and PFEFFER, J-J (1975) *Alauda* **43**, 103-10. LE GRAND, G (1982) *Acoreana* **6**, 195-211; (1983) *Arquipélago* **4**, 85-116; (1986) *Dutch Birding* **8**, 55-7. LEHIKOINEN, E (1979) *Ornis fenn.* **56**, 24-9. LEHIKOINEN, E, and LAAKSONEN, M (1977) *Ornis fenn.* **54**, 133-4. LEHIKOINEN, E and NIEMELÄ, P (1977) *Lintumies* **12**, 33-44. LEHMANN, E von (1952) *Vögel Heimat* **22**, 95-6; (1962) *Orn. Mitt.* **14**, 70-1. LEHMANN, H and MERTENS, R (1969) *Ool. Rec.* **43**, 2-16. LEIBL, F and MELCHIOR, F (1985) *Anz. orn. Ges. Bayern* **24**, 125-33. LEICESTER, M (1959) *Emu* **59**, 295-6. LEIN, M R (1978) *Can. J. Zool.* **56**, 1266-83. LEINONEN, A (1978) *Ornis fenn.* **55**, 182-3. LEKAGUL, B and ROUND, P D (1991) *A guide to the birds of Thailand.* Bangkok. LEKAGUL, B, ROUND, P D, and KOMOLPHALIN, K (1985) *Br. Birds* **78**, 2-39. LENNERSTEDT, I (1964) *Fauna och Flora* **59**, 94-123. LENSINK, R, BIJTEL, H J V VAN DEN, and SCHOLS, R M (1989) *Limosa* **62**, 1-10. LENSINK, R and HUSTINGS, F (1991) *Limosa* **64**, 29-30. LEONOVICH, V V (1976) *Ornitologiya* **12**, 87-94; (1983) *Ornitologiya* **18**, 23-32. LEONTIEV, A N (1965) *Ornitologiya* **7**, 478-9. LEONTIEV, A N and PAVLOV, E I (1963) *Ornitologiya* **6**, 165-72. LEVER, C (1987) *Naturalized birds of the world.* Harlow. LEVIN, A S and GUBIN, B M (1985) *Biologiya ptits intrazonal'nogo lesa.* Alma-Ata. LEWIS, A and POMEROY, D (1989) *A bird atlas of Kenya.* Rotterdam. LEWIS, A D (1989) *Scopus* **13**, 129-31. LEWIS, L R (1985) *Newbury Distr. orn. Club ann. Rep. 1984*, 34-9. LEWIS, S (1920) *Br. Birds* **14**, 26-33. LIEDER, K (1987) *Beitr. Vogelkde.* **33**, 46-8. LIEDEKERKE, R DE (1970) *Aves* **7**, 122; (1979) *Nos Oiseaux* **35**, 288. LIEFF, M R and JORDAN, N P (1950) *Br. Birds* **43**, 56. LIEN, L, ÖSTBYE, E, HAGEN, A, KLEMETSEN, A, and SKAR, H J (1970) *Nytt Mag. Zool.* **18**, 245-51. LIGON, J S (1961) *New Mexico birds and where to find them.* Albuquerque. LILFORD, LORD (1866) *Ibis* (2) **2**, 377-92. LILJA, C (1982) *Growth* **46**, 367-87. LIMBRUNNER, A (1987) *Verh. Orn. Ges. Bayern* **24**, 541-2. LINCOLN, G A, RACEY, R A, SHARP, P J, and KLANDORF, H (1980) *J. Zool. Lond.* **190**, 137-53. LIND, E (1952) *Vår Fågelvärld* **11**, 145-53. LINDGREN, F (1975)

Fauna och Flora **70** (5), 198-210. LINDNER, C (1906) *Orn. Montsschr.* **31**, 46-65, 105-21; (1907) *Orn. Monatsschr.* **32**, 398-410; (1917) *J. Orn.* **65** (2), 161-5. LINDNER, F (1911) *Orn. Monatsschr.* **36**, 62-72. LINDSTRÖM, Å (1987) *Ornis fenn.* **64**, 50-6; (1989) *Auk* **106**, 225-32; (1990) D Phil Thesis. Lund Univ. LINDSTRÖM, Å and ALERSTAM, T (1986) *Behav. Ecol. Sociobiol.* **19**, 417-24. LINDSTRÖM, Å and NILSSON, J-Å (1988) *Ornis scand.* **19**, 165-6. LINDSTRÖM, Å, OTTOSSON, U, and PETTERSSON, J (1984) *Vår Fågelvärld* **43**, 525-30. LINK, R and RITTER, M (1973a) *Orn. Beob.* **70**, 185-6; (1973b) *Orn. Beob.* **70**, 267-72. LINKE, H (1975) *Orn. Mitt.* **27**, 170-1. LINKOLA, P (1960) *Vår Fågelvärld* **19**, 66-7. LINN, H (1984) *Br. Birds* **77**, 489-90. LINSDALE, J M (1928) *Univ. Calif. Publ. Zool.* **30**, 251-392. LINSDALE, J M and SUMNER, E L (1934) *Condor* **36**, 107-12; (1937) *Condor* **39**, 162-63. LINT, A (1964) *Loodus. Selts. aast.* **56**, 167-88; (1971) *Orn. Kogumik* **5**, 132-63. LIPKOVICH, A D (1985) In Sokolov, V E and Sablina, T B (eds) *Izuchenie i okhrana redkikh i ischezayushchikh vidov zhivotnykh fauny SSSR*, 102-5. Moscow; (1986) In Amirkhanov, A M (ed.) *Ekosistemy ekstremal'nykh usloviy sredy v zapovednikakh RSFSR*, 128-34. Moscow. LIPPENS, L (1968) *Gerfaut* **58**, 3-23. LIPPENS, L and WILLE, H (1972) *Atlas des oiseaux de Belgique et d'Europe occidentale.* Tielt; (1976) *Les oiseaux du Zaïre.* Tielt. LI SHI-CHUN, LIU XI-YUO, TAN YAO-KUANG, and SIEN YAO-HUA (1975) *Acta zool. Sin.* **21**, 71-7. LISITSYNA, T Y and NIKOL'SKI, I D (1979) *Ornitologiya* **14**, 216-19. LITTLE, R (1990) *Br. Birds* **83**, 504-6. LITTLEJOHN, A C (1952) *Fair Isle Bird Obs. Bull.* **1** (8), 21-3, 34. LITUN, V I (1986) *Ornitologiya* **21**, 165. LITUN, V I and PLESSKI, P V (1983) *Ornitologiya* **18**, 64-9. LLOYD-EVANS, L (1948) *Br. Birds* **41**, 213. LLOYD-EVANS, L and NAU, B S (1965) *Rye Meads ann. Rep.* **3**, 23-39. LOBB, M G (1981) *J. RAF orn. Soc.* **12**, 25-7. LOBKOV, E G (1986) *Gnezdyashchiesya ptitsy Kamchatki.* Vladivostok. LOCK, J M (1971) In Gooders, J (ed.) *Birds of the world*, 2676-8. London. LOCKIE, J D (1955) *Ibis* **97**, 341-69; (1956a) *Bird Study* **3**, 180-90; (1956b) *J. Anim. Ecol.* **25**, 421-8; (1959) *Br. Birds* **52**, 332-4. LOCKLEY, A K (1992) *J. Orn.* **133**, 77-82. LOCKLEY, R M (1953) *Br. Birds* **46**, 347-8. LØFALDLI, L (1983) *Vår Fuglefauna* **6**, 183-9. LOFTIN, H (1977) *Bird-Banding* **48**, 253-58. LOFTS, B and MARSHALL, A J (1960) *Ibis* **102**, 209-14. LOFTS, B, MURTON, R K, WESTWOOD, N J, and THEARLE, R J P (1973) *Gen. comp. Endocrin.* **21**, 202-9. LOGMINAS, V (1991) *Lietuvos fauna: Paukščiai* **2**. Vilnius. LÖHRL, H (1950) *Z. Tierpsychol.* **7**, 130-3; (1957) *J. Orn.* **98**, 122-3; (1963) *J. Orn.* **104**, 62-8; (1964) *J. Orn.* **105**, 153-81; (1967) *Vogelwelt* **88**, 148-52; (1970) *Anz. orn. Ges. Bayern* **9**, 185-96; (1978a) *Gef. Welt* **102**, 162-5; (1978b) *Vogelwelt* **99**, 121-31; (1982) *Ökol. Vögel* **4**, 81; (1980) *J. Orn.* **121**, 408; (1987a) *Vogelwelt* **108**, 151-2; (1987b) *Vogelwelt* **108**, 189-90. LÖHRL, B and BÖHRINGER, R (1957) *J. Orn.* **98**, 229-40. LOISON, M (1984) *Aves* **21**, 109. LOMAN, J (1975) *Ornis scand.* **6**, 169-78; (1977) *Oikos* **29**, 294-301; (1980a) *Ekol. pol.* **28**, 95-109; (1980b) *Holarctic Ecol.* **3**, 26-35; (1980c) *Ibis* **122**, 494-500; (1980d) Doctoral Thesis. Lund Univ; (1984) *Ornis scand.* **15**, 183-7; (1985) *Ardea* **73**, 61-75. LOMAN, J and TAMM, S (1980) *Amer. Nat.* **115**, 285-9. LOMBARDO, M P, POWER, H W, STOUFFER, P C, ROMAGNANO, L C, and HOFFENBERG, A S (1989) *Behav. Ecol. Sociobiol.* **24**, 217-23. LONDEI, T and GNISCI, R (1988) *Riv. ital. Orn.* **58**, 59-73. LONDEI, T and MAFFIOLI, B (1989) *Riv. ital. Orn.* **59**, 241-58. LONDON NATURAL HISTORY SOCIETY (1957) *The birds of the London* *area since 1900*. London. LONEUX, M (1988) *Cahiers Ethol. appl.* **8**, 337-406. LONG, J L (1981) *Introduced birds of the world*. New York.668efgi. LONGSTAFF, T G (1932) *J. Anim. Ecol.* **1**, 119-42. LÖNNBERG, E (1905) *Arkiv Zool.* **2** (9), 1-23; (1909) *Arkiv Zool.* **5** (9), 1-42; (1918) *Fauna och Flora* **13**, 87-8. LOOFT, V (1965) *Corax* **1** (17), 1-9; (1967) *Corax* **2** (18), 27-31; (1971a) *Corax* **3**, 188-96; (1971b) *Corax* **3**, 196-9. LOPE, F DE, GUERRERO, J, and CRUZ, C DE LA (1984) *Alauda* **52**, 312. LOPE, F DE, GUERRERO, J, CRUZ, C DE LA and SILVA, E DA (1985) *Alauda* **53**, 167-80. LOPE, F DE, GUERRERO, J, GARCÍA, M E, CRUZ, C DE LA CARRETERO, J J, NAVARRO, J A, SILVA, E da, and OTANO, J (1983) *Alytes* **1**, 393-9. LØPPENTHIN, B (1935) *Dansk orn. Foren. Tidsskr.* **29**, 15-22; (1943) *Dansk orn. Foren. Tidsskr.* **37**, 193-214. LORENZ, K S (1931) *J. Orn.* **79**, 67-127; (1932) *J. Orn.* **80**, 50-98; (1940) *Z. Tierpsychol.* **3**, 278-92; (1952) *King Solomon's ring*. London; (1955) *Ik sprak met viervoeters, vogels en vissen*. Amsterdam; (1963) *On aggression*. London. LORENZ, T (1890) *J. Orn.* **38**, 98-100. LOSKOT, V M (1986a) *Trudy Zool. Inst. Akad. Nauk SSSR* **150**, 44-56; (1986b) *Trudy Zool. Inst. Akad. Nauk SSSR* **150**, 147-70; (1991) *Trudy Zool. Inst. Akad. Nauk SSSR* **231**, 43-116. LO VALVO, F and LO VERDE, G (1987) *Riv. ital. Orn.* **57**, 97-110. LOVARI, S (1976a) In Pedrotti, F (ed.) *SOS Fauna*, 189-214. Camerino; (1976b) *Gerfaut* **66**, 207-19; (1978) *Gerfaut* **68**, 163-76; (1979) *Biol. Behav.* **4**, 311-26. LOVATY, F (1985) *Oiseau* **55**, 351-7; (1991) *Nos Oiseaux* **41**, 99-106. LOVE, J A and SUMMERS, R W (1973) *Scott. Birds* **7**, 399-403. LØVENSKIOLD, H L (1947) *Håndbok over Norges fugler*. Oslo; (1963) *Avifauna svalbardensis*. Oslo. LOWERY, G H (1974) *Louisiana birds*. Baton Rouge. LOWERY, G H and MONROE, B L (1968) In *Peters' check-list of birds of the world* **14**. LOWTHER, J K (1961) *Can. J. Zool.* **39**, 281-92. LOWTHER, P E (1979a) *Bird-Banding* **50**, 160-2; (1979b) *Inland Bird Banding* **51**, 23-9; (1988) *J. Fld. Orn.* **59**, 51-4. LOXTON, R G (1968) *Nat. Wales* **11**, 126-30. LUCAS, J V and TOOK, J M E (1974) *Cyprus Orn. Soc.* (*1957*) *Rep.* **19**, 1-72. LUDEWIG, G (1989) *Gef. Welt* **113**, 361-2. LUDWIG, H (1984) *Gef. Welt* **108**, 226-8. LUDLOW, F (1928) *Ibis* (12) **4**, 51-73. LUDLOW, F and KINNEAR, N B (1933) *Ibis* (13) **3**, 658-94. LUKAČ, G (1988) *Orn. Mitt.* **40**, 287-91. LULAV, S (1967) *IUCN Bull.* NS **2**, 11. LUNDBERG, A and EDHOLM, M (1982) *Br. Birds* **75**, 583-5. LUNDBERG, A, MATTSSON, R, NILSSON, B, and WIDÉN, P (1980) *Vår Fågelvärld* **39**, 225-30. LUNDBERG, P (1981) *J. Orn.* **122**, 65-72. LUNDBERG, P and ERIKSSON, L-O (1984) *Ornis scand.* **15**, 105-9. LUNDEVALL, C-F (1950) *Dansk orn. Foren. Tidsskr.* **44**, 30-40; (1952) *Kung. Svenska Vetensk. Akad., Avh. Naturskydd.* **7**, 1-73. LUNDIN, A (1962) *Vår Fågelvärld* **21**, 81-95. LUNIAK, M (1977a) *Acta Orn.* **16**, 213-40; (1977b) *Acta Orn.* **16**, 241-74. LÜPS, P, HAURI, R, HERREN, H, MÄRKI, H, and RYSER, R (1978) *Orn. Beob.* **75** suppl. LÜTGENS, H (1955) *Orn. Mitt.* **7**, 113. LÜTTSCHWAGER, H (1926) *Orn. Monatsber.* **34**, 41-3. LYAISTER, A F and SOSNIN, G V (1942) *Materialy po ornitofaune Armyanskoy SSR (Ornis Armeniaca)*. Yerevan. LYE, R J (1948) *Br. Birds* **41**, 211. LYKHVAR', V P (1983) In Kuchin, A P (ed.) *Ptitsy Sibiri*, 147-8. Gorno-Altaysk. LYNCH, J F, MORTON, E S, and VAN DER VOORT, M E (1985) *Auk* **102**, 714-21. LYNES, H (1924) *Ibis* (11) **6**, 648-719; (1926) *Ibis* (12) **2**, 346-405; (1930) *Diary of Algerian expedition*. Unpubl. MS, Edward Grey Inst., Oxford Univ. LYON, B E (1984) M Sc Thesis. Queen's Univ. Kingston. LYON, B E and MONTGOMERIE, R D (1985) *Behav. Ecol. Sociobiol.* **17**, 279-84; (1987) *Ecology* **68**, 713-22. LYON, B E,

MONTGOMERIE, R D, and HAMILTON, L D (1987) *Behav. Ecol. Sociobiol.* **20**, 377-82.

McATEE, W L (1950) *Auk* **67**, 247. MacBEAN, A F (1949) *Scott. Nat.* **61**, 176-7. McCABE, T T and McCABE, E B (1933) *Condor* **35**, 136-47. McCABE, T T and MILLER, A H (1933) *Condor* **35**, 192-7. McCLURE, H E (1974) *Migration and survival of the birds of Asia*. Bangkok. McCRACKEN, D I (1989) In Bignal, E and Curtis, D J (eds) *Choughs and land-use in Europe*, 52-6. Scottish Chough Study Group. MacDONALD, D (1965) *Scott. Birds* **3**, 235-46; (1968) *Scott. Birds* **5**, 177-8. MACDONALD, J D (1973) *Birds of Australia*. Sydney. MacDONALD, J W (1962) *Bird Study* **9**, 147-67; (1963) *Bird Study* **10**, 91-101. MacDONALD, M (1960) *Birds in my Indian garden*. London. MACDONALD, R A and WHELAN, J (1986) *Ibis* **128**, 540-57. McGILLIVRAY, W B (1980) *J. Fld. Orn.* **51**, 371-2; (1984) *Can. J. Zool.* **62**, 381-5. McGILLIVRAY, W B and JOHNSTON, R F (1987) *Auk* **104**, 681-7. McGREGOR, P K (1980) *Z. Tierpsychol.* **54**, 285-97; (1981) D Phil Thesis. Oxford Univ; (1983) *Z. Tierpsychol.* **62**, 256-60; (1986) *J. Orn.* **127**, 37-42. McGREGOR, P K and THOMPSON, D B A (1988) *Ornis scand.* **19**, 153-9. McGREGOR, P K, WALFORD, V R, and HARPER, D G C (1988) *Bioacoustics* **1**, 107-29. McINTYRE, N (1953) *Br. Birds* **46**, 377-8. MACKE, T (1965) *J. Orn.* **106**, 461-2; (1980) *Charadrius* **16**, 5-13. McKEE, J (1985) *Br. Birds* **78**, 150-1. McKENDRY, W G (1973) *Br. Birds* **66**, 400. McKILLIGAN, N G (1980) *Bird Study* **27**, 93-100. MacKINNON, J (1988) *Field guide to the birds of Java and Bali*. Yogyakarta. MACKINTOSH, D R (1941) *Bull. zool. Soc. Egypt* **3**, 7-29. MACKOWICZ, R, PINOWSKI, J, and WIELOCH, M (1970) *Ekol. Polska* **18**, 465-501. MACKWORTH-PRAED, C W and GRANT, C H B (1960) *Birds of eastern and north eastern Africa* **2**. London; (1963) *Birds of the southern third of Africa* **2**. London; (1973) *Birds of west central and western Africa* **2**. London. McLACHLAN, G R (1963) *Ostrich* **34**, 102-9. McLAREN, I A (1981) *Auk* **98**, 243-57. McLAUGHLIN, R L and MONTGOMERIE, R D (1985) *Auk* **102**, 687-95; (1989a) *Auk* **106**, 738-41; (1989b) *Behav. Ecol. Sociobiol.* **25**, 207-15. MACLEAN, G L (1985) *Roberts' birds of Southern Africa*. Cape Town. MACLEAY, K N G (1960) *Univ. Khartoum nat. Hist. Mus. Bull.* **1**. McLENNAN, J A and MacMILLAN, B W H (1983) *New Zealand J. agric. Res.* **26**, 139-45. MacLEOD, I C (1987) *Scott. Bird News* **7**, 11. McNAIR, D B (1988) *Migrant* **59**, 45-8. McNEIL, R (1982) *J. Fld. Orn.* **53**, 125-32. McVEAN, A and HADDLESEY, P (1980) *Ibis* **122**, 533-6. MADGE, S C, HEARL, G C, HUTCHINGS, S C, AND WILLIAMS, L P (1990) *Br. Birds* **83**, 187-95. MADON, P (1928a) *Les corvidés d'Europe*. Paris; (1928b) *Mém. Soc. Orn. Mamm. France.* **1**. MADSEN, J (1981) In Fox, A D and Stroud, D A (eds) *Rep. Greenland White-fronted Goose Study 1979 Exped. Eqalungmiut Nunât, west Greenland*, 183-7; (1982) *Dansk orn. Foren. Tidsskr.* **76**, 137-45. MADSEN, J J (1990) *Dutch Birding* **12**, 77. MAES, P (1989) *Oriolus* **55**, 66-72. MAESTRI, F and VOLTOLINI, L (1990) *Riv. ital. Orn.* **60**, 99-100. MAESTRI, F, VOLTOLINI, L, and LO VALVO, F (1989) *Riv. ital. Orn.* **59**, 159-71. MAGEE, M J (1928) *Bull. NE Bird-Banding Assoc.* **4**, 149-52; (1930) *Bird-Banding* **1**, 43-5. MAGNENAT, D (1969) *Nos Oiseaux* **30**, 69-70. MAGNÚSSON, K G (1986) *Bliki* **5**, 1-2. MAHABAL, A and VAIDYA, V G (1989) *Proc. Indian Acad. Sci. Anim. Sci.* **98**, 199-209. MAHÉ, E (1985) Thèse. Univ. Sci. Tech. Languedoc. MAHÉO, R (1969) *Ar Vran* **2**, 176-87. MAHER, W J (1964) *Ecology* **45**, 520-8. MAKATSCH, W (1950) *Die Vogelwelt*

Macedoniens. Leipzig; (1955) *Aquila* **59-62**, 347-50; (1963) *Zool. Abh. Staatl. Mus. Tierkde. Dresden* **26**, 135-86; (1971) *Zool. Abh. Staatl. Mus. Tierkde. Dresden* **32**, 17-41; (1976) *Die Eier der Vögel Europas* **2**. Radebeul; (1979) *Vögel Heimat* **49**, 163-6; (1981) *Verzeichnis der Vögel der Deutschen Demokratischen Republik.* Leipzig. MALBRANT, R (1952) *Faune du Centre Africain français.* Paris. MAL'CHEVSKI, A S (1959) *Gnezdovaya zhizn'pevchikh ptits.* St Petersburg. MAL'CHEVSKI, A S and PUKINSKI, Y B (1983) *Ptitsy Leningradskoy oblasti i sopredel'nykh territoriy* **2**. St Petersburg. MALCOLM-COE, Y (1981) *Bull. E. Afr. nat. Hist. Soc.* Jan-Feb, 8-11. MALLALIEU, M (1988) *Birds in Islamabad, Pakistan, 1985-7.* MALLOCH, J R (1922) *Auk* **39**, 569-70. MALMBERG, T (1949) *Vår Fågelvärld* **8**, 121-31; (1971) *Ornis scand.* **2**, 89-117. MAMBETZHUMAEV, A M and ABDREIMOV, T (1972) In Reymov, R (ed) *Ekologiya vazhneyshikh mlekopitayushchikh i ptits Karakalpakii*, 200-12. Tashkent. MANN, P, HERLYN, H, and UNTHEIM, H (1990) *Vogelwelt* **111**, 142-55. MANN, S and HOCHBERG, O (1982) *Israel Land Nat.* **8**, 10-14. MANNICHE, A L V (1910) *Medd. Grønland* **45**. MANSFELD, K (1950) *Nachricht. dtsch. Pflanzenschutzdienst* (NF) **4**, 131-6, 147-54, 164-75. MANSON-BAHR, P (1953) *Br. Birds* **46**, 414-15; (1954) *Br. Birds* **47**, 313. MANWELL, C and BAKER, C M A (1975) *Austral. J. Biol. Sci.* **28**, 545-57. MARCHANT, J (1984) *BTO News* **134**, 7-10. MARCHANT, J and WHITTINGTON, P (1988) *BTO News* **157**, 7-10. MARCHANT, J H, HUDSON, R, CARTER, S P, and WHITTINGTON, P (1990) *Population trends in British breeding birds.* Tring. MARCHANT, S (1962) *Bull. Iraq nat. Hist. Mus.* **2** (1), 1-40; (1963a) *Ibis* **105**, 369-98; (1963b) *Ibis* **105**, 516-57. MARCHANT, S and MACNAB, J W (1962) *Bull. Iraq nat. Hist. Inst.* **2** (3), 1-48. MARDER, J (1973) *Comp. Biochem. Physiol.* **45A**, 421-30. MARÉCHAL, P (1986) *Vogeljaar* **34**, 73-81; (1988) *Vogeljaar* **36**, 88. MARFURT, B (1971) *Orn. Beob.* **68**, 245-9. MARION, L and MARION, P (1975) *Bull. Soc. Sci. nat. Ouest France Suppl.* MARISOVA, I V, GORBAN', I M, and DAVIDOVICH, L I (1990) *Mat. dopov. 5. naradi orn. amat. orn. rukhu Zakhid. Ukraïni*, 29-32. MARJAKANGAS, A (1981) *Ornis. fenn.* **58**, 90-1; (1983) *Ornis fenn.* **60**, 89. MARKGREN, G (1955) *Vår Fågelvärld* **14**, 168-77. MARKGREN, G and LUNDBERG, S (1959) *Vår Fågelvärld* **18**, 185-205. MÄRKI, H (1976) *Orn. Beob.* **73**, 67-88. MÄRKI, H and BIBER, O (1975) *Jahrb. Naturhist. Mus. Bern* **5**, 153-64. MARKKOLA, J and VIERIKKO, E (1983) *Aureola* **8**, 147-50. MARLAN, J (1991) *Birding* **23**, 220-3. MARLE, J G VAN (1949) *Bull. Br. Orn. Club* **69**, 118-19. MARLE, J G VAN and HENS, P A (1938) *Limosa* **11**, 86-92. MARLE, J G VAN and VOOUS, K H (1988) *The birds of Sumatra.* Tring. MARLER, P (1956a) *Behav. Suppl.* **5**; (1956b) *Ibis* **98**, 496-501; (1957) *Behaviour* **11**, 13-39. MARLER, P and BOATMAN, D J (1951) *Ibis* **93**, 90-9. MARLER, P and HAMILTON, W J (1966) *Mechanisms of animal behaviour.* New York. MARLER, P and MUNDINGER, P C (1975) *Ibis* **117**, 1-17. MARPLES, B J and GURR, L (1943) *Emu* **43**, 67-71. MARQUARDT, K (1975) *Vögel Heimat* **46**, 56-9. MARQUISS, M (1980) *Ring. Migr.* **3**, 35-6. MARQUISS, M and BOOTH, C J (1986) *Bird Study* **33**, 190-5. MARQUISS, M, NEWTON, I, and RATCLIFFE, D A (1978) *J. appl. Ecol.* **15**, 129-44. MARR, V and KNIGHT, R L (1982) *Murrelet* **63**, 25. MARSHALL, A J and COOMBS, C J F (1957) *Proc. zool. Soc. Lond.* **128**, 545-88. MARSHALL, D and RAE, R (1981) *Rep. Grampian Ring. Group* **3**, 23-35. MARTEN, J A and JOHNSTON, N K (1986) *Condor* **88**, 409-20. MARTENS, J (1972) *Bonn. zool. Beitr.* **23**, 95-121;

(1979) *Nat. Mus.* **109**, 337-43. MARTI, C D (1974) *Condor* **76**, 229. MARTÍN, A (1985) *Alauda* **53**, 309; (1987) *Atlas de las aves nidificantes en la isla de Tenerife.* Inst. Estud. Canarios Monogr. **32**. MARTÍN, A, BACALLADO, J J, EMMERSON, K W, and BAEZ, M (1984) *Il Reunión Iberoamer. Cons. Zool. Vert.*, 130-9. MARTIN, A J (1990) *Stour Ring. Group Rep. 1990*, 51-64. MARTIN, C E (1939) *Br. Birds* **33**, 108. MARTIN, G R (1986) *J. comp. Physiol.* **159**, 545-57. MARTIN, J-L (1980) *Problèmes de biogéographie insulaire: les cas des oiseaux nicheurs de Corse.* Thèse doct. Univ. Montpellier MARTINS, R P (1989) *Sandgrouse* **11**, 1-41. MARZOCCHI, J F (1990) *Contribution à l'étude de l'avifaune du Cap Corse.* Bastia. MASATOMI, H and KOBAYASHI, S (1982) *J. Yamashina Inst. Orn.* **14**, 306-24. MASCHER, J (1952) *Vår Fågelvärld* **11**, 34-6, 37. MASON, C F (1989) *Bird Study* **36**, 145-6. MASON, C W and MAXWELL-LEFROY, H (1912) *Mem. Dept. Agric. India Ent. Ser.* **3**. MASSA, B (1976) *Ric. Biol. Selvag.* 7 suppl., 427-74; (1984) *Riv. ital. Orn.* **54**, 102-3; (1987) *Bull. Br. Orn. Club* **107**, 118-29; (1989) *Bull. Br. Orn. Club* **109**, 196-8. MASSI, A, MESCHINI, E, and ROSELLI, A (1991) *Riv. ital. Orn.* **61**, 63-5. MASSOTH, K (1981) *Voliere* **4**, 115-21; (1989) *Voliere* **12**, 135-9. MAŠTROVIĆ, A (1942) *Die Vögel des Küstenlandes Kroatiens* **1**. Zagreb. MAT, H A and DAVISON, G W H (1984) *Malaysiana* **13**, 231-8. MATHER, J (1986) *The birds of Yorkshire.* London. MATHEW, K L and NAIK, R M (1986) *Ibis* **128**, 260-5. MATHEWS, G F (1864) *Naturalist* **1**, 49-51, 69-71, 88-90. MATHIASSON, S (1972) *Bull. Br. Orn. Club* **92**, 103-6. MATOUŠEK, B (1968) *Act. Rer. Natur. Mus. Nat. Slov. Bratislava* **14** (2), 101-18; (1969) *Act. Rer. Natur. Mus. Nat. Slov. Bratislava* **15** (1), 59-76; (1971) *Act. Rer. Natur. Mus. Nat. Slov. Bratislava* **17** (1), 155-66. MATOUŠEK, B and JABLONSKI, B (1969) *Annot. zool. bot. Bratislava* **63**, 1-9. MATT, D (1983) *Orn. Mitt.* **35**, 216. MATTES, H (1976) *Orn. Beob.* **73**, 247-8; (1978) *Münster. Geogr. Arb.* **2**. MATTES, H and BÜRKLI, W (1979) *Orn. Beob.* **76**, 317-20. MATTES, H and JENNI, L (1984) *Orn. Beob.* **81**, 303-15. MATTHEWS, M, RECKITT, R, MUŽINIĆ, J, and MIKUSKA, J (1988) *Upoznajmo Ptice.* ICBP Migratory Birds Programme, Zagreb. MATVEJEV, S D (1948) *Godišnjaka Biol. Inst. Sarajevu* **1948**, 75-8; (1955) *Acta Mus. Maced. Sci. Nat.* **4**, 1-22; (1976) *Conspectus Avifaunae Balcanicae* **1**. Belgrade. MATVEJEV, S D and VASIĆ, V F (1973) *Catalogus Faunae Jugoslaviae* **4** (3). Aves. Ljubljana. MATZ, W (1967) *Falke* **14**, 130-3. MAU, K-G (1980) *Gef. Welt* **104**, 171-5, 187-9, 213-15, 234-8; (1982) *Gef. Welt* **106**, 215-19, 246-9. MAUERSBERGER, G (1960) In Stresemann, E and Portenko, L A (eds) *Atlas der Verbreitung Palaearktischer Vögel* **1**. Berlin; (1971) *J. Orn.* **112**, 232-33; (1972) *J. Orn.* **113**, 53-9; (1973) *Beitr. Vogelkde.* **19**, 76-7; (1976) *Falke* **23**, 51-5; (1980) *Mitt. zool. Mus. Berlin* **56**, Suppl. Ann. Orn. **4**, 77- 164; (1982) *Mitt. zool. Mus. Berlin* **58**, Suppl. Ann. Orn. **6**, 101-13; (1983) *Mitt. zool. Mus. Berlin* **59**, Suppl. Ann. Orn. **7**, 47-83. MAUERSBERGER, G and MÖCKEL, R (1987) *Mitt. zool. Mus. Berlin* **63**, Suppl. Ann. Orn. **11**, 97-111. MAUERSBERGER, G and PORTENKO, L A (1971) In Stresemann, E, Portenko, L A, and Mauersberger, G (eds) *Atlas der Verbreitung Palaearktischer Vögel* **3**. Berlin. MAUERSBERGER, G, WAGNER, S, WALLSCHLÄGER, D, and WARTHOLD, R (1982) *Mitt. zool. Mus. Berlin* **58**, 11-74. MAUNDER, J E and MONTEVECCHI, W A (1982) *A field checklist of the birds of insular Newfoundland.* St John's. MAURIZIO, R (1978) *Vögel Heimat* **48**, 74-9, 104-7; (1987) *Orn. Beob.* **84**, 133-4. MAUTSCH, H and RANK, H (1973) *Falke* **20**, 268-72.

MAY, R C (1951) *Br. Birds* **44**, 17. MAYAUD, N (1933*a*) *Alauda* **5**, 192-4; (1933*b*) *Alauda* **5**, 195-220, 345-82; (1933*c*) *Alauda* **5**, 453-99; (1936) *Inventaire des oiseaux de France.* Paris; (1939) *Oiseau* **9**, 486-506; (1941) *Arch. suisses Orn.* (1) 12, 539-43; (1948) *Alauda* **16**, 168-79; (1960) *Alauda* **28**, 287-302 MAYES, W E (1926) *Br. Birds* **20**, 273-4. MAYHOFF, H (1911) *Orn. Monatsschr.* **36**, 72-86; (1915) *Verh. orn. Ges. Bayern* **12**, 109-18. MAYO, A L W (1950) *Br. Birds* **43**, 368. MAYR, E (1926) *J. Orn.* **74**, 571-671; (1927) *J. Orn.* **75**, 596-619; (1949) *Ibis* **91**, 304-6; (1963) *Animal species and evolution.* Cambridge, Mass. MAYR, E and SHORT, L L (1970) *Publ. Nuttall Orn. Club* **9**. MAZZUCCO, K (1974) *Egretta* **17**, 53-9. MEAD, C (1983) *BTO News* **129**, 1; (1986) *BTO News* **145**, 1. MEAD, C J (1975) *Ardeola* **21**, 699-732. MEAD, C J and CLARK, J A (1987) *Ring. Migr.* **8**, 135-200; (1988) *Ring. Migr.* **9**, 169-204; (1989) *Ring. Migr.* **10**, 159-96; (1990) *Ring. Migr.* **11**, 137-76; (1991) *Ring. Migr.* **12**, 139-76; (1993) *Ring. Migr.* in press. MEAD, C J and HUDSON, R (1983) *Ring. Migr.* **4**, 281-319; (1984) *Ring. Migr.* **5**, 153-92; (1985) *Ring. Migr.* **6**, 125-72. MEADE-WALDO, E G (1889*a*) *Ibis* (6) 1, 1-13; (1889*b*) *Ibis* (6) 1, 503-20; (1893) *Ibis* (6) **5**, 185-207. MEARNS, R and MEARNS, B (1989) *Scott. Birds* **15**, 179. MEDWAY, LORD and WELLS, D R (1976) *The birds of the Malay peninsula* **5**. London. MEEK, E R (1984) *Br. Birds* **77**, 160-4. MEESE, R J and FULLER, M R (1989) *Ibis* **131**, 27-32. MEHLUM, F (1978) *Vår Fuglefauna* **1**, 87-9. MEIDELL, O (1943) *Nytt Mag. Naturv.* **84**, 1-91. MEIKLEJOHN, M and MEIKLEJOHN, R F (1938) *Br. Birds* **32**, 194. MEIKLEJOHN, M F M (1948) *Ibis* **90**, 76-86; (1950) *Br. Birds* **43**, 264. MEIKLEJOHN, R F (1930) *Ibis* (12) 6, 560-4; (1936) *Ibis* (13) 6, 377-8. MEINEKE, T.(1979) *Beitr. Naturkde. Niedersachs.* **32**, 86-93. MEINERTZHAGEN, R (1920) *Ibis* (11) 2, 132-95; (1921) *Ibis* (11) 3, 621-71; (1924) *Ibis* (11) 6, 601-25; (1926) *Novit. Zool.* **33**, 57-121; (1927) *Ibis* (12) 3, 363-422; (1930) *Nicoll's birds of Egypt.* London; (1935) *Ibis* (13) 5, 110-51; (1938) *Ibis* (14) 2, 480-520, 671-717; (1939) *Bull. Br. Orn. Club* **59**, 67-8; (1940) *Ibis* (14) 4, 106-36, 187-234; (1947) *Bull. Br. Orn. Club* **67**, 90-8; (1949) *Ibis* **91**, 465-82; (1953) *Bull. Br. Orn. Club* **73**, 41-4; (1954) *Birds of Arabia.* Edinburgh; (1959) *Pirates and predators.* Edinburgh. MEININGER, P L, DUIVEN, P, MARTEIJN, E C L, and SPANJE, T M VAN (1990) *Malimbus* **12**, 19-24. MEININGER, P L, MULLIÉ, W C, and BRUUN, B (1980) *Gerfaut* **70**, 245-50. MEININGER, P L and SØRENSEN, U G (1984) *Bull. Br. Orn. Club* **104**, 54-7. MEISE, K E (1958) *Falke* **5**, 141. MEISE, W (1928) *J. Orn.* **76**, 1-203; (1934*a*) *Zool. Abh. Staatl. Mus. Tierkde. Dresden* 18 (2), 1-86; (1934*b*) *Orn. Monatsber.* **42**, 9-15; (1936) *J. Orn.* **84**, 631-72; (1938) *Compt. Rendu Congr. Orn. Int.* **9**, 233-48. MEKLENBURTSEV, R N (1946) *Byull. Mosk. Obshch. Ispyt. Prir. Otd. Biol.* **5** (1), 87-110. MELCHER, R (1951) *Orn. Beob.* **48**, 122-35. MELCHIOR, E (1975) *Regulus* **11**, 351-63. MELCHIOR, E, MENTGEN, E, PELTZER, R, SCHMITT, R, and WEISS, J. (1987) *Atlas der Brutvögel Luxemburgs.* Luxembourg. MELDE, F and MELDE, M (1976) *Falke* **23**, 88-92. MELDE, M (1984) *Raben- und Nebelkrähe.* Wittenberg Lutherstadt. MELTOFTE, H and FJELDSÅ, J (1989) *Fuglene i Danmark* **2**. Copenhagen. MELVILLE, D S and ROUND, P D (1984) *Bull. Br. Orn. Club* **104**, 127-38. MENDELSSOHN, H (1955) *Sal'it* **2**, 27-36. MENGEL, R M (1963) *Wilson Bull.* **75**, 201-3; (1964) *Living Bird* **3**, 9-43; (1965) *Orn. Monogr. AOU* **3**. MENTGEN, E (1988) *Regulus* **9** suppl., 62-8. MENZBIER, M and SUSCHKIN, P (1913) *Orn. Monatsber.* **21**, 192-3. MENZEL, H (1983)

Beitr. Vogelkde. **29**, 310. MÉRIC, J-D (1973) *Alauda* **41**, 161-3. MERIKALLIO, E (1951) *Proc. int. orn. Congr.* **10**, 484-93. MERKEL, F W (1978) *Luscinia* **43**, 163-81; (1980) *Luscinia* **44**, 133-58. MESSINEO, D J (1985) *Kingbird* **35**, 233-7. MESTER, H (1971) *Bonn. zool. Beitr.* **22**, 28-89. MESTER, H and PRÜNTE, W (1982) *J. Orn.* **123**, 381-99. METZ, E (1981) *Orn. Mitt.* **33**, 272. METZMACHER, M (1984) *Cah. Ethol. appl.* **3**, 191-214; (1986*a*) *Gerfaut* **76**, 131-8; (1986*b*) *Gerfaut* **76**, 317-34; (1986*c*) *Gerfaut* **76**, 335-42; (1986*d*) *Oiseau* **56**, 229-62; (1990) In Pinowski, J and Summers-Smith, J D (eds) *Granivorous birds in the agricultural landscape,* 151-68. MEWALDT, L R (1977) *N. Am. Bird Bander* **2** (4), suppl. MEWALDT, L R and KING, J R (1977) *Condor* **79**, 445-55; (1978) *Auk* **95**, 168-74; (1986) *J. Fld. Orn.* **57**, 155-67. MEWALDT, L R, KIBBY, S S, and MORTON, M L (1968) *Condor* **70**, 14-30. MEWES, W and HOMEYER, E F VON (1886) *Ornis* **2**, 181-288. MEY, E (1988) *Mitt. zool. Mus. Berlin* **64**, *Suppl. Ann. Orn.* **12**, 79-128. MEYER, D and SCHLOSS, W (1968) *Auspicium* **3**, 33-68. MEYER, R M (1990) *Bird Study* **37**, 199-209. MEYER-DEEPEN, H (1954) *Gef. Welt* **78**, 154-5. MEZHENNYI, A A (1964) *Zool. Zh.* **43**, 1679-87; (1979) In Krechmar, A V and Chernyavski, F B (eds) *Ptitsy severo-vostoka Azii,* 64-7. Vladivostok. MEZZAVILLA, F and BATTISTELLA, U (1987) *Riv. ital. Orn.* **57**, 33-40. MICHAELIS, H J (1980) *Falke* **27**, 286. MICHAELSEN, J (1985) *Vår Fuglefauna* **8**, 49-52. MICHEL, C (1986) *Birds of Mauritius.* Mauritius. MICHELS, H (1973) *Orn. Mitt.* **25**, 273. MICHENER, H and MICHENER, J R (1943) *Condor* **45**, 113-16. MIDDLETON, A L A (1969) *Emu* **69**, 145-54; (1970) *Emu* **70**, 12-16. MIKHAYLOV, K E (1984) *Ornitologiya* **19**, 205-6. MIKHAYLOV, K E and FIL'CHAGOV, A V (1984) *Ornitologiya* **19**, 22-9. MIKHEEV, A V (1939) *Zool. Zh.* **18**, 924-38. MIKKOLA, K (1979) *Lintumies* **14**, 125-36. MIKKOLA, K and KOIVUNEN, P (1966) *Orn. fenn.* **43**, 1-12. MIKKONEN, A V (1981) *Ornis scand.* **12**, 194-206; (1983) *Ornis scand.* **14**, 36-47; (1984) *Ornis fenn.* **61**, 33-53; (1985*a*) *Ann. zool. fenn.* **22**, 137-56; (1985*b*) *Acta Univ. Ouluensis* (A) *Sci. Rer. Nat.* **172** *Biol.* **24**. MILCHEV, B (1990) Ph D Thesis. Sofia Univ. MILD, K (1987) *Soviet bird songs* (2 cassettes and booklet). Stockholm; (1990) *Bird songs of Israel and the Middle East* (2 cassettes and booklet). Stockholm. MILDENBERGER, H (1940) *Beitr. Fortpfl. Vögel* **16**, 77-9; (1968) *Bonn. zool. Beitr.* **19**, 322-8; (1984) *Die Vögel des Rheinlandes* **2**. Düsseldorf. MILDENBERGER, H and SCHULZE-HAGEN, K (1973) *Charadrius* **9**, 52-7. MILES, P (1968) *Opera Corcontica* **5**, 201-11. MILLER, A H (1941) *Univ. Calif. Publ. Zool.* **44**, 173-434; (1942) *Condor* **44**, 185-6. MILLER, W DE W (1906) *Bull. Am. Mus. nat. Hist.* **22**, 161-83. MILLINGTON, R and HARRAP, S (1991) *Birding World* **4**, 52-4. MILLS, D G H (1982) *Br. Birds* **75**, 290-1. MILLS, E D and ROGERS, D T (1990) *Wilson Bull.* **102**, 146-50. MILSOM, T P and WATSON, A (1984) *Scott. Birds* **13**, 19-23. MINGOZZI, T (1982) *Riv. ital. Orn.* **52**, 43-5. MINGOZZI, T, BOANO, G, and PULCHER, C (1988) *Atlante degli uccelli nidificanti in Piemonte e Val d'Aosta 1980-84.* Turin. MIRZA, Z B (1973) In Kendeigh, S C and Pinowski, J (eds) *Productivity, population dynamics and systematics of granivorous birds,* 141-50. Warsaw; (1974) *Int. Stud. Sparrows* **7**, 76-87. MIRZA, Z B, KORA, A, SADIK, L S, and DAHNOUS, K (1975) *Int. Stud. Sparrows* **8**, 117-23. MISRA, R K and SHORT, L L (1974) *Condor* **76**, 137-46. MLÍKOVSKÝ, J (1982) *Vogelwarte* **31**, 442-5. MOBERG, A (1949) *Vår Fågelvärld* **8**, 187. MOCCI DEMARTIS, A (1973) *Alauda* **41**, 35-62. MODEL, N and OTREMBA, W (1986) *Anz. orn. Ges. Bayern* **24**,

177-9. Mödlinger, P (1977) *Aquila* 83, 79-89. Moeed, A (1976) *Notornis* 23, 246-9. Moffat, C B (1943) *Irish Nat. J.* 8, 54-5. Mohr, R (1967) *J. Orn.* 108, 484-90; (1974) *J. Orn.* 115, 106-7. Mojsisovics, A (1886) *Mitt. orn. Ver. Wien Schwalbe* 10, 113. Moksnes, A (1973) *Norw. J. Zool.* 21, 113-38. Moksnes, A and Røskaft, E (1987) *Ornis scand.* 18, 168-72. Molamusov, K T (1967) *Ptitsy tsentral'noy chasti Severnogo Kavkaza.* Nal'chik. Molau, U (1985) *Vår Fågelvärld* 44, 5-20. Moll, W (1986) *Vogelkdl. Ber. Niedersachs.* 18, 11-14. Møller, A P (1978a) *Nordjyllands fugle.* Klampenborg; (1978b) *Dansk orn. Foren. Tidsskr.* 72, 61-3; (1978c) *Dansk orn. Foren. Tidsskr.* 72, 197-215; (1979) *Dansk orn. Foren. Tidsskr.* 73, 305-9; (1981a) *Dansk orn. Foren. Tidsskr.* 75, 69-78; (1981b) *Dutch Birding* 3, 148-50; (1982a) *Ornis scand.* 13, 94-100; (1982b) *Ornis scand.* 13, 239-46; (1983a) *J. Orn.* 124, 147-61; (1983b) *Ornis fenn.* 60, 105-11; (1983c) *Ornis scand.* 14, 81-9; (1985) *J. Orn.* 126, 405-19; (1987) *Anim. Behav.* 35, 203-10; (1988) *Ethology* 78, 321-31. Moltoni, E (1950) *Riv. ital. Orn.* 20, 75-8; (1951) *Riv. ital. Orn.* 21, 45-51; (1969) *Riv. ital. Orn.* 39, 128-57; (1973) *Riv. ital. Orn.* 43 suppl. Moltoni, E and Brichetti, P (1978) *Riv. ital. Orn.* 48, 65-142. Monaghan, P (1989) In Bignal and Curtis (1989) 4-8, 63-4. Monaghan, P and Bignal, E (1985) *Bull. Br. ecol. Soc.* 16, 208-10. Monaghan, P, Bignal, E, Bignal, S, Easterbee, N, and McKay, C R (1989) *Scott. Birds* 15, 114-18. Monaghan, P and Thompson, P M (1984) *Bull. Br. ecol. Soc.* 15, 145-6. Monk, J F (1954) *Bird Study* 1, 2-14. Mönke, R (1975) *Beitr. Vogelkde.* 21, 370-1. Monroe, B L, Jr (1968) *Orn. Monogr. AOU* 7. Montell, J (1917) *Acta Soc. Fauna Flora fennica* 44 (7). Montgomerie, R D, Cartar, R V, McLaughlin, R L, and Lyon, B (1983) *Arctic* 36, 65-75. Moody, D C (1954) *Br. Birds* 47, 406. Moore, F R, Kerlinger, P, and Simons, T R (1990) *Wilson Bull.* 102, 487-500. Moore, A S (1991) *Peregrine* 7, 47-53. Moore, H J and Boswell, C (1956) *Iraq nat. Hist. Mus. Publ.* 10; (1957) *Iraq nat. Hist. Mus. Publ.* 12. Moore, N C (1962) *J. Northants. nat. Hist. Soc.* 34, 130-2. Morales, J A G (1969) *Ardeola* 13, 265. Moreau, R E (1931) *Ibis* (13) 1, 204-8; (1960) *Ibis* 102, 298-321, 443-71; (1966) *The bird faunas of Africa and its islands.* London; (1967) *Ibis* 109, 445; (1972) *The Palaearctic-African bird migration systems.* London. Morel, G and Roux, F (1966) *Terre Vie* 113, 19-72, 143-76; (1973) *Terre Vie* 27, 523-50. Morel, G J (1980) *Liste commentée des oiseaux du Sénégal et de la Gambie. Suppl.* 1. Dakar. Morel, G J and Morel, M-Y (1976) *Terre Vie* 30, 493-520; (1978) *Cah. ORSTOM Sér. Biol.* 13, 347-58; (1980) *Proc. int. orn. Congr.* 17, 1150-4; (1990) *Les oiseaux de Sénégambie.* Paris. Morel, M-Y (1964) *Terre Vie* 111, 436-451; (1966) *Ostrich Suppl.* 6, 435-42; (1967) *Terre Vie* 114, 77-82; (1969) Doc. Sci. Nat. Thesis. Rennes Univ; (1973) *Mém. Mus. natn. Hist. nat. Paris* (A) *Zool.* 78, 1-156. Morel, M-Y and Morel, G J (1973a) *Oiseau* 43, 97-118; (1973b) *Oiseau* 43, 314-29. Moreno, J M (1988) *Guía de las aves de las Islas Canarias.* Santa Cruz de Tenerife. Morgan, A (1971) *Br. Birds* 64, 422-3. Morgan, P A and Howse, P E (1973) *Anim. Behav.* 21, 481-91; (1974) *Anim. Behav.* 22, 688-94. Morgan, J and Shirihai, H (1992) *Birding World* 5, 344-7. Moritz, D (1982) *Vogelwelt* 103, 16-18. Morley, A (1943) *Ibis* 85, 132-58. Morozov, V V (1984) *Ornitologiya* 19, 30-40; (1986) *Tez. Dokl.* 1. *S'ezda Vsesoyuz. orn. Obshch.* 9. *Vsesoyuz. orn. Konf.* 2, 83-5; (1987) *Ornitologiya* 22, 134-47. Morphy, M J (1965) *Ibis* 107, 97-100. Morrison, A (1948) *Ibis*

90, 381-7. Morrison, C M (1977) *Scott. Birds* 9, 302. Mortensen, P H and Birkholm-Clausen, F (1963) *Dansk orn. Foren. Tidsskr.* 57, 22-4. Morton, E S (1989) *Wilson Bull.* 101, 460-2. Morton, G A and Morton, M L (1990) *Condor* 92, 813-28. Morton, M L and Welton, D E (1973) *Condor* 75, 184-9. Morton, M L, Horstmann, J L, and Carey, C (1973) *Auk* 90, 83-93. Morton, M L, King, J R, and Farner, D S (1969) *Condor* 71, 376-85. Morton, M L and Morton, G A (1987) *Condor* 89, 197-200. Morton, M L, Orejuela, J E, and Budd, S M (1972) *Condor* 74, 423-30. Morton, M L and Pereyra, M E (1987) *J. Fld. Orn.* 58, 6-21. Mosimann, P (1988) *Orn. Beob.* 85, 179-81. Moss, D (1979) *Wicken Fen Group Rep.* 10, 12-14. Motis, A (1985) Master's Thesis. Barcelona Univ; (1986) *Historia Natural dels Paisos Catalans.* 12: *Ocells. Fundació Enciclopedia Catalana,* 336-415. Barcelona; (1987) *Misc. Zool.* 11, 339-46. Motis, A, Martínez, A, Peris, S, and Ferrer, X (1987) *9th Jorn. Orn. Españolas, Madrid,* poster paper. Motis, A, Mestre, P, and Martínez, A (1983) *Misc. Zool.* 7, 131-7. Mould, J E M (1974) *Bird Study* 21, 157-8. Mountfort, G and Ferguson-Lees, I J (1961a) *Ibis* 103a, 86-109; (1961b) *Ibis* 103a, 443-71. Mountfort, G R (1935) *Br. Birds* 29, 145-8; (1956) *Ibis* 98, 490-5; (1957) *The Hawfinch.* London; (1958) *Portrait of a wilderness.* London; (1962) *Portrait of a river.* London; (1965) *Portrait of a desert.* London. Mountjoy, D J and Lemon, R E (1991) *Behav. Ecol. Sociobiol.* 28, 97-100. Moyer, J W (1930) *Auk* 47, 567-8. Moyson, I Y (1973) *Gerfaut* 63, 257-78. Mühlethaler, F (1952) *Orn. Beob.* 49, 173-82. Muir, R C (1959) *Br. Birds* 52, 434-5. Müller, A K (1953) *Anz. Orn. Ges. Bayern* 4, 76. Müller, H E J (1983) *Falke* 30, 24-31. Müller, H E J and Wernicke, P (1990) *Falke* 37, 83-6. Müller, L (1982) *Falke* 29, 421. Müller, S (1970) *Falke* 17, 199-203; (1973) *Corax* 4, 112-30; (1977) *Falke* 24, 268. Muller, Y (1987) *Acta Oecol.* 8, 185-9. Mulligan, J A and Olsen, K C (1969) In Hinde, R A (ed) *Bird vocalizations,* 165-84. Cambridge. Mullins, J R (1984a) *Br. Birds* 77, 26-7; (1984b) *Br. Birds* 77, 133-5; (1984c) *Br. Birds* 77, 426. Mulvihill, R S and Chandler, C R (1990) *Auk* 107, 490-9; (1991) *Condor* 93, 172-5. Münch, H (1957) *Orn. Beob.* 54, 194-5. Mundinger, P C (1970) *Science NY* 168, 480-2; (1979) *Syst. Zool.* 28, 270-83. Mundy, P J and Cook, A W (1977) *Ostrich* 48, 72-84. Munkejord, A (1987) *Fauna norv.* (C) *Cinclus* 10, 73-80. Munkejord, A, Hauge, F, Folkedal, S, and Kvinnesland, A (1985) *Fauna norv.* (C) *Cinclus* 8, 1-8. Munn, P W (1931) *Novit. Zool.* 37, 53-132. Muñoz-Pulido, R, Bautista, L M, Alonso, J C, and Alonso, J A (1990) *Bird Study* 37, 111-14. Munro, I C (1977) *Scott. Birds* 9, 382. Munro, J H B (1975) *Scott. Birds* 8, 309-14. Munro, J A and Cowan, I M (1947) *British Columbia Prov. Mus. spec. Publ.* 2. Muntaner, J, Ferrer, X, and Martínez-Vilalta, A (1983) *Atlas. dels ocells nidificants de Catalunya i Andorra.* Barcelona. Munteanu, D (1966) *Bull. Br. Orn. Club* 86, 98-100; (1967) *Larus* 19, 179-203. Murphy, M E and King, J R (1984) *Auk* 101, 164-7. Murphy, M E, King, J R, and Lu, J (1988) *Can. J. Zool.* 66, 1403-13. Murphy, R C (1924) *Bull. Amer. Mus. nat. Hist.* 50, 211-78. Murr, F (1957) *Anz. orn. Ges. Bayern* 4, 556-8. Murray, B G (1965) *Wilson Bull.* 77, 122-33; (1989) *Auk* 106, 8-17. Murray, B G and Jehl, J R (1964) *Bird-Banding* 35, 253-63. Murray, R D (1978) *Br. Birds* 71, 318-9. Murton, R K (1971) *Man and birds.* London. Murton, R K and Westwood, N J (1974) *Ibis* 116, 298-313. Mycock, J

(1987) *Orn. Soc. Middle East Bull.* **18**, 1–3. MYLNE, C K (1960) *Br. Birds* **53**, 86–8; (1961) *Br. Birds* **54**, 206–7; (1957) *Br. Birds* **50**, 171–2. MYRCHA, A and PINOWSKI, J (1970) *Condor* **72**, 175–81.

NAGY, E (1943) *Beitr. Fortpfl. Vögel* **19**, 9–13. NAIK, N L and NAIK, R M (1969) *Pavo* **7**, 57–73. NAKAMURA, T, YAMAGUCHI, S, IIJIMA, K, and KAGAWA, T (1968) *Misc. Rep. Yamashina Inst. Orn.* **5**, 313–36. NANKINOV, D N (1974) In Boehme, R L and Flint, V E (eds) *Mat. 6 Vsesoyuz. orn. Konf.* **2**, 92–3. Moscow; (1978) *Acta Orn.* **16**, 285–94; (1984) *Intl. Studies on Sparrows* **11**, 47–70. NARANG, M L and LAMBA, B S (1984) *Rec. zool. Survey India misc. Publ. occ. Pap.* **44**, 1–76. NARDIN, C and NARDIN, G (1985) *Nos Oiseaux* **38**, 113–20. NATORP, O (1931) *J. Orn.* **79**, 338–46; (1940) *Orn. Monatsber.* **48**, 46–8. NAU, B S (1960) *Bird Study* **7**, 185–8. NAUMANN, J A (1900) (ed. Hennicke, C R) *Naturgeschichte der Vögel Mitteleuropas* **3**; (1901) (ed. Hennicke, C R) **4**. NAUROIS, R DE (1969) *Bull. de l'Inst. fond. d'Afrique noire* **31**A, 143–218; (1981) *Bull. de l'Inst. fond. d'Afrique noire* **43**, 202–18; (1986) *Cyanopica* **3**, 533–8; (1988) *Bol. Mus. Mun. Funchal* **40**, 253–73. NAUROIS, R DE and BONNAFFOUX, D (1969) *Alauda* **37**, 93–113. NEALE, J J (1899–1900) *Trans. Cardiff Nat. Soc.* **32**, 1–5; (1901) *Rep. Trans. Cardiff Nat. Soc.* **32**, 49–53. NECHAEV, V A (1969) *Ptitsy yuzhnykh Kuril'skikh ostrovov.* Leningrad; (1974) *Trudy Biol.-pochvenn. Inst. Akad. Nauk SSSR* **17** (120), 120–35; (1975) *Trudy Biol.-pochvenn. Inst. Akad. Nauk SSSR* **29** (132), 114–60; (1977) *Byull. Mosk. Ispyt. Prir. Otd. Biol.* **82** (3), 31–9. NEIN, R (1982) *Beitr. Naturkde. Wetterau* **2**, 154. NELSON, B (1973) *Azraq: desert oasis.* London. NELSON, J B (1950) *Br. Birds* **43**, 293. NELSON, T H (1907) *The birds of Yorkshire* **1**. London. NERO, R W (1951) *Wilson Bull.* **63**, 84–8; (1963) *Sask. nat. Hist. Soc. spec. Publ.* **5**; (1967) *Sask. nat. Hist. Soc. spec. Publ.* **6**. NERO, R W and LEIN, M R (1971) *Sask. Nat. Hist. Soc. Spec. Publ.* **7**. NERUCHEV, V V and MAKAROV, V I (1982) *Ornitologiya* **17**, 125–9. NETHERSOLE-THOMPSON, D (1966) *The Snow Bunting.* Edinburgh; (1975) *Pine Crossbills.* Berkhamsted; (1976) *Scott. Birds* **9**, 147–62. NETHERSOLE-THOMPSON, C and NETHERSOLE-THOMPSON, D (1940) *Br. Birds* **34**, 135; (1943a) *Br. Birds* **37**, 70–4; (1943b) *Br. Birds* **37**, 88–94. NETHERSOLE-THOMPSON, D and WATSON, A (1974) *The Cairngorms.* London. NETHERSOLE-THOMPSON, D and WHITAKER, D (1984) *Scott. Birds* **13**, 87. NEUB, M (1973) *Anz. orn. Ges. Bayern* **12**, 248–55. NEUFELDT, I A (1961) *Zool. Zh.* **40**, 416–26; (1970) *Trudy Zool. Inst. Akad. Nauk SSSR* **47**, 111–81; (1986) *Trudy Zool. Inst. Akad. Nauk SSR* **150**, 7–43. NEUFELDT, I A and LUKINA, E W (1966) *Falke* **13**, 121–5. NEUMANN, O (1907) *Orn. Monatsber.* **15**, 144–6. NEUSCHWANDER, J (1973) *Alauda* **41**, 163–5. NEVES, F I M (1984) *Cyanopica* **3**, 183–96. NEWSTEAD, R (1908) *J. Board Agric. Suppl.* **15** (9). NEWTON, I (1964a) *Bird Study* **11**, 47–68; (1964b) D Phil Thesis. Oxford Univ; (1964c) *J. appl. Ecol.* **1**, 265–79; (1966a) *Br. Birds* **59**, 89–100; (1966b) *Ibis* **108**, 41–67; (1967a) *Ibis* **109**, 33–98; (1967b) *J. Anim. Ecol.* **36**, 721–44; (1967c) *Ibis* **109**, 440–1; (1967d) *Bird Study* **14**, 10–24; (1968a) *Bird Study* **15**, 84–92; (1968b) *Condor* **70**, 323–32; (1969a) *J. Orn.* **110**, 53–61; (1969b) *Physiol. Zoöl* **42**, 96–107; (1970) In Watson, A (ed.) *Animal populations in relation to their food resources*, 337–57; (1972) *The finches.* London. NEWTON, I, DAVIS, P E, and DAVIS, J E (1982) *J. appl. Ecol.* **19**, 681–706. NEWTON, I and EVANS, P R (1966)

Bird Study **13**, 96–8. NICE, M M (1937) *Trans. Linn. Soc. New York* **4**; (1938) *Bird-Banding* **9**, 1–11; (1943) *Trans. Linn. Soc. New York* **6**; (1946) *Condor* **48**, 41–2. NICHOLS, J T (1935) *Bird-Banding* **6**, 11–15; (1945) *Bird-banding* **16**, 29–32; (1953) *Bird-Banding* **24**, 16–7. NICHOLSON, E M (1930) *Ibis* (12) **6**, 280–313, 395–428; (1951) *Birds and men.* London. NICHOLSON, E M and NICHOLSON, B D (1930) *J. Ecol.* **18**, 51–66. NICHT, M (1961) *Zool. Abh. Staatl. Mus. Tierkde. Dresden* **26**, 79–99. NICKELL, W P (1968) *Bird-banding* **39**, 107–16. NICOLAI, B (ed.) (1993) *Atlas der Brutvögel Ostdeutschlands.* Jena. NICOLAI, J (1956) *Z. Tierpsychol.* **13**, 93–132; (1957) *J. Orn.* **98**, 363–71; (1959) *J. Orn.* **100**, 39–46; (1960) *Zool. Jb.* **87**, 317–62; (1964) *Z. Tierpsychol.* **21**, 129–204. NICOLL, M J (1909a) *Bull. Br. Orn. Club* **23**, 99–100; (1909b) *Ibis* (9) **3**, 471–84; (1912) *Ibis* (9) **6**, 405–53; (1922) *Ibis* (11) **4**, 688–701. NIEDRACH, R J and ROCKWELL, R B (1939) *The birds of Denver and mountain parks.* Denver. NIELSEN, O K (1979) *Nátúrufr.* **49**, 204–20; 654ce; (1986) Ph D Thesis. Cornell Univ. NIETHAMMER, G (1936) *Beitr. Fortpfl. Vögel* **12**, 161–2; (1937) *Handbuch der deutschen Vogelkunde* **1**. Leipzig; 647i 658b 665b 651j 653j 658gh 674j 678bcfi 679gh; (1942) *Handbuch der deutschen Vogelkunde* **3**. Leipzig; (1943) *J. Orn.* **91**, 167–238; (1950) *Syllegomena biologica*, 267–86. Leipzig; (1953a) *Bonn. zool. Beitr.* **4**, 73–8; (1953b) *J. Orn.* **94**, 282–9; (1955) *Bonn. zool. Beitr.* **6**, 29–80; (1957a) *Bonn. zool. Beitr.* **8**, 230–47; (1957b) *Bonn. zool. Beitr.* **8**, 275–84; (1957c) *J. Orn.* **98**, 363–71; (1958) *J. Orn.* **99**, 431–7; (1961) *Falke* **8**, 367–70; (1962) *Bonn. zool. Beitr.* **13**, 209–15; (1963) *Bonn. zool. Beitr.* **14**, 129–50; (1967) *Ibis* **109**, 117–8; (1969) *J. Orn.* **110**, 205–8; (1971) *J. Orn.* **112**, 202–26; (1973) *Bonn. zool. Beitr.* **24**, 270–84. NIETHAMMER, G and BAUER, K (1960) *Orn. Beob.* **57**, 241–2. NIETHAMMER, VON G, KRAMER, H, and WOLTERS, H E (1964) *Die Vögel Deutschlands.* Frankfurt. NIETHAMMER, G and THIEDE, W (1962) *J. Orn.* **103**, 289–93. NIETHAMMER, G and WOLTERS, H E (1966) *Bonn. zool. Beitr.* **17**, 157–85; (1969) *Bonn. zool. Beitr.* **20**, 351–4. NIGG, F (1974) *Vögel Heimat* **45**, 38. NIGHTINGALE, B and ALLSOPP, K (1990) *Br. Birds* **83**, 541–8; (1991) *Br. Birds* **84**, 316–28; (1992) *Br. Birds* **85**, 636–47. NIKANDER, P J and JÄNNES, H (1992) *Lintumies* **27**, 275–83. NIKOLAUS, G (1987) *Bonn. zool. Monogr.* **25**. NIKOLAUS, G and PEARSON, D (1991) *Ring. Migr.* **12**, 46–7. NILSSON, I (1957) *Vår Fågelvärld* **16**, 50. NILSSON, R (1971) *Vår Fågelvärld* **30**, 124–5. NILSSON, S G (1979) *Ibis* **121**, 177–85. NISBET, I C T (1970) *Bird-Banding* **41**, 207–40. NISBET, I C T, DRURY, W H, and BAIRD, J (1963) *Bird-Banding* **34**, 107–59. NISBET, I C T and SMOUT, T C (1957) *Br. Birds* **50**, 201–4. NITSCHE, G (1980) *Orn. Mitt.* **32**, 274–5. NOË, A (1983) *Anz. orn. Ges. Bayern* **22**, 110–11. NOESKE, A and HOFFMANN, M (1988) *Limicola* **2**, 22–7. NOGALES, M (1990) Doct. Thesis. La Laguna Univ; (1992) *Ecología* **6**, 215–23. NOGGE, F (1973) *Bonn. zool. Beitr.* **24**, 254–69. NOLAN, V and KETTERSON, E D (1983) *Wilson Bull.* **95**, 603–20. NOLAN, V and KETTERSON, E D (1990) *Wilson Bull.* **102**, 469–79. NOLAN, V, KETTERSON, E D, ZIEGENFUS, C, CULLEN, D P, and CHANDLER, C R (1992) *Condor* **94**, 364–70. NOLL, H (1956) *Ber. St. Gall. naturw. Ges.* **75**, 48–52; 651g. NOLTE, W (1930) *J. Orn.* **78**, 1–19. NOORDEN, B VAN (1991) *Limosa* **64**, 69–71. NORDMEYER, A, OELKE, H, and PLAGEMANN, E (1970) *Int. Stud. Sparrows* **4**, 50–4. NORMAN, D, CROSS, D, and COCKBAIN, R (1981) *BTO News* **114**, 9. NORMAN, E J (1966) *Birds Illustr.* **12**, 74–5. NØHR, H (1984) *Vår Fågelvärld* **43**, 241–3. NØRREVANG, A and HARTOG, J C DEN

(1984) *Cour. Forsch.-Inst. Senckenberg* **68**, 107-34. NORRIS, R A and HIGHT, G L (1957) *Condor* **59**, 40-52. NORRIS, R A and JOHNSTON, D W (1958) *Wilson Bull.* **70**, 114-29. NORTH, C A (1968) Diss. Oklahoma State Univ; (1973) In Hardy, J W and Morton, M L (eds) *Orn. Monogr. AOU* **14**, 79-91. NORTON, W J E (1958) *Ibis* **100**, 179-89. NOSKOV, G A (1975) *Zool. Zh.* **54**, 413-24; (1977) *Zool. Zh.* **56**, 1676-86; (1978) *Ekologiya* **1**, 61-9; (ed.) (1981) *Polevoy vorobey.* Leningrad; (1982) *Orn. Stud. USSR* **2**, 348-63. NOSKOV, G A and GAGINSKAYA, A R (1977) *Ornitologiya* **13**, 190-1. NOSKOV, G A, RYMKEVICH, T A, SHIBDOV, A A, and NANKINOV, D N (1975) *Vestnik Leningradskogo Univ.* **3** (1), 11-16. NOSKOV, G A and SHAMOV, S V (1983) *Soobshch. Pribalt. Kom. Izuch. Migr. Ptits* **14**, 125-9. NOSKOV, G A and SMIRNOV, E N (1979) *Nauch. Dokl. vyssh. Shk. Biol. Nauki* 1970 (3), 38-45. NOSKOV, G A, ZIMIN, V B, REZVYI, S P, RYMKEVICH, T A, LAPSHIN, N V, and GOLOVAN', V I (1981) In Noskov, G A (ed.) *Ekologiya ptits Priladozh'ya*, 3-86. Leningrad. NOTHDURFT, W (1972) *Anz. orn. Ges. Bayern* **11**, 185-9. NOTHDURFT, W, KNOLLE, F, and ZANG, H (1988) *Vogelkdl. Ber. Niedersachs.* **20**, 33-85. NOVAL, A (1971) *Ardeola* spec. vol., 491-507; (1986) *Guía de las aves de Asturias.* Gijon. NOVIKOV, B G (1938) *Trudy Nauch.-issled. Inst. eksper. Morfogen. Mosk. Univ.* **6**, 485-93. NOVIKOV, G A (1949) *Zool. Zh.* **28**, 461-70; (1952) *Trudy Zool. Inst. Akad. Nauk SSSR* **9**, 1133-54; (1972) *Aquilo (Ser. Zool.)* **13**, 95-7. NOVOTNÝ, I (1970) *Acta Sci. Nat. Brno* **4**, 1-57. NUMEROV, A D (1978) *Trudy Okskogo gos. Zapoved.* **14**, 356-7. NÜRNBERGER, F, SIEBOLD, D, and BERGMANN, H-H (1989) *Bioacoustics* **1**, 273-86. W S0006 TNUTTALL, J (1972) *Naturalist* **923**, 140-1. NYBERG, M (1932) *Ornis fenn.* **9**, 83-5. NYHLÉN, V (1950) *Vår Fågelvärld* **9**, 49-63. NYHOLM, N E I and MYHRBERG, H E (1977) *Oikos* **29**, 336-41. NYSTRÖM, B and NYSTRÖM, H (1987) *Vår Fågelvärld* **46**, 119-28; (1991) *Ornis Svecica* **1**, 65-8. NYSTRÖM, E W (1925) *Ornis fenn.* **2**, 8-13.

OAKES, C (1953) *The birds of Lancashire.* Edinburgh. OATES, E W (1883) *A handbook to the birds of British Burmah* **1**; (ed.) (1890) *Hume's nests and eggs of Indian birds.* London. OBERHOLSER, H C (1919) *Auk* **36**, 549-55; (1974) *The bird life of Texas* **2**. Austin. O'CONNOR, R J (1973) *Rep. Rye Meads Ringing Group* **6**, 16-27. O'CONNOR, R J and SHRUBB, M (1986) *Farming and birds.* Cambridge. OEHLER, J (1977) *Wissenschaftl. Z. Humboldt-Univ. Berlin Math.-Nat. R.* **26**, 425-9; (1978) *Biol. Zbl.* **97**, 279-87. OELKE, H (1992) *Beitr. Naturk. Niedersachs.* **45**, 1-17. OESER, R (1975) *Beitr. Vogelkde.* **21**, 475-6. OESER, R E (1984) *Beitr. Vogelkde.* **30**, 162-8. ODDIE, W E (1981) *Br. Birds* **74**, 532-3. ODINZOWA, N P (1967) *Falke* **14**, 414-5. ODUM, E P (1931) *Wilson Bull.* **43**, 316-7; (1949) *Wilson Bull.* **61**, 3-14; (1958) *Bird-Banding* **29**, 105-8. ODUM, E P, CONNELL, C E, and STODDARD, H L (1961) *Auk* **78**, 515-27. ODUM, E P and PERKINSON, J D (1951) *Physiol. Zool.* **24**, 216-30. OGGIER, P A (1986) *Orn. Beob.* **83**, 295-9. OGILVIE, C M (1947) *Br. Birds* **40**, 135-9; (1949) *Br. Birds* **42**, 65-8; (1951) *Br. Birds* **44**, 1-5. OGILVIE, M A and OGILVIE, C C (1984) *Br. Birds* **77**, 368. OGILVIE-GRANT, W R (1901) *Ibis* (8) **1**, 518-21. OGORODNIKOVA, L I and MIRONOVA, V E (1989) In Konstantinov, V M and Klimov, S M (eds.) *Vranovye ptitsy v estestvennykh i antropogennykh landshaftakh* (Materialy II Vsesoyuznogo soveshchaniya) **4**, 12-13. Lipetsk. OJANEN, M and KYLMÄNEN, R (1984) *Ornis fenn.* **61**, 123. OJANEN, M, ORELL,

M, and HIRVELÄ, J (1979) *Holarctic Ecol.* **2**, 81-7. OJANEN, M and LÄHDESMÄKI, P (1984) *Aureola* **9**, 14-21. ÓLAFSSON, E (in press) *Náttúrufr.* OKULEWICZ, J (1989) *Ptaki ŚląSKA* **7**, 1-39; (1991) *Ptaki Śląska* **8**, 1-17. OLIER, A (1958) *Alauda* **26**, 65-66; (1959) *Alauda* **27**, 205-10. OLIOSO, G (1972) *Alauda* **40**, 171-4; (1973) *Alauda* **41**, 227-32; (1974) *Alauda* **42**, 502; (1987) *Bièvre* **9**, 1-8. OLIVIER, G (1949) *Oiseau* **19**, 102-4. OLSEN, K (1984) *Vår Fuglefauna* **7**, 233-4. OLSEN, K M (1987a) *Dansk orn. Foren. Tidsskr.* **81**, 109-20; (1987b) *Fugle* **7** (2), 20-1; (1989) *Dansk orn. Foren. Tidsskr.* **83**, 97-101; (1991a) *Anser* **30**, 29-40; (1991b) *Dansk orn. Foren. Tidsskr.* **85**, 20-34. OLSON, S L (1985) *Bull. Br. Orn. Club* **105**, 29-30. OLSSON, C (1988) *Vår Fågelvärld* **47**, 197-9. OLSSON, M (1954) *Vår Fågelvärld* **13**, 113, 120. OLSSON, V (1960) *Vår Fågelvärld* **19**, 1-19; (1964) *Br. Birds* **57**, 118-23; (1969) *Vogelwarte* **25**, 147-56; (1971) *Br. Birds* **64**, 213-23; (1985) *Vår Fågelvärld* **44**, 269-83. OLSTHOORN, H (1987) *Br. Birds* **80**, 117-18. O'MAHONY, J (1977) *Br. Birds* **70**, 340. O'MALLEY, S L C (1993) D Phil Thesis. Leicester Univ. OORDT, G J VAN, and DAMSTÉ, P H (1939) *Acta Brevia Neerlandica* **9**, 140-3. OREN, D C and SMITH, N J H (1981) *Wilson Bull.* **93**, 281-282. ORFORD, N (1973) *Bird Study* **20**, 50-62, 121-6. ORFORD, N W (1959) *Nature* **184**, 650. ORLOV, P P (1955) *Zool. Zh.* **34**, 950-2. ORNITHOLOGICAL SOCIETY OF JAPAN (1974) *Check-list of Japanese birds.* Tokyo. ORTLIEB, R (1971) *Anz. Orn. Ges. Bayern* **10**, 186-7. OSCAR, P (1986) *Larus* **36-7**, 155-65. OSIECK, E R (1973) *Vogeljaar* **21**, 274-7. OSMASTON, B B (1925) *Ibis* (12) **1**, 663-719; (1927) *J. Bombay Nat. Hist. Soc.* **31**, 975-99. OSTAPENKO, M M, KASHKAROV, D Y, SHERNAZAROV, E, LANOVENKO, E N, and FILATOV, A K (1980) In Abdusalyamov, I A (ed.) *Migratsii ptits v Azii*, 75-97. Dushanbe. OSTAPENKO, V A (1981) *Ornitologiya* **16**, 179-80. OSTASHCHENKO, A N, POPOV, E A, and YABLONKEVICH, M L (1985) *Trudy Zool. Inst. Akad. Nauk SSSR* **137**, 119-28. OSTHAUS, H and SCHLOSS, W (1975) *Auspicium* **6**, 45-89. O'SULLIVAN, O and SMIDDY, P (1988) *Irish Birds* **3**, 609-48; (1989) *Irish Birds* **4**, 79-114; (1991) *Irish Birds* **4**, 423-62. OTTOSSON, U and HAAS, F (1991) *Ornis Svecica* **1**, 113-8. OTTOW, B (1912) *Orn. Monatsber.* **20**, 11. OUBRON, G (1967) *Organisation commune de lutte anti-acridienne et de lutte antiaviare* (Dakar), 707/403.33.21/LAV. OUWENEEL, G L (1970) *Limosa* **43**, 156-8. OWEN, D A L (1985) D Phil Thesis. Univ. Oxford; (1989) In Bignal and Curtis (eds), 57-62. OWEN, D F (1953) *Br. Birds* **46**, 353-64; (1956) *Bird Study* **3**, 257-65; (1959) *Ibis* **101**, 235-9. OWEN, J H (1950) *Br. Birds* **43**, 16.

PAATELA, J E (1938) *Ornis fenn.* **15**, 65-9; (1948) *Ornis fenn.* **25**, 21-8. PACCAUD, O (1954) *Nos Oiseaux* **22**, 214. PACHECO, F, ALBA, F J, GARCÍA DÍAZ, E, and PÉREZ MELLADO, V (1977) *Ardeola* **22**, 55-73. PACKARD, G C (1967a) *Syst. Zool.* **16**, 73-89; (1967b) *Wilson Bull.* **79**, 345-6. PAEVSKI, V A (1968) *Migr. Zhivot.* **5**, 153-60; (1970a) *Mat. 7. Pribalt. orn. Konf.* **1**, 69-75; (1970b) *Zool. Zh.* **49**, 798-9; (1971) *Trudy Zool. Inst. Akad. Nauk SSSR* **50**, 3-110; (1981) *Zool. Zh.* **60**, 109-14; (1985) *Demografiya ptits.* Leningrad. PAEVSKI, V A, and VINOGRADOVA, N V (1974) *Trudy Zool. Inst. Akad. Nauk SSSR* **55**, 186-206. PAEVSKI, V A, VINOGRADOVA, N V, SHAPOVAL, A P, SHUMAKOV, M E, and YABLONKEVICH, M L (1990) *Trudy Zool. Inst. Akad. Nauk SSSR* **210**, 63-72. PAGE, J (1988) *Bat News* **15**, 3. PAIGE, J P (1960) *Ibis* **102**, 520-5. PAIL-

LERETS, DE B DE (1934) *Alauda* 6, 395-6. PAINTIN, A (1991) *Ann. Bull. Soc. Jersiaise* 25, 422-4. PALMER, R S (1949) *Bull. Mus. comp. Zool.* 102. PALMÉR, S and BOSWALL, J (1981) *A field guide to the bird songs of Britain and Europe.* Stockholm. PALMER, W (1894) *Auk* 11, 282-91. PALMGREN, P (1930) *Acta zool. fenn.* 7,; (1935) *Ornis fenn.* 12, 107-21; (1936) *Ornis fenn.* 13, 153-9. PALUDAN, K (1936) *Vidensk. Medd. dansk nat. Foren.* 100, 247-346; (1938) *J. Orn.* 86, 562-638; (1940) *Danish Sci. Invest. Iran* 2, 11-54; (1959) *Vidensk. Medd. dansk nat. Foren.* 122. PANDOLFI, M (1987) *Riv. ital. Orn.* 57, 115-6. PANNACH, D (1983) *Beitr. Vogelkde.* 29, 317-18; (1984) *Beitr. Vogelkde.* 30, 211-12; (1990) *Abh. Ber. Naturkundemus. Görlitz* 63 (3). PANOV, E N (1973a) *Ptitsy yuzhnogo Primor'ya.* Novosibirsk; (1973b) In Vorontsov, N N (ed.) *Problemy Evolyutsii* 3, 261-94; (1989) *Gibridizatsiya i etologicheskaya izolyatsiya u ptits.* Moscow. PANOV, E N and BULATOVA, N S (1972) *Byull. Mosk. Obshch. Ispyt. Prior. Otd. Biol.* 77 (4), 86-94. PANOV, E N and RADZHABLI, S I (1972) In Vorontsov, N N (ed.) *Problemi Evolyutsii* 2, 263-75. Novosibirsk. PAQUET, A (1979) *Aves* 16, 162-3. PARKES, K C (1951) *Wilson Bull.* 63, 5-15; (1952) *Wilson Bull.* 64, 161-2; (1961) *Wilson Bull.* 73, 374-9; (1967) *Wilson Bull.* 79, 456-8; (1978) *Auk* 95, 682-90; (1988) *J. Fld. Orn.* 59, 60-2; (1988b) *Bird Obs. (Mass.)* 16, 324-5; (1990) *Wilson Bull.* 102, 733-4. PARKHURST, R and LACK, D (1946) *Br. Birds* 39, 358-64. PARKIN, D T (1988) *Proc. Int. Orn. Congr.* 19, 1652-7, 1669-73. PARKS, G H (1962) *Bird-Banding* 33, 148-51. PARMELEE, D F and MACDONALD, S D (1960) *Bull. nat. Mus. Canada* 169. PARMELEE, D F and PARMELEE, J M (1988) *Condor* 90, 952. PAROVSHCHIKOV, V Y (1962) *Ornitologiya* 4, 7-10. PARRACK, J D (1973) *The naturalist in Majorca.* Newton Abbot. PARROTT, J, PHILLIPS, J, AND WOOD, J (1987) *Dutch Birding* 9, 17-19. PARROT, C (1907) *Zool. Jahrb.* 25, 1-78. PARRY, P E (1948) *Br. Birds* 41, 344-5. PARSLOW, J L F (1973) *Breeding birds of Britain and Ireland.* Berkhamsted. PARSLOW, J L F and CARTER, M J (1965) *Br. Birds* 58, 208-14. PARSONS, J and BAPTISTA, L F (1980) *Auk* 97, 807-15. PARTRIDGE, L and GREEN, P (1987) *Anim. Behav.* 35, 982-90. PASCUAL, J A (1992) Ph D Thesis, Salamanca Univ. PASCUAL, J A, CALVO, J M, LEHMANN, S, and PERIS, S J (1992) *Congr. nac. Iberoamer. Etología* 4, 381-7. Cáceres. PASHLEY, D N and MARTIN, R P (1988) *Amer. Birds* 42, 1164-76. PASPALEVA, M (1965) *Bull. Inst. Zool. Acad. Sci. Bulg.* 19, 33-7. PASSBURG, R E (1959) *Ibis* 101, 153-69. PASTEUR, G (1956) *Bull. Soc. Sci. nat. Phys. Maroc* 36, 165-84. PATEFF, P (1947) *Ibis* 89, 494-507. PATTERSON, I J (1970) *Br. ecol. Soc. Symp.* 10, 249-52; (1975) In Baerends, G, Beer, C, and Manning, A (ed.) *Function and evolution in behaviour,* 169-83. Oxford; (1980a) *Ardea* 68, 53-62; (1980b) *Br. Birds* 73, 359-60. PATTERSON, I J, CAVALLINI, P, and ROLANDO, A (1991) *Ornis scand.* 22, 79-87. PATTERSON, I J, DUNNET, G M, and FORDHAM, R A (1971) *J. appl. Ecol.* 8, 815-33. PATTERSON, I J and GRACE, E S (1984) *J. Anim. Ecol.* 53, 559-72. PÄTZOLD, R (1975) *Falke* 22, 300-4. PAUDTKE, B (1988) *Limicola* 2, 74. PAYN, W A (1926) *Ool. Rec.* 6, 70; (1947) *Bull. Br. Orn. Club* 67, 41-2. PAYN, W H (1982) *Br. Birds* 75, 134. PAYNE, R B (1972) *Condor* 74, 485-6; (1973) *Orn. Monogr. AOU* 11; (1976) *Auk* 93, 25-38; (1980) *Ibis* 122, 43-56; (1983) *Anim. Behav.* 31, 788-805; (1987) *Occas. Papers Mus. Zool. Univ. Michigan* 714; (1991) In Poole, A, Stettenheim, P and Gill, F B (eds) *Birds of North America* 4. Philadelphia. PAYNE, R B, PAYNE, L L, and DOEHLERT, S M (1988) *Ecology* 69, 104-17. PAYNE, R B,

THOMPSON, W L, FIALA, K L, and SWEANY, L L (1981) *Behaviour* 77, 199-221. PAYNTER, R A (1964) *Condor* 66, 277-81; (1968) In Peters, J L *Check-list of birds of the world* 14; (1970) 13. Cambridge, Mass. PAZ, U (1987) *The birds of Israel.* Tel-Aviv. PAZZUCONI, A (1970) *Riv. ital. Orn.* 40, 458-9. PEABODY, P B (1907) *Auk* 24, 271-8. PEACH, W J, GIBSON, T S H, and FOWLER, J A (1987) *Bird Study* 34, 37-8. PEAKALL, D B (1960) *Bird Study* 7, 94-102. PEAKE, E (1929) *Br. Birds* 22, 174-5. PEARCE, A (1978) *Nat. Wales* 16, 142. PEARSE, T (1940) *Condor* 42, 124-5. PEARSON, D L (1980) In Keast, A and Morton, E S (eds) *Migrant birds in the Neotropics: ecology, behavior, distribution and conservation,* 273-83. Washington DC. PECK, K M (1989) *Biol. Cons.* 48, 41-57. PEDERSEN, B B (1980) *Dansk orn. Foren. Tidsskr.* 74, 127-40. PEDROCCHI, C (1979) *Estudio integrado y multidisciplinario de la dehesa salmantina. 1. Estudio fisiográfico descriptivo.* 3, 221-49. PEDROLI, J-C (1967) *Nos Oiseaux* 29, 164. PEIPONEN, V (1957) *Ornis fenn.* 34, 41-64. PEIPONEN, V A (1962) *Ornis fenn.* 39, 37-60; (1967) *Ann. zool. fenn.* 4, 547-59; (1970) *Proc. Helsinki Sympl. UNESCO,* 281-7; (1974) *Ann. zool. fenn.* 11, 155-65. PEITZMEIER, J (1955) *J. Orn.* 96, 347-8. PEK, L V and FEDYANINA, T F (1961) In Yanushevich, A I (ed.) *Ptitsy Kirgizii* 3, 59-118. Frunze. PELLERIN, M (1981) *Compt. Rend. Acad. Sci. Paris* 293, 713-15; (1982) *Behaviour* 81, 287-95; (1983) *Behav. Proc.* 8, 157-63. PEPPER, S R and KENNEDY, H J (1970) *Ornis fenn.* 47, 35-6. PEPPERBERG, I M (1985) *Auk* 102, 854-64. PERDECK, A C (1967) *Bird Study* 14, 129-52; (1970) *Ardea* 58, 142-70. PÉREZ PADRÓN, F (1981) *Vida Silvestre* 40, 258-63. PERIS, S J (1979) Doct. Thesis. Univ. Complutense de Madrid; (1980a) *Ardeola* 25, 207-40; (1980b) *Doñana Acta Vert.* 7, 249-60; (1981) *Stud. Oec.* 2, 155-69; (1983) *J. Orn.* 124, 78-81; (1984a) *Bol. Real Soc. Española Hist. Nat. (Biol.)* 80, 37-46; (1984b) *Ardeola* 31, 3-16; (1984c) *Reunión Iberoamer. Cons. Zool. Vert.* Cáceres 2, 140-52; (1984d) *Salamanca, Revista de Estudios* 11-12, 175-234; (1988) *Gerfaut* 78, 101-112; (1989) *Misc. Zool.* 13, 217-20; (1991) *Ring. Migr.* 12, 124-5. PERIS, S J, MOTIS, A, and MARTINEZ-VILALTA, A (1987) *Acta VIII BRSE Hist. Nat., Pamplona,* 151-6. PERIS, S, MOTIS, A, MARTINEZ-VILALTA, A, and FERRER, X (1991) *J. Orn.* 132, 445-9. PERSSON, C (1942) *Vår Fågelvärld* 1, 27-8. PESCH, A and GÜTTINGER, H-R (1985) *J. Orn.* 126, 108-10. PESENTI, P G (1945) *Riv. ital. Orn.* 15, 89-97. PESKOV, V N (1990) *Vestnik Zool.* (6), 62-7. PETER, H (1968) *Egretta* 11, 58. PETER, H-U and STEIDEL, G (1990) *Acta XX Congr. Int. Orn. Supp. Abstr.* 703, 384. PETERS, H S and BURLEIGH, T D (1951) *The birds of Newfoundland.* St John's. PETERS, J L (1931) *Auk* 48, 575-87. PETERS, J L and GRISCOM, L (1938) *Bull. Mus. comp. Zool.* 80, 445-78. PETERS, R A (ed.) (1968) *Check-list of birds of the world* 14. Cambridge. PETERSEN, Æ (1985) *Bliki* 4, 57-67; (1989) *Bliki* 8, 56-61. PETERSEN, R, MOUNTFORT, G, and HOLLOM, P A D (1983) *A field guide to the birds of Britain and Europe.* London. PETERSON, R T and CHALIF, E L (1973) *A field guide to Mexican birds.* Boston. PETRETTI, F (1979) *Br. Birds* 72, 290. PETRIDES, G A (1943) *Wilson Bull.* 55, 193-4. PETROV, T (1982) *Isv. Mus. Yuzh. Bŭlg.* 8, 21-41; (1988) *Isv. Mus. Yuzh. Bŭlg.* 14, 25-45. PETTERSSON, Å (1977) *Vår Fågelvärld* 36, 161-73. PETTERSSON, M (1959) *Nature* 184, 649-50. PETTITT, R G and BUTT, D V (1949) *Br. Birds* 42, 327. PÉTURSSON, G and ÓLAFSSON, E (1985) *Bliki* 4, 13-39; (1986) *Bliki* 5, 19-46; (1988) *Bliki* 6, 33-68. PÉTURSSON, G, THRÁINSSON, G, and ÓLAFSSON, E (1991) *Bliki*

10, 15–54; (1992) *Bliki* 11, 31–63. PEUS, F (1954) *Bonn. zool. Beitr.* 5 suppl. 1, 1–50. PFEIFFER, S (1928) *Anz. orn. Ges. Bayern* (12), 142–3. PFEIFER, S (1953) *Vogelwelt* 74, 216–17; (1954) *J. Orn.* 95, 185–6; (1956) *Orn. Mitt.* 8, 148–9; (1966) *Orn. Mitt.* 18, 28–30; (1974) *Orn. Mitt.* 26, 159–60. PFEIFER, S and KEIL, W (1956) *Nachrichtenblatt des Deutschen Pflanzenschutzdienstes* 8 (9), 129–31. PFENNIG, H G (1986) *Charadrius* 22, 221–6; (1988) *Charadrius* 24, 88–91. PFIRTER, A (1975) *Vögel Heimat* 45, 128. PFLUMM, W (1978) *J. Orn.* 119, 308–24; (1984) *J. Orn.* 125, 481–2. PFORR, M and LIMBRUNNER, A (1982) *The breeding birds of Europe* 2. London. PFÜTZNER, W (1988) *Abh. Ber. Naturkundemus: Görlitz* 62 (2). PHILIPSON, W R (1933) *Br. Birds* 27, 66–71; (1939) *Br. Birds* 32, 272. PHILLIPS, A R (1966) *Bull. Br. Orn. Club* 86, 148–59; (1974) *Abstr. Int. Orn. Congr.* 16, 68–9; (1977) *Bird-Banding* 48, 110–7; (1991) *The known birds of North and Middle America*. Part II. Denver. PHILLIPS, A R, MARSHALL, J, and MONSON, G (1964) *The birds of Arizona*. Tucson. PHILLIPS, J C (1915) *Auk* 32, 273–89. PHILLIPS, N J and ROUND, P D (eds) (1975) *Crete Ringing Group Rep. 1973–5*. Aberdeen Univ. PHILLIPS, N R (1982) *Sandgrouse* 4, 37–59. PHILLIPS, W W A (1953) *Ibis* 95, 548–9; (1978) *Annotated checklist of the birds of Ceylon (Sri Lanka)*. Colombo. PICOZZI, N (1975a) *Br. Birds* 68, 409–19; (1975b) *J. Wildl. Mgmt.* 39, 151–5; (1976) *Ibis* 118, 254–7; (1982) *Scott. Birds* 12, 23–4. PIECHOCKI, R (1954) *J. Orn.* 95, 297–305; (1956) *Falke* 3, 10–17; (1959) *Zool. Abh. Staatl. Mus. Tierkde. Dresden* 24, 105–203; (1971a) *Falke* 18, 4–26, 40–57; (1971b) *Falke* 18, 94–100. PIECHOCKI, R and BOLOD, A (1972) *Mitt. zool. Mus. Berlin* 48, 41–175. PIECHOCKI, R, STUBBE, M, UHLENHAUT, K, and SUMJAA, D (1982) *Mitt. zool. Mus. Berlin* 58, Suppl. *Ann. Orn.* 6, 3–53. PIELOWSKI, Z (1963) *Przegląd Zool.* 7, 71–8. PIELOWSKI, Z and PINOWSKI, J (1962) *Bird Study* 9, 116–22. PIERSMA, T and BLOKSMA, N (1987) *Bird Study* 34, 127–8. PIESKER, O (1972) *Beitr. Vogelkde.* 18, 452. PIKULA, J (1973) *Zool. Listy* 22, 155–64; (1989) *Folia Zool.* 38, 167–82. PIKULA, J and FOLK, C (1970) *Zool. Listy* 19, 261–73. PILCHER, C W T (1986) *Sandgrouse* 8, 102–6. PINDER, J M (1991) *Br. Birds* 84, 198. PINEAU, J and GIRAUD-AUDINE, M (1975) *Alauda* 43, 135–41; (1979) *Trav. Inst. Sci. Rabat Sér. Zool.* 38. PINOWSKA, B and PINOWSKI, J (1977) *Int. Stud. Sparrows* 10, 26–41. PINOWSKI, J (1959) *Ekol. Polska A* 7, 435–82; (1965a) *Bird Study* 12, 27–33; (1965b) *Bull. Acad. pol. Sci.* 13, 509–14; (1966) *Ekol. Polska A* 14, 145–72; (1967a) *Ardea* 55, 241–8; (1967b) *Ekol. Polska A* 15, 1–30; (1968) *Ekol. Polska A* 16, 1–58. PINOWSKI, J and KENDEIGH, S C (eds) (1977) *Granivorous birds in ecosystems*. Cambridge. PINOWSKI, J and PINOWSKA, B (1985) *Ring* 124-5, 51–6. PINOWSKI, J, TOMEK, T, and TOMEK, W (1973) In Kendeigh, S C and Pinowski, J (eds) *Productivity, population dynamics and systematics of granivorous birds*, 263–73, Warsaw. PINOWSKI, J and WOJCIK, Z (1969) *Falke* 16, 256–61. PINXTEN, R, EENS, M, VAN ELSACKER, L, and VERHEYEN, R F (1989b) *Bird Study* 36, 45–8. PINXTEN, R, EENS, M, and VERHEYEN, R F (1989a) *Behaviour* 111, 234–56; (1990) *J. Orn.* 131, 141–50; (1991) *Ardea* 79, 15–30. PINXTEN, R, VAN ELSACKER, L, and VERHEYEN, R F (1987) *Ardea* 75, 263–9. PIOTROWSKI, S H (ed.) (1991) *Suffolk Birds* 40, 45–136. PIPER, W H and WILEY, R H (1989) *J. Fld. Orn.* 60, 73–83; (1991) *J. Fld. Orn.* 62, 40–5. PITMAN, C R S (1921) *Ool. Rec.* 1, 15–38, 73–91; (1922) *Ool. Rec.* 2, 49–57; (1961) *Bull. Br. Orn. Club* 81, 148–9. PITT, F (1918) *Br. Birds* 12, 122–31.

PITTAWAY, R (1989) *Ontario Birds* 7, 65–7. PIVAR, G (1965) *Larus* 16-18 (1962–64), 159–280. PLATH, L (1983) *Beitr. Vogelkde.* 29, 53–4; (1984) *Beitr. Vogelkde.* 30, 215; (1987) *Orn. Mitt.* 39, 16; (1988a) *Falke* 35, 27–8; (1988b) *Falke* 35, 358–61; (1989) *Falke* 36, 143–7. PLAYFORD, P F J (1985) *Gwent Bird Rep.* 20, 26. PLESKE, T (1928) *Birds of the Eurasian trundra*. Boston. PLUCINSKI, A (1970) *Orn. Mitt.* 22, 3–4. POCHELON, G (1992) *Alauda* 60, 148. PODARUEVA, V I (1979) In Labutin, Y V (ed.) *Migratsii i ekologiya ptits Sibiri*, 170–2. Yakutsk. PODMORE, R E (1948) *Br. Birds* 41, 272. PODOL'SKI, A L and SADYKOV, O F (1983) *Ornitologiya* 18, 176–8. POHL, H (1971a) *Ibis* 113, 185–93; (1971b) *J. Orn.* 112, 266–78; (1972) *Naturwissenschaften* 59, 518; (1974) *Naturwissenschaften* 61, 406; (1980) *Physiol. Zool.* 53, 186–209; (1989) In Mercer, J B (ed.) *Thermal Physiology*, 713–18. New York. POKORNY, F and PIKULA, J (1986–7) *J. World Pheasant Assoc.* 12, 75–80. POKROVSKAYA, I V (1956) *Zool. Zh.* 35, 96–110; (1963) *Uch. Zap. Leningr. gos. ped. Inst.* 230, 93–102. POLATZEK, J (1909) *Orn. Jahrb.* 20, 1–24. POLIVANOV, V M and POLIVANOVA, N N (1986) *Trudy Teberdinsk. gos. zapoved.* 10, 11–164. POLIVANOVA, N N and POLIVANOV, V M (1977) *Ornitologiya* 13, 82–90. POLOZOV, S A (1990) *Ornitologiya* 24, 132–3. PONOMAREVA, T S (1974) *Ornitologiya* 11, 404–7; (1981) *Ornitologiya* 16, 16–21; (1983) *Ornitologiya* 18, 57–63. POPOV, A V (1959) *Ptitsy Gissaro-Karategina*. Dushanbe. POPOV, V V (1978) *Ptitsy Volzhsko-Kamskogo kraya*. Moscow. POPPE, D (1976) *Wielewaal* 42, 345. PORTENKO, L A (1929) *Ezhegodnik Zool. Muz. Akad. Nauk SSSR* 29, 37–81; (1930) *Auk* 47, 205–7; (1937) *Fauna ptits vnepolyarnoy chasti severnogo Urala*. Moscow; (1939) *Trudy Nauch.-issled. inst. polyarn. zemledeliya zhivotnovod. i promysl. khoz.* 5, 5–211; (1954) *Ptitsy SSSR* 3; Moscow; (1973) *Ptitsy Chukotskogo poluostrova i ostrova Vrangelya* 2. Leningrad. PORTENKO, L A and STÜBS, J (1971) In Stresemann, E, Portenko, L A, and Mauersberger, G (eds) *Atlas der Verbreitung palaearktischer Vögel* 3. Berlin; (1976) In Dathe, H (ed.) *Atlas der Verbreitung palaearktischer Vögel* 5. Berlin. PORTENKO, L A and VIETINGHOFF-SCHEEL, E VON (1974) In Stresemann, E, Portenko, L A, and Mauersberger, G (eds) *Atlas der Verbreitung palaearktischer Vögel* 4. Berlin. PORTER, R F (1983) *Sandgrouse* 5, 45–74. PORTER, S (1941) *Avic. Mag.* (5) 6, 3–8. POSLAVSKI, A N (1974) *Ornitologiya* 11, 238–52. POTAPOV, R L (1966) *Trudy Zool. Inst. Akad. Nauk SSSR* 39, 3–119; (1986) *Trudy Zool. Inst. Akad. Nauk SSSR* 150, 57–73. POTTER, E F (1970) *Auk* 87, 692–713. PARNELL, J F, and TEULINGS, R P (1980) *Birds of the Carolinas*. Chapel Hill. POTTI, J and TELLERÍA, J L (1984) *Stud. Oecol.* 5, 247–58. POUGH, R H (1949) *Audubon bird guide: small land birds*. New York; (1957) *Audubon western bird guide*. New York. POULSEN, H (1949) *Dansk orn. Foren. Tidsskr.* 43, 256–7; (1950) *Dansk orn. Foren. Tidsskr.* 44, 96–9; (1954) *Dansk orn. Foren. Tidsskr.* 48, 32–7; (1956) *Dansk orn. Foren. Tidsskr.* 50, 267–98; (1958) *Dansk orn. Foren. Tidsskr.* 52, 89–105; W SOOO6 T(1959) *Z. Tierpsychol.* 16, 173–8. POUNDS, H E (1950) *Br. Birds* 43, 333; (1972) *Br. Birds* 65, 34. POWER, H W, LITOVICH, E, and LOMBARDO, M P (1981) *Auk* 98, 386–9. POWER, H W, KENNEDY, E D, ROMAGNANO, L C, LOMBARDO, M P, HOFFENBERG, A S, STOUFFER, P C, and McGUIRE, T C (1989) *Condor* 91, 753–65. PRATO, E S DA, and PRATO, S R D DA (1978) *Ring. Migr.* 2, 48–9. PRATT, H D, BRUNER, P L, and BERRETT, D G (1987) *The birds of Hawaii and the tropical Pacific*. Princeton. PRAVOSUDOV, V V (1984) *Zool.*

Zh. 63, 950-3. Praz, J-C (1971) Nos Oiseaux 31, 11-13. Praz, J-C and Oggier, P-A (1973) Nos Oiseaux 32, 109-12; (1976) Bull. Murithienne 93, 29-40. Preble, E A (1908) North American Fauna 27. Preiser, F (1957) Diss. Landwirtschaftliche Hochschule Hohenheim. Preiser, E and Massoth, K H (1990) Voliere 13, 68-71. Preiss, F (1974) Gef. Welt 98, 121-2. Prendergast, E D V (1985) Adjutant 15, 17-18. Prendergast, E D V and Boys, J V (1983) The birds of Dorset. Newton Abbot. Prescott, D R C (1991) Condor 93, 694-700. Prescott, K W (1978) Inland Bird-Banding News 50, 163-83; (1981) Inland Bird-Banding 53, 39-48; (1986) N. American Bird Bander 11, 46-51. Preuss, D (1991a) Gef. Welt 115, 192-6; (1991b) Gef. Welt 115, 340-41. Pricam, R (1957) Nos Oiseaux 24, 160-3. Price, T D (1979) J. Bombay Nat. Hist. Soc. 76, 379-422. Priest, C D (1936) The birds of southern Rhodesia 4. London. Priestley, C F (1947) Br. Birds 40, 176. Prigann, I (1992) Kart. Med. Brutvögel 7, 11-12. Prill, H (1974) Orn. Rundbrief Mecklenburgs N.F. 15, 56-9; (1975) Falke 22, 92-4. Priolo, A (1988) Riv. ital. Orn. 58, 105-24. Prinzinger, R, and Hund, K (1975) Anz. Orn. Ges. Bayern 14, 70-8; (1981) Ökol. Vögel 3, 249-59. Prokofiev, M A (1962) Ornitologiya 4, 333-5. Prokofieva, I V (1963a) Uch. Zap. Leningr. gos. ped. Inst. 230, 57-69; (1963b) Uch. Zap. Leningr. gos. ped. Inst. 230, 71-86. Promptov, A N (1934) Zool. Zh. 13 (3), 523-39. Prozesky, O P M (1974) A field guide to the birds of southern Africa. London. Prŷs-Jones, R P (1976) Bird Study 23, 294; (1977) D Phil Thesis. Oxford Univ; (1984) Gerfaut 74, 15-37. Przygodda, W (1960) Orn. Mitt. 12, 21-5; (1969) Bonn. zool. Beitr. 20, 69-74. Ptushenko, E S and Inozemtsev, A A (1968) Biologiya i khozyaystvennoe znachenie ptits Moskovskoy oblasti i sopredel'nykh territoriy. Moscow. Pukinski, Y B (1969) Vopr. ekol. biotsenol. 9, 62-78. Pulliainen, E (1971) Ann. zool. Fenn. 8, 326-9; (1972) Ann. zool. Fenn. 9, 28-31; (1974) Ann. zool. Fenn. 11, 204-6; (1979a) Aquilo Ser. Zool. 19, 87-96; (1979b) Ornis fenn. 56, 156-62. Pulliainen, E and Hakanen, R (1972) Ornis fenn. 49, 86-90. Pulliainen, E, Kallio, T, and Hallaksela, A-M (1978) Aquilo Ser. Zool. 18, 23-7. Pulliainen, E and Peiponen, V (1981) Ornis fenn. 58, 109-16. Pulliainen, E and Saari, L (1976) Ornis fenn. 53, 46; (1989) Ornis fenn. 66, 161-5. Pulliainen, E and Tuomainen, J (1978) Ornis fenn. 55, 180-2. Pulman, C B (1978) Br. Birds 71, 363. Purchas, T P G (1975) Proc. New Zealand ecol. Soc. 22, 111-12; (1979) New Zealand J. Zool. 6, 321-7; (1980) New Zealand J. Zool. 7, 557-78. Purrmann, P (1973) Falke 20, 102. Pyle, P, Howell, S N G, Yunick, R P, and DeSante, D F (1987) Identification guide to North American passerines. Bolinas.

Quay, W B (1989) Condor 91, 660-70. Quépat, N (1875) Monographie du Cini. Paris. Quilliam, H R (1973) History of the birds of Kingston, Ontario. Kingston.

Rabøl, J (1969) Feltornithologen 11, 123-31. Raboud, C (1988) Orn. Beob. 85, 385-92. Rademacher, B (1951) Z. Pflanzenkrankheiten u. Pflanzenschutz. 58, 416-26. Rademacher, W (1974) Orn. Mitt. 26, 182; (1977) Charadrius 13, 75-8; (1983) Orn. Mitt. 35, 107; (1984) Charadrius 20, 183-5. Radford, A P (1966) Br. Birds 59, 201; (1970a) Br. Birds 63, 138; (1970b) Br. Birds 63, 428-9; (1971) Br. Birds 64, 233-5; (1974) Br. Birds 67, 440; (1983) Br. Birds 76, 580; (1985) Br. Birds 78, 513-14; (1991) Br. Birds 84, 153. Radtke, G A (1986) Voliere 9, 152-4. Radzhabli, S I

and Panov, E N (1972) In Vorontsov, N N (ed.) Problemy Evolyutsii 2, 255-62. Novosibirsk. Radzhabli, S I, Panov, E N, and Bulatova, N S (1970) Zool. Zh. 49, 1857-63. Rae, R and Marquiss, M (1989) Ring. Migr. 10, 133-40. Raevel, P (1981a) Héron (3), 63; (1981b) Héron (4), 1-10; (1983) Héron (4), 105-7. Raevel, P and Deroo, S (1980) Héron (4), 118. Raevel, P and Roussel, F (1983) Héron (4), 104-5. Raffaele, H A (1989) A guide to the birds of Puerto Rico and the Virgin Islands. Princeton. Rahman, M K and Husain, K Z (1988) Bangladesh J. Zool. 16, 155-7. Rahmani, A R and D'Silva, C (1986) J. Bombay nat. Hist. Soc. 82, 657. Raines, R J (1962) Ibis 104, 490-502. Rait Kerr, H (1950) Field 196, 302. Ralph, C J and Pearson, C A (1971) Condor 73, 77-80. Rammner, C (1977) Falke 24, 62-7. Ramos, M A (1988) Proc. int. orn. Congr. 19, 251-93. Rand, A L (1944) Can. Fld.-Nat. 58, 111-25; (1946) Nat. Mus. Canada Bull. 105; (1961a) Wilson Bull. 73, 46-56; (1961b) Bird-Banding 32, 71-9. Rand, A L and Vaurie, C (1955) Bull. Br. Orn. Club 75, 28. Rappe, A (1965) Gerfaut 55, 4-15. Rappe, A and Herroelen, P (1964) Gerfaut 54, 3-11. Rappole, J H (1983) J. Fld. Orn. 54, 152-9. Rappole, J H, Morton, E S, Lovejoy, T E, and Ruos, J L (1983) Nearctic avian migrants in the Neotropics. US Dept. Interior, Washington DC. Rappole, J H, Ramos, M A, Oehlenschlager, R J, Warner, D W, and Barkan, C P (1979) In Drawe, D L (ed.) Proc. First Welder Wildl. Found. Symposium, 199-214. Sinton. Rappole, J H and Warner, D W (1980) In Keast, A and Morton, E S (eds) Migrant birds in the Neotropics: ecology, behavior, distribution, and conservation, 353-93. Washington DC. Rashkevich, N A (1965) Zool. Zh. 44, 1532-7. Raspail, X (1901) Bull. Soc. zool. France 26, 104-9. Ratcliffe, D A (1962) Ibis 104, 13-39. Raveling, D G (1965) Bird-Banding 36, 89-101. Raveling, D G and Warner, D W (1965) Bird-Banding 36, 169-79. Ravkin, Y S (1973) Ptitsy severo-vostochnogo Altaya. Novosibirsk; (1984) Prostranstvennaya organizatsiya naseleniya ptits lesnoy zony. Novosibirsk. Raw, W (1921) Ibis (11) 4, 238-64. Ray, K A (1965) Br. Birds 58, 439. Raymond, T (1988) Bird Obs. Mass. 16, 270. Raynor, J H (1948) Br. Birds 41, 342. Rea, A M (1970) Condor 72, 230-3. Rebecca, G W (1985) Scott. Birds 13, 188; (1986) Scott. Bird News 4, 9. Reber, U (1986) Gef. Welt 110, 153; (1988) Gef. Welt 112, 324. Reboussin, R (1931) Oiseau 1, 283-8. Redman, N (1993) Br. Birds 86, 131-3. Redondo, T (1991) J. Orn. 132, 145-63. Redondo, T and Arias de Reyna, L (1988) Anim. Behav. 36, 653-61. Redondo, T, Arias de Reyna, L, Gonzalez-Arenas, J, Recuerda, P, and Zuniga, J M (1988) Misc. Zool. 10, 287-97. Redondo, T and Carranza, J (1989) Behav. Ecol. Sociobiol. 25, 369-78. Redondo, T and Exposito, F (1990) Ethology 84, 307-18. Redondo, T, Hidalgo de Trucios, S J, and Medina, R (1989) Etología 1, 19-31. Ree, L van and Berg, A B van den (1987) Dutch Birding 9, 108-13. Ree, V (1976) Sterna 15, 179-197; (1977) Sterna 16, 113-202. Reeb, F (1977) Alauda 45, 293-33. Reed, T M (1982) Anim. Behav. 30, 171-81. Reichenow, A (1916) Orn. Monatsber. 24, 154-5. Reid, J C (1979) Bonn. zool. Beitr. 30, 357-66. Reinikainen, A (1937) Ornis fenn. 14, 55-64; (1939) Ornis fenn. 16, 73-95. Reinsch, A (1977) Orn. Mitt. 29, 190. Reinsch, H H (1960) Beitr. Vogelkde. 7, 153-4; (1965) Beitr. Vogelkde. 10, 323-5; (1967) Beitr. Vogelkde. 12, 293-4. Reiser, O (1905) Materialien zu einer Ornis Balcanica 3; (1939) I. Vienna. Reiser, O and Führer, L von (1896) Materialien zu einer Ornis Balcanica 4. Vienna. Rékási, J (1976) Int.

Stud. Sparrows 9, 72-82. RÉKÁSI, J and STERBETZ, I (1975) Aquila 80-1, 215-20. RENDAHL, H (1958) Vogelwarte 19, 199-203; (1959) Ark. Zool. 12, 303-12; (1964) Vogelwarte 22, 229-35; (1968) Ark. Zool. 22, 225-78. RENDELL, L (1947) Br. Birds 40, 49. RENGLIN, P (1975) Vår Fågelvärld 34, 59. RENSSEN, T A (1988) Limosa 61, 137-44; (1991a) Limosa 64, 30; (1991b) In Glandt (1991), 61-6. RESTALL, R L (1975a) Avic. Mag. 81, 107-13; (1975b) Finches and other seed-eating birds. London. RETTIG, K (1985) Beitr. Naturk. Niedersachs. 38, 222-3. RETZ, M (1966a) Diss. Zool. Forschungsinst., Mus. Alexander Koenig (not a Ph. D.); (1966b) Auspicium 2, 231-47; (1968) Auspicium 2, 412-46. REY, E (1907a) Orn. Monatsschr. 32, 205-18; (1907b) Orn. Monatsschr. 32, 235-46; (1908) Orn. Monatsschr. 33, 221-31; (1910) Orn. Monatsschr. 35, 305-13. REYMERS, N F (1954) Zool. Zh. 33, 1358-62; (1959) Zool. Zh. 38, 907-15; (1966) Ptitsy i mlekopitayushchie yuzhnoy taygi sredney Sibiri. Moscow. REYNOLDS, A (1974) Rye Meads Ring. Gr. Rep. 7, 44-52. RHEINWALD, G (1973) Charadrius 9, 58-64; (1982) Brutvogelatlas der Bundesrepublik Deutschland. Bonn; (1992) Die Vögel von Deutschland: Artenliste. Dachverband Deutscher Avifaunisten. RIBAUT, J-P (1954) Nos Oiseaux 22, 225-8. RICHARD, A (1928) Nos Oiseaux 9, 97-104. RICHARDS, C E (1981) Br. Birds 74, 187-8. RICHARDS, D B and THOMPSON, N S (1978) Behaviour 64, 184-203. RICHARDS, M L (1973) Br. Birds 66, 365-6. RICHARDS, P R (1974) Br. Birds 67, 215; (1976a) Bird Study 23, 207-11; (1976b) Bird Study 23, 212. RICHARDSON, C (1989) Oman Bird News 5, 6-7; (1990) The birds of the United Arab Emirates. Warrington. RICHARDSON, S C, PATTERSON, I J, and DUNNET, G M (1979) J. Anim. Ecol. 48, 103-10. RICHFORD, A S (1978) D. Phil. Thesis. Oxford Univ. RICHMOND, C W (1895a) Proc. US. natn. Mus. 18, 451-503; (1895b) Proc. US natn. Mus. 18, 569-91. RICHMOND, W K (1963) Country Life, 3 Jan., 14-15. RICHNER, H (1989a) J. Anim. Ecol. 58, 427-40; (1989b) Anim. Behav. 38, 606-12; (1990) Ibis 132, 105-8. RICHNER, H and MARCLAY, C (1991) Anim. Behav. 41, 433-38. RICHNER, H, SCHNEITER, P, and STIRNIMANN, H (1989) Funct. Ecol. 3, 617-24. RICHTER, H (1958) Abh. Ber. Staatl. Mus. Tierkde. Dresden 23, 219-40. RICHTERS, W (1952) Orn. Mitt. 4, 193-9. RICKLEFS, R E (1979) Auk 96, 10-30; (1984) Auk 101, 319-33. RICKLEFS, R E and PETERS, S (1979) Bird-Banding 50, 338-48. RICKLEFS, R E and HUSSELL, D J T (1984) Ornis scand. 15, 155-61. RIDDIFORD, N and BROOME, T (1983) Br. Birds 76, 174-82. RIDDIFORD, N, HARVEY, P V, and SHEPHERD, K B (1989) Br. Birds 82, 603-12. RIDDOCH, J (1986) Scott. Bird News 3, 2. RIDGELY, R S (1981) A guide to the birds of Panama. Princeton. RIDGELY, R S and TUDOR, G (1989) The birds of South America 1. Oxford. RIDGWAY, R (1869) Proc. Acad. Nat. Sci. Philadelphia 2, 129-33; (1901-11) Bull. US natn. Mus. 50 (1-5). RIGGENBACH, H E (1970) Orn. Beob. 67, 255-69; (1979) Orn. Beob. 76, 153-68. RIMMER, C C (1986) J. Fld. Orn. 57, 114-25; (1988) Condor 90, 141-56. RINGLEBEN, H (1960) Vogelwelt 81, 146-51; (1981) Vogelk. Ber. Niedersachs. 13, 73-8. RINNE, U and BAUCH, J (1970) Luscinia 41, 16-20. RINNHOFER, G (1965) Beitr. Vogelkde. 11, 118-19; (1968) Falke 15, 312-14; (1969) Beitr. Vogelkde. 14, 324-9; (1972) Beitr. Vogelkde. 18, 401-16; (1976) Falke 23, 20-1. RION, P (1990) Aves 27, 119-28. RIS, H (1957) Orn. Beob. 54, 195-7. RISBERG, E L (1968) Vår Fågelvärld 27, 173-4; (1970) Vår Fågelvärld 29, 77-89. RISBERG, L and RISBERG, B (1975) Vår Fågelvärld 34, 139-51. RISING, J D (1969) Comp. Biochem. Physiol. 31,

915-25; (1970a) System. Zool. 19, 315-51; (1973) In Kendeigh, S C and Pinowski, J (eds) Productivity, population dynamics, and systematics of granivorous birds, 327-35. Warsaw; (1973) Can. J. Zool. 51, 1267-73; (1983) Auk 100, 885-97; (1987) Evolution 41, 514-24; (1988) Wilson Bull. 100, 183-203. RISING, J D and SHIELDS, G F (1980) Evolution 34, 654-62. RISTOW, D, WINK, C, and WINK, M (1986) Ric. Biol. Selvaggina Suppl. 10, 285-95. RITTINGHAUS, H (1957) Vogelwarte 19, 90-7. ROBBINS, C A (1932) Auk 49, 159-65. ROBBINS, C S, BRUUN, B, and ZIM, H S (1983) Birds of North America. New York. ROBEL, D (1983) Beitr. Vogelkde. 29, 326-7. ROBERT, G (1975) Gerfaut 65, 168-9; (1977) Gerfaut 67, 101-31. ROBERT, J-C (1979) Héron (4), 75-7. ROBERTS, E L (1955) Br. Birds 48, 91. ROBERTS, J L (1979) Country Life 4 Jan., 32-3. ROBERTS, P (1982) Bird Study 29, 155-61; (1983) Bird Study 30, 67-72; (1985) Br. Birds 78, 217-32. ROBERTS, P J (1989) In Bignal and Curtis (eds), 9-11; (1982) Br. Birds 75, 38-40. ROBERTS, S and LEWIS, J (1988) Gwent Bird Rep. 23 (1987), 7-10. ROBERTS, T J (1992) The birds of Pakistan 2. Karachi. ROBERTSON, A W P (1953) Br. Birds 46, 380-1; (1954) Bird pageant. London. ROBERTSON, C J R (ed.) (1985) Reader's Digest complete book of New Zealand Birds. Sydney. ROBERTSON, H A, WHITAKER, A H, and FITZGERALD, B M (1983) New Zeal. J. Zool. 10, 87-97. ROBERTSON, I S (1975a) Br. Birds 68, 115; (1975b) Br. Birds 68, 453-5. ROER, F (1979) Prachtfinken. Berlin. ROBINSON, J C (1990) Passenger Pigeon 52, 113-18. ROCHE, J (1958) Inst. Rech. Sahar. Univ. Alger 1958, 151-65. RODE, M and LUTZ, K (1991) Corax 14, 95-109. RODEBRAND, S (1975) Calidris 4, 3-12. RÖDER, J (1991) Voliere 14, 204-6. RODRÍGUEZ, L G (1972) Ardeola 16, 215-22. RODRÍGUEZ-TEIJEIRO, J D and CORDERO-TAPIA, P J (1983) Doñana Acta Vert. 10, 77-90. RÖELL, A (1978) Behaviour 64, 1-124; (1979) Ardea 67, 123-9. RÖELL, A and BOSSEMA, I (1982) Behav. Ecol. Sciobiol. 11, 1-6. ROFSTAD, G (1986) J. Zool. Lond. (A) 208, 299-323; (1988) Ornis scand. 19, 27-30. ROFSTAD, G and SANDVIK, J (1985) Ornis scand. 16, 38-44. ROGACHEVA, E V (1988) Ptitsy Sredney Sibiri. Moscow. ROGERS, D T (1965) Bird-Banding 36, 115-6. ROGERS, D T, GARCIA, B J, and RÖGEL, B A (1986a) Wilson Bull. 98, 163-7. ROGERS, D T, HICKS, D L, WISCHUSEN, E W, and PARRISH, J R (1982a) J. Fld. Orn. 53, 133-8. ROGERS, D T, JR and ODUM, E P (1964) Auk 81, 505-13; (1966) Wilson Bull. 78, 415-33. ROGERS, M J (1982) Br. Birds 75, 387; (1984a) Br. Birds 77, 120; (1984b) Isles of Scilly Bird Rep. 1983, 11-62. ROGERS, M J and RARITIES COMMITTEE (1979) Br. Birds 72, 503-49; (1982b) Br. Birds 75, 482-533; (1985) Br. Birds 78, 529-89; (1986b) Br. Birds 79, 526-88; (1988) Br. Birds 81, 535-96; (1989) Br. Birds 82, 505-63; (1990) Br. Birds 83, 439-96; (1991) Br. Birds 84, 449-505; (1992) Br. Birds 85, 507-54. ROGERS, T D (1988) A new list of the birds of Masirah island, Sultanate of Oman. Muscat. ROGGEMAN, W (1983) Gerfaut 73, 451-82. ROHNER, C (1980) Orn. Beob. 77, 103-10; (1981) Orn. Beob. 78, 1-11. ROHWER, S (1986) Auk 103, 281-92. ROHWER, S, KLEIN, W P, and HEARD, S (1983) Wilson Bull. 95, 199-208. ROHWER, S and MANNING, J (1990) Condor 92, 125-40. ROHWER, S, THOMPSON, C W, and YOUNG, B E (1992) Condor 94, 297-300. ROI, O LE (1923) J. Orn. 71, 196-252. ROKITANSKY, G (1934) Falco 30, 6-8; (1969) Gef. Welt 93, 68-9. ROKITANSKY, G and SCHIFTER, H (1971) Ann. naturhist. Mus. Wien 75, 495-538. ROLAND, J (1988) Beih. Veröff. Nat. Land. Bad.-Württ. 53, 93-108. ROLANDO, A and ZUNINO, M (1992)

Ornis scand. **23**, 201-2. ROLFE, R L (1965) *Br. Birds* **58**, 150-1. ROLFE, R (1966) *Bird Study* **13**, 221-36. ROLLIN, N (1958) *Br. Birds* **51**, 290-303. ROLLS, J C (1973) *Br. Birds* **66**, 169. ROMER, M L R (1947) *Br. Birds* **40**, 176-7. ROOKE, K B (1950) *Br. Birds* **43**, 114-15. ROOS, G (1975) *Anser* **14**, 237-46; (1978) *Anser* **17**, 69-89; (1984) *Anser Suppl.* **13**; (1985) *Anser* **24**, 1-28; (1991) *Anser* **30**, 229-58. ROOT, T (1988) *Atlas of wintering North American birds.* Chicago. ROOTSMÄE, L (1990) *Soobshch. Pribalt. Kom. Izuch. Migr. Ptits* **23**, 123-8. ROOTSMÄE, L and VEROMAN, H (1967) *Orn. Kogumik* **4**, 177-99. RÖRIG, G (1900) *Arb. biol. Abt. Land.-Forstwirtsch. kaiserl. Gesundheitsamt Berlin* **1**, 285-400. ROSELAAR, C S (1976) *Bull. zool. Mus. Univ. Amsterdam* **5**, 13-18; (1991) Lijst van alle vogelsoorten van de wereld. In Perrins, C M (ed.) *Geillustreerde encyclopedie van de vogels.* Weert. ROSENBERG, E (1953) *Fåglar i Sverige.* Stockholm. ROSENIUS, P (1929) *Svenska fåglar och fågelbon.* Lund. ROSEVEARE, W L (1951) *Br. Birds* **44**, 16-17. RØSKAFT, E (1978) *Vår Fuglefauna* **1**, 207; (1980a) *Fauna norv.* (C) *Cinclus* **3**, 9-15; (1980b) *Fauna norv.* (C) *Cinclus* **3**, 56-9; (1980c) *Vår Fuglefauna* **3**, 5-10; (1981a) *Fauna norv.* (C) *Cinclus* **5**, 5-9; (1981b) *Fauna norv.* (C) *Cinclus* **4**, 76-81; (1983a) *Ornis scand.* **14**, 175-9; (1983b) *Ornis scand.* **14**, 180-7; (1983c) *Fauna norv.* (C) *Cinclus* **6**, 78-80; (1985) *J. Anim. Ecol.* **54**, 255-60; (1987) *Ornis scand.* **18**, 70-1. RØSKAFT, E and ESPMARK, Y (1982) *Ornis scand.* **13**, 38-46; (1984) *Behav. Proc.* **9**, 223-30. RØSKAFT, E, ESPMARK, Y, and JÄRVI, T (1983) *Ornis scand.* **14**, 169-74. ROSS, H A (1980) *Auk* **97**, 721-32. ROSS, W M (1948) *Scott. Nat.* **60**, 147-56. ROST, F (1992) *Anz. Ver. Thür. Orn.* **1**, 41-2. RÖTHING, H (1969) *Vogelwelt* **90**, 146-7. ROTHSCHILD, M (1955) *Proc. int. orn. Congr.* **11** (1954), 611-617; (1957) *Nos Oiseaux* **24**, 1-6; (1960) *Entomologist*, July 1960, 139-40. ROTHSCHILD, W and HARTERT, E (1911) *Novit. Zool.* **18**, 456-550. ROTHSTEIN, S I (1978) *Auk* **95**, 152-60. ROTHSTEIN, S I, VERNER, J, and STEVENS, E (1980) *Auk* **97**, 253-67. RÖTTLER, G (1985a) *Orn. Mitt.* **37**, 35-6; (1985b) *Orn. Mitt.* **37**, 243-4. ROUX, P (1990) *Oiseau* **60**, 16-38. ROUX, P, CHAKIR, N, and LESNE, L (1990) *Bièvre* **11**, 13-20. RØV, N (1975) *Ornis scand.* **6**, 1-14. ROWELL, C H F (1957) *Bird Study* **4**, 33-50. ROWORTH, P C (1983) *Br. Birds* **76**, 351. ROYALL, W C, JR, GUARINO, J L, DE GRAZIO, J W, and GAMMELL, A (1971) *Condor* **73**, 100-6. RUCNER, D (1973a) *Larus* **24**, 168-70. RUCNER, R (1973b) *Larus* **25**, 27-45. RUDAT, V (1976) *Falke* **23**, 316-17; (1984) *Orn. Jber. Mus. Hein.* **8-9**, 77-85. RUDAT, V and RUDAT, W (1971) *Falke* **18**, 387-9; (1978) *Zool. Jb. Syst.* **105**, 386-98. RUDEBECK, G (1950) *Vår Fågelvärld. Suppl.* **1**. RUELLE, M (1987) *Ornithologue* **55**, 46-57. RUELLE, P J and SEMAILLE, R (1982) *Malimbus* **4**, 27-32. RUF, M (1977) *Vögel Heimat* **47**, 146. RUFINO, R (1989) *Atlas das aves que nidificam em Portugal continental.* Lisbon. RUGE, K (1988) *Beih. Veröff. Nat. Land. Bad.-Württ.* **53**, 21-30. RUNDE, O J (1984) *Sterna* **17**, 129-55. RÜPPELL, G (1970) *Publ. Wiss. Sekt. Biol. E 1508/1969.* Göttingen. RÜPPELL, W (1944) *J. Orn.* **92**, 106-32. RUPPRECHT, A L (1990) *Zool. Abh. Staatl. Mus. Tierkde. Dresden* **45**, 151-4. RUSSELL, D N (1971) *Condor* **73**, 369-72. RUSSELL, S M (1964) *Orn. Monogr. AOU* **1**. RUSTAMOV, A K (1954) *Ptitsy pustyny Kara-Kum.* Ashkhabad; (1958) *Ptitsy Turkmenistana* **2**. Ashkhabad; (1977) In Yurlov, K T (ed.) *Migratsii ptits v Azii*, 202-4. Novosibirsk; (1984) In Konstantinov, V M (ed.) *Ekologiya, biotsenoticheskoe i khozyaystvennoe znachenie vranovykh ptits*, 114-15. Moscow. RUSTAMOW, A K and SOPYEW,

O (1990) *Falke* **37**, 12-15. RUTE, J (1984) *Loodusevaatlusi 1981* (1), 89-94. RUTE, J J and BAUMANIS, J A (1986) In Sokolov, V E and Dobrynina, I N (eds) *Kol'tsevanie i mechenie ptits v SSSR, 1979-82*, 23-9. Moscow. RUTHENBERG, H (1968) *Falke* **15**, 406-13. RUTHKE, P (1939a) *Beitr. Fortpfl. Vögel* **15**, 41-50; (1939b) *Orn. Monatsber.* **47**, 181-2; (1971) *Vogelwelt* **92**, 191. RUTSCHKE, E (ed.) (1983) *Die Vogelwelt Brandenburgs.* Jena. RUTTLEDGE, R F (1954) *Br. Birds* **47**, 447; (1966) *Ireland's birds.* London; (1975) *A list of the birds of Ireland.* Dublin. RUTTLEDGE, W (1965) *Br. Birds* **58**, 442-3. RYALL, C (1986) *New Scientist* **1528**, 48-9; (1990) *Scopus* **14**, 14-16. RYALL, C and REID, C (1987) *Swara* **10** (1), 9-11. RYDZEWSKI, W (1960) *Proc. int. orn. Congr.* **12** (2), 641-4; (1978) *Ring* **96-7**, 218-62. RYMKEVICH, T A (1976) *Zool. Zh.* **55**, 1695-1703; (1977) *Ornitologiya* **13**, 67-73; (1979) *Vestnik Leningrad gos. Univ. 3 Biol.* **1**, 37-47; (ed.) (1990) *Lin'ka vorob'inykh ptits Severo-zapada SSSR.* Leningrad. RYVES, B H (1948) *Bird life in Cornwall.* London. RYVES, B H and RYVES, I N M (1934) *Br. Birds* **28**, 2-26. RYZHANOVSKI, V N (1986) *Zool. Zh.* **65**, 1041-50.

SAARELA, S, KLAPPER, B, and HELDMAIER, G (1988) *Abstr. 10th int. Congr. Photobiol.* **40**. Jerusalem. SABEL, K (1963) *Gef. Welt* **87**, 1-4, 41-5; (1965) *Gef. Welt* **89**, 32-4, 49-51; (1983) *Naturgemässe Finkenzucht.* Bassum. SACARRÃO, G F (1968) *Cyanopica* **1**, 37-46. SACARRÃO, G F and SOARES, A A (1975) *Est. Fauna Portug.* **8**, 1-14; (1976) *Arquivos Mus. Bocage* (2) **6**, 1-13. SACHTLEBEN, H (1918) *Arch. Naturgeschichte* **84** (A) 6, 88-153. SADLER, T S and MYRES, M T (1976) *Prov. Mus. Alberta Nat. Hist. Sect. Occ. Pap.* **1**. SADLIK, J and HAFERLAND, H-J (1981) *Orn. Jber. Mus. Hein.* **5-6**, 77-80. SÆTHER, B-E (1982) *Ornis scand.* **13**, 149-63. SÆTHER, B-E and FONSTAD, T (1981) *Anim. Behav.* **29**, 637-9. SAGE, B L (1957) *Br. Birds* **50**, 353; (1960) *Ardea* **48**, 160-78. SAGE, B and WHITTINGTON, P A (1985) *Bird Study* **32**, 77-81. SAGE, B L and NAU, B S (1963) *Trans. Herts. nat. Hist. Soc.* **25**, 226-44. SAGE, B L and VERNON, J D R (1978) *Bird Study* **25**, 64-86. SAGITOV, A K (1962) *Ornitologiya* **4**, 354-66. SAGITOV, A K and BAKAEV, S (1980) *Ornitologiya* **15**, 142-5. SAGITOV, A K, BELYALOVA, L E, and FUNDUKCHIEV, S E (1990) In Kurochkin, E N (ed.) *Sovremennaya ornitologiya*, 86-97. Moscow. SAINI, H K, DHINDSA, M S, and TOOR, H S (1989) *Gerfaut* **79**, 69-79. SAINO, N and MERIGGI, A (1990) *Ethol. Ecol. Evol.* **2**, 205-14. ST LOUIS, V L and BARLOW, J C (1987) *Wilson Bull.* **99**, 628-41; (1988) *Evolution* **42**, 266-76; (1991) *Wilson Bull.* **103**, 1-12. ST QUINTIN, W H (1907) *Avic. Mag.* **5**, 55-6. SAITO, S-I (1983) *Tori* **32**, 13-20. SALATHÉ, T (1979) *Orn. Beob.* **76**, 247-56; (1987) *Ardea* **75**, 221-9. SALATHÉ, T and RAZUMOVSKY, K (1986) *Terre Vie* **41**, 343-53. SALFELD, D (1963) *Br. Birds* **56**, 221; (1969) *Br. Birds* **62**, 238. SALIKHBAEV, K S and BOGDANOV, A N (1967) *Fauna Uzbekskoy SSR* **2**, Ptitsy **4**. Tashkent. SALMANOVA, L M (1986) *Tez. Dokl. 1 S'ezda Vsesoyuz. orn. Obshch. 9 Vsesoyuz. orn. Konf.* **2**, 221-2. SALMEN, H (1982) *Die Ornis Siebenbürgens* **2**. Köln. SALMON, L (1948) *Br. Birds* **41**, 84. SALOMONSEN, F (1928) *Vidensk. medd. Dansk Naturhist. Foren.* **86**, 123-202; (1930a) *Dansk orn. Foren. Tidsskr.* **24**, 9-101; (1930b) *Dansk orn. Foren. Tidsskr.* **24**, 101-4; (1931a) *Ibis* (13) **1**, 57-70; (1931b) *Orn. Monatsber.* **39**, 112-13; (1935) *Zoology of the Faroes* **64** Aves. Copenhagen; (1947a) *Dansk orn. Foren. Tidsskr.* **41**, 136-40; (1947b) *Dansk orn. Foren. Tidsskr.* **41**, 216-21; (1948) *Dansk orn. Foren. Tidsskr.* **42**, 27-8; (1949) *Dansk* *orn. Foren. Tidsskr.* **43**, 1-45; (1950-1) *Grønlands fugle.* Copenhagen; (1959) *Dansk orn. Foren. Tidsskr.* **53**, 31-9; (1972) *Proc. int. orn. Congr.* **15**, 25-77. SALT, G W (1963) *Proc. int. orn. Congr.* **13**, 905-17. SALT, J R (1984) *Alberta Nat.* **14**, 104. SALT, W R and SALT, J R (1976) *The birds of Alberta.* Edmonton. SALVAN, J (1963) *Oiseau* **33**, 161-2; (1967-9) *Oiseau* **37**, 255-84, **38**, 53-85, 127-50, 249-73, **39**, 38-69. SALZMANN, E (1909) *Orn. Monatsschr.* **34**, 357-67, 400-14; (1911) *Orn. Monatsschr.* **36**, 425-9. SAMCHUK, N D (1971) *Vestnik Zool.* (1), 69-73. SÁNCHEZ-AGUADO, F J (1984) *Ardeola* **31**, 33-45; (1985) *Doñana Acta Vert.* **12**, 197-209; (1986) *Ardeola* **33**, 17-33. SANDBERG, P and WALLENGAARD, R (1987) *Vår Fågelvärld* **46**, 130. SANDEN, W VON (1956) *Orn. Mitt.* **8**, 16. SANTOS JÚNIOR, J R DOS (1968) *Cyanopica* **1**, 9-36. SANTOS, T, SUAREZ, F, and TELLÉRIA, J L (1981) *Proc. 7th int. Conf. Bird Census, 5th Meeting EOAC*, 79-88. León. SAPETINA, I M (1962) *Trudy Okskogo gos. Zapoved.* **4**, 327-36. SAPOZHENKOV, Y F (1962) *Ornitologiya* **5**, 177-82. SAPPINGTON, J N (1977) *Wilson Bull.* **89**, 300-9. SAPSFORD, A (1991) *Calf of Man Bird Obs. Rep. 1990*, 42-61. SARUDNY, N (1904) *Orn. Jahrb.* **15**, 108; (1906) *Orn. Monatsber.* **14**, 47-8; (1907) *Orn. Monatsber.* **15**, 61-3. SARUDNY, N and HARMS, M (1912) *J. Orn.* **60**, 592-619; (1914) *Orn. Monatsber.* **22**, 53-4. SAUNDERS, A A (1951) *A guide to bird songs.* Garden City. SAUNDERS, D and SAUNDERS, S (1992) *Br. Birds* **85**, 337-43. SAUNIER, A (1971) *Nos Oiseaux* **31**, 66-7. SAUROLA, P (1977) *Lintumies* **12**, 118-23; (1979) *Lintumies* **14**, 161-7. SAVAGE, E U (1928) *Br. Birds* **22**, 57. SAVIGNI, G and MASSA, R (1983) *Riv. ital. Orn.* **53**, 3-14. SAWLE, C (1988) *Devon Birds* **41**, 76-7. SAXBY, H L (1874) *The birds of Shetland.* Edinburgh. SCHABER, S (1983) *Nos Oiseaux* **37**, 41-2. SCHAETZEN, R de and JACOB, J P (1985) *Aves* **22**, 127-9. SCHÄFER, E (1938) *J. Orn.* **86** suppl; (1991) *Gef. Welt* **115**, 55. SCHÄFER, K J (1968) *Natur Heimat* **28**, 67-72. SCHÄPPER, R (1986) *Orn. Beob.* **83**, 142-5. SCHARLAU, W (1989) *Kart. med. Brutvögel* **3**, 3-23. SCHAUENSEE, R M DE (1966) *The species of birds of South America and their distribution.* Philadelphia; (1984) *The birds of China.* Oxford. SCHAUENSEE, R M DE and PHELPS, W H (1978) *A guide to the birds of Venezuela.* Princeton. SCHEER, G (1952) *Ornis fenn.* **29**, 77-82; (1953) *Anz. orn. Ges. Bayern* **4**, 70-2. SCHEIFLER, H (1968) *Gef. Welt* **92**, 189-90. SCHEKKERMAN, H (1986a) *Dutch Birding* **8**, 89-97; (1986b) *Graspieper* **6**, 110-15; (1989) *Limosa* **62**, 29-34; (1992) *Dutch Birding* **14**, 7-10. SCHELDE, O (1970) *Feltornithologen* **12**, 188. SCHENK, J (1907) *Aquila* **14**, 252-75; (1919) *Aquila* **26**, 129-31; (1929) *Verh. VI int. Orn.-Kongr.*, 250-64; (1934) *Aquila* **38-41**, 121-53. SCHERNER, E R (1968) *Beitr. Naturkde. Niedersachs.* **20**, 120-1; (1969) *Vogelwelt* **90**, 64-5; (1972a) *Vogelwelt* **93**, 41-68; (1972b) *Angew. Orn.* **4**, 35-42; (1972c) *Orn. Mitt.* **24**, 221; (1974) *Vogelwelt* **95**, 41-60; (1979) *Faun. Mitt. Süd. Niedersachsen* **2**, 11-17. SCHIFFERLI, A (1932) *Orn. Beob.* **29**, 66-84; (1953) *Orn. Beob.* **50**, 65-89; (1963) *Proc. int. orn. Congr.* **13**, 468-74; (1992) *Orn. Beob.* **89**, 48-9. SCHIFFERLI, A, GÉROUDET, P, and WINKLER, R (1982) *Verbreitungsatlas der Brutvögel der Schweiz.* Sempach. SCHIFFERLI, A and LANG, E M (1940a) *J. Orn.* **88**, 550-75; (1940b) *Rev. suisse Zool.* **47**, 217-23; (1946) *Orn. Beob.* **43**, 114-17. SCHIFFERLI, L (1977) *Orn. Beob.* **74**, 71-4; (1978) *Orn. Beob.* **75**, 44-7; (1980) *Avocetta* **4**, 49-62; (1981) *Orn. Beob* **78**, 113-15. SCHIFFERLI, L and FUCHS, E (1981) *Orn. Beob.* **78**, 233-43. SCHIFFERLI, L and SCHIFFERLI, A (1980) *Orn. Beob.* **77**, 21-6. SCHIØLER, E L (1922) *Dansk orn. Foren. Tidsskr.* **16**, 1-55. SCHLEGEL, R (1920) *Z. Ool. Orn.*

25, 29-35. SCHLENKER, R (1976) *Vogelwarte* **28**, 313-14. SCHLEUSSNER, G (1990) *J. Orn.* **131**, 151-5. SCHLEUSSNER, G, DITTAMI, J P, and GWINNER, E (1985) *Physiol. Zool.* **58**, 597-604. SCHLINGER, B A and ADLER, G H (1990) *Wilson Bull.* **102**, 545-50. SCHLÖGEL, N (1987) *Beitr. Vogelkde.* **33**, 65-71. SCHLOSS, W (1984) *Auspicium* **7**, 257-75. SCHMID, H (1991) *Orn. Beob.* **88**, 101-9. SCHMID, T (1974) *Nos Oiseaux* **32**, 274. SCHMID, U (1979) *Vogelkdl. Ber. Niedersachs.* **11**, 45-6. SCHMID, U and GATTER, W (1986) *Vogelwarte* **33**, 335-8. SCHMIDT, E (1960a) *Vogelwarte* **20**, 199-205. SCHMIDT, G (1960b) *Orn. Mitt.* **12**, 3-8; (1962) *Orn. Mitt.* **14**, 33; (1964) *Heimat* **71**, 394-6; (1966) *Vogelwelt* **87**, 154-6. SCHMIDT, G A J (1957) *Orn. Mitt.* **9**, 121-6. SCHMIDT, K (1988) *Beih. Veröff. Nat. Land. Bad.-Württ.* **53**, 191-210. SCHMIDT, O (1991) *Orn. Mitt.* **43**, 3-5. SCHMITT, C and STADLER, H (1914) *Orn. Monatsschr.* **39**, 300-1. SCHMITZ, L (1987) *Aves* **24**, 1-18; (1989) *Aves* **26**, 73-87. SCHNEIDER, W (1957) *Beitr. Vogelkde.* **6**, 43-74; (1964) *Beitr. Vogelkde.* **9**, 455; (1972a) *Der Star.* Wittenberg Lutherstadt; (1972b) *Beitr. Vogelkde.* **18**, 310-46; (1982) *Beitr. Vogelkde.* **28**, 207-21; (1984) *Falke* **31**, 42-3. SCHNELL, F H (1950) *Vogelwelt* **71**, 168. SCHNURRE, O (1959) *Bonn. zool. Beitr.* **10**, 343-50. SCHOENENBERGER, A (1972) *Alauda* **40**, 23-36. SCHÖLL, R W (1959) *J. Orn.* **100**, 439-40; (1960) *Anz. orn. Ges. Bayern* **5**, 591-6. SCHOLS, R (1987) *Limosa* **60**, 119-22. SCHÖLZEL, H (1981) *Orn. Mitt.* **33**, 327-8. SCHÖNBECK, H (1956) *Mitt. Landesmus. Joanneum* 556D, 68-82. SCHÖNFELD, M and BRAUER, P (1972) *Hercynia* (NF) **1**, 40-68. SCHÖNWETTER, M (1984) *Handbuch der Oologie* **3**. SCHOOF, E (1988) *Vogelkdl. Hefte Edertal* **14**, 41-2. SCHOPPE, R (1986) *Beitr. Naturkde. Niedersachs.* **39**, 44-52. SCHOT, W E M VAN DER (1989) *Dutch Birding* **11**, 170-2. SCHRANTZ, F G (1943) *Auk* **60**, 367-87. SCHREIBER, M (1987) *J. Orn.* **128**, 388. SCHREIBER, R W and SCHREIBER, E A (1984) *Bull. Br. Orn. Club* **104**, 62-8. SCHROETER, W (1982a) *Orn. Mitt.* **34**, 127-8; (1982b) *Orn. Mitt.* **34**, 166-8; (1982c) *Orn. Mitt.* **34**, 171. SCHUBERT, G and SCHUBERT, M (1982) *Falke* **29**, 366-72. SCHUBERT, M (1982) *Stimmen der Vögel Zentralasiens.* (2 LP discs). Eterna. SCHUBERT, P (1988) *Beitr. Vogelkde.* **34**, 69-84. SCHUBERT, W (1977) *Anz. orn. Ges. Bayern* **16**, 45-57. SCHUMANN, H (1956) *Beitr. Naturkde. Niedersachs.* **9**, 94. SCHUPHAN, I and HESELER, U (1965) *Vogelwarte* **23**, 77-9. SCHUSTER, L (1921-3) *J. Orn.* **69**, 153-200, 535-70; **71**, 287-361; (1926) *Beitr. Fortpfl. Vögel* **2**, 55-8; (1930) *J. Orn.* **78**, 273-301; (1944) *Beitr. Fortpfl. Vögel* **20**, 132-33; (1950) *Vogelwelt* **71**, 9-17. SCHÜZ, E (1932) *Orn. Monatsber.* **40**, 123; (1941) *Vogelzug* **12**, 152-63; (1959) *Die Vogelwelt des südkaspischen Tieflandes.* Stuttgart. SCHWAB, A (1969) *Orn. Beob.* **66**, 230-1. SCHWAB, R G and MARSH, R E (1967) *Bird-Banding* **38**, 143-7. SCHWABL, H and FARNER, D S (1989) *Condor* **91**, 108-12. SCHWABL, H, SCHWABL-BENZINGER, I, GOLDSMITH, A R, and FARNER, D S (1988) *Gen. comp. Endocrinol.* **71**, 398-405. SCHWARTZ, P (1964) *Living Bird* **3**, 169-84. SCHWEIGER, H (1959) *J. Orn.* **100**, 350-1. SCLATER, W L and MOREAU, R E (1933) *Ibis* (13) **3**, 399-440. SCOTT, D A, HAMADANI, H M, and MIRHOSSEYNI, A A (1975) *Birds of Iran.* Tehran. SCOTT, R E (1959) *Br. Birds* **52**, 388; (1961) *Bird Study* **8**, 152-4; (1965a) *Bull. Br. Orn. Club* **85**, 66-7; (1965b) *Vår Fågelvärld* **24**, 156-71. SCOTTER, G W, CARBYN, L N, NEILY, W P, and HENRY, J D (1985) *Sask. nat. Hist. Soc. spec. Publ.* **15**. SEAGO, M J (ed.) (1986) *Trans. Norfolk Norwich Nat. Soc.* **27** (4), 274-303; (1987) *Trans. Norfolk Norwich Nat. Soc.* **27** (6), 430-62; (1988) *Trans. Norfolk Norwich Nat.*

Soc. **28** (2), 118-51; (1991) *Trans. Norfolk Norwich Nat. Soc.* **29** (2), 123-54. SEALY, S G (1971) *Blue Jay* **29**, 12-16; (1979) *Can. J. Zool.* **57**, 1473-8; (1985) *N. Amer. Bird Bander* **10**, 12-17. SEALY, S G, SEXTON, D A, and COLLINS, K M (1980) *Wilson Bull.* **92**, 114-16. SEARCY, W A (1979) *Condor* **81**, 304-5. SEASTEDT, T R (1980) *Condor* **82**, 232-3. SEASTEDT, T R and MACLEAN, S F (1979) *Auk* **96**, 131-42. SEEL, D C (1960) *Br. Birds* **53**, 303-10; (1964) *Bird Study* **11**, 265-71; (1966) *Bird Study* **13**, 207-9; (1968a) *Ibis* **110**, 129-44; (1968b) *Ibis* **110**, 270-82; (1969) *Ibis* **111**, 36-47; (1970) *Ibis* **112**, 1-14; (1976) *Ibis* **118**, 491-536; (1983) *Bangor Occ. Pap.* **15**. Inst. terr. Ecol., Bangor. SEIFERT, S and SCHÖNFUSS, G (1959) *Beitr. Vogelkde.* **6**, 387-95. SEITZ, E (1964) *Orn. Mitt.* **16**, 212. SELANDER, R K (1967) *Syst. Zool.* **16**, 286-7. SELANDER, R K and GILLER, D R (1960) *Condor* **62**, 202-14. SELANDER, R K and JOHNSTON, R F (1967) *Condor* **69**, 217-58. SELL, M (1984) *Charadrius* **20**, 73-7. SELLERS, R M (1986) *Ring. Migr.* **7**, 99-111. SELLEY, E (1976) *Aquila* **82**, 250. SELLIN, D (1987) *Vogelwelt* **108**, 13-27; (1988) *Beitr. Vogelkde.* **34**, 157-76; (1991) In Glandt (1991), 21-6. SELOUS, F C (1907) *Br. Birds* **1**, 48-51. SEMA, A (1978) *Trudy Inst. Zool. Akad. Nauk Kazakh. SSR Ser. Zool.* **38**, 42-57. SEMENOV-TYAN-SHANSKI, O I and GILYAZOV, A S (1991) *Ptitsy Laplandii.* Moscow. SEMPLE, K R (1971) *Avic. Mag.* **77**, 166-7. SENAR, J C (1983) *Anthropos* **26-7**, 84; (1983) *Misc. Zool.* **7**, 224-6; (1984a) In De Haro, A and Espadaler, X (eds) *Processus d'acquisition précoce. Les communications*, 351-5. Barcelona; (1984b) *Condor* **86**, 213-14; (1985) *Misc. Zool.* **9**, 347-60; (1988) *Ring. Migr.* **9**, 91-2; (1989) *Gerfaut* **79**, 185-7. SENAR, J C and BORRAS, A (1985) *BTO News* **141**, 6. SENAR, J C, BURTON, P J K, and METCALFE, N B (1992) *Ornis scand.* **23**, 63-72. SENAR, J C, CAMERINO, M, and METCALFE, N B (1989) *Behav. Ecol. Sociobiol.* **25**, 141-5. SENAR, J C and COPETE, J L (1990a) *Butll. Gr. Cat. Anell.* **7**, 11-12; (1990b) *Bird Study* **37**, 40-3. SENAR, J C, COPETE, J L, and METCALFE, N B (1990) *Ornis scand.* **21**, 129-32. SENAR, J C, MASÓ, G, and VALLE, M J DEL (1986) *Res. Congr. Nac. Etol.* **8**. Córdoba. SENAR, J C and METCALFE, N B (1988) *Anim. Behav.* **36**, 1549-50. SENGUPTA, S N (1968) *Proc. zool. Soc. Calcutta* **21**, 1-27; (1969) *Auk* **86**, 556; (1973) *J. Bombay nat. Hist. Soc.* **70**, 204-6; (1976) *Proc. Indian natn. Sci. Acad.* **42**, 338-45. SEO, Land Bird Census Committee (1985) Proc. 9th Int. Conf. Bird Census Atlas Work, 117-22. Tring. SERE, D (1986) *Acrocephalus* **7**, 33-4. SEREBRENNIKOV, M K (1931) *J. Orn.* **79**, 29-56. SERLE, W and MOREL, G J (1977) *A field guide to the birds of West Africa.* London. SERMET, E (1967) *Nos Oiseaux* **29**, 17-20; (1973) *Nos Oiseaux* **32**, 113-15. SEUTIN, G, BOAG, P T, WHITE, B N, and RATCLIFFE, L M (1991) *Auk* **108**, 166-70. SEVESI, A (1939) *Riv. ital. Orn.* (2) **9**, 112-13. SHAPOVAL, A P (1989) In Konstantinov, V M and Klimov, S M (eds) *Vranovye ptitsy v estestvennykh i antropogennykh landshaftakh* (1), 76-8. Lipetsk. SHARPE, R B (1885) *Catalogue of the birds in the British Museum* **10**; (1888) **12**; (1890) **13**. London. SHARROCK, J T R (1963) *Br. Birds* **56**, 221; (1974) *Br. Birds* **67**, 356; (1976) *The atlas of breeding birds in Britain and Ireland.* Tring; (1984) *Br. Birds* **77**, 489. SHAUB, B M (1950) *Bird-Banding* **21**, 105-11. SHAUB, B M and SHAUB, M S (1953) *Bird-Banding* **24**, 135-41. SHAW, G (1990) *Bird Study* **37**, 30-5. SHAW, TSEN-HWANG (1935) *Bull. Fan Mem. Inst. Biol. Zool.* **6**, 65-70. SHEPHERD, M (1992) Ph D Thesis. Nottingham Univ. SHIOVITZ, K A and THOMPSON, W L (1970) *Anim. Behav.* **18**, 151-8. SHIRIHAI, H (1987) *Dutch Birding* **9**, 152-7;

(1989) *Br. Birds* **82**, 52-5; (1992) *Br. Birds*, 289; (in press) *Birds of Israel.* SHIRIHAI, H and ALSTRÖM, P (1990) *Br. Birds* **83**, 262-72. SHIRIHAI, H, JONSSON, A, and SEBBA, N (1987) *Dutch Birding* **9**, 120-2. SHKATULOVA, A P (1962) *Ornitologiya* **4**, 176-81; (1979) *Ornitologiya* **14**, 97-107. SHMELEV, A A and BRUNOV, V V (1986) *Tez. Dokl. 1. S'ezda Vsesoyuz. orn. Obshch. 9 Vesesoyuz. orn. Konf.* **2**, 339-41. SHNITNIKOV, V N (1949) *Ptitsy Semirech'ya.* Moscow. SHOEMAKER, H H (1939) *Auk* **56**, 381-406. SHORT, L L (1969) *Auk* **86**, 84-105. SHORT, L L and HORNE, J F M (1981) *Sandgrouse* **3**, 43-61. SHORT, L L and ROBBINS, C S (1967) *Auk* **84**, 534-43. SHORT, L L and SIMON, S W (1965) *Condor* **67**, 438-42. SHRUBB, M (1979) *The birds of Sussex.* Chichester. SHUFORD, W D (1981) *Amer. Birds* **35**, 264-6. SHUKLA, K K (1981) BSc Hons Thesis. Malaya Univ. SHUKUROV, E D (1986) *Ptitsy elovykh lesov Tyan'-Shanya.* Frunze. SHURUPOV, I I (1985) *Ornitologiya* **20**, 201; (1986) *Tez. Dokl. 1. S'ezda Vsesoyuz. orn. Obshch. 9. Vsesoyuz. orn. Konf.* **2**, 348-9. St Petersburg. SHUTOV, S V (1990) *Zool. Zh.* **69** (5), 93-9. SIBLET, J-P (1988) *Les oiseaux du massif de Fontainbleau et des environs.* Paris. SIBLEY, C G (1950) *Univ. Calif. Publ. Zoöl.* **50**, 109-94; (1954) *Evolution* **8**, 252-90; (1970) *Bull. Peabody Mus. nat. Hist.* **32**. SIBLEY, C G and AHLQUIST, J E (1984) *Auk* **101**, 230-43; (1990) *Phylogeny and classification of birds.* New Haven. SIBLEY, C G, CORBIN, K W, AHLQUIST, J E, and FERGUSON, A (1974) In Wright, C A (ed.) *Biochemical and immunological taxonomy of animals*, 89-176. New York. SIBLEY, C G and SHORT, L L (1959) *Auk* **76**, 443-63; (1964) *Condor* **66**, 130-50. SIBLEY, C G and WEST, D A (1958) *Condor* **60**, 85-104; (1959) *Auk* **76**, 326-38. SICK, H (1931) *Mitt. Ver. sächs. Orn.* **3**, 150-4; (1938) *Beitr. Fortpfl. Vögel* **14**, 176-81; (1939) *Orn. Monatsber.* **47**, 65-71; (1957) *Vogelwelt* **78**, 1-18. SIEGFRIED, W R (1968) *Ostrich* **39**, 105-29. SIERING, M (1986) *Verh. orn. Ges. Bayern* **24**, 319-32. SIIVONEN, L (1963) *Sitzber. Finn. Akad. Wiss.* **1962**, 111-25. SILVOLA, T (1966) *Ornis fenn.* **43**, 60-70. SIMEONOV, S D (1964) *Ann. Univ. Sofia* **56**, 239-75; (1970) *Vogelwelt* **91**, 59-67; (1971) Dissertation. Sofia; (1975) *Ekologiya (Sofia)* **1**, 55-63. SIMEONOV, S and DOÏCHEV, R (1973) *Ann. Univ. Sofia Fac. Biol.* **65**, 163-71. SIMEONOV, S and PETROV, T (1977) *Ann. Univ. Sofia Fac. Biol.* **71**, 39-47. SIMMONS, K E L (1952) *Br. Birds* **45**, 323-5; (1954) *Ibis* **96**, 478-81; (1957a) *Br. Birds* **50**, 401-24; (1957b) *Ibis* **99**, 178-81; (1960) *Br. Birds* **53**, 11-15; (1961a) *Ibis* **103a**, 37-49; (1961b) *Avic. Mag.* **67**, 124-32; (1963) *Avic. Mag.* **69**, 148; (1966) *J. Zool. Lond.* **149**, 145-62; (1968) *Br. Birds* **61**, 228-9; (1970) *Br. Birds* **63**, 175-7; (1974) *Br. Birds* **67**, 243; (1984) *Br. Birds* **77**, 121; (1985) In Campbell, B and Lack, E (eds) *A dictionary of birds.* 101-5. Calton; (1986a) *The sunning behaviour of birds.* Bristol; (1986b) *Br. Birds* **79**, 595-6. SIMMS, E (1948) *Br. Birds* **41**, 344; (1962) *Br. Birds* **55**, 1-36; (1971) *Woodland birds.* London; (1975) *Birds of town and suburb.* London. SIMON, P, DELMÉE, E, and DACHY, P (1975) *Gerfaut* **65**, 153-64; (1983) *Gerfaut* **73**, 207-11. SIMON, T (1921) *Orn. Beob.* **19**, 4-6. SIMPSON, T (1970a) *Br. Birds* **63**, 177; (1970b) *Br. Birds* **63**, 254-5. SIMROTH, H (1908) *Orn. Monatsschr.* **33**, 61-71. SIMS, R W (1955) *Bull. Br. Mus. (nat. Hist.)* **2**, 369-93. SIMSON, E C L (1958) *Bull. Jourdain Soc.* **4**, 35-40. SINGER, R and YOM-TOV, Y (1988) *Ornis scand.* **19**, 139-44. SITASUWAN, N and THALER, E (1984) *Gef. Welt* **108**, 12-15, 45-7; (1985) *J. Orn.* **126**, 181-93. SITS, E (1937) *Beitr. Fortpfl. Vögel* **13**, 140-3. SITTERS, H P (1982) *Br. Birds* **75**, 105-8;

(1985) *Bird Study* **32**, 1-10; (1991) M Sc Thesis. Aberdeen Univ. SKEAD, C J (1967) *Ostrich Suppl.* **7**. SKEAD, D M (1974) *Ostrich* **45**, 189-92; (1977) *Ostrich Suppl.* **12**, 117-31. SKEAD, D M and DEAN, W R J (1977) *Ostrich Suppl.* **12**, 3-42. SKARPHÉTHINSSON, K H, NIELSEN, Ó K, THÓRISSON, S, and PETERSEN, I K (1992) *Bliki* **11**, 1-26. SLAGSVOLD, T (1979a) *Fauna norv.* (C) *Cinclus* **2**, 1-6; (1979b) *Fauna norv.* (C) *Cinclus* **2**, 60-4; (1979c) *Fauna norv.* (C) *Cinclus* **2**, 65-9; (1980) *Fauna norv.* (C) *Cinclus* **3**, 16-35; (1981) *Fauna norv.* (C) *Cinclus* **4**, 47-8; (1984) *Fauna norv.* (C) *Cinclus* **7**, 127-31; (1982a) *Ornis scand.* **13**, 141-4; (1982b) *Ornis scand.* **13**, 165-75; (1985) *Fauna norv.* (C) *Cinclus* **8**, 9-17. SLATER, P J B (1981) *Z. Tierpsychol.* **56**, 1-24; (1983) *Anim. Behav.* **31**, 272-81. SLATER, P J B, CLEMENTS, F A, and GOODFELLOW, D J (1984) *Behaviour* **88**, 76-97. SLATER, P J B and INCE, S A (1979) *Behaviour* **71**, 146-66; (1982) *Ibis* **124**, 21-6. SLATER, P J B and SELLAR, P J (1986) *Behaviour* **99**, 46-64. SLESSERS, M (1970) *Auk* **87**, 91-9. SLOAN-CHESSER, S (1937) *Ool. Rec.* **17**, 83-8. SMART, J H (1978) *Br. Birds* **71**, 86. SMETANA, N M (1980) In Kovshar', A F (ed.) *Biologiya ptits Naurzumskogo gosudarstvennogo zapovednika*, 75-104. Alma-Ata. SMETS, F and DRAULANS, D (1982) *Wielewaal* **48**, 55-7. SMIDDY, P (1986) *Br. Birds* **79**, 251-2. SMITH, A E (ed.) (1951) *Gibraltar Point Bird Rep. 1950*; (1953) *Gibraltar Point Bird Rep. 1952*; (1954) *Gibraltar Point Bird Rep. 1953*. SMITH, A E and CORNWALLIS, R K (1953a) In Smith (1953), 8-29; (1953b) *Br. Birds* **46**, 428-30. SMITH, E C (1974) *Ardea* **62**, 226-35. SMITH, F and BORG, S (1976) *Il-Merill* **17**, 25-6. SMITH, F R (1959) *Br. Birds* **52**, 1-9. SMITH, F R and RARITIES COMMITTEE (1967) *Br. Birds* **60**, 309-38; (1973) *Br. Birds* **66**, 331-60. SMITH, J N M and ZACH, R (1979) *Evolution* **33**, 460-7. SMITH, K, WALDON, J, and WILLIAMS, G (1992) *RSPB Conserv. Rev.* **6**, 40-4. SMITH, K D (1955a) *Ibis* **97**, 65-80; (1955b) *Ibis* **97**, 480-507; (1957) *Ibis* **99**, 1-26, 307-37; (1960) *Ibis* **102**, 536-44; (1962-4) Unpubl. diaries of Moroccan expeditions. Edward Grey Inst., Oxford Univ; (1965) *Ibis* **107**, 493-526; (1968) *Ibis* **110**, 452-92. SMITH, K G (1978) *Western Birds* **9**, 79-81. SMITH, C E (1966a) *Bird-Banding* **37**, 49-51. SMITH, P W (1985) *American Birds* **39**, 255-8. SMITH, R (1991) In Stroud, D A and Glue, D (eds) *Britain's birds in 1989/90: the conservation and monitoring review*, 112-13. BTO/NCC; (1992) *Ring. Migr.* **13**, 43-51. SMITH, V W (1966b) *Ibis* **108**, 492-512; (1967) *Bull. Niger. Orn. Soc.* **4** (13-14), 40. SMYTHIES, B E (1960) *The birds of Borneo.* Edinburgh; (1981) *The birds of Borneo.* 3rd edn. Kuala Lumpur; (1986) *The birds of Burma.* Liss. SNIGIREWSKI, S I (1928) *J. Orn.* **76**, 587-607. SNOW, B and SNOW, D (1988) *Birds and berries.* Calton. SNOW, D W (1952) *Ibis* **94**, 473-98; (1953) *Br. Birds* **46**, 379-80. SNOW, D W and MAYER-GROSS, H (1967) *Bird Study* **14**, 43-52. SNOW, D W, OWEN D F, and MOREAU, R E (1955) *Ibis* **97**, 557-71. SNYDER, L L (1957) *Arctic birds of Canada.* Toronto; (1937) *Can. Fld.-Nat.* **51**, 37-9. SOBANSKI, G G (1979) In Labutin, Y V (ed.) *Migratsii i ekologiya ptits Sibiri*, 46. Yakutsk. SOF (SVERIGES ORNITOLOGISKA FÖRENING) (1990) *Vår Fågelvärld Suppl.* **14**. SOKOLOV, E P (1986a) *Trudy Zool. Inst. Akad. Nauk SSSR* **147**, 71-81. SOKOLOV, E P and LOBKOV, E G (1985) *Ornitologiya* **20**, 33-41. SOKOLOV, L V (1986b) *Zool. Zh.* **65**, 1544-51. SOKOŁOWSKI, J (1962) *Acta Orn.* **7**, 33-67. SOLER, J J and SOLER, M (1991) *Ardeola* **38**, 69-89 SOLER, M (1987) *Doñana Acta Vert.* **14**, 67-81; (1988) *Bird Study* **35**, 69-76; (1989a) *Bird Study* **36**, 73-6; (1989b) *Ardeola* **36**, 3-24; (1989c) In

Bignal, E and Curtis, D J (eds) *Choughs and land-use in Europe*, 29-33. Scottish Chough Study Group; (1990) In Pinowski, J and Summers-Smith, J D (eds) *Granivorous birds in the agricultural landscape*, 253-61. Warsaw. SOLER, M, ALCALÁ, N, and SOLER, J J (1990) *Doñana Acta Vert.* **17**, 17-48. SOLER, M and SOLER, J J (1987) *Ardeola* **34**, 3-14; (1990) *Ardeola* **37**, 37-52. SOLLENBERG, P (1959) *Vår Fågelvärld* **18**, 128-31. SOLONEN, T (1985) *Suomen linnusto.* Helsinki. SOMEREN, V D VAN (1958) *A bird watcher in Kenya.* Edinburgh. SOMEREN, V G L VAN (1922) *Novit. Zool.* **29**, 1-246; (1956) *Fieldiana Zool.* **38**. SOMERKOSKI, M (1984) *Lounais-Hameen Luonto* **70**, 47-9. SÖMMER, P (1991) In Glandt (1991), 17-20. SONDBØ, S D (1993) *Birding World* **5**, 458-60. SONDELL, J (1977) *Vår Fågelvärld* **36**, 174-84. SONERUD, G A and BEKKEN, J (1979) *Vår Fuglefauna* **2**, 78-85. SONERUD, G A and FJELD, P E (1987) *Ornis scand.* **18**, 323-5. SOPER, E A (1969) *Br. Birds* **62**, 200-1. SOPER, T (1986) *The bird table book.* Newton Abbot. SOPYEV, O (1965) *Ornitologiya* **7**, 134-41; (1967) *Ornitologiya* **8**, 221-35; (1979) *Izv. Akad. Nauk Turkmen. SSR Ser. Biol. Nauk* (4), 53-7. SORBI, S, ROBBRECHT, G, STEEMAN, C, and WILLE, E (1990) *Aves* **27**, 39-47. SORCI, G, MASSA, B, and CANGIALOSI, G (1971) *Riv. ital. Orn.* **41**, 1-10. SOROKIN, A G (1977) *Ornitologiya* **13**, 210-11. SOUTHERN, H N (1945) *Ibis* **87**, 287. SOUZA, J A de (1991) *Ardeola* **38**, 179-98. SOVON (SAMENWERKENDE ORGANISATIES VOGELONDERZOEK NEDERLAND) (1987) *Atlas van de Nederlandse vogels.* Arnhem. SPAANS, A L (1977) *Ardea* **65**, 83-7. SPAANS, A L, RODENBURG, S, and WOLF, J DE (1982) *Vogeljaar* **30**, 31-5. SPAEPEN, J (1952) *Gerfaut* **42**, 164-214. SPANGENBERG, E P (1965) *Sbor. Trud. zool. Mus. MGU* **9**, 98-202. SPANÒ, S and TRUFFI, G (1986) *Riv. ital. Orn.* **56**, 231-9. SPARKS, J H (1963a) *Ibis* **105**, 558-61; (1963b) *Nature* **200**, 281; (1964) *Anim. Behav.* **12**, 125-36; (1965) *Proc. zool. Soc. Lond.* **145**, 387-403. SPEEK, B J and SPEEK, G (1984) *Thieme's vogeltrekatlas.* Zutphen. SPEICHER, K (1989) *Unser Kanarienvogel.* Stuttgart. SPEIRS, J M (1985) *Birds of Ontario* **2**. Toronto. SPELLMAN, C B, LEMON, R E, and MORRIS, M M J (1987) *Wilson Bull.* **99**, 257-61. SPENCE, B R and CUDWORTH, J (1966) *Br. Birds* **59**, 198-201. SPENCER, K G (1966a) *Naturalist* **91**, 73-80. SPENCER, R (1961) *Br. Birds* **54**, 449-95; (1962) *Br. Birds* **55**, 493-556; (1963) *Br. Birds* **56**, 477-524; (1964) *Br. Birds* **57**, 525-82; (1966b) *Br. Birds* **59**, 441-91; (1967) *Br. Birds* **60**, 429-75; (1969) *Br. Birds* **62**, 393-442; (1972) *Bird Study* **19** Suppl. SPENCER, R and GUSH, G H (1973) *Br. Birds* **66**, 91-9. SPENCER, R and HUDSON, R (1978a) *Ring. Migr.* **1**, 189-252; (1978b) *Ring. Migr.* **2**, 57-104; (1979) *Ring. Migr.* **2**, 161-208; (1980) *Ring. Migr.* **3**, 65-108; (1981) *Ring. Migr.* **3**, 213-56; (1982) *Ring. Migr.* **4**, 65-128. SPENCER, R and RARE BREEDING BIRDS PANEL (1988a) *Br. Birds* **81**, 99-125; (1988b) *Br. Birds* **81**, 417-44; (1991) *Br. Birds* **84**, 349-70, 379-92; (1993) *Br. Birds* **86**, 62-90. SPENNEMANN, A (1926) *Trop. Natuur* **15**, 86-9; (1928) *Beitr. Fortpfl. Vögel* **4**, 112-13; (1937) *Beitr. Fortpfl. Vögel* **13**, 120. SPERL, J (1992) *Falke* **39**, 244-5. SPILLNER, W (1973) *Falke* **20**, 166-9; (1975) *Falke* **22**, 276-9. SPITTLE, R J (1950) *Bull. Raffles Mus.* **21**, 184-204. SPJØTVOLL, Ø (1972) *Sterna* **11**, 201-13. SPRAGUE, R T and WEIR, R D (1984) *The birds of Prince Edward County.* Kingston. SPRAY, C J (1978) Ph D Thesis. Aberdeen Univ. SQUIRES, W A (1976) *New Brunswick Museum Monogr. Ser.* **7**. St John. SSOKOLOW, J J (1932) *Orn. Beob.* **30**, 20-4. STAAV, R (1976) *Fauna och Flora* **71**, 202-7; (1983) *Fauna och Flora* **78**, 265-76. STAAV, R and FRANSSON, T

(1987) *Nordens fåglar*. Stockholm. STACHANOW, W S (1931) *J. Orn.* **79**, 315-17. STADIE, C (1983) *Voliere* **6**, 35. STADLER, H (1926a) *Ber. Ver. schles. Orn.* **12**, 22-38; (1926b) *Ber. Ver. schles. Orn.* **12**, 82-94; (1927) *Ber. Ver. schles. Orn.* **13**, 40-9, 117-25; (1931) *Verh. orn. Ges. Bayern* **19**, 331-59; (1956) *Nachr. naturw. Mus. Aschaffenburg* **51**, 1-39. STAGG, A (1974) *J. RAF orn. Soc.* **9**, 17-38; (1985) *The birds of south-west Saudi Arabia*. Riyadh; (1987) *Birds of the Riyadh region*. Riyadh. STAGG, A and WALKER, F (1982) *A checklist of the birds of Tabuk Kingdom of Saudi Arabia*. STAHLBAUM, G (1957) *Vogelwelt* **78**, 127; (1967) *Beitr. Vogelkde.* **13**, 224. STAINTON, J M (1982) *Br. Birds* **75**, 65-86; (1991) *Br. Birds* **84**, 66-7. STANFORD, J K (1954) *Ibis* **96**, 449-73, 606-24. STANTSCHINSKY, W W (1929) *J. Orn.* **77**, 309-15. STAŠAITIS, J N (1982) *Ornitologiya* **17**, 173. ŠTASTNÝ, K and BEJČEK, V (1991) *Panurus* **3**, 27-36. STEGEMAN, L C (1954) *Auk* **71**, 179-85. STEGEMANN, K-D (1975) *Beitr. Vogelkde.* **21**, 383-4. STEGMANN, B (1928) *Ezhegodnik Zool. Muz. Akad. Nauk SSSR* **28**, 366-90; (1931a) *J. Orn.* **79**, 137-236; (1931b) *Orn. Monatsber.* **39**, 183-4; (1932) *J. Orn.* **80**, 99-114; (1935) *Orn. Monatsber.* **43**, 29-30; (1936) *J. Orn.* **84**, 58-139; (1948) *Zool. Zh.* **27**, 241-4; (1956) *J. Orn.* **97**, 236. STEIN, G (1929) *Orn. Monatsber.* **37**, 7-12. STEINBACHER, F (1930a) *J. Orn.* **78**, 471-87; (1930b) *Aquila* **36-7**, 88-91. STEINBACHER, J (1952) *Bonn. zool. Beitr.* **3**, 23-30; (1954) *Senckenbergiana* **34**, 307-10; (1956) *Senck. biol.* **37**, 213-19. STEINBACHER, J and WOLTERS, H E (1963-5) *Vögel in Käfig and Voliere: Prachtfinken*. Aachen. STEINER, H (1960) *J. Orn.* **101**, 92-112; (1955) *Proc. int. Orn. Congr.* **11**, 350-5. STEINER, H M (1969) *Bonn. zool. Beitr.* **20**, 75-84. STEINFATT, O (1937a) *Beitr. Fortpfl. Vögel* **13**, 210-23; (1937b) *Verh. orn. Ges. Bayern* **21**, 139-54; (1940) *Ber. Ver. schles. Orn.* **25**, 11-22; (1942) *Beitr. Fortpfl. Vögel* **18**, 21-6; (1943) *Beitr. Fortpfl. Vögel* **19**, 68-71; (1944) *Orn. Monatsber.* **52**, 8-16; (1954) *J. Orn.* **95**, 245-62. STEJNEGER, L (1892) *Proc. US natn. Mus.* **15**, 289-359. STENHOUSE, D (1962a) *Ibis* **104**, 250-2; (1962b) *Notornis* **10**, 61-7. STEPANITSKAYA, E V (1987) In Sokolov, V E and Dobrynina, I N (eds) *Kol'tsevanie i mechenie zhivotnykh, 1983-4 gody*, 80-155. Moscow. STEPANOV, E A (1960) *Ornitologiya* **3**, 292-7; (1987) *Ornitologiya* **22**, 118-23. STEPANYAN, L S (1969a) *Nauch. Dokl. vyssh. Shk. biol. Nauki* (2), 22-6; (1969b) *Uchen. zap. Mosk. gos. ped. Inst.* **362**, 176-302; (1970) *Uchen. zap. Mosk. gos. ped. Inst.* **394**, 102-50; (1978) *Sostav i raspredelenie ptits fauny SSSR, Passeriformes*. Moscow; (1983) *Nadvidy i vidy-dvoyniki v avifaune SSSR*. Moscow; (1990) *Konspekt ornitologicheskoy fauny SSSR*. Moscow. STEPANYAN, L S and GALUSHIN, V M (1962) *Ornitologiya* **4**, 200-7. STEPHAN, B (1974) *Falke* **21**, 31; (1982) *Mitt. zool. Mus. Berlin* **58**, Suppl. Ann. Orn. **6**, 91-100; (1984) *Mitt. zool. Mus. Berlin* **60**, Suppl. Ann. Orn. **8**, 89-96; (1986) *Mitt. zool. Mus. Berlin* **62**, Suppl. Ann. Orn. **10**, 25-68. STEPHAN, B and GAVRILOV, E I (1980) *Mitt. zool. Mus. Berlin* **56**, Suppl. Ann. Orn. **4**, 25-8. STERBETZ, I (1964) *Angew. Orn.* **2**, 30-6; (1967) *Aquila* **73-4**, 203; (1968) *Aquila* **75**, 151-7; (1971) *Állattani közl.* **58**, 171-2. STEVENS, C J (1945) *Br. Birds* **38**, 295-6. STEVENSON, H (1866) *The birds of Norfolk* **1**. London. STEVENSON, H M (1957) *Wilson Bull.* **69**, 39-77. STEVENSON, J (1950) *Countryman* **41**, 334-5. STEWART, P A (1937) *Auk* **54**, 324-32. STEWART, R E (1952) *Auk* **69**, 50-9. STEWART, R E and ROBBINS, C S (1958) *North American Fauna* **62**. STEWART, W (1927) *Scott. Nat.* **163**, 104-7. STEWART-HESS, C (1992) *Devon Birds* **45**, 25-6. STIEFEL,

A (1976) *Ruhe und Schlaf bei Vögeln*. Wittenberg Lutherstadt. STIEHL, R B (1978) Ph D Thesis. Portland State Univ. STIEHL, R B and TRAUTWEIN, S N (1991) *Wilson Bull.* **103**, 83-92. STILES, F G and CAMPOS, R G (1983) *Condor* **85**, 254-5. STILES, F G and SKUTCH, A F (1989) *A guide to the birds of Costa Rica*. London. STILL, E, MONAGHAN, P, and BIGNAL, E (1987) *Ibis* **129**, 398-403. STINGELIN, A (1935) *Arch. suisses Orn.* **1**, 251-6. STIRRUP, S A and EVERSHAM, B (1984) *Br. Birds* **77**, 491. STISHOV, M S, PRIDATKO, V I, and BARANYUK, V V (1991) *Ptitsy ostrova Vrangelya*. Novosibirsk. STJERNBERG, T (1979) *Acta zool. Fenn.* **157**; (1985) *Proc. int. orn. Congr.* **18**, 743-53. STOBO, W T and McLAREN, I A (1975) *Proc. Nova Scotia Inst. Sci.* **27**, suppl. 2. STODDARD, H L (1978) *Tall Timbers Res. Stn. Bull.* **21**. STOKOE, R (1949) *Br. Birds* **42**, 359-60. STOLT, B-O (1977) *Zoon* **5**, 57-61; (1987) *Vår Fågelvärld* **46**, 48-53; (1993) *J. Orn.* **134**, 59-68. STOLT, B-O and ÅSTRÖM, G (1975) *Fauna och Flora* **70**, 145-54. STOLT, B O and MASCHER, J W (1971) *Vår Fågelvärld* **30**, 84-90. STONE, W (1937) *Bird Studies at old Cape May* **2**. Philadelphia. STONER, D (1923) *Auk* **40**, 328-30. STORER, R W (1969) *Living Bird* **8**, 127-36. STORER, R W, and ZIMMERMAN, D A (1959) *Occ. Pap. Mus. Zool. Univ. Michigan* **609**, 1-13. STORK, H-J, JÄNICKE, B, and WENDENBURG, U (1976) *Orn. Ber. Berlin* **1**, 295-316. STRACHE, R-R and MADAS, K (1988) *Beitr. Vogelkde.* **34**, 201-2. STRAHM, J (1958) *Nos Oiseaux* **24**, 177-84; (1960) *Nos Oiseaux* **25**, 265-71; (1962) *Nos Oiseaux* **26**, 179-85, 297-303. STRAKA, U (1991) *Egretta* **34**, 34-41. STRAUSS, E (1938a) *Z. Tierpsychol.* **2**, 145-72; (1938b) *Z. Tierpsychol.* **2**, 172-97. STRAUTMAN, F I (1963) *Ptitsy zapadnykh oblastey USSR* **2**. L'vov; (1954) *Ptitsy Sovetskikh Karpat*. Kiev. STRAVINSKI, C and SHCHEPSKI, Y (1972) *Soobshch. Pribalt. Kom. Izuch. Migr. Ptits* **7**, 44-57. STREBEL, S (1991) *Orn. Beob.* **88**, 217-42. STREMKE, D (1990) *Beitr. Vogelkde.* **36**, 10-16. STRESEMANN, E (1910) *Orn. Monatsber.* **18**, 33-9; (1919) *Beitr. Zoogeogr. paläarktischen Region* **1**, 25-56; (1920) *Avifauna Macedonica*. Munich; (1924) *Orn. Monatsber.* **32**, 42-3; (1928a) *Orn. Monatsber.* **36**, 41-2; (1928b) *J. Orn.* **76**, 313-411; (1930) *Orn. Monatsber.* **38**, 17-18; (1935) *Orn. Monatsber.* **43**, 30-1; (1940) *Orn. Monatsber.* **48**, 102-4; (1943a) *Orn. Monatsber.* **51**, 166-8; (1943b) *J. Orn.* **91**, 305-24; (1943c) *J. Orn.* **91**, 448-514; (1956) *J. Orn.* **97**, 44-72. STRESEMANN, E, MEISE, W, and SCHÖNWETTER, M (1937) *J. Orn.* **85**, 375-576. STRESEMANN, E and SCHIEBEL, G (1925) *J. Orn.* **73**, 658-9. STRESEMANN, E and STRESEMANN, V (1966) *J. Orn.* **107** suppl; (1969a) *J. Orn.* **110**, 291-313; (1969b) *J. Orn.* **110**, 475-81; (1970) *Beitr. Vogelkde.* **16**, 386-92. STRIEGLER, R, STRIEGLER, U, and JOST, K-D (1982) *Falke* **29**, 164-70. STROKOV, V V (1962) *Ornitologiya* **5**, 290-9. STRÖM, K (1991) *Ornis Svecica* **1**, 119-20. STRÖMBERG, G (1975) *Fåglar Blekinge* **11**, 232-7. STÜBS, J (1958) *Beitr. Vogelkde.* **5**, 312-14. STUDD, M, MONTGOMERIE, R D, and ROBERTSON, R J (1983) *Can. J. Zool.* **61**, 226-31. STUDER-THIERSCH, A (1969) *Orn. Beob.* **66**, 105-44; (1984) *Orn. Beob.* **81**, 29-44. STYAN, F W (1891) *Ibis* (6) **3**, 316-59. SUDHAUS, W (1969a) *Orn. Mitt.* **21**, 18; (1969b) *Vogelwelt* **90**, 53-9; (1969c) *Vogelwelt* **90**, 234-5. SUDILOVSKAYA, A M (1957) *Byull. Mosk. Obshch. Ispyt. Prir. Otd. Biol.* **57** (3), 19-23. SUEUR, F (1981) *Alauda* **49**, 300-4; (1988) *Oiseau* **58**, 156-8; (1990a) *Oiseau* **60**, 60-2; (1990b) *Oiseau* **60**, 63-5. SUHONEN, J and JOKIMÄKI, J (1988) *Ornis fenn.* **65**, 76-83. SUKHININ, A N (1959) *Trudy Inst. Zool. Parasitol. Akad. Nauk Turkmen. SSR* **4**, 69-124. SULKAVA, S (1969) *Aquilo Ser. Zool.* **7**, 33-7.

SULLIVAN, G A (1976) *Anim. Behav.* **24**, 880-8. SULTANA, J and GAUCI, C (1982) *A new guide to the birds of Malta.* Valletta; (1985*a*) *Il-Merill* **23**, 11; (1985*b*) *Il-Merill* **23**, 32-40; (1988) *Il-Merill* **25**, 41-52. SULTANOV, E G (1987*a*) *Izv. Akad. Nauk Azer. SSR Ser. Biol. Nauk* (1), 43-8; (1987*b*) *Dokl. Akad. Nauk Azer. SSR* **43** (9), 72-5. SULTANOV, E G and GUMBATOVA, S E (1986) *Tez. Dokl. I S'ezda Vsesoyuz. orn. Kongr. 9 Vsesoyuz. orn. Konf.* **2**, 265-6. St Petersburg. SUMMERS, D D B (1979) *Br. Birds* **72**, 249-63; (1982) *J. appl. Ecol.* **19**, 813-19. SUMMERS, R W (1989) *Scott. Birds* **15**, 181. SUMMERS, R W and CROSS, S C (1987) *Ring. Migr.* **8**, 11-18. SUMMERS, R W, WESTLAKE, G E, and FEARE, C J (1987) *Ibis* **129**, 96-102. SUMMERS-SMITH, D (1954*a*) *Ibis* **96**, 116-28; (1954*b*) *Br. Birds* **47**, 249-65; (1955) *Ibis* **97**, 296-305; (1956) *Br. Birds* **49**, 465-88; (1958) *Ibis* **100**, 190-203; (1959) *Br. Birds* **52**, 164-5; (1979) *Il-Merill* **20**, 18-19; (1980) *Il-Merill* **21**, 17-18; (1983) *Br. Birds* **76**, 411; (1984) *Br. Birds* **77**, 25-6. SUMMERS-SMITH, D and LEWIS, L R (1952) *Bird Notes* **25**, 44-8. SUMMERS-SMITH, D and SUMMERS-SMITH, M (1952) *Br. Birds* **45**, 75. SUMMERS-SMITH, D and VERNON, J D R (1972) *Ibis* **114**, 259-62. SUMMERS-SMITH, J D (1963) *The House Sparrow.* London; (1984*a*) *Ostrich* **55**, 141-6; (1984*b*) *Bull. Br. Orn. Club* **104**, 138-42; (1985) In Campbell, B and Lack, E (eds) *A dictionary of birds*, 555-6; (1988) *The sparrows.* Calton; (1989) *Bird Study* **36**, 23-31; (1990) *Phoenix* **7**, 17-19; (1990) In Pinowski, J and Summers-Smith, J D (eds) *Granivorous birds in the agricultural landscape*, 11-29. Warsaw; (1992) *In search of sparrows.* London. SUMMERS-SMITH, M (1951) *Br. Birds* **44**, 16. SUNDERLIN, M A (1978) *Kingbird* **28**, 94. SUNDIN, B (1988) *Vår Fågelvärld* **47**, 15. SUORMALA, K (1938) *Ornis fenn.* **15**, 16-20. SUSCHKIN, P and STEGMANN, B (1929) *J. Orn.* **77**, 386-406. SUSHKIN, P (1913) *Bull. Soc. Imp. Nat. Moskva* **26**, 198-400; (1933) *Ibis* (13) **3**, 55-8. SUSHKIN, P P (1925) *Proc. Boston Soc. nat. Hist.* **38**, 1-55; (1938) *Ptitsy Sovetskogo Altaya* **2**. Moscow. SUTTER, E (1946) *Orn. Beob.* **43**, 81-5; (1948) *Orn. Beob.* **45**, 98-106; (1985) *Proc. int. Orn. Congr.* **18**, 1055. SUTTER, E and AMANN, F (1953) *Orn. Beob.* **50**, 89-90. SUTTON, G M (1932) *Mem. Carnegie Mus. Pittsburgh* **12** (part 2, sect. 2); (1935) *Cranbrook Inst. Sci. Bull.* **3**, 1-36; (1938) *Auk* **55**, 1-6; (1967) *Oklahoma birds.* Norman; (1968) *Oklahoma orn. Soc.* **1**, 1-7. SUTTON, G M and PARMELEE, D F (1954) *Wilson Bull.* **66**, 159-79. SUTTON, R W W and GRAY, J R (1972) In Vittery, A and Squire, J E (eds) *Orn. Soc. Turkey Bird Rep. 1968-9*, 186-205. Sandy. SVÄRDSON, G (1957) *Br. Birds* **50**, 314-43. SVENSSON, B W (1978) *Ornis scand.* **9**, 66-83. SVENSSON, L (1975) *Vår Fågelvärld* **34**, 311-18; (1984*a*) *Soviet birds.* Stockholm (cassette and booklet); (1984*b*) *Identificaion guide to European passerines* 3rd edn. Stockholm; (1991) *Birding World* **4**, 349-52; (1992) *Identification guide to European passerines* 4th edn. Stockholm. SVENSSON, L O (1973) *Vår Fågelvärld* **32**, 46-7. SVENSSON, S (1964) *Vår Fågelvärld* **23**, 43-56; (1988) *Vår Fågelvärld* **47**, 119-20; (1990) *Baltic Birds* **5** (2), 180-91. SVENSSON, S, CARLSSON, U T, and LILJEDAHL, G (1984) *Ann. zool. fenn.* **21**, 339-50. SWANBERG, P O (1951*a*) *Fauna och Flora* **46**, 11-29, 111-36; (1951*b*) *Proc. int. orn. Congr.* **10**, 545-54; (1952) *Br. Birds* **45**, 60-1; (1954) *Vår Fågelvärld* **13**, 213-40; (1956*a*) *Ibis* **98**, 412-19; (1956*b*) In Wingstrand, K G (ed.) *Bertil Hanström: zoological papers in honour of his sixty-fifth birthday*, 278-97. Lund; (1969) *Br. Birds* **62**, 239-40; (1981) *Vår Fågelvärld* **40**, 399-408. SWANN, R L (1954) *Sterna* **14**, 111-12; (1988) *Ring. Migr.* **9**, 1-4.

SWARTH, H S (1920) *Univ. Calif. Publ. Zool.* **21**, 75-224. SWENK, M H (1930) *Wilson Bull.* **42**, 81-95. SWINGLAND, I R (1976) *Anim. Behav.* **24**, 154-8; (1977) *J. Zool. Lond.* **182**, 509-28. SWINHOE, R (1861) *Ibis* (1) **3**, 323-45. SWYNNERTON, C F M (1916) *Ibis* (10) **4**, 264-94. SYKES, T K (1986) *Br. Birds* **79**, 594-5. SYLVESTER, G (1968) *Vogelwelt* **89**, 232. SYMENS, P (1990) *Sandgrouse* **12**, 3-7. SYMENS, D (1991) *Oriolus* **57**, 26-32. SZABÓ, L V (1962) *Aquila* **67-8**, 260-1; (1976) *Aquila* **82**, 145-54. SZABÓ, L V and GYÖRY, J (1962) *Aquila* **67-8**, 141-9. SZCZEPSKI, J P (1970) *Acta Orn.* **12**, 103-75; (1976) *Acta Orn.* **15**, 145-276. SZEMERE, L (1957) *Aquila* **63-4**, 349. SZIJJ, L J (1957) *Aquila* **63-4**, 71-101. SZIVKA, L (1983) *Larus* **33-5**, 141-59. SZYMCZAK, J T (1987*a*) *J. interdiscip. Cycle Res.* **18**, 49-57; (1987*b*) *J. comp. Physiol. (A) Sens. Neural Behav. Physiol.* **161**, 321-7. SZYMCZAK, J T, NAREBSKI, J, and KADZIELA, W (1989) *J. interdiscip. Cycle Res.* **20**, 281-8.

TAAPKEN, J (1976) *Vogeljaar* **24**, 39. TAAPKEN, J, BLOEM, F, and BLOEM, T (1955*a*) *Ardea* **43**, 145-74; (1955*b*) *Ardea* **43**, 286-9; (1957) *Vår Fågelvärld* **16**, 105-12. TACZANOWSKI, L (1873) *J. Orn.* **21**, 81-119. TAHON, J, TORREKENS, C, and GIGOT, J (1978) *EPPO Publ. (B)* **84**, 83-153. TAIT, W C (1924) *The birds of Portugal.* London. TAITT, M J (1973) *Bird Study* **20**, 226-36. TAKEISHI, M (1985) *Orn. Far East Newsl.* **6**, 1-2. TALLMAN, D A and ZUSI, R L (1984) *Auk* **101**, 155-8. TÁLPEANU, M and PASPALEVA, M (1973) *Oiseaux du delta du Danube.* Bucharest; (1979) *Trav. Mus. Hist. nat. Grigore Antipa* **20**, 441-9. TAMM, S (1977) *Behav. Proc.* **2**, 293-9. TARANENKO, L I (1979) *Ornitologiya* **14**, 198-9. TARASHCHUK, V I (1953) *Ptitsy polezashchitnykh nasazhdeniy.* Kiev. TAST, J (1968) *Ann. zool. fenn.* **5**, 159-78; (1970) *Ornis fenn.* **47**, 74-82. TAST, J and RASSI, P (1973) *Ornis fenn.* **50**, 29-45. TATNER, P (1982*a*) *J. Zool. Lond.* **197**, 559-81; (1982*b*) *Bird Study* **29**, 227-34; (1982*c*) *Naturalist* **107**, 47-58; (1983) *Ibis* **125**, 90-107; (1986) *Ring. Migr.* **7**, 112-18. TATSCHL, J L (1968) *Auk* **85**, 514. TAUCHNITZ, H (1972) *Apus* **2**, 245-54. TAVERNER, J H (1958) *Br. Birds* **51**, 126. TAVERNER, P A and SUTTON, G M (1934) *Ann. Carnegie Mus.* **23**. TAYLOR, D W (1980) *Br. Birds* **73**, 39. TAYLOR, D W, DAVENPORT, D L, and FLEGG, J J M (1981) *The birds of Kent.* Meopham. TAYLOR, F (1922) *Br. Birds* **16**, 103-4; (1935) *Br. Birds* **29**, 102-4. TAYLOR, K (1985) *BTO News* **140**, 1; (1986) *BTO News* **142**, 9. TAYLOR, M (1987) *The birds of Sheringham.* North Walsham. TAYLOR, W K (1972) *Bird-Banding* **43**, 15-19; (1974) *Auk* **91**, 485-7; (1976) *Bird-Banding* **47**, 72-3. TEBBUTT, C F (1949) *Br. Birds* **42**, 242. TEIXEIRA, R M (ed) (1979) *Atlas van de Nederlandse broedvogels.* Deventer. TEKKE, M J (1971) *Limosa* **44**, 19-22. TELLERÍA, J L (1981) *La migración de las aves en el Estrecho de Gibraltar* **2**. Madrid. TEMMINCK, C (1835) *Manuel d'Ornithologie ou tableau systématique des oiseaux qui se trouvent en Europe.* Paris. TEMPLE LANG, J and DEVILLERS, P (1975) *Gerfaut* **65**, 137-52. TENOVUO, R (1963) *Ann. zool. Soc. Vanamo* **25** (5). TENOVUO, R and LEMMETYINEN, R (1970) *Ornis fenn.* **47**, 159-66. TERENTIEV, P V (1966) *Trudy Inst. Biol. Sverdlovsk* **51**, 35-55; (1970) *Byull. Mosk. Obshch. Ispyt. Prir. Otd. Biol.* **75** (6), 129-134. TERNE, T (1978) *Vår Fågelvärld* **37**, 255-6. TERNOVSKI, D V (1954) *Byull. Mosk. Obshch. Ispyt. Prir. Otd. Biol.* **59** (1), 37-40. TERRES, J K (1980) *Encyclopedia of North American birds.* New York. TERRY, J H (1986) *Trans. Herts. nat. Hist. Soc.* **29**, 303-12. THALER, E (1977) *Zool. Garten (NF)* **47**, 241-60. THANNER, R VON (1903)

Orn. Jahrb. **14**, 211-17; (1910) *Orn. Jahrb.* **21**, 81-101. THESING, G (1987) *Orn. Mitt.* **39**, 320. THÉVENOT, M (1982) *Oiseau* **52**, 21-86, 97-152. THÉVENOT, M, BEAUBRUN, P, BAOUAB, R E, and BERGIER, P (1982) *Docum. Inst. Sci. Rabat* 7. THÉVENOT, M, BERGIER, P, and BEAUBRUN, P (1981) *Docum. Inst. Sci. Rabat* 6. THIBAULT, J-C (1983) *Les oiseaux de la Corse.* Ajaccio. THIEDE, W (1982) *Ornis fenn.* **59**, 37-8; (1987) *Orn. Mitt.* **39**, 269-75; (1989) *Orn. Mitt.* **41**, 6-11. THIEDE, W and THIEDE, U (1974) *Vogelwelt* **95**, 88-95. THIELCKE, G (1969) In Hinde, R A (ed.) *Bird vocalizations,* 311-39. Cambridge. THIENEMANN, J (1903) *J. Orn.* **51**, 212-23; (1908) *J. Orn.* **56**, 393-470; (1910) *Orn. Monatsber.* **18**, 66. THIENEMANN, J and SCHÜZ, E (1931) *Vogelzug* **2**, 103-10. THIES, H (1990) *Corax* **13**, 281-308. THIOLLAY, J-M (1985) *Malimbus* **7**, 1-59. THOM, V M (1986) *Birds in Scotland.* Calton. THOMAS, A (1989) In Bignal and Curtis (1989), 23-4. THOMAS, M (1982) *Br. Birds* **75**, 36-7. THOMPSON, C F and FLUX, J E C (1988) *Ornis scand.* **19**, 1-6. THOMPSON, D, EVANS, A, and GALBRAITH, C (1992) *BTO News* **178**, 8-9. THOMPSON, D B A and GRIBBIN, S (1986) *Bull. Br. ecol. Soc.* **17**, 69-75. THOMPSON, D B A and NETHERSOLE-THOMPSON, D (1984) *Br. Birds* **77**, 368. THOMPSON, N S (1969a) *Comm. Behav. Biol.* part A, **3**, 1-5; (1969b) *Comm. Behav. Biol.* Part A, **4**, 269-71; (1982) *Behaviour* **80**, 106-17. THOMPSON, W L (1970) *Auk* **87**, 58-71. THOMPSON, W L and COUTLEE, E L (1963) *Wilson Bull.* **75**, 358-72. THOMSEN, P and JACOBSEN, P (1979) *The birds of Tunisia.* Copenhagen. THOMSON, A L (1949) *Bird migration.* London. THOMSON, A L and SPENCER, R (1954) *Br. Birds* **47**, 361-92. THÖNEN, W (1965) *Orn. Beob.* **62**, 196-7. THÓRISSON, S (1981) *Náttúrufr.* **51**, 145-63. THORNEYCROFT, H B (1966) *Science* **154**, 1571-2; (1975) *Evolution* **29**, 611-21. THORPE, W H (1954) *Nature* **173**, 465-9; (1955) *Ibis* **97**, 247-51; (1956) *Br. Birds* **49**, 389-95; (1958) *Ibis* **100**, 535-70. THOUY, P (1976) *Alauda* **44**, 135-51. TIAINEN, J (1979) In Hildén, O, Tiainen, J, and Valjakka, R (eds) *Muuttolinnut,* 264-71. Helsinki. TIAINEN, J and YLIMAUNU, J (1984) *Lintumies* **19**, 26-9. TICEHURST, C B (1910a) *Br. Birds* **3**, 261-2; (1910b) *Br. Birds* **4**, 70-2; (1915) *Ibis* (10) **3**, 662-9; (1922) *Ibis* (11) **4**, 526-72, 605-62; (1926) *J. Bombay nat. Hist. Soc.* **31**, 368-78; (1926-7) *J. Bombay nat. Hist. Soc.* **31**, 687-711, 862-81, **32**, 64-97; (1932) *A history of the birds of Suffolk.* London; (1940) *Ibis* (14) **4**, 523-5. TICEHURST, C B, BUXTON, P A, and CHEESMAN, R E (1921-2) *J. Bombay nat. Hist. Soc.* **28**, 210-50, 381-427, 650-74, 937-56. TICEHURST, C B and CHEESMAN, R E (1925) *Ibis* (12) **1**, 1-31. TICEHURST, C B, COX, P, and CHEESMAN, R E (1926) *J. Bombay nat. Hist. Soc.* **31**, 91-119. TICEHURST, C B and WHISTLER, H (1933) *Ibis* (13) **3**, 97-112; (1938) *Ibis* (14) **2**, 717-46. TICEHURST, N F (1909) *A history of the birds of Kent.* London. TICHON, M (1989) *Aves* **26**, 57. TIETZE, F (1971) *Falke* **18**, 89-93. TIMMERMANN, G (1938) *Beitr. Fortpfl. Vögel* **14**, 201-6; (1938-49) *Die Vögel Islands.* Reykjavik. TIMMIS, W H (1972) *Avic. Mag.* **78**, 9-11; (1973) *Avic. Mag.* **79**, 3-7. TIMOFÉEFF-RESSOVSKY, N W (1940) *J. Orn.* **88**, 334-40. TINBERGEN, J M (1981) *Ardea* **69**, 1-67. TINBERGEN, L (1934) *Ardea* **23**, 99-100; (1946) *Ardea* **34**, 1-213; (1953) *Br. Birds* **46**, 377. TINBERGEN, N (1939) *Trans. Linn. Soc. New York* **5**. TINNING, P C and TINNING, P A (1970) *Br. Birds* **63**, 83. TINTORI, G (1964) *Nos Oiseaux* **27**, 250-1. TIPPER, R P (1987) *Hong Kong Bird Rep. 1986,* 81-21. TISCHLER, F (1931) *Orn. Monatsber.* **39**, 113-15. TODD, W E C (1963) *Birds of the Labrador peninsula and adjacent areas.* Toronto. TODHUNTER, J F

(1987) *Suffolk orn. Group Bull.* **76**, 31-2. TOHMÉ, G and NEUSCHWANDER, J (1978) *Oiseau* **48**, 319-27. TOLSTOY, V A (1986) *Tez. Dokl. 1. S'ezda Vsesoyuz. orn. Obshch. 9. Vsesoyuz. orn. Konf.* **2**, 284-5. TOMBRE-STEEN, I (1991a) *Ornis scand.* **22**, 383-6; (1991b) *Vår Fuglefauna* **14**, 222-3. TOMEK, T and WALIGÓRA, E (1976) *Acta Zool. Cracov.* **21**, 13-30. TOMIAŁOJĆ, L (1967) *Acta Orn.* **10**, 109-56; (1974) *Acta Orn.* **14**, 59-97; (1976a) *Birds of Poland.* Warsaw; (1976b) *Przeglad. Zool.* **20**, 361-4; (1988) *Ring* **12** (134-5), 31; (1990) *Ptaki Polski* 2nd edn. Warsaw. TOMIAŁOJĆ, L and PROFUS, P (1977) *Acta Orn.* **16**, 117-77. TOMIAŁOJĆ, L, WESOŁOWSKI, T, and WALANKIEWICZ, W (1984) *Acta Orn.* **20**, 241-310. TOMKOVICH, P S and MOROZOV, V V (1982) *Ornitologiya* **17**, 173-5. TOMKOVICH, P S and SOROKIN, A G (1983) *Sbor. Trud. Zool. Mus.* **21**, 77-159. TOMLINSON, A G (1917) *J. Bombay nat. Hist. Soc.* **24**, 825-9. TOMPA, F S (1975) *Orn. Beob.* **72**, 181-98; (1976) *Orn. Beob.* **73**, 119-24. TOOK, J M E (1971) *Cyprus orn. Soc.* (1957) *Rep.* **18**, 40-9; (1973) *Common birds of Cyprus.* Nicosia. TORDOFF, H B (1950) *Wilson Bull.* **62**, 3-4; (1952) *Condor* **54**, 200-3; (1954) *Condor* **56**, 346-58. TORDOFF, H B and DAWSON, W R (1965) *Condor* **67**, 416-22. TORDOFF, H B and MENGEL, R M (1956) *Univ. Kansas Publ. Mus. nat. Hist.* **10**, 1-44. TÖRÖK, J (1990) In Pinowski, J and Summers-Smith, J D (eds) *Granivorous birds in the agricultural landscape,* 199-210. Warsaw. TORRES, J A and LEON, A (1985) *Serv. Publ. Univ. Córdoba España.* TOUPS, J A and JACKSON, J A (1987) *Birds and birding on the Mississippi coast.* Jackson. TOWNSEND, C W (1906) *Auk* **23**, 172-9; (1909) *Auk* **26**, 13-19. TRACY, N (1927) *Br. Birds* **21**, 155. TRANSEHE, N VON (1965) *Die Vogelwelt Lettlands.* Hannover. TRAUTMAN, M B (1940) *Univ. Michigan Mus. Zool. Misc. Publ.* **44**. TRAYLOR, M A (1960) *Nat. Hist. Misc. Chicago Acad. Sci.* **175**, 1-2. TRETTAU, W (1964) *J. Orn.* **105**, 475-82. TRETTAU, W and WOLTERS, H E (1967) *Bonn. zool. Beitr.* **18**, 308-20. TRET'YAKOV, G P (1978) In Kashkarov, D Y (ed.) *Migratsii ptits v Azii,* 126-30. Tashkent. TRICOT, J (1968) *Aves* **5**, 146-56. TRISTRAM, H B (1859) *Ibis* (1) **1**, 22-41; (1864) *Proc. zool. Soc. Lond.,* 444; (1868) *Ibis* (2) **4**, 204-15; (1884) *The survey of western Palestine.* London. TROMMER, G (1971) *Orn. Mitt.* **23**, 170-1. TROTMAN, N (1974) *J. Gloucs. Nat. Soc.* **25**, 358-9. TROTTER, W D C (1970) *Oiseau* **40**, 160-70. TROY, D M (1983) *J. Fld. Orn.* **54**, 146-51; (1984) *Can. J. Zool.* **62**, 2302-6; (1985) *Auk* **102**, 82-96. TROY, D M and BRUSH, A H (1983) *Condor* **85**, 443-6. TROY, D M and SHIELDS, G F (1979) *Condor* **81**, 96-7. TRUSCOTT, B (1944) *Br. Birds* **38**, 74. TRYON, P R and MACLEAN, S F (1980) *Auk* **97**, 509-20. TSCHUSI ZU SCHMIDHOFFEN, V VON (1890) *Orn. Jahrb.* **1**, 65-81. TSUNEKI, K (1960) *Jap. J. Ecol.* **10**, 177-89; (1962) *Mem. Fac. Lib. Arts Fukui Univ. Ser. II Nat. Sci.* **12**, 117-78. TSVELYKH, A N (1993) *Russ. orn. Zh.* **2** (1), 94-6. TUAJEW, D G and WASSILJEW, W I (1974) *Falke* **21**, 18-19. TUCKER, B W (ed.) (1950) *Br. Birds* **43**, 114. TUCKER, M and ROWCLIFFE, J P G (1950) *Br. Birds* **43**, 370. TUCKER, N and TUCKER, L A (1978) *Br. Birds* **71**, 363-4. TUCKER, V R (1980a) *Br. Birds* **73**, 538; (1980b) *Devon Birds* **33**, 55-9. TUFTS, H F (1906) *Auk* **23**, 339-40. TUFTS, R W (1961) *Birds of Nova Scotia.* Halifax; (1986) *Birds of Nova Scotia* 3rd edn. Halifax. TURČEK, F J (1948) *Auk* **65**, 297; (1961) *Ökologische Beziehungen der Vögel und Gehölze.* Bratislava. TURČEK, F J and KELSO, L (1968) *Comm. Behav. Biol.* (A) **1**, 277-97. TURNER, B C (1959a) *Br. Birds* **52**, 129-31; (1959b) *Br. Birds* **52**, 388-90. TURNER, D (1983) *Sunday Express* 21 Aug 1983.

Tutman, I (1950) *Larus* 3, 353-60; (1969) *Vogelwelt* 90, 1-8. Tutt, H R (1952) *Ibis* 94, 162-3. Tweedie, M W F (1960) *Common Malayan birds*. London. Tyler, S J (1971) *Br. Birds* 64, 230-1; (1980) *Scopus* 4, 44-5. Tyrberg, T (1987) *Vår Fågelvärld* 46, 375-417; (1988) *Vår Fågelvärld* 47, 378-418; (1990) *Vår Fågelvärld* 49, 389-428; (1991a) *Ornis Svecica* 1, 3-10; (1991b) *Vår Fågelvärld* 50 (6-7), 27-61.

Udvardy, M D F (1956) *Ark. Zool.* (2) 9, 499-505. Uhlig, R (1984) *Beitr. Vogelkde.* 30, 75-6. Ulbricht, H (1975) *Beitr. Vogelkde.* 21, 452-70. Ulfstrand, S (1959) *Vår Fågelvärld* 18, 131-62. Ulfstrand, S and Högstedt, G (1976) *Anser* 15, 1-32. Ulfstrand, S, Roos, G, Alerstam, T, and Österdahl, L (1974) *Vår Fågelvärld* suppl. 8. Ullrich, B (1986) *Orn. Jahresh. Bad.-Württ.* 2, 79-80. Uloth, W (1977) *Falke* 24, 98-9. Umrikhina, G S (1969) *Tez. dokl. V Vsesoyuz. orn. Konf.* 2, 652-5; (1970) *Ptitsy Chuyskoy doliny*. Frunze. Upton, R (1962) *Br. Birds* 55, 592. Urban, E K and Brown, L H (1971) *A checklist of the birds of Ethiopia*. Addis Ababa. Urbánek, B (1959) *Sylvia* 16, 253-61. Uryadova, L P (1986) *Tez. Dokl. 1. S'ezda Vsesoyuz. orn. Obshch 9. Vsesoyuz. Orn. Konf.* 2, 288-90. St Petersburg. Uspenski, S M (1959) *Ornitologiya* 2, 7-15. Ussher, R J (1889) *Zoologist* (3) 13, 180-1.

Vakarenko, V I and Mikhalevich, O A (1986) *Ornitologiya* 21, 150-2. Valeur, P (1946) *Naturen* 9, 270-9. Valverde, J-A (1953) *Nos Oiseaux* 22, 78-82; (1957) *Aves del Sahara Español*. Madrid; (1967) *Monogr. Estac. biol. Doñana* 1, 1-219. Van der Elst, D (1990) *Aves* 27, 73-82. Van der Mueren, E (1980) *Gerfaut* 70, 455-70. Van der Plas, L H W and Wattel, J (1986) *Abstr. Symp. XIX int. orn. Congr. Ottawa*, 516. Van Oss, R M (1950) *Br. Birds* 43, 292-3. Van Tyne, J (1934) *Auk* 51, 529-30. Van Tyne, J and Drury, W H (1959) *Occ. Pap. Mus. Zool. Univ. Michigan* 615. Van Winkel, J (1968) *Wielewaal* 34, 359-60. Vardy, L E (1971) *Condor* 73, 401-14. Varshavski, S N (1977) *Byull. Mosk. Obshch. Ispyt. Prir. Otd. Biol.* 82 (5), 51-7. Varshavski, S N and Shilov, M N (1958) *Zool. Zh.* 37, 1521-30. Vartapetov, L G (1984) *Ptitsy taezhnykh mezhdurechiy Zapadnoy Sibiri*. Novosibirsk. Vásárhelyi, I (1967) *Aquila* 73-4, 196. Vasil'chenko, A A (1982) *Ornitologiya* 17, 130-4. Vaughan, J H (1930) *Ibis* (12) 6, 1-48. Vaughan, R (1953) *Riv. ital. Orn.* 23, 137-42; (1979) *Arctic summer*. Shrewsbury; (1992) *In search of arctic birds*. London. Vauk, G (1961) *Vogelwelt* 82, 179-182; (1964) *Vogelwelt* 85, 113-20; (1968) *Vogelwelt* 89, 142-5; (1970) *Vogelwelt* 91, 11-15; (1972) *Die Vögel Helgolands*. Hamburg; (1973) *Beitr. Vogelkde.* 19, 225-60; (1980) *Z. Jagdwiss.* 26, 93-5. Vaurie, C (1949a) *Amer. Mus. Novit.* 1406; (1949b) *Amer. Mus. Novit.* 1424; (1954a) *Amer. Mus. Novit.* 1658; (1954b) *Amer. Mus. Novit.* 1668; (1954c) *Amer. Mus. Novit.* 1694; (1955) *Amer. Mus. Novit.* 1753; (1956a) *Amer. Mus. Novit.* 1775; (1956b) *Amer. Mus. Novit.* 1786; (1956c) *Amer. Mus. Novit.* 1788; (1956d) *Amer. Mus. Novit.* 1795; (1956e) *Amer. Mus. Novit.* 1805; (1956f) *Amer. Mus. Novit.* 1814; (1957) *Dansk orn. Foren. Tidsskr.* 51, 9-11; (1958a) *Amer. Mus. Novit.* 1898; (1958b) *Ibis* 100, 275-6; (1959) *Birds of the Palearctic fauna: passeriformes*. London; (1972) *Tibet and its birds*. London. Vedum, T V and Tøråsen, A (1988) *Vår Fuglefauna* 11, 83-6. Veger, Z (1968) *Avic. Mag.* 74, 157-9. Veh, M (1988) *Beih. Veröff. Nat. Land. Bad.-Württ.* 53, 75-82.

Veiga, J P (1990) *Anim. Behav.* 39, 496-502. Venables, L S V (1940) *Br. Birds* 33, 334-5; (1949) *Br. Birds* 42, 182. Veprintsev, B N and Zablotskaya, M M (1982) *Akusticheskaya kommunikatsiya tundryanoy chechetki Acanthis hornemanni (Holboell)*. Pushchino. Verbeek, N A M (1972) *J. Orn.* 113, 297-314. Vercauteren, P (1984) *Gerfaut* 74, 327-60. Vere Benson, S (1970) *Birds of Lebanon and the Jordan area*. London. Verheyen, R (1953) *Explor. Parc natn. Upemba Miss. G F de Witte* 19; (1954) *Gerfaut* 44, 324-42; (1955a) *Gerfaut* 45, 5-25; (1955b) *Gerfaut* 45, 173-84; (1956) *Gerfaut* 46, 1-15; (1957a) *Les passereaux de Belgique* 1. Brussels; (1957b) *Gerfaut* 47, 161-70; (1960) *Gerfaut* 50, 101-53. Verheyen, R F (1968) *Gerfaut* 58, 369-93; (1969a) *Gerfaut* 59, 239-59; (1969b) *Gerfaut* 59, 378-84; (1980) In Wright, E N, Inglis, I R, and Feare, C J (eds) *Bird problems in agriculture*, 69-82. Croydon. Veroman, H (1978) *Orn. Kogumik* 8, 253-4. Vertse, A (1943) *Aquila* 50, 142-248. Vertzhutski, B N, Ravkin, Y S, Seryshev, A A, and Verzhutskaya, N V (1979) In Labutin, Y V (ed.) *Migratsii i ekologiya ptits Sibiri*, 127-8. Yakutsk. Vickholm, M, Virolainen, E, and Zetterberg, P (1981) *Ornis fenn.* 58, 133-4. Vidal, A (1991) *Orn. Anz.* 30, 173-5. Vielliard, J (1962) *Oiseau* 32, 74-9. Vierhaus, H and Bruch, A (1963) *J. Orn.* 104, 250. Vieweg, A (1981) *Falke* 28, 205. Víksne, J (1983) *Ptitsy Latvii*. Riga; (1989) *Latvian breeding bird atlas 1980-84*. Riga. Vincent, A W (1949) *Ibis* 91, 660-88. Vines, G (1981) *Ibis* 123, 190-202. Vinicombe, K E (1988) *Br. Birds* 81, 240-1. Vinogradova, N V, Lyuleeva, D S, Paevski, V A, Popov, E A, and Shumakov, M E (1985) *Trudy Zool. Inst. Akad. Nauk SSSR* 137, 138-54. Vinter, S V and Sokolov, E P (1983) *Trudy Zool. Inst. Akad. Nauk SSSR* 116, 61-71. Virkkala, R (1987) *Ann. zool. fenn.* 24, 281-94; (1988) *Ornis fenn.* 65, 104-13; (1989) *Ann. zool. fenn.* 26, 277-85; (1991) *Ornis fenn.* 68, 193-203. Vleugel, D A (1941) *Ardea* 30, 89-106; (1974) *Alauda* 42, 429-35. Vogt, W (1974) *Orn. Beob.* 71, 320. Voigt, A (1961) *Exkursionsbuch zum Studium der Vogelstimmen* 12th edn. Heidelberg. Voipio, P (1961) *Ornis fenn.* 38, 81-92; (1968) *Ornis fenn.* 45, 10-16; (1969) *Ardea* 57, 48-63. Voisin, R (1963) *Nos Oiseaux* 27, 164-71; (1965) *Nos Oiseaux* 28, 28; (1966) *Bull. Murithienne* 83, 107-12; (1968) *Nos Oiseaux* 29, 286-92. Völker, O (1957) *J. Orn.* 98, 210-14. Volsøe, H (1949) *Dansk orn. Foren. Tidsskr.* 43, 237-42; (1951) *Vidensk. Medd. dansk nat. Foren.* 113. Vondráček, J (1988) *Beih. Veröff. Nat. Land. Bad.-Württ.* 53, 66. Vonk, H and IJzendoorn, E J van (1988) *Dutch Birding* 10, 127-30. Voous, K H (1944) *Ardea* 33, 42-50; (1945) *Limosa* 18, 11-22; (1946) *Gerfaut* 36, 199-202; (1947) *Alauda* 15, 172-6; (1949) *Condor* 51, 52-81; (1950) *Limosa* 23, 281-92; (1951a) *Limosa* 24, 81-91; (1951b) *Limosa* 24, 131-3; (1953) *Beaufortia* 2 (30), 1-41; (1960a) *Limosa* 33, 128-34; (1960b) *Atlas of European birds*. London; (1977) *List of recent Holarctic bird species*. London; (1978) *Br. Birds* 71, 3-10. Vorobiev, G P and Likhatski, Y P (1987) *Ornitologiya* 22, 176-7. Vorobiev, K A (1954) *Ptitsy Ussuriyskogo kraya*. Moscow; (1963) *Ptitsy Yakutii*. Moscow; (1980) *Ornitologiya* 15, 194-6. Vorobiev, V N and Kaganova, O Z (1980) *Ornitologiya* 15, 133-7. Voronkova, K A and Ravkin, E S (1974) *Ornitologiya* 11, 364-6. Vowles, G A and Vowles, R S (1987) *Ring. Migr.* 8, 119-20. Vries, R de (1982) *Seevögel* 3 suppl., 27-33. Vronski, I A (1986) *Tez. Dokl. 1 S'ezda Vsesoyuz. orn. Obshch. 9 Vesesoyuz. orn. Konf.* 1, 136. Vtorov, P P (1962) *Ornitologiya* 4, 218-33; (1967) *Orni-*

tologiya **8**, 254–61; (1972) *Ornitologiya* **10**, 242–7. VTOROV, P P and DROZDOV, N N (1960) *Ornitologiya* **3**, 131–8. VUILLEUMIER, F (1977) *Terre Vie* **31**, 459–88.

WADE, V E and RYLANDER, M K (1982) *Bird Study* **29**, 166. WADEWITZ, O (1976) *Falke* **23**, 160–4. WAGNER, U (1981) *Vogelwelt* **102**, 32. WAHLMINO, H and PETERSSON, B (1956) *Vår Fågelvärld* **15**, 61. WAITE, H W (1948) *J. Bombay nat. Hist. Soc.* **48**, 93–117. WAITE, R K (1978) M A Thesis. Keele Univ; (1981) *Z. Tierpsychol.* **57**, 15–36; (1984*a*) *Ornis scand.* **15**, 55–62; (1984*b*) *Behav. Ecol. Sociobiol.* **15**, 55–9; (1985*a*) *Bird Study* **32**, 45–9; (1985*b*) *Proc. int. orn. Congr.* **18**, 1189. Moscow; (1986) *Br. Birds* **79**, 659–60. WALBRIDGE, G (1978) *Br. Birds* **71**, 314–15. WALCHER, A (1918) *Orn. Jahrb.* **29**, 51–5. WALFORD, N T (1930) *Br. Birds* **24**, 51. WALICZKY, Z, MAGYAR, G, and HRASKÓ, G (1983) *Aquila* **90**, 73–9. WALKER, B (1990) *Ontario Birds* **7**, 86–7. WALKER, F J (1981*a*) *Sandgrouse* **2**, 33–55; (1981*b*) *Sandgrouse* **2**, 56–85. WALKINSHAW, L H (1948) *Condor* **50**, 64–70. WALLACE, D I M (1957) *Br. Birds* **50**, 208–9; (1976*a*) *Br. Birds* **69**, 27–33; (1976*b*) *Br. Birds* **69**, 465–73; (1982*a*) *Br. Birds* **75**, 291; (1982*b*) *Sandgrouse* **4**, 77–99; (1983) *Sandgrouse* **5**, 1–18; (1984) *Sandgrouse* **6**, 24–47. WALLACE, D I M, COBB, F K, and TUBBS, C R (1977) *Br. Birds* **70**, 45–9. WALLER, C S (1970) *Br. Birds* **63**, 147–9. WALLGREN, H (1954) *Acta zool. fenn.* **84**; (1956) *Acta Soc. Fauna Flora fenn.* **71** (4). WALLIN, L (1966) *Vår Fågelvärld* **25**, 327–45. WALLIS, H M (1887) *Ibis* (5) **5**, 454–5; (1912) *Bull. Br. Orn. Club* **29**, 83; (1932) *Bull. Br. Orn. Club* **52**, 38–9. WALLSCHLÄGER, D (1983) *Mitt. zool. Mus. Berlin* **59**, *Suppl. Ann. Orn.* **7**, 85–116. WALPOLE-BOND, J (1905) *Countryside* 12 Aug; (1932) *Br. Birds* **25**, 292–300; (1938) *A history of Sussex birds.* London. WALSH, T A (1976) *Br. Birds* **69**, 222. WALTER, H (1965) *J. Orn.* **106**, 81–105. WALTER, H and DEMARTIS, A M (1972) *J. Orn.* **113**, 391–406. WALTERS, J (1988) *Limosa* **61**, 33–40. WALTERS, P M and LAMM, D W (1980) *N. Am. Bird Bander* **5**, 15. WARD, N (1987) *Br. Birds* **80**, 500–2. WARD, P (1977) *Ann. Rep. Inst. terr. Ecol.* Cambridge, 54–6. WARD, P and POH, G E (1968) *Ibis* **110**, 359–63. WARD, P and ZAHAVI, A (1973) *Ibis* **115**, 517–34. WARMBIER, N (1973) *Falke* **20**, 67. WARNCKE, K (1960) *Vogelwelt* **81**, 178–84; (1968) *J. Orn.* **109**, 300–2. WARNES, J M (1983) *Scott. Birds* **12**, 238–46. WARNES, J M and STROUD, D A (1988) In Bignal, E and Curtis, D J (eds) *Choughs and land-use in Europe*, 46–51. Scottish Chough Study Group. WARREN, D R (1974) *Br. Birds* **67**, 440. WARRILOW, G J, FOWLER, J A, and FLEGG, J J M (1978) *Ring. Migr.* **2**, 34–7. WASHINGTON, D (1974) *Br. Birds* **67**, 213–14. WASSENICH, V (1969) *Regulus* **9**, 362–70; (1973) *Regulus* **11**, 55. WASSMANN, R (1990) *Vogelkdl. Ber. Niedersachs.* **22**, 48. WASYLIK, A and PINOWSKI, J (1970) *Bull. Acad. Pol. Sci.* **18**, 29–32. WATERHOUSE, M J (1949) *Ibis* **91**, 1–16. WATSON, A (1957*a*) *Can. Fld.-Nat.* **71**, 87–109; (1957*b*) *Sterna* **2**, 65–99; (1963) *Arctic* **16**, 101–8; (1989) *Scott. Birds* **15**, 178–9; (1992*a*) *Scott. Birds* **16**, 273–5; (1992*b*) *Scott. Birds* **16**, 287. WATSON, A and O'HARE, P J (1980) *Irish Birds* **1**, 487–91. WATSON, A and SMITH, R (1991) *Scott. Birds* **16**, 53–6. WATSON, D (1972) *Birds of moor and mountain.* Edinburgh. WATSON, G E (1960) *Postilla* **52**, 1–15; (1964) Ph D Thesis. Yale Univ. WATT, D J (1986) *J. Fld. Orn.* **57**, 105–13. WATT, D J, RALPH, C J, and ATKINSON, C T (1984) *Auk* **101**, 110–20. WATTEL, J (1971) *Ostrich* **42**, 229. WEAVER, R L (1942) *Wilson Bull.* **54**, 183–91; (1943) *Auk* **60**, 62–74. WEBER, C (1990*a*) *Lim-*

icola **4**, 222–8; (1990*b*) *Limicola* **4**, 276–84; (1991) *Limicola* **5**, 92. WEBER, H (1954) *Orn. Mitt.* **6**, 168–70; (1959) *Beitr. Vogelkde.* **6**, 351–6; (1971–2) *Falke* **18**, 306–14, **19**, 16–27. WECHSLER, B (1988*a*) *Behaviour* **106**, 252–64; (1988*b*) *Behaviour* **107**, 267–77; (1989) *Ethology* **80**, 307–17. WEHRLE, C M (1989) *Orn. Beob.* **86**, 53–68. WEIGOLD, H (1926) *Wiss. Meeresunters.* (NF) **15** (3) article 17. WEINZIERL, H (1961) *Orn. Mitt.* **13**, 153. WEISE, C M (1962) *Auk* **79**, 161–72. WEISE, R (1992) *Vogelwelt* **113**, 47–51. WEISS, I and WIEHE, H (1984) *Orn. Mitt.* **36**, 162. WENDLAND, V (1958) *J. Orn.* **99**, 203–8. WERNICKE, P (1990) *Beitr. Vogelkde.* **36**, 1–9. WERNLI, W (1970) *Vögel der Heimat* **40**, 93–109. WEST, D A (1962) *Auk* **79**, 399–424. WEST, G C, PEYTON, L J, and SAVAGE, S (1968) *Bird-Banding* **39**, 51–5. WEST, M J and KING, A P (1990) *Amer. Sci.* **78**, 107–14. WEST, M J, STROUD, A N, and KING, A P (1983) *Wilson Bull.* **95**, 635–40. WESTERFRÖLKE, P (1958) *Vogelwelt* **79**, 117. WESTERNHAGEN, W VON (1956) *Orn. Mitt.* **8**, 169. WESTERSKOV, K (1953) *Notornis* **5**, 189–91. WESTERTERP, K (1973) *Ardea* **61**, 137–58. WESTERTERP, K, GORTMAKER, W, and WIJNGAARDEN, H (1982) *Ardea* **70**, 153–62. WESTIN, L (1973) *Vår Fågelvärld* **32**, 44. WESTPHAL, D (1976) *J. Orn.* **117**, 70–4; (1981) *Vogelwarte* **31**, 94–101. WETHERBEE, O P (1937) *Bird-Banding* **8**, 1–10. WETMORE, A (1936) *Smithson. Misc. Coll.* **95** (17); (1949) *J. Washington Acad. Sci* **39**, 137–9. WETTON, J H, CARTER, R E, PARKIN, D T, and WALTERS, D (1987) *Nature* **327**, 147–9. WETTON, J H and PARKIN, D T (1991*a*) *Proc. Roy. Soc. Lond. Biol. Sci.* **245**, 227–33; (1991*b*) *Proc. int. orn. Congr.* **20**, 2435–41. WETTSTEIN, O (1959) *J. Orn.* **100**, 103–4. WHISTLER, H (1922*a*) *Ibis* (11) **4**, 259–309; (1922*b*) *J. Bombay nat. Hist. Soc.* **28**, 990–1006; (1923) *Ibis* (11) **5**, 611–29; (1924) *J. Bombay nat. Hist. Soc.* **30**, 177–88; (1941) *Popular handbook of Indian birds.* London; (1945) *J. Bombay nat. Hist. Soc.* **45**, 106–22. WHISTLER, H and HARRISON, J M (1930) *Ibis* (12) **6**, 453–70. WHITAKER, B (1955) *Br. Birds* **48**, 145–6. WHITAKER, J I S (1894) *Ibis* (6) **6**, 78–100; (1898) *Ibis* (7) **4**, 592–610. WHITAKER, L M (1957) *Wilson Bull.* **69**, 195–262. WHITE, C M and WEST, G C (1977) *Oecologia* **27**, 227–38. WHITE, C M N (1960) *Ibis* **102**, 138–9; (1967) *Bull. Br. Orn. Club* **87**, 62–3. WHITE, S J and HINDE, R A (1968) *J. Zool. Lond.* **155**, 145–55. WHITEHEAD, C H T (1909) *Ibis* (9) **3**, 214–84; (1911) *J. Bombay nat. Hist. Soc.* **20**, 776–99. WHITELEY, J D, PRITCHARD, J S, and SLATER, P J B (1990) *Bird Study* **37**, 12–17. WHITNEY, B (1983) *Birding* **15**, 219–22. WHITTAKER, A (1990) *Br. Birds* **83**, 73. WHITTLE, C L (1938) *Bird-Banding* **9**, 196–7. WICKLER, W (1961) *Z. Tierpsychol.* **18**, 320–42; (1982) *Auk* **99**, 590–1. WIECZOREK, P (1975) *Falke* **22**, 282. WIEDENFELD, D A (1991) *Condor* **93**, 712–23. WIEHE, H (1988) *Orn. Mitt.* **40**, 252; (1990) *Orn. Mitt.* **42**, 294–6. WIELOCH, M (1975) *Pol. ecol. Stud.* **1**, 227–42. WIENS, J A and DYER, M I (1977) In Pinowski, J and Kendeigh, S C (eds) *Granivorous birds in ecosystems*, 205–66. Cambridge. WILDASH, P (1968) *Birds of South Vietnam.* Tokyo. WILD BIRD SOCIETY OF JAPAN (1982) *A field guide to the birds of Japan.* Tokyo. WILDE, J (1962) *Br. Birds* **55**, 560–2. WILDER, G D and HUBBARD, H W (1924) *J. N. China Branch Roy. Asiatic Soc.* **55**, 156–239; (1938) *Birds of north-eastern China.* Peking. WILKINSON, D M (1988) *Br. Birds* **81**, 657–8. WILKINSON, R (1975) Ph D Thesis. Southampton Univ; (1980) *Z. Tierpsychol.* **54**, 436–56; (1982) *Ornis scand.* **13**, 117–22; (1990) *Bioacoustics* **2**, 179–97. WILKINSON, R and HOWSE, P E (1975) *Z. Tierpsychol.* **38**, 200–11. WILLCOX, D R C and WILLCOX, B (1978) *Ibis*

120, 329-33. Wille, H-G (1983) Orn. Mitt. 35, 269-73. Williams, C H and Williams, C E (1929) J. Bombay nat. Hist. Soc. 33, 598-613. Williams, J G (1941) Ibis (14) 5, 245-64; (1963) A field guide to the birds of East and Central Africa. London. Williams, J G and Arlott, N (1980) A field guide to the birds of East Africa. London. Williams, L P (1986) Br. Birds 79, 423-6. Williams, M D (1986) (ed.) Rep. Cambridge orn. Exped. China 1985. Williams, M D, Carey, G J, Duff, D G, and Weishu, X (1992) Forktail 7, 3-55. Williams, T D, Reed, T M, and Webb, A (1986) Scott. Birds 14, 57-60. Williams, T S (1946) Br. Birds 39, 149-50. Williams, W M (1989) Devon Birds 42, 49-50. Williamson, F S L and Rausch, R (1956) Condor 58, 165. Williamson, K (1939) Br. Birds 33, 78; (1953) Scott. Nat. 65, 65-94; (1955) Fair Isle Bird Obs. Bull. 2, 327-8; (1956) Dansk orn. Foren. Tidsskr. 50, 125-33; (1957) Scott. Nat. 69, 190-2; (1959) Peregrine 3, 8-14; (1961a) Bird Migr. 1, 235-40; (1961b) Bird Migr. 2, 43-5; (1961c) Br. Birds 54, 238-41; (1962) Br. Birds 55, 130-1; (1963a) Bird Migr. 2, 207-23; (1963b) Bird Migr. 2, 252-60, 329-40; (1965) Fair Isle and its birds. Edinburgh; (undated) Observations of the Chough. Mona's Herald Ltd; (1968) Q. J. Forestry 62, 118-31. Williamson, K and Davis, P (1956) Br. Birds 49, 6-25. Williamson, K and Spencer, R (1960) Bird Migr. 1, 176-81. Williamson, P and Gray, L (1975) Condor 77, 84-9. Willoughby, E J (1992) Condor 94, 295-7. Wilmore, S B (1977) Crows, jays, ravens and their relatives. Newton Abbot. Wilson, C W (1883) Ibis (5) 1, 575-7. Wilson, P R (1965) B Sc Hons Thesis. Victoria Univ; (1973) Ph D Thesis. Victoria Univ. Wilson, R T (1981) African J. Ecol. 19, 285-94. Wiman, C (1943) Vår Fågelvärld 2, 94-5. Windsor, R E (1935) Br. Birds 29, 126. Wingate, D B (1958) Auk 75, 359-60; (1973) A checklist and guide to the birds of Bermuda. Bermuda. Winkelman, J E (1981) Limosa 54, 81-8. Winkler, H (1979a) Egretta 22, 29-30. Winkler, K (1979b) Gef. Welt 103, 201-3; (1992) Gef. Welt 116, 156-7. Winkler, R (1975) Bull. Murithienne (Sion) 92, 48; (1984) Orn. Beob. Suppl. 5. Winkler, R, Daunicht, W D, and Underhill, L G (1988) Orn. Beob. 85, 245-59. Winkler, R and Jenni, L (1987) J. Orn. 128, 243-6. Winkler, R and Winkler, A (1985) Orn. Beob. 82, 55-66; (1986) Orn. Beob. 83, 76. Winterbottom, J M (1962) Ostrich 33 (2), 43-50; (1975) Ostrich 46, 236-50. Wiprächtiger, P (1987) Vögel Heimat 57, 187-8. Wirdheim, A and Carlén, T (1986) Fågelstråk. Halmstad. Witchell, C A (1896) The evolution of bird-song. London. Witherby, H F (1901) Ibis (8) 1, 237-78; (1903) Ibis (8) 3, 501-71; (1908) Bull. Br. Orn. Club 23, 48; (1910) Ibis (9) 4, 491-517; (1913) Br. Birds 7, 126-39; (1915) Bull. Br. Orn. Club 36, 3-4; (1928) Ibis (12) 4, 385-436. Witherby, H F, Jourdain, F C R, Ticehurst, N F, and Tucker, B W (1938) The handbook of British birds 1. London. Witschi, E (1936) Proc. Soc. exp. Biol. Med. 33, 484-6. Witschi, E and Miller, R A (1938) J. exp. Zool. 79, 475-87. Witschi, E and Woods, R P (1936) J. exp. Zool. 73, 445-59. Witt, K (1988) Orn. Ber. Berlin (West) 13, 119-55; (1989) Vogelwelt 110, 142-50. Wittenberg, G (1970) Orn. Mitt. 22, 129-31. Wittenberg, J (1968) Zool. Jb. Syst. 95, 16-146; (1976) Vogelwarte 28, 230-2; (1987) Verh. naturwiss. Ver. Hamburg (NF) 29, 5-49; (1988) Beih. Veröff. Nat. Land. Bad.-Württ. 53, 109-18. Wohl, E (1980) Mitt. Abt. Zool. Landesmus. Joanneum Graz 9, 137-40; (1981) Mitt. Abt. Zool. Landesmus. Joanneum Graz 10, 81-4; (1985) Mitt. Abt. Zool. Landesmus. Joanneum Graz 34, 65-8. Wolfson,

A (1942) Condor 44, 237-63; (1945) Condor 47, 95-127; (1954a) Auk 71, 413-34; (1954b) J. exp. Zool. 125, 353-76; (1954c) Wilson Bull. 66, 112-18. Wolters, H E (1952) Bonn. zool. Beitr. 3, 231-88; (1957) Bonn. zool. Beitr. 8, 90-129; (1958) Bonn. zool. Beitr. 9, 200-7; (1962) Bonn. zool. Beitr. 13, 324-6; (1968) Bonn. zool. Beitr. 19, 157-64. Won, P-o (1961) Avi-mammalian Fauna of Korea 12, 31-139. Won, P-o, Woo, H-c, Chun, M-z, and Ham, K-w (1966) Misc. Rep. Yamashina Inst. Orn. 4, 445-68. Wong, M (1983) Wilson Bull. 95, 287-94. Wontner-Smith, C (1939) Br. Birds 33, 194. Wood, D L and Wood, D S (1972) Bird-Banding 43, 182-90. Wood, H B (1945) Auk 62, 455-6. Woodcock, M (1980) Collins handguide to the birds of the Indian sub-continent. London. Woodford, J and Lovesy, F T (1958) Bird-Banding 29, 109-10. Woods, H E (1950) Br. Birds 43, 82-3; (1975) Bird-Banding 46, 344-6. Woodward, I D (1960a) Nat. Wales 6, 26-7; (1960b) Devon Birds 13, 23-5. Workman, W B (1961) Br. Birds 54, 250-1; (1963) Br. Birds 56, 52-3. Wortelaers, F (1950) Gerfaut 40, 207-12. Wotkyns, D B (1962) Audubon Mag. 64, 235. Wright, M (1972) Br. Birds 65, 260-1. Wunsch, H (1976) Gef. Welt 100, 42-4. Wüst, W (1961) Anz. orn. Ges. Bayern 6, 91-2; (1986) Avifauna Bavariae 2. Munich. Wydoski, R S (1964) Auk 81, 542-50. Wynne-Edwards, V C (1927) Br. Birds 21, 229-30; (1952) Auk 69, 353-91; (1962) Animal dispersion in relation to social behaviour. Edinburgh.

Ximenis, J A (1977) Ardeola 22, 111.

Yablonkevich, M L, Bardin, A V, Bol'shakov, K V, Popov, E A, and Shapoval, A P (1985a) Trudy Zool. Inst. Akad. Nauk SSSR 137, 69-97. Yablonkevich, M L, Bol'shakov, K V, Bulyuk, V N, Eliseev, D O, Efremov, V D, and Shamuradov, A K (1985b) Trudy Zool. Inst. Akad. Nauk SSSR 137, 11-59. Yakobi, V E (1979) Zool. Zh. 58, 136-7. Yamashina, Y (1982) Birds in Japan. Tokyo. Yang, S H and Selander, R K (1968) Syst. Zool. 17, 107-43. Yanushevich, A I, Tyurin, P S, Yakovleva, I D, Kydyraliev, A, and Semenova, N I (1960) Ptitsy Kirgizii 2. Frunze. Yapp, W B (1951) J. Anim. Ecol. 20, 169-72; (1962) Birds and woods. London; (1975) Br. Birds 68, 342. Yarbrough, C G, and Johnston, D W (1965) Wilson Bull. 77, 175-91. Yeates, G K (1932) Br. Birds 26, 30-3; (1934) The life of the Rook. London; (1951) The land of the loon. London. Yeatman, L J (1971) Histoire des oiseaux d'Europe. Paris; (1976) Atlas des oiseaux nicheurs de France. Paris. Yeatman-Berthelot, D (1991) Atlas des oiseaux de France en hiver. Paris. Yeo, P F (1947) Br. Birds 40, 211-12. Yeo, V Y Y (1990) B Sc Hons Thesis. Singapore Univ. Yésou, P (1983) Alauda 51, 161-78. Ylimaunu, J, Ylimaunu, O, and Ylipekkala, J (1986) Lintumies 21, 98-9. Yom-Tov, Y (1974) J. Anim. Ecol. 43, 479-98; (1975a) Bird Study 22, 47-51; (1975b) Auk 92, 778-85; (1976) Behaviour 59, 247-51; (1980a) Ibis 122, 234-7; (1980b) Teva Va'aretz 22, 98-101; (1992) Bird Study 39, 111-14. Yom-Tov, Y and Ar, A (1980) Israel J. Zool. 29, 171-87. Yom-Tov, Y, Ar, A, and Mendelssohn, H (1978) Condor 80, 340-3. Yom-Tov, Y, Dunnet, G M, and Anderson, A (1974) Ibis 116, 87-90. Yosef, R (1991) Sandgrouse 13, 73-9. Yosef, R and Yosef, D (1991) Wilson Bull. 103, 518-20. Young, B E (1991) Condor 93, 236-50. Young, J G (1984) Scott. Birds 13, 88. Young, S (1990) Br. Birds 83, 508-9. Ytreberg, N-J (1972) Norw. J. Zool. 20, 61-89. Yunick, R P (1972) Bird-Banding 43, 38-46; (1976)

Bird-Banding **47**, 276-7; (1977*a*) *N. Amer. Bird Bander* **2**, 12-13; (1977*b*) *N. Amer. Bird Bander* **2**, 155-6; (1981) *N. Amer. Bird Bander* **6**, 97; (1984) *N. Amer. Bird Bander* **9**, 2-4, 6. YURLOV, K T, CHERNYSHOV, V M, KOSHELEV, A I, SAGITOV, R A, TOTUNOV, V M, KHODKOV, G I, and YURLOV, A K (1977) In Yurlov, K T (ed.) *Migratsii ptits v Azii*, 205-9. Novosibirisk.

ZABLOTSKAYA, M M (1975) *Byull. Mosk. Obshch. Ispyt. Prir. Otd. Biol.* **53** (3), 22-38; (1976*a*) *Dokl. Uchast. 2. Vsesoyuz. Konf. Poved. Zhiv.*, 125-7; (1976*b*) *Dokl. Uchast. 2. Vsesoyuz. Konf. Poved. Zhiv.*, 127-9; (1976*c*) *Dokl. Uchast. 2. Vsesoyuz. Konf. Poved. Zhiv.*, 130-1; (1978*a*) *Zool. Zh.* **57**, 105-13; (1978*b*) *Byull. Mosk. Obshch. Ispyt. Prir. Otd. Biol.* **83** (4), 36-54; (1981) *Akusticheskaya kommunikatsiya obyknovennoy chechetki Acanthis flammea flammea* (*L*). Pushchino; (1982) *Akusticheskaya kommunikatsiya konoplyanki Acanthis cannabina* (*L*). Pushchino. ZAHARONI, M (1991) *Israel Bird Ring. Cent. Ringer's Newsl.* **3**, 21. ZARUDNYI. N A (1911) *Orn. Vestnik.* **3-4**, 298-306 (1916) *Nasha Okhota* **20**, 37-8. ZEDLITZ, O VON (1911) *J. Orn.* **59**, 1-92; (1912) *J. Orn.* **60**, 325-64, 529-69; (1921) *Flora och Fauna* **16**, 275-80; (1925) *Flora och Fauna* **20**, 145-73; (1926) *J. Orn.* **74**, 296-308. ZEIDLER, K (1966) *J. Orn.* **107**, 113-53. ZHORDANIA, R G and GOGILASHVILI, G S (1976) *Acta Orn.* **15**, 323-38. ZHURAVLEV, M N and AFONIN, P V (1982) *Ornitologiya* **17**, 182. ZIEGER, R (1967) *Beitr. Vogelkde.* **13**, 117-24. ZIMIN, V B (1981) In Ivanter, E V (ed.) *Ekologiya nazemnykh pozvonochnykh*

Severo-zapada SSSR, 13-31. Petrozavodsk. ZIMKA, J (1968) *Acta Orn.* **11**, 87-102. ZIMMERLI, E (1986) *Vogel Heimat* **56**, 211-12. ZIMMERMAN, J L (1965) *Wilson Bull.* **77**, 55-70. ZIMMERMANN, D (1951) *Orn. Beob.* **48**, 73-111; (1987) *Orn. Beob.* **84**, 66. ZIMMERMANN, R (1907) *Z. Orn. prakt. Geflügelz.* **31**, 2-4, 17-18; (1913) *Orn. Monatsber.* **21**, 112-14; (1931) *Orn. Monatsber.* **39**, 99-102. ZINK, G (1969) *Auspicium* **3**, 195-291; (1981) *Der Zug europäischer Singvögel* **1**; (1985) **4**. Möggingen. ZINK, R M (1982) *Auk* **99**, 632-49. ZINK, R M, DITTMANN, D L, and ROOTES, W L (1991*a*) *Auk* **108**, 578-84. ZINK, R M, DITTMANN, D L, CARDIFF, S W, and RISING, J D (1991*b*) *Condor* **93**, 1016-19. ZINK R M, and KLICKA, J T (1990) *Wilson Bull.* **102**, 514-20. ZINK, R M, ROOTES, W L, and DITTMANN, D L (1991*c*) *Condor* **93**, 318-29. ZINNENLAUF, B (1967) *Orn. Beob.* **64**, 113-18. ZINO, P A (1969) *Bocagiana* **21**, 1-7. ZISWILER, V (1959) *Vjschr. Naturf. Ges. Zürich* **104**, 222-6; (1965) *J. Orn.* **106**, 1-48; (1967*a*) *Orn. Beob.* **64**, 105-10; (1967*b*) *Rev. suisse Zool.* **74** 620-8; (1967*c*) *Zool. Jb. Syst.* **94**, 427-520; (1979) *Rev. suisse Zool.* **86**, 823-31. ZLOTIN, R I (1968) *Ornitologiya* **9**, 158-63. ZONFRILLO, B (1988) *Bull. Br. Orn. Club* **108**, 71-5. ZONOV, G B (1978) In Tolchina, S N (ed.) *Rol' ptits v biotsenozakh Vostochnoy Sibiri*, 168-82. ZUCCHI, H (1975) *Orn. Mitt.* **27**, 171-2. ZUMSTEIN, F (1921) *Verh. orn. Ges. Bayern* **15**, 68-73; (1927) *Beitr. Fortpfl. Vögel* **3**, 181-4. ZUÑIGA, J M (1989) In Bignal, E and Curtis, D J (eds) *Choughs and land-use in Europe*, 65-9. Scottish Chough Study Group. ZYKOVA, L Y and PANOV, E N (1982) *Zool. Zh.* **61**, 1113-16.

ERRATA AND CORRIGENDA

It is regretted that the reproduction of some of the sonagrams in Volumes VI and VII fell short of the high standard to be expected in Birds of the Western Palearctic. The originals supplied to the Press were of the same high quality as for the previous volumes, but unfortunately there were technical problems in the process of reproduction that affected the final printing. The sonagrams in Volume VII have subsequently been reprinted in a separate booklet, and the affected sonagrams from Volumes VI and VII are reproduced here.

VOLUME IV

Picus viridis Green Woodpecker (p. 833)

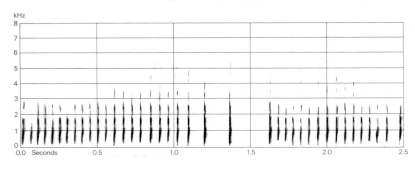

I L Svensson Sweden

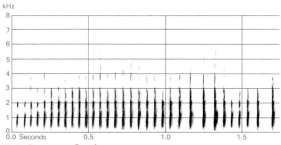

These two figures show the Drumming and some Tapping (Fig I, 1.2–1.7 s) of *Picus viridis*. They replace Fig I in Volume IV, p. 833, which shows the Drumming of *Dendrocopos major*.

II L Svensson Sweden

VOLUME VI

Prinia gracilis Graceful Warbler (p. 37)

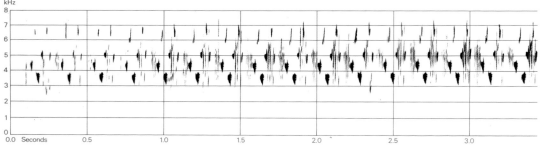

III P A D Hollom Iran April 1972

Locustella fluviatilis River Warbler (pp. 84, 85)

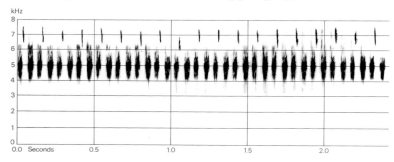

II J Paatela/Sveriges Radio (1972–80)
Finland June 1962

Locustella fluviatilis **River Warbler** (*cont.*)

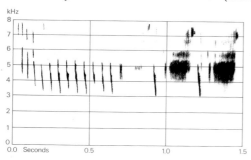

IV J Paatela/Sveriges Radio (1972–80) Finland June 1963

Locustella luscinioides **Savi's Warbler** (p. 97)

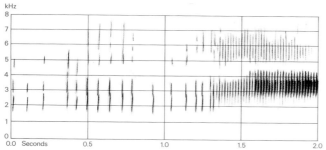

I P Szöke and M Orszag/Sveriges Radio (1972–80) Hungary June 1967

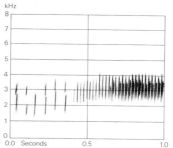

II P A D Hollom Netherlands May 1982

Acrocephalus paludicola **Aquatic Warbler** (p. 127)

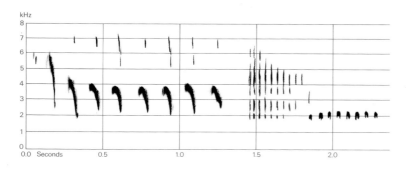

V M Schubert East Germany June 1968

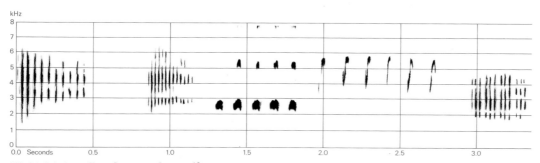

VI M Schubert East Germany June 1968

Acrocephalus schoenobaenus **Sedge Warbler** (p. 141)

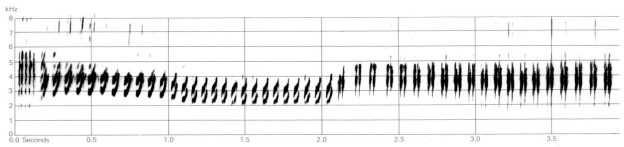

III S Palmér/Sveriges Radio (1972–80) Sweden May 1961

Acrocephalus agricola **Paddyfield Warbler** (pp. 150, 151, 152)

I B N Veprintsev and V V Leonovich
USSR June 1975

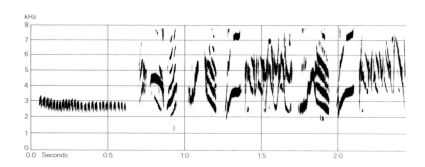

II B N Veprintsev and V V Leonovich
USSR June 1975

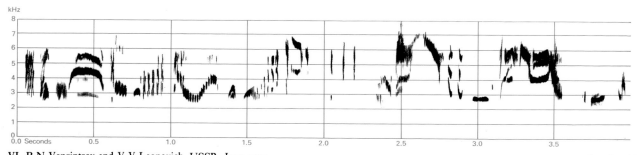

VI B N Veprintsev and V V Leonovich USSR June 1975

(*continued*)

Acrocephalus agricola **Paddyfield Warbler** (*cont.*)

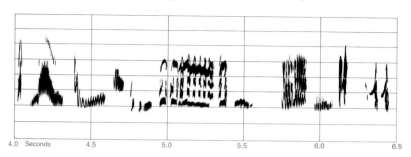

Acrocephalus dumetorum **Blyth's Reed Warbler** (pp. 164, 165)

II P J Sellar Finland June 1966

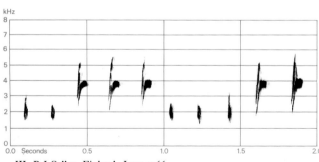

III P J Sellar Finland June 1966

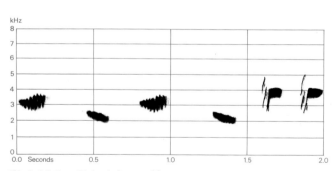

IV P J Sellar Finland June 1966

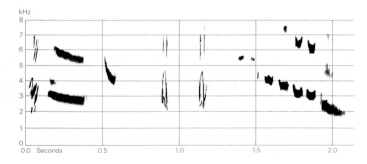

VI N Linnman/Sveriges Radio (1972–80) Finland June 1958

Acrocephalus brevipennis Cape Verde Cane Warbler (p. 170)

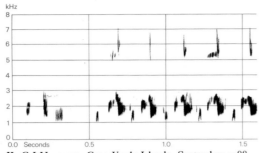 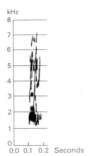

II C J Hazevoet Cape Verde Islands September 1988

IV C J Hazevoet Cape Verde Islands September 1988

Acrocephalus stentoreus Clamorous Reed Warbler (pp. 218, 219)

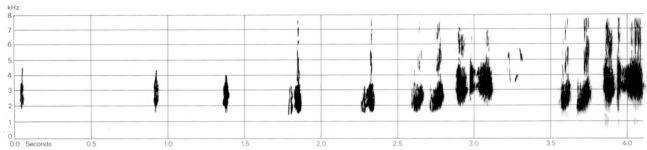

I C Chappuis Afghanistan May 1977

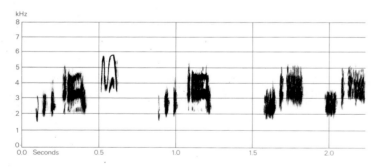

II C Chappuis Afghanistan May 1977

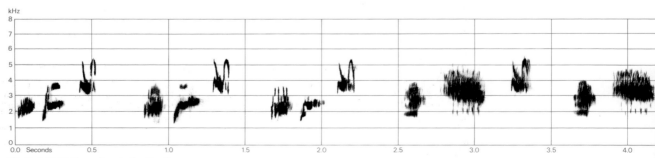

IV P A D Hollom Israel April 1980

Hippolais pallida **Olivaceous Warbler** (pp. 255, 256, 258)

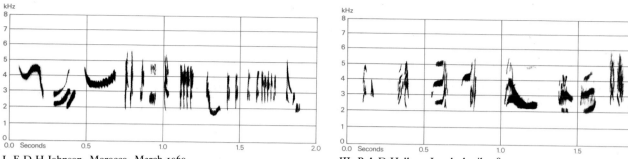

I   E D H Johnson   Morocco   March 1969

III   P A D Hollom   Israel   April 1980

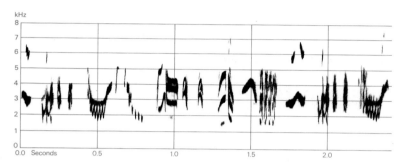

II   E D H Johnson   Algeria   May 1983

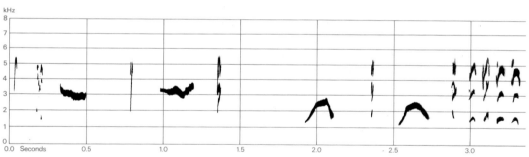

VI   B N Veprintsev and V V Leonovich   USSR   May 1985

Hippolais caligata **Booted Warbler** (p. 268)

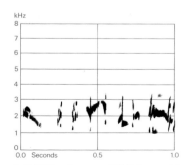

I   B N Veprintsev and V V Leonovich   USSR   June 1975

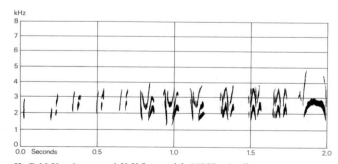

II   B N Veprintsev and V V Leonovich   USSR   April 1974

Hippolais polyglotta **Melodious Warbler** (p. 306)

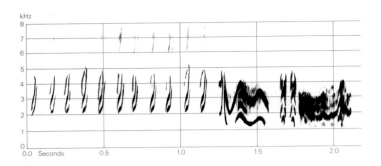

V P A D Hollom France May 1973

Sylvia sarda **Marmora's Warbler** (pp. 313, 314)

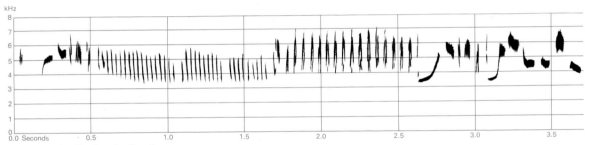

I C Chappuis Corsica April 1969

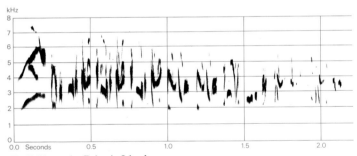

III C Chappuis Balearic Islands 1975

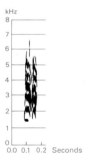

V C Chappuis Balearic Islands 1975

Sylvia undata **Dartford Warbler** (pp. 324, 325)

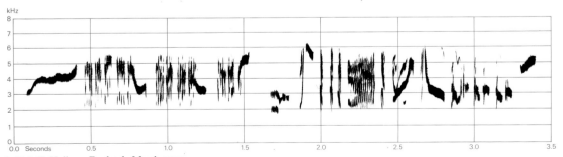

I P A D Hollom England March 1979

Sylvia undata **Dartford Warbler** (*cont.*)

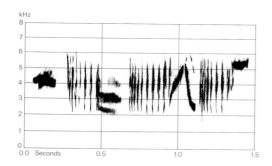

V V C Lewis England August 1976

Sylvia deserticola **Tristram's Warbler** (p. 332)

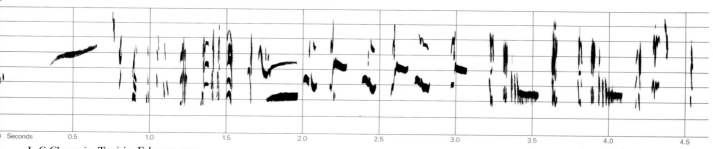

I C Chappuis Tunisia February 1971

Sylvia conspicillata **Spectacled Warbler** (pp. 342, 343)

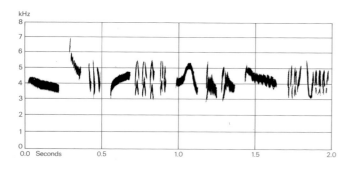

I J-C Roché France June 1958

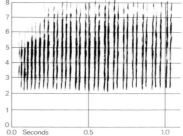

II P A D Hollom Morocco March 1978

IV C Chappuis/Sveriges Radio (1972–80) France April 1965

Sylvia cantillans **Subalpine Warbler** (p. 354)

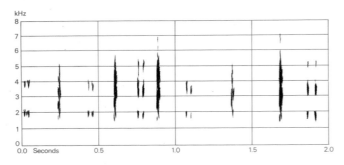

IV E D H Johnson Algeria March 1969

Sylvia mystacea **Ménétries's Warbler** (p. 363)

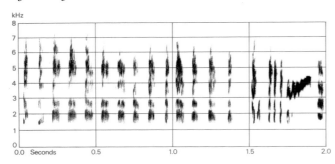

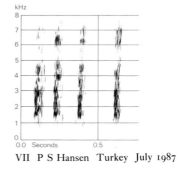

VII P S Hansen Turkey July 1987

Sylvia melanocephala **Sardinian Warbler** (p. 377)

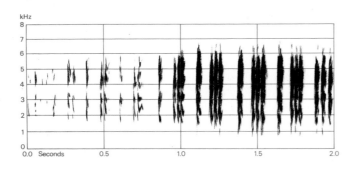

V E D H Johnson Algeria February 1968

Sylvia leucomelaena **Arabian Warbler** (p. 409)

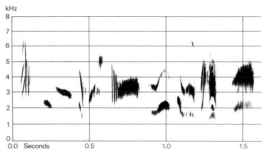

I H Shirihai Israel March 1986

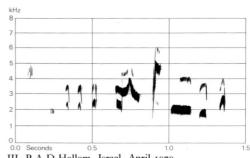

III P A D Hollom Israel April 1979

Sylvia hortensis **Orphean Warbler** (p. 419)

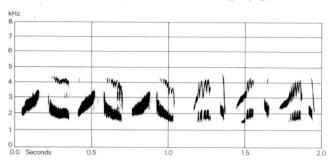

I C Chappuis Spain April 1966

Sylvia nisoria **Barred Warbler** (p. 434)

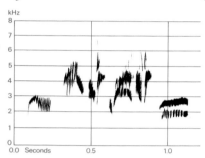

II V Neuvonen Finland June 1982

Sylvia curruca **Lesser Whitethroat** (pp. 452, 453, 454)

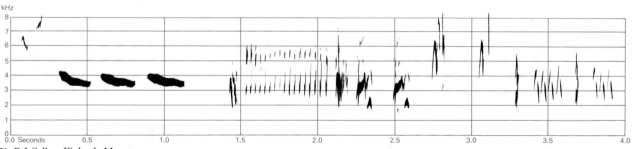

V P J Sellar Finland May 1971

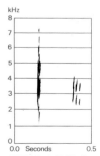

VIII P A D Hollom Egypt March 1984

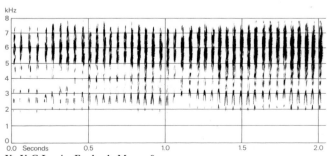

X V C Lewis England May 1987

Sylvia communis Whitethroat (p. 472)

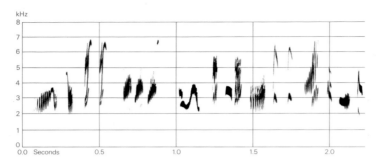

IV P J Sellar England June 1975

Sylvia borin **Garden Warbler** (p. 491)

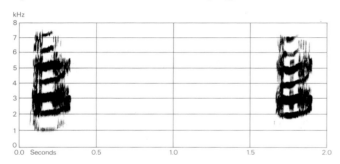

VI V C Lewis England June 1965

Sylvia atricapilla **Blackcap** (p. 509)

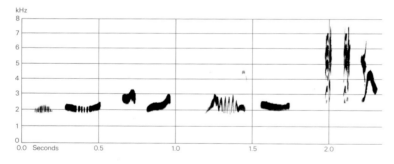

II W T C Seale England May 1985

Phylloscopus trochiloides **Greenish Warbler** (p. 531)

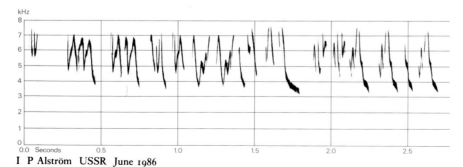

I P Alström USSR June 1986

Phylloscopus trochiloides **Greenish Warbler** (*cont.*)

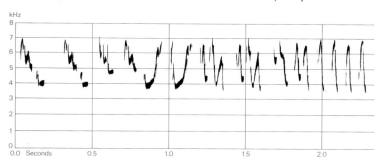

II S Wahlström   Sweden   June 1971

Phylloscopus borealis **Arctic Warbler** (pp. 543, 545)

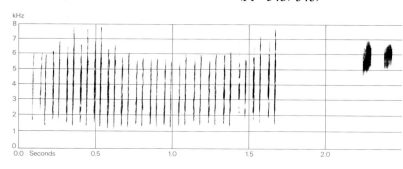

I J-C Roché   Norway   June 1968

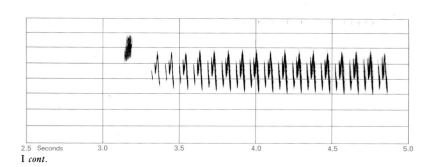

I *cont.*

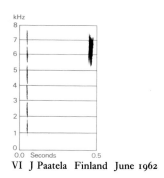

VI J Paatela   Finland   June 1962

Phylloscopus sibilatrix **Wood Warbler** (pp. 599, 600)

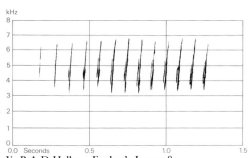

V P A D Hollom   England   June 1987

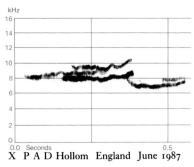

X P A D Hollom   England   June 1987

Phylloscopus sibilatrix **Wood Warbler** (*cont.*)

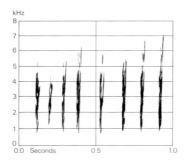

XII   V C Lewis England June 1965

Phylloscopus neglectus **Plain Willow Warbler** (p. 604)

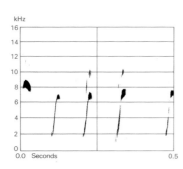

I   B N Veprintsev   USSR   May 1973

Phylloscopus collybita **Chiffchaff** (pp. 631, 632)

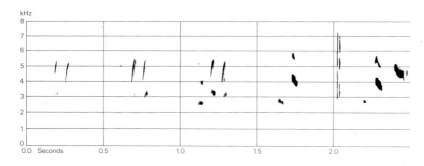

VII   D Wallschläger/BBC Mongolia June 1979

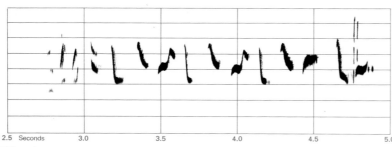

VII *cont.*

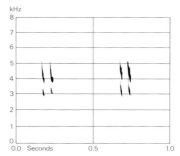

IX   W T C Seale England June 1988

Phylloscopus trochilus **Willow Warbler** (pp. 656, 659, 660)

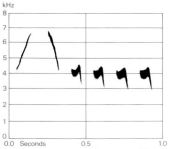

I P J Sellar England May 1988

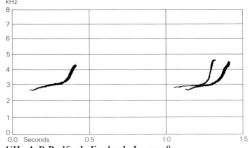

VII A P Radford England June 1985

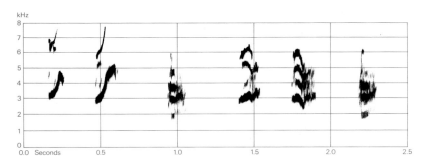

XII V C Lewis England May 1964

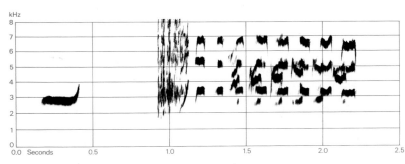

XIII A P Radford England May 1987

Regulus regulus **Goldcrest** (p. 679)

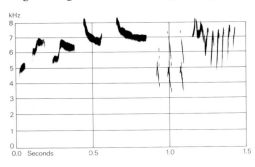

III P J Sellar England March 1963

VOLUME VII

Muscicapa striata Spotted Flycatcher (pp. 22, 23)

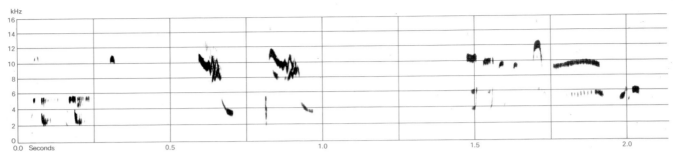

III P A D Hollom England May 1972

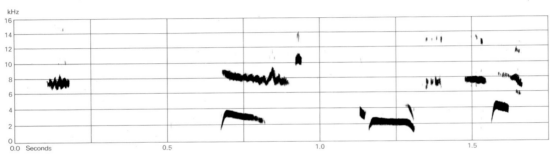

IV P A D Hollom England May 1972

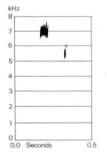

VI P A D Hollom England
June 1975

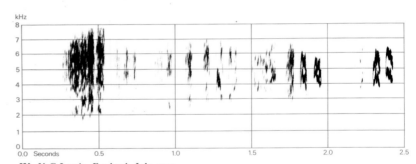

IX V C Lewis England July 1971

Ficedula parva Red-breasted Flycatcher (pp. 35, 36)

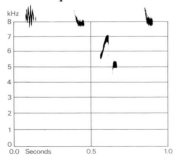

I C Weismann Denmark June 1969

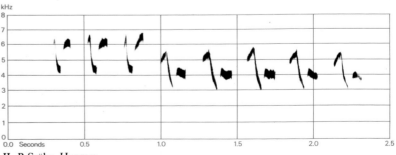

II P Szöke Hungary

Ficedula parva **Red-breasted Flycatcher** (*cont.*)

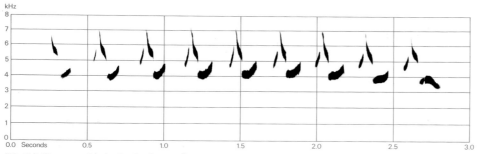

III C J Hazevoet Netherlands June 1981

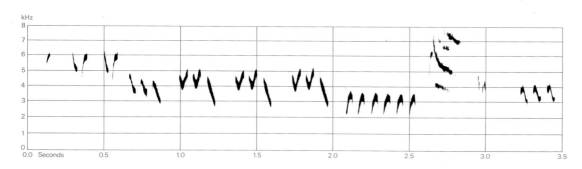

Ficedula semitorquata **Semi-collared Flycatcher** (p. 46)

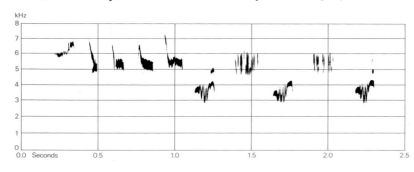

I C Chappuis/Sveriges Radio (1972–80)
Bulgaria May 1967

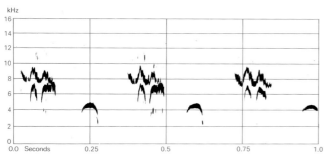

II C Chappuis/Sveriges Radio (1972–80) Bulgaria May
1967

Ficedula albicollis **Collared Flycatcher** (p. 58, 59, 60)

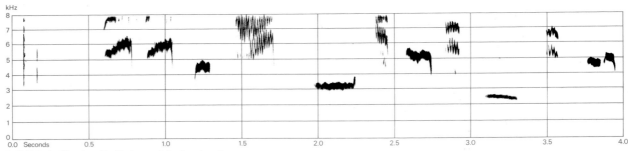

I   S Palmér/Sveriges Radio (1972–80) Sweden June 1966

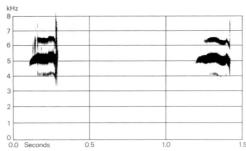

V H P Gelter Sweden April 1983

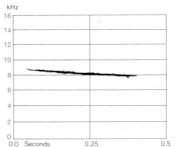

VIII H P Gelter Sweden May 1983

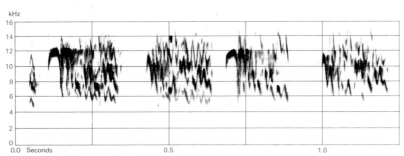

IX H P Gelter Sweden May 1983

Ficedula hypoleuca **Pied Flycatcher** (p. 79)

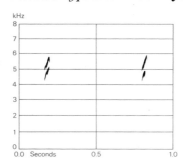

I P J Sellar Sweden May 1970

Aegithalos caudatus Long-tailed Tit (p. 141)

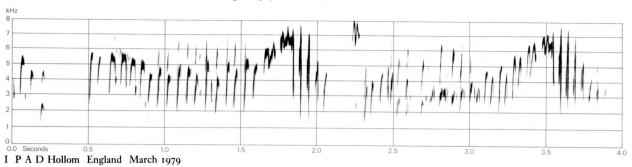

I P A D Hollom England March 1979

Parus cinctus Siberian Tit (p. 192)

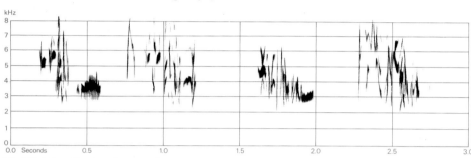

III P J Sellar Sweden August 1963

Parus caeruleus Blue Tit (p. 242)

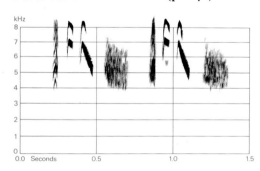

XVI W T C Seale England April 1985

Sitta tephronota Eastern Rock Nuthatch (p. 318)

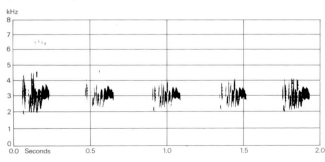

VI C Chappuis Afghanistan May 1977

Tichodroma muraria **Wallcreeper** (p. 340)

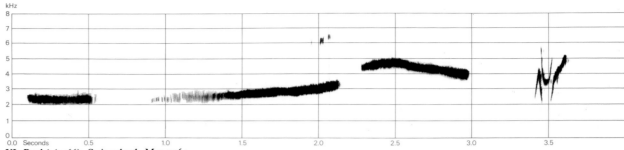

VI Roché (1966) Switzerland May 1964

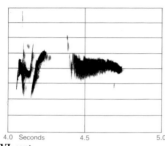

VI *cont.*

Certhia brachydactyla **Short-toed Treecreeper** (p. 372)

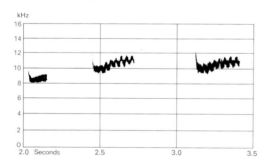

VII P Szöke Hungary

Lanius collurio **Red-backed Shrike** (p. 472)

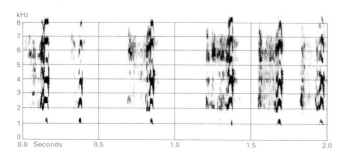

V E D H Johnson Corsica May 1989

CORRECTIONS TO VOLUME IV

Page 540. *Strix aluco* Tawny Owl. **Voice.** Column 2, line 3. Amend to read '. . . 'cher-oooOOooo' . . .'.

Page 540. *Strix aluco* Tawny Owl. **Voice.** Column 2, line 8. Amend to read 'A quiet tremulous tremolo (PAD Hollom).'

Plate 65. *Alcedo atthis* Kingfisher. Amend caption to read '. . . 3 juv, 4 nestling. Nominate *atthis*: 5 ad ♂. (NA)'.

Page 838. *Picus vaillantii* Levaillant's Green Woodpecker. **Voice.** Paragraph 2, lines 9–12. End sentence after '. . . detailed comparison' and delete ', although our recording of that species (Fig I in *P. viridis*) differs from both of *P. vaillantii* in showing accelerando throughout.'

CORRECTIONS TO VOLUME VI

Page 492. *Sylvia borin* Garden Warbler. **Voice.** Line 5. Amend to read '. . . WTC Seale: . . .'.

Page 509. *Sylvia atricapilla* Blackcap. **Voice.** Column 2, last line. Amend '. . . one from Britain (Fig I) apart from . . .' to read '. . . one from France (Fig I) apart from . . .'.

Page 630. *Phylloscopus collybita* Chiffchaff. **Voice.** Column 1, line 20. Amend '. . . Fig VI, see call 7b.' to read '. . . Fig VI, see call 7a.'

Page 630. *Phylloscopus collybita* Chiffchaff. **Voice.** Column 2, line 3 from below. Amend '. . . (sounds occuring at *c.* 2.4 s and *c.* 5.2 s are presumed . . .' to read '. . . (sounds occuring at *c.* 2.0 s and *c.* 4.7 s are presumed . . .'.

CORRECTIONS TO VOLUME VII

The term 'trill' is defined in the Voice Glossary (Vol. V, p. 32) as 'a rapid and regular alternation of two notes, usually with a small pitch interval'. In the accounts of certain species (notably within the Sittidae), this term is used in a more general sense, often to describe an acoustic structure that is not a true trill, but is instead a tremolo, which is defined as a 'rapid reiteration of one note'.

The song-phases of certain species commonly comprise a reiteration of either a single unit-type (e.g. the Sittidae), or a single motif-type (e.g. the Paridae). Sonagrams of these may show extracts from a song-phrase, rather than the whole phrase. The description of such sonagrams as being 'song' or 'a song' should not be taken to indicate that they illustrate a complete song-phrase.

Page 81. *Ficedula hypoleuca* Pied Flycatcher. **Voice.** Column 1, lines 1–2. Delete '(call of *G. passerinum* occurs below 1st call of *F. hypoleuca*)'.

Page 129. *Turdoides fulvus* Fulvous Babbler. **Voice.** Column 1, line 4 from below. Amend to read '. . . terminal ripple in song . . .'.

Page 140. *Aegithalos caudatus* Long-tailed Tit. **Voice.** Line 5 from below. Amend 'simultaneiously' to read 'simultaneously'.

Page 155. *Parus palustris* Marsh Tit. **Voice.** Column 1, line 11. Amend 'in songs illustrated);' to read 'in songs illustrated by Romanowski);'.

Page 165. *Parus lugubris* Sombre Tit. **Voice.** Column 1, lines 10–12. Amend 'Single 'töpp' Steinfatt 1954), 'jup', or 'chip' (WTC Seale: Fig IV) apparently . . .' to read 'Single 'töpp' (Steinfatt 1954), or 'chip' (WTC Seale: Fig IV, 1st unit) apparently . . .'.

Page 180. *Parus montanus* Willow Tit. **Voice.** Column 2, line 9 from below. Amend '. . . 'kett' (Figs V–VI); see . . .' to read '. . . 'kett' (Figs V–VI; narrow vertical lines on Fig V represent bill-tapping during nest excavation); see . . .'.

Page 191. *Parus cinctus* Siberian Tit. **Voice.** Column 2, line 2 from below. Amend to read '. . . (perhaps instrumental rather than vocal: WTC Seale) . . .'.

Page 192. *Parus cinctus* Siberian Tit. **Voice.** Column 1, line 3. Amend to read '. . . (Fig VIII; includes wing-fluttering sounds; WTC Seale), given by . . .'.

Page 195. *Parus cristatus* Crested Tit. **Field characters.** Column 2, paragraph 3, line 4. Amend 'merry trill' to read 'merry tremolo'.

Page 268. *Parus major* Great Tit. **Social pattern and behaviour.** Column 2, lines 24–25. Amend '. . . contact with ♀ (see 6a in Voice), . . .' to read '. . . contact with ♀ (see 7a in Voice), . . .'.

Page 268. *Parus major* Great Tit. **Social pattern and behaviour.** Column 2, line 32. Amend '. . . 'zeedling' call (see 6c in Voice).' to read '. . . 'zeedling' call (see 7c in Voice).'

Page 270. *Parus major* Great Tit. **Voice.** Column 2, lines 3–4. Amend to read '. . . several different song-types (Figs II–III and VI from same individual), number varying . . .'.

Page 271. *Parus major* Great Tit. **Voice.** Column 1, lines 7–8 from below. Amend '. . . not as rough and prolonged; may be . . .' to read '. . . not as rough; may be . . .'.

Page 285. *Sitta krueperi* Krüper's Nuthatch. **Voice.** Column 1, paragraph 2, line 1. Amend 'Trill' to read 'Trill or tremolo'.

Page 290. *Sitta whiteheadi* Corsican Nuthatch. **Voice.** Column 2, paragraph 2, line 2. Amend 'Trill' to read 'Tremolo'.

Page 290. *Sitta whiteheadi* Corsican Nuthatch. **Voice.** Column 2, paragraph 2, lines 14–15. Amend to read '. . .repeated rapidly as a tremolo, at *c.* 10 notes per s; given in 'trilled' form . . .'.

Page 308. *Sitta europaea* Nuthatch. **Voice.** Column 1, lines 5-6. Amend to read '... high tremolo (Fig VII), given by ♀; function unknown. (3g) Whinnying tremolo of much lower pitch ...'.

Page 317. *Sitta tephronota* Eastern Rock Nuthatch. **Voice.** Paragraph 2, line 5. Amend to read 'A trill, tremolo, or succession of ...'.

Page 347. *Certhia familiaris* Treecreeper. **Field characters.** Paragraph 2, line 2. Amend 'cadence' to read 'cascade'.

Page 358. *Certhia familiaris* Treecreeper. **Voice.** Column 2, lines 17-18. Amend '... *Fringilla coelebs* raised a couple of octaves and with ...' to read '*Fringilla coelabs* raised *c.* one octave (WTC Seale) and with ...'.

Page 389. *Remiz pendulinus* Penduline Tit. **Voice.** 2nd part of sonagram Figure II. Amend 'I *cont.*' to read 'II *cont.*'

Page 389. *Remiz pendulinus* Penduline Tit. **Voice.** Lines 1-2 from below. Delete '(after pause: reduced by 0.8 s in Fig II)'.

CORRECTIONS TO VOLUME VIII

Page 411. *Lagonosticta senegala* Red-billed Firefinch. **Field characters.** Column 2, line 11. Amend 'rather feeble twitter' to read 'lively musical phrase'.

INDEXES

Figures in **bold type** refer to plates.
References prefixed by 'S:' refer to
changes in status of the species in the
Western Palearctic

SCIENTIFIC NAMES

ENGLISH NAMES

NOMS FRANÇAIS

DEUTSCHE NAMEN

COMBINED INDEX

Bold type refers to volumes, normal type to pages within volumes.

1. The first entry under each species indicates the volume and starting page of the main **Species account**.
2. References following 'C:' refer to volume and page number of **Corrections** to the main text. Because many of the minor corrections have already been made in reprintings of earlier volumes, the lists of corrections are not all necessarily relevant to any particular set of volumes, other than the first printing of each.
3. Some corrections to references are also given at the end of each set of corrections. It may be necessary to check these.
4. References following 'S:' refer to the **Status** of the species concerned. All amendments to status are in Volume IX (see pp. 372–4 of this volume).
5. References following 'So:' refer to corrections to **Sonagrams**; most of these are ones that printed badly in Volumes VI and VII.

SCIENTIFIC NAMES

aalge (*Uria*), **IV**, 170
abyssinicus (*Coracias*), **IV**, 776
Accipiter badius, **II**, 169, C: **VI**, 719
    brevipes, **II**, 173
    gentilis, **II**, 148
    nisus, **II**, 158
Accipitridae, **II**, 5
Acridotheres tristis, **VIII**, 280
Acrocephalus aedon, **VI**, 244, C: **VIII**, 895
    agricola, **VI**, 146, So: **IX**, 422, 423
    arundinaceus, **VI**, 223
    brevipennis, **VI**, 168, C: **VIII**, 895, So: **IX**, 424
    dumetorum, **VI**, 155, So: **IX**, 423
    melanopogon, **VI**, 106
    paludicola, **VI**, 117, So: **IX**, 421
    palustris, **VI**, 172
    schoenobaenus, **VI**, 130, So: **IX**, 422
    scirpaceus, **VI**, 193
    stentoreus, **VI**, 212, So: **IX**, 424
Actitis hypoleucos, **III**, 594
    macularia, **III**, 605
acuminata (*Calidris*), **III**, 336, C: **VI**, 720
acuta (*Anas*), **I**, 521
acutirostris (*Calandrella*), S: **IX**, 374
adamsii (*Gavia*), **I**, 62, C: **V**, 1053, **VI**, 719
aedon (*Acrocephalus*), **VI**, 244, C: **VIII**, 895
Aegithalidae, **VII**, 132

Aegithalos caudatus, **VII**, 133, C: **VIII**, 895, **IX**, 439, So: **IX**, 437
Aegolius funereus, **IV**, 606
Aegypius monachus, **II**, 89, C: **III**, 907, **IV**, 954
aegyptiacus (*Alopochen*), **I**, 447
aegyptius (*Caprimulgus*), **IV**, 641
aegyptius (*Pluvianus*), **III**, 85, C: **IV**, 954
aereruginosus (*Circus*), **II**, 105
aethereus (*Phaethon*), **I**, 179
Aethia cristatella, **IV**, 229
aethiopica (*Hirundo*), S: **IX**, 374
aethiopicus (*Threskiornis*), **I**, 347
aethiops (*Myrmecocichla*), **V**, 754
affinis (*Apus*), **IV**, 692
affinis (*Aythya*), S: **IX**, 373
africanus (*Phalacrocorax*), **I**, 219
Agelaius phoeniceus, **IX**, 364
agricola (*Acrocephalus*), **VI**, 146, So: **IX**, 422, 423
aguimp (*Motacilla*), **V**, 471
Aix galericulata, **I**, 465
    sponsa, **I**, 465, S: **IX**, 373
Alaemon alaudipes, **V**, 74
Alauda arvensis, **V**, 188, C: **VI**, 721
    gulgula, **V**, 205
    razae, **V**, 207
Alaudidae, **V**, 45
alaudipes (*Alaemon*), **V**, 74
alba (*Calidris*), **III**, 282
alba (*Egretta*), **I**, 297
alba (*Motacilla*), **V**, 454, C: **VI**, 722
alba (*Platalea*), S: **IX**, 372
alba (*Tyto*), **IV**, 432, C: **V**, 1054
albellus (*Mergus*), **I**, 668
albeola (*Bucephala*), **I**, 650
albicilla (*Haliaeetus*), **II**, 48
albicollis (*Ficedula*), **VII**, 49, So: **IX**, 436
albicollis (*Zonotrichia*), **IX**, 94
albifrons (*Anser*), **I**, 403
albifrons (*Sterna*), **IV**, 120
alboniger (*Oenanthe*), **V**, 872
albus (*Corvus*), **VIII**, 195
Alca torda, **IV**, 195
Alcedinidae, **IV**, 700
Alcedininae, **IV**, 710
Alcedo atthis, **IV**, 711, C: **V**, 1055, **VI**, 720, **IX**, 439
alchata (*Pterocles*), **IV**, 269
Alcidae, **IV**, 168
alcyon (*Ceryle*), **IV**, 731
Alectoris barbara, **II**, 469
    chukar, **II**, 452
    graeca, **II**, 458
    rufa, **II**, 463
aleutica (*Sterna*), **IV**, 100
alexandri (*Apus*), **IV**, 652
alexandrinus (*Charadrius*), **III**, 153
alle (*Alle*), **IV**, 219, C: **V**, 1054
Alle alle, **IV**, 219, C: **V**, 1054

alleni (*Porphyrula*), **II**, 588, C: **V**, 1053, **VI**, 719
Alopochen aegyptiacus, **I**, 447
alpestris (*Eremophila*), **V**, 210, C: **VI**, 721
alpina (*Calidris*), **III**, 356
altirostris (*Turdoides*), **VII**, 101
aluco (*Strix*), **IV**, 526, C: **IX**, 439
amandava (*Amandava*), **VIII**, 427
Amandava amandava, **VIII**, 427
americana (*Anas*), **I**, 481, C: **VI**, 719
americana (*Fulica*), **II**, 611, C: **IV**, 954
americana (*Parula*), **IX**, 20
americana (*Spiza*), **IX**, 339
americanus (*Coccyzus*), **IV**, 425
amherstiae (*Chrysolophus*), **II**, 519
Ammodramus sandwichensis, **IX**, 80
Ammomanes cincturus, **V**, 59, C: **VI**, 721
    deserti, **V**, 65, C: **VI**, 721
Ammoperdix griseogularis, **II**, 473
    heyi, **II**, 476
amoena (*Passerina*), **IX**, 352
ampelinus (*Hypocolius*), **V**, 502
anaethetus (*Sterna*), **IV**, 109
Anas acuta, **I**, 521
    americana, **I**, 481, C: **VI**, 719
    capensis, **I**, 504
    clypeata, **I**, 539
    crecca, **I**, 494
    discors, **I**, 537, C: **V**, 1053, **VI**, 719
    erythrorhyncha, S: **IX**, 373
    falcata, **I**, 483
    formosa, **I**, 494, C: **VIII**, 895
    penelope, **I**, 473, C: **IV**, 954
    platyrhynchos, **I**, 505, C: **IV**, 954
    querquedula, **I**, 529, C: **III**, 907
    rubripes, **I**, 519
    smithii, S: **IX**, 373
    strepera, **I**, 485
Anatidae, **I**, 368
Anatinae, **I**, 444
angustirostris (*Marmaronetta*), **I**, 548, **VI**, 719
Anhinga melanogaster, **I**, 223
Anhingidae, **I**, 222
Anous stolidus, **IV**, 163
anser (*Anser*), **I**, 413
Anser albifrons, **I**, 403
    anser, **I**, 413
    brachyrhynchus, **I**, 397
    caerulescens, **I**, 422
    erythropus, **I**, 409
    fabalis, **I**, 391
    indicus, **I**, 422, S: **IX**, 372
    rossii, S: **IX**, 372
Anserinae, **I**, 370
Anthreptes metallicus, **VII**, 401
    platurus, **VII**, 397
Anthropoides virgo, **II**, 631, C: **III**, 907
Anthus berthelotii, **V**, 327
    campestris, **V**, 313

ENGLISH NAMES

NOMS FRANÇAIS

Les nombres en caractères gras se rapportent aux volumes; ceux en caractères normaux se rapportent aux pages de ces volumes.

1. Les deux premiers nombres suivant le nom de chaque espèce se rapportent au volume et à la première page du compte-rendu principal de cette espèce.

2. Les nombres suivant la lettre 'C:' se rapportent au volume et à la page de l'errata du texte principal. Les erreurs qui y sont mentionnées s'appliquent toutes à la première impression de chaque volume, mais certaines ont pu être corrigées dans les impressions subséquentes.

3. Certaines corrections s'appliquant à l'index sont aussi mentionnées à la fin de chaque errata. Il peut donc être nécessaire de vérifier les errata.

4. Les nombres suivant la lettre 'S:' se rapportent au statut de l'espèce en question. Tout changement s'appliquant au statut des espèces se trouve dans le volume IX (voir pages 372-374 du présent volume).

5. Les nombres suivant les lettres 'So:' se rapportent à l'errata des sonogrammes. La plupart de ces corrections s'appliquent aux volumes VI et VII dans lesquels les sonogrammes furent mal reproduits.

Accenteur alpin, **V**, 574, C: **VI**, 722
  à gorge noire, **V**, 568
  montanelle, **V**, 560
  mouchet, **V**, 548
  de Radde, **V**, 565
Agrobate roux, **V**, 586
Aigle de Bonelli, **II**, 258
  botté, **II**, 251
  criard, **II**, 211, C: **VI**, 719
  impérial, **II**, 225
  pêcheur africain, **II**, 44, S: **IX**, 373
  pomarin, **II**, 203
  ravisseur, **II**, 216, C: **VI**, 719
  royal, **II**, 234
  des steppes, **II**, 216, C: **VI**, 719
  de Verreaux, **II**, 245
Aigrette ardoisée, S: **IX**, 372
  bleue, S: **IX**, 372
  garzette, **I**, 290, C: **II**, 689
  Grande, **I**, 297
  intermédiaire, **I**, 296
  neigeuse, S: **IX**, 372
  des récifs, **I**, 286, C: **VI**, 719
  tricolore, S: **IX**, 372
Albatros à bec jaune, **I**, 115
  à cape blanche, S: **IX**, 372

hurleur, **I**, 116
  à pattes noires, **I**, 116
  royal, **I**, 117
  à sourcils noirs, **I**, 113, C: **VI**, 719
  à tête grise, **I**, 115
Alcyon ceinturé, **IV**, 731
  pie, **IV**, 723, C: **V**, 1055
Alouette bilophe, **V**, 225
  calandra, **V**, 93, C: **VI**, 721
  calandrelle, **V**, 123
  des champs, **V**, 188, C: **VI**, 721
  de Clotbey, **V**, 87
  gulgule, **V**, 205
  hausse-col, **V**, 210, C: **VI**, 721
  de Hume, S: **IX**, 374
  leucoptère, **V**, 109, C: **VI**, 721
  lulu, **V**, 173, C: **VI**, 721
  moineau, **V**, 46, C: **VI**, 721
  monticole, **V**, 103
  nègre, **V**, 115
  pispolette, **V**, 135
  de Razo, **V**, 207
Amarante du Sénégal, **VIII**, 411, C: **IX**, 440
Ammomane du désert, **V**, 65, C: **VI**, 721
  de Dunn, **V**, 54
  élégante, **V**, 59, C: **VI**, 721
Anhinga, **I**, 223
Anserelle de Cormandel, S: **IX**, 373
Astrild ondulé, **VIII**, 420
Autour gabar, S: **IX**, 373
  des palombes, **II**, 148
Autour-chanteur sombre, **II**, 144
Autruche, **I**, 37
Avocette, **III**, 48, C: **V**, 1053

Balbuzard pêcheur, **II**, 265
Barge hudsonienne, S: **IX**, 373
  à queue noire, **III**, 458
  rousse, **III**, 473, C: **IV**, 954
Bargette de Terek, **III**, 587, C: **IV**, 954, **VI**, 720
Bartramie à longue queue, **III**, 514, C: **VI**, 720
Bateleur, **II**, 103
Bécasse des bois, **III**, 444, C: **IV**, 954
Bécasseau d'Alaska, **III**, 297, C: **VI**, 719
  de Baird, **III**, 329, C: **VI**, 720
  de Bonaparte, **III**, 325, C: **VI**, 720
  cocorli, **III**, 339
  à col rouge, **III**, 299
  à échasses, **III**, 379, C: **IV**, 954, **V**, 1053, **VI**, 720
  falcinelle, **III**, 372
  Grand, maubèche, **III**, 268
  à long bec, **III**, 441
  à longs doigts, **III**, 319
  maubèche, **III**, 271
  minuscule, **III**, 321
  minute, **III**, 303

  à queue pointue, **III**, 336, C: **VI**, 720
  rousset, **III**, 382, C: **VI**, 720
  roux, **III**, 438, C: **IV**, 954
  sanderling, **III**, 282
  semipalmé, **III**, 293
  spatule, S: **IX**, 373
  tacheté, **III**, 332, C: **V**, 1053
  de Temminck, **III**, 311
  variable, **III**, 356
  violet, **III**, 345, C: **IV**, 954, **V**, 1053
Bécassine double, **III**, 423
  des marais, **III**, 409
  peinte, **III**, 10
  à queue pointue, **III**, 433
  sourde, **III**, 403
  de Swinhoe, **III**, 438
Bec-croisé bifascié, **VIII**, 672
  d'Ecosse, **VIII**, 707
  perroquet, **VIII**, 717
  des sapins, **VIII**, 686
Bec-en-ciseaux d'Afrique, **IV**, 166
Bengali rouge, **VIII**, 427
Bergeronnette citrine, **V**, 433, C: **VI**, 721
  grise, **V**, 454, C: **VI**, 722
  pie, **V**, 471
  printanière, **V**, 413, C: **VI**, 721
  des ruisseaux, **V**, 442, C: **VI**, 721
  d'Yarrell, **V**, 454, C: **VI**, 722
Bernache du Canada, **I**, 424
  à cou roux, **I**, 442, C: **VI**, 719
  cravant, **I**, 436
  nonnette, **I**, 430, C: **V**, 1053
Blongios minute, **I**, 254
  nain, **I**, 255
  de Schrenck, **I**, 260
  de Sturm, **I**, 261
Bondrée apivore, **II**, 13
Bouscarle de Cetti, **VI**, 7
Bouvreuil pivoine, **VIII**, 815
Bruant auréole, **IX**, 264
  à calotte blanche, **IX**, 142
  canelle, **IX**, 204
  cendré, **IX**, 204
  cendrillard, **IX**, 230
  des champs, **IX**, 80
  chanteur, **IX**, 86
  à cou gris, **IX**, 223
  à cou noir, **IX**, 182
  à couronne blanche, **IX**, 90
  fauve, **IX**, 84
  fou, **IX**, 182
  à gorge blanche, **IX**, 94
  jaune, **IX**, 153
  à joues marron, **IX**, 80
  lapon, **IX**, 101
  à longue queue, **IX**, 195
  des marais, **IX**, 90
  masquée, **IX**, 138
  mélanocéphale, **IX**, 313
  nain, **IX**, 250

DEUTSCHE NAMEN

Fett gedruckte Schrift verweist auf die Bandzahl, normal gedruckte Schrift auf die Seiten innerhalb der Bände.

1. Der erste Vermerk nach jeder Artenlistung gibt den Band und die Seitenzahl der allgemeinen **Artenbeschreibung** an.
2. Vermerke, die einem „C:" folgen, beziehen sich auf Band und Seitenzahl von **Korrekturen** im Haupttext. Da etliche der kleineren Korrekturen bereits in den Nachdrucken vorangegangener Bände ausgeführt worden sind, ist die Liste der Korrekturen nicht für jeden Band der gesamten Ausgabe von Bedeutung, sondern nur für den jeweiligen Erstdruck.
3. Einige Korrekturen zu den Vermerken sind auch am Ende jedes korrigierten Abschnitts angegeben. Es empfiehlt sich, diese zu beachten.
4. Hinweise, die einem „S:" folgen, beziehen sich auf den **Status** der betreffenden Art. Alle Berichtigungen zum Status der Art sind in Band IX zu finden (siehe Seite 372-4).
5. Hinweise, die einem „So:" folgen, beziehen sich auf Korrekturen der **Sonagramme**; die meisten davon waren bereits in unzureichender Qualität in Band VI and VII gedruckt worden.

NOMI IN ITALIANO

NOMBRES ESPAÑOLES